Prepare for Your Tests With TEST PREP VIDEO

The Chapter Test Prep Video included in this text will help you prepare for tests and use your study time more efficiently. Step-by-step solutions presented by Elayn Martin-Gay are included for every problem in the Chapter Tests in *Intermediate Algebra A Graphing Approach, Fourth Edition.* Now captioned in English and Spanish.

Good study skills are key to success in mathematics. Use the resources available to help you make the most of your study time and get the practice you need to understand the concepts.

W9-BQV-402

Three Steps to Success

To make the most of this resource when studying for a test, follow these three steps:

1. Take the Chapter Test found at the end of the text chapter.
2. Check your answers in the back of the text.
3. Use this video to review every step of the worked-out solution to those specific questions you answered incorrectly or need to review.

The easy-to-use navigation gives you direct access to the specific questions you need to review.

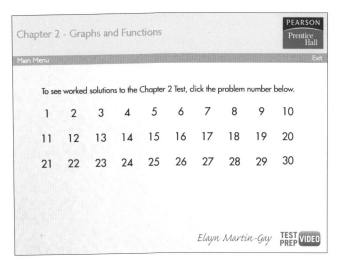

Chapter Test Menu: Lists all questions contained in the Chapter Test. Students select the test questions they wish to review.

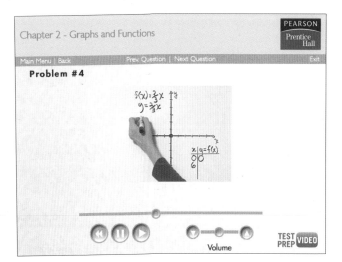

Previous and Next buttons allow easy navigation within tests. Rewind, Pause, and Play buttons let you view the solutions at your own pace.

How to Get Started

1. Insert the Video CD into your CD-ROM drive. The video should start automatically. If it does not, or if you have auto-launch turned off, you will need to open the program manually.

   To open the program manually:

   Windows users

   - Double-click on "My Computer," and locate your CD-ROM drive. It is usually the D: Drive.
   - Double-click on the CD-ROM drive to open the CD Video.
   - Find the file named "Start."
   - Double-click on "Start" to begin the video.

   Macintosh users

   - Locate the CD-ROM icon on your desktop and double-click it to reveal the contents of the CD.
   - Locate "Start," and double-click on it to start the program.

2. The opening screen lists your text chapter titles.

3. Click on the chapter title for the Chapter Test you are working on. You will be taken to the Chapter Test for that chapter.

4. The Chapter Test menu lists all the problems contained in your Chapter Test in the text.

5. Double-click on the problem number(s) to view the solution(s) you need to review. You can also navigate to other problems on the test by using the Previous and Next buttons.

6. If you do not have QuickTime, refer to the ReadMe file included on this Video CD. The ReadMe includes instructions on how to download and install the latest version of QuickTime.

7. **English and Spanish Captioning.** To activate the captioning feature, select the appropriate logo on the main menu, and click the "off" button to "on." Choosing the CC logo will activate English captions, while selecting the ESP logo will activate Spanish captions. Captioning will then be enabled for all video segments on this CD. To disable captioning, click the selected button on the main menu to "off."

Technical Support: To obtain support, please visit us online at http://247pearsoned.custhelp.com where you can search our knowledge base for common solutions, view product alerts, and review all options for additional assistance.

SINGLE PC LICENSE AGREEMENT AND LIMITED WARRANTY

READ THIS LICENSE CAREFULLY BEFORE OPENING THIS PACKAGE. BY OPENING THIS PACKAGE, YOU ARE AGREEING TO THE TERMS AND CONDITIONS OF THIS LICENSE. IF YOU DO NOT AGREE, DO NOT OPEN THE PACKAGE. PROMPTLY RETURN THE UNOPENED PACKAGE AND ALL ACCOMPANYING ITEMS TO THE PLACE YOU OBTAINED THEM [[FOR A FULL REFUND OF ANY SUMS YOU HAVE PAID FOR THE SOFTWARE]]. *THESE TERMS APPLY TO ALL LICENSED SOFTWARE ON THE DISK EXCEPT THAT THE TERMS FOR USE OF ANY SHAREWARE OR FREEWARE ON THE DISKETTES ARE AS SET FORTH IN THE ELECTRONIC LICENSE LOCATED ON THE DISK:*

1. GRANT OF LICENSE AND OWNERSHIP: The enclosed computer programs and data ("Software") are licensed, not sold, to you by Pearson Education, Inc. publishing as Prentice-Hall, Inc. ("We" or the "Company") in consideration of your purchase or adoption of the accompanying Company textbooks and/or other materials, and your agreement to these terms. We reserve any rights not granted to you. You own only the disk(s) but we and/or our licensors own the Software itself. This license allows individuals who have purchased the accompanying Company textbook to use and display their copy of the Software on a single computer (i.e., with a single CPU) at a single location for *academic* use only, so long as you comply with the terms of this Agreement. You may make one copy for back up, or transfer your copy to another CPU, provided that the Software is usable on only one computer.

2. RESTRICTIONS: You may *not* transfer or distribute the Software or documentation to anyone else. Except for backup, you may *not* copy the documentation or the Software. You may *not* network the Software or otherwise use it on more than one computer or computer terminal at the same time. You may *not* reverse engineer, disassemble, decompile, modify, adapt, translate, or create derivative works based on the Software or the Documentation. You may be held legally responsible for any copying or copyright infringement that is caused by your failure to abide by the terms of these restrictions.

3. TERMINATION: This license is effective until terminated. This license will terminate automatically without notice from the Company if you fail to comply with any provisions or limitations of this license. Upon termination, you shall destroy the Documentation and all copies of the Software. All provisions of this Agreement as to limitation and disclaimer of warranties, limitation of liability, remedies or damages, and our ownership rights shall survive termination.

4. LIMITED WARRANTY AND DISCLAIMER OF WARRANTY: Company warrants that for a period of 60 days from the date you purchase this SOFTWARE (or purchase or adopt the accompanying textbook), the Software, when properly installed and used in accordance with the Documentation, will operate in substantial conformity with the description of the Software set forth in the Documentation, and that for a period of 30 days the disk(s) on which the Software is delivered shall be free from defects in materials and workmanship under normal use. The Company does *not* warrant that the Software will meet your requirements or that the operation of the Software will be uninterrupted or error-free. Your only remedy and the Company's only obligation under these limited warranties is, at the Company's option, return of the disk for a refund of any amounts paid for it by you or replacement of the disk. THIS LIMITED WARRANTY IS THE ONLY WARRANTY PROVIDED BY THE COMPANY AND ITS LICENSORS, AND THE COMPANY AND ITS LICENSORS DISCLAIM ALL OTHER WARRANTIES, EXPRESS OR IMPLIED, INCLUDING WITHOUT LIMITATION, THE IMPLIED WARRANTIES OF MERCHANTABILITY AND FITNESS FOR A PARTICULAR PURPOSE. THE COMPANY DOES NOT WARRANT, GUARANTEE OR MAKE ANY REPRESENTATION REGARDING THE ACCURACY, RELIABILITY, CURRENTNESS, USE, OR RESULTS OF USE, OF THE SOFTWARE.

5. LIMITATION OF REMEDIES AND DAMAGES: IN NO EVENT, SHALL THE COMPANY OR ITS EMPLOYEES, AGENTS, LICENSORS, OR CONTRACTORS BE LIABLE FOR ANY INCIDENTAL, INDIRECT, SPECIAL, OR CONSEQUENTIAL DAMAGES ARISING OUT OF OR IN CONNECTION WITH THIS LICENSE OR THE SOFTWARE, INCLUDING FOR LOSS OF USE, LOSS OF DATA, LOSS OF INCOME OR PROFIT, OR OTHER LOSSES, SUSTAINED AS A RESULT OF INJURY TO ANY PERSON, OR LOSS OF OR DAMAGE TO PROPERTY, OR CLAIMS OF THIRD PARTIES, EVEN IF THE COMPANY OR AN AUTHORIZED REPRESENTATIVE OF THE COMPANY HAS BEEN ADVISED OF THE POSSIBILITY OF SUCH DAMAGES. IN NO EVENT SHALL THE LIABILITY OF THE COMPANY FOR DAMAGES WITH RESPECT TO THE SOFTWARE EXCEED THE AMOUNTS ACTUALLY PAID BY YOU, IF ANY, FOR THE SOFTWARE OR THE ACCOMPANYING TEXTBOOK. BECAUSE SOME JURISDICTIONS DO NOT ALLOW THE LIMITATION OF LIABILITY IN CERTAIN CIRCUMSTANCES, THE ABOVE LIMITATIONS MAY NOT ALWAYS APPLY TO YOU.

6. GENERAL: THIS AGREEMENT SHALL BE CONSTRUED IN ACCORDANCE WITH THE LAWS OF THE UNITED STATES OF AMERICA AND THE STATE OF NEW YORK, APPLICABLE TO CONTRACTS MADE IN NEW YORK, AND SHALL BENEFIT THE COMPANY, ITS AFFILIATES AND ASSIGNEES. THIS AGREEMENT IS THE COMPLETE AND EXCLUSIVE STATEMENT OF THE AGREEMENT BETWEEN YOU AND THE COMPANY AND SUPERSEDES ALL PROPOSALS OR PRIOR AGREEMENTS, ORAL, OR WRITTEN, AND ANY OTHER COMMUNICATIONS BETWEEN YOU AND THE COMPANY OR ANY REPRESENTATIVE OF THE COMPANY RELATING TO THE SUBJECT MATTER OF THIS AGREEMENT. If you are a U.S. Government user, this Software is licensed with "restricted rights" as set forth in subparagraphs (a)–(d) of the Commercial Computer-Restricted Rights clause at FAR 52.227-19 or in subparagraphs (c)(1)(ii) of the Rights in Technical Data and Computer Software clause at DFARS 252.227-7013, and similar clauses, as applicable.

Windows System Requirements:
Intel® Pentium® 700-MHz processor
Windows 2000 (Service Pack 4), XP, or Vista
800 χ 600 resolution
8x CD drive
QuickTime 7.x
Sound Card
Internet browser

Macintosh System Requirements:
Power PC® or Intel 300 MHz processor
Mac 10.x
800 χ 600 resolution monitor
8x CD drive
QuickTime 7
Internet browser

Should you have any questions concerning this agreement or if you wish to contact the Company for any reason, please contact in writing:
Director, Media Production
Pearson Education
1 Lake Street
Upper Saddle River, NJ 07458

Intermediate Algebra
A Graphing Approach

Intermediate Algebra
A Graphing Approach

Fourth Edition

Elayn Martin-Gay
University of New Orleans

Margaret Greene
Florida Community College at Jacksonville

PEARSON

Prentice Hall

Upper Saddle River, New Jersey 07458

Library of Congress Cataloging-in-Publication Data

Martin-Gay, K. Elayn and Margaret Greene
  Intermediate algebra: a graphing approach/K. Elayn Martin-Gay—4th ed.
    p. cm.
  Includes index.
  ISBN 0-13-600733-3
  1. Algebra   I. Title

President: *Greg Tobin*
Editor in Chief: *Paul Murphy*
Vice President and Editorial Director: *Christine Hoag*
Sponsoring Editor: *Mary Beckwith*
Assistant Editor: *Christine Whitlock*
Editorial Assistant: *Georgina Brown*
Production Management: *Elm Street Publishing Services*
Senior Managing Editor: *Linda Mihatov Behrens*
Operations Specialist: *Ilene Kahn*
Senior Operations Supervisor: *Diane Peirano*
Vice President and Executive Director of Development: *Carol Trueheart*
Development Editor: *Lisa Collette*
Media Producer: *Audra J. Walsh*
Lead Media Project Manager: *Richard Bretan*
Software Development:
  MyMathLab, Media Producer: Jennifer Sparkes
  MathXL, Software Editor: Janet Szykolony
  TestGen, Software Editor: Ted Hartman
Vice President and Director of Marketing: *Amy Cronin*
Senior Marketing Manager: *Michelle Renda*
Marketing Manager: *Marlana Voerster*
Marketing Assistants: *Jill Kapinus and Nathaniel Koven*
Senior Art Director: *Juan R. López*
Interior Designer: *Mike Fruhbeis*
Cover Art Direction, Design, and Illustration: *Kenny Beck*
AV Project Manager: *Thomas Benfatti*
Director, Image Resource Center: *Melinda Patelli*
Manager, Rights and Permissions: *Zina Arabia*
Manager, Visual Research: *Beth Brenzel*
Image Permission Coordinator: *Craig Jones*
Photo Researcher: *Kathy Ringrose*
Art Studios: *Scientific Illustrators/Laserwords*
Compositor: *ICC Macmillan Inc.*

 © 2009, 2005, 2001, 1997 Pearson Education, Inc.
Pearson Prentice Hall
Pearson Education, Inc.
Upper Saddle River, New Jersey 07458

Pearson Prentice Hall™ is a trademark of Pearson Education, Inc.
Printed in the United States of America

10 9 8 7 6 5 4 3 2 1

ISBN-10 0-13-600733-3
ISBN-13 978-0-13-600733-3

Pearson Education Ltd., London
Pearson Education Singapore, Pte. Ltd.
Pearson Education Canada, Inc.
Pearson Education—Japan
Pearson Education Australia PTY, Limited
Pearson Education North Asia, Ltd., Hong Kong
Pearson Educación de Mexico, S.A. de C.V.
Pearson Education Malaysia, Pte. Ltd.
Pearson Education Upper Saddle River, New Jersey

*To Elizabeth Ashley, Lockwood Ryan, and
Matthew Patrick Greene. You light up my life!*

*To Bailey Frances Martin, Ethan Warren, Avery Blythe, Mia
Adelle Barnes, and Madison Jane Martin . . . ditto.*

Contents

CHAPTER

4

SYSTEMS OF LINEAR EQUATIONS AND INEQUALITIES   267

CHAPTER

5

EXPONENTS, POLYNOMIALS, AND POLYNOMIAL FUNCTIONS   323

CHAPTER

6

RATIONAL EXPRESSIONS   409

CHAPTER

7

RATIONAL EXPONENTS, RADICALS, AND COMPLEX NUMBERS 484

CHAPTER

8

QUADRATIC EQUATIONS AND FUNCTIONS 551

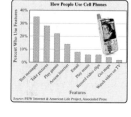

CHAPTER

9

EXPONENTIAL AND LOGARITHMIC FUNCTIONS 624

CHAPTER

10

CONIC SECTIONS   688

CHAPTER

11

SEQUENCES, SERIES, AND THE BINOMIAL THEOREM   723

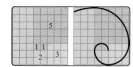

APPENDICES   759

Tools to Help
Students Succeed

Your textbook includes a number of features designed to help you succeed in this math course—as well as the next math course you take. These features include:

Feature	Benefit	Page
Well-crafted Exercise Sets: We learn math by doing math	The exercise sets in your text offer an ample number of exercises carefully ordered so you can master basic mathematical skills and concepts while developing all-important problem solving skills. Exercise sets include Mixed Practice exercises to help you master multiple key concepts, as well as Vocabulary and Readiness Check, Writing, Applications, Concept Check, Concept Extension, and Review and Preview Exercises.	131–137
Study Skills Builders: Maximize your chances for success	Study Skills Builders reinforce the material in *Section 1.1—Tips for Success in Mathematics.* Study Skills Builders are a great resource for study ideas and self-assessment to maximize your opportunity for success in this course. Take your new study skills with you to help you succeed in your next math course.	86
The Bigger Picture: Succeed in this math course and the next one you take	The Bigger Picture focuses on the key concepts of this course—solving equations and inequalities—and asks you to keep an ongoing study guide so you can solve equations and inequalities, and recognize the difference between them. A strong foundation in solving equations and inequalities will help you succeed in this algebra course, as well as the next math course you take.	51
Examples: Step-by-step instruction for you	Examples in the text provide you with clear, concise step-by-step instructions to help you learn. Annotations in the examples provide additional instruction.	47
Helpful Hints: Help where you'll need it most	Helpful Hints provide tips and advice at exact locations where students need it most. Strategically placed where you might have the most difficulty, Helpful Hints will help you work through common trouble spots.	48
Practice Exercises: Immediate reinforcement	New Practice exercises offer immediate reinforcement after every example. Try each Practice exercise after studying the corresponding example to make sure you have a good working knowledge of the concept.	47
Integrated Review: Mid-chapter progress check	To ensure you understand the key concepts covered in the first sections of the chapter, work the exercises in the Integrated Review before you continue with the rest of the chapter.	174
Vocabulary and Readiness Check, Vocabulary Check: Key terms and vocabulary	Use the Vocabulary and Readiness Checks to build your vocabulary and warm-up on concepts in the section. Make sure you understand key terms and vocabulary in each chapter with the end-of-chapter Vocabulary Check.	131, 194
Chapter Highlights: Study smart	Chapter Highlights outline the key concepts of the chapter along with examples to help you focus your studying efforts as you prepare for your test.	194–200
Chapter Test: Take a practice test	In preparation for your classroom test, take this practice test to make sure you understand the key topics in the chapter. Be sure to use the **Chapter Test Prep Video** included with this text to see the author present a fully worked-out solution to each exercise in the Chapter Test.	204
Practice Final Exam: Take a practice final	In preparation for your final, take the practice final exam found in Appendix A.2. **Martin-Gay's Interactive DVD/CD Lecture Series** includes the Practice Final Exam and provides you with full video solutions to each exercise. Overview clips provide a brief overview on how to approach different problem types.	761–763

Martin-Gay's **VIDEO RESOURCES**
Help Students Succeed

MARTIN-GAY'S CHAPTER TEST PREP VIDEO (AVAILABLE WITH THIS TEXT) TEST PREP VIDEO

- Provides students with help during their most "teachable moment"—while they are studying for a test.
- Text author Elayn Martin-Gay presents step-by-step solutions to the exact exercises found in each Chapter Test in the book.
- Easy video navigation allows students to instantly access the worked-out solutions to the exercises they want to review.
- Close-captioned in English and Spanish.

NEW MARTIN-GAY'S INTERACTIVE DVD/CD LECTURE SERIES

Martin-Gay's video series has been comprehensively updated to address the way today's students study and learn. The new videos offer students active learning at their pace, with the following resources and more:

- **A complete lecture** for each section of the text, presented by Elayn Martin-Gay. Students can easily review a section or a specific topic before a homework assignment, quiz, or test. Exercises in the text marked with the 🎥 are worked on the video.
- A **new interface** with menu and navigation features helps students quickly find and focus on the examples and exercises they need to review.
- Martin-Gay's "pop-ups" reinforce key terms and definitions and are a great support for multiple learning styles.
- A new **Practice Final Exam Video** helps students prepare for the final exam. This Practice Final Exam is included in the text in Appendix A.2. At the click of a button, students can watch the full solutions to each exercise on the exam when they need help. Overview clips provide a brief overview on how to approach different problem types—just as they will need to do on a Final Exam.
- **Interactive Concept Checks** allow students to check their understanding of essential concepts. Like the Concept Checks in the text, these multiple choice exercises focus on common misunderstandings. After making their answer selection, students are told whether they're correct or not, and why! Elayn also presents the full solution.
- **Study Skills Builders** help students develop effective study habits and reinforce the advice provided in Section 1.1, Tips for Success in Mathematics, found in the text and video.
- **Close-captioned in Spanish and English**
- Ask your bookstore for information about Martin-Gay's *Intermediate Algebra: A Graphing Approach,* Fourth Edition Interactive DVD/CD Lecture Series or visit www.mypearsonstore.com.

You will find Interactive Concept Checks and Study Skills Builders in the following sections on the Interactive DVD/CD Lecture Series:

Interactive Concept Checks Section		Study Skills Builders Section
1.2	6.5	1.6 Time Management
1.3	6.6	1.7 Doing Your Homework Online
1.4	6.7	1.8 Are You Familiar with the Resources Available with Your Textbook?
1.5	7.1	
2.1	7.2	2.5 Have You Decided to Complete This Course Successfully?
2.2	7.4	
2.3	7.5	2.6 Are You Organized?
2.4	7.6	3.6 How Well Do You Know Your Textbook?
2.7	8.1	
3.1	8.2	4.3 Tips for Studying for an Exam
3.2	8.3	
3.3	8.4	4.5 How Are Your Homework Assignments Going?
3.4	8.6	5.5 Doing Your Homework Online
3.5	9.1	
4.1	9.2	5.8 What to Do the Day of an Exam
4.2	9.3	6.1 Are You Satisfied with Your Performance on a Particular Quiz?
4.4	9.5	
5.1	9.6	6.4 Are You Organized?
5.2	10.1	7.3 Are You Familiar with the Resources Available with Your Textbook?
5.3	10.2	
5.4	10.3	7.7 Have You Decided to Complete This Course Successfully?
5.6	11.1	
5.7	11.2	8.5 How Well Do You Know Your Textbook?
6.2	11.3	8.7 Preparing for Your Final Exam
6.3		
		9.4 Tips for Studying for an Exam
		9.7 Are You Satisfied with Your Performance on a Particular Quiz or Exam?
		10.4 What to Do the Day of an Exam
		11.4 How Are Your Homework Assignments Going?
		11.5 Preparing for Your Final Exam

Additional Resources to Help You Succeed

Student Study Pack
A single, easy-to-use package—available bundled with your textbook or by itself—for purchase through your bookstore. This package contains the following resources to help you succeed:

Student Solutions Manual
- Contains worked-out solutions to odd-numbered exercises from each section exercise set, all Practice exercises, Vocabulary and Readiness Check Exercises, and all exercises found in the Chapter Reviews, Chapter Tests, and Integrated Reviews.

Martin-Gay's Interactive Video Lectures
- Text author Elayn Martin-Gay presents the key concepts from every section of the text with 15–20 minute mini-lectures. Students can easily review a section or a specific topic before a homework assignment, quiz, or test.
- A new interface allows easy navigation through the lesson.
- Includes fully worked-out solutions to exercises marked with an icon in each section. Also includes *Section 1.1, Tips for Success in Mathematics.*
- Close-captioned in English and Spanish.

Pearson Tutor Center

Online Homework and Tutorial Resources

MyMathLab® *MyMathLab*

MyMathLab is a series of text-specific, easily customizable online courses for Pearson Education's textbooks in mathematics and statistics. Powered by CourseCompass™ (our online teaching and learning environment) and MathXL® (our online homework, tutorial, and assessment system), MyMathLab gives you the tools you need to deliver all or a portion of your course online, whether your students are in a lab setting or working from home. MyMathLab provides a rich and flexible set of course materials, featuring free-response exercises that are algorithmically generated for unlimited practice and mastery. Students can also use online tools, such as video lectures, animations, and a multimedia textbook, to independently improve their understanding and performance. Instructors can use MyMathLab's homework and test managers to select and assign online exercises correlated directly to the textbook, and they can also create and assign their own online exercises and import TestGen tests for added flexibility. MyMathLab's online gradebook—designed specifically for mathematics and statistics—automatically tracks students' homework and test results and gives the instructor control over how to calculate final grades. Instructors can also add offline (paper-and-pencil) grades to the gradebook. Includes access to the Pearson Tutor Center. MyMathLab is available to qualified adopters. For more information, visit our website at www.mythlab.com or contact your sales representative.

MathXL® *MathXL* PRACTICE

MathXL® is a powerful online homework, tutorial, and assessment system that accompanies Pearson Education's textbooks in mathematics and statistics. With MathXL, instructors can create, edit, and assign online homework and tests using algorithmically generated exercises correlated at the objective level to the textbook. They can also create and assign their own online exercises and import TestGen tests for added flexibility. All student work is tracked in MathXL's online gradebook. Students can take chapter tests in MathXL and receive personalized study plans based on their test results. The study plan diagnoses weaknesses and links students directly to tutorial exercises for the objectives they need to study and retest. Students can also access supplemental animations and video clips directly from selected exercises. MathXL is available to qualified adopters. For more information, visit our website at www.mathxl.com, or contact your sales representative.

Preface

ABOUT THE BOOK

Intermediate Algebra: A Graphing Approach, Fourth Edition was written to provide a **solid foundation in algebra** for students who might not have had previous experience in algebra. Specific care has been taken to ensure that students have the most **up-to-date and relevant** text preparation for their next mathematics course, as well as to help students succeed in nonmathematical courses that require a grasp of algebraic fundamentals. I have tried to achieve this by writing a user-friendly text that is keyed to objectives and contains many worked-out examples. The basic concepts of graphing and functions are introduced early, and problem solving techniques, real-life and real-data applications, data interpretation, appropriate use of technology, number sense, critical thinking, decision-making, and geometric concepts are emphasized and integrated throughout the book.

The many factors that contributed to the success of the previous editions have been retained. In preparing this edition, I considered the comments and suggestions of colleagues throughout the country, students, and many users of the prior editions. The AMATYC Crossroads in Mathematics: Standards for Introductory College Mathematics before Calculus and the MAA and NCTM standards (plus Addenda), together with advances in technology, also influenced the writing of this text.

Throughout the series, pedagogical features are designed to develop student proficiency in algebra and problem solving, and to prepare students for future courses.

WHAT'S NEW IN THE FOURTH EDITION

New Martin-Gay's Interactive DVD/CD Lecture Series, featuring Elayn Martin-Gay, provides students with active learning at their pace. The new videos offer the following resources and more:

- A complete lecture for each section of the text. A new interface with menu and navigation features allows students to quickly find and focus on the examples and exercises they need to review.
- The new **Practice Final Exam** helps students prepare for an end-of-course final. Students can watch full video solutions to each exercise. Overview clips give students a brief overview on how to approach different problem types.
- **Interactive Concept Check exercises** that allow students to check their understanding of essential concepts.

A New Three-Point System to support students and instructors:

- **Examples** in the text prepare students for class and homework.
- New **Practice Exercises** are paired with each text example and are ready-made for use in the classroom and can be assigned as homework.
- **Classroom Examples,** found only in the Annotated Instructor's Edition, are also paired with each example to give instructors a convenient way to further illustrate skills and concepts.

ENHANCED EXERCISE SETS

- **New Vocabulary and Readiness exercises** appear at the beginning of exercise sets. The **Vocabulary** exercises reinforce a student's understanding of new terms so that forthcoming instructions in the exercise set are clear. The **Readiness** exercises serve as a warm-up on the skills and concepts necessary to complete the exercise set.

- **NEW! Concept Check exercises** have been added to the section exercise sets. These exercises are related to the Concept Check(s) found within the section. They help students measure their understanding of key concepts by focusing on common trouble areas. These exercises may ask students to identify a common error, and/or provide an explanation.

- **New Mixed Review** exercises are included at the end of the Chapter Review. These exercises require students to determine the problem type and strategy needed in order to solve them.

- **New Practice Final Exam**—included in Appendix A.2—helps students prepare for an end-of-course final. Martin-Gay's Interactive DVD/CD Lecture Series provides students with solutions to all exercises on the final. Overview clips give a brief introduction on how to approach different types of problems.

INCREASED EMPHASIS ON STUDY SKILLS AND STUDENT SUCCESS

- **NEW! Study Skills Builders** (formerly Study Skill Reminders) Found at the end of many exercise sets, Study Skills Builders allow instructors to assign exercises that will help students improve their study skills and take responsibility for their part of the learning process. Study Skills Builders reinforce the material found in Section 1.1, "Tips for Success in Mathematics," and serve as an excellent tool for self-assessment.

- **NEW! The Bigger Picture** is a recurring feature, starting in Section 1.5, that focuses on the key concepts of the course—solving equations and inequalities. Students develop an ongoing study guide to help them be able to solve equations and inequalities, and to know the difference between them. By working the exercises and developing this study guide throughout the text, students can begin to transition from thinking "section by section" to thinking about how the mathematics in this course is part of the "bigger picture" of mathematics in general. A completed outline is provided in Appendix A so students have a model for their work. It's great preparation for the Practice Final in A.2.

CONTINUING SUPPORT FOR TEST PREPARATION

- **TEST PREP VIDEO Chapter Test Prep Video** provides students with help during their most "teachable moment"—while they are studying for a test. Included with every copy of the student edition of the text, this video provides fully worked-out solutions by the author to every exercise from each Chapter Test in the text. The easy video navigation allows students to instantly access the solutions to the exercises they want to review. The problems are solved by the author in the same manner as in the text.

- **Chapter Test Files in TestGen®** provide algorithms specific to each exercise from each Chapter Test in the text. Allows for easy replication of Chapter Tests with consistent, algorithmically generated problem types for additional assignments or assessment purposes.

CONTENT CHANGES IN THE FOURTH EDITION

The following content changes are included in the fourth edition:

- Section 1.9 on reading graphs has been deleted. Mean, median, and mode are now covered in the appendix so that they may be introduced at any point in the course. Percent applications are included in problem solving sections.

- New Section 2.7, "Graphing Piecewise-Defined Functions and Shifting and Reflecting Graphs of Functions," has been added to Chapter 2 to better prepare students in this course and subsequent courses. Some formal knowledge on shifting and reflecting of graphs also helps students when using their graphing calculators.

- A new appendix, "Stretching and Compressing Graphs of Absolute Value Functions," has been added to support extended coverage in Section 2.7.

- Section 2.2's discussion of the domain and range of functions has been revised. Set builder notation is now used. All interval notation is formally introduced in Section 3.2.

- A new Section 4.5, "Systems of Linear Inequalities," has been added to Chapter 4. Determinants are now covered in Appendix D.

- There is now a section, 6.4, covering long division and synthetic division. This section contains mixed exercises requiring students to decide what method of division to use.

- Appendix A.2 is a Practice Final Exam that is an excellent review for students. The entire exam is worked by the author on the Interactive DVD/CD Lecture Series.

- All applications have been updated throughout, particularly in the data interpretation sections, 2.6 and 8.7.

KEY PEDAGOGICAL FEATURES

The following key features have been retained and/or updated for the Fourth Edition of the text:

Problem Solving Process   This is formally introduced in Chapter 1 with a four-step process that is integrated throughout the text. The four steps are **Understand, Translate, Solve,** and **Interpret.** The repeated use of these steps in a variety of examples shows their wide applicability. Reinforcing the steps can increase students' comfort level and confidence in tackling problems.

Exercise Sets Revised and Updated   The exercise sets have been carefully examined and extensively revised. Special focus was placed on making sure that even- and odd-numbered exercises are paired.

Each text section ends with an exercise set, usually divided into two parts. Both parts contain graded exercises. The **first part is carefully keyed** to at least one worked example in the text. Once a student has gained confidence in a skill, **the second part contains exercises not keyed to examples.** Exercises and examples marked with a 🄢 have been worked out step-by-step by the author in the interactive videos that accompany this text.

Throughout the text exercises there is an emphasis on data and graphical interpretation via tables, charts, and graphs. The ability to interpret data and read and create a variety of types of graphs is developed gradually so students become comfortable with it. Similarly, throughout the text there is integration of geometric concepts, such as perimeter and area. Exercises and examples marked with a geometry icon ($\triangle$) have been identified for convenience.

Examples   Detailed, step-by-step examples were added, deleted, replaced, or updated as needed. Many of these reflect real life. Additional instructional support is provided in the annotated examples. We have included examples that show algebraic, numerical, and/or graphical approaches to solving and checking solutions. See Section 3.1, page 209.

Helpful Hints   Helpful Hints contain practical advice on applying mathematical concepts. Strategically placed where students are most likely to need immediate reinforcement, Helpful Hints help students avoid common trouble areas and mistakes.

Discover the Concept   These explorations help students recognize patterns or discover a concept on their own immediately before the concept is introduced.

Technology Notes   These notes contain specific suggestions for problem solving with technology.

Concept Checks This feature allows students to gauge their grasp of an idea as it is being presented in the text. Concept Checks stress conceptual understanding at the point-of-use and help suppress misconceived notions before they start. Answers appear at the bottom of the page. Exercises related to Concept Checks are now included in the exercise sets.

Mixed Practice Exercises These exercises combine objectives within a section. They require students to determine the problem type and strategy needed in order to solve them. In doing so, students need to think about key concepts to proceed with a correct method of solving—just as they would need to do on a test.

Concept Extensions These exercises require students to combine several skills or concepts to solve exercises in the section.

Integrated Reviews A unique, mid-chapter exercise set that helps students assimilate new skills and concepts that they have learned separately over several sections. These reviews provide yet another opportunity for students to work with "mixed" exercises as they master the topics.

Vocabulary Check Provides an opportunity for students to become more familiar with the use of mathematical terms as they strengthen their verbal skills. These appear at the end of each chapter before the Chapter Highlights.

Chapter Highlights Found at the end of every chapter, these contain key definitions and concepts with examples to help students understand and retain what they have learned and help them organize their notes and study for tests.

Chapter Review The end of every chapter contains a comprehensive review of topics introduced in the chapter. The Chapter Review offers exercises keyed to every section in the chapter, as well as **Mixed Review (NEW!)** exercises that are not keyed to sections.

Cumulative Review Follows every chapter in the text except Chapter 1. Each odd-numbered exercise contained in the Cumulative Review is an earlier worked example in the text that is referenced in the back of the book along with the answer.

Writing Exercises \ These exercises occur in almost every exercise set and require students to provide a written response to explain concepts or justify their thinking.

Applications Real-world and real-data applications have been thoroughly updated and many new applications are included. These exercises occur in almost every exercise set and show the relevance of mathematics and help students gradually, and continuously, develop their problem solving skills.

Review and Preview Exercises These exercises occur in each exercise set (except Chapter 1) and are keyed to earlier sections. They review concepts learned earlier in the text that will be needed in the next section or chapter.

Exercise Set Resource Icons at the opening of each exercise set remind students of the resources available for extra practice and support:

See Student Resource descriptions, pages xxi–xxii, for details on the individual resources available.

Exercise Icons These icons facilitate the assignment of specialized exercises and let students know what resources can support them.

 Video icon: exercise worked on Martin-Gay's Interactive DVD/CD Lectures.

△ Triangle icon: identifies exercises involving geometric concepts.

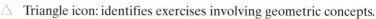

Pencil icon: indicates a written response is needed.

Calculator icons: optional exercises intended to be solved using a scientific or graphing calculator.

Group Activities   Found at the end of each chapter, these activities are for individual or group completion, and are usually hands-on or data-based activities that extend the concepts found in the chapter, allowing students to make decisions and interpretations and to think and write about algebra.

A Word about Textbook Design and Student Success

The design of developmental mathematics textbooks has become increasingly important. As students and instructors have told Pearson in focus groups and market research surveys, these textbooks cannot look "cluttered" or "busy." A "busy" design can distract a student from what is most important in the text. It can also heighten math anxiety.

As a result of the conversations and meetings we have had with students and instructors, we concluded the design of this text should be understated and focused on the most important pedagogical elements. Students and instructors helped us to identify the primary elements that are central to student success. These primary elements include:

- Exercise Sets
- Examples and Practice Problems
- Helpful Hints
- Rules, Property, and Definition boxes

As you will notice in this text, these primary features are the most prominent elements in the design. We have made every attempt to make sure these elements are the features the eye is drawn to. The remaining features, the secondary elements in the design, blend into the "fabric" or "grain" of the overall design. These secondary elements complement the primary elements without becoming distractions.

Pearson's thanks goes to all of the students and instructors (as noted by the author in Acknowledgments) who helped us develop the design of this text. At every step in the design process, their feedback proved valuable in helping us to make the right decisions. Thanks to your input, we're confident the design of this text will be both practical and engaging as it serves its educational and learning purposes.

Sincerely,

Paul Murphy

Editor-in-Chief
Developmental Mathematics
Pearson Arts & Sciences

INSTRUCTOR AND STUDENT RESOURCES

The following resources are available to help instructors and students use this text more effectively.

INSTRUCTOR RESOURCES

Annotated Instructor's Edition

- Answers to all exercises printed on the same text page or in the Graphing Answer Section.
- Teaching Tips throughout the text placed at key points.

- Classroom Examples paired with each text example for use as an additional resource during class.
- General tips and suggestions for classroom or group activities.
- Graphing Answer Section includes graphical answers and answers to the Group Activities.

Instructor's Solutions Manual

- Solutions to the even-numbered exercises
- Solutions to every Vocabulary Check exercise
- Solutions to every Practice exercise
- Solutions to every exercise in the Integrated Reviews, Chapter Reviews, Chapter Tests, and Cumulative Reviews

Instructor's Resource Manual with Tests and Mini-Lectures

- **NEW!** Includes Mini-Lectures for every section from the text
- Additional Exercises now 3 forms per section, to help instructors support students at different skill and ability levels.
- Free-Response Test Forms, Multiple Choice Test Forms, and Group Activities
- Answers to all items

Martin-Gay's Instructor-to-Instructor Videos

- Text author Elayn Martin-Gay presents tips, hints, and suggestions for engaging students and presenting key topics.
- Available as part of the Instructor Resource Kit.
- Contact your sales representative for more information.

Instructor Resource Kit

The Martin-Gay Instructor Resource Kit contains tools and resources to help instructors succeed in the classroom. The kit includes:

- Instructor-to-Instructor CD Videos that offer tips, suggestions, and strategies for engaging students and presenting key topics
- PDF files of the Instructor's Solutions Manual and the Instructor's Resource Manual
- Powerpoint Lecture Slides and Active Learning Questions

MYMATHLAB®

(Instructor Version 0-13-147898-2)

MyMathLab is a series of text-specific, easily customizable, online courses for Pearson Education's textbooks in mathematics and statistics. Powered by CourseCompass™ (our online teaching and learning environment) and MathXL® (our online homework, tutorial, and assessment system), MyMathLab gives you the tools you need to deliver all or a portion of your course online, whether your students are in a lab setting or working from home. MyMathLab provides a rich and flexible set of course materials, featuring free-response exercises that are algorithmically generated for unlimited practice and mastery. Students can also use online tools, such as video lectures, animations, and a multimedia textbook, to independently improve their understanding and performance. Instructors can use MyMathLab's homework and test managers to select and assign online exercises correlated directly to the textbook, and they can also create and assign their own online exercises and import TestGen tests for added flexibility. MyMathLab's online gradebook—designed specifically for mathematics and statistics— automatically tracks students' homework and test results and gives the instructor control over how to calculate final grades. Instructors can also add offline (paper-and-pencil) grades to the gradebook. **Includes access to the Pearson Tutor Center. Students**

can receive tutoring via toll free phone, fax, e-mail, and Internet. MyMathLab is available to qualified adopters. For more information, visit our website at www.mymathlab .com or contact your sales representative.

MATHXL®

(Instructor version 0-13-147895-8)

MathXL® is a powerful online homework, tutorial, and assessment system that accompanies Pearson Education's textbooks in mathematics and statistics. With MathXL, instructors can create, edit, and assign online homework and tests using algorithmically generated exercises correlated at the objective level to the textbook. They can also create and assign their own online exercises and import TestGen tests for added flexibility. All student work is tracked in MathXL's online gradebook. Students can take chapter tests in MathXL and receive personalized study plans based on their test results. The study plan diagnoses weaknesses and links students directly to tutorial exercises for the objectives they need to study and retest. Students can also access supplemental animations and video clips directly from selected exercises. MathXL is available to qualified adopters. For more information, visit our website at www.mathxl.com, or contact your sales representative.

INTERACT MATH TUTORIAL WEBSITE: WWW.INTERACTMATH.COM

Get practice and tutorial help online! This interactive tutorial website provides algorithmically generated practice exercises that correlate directly to the exercises in the textbook. Students can retry an exercise as many times as they like with new values each time for unlimited practice and mastery. Every exercise is accompanied by an interactive guided solution that provides helpful feedback for incorrect answers, and students can also view a worked-out sample problem that steps them through an exercise similar to the one they're working on.

TESTGEN®

TestGen enables instructors to build, edit, print, and administer tests using a computerized bank of questions developed to cover all the objectives of the text. TestGen is algorithmically based, allowing instructors to create multiple but equivalent versions of the same question or test with the click of a button. Instructors can also modify test bank questions or add new questions. Tests can be printed or administered online. The software and test bank are available for download from Pearson Education's online catalog.

STUDENT RESOURCES

Student Solutions Manual

- Solutions to the odd-numbered section exercises
- Solutions to the Practice exercises
- Solutions to the Vocabulary and Readiness Check
- Solutions to every exercise found in the Chapter Reviews, Chapter Tests, Cumulative Reviews, and Integrated Reviews

MARTIN-GAY'S INTERACTIVE DVD/CD LECTURE SERIES

Martin-Gay's video series has been comprehensively updated to address the way today's students study and learn. The new videos offer students active learning at their pace, with the following resources and more:

- **A complete lecture** for each section of the text, presented by Elayn Martin-Gay. Students can easily review a section or a specific topic before a homework assignment, quiz, or test. Exercises in the text marked with the ▒ are worked out on the video.

- A **new interface** with menu and navigation features allows students to quickly find and focus on examples and exercises they need to review.
- Pop-ups reinforce key terms and definitions.
- A new **Practice Final Exam Video** helps students prepare for the final exam. This Practice Final Exam is included in the text in Appendix A.2. At the click of a button, students can watch the full solutions to each exercise on the exam that they need help with. Overview clips provide a brief overview on how to approach different problem types.
- **Interactive Concept Checks** allow students to check their understanding of essential concepts. These multiple choice exercises focus on common misunderstandings. After making their answer selection, students are told whether they're correct or not, and why! Elayn also presents the full solution.
- **Study Skills Builders** help students develop effective study habits and reinforce the advice provided in Section 1.1, "Tips for Success in Mathematics," found in the text and video.
- **Close-captioned in Spanish and English**

Intermediate Algebra: A Graphing Approach, Fourth Edition *Student Study Pack*

The Student Study Pack includes:

- Martin-Gay's Interactive DVD/CD Lecture Series
- Student Solutions Manual
- Pearson Tutor Center

Chapter Test Prep Video CD — Standalone TEST PREP VIDEO

- Includes fully worked-out solutions to every problem from each Chapter Test in the text.

MATHXL® TUTORIALS ON CD

This interactive tutorial CD-ROM provides algorithmically generated practice exercises that are correlated at the objective level to the exercises in the textbook. Every practice exercise is accompanied by an example and a guided solution designed to involve students in the solution process. Selected exercises may also include a video clip to help students visualize concepts. The software provides helpful feedback for incorrect answers and can generate printed summaries of students' progress.

INTERACT MATH TUTORIAL WEBSITE: WWW.INTERACTMATH.COM

Get practice and tutorial help online! This interactive tutorial website provides algorithmically generated practice exercises that correlate directly to the exercises in the textbook. Students can retry an exercise as many times as they like with new values each time for unlimited practice and mastery. Every exercise is accompanied by an interactive guided solution that provides helpful feedback for incorrect answers, and student can also view a worked-out sample problem that steps them through an exercise similar to the one they're working on.

ACKNOWLEDGMENTS

There are many people who helped me develop this text, and I will attempt to thank some of them here. Carrie Green and Edutorial Services were *invaluable* for contributing to the overall accuracy of the text. Suellen Robinson, Lisa Collette, and Kim Lane were *invaluable* for their many suggestions and contributions during the development and writing of this Fourth Edition. Karin Kipp provided guidance throughout the production process.

A special thanks to my Editor-in-Chief, Paul Murphy, for all of his assistance, support, and contributions to this project. A very special thank you goes to my Sponsoring Editor, Mary Beckwith, for being there 24/7/365, as my students say. Last, my thanks to the staff at Pearson Education for all their support: Linda Behrens, Ilene Kahn, Audra Walsh, Juan López, Mike Fruhbeis, Kenny Beck, Richard Bretan, Tom Benfatti, Kate Valentine, Michelle Renda, Chris Hoag, and Greg Tobin.

I would like to thank the following reviewers for their input and suggestions:

Sandi Athanassiou, *University of Missouri—Columbia*
Michelle Beerman, *Pasco Hernandez Community College*
Monika Bender, *Central Texas College*
Bob Hervey, *Hillsborough Community College*
Jorge Romero, *Hillsborough Community College*
Joseph Wakim, *Brevard Community College*
Flo Wilson, *Central Texas College*
Marie Caruso and students, *Middlesex Community College*

I would also like to thank the following dedicated group of instructors who participated in our focus groups, Martin-Gay Summits, and our design review. Their feedback and insights have helped to strengthen this text. These instructors include:

Cedric Atkins, *Mott Community College*
Michelle Beerman, *Pasco Hernandez Community College*
Laurel Berry, *Bryant & Stratton*
John Beyers, *University of Maryland University College*
Lisa and Bob Brown, *Community College Baltimore County–Essex*
Gail Burkett, *Palm Beach Community College*
Cheryl Cantwell, *Seminole Community College*
Jackie Cohen, *Augusta State*
Julie Dewan, *Mohawk Community College*
Janice Ervin, *Central Piedmont Community College*
Karen Estes, *St. Petersburg College*
Cindy Gaddis, *Tyler Junior College*
Pauline Hall, *Iowa State*
Sonya Johnson, *Central Piedmont Community College*
Irene Jones, *Fullerton College*
Paul Jones, *University of Cincinnati*
Nancy Lange, *Inver Hills Community College*
Sandy Lofstock, *St. Petersburg College*
Jean McArthur, *Joliet Junior College*
Marcia Molle, *Metropolitan Community College*
Greg Nguyen, *Fullerton College*
Linda Padilla, *Joliet Junior College*
Rena Petrello, *Moorpark College*
Ena Salter, *Manatee Community College*
Carole Shapero, *Oakton Community College*
Ann Smallen, *Mohawk Community College*
Jennifer Strehler, *Oakton Community College*
Tanomo Taguchi, *Fullerton College*
Sam Tinsley, *Richland College*
Linda Tucker, *Rose State College*
Leigh Ann Wheeler, *Greenville Technical Community College*
Jenny Wilson, *Tyler Junior College*
Valerie Wright, *Central Piedmont Community College*

A special thank you to those students who participated in our design review: Katherine Browne, Mike Bulfin, Nancy Canipe, Ashley Carpenter, Jeff Chojnachi, Roxanne Davis, Mike Dieter, Amy Dombrowski, Kay Herring, Todd Jaycox, Kaleena Levan, Matt Montgomery, Tony Plese, Abigail Polkinghorn, Harley Price, Eli Robinson, Avery Rosen, Robyn Schott, Cynthia Thomas, and Sherry Ward.

Elayn Martin-Gay
Margaret (Peg) Greene

ABOUT THE AUTHORS

Elayn Martin-Gay has taught mathematics at the University of New Orleans for more than 25 years. Her numerous teaching awards include the local University Alumni Association's Award for Excellence in Teaching, and Outstanding Developmental Educator at University of New Orleans, presented by the Louisiana Association of Developmental Educators.

Prior to writing textbooks, Elayn Martin-Gay developed an acclaimed series of lecture videos to support developmental mathematics students in their quest for success. These highly successful videos originally served as the foundation material for her texts. Today, the videos are specific to each book in the Martin-Gay series. The author has also created Chapter Test Prep Videos to help students during their most "teachable moment"—as they prepare for a test—along with Instructor-to-Instructor videos that provide teaching tips, hints, and suggestions for each developmental mathematics course, including basic mathematics, prealgebra, beginning algebra, and intermediate algebra.

Elayn is the author of 12 published textbooks as well as multimedia interactive mathematics, all specializing in developmental mathematics courses. She has participated as an author across the broadest range of educational materials: textbooks, videos, tutorial software, and courseware. This offers an opportunity of various combinations for an integrated teaching and learning package, offering great consistency for the student.

Margaret (Peg) Greene has taught mathematics at Florida Community College at Jacksonville for more than 24 years. Peg has also been an instructor of the College Instructors Training Network, an Ohio State Short course program for teaching college instructors how to enhance their teaching by using technology in the mathematics classroom. A recipient of excellence in teaching awards, Peg has experience that includes being a member of the NASA/AMATYC Coalition: I and II; teaching a telecourse; teaching online courses as well as training instructors to teach via the online environment; developing a video series on using the graphing calculator; conducting workshops; authoring numerous publications; and teaching at the Southeast Center for Cooperative Learning, certified through the University of Minnesota.

Application Index

xxv

Intermediate Algebra
A Graphing Approach

1

Real Numbers, Algebraic Expressions, and Equations

With an increase in population and our increase in purchase and use of electronic products, the percent of energy usage in electric form has quadrupled since 1940. In fact, the Energy Information Administration predicts that national electricity consumption will increase by about 40% by the year 2030. For example,

84% of households have DVD players

76% of households have at least one cell phone

62% of households have digital cameras

32% of households have MP3 players

Unfortunately, our electrical delivery system technology is aging and performing many functions that it was not originally designed to perform. For example,

70% of transmission lines and power transformers are 25 years or older

60% of circuit breakers are more than 30 years old

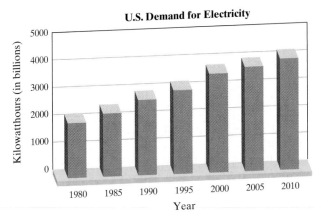

U.S. Demand for Electricity

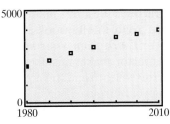

Same electricity data on a graphing calculator screen.

Source: U.S. Department of Energy

Mathematics is a tool for solving problems in such diverse fields as transportation, engineering, economics, medicine, business, and biology. We solve problems using mathematics by modeling real-world phenomena with mathematical equations or inequalities. Our ability to solve problems using mathematics, then, depends in part on our ability to solve equations and inequalities. In this chapter, we review operations on and properties of real numbers. We then solve linear equations and problems that can be modeled by linear equations.

In Section 1.7, Exercise 29, you will have the opportunity to use a table to project future demand for electricity.

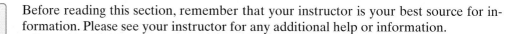

1.1 TIPS FOR SUCCESS IN MATHEMATICS

OBJECTIVES

1 Get ready for this course.

2 Understand some general tips for success.

3 Understand how to use this text.

4 Get help as soon as you need it.

5 Learn how to prepare for and take an exam.

6 Develop good time management.

Before reading this section, remember that your instructor is your best source for information. Please see your instructor for any additional help or information.

OBJECTIVE 1 ▶ Getting ready for this course. Now that you have decided to take this course, remember that a *positive attitude* will make all the difference in the world. Your belief that you can succeed is just as important as your commitment to this course. Make sure you are ready for this course by having the time and positive attitude that it takes to succeed.

Next, make sure you have scheduled your math course at a time that will give you the best chance for success. For example, if you are also working, you may want to check with your employer to make sure that your work hours will not conflict with your course schedule. Also, schedule your class during a time of day when you are more attentive and do your best work.

On the day of your first class period, double-check your schedule and allow yourself extra time to arrive in case of traffic problems or difficulty locating your classroom. Make sure that you bring at least your textbook, paper, and a writing instrument. Are you required to have a lab manual, graph paper, calculator, or some other supply besides this text? If so, also bring this material with you.

OBJECTIVE 2 ▶ General tips for success. Below are some general tips that will increase your chance for success in a mathematics class. Many of these tips will also help you in other courses you may be taking.

Exchange names and phone numbers with at least one other person in class. This contact person can be a great help if you miss an assignment or want to discuss math concepts or exercises that you find difficult.

Choose to attend all class periods and be on time. If possible, sit near the front of the classroom. This way, you will see and hear the presentation better. It may also be easier for you to participate in classroom activities.

Do your homework. You've probably heard the phrase "practice makes perfect" in relation to music and sports. It also applies to mathematics. You will find that the more time you spend solving mathematics problems, the easier the process becomes. Be sure to schedule enough time to complete your assignments before the next class period.

Check your work. Review the steps you made while working a problem. Learn to check your answers to the original problems. You may also compare your answers with the answers to selected exercises section in the back of the book. If you have made a mistake, try to figure out what went wrong. Then correct your mistake. If you can't find what went wrong, don't erase your work or throw it away. Bring your work to your instructor, a tutor in a math lab, or a classmate. It is easier for someone to find where you had trouble if they look at your original work.

Learn from your mistakes and be patient with yourself. Everyone, even your instructor, makes mistakes. (That definitely includes me—Elayn Martin-Gay.) Use your errors to learn and to become a better math student. The key is finding and understanding your errors.

Was your mistake a careless one, or did you make it because you can't read your own math writing? If so, try to work more slowly or write more neatly and make a conscious effort to carefully check your work.

Did you make a mistake because you don't understand a concept? Take the time to review the concept or ask questions to better understand it.

Did you skip too many steps? Skipping steps or trying to do too many steps mentally may lead to preventable mistakes.

Know how to get help if you need it. It's all right to ask for help. In fact, it's a good idea to ask for help whenever there is something that you don't understand. Make sure you know when your instructor has office hours and how to find his or her office. Find out whether math tutoring services are available on your campus. Check out the hours, location, and requirements of the tutoring service. Videotapes and software are available with this text. Learn how to access these resources.

Organize your class materials, including homework assignments, graded quizzes and tests, and notes from your class or lab. All of these items will be valuable references throughout your course especially when studying for upcoming tests and the final exam. Make sure that you can locate these materials when you need them.

Read your textbook before class. Reading a mathematics textbook is unlike leisure reading such as reading a novel or newspaper. Your pace will be much slower. It is helpful to have paper and a pencil with you when you read. Try to work out examples on your own as you encounter them in your text. You should also write down any questions that you want to ask in class. When you read a mathematics textbook, some of the information in a section may be unclear. But when you hear a lecture or watch a videotape on that section, you will understand it much more easily than if you had not read your text beforehand.

Don't be afraid to ask questions. Instructors are not mind readers. Many times we do not know a concept is unclear until a student asks a question. You are not the only person in class with questions. Other students are normally grateful that someone has spoken up.

Hand in assignments on time. This way you can be sure that you will not lose points for being late. Show every step of a problem and be neat and organized. Also be sure that you understand which problems are assigned for homework. You can always double-check this assignment with another student in your class.

OBJECTIVE 3 ▶ Using this text. There are many helpful resources that are available to you in this text. It is important that you become familiar with and use these resources. They should increase your chances for success in this course.

- The main section of exercises in each exercise set is referenced by an example(s). Use this referencing if you have trouble completing an assignment from the exercise set.

- If you need extra help in a particular section, look at the beginning of the section to see what videotapes and software are available.

- Make sure that you understand the meaning of the icons that are beside many exercises. The video icon 🎬 tells you that the corresponding exercise may be viewed on the videotape that corresponds to that section. The pencil icon ✎ tells you that this exercise is a writing exercise in which you should answer in complete sentences. The triangle icon △ tells you that the exercise involves geometry.

- Practice exercises are located immediately after a worked-out example in the text. This exercise is similar to the previous example and is a great way for you to immediately reinforce a learned concept. All answers to the practice exercises are found in the back of the text.

- Integrated Reviews in each chapter offer you a chance to practice—in one place—the many concepts that you have learned separately over several sections.

- There are many opportunities at the end of each chapter to help you understand the concepts of the chapter.

  Chapter Highlights contain chapter summaries and examples.

  Chapter Reviews contain review problems organized by section.

  Chapter Tests are sample tests to help you prepare for an exam.

  Cumulative Reviews are reviews consisting of material from the beginning of the book to the end of that particular chapter.

- *The Bigger Picture.* This feature contains the directions for building an outline to be used throughout the course. The purpose of this outline is to help you make the transition from thinking "section by section" to thinking about how the mathematics in this course is part of a bigger picture.

- *Study Skills Builder.* This feature is found at the end of many exercise sets. In order to increase your chance of success in this course, please read and answer the questions in the Study Skills Builder. For your convenience, the table below contains selected Study Skills Builder titles and their location.

Study Skills Builder Title	*Page of First Occurrence*
Learning New Terms	Page 17
Are You Prepared for a Test on Chapter 2?	Page 200
What to Do the Day of an Exam	Page 333
Have You Decided to Complete This Course Successfully?	Page 76
Organizing a Notebook	Page 63
How Are Your Homework Assignments Going?	Page 86
Are You Familiar with Your Textbook Supplements?	Page 281
How Well Do You Know Your Textbook?	Page 117
Are You Organized?	Page 149
Are You Satisfied with Your Performance on a Particular Quiz or Exam?	Page 309
Tips for Studying for an Exam	Page 193
Are You Getting All the Mathematics Help That You Need?	Page 362
How Are You Doing?	Page 438
Are You Prepared for Your Final Exam?	Page 711

See the preface at the beginning of this text for a more thorough explanation of the features of this text.

OBJECTIVE 4 ▶ Getting help. If you have trouble completing assignments or understanding the mathematics, get help as soon as you need it! This tip is presented as an objective on its own because it is so important. In mathematics, usually the material presented in one section builds on your understanding of the previous section. This means that if you don't understand the concepts covered during a class period, there is a good chance that you will not understand the concepts covered during the next class period. If this happens to you, get help as soon as you can.

Where can you get help? Many suggestions have been made in the section on where to get help, and now it is up to you to do it. Try your instructor, a tutoring center, or a math lab, or you may want to form a study group with fellow classmates. If you do decide to see your instructor or go to a tutoring center, make sure that you have a neat notebook and are ready with your questions.

OBJECTIVE 5 ▶ Preparing for and taking an exam. Make sure that you allow yourself plenty of time to prepare for a test. If you think that you are a little "math anxious," it may be that you are not preparing for a test in a way that will ensure success. The way that you prepare for a test in mathematics is important. To prepare for a test,

1. Review your previous homework assignments. You may also want to rework some of them.
2. Review any notes from class and section-level quizzes you have taken. (If this is a final exam, also review chapter tests you have taken.)
3. Review concepts and definitions by reading the Highlights at the end of each chapter.
4. Practice working out exercises by completing the Chapter Review found at the end of each chapter. (If this is a final exam, go through a Cumulative Review. There is

one found at the end of each chapter except Chapter 1. Choose the review found at the end of the latest chapter that you have covered in your course.) *Don't stop here!*

5. It is important that you place yourself in conditions similar to test conditions to find out how you will perform. In other words, as soon as you feel that you know the material, get a few blank sheets of paper and take a sample test. There is a Chapter Test available at the end of each chapter. During this sample test, do not use your notes or your textbook. Once you complete the Chapter Test, check your answers in the back of the book. If any answer is incorrect, there is a CD available with each exercise of each chapter test worked. Use this CD or your instructor to correct your sample test. Your instructor may also provide you with a review sheet. If you are not satisfied with the results, study the areas that you are weak in and try again.

6. Get a good night's sleep before the exam.

7. On the day of the actual test, allow yourself plenty of time to arrive at where you will be taking your exam.

When taking your test,

1. Read the directions on the test carefully.

2. Read each problem carefully as you take the test. Make sure that you answer the question asked.

3. Pace yourself by first completing the problems you are most confident with. Then work toward the problems you are least confident with. Watch your time so you do not spend too much time on one particular problem.

4. If you have time, check your work and answers.

5. Do not turn your test in early. If you have extra time, spend it double-checking your work.

OBJECTIVE 6 ▶ Managing your time. As a college student, you know the demands that classes, homework, work, and family place on your time. Some days you probably wonder how you'll ever get everything done. One key to managing your time is developing a schedule. Here are some hints for making a schedule:

1. Make a list of all of your weekly commitments for the term. Include classes, work, regular meetings, extracurricular activities, etc. You may also find it helpful to list such things as laundry, regular workouts, grocery shopping, etc.

2. Next, estimate the time needed for each item on the list. Also make a note of how often you will need to do each item. Don't forget to include time estimates for reading, studying, and homework you do outside of your classes. You may want to ask your instructor for help estimating the time needed.

3. In the following exercise set, you are asked to block out a typical week on the schedule grid given. Start with items with fixed time slots like classes and work.

4. Next, include the items on your list with flexible time slots. Think carefully about how best to schedule some items such as study time.

5. Don't fill up every time slot on the schedule. Remember that you need to allow time for eating, sleeping, and relaxing! You should also allow a little extra time in case some items take longer than planned.

6. If you find that your weekly schedule is too full for you to handle, you may need to make some changes in your workload, classload, or in other areas of your life. You may want to talk to your advisor, manager or supervisor at work, or someone in your college's academic counseling center for help with such decisions.

Note: In this chapter, we begin a feature called Study Skills Builder. The purpose of this feature is to remind you of some of the information given in this section and to further expand on some topics in this section.

1.1 EXERCISE SET

PRACTICE **WATCH** **DOWNLOAD** **READ** **REVIEW**

1. What is your instructor's name?

2. What are your instructor's office location and office hours?

3. What is the best way to contact your instructor?

4. What does the ◥ icon mean?

5. What does the ◉ icon mean?

6. What does the △ icon mean?

7. Where are answers located in this text?

8. What Exercise Set answers are available to you in the answers section?

9. What Chapter Review, Chapter Test, and Cumulative Review answers are available to you in the answer section?

10. Search after the solution of any worked example in this text. What are Practice exercises?

11. When might be the best time to work a Practice exercise?

12. Where are the answers to Practice exercises?

13. List any similarities between a Practice exercise and the worked example before it.

14. Go to the Highlights section at the end of this chapter. Describe how this section may be helpful to you when preparing for a test.

15. Do you have the name and contact information of at least one other student in class?

16. Will your instructor allow you to use a calculator in this class?

17. Are videotapes, CDs, and/or tutorial software available to you? If so, where?

18. Is there a tutoring service available? If so, what are its hours?

19. Have you attempted this course before? If so, write down ways that you might improve your chance of success during this second attempt.

20. List some steps that you can take if you begin having trouble understanding the material or completing an assignment.

21. Read or reread objective 6 and fill out the schedule grid below.

22. Study your filled-out grid from Exercise 21. Decide whether you have the time necessary to successfully complete this course and any other courses you may be registered for.

	Monday	Tuesday	Wednesday	Thursday	Friday	Saturday	Sunday
4:00 A.M.							
5:00 A.M.							
6:00 A.M.							
7:00 A.M.							
8:00 A.M.							
9:00 A.M.							
10:00 A.M.							
11:00 A.M.							
12:00 P.M.							
1:00 P.M.							
2:00 P.M.							
3:00 P.M.							
4:00 P.M.							
5:00 P.M.							
6:00 P.M.							
7:00 P.M.							
8:00 P.M.							
9:00 P.M.							
10:00 P.M.							
11:00 P.M.							
Midnight							
1:00 A.M.							
2:00 A.M.							
3:00 A.M.							

1.2 ALGEBRAIC EXPRESSIONS AND SETS OF NUMBERS

OBJECTIVES

1 Identify and evaluate algebraic expressions.

2 Identify natural numbers, whole numbers, integers, and rational and irrational real numbers.

3 Find the absolute value of a number.

4 Find the opposite of a number.

5 Write phrases as algebraic expressions.

TECHNOLOGY NOTE

Throughout this text, we assume that students have access to a graphing utility. Technology notes such as this one will appear often to alert students to possible commands that may be available or to alert students to special considerations that they need to watch for when using a graphing utility.

▶ **Helpful Hint**

Recall that 8443*t* means $8443 \cdot t$.

OBJECTIVE 1 ▶ Evaluating algebraic expressions. Recall that letters that represent numbers are called **variables.** An **algebraic expression** is formed by numbers and variables connected by the operations of addition, subtraction, multiplication, division, raising to powers, and/or taking roots. For example,

$$2x + 3, \quad \frac{x + 5}{6} - \frac{z^2}{y^2}, \quad \text{and} \quad \sqrt{y} - 1.6$$

are algebraic expressions or, more simply, expressions.

Algebraic expressions occur often during problem solving. For example, the B747-400 aircraft costs \$8443 per hour to operate. (*Source: The World Almanac*) The algebraic expression 8443*t* gives the total cost to operate the aircraft for *t* hours. To find the cost to operate the aircraft for 5.2 hours, for example, we replace the variable *t* with 5.2 and perform the indicated operation. This process is called **evaluating** an expression, and the result is called the **value** of the expression for the given replacement value.

Copyright The Boeing Company

In our example, when $t = 5.2$ hours,

$$8443t = 8443(5.2) = 43,903.60$$

Thus, it costs \$43,903.60 to operate the B747-400 aircraft for 5.2 hours.

△ **EXAMPLE 1** Finding the Area of a Tile

The research department of a flooring company is considering a new flooring design that contains parallelograms. The area of a parallelogram with base *b* and height *h* is *bh*. Find the area of a parallelogram with base 10 centimeters and height 8.2 centimeters.

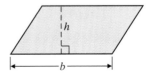

Solution We replace *b* with 10 and *h* with 8.2 in the algebraic expression *bh*.

$$bh = 10 \cdot 8.2 = 82$$

The area is 82 square centimeters. □

PRACTICE

1 The tile edging for a bathroom is in the shape of a triangle. The area of a triangle with base *b* and height *h* is $A = \frac{1}{2}bh$. Find the area of the tile if the base measures 3.5 cm and the height measures 8 cm.

Algebraic expressions simplify to different values depending on replacement values. (Order of operations is needed for simplifying many expressions. We fully review this in Section 1.3.)

EXAMPLE 2 Evaluate: $3x - y$ when $x = 15$ and $y = 4$.

Solution We replace x with 15 and y with 4 in the expression.

$$3x - y = 3 \cdot 15 - 4 = 45 - 4 = 41$$

PRACTICE
2 Evaluate $2p - q$ when $p = 17$ and $q = 3$.

When evaluating an expression to solve a problem, we often need to think about the kind of number that is appropriate for the solution. For example, if we are asked to determine the maximum number of parking spaces for a parking lot to be constructed, an answer of $98\frac{1}{10}$ is not appropriate because $\frac{1}{10}$ of a parking space is not realistic.

OBJECTIVE 2 ▶ Identifying common sets of numbers. Let's review some common sets of numbers and their graphs on a number line. To construct a number line, we draw a line and label a point 0 with which we associate the number 0. This point is called the **origin.** Choose a point to the right of 0 and label it 1. The distance from 0 to 1 is called the **unit distance** and can be used to locate more points. The **positive numbers** lie to the right of the origin, and the **negative numbers** lie to the left of the origin. The number 0 is neither positive nor negative.

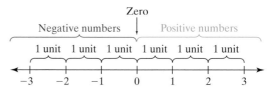

Concept Check ☑

Use the definitions of positive numbers, negative numbers, and zero to describe the meaning of *nonnegative numbers.*

A number is **graphed** on a number line by shading the point on the number line that corresponds to the number. Some common sets of numbers and their graphs include:

Natural numbers: $\{1, 2, 3, \dots\}$	
Whole numbers: $\{0, 1, 2, 3, \dots\}$	
Integers: $\{\dots, -3, -2, -1, 0, 1, 2, 3, \dots\}$	

The symbol ... is used in each set above. This symbol, consisting of three dots, is called an **ellipsis** and means to continue in the same pattern.

The members of a set are called its **elements.** When the elements of a set are listed, such as those displayed in the previous paragraph, the set is written in **roster** form. A set can also be written in **set builder notation,** which describes the members of a set but does not list them. The following set is written in set builder notation.

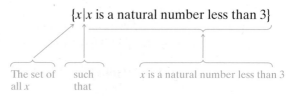

> ▶ **Helpful Hint**
>
> Use { } or ∅ to write the empty set. {∅} is **not** the empty set because it has one element: ∅.

This same set written in roster form is $\{1, 2\}$.

A set that contains *no* elements is called the **empty set** (or **null set**) symbolized by { } or ∅. The set

$$\{x \mid x \text{ is a month with 32 days}\}$$

is ∅ or { } because no month has 32 days. The set has no elements.

EXAMPLE 3 List the elements in each set.

a. $\{x \mid x \text{ is a natural number greater than } 100\}$

b. $\{x \mid x \text{ is a whole number between 1 and 6}\}$

Solution

a. $\{101, 102, 103, \ldots\}$ **b.** $\{2, 3, 4, 5\}$ □

PRACTICE
3 List the elements in each set.

a. $\{x \mid x \text{ is a whole number between 5 and 10}\}$

b. $\{x \mid x \text{ is a natural number greater than 40}\}$

The symbol ∈ is used to denote that an element is in a particular set. The symbol ∈ is read as "is an element of." For example, the true statement

$$3 \text{ is an element of } \{1, 2, 3, 4, 5\}$$

can be written in symbols as

$$3 \in \{1, 2, 3, 4, 5\}$$

The symbol ∉ is read as "is not an element of." In symbols, we write the true statement "p is not an element of $\{a, 5, g, j, q\}$" as

$$p \notin \{a, 5, g, j, q\}$$

EXAMPLE 4 Determine whether each statement is true or false.

a. $3 \in \{x \mid x \text{ is a natural number}\}$ **b.** $7 \notin \{1, 2, 3\}$

Solution

a. True, since 3 is a natural number and therefore an element of the set.

b. True, since 7 is not an element of the set $\{1, 2, 3\}$. □

PRACTICE
4 Determine whether each statement is true or false.

a. $7 \in \{x \mid x \text{ is a natural number}\}$ **b.** $6 \notin \{1, 3, 5, 7\}$

We can use set builder notation to describe three other common sets of numbers.

> **Identifying Numbers**
>
> ***Real Numbers:*** $\{x \mid x \text{ corresponds to a point on the number line}\}$
>
>
> ***Rational numbers:*** $\left\{\dfrac{a}{b} \;\middle|\; a \text{ and } b \text{ are integers and } b \neq 0\right\}$
>
> ***Irrational numbers:*** $\{x \mid x \text{ is a real number and } x \text{ is not a rational number}\}$

> ▶ **Helpful Hint**
>
> Notice from the definition that all real numbers are either rational or irrational.

Every rational number can be written as a decimal that either repeats or terminates. For example,

Rational Numbers

$$\frac{1}{2} = 0.5 \qquad \frac{5}{4} = 1.25$$

$$\frac{2}{3} = 0.6666666\ldots = 0.\overline{6} \qquad \frac{1}{11} = 0.090909\ldots = 0.\overline{09}$$

An irrational number written as a decimal neither terminates nor repeats. For example, π and $\sqrt{2}$ are irrational numbers. Their decimal form neither terminates nor repeats. Decimal approximations of each are below:

Irrational Numbers

$$\pi \approx 3.141592\ldots \qquad \sqrt{2} \approx 1.414213\ldots$$

Notice that every integer is also a rational number since each integer can be written as the quotient of itself and 1:

$$3 = \frac{3}{1}, \quad 0 = \frac{0}{1}, \quad -8 = \frac{-8}{1}$$

Not every rational number, however, is an integer. The rational number $\frac{2}{3}$, for example, is not an integer. Some square roots are rational numbers and some are irrational numbers. For example, $\sqrt{2}$, $\sqrt{3}$, and $\sqrt{7}$ are irrational numbers while $\sqrt{25}$ is a rational number because $\sqrt{25} = 5 = \frac{5}{1}$. The set of rational numbers together with the set of irrational numbers make up the set of real numbers. To help you make the distinction between rational and irrational numbers, here are a few examples of each.

	Real Numbers	
Rational Numbers		**Irrational Numbers** (*Decimal Form Neither Terminates Nor Repeats*)
Numbers	***Equivalent Quotient of Integers, $\frac{a}{b}$*** (*Decimal Form Terminates or Repeats*)	
$-\frac{2}{3}$	$\frac{-2}{3}$ or $\frac{2}{-3}$	$\sqrt{5}$
$\sqrt{36}$	$\frac{6}{1}$	$\frac{\sqrt{6}}{7}$
5	$\frac{5}{1}$	$-\sqrt{13}$
0	$\frac{0}{1}$	π
1.2	$\frac{12}{10}$	$\frac{2}{\sqrt{3}}$
$3\frac{7}{8}$	$\frac{31}{8}$	

Some rational and irrational numbers are graphed below.

Earlier we mentioned that every integer is also a rational number. In other words, all the elements of the set of integers are also elements of the set of rational numbers. When this happens, we say that the set of integers, set Z, is a subset of the set of rational numbers, set Q. In symbols,

$$Z \subseteq Q$$
$$\underbrace{}_{\text{is a subset of}}$$

The natural numbers, whole numbers, integers, rational numbers, and irrational numbers are each a subset of the set of real numbers. The relationships among these sets of numbers are shown in the following diagram.

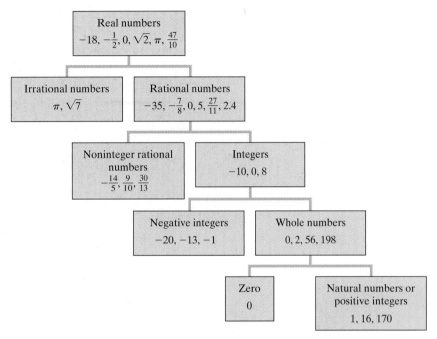

EXAMPLE 5   Determine whether the following statements are true or false.

a. 3 is a real number.

b. $\dfrac{1}{5}$ is an irrational number.

c. Every rational number is an integer.

d. $\{1, 5\} \subseteq \{2, 3, 4, 5\}$

Solution

a. True. Every whole number is a real number.

b. False. The number $\dfrac{1}{5}$ is a rational number, since it is in the form $\dfrac{a}{b}$ with a and b integers and $b \neq 0$.

c. False. The number $\dfrac{2}{3}$, for example, is a rational number, but it is not an integer.

d. False. The element 1 in the first set is not an element of the second set. ☐

PRACTICE
5   Determine whether the following statements are true or false.

a. -5 is a real number.

b. $\sqrt{8}$ is a rational number.

c. Every whole number is a rational number.

d. $\{2, 4\} \subset \{1, 3, 4, 7\}$

OBJECTIVE 3 ▶ Finding the absolute value of a number. The number line can also be used to visualize distance, which leads to the concept of absolute value. The **absolute**

value of a real number a, written as $|a|$, is the distance between a and 0 on the number line. Since distance is always positive or zero, $|a|$ is always positive or zero.

4 units   4 units

-4 -3 -2 -1 $\;0\;$ 1 2 3 4

Using the number line, we see that

$$|4| = 4 \qquad \text{and also} \qquad |-4| = 4$$

Why? Because both 4 and -4 are a distance of 4 units from 0.

An equivalent definition of the absolute value of a real number a is given next.

Absolute Value

The absolute value of a, written as $|a|$, is

$$|a| = \begin{cases} a \text{ if } a \text{ is 0 or a positive number} \\ -a \text{ if } a \text{ is a negative number} \end{cases}$$

the opposite of

EXAMPLE 6 Find each absolute value.

a. $|3|$ **b.** $\left|-\dfrac{1}{7}\right|$ **c.** $-|2.7|$ **d.** $-|-8|$ **e.** $|0|$

Solution

a. $|3| = 3$ since 3 is located 3 units from 0 on the number line.

b. $\left|-\dfrac{1}{7}\right| = \dfrac{1}{7}$ since $-\dfrac{1}{7}$ is $\dfrac{1}{7}$ units from 0 on the number line.

c. $-|2.7| = -2.7$. The negative sign outside the absolute value bars means to take the opposite of the absolute value of 2.7.

d. $-|-8| = -8$. Since $|-8|$ is 8, $-|-8| = -8$.

e. $|0| = 0$ since 0 is located 0 units from 0 on the number line.

TECHNOLOGY NOTE

To find absolute value using a graphing utility, look for a key named ABS or check in a math menu. Check on the use of parentheses with absolute value for your graphing utility.

The screens below show a check using a graphing utility. Many calculators have a subtraction key and a separate negative key. Make sure that you are using the negative key for this exercise.

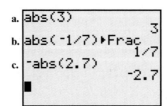

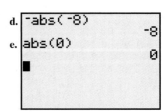

PRACTICE
6 Find each absolute value.

a. $|4|$ **b.** $\left|-\dfrac{1}{2}\right|$ **c.** $|1|$ **d.** $-|6.8|$ **e.** $-|-4|$

Answer to Concept Check:

$|14| = 14$ since the absolute value of a number is the distance between the number and 0 and distance cannot be negative.

Concept Check ☑

Explain how you know that $|14| = -14$ is a false statement.

OBJECTIVE 4 ▶ Finding the opposite of a number. The number line can also help us visualize opposites. Two numbers that are the same distance from 0 on the number line but are on opposite sides of 0 are called **opposites.**

See the definition illustrated on the number lines below.

> ▶ **Helpful Hint**
>
> The opposite of 0 is 0.

The opposite of 6.5 is −6.5.

The opposite of $\frac{2}{3}$ is $-\frac{2}{3}$.

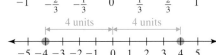

The opposite of −4 is 4.

TECHNOLOGY NOTE

Recall that most calculators have a subtraction key and a separate negative key. If this is so on your calculator, notice the difference in their displays. The subtraction sign is usually a longer horizontal dash and the negative sign is a shorter, raised horizontal dash.

> **Opposite**
> The opposite of a number a is the number $-a$.

Above we state that the opposite of a number a is $-a$. This means that the opposite of −4 is $-(-4)$. But from the number line above, the opposite of −4 is 4. This means that $-(-4) = 4$, and in general, we have the following property.

> **Double Negative Property**
> For every real number a, $-(-a) = a$.

EXAMPLE 7   Write the opposite of each number.

a. 8   **b.** $\frac{1}{5}$ **c.** −9.6

Solution

a. The opposite of 8 is −8.

b. The opposite of $\frac{1}{5}$ is $-\frac{1}{5}$.

c. The opposite of −9.6 is $-(-9.6) = 9.6$. □

PRACTICE
7   Write the opposite of each number.

a. 5.4   **b.** $-\frac{3}{5}$ **c.** 18

OBJECTIVE 5 ▶ Writing phrases as algebraic expressions. Often, solving problems involves translating a phrase to an algebraic expression. The following is a partial list of key words and phrases and their usual direct translations.

Selected Key Words/Phrases and Their Translations			
Addition	**Subtraction**	**Multiplication**	**Division**
sum	difference of	product	quotient
plus	minus	times	divide
added to	subtracted from	multiply	into
more than	less than	twice	ratio
increased by	decreased by	of	
total	less		

EXAMPLE 8 Translate each phrase to an algebraic expression. Use the variable x to represent each unknown number.

a. Eight times a number

b. Three more than eight times a number

c. The quotient of a number and -7

d. One and six-tenths subtracted from twice a number

e. Six less than a number

f. Twice the sum of four and a number

Solution

a. $8 \cdot x$ or $8x$

b. $8x + 3$

c. $x \div -7$ or $\dfrac{x}{-7}$

d. $2x - 1.6$ or $2x - 1\dfrac{6}{10}$

e. $x - 6$

f. $2(4 + x)$

PRACTICE

8 Translate each phrase to an algebraic expression. Use the variable x to represent the unknown number.

a. The product of 3 and a number

b. Five less than twice a number

c. Three and five-eighths more than a number

d. The quotient of a number and 2

e. Fourteen subtracted from a number

f. Five times the sum of a number and ten

VOCABULARY & READINESS CHECK

Use the choices below to fill in each blank. Not all choices will be used.

whole numbers	integers	rational number	a
natural numbers	value	irrational number	$-a$
absolute value	expression	variables	

1. Letters that represent numbers are called _____.

2. Finding the _____ of an expression means evaluating the expression.

3. The _____ of a number is that number's distance from 0 on the number line.

4. A(n) _____ is formed by numbers and variables connected by operations such as addition, subtraction, multiplication, division, raising to powers, and/or taking roots.

5. The _____ are $\{1, 2, 3, \ldots\}$.

6. The _____ are $\{0, 1, 2, 3, \ldots\}$.

7. The _____ are $\{\ldots -3, -2, -1, 0, 1, 2, 3, \ldots\}$.

8. The number $\sqrt{5}$ is a(n) _____.

9. The number $\dfrac{5}{7}$ is a(n) _____.

10. The opposite of a is _____.

1.2 EXERCISE SET

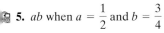

Find the value of each algebraic expression at the given replacement values. See Examples 1 and 2.

1. $5x$ when $x = 7$

2. $3y$ when $y = 45$

3. $9.8z$ when $z = 3.1$

4. $7.1a$ when $a = 1.5$

5. ab when $a = \dfrac{1}{2}$ and $b = \dfrac{3}{4}$

6. yz when $y = \dfrac{2}{3}$ and $z = \dfrac{1}{5}$

7. $3x + y$ when $x = 6$ and $y = 4$

8. $2a - b$ when $a = 12$ and $b = 7$

9. The aircraft B737-400 flies an average speed of 400 miles per hour.

The expression $400t$ gives the distance traveled by the aircraft in t hours. Find the distance traveled by the B737-400 in 5 hours.

10. The algebraic expression $1.5x$ gives the total length of shelf space needed in inches for x encyclopedias. Find the length of shelf space needed for a set of 30 encyclopedias.

11. Employees at Wal-Mart constantly reorganize and reshelve merchandise. In doing so, they calculate floor space needed for displays. The algebraic expression $l \cdot w$ gives the floor space needed in square units for a display that measures length l units and width w units. Calculate the floor space needed for a display whose length is 5.1 feet and whose width is 4 feet.

12. The algebraic expression $\dfrac{x}{5}$ can be used to calculate the distance in miles that you are from a flash of lightning, where x is the number of seconds between the time you see a flash of lightning and the time you hear the thunder. Calculate the distance that you are from the flash of lightning if you hear the thunder 2 seconds after you see the lightning.

13. The B737-400 aircraft costs $2948 dollars per hour to operate. The algebraic expression $2948t$ gives the total cost to operate the aircraft for t hours. Find the total cost to operate the B737-400 for 3.6 hours.

14. Flying the SR-71A jet, Capt. Elden W. Joersz, USAF, set a record speed of 2193.16 miles per hour. At this speed, the algebraic expression $2193.16t$ gives the total distance flown in t hours. Find the distance flown by the SR-71A in 1.7 hours.

List the elements in each set. See Example 3.

15. $\{x \mid x$ is a natural number less than $6\}$

16. $\{x \mid x$ is a natural number greater than $6\}$

17. $\{x \mid x$ is a natural number between 10 and $17\}$

18. $\{x \mid x$ is an odd natural number$\}$

19. $\{x \mid x$ is a whole number that is not a natural number$\}$

20. $\{x \mid x$ is a natural number less than $1\}$

21. $\{x \mid x$ is an even whole number less than $9\}$

22. $\{x \mid x$ is an odd whole number less than $9\}$

Graph each set on a number line.

23. $\{0, 2, 4, 6\}$

24. $\{-1, -2, -3\}$

25. $\left\{\dfrac{1}{2}, \dfrac{2}{3}\right\}$

26. $\{1, 3, 5, 7\}$

27. $\{-2, -6, -10\}$

28. $\left\{\dfrac{1}{4}, \dfrac{1}{3}\right\}$

29. In your own words, explain why the empty set is a subset of every set.

30. In your own words, explain why every set is a subset of itself.

List the elements of the set $\left\{3, 0, \sqrt{7}, \sqrt{36}, \dfrac{2}{5}, -134\right\}$ *that are also elements of the given set. See Example 4.*

31. Whole numbers

32. Integers

33. Natural numbers

34. Rational numbers

35. Irrational numbers

36. Real numbers

Place $\in$ or $\notin$ in the space provided to make each statement true. See Example 4.

37. -11 $\{x \mid x \text{ is an integer}\}$

38. 0 $\{x \mid x \text{ is a positive integer}\}$

39. -6 $\{2, 4, 6, \dots\}$

40. 12 $\{1, 2, 3, \dots\}$

41. 12 $\{1, 3, 5, \dots\}$

42. 0 $\{1, 2, 3, \dots\}$

43. $\dfrac{1}{2}$ $\{x \mid x \text{ is an irrational number}\}$

44. 0 $\{x \mid x \text{ is a natural number}\}$

Determine whether each statement is true or false. See Examples 4 and 5. Use the following sets of numbers.

$$N = \text{set of natural numbers}$$
$$Z = \text{set of integers}$$
$$I = \text{set of irrational numbers}$$
$$Q = \text{set of rational numbers}$$
$$\mathbb{R} = \text{set of real numbers}$$

45. $Z \subseteq \mathbb{R}$

46. $\mathbb{R} \subseteq N$

47. $-1 \in Z$

48. $\dfrac{1}{2} \in Q$

49. $0 \in N$

50. $Z \subseteq Q$

51. $\sqrt{5} \notin I$

52. $\pi \notin \mathbb{R}$

53. $N \subseteq Z$

54. $I \subseteq N$

55. $\mathbb{R} \subseteq Q$

56. $N \subseteq Q$

57. In your own words, explain why every natural number is also a rational number but not every rational number is a natural number.

58. In your own words, explain why every irrational number is a real number but not every real number is an irrational number.

Find each absolute value. See Example 6.

59. $-|2|$

60. $|8|$

61. $|-4|$

62. $|-6|$

63. $|0|$

64. $|-1|$

65. $-|-3|$

66. $-|-11|$

67. Explain why $-(-2)$ and $-|-2|$ simplify to different numbers.

68. The boxed definition of absolute value states that $|a| = -a$ if a is a negative number. Explain why $|a|$ is always nonnegative, even though $|a| = -a$ for negative values of a.

Write the opposite of each number. See Example 7.

69. -6.2

70. -7.8

71. $\dfrac{4}{7}$

72. $\dfrac{9}{5}$

73. $-\dfrac{2}{3}$

74. $-\dfrac{14}{3}$

75. 0

76. 10.3

Write each phrase as an algebraic expression. Use the variable x to represent each unknown number. See Example 8.

77. Twice a number.

78. Six times a number.

79. Five more than twice a number.

80. One more than six times a number.

81. Ten less than a number.

82. A number minus seven.

83. The sum of a number and two.

84. The difference of twenty-five and a number.

85. A number divided by eleven.

86. The quotient of a number and thirteen.

87. Twelve, minus three times a number.

88. Four, subtracted from three times a number.

89. A number plus two and three-tenths.

90. Fifteen and seven-tenths plus a number.

91. A number less than one and one-third.

92. Two and three-fourths less than a number.

93. The quotient of five and the difference of four and a number.

94. The quotient of four and the sum of a number and one.

95. Twice the sum of a number and three.

96. Eight times the difference of a number and nine.

CONCEPT EXTENSIONS

Use the bar graph below to complete the given table by estimating the millions of tourists predicted for each country. (Use whole numbers.)

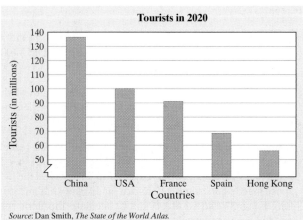

Source: Dan Smith, *The State of the World Atlas.*

97. China	
98. France	
99. Spain	
100. Hong Kong	

101. In your own words, explain how the graphing utility screen suggests that $\sqrt{25}$ is a rational number and $\sqrt{10}$ is an irrational number.

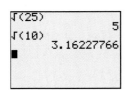

102. In your own words, explain how the graphing utility screen suggests that $\sqrt{\dfrac{4}{9}}$ is a rational number and $\sqrt{7}$ is an irrational number.

📖 STUDY SKILLS BUILDER

Learning New Terms

Many of the terms used in this text may be new to you. It will be helpful to make a list of new mathematical terms and symbols as you encounter them and to review them frequently. Placing these new terms (including page references) on 3 × 5 index cards might help you later when you're preparing for a quiz.

Answer the following.

1. Name one way you might place a word and its definition on a 3 × 5 card.

2. How do new terms stand out in this text so that they can be found?

1.3 OPERATIONS ON REAL NUMBERS

OBJECTIVES

1 Add and subtract real numbers.

2 Multiply and divide real numbers.

3 Evaluate expressions containing exponents.

4 Find roots of numbers.

5 Use the order of operations.

6 Evaluate algebraic expressions.

OBJECTIVE 1 ▶ Adding and subtracting real numbers. When solving problems, we often have to add real numbers. For example, if the New Orleans Saints lose 5 yards in one play, then lose another 7 yards in the next play, their total loss may be described by $-5 + (-7)$.

The addition of two real numbers may be summarized by the following.

> **Adding Real Numbers**
>
> **1.** To add two numbers with the *same sign,* add their absolute values and attach their common sign.
>
> **2.** To add two numbers with *different signs,* subtract the smaller absolute value from the larger absolute value and attach the sign of the number with the larger absolute value.

For example, to add $-5 + (-7)$, first add their absolute values.

$$|-5| = 5, \quad |-7| = 7, \quad \text{and} \quad 5 + 7 = 12$$

Next, attach their common negative sign.

$$-5 + (-7) = -12$$

(This represents a total loss of 12 yards for the New Orleans Saints in the example above.)

To find $(-4) + 3$, first subtract their absolute values.

$$|-4| = 4, |3| = 3, \quad \text{and} \quad 4 - 3 = 1$$

Next, attach the sign of the number with the larger absolute value.

$$(-4) + 3 = -1$$

EXAMPLE 1 Add.

a. $-3 + (-11)$ **b.** $3 + (-7)$ **c.** $-10 + 15$

d. $-8.3 + (-1.9)$ **e.** $-\dfrac{2}{3} + \dfrac{3}{7}$

Solution

a. $-3 + (-11) = -(3 + 11) = -14$ **b.** $3 + (-7) = -4$

c. $-10 + 15 = 5$ **d.** $-8.3 + (-1.9) = -10.2$

e. $-\dfrac{2}{3} + \dfrac{3}{7} = -\dfrac{14}{21} + \dfrac{9}{21} = -\dfrac{5}{21}$

The screens below show a check using a calculator.

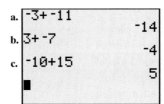
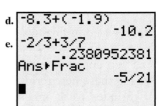

PRACTICE
1   Add.

a. $-6 + (-2)$ **b.** $5 + (-8)$ **c.** $-4 + 9$

d. $(-3.2) + (-4.9)$ **e.** $-\dfrac{3}{5} + \dfrac{2}{3}$

Subtraction of two real numbers may be defined in terms of addition.

> **Subtracting Real Numbers**
> If a and b are real numbers,
> $$a - b = a + (-b)$$

In other words, to subtract a real number, we add its opposite.

EXAMPLE 2 Subtract.

a. $2 - 8$ **b.** $-8 - (-1)$ **c.** $-11 - 5$ **d.** $10.7 - (-9.8)$

e. $\dfrac{2}{3} - \dfrac{1}{2}$ **f.** $1 - 0.06$ **g.** Subtract 7 from 4.

Solution   Add the opposite   Add the opposite

a. $2 - 8 = 2 + (-8) = -6$ **b.** $-8 - (-1) = -8 + (1) = -7$

c. $-11 - 5 = -11 + (-5) = -16$ **d.** $10.7 - (-9.8) = 10.7 + 9.8 = 20.5$

e. $\dfrac{2}{3} - \dfrac{1}{2} = \dfrac{2 \cdot 2}{3 \cdot 2} - \dfrac{1 \cdot 3}{2 \cdot 3} = \dfrac{4}{6} + \left(-\dfrac{3}{6} \right) = \dfrac{1}{6}$

f. $1 - 0.06 = 1 + (-0.06) = 0.94$ **g.** $4 - 7 = 4 + (-7) = -3$

The screen below shows a check of Example 2 **a.–c.** Notice the difference between the negative sign and the subtraction sign.

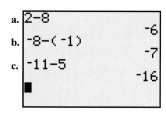

PRACTICE

2   Subtract.

a. $3 - 11$ **b.** $-6 - (-3)$ **c.** $-7 - 5$ **d.** $4.2 - (-3.5)$

e. $\dfrac{5}{7} - \dfrac{1}{3}$ **f.** $3 - 1.2$ **g.** Subtract 9 from 2.

To add or subtract three or more real numbers, add or subtract from left to right.

EXAMPLE 3   Simplify the following expressions.

a. $11 + 2 - 7$ **b.** $-5 - 4 + 2$

Solution

a. $11 + 2 - 7 = 13 - 7 = 6$ **b.** $-5 - 4 + 2 = -9 + 2 = -7$

PRACTICE

3   Simplify the following expressions.

a. $13 + 5 - 6$ **b.** $-6 - 2 + 4$

OBJECTIVE 2 ▶ Multiplying and dividing real numbers. In order to discover sign patterns when you multiply real numbers, recall that multiplication by a positive integer is the same as repeated addition. For example,

$$3(2) = 2 + 2 + 2 = 6$$
$$3(-2) = (-2) + (-2) + (-2) = -6$$

Notice here that $3(-2) = -6$. This illustrates that the product of two numbers with different signs is negative. We summarize sign patterns for multiplying any two real numbers as follows.

> **Multiplying Two Real Numbers**
> The product of two numbers with the *same* sign is positive.
> The product of two numbers with *different* signs is negative.

Also recall that the product of zero and any real number is zero.

$$0 \cdot a = 0$$

> **Product Property of 0**
> $0 \cdot a = 0$ Also $a \cdot 0 = 0$

EXAMPLE 4   Multiply.

a. $(-8)(-1)$ **b.** $-2\left(\dfrac{1}{6}\right)$ **c.** $-1.2(0.3)$ **d.** $0(-11)$

e. $\dfrac{1}{5}\left(-\dfrac{10}{11}\right)$ **f.** $(7)(1)(-2)(-3)$ **g.** $8(-2)(0)$

Solution

a. Since the signs of the two numbers are the same, the product is positive. Thus $(-8)(-1) = +8$, or 8.

b. Since the signs of the two numbers are different or unlike, the product is negative. Thus $-2\left(\dfrac{1}{6}\right) = -\dfrac{2}{6} = -\dfrac{1}{3}$.

c. $-1.2(0.3) = -0.36$

d. $0(-11) = 0$

e. $\dfrac{1}{5}\left(-\dfrac{10}{11}\right) = -\dfrac{10}{55} = -\dfrac{2}{11}$

f. To multiply three or more real numbers, you may multiply from left to right.

$$\begin{aligned} (7)(1)(-2)(-3) &= 7(-2)(-3) \\ &= -14(-3) \\ &= 42 \end{aligned}$$

g. Since zero is a factor, the product is zero.

$$(8)(-2)(0) = 0$$

The screen below shows options for checking parts **b** and **e** where fractions are involved.

```
b. -2*(1/6)
            -.3333333333
   Ans▶Frac
                    -1/3
e. (1/5)*(-10/11)▶F
   rac
                   -2/11
■
```

PRACTICE
4   Multiply.

a. $(-5)(3)$ **b.** $(-7)\left(-\dfrac{1}{14}\right)$ **c.** $5.1(-2)$ **d.** $14(0)$

e. $\left(-\dfrac{1}{4}\right)\left(\dfrac{8}{13}\right)$ **f.** $6(-1)(-2)(3)$ **g.** $5(-2.3)$

▶ **Helpful Hint**

The following sign patterns may be helpful when we are multiplying.

1. An odd number of negative factors gives a negative product.

2. An even number of negative factors gives a positive product.

Recall that $\dfrac{8}{4} = 2$ because $2 \cdot 4 = 8$. Likewise, $\dfrac{8}{-4} = -2$ because $(-2)(-4) = 8$. Also, $\dfrac{-8}{4} = -2$ because $(-2)4 = -8$, and $\dfrac{-8}{-4} = 2$ because $2(-4) = -8$. From these examples, we can see that the sign patterns for division are the same as for multiplication.

TECHNOLOGY NOTE

A calculator will give some type of error message when asked to divide by 0. For example:

```
ERR:DIVIDE BY 0
1█Quit
2:Goto
```

Dividing Two Real Numbers

The quotient of two numbers with the *same* sign is positive.
The quotient of two numbers with *different* signs is negative.

Also recall that division by a nonzero real number b is the same as multiplication by $\frac{1}{b}$. In other words,

$$\frac{a}{b} = a \cdot \frac{1}{b}$$

This means that to simplify $\frac{a}{b}$, we can divide by b or multiply by $\frac{1}{b}$. The nonzero numbers b and $\frac{1}{b}$ are called **reciprocals.** Notice that b *must* be a nonzero number. We do not define division by 0. For example, $5 \div 0$, or $\frac{5}{0}$, is undefined. To see why, recall that if $5 \div 0 = n$, a number, then $n \cdot 0 = 5$. This is not possible since $n \cdot 0 = 0$ for any number n, and never 5. Thus far we have learned that we cannot divide 5 or any other nonzero number by 0.

Can we divide 0 by 0? By the same reasoning, if $0 \div 0 = n$, a number, then $n \cdot 0 = 0$. This is true for any number n so that the quotient $0 \div 0$ would not be a single number. To avoid this, we say that

Division by 0 is undefined.

EXAMPLE 5 Divide.

a. $\dfrac{20}{-4}$ **b.** $\dfrac{-9}{-3}$ **c.** $-\dfrac{3}{8} \div 3$ **d.** $\dfrac{-40}{10}$ **e.** $\dfrac{-1}{10} \div \dfrac{-2}{5}$ **f.** $\dfrac{8}{0}$

Solution

a. Since the signs are different or unlike, the quotient is negative and $\dfrac{20}{-4} = -5$.

b. Since the signs are the same, the quotient is positive and $\dfrac{-9}{-3} = 3$.

c. $-\dfrac{3}{8} \div 3 = -\dfrac{3}{8} \cdot \dfrac{1}{3} = -\dfrac{1}{8}$ **d.** $\dfrac{-40}{10} = -4$

e. $\dfrac{-1}{10} \div \dfrac{-2}{5} = -\dfrac{1}{10} \cdot -\dfrac{5}{2} = \dfrac{1}{4}$ **f.** $\dfrac{8}{0}$ is undefined. □

PRACTICE
5 Divide.

a. $\dfrac{-16}{8}$ **b.** $\dfrac{-15}{-3}$ **c.** $-\dfrac{2}{3} \div 4$

d. $\dfrac{54}{-9}$ **e.** $-\dfrac{1}{12} \div \left(-\dfrac{3}{4}\right)$ **f.** $\dfrac{0}{-7}$

With sign rules for division, we can understand why the positioning of the negative sign in a fraction does not change the value of the fraction. For example,

$$\frac{-12}{3} = -4, \quad \frac{12}{-3} = -4, \quad \text{and} \quad -\frac{12}{3} = -4$$

Since all the fractions equal -4, we can say that

$$\frac{-12}{3} = \frac{12}{-3} = -\frac{12}{3}$$

In general, the following holds true.

> If a and b are real numbers and $b \neq 0$, then $\dfrac{a}{-b} = \dfrac{-a}{b} = -\dfrac{a}{b}$.

OBJECTIVE 3 ▶ Evaluating expressions containing exponents. Recall that when two numbers are multiplied, they are called **factors.** For example, in $3 \cdot 5 = 15$, the 3 and 5 are called factors.

A natural number *exponent* is a shorthand notation for repeated multiplication of the same factor. This repeated factor is called the **base,** and the number of times it is used as a factor is indicated by the **exponent.** For example,

$$\overset{\text{exponent}}{4^{3}} = \underbrace{4 \cdot 4 \cdot 4}_{4 \text{ is a factor } 3 \text{ times.}} = 64$$

base ⟋

TECHNOLOGY NOTE

Most graphing utilities use the symbol $^{\wedge}$ for the exponent key. Some graphing utilities also have an x^2 key for squaring, and cubing may be found in a math menu. However, the $^{\wedge}$ key may be used to raise the base to any power, even 2.

Exponents

If a is a real number and n is a natural number, then the **nth power of a,** or **a raised to the nth power,** written as a^{n}, is the product of n factors, each of which is a.

$$\overset{\text{exponent}}{a^{n}} = \underbrace{a \cdot a \cdot a \cdot a \cdot \cdots \cdot a}_{a \text{ is a factor } n \text{ times.}}$$

base ⟋

It is not necessary to write an exponent of 1. For example, 3 is assumed to be 3^{1}.

EXAMPLE 6   Evaluate each expression.

a. 3^{2} **b.** $\left(\dfrac{1}{2}\right)^{4}$ **c.** -5^{2}

d. $(-5)^{2}$ **e.** -5^{3} **f.** $(-5)^{3}$

Solution

a. $3^{2} = 3 \cdot 3 = 9$ **b.** $\left(\dfrac{1}{2}\right)^{4} = \left(\dfrac{1}{2}\right)\left(\dfrac{1}{2}\right)\left(\dfrac{1}{2}\right)\left(\dfrac{1}{2}\right) = \dfrac{1}{16}$

c. $-5^{2} = -(5 \cdot 5) = -25$ **d.** $(-5)^{2} = (-5)(-5) = 25$

e. $-5^{3} = -(5 \cdot 5 \cdot 5) = -125$ **f.** $(-5)^{3} = (-5)(-5)(-5) = -125$

The screens below show a check using a calculator.

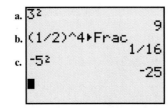

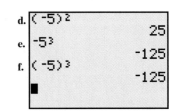

PRACTICE

6   Evaluate each expression.

a. 2^{3} **b.** $\left(\dfrac{1}{3}\right)^{2}$ **c.** -6^{2}

d. $(-6)^{2}$ **e.** -4^{3} **f.** $(-4)^{3}$

> ▶ **Helpful Hint**
> Be very careful when simplifying expressions such as -5^2 and $(-5)^2$.
>
> $$-5^2 = -(5 \cdot 5) = -25 \quad \text{and} \quad (-5)^2 = (-5)(-5) = 25$$
>
> Without parentheses, the base to square is 5, not -5.

TECHNOLOGY NOTE

Some graphing utilities contain a square root key as well as other root options under a menu such as a math menu.

OBJECTIVE 4 ▶ Finding roots of numbers. The opposite of squaring a number is taking the **square root** of a number. For example, since the square of 4, or 4^2, is 16, we say that a square root of 16 is 4. The notation $\sqrt{a}$ is used to denote the **positive,** or **principal, square root** of a nonnegative number a. We then have in symbols that $\sqrt{16} = 4$. The negative square root of 16 is written $-\sqrt{16} = -4$. The square root of a negative number, such as $\sqrt{-16}$ is not a real number. Why? There is no real number that, when squared, gives a negative number.

EXAMPLE 7 Find the square roots.

a. $\sqrt{9}$ **b.** $\sqrt{25}$ **c.** $\sqrt{\dfrac{1}{4}}$ **d.** $-\sqrt{36}$ **e.** $\sqrt{-36}$

Solution

a. $\sqrt{9} = 3$ since 3 is positive and $3^2 = 9$.   **b.** $\sqrt{25} = 5$ since $5^2 = 25$.

c. $\sqrt{\dfrac{1}{4}} = \dfrac{1}{2}$ since $\left(\dfrac{1}{2}\right)^2 = \dfrac{1}{4}$.   **d.** $-\sqrt{36} = -6$

e. $\sqrt{-36}$ is not a real number.

PRACTICE
7 Find the square roots.

a. $\sqrt{49}$ **b.** $\sqrt{\dfrac{1}{16}}$ **c.** $-\sqrt{64}$ **d.** $\sqrt{-64}$ **e.** $\sqrt{100}$

We can find roots other than square roots. Since 2 cubed, written as 2^3, is 8, we say that the **cube root** of 8 is 2. This is written as

$$\sqrt[3]{8} = 2.$$

Also, since $3^4 = 81$ and 3 is positive,

$$\sqrt[4]{81} = 3.$$

EXAMPLE 8 Find the roots.

a. $\sqrt[3]{27}$ **b.** $\sqrt[5]{1}$ **c.** $\sqrt[4]{16}$

Solution

a. $\sqrt[3]{27} = 3$ since $3^3 = 27$.

b. $\sqrt[5]{1} = 1$ since $1^5 = 1$.

c. $\sqrt[4]{16} = 2$ since 2 is positive and $2^4 = 16$.

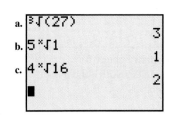

PRACTICE
8 Find the roots.

a. $\sqrt[3]{64}$ **b.** $\sqrt[5]{-1}$ **c.** $\sqrt[4]{10{,}000}$

Of course, as mentioned in Section 1.2, not all roots simplify to rational numbers. We study radicals further in Chapter 7.

OBJECTIVE 5 ▶ Using the order of operations. Expressions containing more than one operation are written to follow a particular agreed-upon **order of operations.** For example, when we write $3 + 2 \cdot 10$, we mean to multiply first, and then add.

Order of Operations

Simplify expressions using the order that follows. If grouping symbols such as parentheses are present, simplify expressions within those first, starting with the innermost set. If fraction bars are present, simplify the numerator and denominator separately.

1. Evaluate exponential expressions, roots, or absolute values in order from left to right.
2. Multiply or divide in order from left to right.
3. Add or subtract in order from left to right.

EXAMPLE 9 Simplify.

a. $20 \div 2 \cdot 10$ **b.** $1 + 2(1 - 4)^2$ **c.** $\dfrac{|-2|^3 + 1}{-7 - \sqrt{4}}$

Solution

a. Be careful! Here, we multiply or divide in order from left to right. Thus, divide, then multiply.

$$20 \div 2 \cdot 10 = 10 \cdot 10 = 100$$

b. Remember order of operations so that you are *not* tempted to add 1 and 2 first.

$$
\begin{aligned}
1 + 2(1 - 4)^2 &= 1 + 2(-3)^2 \quad \text{Simplify inside grouping symbols first.} \\
&= 1 + 2(9) \quad \text{Write } (-3)^2 \text{ as 9.} \\
&= 1 + 18 \quad \text{Multiply.} \\
&= 19 \quad \text{Add.}
\end{aligned}
$$

c. Simplify the numerator and the denominator separately; then divide.

$$
\begin{aligned}
\frac{|-2|^3 + 1}{-7 - \sqrt{4}} &= \frac{2^3 + 1}{-7 - 2} \quad \text{Write } |-2| \text{ as 2 and } \sqrt{4} \text{ as 2.} \\
&= \frac{8 + 1}{-9} \quad \text{Write } 2^3 \text{ as 8.} \\
&= \frac{9}{-9} = -1 \quad \text{Simplify the numerator, then divide.}
\end{aligned}
$$

PRACTICE
9 Simplify.

a. $14 - 3 \cdot 4$ **b.** $3(5 - 8)^2$ **c.** $\dfrac{|-5|^2 + 4}{\sqrt{4} - 3}$

Besides parentheses, other symbols used for grouping expressions are brackets [ ] and braces { }. These other grouping symbols are commonly used when we group expressions that already contain parentheses.

TECHNOLOGY NOTE

To evaluate expressions with a calculator, we sometimes need to insert parentheses that may not be shown in the expression. This is especially true when entering a fraction whose numerator or denominator contains more than one term. (See the screen below.)

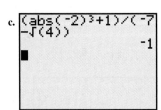

c.

A calculator check for Example 9c. Notice the use of parentheses.

EXAMPLE 10 Simplify: $3 - [(4 - 6) + 2(5 - 9)]$

Solution $3 - [(4 - 6) + 2(5 - 9)] = 3 - [-2 + 2(-4)]$ Simplify within the innermost sets of parentheses.

$$= 3 - [-2 + (-8)]$$
$$= 3 - [-10]$$
$$= 13 \qquad \square$$

**PRACTICE
10** Simplify: $5 - [(3 - 5) + 6(2 - 4)]$.

EXAMPLE 11 Simplify: $\dfrac{-5\sqrt{30 - 5} + (-2)^2}{4^2 + |7 - 10|}$

Solution Here, the fraction bar, radical sign, and absolute value bars serve as grouping symbols. Thus, we simplify within the radical sign and absolute value bars first, remembering to calculate above and below the fraction bar separately.

$$\frac{-5\sqrt{30 - 5} + (-2)^2}{4^2 + |7 - 10|} = \frac{-5\sqrt{25} + (-2)^2}{4^2 + |-3|} = \frac{-5 \cdot 5 + 4}{16 + 3} = \frac{-25 + 4}{16 + 3}$$
$$= \frac{-21}{19} \text{ or } -\frac{21}{19} \qquad \square$$

**PRACTICE
11** Simplify: $\dfrac{-2\sqrt{12 + 4} - (-3)^2}{6^2 + |1 - 9|}$

Concept Check ✓

True or false? If two different people use the order of operations to simplify a numerical expression and neither makes a calculation error, it is not possible that they each obtain a different result. Explain.

OBJECTIVE 6 ▶ **Evaluating algebraic expressions.** Recall from Section 1.2 that an algebraic expression is formed by numbers and variables connected by the operations of addition, subtraction, multiplication, division, raising to powers, and/or taking roots. Also, if numbers are substituted for the variables in an algebraic expression and the operations performed, the result is called the **value of the expression** for the given replacement values. This entire process is called **evaluating an expression.**

EXAMPLE 12 Evaluate each algebraic expression when $x = 2$, $y = -1$, and $z = -3$.

a. $z - y$ **b.** $-2z^2$ **c.** $\dfrac{2x + y}{z}$ **d.** $-x^2 - 4x$

Solution

a. $z - y = -3 - (-1) = -3 + 1 = -2$

b. $-2z^2 = -2(-3)^2$ Let $z = -3$.

$$= -2(9) \qquad \text{Write } (-3)^2 \text{ as } 9.$$
$$= -18 \qquad \text{Multiply.}$$

c. $\dfrac{2x + y}{z} = \dfrac{2(2) + (-1)}{-3} = \dfrac{4 + (-1)}{-3} = \dfrac{3}{-3} = -1$

d. $-x^2 - 4x = -2^2 - 4(2) = -4 - 8 = -12$

Answer to Concept Check:

true; answers may vary

How can we use a calculator to evaluate the expressions in Example 12? One way to evaluate Example 12 d with a calculator is to just enter the numerical expression and evaluate. See the screen below on the left.

Another way to evaluate $-x^2 - 4x$ when $x = 2$ is to store the value 2 in the variable x and then enter and evaluate the algebraic expression, as shown below on the right.

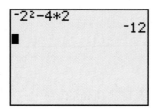

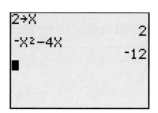

The value of $-x^2 - 4x$ when $x = 2$ is -12. □

PRACTICE
12 Evaluate each expression when $x = 16$ and $y = -5$. Check using your calculator.

a. $2x - 7y$ **b.** $-4y^2$ **c.** $\dfrac{\sqrt{x}}{y} - \dfrac{y}{x}$

Thus far, we have seen that we can evaluate an expression or perform a calculation mentally, using paper and pencil, or using a calculator. In general, how do we decide which method to use?

The following example shows that all three methods are useful.

EXAMPLE 13 Evaluate the expression $3x^2 + 4$ for the following values of x:

a. $x = 2$ **b.** $x = 12$ **c.** $x = 3.91$

Solution

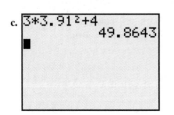

a. Never underestimate the power and speed of mental calculation. Replace x with 2 and mentally calculate x^2 to be 4, multiply by 3 to get 12, and add 4 to get 16. The value of $3x^2 + 4$ when $x = 2$ is 16.

b. Replace x with 12 and mentally calculate the square of 12 to be 144. Next, you might use paper and pencil to evaluate $3 \cdot 144$, which equals 432, and then add 4 to get 436. The value of $3x^2 + 4$ when $x = 12$ is 436.

c. Since the value of x is a decimal, using a calculator is the most appropriate choice. First, let's mentally approximate the value of the expression $3.91 \approx 4$, and the value of $3x^2 + 4$ at 4 is 52. The screen at the left shows that the value of $3x^2 + 4$ when $x = 3.91$ is 49.8643. This is close to our approximation of 52 and thus is reasonable. □

PRACTICE
13 Evaluate the expression $5x^2 + 1$ for the following values of x:

a. $x = 3$ **b.** $x = 11$ **c.** $x = 4.86$

Sometimes variables such as x_1 and x_2 will be used in this book. The small 1 and 2 are called **subscripts**. The variable x_1 can be read as "x sub 1," and the variable x_2 can be read as "x sub 2." The important thing to remember is that they are two different variables. For example, if $x_1 = -5$ and $x_2 = 7$, then

$$x_1 - x_2 = -5 - 7 = -12.$$

EXAMPLE 14   The algebraic expression $\dfrac{5(x-32)}{9}$ represents the equivalent temperature in degrees Celsius when x is the temperature in degrees Fahrenheit. Complete the following table by evaluating this expression at the given values of x.

Degrees Fahrenheit	x	-4	10	32
Degrees Celsius	$\dfrac{5(x-32)}{9}$			

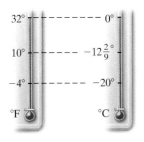

Calculations for Example 14.

Solution   To complete the table, evaluate $\dfrac{5(x-32)}{9}$ at each given replacement value.

When $x = -4$,

$$\frac{5(x-32)}{9} = \frac{5(-4-32)}{9} = \frac{5(-36)}{9} = -20$$

When $x = 10$,

$$\frac{5(x-32)}{9} = \frac{5(10-32)}{9} = \frac{5(-22)}{9} = -\frac{110}{9}$$

When $x = 32$,

$$\frac{5(x-32)}{9} = \frac{5(32-32)}{9} = \frac{5\cdot 0}{9} = 0$$

The completed table is

Degrees Fahrenheit	x	-4	10	32
Degrees Celsius	$\dfrac{5(x-32)}{9}$	-20	$-\dfrac{110}{9}$	0

Thus, $-4°\text{F}$ is equivalent to $-20°\text{C}$, $10°\text{F}$ is equivalent to $-\dfrac{110°}{9}$ C, and $32°\text{F}$ is equivalent to $0°\text{C}$.   □

PRACTICE
14   The algebraic expression $\dfrac{9}{5}x + 32$ represents the equivalent temperature in degrees Fahrenheit when x is the temperature in degrees Celsius. Complete the following table by evaluating this expression at the given values of x.

Degrees Celsius	x	-5	10	25
Degrees Fahrenheit	$\dfrac{9}{5}x + 32$			

VOCABULARY & READINESS CHECK

Choose the fraction(s) equivalent to the given fraction. (There may sometimes be more than one correct choice.)

1. $-\dfrac{1}{7}$ **a.** $\dfrac{-1}{-7}$ **b.** $\dfrac{-1}{7}$ **c.** $\dfrac{1}{-7}$ **d.** $\dfrac{1}{7}$

2. $\dfrac{-x}{y}$ **a.** $\dfrac{x}{-y}$ **b.** $-\dfrac{x}{y}$ **c.** $\dfrac{x}{y}$ **d.** $\dfrac{-x}{-y}$

3. $\dfrac{5}{-(x+y)}$ **a.** $\dfrac{5}{(x+y)}$ **b.** $\dfrac{-5}{(x+y)}$ **c.** $\dfrac{-5}{-(x+y)}$ **d.** $-\dfrac{5}{(x+y)}$

4. $-\dfrac{(y+z)}{3y}$ **a.** $\dfrac{-(y+z)}{3y}$ **b.** $\dfrac{-(y+z)}{-3y}$ **c.** $\dfrac{(y+z)}{3y}$ **d.** $\dfrac{(y+z)}{-3y}$

5. $\dfrac{-9x}{-2y}$ **a.** $\dfrac{-9x}{2y}$ **b.** $\dfrac{9x}{2y}$ **c.** $\dfrac{9x}{-2y}$ **d.** $-\dfrac{9x}{2y}$

6. $\dfrac{-a}{-b}$ **a.** $\dfrac{a}{b}$ **b.** $\dfrac{a}{-b}$ **c.** $\dfrac{-a}{b}$ **d.** $-\dfrac{a}{b}$

Use the choices below to fill in each blank. Some choices may be used more than once and some used not at all.

exponent	undefined	base	1	$\dfrac{-a}{-b}$	$\dfrac{a}{b}$
square root	reciprocal	0	9	$\dfrac{-a}{b}$	$\dfrac{a}{-b}$

7. $0 \cdot a =$ _____ .

8. $\dfrac{0}{4}$ simplifies to _____ while $\dfrac{4}{0}$ is _____ .

9. The _____ of the nonzero number b is $\dfrac{1}{b}$.

10. The fraction $-\dfrac{a}{b} =$ _____ $=$ _____ .

11. A(n) _____ is a shorthand notation for repeated multiplication of the same number.

12. In $(-5)^2$, the 2 is the _____ and the -5 is the _____ .

13. The opposite of squaring a number is taking the _____ of a number.

14. Using order of operations, $9 \div 3 \cdot 3 =$ _____ .

1.3 EXERCISE SET

MyMathLab · PRACTICE · WATCH · DOWNLOAD · READ · REVIEW

Add or subtract as indicated. See Examples 1 through 3

1. $-3 + 8$ **2.** $12 + (-7)$

3. $-14 + (-10)$ **4.** $-5 + (-9)$

5. $-4.3 - 6.7$ **6.** $-8.2 - (-6.6)$

7. $13 - 17$ **8.** $15 - (-1)$

9. $\dfrac{11}{15} - \left(-\dfrac{3}{5}\right)$ **10.** $\dfrac{7}{10} - \dfrac{4}{5}$

11. $19 - 10 - 11$ **12.** $-13 - 4 + 9$

13. $-\dfrac{4}{5} - \left(-\dfrac{3}{10}\right)$ **14.** $-\dfrac{5}{2} - \left(-\dfrac{2}{3}\right)$

15. Subtract 14 from 8.   **16.** Subtract 9 from -3.

Multiply or divide as indicated. See Examples 4 and 5.

17. $-5 \cdot 12$ **18.** $-3 \cdot 8$

19. $-17 \cdot 0$ **20.** $-5 \cdot 0$

21. $\dfrac{0}{-2}$ **22.** $\dfrac{-2}{0}$

23. $\dfrac{-9}{3}$ **24.** $\dfrac{-20}{5}$

25. $\dfrac{-12}{-4}$ **26.** $\dfrac{-36}{-6}$

27. $3\left(-\dfrac{1}{18}\right)$ **28.** $5\left(-\dfrac{1}{50}\right)$

29. $(-0.7)(-0.8)$ **30.** $(-0.9)(-0.5)$

31. $9.1 \div (-1.3)$ **32.** $22.5 \div (-2.5)$

33. $-4(-2)(-1)$ **34.** $-5(-3)(-2)$

Evaluate each expression. See Example 6.

35. -7^2 **36.** $(-7)^2$

37. $(-6)^2$ **38.** -6^2

39. $(-2)^3$ **40.** -2^3

41. $\left(-\dfrac{1}{3}\right)^3$ **42.** $\left(-\dfrac{1}{2}\right)^4$

Find the following roots. See Examples 7 and 8.

43. $\sqrt{49}$ **44.** $\sqrt{81}$

45. $-\sqrt{\dfrac{4}{9}}$ **46.** $-\sqrt{\dfrac{4}{25}}$

47. $\sqrt[3]{64}$ **48.** $\sqrt[5]{32}$

49. $\sqrt[4]{81}$ **50.** $\sqrt[3]{1}$

51. $\sqrt{-100}$ **52.** $\sqrt{-25}$

MIXED PRACTICE

Simplify each expression. See Examples 1 through 11.

53. $3(5 - 7)^4$ **54.** $7(3 - 8)^2$

55. $-3^2 + 2^3$ **56.** $-5^2 - 2^4$

57. $\dfrac{3.1 - (-1.4)}{-0.5}$ **58.** $\dfrac{4.2 - (-8.2)}{-0.4}$

59. $(-3)^2 + 2^3$ **60.** $(-15)^2 - 2^4$

61. $-8 \div 4 \cdot 2$ **62.** $-20 \div 5 \cdot 4$

63. $-8\left(-\dfrac{3}{4}\right) - 8$

64. $-10\left(-\dfrac{2}{5}\right) - 10$

65. $2 - [(7 - 6) + (9 - 19)]$

66. $8 - [(4 - 7) + (8 - 1)]$

67. $\dfrac{(-9 + 6)(-1^2)}{-2 - 2}$

68. $\dfrac{(-1 - 2)(-3^2)}{-6 - 3}$

69. $(\sqrt[3]{8})(-4) - (\sqrt{9})(-5)$

70. $(\sqrt[3]{27})(-5) - (\sqrt{25})(-3)$

71. $25 - [(3 - 5) + (14 - 18)]^2$

72. $10 - [(4 - 5)^2 + (12 - 14)]^4$

73. $\dfrac{(3 - \sqrt{9}) - (-5 - 1.3)}{-3}$

74. $\dfrac{-\sqrt{16} - (6 - 2.4)}{-2}$

75. $\dfrac{|3 - 9| - |-5|}{-3}$

76. $\dfrac{|-14| - |2 - 7|}{-15}$

77. $\dfrac{3(-2 + 1)}{5} - \dfrac{-7(2 - 4)}{1 - (-2)}$

78. $\dfrac{-1 - 2}{2(-3) + 10} - \dfrac{2(-5)}{-1(8) + 1}$

79. $\dfrac{\dfrac{1}{3} \cdot 9 - 7}{3 + \dfrac{1}{2} \cdot 4}$

80. $\dfrac{\dfrac{1}{5} \cdot 20 - 6}{10 + \dfrac{1}{4} \cdot 12}$

81. $3\{-2 + 5[1 - 2(-2 + 5)]\}$

82. $2\{-1 + 3[7 - 4(-10 + 12)]\}$

83. $\dfrac{-4\sqrt{80 + 1} + (-4)^2}{3^3 + |-2(3)|}$

84. $\dfrac{(-2)^4 + 3\sqrt{120 - 20}}{4^3 + |5(-1)|}$

Evaluate each expression when x = 9 and y = −2. See Examples 12 and 13.

85. $9x - 6y$

86. $4x - 10y$

87. $-3y^2$

88. $-7y^2$

89. $\dfrac{\sqrt{x}}{y} - \dfrac{y}{x}$

90. $\dfrac{y}{2x} - \dfrac{\sqrt{x}}{3y}$

91. $\dfrac{3 + 2|x - y|}{x + 2y}$

92. $\dfrac{5 + 2|y - x|}{x + 6y}$

93. $\dfrac{y^3 + \sqrt{x - 5}}{|4x - y|}$

94. $\dfrac{y^2 + \sqrt{x + 7}}{|3x - y|}$

Find the value of each expression when x = 1.4 and y = −6.2. If necessary, round the result to the nearest hundredth. See Examples 12 and 13.

95. $3x - 2y$

96. $5y^2 + x$

97. $\dfrac{|x - y|}{2y}$

98. $\dfrac{\sqrt{x - y}}{2x}$

99. $-3(x^2 + y^2)$

100. $1.6(3x^2 + 1.8)$

For Exercises 101–104, state the expression being evaluated, the values of the variables, and the value of the expression.

101.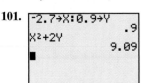

102.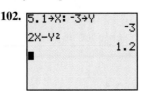

103.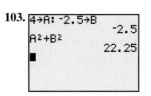

104.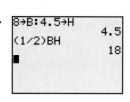

105. Consider the expressions $\left(\dfrac{1}{2}\right)x$ and $\dfrac{1}{2}x$.

   a. Use your calculator to store 2 in x, and then enter and evaluate each expression.

   b. Are the results of part **a** the same or different? Explain why.

106. Consider the expression $\dfrac{12}{(x - 2)}$ and $\dfrac{12}{x} - 2$.

   a. Use your calculator to store 4 in x, and then enter and evaluate each expression.

   b. Are the results of part **a** the same or different? Explain why.

See Example 14.

107. The algebraic expression $8 + 2y$ represents the perimeter of a rectangle with width 4 and length y.

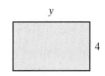

   a. Complete the table that follows by evaluating this expression at the given values of y.

Length	y	5	7	10	100
Perimeter	$8 + 2y$				

   b. Use the results of the table in part **a** to answer the following question. As the width of a rectangle remains the same and the length increases, does the perimeter increase or decrease? Explain how you arrived at your answer.

108. The algebraic expression πr^2 represents the area of a circle with radius r.

   a. Complete the table below by evaluating this expression at given values of r. (Use 3.14 for π.)

Radius	r	2	3	7	10
Area	πr^2				

   b. As the radius of a circle increases (see the table), does its area increase or decrease? Explain your answer.

109. The algebraic expression $\dfrac{100x + 5000}{x}$ represents the cost per bookshelf (in dollars) of producing x bookshelves.

a. Complete the table below.

Number of Bookshelves	x	10	100	1000
Cost per Bookshelf	$\dfrac{100x + 5000}{x}$			

b. As the number of bookshelves manufactured increases (see the table), does the cost per bookshelf increase or decrease? Why do you think that this is so?

110. If c is degrees Celsius, the algebraic expression $1.8c + 32$ represents the equivalent temperature in degrees Fahrenheit.

a. Complete the table below.

Degrees Celsius	c	-10	0	50
Degrees Fahrenheit	$1.8c + 32$			

b. As degrees Celsius increase (see the table), do degrees Fahrenheit increase or decrease?

CONCEPT EXTENSIONS

Find the value of the expression when $x_1 = 2$, $x_2 = 4$, $y_1 = -3$, and $y_2 = 2$.

111. $\dfrac{y_2 - y_1}{x_2 - x_1}$

112. $\sqrt{(x_2 - x_1)^2 + (y_2 - y_1)^2}$

Each circle below represents a whole, or 1. Determine the unknown fractional part of each circle.

113.

114.

115. Most of Mauna Kea, a volcano on Hawaii, lies below sea level. If this volcano begins at 5998 meters below sea level and then rises 10,203 meters, find the height of the volcano above sea level.

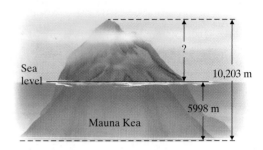

116. The highest point on land on Earth is the top of Mt. Everest in the Himalayas, at an elevation of 29,028 feet above sea level. The lowest point on land is the Dead Sea, between Israel and Jordan, at 1319 feet below sea level. Find the difference in elevations.

Insert parentheses so that each expression simplifies to the given number.

117. $2 + 7 \cdot 1 + 3$; 36

118. $6 - 5 \cdot 2 + 2$; -6

119. Explain why -3^2 and $(-3)^2$ simplify to different numbers.

120. Explain why -3^3 and $(-3)^3$ simplify to the same number.

Use a calculator to approximate each square root. For Exercises 125 and 126, simplify the expression. Round answers to four decimal places.

121. $\sqrt{10}$

122. $\sqrt{273}$

123. $\sqrt{7.9}$

124. $\sqrt{19.6}$

125. $\dfrac{-1.682 - 17.895}{(-7.102)(-4.691)}$

126. $\dfrac{(-5.161)(3.222)}{7.955 - 19.676}$

1.4 PROPERTIES OF REAL NUMBERS

OBJECTIVES

1. Use operation and order symbols to write mathematical sentences.

2. Identify identity numbers and inverses.

3. Identify and use the commutative, associative, and distributive properties.

4. Write algebraic expressions.

5. Simplify algebraic expressions.

OBJECTIVE 1 ▶ Using symbols to write mathematical sentences. In Section 1.2, we used the symbol = to mean "is equal to." All of the following key words and phrases also imply equality.

Equality			
equals	is/was	represents	is the same as
gives	yields	amounts to	is equal to

EXAMPLES Write each sentence as an equation.

1. The sum of x and 5 is 20.

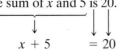

$$x + 5 \quad = \quad 20$$

2. Two times the sum of 3 and y amounts to 4.

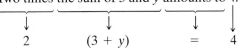

$$2 \qquad (3 + y) \qquad = \qquad 4$$

3. The difference of 8 and x is the same as the product of 2 and x.

$$8 - x \qquad = \qquad 2 \cdot x$$

4. The quotient of z and 9 amounts to 9 plus z.

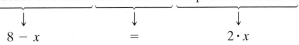

$$z \div 9 \qquad = \qquad 9 + z$$

or $\qquad \dfrac{z}{9} \qquad = \qquad 9 + z$

PRACTICES

1–4 Write each sentence using mathematical symbols.

1. The product of -4 and x is 20.

2. Three times the difference of z and 3 equals 9.

3. The sum of x and 5 is the same as 3 less than twice x.

4. The sum of y and 2 is 4 more than the quotient of z and 8.

If we want to write in symbols that two numbers are not equal, we can use the symbol $\neq$, which means "**is not equal to.**" For example,

$$3 \neq 2$$

Graphing two numbers on a number line gives us a way to compare two numbers. For two real numbers a and b, we say **a is less than b** if on the number line a lies to the left of b. Also, if b is to the right of a on the number line, then **b is greater than a.** The symbol $<$ means "**is less than.**" Since a is less than b, we write

$$a < b$$

> **▶ Helpful Hint**
>
> Notice that if $a < b$, then $b > a$. For example, since $-1 < 7$, then $7 > -1$.

The symbol $>$ means "**is greater than.**" Since b is greater than a, we write

$$b > a$$

EXAMPLE 5   Insert $<$, $>$, or $=$ between each pair of numbers to form a true statement.

a. $-1 \quad -2$ **b.** $\dfrac{12}{4} \quad 3$ **c.** $-5 \quad 0$ **d.** $-3.5 \quad -3.05$

Solution

a. $-1 > -2$ since -1 lies to the right of -2 on the number line.

b. $\dfrac{12}{4} = 3$.

c. $-5 < 0$ since -5 lies to the left of 0 on the number line.

d. $-3.5 < -3.05$ since -3.5 lies to the left of -3.05 on the number line.

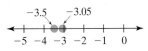

PRACTICE
5   Insert $<$, $>$, or $=$ between each pair of numbers to form a true statement.

a. -6 -5 **b.** $\dfrac{24}{3}$ 8 **c.** 0 -7 **d.** 2.76 2.67

> ▶ **Helpful Hint**
> When inserting the $>$ or $<$ symbol, think of the symbols as arrowheads that "point" toward the smaller number when the statement is true.

In addition to $<$ and $>$, there are the inequality symbols $\leq$ and $\geq$. The symbol

$$\leq \text{ means "\textbf{is less than or equal to}"}$$

and the symbol

$$\geq \text{ means "\textbf{is greater than or equal to}"}$$

For example, the following are true statements.

$$
\begin{array}{lll}
10 \leq 10 & \text{since} & 10 = 10 \\
-8 \leq 13 & \text{since} & -8 < 13 \\
-5 \geq -5 & \text{since} & -5 = -5 \\
-7 \geq -9 & \text{since} & -7 > -9
\end{array}
$$

EXAMPLE 6   Write each sentence using mathematical symbols.

a. The sum of 5 and y is greater than or equal to 7.
b. 11 is not equal to z.
c. 20 is less than the difference of 5 and twice x.

Solution

a. $5 + y \geq 7$ **b.** $11 \neq z$ **c.** $20 < 5 - 2x$ □

PRACTICE
6   Write each sentence using mathematical symbols.

a. The difference of x and 3 is less than or equal to 5.
b. y is not equal to -4.
c. Two is less than the sum of 4 and one-half z.

OBJECTIVE 2 ▶ Identifying identities and inverses. Of all the real numbers, two of them stand out as extraordinary: 0 and 1. Zero is the only number that when *added* to any real number, the result is the same real number. Zero is thus called the **additive identity.** Also, one is the only number that when *multiplied* by any real number, the result is the same real number. One is thus called the **multiplicative identity.**

	Addition	**Multiplication**
Identity Properties	The additive identity is 0. $a + 0 = 0 + a = a$	The multiplicative identity is 1. $a \cdot 1 = 1 \cdot a = a$

In Section 1.2, we learned that a and $-a$ are opposites.

Another name for opposite is **additive inverse.** For example, the additive inverse of 3 is -3. Notice that the sum of a number and its opposite is always 0.

In Section 1.3, we learned that, for a nonzero number, b and $\frac{1}{b}$ are reciprocals. Another name for reciprocal is **multiplicative inverse.** For example, the multiplicative inverse of $-\frac{2}{3}$ is $-\frac{3}{2}$. Notice that the product of a number and its reciprocal is always 1.

	Opposite or Additive Inverse	***Reciprocal or Multiplicative Inverse***
Inverse Properties	For each number a, there is a unique number $-a$ called the **additive inverse** or **opposite** of a such that $$a + (-a) = (-a) + a = 0$$	For each nonzero a, there is a unique number $\frac{1}{a}$ called the **multiplicative inverse** or **reciprocal** of a such that $$a \cdot \frac{1}{a} = \frac{1}{a} \cdot a = 1$$

EXAMPLE 7   Write the additive inverse, or opposite, of each.

a. 4   **b.** $\frac{3}{7}$ **c.** -11.2

Solution

a. The opposite of 4 is -4.

b. The opposite of $\frac{3}{7}$ is $-\frac{3}{7}$.

c. The opposite of -11.2 is $-(-11.2) = 11.2$.   □

PRACTICE
7   Write the additive inverse, or opposite, of each.

a. -7 **b.** 4.7   **c.** $-\frac{3}{8}$

EXAMPLE 8   Write the multiplicative inverse, or reciprocal, of each.

a. 11   **b.** -9 **c.** $\frac{7}{4}$

Solution

a. The reciprocal of 11 is $\frac{1}{11}$.

b. The reciprocal of -9 is $-\frac{1}{9}$.

c. The reciprocal of $\frac{7}{4}$ is $\frac{4}{7}$ because $\frac{7}{4} \cdot \frac{4}{7} = 1$.   □

PRACTICE
8   Write the multiplicative inverse, or reciprocal, of each.

a. $-\frac{5}{3}$ **b.** 14   **c.** -2

▶ **Helpful Hint**
The number 0 has no reciprocal. Why? There is no number that when multiplied by 0 gives a product of 1.

Answer to Concept Check:
no; answers may vary

Concept Check ✓
Can a number's additive inverse and multiplicative inverse ever be the same? Explain.

OBJECTIVE 3 ▶ Using the commutative, associative, and distributive properties. In addition to these special real numbers, all real numbers have certain properties that allow us to write equivalent expressions—that is, expressions that have the same value. These properties will be especially useful in Chapter 2 when we solve equations.

The **commutative properties** state that the order in which two real numbers are added or multiplied does not affect their sum or product.

Commutative Properties

For real numbers a and b,

$$\text{Addition:} \quad a + b = b + a$$
$$\text{Multiplication:} \quad a \cdot b = b \cdot a$$

The **associative properties** state that regrouping numbers that are added or multiplied does not affect their sum or product.

Associative Properties

For real numbers a, b, and c,

$$\text{Addition:} \quad (a + b) + c = a + (b + c)$$
$$\text{Multiplication:} \quad (a \cdot b) \cdot c = a \cdot (b \cdot c)$$

EXAMPLE 9 Use the commutative property of addition to write an expression equivalent to $7x + 5$.

Solution $$7x + 5 = 5 + 7x.$$ □

PRACTICE
9 Use the commutative property of addition to write an expression equivalent to $8 + 13x$.

EXAMPLE 10 Use the associative property of multiplication to write an expression equivalent to $4 \cdot (9y)$. Then simplify this equivalent expression.

Solution $$4 \cdot (9y) = (4 \cdot 9)y = 36y.$$ □

PRACTICE
10 Use the associative property of multiplication to write an expression equivalent to $3 \cdot (11b)$. Then simplify the equivalent expression.

The **distributive property** states that multiplication distributes over addition.

Distributive Property

For real numbers a, b, and c,

$$a(b + c) = ab + ac$$

EXAMPLE 11 Use the distributive property to multiply.

a. $3(2x + y)$ **b.** $-(3x - 1)$ **c.** $0.7a(b - 2)$

Solution

a. $3(2x + y) = 3 \cdot 2x + 3 \cdot y$ Apply the distributive property.

$= 6x + 3y$ Apply the associative property of multiplication.

b. Recall that $-(3x - 1)$ means $-1(3x - 1)$.

$$-1(3x - 1) = -1(3x) + (-1)(-1)$$
$$= -3x + 1$$

c. $0.7a(b - 2) = 0.7a \cdot b - 0.7a \cdot 2 = 0.7ab - 1.4a$

PRACTICE
11   Use the distributive property to multiply.

a. $4(x + 5y)$ **b.** $-(3 - 2z)$ **c.** $0.3x(y - 3)$

Concept Check ☑

Is the statement below true? Why or why not?

$$6(2a)(3b) = 6(2a) \cdot 6(3b)$$

In the following example, a calculator screen illustrates some of the properties.

EXAMPLE 12   State the basic property that is being illustrated on each calculator screen:

a. 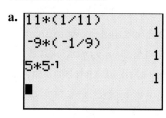   **b.** 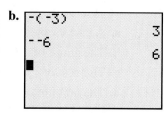   **c.**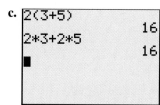

Solution

a. Multiplicative inverses   **b.** Double negative property
c. Distributive property

PRACTICE
12   State the property illustrated.

a. $-(-11.7) = 11.7$ **b.** $7(y + 5) = 7 \cdot y + 7 \cdot 5$
c. $2 + (a + b) = (2 + a) + b$

OBJECTIVE 4 ▶ Writing algebraic expressions. As mentioned earlier, an important step in problem solving is to be able to write algebraic expressions from word phrases. Sometimes this involves a direct translation, but often an indicated operation is not directly stated but rather implied.

EXAMPLE 13   Write each as an algebraic expression.

a. A vending machine contains x quarters. Write an expression for the *value* of the quarters.

b. The number of grams of fat in x pieces of bread if each piece of bread contains 2 grams of fat.

c. The cost of x desks if each desk costs $156.

d. Sales tax on a purchase of x dollars if the tax rate is 9%.

Each of these examples implies finding a product.

Solution

a. The value of the quarters is found by multiplying the value of a quarter (0.25 dollar) by the number of quarters.

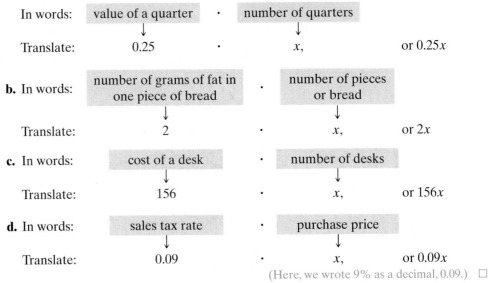

In words: | value of a quarter | · | number of quarters |

Translate: 0.25 · x, or $0.25x$

b. In words: | number of grams of fat in one piece of bread | · | number of pieces or bread |

Translate: 2 · x, or $2x$

c. In words: | cost of a desk | · | number of desks |

Translate: 156 · x, or $156x$

d. In words: | sales tax rate | · | purchase price |

Translate: 0.09 · x, or $0.09x$

(Here, we wrote 9% as a decimal, 0.09.) □

PRACTICE

13 Write each as an algebraic expression.

a. A parking meter contains x dimes. Write an expression for the value of the dimes.

b. The grams of carbohydrates in y cookies if each cookie has 26 g of carbohydrates.

c. The cost of z birthday cards if each birthday card costs $1.75.

d. The amount of money you save on a new cell phone costing t dollars if it has a 15% discount.

Two or more unknown numbers in a problem may sometimes be related. If so, try letting a variable represent one unknown number and then represent the other unknown number or numbers as expressions containing the same variable.

EXAMPLE 14 Write each as an algebraic expression.

a. Two numbers have a sum of 20. If one number is x, represent the other number as an expression in x.

b. The older sister is 8 years older than her younger sister. If the age of the younger sister is x, represent the age of the older sister as an expression in x.

△ **c.** Two angles are complementary if the sum of their measures is 90°. If the measure of one angle is x degrees, represent the measure of the other angle as an expression in x.

d. If x is the first of two consecutive integers, represent the second integer as an expression in x.

Solution

a. If two numbers have a sum of 20 and one number is x, the other number is "the rest of 20."

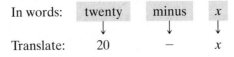

In words: | twenty | minus | x |

Translate: 20 − x

b. The older sister's age is

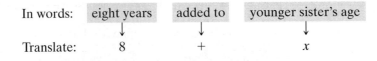

In words: | eight years | added to | younger sister's age |

Translate: 8 + x

c. In words: | ninety | minus | x |

Translate: 90   – x

d. The next consecutive integer is always one more than the previous integer.

In words: | the first integer | plus | one |

Translate: x $+$ 1   ☐

PRACTICE
14   Write each as an algebraic expression.

a. Two numbers have a sum of 16. If one number is x, represent the other number as an expression in x.

b. Two angles are supplementary if the sum of their measures is 180°. If the measure of one angle is x degrees, represent the measure of the other angle as an expression in x.

c. If x is the first of two consecutive even integers, represent the next even integer as an expression in x.

d. One brother is 9 years younger than another brother. If the age of the younger brother is x, represent the age of the older brother as an expression in x.

OBJECTIVE 5 ▶ Simplifying algebraic expressions. Often, an expression may be **simplified** by removing grouping symbols and combining any like terms. The **terms** of an expression are the addends of the expression. For example, in the expression $3x^2 + 4x$, the terms are $3x^2$ and $4x$.

Expression	*Terms*
$-2x + y$	$-2x, y$
$3x^2 - \dfrac{y}{5} + 7$	$3x^2, -\dfrac{y}{5}, 7$

Terms with the same variable(s) raised to the same power are called **like terms.** We can add or subtract like terms by using the distributive property. This process is called **combining like terms.**

EXAMPLE 15   Use the distributive property to simplify each expression.

a. $3x - 5x + 4$ **b.** $7yz + yz$ **c.** $4z + 6.1$

Solution

a. $3x - 5x + 4 = (3 - 5)x + 4$ Apply the distributive property.
                    $= -2x + 4$

b. $7yz + yz = (7 + 1)yz = 8yz$

c. $4z + 6.1$ cannot be simplified further since $4z$ and 6.1 are not like terms.   ☐

PRACTICE
15   Use the distributive property to simplify.

a. $6ab - ab$ **b.** $4x - 5 + 6x$ **c.** $17p - 9$

Let's continue to use properties of real numbers to simplify expressions. Recall that the distributive property can also be used to multiply. For example,

$$-2(x + 3) = -2(x) + (-2)(3) = -2x - 6$$

The associative and commutative properties may sometimes be needed to re-arrange and group like terms when we simplify expressions.

$$-7x^2 + 5 + 3x^2 - 2 = -7x^2 + 3x^2 + 5 - 2$$
$$= (-7 + 3)x^2 + (5 - 2)$$
$$= -4x^2 + 3$$

EXAMPLE 16   Simplify each expression.

a. $3xy - 2xy + 5 - 7 + xy$ **b.** $7x^2 + 3 - 5(x^2 - 4)$

c. $(2.1x - 5.6) - (-x - 5.3)$ **d.** $\frac{1}{2}(4a - 6b) - \frac{1}{3}(9a + 12b - 1) + \frac{1}{4}$

<u>Solution</u>

a. $3xy - 2xy + 5 - 7 + xy = 3xy - 2xy + xy + 5 - 7$ Apply the commutative property.

$$= (3 - 2 + 1)xy + (5 - 7)$$ Apply the distributive property.

$$= 2xy - 2$$ Simplify.

b. $7x^2 + 3 - 5(x^2 - 4) = 7x^2 + 3 - 5x^2 + 20$ Apply the distributive property.

$$= 2x^2 + 23$$ Simplify.

c. Think of $-(-x - 5.3)$ as $-1(-x - 5.3)$ and use the distributive property.

$$(2.1x - 5.6) - 1(-x - 5.3) = 2.1x - 5.6 + 1x + 5.3$$

$$= 3.1x - 0.3$$ Combine like terms.

d. $\frac{1}{2}\left(4a - 6b\right) - \frac{1}{3}\left(9a + 12b - 1\right) + \frac{1}{4}$

$$= 2a - 3b - 3a - 4b + \frac{1}{3} + \frac{1}{4}$$ Use the distributive property.

$$= -a - 7b + \frac{7}{12}$$ Combine like terms.   □

PRACTICE
16   Simplify each expression.

a. $5pq - 2pq - 11 - 4pq + 18$ **b.** $3x^2 + 7 - 2(x^2 - 6)$

c. $(3.7x + 2.5) - (-2.1x - 1.3)$

d. $\frac{1}{5}(15c - 25d) - \frac{1}{2}(8c + 6d + 1) + \frac{3}{4}$

Answer to Concept Check:
$x - 4(x - 5) = x - 4x + 20$
$\qquad\qquad = -3x + 20$

Concept Check ✓
Find and correct the error in the following.

$$x - 4(x - 5) = x - 4x - 20$$
$$= -3x - 20$$

VOCABULARY & READINESS CHECK

Complete the table by filling in the symbols.

	Symbol	Meaning		Symbol	Meaning
1.		is less than	**2.**		is greater than
3.		is not equal to	**4.**		is equal to
5.		is greater than or equal to	**6.**		is less than or equal to

Use the choices below to fill in each blank. Not all choices will be used.

like	terms	distributive	$-a$	commutative
unlike	combining	associative	$\dfrac{1}{a}$	multiplicative inverse
x^{-1}	additive inverse			

7. The opposite or _____ of nonzero number a is _____.

8. The reciprocal or _____ of nonzero number a is _____.

9. The _____ property has to do with "order."

10. The _____ property has to do with "grouping."

11. $a(b + c) = ab + ac$ illustrates the _____ property.

12. Terms with the same variable(s) raised to the same powers are called _____ terms.

13. The _____ of an expression are the addends of the expression.

14. The process of adding or subtracting like terms is called _____ like terms.

15. The _____ key on a calculator is used to find the multiplicative inverse of a number.

1.4 │ EXERCISE SET

MyMathLab   MathXL PRACTICE   WATCH   DOWNLOAD   READ   REVIEW

MIXED PRACTICE

Write each sentence using mathematical symbols. See Examples 1 through 4 and 6.

1. The sum of 10 and x is -12.

2. The difference of y and 3 amounts to 12.

3. Twice x, plus 5, is the same as -14.

4. Three more than the product of 4 and c is 7.

5. The quotient of n and 5 is 4 times n.

6. The quotient of 8 and y is 3 more than y.

7. The difference of z and one-half is the same as the product of z and one-half.

8. Five added to one-fourth q is the same as 4 more than q.

9. The product of 7 and x is less than or equal to -21.

10. 10 subtracted from the reciprocal of x is greater than 0.

11. Twice the difference of x and 6 is greater than the reciprocal of 11.

12. Four times the sum of 5 and x is not equal to the opposite of 15.

13. Twice the difference of x and 6 is -27.

14. 5 times the sum of 6 and y is -35.

Insert $<$, $>$, or $=$ between each pair of numbers to form a true statement. See Example 5.

15. -16 -17

16. -14 -24

17. 7.4 7.40

18. $\dfrac{7}{2}$ $\dfrac{35}{10}$

19. $\dfrac{7}{11}$ $\dfrac{9}{11}$

20. $\dfrac{9}{20}$ $\dfrac{3}{20}$

21. $\dfrac{1}{2}$ $\dfrac{5}{8}$

22. $\dfrac{3}{4}$ $\dfrac{7}{8}$

23. -7.9 -7.09

24. -13.07 -13.7

Fill in the chart. See Examples 7 and 8.

	Number	Opposite	Reciprocal
25.	5		
26.	7		
27.		8	
28.			$-\dfrac{1}{4}$
29.	$-\dfrac{1}{7}$		
30.	$\dfrac{1}{11}$		
31.	0		
32.	1		
33.			$\dfrac{8}{7}$
34.		$\dfrac{23}{5}$	

35. Name the only real number that has no reciprocal, and explain why this is so.

36. Name the only real number that is its own opposite, and explain why this is so.

Use a commutative property to write an equivalent expression. See Example 9.

37. $7x + y$

38. $3a + 2b$

39. $z \cdot w$

40. $r \cdot s$

41. $\dfrac{1}{3} \cdot \dfrac{x}{5}$

42. $\dfrac{x}{2} \cdot \dfrac{9}{10}$

43. Is subtraction commutative? Explain why or why not.

44. Is division commutative? Explain why or why not.

Use an associative property to write an equivalent expression. See Example 10.

45. $5 \cdot (7x)$

46. $3 \cdot (10z)$

47. $(x + 1.2) + y$

48. $5q + (2r + s)$

49. $(14z) \cdot y$

50. $(9.2x) \cdot y$

51. Evaluate $12 - (5 - 3)$ and $(12 - 5) - 3$. Use these two expressions and discuss whether subtraction is associative.

52. Evaluate $24 \div (6 \div 3)$ and $(24 \div 6) \div 3$. Use these two expressions and discuss whether division is associative.

Use the distributive property to find the product. See Example 11.

53. $3(x + 5)$

54. $7(y + 2)$

55. $-(2a + b)$

56. $-(c + 7d)$

57. $2(6x + 5y + 2z)$

58. $5(3a + b + 9c)$

59. $-4(x - 2y + 7)$

60. $-10(2a - 3b - 4)$

61. $0.5x(6y - 3)$

62. $1.2m(9n - 4)$

Complete the statement to illustrate the given property. See Example 12.

63. $3x + 6 = $ _____ Commutative property of addition

64. $8 + 0 = $ _____ Additive identity property

65. $\dfrac{2}{3} + \left(-\dfrac{2}{3}\right) = $ _____ Additive inverse property

66. $4(x + 3) = $ _____ Distributive property

67. $7 \cdot 1 = $ _____ Multiplicative identity property

68. $0 \cdot (-5.4) = $ _____ Multiplication property of zero

69. $10(2y) = $ _____ Associative property

70. $9y + (x + 3z) = $ _____ Associative property

Write the property illustrated on each screen. See Example 12.

71.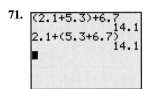

72.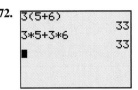

73.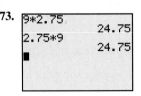

74.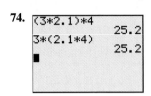

75. To demonstrate the distributive property geometrically, represent the area of the larger rectangle in two ways: First as length a times width $b + c$, and second as the sum of the areas of the smaller rectangles.

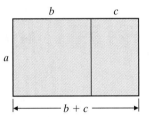

Write each of the following as an algebraic expression. See Examples 13 and 14.

76. Write an expression for the amount of money (in dollars) in n nickels.

77. Write an expression for the amount of money (in dollars) in d dimes.

78. Two numbers have a sum of 25. If one number is x, represent the other number as an expression in x.

79. Two numbers have a sum of 112. If one number is x, represent the other number as an expression in x.

80. Two angles are supplementary if the sum of their measures is $180°$. If the measure of one angle is x degrees, represent the measure of the other angle as an expression in x.

81. If the measure of an angle is $5x$ degrees, represent the measure of its complement as an expression in x.

82. The cost of x compact discs if each compact disc costs $6.49.

83. The cost of y books if each book costs $35.61.

84. If x is an odd integer, represent the next odd integer as an expression in x.

85. If $2x$ is an even integer, represent the next even integer as an expression in x.

MIXED PRACTICE

Simplify each expression. See Examples 11, 15, and 16.

86. $-9 + 4x + 18 - 10x$

87. $5y - 14 + 7y - 20y$

88. $5k - (3k - 10)$

89. $-11c - (4 - 2c)$

90. $(3x + 4) - (6x - 1)$

91. $(8 - 5y) - (4 + 3y)$

92. $3(xy - 2) + xy + 15 - x^2$

93. $-4(yz + 3) - 7yz + 1 + y^2$

94. $-(n + 5) + (5n - 3)$

95. $-(8 - t) + (2t - 6)$

96. $4(6n^2 - 3) - 3(8n^2 + 4)$

97. $5(2z^3 - 6) + 10(3 - z^3)$

98. $3x - 2(x - 5) + x$

99. $7n + 3(2n - 6) - 2$

100. $1.5x + 2.3 - 0.7x - 5.9$

101. $6.3y - 9.7 + 2.2y - 11.1$

102. $\frac{3}{4}b - \frac{1}{2} + \frac{1}{6}b - \frac{2}{3}$

103. $\frac{7}{8}a - \frac{11}{12} - \frac{1}{2}a + \frac{5}{6}$

104. $2(3x + 7)$

105. $4(5y + 12)$

106. $\frac{1}{4}(8x - 4) - \frac{1}{5}(20x - 6y)$

107. $\frac{1}{2}(10x - 2) - \frac{1}{6}(60x - 5y)$

108. $\frac{1}{6}(24a - 18b) - \frac{1}{7}(7a - 21b - 2) - \frac{1}{5}$

109. $\frac{1}{3}(6x - 33y) - \frac{1}{8}(24x - 40y + 1) - \frac{1}{3}$

CONCEPT EXTENSIONS

In each statement, a property of real numbers has been incorrectly applied. Correct the right-hand side of each statement. See the second Concept Check in this section.

110. $3(x + 4) = 3x + 4$　　　**111.** $5(7y) = (5 \cdot 7)(5 \cdot y)$

Simplify each expression.

112. $8.1z + 7.3(z + 5.2) - 6.85$

113. $6.5y - 4.4(1.8x - 3.3) + 10.95$

△ **114.** Do figures with the same surface area always have the same volume? To see, take two $8\frac{1}{2}$-by-11-inch sheets of paper and construct two cylinders using the following figures as a guide. Working with a partner, measure the height and the radius of each resulting cylinder and use the expression $\pi r^2 h$ to approximate each volume to the nearest tenth of a cubic inch. Explain your results.

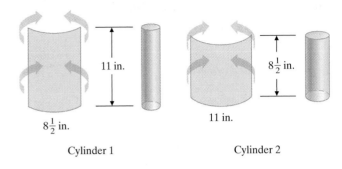

Cylinder 1　　　　　　　　Cylinder 2

△ **115.** Use the same idea as in Exercise 114, work with a partner, and discover whether two rectangles with the same perimeter always have the same area. Explain your results.

The following graph is called a broken-line graph, or simply a line graph. This particular graph shows the past, present, and future predicted U.S. population over 65. Just as with a bar graph, to find the population over 65 for a particular year, read the height of the corresponding point. To read the height, follow the point horizontally to the left until you reach the vertical axis. Use this graph to answer Exercises 116 through 121.

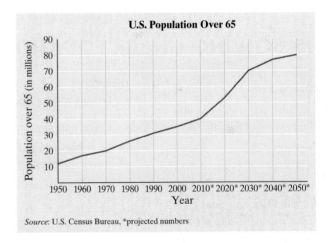

Source: U.S. Census Bureau, *projected numbers

116. Estimate the population over 65 in the year 1970.

117. Estimate the predicted population over 65 in the year 2050.

118. Estimate the predicted population over 65 in the year 2030.

119. Estimate the population over 65 in the year 2000.

120. Is the population over 65 increasing as time passes or decreasing? Explain how you arrived at your answer.

121. The percent of Americans over 65 in 1950 was 8.1%. The percent of Americans over 65 in 2050 is expected to be 2.5 times the percent over 65 in 1950. Estimate the percent of Americans expected to be over age 65 in 2050.

INTEGRATED REVIEW ALGEBRAIC EXPRESSIONS, OPERATIONS ON REAL NUMBERS, AND PROPERTIES

Find the value of each expression when $x = -1$, $y = 3$, and $z = -4$.

1. z^2

2. $-z^2$

3. $\dfrac{4x - z}{2y}$

4. $x(y - 2z)$

Perform indicated operations.

5. $-7 - (-2)$

6. $\dfrac{9}{10} - \dfrac{11}{12}$

7. $\dfrac{-13}{2 - 2}$

8. $(1.2)^2 - (2.1)^2$

9. $\sqrt{64} - \sqrt[3]{64}$

10. $-5^2 - (-5)^2$

11. $9 + 2[(8 - 10)^2 + (-3)^2]$

12. $8 - 6[\sqrt[3]{8}(-2) + \sqrt{4}(-5)]$

Write each phrase as an algebraic expression. Use x to represent each unknown number.

13. Subtract twice a number from -15.

14. Five more than three times a number.

15. Name the whole number that is not a natural number.

16. True or false: A real number is either a rational number or an irrational number, but never both.

Use properties of real numbers to simplify each expression.

17. $-5(9x)$

18. $(3x - 7) - (4x + 1)$

19. $8.6a + 2.3b - a + 4.9b$

20. $\dfrac{2}{3}y - \dfrac{2}{3} + y - \dfrac{1}{9}y + \dfrac{9}{10}$

1.5 SOLVING LINEAR EQUATIONS ALGEBRAICALLY

OBJECTIVES

1 Solve linear equations using properties of equality.

2 Solve linear equations that can be simplified by combining like terms.

3 Solve linear equations containing fractions or decimals.

4 Recognize identities and equations with no solution.

OBJECTIVE 1 ▶ Solving linear equations using properties of equality. Linear equations model many real-life problems. For example, we can use a linear equation to calculate the increase in the percent of households with digital cameras.

With the help of your computer, digital cameras allow you to see your pictures and make copies immediately, send them in e-mail, or use them on a Web page. Percentage of households with these cameras is shown in the graph below.

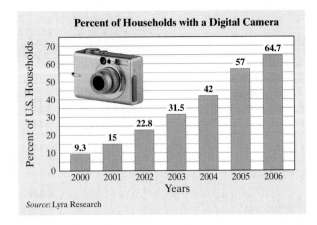

Percent of Households with a Digital Camera

Source: Lyra Research

To find the increase in percent of households from 2005 to 2006, for example, we can use the equation below:

In words:	increase in percent of households	is	percent of house-holds in 2006	minus	percent of house-holds in 2005
	↓	↓	↓	↓	↓
Translate:	x	$=$	64.7	$-$	57

Since our variable x (increase in percent of households) is by itself on one side of the equation, we can find the value of x by simplifying the right side.

$$x = 7.7$$

The increase in the households with digital cameras from 2005 to 2006 was 7.7%.

The **equation** $x = 64.7 - 57$, like every other equation, is a statement that two expressions are equal. Oftentimes, the unknown variable is not by itself on one side of the equation. In these cases, we will use properties of equality to write equivalent equations so that a solution may be found. This is called **solving the equation.** In this section, we concentrate on solving equations such as this one, called **linear equations** in one **variable.** Linear equations are also called **first-degree equations** since the exponent on the variable is 1.

Linear Equations in One Variable

$$3x = -15 \qquad 7 - y = 3y \qquad 4n - 9n + 6 = 0 \qquad z = -2$$

Linear Equations in One Variable

A linear equation in one variable is an equation that can be written in the form

$$ax + b = c$$

where $a, b,$ and c are real numbers and $a \neq 0$.

When a variable in an equation is replaced by a number and the resulting equation is true, then that number is called a **solution** of the equation. For example, 1 is a solution of the equation $3x + 4 = 7$, since $3(1) + 4 = 7$ is a true statement. But 2 is not a solution of this equation, since $3(2) + 4 = 7$ is not a true statement. The **solution set** of an equation is the set of solutions of the equation. For example, the solution set of $3x + 4 = 7$ is $\{1\}$.

To **solve an equation** is to find the solution set of an equation. Equations with the same solution set are called **equivalent equations.** For example,

$$3x + 4 = 7 \qquad 3x = 3 \qquad x = 1$$

are equivalent equations because they all have the same solution set, namely $\{1\}$. To solve an equation in x, we start with the given equation and write a series of simpler equivalent equations until we obtain an equation of the form

$$x = \textbf{number}$$

Two important properties are used to write equivalent equations.

The Addition and Multiplication Properties of Equality

If $a, b,$ and $c,$ are real numbers, then

$$a = b \quad \text{and} \quad a + c = b + c \text{ are equivalent equations.}$$
$$\text{Also, } a = b \quad \text{and} \quad ac = bc \text{ are equivalent equations as long as } c \neq 0.$$

The **addition property of equality** guarantees that the same number may be added to both sides of an equation, and the result is an equivalent equation. The **multiplication property of equality** guarantees that both sides of an equation may be multiplied by the same nonzero number, and the result is an equivalent equation. Because we define subtraction in terms of addition $(a - b = a + (-b))$, and division in terms of multiplication $\left(\dfrac{a}{b} = a \cdot \dfrac{1}{b} \right)$, these properties also guarantee that we may *subtract* the same number from both sides of an equation, or *divide* both sides of an equation by the same nonzero number, and the result is an equivalent equation.

For example, to solve $2x + 5 = 9$, use the addition and multiplication properties of equality to isolate x—that is, to write an equivalent equation of the form

$$x = \textbf{number}$$

We will do this in the next example.

EXAMPLE 1 Solve for x: $2x + 5 = 9$.

Solution First, use the addition property of equality and subtract 5 from both sides. We do this so that our only variable term, $2x$, is by itself on one side of the equation.

$$2x + 5 = 9$$
$$2x + 5 - 5 = 9 - 5 \quad \text{Subtract 5 from both sides.}$$
$$2x = 4 \quad \text{Simplify.}$$

Now that the variable term is isolated, we can finish solving for x by using the multiplication property of equality and dividing both sides by 2.

$$\frac{2x}{2} = \frac{4}{2} \quad \text{Divide both sides by 2.}$$
$$x = 2 \quad \text{Simplify.}$$

Check: To see that 2 is the solution, replace x in the original equation with 2 and see that a true statement results using any method below.

$$2x + 5 = 9 \quad \text{Original equation}$$

$$2(2) + 5 = 9 \quad \text{Let } x = 2.$$
$$4 + 5 = 9$$
$$9 = 9 \quad \text{True}$$

Let $x = 2$.

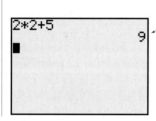

The left side of the equation evaluates to 9, the value of the right side.

Store 2 in x and evaluate the left side of the equation.

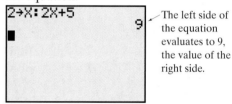

The left side of the equation evaluates to 9, the value of the right side.

Since we arrive at a true statement, 2 is the solution, or the solution set is $\{2\}$. □

PRACTICE

1 Solve for x: $3x + 7 = 22$.

EXAMPLE 2 Solve: $0.6 = 2 - 3.5c$.

Solution We use both the addition property and the multiplication property of equality.

$$0.6 = 2 - 3.5c$$
$$0.6 - 2 = 2 - 3.5c - 2 \quad \text{Subtract 2 from both sides.}$$
$$-1.4 = -3.5c \quad \text{Simplify. The variable term is now isolated.}$$
$$\frac{-1.4}{-3.5} = \frac{-3.5c}{-3.5} \quad \text{Divide both sides by } -3.5.$$
$$0.4 = c \quad \text{Simplify } \frac{-1.4}{-3.5}.$$

> **Helpful Hint**
>
> Don't forget that
>
> $0.4 = c$ and $c = 0.4$ are equivalent equations.
>
> We may solve an equation so that the variable is alone on either side of the equation.

Check:

$$0.6 = 2 - 3.5c$$
$$0.6 \stackrel{?}{=} 2 - 3.5(0.4) \quad \text{Replace } c \text{ with } 0.4.$$
$$0.6 \stackrel{?}{=} 2 - 1.4 \quad \text{Multiply.}$$
$$0.6 = 0.6 \quad \text{True}$$

The solution is 0.4, or the solution set is $\{0.4\}$. □

PRACTICE

2 Solve: $2.5 = 3 - 2.5t$.

OBJECTIVE 2 ▶ Solving linear equations that can be simplified by combining like terms.
Often, an equation can be simplified by removing any grouping symbols and combining
any like terms.

EXAMPLE 3   Solve: $-4x - 1 + 5x = 9x + 3 - 7x$.

Solution   First we simplify both sides of this equation by combining like terms. Then,
let's get variable terms on the same side of the equation by using the addition property
of equality to subtract $2x$ from both sides. Next, we use this same property to add 1 to
both sides of the equation.

$$-4x - 1 + 5x = 9x + 3 - 7x$$
$$x - 1 = 2x + 3 \qquad \text{Combine like terms.}$$
$$x - 1 - 2x = 2x + 3 - 2x \quad \text{Subtract } 2x \text{ from both sides.}$$
$$-x - 1 = 3 \qquad \text{Simplify.}$$
$$-x - 1 + 1 = 3 + 1 \qquad \text{Add 1 to both sides.}$$
$$-x = 4 \qquad \text{Simplify.}$$

Notice that this equation is not solved for x since we have $-x$ or $-1x$, not x. To solve
for x, we divide both sides by -1.

$$\frac{-x}{-1} = \frac{4}{-1} \quad \text{Divide both sides by } -1.$$
$$x = -4 \quad \text{Simplify.}$$

Check: Let $x = -4$ in the original equation and check that a true statement results.

$$-4x - 1 + 5x = 9x + 3 - 7x \quad \text{Original equation}$$

To use a calculator to check, store -4 in x and evaluate the left side and the right side
of the equation.

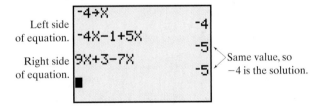

Left side of equation.

Right side of equation.

Same value, so -4 is the solution.

The solution is -4. ☐

**PRACTICE
3**   Solve: $-8x - 4 + 6x = 5x + 11 - 4x$.

If an equation contains parentheses, use the distributive property to remove
them.

EXAMPLE 4   Solve: $2(x - 3) = 5x - 9$.

Solution   First, use the distributive property.

$$2(x - 3) = 5x - 9$$
$$2x - 6 = 5x - 9 \quad \text{Use the distributive property.}$$

```
1→X
                    1
2(X-3)
                   -4
5X-9
                   -4
■
```

A calculator check for
Example 4.

Next, get variable terms on the same side of the equation by subtracting $5x$ from both sides.

$$2x - 6 - 5x = 5x - 9 - 5x \quad \text{Subtract } 5x \text{ from both sides.}$$
$$-3x - 6 = -9 \qquad\qquad \text{Simplify.}$$
$$-3x - 6 + 6 = -9 + 6 \qquad \text{Add 6 to both sides.}$$
$$-3x = -3 \qquad\qquad \text{Simplify.}$$
$$\frac{-3x}{-3} = \frac{-3}{-3} \qquad\qquad \text{Divide both sides by } -3.$$
$$x = 1$$

Let $x = 1$ in the original equation to see that 1 is the solution. A calculator check is shown in the margin. □

PRACTICE
4 Solve: $3(x - 5) = 6x - 3$.

OBJECTIVE 3 ▶ Solving linear equations containing fractions or decimals. If an equation contains fractions, we first clear the equation of fractions by multiplying both sides of the equation by the *least common denominator* (LCD) of all fractions in the equation.

EXAMPLE 5 Solve for y: $\dfrac{y}{3} - \dfrac{y}{4} = \dfrac{1}{6}$.

Solution First, clear the equation of fractions by multiplying both sides of the equation by 12, the LCD of denominators 3, 4, and 6.

$$\frac{y}{3} - \frac{y}{4} = \frac{1}{6}$$

$$12\left(\frac{y}{3} - \frac{y}{4}\right) = 12\left(\frac{1}{6}\right) \quad \text{Multiply both sides by the LCD, 12.}$$

$$12\left(\frac{y}{3}\right) - 12\left(\frac{y}{4}\right) = 2 \quad \text{Apply the distributive property.}$$

$$4y - 3y = 2 \qquad\qquad \text{Simplify.}$$

$$y = 2 \qquad\qquad \text{Simplify.}$$

Check: To check, let $y = 2$ in the original equation.

$$\frac{y}{3} - \frac{y}{4} = \frac{1}{6} \quad \text{Original equation.}$$

$$\frac{2}{3} - \frac{2}{4} \stackrel{?}{=} \frac{1}{6} \quad \text{Let } y = 2.$$

$$\frac{8}{12} - \frac{6}{12} \stackrel{?}{=} \frac{1}{6} \quad \text{Write fractions with the LCD.}$$

$$\frac{2}{12} \stackrel{?}{=} \frac{1}{6} \quad \text{Subtract.}$$

$$\frac{1}{6} = \frac{1}{6} \quad \text{Simplify.}$$

```
2→Y:Y/3-Y/4
          .1666666667
Ans▶Frac
                  1/6
■
```

A calculator check for
Example 5.

This is a true statement, so the solution is 2. A calculator check is shown in the margin. □

PRACTICE
5 Solve for y: $\dfrac{y}{2} - \dfrac{y}{5} = \dfrac{1}{4}$.

As a general guideline, the following steps may be used to solve a linear equation in one variable.

> **Solving a Linear Equation in One Variable**
>
> **STEP 1.** Clear the equation of fractions by multiplying both sides of the equation by the least common denominator (LCD) of all denominators in the equation.
>
> **STEP 2.** Use the distributive property to remove grouping symbols such as parentheses.
>
> **STEP 3.** Combine like terms on each side of the equation.
>
> **STEP 4.** Use the addition property of equality to rewrite the equation as an equivalent equation with variable terms on one side and numbers on the other side.
>
> **STEP 5.** Use the multiplication property of equality to isolate the variable.
>
> **STEP 6.** Check the proposed solution in the original equation.

EXAMPLE 6   Solve for x: $\dfrac{x + 5}{2} + \dfrac{1}{2} = 2x - \dfrac{x - 3}{8}$.

Solution   Multiply both sides of the equation by 8, the LCD of 2 and 8.

$$8\left(\frac{x + 5}{2} + \frac{1}{2}\right) = 8\left(2x - \frac{x - 3}{8}\right) \quad \text{Multiply both sides by 8.}$$

> ▶ **Helpful Hint**
>
> When we multiply both sides of an equation by a number, the distributive property tells us that each term of the equation is multiplied by the number.

$$8\left(\frac{x + 5}{2}\right) + 8 \cdot \frac{1}{2} = 8 \cdot 2x - 8\left(\frac{x - 3}{8}\right) \quad \text{Apply the distributive property.}$$

$$4(x + 5) + 4 = 16x - (x - 3) \quad \text{Simplify.}$$

$$4x + 20 + 4 = 16x - x + 3 \quad \text{Use the distributive property to remove parentheses.}$$

$$4x + 24 = 15x + 3 \quad \text{Combine like terms.}$$

$$-11x + 24 = 3 \quad \text{Subtract } 15x \text{ from both sides.}$$

$$-11x = -21 \quad \text{Subtract 24 from both sides.}$$

$$\frac{-11x}{-11} = \frac{-21}{-11} \quad \text{Divide both sides by } -11.$$

$$x = \frac{21}{11} \quad \text{Simplify.}$$

Check: To check, verify that replacing x with $\dfrac{21}{11}$ makes the original equation true. If you use a calculator to verify, make sure that parentheses are placed about a numerator or denominator that contains more than one term, as shown to the left. The solution is $\dfrac{21}{11}$.   □

```
21/11→X
        1.909090909
(X+5)/2+1/2
        3.954545455
2X-(X-3)/8
        3.954545455
■
```

A calculator check for Example 6.

PRACTICE
6   Solve for x: $x - \dfrac{x - 2}{12} = \dfrac{x + 3}{4} + \dfrac{1}{4}$.

If an equation contains decimals, you may want to first clear the equation of decimals.

EXAMPLE 7   Solve: $0.3x + 0.1 = 0.27x - 0.02$.

Solution   To clear this equation of decimals, we multiply both sides of the equation by 100. Recall that multiplying a number by 100 moves its decimal point two places to the right.

$$100(0.3x + 0.1) = 100(0.27x - 0.02)$$

$$100(0.3x) + 100(0.1) = 100(0.27x) - 100(0.02) \quad \text{Use the distributive property.}$$

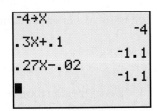

A calculator check for
Example 7.

$$30x + 10 = 27x - 2 \qquad \text{Multiply.}$$
$$30x - 27x = -2 - 10 \qquad \text{Subtract } 27x \text{ and } 10 \text{ from both sides.}$$
$$3x = -12 \qquad \text{Simplify.}$$
$$\frac{3x}{3} = \frac{-12}{3} \qquad \text{Divide both sides by 3.}$$
$$x = -4 \qquad \text{Simplify.}$$

Check to see that the solution is -4. A calculator check is shown in the margin. □

PRACTICE
7   Solve: $0.15x - 0.03 = 0.2x + 0.12$.

Concept Check ☑

Explain what is wrong with the following:

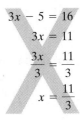

$$3x - 5 = 16$$
$$3x = 11$$
$$\frac{3x}{3} = \frac{11}{3}$$
$$x = \frac{11}{3}$$

OBJECTIVE 4 ▶ Recognizing identities and equations with no solution. So far, each linear equation that we have solved has had a single solution. A linear equation in one variable that has exactly one solution is called a **conditional equation.** We will now look at two other types of equations: contradictions and identities.

An equation in one variable that has no solution is called a **contradiction,** and an equation in one variable that has every number (for which the equation is defined) as a solution is called an **identity.** For review: A linear equation in one variable with

No solution	Is a	Contradiction
Every real number as a solution (as long as the equation is defined)	Is an	Identity

The next examples show how to recognize contradictions and identities.

EXAMPLE 8   Solve for x: $3x + 5 = 3(x + 2)$.

Solution   First, use the distributive property and remove parentheses.

$$3x + 5 = 3(x + 2)$$
$$3x + 5 = 3x + 6 \qquad \text{Apply the distributive property.}$$
$$3x + 5 - 3x = 3x + 6 - 3x \qquad \text{Subtract } 3x \text{ from both sides.}$$
$$5 = 6$$

The equation $5 = 6$ is a false statement no matter what value the variable x might have. Thus, the original equation has no solution. Its solution set is written either as $\{\ \}$ or $\varnothing$. This equation is a contradiction. □

PRACTICE
8   Solve for x: $4x - 3 = 4(x + 5)$.

▶ **Helpful Hint**

A solution set of $\{0\}$ and a solution set of $\{\ \}$ are not the same. The solution set $\{0\}$ means 1 solution, 0. The solution set $\{\ \}$ means *no* solution.

Answer to Concept Check:
$$3x - 5 = 16$$
$$3x = 21$$
$$x = 7$$
Therefore, the correct solution set is $\{7\}$.

EXAMPLE 9   Solve for x: $6x - 4 = 2 + 6(x - 1)$.

Solution   First, use the distributive property and remove parentheses.

$$6x - 4 = 2 + 6(\overset{\frown}{x - 1})$$

$$6x - 4 = 2 + 6x - 6 \qquad \text{Apply the distributive property.}$$

$$6x - 4 = 6x - 4 \qquad \text{Combine like terms.}$$

At this point we might notice that both sides of the equation are the same, so replacing x by any real number gives a true statement. Thus the solution set of this equation is the set of real numbers, and the equation is an identity. Continuing to "solve" $6x - 4 = 6x - 4$, we eventually arrive at the same conclusion.

$$6x - 4 + 4 = 6x - 4 + 4 \quad \text{Add 4 to both sides.}$$

$$6x = 6x \qquad\qquad \text{Simplify.}$$

$$6x - 6x = 6x - 6x \qquad \text{Subtract } 6x \text{ from both sides.}$$

$$0 = 0 \qquad\qquad \text{Simplify.}$$

Since $0 = 0$ is a true statement for every value of x, all real numbers are solutions. The solution set is the set of all real numbers, $\mathbb{R}$, or $\{x \mid x \text{ is a real number}\}$, and the equation is called an identity. □

PRACTICE
9   Solve for x: $5x - 2 = 3 + 5(x - 1)$.

> ▶ **Helpful Hint**
>
> For linear equations, *any* false statement such as $5 = 6, 0 = 1$, or $-2 = 2$ informs us that the original equation has no solution. Also, *any* true statement such as $0 = 0, 2 = 2$, or $-5 = -5$ informs us that the original equation is an identity.

VOCABULARY & READINESS CHECK

Use the choices below to fill in the blanks. Not all choices will be used.

multiplication	value	like
addition	solution	equivalent

1. Equations with the same solution set are called _____ equations.

2. A value for the variable in an equation that makes the equation a true statement is called a _____ of the equation.

3. By the _____ property of equality, $y = -3$ and $y - 7 = -3 - 7$ are equivalent equations.

4. By the _____ property of equality, $2y = -3$ and $\dfrac{2y}{2} = \dfrac{-3}{2}$ are equivalent equations.

Identify each as an equation or an expression.

5. $\dfrac{1}{3}x - 5$ _____
6. $2(x - 3) = 7$ _____
7. $\dfrac{5}{9}x + \dfrac{1}{3} = \dfrac{2}{9} - x$ _____
8. $\dfrac{5}{9}x + \dfrac{1}{3} - \dfrac{2}{9} - x$ _____

By inspection, decide which equations have no solution and which equations have all real numbers as solutions.

9. $2x + 3 = 2x + 3$
10. $2x + 1 = 2x + 3$
11. $5x - 2 = 5x - 7$
12. $5x - 3 = 5x - 3$

1.5 EXERCISE SET

MathXL PRACTICE · WATCH · DOWNLOAD · READ · REVIEW

Solve each equation and check. See Examples 1 and 2.

1. $-5x = -30$

2. $-2x = 18$

3. $-10 = x + 12$

4. $-25 = y + 30$

5. $x - 2.8 = 1.9$

6. $y - 8.6 = -6.3$

7. $5x - 4 = 26 + 2x$

8. $5y - 3 = 11 + 3y$

9. $-4.1 - 7z = 3.6$

10. $10.3 - 6x = -2.3$

11. $5y + 12 = 2y - 3$

12. $4x + 14 = 6x + 8$

Solve each equation and check. See Examples 3 and 4.

13. $3x - 4 - 5x = x + 4 + x$

14. $13x - 15x + 8 = 4x + 2 - 24$

15. $8x - 5x + 3 = x - 7 + 10$

16. $6 + 3x + x = -x + 8 - 26 + 24$

17. $5x + 12 = 2(2x + 7)$

18. $2(4x + 3) = 7x + 5$

19. $3(x - 6) = 5x$

20. $6x = 4(x - 5)$

21. $-2(5y - 1) - y = -4(y - 3)$

22. $-4(3n - 2) - n = -11(n - 1)$

Solve each equation and check. See Examples 5 through 7.

23. $\dfrac{x}{2} + \dfrac{x}{3} = \dfrac{3}{4}$

24. $\dfrac{x}{2} + \dfrac{x}{5} = \dfrac{5}{4}$

25. $\dfrac{3t}{4} - \dfrac{t}{2} = 1$

26. $\dfrac{4r}{5} - \dfrac{r}{10} = 7$

27. $\dfrac{n-3}{4} + \dfrac{n+5}{7} = \dfrac{5}{14}$

28. $\dfrac{2+h}{9} + \dfrac{h-1}{3} = \dfrac{1}{3}$

29. $0.6x - 10 = 1.4x - 14$

30. $0.3x + 2.4 = 0.1x + 4$

31. $\dfrac{3x-1}{9} + x = \dfrac{3x+1}{3} + 4$

32. $\dfrac{2z+7}{8} - 2 = z + \dfrac{z-1}{2}$

33. $1.5(4 - x) = 1.3(2 - x)$

34. $2.4(2x + 3) = -0.1(2x + 3)$

Solve each equation. See Examples 8 and 9.

35. $4(n + 3) = 2(6 + 2n)$

36. $6(4n + 4) = 8(3 + 3n)$

37. $3(x + 1) + 5 = 3x + 2$

38. $4(x + 2) + 4 = 4x - 8$

39. $2(x - 8) + x = 3(x - 6) + 2$

40. $5(x - 4) + x = 6(x - 2) - 8$

41. $4(x + 5) = 3(x - 4) + x$

42. $9(x - 2) = 8(x - 3) + x$

Each screen shows a calculator check of a proposed solution to an equation. Write the equation and the verified solution.

43.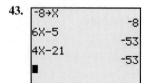

44.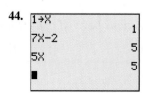

45.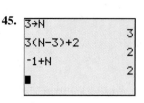

46.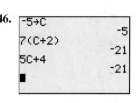

MIXED PRACTICE

Solve each equation. See Examples 1 through 9.

47. $\dfrac{3}{8} + \dfrac{b}{3} = \dfrac{5}{12}$

48. $\dfrac{a}{2} + \dfrac{7}{4} = 5$

49. $x - 10 = -6x - 10$

50. $4x - 7 = 2x - 7$

51. $5(x - 2) + 2x = 7(x + 4) - 38$

52. $3x + 2(x + 4) = 5(x + 1) + 3$

53. $y + 0.2 = 0.6(y + 3)$

54. $-(w + 0.2) = 0.3(4 - w)$

55. $\dfrac{1}{4}(a + 2) = \dfrac{1}{6}(5 - a)$

56. $\dfrac{1}{3}(8 + 2c) = \dfrac{1}{5}(3c - 5)$

57. $2y + 5(y - 4) = 4y - 2(y - 10)$

58. $9c - 3(6 - 5c) = c - 2(3c + 9)$

59. $6x - 2(x - 3) = 4(x + 1) + 4$

60. $10x - 2(x + 4) = 8(x - 2) + 6$

61. $\dfrac{m-4}{3} - \dfrac{3m-1}{5} = 1$

62. $\dfrac{n+1}{8} - \dfrac{2-n}{3} = \dfrac{5}{6}$

63. $8x - 12 - 3x = 9x - 7$

64. $10y - 18 - 4y = 12y - 13$

65. $-(3x - 5) - (2x - 6) + 1 = -5(x - 1) - (3x + 2) + 3$

66. $-4(2x - 3) - (10x + 7) - 2 = -(12x - 5) - (4x + 9) - 1$

67. $\dfrac{1}{3}(y + 4) + 6 = \dfrac{1}{4}(3y - 1) - 2$

68. $\dfrac{1}{5}(2y - 1) - 2 = \dfrac{1}{2}(3y - 5) + 3$

69. $2[7 - 5(1 - n)] + 8n = -16 + 3[6(n + 1) - 3n]$

70. $3[8 - 4(n - 2)] + 5n = -20 + 2[5(1 - n) - 6n]$

CONCEPT EXTENSIONS

Find the error for each proposed solution. Then correct the proposed solution. See the Concept Check in this section.

71. $2x + 19 = 13$

$2x = 32$

$\dfrac{2x}{2} = \dfrac{32}{2}$

$x = 16$

72. $-3(x - 4) = 10$

$-3x - 12 = 10$

$-3x = 22$

$\dfrac{-3x}{-3} = \dfrac{22}{-3}$

$x = -\dfrac{22}{3}$

73. $9x + 1.6 = 4x + 0.4$

$$5x = 1.2$$
$$\frac{5x}{5} = \frac{1.2}{5}$$
$$x = 0.24$$

74. $\dfrac{x}{3} + 7 = \dfrac{5x}{3}$

$$x + 7 = 5x$$
$$7 = 4x$$
$$\frac{7}{4} = \frac{4x}{4}$$
$$\frac{7}{4} = x$$

75. a. Simplify the expression $4(x + 1) + 1$.

   b. Solve the equation $4(x + 1) + 1 = -7$.

   c. Explain the difference between solving an equation for a variable and simplifying an expression.

76. Explain why the multiplication property of equality does not include multiplying both sides of an equation by 0. (*Hint:* Write down a false statement and then multiply both sides by 0. Is the result true or false? What does this mean?)

77. In your own words, explain why the equation $x + 7 = x + 6$ has no solution while the solution set of the equation $x + 7 = x + 7$ contains all real numbers.

78. In your own words, explain why the equation $x = -x$ has one solution—namely, 0—while the solution set of the equation $x = x$ is all real numbers.

Find the value of K such that the equations are equivalent.

79. $3.2x + 4 = 5.4x - 7$

$$3.2x = 5.4x + K$$

80. $-7.6y - 10 = -1.1y + 12$

$$-7.6y = -1.1y + K$$

81. $\dfrac{7}{11}x + 9 = \dfrac{3}{11}x - 14$

$$\frac{7}{11}x = \frac{3}{11}x + K$$

82. $\dfrac{x}{6} + 4 = \dfrac{x}{3}$

$$x + K = 2x$$

83. Write a linear equation in x whose only solution is 5.

84. Write an equation in x that has no solution.

Solve the following.

85. $x(x - 6) + 7 = x(x + 1)$

86. $7x^2 + 2x - 3 = 6x(x + 4) + x^2$

87. $3x(x + 5) - 12 = 3x^2 + 10x + 3$

88. $x(x + 1) + 16 = x(x + 5)$

Solve and check.

89. $2.569x = -12.48534$

90. $-9.112y = -47.537304$

91. $2.86z - 8.1258 = -3.75$

92. $1.25x - 20.175 = -8.15$

THE BIGGER PICTURE   SOLVING EQUATIONS AND INEQUALITIES

This is a special feature that will be repeated and expanded throughout this text. It is very important for you to be able to recognize and solve different types of equations and inequalities. To help you do this, we will begin an outline below and continually expand this outline as different equations (and inequalities) are introduced. Although suggestions will be given, this outline should be in your own words and you should include at least "how to recognize" and "how to begin to solve" under each letter heading.

For example:

Solving Equations and Inequalities

I. Equations

   A. Linear equations: Power on variable is 1 and there are no variables in denominator. (Section 1.5)

$$5(x - 2) = \frac{4(2x + 1)}{3}$$

$$3 \cdot 5(x - 2) = \cancel{3} \cdot \frac{4(2x + 1)}{\cancel{3}} \qquad \text{Multiply both sides by the LCD, 3.}$$

$$15(x - 2) = 4(2x + 1) \qquad \text{Simplify.}$$

$$15x - 30 = 8x + 4 \qquad \text{Multiply.}$$

$$7x = 34 \qquad \text{Add 30 and subtract } 8x \text{ from both sides.}$$

$$x = \frac{34}{7} \qquad \text{Divide both sides by 7.}$$

II. Inequalities

Solve.

1. $3x - 4 = 3(2x - 1) + 7$

2. $5 + 2x = 5(x + 1)$

3. $\dfrac{x + 3}{2} = 1$

4. $\dfrac{x - 2}{2} - \dfrac{x - 4}{3} = \dfrac{5}{6}$

5. $\dfrac{7}{5} + \dfrac{y}{10} = 2$

6. $5 + 2x = 2(x + 1)$

7. $4(x - 2) + 3x = 9(x - 1) - 2$

8. $6(x + 1) - 2 = 6x + 4$

1.6 AN INTRODUCTION TO PROBLEM SOLVING

OBJECTIVE 1 ▶ Writing and simplifying algebraic expressions. In order to prepare for problem solving, we practice writing algebraic expressions that can be simplified.

Our first example involves consecutive integers and perimeter. Recall that *consecutive integers* are integers that follow one another in order. Study the examples of consecutive, even, and odd integers and their representations.

Consecutive Integers: **Consecutive Even Integers:**   **Consecutive Odd Integers:**

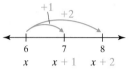   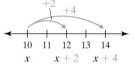   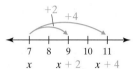

EXAMPLE 1   Write the following as algebraic expressions. Then simplify.

a. The sum of three consecutive integers, if x is the first consecutive integer.

△ **b.** The perimeter of the triangle with sides of length $x, 5x$, and $6x - 3$.

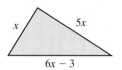

Solution

a. Recall that if x is the first integer, then the next consecutive integer is 1 more, or $x + 1$, and the next consecutive integer is 1 more than $x + 1$, or $x + 2$.

In words:	first integer	plus	next consecutive integer	plus	next consecutive integer
	↓	↓	↓	↓	↓
Translate:	x	$+$	$(x + 1)$	$+$	$(x + 2)$

Then $x + (x + 1) + (x + 2) = x + x + 1 + x + 2$

$= 3x + 3$ Simplify by combining like terms.

b. The perimeter of a triangle is the sum of the lengths of the sides.

In words:	side	+	side	+	side
	↓		↓		↓
Translate:	x	$+$	$5x$	$+$	$(6x - 3)$

Then $x + 5x + (6x - 3) = x + 5x + 6x - 3$

$= 12x - 3$ Simplify.   □

PRACTICE

1   Write the following algebraic expressions. Then simplify.

a. The sum of three consecutive odd integers, if x is the first consecutive odd integer

b. The perimeter of a trapezoid with bases x and $2x$, and sides of $x + 2$ and $2x - 3$

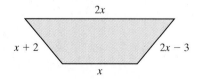

EXAMPLE 2   The three busiest airports in the United States are in Chicago, Atlanta, and Los Angeles. The airport in Atlanta has 13.3 million more arrivals and departures than the Los Angeles airport. The Chicago airport has 1.3 million more arrivals and departures than the Los Angeles airport. Write the sum of the arrivals and departures from these three cities as a simplified algebraic expression. Let x be the number of arrivals and departures at the Los Angeles airport. (*Source:* U.S. Department of Transportation)

Solution   If x = millions of arrivals and departures at the Los Angeles airport, then

$x + 13.3$ = millions of arrivals and departures at the Atlanta airport and

$x + 1.3$ = millions of arrivals and departures at the Chicago airport.

Since we want their sum, we have

In words:	arrivals and departures at Los Angeles	+	arrivals and departures at Atlanta	+	arrivals and departures at Chicago
Translate:	x	+	$(x + 13.3)$	+	$(x + 1.3)$

Then
$$x + (x + 13.3) + (x + 1.3) = x + x + 13.3 + x + 1.3$$
$$= 3x + 14.6 \quad \text{Combine like terms.}$$

In Exercise 57, we will find the actual number of arrivals and departures at these airports.   □

PRACTICE
2   The three busiest airports in Europe are in London, England; Paris, France; and Frankfurt, Germany. The airport in London has 15.7 million more arrivals and departures than the Frankfurt airport. The Paris airport has 1.6 million more arrivals and departures than the Frankfurt airport. Write the sum of the arrivals and departures from these three cities as a simplified algebraic expression. Let x be the number of arrivals and departures at the Frankfurt airport. (*Source:* Association of European Airlines)

▶ **Helpful Hint**

You may want to begin this section by studying key words and phrases and their translations in Section 1.2 Objective 5 and Section 1.4 Objective 4.

OBJECTIVE 2 ▶ **Applying steps for problem solving.** Our main purpose for studying algebra is to solve problems. The following problem-solving strategy will be used throughout this text and may also be used to solve real-life problems that occur outside the mathematics classroom.

General Strategy for Problem Solving

1. UNDERSTAND the problem. During this step, become comfortable with the problem. Some ways of doing this are:

   Read and reread the problem.

   Propose a solution and check. Pay careful attention to how you check your proposed solution. This will help when writing an equation to model the problem.

   Construct a drawing.

   Choose a variable to represent the unknown. (Very important part)

2. TRANSLATE the problem into an equation.

3. SOLVE the equation.

4. INTERPRET the results: *Check* the proposed solution in the stated problem and *state* your conclusion.

Let's review this strategy by solving a problem involving unknown numbers.

EXAMPLE 3 Finding Unknown Numbers

Find three numbers such that the second number is 3 more than twice the first number, and the third number is four times the first number. The sum of the three numbers is 164.

Solution

1. UNDERSTAND the problem. First let's read and reread the problem and then propose a solution. For example, if the first number is 25, then the second number is 3 more than twice 25, or 53. The third number is four times 25, or 100. The sum of 25, 53, and 100 is 178, not the required sum, but we have gained some valuable information about the problem. First, we know that the first number is less than 25 since our guess led to a sum greater than the required sum. Also, we have gained some information as to how to model the problem.

> ▶ **Helpful Hint**
>
> The purpose of guessing a solution is not to guess correctly but to gain confidence and to help understand the problem and how to model it.

Next let's assign a variable and use this variable to represent any other unknown quantities. If we let

$$x = \text{the first number, then}$$
$$2x + 3 = \text{the second number}$$

3 more than
twice the second number

$$4x = \text{the third number}$$

2. TRANSLATE the problem into an equation. To do so, we use the fact that the sum of the numbers is 164. First let's write this relationship in words and then translate to an equation.

In words:	first number	added to	second number	added to	third number	is	164
Translate:	x	$+$	$(2x + 3)$	$+$	$4x$	$=$	164

3. SOLVE the equation.

$$x + (2x + 3) + 4x = 164$$
$$x + 2x + 4x + 3 = 164 \quad \text{Remove parentheses.}$$
$$7x + 3 = 164 \quad \text{Combine like terms.}$$
$$7x = 161 \quad \text{Subtract 3 from both sides.}$$
$$x = 23 \quad \text{Divide both sides by 7.}$$

4. INTERPRET. Here, we *check* our work and *state* the solution. Recall that if the first number $x = 23$, then the second number $2x + 3 = 2 \cdot 23 + 3 = 49$ and the third number $4x = 4 \cdot 23 = 92$.

Check: Is the second number 3 more than twice the first number? Yes, since 3 more than twice 23 is $46 + 3$, or 49. Also, their sum, $23 + 49 + 92 = 164$, is the required sum.

State: The three numbers are 23, 49, and 92. ☐

PRACTICE

3 Find three numbers such that the second number is 8 less than triple the first number, the third number is five times the first number, and the sum of the three numbers is 118.

Many of today's rates and statistics are given as percents. Interest rates, tax rates, nutrition labeling, and percent of households in a given category are just a few examples.

Before we practice solving problems containing percents, let's briefly take a moment and review the meaning of percent and how to find a percent of a number.

The word *percent* means "per hundred," and the symbol % is used to denote percent. This means that 23% is 23 per hundred, or $\frac{23}{100}$. Also,

$$41\% = \frac{41}{100} = 0.41$$

To find a percent of a number, we multiply.

$$16\% \text{ of } 25 = 16\% \cdot 25 = 0.16 \cdot 25 = 4$$

Thus, 16% of 25 is 4.

Study the table below. It will help you become more familiar with finding percents.

Percent	*Meaning/Shortcut*	*Example*
50%	$\frac{1}{2}$ or half of a number	50% of 60 is 30.
25%	$\frac{1}{4}$ or a quarter of a number	25% of 60 is 15.
10%	0.1 or $\frac{1}{10}$ of a number (move the decimal point 1 place to the left)	10% of 60 is 6.0 or 6.
1%	0.01 or $\frac{1}{100}$ of a number (move the decimal point 2 places to the left)	1% of 60 is 0.60 or 0.6.
100%	1 or all of a number	100% of 60 is 60.
200%	2 or double a number	200% of 60 is 120.

Concept Check ☑

Suppose you are finding 112% of a number x. Which of the following is a correct description of the result? Explain.

a. The result is less than x.   **b.** The result is equal to x.   **c.** The result is greater than x.

Next, we solve a problem containing a percent.

EXAMPLE 4   Finding the Original Price of a Computer

Suppose that a computer store just announced an 8% decrease in the price of a particular computer model. If this computer sells for $2162 after the decrease, find the original price of this computer.

Solution

1. UNDERSTAND. Read and reread the problem. Recall that a percent decrease means a percent of the original price. Let's guess that the original price of the computer is $2500. The amount of decrease is then 8% of $2500, or $(0.08)(\$2500) = \$200$. This means that the new price of the computer is the original price minus the decrease, or $2500 − $200 = $2300. Our guess is incorrect, but we now have an idea of how to model this problem. In our model, we will let $x =$ the original price of the computer.

2. TRANSLATE.

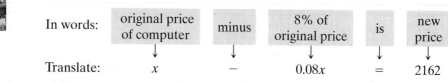

In words:	original price of computer	minus	8% of original price	is	new price
Translate:	x	−	$0.08x$	=	2162

3. SOLVE the equation.

$$x - 0.08x = 2162$$
$$0.92x = 2162 \qquad \text{Combine like terms.}$$
$$x = \frac{2162}{0.92} = 2350 \quad \text{Divide both sides by 0.92.}$$

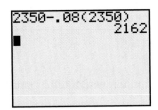

4. INTERPRET.

Check: If the original price of the computer was \$2350, the new price is

$$\$2350 - (0.08)(\$2350) = \$2350 - \$188$$
$$= \$2162 \qquad \text{The given new price}$$

State: The original price of the computer was \$2350.   ☐

PRACTICE

4   At the end of the season, the cost of a snowboard was reduced by 40%. If the snowboard sells for \$270 after the decrease, find the original price of the board.

◢ **EXAMPLE 5**   **Finding the Lengths of a Triangle's Sides**

A pennant in the shape of an isosceles triangle is to be constructed for the Slidell High School Athletic Club and sold at a fundraiser. The company manufacturing the pennant charges according to perimeter, and the athletic club has determined that a perimeter of 149 centimeters should make a nice profit. If each equal side of the triangle is twice the length of the third side, increased by 12 centimeters, find the lengths of the sides of the triangular pennant.

Solution

1. UNDERSTAND. Read and reread the problem. Recall that the perimeter of a triangle is the distance around. Let's guess that the third side of the triangular pennant is 20 centimeters. This means that each equal side is twice 20 centimeters, increased by 12 centimeters, or $2(20) + 12 = 52$ centimeters.

This gives a perimeter of $20 + 52 + 52 = 124$ centimeters. Our guess is incorrect, but we now have a better understanding of how to model this problem.

Now we let the third side of the triangle $= x$

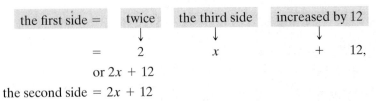

the first side =	twice	the third side	increased by 12
	↓	↓	↓
=	2	x	+ 12,

or $2x + 12$

the second side $= 2x + 12$

2. TRANSLATE.

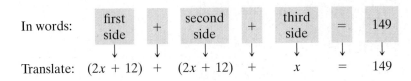

In words:	first side	+	second side	+	third side	=	149
	↓	↓	↓	↓	↓	↓	↓
Translate:	$(2x + 12)$	+	$(2x + 12)$	+	x	=	149

3. SOLVE the equation.

$$(2x + 12) + (2x + 12) + x = 149$$

$$2x + 12 + 2x + 12 + x = 149 \quad \text{Remove parentheses.}$$

$$5x + 24 = 149 \quad \text{Combine like terms.}$$

$$5x = 125 \quad \text{Subtract 24 from both sides.}$$

$$x = 25 \quad \text{Divide both sides by 5.}$$

4. INTERPRET. If the third side is 25 centimeters, then the first side is $2(25) + 12 = 62$ centimeters and the second side is 62 centimeters also.

Check: The first and second sides are each twice 25 centimeters increased by 12 centimeters, or 62 centimeters. Also, the perimeter is $25 + 62 + 62 = 149$ centimeters, the required perimeter.

State: The lengths of the sides of the triangle are 25 centimeters, 62 centimeters, and 62 centimeters.   ☐

PRACTICE

5   For its 40th anniversary, North Campus Community College is placing rectangular banners on all the light poles on campus. The perimeter of these banners is 160 inches. If the longer side of each banner is 16″ less than double the width, find the dimensions of the banners.

EXAMPLE 6   **Finding Consecutive Integers**

Kelsey Ohleger was helping her friend Benji Burnstine study for an algebra exam. Kelsey told Benji that her three latest art history quiz scores are three consecutive even integers whose sum is 264. Help Benji find the scores.

Solution

1. UNDERSTAND. Read and reread the problem. Since we are looking for consecutive even integers, let

$$x = \text{the first integer. Then}$$
$$x + 2 = \text{the second consecutive even integer.}$$
$$x + 4 = \text{the third consecutive even integer.}$$

2. TRANSLATE.

In words:	first integer	+	second even integer	+	third even integer	=	264
Translate:	x	+	$(x + 2)$	+	$(x + 4)$	=	264

3. SOLVE.

$$x + (x + 2) + (x + 4) = 264$$

$$3x + 6 = 264 \quad \text{Combine like terms.}$$

$$3x = 258 \quad \text{Subtract 6 from both sides.}$$

$$x = 86 \quad \text{Divide both sides by 3.}$$

4. INTERPRET. If $x = 86$, then $x + 2 = 86 + 2$ or 88, and $x + 4 = 86 + 4$ or 90.

Check: The numbers 86, 88, and 90 are three consecutive even integers. Their sum is 264, the required sum.

State: Kelsey's art history quiz scores are 86, 88, and 90.   ☐

PRACTICE

6   Find three consecutive odd integers whose sum is 81.

VOCABULARY & READINESS CHECK

Fill in each blank with $<$, $>$, *or* $=$. *(Assume that the unknown number is a positive number.)*

1. 130% of a number _____ the number.

2. 70% of a number _____ the number.

3. 100% of a number _____ the number.

4. 200% of a number _____ the number.

Complete the table. The first row has been completed for you.

	First Integer	All Described Integers
Three consecutive integers	18	18, 19, 20
5. Four consecutive integers	31	
6. Three consecutive odd integers	31	
7. Three consecutive even integers	18	
8. Four consecutive even integers	92	
9. Three consecutive integers	y	
10. Three consecutive even integers	z (z is even)	
11. Four consecutive integers	p	
12. Three consecutive odd integers	s (s is odd)	

1.6 EXERCISE SET

MyMathLab® *Powered by CourseCompass™ and MathXL®*

MathXL PRACTICE | WATCH | DOWNLOAD | READ | REVIEW

Write the following as algebraic expressions. Then simplify. See Examples 1 and 2.

△ **1.** The perimeter of the square with side length y.

y

△ **2.** The perimeter of the rectangle with length x and width $x - 5$.

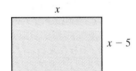
x

$x - 5$

3. The sum of three consecutive integers if the first is z.

4. The sum of three consecutive odd integers if the first integer is x.

5. The total amount of money (in cents) in x nickels, $(x + 3)$ dimes, and $2x$ quarters. (*Hint:* The value of a nickel is 5 cents, the value of a dime is 10 cents, and the value of a quarter is 25 cents.)

6. The total amount of money (in cents) in y quarters, $7y$ dimes, and $(2y - 1)$ nickels. (Use the hint for Exercise 5.)

△ **7.** A piece of land along Bayou Liberty is to be fenced and subdivided as shown so that each rectangle has the same dimensions. Express the total amount of fencing needed as an algebraic expression in x.

x

$2x + 1$ | ? | ?

8. A flooded piece of land near the Mississippi River in New Orleans is to be surveyed and divided into 4 rectangles of equal dimension. Express the total amount of fencing needed as an algebraic expression in x.

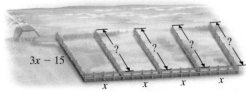
$3x - 15$
? ? ? ?
x x x x

△ **9.** Write the perimeter of the floor plan shown as an algebraic expression in x.

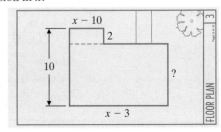
$x - 10$
2
10
?
$x - 3$
FLOOR PLAN

10. Write the perimeter of the floor plan shown as an algebraic expression in x.

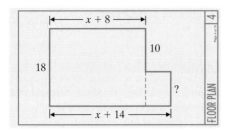

Solve. See Example 3.

11. Four times the difference of a number and 2 is the same as 2 increased by four times the number plus twice the number. Find the number.

12. Twice the sum of a number and 3 is the same as five times the number minus 1 minus four times the number. Find the number.

13. A second number is five times a first number. A third number is 100 more than the first number. If the sum of the three numbers is 415, find the numbers.

14. A second number is 6 less than a first number. A third number is twice the first number. If the sum of the three numbers is 306, find the numbers.

Solve. See Example 4.

15. The United States consists of 2271 million acres of land. Approximately 29% of this land is federally owned. Find the number of acres that are not federally owned. (*Source:* U.S. General Services Administration)

16. The state of Nevada contains the most federally owned acres of land in the United States. If 90% of the state's 70 million acres of land is federally owned, find the number of acres that are not federally owned. (*Source:* U.S. General Services Administration)

17. In 2006, a total of 2748 earthquakes occurred in the United States. Of these, 85.3% were minor tremors with magnitudes of 3.9 or less on the Richter scale. How many minor earthquakes occurred in the United States in 2006? Round to the nearest whole number. (*Source:* U.S. Geological Survey National Earthquake Information Center)

18. Of the 958 tornadoes that occurred in the United States during 2006, 25.5% occurred during the month of April. How many tornadoes occurred during April 2006? Round to the nearest whole number. (*Source:* Storm Prediction Center)

19. In a recent survey, 15% of online shoppers in the United States say that they prefer to do business only with large, well-known retailers. In a group of 1500 online shoppers, how many are willing to do business with any size retailers? (*Source:* Inc.com)

20. In 2006, the restaurant and food service industry employed 9.1% of the total employees in the state of California. In 2006, California had approximately 17,029,300 employed workers. How many people worked in the restaurant and food service industry in California? Round to the nearest whole number. (*Source:* California Employment Development Center)

The following graph is called a circle graph or a pie chart. The circle represents a whole, or in this case, 100%. This particular graph shows the number of minutes per day that people use e-mail at work. Use this graph to answer Exercises 21 through 24.

Time Spent on E-Mail at Work

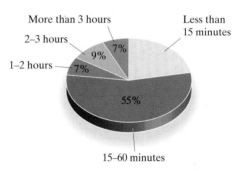

Source: Pew Internet & American Life Project

21. What percent of e-mail users at work spend less than 15 minutes on e-mail per day?

22. Among e-mail users at work, what is the most common time spent on e-mail per day?

23. If it were estimated that a large company has 5957 employees, how many of these would you expect to be using e-mail more than 3 hours per day? Round to the nearest whole employee.

24. If it were estimated that a medium-size company has 278 employees, how many of these would you expect to be using e-mail between 2 and 3 hours per day? Round to the nearest whole employee.

25. In 2006, the population of Canada was 31.6 million. This represented an increase in population of 5.4% since 2001. What was the population of Canada in 2001? (Round to the nearest hundredth of a million.) (*Source:* Statistics Canada)

26. In 2006, the cost of an average hotel room per night was $96.73. This was an increase of 6.8% over the average cost in 2005. Find the average hotel room cost in 2005. (*Source:* Smith Travel Research)

Use the diagrams to find the unknown measures of angles or lengths of sides. Recall that the sum of the angle measures of a triangle is 180°. See Example 5.

27.

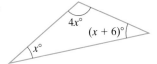

28.

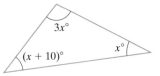

29.

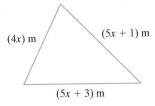

Perimeter is 102 meters.

30.

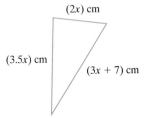

Perimeter is 75 centimeters.

31.

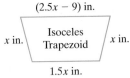

Perimeter is 99 inches.

32.

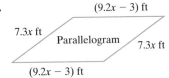

Perimeter is 324 feet.

Solve. See Example 6.

33. The sum of three consecutive integers is 228. Find the integers.

34. The sum of three consecutive odd integers is 327. Find the integers.

35. The zip codes of three Nevada locations—Fallon, Fernley, and Gardnerville Ranchos—are three consecutive even integers. If twice the first integer added to the third is 268,222, find each zip code.

36. During a recent year, the average SAT scores in math for the states of Alabama, Louisiana, and Michigan were 3 consecutive integers. If the sum of the first integer, second integer, and three times the third integer is 2637, find each score.

MIXED PRACTICE

Solve. See Examples 1 through 6. Many companies and government agencies predict the growth or decline of occupations. The following data are based on information from the U.S. Department of Labor. Notice that the first table is in increase in number of jobs (in thousands) and the second table is in percent increase in number of jobs.

37. Use the table to find the actual increase in number of jobs for each occupation.

Occupation	Increase in Number of Jobs (in thousands) from 2000 to 2012
Security guards	$2x - 51$
Home health aides	$\frac{3}{2}x + 3$
Computer systems analysts	x
Total	780 thousand

38. Use the table to find the actual percent increase in number of jobs for each occupation.

Occupation	Percent Increase in Number of Jobs from 2000 to 2012
Computer software engineers	$\frac{3}{2}x + 1$
Management analysts	x
Receptionist and information clerks	$x - 1$
Total	105%

39. The occupations of postsecondary teachers, registered nurses, and medical assistants are among the ten with the largest growth from 2000 to 2012. (See the Chapter 2 opener.) The number of postsecondary teacher jobs will grow 173 thousand more than twice the number of medical assistant jobs. The number of registered nurse jobs will grow 22 thousand less than three times the number of medical assistant jobs. If the total growth of these three jobs is predicted to be 1441 thousand, find the predicted growth of each job.

40. The occupations of telephone operators, fishers, and sewing machine operators are among the ten with the largest job decline from 2000 to 2012, according to the U.S. Department of Labor. The number of telephone operators will decline 8 thousand more than twice the number of fishers. The number of sewing machine operators will decline 1 thousand less than

10 times the number of fishers. If the total decline of these three jobs is predicted to be 137 thousand, find the predicted decline of each job.

41. America West Airlines has 3 models of Boeing aircraft in their fleet. The 737-300 contains 21 more seats than the 737-200. The 757-200 contains 36 less seats than twice the number of seats in the 737-200. Find the number of seats for each aircraft if the total number of seats for the 3 models is 437. (*Source:* America West Airlines)

42. The new governor of California makes $27,500 more than the governor of New York, and $120,724 more than the governor of Alaska. If the sum of these 3 salaries is $471,276, find the salary of each governor. (*Source:* State Web sites)

43. A new fax machine was recently purchased for an office in Hopedale for $464.40 including tax. If the tax rate in Hopedale is 8%, find the price of the fax machine before tax.

44. A premedical student at a local university was complaining that she had just paid $158.60 for her human anatomy book, including tax. Find the price of the book before taxes if the tax rate at this university is 9%.

45. In 2006, the population of South Africa was 44.2 million people. From 2006 to 2050, South Africa's population is expected to decrease by 5.6%. Find the expected population of South Africa in 2050. Round to the nearest tenth of a million. (*Source:* Population Reference Bureau)

46. In 2006, the population of Morocco was 33.2 million. This represented an increase in population of 1.5% from a year earlier. What was the population of Morocco in 2005? Round to the nearest tenth of a million. (*Source:* Population Reference Bureau)

Recall that two angles are complements of each other if their sum is 90°. Two angles are supplements of each other if their sum is 180°. Find the measure of each angle.

47. One angle is three times its supplement increased by 20°. Find the measures of the two supplementary angles.

48. One angle is twice its complement increased by 30°. Find the measure of the two complementary angles.

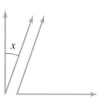

Recall that the sum of the angle measures of a triangle is 180°.

49. Find the measures of the angles of a triangle if the measure of one angle is twice the measure of a second angle and the third angle measures 3 times the second angle decreased by 12.

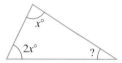

50. Find the angles of an isoceles triangle whose two base angles are equal and whose third angle is 10° less than three times a base angle.

51. Two frames are needed with the same perimeter: one frame in the shape of a square and one in the shape of an equilateral triangle. Each side of the triangle is 6 centimeters longer than each side of the square. Find the dimensions of each frame. (An equilateral triangle has sides that are the same length.)

52. Two frames are needed with the same perimeter: one frame in the shape of a square and one in the shape of a regular pentagon. Each side of the square is 7 inches longer than each side of the pentagon. Find the dimensions of each frame. (A regular polygon has sides that are the same length.)

53. The sum of the first and third of three consecutive even integers is 156. Find the three even integers.

54. The sum of the second and fourth of four consecutive integers is 110. Find the four integers.

△ **55.** The perimeter of the triangle in Example 1b in this section is 483 feet. Find the length of each side.

△ **56.** The perimeter of the trapezoid in Practice 1b in this section is 110 meters. Find the lengths of its sides and bases.

57. The airports in Chicago, Atlanta, and Los Angeles have a total of 197.6 million annual arrivals and departures. Use this information and Example 2 in this section to find the number from each individual airport.

58. The airports in London, Paris, and Frankfurt have a total of 173.9 million annual arrivals and departures. Use this information and Practice 2 in this section to find the number from each airport.

59. Incandescent, fluorescent, and halogen bulbs are lasting longer today than ever before. On average, the number of bulb hours for a fluorescent bulb is 25 times the number of bulb hours for a halogen bulb. The number of bulb hours for an incandescent bulb is 2500 less than the halogen bulb. If the total number of bulb hours for the three types of bulbs is 105,500, find the number of bulb hours for each type. (*Source: Popular Science Magazine*)

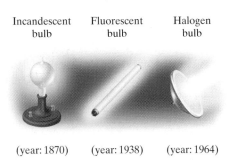

Incandescent   Fluorescent   Halogen
bulb   bulb   bulb

(year: 1870)   (year: 1938)   (year: 1964)

60. The three tallest hospitals in the world are Guy's Tower in London, Queen Mary Hospital in Hong Kong, and Galter Pavilion in Chicago. These buildings have a total height of 1320 feet. Guy's Tower is 67 feet taller than Galter Pavilion and the Queen Mary Hospital is 47 feet taller than Galter Pavilion. Find the heights of the three hospitals.

△ **61.** The official manual for traffic signs is the *Manual on Uniform Traffic Control Devices* published by the Government Printing Office. The rectangular sign below has a length 12 inches more than twice its height. If the perimeter of the sign is 312 inches, find its dimensions.

62. INVESCO Field at Mile High, home to the Denver Broncos, has 11,675 more seats than Heinz Field, home to the Pittsburgh Steelers. Together, these two stadiums can seat a total of 140,575 NFL fans. How many seats does each stadium have? (*Sources:* Denver Broncos, Pittsburgh Steelers)

63. In 2004, sales in the athletic footwear market in the United States were $4.7 billion. (*Source:* NPD Group)
   a. There was a 5% increase in the athletic footwear sales from 2004 to 2006. Find the total sales for the athletic footwear market in 2006. (Round to the nearest tenth of a billion.)
   b. Twenty-five percent of the sales of footwear were running shoes. Find the amount of money spent on running shoes in 2006. (Round to the nearest hundredth of a billion.)

64. The number of deaths caused by tornadoes decreased 59.2% from the 1950s to the 1990s. There were 579 deaths from tornadoes in the 1990s. (*Source:* National Weather Service)
   a. Find the number of deaths caused by tornadoes in the 1950s. Round to the nearest whole number.
   ✎ **b.** In your own words, explain why you think that the number of tornado-related deaths has decreased so much since the 1950s.

65. China, the United States, and Russia are the countries with the most cellular subscribers in the world. Together, the three countries have 34.8% of the world's cellular subscribers. If the percent of world subscribers in China is 3.1 less than 4 times the percent of world subscribers in Russia, and the percent of world subscribers in the United States is 4.3% more than the percent of world subscribers in Russia, find the percent of world subscribers for each country. (*Source:* Computer Industry of America)

66. In 2006, 74.2 million tax returns were filed electronically (online filing, professional electronic filing, and TeleFile). This represents a 108.8% increase over the year 2005. How many income tax returns were filed electronically in 2005? Round to the nearest tenth of a million. (*Source:* IRS)

67. The popularity of the Harry Potter series of books is an unmatched phenomenon among readers of all ages. To satisfy their desires for Harry Potter adventures, author J. K. Rowling increased the number of pages in the later books of the Harry Potter adventures. The final adventure of the series, *Harry Potter and the Deathly Hallows,* boasts 784 pages in its hardcover edition. This is a 154% increase over the number of pages in the first Harry Potter adventure. How many pages are in *Harry Potter and the Sorcerer's Stone,* the first adventure? (Round to the nearest whole page.) (*Source:* Amazon.com)

68. During the 2006 Major League Baseball season, the number of home runs hit by Alfonso Soriano of the Washington Nationals, Lance Berkman of the Houston Astros, and Jermaine Dye of the Chicago White Sox were three consecutive integers. Of these three players, Soriano hit the most home runs and Dye hit the least. The total number of home runs hit by these three players over the course of the season was 135. How many home runs did each player hit during the 2006 season? (*Source:* Major League Baseball)

69. During the 2006 Winter Olympic Games in Torino, Italy, the total number of gold medals won by Germany, Canada, and the United States were three consecutive odd integers. Of these three countries, Germany won the most gold medals and Canada won the fewest. If the sum of the first integer, twice the second integer, and four times the third integer is 69, find the number of gold medals won by each country. (*Source: Sports Illustrated Almanac*)

CONCEPT EXTENSIONS

70. For Exercise 38, the percents have a sum of 105%. Is this possible? Why or why not?

71. In your own words, explain the differences in the tables for Exercises 37 and 38.

72. Choose two occupations from the six given in Exercises 37 and 38. Define these occupations and include the education needed for each.

73. Find an angle such that its supplement is equal to twice its complement increased by 50°.

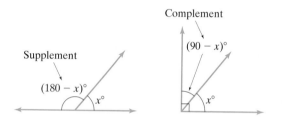

74. Newsprint is either discarded or recycled. Americans recycle about 27% of all newsprint, but an amount of newsprint equivalent to 30 million trees is discarded every year. About how many trees' worth of newsprint is *recycled* in the United States each year? (*Source:* The Earth Works Group)

75. The average annual number of cigarettes smoked by an American adult continues to decline. For the years 1997–2006, the equation $y = -49.4x + 1756.8$ approximates this data. Here, x is the number of years since 1997 and y is the average annual number of cigarettes smoked. (*Source:* Centers for Disease Control)

a. If this trend continues, find the year in which the average annual number of cigarettes smoked is 0. To do this, let $y = 0$ and solve for x.

b. Predict the average annual number of cigarettes smoked by an American adult in 2012. To do so, let $x = 15$ (since $2012 - 1997 = 15$) and find y.

c. Use the result of part **b** to predict the average *daily* number of cigarettes smoked by an American adult in 2012. Round to the nearest whole cigarette. Do you think this number represents the average daily number of cigarettes smoked by an adult smoker? Why or why not?

To break even in a manufacturing business, income or revenue R must equal the cost of production C. Use this information to answer Exercises 76 through 79.

76. The cost C to produce x number of skateboards is $C = 100 + 20x$. The skateboards are sold wholesale for $24 each, so revenue R is given by $R = 24x$. Find how many skateboards the manufacturer needs to produce and sell to break even. (*Hint:* Set the cost expression equal to the revenue expression and solve for x.)

77. The revenue R from selling x number of computer boards is given by $R = 60x$, and the cost C of producing them is given by $C = 50x + 5000$. Find how many boards must be sold to break even. Find how much money is needed to produce the break-even number of boards.

78. In your own words, explain what happens if a company makes and sells fewer products than the break-even number.

79. In your own words, explain what happens if more products than the break-even number are made and sold.

80. In 2007, 76.8 million tax returns were filed electronically. From 2008 to 2011, the average annual increase in electronic filing is projected to be 4.2% of each previous year. If this holds true, what will the number of electronic tax returns be for the years 2008–2011? Round to the nearest tenth of a million. (*Source:* IRS)

 STUDY SKILLS BUILDER

Organizing a Notebook

It's never too late to get organized. If you need ideas about organizing a notebook for your mathematics course, try some of these:

- Use a spiral or ring binder notebook with pockets and use it for mathematics only.

- Start each page by writing the book's section number you are working on at the top.

- When your instructor is lecturing, take notes. *Always* include any examples your instructor works for you.

- Place your worked-out homework exercises in your notebook immediately after the lecture notes from that section. This way, a section's worth of material is together.

- Homework exercises: Attempt all assigned homework. For odd-numbered exercises, you are not through until you check your answers against the back of the book. Correct any exercises with incorrect answers. You may want to place a "?" by any homework exercises or notes that you need to ask questions about. Also, consider placing a "!" by any notes or exercises you feel are important.

- Place graded quizzes in the pockets of your notebook. If you are using a binder, you can place your quizzes in a special section of your binder.

Let's check your notebook organization by answering the following questions.

1. Do you have a spiral or ring binder notebook for your mathematics course only?

2. Have you ever had to flip through several sheets of notes and work in your mathematics notebook to determine what section's work you are in?

3. Are you now writing the textbook's section number at the top of each notebook page?

4. Have you ever lost or had trouble finding a graded quiz or test?

5. Are you now placing all your graded work in a dedicated place in your notebook?

6. Are you attempting all of your homework and placing all of your work in your notebook?

7. Are you checking and correcting your homework in your notebook? If not, why not?

8. Are you writing in your notebook the examples your instructor works for you in class?

1.7 A NUMERICAL APPROACH: MODELING WITH TABLES

OBJECTIVE

1 Use tables to solve problems.

In this section, we introduce a numerical approach to problem solving. This approach may be used instead of or in addition to an algebraic approach. A numerical approach is invaluable for developing estimation skills and number sense.

In earlier sections, we introduced methods for using a calculator to evaluate an algebraic expression at given values of the variable. Often we organized our results in tables such as the one below.

Time (seconds)	x	2	3	4	5	6
Speed (feet per second)	$32x$	64	96	128	160	192

Notice for this particular table that the minimum x-value is 2, and the x-values increase by 1. Another method for completing a table such as this is the table feature of your calculator.

When generating a **table** on a calculator, we need to give the calculator specific instructions as to a minimum x-value to *start* with and the *increment* or change in the x-values. (For many calculators, the Greek letter delta, Δ, is used in the table setup menu. Here, Δ indicates the change in x.) In this text, we will use the term *table start* to indicate the first x-value to appear in the table and the term *table increment* to indicate the increment or change in x-values.

To display the table above on your calculator, go to the Y= editor and enter the expression $32x$. Next, go to the table setup menu. There, let table start = 2 and table increment = 1 as shown below to the left. Then the displayed table should look like the table below to the right.

TECHNOLOGY NOTE

On some graphing utilities, table increment is designated by Δ Tbl or Pitch.

OBJECTIVE 1 ▶ Using tables to solve problems. Next, we generate tables to help us solve problems.

EXAMPLE 1   Finding Commission

Bette Meish is considering a job as a salesperson for a computer software company. She is offered monthly gross pay of $2500 plus 5% commission on sales. An expression that models her gross pay is

In words:	$2500	plus	5% of sales
	↓	↓	↓
Translate:	2500	+	$0.05x$

To help her decide about the job, she considers her monthly pay for sales amounts of $1000 to $9000 in $1000 increments.

Sales, x	1000	2000	3000	4000	5000	6000	7000	8000	9000
Gross pay $2500 + 0.05x$									

a. Complete the table.

b. The company informs her that the average sales per month for its sales force is $6000. Find her annual gross pay if she maintains this average.

c. Look for a pattern in the table. Complete the following sentence. For each additional $1000 in sales, Bette's gross pay is increased by _____.

d. What could Bette expect her annual gross pay to be if she averaged $5000 in sales per month?

e. If Bette decides that she needs to earn a gross pay of $2900 per month, how much must she average in sales per month?

Solution

a. To complete the table, go to the Y= editor and enter $y_1 = 2500 + 0.05x$. Notice that the sales values in the given table start at 1000 and the change or increment is 1000. In the table setup menu, let table start = 1000 and table increment = 1000.

TECHNOLOGY NOTE

When using the table feature, it is necessary to use x as the variable. We enter the expression to be evaluated in y_1 and then indicate where the table is to start, and the increment, or change in x, under the table setup menu. To view additional table values, we can scroll up or down in the x column and down in the y_1 column.

X	Y1
1000	2550
2000	2600
3000	2650
4000	2700
5000	2750
6000	2800
7000	2850

Y1⊟2500+.05X

X	Y1
4000	2700
5000	2750
6000	2800
7000	2850
8000	2900
9000	2950
10000	3000

X=9000

From the screens above we can complete the table as follows:

Sales, x	1000	2000	3000	4000	5000	6000	7000	8000	9000
Gross pay **$2500 + 0.05x$**	2550	2600	2650	2700	2750	2800	2850	2900	2950

b. To find her annual gross pay if she maintains $6000 in sales per month, find 6000 under x, and the corresponding y_1 entry is 2800. Therefore, if she averages $6000 worth of sales per month, she will average $2800 gross pay per month. To find her annual gross pay, find 12($2800) = $33,600.

c. Observing the table entries, we see that for each additional $1000 sales, Bette's gross pay is increased by $50.

d. To find her annual gross pay if she averages $5000 in sales monthly, scroll to the 5000 entry in the x column. Read the corresponding y_1-value to find a monthly gross pay of $2750. The annual gross pay is 12($2750) = $33,000.

e. Find the 2900 entry in the y_1 (gross pay) column and read the corresponding x-value, which is sales. She must average $8000 in sales per month to earn a gross pay of $2900 per month.

1 Elizabeth Ashley is considering a job offer that has a base salary of $2000 per month plus a 6% sales commission. Write an expression that models gross pay; then complete the table. How much must she average in sales per month to have her gross pay be $2540?

Sales, x	1000	3000	5000	7000
Gross pay				

EXAMPLE 2 Calculating Costs

Tom Sabo, a licensed electrician, charges $45 per house visit and $30 per hour.

a. Write an equation for Tom's total charge given the number of hours, x, on the job.

b. Use a table to find Tom's total charge for a job that could take from 1 to 6 hours.

Solution

a. To model this problem, recall that we are given that x is the number of hours.

In words:	Total Charge	is	$45	plus	$30 per hour
	↓	↓	↓	↓	↓
Translate:	Total Charge	=	45	+	$30x$

b. To produce a table to model what Tom charges, go to the Y= editor and enter $y_1 = 45 + 30x$. Notice that the x-values (hours) we are interested in start at 1. In the table setup menu, set table start at 1 and table increment at 1.

From the table in the margin, we read that Tom charges $75 for 1 hour, $105 for 2 hours, $135 for 3 hours, and so on, up to $255 for 7 hours. □

2 Patrick needs to hire a plumber. One plumber charges $35 per house visit and $25 per hour. The second plumber charges no house visit fee but $35 per hour. Patrick estimates the job to take 5 hrs. Which plumber charges less for 5 hours? How much less?

X	Y1
1	75
2	105
3	135
4	165
5	195
6	225
7	255

Y1◻45+30X

Table for Example 2b.

X	Y1
0	45
.25	52.5
.5	60
.75	67.5
1	75
1.25	82.5
1.5	90

Y1◻45+30X

There are numerous advantages to producing a table rather than evaluating the algebraic expression for each value of x, but one of the most advantageous situations occurs when you want to change the increment slightly. Let's say that Tom decides to calculate his fee every 15 minutes, or in 15-minute increments. In this case, let's start the table at 0 and have the table increment be 0.25, since every 15 minutes is a quarter of an hour.

From the second table in the margin, we see that it would cost us $45 just for having Tom show up. We can look down the table (scrolling if necessary) to find other charges. For example, we see that the charge is $82.50 to hire him for 1 hour and 15 minutes. This is a reasonable number since we found earlier that the charge is $75 for 1 hour and $105 for 2 hours.

Recall that a *mathematical model* is a graph, table, list, equation, or inequality that describes a situation. We call the equations that model Examples 1 and 2 **linear models** because they can be written in the form $y = mx + b$.

	Equation	$y = mx + b$
Example 1	$y = 2500 + 0.05x$	$y = 0.05x + 2500$
Example 2	$y = 45 + 30x$	$y = 30x + 45$

For linear models, as the x-values increase, the y-values or expressions in x always increase, always decrease, or always stay the same. Check the tables in Examples 1 and 2 to see that this is true. We study linear models and the equation $y = mx + b$ further in Chapter 2.

It is possible to enter more than one expression in the Y = editor.

EXAMPLE 3   **Finding Discounted Price, Sales Tax, and Total Cost**

Matthew Kramer wants to buy audio equipment at a store that has a 20% off the original price sale. There is also a 7% sales tax. Each of the items' original price, x, is shown in the table below.

a. Find the discounted price of each item and the total cost if he buys all the items during the sale.

b. Find the total cost if he does not purchase them during the sale.

c. How much does he save if he purchases them during the sale?

Item	Original Price	Discounted Price	Tax	Total Discounted Cost
MP3 Ready Stereo Receiver	199.95			
80GB MP3 Player	249.95			
DVD/CD Player	149.99			
Home Theater Speaker System	499.50			
Totals (for Discounted Prices)				

Solution

a. To model this problem, recall that we are given that x is the original price. Since we are looking for the discounted price and the sale is 20% off the original price, we have that

In words: Discounted price   is   80%   of   original cost

Translate: Discounted price $=$ 0.80 $\cdot$ x

We define $y_1 = 0.80x$ to represent the discounted price.

To calculate sales tax, we know that

In words: Sales tax   is   7%   of   discounted price

Translate: Sales tax $=$ 0.07 $\cdot$ $(0.80x)$

We define $y_2 = 0.07(0.80x)$ to be the tax, and $y_3 = y_1 + y_2$ to be the total cost of each item. To access Y1 and Y2, enter VARS, YVars, Function, and the appropriate Y=.

Notice that there is no constant change in the original price, x, of the four items. To keep from scrolling over a large span of numbers in a table, we introduce the ASK feature. This feature allows us to enter x-values one at a time and see the corresponding y-values. Access the ASK feature of the table and enter the tag prices in the x-column. The corresponding y_1 entry represents the discounted price, y_2 the tax, and y_3 the total discounted cost.

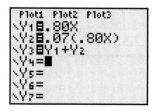

Fill in the corresponding entries on the table and find the sum of each column below.

Item	Original Price	Discounted Price	Tax	Total Discounted Cost
MP3 Ready Stereo Receiver	199.95	159.96	11.20	171.16
80GB MP3 Player	249.95	199.96	14.00	213.96
DVD/CD Player	149.99	119.99	8.40	128.39
Home Theater Speaker System	499.50	399.60	27.97	427.57
Totals (for Discounted Prices)		879.51	61.57	941.08

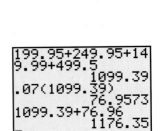

Calculations for Example 3b and c.

The discounted price with tax is $941.08.

b. To find the total cost if he does not purchase the items during the sale, we can calculate separately.

Total of original prices = 199.95 + 249.95 + 149.99 + 499.50 = $1099.39

Tax of total = 0.07($1099.39) = $76.96

Total Original Cost (if not discounted) = $1099.39 + $76.96 = $1176.35

c. His savings = Total Original Cost − Total Discounted Cost
   = $1176.35 − $941.08 = $235.27

He saves $235.27.

To break even in a manufacturing business, income or revenue R must equal the cost of production C. In other words the break-even point, sometimes called the equilibrium point in economics, is where the Revenue = Cost.

PRACTICE

3   Gabrielle went to a 40% off sale and purchased sale items whose original prices were $79.99, $16.49, $32.79, and $15.75. What was the total price paid for the discounted items before tax? The store charged an additional 6.5% tax. What was her discounted cost of items including tax?

EXAMPLE 4   Break-Even Point

The cost to produce a certain type of recycled tire is $1400 in fixed costs plus $50 per tire. The tires sell for $85 per tire. The revenue to produce x tires is $R = 85x$. Complete the table below to find the break-even point

No. of Tires	10	20	30	40	50
Cost					
Revenue					

Solution

1. UNDERSTAND. Read and reread the problem. Recall that the break-even point is where the cost and the revenue are equal. Let's first guess what the cost would be if the company produced 10 tires. The fixed cost is $1400 and the cost per tire is $50, so the total cost is 1400 + 50(10) = 1400 + 500 = 1900. The revenue brought in by the sale of 10 tires would be $R = 85(10) = 850$ since each tire sells for $85 each. The cost in this case is $1900 and the revenue is $850, which are not equal. Our guess is not correct, but we have a better understanding of the problem and see that there is a need to produce more tires to break even.

2. TRANSLATE.

In words: | Cost | = | Revenue |

Translate: $1400 + 50x$ = $85x$

3. COMPLETE THE TABLE. Enter the cost in y_1 as $y_1 = 1400 + 50x$, and the revenue as $y_2 = 85x$. Set the table to start at 10 and the increment (Δ table) to be 10.

No. of Tires, x	10	20	30	40	50
Cost (x Tires)	1900	2400	2900	3400	3900
Revenue (x Tires)	850	1700	2550	3400	4250

4. INTERPRET. Since the table entries for 40 tires are the same for both the cost and the revenue, the sale and production of 40 tires produces the break-even point.   ☐

PRACTICE

4   The cost to produce an item is $5600 in fixed costs plus $4 an item. The item sells for $11 per item. Find an expression for cost and one for revenue, and complete the table to find the break-even point.

No. of items	600	700	800	900	1000
Cost (x items)					
Revenue (x items)					

> **▶ Helpful Hint**
>
> Of course, to solve Example 4 algebraically (without tables), just solve the equation in blue above,
>
> $1400 + 50x = 85x.$
>
> For this section, though, our objective is to become familiar with the table feature, so we continue to use tables to solve.

Thus far, we have studied linear models only. Another common model is a quadratic model. A quadratic model has the form $y = ax^2 + bx + c, a \neq 0$. For quadratic models, as x-values increase, y-values increase and then decrease, or decrease and then increase. We study quadratic models further in Chapters 5 and 8.

EXAMPLE 5   **Height of a Rocket**

A small rocket is fired straight up from ground level on a campus parking lot with a velocity of 80 feet per second. Neglecting air resistance, the rocket's height y feet at time x seconds is given by the equation

$$y = -16x^2 + 80x.$$

If you are standing at a window that is 64 feet high,

a. In how many seconds will you see the rocket pass by the window?

b. When will you see the rocket pass by the window again?

c. From firing, how many seconds does it take the rocket to hit the ground?

d. Approximate to the nearest tenth of a second when the rocket reaches its maximum height. Approximate the maximum height to the nearest foot.

Solution   Enter $y_1 = -16x^2 + 80x$ in the Y=editor. Using the table feature, let table start = 0 and table increment = 1.

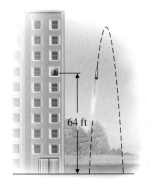

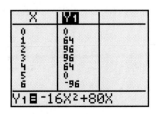

a. To find the number of seconds at which the rocket passes the window, recall that the window height is 64 feet and look for a y_1 value of 64. From the table we can see that it will first pass the window 1 second after it is fired.

b. Scrolling down the y_1 column in the table we see that at 4 seconds the rocket again passes the window on the way down.

c. We see that at 5 seconds the height is again 0, indicating that the rocket has hit the ground.

d. From the table, the rocket appears to reach its maximum height somewhere between 2 and 3 seconds. To get a better approximation, start the table at 2 and set the table increment to be 0.1 to find one-tenth of a second intervals of time.

X	Y₁
2	96
2.1	97.44
2.2	98.56
2.3	99.36
2.4	99.84
2.5	100
2.6	99.84

Y₁=100

The maximum height, 100 feet, occurs 2.5 seconds after the rocket is launched. □

Notice that the table above displays a relation between two values. For example, the *x*-value 2 is paired with the y-value 96 and so on. We study relations and paired data further in Chapter 2.

PRACTICE

5 A small rocket with a velocity of 112 feet per second is fired straight up from ground level on a parking lot by a multistory building. Neglecting air resistance, the rocket's height *y* feet at time *x* seconds is given by the equation

$$y = -16x^2 + 112x.$$

If you are standing at a window that is 160 feet high, answer the same questions as Example 5 parts a–d.

VOCABULARY & READINESS CHECK

For each table below, the y_1 column is generated by evaluating an expression in x for corresponding values in the x column. Match each table with the expression in x that generates the y_1 column.

a. $5x$ **b.** $3x + 1$ **c.** x^2

d. $x^2 + 1$ **e.** $-x + 5$ **f.** $-2x$

1.

X	Y₁
1	-2
2	-4
3	-6
4	-8
5	-10
6	-12
7	-14

X=7

2.

X	Y₁
1	5
2	10
3	15
4	20
5	25
6	30
7	35

X=7

3.

X	Y₁
1	4
2	7
3	10
4	13
5	16
6	19
7	22

X=7

4.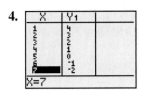

5.

6.

1.7 | EXERCISE SET

MyMathLab *Powered by CourseCompass™ and MathXL®*

Math XL PRACTICE · WATCH · DOWNLOAD · READ · REVIEW

MIXED PRACTICE

Use the table feature to complete each table for the given expression. If necessary, round results to two decimal places. See Examples 1 through 5.

1.

Side Length of a Cube	x	7	8	9	10	11
Volume	x^3					

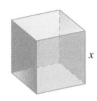

a. From the table, give the volume of a cube whose side measures 9 centimeters. Use correct units.

b. From the table, give the volume of a cube whose side measures 11 feet. Use correct units.

c. If a cube has volume 343 cubic inches, what is the length of its side?

2.

Radius of Cylinder	x	2	3	4	5	6
Volume (If height is 3 units)	$3\pi x^2$					

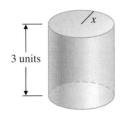

a. Give the volume of a cylinder whose height is 3 inches and whose radius is 5 inches. Use correct units.

b. Give the volume of a cylinder whose height is 3 meters and whose radius is 4 meters. Use correct units.

c. If a cylinder with height 3 kilometers has a volume of 150.8 cubic kilometers, approximate the radius of the cylinder.

3.

Hours Worked	x	39	39.5	40	40.5
Gross Pay (Dollars)	$25 + 7x$				

a. Scroll down the table on your calculator to find how many hours worked gives a gross pay of $333.

b. Scroll up the table to find how many hours worked gives a gross pay of $270.

c. If gross pay is $284, how many hours were worked?

4.

Minutes on Phone	x	7	7.5	8	8.5
Total Charge (Dollars)	$1.30 + 0.15x$				

a. Scroll on your calculator to find how long a person with $3.25 can talk on a pay phone.

b. If a person wants to spend no more than $4 for a single phone call, how long can that person stay on the phone?

Solve. If necessary, round amounts to two decimal places.

5. Emily Keaton has a job typing term papers for students on her computer. She charges $15.25 per hour for a rough draft and $17 per hour for a final bound manuscript. Her price increases in 15-minute intervals with a 1-hour minimum.

a. Let x = number of hours and write an algebraic representation for y_1, the cost of preparing a rough draft.

b. Let x = number of hours and write an algebraic representation for y_2, the cost of preparing a final manuscript.

c. Using a table, complete the following chart for the costs of jobs ranging from 1 to 3 hours.

Hours	1	1.25	1.5	1.75	2	2.25	2.5	2.75	3
Rough Draft									
Manu- script									

d. If Emily can type at a rate of 2 pages per 15 minutes (0.25 of an hour), how much does she charge for a 10-page rough draft?

e. If a job that consists of typing a bound manuscript takes Emily 10.75 hours, how much does she charge?

6. The cost of tuition at the local community college is $69.75 per credit hour for in-state students and $263.25 per credit hour for out-of-state students.

a. Write an algebraic representation, y_1, for the cost of in-state tuition dependent on x, the number of credit hours.

b. Write an algebraic representation, y_2, for the cost of out-of-state tuition dependent on x, the number of credit hours.

c. Complete the following table given the number of credit hours.

Credit Hours	2	3	4	5	6
In-State Cost (Dollars)					
Out-of-State Cost (Dollars)					

d. What is the total tuition due if Mandy, an in-state student, is taking a 3-hour math course, a 4-hour science course, a 2-hour study skills course, a 3-hour history course, and a 3-hour Spanish course? The science course has an additional $15 lab fee.

Use the table and the ask feature of your calculator to complete each table for the given expression. If necessary, round to two decimal places. See Example 3.

7.

Radius of Circle	x	2	5	21	94.2
Circumference	$2\pi x$				
Area	πx^2				

a. Find the circumference and the area of a circle whose radius is 30 yards. Use correct units.

b. Find the circumference and the area of a circle whose radius is 78.5 millimeters. Use correct units.

8.

Radius of Sphere	x	3	7.1	43	50
Volume	$\frac{4}{3}\pi x^3$				
Surface Area	$4\pi x^2$				

a. Find the volume and surface area of a sphere whose radius is 1.7 centimeters. Use correct units.

b. Find the volume and surface area of a sphere whose radius is 25 feet. Use correct units.

9. Krista Handefeld is considering a job as a real estate agent. She is offered a monthly salary of $500 plus 4% commission on sales of homes. She would like to know her possible gross pay if the homes in the area range from a sales price of $125,000 to $400,000 and she averages selling one home per month.

a. Complete the table below.

Sales (In Thousands)	125	200	250	350	400
Gross Pay					

b. If she sells at least one home a month, her gross salary would be between _____ .

10. Elizabeth Gocek is considering two part-time job offers. Company A pays a starting salary of $23,500 with raises of $1000 per year. Company B pays a starting salary of $25,000 with raises of $475 each year.

a. Make a table showing her salary during each year if she plans to stay with the company for 5 years.

Years	1	2	3	4	5
Gross Pay A					
Gross Pay B					

b. Which job do you think pays the most money if she stays for five years?

11. Kinsley Water Service charges a basic water fee of $12.96 per month plus $1.10 per thousand gallons of water used.

a. Write an algebraic representation for the total cost, y_1, in terms of the number of thousands of gallons of water used, x.

b. Make a table for the cost each month of 0 to 6 thousand gallons of water used.

Gallons of Water Used (Thousands)	0	1	2	3	4	5	6
Monthly Cost (Dollars)							

12. Jewell Water Service charges a basic water fee of $6.00 per month plus $1.72 per thousand gallons of water used.

a. Write an algebraic representation for the total cost, y_1, in terms of the number of thousands of gallons of water used, x.

b. Make a table for the cost each month of 0 to 6 thousand gallons of water.

Gallons of Water Used (Thousands)	0	1	2	3	4	5	6
Monthly Cost (Dollars)							

13. A company that has 500 computers has been infected by a computer virus. The virus is infecting the computers at a rate of 8 per minute.

  a. Write an algebraic expression representation, y_1, for the number of computers that have not been infected. Let x be the number of minutes that have elapsed since the first computer has been infected.

  b. Complete the table for the number of computers not infected depending on the time elapsed.

No. of Minutes	0	10	20	30	40	50	60	70
No. of Computers Not Infected								

  c. When will there be 260 computers that are not infected?

  d. How long will it take for all the computers to be infected?

14. Mark Cook is the general manager of the Acme Used Car lot. At the beginning of the month, he has 850 cars on the lot and his sales team has been selling the cars at the rate of 25 per day.

  a. Write an algebraic representation for the number of cars on the lot, y_1, in terms of the number of days that have passed, x. (Assume no new cars are delivered.)

  b. Complete the table for the number of cars remaining on the lot given the number of days that have elapsed.

Days Ellapsed	0	10	15	20	25	30
Cars Remaining						

  c. If Mark does not replace any of the sold inventory, when will his lot be completely empty of cars?

15. Kelsey is considering two job offers. The first job pays a starting salary of $14,500, with raises of $1000 every year. The second job pays a starting salary of $20,000 with raises of $575 each year.

  a. Make a chart showing both salaries during 5, 10, 15, and 20 years.

Year	5	10	15	20
First Job				
Second Job				

  b. Which plan do you think is more advantageous for Kelsey? Why do you think so?

16. Rory Baruch considers two different salary plans that his company offers. Plan A starts at $35,000, with raises of $650 per year. Plan B starts at $32,000, with raises of $1000 per year.

  a. If Rory plans to stay with the company less than 6 years, make a chart showing his salary for both plans during years 1 to 6.

  b. Which plan looks more advantageous for Rory? Why do you think so?

Year	1	2	3	4	5	6
Plan A						
Plan B						

17. The Jackson Electric Authority charges a basic fee of $15.00 per month plus $0.08924 per kilowatt-hour of electricity used.

  a. Write an algebraic representation for the total cost, y_1, in terms of the number of kilowatt-hours used, x.

  b. Complete the table for the cost each month of 0 to 2000 kilowatt-hours.

Kilowatt-Hours	0	500	1000	1500	2000
Monthly Cost (Dollars)					

18. The First Coast Electric charges a basic fee of $10.00 per month plus $0.09850 per kilowatt-hour of electricity used.

  a. Write an algebraic representation for the total cost, y_1, in terms of the number of kilowatt-hours used, x.

  b. Complete the table for the cost each month of 0 to 2000 kilowatt-hours.

Kilowatt-Hours	0	500	1000	1500	2000
Monthly Cost (Dollars)					

19. Kara White, a nurse, is starting a 1000 ml IV on one of her hospital patients. It has a drip rate of 2.5 ml per min.

  a. Write an algebraic expression for the number of ml of solution left in the bag, y_1, after x minutes has elapsed since she started the IV.

  b. Complete the table below to find how many ml are left in the bag after x minutes.

  c. At what time should Kara expect to return to the room if she starts the drip at 10 a.m. and wants to put a new bag on when there is at least 50 ml remaining in the bag?

Minutes	0	60	120	180	240	300	360
ml Solution Remaining							

20. Cathy Grooms, a nurse, is starting a 1000 ml IV on one of her hospital patients. It has a drip rate of 1.5 ml per min.

  a. Write an algebraic expression for the number of ml of solution left in the bag, y_1, after x minutes have elapsed since she started the IV.

  b. Complete the table below to find how many ml are left in the bag after x minutes.

  c. At what time should Cathy expect to return to the room if she starts the drip at 1 p.m. and wants to put a new bag on when there is at least 10 ml remaining in the bag?

Minutes	0	60	120	180	240	300	360
ml Solution Remaining							

21. Cara D'Alesio bought a new Sports Utility Vehicle (SUV) that has a 24-gallon gas tank. The salesperson told her that the car averages 15 miles per gallon (mpg) in the city and 25 mpg for highway travel. (If needed, round answers to one decimal place.)

   a. Write an algebraic representation for the amount of gas left in the tank, y_1, in terms of miles driven, x, for city driving.

   b. Write an algebraic representation for the amount of gas left in the tank, y_2, in terms of miles driven, x, for highway driving.

   c. Complete the table.

Miles	0	100	200	300	400	500	600
City							
Highway							

   d. How much gas is left in the tank after the first 100 miles of city driving?

   e. How much gas is left in the tank after the first 100 miles of highway driving?

   f. How many miles can she drive before her tank is empty if she is only driving on a highway?

   g. How many miles can she drive before her tank is empty if she is only driving in the city?

   h. Write a sentence telling what the negative entries in the table represent.

22. Jesus Thornbury has an economy car that has a 14-gallon tank. His car averages 24 miles per gallon (mpg) in the city and 30 mpg when traveling on a highway. (If needed, round answers to one decimal place.)

   a. Write an algebraic representation for the amount of gas left in the tank, y_1, in terms of miles driven, x, for city driving.

   b. Write an algebraic representation for the amount of gas left in the tank, y_2, in terms of miles driven, x, for highway driving.

   c. Complete the table.

Miles	0	50	100	150	200	250	300	350	400
City									
Highway									

   d. How much gas is left in the tank after the first 300 miles of city driving?

   e. How much gas is left in the tank after the first 300 miles of highway driving?

   f. How many miles can he drive before his tank is empty if he is only driving on a highway?

   g. How many miles can he drive before his tank is empty if he is only driving in the city?

   h. Write a sentence telling what the negative entries in the table represent.

23. Judy Martinez has been shopping for school clothes and has picked out several outfits. She knows that the following Wednesday there will be a Super Wednesday sale and everything in the store will be discounted 35%.

   a. Write an algebraic representation for the sale price, y_1, in terms of the price, x, on the price tag.

   b. Complete the table and find the cost of each item on sale, given the price on the price tag. Round dollar amounts to the nearest cent.

Item	Blouse	Skirt	Shorts	Shoes	Purse	Earrings	Backpack
Price Tag	$29.95	$35.95	$19.25	$39.95	$17.95	$9.95	$25.75
Sale Price							

   c. If she purchases all the items when on sale, what will her total bill be before taxes?

   d. What is the total after taxes if a 7% tax is added to the bill?

   e. How much did she save before taxes by waiting for the sale?

24. Saumil has been shopping for tools and has picked out several at Sears. Next month, Sears is having a tool sale and all tools in the store will be discounted 30%.

   a. Write an algebraic representation for the sale price, y_1, in terms of the price, x, on the price tag.

   b. Complete the table and find the cost of each item on sale, given the price on the price tag.

Item	Hammer	Drill	Sander	Glue gun	Screwdriver set	Socket wrench set
Price Tag	$9.95	$29.95	$49.75	$19.95	$27.95	$79.95
Sale Price						

   c. If he purchases all the items when on sale, what will his total bill be before taxes?

   d. What is the total after taxes if a 7% tax is added to the bill?

   e. How much did he save before taxes by waiting for the sale?

25. The cost of producing x calculators is $3500 in fixed costs and $55 per calculator. The calculators sell for $75 each. Find the break-even point by completing the following table.

x Calculators	100	125	150	175	200
Cost					
Revenue					

26. The cost of producing x books of digital pictures is $1750 in fixed costs and $21 per book. The books sell for $35 each wholesale. Find the break-even point by completing the following table.

x Books	25	50	75	100	125
Cost					
Revenue					

27. In Exercise 25, write in words what the entries for the cost and revenue tell you about the profit when 200 calculators are produced and sold.

28. In Exercise 26, write in words what the entries for the cost and revenue tell you about the profit when 100 books have been produced and sold.

29. From 2010 to 2030, the U.S. demand for electricity is projected to increase according to the equation $y = 66x + 4050$ where x is the number of years since 2010 and y is billions of kilowatt hours. (*Source:* U.S. Dept. of Energy; Energy Information Admin.)

a. Use the given equation to fill in the table.

Years since 2010	0	5	10	15	20
Projected Billions of Kilowatt Hours					

b. Use a sentence to describe the projected amount of energy expected in the year 2025.

30. GROUP ACTIVITY. The owner of Descartes Pizzaria is in the process of changing his pizza prices. He is considering two methods for determining the new price for a pizza. One method is to charge 5¢ per square inch of area of pizza. Another method is to charge 25¢ per inch of circumference. Fill in the table below and compare the costs of pizzas for the given *diameters* using the two methods. Round costs to 2 decimal places. Which price do you think the owner should choose for each size of pizza? Why?

Diameter (Inches)	Area	Circumference	Cost (Area)	Cost (Circumference)
6				
12				
18				
24				
30				
36				

Diameter

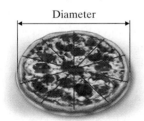

31. A rocket with an initial velocity of 85 feet per second is launched from the top of a mountain 1500 feet above sea level. Neglecting air resistance, the height of the rocket above sea level at the time x seconds is given by the equation $y = -16x^2 + 85x + 1500$ where y is measured in feet.

a. Find its height for the first 5 seconds in one-second intervals.

Seconds	1	2	3	4	5
Height (Feet)					

b. Find the rocket's maximum height to the nearest foot.

32. A firecracker rocket is fired at ground level with an initial velocity of 100 feet per second. Neglecting air resistance, the height of the rocket at time x seconds is given by the equation $y = -16x^2 + 100x$ where y is measured in feet.

a. Find the height of the rocket for the first five seconds in one-second intervals.

Seconds	1	2	3	4	5
Height (Feet)					

b. Explain why the rocket heights increase and then decrease.

c. If it doesn't explode, determine to the nearest tenth of a second when the rocket hits the ground.

33. Anne Holloway rents a car for $25 per day plus $0.08 per mile. Michelle Goods rents a similar car from a different rental agency for $35 per day plus $0.05 per mile.

a. Find both rental charges for one day if they each drive 300 miles.

b. For what distance is Anne's rental agreement cheaper than Michelle's for one day?

34. Matthew Aires repairs computers and charges $35 per house visit plus $25 per hour.

a. Write an expression for Matthew's total charge given the number of hours, x, on the job.

b. Complete the following table for the total charges given the number of hours on the job.

Hours on Job	1.5	2	2.5	3
Cost (Dollars)				

35. William Kramer, a lawyer, charges an initial consultation fee of $90 plus $75 per hour after this consultation.

a. Write an expression for William's total charges given the number of hours, x, on the job after the consultation.

b. Complete the table below for the total charges given the number of hours *after* the initial consultation.

No. of Hours	2	4	6	8
Total Fee (Dollars)				

36. An object is *dropped* from the top of the Nations Bank Tower in Atlanta, Georgia. Neglecting air resistance, the height of the object at time x seconds is given by

$$y = -16x^2 + 1050$$

where y is measured in feet. Use a table to determine the following.

a. What is the object's maximum height? How did you arrive at your answer?

b. To the nearest tenth of a second, determine when the object hits the ground.

37. The path of some leaping animals can be described by a quadratic model. A particular leap of a frog can be modeled by the equation

$$y = -0.14x^2 + 1.4x$$

where x is the number of feet horizontally from a starting point and y is the vertical height of the frog in feet. Use a table to determine the following.

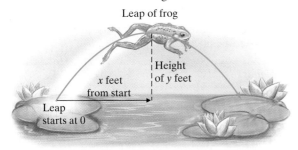

Leap of frog

Height of y feet

x feet from start

Leap starts at 0

a. Determine to the nearest tenth of a foot the maximum height reached by the frog.

b. What is the frog's horizontal distance from the starting point when the maximum height is reached?

c. How far is the frog from the starting point when it lands? (By the way, the longest frog leap on record is 21 feet $5\frac{3}{4}$ inches.)

CONCEPT EXTENSIONS

38. A gardener has 52 feet of garden edging in 1-foot sections. A rectangular garden is to be formed using the edging for 3 sides and the side of a patio for the fourth side. Use a table to find the dimensions that will form a garden of largest possible area.

39. Answer Exercise 38 using 92 feet of 1-foot sections of garden edging instead of 52.

40. The high temperature for a particular week in Jackson, Mississippi, can be described by the quadratic model

$$y = 1.05x^2 - 8.38x + 89.43$$

where x is the day of the week (Sunday, $x = 1$; Monday, $x = 2$; and so on) and y is the temperature in degrees Fahrenheit.

a. Find the day of the week with the lowest temperature.

b. What is the lowest temperature for the week rounded to the nearest tenth?

c. Do you think that this equation models the temperature for much longer than a week, say, 2 weeks? Why or why not?

41. The sunrise times in Valdivia, Chile, for the months of April through October can be described by the quadratic model

$$y = -0.14x^2 + 1.09x + 5.11$$

where $x = 1$ is the first day of April, $x = 2$ is the first day of May, $x = 3$ is the first day of June, and so on, and y is the time of sunrise in hours a.m.

a. Approximate the time of sunrise on May 1. Give the time in hundredths of an hour and also in hours and minutes.

b. For what integer value of x is the sunrise the latest? What calendar date does this correspond to?

c. Give this latest sunrise time in hours and minutes.

d. Approximate the sunrise time in hours and minutes for September 15.

e. Do you think that this equation models the sunrise times for all the months of the year? Why or why not?

South America

Valdivia

📖 STUDY SKILLS BUILDER

Have You Decided to Complete This Course Successfully?

Ask yourself if one of your current goals is to complete this course successfully.

If it is not a goal of yours, ask yourself why? One common reason is fear of failure. Amazingly enough, fear of failure alone can be strong enough to keep many of us from doing our best in any endeavor.

Another common reason is that you simply haven't taken the time to make successfully completing this course one of your goals. How do you do this? Start by writing this goal in your mathematics notebook. Then list steps you will take to ensure success. A great first step is to read or reread Section 1.1 and make a commitment to try the suggestions in that section.

Good luck, and don't forget that a positive attitude will make a big difference.

Let's see how you are doing.

1. Have you decided to make "successfully completing this course" a goal of yours? If not, please list reasons why this has not happened. Study your list and talk to your instructor about this.

2. If your answer to question 1 is yes, take a moment and list in your notebook further specific goals that will help you achieve this major goal of successfully completing this course. (For example, "My goal this semester is not to miss any of my mathematics classes.")

3. Rate your commitment to this course with a number between 1 and 5. Use the diagram below to help.

High Commitment	Average Commitment	Not Committed at All
5	4	3	2	1

4. If you have rated your personal commitment level (from the exercise above) as a 1, 2, or 3, list the reasons why this is so. Then determine whether it is possible to increase your commitment level to a 4 or 5.

1.8 FORMULAS AND PROBLEM SOLVING

OBJECTIVE 1 ▶ Solving a formula for a specified variable. Solving problems that we encounter in the real world sometimes requires us to express relationships among measured quantities. An equation that describes a known relationship among quantities, such as distance, time, volume, weight, money, and gravity, is called a **formula.** Some examples of formulas are

Formula	*Meaning*
$I = PRT$	Interest = principal · rate · time
$A = lw$	Area of a rectangle = length · width
$d = rt$	Distance = rate · time
$C = 2\pi r$	Circumference of a circle = $2 \cdot \pi \cdot$ radius
$V = lwh$	Volume of a rectangular solid = length · width · height

Other formulas are listed in the front cover of this text. Notice that the formula for the volume of a rectangular solid, $V = lwh$, is solved for V since V is by itself on one side of the equation with no V's on the other side of the equation. Suppose that the volume of a rectangular solid is known as well as its width and its length, and we wish to find its height. One way to find its height is to begin by solving the **formula** $V = lwh$ for h.

 EXAMPLE 1   Solve: $V = lwh$ for h.

Solution   To solve $V = lwh$ for h, isolate h on one side of the equation. To do so, divide both sides of the equation by lw.

$$V = lwh$$
$$\frac{V}{lw} = \frac{lwh}{lw} \quad \text{Divide both sides by } l\,w.$$
$$\frac{V}{lw} = h \quad \text{Simplify.}$$

Then to find the height of a rectangular solid, divide the volume by the product of its length and its width. ☐

PRACTICE
1   Solve: $I = Prt$ for t.

The following steps may be used to solve formulas and equations in general for a specified variable.

Solving Equations for a Specified Variable

STEP 1. Clear the equation of fractions by multiplying each side of the equation by the least common denominator.

STEP 2. Use the distributive property to remove grouping symbols such as parentheses.

STEP 3. Combine like terms on each side of the equation.

STEP 4. Use the addition property of equality to rewrite the equation as an equivalent equation with terms containing the specified variable on one side and all other terms on the other side.

STEP 5. Use the distributive property and the multiplication property of equality to isolate the specified variable.

EXAMPLE 2 Solve: $3y - 2x = 7$ for y.

Solution This is a linear equation in two variables. Often an equation such as this is solved for y in order to reveal some properties about the graph of this equation, which we will learn more about in Chapter 3. Since there are no fractions or grouping symbols, we begin with Step 4 and isolate the term containing the specified variable, y, by adding $2x$ to both sides of the equation.

$$3y - 2x = 7$$
$$3y - 2x + 2x = 7 + 2x \quad \text{Add } 2x \text{ to both sides.}$$
$$3y = 7 + 2x$$

To solve for y, divide both sides by 3.

$$\frac{3y}{3} = \frac{7 + 2x}{3} \quad \text{Divide both sides by 3.}$$
$$y = \frac{2x + 7}{3} \quad \text{or} \quad y = \frac{2x}{3} + \frac{7}{3}$$

□

PRACTICE
2 Solve: $7x - 2y = 5$ for y.

EXAMPLE 3 Solve: $A = \frac{1}{2}(B + b)h$ for b.

Solution Since this formula for finding the area of a trapezoid contains fractions, we begin by multiplying both sides of the equation by the LCD, 2.

$$A = \frac{1}{2}(B + b)h$$
$$2 \cdot A = 2 \cdot \frac{1}{2}(B + b)h \quad \text{Multiply both sides by 2.}$$
$$2A = (B + b)h \quad \text{Simplify.}$$

Next, use the distributive property and remove parentheses.

$$2A = (B + b)h$$
$$2A = Bh + bh \quad \text{Apply the distributive property.}$$
$$2A - Bh = bh \quad \text{Isolate the term containing } b \text{ by}$$
$$\qquad\qquad\qquad\quad \text{subtracting } Bh \text{ from both sides.}$$
$$\frac{2A - Bh}{h} = \frac{bh}{h} \quad \text{Divide both sides by } h.$$
$$\frac{2A - Bh}{h} = b \quad \text{or} \quad b = \frac{2A - Bh}{h}$$

□

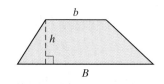

▶ **Helpful Hint**

Remember that we may isolate the specified variable on either side of the equation.

PRACTICE
3 Solve: $A = P + Prt$ for r.

OBJECTIVE 2 ▶ **Using formulas to solve problems.** In this section, we also solve problems that can be modeled by known formulas. We use the same problem-solving steps that were introduced in the previous section.

Formulas are very useful in problem solving. For example, the compound interest formula

$$A = P\left(1 + \frac{r}{n}\right)^{nt}$$

is used by banks to compute the amount A in an account that pays compound interest. The variable P represents the principal or amount invested in the account, r is the annual rate of interest, t is the time in years, and n is the number of times compounded per year.

EXAMPLE 4   Finding the Amount in a Savings Account

Karen Estes just received an inheritance of $10,000 and plans to place all the money in a savings account that pays 5% compounded quarterly to help her son go to college in 3 years. How much money will be in the account in 3 years?

Solution

1. UNDERSTAND. Read and reread the problem. The appropriate formula needed to solve this problem is the compound interest formula

$$A = P\left(1 + \frac{r}{n}\right)^{nt}$$

   Make sure that you understand the meaning of all the variables in this formula.

$$A = \text{amount in the account after } t \text{ years}$$
$$P = \text{principal or amount invested}$$
$$t = \text{time in years}$$
$$r = \text{annual rate of interest}$$
$$n = \text{number of times compounded per year}$$

2. TRANSLATE. Use the compound interest formula and let $P = \$10,000$, $r = 5\% = 0.05$, $t = 3$ years, and $n = 4$ since the account is compounded quarterly, or 4 times a year.

   Formula: $A = P\left(1 + \frac{r}{n}\right)^{nt}$

   Substitute: $A = 10,000\left(1 + \frac{0.05}{4}\right)^{4 \cdot 3}$

3. SOLVE. We use our calculator and simplify the right side of the equation.

```
10000(1+(.05/4))
^(4*3)
          11607.54518
```

4. INTERPRET.

Check: Repeat your calculations to make sure that no error was made. Notice that $11,607.55 is a reasonable amount to have in the account after 3 years.

State: In 3 years, the account will contain $11,607.55.   □

PRACTICE

4   Russ placed $8000 into his credit union account paying 6% compounded semiannually (twice a year). How much will be in Russ's account in 4 years?

EXAMPLE 5 **Finding Cycling Time**

The fastest average speed by a cyclist across the continental United States is 15.4 mph, by Pete Penseyres. If he traveled a total distance of about 3107.5 miles at this speed, find his time cycling. Write the time in days, hours, and minutes. (*Source: The Guinness Book of World Records*)

Solution

1. UNDERSTAND. Read and reread the problem. The appropriate formula needed is the distance formula

$$d = rt \quad \text{where}$$
$$d = \text{distance traveled} \quad r = \text{rate} \quad \text{and} \quad t = \text{time}$$

2. TRANSLATE. Use the distance formula and let $d = 3107.5$ miles and $r = 15.4$ mph.

    Check: $d = rt$

    State: $3107.5 = 15.4t$

3. SOLVE.

$$\frac{3107.5}{15.4} = \frac{15.4t}{15.4} \quad \text{Divide both sides by 15.4.}$$

$$201.79 \approx t$$

The time is approximately 201.79 hours. Since there are 24 hours in a day, we divide 201.79 by 24 and find that the time is approximately 8.41 days. Now, let's convert the decimal part of 8.41 days back to hours. To do this, multiply 0.41 by 24 and the result is 9.84 hours. Next, we convert the decimal part of 9.84 hours to minutes by multiplying by 60 since there are 60 minutes in an hour. We have $0.84 \cdot 60 \approx 50$ minutes rounded to the nearest whole. The time is then approximately

8 days, 9 hours, 50 minutes.

4. INTERPRET.

Check: Repeat your calculations to make sure that an error was not made.

State: Pete Penseyres's cycling time was approximately 8 days, 9 hours, 50 minutes. □

PRACTICE
5 Nearly 4800 cyclists from 36 U.S. states and 6 countries recently rode in the Pan-Massachusetts Challenge to raise money for cancer research and treatment. If the riders of a certain team traveled their 192-mile route at an average speed of 7.5 miles per hour, find the time they spent cycling. Write the answer in hours and minutes.

EXAMPLE 6 **Finding Degrees Celsius**

The formula $C = \dfrac{5}{9}(F - 32)$ converts degrees Fahrenheit to degrees Celsius. Use this formula and the table feature of your calculator to complete the given table. If necessary, round values to the nearest hundredth.

Fahrenheit	−4	10	32	70	100
Celsius					

Solution

Let x = degrees Fahrenheit. Then enter $y_1 = \dfrac{5}{9}(x - 32)$ to find the corresponding degrees Celsius. Notice that there is no constant increment change in the Fahrenheit values in the given table. To keep from scrolling over a large span of numbers in a table (not a good use of time), we activate the ask feature in table setup.

Enter each x-value from the table. The corresponding y_1-values are shown to the left. The completed table is

Fahrenheit	x	-4	10	32	70	100
Celsius	$\dfrac{5}{9}(x - 32)$	-20	-12.22	0	21.11	37.78

PRACTICE

6 Solve the formula $C = \dfrac{5}{9}(F - 32)$ for F and complete the table.

Celsius	-40	-30	10	30
Fahrenheit				

DISCOVER THE CONCEPT

Finding the Maximum Area. While this discovery can be done by an individual, we suggest working in groups of 2 or 3 if feasible. Each group/person begins with 12 toothpicks.

Suppose that each toothpick represents a 1-foot piece of fencing. Use all the toothpicks (without bending or breaking them) to form a rectangular fence. First, form a rectangular fence of width 1 foot, next, a rectangular fence of width 2 feet, then 3 feet, and so on. For each rectangular fence constructed, record the width, length, area, and perimeter in a table such as the one below.

Width	Length	Area	Perimeter
1			
2			
3			
4			
5			
6			

a. Find the dimensions of the rectangle with the largest area.

b. Compare the areas of the rectangles with their perimeters. What patterns do you notice?

c. Discuss what happens when the width is 6.

d. Look for other patterns in the table. Let the width be represented by x. Write expressions for the length and area in terms of x.

From the discovery above, we see that the rectangle with largest area has dimensions 3 feet by 3 feet and an area of 9 square feet. Notice that although a change in dimensions results in a change in area, the perimeter remains constant.

EXAMPLE 7 **Finding Maximum Area**

Sylvia Daschle has purchased 60 feet of fencing in 1-foot sections. She wants to enclose the largest possible rectangular garden using a portion of a side of her house as a side of the rectangle. Use a table to find the dimensions that will give Sylvia the largest area.

1. UNDERSTAND. Read and reread the problem and then try some dimensions in particular. For example, if the width is 1 foot, the length is $60 - 2(1)$ or 58 feet. The area of the rectangle is $w \cdot l$ or $1(58) = 58$ square feet. Next, let the width be 2 feet and continue until a pattern is found. A table can be useful when looking for a pattern.

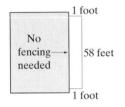

1 foot

No fencing→ needed 58 feet

1 foot

Solution

Width	Length	Area
1	$60 - 2(1)$ or 58	$1 \cdot 58 = 58$
2	$60 - 2(2)$ or 56	$2 \cdot 56 = 112$
3	$60 - 2(3)$ or 54	$3 \cdot 54 = 162$

2. TRANSLATE. Let $x =$ the width of the garden. From the table, we can see that $60 - 2x =$ the length of the garden

In words: Area = width · length or

Translate: Area = x · $(60 - 2x)$

3. SOLVE. Since Sylvia has fencing in 1-foot sections, we know that the width and the length are natural numbers. To solve, we can use a table. We can enter the expression for length in y_1 and the area in y_2.

$$y_1 = 60 - 2x \qquad \text{Length}$$

$$y_2 = x(60 - 2x) \qquad \text{Area}$$

Set table start to be 1 and table increment to be 1. Scroll down the y_2 or area column to find the largest area. Notice that equation $y_2 = x(60 - 2x)$ is a quadratic model, so for increasing values of x, the y_2 values will increase and then decrease, or decrease and then increase.

X	Y1	Y2
1	58	58
2	56	112
3	54	162
4	52	208
5	50	250
6	48	288
7	46	322

Y1◼60-2X

X	Y1	Y2
10	40	400
11	38	418
12	36	432
13	34	442
14	32	448
15	30	450
16	28	448

Y2=450

The largest area is 450 square feet.

4. INTERPRET. *Check* the solution and *state*. From the table, we see that the dimensions that give the largest area are 15 feet by 30 feet. The largest area is 450 square feet. □

PRACTICE
7 A do-it-yourselfer has purchased 88 feet of fencing in 1-foot sections. He wants to form a pen for his dogs of largest possible area in the shape of a rectangle. He is using the side of his large shed as one side of the rectangle. Use a table to find the dimensions that will give him the largest area.

1.8 | EXERCISE SET

 MyMathLab Powered by CourseCompass™ and MathXL®

 Math XL PRACTICE   WATCH   DOWNLOAD READ   REVIEW

Solve each equation for the specified variable. See Examples 1–3.

1. $D = rt$; for t

2. $W = gh$; for g

3. $I = PRT$; for R

△ **4.** $V = lwh$; for l

5. $9x - 4y = 16$; for y

6. $2x + 3y = 17$; for y

△ **7.** $P = 2L + 2W$; for W

8. $A = 3M - 2N$; for N

9. $J = AC - 3$; for A

10. $y = mx + b$; for x

11. $W = gh - 3gt^2$; for g

12. $A = Prt + P$; for P

13. $T = C(2 + AB)$; for B

14. $A = 5H(b + B)$; for b

15. $C = 2\pi r$; for r

16. $S = 2\pi r^2 + 2\pi rh$; for h

17. $E = I(r + R)$; for r

18. $A = P(1 + rt)$; for t

19. $s = \dfrac{n}{2}(a + L)$; for L

20. $C = \dfrac{5}{9}(F - 32)$; for F

21. $N = 3st^4 - 5sv$; for v

22. $L = a + (n - 1)d$; for d

23. $S = 2LW + 2LH + 2WH$; for H

24. $T = 3vs - 4ws + 5vw$; for v

In this exercise set, round all dollar amounts to two decimal places. Solve. See Example 4.

25. Complete the table and find the balance A if $3500 is invested at an annual percentage rate of 3% for 10 years and compounded n times a year.

n	1	2	4	12	365
A					

26. Complete the table and find the balance A if $5000 is invested at an annual percentage rate of 6% for 15 years and compounded n times a year.

n	1	2	4	12	365
A					

27. A principal of $6000 is invested in an account paying an annual percentage rate of 4%. Find the amount in the account after 5 years if the account is compounded

   a. semiannually **b.** quarterly

   c. monthly

28. A principal of $25,000 is invested in an account paying an annual percentage rate of 5%. Find the amount in the account after 2 years if the account is compounded

   a. semiannually **b.** quarterly

   c. monthly

MIXED PRACTICE

Solve. See Examples 4 through 7.

29. The day's high temperature in Phoenix, Arizona, was recorded as 104°F. Write 104°F as degrees Celsius. [Use the formula $C = \dfrac{5}{9}(F - 32)$.]

30. The annual low temperature in Nome, Alaska, was recorded as $-15°C$. Write $-15°C$ as degrees Fahrenheit. [Use the formula $F = \dfrac{9}{5}C + 32$.]

31. Omaha, Nebraska, is about 90 miles from Lincoln, Nebraska. Irania must go to the law library in Lincoln to get a document for the law firm she works for. Find how long it takes her to drive round-trip if she averages 50 mph.

32. It took the Selby family $5\dfrac{1}{2}$ hours round-trip to drive from their house to their beach house 154 miles away. Find their average speed.

△ **33.** A package of floor tiles contains 24 1-foot-square tiles. Find how many packages should be bought to cover a square ballroom floor whose side measures 64 feet.

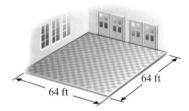

64 ft · 64 ft

△ **34.** One-foot-square ceiling tiles are sold in packages of 50. Find how many packages must be bought for a rectangular ceiling 18 feet by 12 feet.

35. If the area of a triangular kite is 18 square feet and its base is △ 4 feet, find the height of the kite.

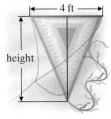

← 4 ft → height

36. Bryan, Eric, Mandy, and Melissa would like to go to Disneyland in 3 years. Their total cost should be $4500. If each invests $1000 in a savings account paying 5.5% interest, compounded semiannually, will they have enough in 3 years?

△ **37.** A gallon of latex paint can cover 500 square feet. Find how many gallon containers of paint should be bought to paint two coats on each wall of a rectangular room whose dimensions are 14 feet by 16 feet (assume 8-foot ceilings).

△ **38.** A gallon of enamel paint can cover 300 square feet. Find how many gallon containers of paint should be bought to paint three coats on a wall measuring 21 feet by 8 feet.

△ **39.** A portion of the external tank of the Space Shuttle *Endeavour* is a liquid hydrogen tank. If the ends of the tank are hemispheres, find the volume of the tank. To do so, answer parts **a** through **c**.

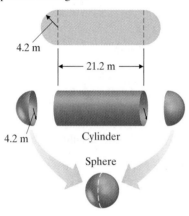

4.2 m

21.2 m

4.2 m Cylinder

Sphere

a. Find the volume of the cylinder shown. Round to 2 decimal places.

b. Find the volume of the sphere shown. Round to 2 decimal places.

c. Add the results of parts **a** and **b**. This sum is the approximate volume of the tank.

△ **40.** In 1945, Arthur C. Clarke, a scientist and science-fiction writer, predicted that an artificial satellite placed at a height of 22,248 miles directly above the equator would orbit the globe at the same speed with which the Earth was rotating. This belt along the equator is known as the Clarke belt. Use the formula for circumference of a circle and find the "length" of the Clarke belt. (*Hint:* Recall that the radius of the Earth is approximately 4000 miles. Round to the nearest whole mile.)

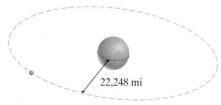

22,248 mi

41. Amelia Earhart was the first woman to fly solo nonstop coast to coast, setting the women's nonstop transcontinental speed record. She traveled 2447.8 miles in 19 hours 5 minutes. Find the average speed of her flight in miles per hour. (Change 19 hours 5 minutes into hours and use the formula $d = rt$.) Round to the nearest tenth of a mile per hour.

△ **42.** The Space Shuttle *Endeavour* has a cargo bay that is in the shape of a cylinder whose length is 18.3 meters and whose diameter is 4.6 meters. Find its volume.

△ **43.** The deepest hole in the ocean floor is beneath the Pacific Ocean and is called Hole 504B. It is located off the coast of Ecuador. Scientists are drilling it to learn more about the Earth's history. Currently, the hole is in the shape of a cylinder whose volume is approximately 3800 cubic feet and whose length is 1.3 miles. Find the radius of the hole to the nearest hundredth of a foot. (*Hint:* Make sure the same units of measurement are used.)

44. The deepest man-made hole is called the Kola Superdeep Borehole. It is approximately 8 miles deep and is located near a small Russian town in the Arctic Circle. If it takes 7.5 hours to remove the drill from the bottom of the hole, find the rate that the drill can be retrieved in feet per second. Round to the nearest tenth. (*Hint:* Write 8 miles as feet, 7.5 hours as seconds, then use the formula $d = rt$.)

△ **45.** Eartha is the world's largest globe. It is located at the headquarters of DeLorme, a mapmaking company in Yarmouth, Maine. Eartha is 41.125 feet in diameter. Find its exact circumference (distance around) and then approximate its circumference using 3.14 for π. (*Source:* DeLorme)

△ **46.** Eartha is in the shape of a sphere. Its radius is about 20.6 feet. Approximate its volume to the nearest cubic foot. (*Source:* DeLorme)

47. A gardener has 52 feet of garden edging in 1-foot sections. A rectangular garden is to be formed using the edging for 3 sides and the side of a patio for the fourth side. Use a table to find the dimensions that will form a garden of largest possible area.

48. Answer Exercise 47 using 92 feet of 1-foot sections of garden edging instead of 52.

The calorie count of a serving of food can be computed based on its composition of carbohydrate, fat, and protein. The calorie count C for a serving of food can be computed using the formula $C = 4h + 9f + 4p$, where h is the number of grams of carbohydrate contained in the serving, f is the number of grams of fat contained in the serving, and p is the number of grams of protein contained in the serving.

49. Solve this formula for f, the number of grams of fat contained in a serving of food.

50. Solve this formula for h, the number of grams of carbohydrate contained in a serving of food.

51. A serving of cashews contains 14 grams of fat, 7 grams of carbohydrate, and 6 grams of protein. How many calories are in this serving of cashews?

52. A serving of chocolate candies contains 9 grams of fat, 30 grams of carbohydrate, and 2 grams of protein. How many calories are in this serving of chocolate candies?

53. A serving of raisins contains 130 calories and 31 grams of carbohydrate. If raisins are a fat-free food, how much protein is provided by this serving of raisins?

54. A serving of yogurt contains 120 calories, 21 grams of carbohydrate, and 5 grams of protein. How much fat is provided by this serving of yogurt? Round to the nearest tenth of a gram.

CONCEPT EXTENSIONS

55. Solar system distances are so great that units other than miles or kilometers are often used. For example, the astronomical unit (AU) is the average distance between Earth and the Sun, or 92,900,000 miles. Use this information to convert each planet's distance in miles from the Sun to astronomical units. Round to three decimal places. (*Source:* National Space Science Data Center)

Planet	Miles from the Sun	AU from the Sun	Planet	Miles from the Sun	AU from the Sun
Mercury	36 million		Saturn	886.1 million	
Venus	67.2 million		Uranus	1783 million	
Earth	92.9 million		Neptune	2793 million	
Mars	141.5 million		Pluto	3670 million	
Jupiter	483.3 million				

56. An orbit such as Clarke's belt in Exercise 40 is called a geostationary orbit. In your own words, why do you think that communications satellites are placed in geostationary orbits?

57. How much do you think it costs each American to build a space shuttle? Write down your estimate. The Space Shuttle *Endeavour* was completed in 1992 and cost approximately $1.7 billion. If the population of the United States in 1992 was 250 million, find the cost per person to build the *Endeavour*. How close was your estimate?

58. If you are investing money in a savings account paying a rate of r, which account should you choose—an account compounded 4 times a year or 12 times a year? Explain your choice.

59. To borrow money at a rate of r, which loan plan should you choose—one compounding 4 times a year or 12 times a year? Explain your choice.

60. The Drake Equation is a formula used to estimate the number of technological civilizations that might exist in our own Milky Way Galaxy. The Drake Equation is given as $N = R^* \times f_p \times n_e \times f_l \times f_i \times f_c \times L$. Solve the Drake Equation for the variable n_e. (*Note:* Descriptions of the meaning of each variable in this equation, as well as Drake Equation calculators, exist online. For more information, try doing a Web search on "Drake Equation.")

61. On April 1, 1985, *Sports Illustrated* published an April Fool's story by writer George Plimpton. He wrote that the New York Mets had discovered a man who could throw a 168-miles-per-hour fast ball. If the distance from the pitcher's mound to the plate is 60.5 feet, how long would it take for a ball thrown at that rate to travel that distance? (*Hint:* Write the rate 168 miles per hour in feet per second. Then use the formula $d = r \cdot t$.)

$$168 \text{ miles per hour} = \frac{168 \text{ miles}}{1 \text{ hour}}$$
$$= \frac{\_\text{feet}}{\_\text{seconds}}$$
$$= \frac{\_\text{feet}}{1 \text{ second}}$$
$$= \_\text{feet per second}$$

*The measure of the chance or likelihood of an event occurring is its **probability**. A formula basic to the study of probability is the formula for the probability of an event when all the outcomes are equally likely. This formula is*

$$\text{Probability of an event} = \frac{\text{number of ways that the event can occur}}{\text{number of possible outcomes}}$$

For example, to find the probability that a single spin on the spinner will result in red, notice first that the spinner is divided into 8 parts, so there are 8 possible outcomes. Next, notice that there is only one sector of the spinner colored red, so the number of ways that the spinner can land on red is 1. Then this probability denoted by P(red) is

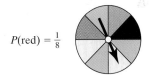

$$P(\text{red}) = \frac{1}{8}$$

Find each probability in simplest form.

62. $P(\text{green})$

63. $P(\text{yellow})$

64. $P(\text{black})$

65. $P(\text{blue})$

66. $P(\text{green or blue})$

67. $P(\text{black or yellow})$

68. $P(\text{red, green, or black})$

69. $P(\text{yellow, blue, or black})$

70. $P(\text{white})$

71. $P(\text{red, yellow, green, blue, or black})$

72. From the previous probability formula, what do you think is always the probability of an event that is impossible occurring?

73. What do you think is always the probability of an event that is sure to occur?

 STUDY SKILLS BUILDER

How Are Your Homework Assignments Going?

It is very important in mathematics to keep up with homework. Why? Many concepts build on each other. Often your understanding of a day's concepts depends on an understanding of the previous day's material.

Remember that completing your homework assignment involves a lot more than attempting a few of the problems assigned.

To complete a homework assignment, remember these four things:

- Attempt all of it.
- Check it.
- Correct it.
- If needed, ask questions about it.

Take a moment and review your completed homework assignments. Answer the questions below based on this review.

1. Approximate the fraction of your homework you have attempted.

2. Approximate the fraction of your homework you have checked (if possible).

3. If you are able to check your homework, have you corrected it when errors have been found?

4. When working homework, if you do not understand a concept, what do you do?

CHAPTER 1 GROUP ACTIVITY

Analyzing Newspaper Circulation

The number of daily newspapers in business in the United States has declined steadily recently, continuing a trend that started in the mid-1970s. In 1900, there were roughly 2300 daily newspapers in operation. However, by 2003, that number had dropped to only 1456 newspapers in existence. Average overall daily newspaper circulation also continues to decline, from 60.2 million in 1992 to 55.2 million in 2003.

The table gives data about daily newspaper circulation for New York City–based newspapers for the years 2005 and 2006. (*Source: Burrels Luce Media Database*) In this activity you will have the opportunity to analyze this data. This activity may be completed by working in groups or individually.

New York–Based Newspaper	2005 Daily Circulation	2006 Daily Circulation	Daily Edition Newsstand Price
Wall Street Journal	2,070,498	2,049,786	$1.00
New York Times	1,121,623	1,142,464	$1.00
New York Daily News	708,773	708,477	$0.50
New York Post	643,056	673,379	$0.35

1. Find the change in circulation from 2005 to 2006 for each newspaper. Did any newspaper gain circulation? If so, which one(s) and by how much?

2. Construct a bar graph showing the change in circulation from 2005 to 2006 for each newspaper. Which newspaper experienced the greatest change in circulation?

3. What was the total daily circulation of these New York City–based newspapers in 2005? 2006? Did total circulation increase or decrease from 2005 to 2006? By how much?

4. Discuss factors that may have contributed to the overall change in daily newspaper circulation.

5. The population of New York City is approximately 8,160,000. Find the number of New York City–based newspapers sold per person in 2006 for New York City. (*Source: CIA Fact Book*)

6. The population of the New York City metropolitan area is approximately 18,700,000. Find the number of New York City–based newspapers sold per person in 2006 for the New York City metropolitan area. (*Source: CIA Fact Book*)

7. Which of the figures found in Questions 5 and 6 do you think is more meaningful? Why might neither of these figures be capable of describing the full circulation situation?

8. Assuming that each copy was sold from a newsstand in the New York City metropolitan area, use the daily edition prices given in the table to approximate the total amount spent each day on these New York City–based newspapers in 2006. Find the total amount spent annually on these daily newspapers in 2006.

9. How accurate do you think the figures you found in Question 8 are? Explain your reasoning.

CHAPTER 1 VOCABULARY CHECK

Fill in each blank with one of the words or phrases listed below.

identity	distributive	real	reciprocals	contradiction
solution	absolute value	opposite	associative	linear equation in one variable
formula	inequality	commutative	whole	algebraic expression
exponent	variable	consecutive integers		

1. A(n) _____ is formed by numbers and variables connected by the operations of addition, subtraction, multiplication, division, raising to powers, and/or taking roots.

2. The _____ of a number a is $-a$.

3. $3(x - 6) = 3x - 18$ by the _____ property.

4. The _____ of a number is the distance between that number and 0 on the number line.

5. A(n) _____ is a shorthand notation for repeated multiplication of the same factor.

6. A letter that represents a number is called a _____ .

7. The symbols $<$ and $>$ are called _____ symbols.

8. If a is not 0, then a and $1/a$ are called _____ .

9. $A + B = B + A$ by the _____ property.

10. $(A + B) + C = A + (B + C)$ by the _____ property.

11. The numbers $0, 1, 2, 3, \ldots$ are called _____ numbers.

12. If a number corresponds to a point on the number line, we know that number is a ____ number.

13. An equation in one variable that has no solution is called a(n) _____ .

14. An equation in one variable that has every number (for which the equation is defined) as a solution is called a(n) _____ .

15. The equation $d = rt$ is also called a(n) _____ .

16. When a variable in an equation is replaced by a number and the resulting equation is true, then that number is called a(n) _____ of the equation.

17. The integers $17, 18, 19$ are examples of _____ .

18. The statement $5x - 0.2 = 7$ is an example of a(n) _____ .

> ▶ **Helpful Hint**
>
> Are you preparing for your test? Don't forget to take the Chapter 1 Test on page 96. Then check your answers at the back of the text and use the Chapter Test Prep Video CD to see the fully worked-out solutions to any of the exercises you want to review.

CHAPTER 1 HIGHLIGHTS

DEFINITIONS AND CONCEPTS	EXAMPLES

SECTION 1.2 ALGEBRAIC EXPRESSIONS AND SETS OF NUMBERS

Letters that represent numbers are called **variables.**	Examples of variables are $$x, a, m, y$$
An **algebraic expression** is formed by numbers and variables connected by the operations of addition, subtraction, multiplication, division, raising to powers, and/or taking roots.	Examples of algebraic expressions are $$7y, -3, \frac{x^2 - 9}{-2} + 14x, \sqrt{3} + \sqrt{m}$$
To **evaluate** an algebraic expression containing variables, substitute the given numbers for the variables and simplify. The result is called the **value** of the expression.	Evaluate $2.7x$ if $x = 3$. $$2.7x = 2.7(3)$$ $$= 8.1$$

(continued)

DEFINITIONS AND CONCEPTS	**EXAMPLES**

SECTION 1.2 ALGEBRAIC EXPRESSIONS AND SETS OF NUMBERS (continued)

Natural numbers: $\{1, 2, 3, \dots\}$
Whole numbers: $\{0, 1, 2, 3, \dots\}$
Integers: $\{\dots, -3, -2, -1, 0, 1, 2, 3, \dots\}$
Each listing of three dots above is called an **ellipsis,** which means the pattern continues.
The members of a set are called its **elements.**
Set builder notation describes the elements of a set but does not list them.
Real numbers: $\{x \mid x$ corresponds to a point on the number line$\}$.
Rational numbers: $\left\{\dfrac{a}{b} \;\middle|\; a \text{ and } b \text{ are integers and } b \neq 0\right\}$.
Irrational numbers: $\{x \mid x$ is a real number and x is not a rational number$\}$.
If 3 is an element of set A, we write $3 \in A$.
If all the elements of set A are also in set B, we say that set A is a **subset** of set B, and we write $A \subseteq B$.
Absolute value:

$$|a| = \begin{cases} a & \text{if } a \text{ is } 0 \text{ or a positive number} \\ -a & \text{if } a \text{ is a negative number} \end{cases}$$

The opposite of a number a is the number $-a$.

Given the set $\left\{-9.6, -5, -\sqrt{2}, 0, \dfrac{2}{5}, 101\right\}$ list the elements that belong to the set of

Natural numbers 101
Whole numbers $0, 101$
Integers $-5, 0, 101$
Real numbers $-9.6, -5, -\sqrt{2}, 0, \dfrac{2}{5}, 101$
Rational numbers $-9.6, -5, 0, \dfrac{2}{5}, 101$
Irrational numbers $-\sqrt{2}$

List the elements in the set
$\{x \mid x$ is an integer between -2 and $5\}$.

```
abs(3)
              3
abs(0)
              0
abs(-7.2)
            7.2
■
```

$\{-1, 0, 1, 2, 3, 4\}$
$\{1, 2, 4\} \subseteq \{1, 2, 3, 4\}$
$|3| = 3, |0| = 0, |-7.2| = 7.2$

The opposite of 5 is -5. The opposite of -11 is 11.

SECTION 1.3 OPERATIONS ON REAL NUMBERS

Adding real numbers:

1. To add two numbers with the same sign, add their absolute values and attach their common sign.
2. To add two numbers with different signs, subtract the smaller absolute value from the larger absolute value and attach the sign of the number with the larger absolute value.

Subtracting real numbers:

$$a - b = a + (-b)$$

Multiplying and dividing real numbers:

The product or quotient of two numbers with the same sign is positive.
The product or quotient of two numbers with different signs is negative.

A natural number **exponent** is a shorthand notation for repeated multiplication of the same factor.

The notation $\sqrt{a}$ is used to denote the **positive**, or **principal, square root** of a nonnegative number a.

$$\sqrt{a} = b \text{ if } b^2 = a \text{ and } b \text{ is positive}.$$

Also,

$$\sqrt[3]{a} = b \text{ if } b^3 = a$$

$$\sqrt[4]{a} = b \text{ if } b^4 = a \text{ and } b \text{ is positive}.$$

$20.8 + (-10.2) = 10.6$

$-18 + 6 = -12$

$\dfrac{2}{7} + \dfrac{1}{7} = \dfrac{3}{7}$

$-5 + (-2.6) = -7.6$

```
-18+6
              -12
2/7+1/7▸Frac
              3/7
-5+(-2.6)
             -7.6
■
```

$18 - 21 = 18 + (-21) = -3$

$(-8)(-4) = 32 \qquad \dfrac{-8}{-4} = 2$

$8 \cdot 4 = 32 \qquad \dfrac{8}{4} = 2$

$-17 \cdot 2 = -34 \qquad \dfrac{-14}{2} = -7$

$4(-1.6) = -6.4 \qquad \dfrac{22}{-2} = -11$

$3^4 = 3 \cdot 3 \cdot 3 \cdot 3 = 81$

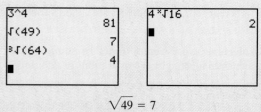

```
3^4
              81
√(49)
               7
3√(64)
               4
■
```
```
4*√16
               2
■
```

$\sqrt{49} = 7$
$\sqrt[3]{64} = 4$
$\sqrt[4]{16} = 2$

DEFINITIONS AND CONCEPTS	**EXAMPLES**

Order of Operations

Simplify expressions using the order that follows. If grouping symbols such as parentheses are present, simplify expressions within those first, starting with the innermost set. If fraction bars are present, simplify the numerator and denominator separately.

1. Raise to powers or take roots in order from left to right.
2. Multiply or divide in order from left to right.
3. Add or subtract in order from left to right.

Simplify $\dfrac{42 - 2(3^2 - \sqrt{16})}{-8}$.

$$\frac{42 - 2(3^2 - \sqrt{16})}{-8} = \frac{42 - 2(9 - 4)}{-8}$$
$$= \frac{42 - 2(5)}{-8}$$
$$= \frac{42 - 10}{-8}$$
$$= \frac{32}{-8} = -4$$

To **evaluate** an algebraic expression containing variables, substitute the given numbers for the variables and simplify. The result is called the **value** of the expression.

Evaluate $-2.7x$ if $x = 3$.
$$-2.7x = -2.7(3)$$
$$= -8.1$$

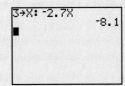

```
3→X: -2.7X
              -8.1
■
```

Symbols: $=$ is equal to
$\neq$ is not equal to
$>$ is greater than
$<$ is less than
$\geq$ is greater than or equal to
$\leq$ is less than or equal to

$$-5 = -5$$
$$-5 \neq -3$$
$$1.7 > 1.2$$
$$-1.7 < -1.2$$
$$\frac{5}{3} \geq \frac{5}{3}$$
$$-\frac{1}{2} \leq \frac{1}{2}$$

Identity:

$$a + 0 = a \qquad 0 + a = a$$
$$a \cdot 1 = a \qquad 1 \cdot a = a$$

$$3 + 0 = 3 \qquad 0 + 3 = 3$$
$$-1.8 \cdot 1 = -1.8 \qquad 1 \cdot -1.8 = -1.8$$

Inverse:

$$a + (-a) = 0 \qquad -a + a = 0$$
$$a \cdot \frac{1}{a} = 1 \qquad \frac{1}{a} \cdot a = 1, a \neq 0$$

$$7 + (-7) = 0 \qquad -7 + 7 = 0$$
$$5 \cdot \frac{1}{5} = 1 \qquad \frac{1}{5} \cdot 5 = 1$$

Commutative:

$$a + b = b + a$$
$$a \cdot b = b \cdot a$$

$$x + 7 = 7 + x$$
$$9 \cdot y = y \cdot 9$$

Associative:

$$(a + b) + c = a + (b + c)$$
$$(a \cdot b) \cdot c = a \cdot (b \cdot c)$$

$$(3 + 1) + 10 = 3 + (1 + 10)$$
$$(3 \cdot 1) \cdot 10 = 3(1 \cdot 10)$$

Distributive:

$$a(b + c) = ab + ac$$

$$6(x + 5) = 6 \cdot x + 6 \cdot 5$$
$$= 6x + 30$$

DEFINITIONS AND CONCEPTS	**EXAMPLES**

SECTION 1.5 SOLVING LINEAR EQUATIONS ALGEBRAICALLY

An **equation** is a statement that two expressions are equal.

Equations:

$$5 = 5 \qquad 7x + 2 = -14 \qquad 3(x - 1)^2 = 9x^2 - 6$$

A **linear equation in one variable** is an equation that can be written in the form $ax + b = c$, where a, b, and c are real numbers and a is not 0.

Linear equations:

$$7x + 2 = -14 \qquad x = -3$$
$$5(2y - 7) = -2(8y - 1)$$

A **solution** of an equation is a value for the variable that makes the equation a true statement.

Check to see that -1 is a solution of

$$3(x - 1) = 4x - 2.$$

$$3(-1 - 1) = 4(-1) - 2$$
$$3(-2) = -4 - 2$$
$$-6 = -6 \qquad \text{True}$$

Thus, -1 is a solution.

Equivalent equations have the same solution.

$x - 12 = 14$ and $x = 26$ are equivalent equations.

The **addition property of equality** guarantees that the same number may be added to (or subtracted from) both sides of an equation, and the result is an equivalent equation.

Solve for x: $-3x - 2 = 10$.

$$-3x - 2 + 2 = 10 + 2 \quad \text{Add 2 to both sides.}$$
$$-3x = 12$$

The **multiplication property of equality** guarantees that both sides of an equation may be multiplied by (or divided by) the same nonzero number, and the result is an equivalent equation.

$$\frac{-3x}{-3} = \frac{12}{-3} \qquad \text{Divide both sides by } -3.$$
$$x = -4$$

To solve linear equations in one variable:

Solve for x:

$$x - \frac{x - 2}{6} = \frac{x - 7}{3} + \frac{2}{3}$$

1. Clear the equation of fractions.

1.
$$6\left(x - \frac{x - 2}{6}\right) = 6\left(\frac{x - 7}{3} + \frac{2}{3}\right) \quad \begin{array}{l}\text{Multiply}\\\text{both}\\\text{sides by 6.}\end{array}$$
$$6x - (x - 2) = 2(x - 7) + 2(2)$$

2. Remove grouping symbols such as parentheses.

2. $\qquad 6x - x + 2 = 2x - 14 + 4 \qquad \begin{array}{l}\text{Remove}\\\text{grouping}\\\text{symbols.}\end{array}$

3. Simplify by combining like terms.

3. $\qquad 5x + 2 = 2x - 10$

4. Write variable terms on one side and numbers on the other side using the addition property of equality.

4. $\qquad 5x + 2 - 2 = 2x - 10 - 2 \qquad \text{Subtract 2.}$
$$5x = 2x - 12$$
$$5x - 2x = 2x - 12 - 2x \qquad \text{Subtract } 2x.$$
$$3x = -12$$

5. Isolate the variable using the multiplication property of equality.

5. $\qquad \dfrac{3x}{3} = \dfrac{-12}{3} \qquad \text{Divide by 3.}$
$$x = -4$$

6. Use your calculator to check the proposed solution in the original equation.

6.
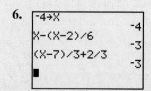

DEFINITIONS AND CONCEPTS	**EXAMPLES**

Problem-Solving Strategy

Colorado is shaped like a rectangle whose length is about 1.3 times its width. If the perimeter of Colorado is 2070 kilometers, find its dimensions.

1. UNDERSTAND the problem.

1. Read and reread the problem. Guess a solution and check your guess.
Let x = width of Colorado in kilometers. Then $1.3x$ = length of Colorado in kilometers.

x

1.3x

2. TRANSLATE the problem.

2. In words:

$$\underbrace{\text{twice the length}} + \underbrace{\text{twice the width}} = \underbrace{\text{perimeter}}$$

Translate: $2(1.3x) + 2x = 2070$

3. SOLVE the equation.

3. $2.6x + 2x = 2070$
$4.6x = 2070$
$x = 450$

4. INTERPRET the results.

4. If $x = 450$ kilometers, then $1.3x = 1.3(450) = 585$ kilometers. *Check:* The perimeter of a rectangle whose width is 450 kilometers and whose length is 585 kilometers is $2(450) + 2(585) = 2070$ kilometers, the required perimeter. *State:* The dimensions of Colorado are 450 kilometers by 585 kilometers.

To generate a table on a calculator, a **table start** x-value and a **table increment**, or change in x-values, must be given.

Use your calculator to complete the given table.

Hours Worked	x	8	10	12	14	16
Gross Pay	$8 + 6.25x$					

Go to the **Y** = editor and enter $y_1 = 8 + 6.25x$. In the table setup menu, let table start = 8 and table increment = 2 since the change in x-values in the table is 2. The table generated is shown below.

X	Y1
8	58
10	70.5
12	83
14	95.5
16	108
18	120.5
20	133

Y1⯀8+6.25X

DEFINITIONS AND CONCEPTS	EXAMPLES

SECTION 1.8 FORMULAS AND PROBLEM SOLVING

An equation that describes a known relationship among quantities is called a **formula**.

Formulas:

$A = \pi r^2$ (area of a circle)

$I = PRT$ (interest = principal · rate · time)

To solve a formula for a specified variable, use the steps for solving an equation. Treat the specified variable as the only variable of the equation.

Solve $A = 2HW + 2LW + 2LH$ for H

$A - 2LW = 2HW + 2LH$ Subtract $2LW$.

$A - 2LW = H(2W + 2L)$ Factor out H.

$$\frac{A - 2LW}{2W + 2L} = \frac{H(2W + 2L)}{2W + 2L}$$ Divide by $2W + 2L$.

$$\frac{A - 2LW}{2W + 2L} = H$$ Simplify.

CHAPTER 1 REVIEW

(1.2) *Find the value of each algebraic expression at the given replacement values.*

1. $7x$ when $x = 3$

2. st when $s = 1.6$ and $t = 5$

3. The hummingbird has an average wing speed of 90 beats per second. The expression $90t$ gives the number of wing beats in t seconds. Calculate the number of wing beats in *1 hour* for the hummingbird.

List the elements in each set.

4. $\{x | x \text{ is an odd integer between } -2 \text{ and } 4\}$

5. $\{x | x \text{ is an even integer between } -3 \text{ and } 7\}$

6. $\{x | x \text{ is a negative whole number}\}$

7. $\{x | x \text{ is a natural number that is not a rational number}\}$

8. $\{x | x \text{ is a whole number greater than } 5\}$

9. $\{x | x \text{ is an integer less than } 3\}$

Determine whether each statement is true or false if $A = \{6, 10, 12\}$, $B = \{5, 9, 11\}$, $C = \{\dots, -3, -2, -1, 0, 1, 2, 3, \dots\}$, $D = \{2, 4, 6, \dots, 16\}$ $E = \{x | x \text{ is a rational number}\}$, $F = \{\ \}$, $G = \{x | x \text{ is an irrational number}\}$, and $H = \{x | x \text{ is a real number}\}$.

10. $10 \in D$

11. $B \in 9$

12. $\sqrt{169} \notin G$

13. $0 \notin F$

14. $\pi \in E$

15. $\pi \in H$

16. $\sqrt{4} \in G$

17. $-9 \in E$

18. $A \subseteq D$

19. $C \nsubseteq B$

20. $C \nsubseteq E$

21. $F \subseteq H$

22. $B \subseteq B$

23. $D \subset C$

24. $C \subseteq H$

25. $G \subseteq H$

26. $\{5\} \in B$

27. $\{5\} \subseteq B$

List the elements of the set $\left\{5, -\dfrac{2}{3}, \dfrac{8}{2}, \sqrt{9}, 0.3, \sqrt{7}, 1\dfrac{5}{8}, -1, \pi\right\}$ that are also elements of each given set.

28. Whole numbers

29. Natural numbers

30. Rational numbers

31. Irrational numbers

32. Real numbers

33. Integers

Find the opposite.

34. $-\dfrac{3}{4}$　　　　　　**35.** 0.6

36. 0　　　　　　　　**37.** 1

Find the reciprocal.

38. $-\dfrac{3}{4}$　　　　　　**39.** 0.6

40. 0　　　　　　　　**41.** 1

(1.3) Simplify.

42. $-7 + 3$　　　　　　**43.** $-10 + (-25)$

44. $5(-0.4)$　　　　　　**45.** $(-3.1)(-0.1)$

46. $-7 - (-15)$　　　　　**47.** $9 - (-4.3)$

48. $(-6)(-4)(0)(-3)$　　　**49.** $(-12)(0)(-1)(-5)$

50. $(-24) \div 0$　　　　　**51.** $0 \div (-45)$

52. $(-36) \div (-9)$　　　　**53.** $60 \div (-12)$

54. $\left(-\dfrac{4}{5}\right) - \left(-\dfrac{2}{3}\right)$　　　**55.** $\left(\dfrac{5}{4}\right) - \left(-2\dfrac{3}{4}\right)$

56. Determine the unknown fractional part.

Simplify.

57. $-5 + 7 - 3 - (-10)$　　**58.** $8 - (-3) + (-4) + 6$

59. $3(4 - 5)^4$　　　　　**60.** $6(7 - 10)^2$

61. $\left(-\dfrac{8}{15}\right) \cdot \left(-\dfrac{2}{3}\right)^2$　　**62.** $\left(-\dfrac{3}{4}\right)^2 \cdot \left(-\dfrac{10}{21}\right)$

63. $\dfrac{-\dfrac{6}{15}}{\dfrac{8}{25}}$　　　　　**64.** $\dfrac{\dfrac{4}{9}}{-\dfrac{8}{45}}$

65. $-\dfrac{3}{8} + 3(2) \div 6$　　**66.** $5(-2) - (-3) - \dfrac{1}{6} + \dfrac{2}{3}$

67. $|2^3 - 3^2| - |5 - 7|$　　**68.** $|5^2 - 2^2| + |9 \div (-3)|$

69. $(2^3 - 3^2) - (5 - 7)$　　**70.** $(5^2 - 2^4) + [9 \div (-3)]$

71. $\dfrac{(8 - 10)^3 - (-4)^2}{2 + 8(2) \div 4}$　　**72.** $\dfrac{(2 + 4)^2 + (-1)^5}{12 \div 2 \cdot 3 - 3}$

73. $\dfrac{(4 - 9) + 4 - 9}{10 - 12 \div 4 \cdot 8}$　　**74.** $\dfrac{3 - 7 - (7 - 3)}{15 + 30 \div 6 \cdot 2}$

75. $\dfrac{\sqrt{25}}{4 + 3 \cdot 7}$　　　　**76.** $\dfrac{\sqrt{64}}{24 - 8 \cdot 2}$

Find the value of each expression when $x = 0$, $y = 3$, and $z = -2$.

77. $x^2 - y^2 + z^2$　　　**78.** $\dfrac{5x + z}{2y}$

79. $\dfrac{-7y - 3z}{-3}$　　　　**80.** $(x - y + z)^2$

△ **81.** The algebraic expression $2\pi r$ represents the circumference of (distance around) a circle of radius r.

a. Complete the table below by evaluating the expression at given values of r. (Use 3.14 for π.)

Radius	r	1	10	100
Circumference	$2\pi r$			

b. As the radius of a circle increases, does the circumference of the circle increase or decrease?

(1.4) Simplify each expression.

82. $5xy - 7xy + 3 - 2 + xy$

83. $4x + 10x - 19x + 10 - 19$

84. $6x^2 + 2 - 4(x^2 + 1)$

85. $-7(2x^2 - 1) - x^2 - 1$

86. $(3.2x - 1.5) - (4.3x - 1.2)$

87. $(7.6x + 4.7) - (1.9x + 3.6)$

Write each statement using mathematical symbols.

88. Twelve is the product of x and negative 4.

89. The sum of n and twice n is negative fifteen.

90. Four times the sum of y and three is -1.

91. The difference of t and five, multiplied by six is four.

92. Seven subtracted from z is six.

93. Ten less than the product of x and nine is five.

94. The difference of x and 5 is at least 12.

95. The opposite of four is less than the product of y and seven.

96. Two-thirds is not equal to twice the sum of n and one-fourth.

97. The sum of t and six is not more than negative twelve.

Name the property illustrated.

98. $(M + 5) + P = M + (5 + P)$

99. $5(3x - 4) = 15x - 20$

100. $(-4) + 4 = 0$

101. $(3 + x) + 7 = 7 + (3 + x)$

102. $(XY)Z = (YZ)X$

103. $\left(-\dfrac{3}{5}\right) \cdot \left(-\dfrac{5}{3}\right) = 1$

104. $T \cdot 0 = 0$

105. $(ab)c = a(bc)$

106. $A + 0 = A$

107. $8 \cdot 1 = 8$

Complete the equation using the given property.

108. $5x - 15z =$ _____ Distributive property

109. $(7 + y) + (3 + x) =$ _____ Commutative property

110. $0 =$ ____ Additive inverse property

111. $1 =$ ____ Multiplicative inverse property

112. $[(3.4)(0.7)]5 =$ _____ Associative property

113. $7 =$ _____ Additive identity property

Insert $<$, $>$, or $=$ to make each statement true.

114. -9 ___ -12

115. -3 ___ -1

116. 7 ___ $|-7|$

117. -5 ___ $-(-5)$

118. $-(-2)$ ___ -2

(1.5) *Solve each linear equation.*

119. $4(x - 5) = 2x - 14$

120. $x + 7 = -2(x + 8)$

121. $3(2y - 1) = -8(6 + y)$

122. $-(z + 12) = 5(2z - 1)$

123. $n - (8 + 4n) = 2(3n - 4)$

124. $4(9v + 2) = 6(1 + 6v) - 10$

125. $0.3(x - 2) = 1.2$

126. $1.5 = 0.2(c - 0.3)$

127. $-4(2 - 3x) = 2(3x - 4) + 6x$

128. $6(m - 1) + 3(2 - m) = 0$

129. $6 - 3(2g + 4) - 4g = 5(1 - 2g)$

130. $20 - 5(p + 1) + 3p = -(2p - 15)$

131. $\dfrac{x}{3} - 4 = x - 2$

132. $\dfrac{9}{4}y = \dfrac{2}{3}y$

133. $\dfrac{3n}{8} - 1 = 3 + \dfrac{n}{6}$

134. $\dfrac{z}{6} + 1 = \dfrac{z}{2} + 2$

135. $\dfrac{y}{4} - \dfrac{y}{2} = -8$

136. $\dfrac{2x}{3} - \dfrac{8}{3} = x$

137. $\dfrac{b - 2}{3} = \dfrac{b + 2}{5}$

138. $\dfrac{2t - 1}{3} = \dfrac{3t + 2}{15}$

139. $\dfrac{2(t + 1)}{3} = \dfrac{2(t - 1)}{3}$

140. $\dfrac{3a - 3}{6} = \dfrac{4a + 1}{15} + 2$

(1.6) *Solve.*

141. Twice the difference of a number and 3 is the same as 1 added to three times the number. Find the number.

142. One number is 5 more than another number. If the sum of the numbers is 285, find the numbers.

143. In 2000, a record number of music CDs were sold by manufacturers in the United States. By 2005, this number had decreased to 705.4 million music CDs. If this represented a decrease of 25%, find the number of music CDs sold by U.S. manufacturers in 2000. (*Source:* Recording Industry Association of America)

144. Find four consecutive integers such that twice the first subtracted from the sum of the other three integers is 16.

145. Determine whether there are two consecutive odd integers such that 5 times the first exceeds 3 times the second by 54.

146. The length of a rectangular playing field is 5 meters less than twice its width. If 230 meters of fencing goes around the field, find the dimensions of the field.

x

$2x - 5$

147. A car rental company charges $19.95 per day for a compact car plus 12 cents per mile for every mile over 100 miles driven per day. If Mr. Woo's bill for 2 days' use is $46.86, find how many miles he drove.

(1.7) *Use your calculator and the table feature to complete each table. If necessary, round any amounts to 2 decimal places.*

148.

Radius of Cone	x	1	1.5	2	2.5	3
Volume (If Height is 10 Inches)	$\dfrac{10}{3}\pi x^2$					

10

x

149.

Width of Rectangle	x	2	4.68	9.5	12.68
Perimeter (If Length is 15 Units)	$30 + 2x$				

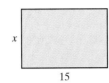

150. Gina Saltalamacchia works as a hostess at a restaurant and makes $5.75 per hour. Complete the table and find her gross pay for the hours given.

Hours	5	10	15	20	25	30
Gross Pay (Dollars)						

151. Coast Waterworks, Inc. charges $8.00 for the first 4 thousand gallons of water used and then $1.10 per thousand gallons used over 4 thousand gallons. Cross Gates Water Company charges $12.00 for the first 5 thousand gallons and $1.50 per thousand gallons used over 5 thousand gallons. Complete the table and find the total charges for the gallons used.

Gallons (In Thousands Used)	5	6	7	8
Coast Charge (Dollars)				
Cross Gates Charge (Dollars)				

152. A rock is thrown upward from a bridge. Neglecting air resistance, the height of the rock at time x seconds is given by

$$y = -16x^2 + 40x + 500$$

where y is measured in feet.

a. Complete the table to find the height of the rock at the given times.

Seconds	0	1	2	3	4	5
Height in Feet						

b. Find the maximum height of the rock to the nearest foot.

c. Find to the nearest tenth of a second when the rock strikes the ground.

(1.8) Solve each equation for the specified variable.

153. $V = lwh; w$

154. $C = 2\pi r; r$

155. $5x - 4y = -12; y$

156. $5x - 4y = -12; x$

157. $y - y_1 = m(x - x_1); m$

158. $y - y_1 = m(x - x_1); x$

159. $E = I(R + r); r$

160. $S = vt + gt^2; g$

△ **161.** $T = gr + gvt; g$

△ **162.** $I = Prt + P; P$

163. $A = \dfrac{h}{2}(B + b); B$

164. $V = \dfrac{1}{3}\pi r^2 h; h$

165. $R = \dfrac{r_1 + r_2}{2}; r_1$

166. $\dfrac{V_1}{T_1} = \dfrac{V_2}{T_2}; T_2$

Solve.

167. A principal of $3000 is invested in an account paying an annual percentage rate of 3%. Find the amount in the account after 7 years if the interest is compounded

a. semiannually

b. weekly

(Approximate to the nearest cent.)

168. Angie has a photograph in which the length is 2 inches longer than the width. If she increases each dimension by 4 inches, the area is increased by 88 square inches. Find the original dimensions.

169. The formula for converting Celsius temperatures to Fahrenheit temperatures is $F = \dfrac{9C + 160}{5}$.

a. Complete the following table to find Fahrenheit temperatures given the Celsius temperatures.

Celsius	−40	−15	10	60
Fahrenheit				

b. If the boiling point of water is 100 degrees Celsius, what is this in degrees Fahrenheit?

c. If the freezing point of water is 0 degrees Celsius, what is this in degrees Fahrenheit?

170. The formula for finding the distance traveled at the rate of 55 miles per hour for x hours is given by $y = 55x$, where y is measured in miles.

a. Find the distance traveled in 15-minute increments for 0 to 1.5 hours.

Hours	0	0.25	0.50	0.75	1	1.25	1.5
Miles							

b. If the distance traveled was 192.5 miles, how many hours did it take to travel this distance at 55 mph?

171. One-square-foot floor tiles come 24 to a package. Find how many packages are needed to cover a rectangular floor 18 feet by 21 feet.

172. Determine which container holds more ice cream, an 8 inch $\times$ 5 inch $\times$ 3 inch box or a cylinder with radius 3 inches and height 6 inches.

MIXED REVIEW

Complete the table.

	Number	Opposite of Number	Reciprocal of Number
173.	$-\dfrac{3}{4}$		
174.		-5	

Simplify. If necessary, write answers with positive exponents only.

175. $-2\left(5x + \dfrac{1}{2}\right) + 7.1$ **176.** $\sqrt{36} \div 2 \cdot 3$

177. $-\dfrac{7}{11} - \left(-\dfrac{1}{11}\right)$ **178.** $10 - (-1) + (-2) + 6$

179. $\left(-\dfrac{2}{3}\right)^3 \div \dfrac{10}{9}$ **180.** $\dfrac{(3 - 5)^2 + (-1)^3}{1 + 2(3 - (-1))^2}$

Solve.

181. $\dfrac{x - 2}{5} + \dfrac{x + 2}{2} = \dfrac{x + 4}{3}$

182. $\dfrac{2z - 3}{4} - \dfrac{4 - z}{2} = \dfrac{z + 1}{3}$

183. China, the United States, and France are predicted to be the top tourist destinations by 2020. In this year, the United States is predicted to have 9 million more tourists than France, and China is predicted to have 44 million more tourists than France. If the total number of tourists predicted for these three countries is 332 million, find the number predicted for each country in 2020.

△ **184.** $A = \dfrac{h}{2}(B + b)$ for B

△ **185.** $V = \dfrac{1}{3}\pi r^2 h$ for h

The volume of a cone can be found using the formula $V = \dfrac{1}{3}\pi r^2 h$.

Find the volumes of the cones with the following heights and radii. Round to two decimal places.

186.

Height h	6	6	6	6	6
Radius x	1	1.5	2	2.25	3
Volume					

187.

Height h	10	10	10	10	10
Radius x	1.5	2.1	2.75	3	3.5
Volume					

CHAPTER 1 TEST

Remember to use the Chapter Test Prep Video CD to see the fully worked-out solutions to any of the exercises you want to review.

Determine whether each statement is true or false.

1. $-2.3 > -2.33$ **2.** $-6^2 = (-6)^2$

3. $-5 - 8 = -(5 - 8)$ **4.** $(-2)(-3)(0) = \dfrac{-4}{0}$

5. All natural numbers are integers.

6. All rational numbers are integers.

Simplify.

7. $5 - 12 \div 3(2)$ **8.** $5^2 - 3^4$

9. $(4 - 9)^3 - |-4 - 6|^2$ **10.** $12 + \{6 - [5 - 2(-5)]\}$

11. $\dfrac{6(7 - 9)^3 + (-2)}{(-2)(-5)(-5)}$ **12.** $\dfrac{(4 - \sqrt{16}) - (-7 - 20)}{-2(1 - 4)^2}$

Evaluate each expression when $q = 4$, $r = -2$, and $t = 1$.

13. $q^2 - r^2$ **14.** $\dfrac{5t - 3q}{3r - 1}$

15. The algebraic expression $5.75x$ represents the total cost for x adults to attend the theater.

a. Complete the table that follows.

b. As the number of adults increases, does the total cost increase or decrease?

Adults	x	1	3	10	20
Total Cost	$5.75x$				

Write each statement using mathematical symbols.

16. Twice the sum of x and five is 30.

17. The square of the difference of six and y, divided by seven, is less than -2.

18. The product of nine and z, divided by the absolute value of -12, is not equal to 10.

19. Three times the quotient of n and five is the opposite of n.

20. Twenty is equal to 6 subtracted from twice x.

21. Negative two is equal to x divided by the sum of x and five.

Name each property illustrated.

22. $6(x - 4) = 6x - 24$

23. $(4 + x) + z = 4 + (x + z)$

24. $(-7) + 7 = 0$

25. $(-18)(0) = 0$

26. Write an expression for the total amount of money (in dollars) in n nickels and d dimes.

Simplify each expression.

27. $-2(3x + 7)$

28. $\frac{1}{3}a - \frac{3}{8} + \frac{1}{6}a - \frac{3}{4}$

29. $4y + 10 - 2(y + 10)$

30. $(8.3x - 2.9) - (9.6x - 4.8)$

31. Evaluate the expression $0.2x^3 + 5x^2 - 6.2x + 3$ if $x = -3.1$.

32. The circumference of a circle can be found by multiplying π times the diameter or using the formula $C = \pi d$. The area of a circle is found by multiplying the radius squared times π or using the formula $A = \pi r^2$. Complete the table below to find the circumference and area, given the following diameters. If necessary, round answers to the nearest hundredth.

Diameter	d	2	3.8	10	14.9
Radius	r				
Circumference	πd				
Area	πr^2				

Solve each equation.

33. $8x + 14 = 5x + 44$

34. $3(x + 2) = 11 - 2(2 - x)$

35. $3(y - 4) + y = 2(6 + 2y)$

36. $7n - 6 + n = 2(4n - 3)$

37. $\frac{7w}{4} + 5 = \frac{3w}{10} + 1$

Solve each equation for the specified variable.

38. $3x - 4y = 8$; y

39. $S = gt^2 + gvt$; g

40. $F = \frac{9}{5}C + 32$; C

Solve.

41. In 2014, the number of people employed as network systems and data communications analysts is expected to be 357,000 in the United States. This represents a 55% increase over the number of people employed in these fields in 2004. Find the number of network systems and data communications analysts employed in 2004. (*Source:* Bureau of Labor Statistics)

△ **42.** A circular dog pen has a circumference of 78.5 feet. Approximate π by 3.14 and estimate how many hunting dogs could be safely kept in the pen if each dog needs at least 60 square feet of room.

43. Find the amount of money in an account after 10 years if a principal of $2500 is invested at 3.5% interest compounded quarterly. (Round to the nearest cent.)

44. The most populous city in the United States is New York, although it is only the third most populous city in the world. Tokyo is the most populous city in the world. Second place is held by Seoul, Korea. Seoul's population is 1.3 million more than New York's, and Tokyo's is 10.2 million less than twice the population of New York. If the sum of the populations of these three cities is 78.3 million, find the population of each city.

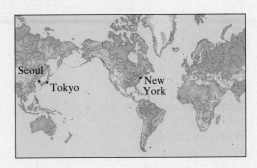

45. Ann-Margaret Tober is deciding whether to accept a part-time sales position at Campo Electronics. She is offered a gross monthly pay of $1500 plus 5% commission on her sales.

a. Complete the gross monthly pay table for the given amounts of sales.

Sales (Dollars)	8000	9000	10,000	11,000	12,000
Gross Monthly Pay (Dollars)					

b. If she has sales of $11,000 per month, what is her gross annual pay?

c. If Ann-Margaret decides she needs a gross monthly pay of $2200, how much must she sell every month?

46. The fireworks display for a small community is launched from a platform that is 20 feet from ground level. A particular rocket is launched with a velocity of 80 feet per second and explodes 0.5 second after it reaches its maximum height. The height of the rocket at time x seconds is given by

$$y = -16x^2 + 80x + 20$$

where y is measured in feet from the ground.

a. When is the height of the rocket 116 feet?

b. What is the rocket's maximum height?

c. When does the rocket reach its maximum height?

d. When does the rocket explode?

2

Graphs and Functions

The linear equations we explored in Chapter 1 are statements about a single variable. This chapter examines statements about two variables: linear equations in two variables. We focus particularly on graphs of those equations which lead to the notion of relation and to the notion of function, perhaps the single most important and useful concept in all of mathematics.

Over the past few years, diamonds have gained much higher visibility by increased media advertising. Strong consumer demand has caused the industry to increase production and is the basis for the bar graph below. By a method called least squares (Section 2.2), the function $f(x) = 0.42x + 10.5$ approximates the data below where $f(x)$ is world diamond production value (in billions of dollars) and where x is the number of years past 2000. In Section 2.2, Exercises 101 and 102, page 135, we will use this linear equation to predict diamond production.

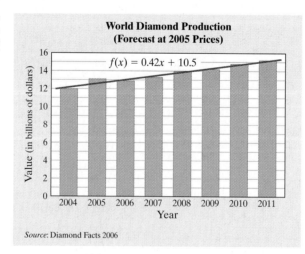

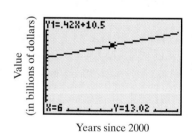

Years since 2000

2.1 GRAPHING EQUATIONS

OBJECTIVES

1 Plot ordered pairs.

2 Introduce the window setting capacity of a graphing utility.

3 Determine whether an ordered pair of numbers is a solution to an equation in two variables.

4 Graph linear equations.

5 Graph nonlinear equations.

OBJECTIVE 1 ▶ Plotting ordered pairs. Graphs are widely used today in newspapers, magazines, and all forms of newsletters. A few examples of graphs are shown here.

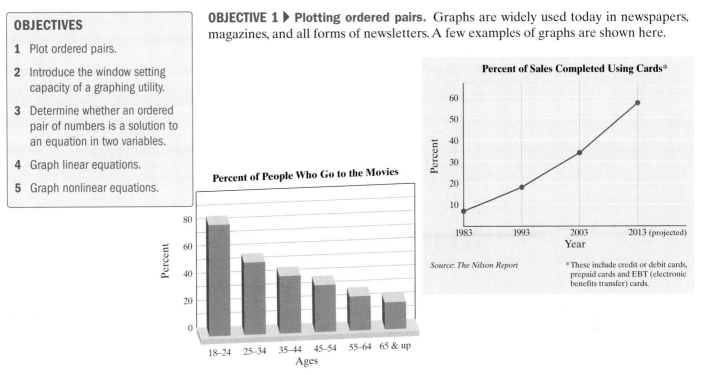

Percent of People Who Go to the Movies

Source: TELENATION/Market Facts, Inc.

Percent of Sales Completed Using Cards*

Source: The Nilson Report

*These include credit or debit cards, prepaid cards and EBT (electronic benefits transfer) cards.

To review how to read these graphs, we review their origin—the rectangular coordinate system. One way to locate points on a plane is by using a **rectangular coordinate system,** which is also called a **Cartesian coordinate system** after its inventor, René Descartes (1596–1650).

A rectangular coordinate system consists of two number lines that intersect at right angles at their 0 coordinates. We position these axes on paper such that one number line is horizontal and the other number line is then vertical. The horizontal number line is called the **x-axis** (or the axis of the **abscissa**), and the vertical number line is called the **y-axis** (or the axis of the **ordinate**). The point of intersection of these axes is named the **origin.**

Notice in the left figure below that the axes divide the plane into four regions. These regions are called **quadrants.** The top-right region is quadrant I. Quadrants II, III, and IV are numbered counterclockwise from the first quadrant as shown. The x-axis and the y-axis are not in any quadrant.

Each point in the plane can be located, or **plotted,** or graphed by describing its position in terms of distances along each axis from the origin. An **ordered pair,** represented by the notation (x, y), records these distances.

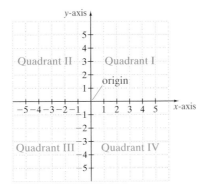

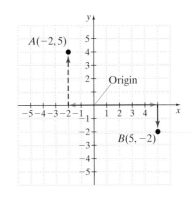

For example, the location of point A in the figure on the right on the previous page is described as 2 units to the left of the origin along the x-axis and 5 units upward parallel to the y-axis. Thus, we identify point A with the ordered pair $(-2, 5)$. Notice that the order of these numbers is *critical*. The x-value -2 is called the **x-coordinate** and is associated with the x-axis. The y-value 5 is called the **y-coordinate** and is associated with the y-axis.

Compare the location of point A with the location of point B, which corresponds to the ordered pair $(5, -2)$. Can you see that the order of the coordinates of an ordered pair matters? Also, two ordered pairs are considered equal and correspond to the same point if and only if their x-coordinates are equal and their y-coordinates are equal.

Keep in mind that **each ordered pair corresponds to exactly one point in the real plane and that each point in the plane corresponds to exactly one ordered pair.** Thus, we may refer to the ordered pair (x, y) as the point (x, y).

EXAMPLE 1 Plot each ordered pair on a Cartesian coordinate system and name the quadrant or axis in which the point is located.

a. $(2, -1)$ **b.** $(0, 5)$ **c.** $(-3, 5)$ **d.** $(-2, 0)$ **e.** $\left(-\frac{1}{2}, -4\right)$ **f.** $(1.5, 1.5)$

Solution The six points are graphed as shown.

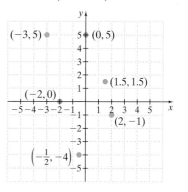

a. $(2, -1)$ lies in quadrant IV.

b. $(0, 5)$ is on the y-axis.

c. $(-3, 5)$ lies in quadrant II.

d. $(-2, 0)$ is on the x-axis.

e. $\left(-\frac{1}{2}, -4\right)$ is in quadrant III.

f. $(1.5, 1.5)$ is in quadrant I.

PRACTICE

1 Plot each ordered pair on a Cartesian coordinate system and name the quadrant or axis in which the point is located.

a. $(3, -4)$ **b.** $(0, -2)$ **c.** $(-2, 4)$ **d.** $(4, 0)$ **e.** $\left(-1\frac{1}{2}, -2\right)$ **f.** $(2.5, 3.5)$

Notice that the y-coordinate of any point on the x-axis is 0. For example, the point with coordinates $(-2, 0)$ lies on the x-axis. Also, the x-coordinate of any point on the y-axis is 0. For example, the point with coordinates $(0, 5)$ lies on the y-axis. These points that lie on the axes do not lie in any quadrants.

Concept Check ✓

Which of the following correctly describes the location of the point $(3, -6)$ in a rectangular coordinate system?

a. 3 units to the left of the y-axis and 6 units above the x-axis

b. 3 units above the x-axis and 6 units to the left of the y-axis

c. 3 units to the right of the y-axis and 6 units below the x-axis

d. 3 units below the x-axis and 6 units to the right of the y-axis

Answer to Concept Check: c

Many types of real-world data occur in pairs. The graph below was shown at the beginning of this section. Notice the paired data $(2013, 57)$ and the corresponding plotted point, both in blue.

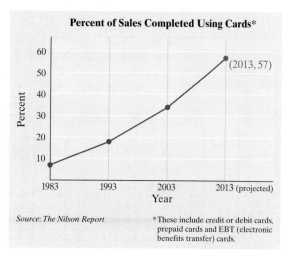

Percent of Sales Completed Using Cards*

Source: The Nilson Report

*These include credit or debit cards, prepaid cards and EBT (electronic benefits transfer) cards.

This paired data point, $(2013, 57)$, means that in the year 2013, it is predicted that 57% of sales will be completed using some type of card (credit, debit, etc.).

OBJECTIVE 2 ▶ Introducing window settings. The rectangular coordinate system extends infinitely in all directions with the number lines that form it. A graphing utility can be used to display a portion of the system. The portion being viewed is called a **viewing window** or simply a **window.**

Below, the rectangular coordinate system is shown with a corresponding viewing window indicated.

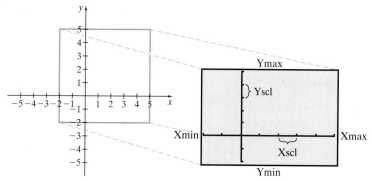

Yscl indicates the number of units per tick mark on the y-axis.
Xscl indicates the number of units per tick mark on the x-axis.

The dimensions of a window are selected by entering the minimum value, maximum value, and scale for both x and y under the window setting.

Several viewing windows can be accessed automatically on most graphing utilities. The screen below illustrates viewing the rectangular coordinate system from -10 to 10 on both the x-axis and the y-axis, with each tick mark representing 1 unit. This particular window setting is called the **standard window** on some graphing utilities.

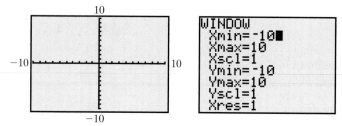

Standard Window

To refer to the window dimensions throughout this text, we will use the following notation. $[-10, 10, 1]$ by $[-10, 10, 1]$ refers to [Xmin, Xmax, Xscl] by [Ymin, Ymax, Yscl].

If we select the same standard window setting but change the Xmin to -20, the Xmax to 20, and the Xscl to 5, observe to the left below how the window has been changed. Next, select the standard window setting but change the Ymin to -50, the Ymax to 50, and the Yscl to 10 and observe the changes in the window to the right.

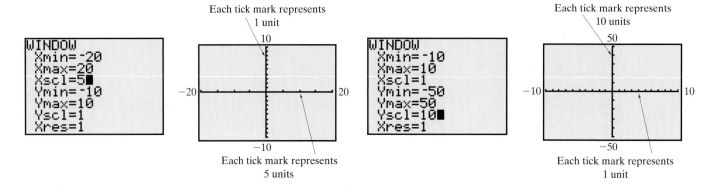

Many graphing utilities have the ability to show placement (via coordinates) and movement about the viewing window by means of a **cursor.**

DISCOVER THE CONCEPT

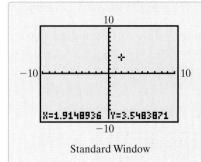

Standard Window

Clear any equations from the **Y=** editor and access the **standard window.** The screen shows the standard window with the cursor located at the ordered pair (1.9148936, 3.5483871).

Use the arrow keys and move the cursor around the rectangular coordinate system. Using the arrow keys, move to some points in quadrant I. What do you observe about the signs of the x- and y-coordinates in this quadrant? Repeat this process in quadrants II, III, and IV. What observations can you make about the signs of the coordinates in each quadrant?

From the discovery above, we see that the signs of the coordinates of an ordered pair depend on the quadrant in which it is located.

Quadrant I:	$(+, +)$		Quadrant III:	$(-, -)$
Quadrant II:	$(-, +)$		Quadrant IV:	$(+, -)$
x-axis:	$(x, 0)$		y-axis:	$(0, y)$

An **integer window** is another window that you may be able to automatically access on your graphing utility. In an integer window the x-coordinates are integers as you move the cursor around the coordinate system.

Access an integer window on your graphing utility and observe how the x- and y-coordinates are different from the x- and y-coordinates in a standard window. The screen below shows an integer window with the ordered pair $(10, 7)$ indicated.

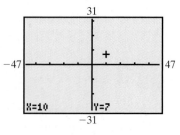

Integer Window

Another common window on many graphing utilities is a **decimal window.** Access this window and then move the cursor around the screen. Notice that the *x*-coordinates are incrementing or changing by 0.1 as you move from one point to another. The screen below shows an example of a decimal window on a graphing utility and its window settings.

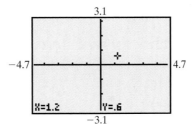

Decimal Window

TECHNOLOGY NOTE

For many graphing utilities, both the standard window and the decimal window are centered at the origin. However, your graphing utility may allow you to center an integer window anywhere in the rectangular coordinate system. The center may depend on where you last left a cursor before selecting the integer setting. Check your graphing utility manual to see what window options are available to you.

Experiment on your own by changing the window settings and predicting the resulting windows.

▶ **Helpful Hint**

Remember that each point in the rectangular coordinate system corresponds to a unique ordered pair (x, y), and that each ordered pair corresponds to a unique point.

EXAMPLE 2   Display a window of the rectangular coordinate system with a setting of $[-20, 40, 5]$ by $[-30, 30, 10]$.

Solution   The window setting and the resulting window are shown next.

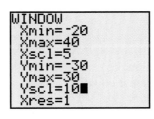   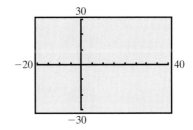

PRACTICE
2   Display a window of the rectangular coordinate system with a setting of $[-10, 30, 10]$ by $[-25, 25, 5]$.

OBJECTIVE 3 ▶ Determining whether an ordered pair is a solution. Solutions of equations in two variables consist of two numbers that form a true statement when substituted into the equation. A convenient notation for writing these numbers is as ordered pairs. A solution of an equation containing the variables x and y is written as a pair of numbers in the order (x, y). If the equation contains other variables, we will write ordered pair solutions in alphabetical order.

EXAMPLE 3   Determine whether $(0, -12), (1, 9),$ and $(2, -6)$ are solutions of the equation $3x - y = 12$.

Solution   To check each ordered pair, replace x with the x-coordinate and y with the y-coordinate and see whether a true statement results.

Let $x = 0$ and $y = -12$. Let $x = 1$ and $y = 9$. Let $x = 2$ and $y = -6$.

$$3x - y = 12 \qquad\qquad 3x - y = 12 \qquad\qquad 3x - y = 12$$
$$3(0) - (-12) \overset{?}{=} 12 \qquad 3(1) - 9 \overset{?}{=} 12 \qquad 3(2) - (-6) \overset{?}{=} 12$$
$$0 + 12 \overset{?}{=} 12 \qquad\qquad 3 - 9 \overset{?}{=} 12 \qquad\qquad 6 + 6 \overset{?}{=} 12$$
$$12 = 12 \quad \text{True} \qquad -6 = 12 \quad \text{False} \qquad 12 = 12 \quad \text{True}$$

Thus, $(1, 9)$ is not a solution of $3x - y = 12$, but both $(0, -12)$ and $(2, -6)$ are solutions. ☐

PRACTICE
3   Determine whether $(1, 4), (0, 6),$ and $(3, -4)$ are solutions of the equation $4x + y = 8$.

OBJECTIVE 4 ▶ Graphing linear equations. The equation $3x - y = 12$, from Example 3, actually has an infinite number of ordered pair solutions. Since it is impossible to list all solutions, we visualize them by graphing.

A few more ordered pairs that satisfy $3x - y = 12$ are $(4, 0), (3, -3), (5, 3),$ and $(1, -9)$. These ordered pair solutions along with the ordered pair solutions from Example 3 are plotted on the following graph. The graph of $3x - y = 12$ is the single line containing these points. Every ordered pair solution of the equation corresponds to a point on this line, and every point on this line corresponds to an ordered pair solution.

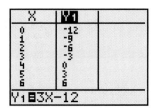

Using the table feature on a graphing utility to verify the ordered pairs solutions to the right.

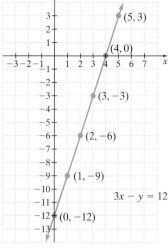

x	y	$3x - y = 12$
5	3	$3 \cdot 5 - 3 = 12$
4	0	$3 \cdot 4 - 0 = 12$
3	-3	$3 \cdot 3 - (-3) = 12$
2	-6	$3 \cdot 2 - (-6) = 12$
1	-9	$3 \cdot 1 - (-9) = 12$
0	-12	$3 \cdot 0 - (-12) = 12$

The equation $3x - y = 12$ is called a linear equation in two variables, and **the graph of every linear equation in two variables is a line.**

> **Linear Equation in Two Variables**
> A linear equation in two variables is an equation that can be written in the form
> $$Ax + By = C$$
> where A and B are not both 0. This form is called **standard form.**

Some examples of equations in standard form:

$$3x - y = 12$$
$$-2.1x + 5.6y = 0$$

▶ **Helpful Hint**

Remember: A linear equation is written in standard form when all of the variable terms are on one side of the equation and the constant is on the other side.

Many real-life applications are modeled by linear equations. Suppose you have a part-time job at a store that sells office products.

Your pay is $3000 plus 20% or $\frac{1}{5}$ of the price of the products you sell. If we let x represent products sold and y represent monthly salary, the linear equation that models your salary is

$$y = 3000 + \frac{1}{5}x$$

(Although this equation is not written in standard form, it is a linear equation. To see this, subtract $\frac{1}{5}x$ from both sides.)

Some ordered pair solutions of this equation are below.

Products Sold	x	0	1000	2000	3000	4000	10,000
Monthly Salary	y	3000	3200	3400	3600	3800	5000

For example, we say that the ordered pair $(1000, 3200)$ is a solution of the equation $y = 3000 + \frac{1}{5}x$ because when x is replaced with 1000 and y is replaced with 3200, a true statement results.

$$y = 3000 + \frac{1}{5}x$$

$$3200 \stackrel{?}{=} 3000 + \frac{1}{5}(1000) \quad \text{Let } x = 1000 \text{ and } y = 3200.$$

$$3200 \stackrel{?}{=} 3000 + 200$$

$$3200 = 3200 \qquad\qquad\qquad \text{True}$$

A portion of the graph of $y = 3000 + \frac{1}{5}x$ is shown in the next example.

Since we assume that the smallest amount of product sold is none, or 0, then x must be greater than or equal to 0. Therefore, only the part of the graph that lies in quadrant I is shown. Notice that the graph gives a visual picture of the correspondence between products sold and salary.

▶ **Helpful Hint**

A line contains an infinite number of points and each point corresponds to an ordered pair that is a solution of its corresponding equation.

EXAMPLE 4   Use the graph of $y = 3000 + \frac{1}{5}x$ to answer the following questions.

a. If the salesperson has $8000 of products sold for a particular month, what is the salary for that month?

b. If the salesperson wants to make more than $5000 per month, what must be the total amount of products sold?

Solution

a. Since *x* is products sold, find 8000 along the *x*-axis and move vertically up until you reach a point on the line. From this point on the line, move horizontally to the left until you reach the *y*-axis. Its value on the *y*-axis is 4600, which means if $8000 worth of products is sold, the salary for the month is $4600.

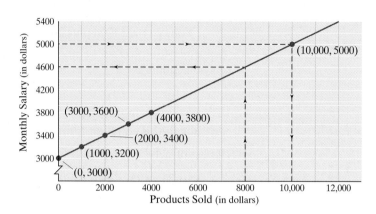

b. Since *y* is monthly salary, find 5000 along the *y*-axis and move horizontally to the right until you reach a point on the line. Either read the corresponding *x*-value from the labeled ordered pair, or move vertically downward until you reach the *x*-axis. The corresponding *x*-value is 10,000. This means that $10,000 worth of products sold gives a salary of $5000 for the month. For the salary to be greater than $5000, products sold must be greater that $10,000. □

PRACTICE
4   Use the graph in Example 4 to answer the following questions.

a. If the salesperson has $6000 of products sold for a particular month, what is the salary for that month?

b. If the salesperson wants to make more than $4800 per month, what must be the total amount of products sold?

Recall from geometry that a line is determined by two points. This means that to graph a linear equation in two variables, just two solutions are needed. We will find a third solution, just to check our work. To find ordered pair solutions of linear equations in two variables, we can choose an *x*-value and find its corresponding *y*-value, or we can choose a *y*-value and find its corresponding *x*-value. The number 0 is often a convenient value to choose for *x* and also for *y*.

EXAMPLE 5   Graph the equation $y = -2x + 3$. Then use a graphing utility to check.

Solution   This is a linear equation. (In standard form it is $2x + y = 3$.) Find three ordered pair solutions, and plot the ordered pairs. The line through the plotted points is the graph. Since the equation is solved for *y*, let's choose three *x*-values. We'll choose 0, 2, and then −1 for *x* to find our three ordered pair solutions.

Let $x = 0$ Let $x = 2$ Let $x = -1$
$y = -2x + 3$ $y = -2x + 3$ $y = -2x + 3$
$y = -2 \cdot 0 + 3$ $y = -2 \cdot 2 + 3$ $y = -2(-1) + 3$
$y = 3$ Simplify. $y = -1$ Simplify. $y = 5$ Simplify.

The three ordered pairs $(0, 3)$, $(2, -1)$ and $(-1, 5)$ are listed in the table and the graph is shown.

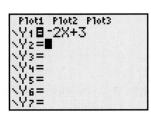

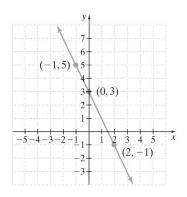

To check, we graph $y = -2x + 3$ using a graphing utility. To do so, enter it into the Y= editor.

Define $y_1 = -2x + 3$ and select the integer window. The screen below illustrates the graph of the linear equation. Access a trace feature to locate the ordered pair solutions in the table above. Trace to verify that the point $(0, 3)$ is a solution for the equation as well as $(2, -1)$ and $(-1, 5)$.

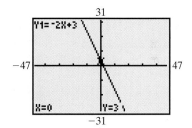

PRACTICE

5   Graph the equation $y = -3x - 2$. Then use a graphing utility to check.

Notice that the graph crosses the y-axis at the point $(0, 3)$. This point is called the **y-intercept.** (You may sometimes see just the number 3 called the y-intercept.) This graph also crosses the x-axis at the point $\left(\frac{3}{2}, 0\right)$. This point is called the **x-intercept.** (You may also see just the number $\frac{3}{2}$ called the x-intercept.)

Since every point on the y-axis has an x-value of 0, we can find the y-intercept of a graph by letting $x = 0$ and solving for y. Also, every point on the x-axis has a y-value of 0. To find the x-intercept, we let $y = 0$ and solve for x.

> **Finding x- and y-Intercepts**
>
> To find an x-intercept, let $y = 0$ and solve for x.
> To find a y-intercept, let $x = 0$ and solve for y.

We will study intercepts further in Section 2.3.

EXAMPLE 6 Graph the linear equation $y = \frac{1}{3}x$.

Solution To graph, we find ordered pair solutions, plot the ordered pairs, and draw a line through the plotted points. We will choose x-values and substitute in the equation. To avoid fractions, we choose x-values that are multiples of 3. To find the y-intercept, we let $x = 0$.

> ▶ **Helpful Hint**
> Notice that by using multiples of 3 for x, we avoid fractions.

> ▶ **Helpful Hint**
> Since the equation $y = \frac{1}{3}x$ is solved for y, we choose x-values for finding points. This way, we simply need to evaluate an expression to find the y-value, as shown.

$$y = \frac{1}{3}x$$

If $x = 0$, then $y = \frac{1}{3}(0)$, or 0.

If $x = 6$, then $y = \frac{1}{3}(6)$, or 2.

If $x = -3$, then $y = \frac{1}{3}(-3)$, or -1.

x	y
0	0
6	2
-3	-1

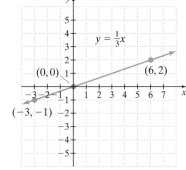

This graph crosses the x-axis at $(0, 0)$ and the y-axis at $(0, 0)$. This means that the x-intercept is $(0, 0)$ and that the y-intercept is $(0, 0)$.

PRACTICE
6 Graph the linear equation $y = -\frac{1}{2}x$.

EXAMPLE 7 Graph $5x - 3y = 10$ in an integer window centered at the origin. Find the y-coordinate that makes the ordered pair $(8, ?)$ a solution for the given equation.

Solution To enter the equation in the Y= editor, first solve the equation for y.

$$5x - 3y = 10$$
$$5x - 3y - 5x = -5x + 10 \quad \text{Subtract } 5x \text{ from both sides.}$$
$$-3y = -5x + 10$$
$$y = \frac{-5x + 10}{-3} \quad \text{Divide both sides by } -3.$$

Define $y_1 = \dfrac{-5x + 10}{-3}$ and graph in an integer window centered at $(0, 0)$. Trace to locate the point with x-coordinate 8 and find the corresponding y-value to be 10.

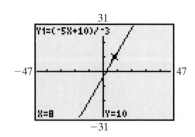

The ordered pair $(8, 10)$ is a solution of the given equation. To check, let $x = 8$ and $y = 10$ in the equation $5x - 3y = 10$ and see that a true statement results.

PRACTICE
7 Graph $7x - 2y = 21$ in an integer window centered at the origin. Find the y-coordinate that makes the ordered pair $(5, ?)$ a solution for the given equation.

OBJECTIVE 5 ▶ Graphing nonlinear equations. Not all equations in two variables are linear equations, and not all graphs of equations in two variables are lines.

EXAMPLE 8   Graph $y = x^2$. Use a graphing utility to check.

Solution   This equation is not linear because the x^2 term does not allow us to write it in the form $Ax + By = C$. Its graph is not a line. We begin by finding ordered pair solutions. Because this graph is solved for y, we choose x-values and find corresponding y-values.

If $x = -3$, then $y = (-3)^2$, or 9.

If $x = -2$, then $y = (-2)^2$, or 4.

If $x = -1$, then $y = (-1)^2$, or 1.

If $x = 0$, then $y = 0^2$, or 0.

If $x = 1$, then $y = 1^2$, or 1.

If $x = 2$, then $y = 2^2$, or 4.

If $x = 3$, then $y = 3^2$, or 9.

x	y
-3	9
-2	4
-1	1
0	0
1	1
2	4
3	9

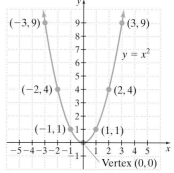

Study the table a moment and look for patterns. Notice that the ordered pair solution $(0, 0)$ contains the smallest y-value because any other x-value squared will give a positive result. This means that the point $(0, 0)$ will be the lowest point on the graph. Also notice that all other y-values correspond to two different x-values. For example, $3^2 = 9$ and also $(-3)^2 = 9$. This means that the graph will be a mirror image of itself across the y-axis. Connect the plotted points with a smooth curve to sketch the graph.

To check, graph $y_1 = x^2$ in an integer window. Compare the graph and table shown with the calculations done by hand.

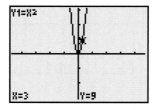

PRACTICE
8   Graph $y = 2x^2$. Use a graphing utility to check.

This curve is given a special name called a **parabola.** It can be shown that the graph of a quadratic equation of the form $y = ax^2 + bx + c$ is a parabola. We will study more about parabolas in later chapters.

The graphing utility finds and plots various ordered pair solutions in the same manner that we find and plot ordered pair solutions. The advantage of a graphing utility is that it can perform this process much faster and with greater accuracy than we can.

DISCOVER THE CONCEPT

a. Compare the graphs of $y = x^2$ and $y = -x^2$ by graphing each in a standard window.

b. How can the graph of $y = x^2$ be transformed to look like the graph of $y = -x^2$?

In the above discovery, we see that if the coefficient of x^2 is positive, the parabola opens upward, and if the coefficient is negative, the parabola opens downward. Compare this fact with the table of values below. We defined $y_1 = x^2$ and $y_2 = -x^2$.

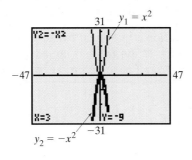

For $y = -x^2$, the y-value is always negative or 0 and the parabola opens downward.

For $y = x^2$, the y-value is always positive or 0 and the parabola opens upward.

We say the graph of $y = x^2$ is reflected across the x-axis to obtain the graph of $y = -x^2$.

Another nonlinear equation is a third-degree or cubic equation.

EXAMPLE 9   Use a graphing utility to graph $y = x^3$ and examine a table of values with x equal to the integers from -3 to 3.

Solution   Define $y_1 = x^3$ and graph the equation in a standard window. The figure below shows the graph of $y = x^3$ with the trace cursor on $(2, 8)$. The table of values shows ordered pair solutions for $x = -3, -2, -1, 0, 1, 2,$ and 3.

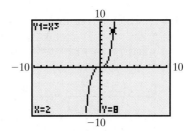

Notice in the table that when x is negative, y is also negative. Thus, these points lie in the third quadrant. When x is positive, y is positive; thus, this portion of the graph lies in the first quadrant. When $x = 0$, then $y = 0$, so the graph crosses the axes at the origin.   □

PRACTICE
9   Use a graphing utility to graph $y = x^3 + 1$ and examine a table of values with x equal to the integers from -3 to 3. Compare this table with the table from Example 9.

DISCOVER THE CONCEPT

How do you think the graph of $y = x^3$ is affected when we place a negative sign in front of the x^3? Test your prediction by graphing $y = x^3$ and $y = -x^3$ separately in a standard window. State in words how you can obtain the graph of $y = -x^3$ from the graph of $y = x^3$.

We next graph the basic absolute value equation, $y = |x|$. Begin by finding and plotting ordered pair solutions and analyze any patterns.

EXAMPLE 10 Graph the equation $y = |x|$. Use a graphing utility to check.

Solution This is not a linear equation since it cannot be written in the form $Ax + By = C$. Its graph is not a line. Because we do not know the shape of this graph, we find many ordered pair solutions. We will choose x-values and substitute to find corresponding y-values.

If $x = -3$, then $y = |-3|$, or 3.

If $x = -2$, then $y = |-2|$, or 2.

If $x = -1$, then $y = |-1|$, or 1.

If $x = 0$, then $y = |0|$, or 0.

If $x = 1$, then $y = |1|$, or 1.

If $x = 2$, then $y = |2|$, or 2.

If $x = 3$, then $y = |3|$, or 3.

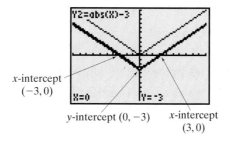

x	y
-3	3
-2	2
-1	1
0	0
1	1
2	2
3	3

Again, study the table of values for a moment and notice any patterns.

From the plotted ordered pairs, we see that the graph of this absolute value equation is V-shaped. The graph of $y = |x|$ using a standard window is in the margin. □

PRACTICE

10   Graph $y = -|x|$. Use a graphing utility to check.

EXAMPLE 11   Graph the equation $y = |x| - 3$ and examine a table of values for x equal to the integers from -3 to 3. Compare the graph and the table to the graph and table for $y = |x|$.

Solution   Graph both $y_1 = |x|$ and $y_2 = |x| - 3$ using a standard window. Notice that compared with the equation $y = |x|$, the equation $y = |x| - 3$ decreases the y-values by 3 units for the same x-values. This means that the graph of $y = |x| - 3$ is the same as the graph of $y = |x|$ lowered by 3 units.

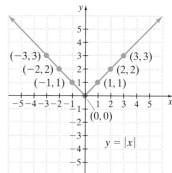

Indicates two x-intercepts $(-3, 0)$ and $(3, 0)$

x-intercept $(-3, 0)$

y-intercept $(0, -3)$ x-intercept $(3, 0)$

This graph of $y = |x| - 3$ shows us that it is possible to have more than one x- or y-intercept. Notice that this graph has one y-intercept $(0, -3)$ and two x-intercepts $(-3, 0)$ and $(3, 0)$. □

PRACTICE

11   Graph the equation $y = |x| + 5$. Compare this graph and table to those for $y = |x|$.

DISCOVER THE CONCEPT

a. Predict how the graph of $y = |x| + 7$ can be obtained from the graph of $y = |x|$. Then check your prediction by graphing both equations using a standard window.

b. Predict how the graph of $y = x^2 - 3$ can be obtained from the graph of $y = x^2$. Use a graphing utility to check your prediction. Next, predict how the graph of $y = x^2 + 5$ can be obtained from $y = x^2$.

c. In general, can you predict how the graph of a basic equation with a constant added can be obtained from the graph of the basic equation? Test your prediction with the basic cubic equation $y = x^3$.

If we add a constant K to a basic graph, it is said to cause a **vertical translation** of the graph. That is, if the constant K is positive, the basic graph is moved or translated K units upward. Similarly, if the constant K is negative, the basic graph is moved, or translated, $|K|$ units downward. We study the translations more in Section 2.7.

To summarize, the graphs of the basic equations we have discussed so far are shown below.

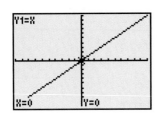 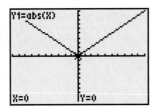 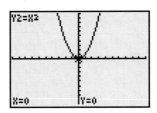 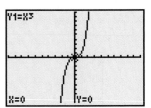

Standard Window

Basic linear Equation
$y = x$

Absolute Value Equation
$y = |x|$

Quadratic Equation (Parabola)
$y = x^2$

Cubic Equation
$y = x^3$

VOCABULARY & READINESS CHECK

Determine the coordinates of each point on the graph.

1. Point A
2. Point B
3. Point C
4. Point D
5. Point E
6. Point F
7. Point G
8. Point H

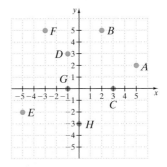

Without graphing, visualize the location of each point. Then give its location by quadrant or x- or y-axis.

9. $(2, 3)$ **10.** $(0, 5)$ **11.** $(-2, 7)$
12. $(-3, 0)$ **13.** $(-1, -4)$ **14.** $(4, -2)$
15. $(0, -100)$ **16.** $(10, 30)$ **17.** $(-10, -30)$
18. $(0, 0)$ **19.** $(-87, 0)$ **20.** $(-42, 17)$

2.1 EXERCISE SET

PRACTICE   WATCH   DOWNLOAD   READ   REVIEW

Plot each point and name the quadrant or axis in which the point lies. See Example 1.

1. $(3, 2)$ **2.** $(2, -1)$
3. $(-5, 3)$ **4.** $(-3, -1)$
5. $\left(5\frac{1}{2}, -4\right)$ **6.** $\left(-2, 6\frac{1}{3}\right)$
7. $(0, 3.5)$ **8.** $(-5.2, 0)$
9. $(-2, -4)$ **10.** $(-4.2, 0)$

Given that x is a positive number and that y is a positive number, determine the quadrant or axis in which each point lies. See Example 1.

11. $(x, -y)$ **12.** $(-x, y)$
13. $(x, 0)$ **14.** $(0, -y)$
15. $(-x, -y)$ **16.** $(0, 0)$

Determine the coordinates of the point shown and name the quadrant in which the point lies.

17.
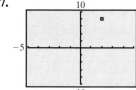

18.
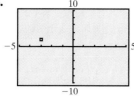

19.
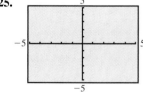

20.
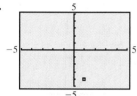

Determine a window setting so that the three given points will lie in that window. Report your answer in the [Xmin, Xmax, Xscl] by [Ymin, Ymax, Yscl] format. See Example 2.

 21. $(-5, 2), (3, 8), (10, 15)$

22. $(-10, 12), (0, 9), (6, -15)$

23. $(-25, 0), (5, 7), (20, 100)$

24. $(0, 50), (25, 75), (50, 150)$

Match the following screens with the window settings listed below.

25.
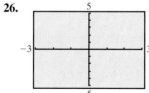

26.
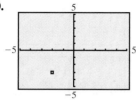

27.
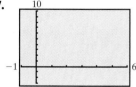

28.

A.
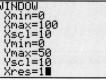

B.
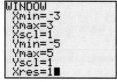

C.
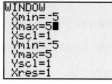
```
WINDOW
 Xmin=-5
 Xmax=5■
 Xscl=1
 Ymin=-5
 Ymax=5
 Yscl=1
 Xres=1
```

D.
```
WINDOW
 Xmin=-1
 Xmax=6
 Xscl=1
 Ymin=-3
 Ymax=10
 Yscl=1■
 Xres=1
```

Use a table or a graph to determine whether each ordered pair is a solution of the given equation. See Example 3.

29. $y = 3x - 5; (0, 5), (-1, -8)$

30. $y = -2x + 7; (1, 5), (-2, 3)$

31. $-6x + 5y = -6; (1, 0), \left(2, \dfrac{6}{5}\right)$

32. $5x - 3y = 9; (0, 3), \left(\dfrac{12}{5}, -1\right)$

33. $y = 2x^2; (1, 2), (3, 18)$

34. $y = 2|x|; (-1, 2), (0, 2)$

MIXED PRACTICE

Determine whether each equation is linear or nonlinear and tell the basic shape of the graph (line, parabola, cubic, V-shaped). See Examples 4 through 11.

35.

Equation	Linear or nonlinear	Shape (Line, Parabola, Cubic, V-shaped)
$y - x = 8$		
$y = 6x$		
$y = x^2 + 3$		
$y = 6x - 5$		
$y = -|x| + 2$		
$y = 3x^2$		
$y = -4x + 2$		
$y = -|x|$		
$y = x^3$		

36.

Equation	Linear or nonlinear	Shape (Line, Parabola, Cubic, V-shaped)
$x + y = 3$		
$y = 4 - x$		
$y = 2x^2 - 5$		
$y = -8x + 6$		
$y = |x - 3|$		
$y = 7x^2$		
$2x - y = 5$		
$y = -|x - 1|$		
$y = x^3 - 2$		

Rewrite each equation as an equivalent equation that can be entered in the Y= editor of your graphing utility. See Example 7.

37. $2x + y = 10$

38. $-6x + y = 2$

39. $-7x - 3y = 4$

40. $5x - 11y = -1.2$

41. The graph of $y = 2x^2 + 1.2x - 5.6$ has two x-intercepts and one y-intercept. Use your graphing utility and graph this equation using a decimal window. Find the coordinates of the intercepts.

42. The graph of $y = x^2 + 7x - 6$ has two ordered pair solutions whose y-value is 12. Use your graphing utility and graph this equation using an integer window. Find the coordinates of these points.

43. If you trace along the graph of $y = 1.5x - 6$ using an integer window, which of the following coordinates would not be displayed?

  a. $(0, -6)$ **b.** $(1, -4.5)$

  c. $(4, 0)$ **d.** $(1.2, -4.2)$

44. If you trace along the graph of $y = x + 2.3$ using an integer window, which of the following coordinates would not be displayed?

  a. $(1.7, 4)$ **b.** $(0, 2.3)$

  c. $(2, 4.3)$ **d.** $(-3, -0.7)$

Match each equation with its graph without using a graphing utility. Then check your answers by graphing each equation using a standard window.

45. $y = x^2 - 4$ **46.** $y = x + 5$

47. $y = x^3 + x^2 - 4$ **48.** $y = |x| + 3$

A. 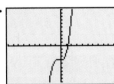   **B.**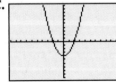

C. 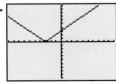   **D.**

Match each equation with its graph without using a graphing utility. Then check your answers by graphing each equation using a standard window.

49. $y = x - 5$ **50.** $y = -x^2 + 4$

51. $y = -x^3 + x^2 + 2x - 3$ **52.** $y = |x + 3|$

A. 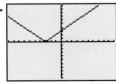   **B.**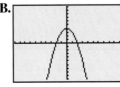

C. 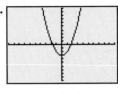   **D.**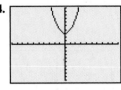

Match each graph with its equation without using a graphing utility. Then graph using a standard window to check your answers.

53. 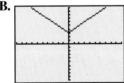   **54.**

55. 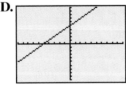   **56.**

  A. $y = x^2 - 3$ **B.** $y = x^2 + 3$

  C. $y = |x| + 5$ **D.** $y = |x| - 3$

Sketch the graphs of the three equations by hand on the same coordinate plane. Then use your graphing utility to check your results. Describe the patterns you see in these graphs.

57. $y = x$

   $y = x + 2$

   $y = x - 3$

58. $y = x^2$

   $y = x^2 + 1$

   $y = x^2 - 3$

59. $y = -x^2$

   $y = -x^2 + 1$

   $y = -x^2 - 2$

60. $y = -|x|$

   $y = -|x| + 2$

   $y = -|x| - 4$

MIXED PRACTICE

Graph each equation by plotting ordered pair solutions by hand. If the equation is not linear, suggested x-values have been given for generating ordered pair solutions. Check the graph using a graphing utility. See Examples 4 through 11.

61. $x + y = 3$ **62.** $y - x = 8$

63. $y = 4x$ **64.** $y = 6x$

65. $y = 4x - 2$ **66.** $y = 6x - 5$

67. $y = |x| + 3$

   Let $x = -3, -2, -1, 0, 1, 2, 3$.

68. $y = |x| + 2$

   Let $x = -3, -2, -1, 0, 1, 2, 3$.

69. $2x - y = 5$

70. $4x - y = 7$

71. $y = 2x^2$

   Let $x = -3, -2, -1, 0, 1, 2, 3$.

72. $y = 3x^2$

   Let $x = -3, -2, -1, 0, 1, 2, 3$.

73. $y = x^2 - 3$
Let $x = -3, -2, -1, 0, 1, 2, 3$.

74. $y = x^2 + 3$
Let $x = -3, -2, -1, 0, 1, 2, 3$.

75. $y = -2x$

76. $y = -3x$

77. $y = -2x + 3$

78. $y = -3x + 2$

79. $y = |x + 2|$
Let $x = -4, -3, -2, -1, 0, 1$.

80. $y = |x - 1|$
Let $x = -1, 0, 1, 2, 3, 4$.

81. $y = x^3$
Let $x = -3, -2, -1, 0, 1, 2$.

82. $y = x^3 - 2$
Let $x = -3, -2, -1, 0, 1, 2$.

83. $y = -|x|$
Let $x = -3, -2, -1, 0, 1, 2, 3$.

84. $y = -x^2$
Let $x = -3, -2, -1, 0, 1, 2, 3$.

85. Given the following table of values for ordered pairs satisfying an absolute value equation, plot the points and join them to complete the graph. Then write an equation of the graph.

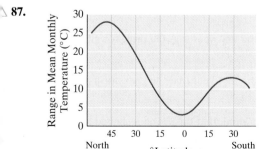

86. Given the following table of values for ordered pairs satisfying a second-degree equation, plot the points and join them to complete the graph. Then write an equation of the graph.

For Exercises 87 through 90, fill in the blank with the word "line" or "parabola."

87.

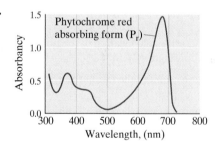

a. The shape of the graph from 55°N to 15°N only resembles a _____ .

b. The shape of the graph from 30°N to 15°N resembles a _____ .

88.

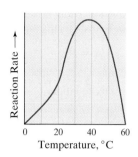

a. The shape of the graph from 20°C to 60°C resembles a _____ .

b. The shape of the graph from 0°C to 20°C resembles a _____ .

89.

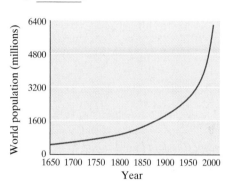

a. The shape of the graph from 1650 to 1800 resembles a _____ .

b. The shape of the graph from 1900 to 2000 resembles a _____ .

90.

a. The shape of the graph from 600 nm to 700 nm resembles a _____ .

b. The shape of the graph from 550 nm to 625 nm resembles a _____ .

Solve See Example 4.

91. The perimeter y of a rectangle whose width is a constant 3 inches and whose length is x inches is given by the equation

$$y = 2x + 6$$

a. Draw a graph of this equation.
b. Read from the graph the perimeter y of a rectangle whose length x is 4 inches.

92. The distance y traveled in a train moving at a constant speed of 50 miles per hour is given by the equation

$$y = 50x$$

where x is the time in hours traveled.

a. Draw a graph of this equation.

b. Read from the graph the distance y traveled after 6 hours.

REVIEW AND PREVIEW

Solve the following equations. See Section 1.5.

93. $3(x - 2) + 5x = 6x - 16$

94. $5 + 7(x + 1) = 12 + 10x$

95. $3x + \dfrac{2}{5} = \dfrac{1}{10}$

96. $\dfrac{1}{6} + 2x = \dfrac{2}{3}$

CONCEPT EXTENSIONS

Solve. See the Concept Check in this section.

97. Which correctly describes the location of the point $(-1, 5.3)$ in a rectangular coordinate system?

a. 1 unit to the right of the y-axis and 5.3 units above the x-axis

b. 1 unit to the left of the y-axis and 5.3 units above the x-axis

c. 1 unit to the left of the y-axis and 5.3 units below the x-axis

d. 1 unit to the right of the y-axis and 5.3 units below the x-axis

98. Which correctly describes the location of the point $\left(0, -\dfrac{3}{4}\right)$ in a rectangular coordinate system?

a. on the x-axis and $\dfrac{3}{4}$ unit to the left of the y-axis

b. on the x-axis and $\dfrac{3}{4}$ unit to the right of the y-axis

c. on the y-axis and $\dfrac{3}{4}$ unit above the x-axis

d. on the y-axis and $\dfrac{3}{4}$ unit below the x-axis

For Exercises 99 through 102, match each description with the graph that best illustrates it.

99. Moe worked 40 hours per week until the fall semester started. He quit and didn't work again until he worked 60 hours a week during the holiday season starting mid-December.

100. Kawana worked 40 hours a week for her father during the summer. She slowly cut back her hours to not working at all during the fall semester. During the holiday season in December, she started working again and increased her hours to 60 hours per week.

101. Wendy worked from July through February, never quitting. She worked between 10 and 30 hours per week.

102. Bartholomew worked from July through February. During the holiday season between mid-November and the beginning of January, he worked 40 hours per week. The rest of the time, he worked between 10 and 40 hours per week.

a.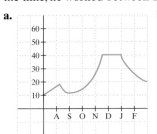

b.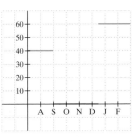

c.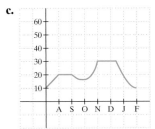

d.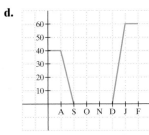

The graph below shows first-class postal rates and the years it increased. Use this graph for Exercises 103 through 106.

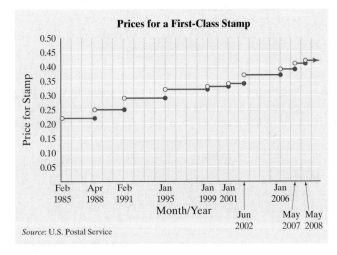

Source: U.S. Postal Service

103. What was the first year that the price for a first-class stamp rose above $0.25?

104. What was the first year that the price for a first-class stamp rose above $0.30?

105. Why do you think that this graph is shaped the way it is?

106. The U.S. Postal Service issued first-class stamps as far back as 1885. The cost for a first-class stamp then was $0.02. By how much had it increased by June 2007?

Group Activity. *For income tax purposes, Jason Verges, owner of Copy Services, uses a method called* **straight-line depreciation** *to show the loss in value of a copy machine he recently purchased. Jason assumes that he can use the machine for 7 years. The following graph shows the value of the machine over the years. Use this graph to answer Exercises 107 through 112.*

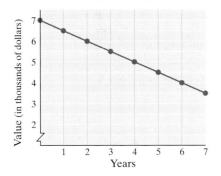

107. What was the purchase price of the copy machine?

108. What is the depreciated value of the machine in 7 years?

109. What loss in value occurred during the first year?

110. What loss in value occurred during the second year?

111. Why do you think that this method of depreciating is called straight-line depreciation?

112. Why is the line tilted downward?

Write each statement as an equation in two variables. Then graph each equation.

113. The *y*-value is 5 more than three times the *x*-value.

114. The *y*-value is −3 decreased by twice the *x*-value.

115. The *y*-value is 2 more than the square of the *x*-value.

116. The *y*-value is 5 decreased by the square of the *x*-value.

📖 STUDY SKILLS BUILDER

How Well Do You Know Your Textbook?

The questions below will determine whether you are familiar with your textbook. For help, see Section 1.1 in this text.

1. What does the 🖩 icon mean?
2. What does the ＼ icon mean?
3. What does the △ icon mean?
4. Which exercise set answers are given in the back of your text?
5. Where can you find a review for each chapter? What answers to this review can be found in the back of your text?
6. Each chapter contains an overview of the chapter along with examples. What is this feature called?
7. Each chapter contains a review of vocabulary. What is this feature called?
8. There is a CD in your text. What content is contained on this CD?
9. What is the location of the section that is entirely devoted to study skills?
10. There is a Practice exercise located after each worked example in the text. What are they and how can they be used?
11. There is a Practice Final Exam in your text. Where is it located?

2.2 INTRODUCTION TO FUNCTIONS

OBJECTIVES

1. Define relation, domain, and range.
2. Identify functions.
3. Use the vertical line test for functions.
4. Find the domain and range of a function.
5. Use function notation.

OBJECTIVE 1 ▶ Defining relation, domain, and range. Recall our example from the last section about products sold and monthly salary. We modeled the data given by the equation $y = 3000 + \frac{1}{5}x$. This equation describes a relationship between x-values and y-values. For example, if $x = 1000$, then this equation describes how to find the y-value related to $x = 1000$. In words, the equation $y = 3000 + \frac{1}{5}x$ says that 3000 plus $\frac{1}{5}$ of the x-value gives the corresponding y-value. The x-value of 1000 corresponds to the y-value of $3000 + \frac{1}{5} \cdot 1000 = 3200$ for this equation, and we have the ordered pair $(1000, 3200)$.

There are other ways of describing relations or correspondences between two numbers or, in general, a first set (sometimes called the set of *inputs*) and a second set (sometimes called the set of *outputs*). For example,

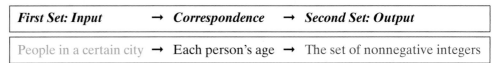

A few examples of ordered pairs from this relation might be (Ana, 4); (Bob, 36); (Trey, 21); and so on.

Below are just a few other ways of describing relations between two sets and the ordered pairs that they generate.

Correspondence

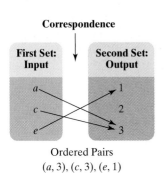

Ordered Pairs
$(a, 3), (c, 3), (e, 1)$

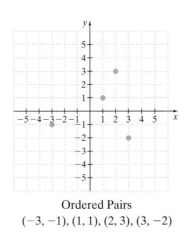

Ordered Pairs
$(-3, -1), (1, 1), (2, 3), (3, -2)$

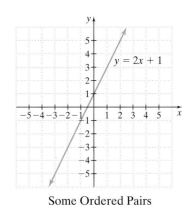

Some Ordered Pairs
$(1, 3), (0, 1)$, and so on

Relation, Domain, and Range

A **relation** is a set of ordered pairs.
The **domain** of the relation is the set of all first components of the ordered pairs.
The **range** of the relation is the set of all second components of the ordered pairs.

For example, the domain for our relation as shown above to the left is $\{a, c, e\}$ and the range is $\{1, 3\}$. Notice that the range does not include the element 2 of the second set. This is because no element of the first set is assigned to this element. If a relation is defined in terms of x- and y-values, we will agree that the domain corresponds to x-values and that the range corresponds to y-values that have x-values assigned to them.

▶ **Helpful Hint**

Remember that the range only includes elements that are paired with domain values. For the correspondence below, the range is $\{a\}$.

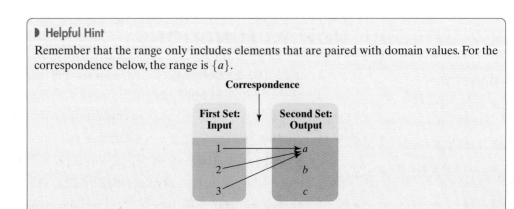

EXAMPLE 1   Determine the domain and range of each relation.

a. $\{(2, 3), (2, 4), (0, -1), (3, -1)\}$

b.

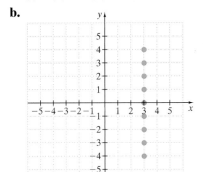

c.

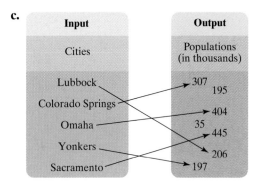

Solution

a. The domain is the set of all first coordinates of the ordered pairs, $\{2, 0, 3\}$. The range is the set of all second coordinates, $\{3, 4, -1\}$.

b. Ordered pairs are not listed here, but are given in graph form. The relation is $\{(-4, 1), (-3, 1), (-2, 1), (-1, 1), (0, 1), (1, 1), (2, 1), (3, 1)\}$. The domain is $\{-4, -3, -2, -1, 0, 1, 2, 3\}$. The range is $\{1\}$.

c. The domain is the set of inputs, {Lubbock, Colorado Springs, Omaha, Yonkers, Sacramento}. The range is the numbers in the set of outputs that correspond to elements in the set of inputs {307, 404, 445, 206, 197}.

▶ **Helpful Hint**

Domain or range elements that occur more than once need only to be listed once.

PRACTICE

1   Determine the domain and range of each relation.

a. $\{(4, 1)(4, -3)(5, -2)(5, 6)\}$

b.

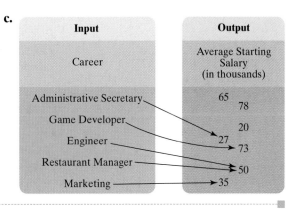

c.

OBJECTIVE 2 ▶ **Identifying functions.** Now we consider a special kind of relation called a function.

Function

A **function** is a relation in which each first component in the ordered pairs corresponds to *exactly* one second component.

> ▶ **Helpful Hint**
> A function is a special type of relation, so all functions are relations, but not all relations are functions.

EXAMPLE 2 Which of the following relations are also functions?

a. $\{(-2, 5), (2, 7), (-3, 5), (9, 9)\}$

b.

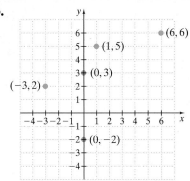

c.

Input	Correspondence	Output
People in a certain city	Each person's age	The set of nonnegative integers

Solution

a. Although the ordered pairs $(-2, 5)$ and $(-3, 5)$ have the same y-value, each x-value is assigned to only one y-value, so this set of ordered pairs is a function.

b. The x-value 0 is assigned to two y-values, -2 and 3, in this graph so this relation does not define a function.

c. This relation is a function because although two different people may have the same age, each person has only one age. This means that each element in the first set is assigned to only one element in the second set. □

PRACTICE
2 Which of the following relations are also functions?

a. $\{(3, 1), (-3, -4), (8, 5), (9, 1)\}$

b.

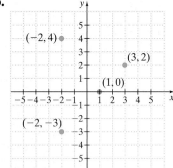

c.

Input	Correspondence	Output
People in a certain city	Birth date (day of month)	Set of positive integers

Concept Check ☑

Explain why a function can contain both the ordered pairs $(1, 3)$ and $(2, 3)$ but not both $(3, 1)$ and $(3, 2)$.

Answer to Concept Check:
Two different ordered pairs can have the same y-value, but not the same x-value in a function.

We will call an equation such as $y = 2x + 1$ a **relation** since this equation defines a set of ordered pair solutions.

EXAMPLE 3 Is the relation $y = 2x + 1$ also a function?*

Solution The relation $y = 2x + 1$ is a function if each x-value corresponds to just one y-value. For each x-value substituted in the equation $y = 2x + 1$, the multiplication and addition performed on each gives a single result, so only one y-value will be associated with each x-value. Thus, $y = 2x + 1$ is a function.

*For further discussion including the graph, see Objective 3.

PRACTICE

3 Is the relation $y = -3x + 5$ also a function?

EXAMPLE 4 Is the relation $x = y^2$ also a function?*

Solution In $x = y^2$, if $y = 3$, then $x = 9$. Also, if $y = -3$, then $x = 9$. In other words, we have the ordered pairs $(9, 3)$ and $(9, -3)$. Since the x-value 9 corresponds to two y-values, 3 and -3, $x = y^2$ is not a function.

*For further discussion including the graph, see Objective 3.

PRACTICE

4 Is the relation $y = -x^2$ also a function?

OBJECTIVE 3 ▶ Using the vertical line test. As we have seen so far, not all relations are functions. Consider the graphs of $y = 2x + 1$ and $x = y^2$ shown next. For the graph of $y = 2x + 1$, notice that each x-value corresponds to only one y-value. Recall from Example 3 that $y = 2x + 1$ is a function.

Graph of Example 3:
$y = 2x + 1$

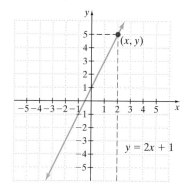

Graph of Example 4:
$x = y^2$

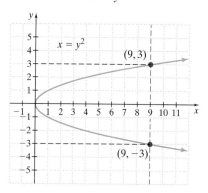

For the graph of $x = y^2$ the x-value 9, for example, corresponds to two y-values, 3 and -3, as shown by the vertical line. Recall from Example 4 that $x = y^2$ is not a function.

Graphs can be used to help determine whether a relation is also a function by the following vertical line test.

Vertical Line Test

If no vertical line can be drawn so that it intersects a graph more than once, the graph is the graph of a function.

EXAMPLE 5   Which of the following graphs are graphs of functions?

a.
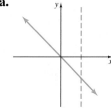

b.
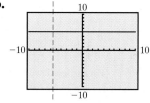

c.
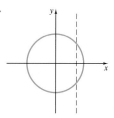

Solution

Yes, this is the graph of a function since no vertical line will intersect this graph more than once.

Yes, this is the graph of a function.

No, this is not the graph of a function. Note that vertical lines can be drawn that intersect the graph in two points.

d.
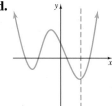

e.
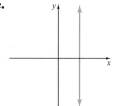

f.
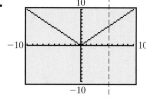

Solution

Yes, this is the graph of a function.

No, this is not the graph of a function. A vertical line can be drawn that intersects this line at every point.

Yes, this is the graph of a function.   □

PRACTICE
5   Which of the following graphs are graphs of functions?

a.
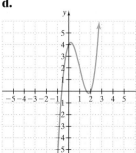

b.
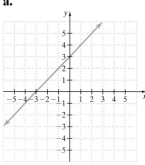

c.
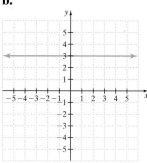

d.
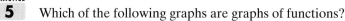

e.
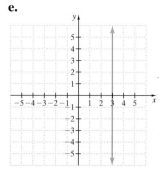

f.
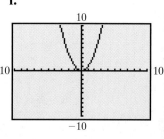

Recall that the graph of a linear equation in two variables is a line, and a line that is not vertical will pass the vertical line test. Thus, **all linear equations are functions except those whose graph is a vertical line.** For now, we will use set builder notation to write domains and ranges.

For ease of writing, if the domain or range is all real numbers, we simply write "all real numbers" instead of $\{x \mid x \text{ is a real number}\}$ or $\{y \mid y \text{ is a real number}\}$.

In Section 3.2, we will learn and use a new notation called interval notation.

Concept Check ☑

Determine which equations represent functions. Explain your answer.

a. $y = |x|$ **b.** $y = x^2$ **c.** $x + y = 6$

OBJECTIVE 4 ▶ Finding the domain and range of a function. Next, we practice finding the domain and range of a relation from its graph.

EXAMPLE 6 Find the domain and range of each relation. Determine whether the relation is also a function.

a.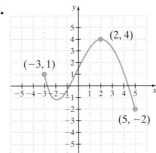

b.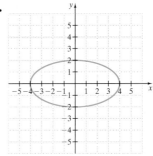

c.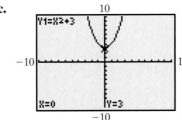

d.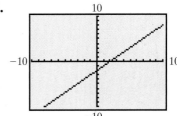

e.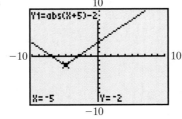

Solution By the vertical line test, graphs **a**, **c**, **d**, and **e** are graphs of functions. The domain is the set of values of x and the range is the set of values of y. We read these values from each graph.

a.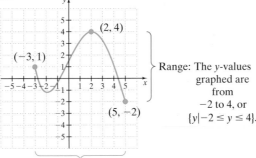

Range: The y-values graphed are from -2 to 4, or $\{y \mid -2 \le y \le 4\}$.

Domain: The x-values graphed are from -3 to 5, or $\{x \mid -3 \le x \le 5\}$.

b.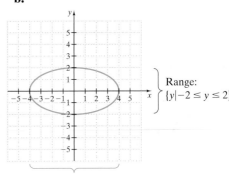

Range: $\{y \mid -2 \le y \le 2\}$

Domain: $\{x \mid -4 \le x \le 4\}$

c.   Domain: all real numbers

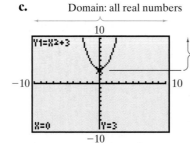

Range: $\{y | y \geq 3\}$

d.   Domain: all real numbers

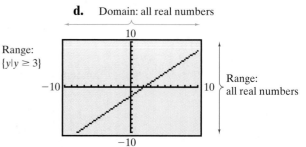

Range: all real numbers

e.   Domain: all real numbers

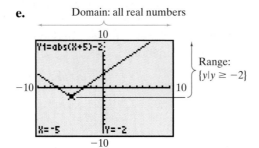

Range: $\{y | y \geq -2\}$

PRACTICE

6   Find the domain and range of each relation. Determine whether each relation is also a function.

a.

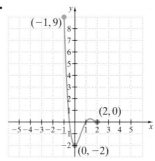

b.

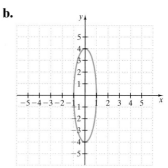

c.

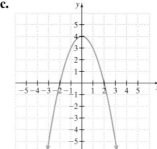

d.

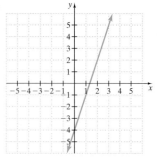

OBJECTIVE 5 ▶ Using function notation. Many times letters such as f, g, and h are used to name functions.

Function Notation

To denote that y is a function of x, we can write

$$y = \underbrace{f(x)}_{\text{Function Notation}} \quad (\text{Read "} f \text{ of } x.\text{"})$$

This notation means that **y is a function of x** or that y *depends on x*. For this reason, y is called the **dependent variable** and x the **independent variable.**

For example, to use function notation with the function $y = 4x + 3$, we write $f(x) = 4x + 3$. The notation $f(1)$ means to replace x with 1 and find the resulting y or function value. Since

$$f(x) = 4x + 3$$

then

$$f(1) = 4(1) + 3 = 7$$

This means that when $x = 1$, y or $f(x) = 7$. The corresponding ordered pair is $(1, 7)$. Here, the input is 1 and the output is $f(1)$ or 7. Now let's find $f(2)$, $f(0)$, and $f(-1)$.

$$f(x) = 4x + 3 \qquad\qquad f(x) = 4x + 3 \qquad\qquad f(x) = 4(x) + 3$$
$$f(2) = 4(2) + 3 \qquad\qquad f(0) = 4(0) + 3 \qquad\qquad f(-1) = 4(-1) + 3$$
$$= 8 + 3 \qquad\qquad\qquad = 0 + 3 \qquad\qquad\qquad\qquad = -4 + 3$$
$$= 11 \qquad\qquad\qquad\qquad = 3 \qquad\qquad\qquad\qquad\qquad = -1$$

▶ Helpful Hint

Make sure you remember that $f(2) = 11$ corresponds to the ordered pair $(2, 11)$.

Ordered Pairs:

$(2, 11)$ $\qquad\qquad\qquad\qquad (0, 3) \qquad\qquad\qquad\qquad (-1, -1)$

There are many ways of evaluating function values using a graphing utility. One way is to use the store feature. For example, to find $f(2)$ when $f(x) = 4x + 3$, store the number 2, enter the expression, and calculate. To find $f(0)$ and $f(-1)$ for the same function and save time, have your graphing utility replay the last entry and then edit it.

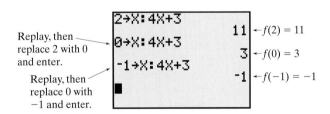

Replay, then replace 2 with 0 and enter.

Replay, then replace 0 with −1 and enter.

We can also use the graph to evaluate a function at a given value. The graph of $y_1 = 4x + 3$ in an integer window is shown below. Move the cursor to the point with x-coordinate 2 to find $f(2)$. Continue in this manner to find $f(0)$ and $f(-1)$.

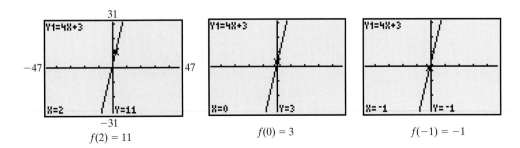

$f(2) = 11$ $\qquad\qquad\qquad\qquad f(0) = 3 \qquad\qquad\qquad\qquad f(-1) = -1$

A third method for finding function values is by using a table.

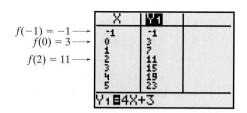

$$f(-1) = -1 \longrightarrow$$
$$f(0) = 3 \longrightarrow$$
$$f(2) = 11 \longrightarrow$$

The method you use to find function values depends on the particular situation. For instance, in Example 7, we evaluate several different functions for various values of x, so we calculate by hand and use the store feature to check. However, in Example 8 we evaluate the same function, but for different values of x, so we choose to look at the graph of the function. These are just a few methods that can be used to evaluate functions at given values.

> ▶ **Helpful Hint**
>
> Note that $f(x)$ is a special symbol in mathematics used to denote a function. The symbol $f(x)$ is read "f of x." It does *not* mean $f \cdot x$ (f times x).

EXAMPLE 7   If $f(x) = 7x^2 - 3x + 1, g(x) = 3x - 2$, and $h(x) = x^2$ find the following.

a. $f(1)$ **b.** $g(3)$ **c.** $h(-2)$

Solution

a. Substitute 1 for x in $f(x) = 7x^2 - 3x + 1$ and simplify.

$$f(x) = 7x^2 - 3x + 1$$
$$f(1) = 7(1)^2 - 3(1) + 1 = 5$$

b. $g(x) = 3x - 2$
$$g(3) = 3(3) - 2 = 7$$

c. $h(x) = x^2$
$$h(-2) = (-2)^2 = 4$$

A calculator check is in the margin.

```
1→X:7X²-3X+1
                5 ← f(1) = 5
3→X:3X-2
                7 ← g(3) = 7
-2→X:X²
                4 ← h(-2) = 4
■
```

PRACTICE

7   If $f(x) = 3x - 2$ and $g(x) = 5x^2 + 2x - 1$, find the following.

a. $f(1)$ **b.** $g(1)$ **c.** $f(0)$ **d.** $g(-2)$

Concept Check ☑

Suppose $y = f(x)$ and we are told that $f(3) = 9$. Which is not true?

a. When $x = 3, y = 9$.

b. A possible function is $f(x) = x^2$.

c. A point on the graph of the function is $(3, 9)$.

d. A possible function is $f(x) = 2x + 4$.

If it helps, think of a function, f, as a machine that has been programmed with a certain correspondence or rule. An input value (a member of the domain) is then fed into the machine, the machine does the correspondence or rule, and the result is the output (a member of the range).

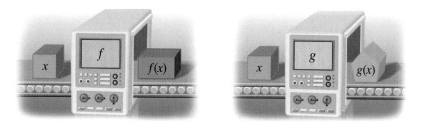

EXAMPLE 8   If $f(x) = 0.5x - 25$, find

a. $f(-10)$ **b.** $f(18)$ **c.** $f(11)$ **d.** $f(0)$

Solution

a. We choose to use a graphical method to evaluate. Define $y_1 = 0.5x - 25$ and graph using an integer window. Move the cursor to find $x = -10$. In the screen shown below, we see that -10 is paired with $y = -30$; therefore $f(-10) = -30$.

b. Move the cursor to $x = 18$ and see that it is paired with -16 and thus $f(18) = -16$.

c. Move the cursor to $x = 11$ and see that $y = -19.5$, so $f(11) = -19.5$.

d. Move the cursor to $x = 0$ and see that $y = -25$, so $f(0) = -25$.

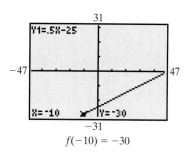

$$f(-10) = -30$$

PRACTICE
8   If $f(x) = 2.6x^2 - 4$, find:

a. $f(2)$ **b.** $f(-4)$ **c.** $f(0)$ **d.** $f(0.7)$

Many formulas that are familiar to you describe functions. For example, we have used the formula for finding the area of a circle, $A = \pi r^2$. The area of the circle is actually a function of the length of the radius. Using this function notation, we write

$$A(r) = \pi r^2$$

$A(r)$ can be read as the area with respect to r. To find the area of the circle whose radius is 3 cm, we write

$$A(r) = \pi r^2$$
$$A(3) = \pi(3)^2 = 9\pi \text{ square centimeters}$$

An approximation to two decimal places is

$$9\pi \text{ sq cm} \approx 28.27 \text{ sq cm}$$

EXAMPLE 9 Given the graphs of the functions f and g, find each function value by inspecting the graphs.

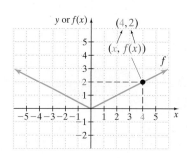 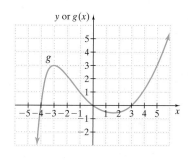

a. $f(4)$ **b.** $f(-2)$ **c.** $g(5)$ **d.** $g(0)$

e. Find all x-values such that $f(x) = 1$.

f. Find all x-values such that $g(x) = 0$.

Solution

a. To find $f(4)$, find the y-value when $x = 4$. We see from the graph that when $x = 4$, y or $f(x) = 2$. Thus, $f(4) = 2$.

b. $f(-2) = 1$ from the ordered pair $(-2, 1)$.

c. $g(5) = 3$ from the ordered pair $(5, 3)$.

d. $g(0) = 0$ from the ordered pair $(0, 0)$.

e. To find x-values such that $f(x) = 1$, we are looking for any ordered pairs on the graph of f whose $f(x)$ or y-value is 1. They are $(2, 1)$ and $(-2, 1)$. Thus $f(2) = 1$ and $f(-2) = 1$. The x-values are 2 and -2.

f. Find ordered pairs on the graph of g whose $g(x)$ or y-value is 0. They are $(3, 0)$ $(0, 0)$, and $(-4, 0)$. Thus $g(3) = 0$, $g(0) = 0$, and $g(-4) = 0$. The x-values are 3, 0, and -4. ☐

PRACTICE
9 Given the graphs of the functions f and g, find each function value by inspecting the graphs.

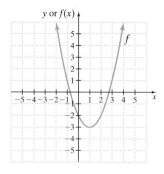 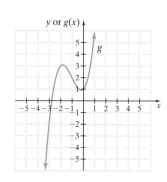

a. $f(1)$ **b.** $f(0)$ **c.** $g(-2)$ **d.** $g(0)$

e. Find all x-values such that $f(x) = 1$.

f. Find all x-values such that $g(x) = -2$.

When a store manager is setting the retail price of an article, the price may be dependent on the wholesale price of the article. When this happens, we say that the retail price is a function of the wholesale price.

EXAMPLE 10 Finding Retail Prices

Elizabeth Lockwood manages the college bookstore and purchases the books from several wholesale companies. She finds that she needs to mark up the wholesale cost by 25%.

a. Write the retail price as a function of the wholesale cost.

b. Find the retail price of the following books given the wholesale cost.

Wholesale Cost	$15.00	$22.75	$38.50	$53.00
Retail Price				

Solution

a. Let w = the wholesale cost of a book. To denote that the retail price is a function of wholesale cost, we define the function $R(w)$.

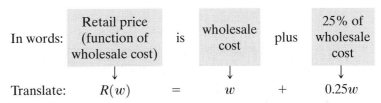

In words:	Retail price (function of wholesale cost)	is	wholesale cost	plus	25% of wholesale cost
	↓		↓		↓
Translate:	$R(w)$	=	w	+	$0.25w$

b. Here we evaluate $R(w) = w + 0.25w$ for the given values of w.

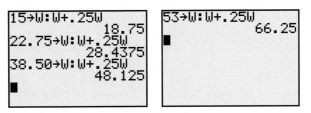

Completing the table, we have

Wholesale Cost	w	$15.00	$22.75	$38.50	$53.00
Retail Price	$R(w)$	$18.75	$28.44	$48.13	$66.25

☐

PRACTICE

10 Find the cost of each book in Example 10 if the mark-up is 15%.

Many types of real-world paired data form functions. The following broken-line graph shows the research and development spending by the Pharmaceutical Manufacturers Association.

EXAMPLE 11 The following graph shows the research and development expenditures by the Pharmaceutical Manufacturers Association as a function of time.

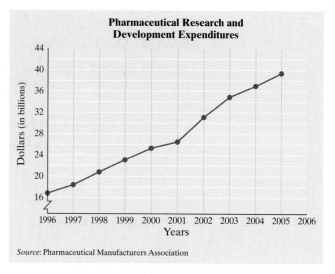

Source: Pharmaceutical Manufacturers Association

a. Approximate the money spent on research and development in 2002.

b. In 1958, research and development expenditures were $200 million. Find the increase in expenditures from 1958 to 2004.

Solution

a. Find the year 2002 and move upward until you reach the graph. From the point on the graph move horizontally, to the left, until the other axis is reached. In 2002, approximately $31 billion was spent.

b. In 2004, approximately $37 billion, or $37,000 million was spent. The increase in spending from 1958 to 2004 is $37,000 − $200 = $36,800 million or $36.8 billion. □

PRACTICE
11 Use the graph in Example 11 and approximate the money spent in 2003.

Notice that the graph in Example 11 is the graph of a function since for each year there is only one total amount of money spent by the Pharmaceutical Manufacturers Association on research and development. Also notice that the graph resembles the

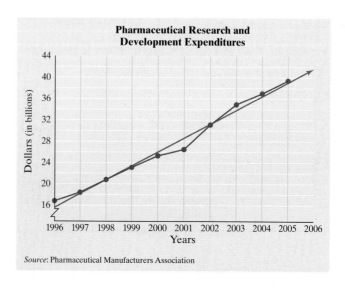

Source: Pharmaceutical Manufacturers Association

graph of a line. Often, businesses depend on equations that "closely fit" data-defined functions like this one in order to model the data and predict future trends. For example, by a method called **least squares,** the function $f(x) = 2.602x - 5178$ approximates the data shown. For this function, x is the year and $f(x)$ is total money spent. Its graph and the actual data function are shown at bottom of previous page.

EXAMPLE 12   Use the function $f(x) = 2.602x - 5178$ to predict the amount of money that will be spent by the Pharmaceutical Manufacturers Association on research and development in 2014.

Solution   To predict the amount of money that will be spent in the year 2014 we use $f(x) = 2.602x - 5178$ and find $f(2014)$.

$$f(x) = 2.602x - 5178$$
$$f(2014) = 2.602(2014) - 5178 \quad \text{See the graphing calculator}$$
$$= 62.428 \qquad\qquad\quad \text{screen in the margin.}$$

We predict that in the year 2014, $62.428 billion dollars will be spent on research and development by the Pharmaceutical Manufacturers Association.

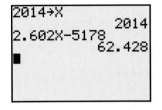

A calculator check for Example 12.

PRACTICE
12   Use $f(x) = 2.602x - 5178$ to approximate the money spent in 2012.

VOCABULARY & READINESS CHECK

Use the choices below to fill in each blank. Some choices may not be used. These exercises have to do with functions and the rectangular coordinate system (Sections 2.1 and 2.2).

x	domain	vertical	relation	$(1.7, -2)$	line	parabola
y	range	horizontal	function	$(-2, 1.7)$	origin	V-shaped

1. The intersection of the x-axis and y-axis is a point, called the _____.

2. To find an x-intercept, let _____ = 0 and solve for _____.

3. To find a y-intercept, let _____ = 0 and solve for _____.

4. The graph of $Ax + By = C$, where A and B are not both 0, is a _____.

5. The graph of $y = |x|$ looks _____.

6. The graph of $y = x^2$ is a _____.

7. A _____ is a set of ordered pairs.

8. The _____ of a relation is the set of all second components of the ordered pairs.

9. The _____ of a relation is the set of all first components of the ordered pairs.

10. A _____ is a relation in which each first component in the ordered pairs corresponds to *exactly* one second component.

11. By the vertical line test, all linear equations are functions except those whose graphs are _____ lines.

12. If $f(-2) = 1.7$, the corresponding ordered pair is _____.

2.2 EXERCISE SET

Find the domain and the range of each relation. Also determine whether the relation is a function. See Examples 1 and 2.

1. $\{(-1, 7), (0, 6), (-2, 2), (5, 6)\}$

2. $\{(4, 9), (-4, 9), (2, 3), (10, -5)\}$

3. $\{(-2, 4), (6, 4), (-2, -3), (-7, -8)\}$

4. $\{(6, 6), (5, 6), (5, -2), (7, 6)\}$

5. $\{(1, 1), (1, 2), (1, 3), (1, 4)\}$

6. $\{(1, 1), (2, 1), (3, 1), (4, 1)\}$

7. $\left\{\left(\dfrac{3}{2}, \dfrac{1}{2}\right), \left(1\dfrac{1}{2}, -7\right), \left(0, \dfrac{4}{5}\right)\right\}$

8. $\{(\pi, 0), (0, \pi), (-2, 4), (4, -2)\}$

9. $\{(-3, -3), (0, 0), (3, 3)\}$

10. $\left\{\left(\dfrac{1}{2}, \dfrac{1}{4}\right), \left(0, \dfrac{7}{8}\right), (0.5, \pi)\right\}$

11.

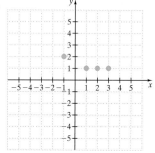

12.

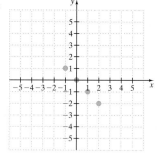

13.

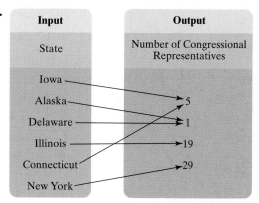

14.

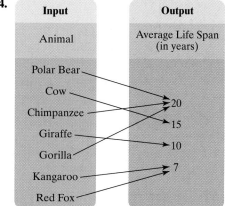

15.

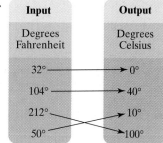

16.

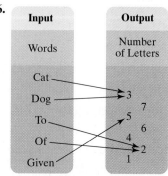

17.

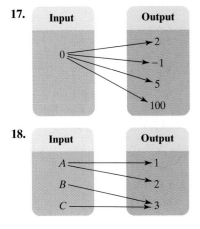

18.

In Exercises 19 through 22, determine whether the relation is a function. See Example 2.

First Set: Input	Correspondence	Second Set: Output
19. Class of algebra students	Final grade average	nonnegative numbers
20. People who live in Cincinnati, Ohio	Birth date	days of the year
21. blue, green, brown	Eye color	People who live in Cincinnati, Ohio
22. Whole numbers from 0 to 4	Number of children	50 Women in a water aerobics class

Use the vertical line test to determine whether each graph is the graph of a function. See Example 5.

23.

24.

25.

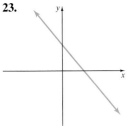

26.

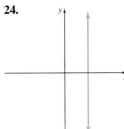

27.

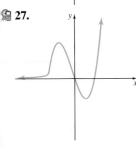

28.

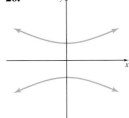
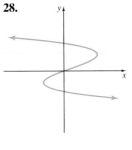

29.

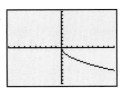

30.

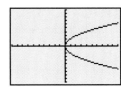

31.

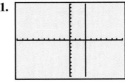

32.

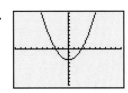

Find the domain and the range of each relation. Use the vertical line test to determine whether each graph is the graph of a function. See Example 6.

33.

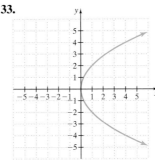

34.

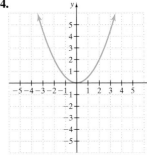

35.

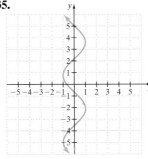

36.

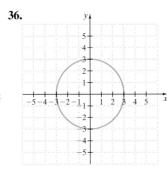

37.

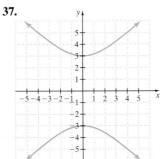

38.

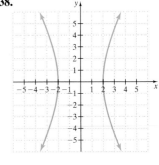

39.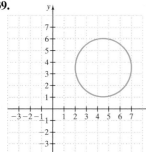

40.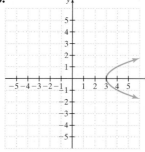

53. $y - x = 7$

54. $2x - 3y = 9$

55. $y = \dfrac{1}{x}$

56. $y = \dfrac{1}{x - 3}$

57. $y = 5x - 12$

58. $y = \dfrac{1}{2}x + 4$

59. $x = y^2$

60. $x = |y|$

If $f(x) = 3x + 3$, $g(x) = 4x^2 - 6x + 3$, and $h(x) = 5x^2 - 7$, find the following. See Examples 7 and 8.

61. $f(4)$ **62.** $f(-1)$

63. $h(-3)$ **64.** $h(0)$

65. $g(2)$ **66.** $g(1)$

67. $g(0)$ **68.** $h(-2)$

41.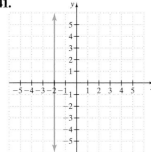

42.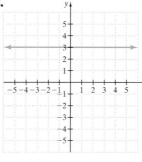

Given the following functions, find the indicated values. See Examples 7 and 8.

69. $f(x) = \dfrac{1}{2}x$;

    a. $f(0)$ **b.** $f(2)$ **c.** $f(-2)$

70. $g(x) = -\dfrac{1}{3}x$;

    a. $g(0)$ **b.** $g(-1)$ **c.** $g(3)$

71. $g(x) = 2x^2 + 4$;

    a. $g(-11)$ **b.** $g(-1)$ **c.** $g\left(\dfrac{1}{2}\right)$

72. $h(x) = -x^2$;

    a. $h(-5)$ **b.** $h\left(-\dfrac{1}{3}\right)$ **c.** $h\left(\dfrac{1}{3}\right)$

73. $f(x) = -5$;

    a. $f(2)$ **b.** $f(0)$ **c.** $f(606)$

74. $h(x) = 7$;

    a. $h(7)$ **b.** $h(542)$ **c.** $h\left(-\dfrac{3}{4}\right)$

43.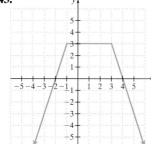

44.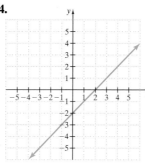

75. $f(x) = 1.3x^2 - 2.6x + 5.1$

    a. $f(2)$ **b.** $f(-2)$ **c.** $f(3.1)$

76. $g(x) = 2.7x^2 + 6.8x - 10.2$

    a. $g(1)$ **b.** $g(-5)$ **c.** $g(7.2)$

77. Given the following table of values for $f(x) = |2x - 5| + 8$, find the indicated values.

    a. $f(-5)$ **b.** $f(10)$ **c.** $f(15)$

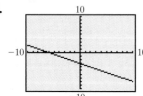

45.

46.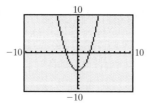

78. Given the following table of values for $f(x) = -2x^3 + x$, find the indicated values.

    a. $f(-4)$ **b.** $f(2)$ **c.** $f(6)$

47. In your own words define **(a)** function; **(b)** domain; **(c)** range.

48. Explain the vertical line test and how it is used.

MIXED PRACTICE

Decide whether each is a function. See Examples 3 through 6.

49. $y = x + 1$ **50.** $y = x - 1$

51. $x = 2y^2$ **52.** $y = x^2$

79. Given the graph of the function $f(x) = x^2 + x + 1$, find the value of $f(2)$.

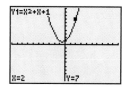

80. Given the graph of the function $f(x) = x^3 + x^2 + 1$, find $f(-2)$.

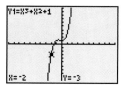

Use the graph of the functions below to answer Exercises 81 through 92. See Example 9.

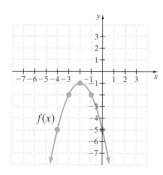
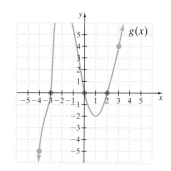

81. If $f(1) = -10$, write the corresponding ordered pair.

82. If $f(-5) = -10$, write the corresponding ordered pair.

83. If $g(4) = 56$, write the corresponding ordered pair.

84. If $g(-2) = 8$, write the corresponding ordered pair.

85. Find $f(-1)$.   **86.** Find $f(-2)$.

87. Find $g(2)$.   **88.** Find $g(-4)$.

89. Find all values of x such that $f(x) = -5$.

90. Find all values of x such that $f(x) = -2$.

91. Find all values of x such that $g(x) = -5$

92. Find all values of x such that $g(x) = 0$.

93. What is the greatest number of x-intercepts that a function may have? Explain your answer.

94. What is the greatest number of y-intercepts that a function may have? Explain your answer.

Solve. See Examples 10 and 11.

95. Julie needs to hire a plumber. The cost of hiring a plumber, C, in dollars is a function of the time spent on the job, t, in hours. If the plumber charges a one-time fee of $20 plus $35 per hour, we can write this in function notation as $C(t) = 20 + 35t$. Complete the table.

Time in Hours	t	1	2	3	4	5
Total Cost	$C(t)$					

96. The cost of hiring a secretary to type a term paper is $10 plus $5.25 per hour. Therefore, the cost, C, in dollars is a function of time, t, in hours, or $C(t) = 10 + 5.25t$. Complete the table.

Time in Hours	t	0.5	1	1.5	2.5	3
Total Cost	$C(t)$					

97. The height h of a firecracker being shot into the air from ground level is a function of time t and can be represented using the following formula from physics: $h = -16t^2 + vt$, where v is initial velocity. If the velocity is 80 feet per second, the height in terms of time is $h(t) = -16t^2 + 80t$. Find the following:

  a. $h(0.5)$ **b.** $h(1)$ **c.** $h(1.5)$

  d. $h(2)$ **e.** $h(2.5)$ **f.** $h(5)$

98. The height h of a rocket launched from a 200-foot-high building with velocity of 120 feet per second can be expressed as a function of time t using the formula $h = -16t^2 + vt + h_0$, where h_0 stands for the initial height. If $h(t) = -16t^2 + 120t + 200$, find the following:

  a. $h(0.2)$ **b.** $h(0.6)$ **c.** $h(2.25)$

  d. $h(3)$ **e.** $h(4)$

Use the graph in Example 11 to answer the following. Also see Example 12.

99. **a.** Use the graph to approximate the money spent on research and development in 1996.

  b. Recall that the function $f(x) = 2.602x - 5178$ approximates the graph in Example 11. Use this equation to approximate the money spent on research and development in 1996.

100. **a.** Use the graph to approximate the money spent on research and development in 1999.

  b. Use the function $f(x) = 2.602x - 5178$ to approximate the money spent on research and development in 1999.

The function $f(x) = 0.42x + 10.5$, can be used to predict diamond production. For this function, x is the number of years after 2000, and $f(x)$ is the value (in billions of dollars) of the year's diamond production. (See the Chapter 2 opener.)

101. Use the function in the directions above to predict diamond production in 2012.

102. Use the function in the directions above to predict diamond production in 2015.

103. Since $y = x + 7$ describes a function, rewrite the equation using function notation.

104. In your own words, explain how to find the domain of a function given its graph.

The function $A(r) = \pi r^2$ may be used to find the area of a circle if we are given its radius.

△ **105.** Find the area of a circle whose radius is 5 centimeters. (Do not approximate π.)

△ **106.** Find the area of a circular garden whose radius is 8 feet. (Do not approximate π.)

The function $V(x) = x^3$ may be used to find the volume of a cube if we are given the length x of a side.

107. Find the volume of a cube whose side is 14 inches.

108. Find the volume of a die whose side is 1.7 centimeters.

Forensic scientists use the following functions to find the height of a woman if they are given the height of her femur bone f or her tibia bone t in centimeters.

$$H(f) = 2.59f + 47.24$$
$$H(t) = 2.72t + 61.28$$

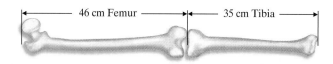

109. Find the height of a woman whose femur measures 46 centimeters.

110. Find the height of a woman whose tibia measures 35 centimeters.

The dosage in milligrams D of Ivermectin, a heartworm preventive, for a dog who weighs x pounds is given by

$$D(x) = \frac{136}{25}x$$

111. Find the proper dosage for a dog that weighs 30 pounds.

112. Find the proper dosage for a dog that weighs 50 pounds.

113. The per capita consumption (in pounds) of all poultry in the United States is approximated by the function $C(x) = 2.28x + 94.86$, where x is the number of years since 2001. (*Source*: Based on actual and estimated data from the Economic Research Service, U.S. Department of Agriculture)

    a. Find and interpret $C(5)$.

    b. Estimate the per capita consumption of all poultry in the United States in 2007.

114. The average length of U.S. hospital stays has been decreasing, following the equation $y = -0.09x + 8.02$, where x is the number of years since 1970 and y is the length of the average stay in days. (*Source:* National Center for Health Statistics)

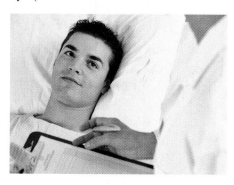

    a. What was the length of the average hospital stay in 1995?

    b. If this trend continues, what will the average length be in 2011?

REVIEW AND PREVIEW

Complete the given table and use the table to graph the linear equation. See Section 2.1.

115. $x - y = -5$

x	0		1
y		0	

116. $2x + 3y = 10$

x	0		
y		0	2

117. $7x + 4y = 8$

x	0		
y		0	-1

118. $5y - x = -15$

x	0		-2
y		0	

119. $y = 6x$

x	0		-1
y		0	

120. $y = -2x$

x	0		-2
y		0	

△ **121.** Is it possible to find the perimeter of the following geometric figure? If so, find the perimeter.

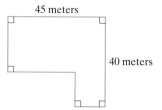

45 meters

40 meters

CONCEPT EXTENSIONS

For Exercises 122 through 125, suppose that $y = f(x)$ and it is true that $f(7) = 50$. Determine whether each is true or false. See the second Concept Check in this section.

122. An ordered pair solution of the function is $(7, 50)$.

123. When x is 50, y is 7.

124. A possible function is $f(x) = x^2 + 1$.

125. A possible function is $f(x) = 10x - 20$.

Given the following functions, find the indicated values.

126. $f(x) = 2x + 7$;

    a. $f(2)$ **b.** $f(a)$

127. $g(x) = -3x + 12$;

    a. $g(s)$ **b.** $g(r)$

128. $h(x) = x^2 + 7$;

    a. $h(3)$ **b.** $h(a)$

129. $f(x) = x^2 - 12$;

    a. $f(12)$ **b.** $f(a)$

130. Describe a function whose domain is the set of people in your hometown.

131. Describe a function whose domain is the set of people in your algebra class.

2.3 GRAPHING LINEAR FUNCTIONS

OBJECTIVES

1 Graph linear functions.

2 Graph linear functions by finding intercepts.

3 Graph vertical and horizontal lines.

4 Use graphs to solve problems.

OBJECTIVE 1 ▶ Graphing linear functions. In this section, we identify and graph linear functions. By the vertical line test, we know that all linear equations except those whose graphs are vertical lines are functions. For example, we know from Section 2.1 that $y = 2x$ is a linear equation in two variables. Its graph is shown.

x	$y = 2x$
1	2
0	0
-1	-2

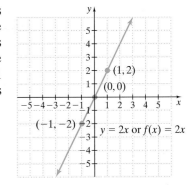

Because this graph passes the vertical line test, we know that $y = 2x$ is a function. If we want to emphasize that this equation describes a function, we may write $y = 2x$ as $f(x) = 2x$.

    A graphing utility is a very versatile tool for exploring the graphs of functions.

DISCOVER THE CONCEPT

a. On your graphing utility, graph both $f(x) = 2x$ as $y_1 = 2x$, and $g(x) = 2x + 10$ as $y_2 = 2x + 10$ using an integer window.

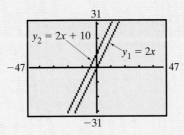

Integer Window

b. Trace on the graph of $y_1 = 2x$ to find the point whose ordered pair has x-coordinate 3 and find the corresponding y-coordinate. Now, press the down arrow key and find the corresponding y-coordinate on the graph of $y_2 = 2x + 10$. Compare the two y-coordinates for the same x-coordinate.

c. Trace to another point on the graph of y_1 and repeat this process. Compare the two lines, and see if you can state how we could sketch the graph of $g(x) = 2x + 10$ from the graph of $f(x) = 2x$.

d. Predict how the graph of $h(x) = 2x - 15$ could be drawn. Graph $y_3 = 2x - 15$ and see if your prediction is correct.

In the discovery on the previous page, we confirmed what we discovered on page 111. The y-values for the graph of $g(x)$ or $y = 2x + 10$ are obtained by adding 10 to the y-value of each corresponding point of the graph of $f(x)$ or $y = 2x$. The graph of $g(x) = 2x + 10$ is the same as the graph of $f(x) = 2x$ shifted upward 10 units.

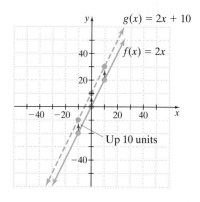

x	-1	0	1
$f(x) = 2x$	-2	0	2
$g(x) = 2x + 10$	8	10	12

add 10

Also, the graph of $h(x)$ or $y = 2x - 15$ is obtained by subtracting 15 from the y-value of each corresponding point of the graph of $f(x)$ or $y = 2x$. Thus, the graph of $h(x) = 2x - 15$ is the same as the graph of $f(x) = 2x$ shifted downward 15 units.

The functions $f(x) = 2x$, $g(x) = 2x + 10$, and $h(x) = 2x - 15$ are called linear functions—"linear" because each graph is a line, and "function" because each graph passes the vertical line test.

In general, a **linear function** is a function that can be written in the form $f(x) = mx + b$. For example, $g(x) = 2x + 10$ is in this form, with $m = 2$ and $b = 10$.

Note: Ordered pairs may be listed in a variety of ways, including tables. In Section 2.1, the tables were written vertically. Above, we see a table written horizontally. Throughout this section, you will see ordered-pair tables written vertically and horizontally. Both ways of writing tables are useful and have advantages.

EXAMPLE 1 Graph $g(x) = 2x + 1$. Compare this graph with the graph of $f(x) = 2x$.

Solution To graph $g(x) = 2x + 1$, find three ordered pair solutions.

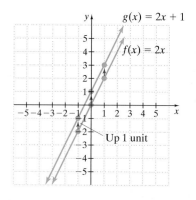

x	$f(x) = 2x$	$g(x) = 2x + 1$
0	0	1
-1	-2	-1
1	2	3

add 1

Notice that y-values for the graph of $g(x) = 2x + 1$ are obtained by adding 1 to each y-value of each corresponding point of the graph of $f(x) = 2x$. The graph of $g(x) = 2x + 1$ is the same as the graph of $f(x) = 2x$ shifted upward 1 unit. ☐

PRACTICE

1 Graph $g(x) = 4x - 3$ and $f(x) = 4x$ on the same axes.

Notice that $g(x) = 2x + 1$ above is a **linear function** in the form $g(x) = mx + b$. For this function, $g(x) = 2x + 1$, $m = 2$ and $b = 1$.

EXAMPLE 2   Graph both linear functions $f(x) = 0.5x$ and $g(x) = 0.5x + 8$ using the same integer window.

a. Use the graphs to complete the following table of solution pairs.

x	-6	0	5	13	18
$f(x) = 0.5x$					
$g(x) = 0.5x + 8$					

b. Complete the following sentence. The graph of $g(x) = 0.5x + 8$ can be obtained from the graph of $f(x) = 0.5x$ by _____.

Solution

a. Define $y_1 = 0.5x$ and $y_2 = 0.5x + 8$ and graph in the integer window. The screens below show the values of $f(x)$ and $g(x)$ when $x = 4$. In other words, $f(4) = 2$ and $g(4) = 10$.

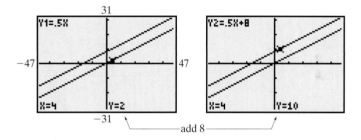

Trace along each graph to complete the table as follows:

x	-6	0	5	13	18
$f(x) = 0.5x$	-3	0	2.5	6.5	9
$g(x) = 0.5x + 8$	5	8	10.5	14.5	17

b. The graph of $g(x) = 0.5x + 8$ can be obtained from the graph of $f(x) = 0.5x$ by shifting it upward 8 units. □

PRACTICE

2   Graph both linear functions $f(x) = -3x$ and $g(x) = -3x + 2$. Complete the sentence. The graph of $g(x) = -3x + 2$ can be obtained from the graph of $f(x) = -3x$ by _____.

EXAMPLE 3   Graph both linear functions $f(x) = -x$ and $g(x) = -x - 16$ using an integer window. Complete the following statement. The graph of $g(x) = -x - 16$ can be obtained from the graph of $f(x) = -x$ by _____.

Solution   Graph $y_1 = -x$ and $y_2 = -x - 16$ using an integer window as shown on the next page. We can complete the statement as follows. The graph of $g(x) = -x - 16$ can be obtained from the graph of $f(x)$ by shifting it downward 16 units. To further illustrate this fact, see the table on the next page. And compare the y_1 and y_2 values for the same x-value.

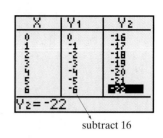

subtract 16

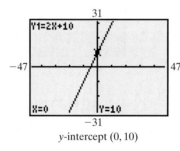

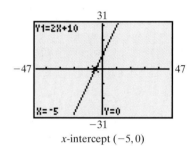

PRACTICE

3   Complete the sentence. The graph of $y_2 = -x - 5$ can be obtained from the graph of $y = -x$ by _____.

In general, for any function $f(x)$, the graph of $y = f(x) + K$ is the same as the graph of $y = f(x)$ shifted $|K|$ units upward if K is positive and downward if K is negative.

OBJECTIVE 2 ▶ Graphing linear functions using intercepts. The graph of $y = 2x + 10$ is shown in both screens below. Notice that this graph crosses both the x-axis and the y-axis. Recall that a point where a graph crosses the x-axis is called the ***x*-intercept,** and a point where a graph crosses the y-axis is called the **y-intercept.**

y-intercept $(0, 10)$

x-intercept $(-5, 0)$

By tracing along the graph, we can see that the y-intercept of the graph of $y = 2x + 10$ is $(0, 10)$. Also the x-intercept of the graph is $(-5, 0)$.

One way to find the y-intercept of the graph of an equation is to let $x = 0$, since a point on the y-axis has an x-coordinate of 0. To find the x-intercept, let $y = 0$ or $f(x) = 0$, since a point on the x-axis has a y-coordinate of 0.

Finding x- and y-Intercepts

To find an x-intercept, let $y = 0$ or $f(x) = 0$ and solve for x.
To find a y-intercept, let $x = 0$ and solve for y.

Intercepts are usually easy to find and plot since one coordinate is 0.

In the next example, we sketch a linear function by plotting x- and y-intercepts.

EXAMPLE 4   Graph $x - 3y = 6$ by plotting intercepts. Check using a graphing utility.

Solution   Let $y = 0$ to find the x-intercept and $x = 0$ to find the y-intercept.

$$\text{If } y = 0 \quad \text{then} \qquad \text{If } x = 0 \quad \text{then}$$

$$x - 3(0) = 6 \qquad\qquad 0 - 3y = 6$$

$$x - 0 = 6 \qquad\qquad\quad -3y = 6$$

$$x = 6 \qquad\qquad\qquad y = -2$$

The x-intercept is $(6, 0)$ and the y-intercept is $(0, -2)$. We find a third ordered pair solution to check our work. If we let $y = -1$, then $x = 3$. Plot the points $(6, 0)$, $(0, -2)$, and $(3, -1)$. The graph of $x - 3y = 6$ is the line drawn through these points, as shown.

x	y
x-intercept → 6	0
0	-2 ← y-intercept
3	-1

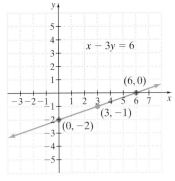

To check using a graphing utility, solve for y to enter the equation in the Y= editor.

$$x - 3y = 6$$
$$-3y = -x + 6$$
$$y = \frac{-x + 6}{-3}$$

Define $y_1 = \dfrac{-x + 6}{-3}$ and graph in an integer window as shown below.

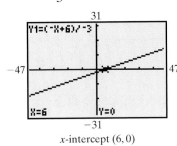

x-intercept $(6, 0)$

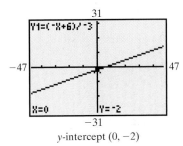

y-intercept $(0, -2)$

PRACTICE

4   Graph $4x - 5y = -20$ by plotting intercepts. Check using a graphing utility.

Notice that the equation $x - 3y = 6$ describes a linear function—"linear" because its graph is a line and "function" because the graph passes the vertical line test.

If we want to emphasize that the equation $x - 3y = 6$ from Example 4 describes a function, we first solve the equation for y.

$$x - 3y = 6$$
$$-3y = -x + 6 \qquad \text{Subtract } x \text{ from both sides.}$$
$$\frac{-3y}{-3} = \frac{-x}{-3} + \frac{6}{-3} \qquad \text{Divide both sides by } -3.$$
$$y = \frac{1}{3}x - 2 \qquad \text{Simplify.}$$

Next, let

$$y = f(x).$$

$$f(x) = \frac{1}{3}x - 2$$

> ▶ **Helpful Hint**
>
> Any linear equation that describes a function can be written using function notation. To do so,
>
> **1.** Solve the equation for y and then
> **2.** Replace y with $f(x)$, as we did above.

> ▶ **Helpful Hint**
>
> Recall that when generating a graph using a graphing utility, we first solve for the variable y. When a function is given in the $f(x)$ notation, it is already solved for y and can be entered in the Y= editor directly.

DISCOVER THE CONCEPT

a. Use an integer setting and graph each linear function.

$$y_1 = x + 15 \qquad y_2 = x \qquad y_3 = x - 11$$

b. Decide how the y-intercept of the graph of the equation compares with the equation.

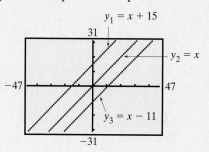

From the discovery above, we found the following:

Equation	y-intercept
$y = x + 15$	$(0, 15)$
$y = x$	$(0, 0)$
$y = x - 11$	$(0, -11)$

In general,

$$y = mx + b \qquad (0, b)$$

Notice that the y-intercept of the graph of the equation of the form $y = mx + b$ is $(0, b)$ each time. This is because we find the y-intercept of the graph of an equation by letting $x = 0$. Thus,

$$y = mx + b$$
$$y = m \cdot 0 + b \quad \text{Let } x = 0.$$
$$y = b$$

The intercept is $(0, b)$.

EXAMPLE 5   Find the y-intercept of the graph of each equation.

a. $f(x) = \dfrac{1}{2}x + \dfrac{3}{7}$

b. $y = -2.5x - 3.2$

Solution

a. The y-intercept of $f(x) = \dfrac{1}{2}x + \dfrac{3}{7}$ is $\left(0, \dfrac{3}{7}\right)$.

b. The y-intercept of $y = -2.5x - 3.2$ is $(0, -3.2)$.

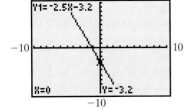

A calculator check of
Example 5b.

PRACTICE
5   Find the y-intercept of the graph of each equation.

a. $f(x) = \dfrac{3}{4}x - \dfrac{2}{5}$

b. $y = 2.6x + 4.1$

EXAMPLE 6   Graph $x = -2y$ by plotting intercepts.

Solution   Let $y = 0$ to find the x-intercept and $x = 0$ to find the y-intercept.

$$
\begin{array}{ll}
\text{If } y = 0 \quad \text{then} & \text{If } x = 0 \quad \text{then} \\
x = -2(0) \quad \text{or} & 0 = -2y \quad \text{or} \\
x = 0 & 0 = y \\
(0, 0) & (0, 0)
\end{array}
$$

Ordered pairs   Both the x-intercept and y-intercept are $(0, 0)$. This happens when the graph passes through the origin. Since two points are needed to determine a line, we must find at least one more ordered pair that satisfies $x = -2y$. Let $y = -1$ to find a second ordered pair solution and let $y = 1$ as a check point.

$$
\begin{array}{ll}
\text{If } y = -1 \quad \text{then} & \text{If } y = 1 \quad \text{then} \\
x = -2(-1) \quad \text{or} & x = -2(1) \quad \text{or} \\
x = 2 & x = -2
\end{array}
$$

The ordered pairs are $(0, 0)$, $(2, -1)$, and $(-2, 1)$. Plot these points to graph $x = -2y$.

x	y
0	0
2	-1
-2	1

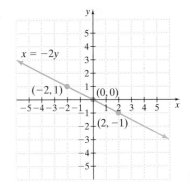

PRACTICE
6   Graph $y = -3x$ by plotting intercepts.

OBJECTIVE 3 ▶ Graphing vertical and horizontal lines. The equations $x = c$ and $y = c$, where c is a real number constant, are both linear equations in two variables. Why? Because $x = c$ can be written as $x + 0y = c$ and $y = c$ can be written as $0x + y = c$. We graph these two special linear equations on the next page.

EXAMPLE 7 Graph $x = 2$.

Solution The equation $x = 2$ can be written as $x + 0y = 2$. For any y-value chosen, notice that x is 2. No other value for x satisfies $x + 0y = 2$. Any ordered pair whose x-coordinate is 2 is a solution to $x + 0y = 2$ because 2 added to 0 times any value of y is $2 + 0$, or 2. We will use the ordered pairs $(2, 3)$, $(2, 0)$, and $(2, -3)$ to graph $x = 2$.

TECHNOLOGY NOTE

Since the graph of $x = c$ is a vertical line and is not a function, we cannot enter this equation in the Y= editor.

However, most graphing utilities have a draw feature that allows you to draw the vertical line, but you cannot use trace on it since it is not a function. See if your graphing utility has this feature.

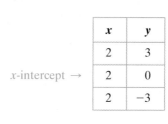

x	y
2	3
x-intercept → 2	0
2	-3

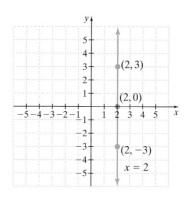

The graph is a vertical line with x-intercept $(2, 0)$. Notice that this graph is not the graph of a function, and it has no y-intercept because x is never 0. ☐

PRACTICE
7 Graph $x = -4$.

EXAMPLE 8 Graph $y = -3$.

Solution The equation $y = -3$ can be written as $0x + y = -3$. For any x-value chosen, y is -3. If we choose 4, 0, and -2 as x-values, the ordered pair solutions are $(4, -3)$, $(0, -3)$, and $(-2, -3)$. We will use these ordered pairs to graph $y = -3$.

x	y
4	-3
0	-3
-2	-3

← y-intercept

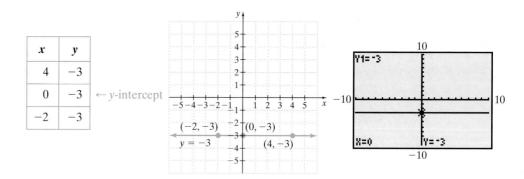

The graph is a horizontal line with y-intercept $(0, -3)$ and no x-intercept. Notice that this graph is the graph of a function. Above to the right is the graph of $y = -3$ using a graphing utility. Trace along the line to see that all ordered pair solutions have a y-coordinate of -3. ☐

PRACTICE
8 Graph $y = 4$.

From Examples 7 and 8, we have the following generalization.

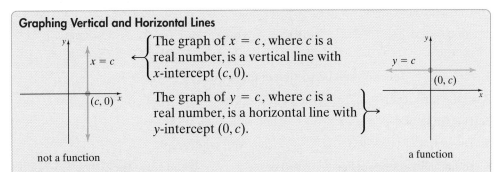

Graphing Vertical and Horizontal Lines

The graph of $x = c$, where c is a real number, is a vertical line with x-intercept $(c, 0)$.

The graph of $y = c$, where c is a real number, is a horizontal line with y-intercept $(0, c)$.

not a function

a function

Notice the graph of $x = c$ is not a function, so it cannot be graphed in function mode on the calculator. The graph of $y = c$ is a function, so it can be graphed on the calculator.

OBJECTIVE 4 ▶ Using graphs to solve problems. We can use the graph of a linear equation to solve problems.

EXAMPLE 9   Cost of Renting a Car

The cost of renting a car for a day is given by the linear function $C(x) = 35 + 0.15x$, where $C(x)$ represents the cost in dollars and x is the number of miles driven. Use the graph of the function to complete the table below in dollars, and find the cost $C(x)$ for the given number of miles.

No. of Miles	x	150	200	325	500
Cost	$C(x) = 35 + 0.15x$				

Solution   Define $y_1 = 35 + 0.15x$ and graph in a $[0, 600, 100]$ by $[0, 200, 50]$ window. We choose $0 \le x \le 600$ since the values in the table are in this interval. We choose $0 \le y \le 200$ as we are estimating our cost to be less than \$200. The graph to the right illustrates how to obtain the cost for 150 miles.

Finding the remaining values from the graph, we can complete the table.

No. of Miles	x	150	200	325	500
Cost	$C(x) = 35 + 0.15x$	57.5	65	83.75	110

Notice from the graph that the cost $C(x)$ increases as the number of miles, x, increases.

PRACTICE
9   The cost of renting a premium car for a day is given by the linear function $C(x) = 79 + 0.10x$, where x is the number of miles driven. Use this function to complete the table.

No. of Miles	x	50	175	230	450
Cost	$C(x)$				

VOCABULARY & READINESS CHECK

Use the choices below to fill in each blank. Some choices may be used more than once and some not at all.

horizontal	y	$(c, 0)$	$(b, 0)$	$(m, 0)$	linear
vertical	x	$(0, c)$	$(0, b)$	$(0, m)$	$f(x)$

1. A _____ function can be written in the form $f(x) = mx + b$.

2. In the form $f(x) = mx + b$, the y-intercept is _____.

3. The graph of $x = c$ is a _____ line with x-intercept _____.

4. The graph of $y = c$ is a _____ line with y-intercept _____.

5. To find an x-intercept, let ____ = 0 or _____ = 0 and solve for ____.

6. To find a y-intercept, let ____ = 0 and solve for ____.

2.3 EXERCISE SET

PRACTICE   WATCH   DOWNLOAD   READ   REVIEW

Graph each linear function. See Examples 1 and 2.

1. $f(x) = -2x$ **2.** $f(x) = 2x$

3. $f(x) = -2x + 3$ **4.** $f(x) = 2x + 6$

5. $f(x) = \dfrac{1}{2}x$ **6.** $f(x) = \dfrac{1}{3}x$

7. $f(x) = \dfrac{1}{2}x - 4$ **8.** $f(x) = \dfrac{1}{3}x - 2$

The graph of $f(x) = 5x$ in a standard window follows. Use this graph to match each linear function with its graph. See Examples 1 and 2.

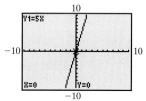

A.    **B.**

C. 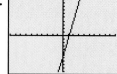   **D.**

9. $f(x) = 5x - 6$ **10.** $f(x) = 5x - 2$
11. $f(x) = 5x + 7$ **12.** $f(x) = 5x + 3$

Graph each linear function by finding x- and y-intercepts. Then write each equation using function notation. See Examples 4 and 5.

13. $x - y = 3$

14. $x - y = -4$

15. $x = 5y$

16. $2x = y$

17. $-x + 2y = 6$

18. $x - 2y = -8$

19. $2x - 4y = 8$

20. $2x + 3y = 6$

21. In your own words, explain how to find x- and y-intercepts.

22. Explain why it is a good idea to use three points to graph a linear equation.

Graph each linear equation. See Examples 6 and 7.

23. $x = -1$

24. $y = 5$

25. $y = 0$

26. $x = 0$

27. $y + 7 = 0$

28. $x - 3 = 0$

Match each equation below with its graph.

A

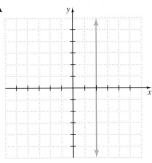

B

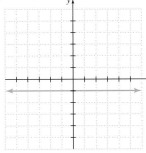

C

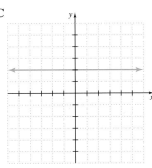

D

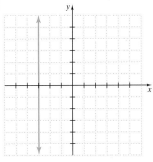

29. $y = 2$

30. $x = -3$

31. $x - 2 = 0$

32. $y + 1 = 0$

33. Discuss whether a vertical line ever has a y-intercept.

34. Discuss whether a horizontal line ever has an x-intercept.

MIXED PRACTICE

Graph each linear equation. See Examples 1 through 7.

35. $x + 2y = 8$

36. $x - 3y = 3$

37. $3x + 5y = 7$

38. $3x - 2y = 5$

39. $x + 8y = 8$

40. $x - 3y = 9$

41. $5 = 6x - y$

42. $4 = x - 3y$

43. $-x + 10y = 11$

44. $-x + 9 = -y$

45. $y = \dfrac{3}{2}$

46. $x = \dfrac{3}{2}$

47. $2x + 3y = 6$

48. $4x + y = 5$

49. $x + 3 = 0$

50. $y - 6 = 0$

51. $f(x) = \dfrac{3}{4}x + 2$

52. $f(x) = \dfrac{4}{3}x + 2$

53. $f(x) = x$

54. $f(x) = -x$

55. $f(x) = \dfrac{1}{2}x$

56. $f(x) = -2x$

57. $f(x) = 4x - \dfrac{1}{3}$

58. $f(x) = -3x + \dfrac{3}{4}$

59. $x = -3$

60. $f(x) = 3$

61. Given the graph of $f(x)$, which of the following statements are true?

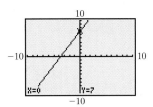

a. $f(7) = 0$ **b.** $f(0) = 7$

c. The x-intercept is $(7, 0)$.   **d.** The y-intercept is $(0, 7)$.

62. Given the graph of $g(x)$, which of the following statements are true?

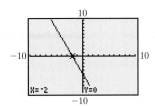

a. $g(-2) = 0$

b. $g(0) = -2$

c. The x-intercept is $(-2, 0)$.

d. The y-intercept is $(0, -2)$.

Solve. See Example 9.

63. Kevin Elliott works in the Human Resource department at a rate of \$15.75 per hour for a 40-hour work week. He makes time and a half for overtime. His salary, before any deductions, can be represented by the function $S(x) = 15.75(40) + 1.5(15.75)x$, where x is the number of overtime hours. Graph the function and use it to complete the table below for the salary $S(x)$ in dollars.

No. of Over-time Hours	x	2	3.25	7.5	9.75
Salary in Dollars	$S(x)$				

64. The average hourly earnings of someone who works in retail trade is \$8.34. If the normal hourly rate for overtime work (work over 40 hours per week) is 1.5 (\$8.34), the salary, before deductions, can be represented by $S(x) = 8.34(40) + 1.5(8.34)x$, where x is the number of overtime hours. Graph the function $S(x)$ and use the graph to complete the table below. (*Source:* U.S. Bureau of Labor Statistics)

No. of Overtime Hours	2	10	12.5	15.75
Salary in Dollars				

REVIEW AND PREVIEW

Simplify. See Section 1.3.

65. $\dfrac{-6 - 3}{2 - 8}$

66. $\dfrac{4 - 5}{-1 - 0}$

67. $\dfrac{-8 - (-2)}{-3 - (-2)}$

68. $\dfrac{12 - 3}{10 - 9}$

69. $\dfrac{0 - 6}{5 - 0}$

70. $\dfrac{2 - 2}{3 - 5}$

CONCEPT EXTENSIONS

Solve.

71. Broyhill Furniture found that it takes 2 hours to manufacture each table for one of its special dining room sets. Each chair takes 3 hours to manufacture. A total of 1500 hours is available to produce tables and chairs of this style. The linear equation that models this situation is $2x + 3y = 1500$, where x represents the number of tables produced and y the number of chairs produced.

   a. Complete the ordered pair solution $(0, \ )$ of this equation. Describe the manufacturing situation this solution corresponds to.

   b. Complete the ordered pair solution $(\ , 0)$ for this equation. Describe the manufacturing situation this solution corresponds to.

   c. If 50 tables are produced, find the greatest number of chairs the company can make.

72. While manufacturing two different camera models, Kodak found that the basic model costs \$55 to produce, whereas the deluxe model costs \$75. The weekly budget for these two models is limited to \$33,000 in production costs. The linear equation that models this situation is $55x + 75y = 33,000$, where x represents the number of basic models and y the number of deluxe models.

   a. Complete the ordered pair solution $(0, \ )$ of this equation. Describe the manufacturing situation this solution corresponds to.

   b. Complete the ordered pair solution $(\ , 0)$ of this equation. Describe the manufacturing situation this solution corresponds to.

   c. If 350 deluxe models are produced, find the greatest number of basic models that can be made in one week.

73. The cost of renting a car for a day is given by the linear function $C(x) = 0.2x + 24$, where $C(x)$ is in dollars and x is the number of miles driven.

   a. Find the cost of driving the car 200 miles.

   b. Graph $C(x) = 0.2x + 24$.

   c. How can you tell from the graph of $C(x)$ that as the number of miles driven increases, the total cost increases also?

74. The cost of renting a piece of machinery is given by the linear function $C(x) = 4x + 10$, where $C(x)$ is in dollars and x is given in hours.

   a. Find the cost of renting the piece of machinery for 8 hours.

   b. Graph $C(x) = 4x + 10$.

   c. How can you tell from the graph of $C(x)$ that as the number of hours increases, the total cost increases also?

75. The yearly cost of tuition (in-state) and required fees for attending a public two-year college full time can be estimated by the linear function $f(x) = 107.3x + 1245.62$, where x is the number of years after 2000 and $f(x)$ is the total cost. (*Source:* U.S. National Center for Education Statistics)

   a. Use this function to approximate the yearly cost of attending a two-year college in the year 2015. [*Hint:* Find $f(15)$.]

   b. Use the given function to predict in what year the yearly cost of tuition and required fees will exceed \$2500. [*Hint:* Let $f(x) = 2500$, solve for x, then round your solution up to the next whole year.]

   c. Use this function to approximate the yearly cost of attending a two-year college in the present year. If you attend a two-year college, is this amount greater than or less than the amount that is currently charged by the college you attend?

76. The yearly cost of tuition (in-state) and required fees for attending a public four-year college full time can be estimated by the linear function $f(x) = 291.5x + 2944.05$, where x is the number of years after 2000 and $f(x)$ is the total cost in dollars. (*Source:* U.S. National Center for Education Statistics)

   a. Use this function to approximate the yearly cost of attending a four-year college in the year 2015. [*Hint:* Find $f(15)$.]

   b. Use the given function to predict in what year the yearly cost of tuition and required fees will exceed \$6000. [*Hint:* Let $f(x) = 6000$, solve for x, then round your solution up to the next whole year.]

   c. Use this function to approximate the yearly cost of attending a four-year college in the present year. If you attend a four-year college, is this amount greater than or less than the amount that is currently charged by the college you attend?

77. The graph of $f(x)$ or $y = -4x$ is given below. Without actually graphing, describe the shape and location of

   a. $y = -4x + 2$ **b.** $y = -4x - 5$

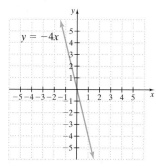

It is true that for any function $f(x)$, the graph of $f(x) + K$ is the same as the graph of $f(x)$ shifted K units up if K is positive and $|K|$ units down if K is negative. (We study this further in Section 2.7.)

The graph of $y = |x|$ is

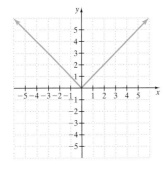

Without actually graphing, match each equation with its graph.

a. $y = |x| - 1$ **b.** $y = |x| + 1$

c. $y = |x| - 3$ **d.** $y = |x| + 3$

78.

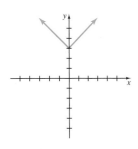

79.

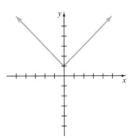

80.

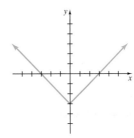

81.

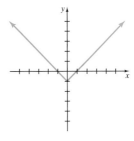

📖 STUDY SKILLS BUILDER

Are You Organized?

When it's time to study for a test, are your notes neat and organized? Have you ever had trouble reading your own mathematics handwriting? (Be honest—I have.)

When any of these things happen, it's time to get organized. Here are a few suggestions:

Write your notes and complete your homework assignment in a notebook with pockets (spiral or ring binder). When you receive graded papers or handouts, place them in the notebook pocket so that you will not lose them.

Remember to mark (possibly with an exclamation point) any note(s) that seem extra important to you. Also mark (possibly with a question mark) any notes or homework that you are having trouble with.

Also, if you are having trouble reading your own handwriting, *slow down* and write your mathematics work clearly!

Exercises

1. Have you been completing your assignments on time?
2. Have you been correcting any exercises you may be having difficulty with?
3. If you are having trouble with a mathematical concept or correcting any homework exercises, have you visited your instructor, a tutor, or your campus math lab?
4. Are you taking lecture notes in your mathematics course? (By the way, these notes should include all worked-out examples solved by your instructor.)
5. Is your mathematics course material (handouts, graded papers, lecture notes) organized?
6. If your answer to Exercise 5 is no, take a moment and review your course material. List at least two ways that you might better organize it. Then read the Study Skills Builder on organizing a notebook in Chapter 1.

2.4 THE SLOPE OF A LINE

OBJECTIVE 1 ▶ **Finding slope given two points.** You may have noticed by now that different lines often tilt differently.

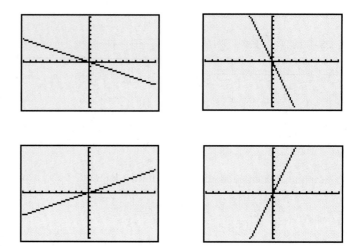

It is very important in many fields to be able to measure and compare the tilt, or **slope,** of lines. For example, a wheelchair ramp with a slope of $\frac{1}{12}$ means that the ramp rises 1 foot for every 12 horizontal feet. A road with a slope or grade of 11% $\left(\text{or } \frac{11}{100}\right)$ means that the road rises 11 feet for every 100 horizontal feet.

We measure the slope of a line as a ratio of **vertical change** to **horizontal change.** Slope is usually designated by the letter m.

Suppose that we want to measure the slope of the following line.

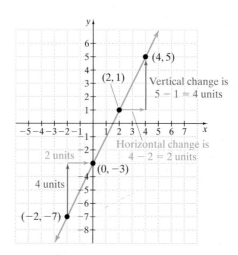

The vertical change between *both* pairs of points on the line is 4 units per horizontal change of 2 units. Then

$$\text{slope } m = \frac{\text{change in } y \text{ (vertical change)}}{\text{change in } x \text{ (horizontal change)}} = \frac{4}{2} = 2$$

We can also think of slope as a **rate of change** between points. A slope of 2 or $\frac{2}{1}$ means that between pairs of points on the line, the rate of change is a vertical change of 2 units per horizontal change of 1 unit.

Consider the line in the box below, which passes through the points (x_1, y_1) and (x_2, y_2). (The notation x_1 is read "x-sub-one.") The vertical change, or *rise*, between these points is the difference of the y-coordinates: $y_2 - y_1$. The horizontal change, or *run*, between the points is the difference of the x-coordinates: $x_2 - x_1$.

Slope of a Line

Given a line passing through points (x_1, y_1) and (x_2, y_2), the **slope** m of the line is

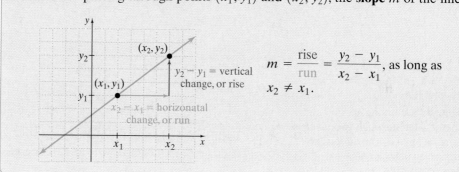

$$m = \frac{\text{rise}}{\text{run}} = \frac{y_2 - y_1}{x_2 - x_1}, \text{ as long as } x_2 \neq x_1.$$

Concept Check ☑

In the definition of slope, we state that $x_2 \neq x_1$. Explain why.

EXAMPLE 1   Find the slope of the line containing the points $(0, 3)$ and $(2, 5)$. Graph the line.

Solution   We use the slope formula. It does not matter which point we call (x_1, y_1) and which point we call (x_2, y_2). We'll let $(x_1, y_1) = (0, 3)$ and $(x_2, y_2) = (2, 5)$.

$$m = \frac{y_2 - y_1}{x_2 - x_1}$$

$$= \frac{5 - 3}{2 - 0} = \frac{2}{2} = 1$$

Notice in this example that the slope is *positive* and that the graph of the line containing $(0, 3)$ and $(2, 5)$ moves *upward*, or increases, as we go from left to right.   □

PRACTICE
1   Find the slope of the line containing the points $(4, 0)$ and $(-2, 3)$. Graph the line.

Answer to Concept Check:
So that the denominator is not 0

> ▶ **Helpful Hint**
>
> The slope of a line is the same no matter which 2 points of a line you choose to calculate slope. The line in Example 1 also contains the point $(-3, 0)$. Below, we calculate the slope of the line using $(0, 3)$ as (x_1, y_1) and $(-3, 0)$ as (x_2, y_2).
>
> $$m = \frac{y_2 - y_1}{x_2 - x_1} = \frac{0 - 3}{-3 - 0} = \frac{-3}{-3} = 1 \quad \text{Same slope as found in Example 1.}$$

EXAMPLE 2 Find the slope of the line containing the points $(5, -4)$ and $(-3, 3)$. Graph the line.

Solution We use the slope formula, and let $(x_1, y_1) = (5, -4)$ and $(x_2, y_2) = (-3, 3)$.

$$m = \frac{y_2 - y_1}{x_2 - x_1}$$

$$= \frac{3 - (-4)}{-3 - 5} = \frac{7}{-8} = -\frac{7}{8}$$

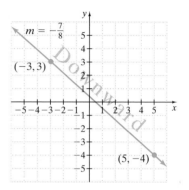

Notice in this example that the slope is negative and that the graph of the line through $(5, -4)$ and $(-3, 3)$ moves downward, or decreases, as we go from left to right. ☐

> ▶ **Helpful Hint**
>
> When we are trying to find the slope of a line through two given points, it makes no difference which given point is called (x_1, y_1) and which is called (x_2, y_2). Once an x-coordinate is called x_1, however, make sure its corresponding y-coordinate is called y_1.

PRACTICE

2 Find the slope of the line containing the points $(-5, -4)$ and $(5, 2)$. Graph the line.

Concept Check ☑

Find and correct the error in the following calculation of slope of the line containing the points $(12, 2)$ and $(4, 7)$.

$$m = \frac{12 - 4}{2 - 7} = \frac{8}{-5} = -\frac{8}{5}$$

To give meaning to the fact that the slope of a line is constant, see the following.

$$y = 2x + 3$$

x	y
-2	-1
-1	1
0	3
1	5
2	7

$+1($ $)+2$
$+1($ $)+2$
$+1($ $)+2$
$+1($ $)+2$

$$m = \frac{2}{1} = 2.$$

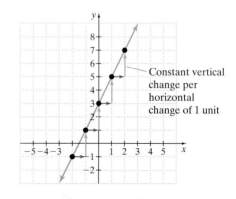

Constant vertical change per horizontal change of 1 unit

Answer to Concept Check:

$$m = \frac{2 - 7}{12 - 4} = \frac{-5}{8} = -\frac{5}{8}$$

For linear equations, there is a constant vertical change per constant horizontal change. This means that the slope is the same between any two points of the graph.

OBJECTIVE 2 ▶ **Finding slope given an equation.** The slope of a line indicates the direction of the line, that is, whether it slants upward or downward from left to right, as well as the amount of tilt to the line. In the following discovery we investigate how the equation itself can indicate direction and amount of tilt.

DISCOVER THE CONCEPT

a. Graph $y_1 = x$, $y_2 = 3x$, and $y_3 = 0.5x$ using a standard window. Examine the graphs. What observation can you make with regard to the slant of the lines? What observation can you make about each line with respect to the amount of tilt and the coefficient of x in the equation?

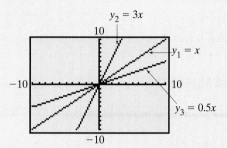

b. Can you change the equations so that the graphs slant downward instead of slanting upward from left to right?

c. Graph each equation separately. Predict the direction of slant and the amount of tilt of the line before seeing its graph on a graphing utility.

   i. $y = x + 3$ **ii.** $y = 2x + 3$ **iii.** $y = -2x + 3$ **iv.** $y = 0.5x + 3$

In part **a** of the discovery, we see that each line moves upward from left to right. Also, notice that the greater the positive coefficient of x is, the steeper the tilt of the line. From part **b**, we discover that if the coefficient of x is a negative number, the line moves downward from left to right.

As we have seen, the slope of a line is defined by two points on the line. Thus, if we know an equation of a line, we can find its slope.

DISCOVER THE CONCEPT

Let's see if we can discover a relationship between an equation written in the form $y = ax + b$ and its slope. We'll use the equations from the previous Discover the Concept box.

a. Find the slope of the graph of each equation below. To do so, use two ordered pair solutions and the slope formula. (We will use the y-intercept as one ordered pair solution.)

$$y = x + 3 \qquad\qquad y = 2x + 3$$
$$y = -2x + 3 \qquad\qquad y = 0.5x + 3$$

Equation	y-intercept	Any second ordered pair solution	Slope
$y = x + 3$ or $y = 1x + 3$	$(0, 3)$	$(4, 7)$	$m = \dfrac{7 - 3}{4 - 0} = 1$
$y = 2x + 3$	$(0, 3)$	$(1, 5)$	$m = \dfrac{5 - 3}{1 - 0} = 2$
$y = -2x + 3$	$(0, 3)$	$(1, 1)$	$m = \dfrac{1 - 3}{1 - 0} = -2$
$y = 0.5x + 3$	$(0, 3)$	$(2, 4)$	$m = \dfrac{4 - 3}{2 - 0} = 0.5$

b. For the equations in the general form $y = ax + b$, compare the slope of each equation above with a and compare the y-intercept with b.

We discovered that when a linear equation is written in the form $f(x) = ax + b$ or $y = ax + b$, a is the slope of the line and $(0, b)$ is its y-intercept. The form $y = mx + b$ is appropriately called the **slope-intercept form.** (Instead of a, we now use m to denote slope.)

> **Slope-Intercept Form**
>
> When a linear equation in two variables is written in slope-intercept form,
>
> $$\overset{\displaystyle \text{slope}}{\underset{\displaystyle \downarrow}{}} \quad \overset{\displaystyle y\text{-intercept is } (0, b)}{\underset{\displaystyle \downarrow}{}}$$
>
> $$y = mx + b$$
>
> then m is the slope of the line and $(0, b)$ is the y-intercept of the line.

EXAMPLE 3 Find the slope and the y-intercept of the line $3x - 4y = 4$.

Solution We write the equation in slope-intercept form by solving for y.

$$3x - 4y = 4$$

$$-4y = -3x + 4 \qquad \text{Subtract } 3x \text{ from both sides.}$$

$$\frac{-4y}{-4} = \frac{-3x}{-4} + \frac{4}{-4} \qquad \text{Divide both sides by } -4.$$

$$y = \frac{3}{4}x - 1 \qquad \text{Simplify.}$$

The coefficient of x, $\frac{3}{4}$, is the slope, and the y-intercept is $(0, -1)$.

To check, we graph the equation $y_1 = \dfrac{(-3x + 4)}{-4}$ in a decimal window. Since the slope is positive, the line does indeed slant upward from left to right (as it should) and the y-intercept is $(0, -1)$.

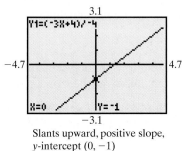

Slants upward, positive slope,
y-intercept $(0, -1)$

PRACTICE

3 Find the slope and the y-intercept of the line $2x - 3y = 9$.

OBJECTIVE 3 ▶ Interpreting slope-intercept form. On the following page is the graph of one-day ticket prices at Disney World for the years shown.

Notice that the graph resembles the graph of a line. Recall that businesses often depend on equations that "closely fit" graphs like this one to model the data and to predict future trends. By the **least squares** method, the linear function $f(x) = 2.7x + 38.64$ approximates the data shown, where x is the number of years since 1996 and $f(x)$ or y is the ticket price for that year.

> **▶ Helpful Hint**
>
> The notation $0 \leftrightarrow 1996$ below the graph on the next page means that the number 0 corresponds to the year 1996, 1 corresponds to the year 1997, and so on.

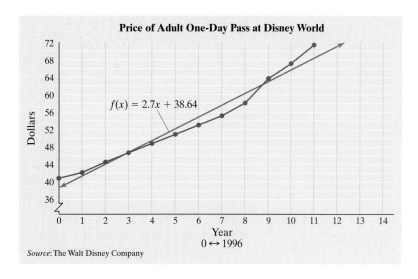

Price of Adult One-Day Pass at Disney World

$f(x) = 2.7x + 38.64$

Source: The Walt Disney Company

EXAMPLE 4   Predicting Future Prices

The adult one-day pass price $f(x)$ for Disney World is given by

$$f(x) = 2.7x + 38.64$$

where x is the number of years since 1996

a. Use this equation to predict the ticket price for the year 2010.

b. What does the slope of this equation mean?

c. What does the y-intercept of this equation mean?

Solution

a. To predict the price of a pass in 2010, we need to find $f(14)$. (Since year 1996 corresponds to $x = 0$, year 2010 corresponds to $x = 14$.)

$$f(x) = 2.7x + 38.64$$
$$f(14) = 2.7(14) + 38.64 \quad \text{Let } x = 14.$$
$$= 76.44$$

We predict that in the year 2010 the price of an adult one-day pass to Disney World will be about $76.44.

b. The slope of $f(x) = 2.7x + 38.64$ is 2.7. We can think of this number as $\dfrac{\text{rise}}{\text{run}}$ or $\dfrac{2.7}{1}$. This means that the ticket price increases on the average by $2.70 every 1 year.

c. The y-intercept of $f(x) = 2.7x + 38.64$ is $(0, 38.64)$.
$\qquad\qquad\qquad\qquad\qquad\qquad\quad\;\uparrow\quad\;\uparrow$
$\qquad\qquad\qquad\qquad\qquad\qquad\;\;\text{year price}$

This means that at year 0, or 1996, the ticket price was about $38.64. □

PRACTICE

4   Use the equation from Example 4 to predict the ticket price for the year 2012.

OBJECTIVE 4 ▶ Finding slopes of horizontal and vertical lines. Next we find the slopes of two special types of lines: vertical lines and horizontal lines.

EXAMPLE 5   Find the slope of the line $x = -5$.

Solution   Recall that the graph of $x = -5$ is a vertical line with x-intercept $(-5, 0)$. To find the slope, we find two ordered pair solutions of $x = -5$. Of course, solutions of $x = -5$ must have an x-value of -5. We will let $(x_1, y_1) = (-5, 0)$ and $(x_2, y_2) = (-5, 4)$.

Then

$$m = \frac{y_2 - y_1}{x_2 - x_1}$$

$$= \frac{4 - 0}{-5 - (-5)}$$

$$= \frac{4}{0}$$

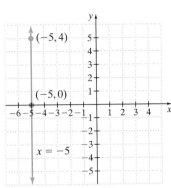

Since $\frac{4}{0}$ is undefined, we say that the slope of the vertical line $x = -5$ is undefined. □

PRACTICE
5   Find the slope of the line $x = 4$.

EXAMPLE 6   Find the slope of the line $y = 2$.

Solution   The graph of $y = 2$ is shown. Trace to find two ordered pair solutions. We select $(0, 2)$ and $(1, 2)$. Use these points to find the slope.

$$m = \frac{2 - 2}{1 - 0}$$

$$= \frac{0}{1}$$

$$= 0$$

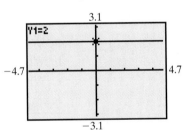

The slope of the horizontal line $y = 2$ is 0. □

PRACTICE
6   Find the slope of the line $y = -3$.

From the previous two examples, we have the following generalization.

> The slope of any vertical line is undefined.
> The slope of any horizontal line is 0.

> ▶ **Helpful Hint**
> Slope of 0 and undefined slope are not the same. Vertical lines have undefined slope, whereas horizontal lines have slope of 0.

The following four graphs summarize the overall appearance of lines with positive, negative, zero, or undefined slopes.

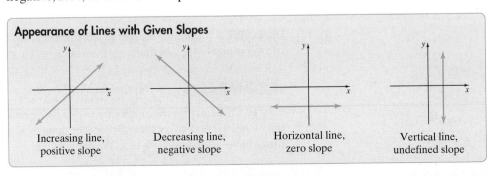

Appearance of Lines with Given Slopes

| Increasing line, positive slope | Decreasing line, negative slope | Horizontal line, zero slope | Vertical line, undefined slope |

The appearance of a line can give us further information about its slope.

The graphs of $y = \dfrac{1}{2}x + 1$ and $y = 5x + 1$ are shown to the right. Recall that the graph of $y = \dfrac{1}{2}x + 1$ has a slope of $\dfrac{1}{2}$ and that the graph of $y = 5x + 1$ has a slope of 5.

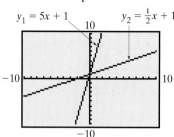

Notice that the line with the slope of 5 is steeper than the line with the slope of $\dfrac{1}{2}$. This is true in general for positive slopes.

> For a line with positive slope m, as m increases, the line becomes steeper.

To see why this is so, compare the slopes from above.

$\dfrac{1}{2}$ means a vertical change of 1 unit per horizontal change of 2 units.

5 or $\dfrac{10}{2}$ means a vertical change of 10 units per horizontal change of 2 units.

For larger positive slopes, the vertical change is greater for the same horizontal change. Thus, larger positive slopes mean steeper lines.

OBJECTIVE 5 ▶ Comparing slopes of parallel and perpendicular lines. Slopes of lines can help us determine whether lines are parallel or perpendicular. Recall that parallel lines are distinct lines with the same tilt that do not meet, and perpendicular lines are lines that intersect to form right angles.

TECHNOLOGY NOTE

Although the graphs of the equations shown below to the left are perpendicular lines, they do not appear to be so when graphed using a standard viewing window.

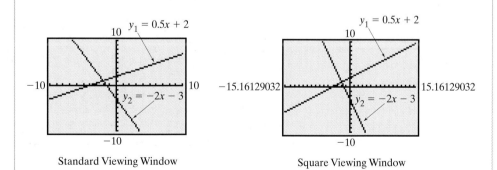

Standard Viewing Window   Square Viewing Window

This is because a graphing calculator screen is rectangular, so the tick marks along the x-axis are farther apart than the tick marks along the y-axis. The graphs of the same lines at the right do appear to be perpendicular because the viewing window has been selected so that there is equal spacing between tick marks on both axes. We say that the lines to the right are graphed using a **square viewing window.** Some graphing utilities can automatically provide a square setting. (This technology note shows the necessity of being able to algebraically determine whether lines are perpendicular. A graphing utility can then be used to partially verify the result.)

DISCOVER THE CONCEPT

Graph each pair of equations in a square window, such as an integer window. Do the lines appear to be parallel, perpendicular, or neither? Find the relationship between the slopes of the lines and whether they appear parallel, perpendicular, or neither.

a. $y = 3x - 5$ and $y = 3x + 7$ **b.** $y = 2x + 3$ and $y = 4x - 5$

c. $y = 0.5x - 9$ and $y = 2x - 3$ **d.** $y = -5x - 6$ and $y = -5x - 2$

e. $y = -\dfrac{1}{3}x - 7$ and $y = 3x + 5$

In the previous discovery, we see that parallel lines have the same tilt and therefore have the same slope.

Parallel Lines

Two nonvertical lines are parallel if they have the same slope and different y-intercepts.

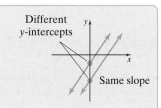

Different y-intercepts

Same slope

How do the slopes of perpendicular lines compare? (Two lines intersecting at right angles are called **perpendicular lines.**) Suppose that a line has a slope of $\dfrac{a}{b}$. If the line is rotated 90°, the rise and run are now switched, except that the run is now negative. This means that the new slope is $-\dfrac{b}{a}$. Notice that

$$\left(\frac{a}{b}\right) \cdot \left(-\frac{b}{a}\right) = -1$$

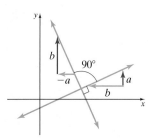

This is how we tell whether two lines are perpendicular.

Perpendicular Lines

Two nonvertical lines are perpendicular if the product of their slopes is -1.

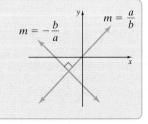

$m = -\dfrac{b}{a}$ $m = \dfrac{a}{b}$

In other words, two nonvertical lines are perpendicular if the slope of one is the negative reciprocal of the slope of the other.

EXAMPLE 7 Given the following pairs of equations, determine whether their graphs are parallel lines, perpendicular lines, or neither. Then use a graphing utility and a square window setting to visualize your results.

a. $3x + 7y = 21$ **b.** $-x + 3y = 2$ **c.** $2x - 3y = 12$

      $6x + 14y = 7$ $2x + 6y = 5$ $6x + 4y = 16$

Solution Find the slope of each line by solving each equation for y.

a. $3x + 7y = 21$

$7y = -3x + 21$

$\dfrac{7y}{7} = \dfrac{-3x}{7} + \dfrac{21}{7}$

$y = -\dfrac{3}{7}x + 3$

$6x + 14y = 7$

$14y = -6x + 7$

$\dfrac{14y}{14} = \dfrac{-6x}{14} + \dfrac{7}{14}$

$y = -\dfrac{3}{7}x + \dfrac{1}{2}$

The slope of both lines is $-\dfrac{3}{7}$. The y-intercept of one line is 3, whereas the y-intercept of the other line is $\dfrac{1}{2}$. Since these lines have the same slope and different y-intercepts, the lines are parallel.

b. Solve each equation for y.

$$-x + 3y = 2 \quad \text{and} \quad 2x + 6y = 5$$

$$y = \dfrac{1}{3}x + \dfrac{2}{3} \qquad y = -\dfrac{1}{3}x + \dfrac{5}{6}$$

The slopes $\dfrac{1}{3}$ and $-\dfrac{1}{3}$ are not equal, nor are they negative reciprocals of each other, so they are simply intersecting lines.

c. Solve each equation for y.

$$2x - 3y = 12 \quad \text{and} \quad 6x + 4y = 16$$

$$y = \dfrac{2}{3}x - 4 \qquad y = -\dfrac{3}{2}x + 4$$

The slopes of the lines are $\dfrac{2}{3}$ and $-\dfrac{3}{2}$, which are negative reciprocals of each other, so the lines are perpendicular.

To visualize the results of parts **a**, **b**, and **c**, see the graphs below using square window settings.

$$y_1 = -\dfrac{3}{7}x + 3 \qquad y_1 = \dfrac{1}{3}x + \dfrac{2}{3} \qquad y_1 = \dfrac{2}{3}x - 4$$

$$y_2 = -\dfrac{3}{7}x + \dfrac{1}{2} \qquad y_2 = -\dfrac{1}{3}x + \dfrac{5}{6} \qquad y_2 = -\dfrac{3}{2}x + 4$$

Parallel lines   Neither parallel nor perpendicular   Perpendicular lines

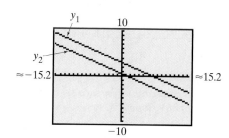

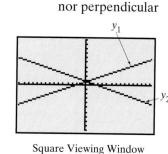

Square Viewing Window

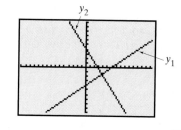

PRACTICE

7 Are the following pairs of lines parallel, perpendicular, or neither?

a. $x - 2y = 3$

$2x + y = 3$

b. $4x - 3y = 2$

$-8x + 6y = -6$

Answer to Concept Check:
y-intercepts are different

Concept Check ☑

What is different *about the equations of two parallel lines?*

VOCABULARY & READINESS CHECK

Use the choices below to fill in each blank. Some choices may be used more than once and some not at all.

horizontal	the same	-1	y-intercepts	$(0, b)$	slope
vertical	different	m	x-intercepts	$(b, 0)$	slope-intercept

1. The measure of the steepness or tilt of a line is called _____.
2. The slope of a line through two points is measured by the ratio of _____ change to _____ change.
3. If a linear equation is in the form $y = mx + b$, or $f(x) = mx + b$, the slope of the line is _____ and the y-intercept is _____.
4. The form $y = mx + b$ or $f(x) = mx + b$ is the _____ form.
5. The slope of a _____ line is 0.
6. The slope of a _____ line is undefined.
7. Two nonvertical perpendicular lines have slopes whose product is _____.
8. Two nonvertical lines are parallel if they have _____ slope and different _____.

Decide whether a line with the given slope slants upward or downward from left to right, or is horizontal or vertical.

9. $m = \dfrac{7}{6}$ 10. $m = -3$ 11. $m = 0$ 12. m is undefined

---

2.4 EXERCISE SET

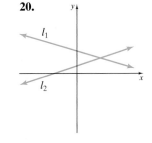

Find the slope of the line that goes through the given points. See Examples 1 and 2.

1. $(3, 2), (8, 11)$ 2. $(1, 6), (7, 11)$

3. $(-2, 8), (4, 3)$ 4. $(3, 7), (-2, 11)$

5. $(-2, -6), (4, -4)$ 6. $(-3, -4), (-1, 6)$

7. $(-2, 5), (3, 5)$ 8. $(4, 2), (4, 0)$

9. $(-1, 1), (-1, -5)$ 10. $(-2, -5), (3, -5)$

11. $(-1, 2), (-3, 4)$ 12. $(3, -2), (-1, -6)$

A linear function is graphed below. Using the trace feature, we found two points on the line. Give the slope of the line.

13.
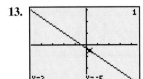
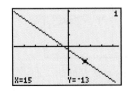

A linear function is graphed below. Using the trace feature we found two points on the line. Give the slope of the line.

14.
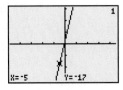
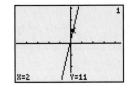

Examine the table of values for a linear function. Select two ordered pairs and find the slope of the line joining these points.

15.
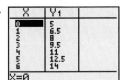

16.
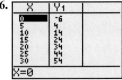

17.
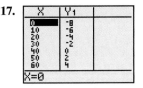

18.
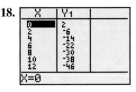

Two lines are graphed on each set of axes. Decide whether l_1 or l_2 has the greater slope.

19.
20.
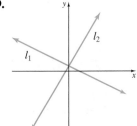

21.
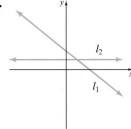

22.
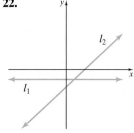

23.
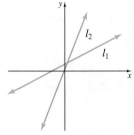

24.
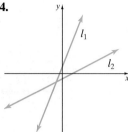

Find the slope and the y-intercept of each line. See Example 3.

25. $f(x) = 5x - 2$

26. $f(x) = -2x + 6$

27. $2x + y = 7$

28. $-5x + y = 10$

29. $2x - 3y = 10$

30. $-3x - 4y = 6$

31. $f(x) = \dfrac{1}{2}x$

32. $f(x) = -\dfrac{1}{4}x$

Match each graph with its equation.

A
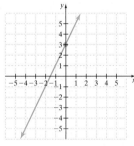

B
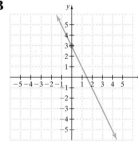

C
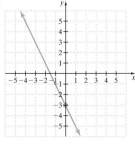

D
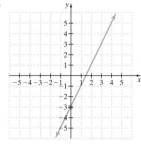

33. $f(x) = 2x + 3$

34. $f(x) = 2x - 3$

35. $f(x) = -2x + 3$

36. $f(x) = -2x - 3$

Find the slope of each line. See Examples 5 and 6.

37. $x = 1$

38. $y = -2$

39. $y = -3$

40. $x = 4$

41. $x + 2 = 0$

42. $y - 7 = 0$

43. Explain how merely looking at a line can tell us whether its slope is negative, positive, undefined, or zero.

44. Explain why the graph of $y = b$ is a horizontal line.

MIXED PRACTICE

Find the slope and the y-intercept of each line. See Examples 3 through 6.

45. $f(x) = -x + 5$

46. $f(x) = x + 2$

47. $-6x + 5y = 30$

48. $4x - 7y = 28$

49. $3x + 9 = y$

50. $2y - 7 = x$

51. $y = 4$

52. $x = 7$

53. $f(x) = 7x$

54. $f(x) = \dfrac{1}{7}x$

55. $6 + y = 0$

56. $x - 7 = 0$

57. $2 - x = 3$

58. $2y + 4 = -7$

Determine whether the lines are parallel, perpendicular, or neither. See Example 7.

59. $f(x) = -3x + 6$
  $g(x) = 3x + 5$

60. $f(x) = 5x - 6$
  $g(x) = 5x + 2$

61. $-4x + 2y = 5$
  $2x - y = 7$

62. $2x - y = -10$
  $2x + 4y = 2$

63. $-2x + 3y = 1$
  $3x + 2y = 12$

64. $x + 4y = 7$
  $2x - 5y = 0$

65. Explain whether two lines, both with positive slopes, can be perpendicular.

66. Explain why it is reasonable that nonvertical parallel lines have the same slope.

Determine the slope of each line.

67.
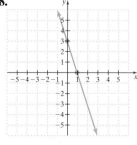

68.
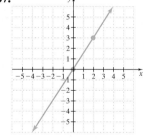

69.
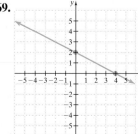

70.
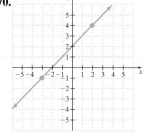

Find each slope. See Examples 1 and 2.

71. Find the pitch, or slope, of the roof shown.

72. Upon takeoff, a Delta Airlines jet climbs to 3 miles as it passes over 25 miles of land below it. Find the slope of its climb.

73. Driving down Bald Mountain in Wyoming, Bob Dean finds that he descends 1600 feet in elevation by the time he is 2.5 miles (horizontally) away from the high point on the mountain road. Find the slope of his descent rounded to two decimal places (1 mile = 5280 feet).

74. Find the grade, or slope, of the road shown.

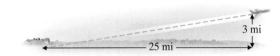

Solve. See Example 4.

75. The annual average income y of an American man over 25 with an associate's degree is approximated by the linear equation $y = 694.9x + 43,884.9$, where x is the number of years after 2000. (*Source:* Based on data from the U.S. Bureau of the Census)

  a. Predict the average income of an American man with an associate's degree in 2009.

  b. Find and interpret the slope of the equation.

  c. Find and interpret the y-intercept of the equation.

76. The annual average income of an American woman over 25 with a bachelor's degree is given by the linear equation $y = 1059.6x + 36,827.4$, where x is the number of years after 2000. (*Source:* Based on data from the U.S. Bureau of the Census)

  a. Find the average income of an American woman with a bachelor's degree in 2009.

  b. Find and interpret the slope of the equation.

  c. Find and interpret the y-intercept of the equation.

77. With wireless Internet (WiFi) gaining popularity, the number of public wireless Internet access points (in thousands) was projected to grow from 2003 to 2008 according to the equation

$$-66x + 2y = 84$$

where x is the number of years after 2003.

  a. Find the slope and y-intercept of the linear equation.

  b. What does the slope mean in this context?

  c. What does the y-intercept mean in this context?

78. One of the faster growing occupations over the next few years is expected to be nursing. The number of people y in thousands employed in nursing in the United States can be estimated by the linear equation $-266x + 10y = 27,409$, where x is the number of years after 2000. (*Source:* Based on data from American Nurses Association)

  a. Find the slope and y-intercept of the linear equation.

  b. What does the slope mean in this context?

  c. What does the y-intercept mean in this context?

79. In an earlier section, it was given that the yearly cost of tuition and required fees for attending a public four-year college full time can be estimated by the linear function

$$f(x) = 291.5x + 2944.05$$

where x is the number of years after 2000 and $f(x)$ is the total cost. (*Source:* U.S. National Center for Education Statistics)

  a. Find and interpret the slope of this equation.

  b. Find and interpret the y-intercept of this equation.

80. In an earlier section, it was given that the yearly cost of tuition and required fees for attending a public two-year college full time can be estimated by the linear function

$$f(x) = 107.3x + 1245.62$$

where x is the number of years after 2000 and $f(x)$ is the total cost. (*Source:* U.S. National Center for Education Statistics)

a. Find and interpret the slope of this equation.

b. Find and interpret the y-intercept of this equation.

REVIEW AND PREVIEW

Simplify and solve for y. See Section 2.1.

81. $y - 2 = 5(x + 6)$

82. $y - 0 = -3[x - (-10)]$

83. $y - (-1) = 2(x - 0)$

84. $y - 9 = -8[x - (-4)]$

CONCEPT EXTENSIONS

Each slope calculation is incorrect. Find the error and correct the calculation. See the second Concept Check in this section.

85. $(-2, 6)$ and $(7, -14)$

$$m = \frac{-14 - 6}{7 - 2} = \frac{-20}{5} = -4$$

86. $(-1, 4)$ and $(-3, 9)$

$$m = \frac{9 - 4}{-3 - 1} = \frac{5}{-4} \text{ or } -\frac{5}{4}$$

87. $(-8, -10)$ and $(-11, -5)$

$$m = \frac{-10 - 5}{-8 - 11} = \frac{-15}{-19} = \frac{15}{19}$$

88. $(0, -4)$ and $(-6, -6)$

$$m = \frac{0 - (-6)}{-4 - (-6)} = \frac{6}{2} = 3$$

89. Find the slope of a line parallel to the line $f(x) = -\dfrac{7}{2}x - 6$.

△**90.** Find the slope of a line parallel to the line $f(x) = x$.

△**91.** Find the slope of a line perpendicular to the line

$$f(x) = -\frac{7}{2}x - 6.$$

△**92.** Find the slope of a line perpendicular to the line $f(x) = x$.

△**93.** Find the slope of a line parallel to the line $5x - 2y = 6$.

△**94.** Find the slope of a line parallel to the line $-3x + 4y = 10$.

Determine whether the graphs of y_1 and y_2 are parallel, perpendicular, or neither.

95.

96.

97.

98.

99. The following graph shows the altitude of a seagull in flight over a time period of 30 seconds.

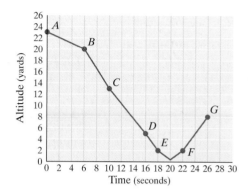

a. Find the coordinates of point B.

b. Find the coordinates of point C.

c. Find the rate of change of altitude between points B and C. (Recall that the rate of change between points is the slope between points. This rate of change will be in yards per second.)

d. Find the rate of change of altitude (in yards per second) between points F and G.

100. Each line below has negative slope.

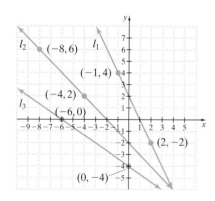

a. Find the slope of each line.

b. Use the result of part **a** to fill in the blank. For lines with negative slopes, the steeper line has the _____ (greater/lesser) slope.

2.5 EQUATIONS OF LINES

OBJECTIVE 1 ▶ Using slope-intercept form to write equations of lines. In the last section, we learned that the slope-intercept form of a linear equation is $y = mx + b$. When a linear equation is written in this form, the slope of the line is the same as the coefficient m of x. Also, the y-intercept of the line is $(0, b)$. For example, the slope of the line defined by $y = 2x + 3$ is 2, and its y-intercept is $(0, 3)$.

We may also use the slope-intercept form to write the equation of a line given its slope and y-intercept. The equation of a line is a linear equation in 2 variables that, if graphed, would produce the line described.

EXAMPLE 1 Write an equation of the line with y-intercept $(0, -3)$ and slope of $\frac{1}{4}$.

Solution We want to write a linear equation in 2 variables that describes the line with y-intercept $(0, -3)$ and has a slope of $\frac{1}{4}$. We are given the slope and the y-intercept. Let $m = \frac{1}{4}$ and $b = -3$, and write the equation in slope-intercept form, $y = mx + b$.

$$y = mx + b$$
$$y = \frac{1}{4}x + (-3) \quad \text{Let } m = \frac{1}{4} \text{ and } b = -3.$$
$$y = \frac{1}{4}x - 3 \quad \text{Simplify.}$$

PRACTICE
1 Write an equation of the line with y-intercept $(0, 4)$ and slope of $-\frac{3}{4}$.

Concept Check ✓

What is wrong with the following equation of a line with y-intercept $(0, 4)$ and slope 2?

$$y = 4x + 2$$

OBJECTIVE 2 ▶ Graph a line using slope and y-intercept. Given the slope and y-intercept of a line, we may graph the line as well as write its equation. Let's graph the line from Example 1.

EXAMPLE 2 Graph $y = \frac{1}{4}x - 3$.

Solution Recall that the slope of the graph of $y = \frac{1}{4}x - 3$ is $\frac{1}{4}$ and the y-intercept is $(0, -3)$. To graph the line, we first plot the y-intercept $(0, -3)$. To find another point on the line, we recall that slope is $\frac{\text{rise}}{\text{run}} = \frac{1}{4}$. Another point may then be plotted by starting at $(0, -3)$, rising 1 unit up, and then running 4 units to the right. We are now at the point $(4, -2)$. The graph is the line through these two points.

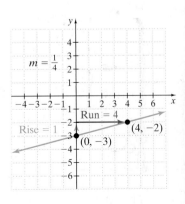

Notice that the line does have a y-intercept of $(0, -3)$ and a slope of $\frac{1}{4}$.   □

PRACTICE
2   Graph $y = \frac{3}{4}x + 2$.

EXAMPLE 3   Graph $2x + 3y = 12$.

Solution   First, we solve the equation for y to write it in slope-intercept form. In slope-intercept form, the equation is $y = -\frac{2}{3}x + 4$. Next we plot the y-intercept $(0, 4)$. To find another point on the line, we use the slope $-\frac{2}{3}$, which can be written as $\frac{\text{rise}}{\text{run}} = \frac{-2}{3}$. We start at $(0, 4)$ and move down 2 units since the numerator of the slope is -2; then we move 3 units to the right since the denominator of the slope is 3. We arrive at the point $(3, 2)$. The line through these points is the graph, shown below to the left.

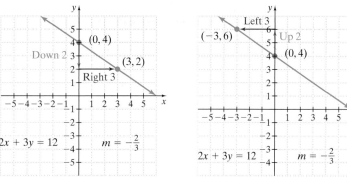

The slope $\frac{-2}{3}$ can also be written as $\frac{2}{-3}$, so to find another point in Example 3 we could start at $(0, 4)$ and move up 2 units and then 3 units to the left. We would arrive at the point $(-3, 6)$. The line through $(-3, 6)$ and $(0, 4)$ is the same line as shown previously through $(3, 2)$ and $(0, 4)$. See the graph above to the right.   □

PRACTICE
3   Graph $x + 2y = 6$.

OBJECTIVE 3 ▶ Using point-slope form to write equations of lines. When the slope of a line and a point on the line are known, the equation of the line can also be found. To do this, use the slope formula to write the slope of a line that passes through points (x_1, y_1) and (x, y). We have

$$m = \frac{y - y_1}{x - x_1}$$

Multiply both sides of this equation by $x - x_1$ to obtain

$$y - y_1 = m(x - x_1)$$

This form is called the **point-slope form** of the equation of a line.

Point-Slope Form of the Equation of a Line
The **point-slope form** of the equation of a line is

$$\underset{\underset{\text{point}}{\underbrace{\qquad}}}{y - y_1 = \overset{\overset{\text{slope}}{\downarrow}}{m}(x - x_1)}$$

where m is the slope of the line and (x_1, y_1) is a point on the line.

EXAMPLE 4 Find an equation of the line with slope -3 containing the point $(1, -5)$. Write the equation in slope-intercept form, $y = mx + b$.

Solution Because we know the slope and a point of the line, we use the point-slope form with $m = -3$ and $(x_1, y_1) = (1, -5)$.

$$y - y_1 = m(x - x_1) \quad \text{Point-slope form}$$

$$y - (-5) = -3(x - 1) \quad \text{Let } m = -3 \text{ and } (x_1, y_1) = (1, -5).$$

$$y + 5 = -3x + 3 \quad \text{Apply the distributive property.}$$

$$y = -3x - 2 \quad \text{Write in slope-intercept form.}$$

In slope-intercept form, the equation is $y = -3x - 2$.

To check, define $y_1 = -3x - 2$ and graph in an integer window. From the equation we see that $m = -3$, so the slope is -3. Next, check on the graph to see that it contains the point $(1, -5)$ as in the screen below to the left.

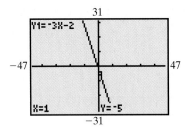
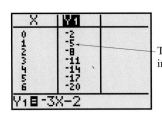

The point $(1, -5)$ is included in the solution pairs.

A second way to see that $(1, -5)$ is an ordered pair solution is to examine a table of values as shown above to the right. □

PRACTICE
4 Find an equation of the line with slope -4 containing the point $(-2, 5)$. Write the equation in slope-intercept form, $y = mx + b$.

> ▶ **Helpful Hint**
> Remember, "slope-intercept form" means the equation is "solved for y."

EXAMPLE 5 Find an equation of the line through points $(4, 0)$ and $(-4, -5)$. Write the equation using function notation. Then use a graphing utility to check.

Solution First, find the slope of the line.

$$m = \frac{-5 - 0}{-4 - 4} = \frac{-5}{-8} = \frac{5}{8}$$

Next, make use of the point-slope form. Replace (x_1, y_1) by either $(4, 0)$ or $(-4, -5)$ in the point-slope equation. We will choose the point $(4, 0)$. The line through $(4, 0)$ with slope $\frac{5}{8}$ is

$$y - y_1 = m(x - x_1) \quad \text{Point-slope form}$$

$$y - 0 = \frac{5}{8}(x - 4) \quad \text{Let } m = \frac{5}{8} \text{ and } (x_1, y_1) = (4, 0).$$

$$8y = 5(x - 4) \quad \text{Multiply both sides by 8.}$$

$$8y = 5x - 20 \quad \text{Apply the distributive property.}$$

To write the equation using function notation, we solve for y, then replace y with $f(x)$.

$$8y = 5x - 20$$

$$y = \frac{5}{8}x - \frac{20}{8} \quad \text{Divide both sides by 8.}$$

$$f(x) = \frac{5}{8}x - \frac{5}{2} \quad \text{Write using function notation.}$$

To check, graph $y_1 = \dfrac{5}{8}x - \dfrac{5}{2}$ in an integer window. Trace to find $(4, 0)$ and $(-4, -5)$ on the graph of the line. We can also examine a table of ordered pair solutions to check.

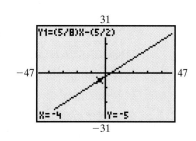
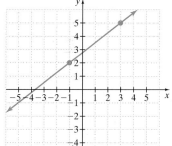

PRACTICE

5   Find an equation of the line through points $(-1, 2)$ and $(2, 0)$. Write the equation using function notation.

> **Helpful Hint**
>
> If two points of a line are given, either one may be used with the point-slope form to write an equation of the line.

EXAMPLE 6   Find an equation of the line graphed. Write the equation in standard form.

Solution   First, find the slope of the line by identifying the coordinates of the noted points on the graph.

The points have coordinates $(-1, 2)$ and $(3, 5)$.

$$m = \frac{5 - 2}{3 - (-1)} = \frac{3}{4}$$

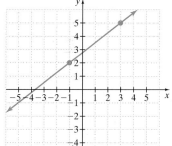

Next, use the point-slope form. We will choose $(3, 5)$ for (x_1, y_1), although it makes no difference which point we choose. The line through $(3, 5)$ with slope $\dfrac{3}{4}$ is

$$y - y_1 = m(x - x_1) \quad \text{Point-slope form}$$

$$y - 5 = \frac{3}{4}(x - 3) \quad \text{Let } m = \frac{3}{4} \text{ and } (x_1, y_1) = (3, 5).$$

$$4(y - 5) = 3(x - 3) \quad \text{Multiply both sides by 4.}$$

$$4y - 20 = 3x - 9 \quad \text{Apply the distributive property.}$$

To write the equation in standard form, move x- and y-terms to one side of the equation and any numbers (constants) to the other side.

$$4y - 20 = 3x - 9$$

$$-3x + 4y = 11 \quad \text{Subtract } 3x \text{ from both sides and add 20 to both sides.}$$

The equation of the graphed line is $-3x + 4y = 11$.

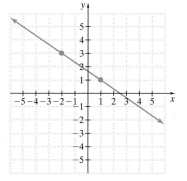

PRACTICE

6   Find an equation of the line graphed. Write the equation in standard form.

The point-slope form of an equation is very useful for solving real-world problems.

EXAMPLE 7 **Predicting Sales**

Southern Star Realty is an established real estate company that has enjoyed constant growth in sales since 2000. In 2002 the company sold 200 houses, and in 2007 the company sold 275 houses. Use these figures to predict the number of houses this company will sell in the year 2016.

Solution

1. UNDERSTAND. Read and reread the problem. Then let

x = the number of years after 2000 and

y = the number of houses sold in the year corresponding to x.

The information provided then gives the ordered pairs $(2, 200)$ and $(7, 275)$. To better visualize the sales of Southern Star Realty, we graph the linear equation that passes through the points $(2, 200)$ and $(7, 275)$.

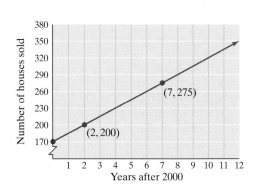

2. TRANSLATE. We write a linear equation that passes through the points $(2, 200)$ and $(7, 275)$. To do so, we first find the slope of the line.

$$m = \frac{275 - 200}{7 - 2} = \frac{75}{5} = 15$$

Then, using the point-slope form and the point $(2, 200)$ to write the equation, we have

$$y - y_1 = m(x - x_1)$$
$$y - 200 = 15(x - 2) \quad \text{Let } m = 15 \text{ and } (x_1, y_1) = (2, 200).$$
$$y - 200 = 15x - 30 \quad \text{Multiply.}$$
$$y = 15x + 170 \quad \text{Add 200 to both sides.}$$

3. SOLVE. To predict the number of houses sold in the year 2016, we use $y = 15x + 170$ and complete the ordered pair $(16, \quad)$, since $2016 - 2000 = 16$.

$$y = 15(16) + 170 \quad \text{Let } x = 16.$$
$$y = 410$$

4. INTERPRET.

Check: Verify that the point $(16, 410)$ is a point on the line graphed in step 1.

State: Southern Star Realty should expect to sell 410 houses in the year 2016. □

PRACTICE
7 Southwest Florida, including Fort Myers and Cape Coral, has been a growing real estate market in past years. In 2002, there were 7513 house sales in the area, and in 2006, there were 9198 house sales. Use these figures to predict the number of house sales there will be in 2014.

OBJECTIVE 4 ▶ Writing equations of vertical and horizontal lines. A few special types of linear equations are linear equations whose graphs are vertical and horizontal lines.

EXAMPLE 8 Find an equation of the horizontal line containing the point $(2, 3)$.

Solution Recall that a horizontal line has an equation of the form $y = b$. Since the line contains the point $(2, 3)$, the equation is $y = 3$. Below is a graph of $y = 3$ with the point $(2, 3)$ indicated.

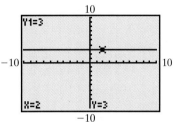

PRACTICE
8 Find the equation of the horizontal line containing the point $(6, -2)$.

EXAMPLE 9 Find an equation of the line containing the point $(2, 3)$ with undefined slope.

Solution Since the line has undefined slope, the line must be vertical. A vertical line has an equation of the form $x = c$. Since the line contains the point $(2, 3)$, the equation is $x = 2$, as shown to the right.

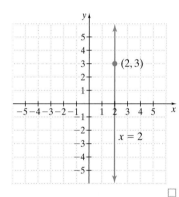

TECHNOLOGY NOTE

Recall that a vertical line does not represent the graph of a function and cannot be graphed in the Y= editor in function mode. Vertical lines may be drawn using the draw menu. Below is an example of using a draw vertical feature.

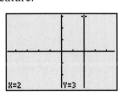

PRACTICE
9 Find an equation of the line containing the point $(6, -2)$ with undefined slope.

OBJECTIVE 5 ▶ Finding equations of parallel and perpendicular lines. Next, we find equations of parallel and perpendicular lines.

△ **EXAMPLE 10** Find an equation of the line containing the point $(4, 4)$ and parallel to the line $2x + 3y = -6$. Write the equation in standard form.

Solution Because the line we want to find is *parallel* to the line $2x + 3y = -6$, the two lines must have equal slopes. Find the slope of $2x + 3y = -6$ by writing it in the form $y = mx + b$. In other words, solve the equation for y.

$$2x + 3y = -6$$

$$3y = -2x - 6 \quad \text{Subtract } 2x \text{ from both sides.}$$

$$y = \frac{-2x}{3} - \frac{6}{3} \quad \text{Divide by 3.}$$

$$y = -\frac{2}{3}x - 2 \quad \text{Write in slope-intercept form.}$$

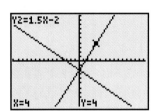

The graphs of parallel lines $2x + 3y = -6$ and $2x + 3y = 20$.

> **Helpful Hint**
> Multiply both sides of the equation $2x + 3y = 20$ by -1 and it becomes $-2x - 3y = -20$. Both equations are in standard form, and their graphs are the same line.

The slope of this line is $-\frac{2}{3}$. Thus, a line parallel to this line will also have a slope of $-\frac{2}{3}$. The equation we are asked to find describes a line containing the point $(4, 4)$ with a slope of $-\frac{2}{3}$. We use the point-slope form.

$$y - y_1 = m(x - x_1)$$
$$y - 4 = -\frac{2}{3}(x - 4) \quad \text{Let } m = -\frac{2}{3}, x_1 = 4, \text{ and } y_1 = 4.$$
$$3(y - 4) = -2(x - 4) \quad \text{Multiply both sides by 3.}$$
$$3y - 12 = -2x + 8 \quad \text{Apply the distributive property.}$$
$$2x + 3y = 20 \quad \text{Write in standard form.} \qquad \square$$

PRACTICE
10 Find an equation of the line containing the point $(8, -3)$ and parallel to the line $3x + 4y = 1$. Write the equation in standard form.

EXAMPLE 11 Write a function that describes the line containing the point $(4, 4)$ and is perpendicular to the line $2x + 3y = -6$.

Solution In the previous example, we found that the slope of the line $2x + 3y = -6$ is $-\frac{2}{3}$. A line perpendicular to this line will have a slope that is the negative reciprocal of $-\frac{2}{3}$, or $\frac{3}{2}$. From the point-slope equation, we have

$$y - y_1 = m(x - x_1)$$
$$y - 4 = \frac{3}{2}(x - 4) \quad \text{Let } x_1 = 4, y_1 = 4, \text{ and } m = \frac{3}{2}.$$
$$2(y - 4) = 3(x - 4) \quad \text{Multiply both sides by 2.}$$
$$2y - 8 = 3x - 12 \quad \text{Apply the distributive property.}$$
$$2y = 3x - 4 \quad \text{Add 8 to both sides.}$$
$$y = \frac{3}{2}x - 2 \quad \text{Divide both sides by 2.}$$
$$f(x) = \frac{3}{2}x - 2 \quad \text{Write using function notation.} \qquad \square$$

The graphs of perpendicular lines $2x + 3y = -6$ and $f(x) = \frac{3}{2}x - 2$.

PRACTICE
11 Write a function that describes the line containing the point $(8, -3)$ and is perpendicular to the line $3x + 4y = 1$.

Forms of Linear Equations

$Ax + By = C$	**Standard form** of a linear equation A and B are not both 0.
$y = mx + b$	**Slope-intercept form** of a linear equation The slope is m, and the y-intercept is $(0, b)$.
$y - y_1 = m(x - x_1)$	**Point-slope form** of a linear equation The slope is m, and (x_1, y_1) is a point on the line.
$y = c$	**Horizontal line** The slope is 0, and the y-intercept is $(0, c)$.
$x = c$	**Vertical line** The slope is undefined, and the x-intercept is $(c, 0)$.

Parallel and Perpendicular Lines
Nonvertical parallel lines have the same slope. The product of the slopes of two nonvertical perpendicular lines is -1.

VOCABULARY & READINESS CHECK

State the slope and the y-intercept of each line with the given equation.

1. $y = -4x + 12$

2. $y = \dfrac{2}{3}x - \dfrac{7}{2}$

3. $y = 5x$

4. $y = -x$

5. $y = \dfrac{1}{2}x + 6$

6. $y = -\dfrac{2}{3}x + 5$

Decide whether the lines are parallel, perpendicular, or neither.

7. $y = 12x + 6$
$y = 12x - 2$

8. $y = -5x + 8$
$y = -5x - 8$

9. $y = -9x + 3$
$y = \dfrac{3}{2}x - 7$

10. $y = 2x - 12$
$y = \dfrac{1}{2}x - 6$

2.5 EXERCISE SET

Use the slope-intercept form of the linear equation to write the equation of each line with the given slope and y-intercept. See Example 1.

1. Slope -1; y-intercept $(0, 1)$

2. Slope $\dfrac{1}{2}$; y-intercept $(0, -6)$

3. Slope 2; y-intercept $\left(0, \dfrac{3}{4}\right)$

4. Slope -3; y-intercept $\left(0, -\dfrac{1}{5}\right)$

5. Slope $\dfrac{2}{7}$; y-intercept $(0, 0)$

6. Slope $-\dfrac{4}{5}$; y-intercept $(0, 0)$

Graph each linear equation. See Examples 2 and 3.

7. $y = 5x - 2$

8. $y = 2x + 1$

9. $4x + y = 7$

10. $3x + y = 9$

11. $-3x + 2y = 3$

12. $-2x + 5y = -16$

Find an equation of the line with the given slope and containing the given point. Write the equation in slope-intercept form. See Example 4.

13. Slope 3; through $(1, 2)$

14. Slope 4; through $(5, 1)$

15. Slope -2; through $(1, -3)$

16. Slope -4; through $(2, -4)$

17. Slope $\dfrac{1}{2}$; through $(-6, 2)$

18. Slope $\dfrac{2}{3}$; through $(-9, 4)$

19. Slope $-\dfrac{9}{10}$; through $(-3, 0)$

20. Slope $-\dfrac{1}{5}$; through $(4, -6)$

Find an equation of the line passing through the given points. Use function notation to write the equation. See Example 5.

21. $(2, 0), (4, 6)$

22. $(3, 0), (7, 8)$

23. $(-2, 5), (-6, 13)$

24. $(7, -4), (2, 6)$

25. $(-2, -4), (-4, -3)$

26. $(-9, -2), (-3, 10)$

27. $(-3, -8), (-6, -9)$

28. $(8, -3), (4, -8)$

29. $\left(\dfrac{3}{5}, \dfrac{4}{10}\right)$ and $\left(-\dfrac{1}{5}, \dfrac{7}{10}\right)$

30. $\left(\dfrac{1}{2}, -\dfrac{1}{4}\right)$ and $\left(\dfrac{3}{2}, \dfrac{3}{4}\right)$

Find an equation of each line graphed. Write the equation in standard form. See Example 6.

31.

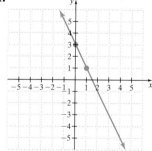

32.

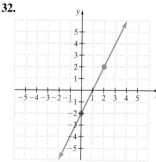

33.

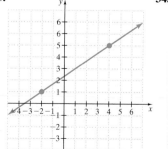

34.

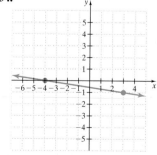

Use the graph of the following function f(x) to find each value.

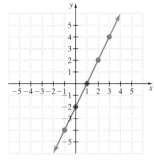

35. $f(0)$ **36.** $f(-1)$

37. $f(2)$ **38.** $f(1)$

39. Find x such that $f(x) = -6$.

40. Find x such that $f(x) = 4$.

Write an equation of each line. See Examples 8 and 9.

41. Slope 0; through $(-2, -4)$

42. Horizontal; through $(-3, 1)$

43. Vertical; through $(4, 7)$

44. Vertical; through $(2, 6)$

45. Horizontal; through $(0, 5)$

46. Undefined slope; through $(0, 5)$

Find an equation of each line. Write the equation using function notation. See Examples 10 and 11.

47. Through $(3, 8)$; parallel to $f(x) = 4x - 2$

48. Through $(1, 5)$; parallel to $f(x) = 3x - 4$

49. Through $(2, -5)$; perpendicular to $3y = x - 6$

50. Through $(-4, 8)$; perpendicular to $2x - 3y = 1$

51. Through $(-2, -3)$; parallel to $3x + 2y = 5$

52. Through $(-2, -3)$; perpendicular to $3x + 2y = 5$

MIXED PRACTICE

Find the equation of each line. Write the equation in standard form unless indicated otherwise. See Examples 1, 4, 5 and 8 through 11.

53. Slope 2; through $(-2, 3)$

54. Slope 3; through $(-4, 2)$

55. Through $(1, 6)$ and $(5, 2)$; use function notation.

56. Through $(2, 9)$ and $(8, 6)$

57. With slope $-\frac{1}{2}$; y-intercept 11

58. With slope -4; y-intercept $\frac{2}{9}$; use function notation.

59. Through $(-7, -4)$ and $(0, -6)$

60. Through $(2, -8)$ and $(-4, -3)$

61. Slope $-\frac{4}{3}$; through $(-5, 0)$

62. Slope $-\frac{3}{5}$; through $(4, -1)$

63. Vertical line; through $(-2, -10)$

64. Horizontal line; through $(1, 0)$

65. Through $(6, -2)$; parallel to the line $2x + 4y = 9$

66. Through $(8, -3)$; parallel to the line $6x + 2y = 5$

67. Slope 0; through $(-9, 12)$

68. Undefined slope; through $(10, -8)$

69. Through $(6, 1)$; parallel to the line $8x - y = 9$

70. Through $(3, 5)$; perpendicular to the line $2x - y = 8$

71. Through $(5, -6)$; perpendicular to $y = 9$

72. Through $(-3, -5)$; parallel to $y = 9$

73. Through $(2, -8)$ and $(-6, -5)$; use function notation.

74. Through $(-4, -2)$ and $(-6, 5)$; use function notation.

Solve. See Example 7.

75. Del Monte Fruit Company recently released a new applesauce. By the end of its first year, profits on this product amounted to $30,000. The anticipated profit for the end of the fourth year is $66,000. The ratio of change in time to change in profit is constant. Let x be years and P be profit.

   a. Write a linear function $P(x)$ that expresses profit as a function of time.

   b. Use this function to predict the company's profit at the end of the seventh year.

   c. Predict when the profit should reach $126,000.

76. The value of a computer bought in 2003 depreciates, or decreases, as time passes. Two years after the computer was bought, it was worth $2000; 4 years after it was bought, it was worth $800.

   a. If this relationship between number of years past 2003 and value of computer is linear, write an equation describing this relationship. [Use ordered pairs of the form (years past 2003, value of computer).]

   b. Use this equation to estimate the value of the computer in the year 2008.

77. The Pool Fun Company has learned that, by pricing a newly released Fun Noodle at $3, sales will reach 10,000 Fun Noodles per day during the summer. Raising the price to $5 will cause the sales to fall to 8000 Fun Noodles per day.

   a. Assume that the relationship between sales price and number of Fun Noodles sold is linear and write an equation describing this relationship.

   b. Predict the daily sales of Fun Noodles if the price is $3.50.

78. The value of a building bought in 1990 appreciates, or increases, as time passes. Seven years after the building was bought, it was worth $165,000; 12 years after it was bought, it was worth $180,000.

   a. If this relationship between number of years past 1990 and value of building is linear, write an equation describing this relationship. [Use ordered pairs of the form (years past 1990, value of building).]

   b. Use this equation to estimate the value of the building in the year 2010.

79. In 2006, the median price of an existing home in the United States was approximately $222,000. In 2001, the median price of an existing home was $150,900. Let y be the median price of an existing home in the year x, where $x = 0$ represents 2001. (*Source:* National Association of REALTORS®)

   a. Write a linear equation that models the median existing home price in terms of the year x. [*Hint:* The line must pass through the points $(0, 150,900)$ and $(5, 222,000)$.]

b. Use this equation to predict the median existing home price for the year 2010.

c. Interpret the slope of the equation found in part **a**.

80. The number of births (in thousands) in the United States in 2000 was 4060. The number of births (in thousands) in the United States in 2004 was 4116. Let y be the number of births (in thousands) in the year x, where $x = 0$ represents 2000. (*Source:* National Center for Health Statistics)

a. Write a linear equation that models the number of births (in thousands) in terms of the year x. (See hint for Exercise 79a.)

b. Use this equation to predict the number of births in the United States for the year 2013.

c. Interpret the slope of the equation in part **a**.

81. The number of people employed in the United States as medical assistants was 387 thousand in 2004. By the year 2014, this number is expected to rise to 589 thousand. Let y be the number of medical assistants (in thousands) employed in the United States in the year x, where $x = 0$ represents 2004. (*Source:* Bureau of Labor Statistics)

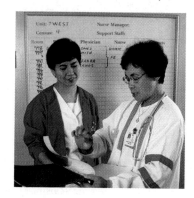

a. Write a linear equation that models the number of people (in thousands) employed as medical assistants in the year x. (See hint for Exercise 79a.)

b. Use this equation to estimate the number of people who will be employed as medical assistants in the year 2013.

82. The number of people employed in the United States as systems analysts was 487 thousand in 2004. By the year 2014, this number is expected to rise to 640 thousand. Let y be the number of systems analysts (in thousands) employed in the United States in the year x, where $x = 0$ represents 2004. (*Source:* Bureau of Labor Statistics)

a. Write a linear equation that models the number of people (in thousands) employed as systems analysts in the year x. (See hint for Exercise 79a.)

b. Use this equation to estimate the number of people who will be employed as systems analysts in the year 2012.

REVIEW AND PREVIEW

Solve. See Section 1.5.

83. $2x - 7 = 21$

84. $-3x + 1 = 0$

85. $5(x - 2) = 3(x - 1)$

86. $-2(x + 1) = -x + 10$

87. $\dfrac{x}{2} + \dfrac{1}{4} = \dfrac{1}{8}$

88. $\dfrac{x}{5} - \dfrac{3}{10} = \dfrac{x}{2} - 1$

CONCEPT EXTENSIONS

Answer true or false.

89. A vertical line is always perpendicular to a horizontal line.

90. A vertical line is always parallel to a vertical line.

Example:

Find an equation of the perpendicular bisector of the line segment whose endpoints are (2, 6) and (0, −2).

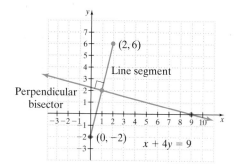

Solution:

A perpendicular bisector is a line that contains the midpoint of the given segment and is perpendicular to the segment.

Step 1: The midpoint of the segment with end points $(2, 6)$ and $(0, -2)$ is $(1, 2)$.

Step 2: The slope of the segment containing points $(2, 6)$ and $(0, -2)$ is 4.

Step 3: A line perpendicular to this line segment will have slope of $-\dfrac{1}{4}$.

Step 4: The equation of the line through the midpoint $(1, 2)$ with a slope of $-\dfrac{1}{4}$ will be the equation of the perpendicular bisector. This equation in standard form is $x + 4y = 9$.

Find an equation of the perpendicular bisector of the line segment whose end points are given. See the previous example.

△ **91.** $(3, -1); (-5, 1)$

△ **92.** $(-6, -3); (-8, -1)$

△ **93.** $(-2, 6); (-22, -4)$

△ **94.** $(5, 8); (7, 2)$

△ **95.** $(2, 3); (-4, 7)$

△ **96.** $(-6, 8); (-4, -2)$

✎ **97.** Describe how to check to see if the graph of $2x - 4y = 7$ passes through the points $(1.4, -1.05)$ and $(0, -1.75)$. Then follow your directions and check these points.

INTEGRATED REVIEW LINEAR EQUATIONS IN TWO VARIABLES

Sections 2.1–2.5

Below is a review of equations of lines.

Forms of Linear Equations

$Ax + By = C$	**Standard form** of a linear equation A and B are not both 0.
$y = mx + b$	**Slope-intercept form** of a linear equation. The slope is m, and the y-intercept is $(0, b)$.
$y - y_1 = m(x - x_1)$	**Point-slope form** of a linear equation. The slope is m, and (x_1, y_1) is a point on the line.
$y = c$	**Horizontal line** The slope is 0, and the y-intercept is $(0, c)$.
$x = c$	**Vertical line** The slope is undefined and the x-intercept is $(c, 0)$.

Parallel and Perpendicular Lines

Nonvertical parallel lines have the same slope. The product of the slopes of two nonvertical perpendicular lines is -1.

Graph each linear equation.

1. $y = -2x$ **2.** $3x - 2y = 6$ **3.** $x = -3$ **4.** $y = 1.5$

Find the slope of the line containing each pair of points.

5. $(-2, -5), (3, -5)$ **6.** $(5, 2), (0, 5)$

Find the slope and y-intercept of each line.

7. $y = 3x - 5$ **8.** $5x - 2y = 7$

Determine whether each pair of lines is parallel, perpendicular, or neither.

9. $y = 8x - 6$ **10.** $y = \dfrac{2}{3}x + 1$
    $y = 8x + 6$ $2y + 3x = 1$

Find the equation of each line. Write the equation in the form $x = a$, $y = b$, or $y = mx + b$. For Exercises 14 through 17, write the equation in the form $f(x) = mx + b$.

11. Through $(1, 6)$ and $(5, 2)$

12. Vertical line; through $(-2, -10)$

13. Horizontal line; through $(1, 0)$

14. Through $(2, -9)$ and $(-6, -5)$

15. Through $(-2, 4)$ with slope -5

16. Slope -4; y-intercept $\left(0, \dfrac{1}{3}\right)$

17. Slope $\dfrac{1}{2}$; y-intercept $(0, -1)$

18. Through $\left(\dfrac{1}{2}, 0\right)$ with slope 3

19. Through $(-1, -5)$; parallel to $3x - y = 5$

20. Through $(0, 4)$; perpendicular to $4x - 5y = 10$

21. Through $(2, -3)$; perpendicular to $4x + y = \dfrac{2}{3}$

22. Through $(-1, 0)$; parallel to $5x + 2y = 2$

23. Undefined slope; through $(-1, 3)$

24. $m = 0$; through $(-1, 3)$

2.6 INTERPRETING DATA: LINEAR MODELS

OBJECTIVE

1 Use a graphing utility to find a linear equation that models data.

The graph below shows the number of passengers who rode public transportation such as a train, bus, or subway during the years shown.

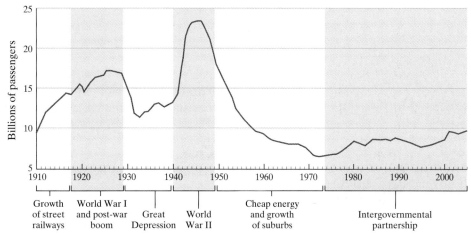

Source: American Public Transit Association

We see that different periods of history affect the number of passengers. For this complex graph there is not one model that best describes the number of passengers. Instead, we can take different periods of time and model each period to study it more closely. Once we get an equation (function) that best models a period of time, we can further study trends by evaluating the function within that period of time, or domain. It is important to remember that if a model is used to predict results outside its domain, this prediction is accurate only if the trend continues in the same manner. For example, if we had modeled the data from 1910 to 1925 and used it to predict the number of passengers for the next decade, we would have expected the number to continue to increase rapidly. Instead, the Great Depression occurred and our prediction would have been very inaccurate.

The table below shows different periods of time for our graph above and possible types of models.

Period of Time	*Approximate Shape of Graph*	*Type of Model (Equation)*
1910–1925	Line	Linear
1929–1945	Upward parabola	Quadratic
1940–1950	Downward parabola	Quadratic
1995–2005	Line	Linear

OBJECTIVE 1 ▶ Using graphing utilities to find linear equations. Regression analysis is the process of fitting a line or a curve to a set of data points. The equation of this line or curve can then be used to further study the graph itself or to predict something in the future. Your calculator has many regression equations to choose from, but in this section we will concentrate on **linear regression.** In other words, we will study data that resemble a line and model it with a linear equation that best fits the data. (In Section 8.7 we will study another regression equation.)

Below is public transit data from the years 1995–2005.

x-List Year (1995 = 0)	y-List Billions of Riders
0	7.8
2	8.4
4	9.2
6	9.7
8	9.4
10	9.8

First we will draw a scatter plot of the ordered pairs and examine it. A scatter plot is a quick way to visually see if there is a relationship, perhaps linear, between the two sets of data.

Plotting the Data
To plot the data, see the steps below.

> ▶ Helpful Hint
>
> It is a good idea to first clear or deselect any equations in the Y= editor so that only the plotted data points are graphed.

1. Enter the data. To do so, use the EDIT menu found in the STAT feature. Enter years as L1 and billions of passengers as L2.

2. Use Stat Plot feature to indicate the type of graph as shown in the screens to the right.

3. Find an appropriate window. In this case $0 \leq x \leq 10$, so we choose $[-2, 15, 1]$ for the x-window; and $7.8 \leq y \leq 9.8$, so we choose $[6, 12, 1]$ for the y-window. You can choose various windows. Just make sure the domain and range are both included in your choice. Press GRAPH to see the individual points plotted.

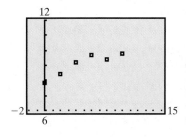

Fitting a Line to the Data

Since this data resemble a line, let's model it with a linear equation. Your calculator will use a method beyond the scope of this course to find a line that best fits this data. Under the Stat Calc menu select LinReg $(ax + b)$. On the home screen indicate the x-list (L1), y-list (L2), and the Y= position where you want the regression equation to be stored (Y1). (See the second screen.)

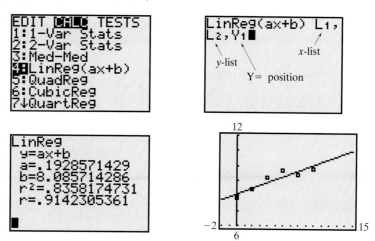

Notice in the above left screen that the calculator gives the slope, a, and the y-intercept, b, for the linear equation of the form $y = ax + b$. It is up to us to insert these values in this form. Thus, the linear regression equation is $y = 0.1928571429x + 8.085714286$ where x is the number of years since 1995 and y is the number of passengers in the billions.

The r-value shown in the third screen is called the **correlation coefficient.** The values r and r^2 indicate how well the regression equation fits the data. If the diagnostic feature of your calculator is turned on, these values are shown. Check with your instructor to see whether you are to make use of this feature.

EXAMPLE 1   **Using a Linear Regression Equation**

Use the equation $y = 0.1928571429x + 8.085714286$ from the above data and estimate the number of passengers in the year 2000.

Solution   Trace on the above graph of the regression equation to evaluate the function at $x = 5$. Another alternative method is to evaluate the function in Y1 for 5 as shown next.

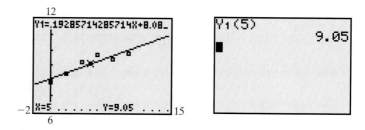

Using either method, the estimate of the number of passengers in 2000 is approximately 9.05 billion.   □

PRACTICE

1   Use the equation in Example 1 and estimate the number of passengers in 2004.

You may want to use the following steps to find a linear regression model.

Finding a Linear Regression Model

1. Enter the given data into lists.
2. Draw a scatter plot of the ordered pairs to see if the data appear linear.
3. Find the linear regression equation.
4. Graph the linear regression equation and use it to answer questions.

EXAMPLE 2   **Cell Phone Usage Increases**

According to the Bureau of Labor Statistics the annual amount consumers spend for cellular phone services increased rapidly between 2001 and 2006. The table below shows the annual average amount consumers spend per customer unit.

Years after 2000 (x)	1	2	3	4	5	6
Annual Money Spent for Cell Phone Services (y)	$210	$294	$316	$378	$455	$524

a. Plot the data points.
b. Use linear regression to fit a line to the data.
c. What does the slope of this line indicate in this situation?
d. If the amount spent continues to increase at the same rate, predict the amount spent per cell phone in the year 2016. Write your answer in a sentence including the rate of increase.

Solution

a. Enter the data in L1 and L2. Turn the Stat Plot on and graph in an appropriate window. We used the window $[-0.5, 7, 1]$ by $[200, 600, 100]$.

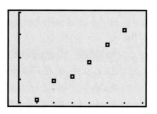

b.

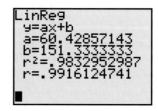
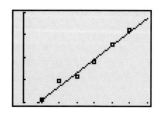

If *a* and *b* are rounded to two decimal places, the linear regression equation is $y = 60.43x + 151.33$.

c. The slope, 60.43, indicates that the average rate of increase that consumers spend is $60.43 per unit per year.

d. Since the equation is stored in Y1 we evaluate the equation for the year 2016 as Y1(16).

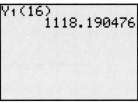

If the trend continues to increase at the rate of $60.43 per year, we predict an annual amount that consumers spend of $1118.19 per unit in the year 2016.   □

PRACTICE

2   For the years 2001–2006, the average annual money spent for phone cards, pagers and similar services per customer unit is listed in the table below. If the trend continues at the same rate, predict the average amount spent per customer unit in the year 2016. Write a sentence explaining the meaning of slope in this situation.

Years after 2000 (x)	1	2	3	4	5	6
Annual Money Spent for Phone Cards, Pagers, etc. (y)	$19	$20	$20	$21	$21	$22

EXAMPLE 3   **Number of Injured Workers Decreases**

The number of cases of injuries and sickness suffered in the workplace continues to decrease according to the Bureau of Labor and Statistics.

Year	1997	1998	1999	2000	2001	2002	2006
Injuries and Sickness per 100 Workers	7.1	6.5	6.3	6.1	5.8	5.3	4.1

a. Plot the data points. (Let $x = 0$ represent the number of years since 1990.)

b. Find the regression equation that best fits the data. Then indicate whether the line is increasing or decreasing.

c. What is the rate at which the number of sick or injured workers is increasing or decreasing?

d. If the trend continues at the same rate, predict the number of sick or injured workers out of 100 to expect in the year 2015.

Solution

a. Enter the years in L1 with 1990 represented by $x = 0$ and the corresponding number of cases per 100 workers in L2. Find an appropriate window. We choose $[6, 17, 1]$ by $[4, 8, 1]$.

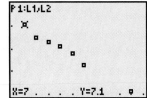

b.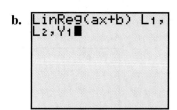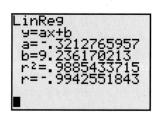

If *a* and *b* are rounded to three decimal places, the linear regression equation is $y = -0.321x + 9.236$. Since the slope is negative (-0.321), the line is decreasing.

c.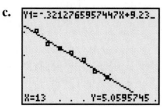

The number of injured or sick workers is decreasing at the rate of 0.321 per 100 workers per year.

d. Since the equation is stored in Y1, we evaluate the equation for the year 2015 by finding Y1(25).

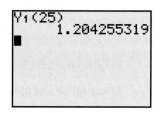

Y1(25) is approximately 1.2, so approximately 1.2 workers per 100 in the year 2015. □

PRACTICE

3 The safety committee of a large construction company is studying the increase in the number of injuries and sickness for their company.

Year	1999	2001	2003	2005	2007
Injuries and Sickness per 100 Workers	4.7	4.9	5.6	5.7	6.8

Answer Example 3, parts b, c, and d. (Let $x = 0$ represent the number of years since 1990.)

EXAMPLE 4 **A Linear Model for Predicting Selling Price**

The following houses were sold in an Orange Park neighborhood in the past month. Listed below are each house's selling price and the number of square feet of living area in the house.

Total Square Feet	2150	2273	2474	2416
Selling Price in Dollars	120,000	134,500	158,900	149,800

a. Plot the data points.

b. Use linear regression to fit a line to the data.

c. Predict the approximate selling price for a house in the same neighborhood that has 2350 square feet of living area.

Solution

a. Enter the total number of square feet in a first list, L1, and the corresponding selling price in a second list, L2. Find an appropriate window and graph as shown.

b.

162,000

| P1:L1,L2 | ☒ |

2100 |X=2474   .Y=158900 | 2500
110,000

LinReg(ax+b) L1,
L2,Y1■

LinReg
y=ax+b
a=117.3537195
b=-132428.7974
r²=.9969084839
r=.9984530454
■

162,000
Y1=117.35371949427X+-13_־

2100 |X=2350   .Y=143352.44 .| 2500
110,000

If a and b are rounded to 2 decimal places, the linear regression equation is $y = 117.35x - 132{,}428.80$.

c. Since this equation is stored in Y1, we evaluate the equation for 2350 square feet, or Y1(2350).

Y1(2350)
       143352.4434
■

We predict the selling price for a house in the same neighborhood that has 2350 square feet of living area to be $143,352.   □

PRACTICE
4   The table below shows houses that were sold in two adjoining subdivisions.

Total Square Feet (living area)	1750	1810	1536	1975
Selling Price (in dollars)	420,000	405,000	315,000	399,000

a. Use linear regression to fit a line to the data. Round numbers to two decimal places.

b. Use part a to predict the selling price for a house with 1662 square feet of living area.

EXAMPLE 5   Adults Working Longer

The percent of adults who are 55 and older who are still working has risen since 1990. The table below shows the percent of the workforce ages 55 and older for the years 1990–2007.

Year	1990	1995	2000	2005	2007
% 55 or older	30.2%	30.8%	32.1%	38%	38.6%

a. Let $x = 0$ represent the numbers of years since 1990. Use linear regression to fit a line to the data. Round decimals to three places.

b. Find the rate at which the number of workers ages 55 and older is increasing. Round to 2 decimal places.

c. If the trend continues to increase at the same linear rate, find the percent of workers ages 55 and older in the year 2015. Round to 2 decimal places.

Solution

a. Enter the data into L1 and L2. Turn the Stat Plot on and graph in an appropriate window. We chose the window $[-2, 19, 1]$ by $[29, 40, 2]$. The linear regression equation is $y = 0.538x + 28.886$.

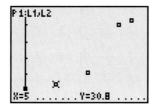 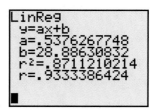 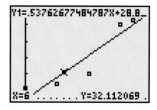

b. The number of workers ages 55 and older is increasing at the rate of 0.54% per year.

c. We stored the equation in Y1, thus we evaluate the equation for the year 2015 by finding Y1(25).

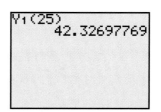

If the trend continues we can expect to see about 42.33% of the workforce being ages 55 years and older in the year 2015. ☐

PRACTICE

5 If the trend given in Example 5 continues, find the number of workers ages 55 and older expected in the year 2020. Round to 2 decimal places.

VOCABULARY & READINESS CHECK

Use the choices below to fill in each blank.

domain	regression analysis	linear regression equation	Stat Plot
data	scatter plot	range	lists

1. _____ is the process of fitting a line or a curve to a set of data points.

2. Enter the _____ into the calculator using _____.

3. Use the _____ feature to indicate the _____ as the type of graph.

4. When finding an appropriate window, make sure the _____ and _____ are both included in your choice.

5. Use a _____ to find the line of best fit.

2.6 EXERCISE SET

Solve. See Examples 1 through 5.

1. Use the function from Section 2.2, Example 12, $f(x) = 2.602x - 5178$, to predict the amount of money that will be spent by the Pharmaceutical Manufacturers Association on research and development in 2015. (*Hint: x* corresponds to the year and $f(x)$ represents amount of money in billions of dollars.

2. Use the function from Exercise 1 to predict the year that $34.9 billion was spent on research and development. (*Hint: x* corresponds to the year and $f(x)$ represents amount of money in billions of dollars.

3. In Section 2.4, Example 4, the adult one-day pass price $f(x)$ for Disney World is given by $f(x) = 2.7x + 38.64$ where x is the number of years since 1996. Find what year the price was approximately $50.

4. Using the function in Exercise 3, predict the price of a one-day pass to Disney World in 2013.

MIXED PRACTICE

Solve. Round the values in each linear regression equation to three decimal places. Predictions using linear regression equations may differ slightly depending on different rounding of the coefficients in the equation. See Examples 1 through 5.

5. The annual expenditures per residential phone services decreased for the years 2001–2006. The table below shows the annual average expenditures per customer unit.

Year	2001	2002	2003	2004	2005	2006
Avg $ Spent per Residential Phone	$686	$641	$620	$592	$570	$542

Let x = the number of years since 2000.

   a. Use linear regression to fit a line to the data. Round your answer to three places.

   b. Find the rate at which the amount of expenditures is decreasing each year.

   c. If the trend continues to decrease at this rate, find the expected amount of expenditures for residential phones in the year 2016.

   d. What does the y-intercept indicate in this context?

6. The average total expenditures for all phone services increased for the years 2001–2006. The table below shows the annual average expenditures per consumer unit.

Year	2001	2002	2003	2004	2005	2006
Avg $ Spent per Phone Services	$914	$957	$956	$990	$1048	$1087

Let x = the number of years since 2000.

   a. Use linear regression to fit a line to the data. Round your answer to three places.

   b. Find the rate at which the amount of expenditures is increasing each year.

   c. If the trend continues to increase at this rate, find the expected amount of expenditures for residential phones in the year 2016.

   d. What does the y-intercept indicate in this context?

7. The percent of male smokers 18 years and older has decreased from 1965 to 2004.

   a. Use linear regression to fit a line to the data in the table.

Years since 1960	5	11	25	30	35	40	44
Percent of Male Smokers	51.9	43.1	32.6	28.4	27	25.7	23.4

   b. Predict the percent of male smokers in the year 2013 if the trend continues to decrease at the same rate. Round to the nearest tenth of a percent.

   c. Predict the percent at which male smokers are decreasing per year.

8. The percent of female smokers 18 years and older has decreased from 1965 to 2004.

   a. Use linear regression to fit a line to the data in the table.

Years since 1960	5	11	25	30	35	40	44
Percent of Female Smokers	33.9	32.1	27.9	22.8	22.6	21	18.5

   b. Predict the percent of female smokers in the year 2013 if the trend continues to decrease at the same rate. Round to the nearest tenth of a percent.

   c. Predict the percent at which female smokers are decreasing per year.

9. The number of sentenced male prisoners in state and federal prisons rose from 1980 to 2004.

   a. Using the data in the table, find the linear regression equation that best fits the data.

Years since 1980	0	5	10	15	20	24
Number of Male Prisoners (in thousands)	304	459	699	1021	1246	1338

b. Predict the number of male prisoners in the year 2014 if the trend continues to increase at the same rate. Round to the nearest thousand.

c. Predict the rate at which male prisoners are increasing. Round to the nearest thousand.

10. The number of sentenced female prisoners in state and federal prisons rose from 1980 to 2004.

a. Using the data in the table, find the regression equation that best fits the data.

Years since 1980	0	5	10	15	20	24
Number of Female Prisoners (in thousands)	12	21	41	64	85	96

b. Predict the number of female prisoners in the year 2014 if the trend continues to increase at the same rate. Round to the nearest thousand.

c. Predict the rate at which female prisoners are increasing. Round to the nearest thousand.

11. The percent of adults in the labor force ages 65 and older who are still working has risen since 1990. The table below shows the data from 1990 to 2007.

Year	1990	1995	2000	2005	2007
% of Labor Force	12.2%	13%	14%	15.2%	16%

Let x = the number of years since 1990.

a. Plot the data points.

b. Find the linear regression equation of the line of best fit for the data.

c. What does the slope of this line indicate in this situation?

d. If the number continues to increase at the same rate, predict the percent of the labor force that is 65 and older in the year 2015.

12. The number of high school girls softball teams since Title IX was enacted has grown rapidly. From 1972 to 1974, the number of teams rose from 373 to 6000. Since 1974, the growth of the number of teams has been more steady. The table below shows the increase between 1974 and 2004.

a. Plot the data in the table and find the linear regression equation that best fits the data. (Let $x = 0$ represent the number of years since 1970.)

Year	1974	1984	1990	1994	2001	2004
No. of Teams	6000	8700	8200	12,000	13,001	14,493

b. Predict the number of girls softball teams in 2012 if this trend continues. Round to the nearest whole.

c. Find the rate at which the number of teams is rising annually.

Given the following data for four houses sold in comparable neighborhoods and their corresponding number of square feet, draw a scatter plot and find a linear regression equation representing a relationship between the number of square feet and the selling price of the house.

13.

Square Feet	1519	2593	3005	3016
Selling Price	$91,238	$220,000	$254,000	$269,000

14.

Square Feet	1754	2151	2587	2671
Selling Price	$91,238	$130,000	$155,000	$172,500

15. Use the equation found in Exercise 13 to predict the selling price of a house with 2200 square feet.

16. Use the equation found in Exercise 14 to predict the selling price of a house with 2400 square feet.

A salesperson's salary depends on the amount of sales for the week. Given the following amounts of sales and the corresponding weekly salaries, plot the data points and find a linear regression equation that describes the salary in terms of the amount of sales.

17.

Sales (in dollars)	1000	3000	5000	8000
Salary (in dollars)	780	940	1100	1340

18.

Sales (in dollars)	5000	10,000	25,000	30,000
Salary (in dollars)	550	800	1550	1800

19. Life expectancy at birth of males was rising in the U.S. between 1900 and 2004 according to the table below.

Year	1900	1930	1950	1970	1990	1999	2004
Life Expectancy	46.3	57.5	65.6	67.1	71.8	73.9	75.2

a. Using the data in the table, find the regression equation that best fits this data. (Let $x = 0$ represent the number of years since 1900.)

b. What is the rate at which the life expectancy at birth of males is rising per year?

c. Predict the life expectancy of a male born in the year 2015 if this trend continues.

20. Life expectancy at birth of females was rising in the U.S. between 1900 and 2004 according to the table below.

Year	1900	1930	1950	1970	1990	1999	2004
Life Expectancy	48.3	59.8	71.1	74.7	78.8	79.4	80.4

a. Using the data in the table, find the regression equation that best fits this data. (Let $x = 0$ represent the number of years since 1900.)

b. What is the rate at which the life expectancy at birth of females is rising?

c. Predict the life expectancy of a female born in the year 2015 if the trend continues.

21. The number of people receiving care in PPOs has increased between the years of 1992 and 2001 according to the American Association of Health Plans.

Year	1992	1994	1996	2001
Number of PPO Members (in millions)	50.4	79.2	96.1	127.8

a. Using the data in the table, find the regression equation that best fits the data. (Let $x = 0$ represent the number of years since 1990.)

b. Predict the number of people receiving care in PPOs in the year 2010 if the trend continues to increase at the same rate.

c. If this trend continues, predict the year that there will be over 240 million members.

d. What does the slope of the linear regression equation represent?

22. The number of people receiving care in HMOs has increased between the years of 1988 and 2004 according to the American Association of Health Plans.

Year	1988	1990	1992	1994	1996	2004
Number of HMO Members (in millions)	32.7	36.5	41.4	51.1	61.8	67.7

a. Using the data in the table, find the regression equation that best fits the data. (Let $x = 0$ represent the number of years since 1980.)

b. Predict the number of people receiving care in HMOs in the year 2012 if the trend continues to increase at the same rate.

c. If this trend continues, predict the year that there will be over 100 million members.

d. What does the slope of the linear regression equation represent?

23. The average top ticket price for Broadway musicals increased dramatically between 1975 and 2003.

Year	1975	1985	1998	2000	2003
Average Ticket Price	13.76	45.26	73.03	118.89	130.50

a. Using the data in the table, find the regression equation that best fits the data. (Let $x = 0$ represent the number of years since 1970.)

b. Predict the average top ticket price for Broadway musicals in the year 2013 if the trend continues to increase at the same rate.

c. Find the rate at which the cost is rising.

24. The average top ticket price for Broadway plays increased dramatically between 1975 and 2003.

Year	1975	1985	1998	2000	2003
Average Ticket Price	10.76	35.29	56.35	91.35	100.05

a. Using the data in the table, find the regression equation that best fits the data. (Let $x = 0$ represent the number of years since 1970.)

b. Predict the average top ticket price for Broadway plays in the year 2013 if the trend continues to increase at the same rate.

c. Find the rate at which the cost is rising.

25. Super Bowl Commercials. The Super Bowl is the most widely watched event on television. The table below gives the average cost for a 30-second commercial for the years 1991 through 2007.

Years since 1990	1	3	5	7	9	11	13	15	17
Super Bowl Commercial Cost (in millions of dollars)	0.8	0.85	1	1.2	1.6	2.05	2.1	2.4	2.6

Source: Advertising Age

a. Plot the data points.

b. Use a linear regression to fit a line to the data.

c. Find the rate at which the cost of a 30-second commercial during the Super Bowl is increasing per year.

d. If the rate continues to increase in the same manner, use the approximate equation in part **b** and predict the cost of a 30-second commercial during the Super Bowl in the year 2013.

26. Academy Award Commercials. The Academy Awards is one of the most watched television shows. The table below gives the average cost of a 30-second commercial slot for the years 1997 through 2007.

Years since 1990	7	9	11	13	15	17
Academy Award Commercial Cost (in millions of dollars)	0.85	1.0	1.45	1.35	1.5	1.67

a. Plot the data points.

b. Use a linear regression to fit a line to the data.

c. Find the rate at which the cost of a 30-second commercial on the Academy Awards presentation is increasing per year.

d. If the rate continues to increase in the same manner, use the approximate equation in part **b** and predict the cost of a 30-second commercial in the year 2014.

REVIEW AND PREVIEW

Solve each equation algebraically. See Section 1.5.

27. $7x + 2 = 9x - 14$

28. $5(x - 2) = 4(x + 7)$

29. $\dfrac{y}{7} + \dfrac{y}{3} = \dfrac{1}{21}$

30. $y + 0.8 = 0.3(y - 2)$

CONCEPT EXTENSIONS

31. In the United States, the revenue (money taken in from sales) at a "full service" restaurant is increasing at a faster rate than the revenue at a "fast food" restaurant. The data below represent the annual revenue in billions of dollars for each type of restaurant.

Year	1995	1999	2000	2001	2002	2003	2004
Full Service	99	126	134	141	148	155	164
Fast Food	103	120	128	133	138	147	159

Source: U.S. Census Bureau

Let $x = 0$ represent the number of years since 1990.

a. Write a linear regression equation for "full service" and one for "fast food." Round the coefficients to three decimal places.

b. Use the equations and their graphs to approximate the year that the revenue from the two types of restaurants was the same. Round up to the nearest whole year.

32. GROUP ACTIVITY. Divide into groups of 5–6 students to see if a relationship exists between the length of a person's forearm (in cm) and their height (in cm).

a. Complete a table recording the data of each student in the group by measuring the forearm length in cm in L1 and the height in cm in L2.

b. Make a scatter plot of the data. What type of graph does the data seem to best fit?

c. Find the regression equation that best fits the data.

d. Interpret the slope and the y-intercept of the regression equation.

33. GROUP ACTIVITY. Complete a similar table comparing the circumference of a student's neck to the circumference of his or her wrist.

2.7 GRAPHING PIECEWISE-DEFINED FUNCTIONS AND SHIFTING AND REFLECTING GRAPHS OF FUNCTIONS

OBJECTIVES

1 Graph piecewise-defined functions.

2 Vertical and horizontal shifts.

3 Reflect graphs.

OBJECTIVE 1 ▶ Graphing piecewise-defined functions. Throughout Chapter 2, we have graphed functions. There are many special functions. In this objective, we study functions defined by two or more expressions. The expression used to complete the function varies with, and depends upon, the value of x. Before we actually graph these piecewise-defined functions, let's practice finding function values. (For this objective, the graphing utility is not mentioned, although it is possible to use it to check separate pieces of your graph.)

EXAMPLE 1 Evaluate $f(2), f(-6)$, and $f(0)$ for the function

$$f(x) = \begin{cases} 2x + 3 & \text{if } x \le 0 \\ -x - 1 & \text{if } x > 0 \end{cases}$$

Then write your results in ordered pair form.

Solution Take a moment and study this function. It is a single function defined by two expressions depending on the value of x. From above, if $x \le 0$, use $f(x) = 2x + 3$. If $x > 0$, use $f(x) = -x - 1$. Thus

$f(2) = -(2) - 1$	$f(-6) = 2(-6) + 3$	$f(0) = 2(0) + 3$
$\quad = -3 \text{ since } 2 > 0$	$\quad = -9 \text{ since } -6 \le 0$	$\quad = 3 \text{ since } 0 \le 0$
$f(2) = -3$	$f(-6) = -9$	$f(0) = 3$
Ordered pairs: $(2, -3)$	$(-6, -9)$	$(0, 3)$

1   Evaluate $f(4)$, $f(-2)$, and $f(0)$ for the function

$$f(x) = \begin{cases} -4x - 2 & \text{if } x \le 0 \\ x + 1 & \text{if } x > 0. \end{cases}$$

Now, let's graph a piecewise-defined function.

EXAMPLE 2   Graph $f(x) = \begin{cases} 2x + 3 & \text{if } x \le 0 \\ -x - 1 & \text{if } x > 0 \end{cases}$

Solution   Let's graph each piece.

If $x \le 0$,   If $x > 0$,

$f(x) = 2x + 3$ $f(x) = -x - 1$

Values ≤ 0

x	$f(x) = 2x + 3$
0	3 Closed circle
−1	1
−2	−1

Values > 0

x	$f(x) = -x - 1$
1	−2
2	−3
3	−4

The graph of the first part of $f(x)$ listed will look like a ray with a closed-circle end point at $(0, 3)$. The graph of the second part of $f(x)$ listed will look like a ray with an open-circle end point. To find the exact location of the open-circle end point, use $f(x) = -x - 1$ and find $f(0)$. Since $f(0) = -0 - 1 = -1$, we graph the second table and place an open circle at $(0, -1)$.

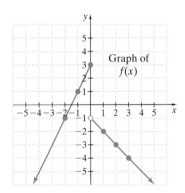

Notice that this graph is the graph of a function because it passes the vertical line test. The domain of this function is all real numbers and the range is $\{y \mid y \le 3\}$.

2   Graph

$$f(x) = \begin{cases} -4x - 2 & \text{if } x \le 0 \\ x + 1 & \text{if } x > 0 \end{cases}$$

OBJECTIVE 2 ▶ Vertical and horizontal shifting.

Review of Common Graphs

We now take common graphs and learn how more complicated graphs are actually formed by shifting and reflecting these common graphs. These shifts and reflections are called transformations, and it is possible to combine transformations. A knowledge of these transformations will help you simplify future graphs.

Let's begin with a review of the graphs of three common functions. We graphed these functions in Section 2.1.

Common Graphs from Section 2.1

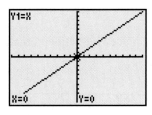
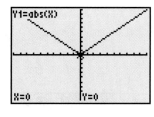
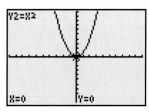

Standard Window

| Basic linear Equation | Absolute Value Equation | Quadratic Equation (Parabola) |
| $y = x$ | $y = |x|$ | $y = x^2$ |

Let's graph by hand a fourth common function, $f(x) = \sqrt{x}$ or $y = \sqrt{x}$. For this graph, you need to recall basic facts about square roots and use your calculator to approximate some square roots to help locate points. Recall also that the square root of a negative number is not a real number, so be careful when finding your domain.

Now **let's graph the square root function $f(x) = \sqrt{x}$, or $y = \sqrt{x}$.**

To graph, we identify the domain, evaluate the function for several values of x, plot the resulting points, and connect the points with a smooth curve. Since $\sqrt{x}$ represents the nonnegative square root of x, the domain of this function is the set of all nonnegative numbers, $\{x \mid x \geq 0\}$. We have approximated $\sqrt{3}$ below to help us locate the point corresponding to $(3, \sqrt{3})$.

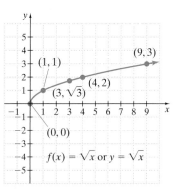

If $x = 0$, then $y = \sqrt{0}$, or 0.

If $x = 1$, then $y = \sqrt{1}$, or 1.

If $x = 3$, then $y = \sqrt{3}$, or 1.7.

If $x = 4$, then $y = \sqrt{4}$, or 2.

If $x = 9$, then $y = \sqrt{9}$, or 3.

x	$f(x) = \sqrt{x}$
0	0
1	1
3	$\sqrt{3} \approx 1.7$
4	2
9	3

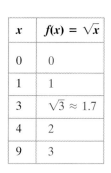

A calculator check: square root equation

$y = \sqrt{x}$

Notice that the graph of this function also passes the vertical line test, as expected.

Take a moment and review all four common graphs. Your success in the rest of this section depends on your knowledge of these graphs. Although we have studied some simple vertical transformations in earlier sections, we now formally study vertical, horizontal, and other transformations. These graphs are hand-drawn, but feel free to check your work with your calculator.

Your knowledge of the slope-intercept form, $f(x) = mx + b$, will help you understand simple shifting of transformations such as vertical shifts. For example, what is the difference between the graphs of $f(x) = x$ and $g(x) = x + 3$?

$f(x) = x$

slope, $m = 1$

y-intercept is $(0, 0)$

$g(x) = x + 3$

slope, $m = 1$

y-intercept is $(0, 3)$

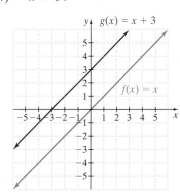

Notice that the graph of $g(x) = x + 3$ is the same as the graph of $f(x) = x$, but moved upward 3 units. This is an example of a **vertical shift** and is true for graphs in general.

Vertical Shifts (Upward and Downward)
Let k be a Positive Number

Graph of	Same As	Moved
$g(x) = f(x) + k$	$f(x)$	k units upward
$g(x) = f(x) - k$	$f(x)$	k units downward

EXAMPLES   Without plotting points, sketch the graph of each pair of functions on the same set of axes.

3. $f(x) = x^2$ and $g(x) = x^2 + 2$

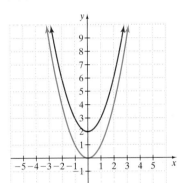

4. $f(x) = \sqrt{x}$ and $g(x) = \sqrt{x} - 3$

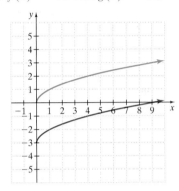

PRACTICES
3–4   Without plotting points, sketch the graphs of each pair of functions on the same set of axes.

3. $f(x) = x^2$ and $g(x) = x^2 - 3$

4. $f(x) = \sqrt{x}$ and $g(x) = \sqrt{x} + 1$

A horizontal shift to the left or right may be slightly more difficult to understand. Let's graph $g(x) = |x - 2|$ and compare it with $f(x) = |x|$.

EXAMPLE 5   Sketch the graphs of $f(x) = |x|$ and $g(x) = |x - 2|$ on the same set of axes.

Solution   Study the table to the left to understand the placement of both graphs.

x	$f(x) = |x|$	$g(x) = |x - 2|$
-3	3	5
-2	2	4
-1	1	3
0	0	2
1	1	1
2	2	0
3	3	1

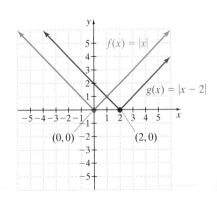

PRACTICE
5   Sketch the graphs of $f(x) = |x|$ and $g(x) = |x - 3|$ on the same set of axes.

The graph of $g(x) = |x - 2|$ is the same as the graph of $f(x) = |x|$, but moved 2 units to the right. This is an example of a **horizontal shift** and is true for graphs in general.

Horizontal Shift (To the Left or Right)
Let h be a Positive Number

Graph of	*Same as*	*Moved*
$g(x) = f(x - h)$	$f(x)$	h units to the right
$g(x) = f(x + h)$	$f(x)$	h units to the left

> ▶ **Helpful Hint**
>
> Notice that $f(x - h)$ corresponds to a shift to the right and $f(x + h)$ corresponds to a shift to the left.

Vertical and horizontal shifts can be combined.

EXAMPLE 6 Sketch the graphs of $f(x) = x^2$ and $g(x) = (x - 2)^2 + 1$ on the same set of axes.

Solution The graph of $g(x)$ is the same as the graph of $f(x)$ shifted 2 units to the right and 1 unit up.

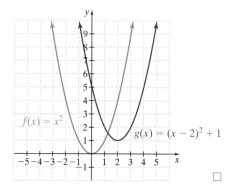

PRACTICE
6 Sketch the graphs of $f(x) = |x|$ and $g(x) = |x - 2| + 3$ on the same set of axes.

OBJECTIVE 3 ▶ Reflecting graphs. Another type of transformation is called a **reflection.** In this section, we will study reflections (mirror images) about the x-axis only. For example, take a moment and study these two graphs. The graph of $g(x) = -x^2$ can be verified, as usual, by plotting points.

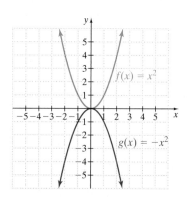

Reflection about the x-axis
The graph of $g(x) = -f(x)$ is the graph of $f(x)$ reflected about the x-axis.

EXAMPLE 7 Sketch the graph of $h(x) = -|x - 3| + 2$.

Solution The graph of $h(x) = -|x - 3| + 2$ is the same as the graph of $f(x) = |x|$ reflected about the x-axis, then moved three units to the right and two units upward.

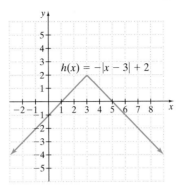

PRACTICE
7 Sketch the graph of $h(x) = -(x + 2)^2 - 1$.

There are other transformations, such as stretching, that won't be covered in this section. For a review of this transformation, see the Appendix.

VOCABULARY & READINESS CHECK

Match each equation with its graph.

1. $y = \sqrt{x}$

A

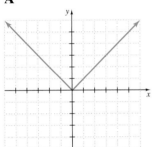

2. $y = x^2$

B

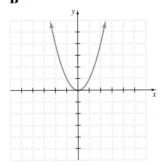

3. $y = x$

C

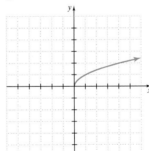

4. $y = |x|$

D

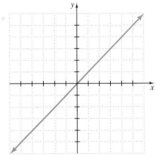

2.7 EXERCISE SET

 MyMathLab Math XL PRACTICE WATCH DOWNLOAD READ REVIEW

Graph each piecewise-defined function. See Examples 1 and 2.

1. $f(x) = \begin{cases} 2x & \text{if} \quad x < 0 \\ x + 1 & \text{if} \quad x \geq 0 \end{cases}$

2. $f(x) = \begin{cases} 3x & \text{if} \quad x < 0 \\ x + 2 & \text{if} \quad x \geq 0 \end{cases}$

3. $f(x) = \begin{cases} 4x + 5 & \text{if} \quad x \leq 0 \\ \dfrac{1}{4}x + 2 & \text{if} \quad x > 0 \end{cases}$

4. $f(x) = \begin{cases} 5x + 4 & \text{if} \quad x \leq 0 \\ \dfrac{1}{3}x - 1 & \text{if} \quad x > 0 \end{cases}$

5. $g(x) = \begin{cases} -x & \text{if} \quad x \leq 1 \\ 2x + 1 & \text{if} \quad x > 1 \end{cases}$

6. $g(x) = \begin{cases} 3x - 1 & \text{if} \quad x \leq 2 \\ -x & \text{if} \quad x > 2 \end{cases}$

7. $f(x) = \begin{cases} 5 & \text{if } x < -2 \\ 3 & \text{if } x \geq -2 \end{cases}$

8. $f(x) = \begin{cases} 4 & \text{if } x < -3 \\ -2 & \text{if } x \geq -3 \end{cases}$

C 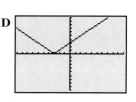 D

MIXED PRACTICE

(Sections 2.2, 2.7) Graph each piecewise-defined function. Use the graph to determine the domain and range of the function. See Examples 1 and 2.

9. $f(x) = \begin{cases} -2x & \text{if } x \leq 0 \\ 2x + 1 & \text{if } x > 0 \end{cases}$

10. $g(x) = \begin{cases} -3x & \text{if } x \leq 0 \\ 3x + 2 & \text{if } x > 0 \end{cases}$

11. $h(x) = \begin{cases} 5x - 5 & \text{if } x < 2 \\ -x + 3 & \text{if } x \geq 2 \end{cases}$

12. $f(x) = \begin{cases} 4x - 4 & \text{if } x < 2 \\ -x + 1 & \text{if } x \geq 2 \end{cases}$

13. $f(x) = \begin{cases} x + 3 & \text{if } x < -1 \\ -2x + 4 & \text{if } x \geq -1 \end{cases}$

14. $h(x) = \begin{cases} x + 2 & \text{if } x < 1 \\ 2x + 1 & \text{if } x \geq 1 \end{cases}$

15. $g(x) = \begin{cases} -2 & \text{if } x \leq 0 \\ -4 & \text{if } x \geq 1 \end{cases}$

16. $f(x) = \begin{cases} -1 & \text{if } x \leq 0 \\ -3 & \text{if } x \geq 2 \end{cases}$

For Exercises 17 through 20, match each equation to its graph.

17. $f(x) = |x| + 3$

18. $f(x) = |x| - 2$

19. $f(x) = \sqrt{x} - 2$

20. $f(x) = \sqrt{x} + 3$

A 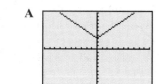 B

C 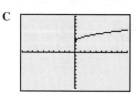 D

For Exercises 21 through 24, match each equation to its graph.

21. $f(x) = |x - 4|$

22. $f(x) = |x + 3|$

23. $f(x) = \sqrt{x + 2}$

24. $f(x) = \sqrt{x - 2}$

A 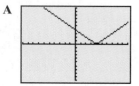 B

Sketch the graph of function. See Examples 3 through 6.

25. $y = (x - 4)^2$

26. $y = (x + 4)^2$

27. $f(x) = x^2 + 4$

28. $f(x) = x^2 - 4$

29. $f(x) = \sqrt{x - 2} + 3$

30. $f(x) = \sqrt{x - 1} + 3$

31. $f(x) = |x - 1| + 5$

32. $f(x) = |x - 3| + 2$

33. $f(x) = \sqrt{x + 1} + 1$

34. $f(x) = \sqrt{x + 3} + 2$

35. $f(x) = |x + 3| - 1$

36. $f(x) = |x + 1| - 4$

37. $g(x) = (x - 1)^2 - 1$

38. $h(x) = (x + 2)^2 + 2$

39. $f(x) = (x + 3)^2 - 2$

40. $f(x) = (x + 2)^2 + 4$

Sketch the graph of each function. See Examples 3 through 7.

41. $f(x) = -(x - 1)^2$

42. $g(x) = -(x + 2)^2$

43. $h(x) = -\sqrt{x} + 3$

44. $f(x) = -\sqrt{x + 3}$

45. $h(x) = -|x + 2| + 3$

46. $g(x) = -|x + 1| + 1$

47. $f(x) = (x - 3) + 2$

48. $f(x) = (x - 1) + 4$

REVIEW AND PREVIEW

Match each equation with its graph. See Section 2.3.

49. $y = -1$

50. $x = -1$

51. $x = 3$

52. $y = 3$

A 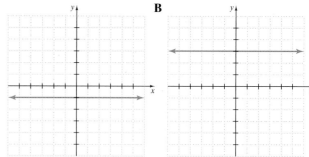 B

C 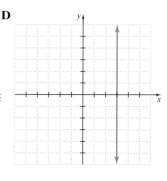 D

CONCEPT EXTENSIONS

53. Draw a graph whose domain is $\{x \mid x \le 5\}$ and whose range is $\{y \mid y \ge 2\}$.

54. In your own words, describe how to graph a piecewise-defined function.

55. Graph: $f(x) = \begin{cases} -\dfrac{1}{2}x & \text{if} \quad x \le 0 \\ x + 1 & \text{if} \quad 0 < x \le 2 \\ 2x - 1 & \text{if} \quad x > 2 \end{cases}$

56. Graph: $f(x) = \begin{cases} -\dfrac{1}{3}x & \text{if} \quad x \le 0 \\ x + 2 & \text{if} \quad 0 < x \le 4 \\ 3x - 4 & \text{if} \quad x > 4 \end{cases}$

Write the domain and range of the following exercises.

57. Exercise 29

58. Exercise 30

59. Exercise 45

60. Exercise 46

Without graphing, find the domain of each function.

61. $f(x) = 5\sqrt{x - 20} + 1$

62. $g(x) = -3\sqrt{x + 5}$

63. $h(x) = 5|x - 20| + 1$

64. $f(x) = -3|x + 5.7|$

65. $g(x) = 9 - \sqrt{x + 103}$

66. $h(x) = \sqrt{x - 17} - 3$

Sketch the graph of each piecewise-defined function. Write the domain and range of each function.

67. $f(x) = \begin{cases} |x| & \text{if} \quad x \le 0 \\ x^2 & \text{if} \quad x > 0 \end{cases}$

68. $f(x) = \begin{cases} x^2 & \text{if} \quad x < 0 \\ \sqrt{x} & \text{if} \quad x \ge 0 \end{cases}$

69. $g(x) = \begin{cases} |x - 2| & \text{if} \quad x < 0 \\ -x^2 & \text{if} \quad x \ge 0 \end{cases}$

70. $g(x) = \begin{cases} -|x + 1| - 1 & \text{if} \quad x < -2 \\ \sqrt{x + 2} - 4 & \text{if} \quad x \ge -2 \end{cases}$

📖 STUDY SKILLS BUILDER

Tips for Studying for an Exam

To prepare for an exam, try the following study techniques:

- Start the study process days before your exam.
- Make sure that you are up-to-date on your assignments.
- If there is a topic that you are unsure of, use one of the many resources that are available to you. For example,

    See your instructor.

    Visit a learning resource center on campus.

    Read the textbook material and examples on the topic.

    View a video on the topic.

- Reread your notes and carefully review the Chapter Highlights at the end of any chapter.
- Work the review exercises at the end of the chapter. Check your answers and correct any mistakes. If you have trouble, use a resource listed above.
- Find a quiet place to take the Chapter Test found at the end of the chapter. Do not use any resources when taking this sample test. This way, you will have a clear

indication of how prepared you are for your exam. Check your answers and make sure that you correct any missed exercises.

- Get lots of rest the night before the exam. It's hard to show how well you know the material if your brain is foggy from lack of sleep.

Good luck and keep a positive attitude.

Let's see how you did on your last exam.

1. How many days before your last exam did you start studying for that exam?

2. Were you up-to-date on your assignments at that time or did you need to catch up on assignments?

3. List the most helpful text supplement (if you used one).

4. List the most helpful campus supplement (if you used one).

5. List your process for preparing for a mathematics test.

6. Was this process helpful? In other words, were you satisfied with your performance on your exam?

7. If not, what changes can you make in your process that will make it more helpful to you?

CHAPTER 2 GROUP ACTIVITY

Modeling Real Data

The number of children who live with only one parent has been steadily increasing in the United States since the 1960s. According to the U.S. Bureau of the Census, the percent of children living with both parents is declining. The following table shows the percent of children (under age 18) living with *both* parents during selected years from 1980 to 2005. In this project, you will have the opportunity to use the data in the table to find a linear function $f(x)$ that represents the data, reflecting the change in living arrangements for children. This project may be completed by working in groups or individually.

Percent of U.S. Children Who Live with Both Parents

Year	1980	1985	1990	1995	2000	2005
x	0	5	10	15	20	25
Percent, y	77	74	73	69	67	68

Source: U.S. Bureau of the Census

1. Plot the data given in the table as ordered pairs.
2. Use a straight edge to draw on your graph what appears to be the line that "best fits" the data you plotted.
3. Estimate the coordinates of two points that fall on your best fitting line. Use these points to find a linear function $f(x)$ for the line.
4. What is the slope of your line? Interpret its meaning. Does it make sense in the context of this situation?
5. Find the value of $f(50)$. Write a sentence interpreting its meaning in context.
6. Compare your linear function with that of another student or group. Are they different? If so, explain why.
7. Enter the data from the table into a graphing calculator. Use the linear regression feature of the calculator to find a linear function for the data. Compare this function to the one you found in Question 3. How are they alike or different? Find the value of $f(50)$ using the model you found with the graphing calculator. Compare it to the value of $f(50)$ you found in Question 5.

CHAPTER 2 VOCABULARY CHECK

Fill in each blank with one of the words or phrases listed below.

relation   standard   slope-intercept   range   point-slope
line   slope   *x*   parallel   perpendicular
function   domain   *y*   linear function

1. A _____ is a set of ordered pairs.
2. The graph of every linear equation in two variables is a _____ .
3. The equation $y - 8 = -5(x + 1)$ is written in _____ form.
4. _____ form of a linear equation in two variables is $Ax + By = C$.
5. The _____ of a relation is the set of all second components of the ordered pairs of the relation.
6. _____ lines have the same slope and different y-intercepts.
7. _____ form of a linear equation in two variables is $y = mx + b$.
8. A _____ is a relation in which each first component in the ordered pairs corresponds to exactly one second component.
9. In the equation $y = 4x - 2$, the coefficient of x is the _____ of its corresponding graph.
10. Two lines are _____ if the product of their slopes is -1.
11. To find the x-intercept of a linear equation, let _____ = 0 and solve for the other variable.
12. The _____ of a relation is the set of all first components of the ordered pairs of the relation.
13. A _____ is a function that can be written in the form $f(x) = mx + b$.
14. To find the y-intercept of a linear equation, let _____ = 0 and solve for the other variable.

▶ **Helpful Hint**

Are you preparing for your test? Don't forget to take the Chapter 2 Test on page 204. Then check your answers at the back of the text and use the Chapter Test Prep Video CD to see the fully worked-out solutions to any of the exercises you want to review.

CHAPTER 2 HIGHLIGHTS

DEFINITIONS AND CONCEPTS	EXAMPLES

SECTION 2.1   GRAPHING EQUATIONS

The **rectangular coordinate system,** or **Cartesian coordinate system,** consists of a vertical and a horizontal number line intersecting at their 0 coordinate. The vertical number line is called the **y-axis,** and the horizontal number line is called the **x-axis.** The point of intersection of the axes is called the **origin.** The axes divide the plane into four regions called **quadrants.**

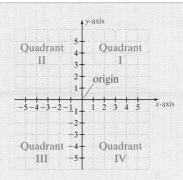

To **plot** or **graph** an ordered pair means to find its corresponding point on a rectangular coordinate system.

To plot or graph the ordered pair $(-2, 5)$, start at the origin. Move 2 units to the left along the x-axis, then 5 units upward parallel to the y-axis.

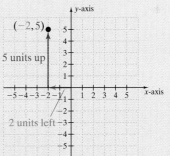

A graphing utility can be used to display a portion of the rectangular coordinate system. The portion being viewed is called a **viewing window** or simply **window.**

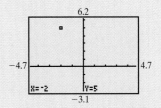

Ordered pair $(-2, 5)$ plotted

To graph an equation using a graphing utility, solve the equation for y and enter it into the Y= editor.

Graph $y = x^3 + 2$.

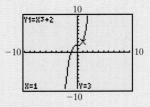

An ordered pair is a **solution** of an equation in two variables if replacing the variables by the corresponding coordinates results in a true statement.

Determine whether $(-2, 3)$ is a solution of

$$3x + 2y = 0$$
$$3(-2) + 2(3) = 0$$
$$-6 + 6 = 0$$
$$0 = 0 \quad \text{True}$$

$(-2, 3)$ is a solution.

(continued)

| **DEFINITIONS AND CONCEPTS** | **EXAMPLES** |

SECTION 2.1   GRAPHING EQUATIONS (continued)

A **linear equation in two variables** is an equation that can be written in the form $Ax + By = C$, where A, B, and C are real numbers and A and B are not both 0. The form $Ax + By = C$ is called **standard form.**

Linear Equations in Two Variables

$$y = -2x + 5, \quad x = 7$$
$$y - 3 = 0, \quad 6x - 4y = 10$$

$6x - 4y = 10$ is in standard form.

The graph of a linear equation in two variables is a line. To graph a linear equation in two variables, find three ordered pair solutions. Plot the solution points, and draw the line connecting the points.

Graph $3x + y = -6$.

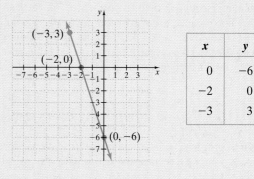

x	y
0	−6
−2	0
−3	3

To graph an equation that is not linear, find a sufficient number of ordered pair solutions so that a pattern may be discovered.

Graph $y = x^3 + 2$.

x	y
−2	−6
−1	1
0	2
1	3
2	10

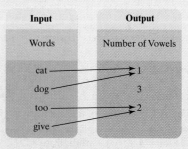

SECTION 2.2   INTRODUCTION TO FUNCTIONS

A **relation** is a set of ordered pairs. The **domain** of the relation is the set of all first components of the ordered pairs. The **range** of the relation is the set of all second components of the ordered pairs.

Relation

Input	Output
Words	Number of Vowels

cat → 1
dog → 1
too → 2
give → 2
(3)

Domain: {cat, dog, too, give}
Range: {1, 2}

A **function** is a relation in which each element of the first set corresponds to exactly one element of the second set.

The previous relation is a function. Each word contains exactly one number of vowels.

DEFINITIONS AND CONCEPTS	**EXAMPLES**

Vertical Line Test

If no vertical line can be drawn so that it intersects a graph more than once, the graph is the graph of a function.

Find the domain and the range of the relation. Also determine whether the relation is a function.

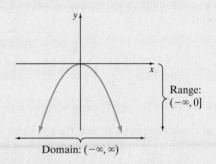

By the vertical line test, this graph is the graph of a function.

The symbol $f(x)$ means **function of x** and is called **function notation.**

If $f(x) = 2x^2 - 5$, find $f(-3)$.

$$f(-3) = 2(-3)^2 - 5 = 2(9) - 5 = 13$$

SECTION 2.3   GRAPHING LINEAR FUNCTIONS

A **linear function** is a function that can be written in the form $f(x) = mx + b$.

To graph a linear function, find three ordered pair solutions. (Use the third ordered pair to check.) Graph the solutions and draw a line through the plotted points.

Linear Functions

$$f(x) = -3, g(x) = 5x, h(x) = -\frac{1}{3}x - 7$$

Graph $f(x) = -2x$.

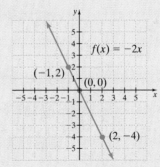

x	y or $f(x)$
-1	2
0	0
2	-4

The graph of $y = mx + b$ is the same as the graph of $y = mx$, but shifted b units up if b is positive and b units down if b is negative.

Graph $g(x) = -2x + 3$.

This is the same as the graph of $f(x) = -2x$ shifted 3 units up.

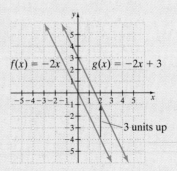

(continued)

DEFINITIONS AND CONCEPTS	EXAMPLES

The x-coordinate of a point where a graph crosses the x-axis is called an **x-intercept.** The y-coordinate of a point where a graph crosses the y-axis is called a **y-intercept.**

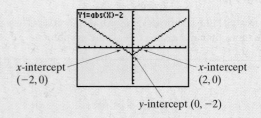

x-intercept $(-2, 0)$ x-intercept $(2, 0)$

y-intercept $(0, -2)$

The x-intercepts of the graph are -2 and 2. The y-intercept is -2.

To find an x-intercept, let $y = 0$ or $f(x) = 0$ and solve for x.

To find a y-intercept, let $x = 0$ and solve for y.

Graph $5x - y = -5$ by finding intercepts.

If $x = 0$, then	If $y = 0$, then
$5x - y = -5$	$5x - y = -5$
$5 \cdot 0 - y = -5$	$5x - 0 = -5$
$-y = -5$	$5x = -5$
$y = 5$	$x = -1$
$(0, 5)$	$(-1, 0)$

Ordered pairs are $(0, 5)$ and $(-1, 0)$.

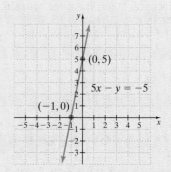

The graph of $x = c$ is a vertical line with x-intercept $(c, 0)$.

The graph of $y = c$ is a horizontal line with y-intercept $(0, c)$.

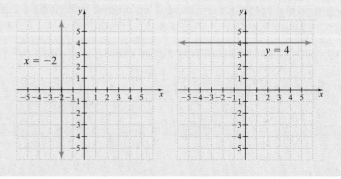

DEFINITIONS AND CONCEPTS	**EXAMPLES**

The **slope** m of the line through (x_1, y_1) and (x_2, y_2) is given by

$$m = \frac{y_2 - y_1}{x_2 - x_1} \text{ as long } x_2 \neq x_1$$

The **slope-intercept form** of a linear equation is $y = mx + b$, where m is the slope of the line and b is the y-intercept.

Find the slope of the line through $(-1, 7)$ and $(-2, -3)$.

$$m = \frac{y_2 - y_1}{x_2 - x_1} = \frac{-3 - 7}{-2 - (-1)} = \frac{-10}{-1} = 10$$

Find the slope and y-intercept of $-3x + 2y = -8$.

$$2y = 3x - 8$$
$$\frac{2y}{2} = \frac{3x}{2} - \frac{8}{2}$$
$$y = \frac{3}{2}x - 4$$

The slope the line is $\frac{3}{2}$, and the y-intercept is $(0, -4)$.

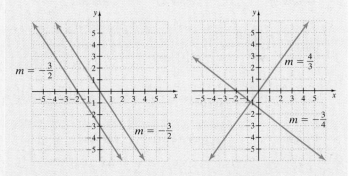

Nonvertical parallel lines have the same slope.

If the product of the slopes of two lines is -1, then the lines are perpendicular.

The slope of a horizontal line is 0.

The slope of a vertical line is undefined.

The slope of $y = -2$ is 0.

The slope of $x = 5$ is undefined.

We can use the slope-intercept form to write an equation of a line given its slope and y-intercept.

Write an equation of the line with y-intercept $(0, -1)$ and slope $\frac{2}{3}$.

$$y = mx + b$$
$$y = \frac{2}{3}x - 1$$

The point-slope form of the equation of a line is $y - y_1 = m(x - x_1)$, where m is the slope of the line and (x_1, y_1) is a point on the line.

Find an equation of the line with slope 2 containing the point $(1, -4)$. Write the equation in standard form: $Ax + By = C$.

$$y - y_1 = m(x - x_1)$$
$$y - (-4) = 2(x - 1)$$
$$y + 4 = 2x - 2$$
$$-2x + y = -6 \quad \text{Standard form}$$

DEFINITIONS AND CONCEPTS	**EXAMPLES**

SECTION 2.6 INTERPRETING DATA: LINEAR MODELS

Regression analysis is the process of fitting a line or a curve to a set of data points.

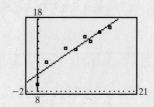

SECTION 2.7 GRAPHING PIECEWISE-DEFINED FUNCTIONS AND SHIFTING AND REFLECTING GRAPHS OF FUNCTIONS

Vertical shifts (upward and downward):
Let k be a positive number.

Graph of	*Same as*	*Moved*
$g(x) = f(x) + k$	$f(x)$	k units upward
$g(x) = f(x) + (-k)$	$f(x)$	k units downward

Horizontal shift (to the left or right):
Let h be a positive number.

Graph of	*Same as*	*Moved*
$g(x) = f(x - h)$	$f(x)$	h units to the right
$g(x) = f(x + h)$	$f(x)$	h units to the left

Reflection about the x-axis:

The graph of $g(x) = -f(x)$ is the graph of $f(x)$ reflected about the x-axis.

The graph of $h(x) = -|x - 3| + 1$ is the same as the graph of $f(x) = |x|$ reflected about the x-axis, shifted 3 units right, then 1 unit up.

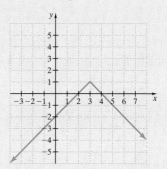

📖 STUDY SKILLS BUILDER

Are You Prepared for a Test on Chapter 2?

Below I have listed some common trouble areas for students in Chapter 2. After studying for your test—but before taking your test—read these.

- Don't forget that the graph of an ordered pair is a *single* point in the rectangular coordinate plane.
- Remember that the slope of a horizontal line is 0 while a vertical line has undefined slope or no slope.
- For a linear equation such as $2y = 3x - 6$, the slope is not the coefficient of x unless the equation is solved for y. Solving this equation for y, we have $y = \frac{3}{2}x - 3$. The slope is $\frac{3}{2}$ and the y-intercept is $(0, -3)$.

Slope	*Parallel line*	*Perpendicular line*
$m = 6$	$m = 6$	$m = -\dfrac{1}{6}$
$m = -\dfrac{2}{3}$	$m = -\dfrac{2}{3}$	$m = \dfrac{3}{2}$

- Parallel lines have the same slope while perpendicular lines have negative reciprocal slopes.
- Don't forget that the statement $f(2) = 3$ corresponds to the ordered pair $(2, 3)$.

Remember: This is simply a checklist of common trouble areas. For a review of Chapter 2, see the Highlights and Chapter Review at the end of this chapter.

CHAPTER 2 REVIEW

(2.1) *Plot the points and name the quadrant or axis in which each point lies.*

1. $A(2, -1), B(-2, 1), C(0, 3), D(-3, -5)$

2. $A(-3, 4), B(4, -3), C(-2, 0), D(-4, 1)$

Determine whether each ordered pair is a solution to the given equation.

3. $7x - 8y = 56; (0, 56), (8, 0)$

4. $-2x + 5y = 10; (-5, 0), (1, 1)$

5. $x = 13; (13, 5), (13, 13)$

6. $y = 2; (7, 2), (2, 7)$

Determine whether each equation is linear or not. Then graph the equation.

7. $y = 3x$ **8.** $y = 5x$

9. $3x - y = 4$ **10.** $x - 3y = 2$

11. $y = |x| + 4$ **12.** $y = x^2 + 4$

13. $y = -\dfrac{1}{2}x + 2$ **14.** $y = -x + 5$

Match each graph with its equation.

15.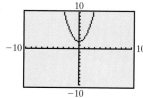

16.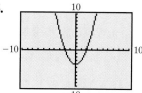

17.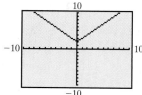

18.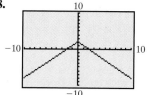

a. $y = x^2 - 4$ **b.** $y = -|x| + 2$

c. $y = |x| + 2$ **d.** $y = x^2 + 2$

(2.2) *Find the domain and range of each relation. Also determine whether the relation is a function.*

19. $\left\{ \left(-\dfrac{1}{2}, \dfrac{3}{4} \right), (6, 0.75), (0, -12), (25, 25) \right\}$

20. $\left\{ \left(\dfrac{3}{4}, -\dfrac{1}{2} \right), (0.75, 6), (-12, 0), (25, 25) \right\}$

21.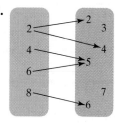

22.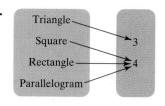

23.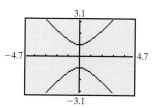

24.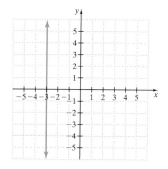

25.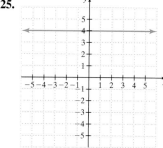

26.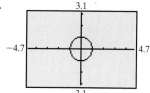

If $f(x) = x - 5$, $g(x) = -3x$, and $h(x) = 2x^2 - 6x + 1$, find the following.

27. $f(2)$ **28.** $g(0)$

29. $g(-6)$ **30.** $h(-1)$

31. $h(1)$ **32.** $f(5)$

The function $J(x) = 2.54x$ may be used to calculate the weight of an object on Jupiter J given its weight on Earth x.

33. If a person weighs 150 pounds on Earth, find the equivalent weight on Jupiter.

34. A 2000-pound probe on Earth weighs how many pounds on Jupiter?

Use the graph of the function below to answer Exercises 35 through 38.

35. Find $f(-1)$.

36. Find $f(1)$.

37. Find all values of x such that $f(x) = 1$.

38. Find all values of x such that $f(x) = -1$.

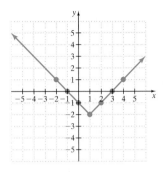

(2.3) Graph each linear function.

39. $f(x) = x$

40. $f(x) = -\dfrac{1}{3}x$

41. $g(x) = 4x - 1$

The graph of $f(x) = 3x$ is sketched below. Use this graph to match each linear function with its graph.

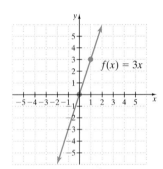

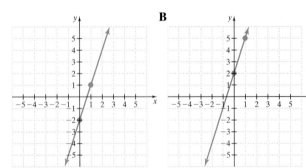

42. $f(x) = 3x + 1$

43. $f(x) = 3x - 2$

44. $f(x) = 3x + 2$

45. $f(x) = 3x - 5$

Graph each linear equation by finding intercepts if possible.

46. $4x + 5y = 20$

47. $3x - 2y = -9$

48. $4x - y = 3$

49. $2x + 6y = 9$

50. $y = 5$

51. $x = -2$

Graph each linear equation.

52. $x - 2 = 0$

53. $y + 3 = 0$

54. The cost C, in dollars, of renting a minivan for a day is given by the linear function $C(x) = 0.3x + 42$, where x is number of miles driven.

　a. Find the cost of renting the minivan for a day and driving it 150 miles.

　b. Graph $C(x) = 0.3x + 42$.

(2.4) Find the slope of the line through each pair of points.

55. $(2, 8)$ and $(6, -4)$

56. $(-3, 9)$ and $(5, 13)$

57. $(-7, -4)$ and $(-3, 6)$

58. $(7, -2)$ and $(-5, 7)$

Find the slope and y-intercept of each line.

59. $6x - 15y = 20$

60. $4x + 14y = 21$

Find the slope of each line.

61. $y - 3 = 0$

62. $x = -5$

Two lines are graphed on each set of axes. Decide whether l_1 or l_2 has the greater slope.

63.

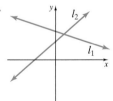

64.

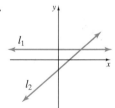

65.

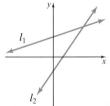

66.

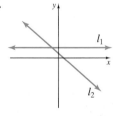

67. Recall from Exercise 54 that the cost C, in dollars, of renting a minivan for a day is given by the linear equation $y = 0.3x + 42$, where x is number of miles driven.

　a. Find and interpret the slope of this equation.

　b. Find and interpret the y-intercept of this equation.

Decide whether the lines are parallel, perpendicular, or neither.

△ **68.** $f(x) = -2x + 6$
$g(x) = 2x - 1$

△ **69.** $-x + 3y = 2$
$6x - 18y = 3$

(2.5) *Graph each linear equation using the slope and y-intercept.*

70. $y = -x + 1$

71. $y = 4x - 3$

72. $3x - y = 6$

73. $y = -5x$

Find an equation of the line satisfying the conditions given.

74. 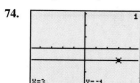 Horizontal, through $(3, -1)$

75. 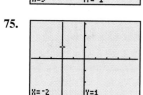 Vertical, through $(-2, 1)$

△ **76.** Parallel to the line $x = 6$; through $(-4, -3)$

77. Slope 0; through $(2, 5)$

Find the standard form equation of each line satisfying the given conditions.

78. Through $(-3, 5)$; slope 3

79. Slope 2; through $(5, -2)$

80. Through $(-6, -1)$ and $(-4, -2)$

81. Through $(-5, 3)$ and $(-4, -8)$

△ **82.** Through $(-2, 3)$; perpendicular to $x = 4$

△ **83.** Through $(-2, -5)$; parallel to $y = 8$

Find the equation of each line satisfying the given conditions. Write each equation using function notation.

84. Slope $-\dfrac{2}{3}$; y-intercept $(0, 4)$

85. Slope -1; y-intercept $(0, -2)$

△ **86.** Through $(2, -6)$; parallel to $6x + 3y = 5$

△ **87.** Through $(-4, -2)$; parallel to $3x + 2y = 8$

△ **88.** Through $(-6, -1)$; perpendicular to $4x + 3y = 5$

△ **89.** Through $(-4, 5)$; perpendicular to $2x - 3y = 6$

90. In 2005, the percent of U.S. drivers wearing seat belts was 82%. The number of drivers wearing seat belts in 2000 was 71%. Let y be the number of drivers wearing seat belts in

the year x, where $x = 0$ represents 2000. (*Source:* Strategis Group for Personal Communications Asso.)

a. Write a linear equation that models the percent of U.S. drivers wearing seat belts in terms of the year x. [*Hint:* Write 2 ordered pairs of the form (years past 2000, percent of drivers).]

b. Use this equation to predict the number of U.S. drivers wearing seat belts in the year 2009. (Round to the nearest percent.)

91. In 1998, the number of people (in millions) reporting arthritis was 43. The number of people (in millions) predicted to be reporting arthritis in 2020 is 60. Let y be the number of people (in millions) reporting arthritis in the year x, where $x = 0$ represents 1998. (*Source:* Arthritis Foundation)

a. Write a linear equation that models the number of people (in millions) reporting arthritis in terms of the year x. (See the hint for Exercise 90a.)

b. Use this equation to predict the number of people reporting arthritis in 2010. (Round to the nearest million.)

(2.6)

92. The U.S. population in thousands according to the U.S. Bureau of the Census is as given in the table below.

Year	1970	1980	1990	2000
Population (in Thousands)	203,302	226,548	248,710	275,306

(Let x represent the number of years since 1970.)

a. Plot the data using the statistical plotting feature of your graphing utility.

b. Use the data for the years given to find a linear regression equation that models the data. Graph this line with the data.

c. Use this equation to predict the population of the United States in the year 2010.

(2.7) *Graph each function.*

93. $g(x) = \begin{cases} -\dfrac{1}{5}x & \text{if } x \le -1 \\ -4x + 2 & \text{if } x > -1 \end{cases}$

94. $f(x) = \begin{cases} -3x & \text{if } x < 0 \\ x - 3 & \text{if } x \ge 0 \end{cases}$

Graph each function.

95. $f(x) = \sqrt{x - 4}$

96. $y = \sqrt{x} - 4$

97. $h(x) = -(x + 3)^2 - 1$

98. $g(x) = |x - 2| - 2$

MIXED REVIEW

Graph each linear equation.

99. $x = -4y$

100. $3x - 2y = -9$

Write an equation of the line satisfying each set of conditions. If possible, write the equation in the form $y = mx + b$.

101. Vertical; through $\left(-7, -\dfrac{1}{2}\right)$

102. Slope 0; through $\left(-4, \dfrac{9}{2}\right)$

103. Slope $\dfrac{3}{4}$; through $(-8, -4)$

104. Through $(-3, 8)$ and $(-2, 3)$

105. Through $(-6, 1)$; parallel to $y = -\dfrac{3}{2}x + 11$

106. Through $(-5, 7)$; perpendicular to $5x - 4y = 10$

Graph each piecewise-defined function.

107. $f(x) = \begin{cases} x - 2 & \text{if } x \le 0 \\ -\dfrac{x}{3} & \text{if } x \ge 3 \end{cases}$

108. $g(x) = \begin{cases} 4x - 3 & \text{if } x \le 1 \\ 2x & \text{if } x > 1 \end{cases}$

Graph each function.

109. $f(x) = \sqrt{x - 2}$

110. $f(x) = |x + 1| - 3$

CHAPTER 2 TEST TEST PREP VIDEO

Remember to use the Chapter Test Prep Video CD to see the fully worked-out solutions to any of the exercises you want to review.

1. Plot the points, and name the quadrant or axis in which each is located: $A(6, -2), B(4, 0), C(-1, 6)$.

Graph each line.

2. $2x - 3y = -6$

3. $4x + 6y = 7$

4. $f(x) = \dfrac{2}{3}x$

5. $y = -3$

6. Find the slope of the line that passes through $(5, -8)$ and $(-7, 10)$.

7. Find the slope and the y-intercept of the line $3x + 12y = 8$.

Find an equation of each line satisfying the given conditions. Write Exercises 8–12 in standard form. Write Exercises 13–15 using function notation.

8. Horizontal; through $(2, -8)$

9. Vertical; through $(-4, -3)$

△ **10.** Perpendicular to $x = 5$; through $(3, -2)$

11. Through $(4, -1)$; slope -3

12. Through $(0, -2)$; slope 5

13. Through $(4, -2)$ and $(6, -3)$

△ **14.** Through $(-1, 2)$; perpendicular to $3x - y = 4$

△ **15.** Parallel to $2y + x = 3$; through $(3, -2)$

△ **16.** Line L_1 has the equation $2x - 5y = 8$. Line L_2 passes through the points $(1, 4)$ and $(-1, -1)$. Determine whether these lines are parallel lines, perpendicular lines, or neither.

Match each graph with its equation.

17.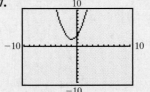

18.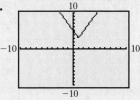

19.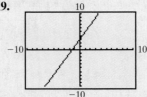

20.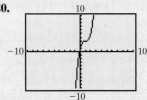

a. $y = 2|x - 1| + 3$

b. $y = x^2 + 2x + 3$

c. $y = 2(x - 1)^3 + 3$

d. $y = 2x + 3$

Find the domain and range of each relation. Also determine whether the relation is a function.

21.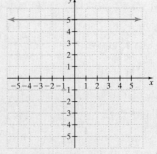

22.

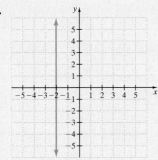

23.

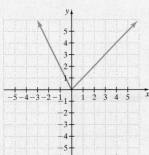

24.

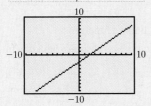

25. The average yearly earnings for high school graduates age 18 and older is given by the linear function

$$f(x) = 1031x + 25{,}193$$

where x is the number of years since 2000 that a person graduated. (*Source:* U.S. Census Bureau)

a. Find the average earnings in 2000 for high school graduates.

b. Find the average earnings for high school graduates in the year 2007.

c. Predict the first whole year that the average earnings for high school graduates will be greater than $40,000.

d. Find and interpret the slope of this equation.

e. Find and interpret the y-intercept of this equation.

Graph each function. For Exercises 26 and 29, state the domain and the range of the function.

26. $f(x) = \begin{cases} -\dfrac{1}{2}x & \text{if } x \le 0 \\ 2x - 3 & \text{if } x > 0 \end{cases}$

27. $f(x) = (x - 4)^2$

28. $g(x) = |x| + 2$

29. $g(x) = -|x + 2| - 1$

30. $h(x) = \sqrt{x} - 1$

CHAPTER 2 CUMULATIVE REVIEW

1. Evaluate: $3x - y$ when $x = 15$ and $y = 4$.

2. Add.

  a. $-4 + (-3)$

  b. $\dfrac{1}{2} - \left(-\dfrac{1}{3}\right)$

  c. $7 - 20$

3. Determine whether the following statements are true or false.

  a. 3 is a real number.

  b. $\frac{1}{5}$ is an irrational number.

  c. Every rational number is an integer.

  d. $\{1, 5\} \subseteq \{2, 3, 4, 5\}$

4. Write the opposite of each.

  a. -7

  b. 0

  c. $\dfrac{1}{4}$

5. Subtract.

  a. $2 - 8$

  b. $-8 - (-1)$

  c. $-11 - 5$

  d. $10.7 - (-9.8)$

  e. $\dfrac{2}{3} - \dfrac{1}{2}$

  f. $1 - 0.06$

  g. Subtract 7 from 4.

6. Multiply or divide.

  a. $\dfrac{-42}{-6}$

  b. $\dfrac{0}{14}$

  c. $-1(-5)(-2)$

7. Simplify each expression.

   a. 3^2

   b. $\left(\dfrac{1}{2}\right)^4$

   c. -5^2

   d. $(-5)^2$

   e. -5^3

   f. $(-5)^3$

8. Which property is illustrated?

   a. $5(x + 7) = 5 \cdot x + 5 \cdot 7$

   b. $5(x + 7) = 5(7 + x)$

9. Insert $<, >$, or $=$ between each pair of numbers to form a true statement.

   a. $-1 \qquad -2$

   b. $\dfrac{12}{4} \qquad 3$

   c. $-5 \qquad 0$

   d. $-3.5 \qquad -3.05$

10. Evaluate $2x^2$ for

   a. $x = 7$

   b. $x = -7$

11. Write the multiplicative inverse, or reciprocal, of each.

   a. 11

   b. -9

   c. $\dfrac{7}{4}$

12. Simplify $-2 + 3[5 - (7 - 10)]$.

13. Solve: $0.6 = 2 - 3.5c$

14. Solve: $2(x - 3) = -40$.

15. Solve for x: $3x + 5 = 3(x + 2)$

16. Solve: $5(x - 7) = 4x - 35 + x$.

17. Write the following as algebraic expressions. Then simplify.

   a. The sum of three consecutive integers, if x is the first consecutive integer.

   b. The perimeter of a triangle with sides of length x, $5x$, and $6x - 3$.

18. Find 25% of 16.

19. Kelsey Ohleger was helping her friend Benji Burnstine study for an algebra exam. Kelsey told Benji that her three latest art history quiz scores are three consecutive even integers whose sum is 264. Help Benji find the scores.

20. Find 3 consecutive odd integers whose sum is 213.

21. Solve $V = lwh$ for h.

22. Solve $7x + 3y = 21$ for y

23. Name the quadrant in which the point is located.

   a. $(2, -1)$ **b.** $(0, 5)$

   c. $(-3, 5)$ **d.** $(-2, 0)$

   e. $\left(-\dfrac{1}{2}, -4\right)$

   f. $(1.5, 1.5)$

24. Name the quadrant or axis in which each point is located.

   a. $(0, -2)$ **b.** $(-1, -2.5)$

   c. $\left(\dfrac{1}{2}, 0\right)$ **d.** $(4, -0.5)$

25. Determine whether $(0, -12), (1, 9),$ and $(2, -6)$ are solutions of the equation $3x - y = 12$.

26. Find the slope and y-intercept of $7x + 2y = 10$.

27. Is the relation $y = 2x + 1$ also a function?

28. Determine whether the graph below is the graph of a function.

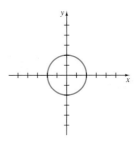

29. Find the y-intercept of the graph of each equation.

   a. $f(x) = \dfrac{1}{2}x + \dfrac{3}{7}$

   b. $y = -2.5x - 3.2$

30. Find the slope of the line through $(-1, 6)$ and $(0, 9)$.

31. Find the slope and y-intercept of the line $3x - 4y = 4$.

32. Find an equation of the vertical line through $\left(-2, -\dfrac{3}{4}\right)$.

33. Write an equation of the line with y-intercept $(0, -3)$ and slope of $\dfrac{1}{4}$.

34. Find an equation of the horizontal line through $\left(-2, -\dfrac{3}{4}\right)$.

3 Equations and Inequalities

Mathematics is a tool for solving problems in such diverse fields as transportation, engineering, economics, medicine, business, and biology. We solve problems using mathematics by modeling real-world phenomena with mathematical equations or inequalities. Our ability to solve problems using mathematics, then, depends in part on our ability to solve equations and inequalities. In this chapter, we solve linear equations and inequalities and graph their solutions.

The federal Bureau of Labor Statistics (BLS) issued its projections for job growth in the United States for the Civilian Labor Force. The table and graph below show the rise in the civilian labor force between the years 1998 and 2008. In Exercise Set 3.2, Exercise 80, you will use the equation calculated below to predict future civilian labor force. (*Source*: Bureau of Labor Statistics)

Years after 1990 (x) Civilian Labor Force (y, in thousands)

x y

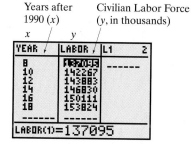

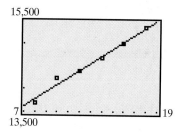

3.1 SOLVING LINEAR EQUATIONS GRAPHICALLY

In Chapter 1, we solved linear equations algebraically and checked the solution numerically with a calculator. Recall that a solution of an equation is a value for the variable that makes the equation a true statement. In this section we solve equations algebraically and graphically. Then we use various methods to check the solution. We need to stress the importance of the check. Technology allows us to reduce the time spent on solving and aids us in verifying solutions. Consequently, by using technology we can increase the accuracy of our work.

OBJECTIVE 1 ▶ Solving equations using the intersection-of-graphs method. There are two methods we will use to graphically obtain a solution of an equation. The first method we refer to as the **intersection-of-graphs method.** For this method, we graph

$$y_1 = \text{left side of equation and}$$
$$y_2 = \text{right side of equation}.$$

Recall that when a solution is substituted for a variable in an equation, the left side of the equation is equal to the right side of the equation. This means that graphically a solution occurs when $y_1 = y_2$ or where the graphs of y_1 and y_2 intersect.

TECHNOLOGY NOTE

Most graphing utilities have the ability to select and deselect graphs, graph equations simultaneously, and define different graph styles. See your graphing utility manual to check its capabilities. Also, find the instructions for using the intersection and root, or zero, features.

DISCOVER THE CONCEPT

Consider the equation $2x - 5 = 27$ and graph $y_1 = 2x - 5$ and $y_2 = 27$ in an integer window.

a. Use the trace feature to estimate the point of intersection of the two graphs.

b. Solve the equation algebraically and compare the x-coordinate of the point of intersection found in part **a** to the algebraic solution of the equation.

c. Locate the intersect feature on your graphing utility, sometimes found on the calculate menu. Use this feature to find the point of intersection of the two graphs.

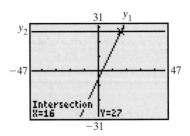

The above discovery indicates that the solution of the equation $2x - 5 = 27$ is 16, the x-coordinate of the point of intersection, as shown above.

To check this solution, we may replace x with 16 in the original equation.

$$2x - 5 = 27$$
$$2 \cdot 16 - 5 = 27 \quad \text{Let } x = 16.$$
$$32 - 5 = 27$$
$$27 = 27 \quad \text{True}$$

Recall that 27 is the y-coordinate of the point of intersection. Why? Because it is the value of both the left side and the right side of the equation when x is replaced with the solution.

The steps below may be used to solve an equation by the intersection-of-graphs method.

Intersection-of-Graphs Method for Solving an Equation

STEP 1. Graph $y_1 = $ left side of the equation and $y_2 = $ right side of the equation.

STEP 2. Find the point(s) of intersection of the two graphs.

STEP 3. The x-coordinate of a point of intersection is a solution to the equation.

STEP 4. The y-coordinate of the point of intersection is the value of both the left side and the right side of the original equation when x is replaced with the solution.

EXAMPLE 1   Solve the equation $5(x - 2) + 15 = 20$.

Algebraic Solution:

$$5(x - 2) + 15 = 20$$
$$5x - 10 + 15 = 20 \quad \text{Use the distributive property.}$$
$$5x + 5 = 20 \quad \text{Combine like terms.}$$
$$5x = 15 \quad \text{Subtract 5 from both sides.}$$
$$x = 3 \quad \text{Divide both sides by 5.}$$

To check, we replace x with 3 and see that a true statement results.

$$5(x - 2) + 15 = 20$$
$$5(3 - 2) + 15 = 20$$
$$5(1) + 15 = 20$$
$$20 = 20 \quad \text{True}$$

The solution is 3.

Graphical Solution:

$$\text{Graph } y_1 = 5(x - 2) + 15 \quad \text{Left side of equation}$$
$$y_2 = 20 \quad \text{Right side of equation}$$

Since the graph of the equation $y = 20$ is a horizontal line with y-intercept 20, use the window $[-25, 25, 5]$ by $[-25, 25, 5]$.

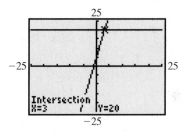

The x-coordinate of the point of intersection is 3. Thus, the solution is 3.

We can also use a table feature to check the solution. If $y_1 = 5(x - 2) + 15$ and $y_2 = 20$, scroll to $x = 3$ and see that y_1 and y_2 are both 20.   ☐

PRACTICE

1   Solve the equation $2(x - 1) + 6 = 8$.

When solving equations graphically, we often have no indication where the intersection lies before looking at the graphs of the two equations. The zoom feature of your graphing utility may be used to quickly look for an appropriate window. If the graphs are present but the intersection point is missing, you can assess the situation and decide which component to adjust on the window setting. You will quickly get used to this assessment process as you solve more equations.

EXAMPLE 2 Graphically solve the equation

$$2(x - 30) + 6(x - 10) - 70 = 35 - x.$$

Solution Graph $y_1 = 2(x - 30) + 6(x - 10) - 70$ and $y_2 = 35 - x$ in an integer window.

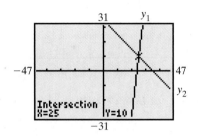

The point of intersection $(25, 10)$ indicates that the solution is 25. To check numerically with a calculator, see the screen below.

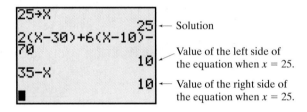

Solution ←

Value of the left side of the equation when $x = 25$.

Value of the right side of the equation when $x = 25$.

PRACTICE
2 Solve graphically: $3(x - 10) + 5(x + 2) - 25 = 9 - x$.

In Example 2, recall that the intersection point $(25, 10)$ indicates that if $x = 25$, then the value of the expression on the left side of the equation $2(x - 30) + 6(x - 10) - 70$ is 10, and the value of the expression on the right side of the equation, $35 - x$, is also 10.

> **▶ Helpful Hint**
> The integer window allows the cursor to move along a graph with integer x-values only. Using this window does *not* mean that the calculator will give only integer values when calculating a point of intersection.

> **▶ Helpful Hint**
> In general, when using the intersection method, the x-coordinate of the point of intersection is a solution of the equation, and the y-coordinate of the point of intersection is the value of each side of the original equation when the variable is replaced with the solution.

EXAMPLE 3 Solve the equation using the intersection-of-graphs method.

$$5.1x + 3.78 = x + 4.7$$

Solution Define $y_1 = 5.1x + 3.78$ and $y_2 = x + 4.7$ and graph in a standard window. The screen below shows the intersection point.

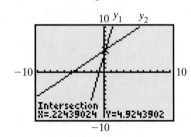

Notice the coordinates of the intersection point. When this happens, we will approximate the solution. For this example, the solution rounded to four decimal places is 0.2244.

To check when the solution is an approximation, note that the value of the left side of the equation and the value of the right side may differ slightly because of rounding. In the screen below we see a check of the equation for $x \approx 0.2244$.

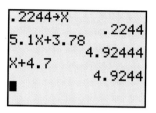

The solution is approximately ($\approx$) 0.2244.

PRACTICE

3 Solve the equation using the intersection-of-graphs method. Round the solution to 3 decimal places. $3.51x + 5.728 = x - 3.41$

OBJECTIVE 2 ▶ Solving applications using graphing methods. Next, we use the intersection-of-graphs method to solve a problem.

EXAMPLE 4 **Purchasing a Refrigerator**

The McDonalds are purchasing a new refrigerator. They find two models that fit their needs. Model 1 sells for $575 and costs $0.07 per hour to run. Model 2 is the energy efficient model that sells for $825, but only costs $0.04 per hour to run. If x represents number of hours, then the costs to purchase and run the refrigerators are modeled by the equations

$$C_1(x) = 575 + 0.07x \quad \text{Cost to buy and run Model 1}$$
$$C_2(x) = 825 + 0.04x \quad \text{Cost to buy and run Model 2}$$

a. If the McDonalds buy Model 2, how many hours must it run before they save money?

b. How many days must it run before they save money?

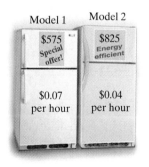

Solution Graph $y_1 = 575 + 0.07x$ and $y_2 = 825 + 0.04x$. Here, the ordered pairs represent (x, y).

hours   cost

a. Find the point of intersection of the graphs.

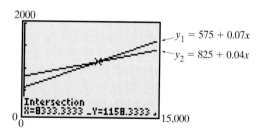

The point of intersection is $(8333.\overline{3}, 1158.\overline{3})$. This means that in approximately 8333.33 hours, both machines cost a total of approximately $1158.33 to buy and run. Before that point of intersection, the graph of y_1 is below the graph of y_2. This means that the cost of Model 1 is less than the cost of Model 2. Likewise, after the point of intersection, the graph of y_2 is below the graph of y_1. This means that the cost of Model 2 is less than the cost of Model 1. Thus, in approximately 8333.33 hours, the McDonalds start saving money.

b. To find the approximate number of days, find

$$\frac{8333.33}{24} \approx 347.22 \text{ days}.$$ ☐

PRACTICE

4   Using the data from Example 4, find how much more its costs (original purchase price and use) for the $575 refrigerator at the end of 2 years than the $825 model.

EXAMPLE 5   Cell Phone Usage vs. Residential Phone Usage

The annual expenditure for cell phone usage can be modeled using the equation $C(x) = 60.429x + 151.333$. The annual expenditure for residential (landline) phone usage can be modeled by the equation $R(x) = -27.457x + 704.6$. In both cases $x =$ the number of years since 2000. Using these models, find what year we can expect the amount of money spent for cell phone usage to equal the amount spent for residential (landline) usage, or find when $C(x) = R(x)$.

Solution   Graph

$$y_1 = 60.429x + 151.333 \text{ and}$$
$$y_2 = -27.457x + 704.6$$

where the ordered pairs (x, y) represent (years, cost). Find the point of intersection of the two graphs. Let's first find an appropriate window. We'll start by looking at the years between 2000 and 2015 (x-values) and considering annual costs from $0 to $1000 ($y$-values). In other words, we'll start with the window [0, 15, 1) by [0, 1000, 100].

 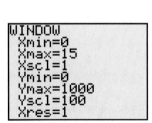 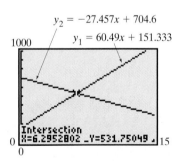

The point of intersection is approximately (6.295, 531.750). This means that during the year 2006 (2000 + 6) the annual usage cost per unit for cell phones equaled the cost per unit for residential (landline) phones and that average cost was about $531.75. ☐

PRACTICE

5   Use the graphs in Example 5 to predict the average annual expenditure for cell phone usage and residential (landline) phone usage in the year 2012 (if the trend continues at the same rates.) Approximate the answers to 2 decimal places.

In the Exercise set for this section, we will revisit some of the exercises from Section 1.7 where we solved equations numerically using tables. Now we look at another option for solving the same situations graphically.

OBJECTIVE 3 ▶ Solving an identity or a contradiction. Recall in Section 1.5 that we solved two special types of equations: contradictions and identities. An equation in one variable that has no solution is called a contradiction and an equation in one variable that has every number (for which the equation is defined) as a solution is called an identity. We now look at these two special types of equations graphically.

EXAMPLE 6   Solve: $4(x + 6) - 2(x - 3) = 2x + 30$

Algebraic Solution:

$$4(x + 6) - 2(x - 3) = 2x + 30$$
$$4x + 24 - 2x + 6 = 2x + 30 \quad \text{Multiply.}$$
$$2x + 30 = 2x + 30 \quad \text{Simplify.}$$

Since both sides are the same, we see that replacing x with any real number will result in a true statement. This equation is an identity and all real numbers are solutions. Using set notation, the solution set is $\{x \mid x \text{ is a real number}\}$.

Graphical Solution:

Graph $y_1 = 4(x + 6) - 2(x - 3)$ and
$$y_2 = 2x + 30$$

The graphs shown below appear to be the same since there appears to be a single line only. When we trace along y_1, we have the same ordered pairs as when we trace along y_2. (If 2 points of y_1 are the same as 2 points of y_2, we know the lines are identical because 2 points uniquely determine a line.)

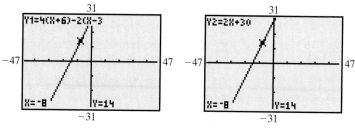

Thus, this equation is an identity. All real numbers are solutions, or the solution set is $\{x \mid x \text{ is a real number}\}$.

Once again, both solutions show that the equation is an identity and the solution set is the set of real numbers or $\{x \mid x \text{ is a real number}\}$.   ☐

PRACTICE
6   Solve: $5(x - 1) - 3(x + 7) = 2x - 26$.

EXAMPLE 7   Solve: $3x - 8 = 5(x - 1) - 2(x + 6)$.

Algebraic Solution:

$$3x - 8 = 5(x - 1) - 2(x + 6)$$
$$3x - 8 = 5x - 5 - 2x - 12 \quad \text{Multiply.}$$
$$3x - 8 = 3x - 17 \quad \text{Simplify.}$$
$$3x - 8 - 3x = 3x - 17 - 3x \quad \text{Subtract } 3x.$$
$$-8 = -17$$

This equation is a false statement no matter what value the variable x might have. The equation has no solution and is a contradiction. The solution set is $\{\ \}$ or $\varnothing$.

Graphical Solution:

Graph $y_1 = 3x - 8$ and
$$y_2 = 5(x - 1) - 2(x + 6)$$

The graph to the right shows that the lines appear to be parallel and, therefore, will never intersect.

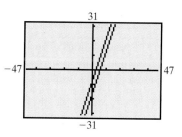

To be sure, check the slope of each graph.

$$y_1 = 3x - 8 \qquad y_2 = 5(x - 1) - 2(x + 6)$$
$$y_2 = 3x - 17$$

The lines have the same slope, 3, and their graphs are distinct, so they are indeed parallel. Since the lines do not intersect, there is no solution. The equation is a contradiction and the solution set is $\{\ \}$ or $\varnothing$.

Both solutions show that the equation is a contradiction and the solution set is $\{\ \}$ or $\varnothing$. □

PRACTICE

7 Solve: $2x - 7 = 7(x + 1) - 5(x - 2)$.

OBJECTIVE 4 ▶ **Solving equations using the x-intercept method.** We now look at another method for solving equations graphically, called the **x-intercept method.** It is actually a form of the intersection-of-graphs method, but for this method we write an equivalent equation with one side of the equation 0. Since $y = 0$ is the x-axis, we look for points where the graph intersects the x-axis or x-intercepts. Recall that an x-intercept is of the form $(x, 0)$. (An x-intercept of $(a, 0)$ is sometimes simply called an x-intercept of a.) Thus, the solutions that lie on the x-axis are called the **zeros** of the equation since this is where $y = 0$. They are also referred to as **roots** of the equation. The built-in feature on graphing utilities to find these solutions is referred to as root or zero on different graphing utilities.

---

x-Intercept Method (or Zeros Method) for Solving an Equation

STEP 1. Write the equation so that one side is 0.

STEP 2. Graph y_1 = the nonzero side of the equation.

STEP 3. Find an x-intercept or zero of the graph.

STEP 4. The x-coordinate of an x-intercept is a solution of the equation.

▶ **Helpful Hint**

Don't forget: One way to check a solution is to substitute it back into the *original equation* and see that a true statement results.

EXAMPLE 8 Solve the following equation using the x-intercept method.

$$-3.1(x + 1) + 8.3 = -x + 12.4$$

Solution We begin by writing the equation so that one side equals 0 by adding x to both sides and subtracting 12.4 from both sides.

$$-3.1(x + 1) + 8.3 = -x + 12.4$$
$$-3.1(x + 1) + 8.3 + x - 12.4 = 0$$

Define $y_1 = -3.1(x + 1) + 8.3 + x - 12.4$ and graph in a standard window. Choose the root, or zero, feature to solve. You will be prompted to indicate a left bound—an x-value less than the x-intercept and a right bound—an x-value greater than the x-intercept. The guess portion of the prompting asks you to move the cursor close to the x-intercept.

TECHNOLOGY NOTE

The real solutions of an equation in the form $f(x) = 0$ occur at the x-intercepts of the graph $y = f(x)$. Because of this, most graphing utilities have a root or zero feature to find the x-intercepts.

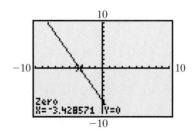

The x-intercept is approximately $(-3.428571, 0)$, which means that the solution to the equation is approximately -3.428571.

To check, replace x with -3.428571 and see that the left side of the equation is approximately the right side. Rounded to two decimal places, $x \approx -3.43$, or the solution is approximately -3.43.

PRACTICE
8   Solve the equation using the x-intercept method.
$$2.5(x - 1) + 3.4 = 2x - 1$$

VOCABULARY & READINESS CHECK

Fill in each blank using the choices below.

x-intercept	x	zero	intersection-of-graphs	y	root

1. To solve an equation by the _____ method, graph the left-hand side of the equation in y_1, the right-hand side of the equation in y_2, and find the intersection point of the two graphs.

2. To solve an equation by the _____ method, first write the equation as an equivalent equation with one side 0.

3. For the intersection-of-graphs method, the __-coordinate of the point of intersection is the solution of the equation.

4. For the intersection-of-graphs method, the __-coordinate of the point of intersection is the value of each side of the equation when the solution is substituted into the equation.

5. List two other terms for the solution of an equation: _____ , _____ .

Mentally solve the following equations.

6. $4x = 24$

7. $6x = -12$

8. $2x + 10 = 20$

9. $5x + 25 = 30$

10. $-3x = 0$

11. $-2x = -14$

12. $2x + 3 = 2x - 1$

13. $2(x - 5) = 2x - 10$

3.1 | EXERCISE SET   MyMathLab   PRACTICE   WATCH   DOWNLOAD   READ   REVIEW

Solve each equation algebraically and graphically. See Examples 1 through 3 and 8.

1. $5x + 2 = 3x + 6$

2. $2x + 9 = 3x + 7$

3. $9 - x = 2x + 12$

4. $3 - 4x = 2 - 3x$

5. $8 - (2x - 1) = 13$

6. $3(2 - x) + 4 = 2x + 3$

For each given screen, write an equation in x and its solution.

7.
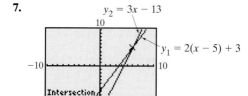
$y_2 = 3x - 13$
$y_1 = 2(x - 5) + 3$
Intersection
X=6   Y=5

8.
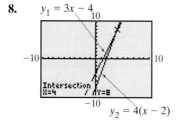
$y_1 = 3x - 4$
Intersection
X=4   Y=8
$y_2 = 4(x - 2)$

9.

$y_1 = -(2x + 3) + 2$
Same Line
$y_2 = 5x - 1 - 7x$

10.
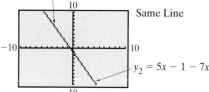
$y_1 = 5x + 2$
Parallel Lines
$y_2 = 2(x - 1) + 3x$

Solve each equation algebraically and graphically. See Examples 6 and 7.

11. $7(x - 6) = 5(x + 2) + 2x$

12. $6x - 9 = 6(x - 3)$

🔧 **13.** $3x - (6x + 2) = -(3x + 2)$

14. $8x - (2x + 3) = 6(x + 5)$

15. $5(x + 1) - 3(x - 7) = 2(x + 4) - 3$

16. $(5x + 8) - 2(x + 3) = (7x - 4) - (4x + 6)$

17. $3(x + 2) - 6(x - 5) = 36 - 3x$

18. $2(x + 3) - 7 = 4 + 2x$

MIXED PRACTICE

Solve each equation. If necessary, round solutions to two decimal places.

🔧 **19.** $(x + 2.1) - (0.5x + 3) = 12$

20. $2(x + 1.3) - 4(5 - x) = 15$

21. $5(a - 12) + 2(a + 15) = a - 9$

22. $8(b + 2) - 4(b - 5) = b + 12$

23. $8(p - 4) - 5(2p + 3) = 3.5(2p - 5)$

24. $3(x + 2) - 5(x - 7) = x + 3$

🔧 **25.** $5(x - 2) + 2x = 7(x + 4)$

26. $3x + 2(x + 4) = 5(x + 1) + 3$

27. $y + 0.2 = 0.6(y + 3)$

28. $-(w + 0.2) = 0.3(4 - w)$

29. $2y + 5(y - 4) = 4y - 2(y - 10)$

30. $9c - 3(6 - 5c) = c - 2(3c + 9)$

31. $2(x - 8) + x = 3(x - 6) + 2$

32. $4(x + 5) = 3(x - 4) + x$

33. $\dfrac{5x - 1}{6} - 3x = \dfrac{1}{3} + \dfrac{4x + 3}{9}$

34. $\dfrac{2r - 5}{3} - \dfrac{r}{5} = 4 - \dfrac{r + 8}{10}$

35. $-2(b - 4) - (3b - 1) = 5b + 3$

36. $4(t - 3) - 3(t - 2) = 2t + 8$

37. $1.5(4 - x) = 1.3(2 - x)$

38. $2.4(2x + 3) = -0.1(2x + 3)$

39. $\dfrac{1}{4}(a + 2) = \dfrac{1}{6}(5 - a)$

40. $\dfrac{1}{3}(8 + 2c) = \dfrac{1}{5}(3c - 5)$

Use a graphing method to solve. See Examples 4 and 5.

41. Acme Mortgage Company wants to hire a student computer consultant. A first consultant charges an initial fee of $30 plus $20 per hour. A second consultant charges a flat fee of $25 per hour. If x represents number of hours, then the costs of the consultants are modeled by

$C_1(x) = 30 + 20x$ First consultant's total cost

$C_2(x) = 25x$ Second consultant's total cost

  a. Find which consultant's cost is lower if the job takes 2 hours.

  b. Find which consultant's cost is lower if the job takes 8 hours.

  c. When is the cost of hiring each consultant the same?

42. Rod Pasch needs to hire a graphic artist. A first graphic artist charges $50 plus $35 per hour. A second one charges $50 per hour. If x represents number of hours, then the costs of the graphic artists can be modeled by

$G_1(x) = 50 + 35x$ First graphic artist's total cost

$G_2(x) = 50x$ Second graphic artist's total cost

  a. Which graphic artist costs less if the job takes 3 hours?

  b. Which graphic artist costs less if the job takes 5 hours?

  c. When is the cost of hiring each artist the same?

43. One car rental agency charges $25 a day plus $0.30 a mile. A second car rental agency charges $28 a day plus $0.25 a mile. If x represents the number of miles driven, then the costs of the car rental agencies for a 1-day rental can be modeled by

$R_1(x) = 25 + 0.30x$ First agency's cost

$R_2(x) = 28 + 0.25x$ Second agency's cost

  a. Which agency's cost is lower if 50 miles are driven?

  b. Which agency's cost is lower if 100 miles are driven?

  c. When is the cost of using each agency the same?

44. Copycat Printing charges $18 plus $0.03 per page for making black and white copies. Duplicate, Inc. charges $0.05 per black and white page copied. If x represents the number of pages copied, then the charges for the companies can be modeled by

$C_1(x) = 18 + 0.03x$ Copycat

$C_2(x) = 0.05x$ Duplicate, Inc.

  a. For 500 copies, which company charges less?

  b. For 1000 copies, which company charges less?

  c. When is the cost of using each company the same?

45. Tara Outzen is considering a job in pharmaceutical sales. She is offered a monthly salary of $1000 plus 4% commission with one company or an 8% commission-only position. How much must she sell per month, in order that the two plans give her the same gross pay?

46. Cheryl Brooks is a Certified Public Accountant and has clients who she charges a flat $500 fee for doing tax returns and others she charges an initial consultation fee of $150 and then $125 per hour. How many hours would she work that both plans would be the same charge to the client?

🔧 **47.** The cost of producing x calculators is $125,000 in fixed costs and $75 per calculator. The calculators sell for $150 each. Find the break-even point.

48. The local college finds that the cost of producing CDs for their distance learning program is $550 and $2 per CD and there are 10 CDs in each course packet. The CD packet sells for $45 each. Find how many students must purchase the packet in order for the college to break even.

REVIEW AND PREVIEW

Determine which numbers in the set $\{-3, -2, -1, 0, 1, 2, 3\}$ *are solutions of each inequality. See Section 1.4.*

49. $x < 0$ **50.** $x > 1$

51. $x + 5 \leq 6$ **52.** $x - 3 \geq -7$

53. In your own words, explain what real numbers are solutions of $x < 0$.

54. In your own words, explain what real numbers are solutions of $x > 1$.

CONCEPT EXTENSIONS

The given screen shows the graphs of y_1 *and* y_2 *and their intersection. Use this screen to answer the questions below.*

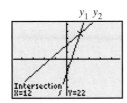

55. Complete the ordered pair for the graphs of both y_1 and y_2:
(12,   ).

56. If x is less than 12, is y_1 less than, greater than, or equal to y_2?

57. If x is greater than 12, is y_1 less than, greater than, or equal to y_2?

58. True or false? If x is 45, $y_1 < y_2$.

Solve each equation graphically and check by a method of your choice. Round solutions to the nearest hundredth.

59. $1.75x - 2.5 = 0$ **60.** $3.1x + 5.6 = 0$

61. $2.5x + 3 = 7.8x - 5$ **62.** $4.8x - 2.3 = 6.8x + 2.7$

63. $3x + \sqrt{5} = 7x - \sqrt{2}$ **64.** $0.9x + \sqrt{3} = 2.5x - \sqrt{5}$

65. $2\pi x - 5.6 = 7(x - \pi)$ **66.** $-\pi x + 1.2 = 0.3(x - 5)$

67. If the intersection-of-graphs method leads to parallel lines, explain what this means in terms of the solution of the original equation.

68. If the intersection-of-graphs method leads to the same line, explain what this means in terms of the solution of the original equation.

3.2 LINEAR INEQUALITIES AND PROBLEM SOLVING

OBJECTIVES

1 Use interval notation.

2 Solve linear inequalities using the addition property of inequality.

3 Solve linear inequalities using the multiplication and the addition properties of inequality.

4 Solve problems that can be modeled by linear inequalities.

Relationships among measurable quantities are not always described by equations. For example, suppose that a salesperson earns a base of $600 per month plus a commission of 20% of sales. Suppose we want to find the minimum amount of sales needed to receive a total income of *at least* $1500 per month. Here, the phrase "at least" implies that an income of $1500 *or more* is acceptable. In symbols, we can write

$$\text{income} \geq 1500$$

This is an example of an inequality, and we will solve this problem in Example 8.

A **linear inequality** is similar to a linear equation except that the equality symbol is replaced with an inequality symbol, such as $<$, $>$, $\leq$, or $\geq$.

Linear Inequalities in One Variable

$3x + 5 \geq 4$	$2y < 0$	$3(x - 4) > 5x$	$\dfrac{x}{3} \leq 5$
↑	↑	↑	↑
is greater than or equal to	is less than	is greater than	is less than or equal to

> **Linear Inequality in One Variable**
>
> A linear inequality in one variable is an inequality that can be written in the form
>
> $$ax + b < c$$
>
> where a, b, and c are real numbers and $a \neq 0$.

In this section, when we make definitions, state properties, or list steps about an inequality containing the symbol $<$, we mean that the definition, property, or steps apply to inequalities containing the symbols $>$, $\leq$, and $\geq$ also.

OBJECTIVE 1 ▶ Using interval notation. A **solution** of an inequality is a value of the variable that makes the inequality a true statement. The **solution set** of an inequality is the set of all solutions. Notice that the solution set of the inequality $x > 2$, for example, contains all numbers greater than 2. Its graph is an interval on the number line since an infinite number of values satisfy the variable. If we use open/closed-circle notation, the graph of $\{x \mid x > 2\}$ looks like the following.

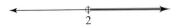

In this text **interval notation** will be used to write solution sets of inequalities. To help us understand this notation, a different graphing notation will be used. Instead of an open circle, we use a parenthesis. With this new notation, the graph of $\{x \mid x > 2\}$ now looks like

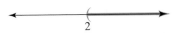

and can be represented in interval notation as $(2, \infty)$. The symbol ∞ is read "infinity" and indicates that the interval includes *all* numbers greater than 2. The left parenthesis indicates that 2 *is not* included in the interval.

In the case where 2 *is* included in the interval, we use a bracket. The graph of $\{x \mid x \geq 2\}$ is below

and can be represented as $[2, \infty)$.

The following table shows three equivalent ways to describe an interval: in set notation, as a graph, and in interval notation.

Set Notation	Graph	Interval Notation
$\{x \mid x < a\}$		$(-\infty, a)$
$\{x \mid x > a\}$		(a, ∞)
$\{x \mid x \leq a\}$		$(-\infty, a]$
$\{x \mid x \geq a\}$		$[a, \infty)$
$\{x \mid a < x < b\}$		(a, b)
$\{x \mid a \leq x \leq b\}$		$[a, b]$
$\{x \mid a < x \leq b\}$		$(a, b]$
$\{x \mid a \leq x < b\}$		$[a, b)$
$\{x \mid x \text{ is a real number}\}$		$(-\infty, \infty)$

▶ **Helpful Hint**

Notice that a parenthesis is always used to enclose ∞ and $-\infty$.

Concept Check ✓

Explain what is wrong with writing the interval $(5, \infty]$.

EXAMPLE 1   Graph each set on a number line and then write in interval notation.

a. $\{x \mid x \geq 2\}$ **b.** $\{x \mid x < -1\}$ **c.** $\{x \mid 0.5 < x \leq 3\}$

Solution

a. $[2, \infty)$

b. $(-\infty, -1)$

c. $(0.5, 3]$

PRACTICE
1   Graph each set on a number line and then write in interval notation.

a. $\{x \mid x < 3.5\}$
b. $\{x \mid x \geq -3\}$
c. $\{x \mid -1 \leq x < 4\}$

OBJECTIVE 2 ▶ Solving linear inequalities using the addition property. We will use interval notation to write solutions of linear inequalities. To solve a linear inequality, we use a process similar to the one used to solve a linear equation. We use properties of inequalities to write equivalent inequalities until the variable is isolated.

Addition Property of Inequality
If a, b, and c are real numbers, then

$$a < b \quad \text{and} \quad a + c < b + c$$

are equivalent inequalities.

In other words, we may add the same real number to both sides of an inequality and the resulting inequality will have the same solution set. This property also allows us to subtract the same real number from both sides.

EXAMPLE 2   Solve algebraically: $x - 2 < 5$. Graph the solution set on a number line and write it in interval notation.

Solution
$$x - 2 < 5$$
$$x - 2 + 2 < 5 + 2 \quad \text{Add 2 to both sides.}$$
$$x < 7 \quad \text{Simplify.}$$

The solution set is $\{x \mid x < 7\}$, which in interval notation is $(-\infty, 7)$. The number-line graph of the solution set is

PRACTICE
2   Solve: $x + 5 > 9$. Graph the solution set on a number line and write it in interval notation.

▶ **Helpful Hint**

In Example 2, the solution set is $\{x \mid x < 7\}$. This means that *all* numbers less than 7 are solutions. For example, $6.9, 0, -\pi, 1$, and -56.7 are solutions, just to name a few. To see this, replace x in $x - 2 < 5$ with each of these numbers and see that the result is a true inequality.

EXAMPLE 3   Solve algebraically: $3x + 4 \geq 2x - 6$. Graph the solution set on a number line and write it in interval notation.

Solution

$$3x + 4 \geq 2x - 6$$
$$3x + 4 - 2x \geq 2x - 6 - 2x \quad \text{Subtract } 2x \text{ from both sides.}$$
$$x + 4 \geq -6 \quad \text{Combine like terms.}$$
$$x + 4 - 4 \geq -6 - 4 \quad \text{Subtract 4 from both sides.}$$
$$x \geq -10 \quad \text{Simplify.}$$

The solution set is $\{x \mid x \geq -10\}$, which in interval notation is $[-10, \infty)$. The number-line graph of the solution set is

$$\xleftarrow{\quad} \underset{-11\ -10\ -9\ -8\ -7\ -6}{\rule{3cm}{0pt}} \xrightarrow{\quad}$$

□

PRACTICE

3   Solve: $8x + 21 \leq 2x - 3$. Graph the solution set on a number line and write it in interval notation.

DISCOVER THE CONCEPT

Let's use a graphing utility to check the solution of $3x + 4 \geq 2x - 6$ from Example 3. Let $y_1 = 3x + 4$ and $y_2 = 2x - 6$ and we can now think of the inequality as $y_1 \geq y_2$.

a. Graph y_1 and y_2 using the window $[-15, 20, 5]$ by $[-40, 30, 10]$.

b. Find the point of intersection of the graphs. What does this point represent?

c. Determine where the graph of y_1 is above the graph of y_2.

d. Now find the x-values for which $y_1 \geq y_2$.

e. Write the solution set of the inequality $y_1 \geq y_2$.

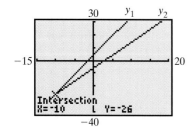

The graphs of y_1 and y_2 are shown to the left. From the above discovery, we find that

b. The point of intersection is $(-10, -26)$. This means that the solution of the equation $y_1 = y_2$ is -10.

c. The graph of y_1 is above the graph of y_2 to the right of the point of intersection, $(-10, -26)$.

d. The x-values for which $y_1 \geq y_2$ are -10 and those x-values to the right of -10 on the x-axis.

e. The solution set is $[-10, \infty)$.

Since there are an infinite number of solutions, checking solution after solution by hand is impossible. Here, the ability to check by graphing is a huge advantage.

OBJECTIVE 3 ▶ **Solving linear inequalities using the multiplication and addition properties.** Next, we introduce and use the multiplication property of inequality to solve linear inequalities. To understand this property, let's start with the true statement $-3 < 7$ and multiply both sides by 2.

$$-3 < 7$$
$$-3(2) < 7(2) \quad \text{Multiply by 2.}$$
$$-6 < 14 \quad \text{True}$$

The statement remains true.

Notice what happens if both sides of $-3 < 7$ are multiplied by -2.

$$-3 < 7$$
$$-3(-2) < 7(-2) \quad \text{Multiply by } -2.$$
$$6 < -14 \quad \text{False}$$

The inequality $6 < -14$ is a false statement. However, **if the direction of the inequality sign is reversed,** the result is true.

$$6 > -14 \quad \text{True}$$

These examples suggest the following property.

Multiplication Property of Inequality

If a, b, and c are real numbers and c is **positive**, then

$$a < b \quad \text{and} \quad ac < bc$$

are equivalent inequalities.

If a, b, and c are real numbers and c is **negative**, then

$$a < b \quad \text{and} \quad ac > bc$$

are equivalent inequalities.

In other words, we may multiply both sides of an inequality by the same positive real number and the result is an equivalent inequality.

We may also multiply both sides of an inequality by the same **negative number** and **reverse the direction of the inequality symbol**, and the result is an equivalent inequality. The multiplication property holds for division also, since division is defined in terms of multiplication.

▶ **Helpful Hint**

Whenever both sides of an inequality are multiplied or divided by a negative number, the direction of the inequality symbol **must be** reversed to form an equivalent inequality.

EXAMPLE 4   Solve algebraically and graph the solution set on a number line. Write the solution set in interval notation.

a. $\dfrac{1}{4}x \le \dfrac{3}{8}$ **b.** $-2.3x < 6.9$

Solution

a.

$$\frac{1}{4}x \le \frac{3}{8}$$

$$4 \cdot \frac{1}{4}x \le 4 \cdot \frac{3}{8} \quad \text{Multiply both sides by 4.}$$

$$x \le \frac{3}{2} \quad \text{Simplify.}$$

▶ **Helpful Hint**

The inequality symbol is the same since we are multiplying by a *positive* number.

The solution set is $\left\{ x \mid x \le \dfrac{3}{2} \right\}$, which in interval notation is $\left(-\infty, \dfrac{3}{2} \right]$. The number-line graph of the solution set is

b.

$$-2.3x < 6.9$$

$$\dfrac{-2.3x}{-2.3} > \dfrac{6.9}{-2.3} \qquad \text{Divide both sides by } -2.3 \text{ and reverse the inequality symbol.}$$

$$x > -3 \qquad \text{Simplify.}$$

> ▶ **Helpful Hint**
> The inequality symbol is *reversed* since we divided by a *negative* number.

The solution set is $\{ x \mid x > -3 \}$, which is $(-3, \infty)$ in interval notation. The number-line graph of the solution set is

PRACTICE

4 Solve and graph the solution set on a number line. Write the solution set in interval notation.

a. $\dfrac{2}{5}x \ge \dfrac{4}{15}$

b. $-2.4x < 9.6$

Concept Check ✓

In which of the following inequalities must the inequality symbol be reversed during the solution process?

a. $-2x > 7$　　　　　　　　　**b.** $2x - 3 > 10$

c. $-x + 4 + 3x < 7$　　　　　**d.** $-x + 4 < 5$

To solve linear inequalities in general, we follow steps similar to those for solving linear equations.

> **Solving a Linear Inequality in One Variable**
>
> **STEP 1.** Clear the inequality of fractions by multiplying both sides of the inequality by the least common denominator (LCD) of all fractions in the inequality.
>
> **STEP 2.** Use the distributive property to remove grouping symbols such as parentheses.
>
> **STEP 3.** Combine like terms on each side of the inequality.
>
> **STEP 4.** Use the addition property of inequality to write the inequality as an equivalent inequality with variable terms on one side and numbers on the other side.
>
> **STEP 5.** Use the multiplication property of inequality to isolate the variable.

EXAMPLE 5 Solve algebraically: $-(x - 3) + 2 \le 3(2x - 5) + x$ and check graphically.

Solution

$$-(x - 3) + 2 \le 3(2x - 5) + x$$

$$-x + 3 + 2 \le 6x - 15 + x \qquad \text{Apply the distributive property.}$$

$$5 - x \le 7x - 15 \qquad \text{Combine like terms.}$$

$$5 - x + x \le 7x - 15 + x \qquad \text{Add } x \text{ to both sides.}$$

$$5 \le 8x - 15 \qquad \text{Combine like terms.}$$

Answer to Concept Check:
a, d

$$5 + 15 \leq 8x - 15 + 15 \qquad \text{Add 15 to both sides.}$$
$$20 \leq 8x \qquad\qquad \text{Combine like terms.}$$
$$\frac{20}{8} \leq \frac{8x}{8} \qquad\qquad \text{Divide both sides by 8.}$$
$$\frac{5}{2} \leq x, \quad \text{or} \quad x \geq \frac{5}{2} \qquad \text{Simplify.}$$

The solution set written in interval notation is $\left[\dfrac{5}{2}, \infty\right)$.

To check, graph $y_1 = -(x - 3) + 2$ and $y_2 = 3(2x - 5) + x$ and find x-values such that $y_1 \leq y_2$.

The point of intersection is $(2.5, 2.5)$. The graph of y_1 is equal to or below the graph of y_2 for all x-values equal to or greater than 2.5, or for the interval $\left[\dfrac{5}{2}, \infty\right)$.

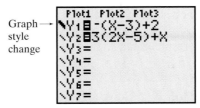

Graph style change →

To change the graph style, go to the icon in front of the **Y=** and press Enter until you get the desired style.

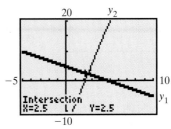

A calculator check of Example 5.   □

PRACTICE

5   Solve: $-(4x + 6) \leq 2(5x + 9) + 2x$. Check graphically and write the solution set in interval notation.

EXAMPLE 6   Solve algebraically: $\dfrac{2}{5}(x - 6) \geq x - 1$ and check graphically.

**Solution**

$$\frac{2}{5}(x - 6) \geq x - 1$$

$$5\left[\frac{2}{5}(x - 6)\right] \geq 5(x - 1) \qquad \text{Multiply both sides by 5 to eliminate fractions.}$$

$$2(x - 6) \geq 5(x - 1)$$

$$2x - 12 \geq 5x - 5 \qquad \text{Apply the distributive property.}$$

$$-3x - 12 \geq -5 \qquad \text{Subtract } 5x \text{ from both sides.}$$

$$-3x \geq 7 \qquad\qquad \text{Add 12 to both sides.}$$

$$\frac{-3x}{-3} \leq \frac{7}{-3} \qquad\qquad \text{Divide both sides by } -3 \text{ and reverse the inequality symbol.}$$

$$x \leq -\frac{7}{3} \qquad\qquad \text{Simplify.}$$

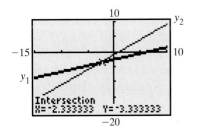

The intersection of y_1 and y_2 is the point $\left(-\dfrac{7}{3}, -\dfrac{10}{3}\right)$. $y_1 \geq y_2$ for all x-values in the interval $\left(-\infty, -\dfrac{7}{3}\right]$ since this is where y_1 is equal to or above y_2.

The solution set written in interval notation is $\left(-\infty, -\dfrac{7}{3}\right]$. The graphing utility check is shown in the margin.   □

PRACTICE

6   Solve: $\dfrac{3}{5}(x - 3) \geq x - 7$. Check graphically and write the solution set in interval notation.

Recall that when the graph of the left side and the right side of an equation do not intersect, the equation has no solution. Let's see what a similar situation means for inequalities.

DISCOVER THE CONCEPT

a. Use a graphing utility to solve $2(x + 3) > 2x + 1$. To do so, graph $y_1 = 2(x + 3)$ and $y_2 = 2x + 1$ and find the x-values for which the graph of y_1 is above the graph of y_2.

b. Next, solve $2(x + 3) < 2x + 1$. Notice that the graphs of y_1 and y_2 found in part **a** can be used. Find the x-values for which the graph of y_1 is below the graph of y_2.

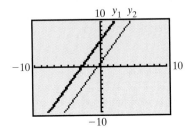

The graphs of parallel lines y_1 and y_2 from the discovery above are shown at the left. Since the graph of y_1 is *always* above the graph of y_2, all x-values satisfy the inequality $y_1 > y_2$ or $2(x + 3) > 2x + 1$. The solution set is all real numbers or $(-\infty, \infty)$.

Also, notice that the graph of y_2 is *never* above the graph of y_1. This means that no x-values satisfy $y_1 < y_2$ or $2(x + 3) < 2x + 1$. The solution set is the empty set or $\emptyset$.

These results can also be obtained algebraically as we see in Example 7.

EXAMPLE 7 Solve algebraically:

a. $2(x + 3) > 2x + 1$ **b.** $2(x + 3) < 2x + 1$

Solution

a.

$$2(x + 3) > 2x + 1$$

$2x + 6 > 2x + 1$ Distribute on the left side.

$2x + 6 - 2x > 2x + 1 - 2x$ Subtract $2x$ from both sides.

$6 > 1$ Simplify.

$6 > 1$ is a true statement for all values of x, so this inequality and the original inequality are true for all numbers. The solution set is $\{x \,|\, x \text{ is a real number}\}$, or $(-\infty, \infty)$ in interval notation.

b. Solving $2(x + 3) < 2x + 1$ in a similar fashion leads us to the statement $6 < 1$. This statement, as well as the original inequality, is false for all values of x. This means that the solution set is the empty set, or $\emptyset$.   □

PRACTICE

7   Solve algebraically: **a.** $4(x - 2) < 4x + 5$ **b.** $4(x - 2) > 4x + 5$. Write the solution set in interval notation.

OBJECTIVE 4 ▶ Solving problems modeled by linear inequalities. Application problems containing words such as "at least," "at most," "between," "no more than," and "no less than" usually indicate that an inequality be solved instead of an equation. In solving applications involving linear inequalities, we use the same procedure as when we solved applications involving linear equations.

EXAMPLE 8   **Calculating Income with Commission**

A salesperson earns $600 per month plus a commission of 20% of sales. Find the minimum amount of sales needed to receive a total income of at least $1500 per month.

Solution

1. UNDERSTAND. Read and reread the problem. Let $x =$ amount of sales.

2. TRANSLATE. As stated in the beginning of this section, we want the income to be at least, or greater than or equal to, $1500. To write an inequality, notice that the salesperson's income consists of $600 plus a commission (20% of sales).

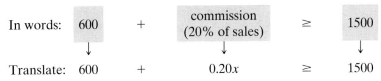

In words: 600   +   commission (20% of sales)   ≥   1500

Translate: 600   +   0.20x   ≥   1500

3. SOLVE the inequality for x.

First, let's decide on a window.

For sales of $2000, gross pay = $600 + 20%($2000) = $1000. Similarly, for sales of $5000, gross pay is $1600. This means that a sales amount between $2000 and $5000 will give a gross pay of $1500. Although many windows are acceptable, for this example, we graph $y_1 = 600 + 0.20x$ and $y_2 = 1500$ on a [2000, 5000, 1000] by [1000, 2000, 100] window.

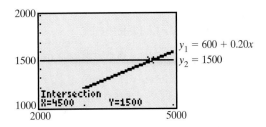

From the graph we see that $y_1 \geq y_2$ for x-values greater than or equal to 4500, or for x-values in the interval $[4500, \infty)$.

4. INTERPRET.

Check: The income for sales of $4500 is

$$600 + 0.20(4500), \text{ or } 1500.$$

Thus, if sales are greater than or equal to $4500, income is greater than or equal to $1500.

State: The minimum amount of sales needed for the salesperson to earn at least $1500 per month is $4500 per month.   □

8   A salesperson earns $900 a month plus a commission of 15% of sales. Find the minimum amount of sales needed to receive a total income of at least $2400 per month.

EXAMPLE 9   **Finding the Annual Consumption**

In the United States, the annual consumption of cigarettes is declining. The consumption c in billions of cigarettes per year since the year 1990 can be approximated by the formula

$$c = -9.2t + 527.33$$

where t is the number of years after 1990. Use this formula to predict the years that the consumption of cigarettes will be less than 200 billion per year.

Solution

1. UNDERSTAND. Read and reread the problem. To become familiar with the given formula, let's find the cigarette consumption after 20 years, which would be the year 1990 + 20, or 2010. To do so, we substitute 20 for t in the given formula.

$$c = -9.2(20) + 527.33 = 343.33$$

Thus, in 2010, we predict cigarette consumption to be about 343.3 billion.

Variables have already been assigned in the given formula. For review, they are
c = the annual consumption of cigarettes in the United States in billions of cigarettes

t = the number of years after 1990

2. TRANSLATE. We are looking for the years that the consumption of cigarettes c is less than 200. Since we are finding years t, we substitute the expression in the formula given for c, or

$$-9.2t + 527.33 < 200$$

3. SOLVE the inequality.

$$-9.2t + 527.33 < 200$$
$$-9.2t < -327.33$$
$$t > \text{approximately } 35.58$$

4. INTERPRET.

Check: Substitute a number greater than 35.58 and see that c is less than 200.

State: The annual consumption of cigarettes will be less than 200 billion more than 35.58 years after 1990, or in approximately $36 + 1990 = 2026$. □

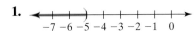

$y_1 = -9.2x + 527.33$

Intersection
X=35.579348 Y=200

A calculator solution for Example 9.

PRACTICE
9 Use the formula given in Example 9 to predict when the consumption of cigarettes will be less than 250 billion per year.

VOCABULARY & READINESS CHECK

Match each graph with the interval notation that describes it.

1.

$$\begin{array}{cccccccccc} & &) & & & & & & & \\ \hline -7 & -6 & -5 & -4 & -3 & -2 & -1 & 0 & & \end{array}$$

a. $(-5, \infty)$ **b.** $(-5, -\infty)$

c. $(\infty, -5)$ **d.** $(-\infty, -5)$

2.

$$\begin{array}{cccccc} & & [& & & \\ \hline -13 & -12 & -11 & -10 & -9 & -8 \end{array}$$

a. $(-\infty, -11]$ **b.** $(-11, \infty)$

c. $[-11, \infty)$ **d.** $(-\infty, -11)$

3.

$$\begin{array}{c} -2.5 \qquad \frac{7}{4} \\ \hline (\qquad] \\ -3 \ -2 \ -1 \ \ 0 \ \ 1 \ \ 2 \ \ 3 \ \ 4 \end{array}$$

a. $\left[\frac{7}{4}, -2.5\right)$ **b.** $\left(-2.5, \frac{7}{4}\right]$

c. $\left[-2.5, \frac{7}{4}\right)$ **d.** $\left(\frac{7}{4}, -2.5\right)$

4.

$$\begin{array}{c} -\frac{10}{3} \qquad 0.2 \\ \hline [\qquad] \\ -4 \ -3 \ -2 \ -1 \ \ 0 \ \ 1 \ \ 2 \ \ 3 \end{array}$$

a. $\left[-\frac{10}{3}, 0.2\right)$ **b.** $\left(0.2, -\frac{10}{3}\right]$

c. $\left(-\frac{10}{3}, 0.2\right]$ **d.** $\left[0.2, -\frac{10}{3}\right)$

Use the choices below to fill in each blank.

$(-\infty, -0.4)$ $\qquad$ $(-\infty, -0.4]$ $\qquad$ $[-0.4, \infty)$ $\qquad$ $(-0.4, \infty)$ $\qquad$ $(\infty, -0.4]$

5. The set $\{x \mid x \geq -0.4\}$ written in interval notation is _____.

6. The set $\{x \mid x < -0.4\}$ written in interval notation is _____.

7. The set $\{x \mid x \leq -0.4\}$ written in interval notation is _____.

8. The set $\{x \mid x > -0.4\}$ written in interval notation is _____.

Each inequality below is solved by dividing both sides by the coefficient of x. In which inequality will the inequality symbol be reversed during this solution process?

9. $3x > -14$ $\qquad$ **10.** $-3x \leq 14$ $\qquad$ **11.** $-3x < -14$ $\qquad$ **12.** $-x \geq 23$

3.2 | EXERCISE SET

Graph the solution set of each inequality on a number line and write it in interval notation. See Example 1.

1. $\{x \mid x < -3\}$

2. $\{x \mid x > 5\}$

3. $\{x \mid x \geq 0.3\}$

4. $\{x \mid x < -0.2\}$

5. $\{x \mid -2 < x < 5\}$

6. $\{x \mid -5 \leq x \leq -1\}$

7. $\{x \mid 5 \geq x > -1\}$

8. $\{x \mid -3 > x \geq -7\}$

Solve. Graph the solution set on a number line and write it in interval notation. See Examples 2 through 4.

9. $x - 7 \geq -9$

10. $x + 2 \leq -1$

11. $8x - 7 \leq 7x - 5$

12. $7x - 1 \geq 6x - 1$

13. $\dfrac{3}{4}x \geq 6$

14. $\dfrac{5}{6}x \geq 5$

15. $-5x > 23.5$

16. $-4x < 11.2$

Use the given screen to solve each inequality. Write the solution in interval notation.

17. $y_1 < y_2$

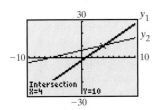

18. $y_1 \geq 0$

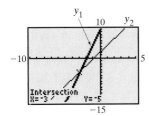

19. $y_1 \geq y_2$

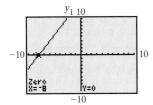

20. $y_1 > y_2$

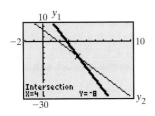

Use the graph of parallel lines y_1 and y_2 to solve each inequality. Write the solution in interval notation.

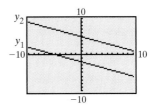

21. $y_1 > y_2$

22. $y_1 < y_2$

Solve. Write the solution set using interval notation. See Examples 5 through 7.

23. $-2x + 7 \geq 9$

24. $8 - 5x \leq 23$

25. $15 + 2x \geq 4x - 7$

26. $20 + x < 6x - 15$

27. $4(2x + 1) > 4$

28. $6(2 - 3x) \geq 12$

29. $3(x - 5) < 2(2x - 1)$

30. $5(x + 4) \leq 4(2x + 3)$

31. $\dfrac{5x + 1}{7} - \dfrac{2x - 6}{4} \geq -4$

32. $\dfrac{1 - 2x}{3} + \dfrac{3x + 7}{7} > 1$

33. $-3(2x - 1) < -4[2 + 3(x + 2)]$

34. $-2(4x + 2) > -5[1 + 2(x - 1)]$

MIXED PRACTICE

Solve. Write the solution set using interval notation. See Examples 1 through 7. Check solutions graphically.

35. $x + 9 < 3$

36. $x - 9 < -12$

37. $-x < -4$

38. $-x > -2$

39. $-7x \leq 3.5$

40. $-6x \leq 4.2$

41. $\dfrac{1}{2} + \dfrac{2}{3} \geq \dfrac{x}{6}$

42. $\dfrac{3}{4} - \dfrac{2}{3} \geq \dfrac{x}{6}$

43. $-5x + 4 \leq -4(x - 1)$

44. $-6x + 2 < -3(x + 4)$

45. $\dfrac{3}{4}(x - 7) \geq x + 2$

46. $\dfrac{4}{5}(x + 1) \leq x + 1$

47. $0.8x + 0.6x \geq 4.2$

48. $0.7x - x > 0.45$

49. $4(x - 6) + 2x - 4 \geq 3(x - 7) + 10x$

50. $7(2x + 3) + 4x \leq 7 + 5(3x - 4) + x$

51. $14 - (5x - 6) \geq -6(x + 1) - 5$

52. $13y - (9y + 2) \leq 5(y - 6) + 10$

53. $\dfrac{1}{2}(3x - 4) \leq \dfrac{3}{4}(x - 6) + 1$

54. $\dfrac{2}{3}(x + 3) < \dfrac{1}{6}(2x - 8) + 2$

55. $\dfrac{-x + 2}{2} - \dfrac{1 - 5x}{8} < -1$

56. $\dfrac{3 - 4x}{6} - \dfrac{1 - 2x}{12} \leq -2$

57. $\dfrac{x + 5}{5} - \dfrac{3 + x}{8} \geq -\dfrac{3}{10}$

58. $\dfrac{x - 4}{2} - \dfrac{x - 2}{3} > \dfrac{5}{6}$

59. $\dfrac{x + 3}{12} + \dfrac{x - 5}{15} < \dfrac{2}{3}$

60. $\dfrac{3x + 2}{18} - \dfrac{1 + 2x}{6} \leq -\dfrac{1}{2}$

61. $0.4(4x - 3) < 1.2(x + 2)$

62. $0.2(8x - 2) < 1.2(x - 3)$

63. $\dfrac{2}{5}x - \dfrac{1}{4} \leq \dfrac{3}{10}x - \dfrac{4}{5}$

64. $\dfrac{7}{12}x - \dfrac{1}{3} \leq \dfrac{3}{8}x - \dfrac{5}{6}$

65. $4(x - 1) \geq 4x - 8$

66. $3x + 1 < 3(x - 2)$

67. $7x < 7(x - 2)$

68. $8(x + 3) \leq 7(x + 5) + x$

Solve. See Examples 8 and 9. For Exercises 69 through 76, **a.** *answer with an inequality, and* **b.** *in your own words, explain the meaning of your answer to part* **a.**

69. Shureka Washburn has scores of 72, 67, 82, and 79 on her algebra tests.

    a. Use an inequality to find the scores she must make on the final exam to pass the course with an average of 77 or higher, given that the final exam counts as two tests.

    b. In your own words explain the meaning of your answer to part **a.**

70. In a Winter Olympics 5000-meter speed-skating event, Hans Holden scored times of 6.85, 7.04, and 6.92 minutes on his first three trials.

    a. Use an inequality to find the times he can score on his last trial so that his average time is under 7.0 minutes.

    b. In your own words, explain the meaning of your answer to part **a.**

71. A small plane's maximum takeoff weight is 2000 pounds or less. Six passengers weigh an average of 160 pounds each. Use an inequality to find the luggage and cargo weights the plane can carry.

72. A shopping mall parking garage charges $1 for the first half-hour and 60 cents for each additional half-hour. Use an inequality to find how long you can park if you have only $4.00 in cash.

73. A clerk must use the elevator to move boxes of paper. The elevator's maximum weight limit is 1500 pounds. If each box of paper weighs 66 pounds and the clerk weighs 147 pounds, use an inequality to find the number of whole boxes she can move on the elevator at one time.

74. As of May 2008, the cost to mail an envelope first class is 42 cents for the first ounce and 17 cents per ounce for each additional ounce. Use an inequality to find the number of whole ounces that can be mailed for no more than $2.50.

75. Northeast Telephone Company offers two billing plans for local calls.

    Plan 1: $25 per month for unlimited calls

    Plan 2: $13 per month plus $0.06 per call

    Use an inequality to find the number of monthly calls for which plan 1 is more economical than plan 2.

76. A car rental company offers two subcompact rental plans.

    Plan A: $36 per day and unlimited mileage

    Plan B: $24 per day plus $0.15 per mile

    Use an inequality to find the number of daily miles for which plan A is more economical than plan B.

77. At room temperature, glass used in windows actually has some properties of a liquid. It has a very slow, viscous flow. (Viscosity is the property of a fluid that resists internal flow. For example, lemonade flows more easily than fudge syrup. Fudge syrup has a higher viscosity than lemonade.) Glass does not become a true liquid until temperatures are greater than or equal to 500°C. Find the Fahrenheit temperatures for which glass is a liquid. (Use the formula $F = \dfrac{9}{5}C + 32$.)

78. Stibnite is a silvery white mineral with a metallic luster. It is one of the few minerals that melts easily in a match flame or at temperatures of approximately 977°F or greater. Find the Celsius temperatures for which stibnite melts. [Use the formula $C = \dfrac{5}{9}(F - 32)$.]

79. Although beginning salaries vary greatly according to your field of study, the equation

$$s = 651.2t + 28,472$$

can be used to approximate and to predict average beginning salaries for candidates with bachelor's degrees. The variable s is the starting salary and t is the number of years after 1990.

a. Approximate when beginning salaries for candidates will be greater than $42,000.

b. Determine the year you plan to graduate from college. Use this year to find the corresponding value of t and approximate your beginning salary.

80. The equation $y = 1573x + 125,217$ can be used to approximate the civilian labor force. Here, x is the number of years after 1990 and y is the civilian labor force (in thousands). (See Chapter 3 opener.)

a. Assume this trend continues and predict the year when the civilian labor force will be greater than 160,000 thousand. Round up to the nearest year.

b. Determine the year you plan to graduate from college. Use this year to predict the corresponding civilian labor force.

The average consumption per person per year of whole milk w in gallons can be approximated by the equation

$$y = -0.19t + 7.6$$

where t is the number of years after 2000. The average consumption of nonfat milk s per person per year can be approximated by the equation

$$y = -0.07t + 3.5$$

where t is the number of years after 2000. The consumption of whole milk is shown on the graph in red and the consumption of nonfat milk is shown on the graph in blue. Use this information to answer Exercises 81–88.

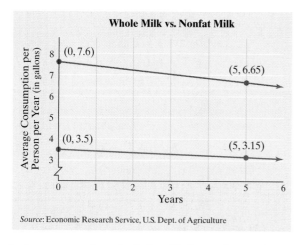

Whole Milk vs. Nonfat Milk

Average Consumption per Person per Year (in gallons)

(0, 7.6) (5, 6.65) (0, 3.5) (5, 3.15)

Years

Source: Economic Research Service, U.S. Dept. of Agriculture

81. Is the consumption of whole milk increasing or decreasing over time? Explain how you arrived at your answer.

82. Is the consumption of nonfat milk increasing or decreasing over time? Explain how you arrived at your answer.

83. Predict the consumption of whole milk in the year 2010. (*Hint:* Find the value of t that corresponds to the year 2010.)

84. Predict the consumption of nonfat milk in the year 2010. (*Hint:* Find the value of t that corresponds to the year 2010.)

85. Determine when the consumption of whole milk was first less than 6 gallons per person per year.

86. Determine when the consumption of nonfat milk was first less than 3 gallons per person per year.

87. For 2000 through 2005 the consumption of whole milk was greater than the consumption of nonfat milk. Explain how this can be determined from the graph.

88. How will the two lines in the graph appear if the consumption of whole milk is the same as the consumption of nonfat milk?

REVIEW AND PREVIEW

List or describe the integers that make both inequalities true. See Section 1.4.

89. $x < 5$ and $x > 1$

90. $x \geq 0$ and $x \leq 7$

91. $x \geq -2$ and $x \geq 2$

92. $x < 6$ and $x < -5$

Solve each equation for x. See Section 1.5.

93. $2x - 6 = 4$

94. $3x - 12 = 3$

95. $-x + 7 = 5x - 6$

96. $-5x - 4 = -x - 4$

CONCEPT EXTENSIONS

Each row of the table shows three equivalent ways of describing an interval. Complete this table by filling in the equivalent descriptions. The first row has been completed for you.

	Set Notation	Graph	Interval Notation
	$\{x \mid x < -3\}$	−3	$(-\infty, -3)$
97.		2	
98.		−4	
99.	$\{x \mid x < 0\}$		
100.	$\{x \mid x \leq 5\}$		
101.			$(-2, 1.5]$
102.			$[-3.7, 4)$

103. Solve: $2x - 3 = 5$.

104. Solve: $2x - 3 < 5$.

105. Solve: $2x - 3 > 5$.

106. Read the equations and inequalities for Exercises 103, 104, and 105 and their solutions. In your own words, write down your thoughts.

107. When graphing the solution set of an inequality, explain how you know whether to use a parenthesis or a bracket.

108. Explain what is wrong with the interval notation $(-6, -\infty)$.

109. Explain how solving a linear inequality is similar to solving a linear equation.

110. Explain how solving a linear inequality is different from solving a linear equation.

111. Write an inequality whose solution set is $\{x \mid x \le 2\}$.

THE BIGGER PICTURE   SOLVING EQUATIONS AND INEQUALITIES

We now continue the outline started in Section 1.5. Although suggestions will be given, this outline should be in your own words. Once you complete this new portion, try the exercises below.

Solving Equations and Inequalities

I. Equations

   A. Linear equations: Power on variable is 1 and there are no variables in denominator. (Sections 1.5 and 3.1)

II. Inequalities

   A. Linear Inequalities: Inequality sign and power on x is 1 and there are no variables in denominator. (Section 3.2)

$$-3(x + 2) \ge 6 \quad \text{Linear inequality}$$
$$-3x - 6 \ge 6 \quad \text{Multiply.}$$
$$-3x \ge 12 \quad \text{Add 6 to both sides.}$$
$$\frac{-3x}{-3} \le \frac{12}{-3} \quad \begin{array}{l}\text{Divide both sides by } -3, \text{ and } \textit{change}\\ \text{the direction of the inequality sign.}\end{array}$$
$$x \le -4 \text{ or } (-\infty, -4]$$

Solve. Write inequality solutions in interval notation.

1. $7x - 2 = 5(2x + 1) + 3$

2. $5 + 9x = 5(x + 1)$

3. $\dfrac{x + 3}{2} > 1$

4. $\dfrac{x + 2}{2} + \dfrac{x + 4}{3} = \dfrac{29}{6}$

5. $\dfrac{7}{5} - \dfrac{y}{10} = 2$

6. $5 + 9x = 9(x + 1)$

7. $4(x - 2) + 3x \ge 9(x - 1) - 2$

8. $8(x + 1) - 2 = 8x + 6$

INTEGRATED REVIEW   LINEAR EQUATIONS AND INEQUALITIES

Sections 3.1– 3.2

Solve each equation or inequality. For inequalities, write the solution set in interval notation.

1. $-4x = 20$

2. $-4x < 20$

3. $\dfrac{3x}{4} \ge 2$

4. $5x + 3 \ge 2 + 4x$

5. $6(y - 4) = 3(y - 8)$

6. $-4x \le \dfrac{2}{5}$

7. $-3x \ge \dfrac{1}{2}$

8. $5(y + 4) = 4(y + 5)$

9. $7x < 7(x - 2)$

10. $\dfrac{-5x + 11}{2} \le 7$

11. $-5x + 1.5 = -19.5$

12. $-5x + 4 = -26$

13. $5 + 2x - x = -x + 3 - 14$

14. $12x + 14 < 11x - 2$

15. $\dfrac{x}{5} - \dfrac{x}{4} = \dfrac{x - 2}{2}$

16. $12x - 12 = 8(x - 1)$

17. $2(x - 3) > 70$

18. $-3x - 4.7 = 11.8$

19. $-2(b - 4) - (3b - 1) = 5b + 3$

20. $8(x + 3) < 7(x + 5) + x$

21. $\dfrac{3t + 1}{8} = \dfrac{5 + 2t}{7} + 2$

22. $4(x - 6) - x = 8(x - 3) - 5x$

23. $\dfrac{x}{6} + \dfrac{3x - 2}{2} < \dfrac{2}{3}$

24. $\dfrac{y}{3} + \dfrac{y}{5} = \dfrac{y + 3}{10}$

25. $5(x - 6) + 2x > 3(2x - 1) - 4$

26. $14(x - 1) - 7x \le 2(3x - 6) + 4$

27. $\dfrac{1}{4}(3x + 2) - x \ge \dfrac{3}{8}(x - 5) + 2$

28. $\dfrac{1}{3}(x - 10) - 4x > \dfrac{5}{6}(2x + 1) - 1$

3.3 COMPOUND INEQUALITIES

Two inequalities joined by the words **and** or **or** are called **compound inequalities.**

Compound Inequalities

$$x + 3 < 8 \quad \text{and} \quad x > 2$$

$$\frac{2x}{3} \geq 5 \quad \text{or} \quad -x + 10 < 7$$

OBJECTIVE 1 ▶ Finding the intersection of two sets. The solution set of a compound inequality formed by the word **and** is the **intersection** of the solution sets of the two inequalities. We use the symbol $\cap$ to represent "intersection."

Intersection of Two Sets

The intersection of two sets, A and B, is the set of all elements common to both sets. A intersect B is denoted by

---

EXAMPLE 1   If $A = \{x \mid x \text{ is an even number greater than 0 and less than 10}\}$ and $B = \{3, 4, 5, 6\}$, find $A \cap B$.

Solution   Let's list the elements in set A.

$$A = \{2, 4, 6, 8\}$$

The numbers 4 and 6 are in sets A and B. The intersection is $\{4, 6\}$. □

PRACTICE

1   If $A = \{x \mid x \text{ is an odd number greater than 0 and less than 10}\}$ and $B = \{1, 2, 3, 4\}$, find $A \cap B$.

OBJECTIVE 2 ▶ Solving compound inequalities containing "and." A value is a solution of a compound inequality formed by the word **and** if it is a solution of *both* inequalities. For example, the solution set of the compound inequality $x \leq 5$ and $x \geq 3$ contains all values of x that make the inequality $x \leq 5$ a true statement **and** the inequality $x \geq 3$ a true statement. The first graph shown below is the graph of $x \leq 5$, the second graph is the graph of $x \geq 3$, and the third graph shows the intersection of the two graphs. The third graph is the graph of $x \leq 5$ **and** $x \geq 3$.

$\{x \mid x \leq 5\}$	number line graph from -1 to 6, shaded $(-\infty, 5]$	$(-\infty, 5]$
$\{x \mid x \geq 3\}$	number line graph from -1 to 6, shaded $[3, \infty)$	$[3, \infty)$
$\{x \mid x \leq 5 \text{ and } x \geq 3\}$ also $\{x \mid 3 \leq x \leq 5\}$ (see below)	number line graph from -1 to 6, shaded $[3, 5]$	$[3, 5]$

Since $x \geq 3$ is the same as $3 \leq x$, the compound inequality $3 \leq x$ and $x \leq 5$ can be written in a more compact form as $3 \leq x \leq 5$. The solution set $\{x \mid 3 \leq x \leq 5\}$ includes all numbers that are greater than or equal to 3 and at the same time less than or equal to 5.

In interval notation, the set $\{x \mid x \leq 5 \text{ and } x \geq 3\}$ or $\{x \mid 3 \leq x \leq 5\}$ is written as $[3, 5]$.

▶ **Helpful Hint**

Don't forget that some compound inequalities containing "and" can be written in a more compact form.

Compound Inequality	Compact Form	Interval Notation
$2 \le x$ and $x \le 6$	$2 \le x \le 6$	$[2, 6]$
Graph:		

EXAMPLE 2   Solve: $x - 7 < 2$ and $2x + 1 < 9$

Solution   First we solve each inequality separately.

$$x - 7 < 2 \quad and \quad 2x + 1 < 9$$
$$x < 9 \quad and \quad 2x < 8$$
$$x < 9 \quad and \quad x < 4$$

Now we can graph the two intervals on two number lines and find their intersection. Their intersection is shown on the third number line.

$\{x \mid x < 9\}$ $(-\infty, 9)$

$\{x \mid x < 4\}$ $(-\infty, 4)$

$\{x \mid x < 9 \text{ and } x < 4\} = \{x \mid x < 4\}$ $(-\infty, 4)$

The solution set is $(-\infty, 4)$.

PRACTICE
2   Solve: $x + 3 < 8$ and $2x - 1 < 3$. Write the solution set in interval notation.

EXAMPLE 3   Solve: $2x \ge 0$ and $4x - 1 \le -9$.

Solution   First we solve each inequality separately.

$$2x \ge 0 \quad and \quad 4x - 1 \le -9$$
$$x \ge 0 \quad and \quad 4x \le -8$$
$$x \ge 0 \quad and \quad x \le -2$$

Now we can graph the two intervals on number lines and find their intersection.

$\{x \mid x \ge 0\}$ $[0, \infty)$

$\{x \mid x \le -2\}$ $(-\infty, -2]$

$\{x \mid x \ge 0 \text{ and } x \le -2\} = \varnothing$ $\varnothing$

There is no number that is greater than or equal to 0 *and* less than or equal to -2. The solution set is $\varnothing$.

PRACTICE
3   Solve: $4x \le 0$ and $3x + 2 > 8$. Write the solution set in interval notation.

▶ **Helpful Hint**

Example 3 shows that some compound inequalities have no solution. Also, some have all real numbers as solutions.

To solve a compound inequality written in a compact form, such as $2 < 4 - x < 7$, we get x alone in the "middle part." Since a compound inequality is really two inequalities in one statement, we must perform the same operations on all three parts of the inequality.

EXAMPLE 4   Solve: $2 < 4 - x < 7$

Solution   To get x alone, we first subtract 4 from all three parts.

$$2 < 4 - x < 7$$
$$2 - 4 < 4 - x - 4 < 7 - 4 \quad \text{Subtract 4 from all three parts.}$$
$$-2 < -x < 3 \quad\quad\quad\quad \text{Simplify.}$$
$$\frac{-2}{-1} > \frac{-x}{-1} > \frac{3}{-1} \quad\quad \text{Divide all three parts by } -1 \text{ and}$$
$$\quad\quad\quad\quad\quad\quad\quad\quad\quad \text{reverse the inequality symbols.}$$
$$2 > x > -3$$

> **Helpful Hint**
> Don't forget to reverse both inequality symbols.

This is equivalent to $-3 < x < 2$.

The solution set in interval notation is $(-3, 2)$, and its number-line graph is shown.

$$\xleftarrow{\quad}\overset{\displaystyle (\quad\quad\quad\quad\quad\quad)}{\underset{-4\ -3\ -2\ -1\ \ \ 0\ \ \ 1\ \ \ 2\ \ \ 3}{\vert\ \vert\ \vert\ \vert\ \vert\ \vert\ \vert\ \vert}}\xrightarrow{\quad}$$

PRACTICE
4   Solve: $3 < 5 - x < 9$. Write the solution set in interval notation.

DISCOVER THE CONCEPT

For the compound inequality $2 < 4 - x < 7$, in Example 4, graph $y_1 = 2$, $y_2 = 4 - x$, and $y_3 = 7$. With this notation, we can think of our inequality as $y_1 < y_2 < y_3$.

a. Find the point of intersection of y_1 and y_2.

b. Find the point of intersection of y_2 and y_3.

c. Determine where the graph of y_2 is between the graphs of y_1 and y_3.

d. Find the x-values for which $y_1 < y_2 < y_3$.

e. Write the solution set of $y_1 < y_2 < y_3$.

In the discovery above, we find the points of intersection as shown.

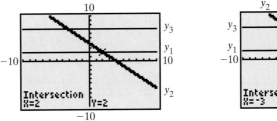
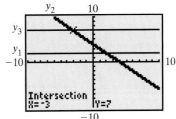

The solution of $y_1 < y_2 < y_3$, or $2 < 4 - x < 7$ contains the x-values where the graph of $y_2 = 4 - x$ is between $y_1 = 2$ and $y_3 = 7$. These x-values in interval notation are $(-3, 2)$. (Parentheses are used because of the inequality symbols $<$.)

EXAMPLE 5 Solve algebraically: $-1 \le \dfrac{2x}{3} + 5 \le 2$. Check graphically.

Solution First, clear the inequality of fractions by multiplying all three parts by the LCD of 3.

$$-1 \le \frac{2x}{3} + 5 \le 2$$

$$3(-1) \le 3\left(\frac{2x}{3} + 5\right) \le 3(2) \qquad \text{Multiply all three parts by the LCD of 3.}$$

$$-3 \le 2x + 15 \le 6 \qquad \text{Use the distributive property and multiply.}$$

$$-3 - 15 \le 2x + 15 - 15 \le 6 - 15 \qquad \text{Subtract 15 from all three parts.}$$

$$-18 \le 2x \le -9 \qquad \text{Simplify.}$$

$$\frac{-18}{2} \le \frac{2x}{2} \le \frac{-9}{2} \qquad \text{Divide all three parts by 2.}$$

$$-9 \le x \le -\frac{9}{2} \qquad \text{Simplify.}$$

The number-line graph of the solution is shown.

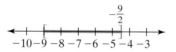

The solution set in interval notation is $\left[-9, -\dfrac{9}{2}\right]$.

To check, graph $y_1 = -1$, $y_2 = \dfrac{2x}{3} + 5$, and $y_3 = 2$ in a $[-15, 5, 3]$ by $[-5, 5, 1]$ window and solve $y_1 \le y_2 \le y_3$.

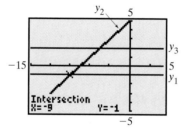
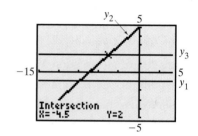

The solution of $y_1 \le y_2 \le y_3$ consists of the x-values where the graph of y_2 is between or equal to the graphs of y_1 and y_3. The solution set in interval notation is $[-9, -4.5]$ or $\left[-9, -\dfrac{9}{2}\right]$. □

PRACTICE
5 Solve algebraically: $-4 \le \dfrac{x}{2} - 1 \le 3$. Check graphically and write the solution set in interval notation.

OBJECTIVE 3 ▶ Finding the union of two sets. The solution set of a compound inequality formed by the word **or** is the **union** of the solution sets of the two inequalities. We use the symbol $\cup$ to denote "union."

▶ **Helpful Hint**
The word "either" in this definition means "one or the other or both."

Union of Two Sets

The **union** of two sets, A and B, is the set of elements that belong to _either_ of the sets. A union B is denoted by

$$A \cup B$$

EXAMPLE 6   If $A = \{x \mid x \text{ is an even number greater than 0 and less than 10}\}$ and $B = \{3, 4, 5, 6\}$, find $A \cup B$.

Solution   Recall from Example 1 that $A = \{2, 4, 6, 8\}$. The numbers that are in either set or both sets are $\{2, 3, 4, 5, 6, 8\}$. This set is the union. □

PRACTICE
6   If $A = \{x \mid x \text{ is an odd number greater than 0 and less than 10}\}$ and $B = \{2, 3, 4, 5, 6\}$. Find $A \cup B$.

OBJECTIVE 4 ▶ Solving compound inequalities containing "or." A value is a solution of a compound inequality formed by the word **or** if it is a solution of **either** inequality. For example, the solution set of the compound inequality $x \leq 1$ **or** $x \geq 3$ contains all numbers that make the inequality $x \leq 1$ a true statement **or** the inequality $x \geq 3$ a true statement.

$$\{x \mid x \leq 1\} \qquad \qquad \qquad (-\infty, 1]$$

$$\{x \mid x \geq 3\} \qquad \qquad \qquad [3, \infty)$$

$$\{x \mid x \leq 1 \text{ or } x \geq 3\} \qquad \qquad (-\infty, 1] \cup [3, \infty)$$

In interval notation, the set $\{x \mid x \leq 1 \text{ or } x \geq 3\}$ is written as $(-\infty, 1] \cup [3, \infty)$.

EXAMPLE 7   Solve: $5x - 3 \leq 10 \text{ or } x + 1 \geq 5$.

Solution   First we solve each inequality separately.

$$\begin{array}{rcl}
5x - 3 \leq 10 & \text{or} & x + 1 \geq 5 \\
5x \leq 13 & \text{or} & x \geq 4 \\
x \leq \dfrac{13}{5} & \text{or} & x \geq 4
\end{array}$$

Now we can graph each interval on a number line and find their union.

$$\left\{ x \mid x \leq \frac{13}{5} \right\} \qquad \qquad \left(-\infty, \frac{13}{5} \right]$$

$$\{x \mid x \geq 4\} \qquad \qquad [4, \infty)$$

$$\left\{ x \mid x \leq \frac{13}{5} \text{ or } x \geq 4 \right\} \qquad \left(-\infty, \frac{13}{5} \right] \cup [4, \infty)$$

The solution set is $\left(-\infty, \dfrac{13}{5} \right] \cup [4, \infty)$. □

PRACTICE
7   Solve: $8x + 5 \leq 8 \text{ or } x - 1 \geq 2$. Write the solution set in interval notation.

EXAMPLE 8   Solve: $-2x - 5 < -3 \text{ or } 6x < 0$.

Solution   First we solve each inequality separately.

$$\begin{array}{rcl}
-2x - 5 < -3 & \text{or} & 6x < 0 \\
-2x < 2 & \text{or} & x < 0 \\
x > -1 & \text{or} & x < 0
\end{array}$$

Now we can graph each interval on a number line and find their union.

$\{x|x > -1\}$ 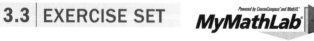 $(-1, \infty)$

$\{x|x < 0\}$ $(-\infty, 0)$

$\{x|x > -1 \ or \ x < 0\}$ $(-\infty, \infty)$
= all real numbers

The solution set is $(-\infty, \infty)$.

PRACTICE
8 Solve: $-3x - 2 > -8 \ or \ 5x > 0$. Write the solution set in interval notation.

Concept Check ✓

Which of the following is *not* a correct way to represent the set of all numbers between -3 and 5?

a. $\{x| -3 < x < 5\}$ **b.** $-3 < x \ or \ x < 5$

c. $(-3, 5)$ **d.** $x > -3$ and $x < 5$

Answer to Concept Check:
b is not correct

VOCABULARY & READINESS CHECK

Use the choices below to fill in each blank. Some choices may be used more than once.

or	∪	∅
and	∩	compound

1. Two inequalities joined by the words "and" or "or" are called _____ inequalities.
2. The word _____ means intersection.
3. The word _____ means union.
4. The symbol _____ represents intersection.
5. The symbol _____ represents union.
6. The symbol _____ is the empty set.
7. The inequality $-2 \le x < 1$ means $-2 \le x$ _____ $x < 1$.
8. $\{x|x < 0 \ and \ x > 0\} = $ _____

3.3 | EXERCISE SET

 MyMathLab   Math XL PRACTICE | WATCH | DOWNLOAD | READ | REVIEW

MIXED PRACTICE

If $A = \{x|x \text{ is an even integer}\}$, $B = \{x|x \text{ is an odd integer}\}$, $C = \{2, 3, 4, 5\}$, and $D = \{4, 5, 6, 7\}$, list the elements of each set. See Examples 1 and 6.

 1. $C \cup D$ **2.** $C \cap D$
 3. $A \cap D$ **4.** $A \cup D$
     5. $A \cup B$ **6.** $A \cap B$
     7. $B \cap D$ **8.** $B \cup D$
     9. $B \cup C$ **10.** $B \cap C$
     11. $A \cap C$ **12.** $A \cup C$

Solve each compound inequality. Graph the solution set on a number line and write it in interval notation. See Examples 2 and 3.

13. $x < 1 \text{ and } x > -3$ **14.** $x \le 0 \text{ and } x \ge -2$
15. $x \le -3 \text{ and } x \ge -2$ **16.** $x < 2 \text{ and } x > 4$
17. $x < -1 \text{ and } x < 1$ **18.** $x \ge -4 \text{ and } x > 1$

Solve each compound inequality. Write solutions in interval notation. See Examples 2 and 3.

19. $x + 1 \ge 7 \text{ and } 3x - 1 \ge 5$
20. $x + 2 \ge 3 \text{ and } 5x - 1 \ge 9$
21. $4x + 2 \le -10 \text{ and } 2x \le 0$
22. $2x + 4 > 0 \text{ and } 4x > 0$

23. $-2x < -8$ and $x - 5 < 5$

24. $-7x \le -21$ and $x - 20 \le -15$

Solve each compound inequality. See Examples 4 and 5.

25. $5 < x - 6 < 11$

26. $-2 \le x + 3 \le 0$

27. $-2 \le 3x - 5 \le 7$

28. $1 < 4 + 2x < 7$

29. $1 \le \dfrac{2}{3}x + 3 \le 4$

30. $-2 < \dfrac{1}{2}x - 5 < 1$

31. $-5 \le \dfrac{-3x + 1}{4} \le 2$

32. $-4 \le \dfrac{-2x + 5}{3} \le 1$

Solve each compound inequality. Graph the solution set on a number line and write it in interval notation. See Examples 7 and 8.

33. $x < 4$ or $x < 5$

34. $x \ge -2$ or $x \le 2$

35. $x \le -4$ or $x \ge 1$

36. $x < 0$ or $x < 1$

37. $x > 0$ or $x < 3$

38. $x \ge -3$ or $x \le -4$

Solve each compound inequality. Write solutions in interval notation. See Examples 7 and 8.

39. $-2x \le -4$ or $5x - 20 \ge 5$

40. $-5x \le 10$ or $3x - 5 \ge 1$

41. $x + 4 < 0$ or $6x > -12$

42. $x + 9 < 0$ or $4x > -12$

43. $3(x - 1) < 12$ or $x + 7 > 10$

44. $5(x - 1) \ge -5$ or $5 - x \le 11$

MIXED PRACTICE

Solve each compound inequality. Write solutions in interval notation. See Examples 1 through 8.

45. $x < \dfrac{2}{3}$ and $x > -\dfrac{1}{2}$

46. $x < \dfrac{5}{7}$ and $x < 1$

47. $x < \dfrac{2}{3}$ or $x > -\dfrac{1}{2}$

48. $x < \dfrac{5}{7}$ or $x < 1$

49. $0 \le 2x - 3 \le 9$

50. $3 < 5x + 1 < 11$

51. $\dfrac{1}{2} < x - \dfrac{3}{4} < 2$

52. $\dfrac{2}{3} < x + \dfrac{1}{2} < 4$

53. $x + 3 \ge 3$ and $x + 3 \le 2$

54. $2x - 1 \ge 3$ and $-x > 2$

55. $3x \ge 5$ or $-\dfrac{5}{8}x - 6 > 1$

56. $\dfrac{3}{8}x + 1 \le 0$ or $-2x < -4$

57. $0 < \dfrac{5 - 2x}{3} < 5$

58. $-2 < \dfrac{-2x - 1}{3} < 2$

59. $-6 < 3(x - 2) \le 8$

60. $-5 < 2(x + 4) < 8$

61. $-x + 5 > 6$ and $1 + 2x \le -5$

62. $5x \le 0$ and $-x + 5 < 8$ **63.** $3x + 2 \le 5$ or $7x > 29$

64. $-x < 7$ or $3x + 1 < -20$

65. $5 - x > 7$ and $2x + 3 \ge 13$

66. $-2x < -6$ or $1 - x > -2$

67. $-\dfrac{1}{2} \le \dfrac{4x - 1}{6} < \dfrac{5}{6}$

68. $-\dfrac{1}{2} \le \dfrac{3x - 1}{10} < \dfrac{1}{2}$

69. $\dfrac{1}{15} < \dfrac{8 - 3x}{15} < \dfrac{4}{5}$

70. $-\dfrac{1}{4} < \dfrac{6 - x}{12} < -\dfrac{1}{6}$

71. $0.3 < 0.2x - 0.9 < 1.5$

72. $-0.7 \le 0.4x + 0.8 < 0.5$

Solve each compound inequality using the graphing utility screens.

73. a. $y_1 < y_2 < y_3$

 b. $y_2 < y_1$ or $y_2 > y_3$

74. a. $y_1 < y_2 < y_3$

 b. $y_2 < y_1$ or $y_2 > y_3$

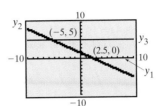

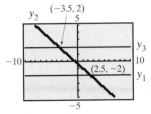

75. a. $y_1 \le y_2 \le y_3$

 b. $y_2 \le y_1$ or $y_2 \ge y_3$

76. a. $y_1 \le y_2 \le y_3$

 b. $y_2 \le y_1$ or $y_2 \ge y_3$

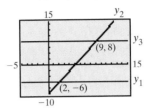

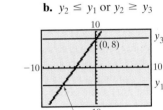

REVIEW AND PREVIEW

Evaluate the following. See Sections 1.2 and 1.3.

77. $|-7| - |19|$

78. $|-7 - 19|$

79. $-(-6) - |-10|$

80. $|-4| - (-4) + |-20|$

Find by inspection all values for x that make each equation true.

81. $|x| = 7$

82. $|x| = 5$

83. $|x| = 0$

84. $|x| = -2$

CONCEPT EXTENSIONS

Use the graph to answer Exercises 85 and 86.

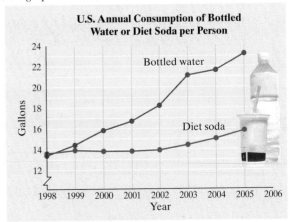

U.S. Annual Consumption of Bottled Water or Diet Soda per Person

85. For what years was the consumption of bottled water greater than 20 gallons per person *and* the consumption of diet soda greater than 14 gallons per person?

86. For what years was the consumption of bottled water less than 15 gallons per person *or* the consumption of diet soda greater than 14 gallons per person?

The formula for converting Fahrenheit temperatures to Celsius temperatures is $C = \dfrac{5}{9}(F - 32)$. Use this formula for Exercises 87 and 88.

87. During a recent year, the temperatures in Chicago ranged from $-29°$ to $35°C$. Use a compound inequality to convert these temperatures to Fahrenheit temperatures.

88. In Oslo, the average temperature ranges from $-10°$ to $18°$ Celsius. Use a compound inequality to convert these temperatures to the Fahrenheit scale.

Solve.

89. Christian D'Angelo has scores of 68, 65, 75, and 78 on his algebra tests. Use a compound inequality to find the scores he can make on his final exam to receive a C in the course. The final exam counts as two tests, and a C is received if the final course average is from 70 to 79.

90. Wendy Wood has scores of 80, 90, 82, and 75 on her chemistry tests. Use a compound inequality to find the range of scores she can make on her final exam to receive a B in the course. The final exam counts as two tests, and a B is received if the final course average is from 80 to 89.

*Solve each compound inequality for x. See the example below. To solve $x - 6 < 3x < 2x + 5$, notice that this inequality contains a variable not only in the middle, but also on the left and the right. When this occurs, we solve by rewriting the inequality using the word **and**.*

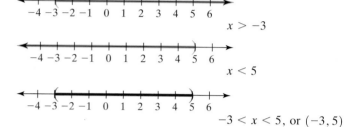

$$x - 6 < 3x \quad \text{and} \quad 3x < 2x + 5$$
$$-6 < 2x \quad \text{and} \quad x < 5$$
$$-3 < x$$
$$x > -3 \quad \text{and} \quad x < 5$$

$x > -3$

$x < 5$

$-3 < x < 5$, or $(-3, 5)$

91. $2x - 3 < 3x + 1 < 4x - 5$

92. $x + 3 < 2x + 1 < 4x + 6$

93. $-3(x - 2) \leq 3 - 2x \leq 10 - 3x$

94. $7x - 1 \leq 7 + 5x \leq 3(1 + 2x)$

95. $5x - 8 < 2(2 + x) < -2(1 + 2x)$

96. $1 + 2x < 3(2 + x) < 1 + 4x$

THE BIGGER PICTURE SOLVING EQUATIONS AND INEQUALITIES

We now continue the outline from Sections 1.5 and 3.2. Although suggestions will be given, this outline should be in your own words. Once you complete this new portion, try the exercises below.

Solving Equations and Inequalities

I. Equations

    A. Linear equations (Sections 1.5 and 3.1)

II. Inequalities

    A. Linear Inequalities (Section 3.2)

    B. Compound Inequalities: Two inequality signs or 2 inequalities separated by "and" or "or." *Or* means *union* and *and* means *intersection*. (Section 3.3)

Solve. Write inequality solutions in interval notation.

1. $x - 2 \leq 1$ and $3x - 1 \geq -4$

2. $-2 < x - 1 < 5$

3. $-2x + 2.5 = -7.7$

4. $-5x > 20$

5. $x \leq -3$ or $x \leq -5$

6. $5x < -10$ or $3x - 4 > 2$

7. $\dfrac{5t}{2} - \dfrac{3t}{4} = 7$

8. $5(x - 3) + x + 2 \geq 3(x + 2) + 2x$

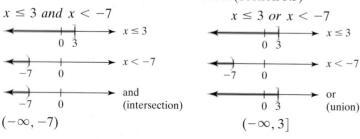

$x \leq 3$ and $x < -7$

$x \leq 3$ or $x < -7$

$(-\infty, -7)$

$(-\infty, 3]$

3.4 ABSOLUTE VALUE EQUATIONS

OBJECTIVE

1 Solve absolute value equations.

OBJECTIVE 1 ▶ Solving absolute equations. In Chapter 1, we defined the absolute value of a number as its distance from 0 on a number line.

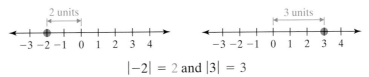

$$|-2| = 2 \text{ and } |3| = 3$$

In this section, we concentrate on solving equations containing the absolute value of a variable or a variable expression. Examples of absolute value equations are

$$|x| = 3 \qquad -5 = |2y + 7| \qquad |z - 6.7| = |3z + 1.2|$$

Since distance and absolute value are so closely related, absolute value equations and inequalities (see Section 3.5) are extremely useful in solving distance-type problems, such as calculating the possible error in a measurement.

For the absolute value equation $|x| = 3$, its solution set will contain all numbers whose distance from 0 is 3 units. Two numbers are 3 units away from 0 on the number line: 3 and -3.

Thus, the solution set of the equation $|x| = 3$ is $\{3, -3\}$. This suggests the following:

> **Solving Equations of the Form $|X| = a$**
>
> If a is a positive number, then $|X| = a$ is equivalent to $X = a$ or $X = -a$.

EXAMPLE 1 Solve: $|p| = 2$.

Solution Since 2 is positive, $|p| = 2$ is equivalent to $p = 2$ or $p = -2$.

To check, let $p = 2$ and then $p = -2$ in the original equation.

$|p| = 2$	Original equation		$|p| = 2$	Original equation
$|2| = 2$	Let $p = 2$.		$|-2| = 2$	Let $p = -2$.
$2 = 2$	True		$2 = 2$	True

The solutions are 2 and -2 or the solution set is $\{2, -2\}$.

To visualize the solution, we solve $|x| = 2$ by the intersection-of-graphs method. Graph $y_1 = |x|$ and $y_2 = 2$.

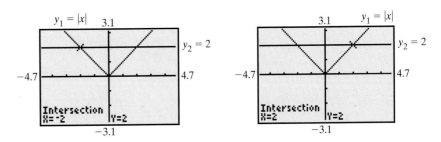

The graphs of $y_1 = |x|$ and $y_2 = 2$ intersect at $(-2, 2)$ and $(2, 2)$. The solutions are thus the x-values -2 and 2.   □

PRACTICE

1 Solve: $|q| = 7$.

TECHNOLOGY NOTE

Remember, absolute value can be found under the Math Num menu or under Catolog. Check your graphing calculator manual to find the location on your calculator.

If the expression inside the absolute value bars is more complicated than a single variable, we can still apply the absolute value property.

> ▶ **Helpful Hint**
> For the equation $|X| = a$ in the box above, X can be a single variable or a variable expression.

EXAMPLE 2 Solve algebraically: $|5w + 3| = 7$ and check graphically.

Solution Here the expression inside the absolute value bars is $5w + 3$. If we think of the expression $5w + 3$ as X in the absolute value property, we see that $|X| = 7$ is equivalent to

$$X = 7 \quad \text{or} \quad X = -7$$

Then substitute $5w + 3$ for X, and we have

$$5w + 3 = 7 \quad \text{or} \quad 5w + 3 = -7$$

Solve these two equations for w.

$$
\begin{aligned}
5w + 3 &= 7 & \text{or} \quad 5w + 3 &= -7 \\
5w &= 4 & \text{or} \quad 5w &= -10 \\
w &= \frac{4}{5} & \text{or} \quad w &= -2
\end{aligned}
$$

Check: To check, graph $y_1 = |5x + 3|$ and $y_2 = 7$ in a standard window as shown.

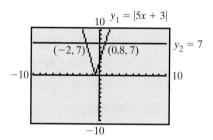

The intersections of the two graphs are the points $(-2, 7)$ and $(0.8, 7)$. Therefore, the solutions to the equation $|5x + 3| = 7$ are -2 and 0.8, or $\frac{4}{5}$.

Both solutions check, and the solutions are -2 and $\frac{4}{5}$ or the solution set is $\left\{ -2, \frac{4}{5} \right\}$. □

PRACTICE
2 Solve algebraically: $|2x - 3| = 5$ and check graphically.

EXAMPLE 3 Solve: $\left| \dfrac{x}{2} - 1 \right| = 11$.

Solution $\left| \dfrac{x}{2} - 1 \right| = 11$ is equivalent to

$$
\begin{aligned}
\frac{x}{2} - 1 &= 11 & \text{or} \quad \frac{x}{2} - 1 &= -11 \\
2\left(\frac{x}{2} - 1 \right) &= 2(11) & \text{or} \quad 2\left(\frac{x}{2} - 1 \right) &= 2(-11) & \text{Clear fractions.} \\
x - 2 &= 22 & \text{or} \quad x - 2 &= -22 & \text{Apply the distributive property.} \\
x &= 24 & \text{or} \quad x &= -20
\end{aligned}
$$

The solutions are 24 and -20. □

PRACTICE
3 Solve: $\left| \dfrac{x}{5} + 1 \right| = 15$.

To apply the absolute value rule, first make sure that the absolute value expression is isolated.

> ▶ **Helpful Hint**
>
> If the equation has a single absolute value expression containing variables, isolate the absolute value expression first.

EXAMPLE 4 Solve: $|2x| + 5 = 7$.

Solution We want the absolute value expression alone on one side of the equation, so begin by subtracting 5 from both sides. Then apply the absolute value property.

$$|2x| + 5 = 7$$
$$|2x| = 2 \qquad \text{Subtract 5 from both sides.}$$
$$2x = 2 \quad \text{or} \quad 2x = -2$$
$$x = 1 \quad \text{or} \quad x = -1$$

The solutions are -1 and 1. ☐

PRACTICE
4 Solve: $|3x| + 8 = 14$.

EXAMPLE 5 Solve: $|y| = 0$.

Solution We are looking for all numbers whose distance from 0 is zero units. The only number is 0. The solution is 0. ☐

PRACTICE
5 Solve: $|z| = 0$.

The next two examples illustrate a special case for absolute value equations. This special case occurs when an isolated absolute value is equal to a negative number.

EXAMPLE 6 Solve $2|x| + 25 = 23$ algebraically and graphically.

Algebraic Solution:

First, isolate the absolute value.

$$2|x| + 25 = 23$$
$$2|x| = -2 \qquad \text{Subtract 25 from both sides.}$$
$$|x| = -1 \qquad \text{Divide both sides by 2.}$$

The absolute value of a number is never negative, so this equation has no solution.

Graphical Solution:

Graph $y_1 = 2|x| + 25$ and $y_2 = 23$ in a $[-47, 47, 10]$ by $[-30, 50, 10]$ window.

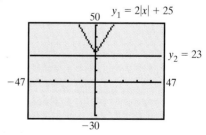

Since there is no point of intersection, there is no solution.

The solution set is $\{\ \}$ or $\varnothing$. ☐

PRACTICE
6 Solve $3|z| + 9 = 7$ algebraically and graphically.

EXAMPLE 7   Solve: $\left|\dfrac{3x + 1}{2}\right| = -2$.

Solution   Again, the absolute value of any expression is never negative, so no solution exists. The solution set is $\{\ \}$ or $\varnothing$.   □

PRACTICE
7   Solve: $\left|\dfrac{5x + 3}{4}\right| = -8$.

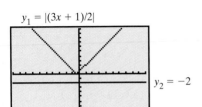

$y_1 = |(3x + 1)/2|$

$y_2 = -2$

A calculator check for Example 7.

Given two absolute value expressions, we might ask, when are the absolute values of two expressions equal? To see the answer, notice that

$$|2| = |2|, \quad |-2| = |-2|, \quad |-2| = |2|, \quad \text{and} \quad |2| = |-2|$$

same   same   opposites   opposites

Two absolute value expressions are equal when the expressions inside the absolute value bars are equal to or are opposites of each other.

EXAMPLE 8   Solve $|3x + 2| = |5x - 8|$ algebraically and check graphically.

Solution   This equation is true if the expressions inside the absolute value bars are equal to or are opposites of each other.

$$3x + 2 = 5x - 8 \quad \text{or} \quad 3x + 2 = -(5x - 8)$$

Next, solve each equation.

$$\begin{aligned} 3x + 2 &= 5x - 8 & \text{or} & \quad 3x + 2 = -5x + 8 \\ -2x + 2 &= -8 & \text{or} & \quad 8x + 2 = 8 \\ -2x &= -10 & \text{or} & \quad 8x = 6 \\ x &= 5 & \text{or} & \quad x = \frac{3}{4} \end{aligned}$$

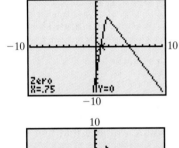

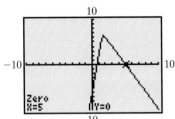

Using the x-intercept method to check, rewrite the equation as

$$|3x + 2| - |5x - 8| = 0$$

and graph $y_1 = |3x + 2| - |5x - 8|$. The x-intercepts are $x = 0.75$ and $x = 5$ as shown in the margin. The solutions are $\dfrac{3}{4}$ and 5.   □

PRACTICE
8   Solve $|2x + 4| = |3x - 1|$ algebraically and check graphically.

EXAMPLE 9   Solve: $|x - 3| = |5 - x|$.

Solution

$$\begin{aligned} x - 3 &= 5 - x & \text{or} & \quad x - 3 = -(5 - x) \\ 2x - 3 &= 5 & \text{or} & \quad x - 3 = -5 + x \\ 2x &= 8 & \text{or} & \quad x - 3 - x = -5 + x - x \\ x &= 4 & \text{or} & \quad -3 = -5 \quad \text{False} \end{aligned}$$

Recall from Section 1.5 that when an equation simplifies to a false statement, the equation has no solution. Thus, the only solution for the original absolute value equation is 4.   □

PRACTICE
9   Solve: $|x - 2| = |8 - x|$.

Answer to Concept Check:
false; answers may vary

Concept Check ☑

True or false? Absolute value equations always have two solutions. Explain your answer.

The following box summarizes the methods shown for solving absolute value equations.

Absolute Value Equations

$|X| = a$
$$\begin{cases} \text{If } a \text{ is positive, then solve } X = a \text{ or } X = -a. \\ \text{If } a \text{ is 0, solve } X = 0. \\ \text{If } a \text{ is negative, the equation } |X| = a \text{ has no solution.} \end{cases}$$

$|X| = |Y|$ Solve $X = Y$ or $X = -Y$.

VOCABULARY & READINESS CHECK

Match each absolute value equation with an equivalent statement.

1. $|x - 2| = 5$
2. $|x - 2| = 0$
3. $|x - 2| = |x + 3|$
4. $|x + 3| = 5$
5. $|x + 3| = -5$

A. $x - 2 = 0$
B. $x - 2 = x + 3$ or $x - 2 = -(x + 3)$
C. $x - 2 = 5$ or $x - 2 = -5$
D. $\varnothing$
E. $x + 3 = 5$ or $x + 3 = -5$

3.4 EXERCISE SET

PRACTICE   WATCH   DOWNLOAD   READ   REVIEW

Solve each absolute value equation. See Examples 1 through 7.

1. $|x| = 7$
2. $|y| = 15$
3. $|3x| = 12.6$
4. $|6n| = 12.6$
5. $|2x - 5| = 9$
6. $|6 + 2n| = 4$
7. $\left|\dfrac{x}{2} - 3\right| = 1$
8. $\left|\dfrac{n}{3} + 2\right| = 4$
9. $|z| + 4 = 9$
10. $|x| + 1 = 3$
11. $|3x| + 5 = 14$
12. $|2x| - 6 = 4$
13. $|2x| = 0$
14. $|7z| = 0$
15. $|4n + 1| + 10 = 4$
16. $|3z - 2| + 8 = 1$
17. $|5x - 1| = 0$
18. $|3y + 2| = 0$

19. Write an absolute value equation representing all numbers x whose distance from 0 is 5 units.

20. Write an absolute value equation representing all numbers x whose distance from 0 is 2 units.

Solve. See Examples 8 and 9.

21. $|5x - 7| = |3x + 11|$
22. $|9y + 1| = |6y + 4|$
23. $|z + 8| = |z - 3|$
24. $|2x - 5| = |2x + 5|$

25. Describe how solving an absolute value equation such as $|2x - 1| = 3$ is similar to solving an absolute value equation such as $|2x - 1| = |x - 5|$.

26. Describe how solving an absolute value equation such as $|2x - 1| = 3$ is different from solving an absolute value equation such as $|2x - 1| = |x - 5|$.

Use the given graphing utility screens to solve the equations shown. Write the solution set of the equation.

27. $|2x - 3| = 5$, where $y_1 = |2x - 3|$ and $y_2 = 5$

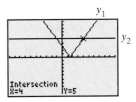

28. $|3x - 4| = 14$, where $y_1 = |3x - 4|$ and $y_2 = 14$

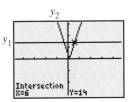

29. $|x - 4| = |1 - x|$, where $y_1 = |x - 4|$ and $y_2 = |1 - x|$

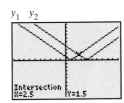

30. $|x + 2| = |3 - x|$, where $y_1 = |x + 2|$ and $y_2 = |3 - x|$

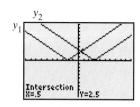

MIXED PRACTICE

Solve each absolute value equation. See Examples 1 through 9.

31. $|x| = 4$

32. $|x| = 1$

33. $|z| = -2$

34. $|y| = -9$

35. $|7 - 3x| = 7$

36. $|4m + 5| = 5$

37. $|6x| - 1 = 11$

38. $|7z| + 1 = 22$

39. $|x - 3| + 3 = 7$

40. $|x + 4| - 4 = 1$

41. $\left|\dfrac{z}{4} + 5\right| = -7$

42. $\left|\dfrac{c}{5} - 1\right| = -2$

43. $|9v - 3| = -8$

44. $|1 - 3b| = -7$

45. $|8n + 1| = 0$

46. $|5x - 2| = 0$

47. $|1 + 6c| - 7 = -3$

48. $|2 + 3m| - 9 = -7$

49. $|5x + 1| = 11$

50. $|8 - 6c| = 1$

51. $|4x - 2| = |-10|$

52. $|3x + 5| = |-4|$

53. $|5x + 1| = |4x - 7|$

54. $|3 + 6n| = |4n + 11|$

55. $|6 + 2x| = -|-7|$

56. $|4 - 5y| = -|-3|$

57. $|2x - 6| = |10 - 2x|$

58. $|4n + 5| = |4n + 3|$

59. $\left|\dfrac{2x - 5}{3}\right| = 7$

60. $\left|\dfrac{1 + 3n}{4}\right| = 4$

61. $2 + |5n| = 17$

62. $8 + |4m| = 24$

63. $\left|\dfrac{2x - 1}{3}\right| = |-5|$

64. $\left|\dfrac{5x + 2}{2}\right| = |-6|$

65. $|2y - 3| = |9 - 4y|$

66. $|5z - 1| = |7 - z|$

67. $\left|\dfrac{3n + 2}{8}\right| = |-1|$

68. $\left|\dfrac{2r - 6}{5}\right| = |-2|$

69. $|x + 4| = |7 - x|$

70. $|8 - y| = |y + 2|$

71. $\left|\dfrac{8c - 7}{3}\right| = -|-5|$

72. $\left|\dfrac{5d + 1}{6}\right| = -|-9|$

73. Explain why some absolute value equations have two solutions.

74. Explain why some absolute value equations have one solution.

REVIEW AND PREVIEW

The circle graph shows the U.S. cheese consumption for 2001. Use this graph to answer Exercises 75–77. See Section 1.6.

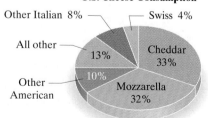

U.S. Cheese Consumption

75. What percent of cheese consumption came from cheddar cheese?

76. A circle contains $360°$. Find the number of degrees in the 4% sector for swiss cheese.

77. If a family consumed 120 pounds of cheese in 2001, find the amount of mozzarella we might expect they consumed.

List five integer solutions of each inequality. See Sections 1.2 and 1.4.

78. $|x| \leq 3$

79. $|x| \geq -2$

80. $|y| > -10$

81. $|y| < 0$

CONCEPT EXTENSIONS

82. Write an absolute value equation representing all numbers x whose distance from 1 is 5 units.

83. Write an absolute value equation representing all numbers x whose distance from 3 is 2 units.

Write each as an equivalent absolute value.

84. $x = 6$ or $x = -6$

85. $2x - 1 = 4$ or $2x - 1 = -4$

86. $x - 2 = 3x - 4$ or $x - 2 = -(3x - 4)$

87. For what value(s) of c will an absolute value equation of the form $|ax + b| = c$ have

   a. one solution?

   b. no solution?

   c. two solutions?

Use a graphical approach to approximate the solutions of each equation. Round the solutions to two decimal places.

88. $|2.3x - 1.5| = 5$

89. $|-7.6x + 2.6| = 1.9$

90. $3.6 - |4.1x - 2.6| = |x - 1.4|$

91. $-1.2 + |5x + 12.1| = -|x + 7.3| + 10$

3.5 ABSOLUTE VALUE INEQUALITIES

OBJECTIVE 1 ▶ Solving absolute value inequalities of the form $|x| < a$. The solution set of an absolute value inequality such as $|x| < 2$ contains all numbers whose distance from 0 is less than 2 units, as shown below.

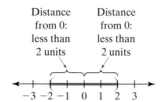

The solution set is $\{x \mid -2 < x < 2\}$, or $(-2, 2)$ in interval notation.

EXAMPLE 1   Solve: $|x| \le 3$.

Solution   The solution set of this inequality contains all numbers whose distance from 0 is less than or equal to 3. Thus 3, -3, and all numbers between 3 and -3 are in the solution set.

The solution set is $[-3, 3]$.

PRACTICE

1   Solve: $|x| < 2$ and graph the solution set on a number line.

In general, we have the following.

> **Solving Absolute Value Inequalities of the Form $|X| < a$**
> If a is a positive number, then $|X| < a$ is equivalent to $-a < X < a$.

This property also holds true for the inequality symbol $\le$.

DISCOVER THE CONCEPT

a. Solve the equation $|x| = 7$ by graphing $y_1 = |x|$, $y_2 = 7$, and finding the point(s) of intersection.
b. Move the trace cursor along y_1 from the left-hand point of intersection of the two graphs to the right-hand point of intersection. Observe the y-values of the points as you move the cursor.
c. Find the x-values for which $|x| < 7$ (or $y_1 < y_2$).
d. Write the solution set of $|x| < 7$ (or $y_1 < y_2$).

In the discovery above, we see that the solutions of $|x| = 7$ are 7 and -7, the x-values of the points of intersection of y_1 and y_2. The solutions of $|x| < 7$ consist of all x-values for which $y_1 < y_2$ or for which the graph of y_1 is below the graph of y_2. These x-values are between -7 and 7, or $(-7, 7)$.

Also notice that the solutions of $|x| > 7$ consist of all x-values for which $y_1 > y_2$ or for which the graph of y_1 is above the graph of y_2. These x-values are less than -7 or greater than 7, or $(-\infty, -7) \cup (7, \infty)$.

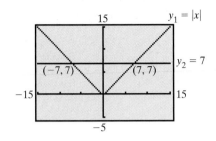

The solution set of the inequality $|x| < 7$ consists of the x-values where $y_1 < y_2$, or in interval notation $(-7, 7)$.

The solution set of the inequality $|x| > 7$ consists of the x-values where $y_1 > y_2$, or in interval notation $(-\infty, -7) \cup (7, \infty)$.

EXAMPLE 2

Solve algebraically and graphically for m: $|m - 6| < 2$.

Algebraic Solution:

From the preceding property, we see that

$$|m - 6| < 2 \text{ is equivalent to } -2 < m - 6 < 2$$

Solve this compound inequality for m by adding 6 to all three sides.

$$-2 < m - 6 < 2$$

$-2 + 6 < m - 6 + 6 < 2 + 6$ Add 6 to all three sides.
$4 < m < 8$ Simplify.

The solution set is $(4, 8)$.

Graphical Solution:

In the inequality $|m - 6| < 2$, replace m with x and proceed as usual. Graph $y_1 = |x - 6|$, $y_2 = 2$, and find x-values for which $y_1 < y_2$.

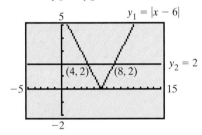

y_1 is below y_2 for x-values between 4 and 8. Thus, the solution set of $|x - 6| < 2$ in interval notation is $(4, 8)$.

The solution set is $(4, 8)$, and its number-line graph is shown.

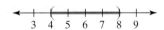

PRACTICE
2　Solve algebraically and graphically for b: $|b + 1| < 3$. Graph the solution set on a number line.

▶ **Helpful Hint**
Before using an absolute value inequality property, isolate the absolute value expression on one side of the inequality.

EXAMPLE 3　Solve algebraically for x: $|5x + 1| + 1 \le 10$.

Solution　First, isolate the absolute value expression by subtracting 1 from both sides.

$$|5x + 1| + 1 \le 10$$

$$|5x + 1| \le 10 - 1 \quad \text{Subtract 1 from both sides.}$$

$$|5x + 1| \le 9 \quad \text{Simplify.}$$

Since 9 is positive, we apply the absolute value property for $|X| \leq a$.

$$-9 \leq 5x + 1 \leq 9$$

$$-9 - 1 \leq 5x + 1 - 1 \leq 9 - 1 \qquad \text{Subtract 1 from all three parts.}$$

$$-10 \leq 5x \leq 8 \qquad \text{Simplify.}$$

$$-2 \leq x \leq \frac{8}{5} \qquad \text{Divide all three parts by 5.}$$

The solution set is $\left[-2, \dfrac{8}{5}\right]$, and the number-line graph is shown above.   □

PRACTICE
3   Solve for x: $|3x - 2| + 5 \leq 9$. Graph the solution set on a number line.

EXAMPLE 4   Solve for x: $\left|2x - \dfrac{1}{10}\right| < -13$. Check graphically.

Solution   The absolute value of a number is always nonnegative and can never be less than -13. Thus this absolute value inequality has no solution. The solution set is $\{ \ \}$ or $\varnothing$.   □

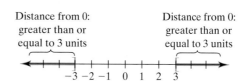

A calculator check for Example 4. Here, the absolute value graph is always above or greater than the horizontal line $y = -13$ so the solution set is $\{ \ \}$ or $\varnothing$.

PRACTICE
4   Solve for x: $\left|3x + \dfrac{5}{8}\right| < -4$. Check graphically.

OBJECTIVE 2 ▶ Solving absolute value inequalities of the form $|x| > a$. Let us now solve an absolute value inequality of the form $|X| > a$, such as $|x| \geq 3$. The solution set contains all numbers whose distance from 0 is 3 or more units. Thus the graph of the solution set contains 3 and all points to the right of 3 on the number line or -3 and all points to the left of -3 on the number line.

Distance from 0: greater than or equal to 3 units   Distance from 0: greater than or equal to 3 units

This solution set is written as $\{x \mid x \leq -3 \text{ or } x \geq 3\}$. In interval notation, the solution is $(-\infty, -3] \cup [3, \infty)$, since "or" means "union." In general, we have the following.

Solving Absolute Value Inequalities of the Form $|X| > a$

If a is a positive number, then $|X| > a$ is equivalent to $X < -a$ or $X > a$.

This property also holds true for the inequality symbol $\geq$.

EXAMPLE 5 Solve algebraically and graphically for x: $|x - 3| \geq 7$.

Algebraic Solution:

Since 7 is positive,

$$|x - 3| \geq 7 \text{ is equivalent to}$$
$$x - 3 \leq -7 \text{ or } x - 3 \geq 7$$

Next, solve the compound inequality.

$$x - 3 \leq -7 \qquad \text{or} \qquad x - 3 \geq 7$$
$$x - 3 + 3 \leq -7 + 3 \quad \text{or} \quad x - 3 + 3 \geq 7 \; + 3$$
$$x \leq -4 \qquad \text{or} \qquad x \geq 10$$

The solution set is $(-\infty, -4] \cup [10, \infty)$, and its number-line graph is shown.

Graphical Solution:

Graph $y_1 = |x - 3|$ and $y_2 = 7$. Find the x-values for which $y_1 \geq y_2$.

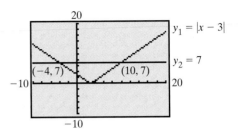

y_1 intersects or is above y_2 for x-values less than or equal to -4 and also x-values greater than or equal to 10, or $(-\infty, -4] \cup [10, \infty)$. □

PRACTICE
5 Solve algebraically and graphically for y: $|y + 4| \geq 6$. Graph the solution set on a number line.

Examples 6 and 8 illustrate special cases of absolute value inequalities. These special cases occur when an isolated absolute value expression is less than, less than or equal to, greater than, or greater than or equal to a negative number or 0.

EXAMPLE 6 Solve $|2x + 9| + 5 > 3$ algebraically and check graphically.

Solution First isolate the absolute value expression by subtracting 5 from both sides.

$$|2x + 9| + 5 > 3$$
$$|2x + 9| + 5 - 5 > 3 - 5 \quad \text{Subtract 5 from both sides.}$$
$$|2x + 9| > -2 \qquad \text{Simplify.}$$

The absolute value of any number is always nonnegative and thus is always greater than -2. This inequality and the original inequality are true for all values of x. The solution set is $\{x \mid x \text{ is a real number}\}$ or $(-\infty, \infty)$.

To check graphically, see the screen to the left. Notice that $y_1 > y_2$ or $|2x + 9| + 5 > 3$ for all real numbers.

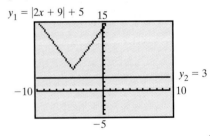

PRACTICE
6 Solve $|4x + 3| + 5 > 3$ algebraically and check graphically.

Concept Check ☑

Without taking any solution steps, how do you know that the absolute value inequality $|3x - 2| > -9$ has a solution? What is its solution?

Answer to Concept Check:
$(-\infty, \infty)$ since the absolute value is always nonnegative

EXAMPLE 7   Use a graphical approach to solve $\left|\dfrac{x}{3} - 1\right| - 2 \geq 0$.

Solution   Graph $y_1 = \left|\dfrac{x}{3} - 1\right| - 2$ and use the graph of y_1 to solve $y_1 \geq 0$.
Find x-values where the graph of y_1 is above the x-axis.

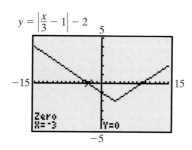 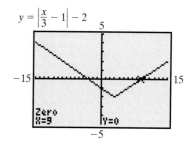

The graph of $y = \left|\dfrac{x}{3} - 1\right| - 2$ is above or on the x-axis for x-values less than or equal to -3 or greater than or equal to 9.

The solution set is $(-\infty, -3] \cup [9, \infty)$.

PRACTICE 7   Use a graphical approach to solve $\left|\dfrac{x}{2} - 3\right| - 3 > 0$.

EXAMPLE 8   Solve for x: $\left|\dfrac{2(x + 1)}{3}\right| \leq 0$.

Solution   Recall that "$\leq$" means "less than or equal to." The absolute value of any expression will never be less than 0, but it may be equal to 0. Thus, to solve $\left|\dfrac{2(x + 1)}{3}\right| \leq 0$ we solve $\left|\dfrac{2(x + 1)}{3}\right| = 0$

$$\frac{2(x + 1)}{3} = 0$$

$$3\left[\frac{2(x + 1)}{3}\right] = 3(0) \quad \text{Clear the equation of fractions.}$$

$$2x + 2 = 0 \quad \text{Apply the distributive property.}$$

$$2x = -2 \quad \text{Subtract 2 from both sides.}$$

$$x = -1 \quad \text{Divide both sides by 2.}$$

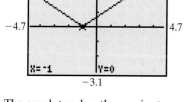

The graph touches the x-axis at $x = -1$ and is never below the x-axis. Therefore, the only point that satisfies the inequality has x-value -1.

The solution set is $\{-1\}$. See the screen to the left to check this solution set graphically.

PRACTICE 8   Solve for x: $\left|\dfrac{3(x - 2)}{5}\right| \leq 0$.

The following box summarizes the types of absolute value equations and inequalities.

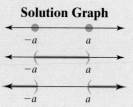

Solving Absolute Value Equations and Inequalities with $a > 0$

Algebraic Solution

$|X| = a$ is equivalent to $X = a$ or $X = -a$.

$|X| < a$ is equivalent to $-a < X < a$.

$|X| > a$ is equivalent to $X < -a$ or $X > a$.

Solution Graph

VOCABULARY & READINESS CHECK

Match each absolute value statement with an equivalent statement.

1. $|2x + 1| = 3$
2. $|2x + 1| \le 3$
3. $|2x + 1| < 3$
4. $|2x + 1| \ge 3$
5. $|2x + 1| > 3$

A. $2x + 1 > 3$ or $2x + 1 < -3$
B. $2x + 1 \ge 3$ or $2x + 1 \le -3$
C. $-3 < 2x + 1 < 3$
D. $2x + 1 = 3$ or $2x + 1 = -3$
E. $-3 \le 2x + 1 \le 3$

3.5 EXERCISE SET

Solve each inequality. Then graph the solution set on a number line and write it in interval notation. See Examples 1 through 4.

1. $|x| \le 4$
2. $|x| < 6$
3. $|x - 3| < 2$
4. $|y - 7| \le 5$
5. $|x + 3| < 2$
6. $|x + 4| < 6$
7. $|2x + 7| \le 13$
8. $|5x - 3| \le 18$
9. $|x| + 7 \le 12$
10. $|x| + 6 \le 7$
11. $|3x - 1| < -5$
12. $|8x - 3| < -2$
13. $|x - 6| - 7 \le -1$
14. $|z + 2| - 7 < -3$

Solve each inequality. Graph the solution set on a number line and write it in interval notation. See Examples 5 through 7.

15. $|x| > 3$
16. $|y| \ge 4$
17. $|x + 10| \ge 14$
18. $|x - 9| \ge 2$
19. $|x| + 2 > 6$
20. $|x| - 1 > 3$
21. $|5x| > -4$
22. $|4x - 11| > -1$
23. $|6x - 8| + 3 > 7$
24. $|10 + 3x| + 1 > 2$

Solve each inequality. Graph the solution set on a number line and write it in interval notation. See Example 8.

25. $|x| \le 0$
26. $|x| \ge 0$
27. $|8x + 3| > 0$
28. $|5x - 6| < 0$

MIXED PRACTICE

Solve each inequality. Graph the solution set on a number line and write it in interval notation. See Examples 1 through 8.

29. $|x| \le 2$
30. $|z| < 8$
31. $|y| > 1$
32. $|x| \ge 10$
33. $|x - 3| < 8$
34. $|-3 + x| \le 10$
35. $|0.6x - 3| > 0.6$
36. $|1 + 0.3x| \ge 0.1$
37. $5 + |x| \le 2$
38. $8 + |x| < 1$
39. $|x| > -4$
40. $|x| \le -7$
41. $|2x - 7| \le 11$
42. $|5x + 2| < 8$
43. $|x + 5| + 2 \ge 8$
44. $|-1 + x| - 6 > 2$
45. $|x| > 0$
46. $|x| < 0$
47. $9 + |x| > 7$
48. $5 + |x| \ge 4$
49. $6 + |4x - 1| \le 9$
50. $-3 + |5x - 2| \le 4$
51. $\left|\dfrac{2}{3}x + 1\right| > 1$
52. $\left|\dfrac{3}{4}x - 1\right| \ge 2$
53. $|5x + 3| < -6$
54. $|4 + 9x| \ge -6$
55. $\left|\dfrac{8x - 3}{4}\right| \le 0$
56. $\left|\dfrac{5x + 6}{2}\right| \le 0$
57. $|1 + 3x| + 4 < 5$
58. $|7x - 3| - 1 \le 10$
59. $\left|\dfrac{x + 6}{3}\right| > 2$
60. $\left|\dfrac{7 + x}{2}\right| \ge 4$
61. $-15 + |2x - 7| \le -6$

62. $-9 + |3 + 4x| < -4$

63. $\left|2x + \dfrac{3}{4}\right| - 7 \le -2$

64. $\left|\dfrac{3}{5} + 4x\right| - 6 < -1$

MIXED PRACTICE

Solve each equation or inequality for x. (See Sections 3.4 and 3.5.)

65. $|2x - 3| < 7$

66. $|2x - 3| > 7$

67. $|2x - 3| = 7$

68. $|5 - 6x| = 29$

69. $|x - 5| \ge 12$

70. $|x + 4| \ge 20$

71. $|9 + 4x| = 0$

72. $|9 + 4x| \ge 0$

73. $|2x + 1| + 4 < 7$

74. $8 + |5x - 3| \ge 11$

75. $|3x - 5| + 4 = 5$

76. $|5x - 3| + 2 = 4$

77. $|x + 11| = -1$

78. $|4x - 4| = -3$

79. $\left|\dfrac{2x - 1}{3}\right| = 6$

80. $\left|\dfrac{6 - x}{4}\right| = 5$

81. $\left|\dfrac{3x - 5}{6}\right| > 5$

82. $\left|\dfrac{4x - 7}{5}\right| < 2$

Use the given graphing utility screen to solve each equation or inequality.

83. a. $|x - 3| - 2 = 6$
   b. $|x - 3| - 2 < 6$
   c. $|x - 3| - 2 \ge 6$

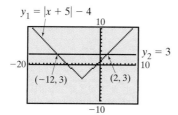

84. a. $|x + 5| - 4 = 3$
   b. $|x + 5| - 4 \le 3$
   c. $|x + 5| - 4 > 3$

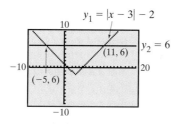

85. a. $|x + 2| - 10 = -4$
   b. $|x + 2| - 10 \le -4$
   c. $|x + 2| - 10 > -4$

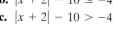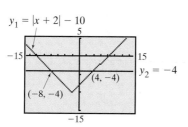

86. a. $|x + 2| + 1 = -5$
   b. $|x + 2| + 1 < -5$
   c. $|x + 2| + 1 > -5$

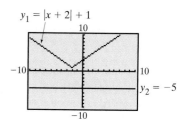

REVIEW AND PREVIEW

Recall the formula:

$$\text{Probability of an event} = \frac{\begin{array}{c}\text{number of ways that}\\ \text{the event can occur}\end{array}}{\begin{array}{c}\text{number of possible}\\ \text{outcomes}\end{array}}$$

Find the probability of rolling each number on a single toss of a die. (Recall that a die is a cube with each of its six sides containing 1, 2, 3, 4, 5, and 6 black dots, respectively.) See Section 1.8.

87. $P(\text{rolling a 2})$ **88.** $P(\text{rolling a 5})$

89. $P(\text{rolling a 7})$ **90.** $P(\text{rolling a 0})$

91. $P(\text{rolling a 1 or 3})$

92. $P(\text{rolling a 1, 2, 3, 4, 5, or 6})$

Consider the equation $3x - 4y = 12$. For each value of x or y given, find the corresponding value of the other variable that makes the statement true. See Section 1.8.

93. If $x = 2$, find y **94.** If $y = -1$, find x

95. If $y = -3$, find x **96.** If $x = 4$, find y

CONCEPT EXTENSIONS

97. Write an absolute value inequality representing all numbers x whose distance from 0 is less than 7 units.

98. Write an absolute value inequality representing all numbers x whose distance from 0 is greater than 4 units.

99. Write $-5 \le x \le 5$ as an equivalent inequality containing an absolute value.

100. Write $x > 1$ or $x < -1$ as an equivalent inequality containing an absolute value.

101. Describe how solving $|x - 3| = 5$ is different from solving $|x - 3| < 5$.

102. Describe how solving $|x + 4| = 0$ is similar to solving $|x + 4| \le 0$.

The expression $|x_T - x|$ is defined to be the absolute error in x, where x_T is the true value of a quantity and x is the measured value or value as stored in a computer.

103. If the true value of a quantity is 3.5 and the absolute error must be less than 0.05, find the acceptable measured values.

104. If the true value of a quantity is 0.2 and the approximate value stored in a computer is $\dfrac{51}{256}$, find the absolute error.

THE BIGGER PICTURE   SOLVING EQUATIONS AND INEQUALITIES

We now continue the outline from Sections 1.5, 3.2, and 3.3. Although suggestions will be given, this outline should be in your own words and you should include at least "how to recognize" and "how to begin to solve" under each letter heading.
   For example:

Solving Equations and Inequalities

I. Equations

  A. Linear equations (Sections 1.5 and 3.1)

  B. Absolute Value Equations: Equation contains the absolute value of a variable expression. (Section 3.4)

$$|3x - 1| - 12 = -4 \quad \text{Absolute value equation.}$$
$$|3x - 1| = 8 \quad \text{Isolate absolute value.}$$
$$3x - 1 = 8 \quad \text{or} \quad 3x - 1 = -8$$
$$3x = 9 \quad \text{or} \quad 3x = -7$$
$$x = 3 \quad \text{or} \quad x = -\frac{7}{3}$$

$$|x - 5| = |x + 1| \quad \text{Absolute value equation.}$$
$$x - 5 = x + 1 \quad \text{or} \quad x - 5 = -(x + 1)$$
$$\underbrace{-5 = 1}_{\text{No solution}} \quad \text{or} \quad x - 5 = -x - 1$$
$$\text{or} \quad 2x = 4$$
$$x = 2$$

II. Inequalities

  A. Linear Inequalities (Section 3.2)

  B. Compound Inequalities (Section 3.3)

  C. Absolute Value Inequalities: Inequality with absolute value bars about variable expression. (Section 3.5)

$|x - 5| - 8 < -2$	$|2x + 1| \geq 17$
$|x - 5| < 6$	$2x + 1 \geq 17$ or $2x + 1 \leq -17$
$-6 < x - 5 < 6$	$2x \geq 16$ or $\quad 2x \leq -18$
$-1 < x < 11$	$x \geq 8$ or $\quad x \leq -9$
$(-1, 11)$	$(-\infty, -9] \cup [8, \infty)$

Solve. If an inequality, write your solutions in interval notation.

1. $9x - 14 = 11x + 2$

2. $|x - 4| = 17$

3. $x - 1 \leq 5$ or $3x - 2 \leq 10$

4. $-x < 7$ and $4x \leq 20$

5. $|x - 2| = |x + 15|$

6. $9y - 6y + 1 = 4y + 10 - y + 3$

7. $1.5x - 3 = 1.2x - 18$

8. $\dfrac{7x + 1}{8} - 3 = x + \dfrac{2x + 1}{4}$

9. $|5x + 2| - 10 \leq -3$

10. $|x + 11| > 2$

11. $|9x + 2| - 1 = 24$

12. $\left|\dfrac{3x - 1}{2}\right| = |2x + 5|$

3.6 GRAPHING LINEAR INEQUALITIES IN TWO VARIABLES

OBJECTIVES

1 Graph linear inequalities.

2 Graph the intersection or union of two linear inequalities.

OBJECTIVE 1 ▶ Graphing linear inequalities. Recall that the graph of a linear equation in two variables is the graph of all ordered pairs that satisfy the equation, and we determined that the graph is a line. Here we graph **linear inequalities** in two variables; that is, we graph all the ordered pairs that satisfy the inequality.
   If the equal sign in a linear equation in two variables is replaced with an inequality symbol, the result is a linear inequality in two variables.

Examples of Linear Inequalities in Two Variables

$$3x + 5y \geq 6 \qquad 2x - 4y < -3$$
$$4x > 2 \qquad\qquad y \leq 5$$

DISCOVER THE CONCEPT

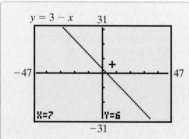

The graph of $y = 3 - x$ and a point $(7, 6)$ above the line.

a. Graph $x + y = 3$ by graphing $y_1 = 3 - x$ in an integer window.

b. Access the cursor that results from using the arrow key pad (not the trace cursor). Move the cursor to points above the line and notice the sum of the x- and y-coordinates.

c. Move the cursor to points below the line and notice the sum of the x- and y-coordinates.

d. Describe the set of points that satisfies $x + y < 3$ and the set of points that satisfies $x + y > 3$.

To graph the linear inequality $x + y < 3$, we first graph the related **boundary** equation $x + y = 3$. The resulting boundary line contains all ordered pairs the sum of whose coordinates is 3. This line separates the plane into two **half-planes.** All points "above" the boundary line $x + y = 3$ have coordinates that satisfy the inequality $x + y > 3$, and all points "below" the line have coordinates that satisfy the inequality $x + y < 3$.

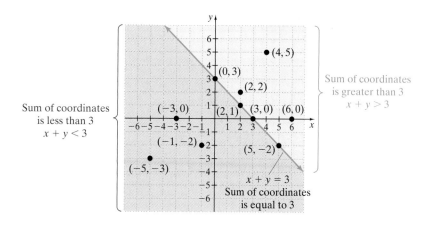

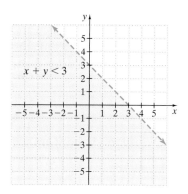

The graph, or **solution region,** for $x + y < 3$, then, is the half-plane below the boundary line and is shown shaded in the graph on the left. The boundary line is shown dashed since it is not a part of the solution region. These ordered pairs on this line satisfy $x + y = 3$ and not $x + y < 3$.

The following steps may be used to graph linear inequalities in two variables.

Graphing a Linear Inequality in Two Variables

STEP 1. Graph the boundary line found by replacing the inequality sign with an equal sign. If the inequality sign is $<$ or $>$, graph a dashed line indicating that points on the line are not solutions of the inequality. If the inequality sign is $\leq$ or $\geq$, graph a solid line indicating that points on the line are solutions of the inequality.

STEP 2. Choose a **test point not on the boundary line** and substitute the coordinates of this test point into the **original inequality.**

STEP 3. If a true statement is obtained in Step 2, shade the half-plane that contains the test point. If a false statement is obtained, shade the half-plane that does not contain the test point.

EXAMPLE 1   Graph $2x - y < 6$.

Solution   First, the boundary line for this inequality is the graph of $2x - y = 6$. Graph a dashed boundary line because the inequality symbol is $<$. Next, choose a test point on either side of the boundary line. The point $(0, 0)$ is not on the boundary line, so we use this point. Replacing x with 0 and y with 0 in the *original inequality* $2x - y < 6$ leads to the following:

$$2x - y < 6$$

$$2(0) - 0 < 6 \quad \text{Let } x = 0 \text{ and } y = 0.$$

$$0 < 6 \quad \text{True}$$

To shade a half-plane, go to the icon in front of Y= and press Enter to see the possible styles.

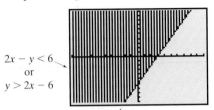

$2x - y < 6$
or
$y > 2x - 6$

On most graphing utilities, the boundary line will always appear solid. Thus, simply note that the boundary line should be dashed.

Because $(0, 0)$ satisfies the inequality, so does every point on the same side of the boundary line as $(0, 0)$. Shade the half-plane that contains $(0, 0)$. The half-plane graph of the inequality is shown below. The graph generated by a graphing utility is in the margin.

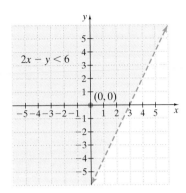

Every point in the shaded half-plane satisfies the original inequality. Notice that the inequality $2x - y < 6$ does not describe a function since its graph does not pass the vertical line test. ☐

PRACTICE

1 Graph $3x + y < 8$.

In general, linear inequalities of the form $Ax + By \leq C$, where A and B are not both 0, do not describe functions.

EXAMPLE 2 Graph $3x \geq y$.

Solution First, graph the boundary line $3x = y$. Graph a solid boundary line because the inequality symbol is $\geq$. Test a point not on the boundary line to determine which half-plane contains points that satisfy the inequality. We choose $(0, 1)$ as our test point.

$$3x \geq y$$
$$3(0) \geq 1 \quad \text{Let } x = 0 \text{ and } y = 1.$$
$$0 \geq 1 \quad \text{False}$$

This point does not satisfy the inequality, so the correct half-plane is on the opposite side of the boundary line from $(0, 1)$. The graph of $3x \geq y$ is the boundary line together with the shaded region shown.

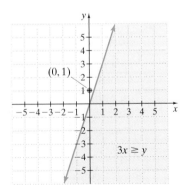

PRACTICE

2 Graph $x \geq 3y$.

Concept Check ☑

If a point on the boundary line is included in the solution of an inequality in two variables, should the graph of the boundary line be solid or dashed?

OBJECTIVE 2 ▶ Graphing intersections or unions of linear inequalities. The intersection and the union of linear inequalities can also be graphed, as shown in the next two examples.

EXAMPLE 3   Graph the intersection of $x \geq 1$ and $y \geq 2x - 1$.

Solution   Graph each inequality. The intersection of the two graphs is all points common to both regions, as shown by the dark pink shading in the third graph.

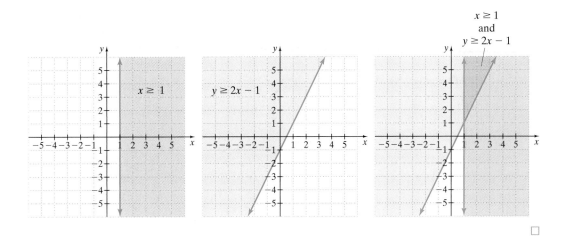

PRACTICE
3   Graph the intersection of $x \leq 3$ and $y \leq x - 2$.

EXAMPLE 4   Graph the union of $x + \frac{1}{2}y \geq -4$ or $y \leq -2$.

Solution   Graph each inequality. The union of the two inequalities is both shaded regions, including the solid boundary lines shown in the third graph.

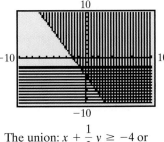

The union: $x + \frac{1}{2}y \geq -4$ or $y \leq -2$ using $y_1 = -2x - 4$, $y_2 = -2$, and the appropriate icons in front of each equation in the Y= screen.

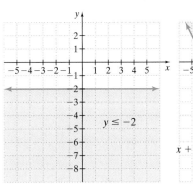

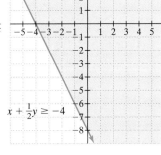

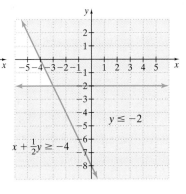

Answer to Concept Check:
Solid

PRACTICE
4   Graph the union of $2x - 3y \leq -2$ or $y \geq 1$.

3.6 | EXERCISE SET

PRACTICE   WATCH   DOWNLOAD   READ   REVIEW

Graph each inequality. See Examples 1 and 2.

1. $x < 2$

2. $x > -3$

3. $x - y \geq 7$

4. $3x + y \leq 1$

5. $3x + y > 6$

6. $2x + y > 2$

7. $y \leq -2x$

8. $y \leq 3x$

9. $2x + 4y \geq 8$

10. $2x + 6y \leq 12$

11. $5x + 3y > -15$

12. $2x + 5y < -20$

13. Explain when a dashed boundary line should be used in the graph of an inequality.

14. Explain why, after the boundary line is sketched, we test a point on either side of this boundary in the original inequality.

Graph each union or intersection. See Examples 3 and 4.

15. The intersection of $x \geq 3$ and $y \leq -2$

16. The union of $x \geq 3$ or $y \leq -2$

17. The union of $x \leq -2$ or $y \geq 4$

18. The intersection of $x \leq -2$ and $y \geq 4$

19. The intersection of $x - y < 3$ and $x > 4$

20. The intersection of $2x > y$ and $y > x + 2$

21. The union of $x + y \leq 3$ or $x - y \geq 5$

22. The union of $x - y \leq 3$ or $x + y > -1$

MIXED PRACTICE

Graph each inequality.

23. $y \geq -2$

24. $y \leq 4$

25. $x - 6y < 12$

26. $x - 4y < 8$

27. $x > 5$

28. $y \geq -2$

29. $-2x + y \leq 4$

30. $-3x + y \leq 9$

31. $x - 3y < 0$

32. $x + 2y > 0$

33. $3x - 2y \leq 12$

34. $2x - 3y \leq 9$

35. The union of $x - y > 2$ or $y < 5$

36. The union of $x - y < 3$ or $x > 4$

37. The intersection of $x + y \leq 1$ and $y \leq -1$

38. The intersection of $y \geq x$ and $2x - 4y \geq 6$

39. The union of $2x + y > 4$ or $x \geq 1$

40. The union of $3x + y < 9$ or $y \leq 2$

41. The intersection of $x \geq -2$ and $x \leq 1$

42. The intersection of $x \geq -4$ and $x \leq 3$

43. The union of $x + y \leq 0$ or $3x - 6y \geq 12$

44. The intersection of $x + y \leq 0$ and $3x - 6y \geq 12$

45. The intersection of $2x - y > 3$ and $x > 0$

46. The union of $2x - y > 3$ or $x > 0$

Match each inequality with its graph.

47. $y \leq 2x + 3$

48. $y < 2x + 3$

49. $y > 2x + 3$

50. $y \geq 2x + 3$

A

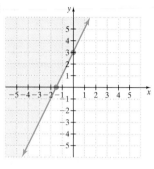

B

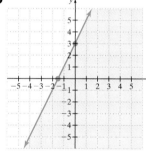

C

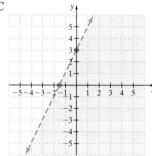

D

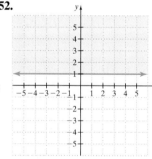

Write the inequality whose graph is given.

51.

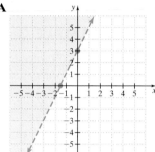

52.

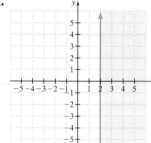

53.

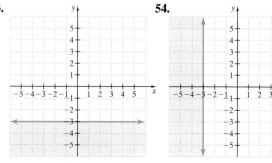

54.

55.

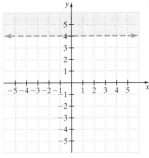

56.

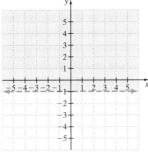

68.

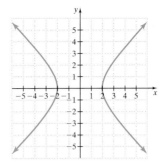

57.

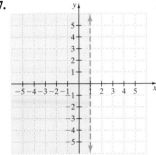

58.

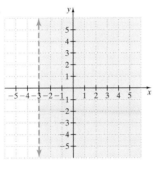

CONCEPT EXTENSIONS

Solve.

69. Rheem Abo-Zahrah decides that she will study at most 20 hours every week and that she must work at least 10 hours every week. Let x represent the hours studying and y represent the hours working. Write two inequalities that model this situation and graph their intersection.

70. The movie and TV critic for the *New York Times* spends between 2 and 6 hours daily reviewing movies and fewer than 5 hours reviewing TV shows. Let x represent the hours watching movies and y represent the time spent watching TV. Write two inequalities that model this situation and graph their intersection.

71. Chris-Craft manufactures boats out of Fiberglas and wood. Fiberglas hulls require 2 hours work, whereas wood hulls require 4 hours work. Employees work at most 40 hours a week. The following inequalities model these restrictions, where x represents the number of Fiberglas hulls produced and y represents the number of wood hulls produced.

$$\begin{cases} x \geq 0 \\ y \geq 0 \\ 2x + 4y \leq 40 \end{cases}$$

Graph the intersection of these inequalities.

REVIEW AND PREVIEW

Evaluate each expression. See Sections 1.3 and 1.4.

59. 2^3 **60.** 3^2 **61.** -5^2

62. $(-5)^2$ **63.** $(-2)^4$ **64.** -2^4

65. $\left(\dfrac{3}{5}\right)^3$ **66.** $\left(\dfrac{2}{7}\right)^2$

Find the domain and the range of each relation. Determine whether the relation is also a function. See Section 2.2.

67.

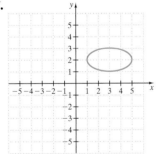

CHAPTER 3 GROUP ACTIVITY

Analyzing Municipal Budgets

Nearly all cities, towns, and villages operate with an annual budget. Budget items might include expenses for fire and police protection as well as for street maintenance and parks. No matter how big or small the budget, city officials need to know if municipal spending is over or under budget. In this project, you will have the opportunity to analyze a municipal budget and make budgetary recommendations. This project may be completed by working in groups or individually.

Suppose that each year your town creates a municipal budget. The next year's annual municipal budget is submitted for approval by the town's citizens at the annual town meeting. This year's budget was printed in the town newspaper earlier in the year.

You have joined a group of citizens who are concerned about your town's budgeting and spending processes. Your group plans to analyze this year's budget along with what was actually spent by the town this year. You hope to present your findings at the annual town meeting and make some budgetary recommendations for next year's budget. The municipal budget contains many different areas of spending. To help focus your group's analysis, you have decided to research spending habits only for categories in which the actual expenses differ from the budgeted amount by more than 12% of the budgeted amount.

1. For each category in the budget, write a specific absolute value inequality that describes the condition that must be met before your group will research spending habits for that category. In each case, let the variable x represent the actual expense for a budget category.

2. For each category in the budget, write an equivalent compound inequality for the condition described in Question 1. Again, let the variable x represent the actual expense for a budget category.

3. Below is a listing of the actual expenditures made this year for each budget category. Use the inequalities from either

Question 1 or Question 2 to complete the Budget Worksheet given at the end of this project. (The first category has been filled in.) From the Budget Worksheet, decide which categories must be researched.

4. Can you think of possible reasons why spending in the categories that must be researched were over or under budget?

5. Based on this year's municipal budget and actual expenses, what recommendations would you make for next year's budget? Explain your reasoning.

6. (Optional) Research the annual budget used by your own town or your college or university. Conduct a similar analysis of the budget with respect to actual expenses. What can you conclude?

	Department/Program	Actual Expenditure
I.	**Board of Health**	
	Immunization Programs	$14,800
	Inspections	$41,900
II.	**Fire Department**	
	Equipment	$375,000
	Salaries	$268,500
III.	**Libraries**	
	Book/Periodical Purchases	$107,300
	Equipment	$29,000
	Salaries	$118,400
IV.	**Parks and Recreation**	
	Maintenance	$82,500
	Playground Equipment	$45,000
	Salaries	$118,000
	Summer Programs	$96,200
V.	**Police Department**	
	Equipment	$328,000
	Salaries	$405,000
VI.	**Public Works**	
	Recycling	$48,100
	Sewage	$92,500
	Snow Removal & Road Salt	$268,300
	Street Maintenance	$284,000
	Water Treatment	$94,100
	TOTAL	$2,816,600

THE TOWN CRIER
Annual Budget Set at Town Meeting
ANYTOWN, USA (MG)—This year's annual budget is as follows:

	Amount Budgeted
BOARD OF HEALTH	
Immunization Programs	$15,000
Inspections	$50,000
FIRE DEPARTMENT	
Equipment	$450,000
Salaries	$275,000
LIBRARIES	
Book/Periodical Purchases	$90,000
Equipment	$30,000
Salaries	$120,000
PARKS AND RECREATION	
Maintenance	$70,000
Playground Equipment	$50,000
Salaries	$140,000
Summer Programs	$80,000
POLICE DEPARTMENT	
Equipment	$300,000
Salaries	$400,000
PUBLIC WORKS	
Recycling	$50,000
Sewage	$100,000
Snow Removal & Road Salt	$200,000
Street Maintenance	$250,000
Water Treatment	$100,000
TOTAL	**$2,770,000**

BUDGET WORKSHEET

Budget category	Budgeted amount	Minimum allowed	Actual expense	Maximum allowed	Within budget?	Amt over/ under budget
Immunization Programs	$15,000	$13,200	$14,800	$16,800	Yes	Under $200

STUDY SKILLS BUILDER

Are You Preparing for a Test on Chapter 3?

Below are listed some common trouble areas for students in Chapter 3. After studying for your test—but before taking your test—read these.

- Remember to solve equations both algebraically and graphically to ensure accuracy.

- Remember to reverse the direction of the inequality symbol when multiplying or dividing both sides of an inequality by a negative number.

$$-11x < 33 \quad \text{Direction of arrow is reversed.}$$
$$\frac{-11x}{-11} > \frac{33}{-11}$$
$$x > -3$$

- Remember the differences when solving absolute value equations and inequalities.

$$|x + 1| = 3$$
$$x + 1 = 3 \quad \text{or} \quad x + 1 = -3$$
$$x = 2 \quad \text{or} \quad x = -4$$
$$\{2, -4\}$$

$$|x + 1| < 3$$
$$-3 < x + 1 < 3$$
$$-3 - 1 < x < 3 - 1$$
$$-4 < x < 2$$
$$(-4, 2)$$

$$|x + 1| > 3$$
$$x + 1 < -3 \quad \text{or} \quad x + 1 > 3$$
$$x < -4 \quad \text{or} \quad x > 2$$
$$(-\infty, -4) \cup (2, \infty)$$

- Remember that an equation is not solved for a specified variable unless the variable is alone on one side of an equation *and* the other side contains *no* specified variables.

$$y = 10x + 6 - y \quad \text{Equation is not solved for } y.$$
$$2y = 10x + 6 \quad \text{Add } y \text{ to both sides.}$$
$$y = 5x + 3 \quad \text{Divide both sides by 2.}$$

Remember: This is simply a checklist of common trouble areas. For a review of Chapter 3, see the Highlights and Chapter Review at the end of this chapter.

CHAPTER 3 VOCABULARY CHECK

Fill in each blank with one of the words or phrases listed below.

contradiction   absolute value   linear inequality in one variable
linear equation in one variable   compound inequality   identity
intersection   solution   union

1. Two inequalities joined by the words "and" or "or" is called a(n) _____.

2. An equation in one variable that has no solution is called a(n) _____.

3. The _____ of two sets is the set of all elements common to both sets.

4. The _____ of two sets is the set of all elements that belong to either of the sets.

5. An equation in one variable that has every number (for which the equation is defined) as a solution is called a(n) _____.

6. A number's distance from 0 is called its _____.

7. When a variable in an equation is replaced by a number and the resulting equation is true, then that number is called a(n) _____ of the equation.

8. The statement $5x - 0.2 < 7$ is an example of a(n) _____.

9. The statement $5x - 0.2 = 7$ is an example of a(n) _____.

▶ **Helpful Hint**
Are you preparing for your test? Don't forget to take the Chapter 3 Test on page 265. Then check your answers at the back of the text and use the Chapter Test Prep Video CD to see the fully worked-out solutions to any of the exercises you want to review.

CHAPTER 3 HIGHLIGHTS

DEFINITIONS AND CONCEPTS	EXAMPLES

SECTION 3.1 SOLVING LINEAR EQUATIONS GRAPHICALLY

To solve an equation graphically by the **intersection-of-graphs method:**

- Graph the left side of the equation as y_1.
- Graph the right side of the equation as y_2.
- Find any points of intersection, or where $y_1 = y_2$.
- The x-coordinate of an intersection point is a solution.
- The y-coordinate of an intersection point is the value of each side of the original equation when the variable is replaced with the solution.

Solve $5(x - 2) + 1 = 2(x - 1) + 2$

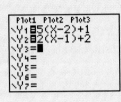
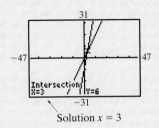

Solution $x = 3$

To solve an equation using the **x-intercept method:**

- Write the equation so that one side is 0.
- For the equation $y_1 = 0$, graph y_1.
- The x-intercepts of the graph are solutions of $y_1 = 0$.

Solve $5(x - 2) + 1 = 2(x - 1) + 2$

$5(x - 2) + 1 - 2(x - 1) - 2 = 0$

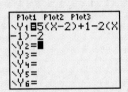
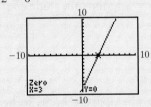

SECTION 3.2 LINEAR INEQUALITIES AND PROBLEM SOLVING

A **linear inequality in one variable** is an inequality that can be written in the form $ax + b < c$, where a, b, and c are real numbers and $a \neq 0$. (The inequality symbols $\leq$, $>$, and $\geq$ also apply here.)

The **addition property of inequality** guarantees that the same number may be added to (or subtracted from) both sides of an inequality, and the resulting inequality will have the same solution set.

The **multiplication property of inequality** guarantees that both sides of an inequality may be multiplied by (or divided by) the same **positive** number, and the resulting inequality will have the same solution set. We may also multiply (or divide) both sides of an inequality by the same **negative** number and **reverse the direction of the inequality symbol,** and the result is an inequality with the same solution set.

Linear inequalities:

$$5x - 2 \leq -7 \qquad 3y > 1 \qquad \frac{z}{7} < -9(z - 3)$$

$$x - 9 \leq -16$$
$$x - 9 + 9 \leq -16 + 9 \quad \text{Add 9.}$$
$$x \leq -7$$

Solve.

$$6x < -66$$

$$\frac{6x}{6} < \frac{-66}{6} \quad \text{Divide by 6. Do not reverse direction of inequality symbol.}$$

$$x < -11$$

Solve.

$$-6x < -66$$

$$\frac{-6x}{-6} > \frac{-66}{-6} \quad \text{Divide by } -6. \text{ Reverse direction of inequality symbol.}$$

$$x > 11$$

DEFINITIONS AND CONCEPTS	**EXAMPLES**

To solve a linear inequality in one variable:	Solve for x:
	$$\frac{3}{7}(x - 4) \geq x + 2$$
1. Clear the equation of fractions.	**1.** $7\left[\dfrac{3}{7}(x - 4)\right] \geq 7(x + 2)$ Multiply by 7.
	$3(x - 4) \geq 7(x + 2)$
2. Remove grouping symbols such as parentheses.	**2.** $3x - 12 \geq 7x + 14$ Apply the distributive property.
3. Simplify by combining like terms.	
4. Write variable terms on one side and numbers on the other side using the addition property of inequality.	**4.** $-4x - 12 \geq 14$ Subtract $7x$.
	$ -4x \geq 26$ Add 12.
	$\dfrac{-4x}{-4} \leq \dfrac{26}{-4}$ Divide by -4. Reverse direction of inequality symbol.
5. Isolate the variable using the multiplication property of inequality.	$x \leq -\dfrac{13}{2}$

Two inequalities joined by the words **and** or **or** are called **compound inequalities.**	Compound inequalities:
	$\quad x - 7 \leq 4 \quad$ and $\quad x \geq -21$
	$2x + 7 > x - 3 \quad$ or $\quad 5x + 2 > -3$
The solution set of a compound inequality formed by the word **and** is the **intersection** $\cap$ of the solution sets of the two inequalities.	Solve for x:
	$x < 5$ and $x < 3$
	$\{x \mid x < 5\}$ $(-\infty, 5)$
	$\{x \mid x < 3\}$ $(-\infty, 3)$
	$\{x \mid x < 3$ and $x < 5\}$ $(-\infty, 3)$
The solution set of a compound inequality formed by the word **or** is the **union,** $\cup$, of the solution sets of the two inequalities.	Solve for x:
	$\quad x - 2 \geq -3 \quad$ or $\quad 2x \leq -4$
	$\quad\quad x \geq -1 \quad$ or $\quad\quad x \leq -2$
	$\{x \mid x \geq -1\}$ $[-1, \infty)$
	$\{x \mid x \leq -2\}$ $(-\infty, -2]$
	$\{x \mid x \leq -2$ or $x \geq -1\}$ $(-\infty, -2] \cup [-1, \infty)$

(continued)

| **DEFINITIONS AND CONCEPTS** | **EXAMPLES** |

SECTION 3.3   COMPOUND INEQUALITIES (continued)

To solve a compound inequality $y_1 < y_2 < y_3$ graphically,

1. Graph separately each of the three parts y_1, y_2, and y_3, respectively, in an appropriate window.

2. Observe where the graph of y_2 is between the graphs of y_1 and y_3.

3. Find the x-coordinates of the points of intersection and determine the appropriate interval of the solution.

Solve for x: $-13 < 3x - 4 \leq 8$

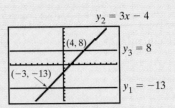

The solution in interval notation is $(-3, 4]$.

SECTION 3.4   ABSOLUTE VALUE EQUATIONS

If a is a positive number, then $|x| = a$ is equivalent to $x = a$ or $x = -a$.

Solve for y:

$$|5y - 1| - 7 = 4$$

$$|5y - 1| = 11$$

$$5y - 1 = 11 \quad \text{or} \quad 5y - 1 = -11 \quad \text{Add 7.}$$

$$5y = 12 \quad \text{or} \qquad 5y = -10 \quad \text{Add 1.}$$

$$y = \frac{12}{5} \qquad \text{or} \quad y = -2 \quad \text{Divide by 5.}$$

The solutions are -2 and $\frac{12}{5}$.

If a is negative, then $|x| = a$ has no solution.

Solve for x:

$$\left|\frac{x}{2} - 7\right| = -1$$

The solution set is $\{ \ \}$ or $\varnothing$.

If an absolute value equation is of the form $|x| = |y|$, solve $x = y$ or $x = -y$.

Solve for x:

$$|x - 7| = |2x + 1|$$

$$x - 7 = 2x + 1 \quad \text{or} \quad x - 7 = -(2x + 1)$$

$$x = 2x + 8 \qquad\qquad x - 7 = -2x - 1$$

$$-x = 8 \qquad\qquad\qquad x = -2x + 6$$

$$x = -8 \qquad \text{or} \qquad 3x = 6$$

$$x = 2$$

The solutions are -8 and 2.

SECTION 3.5   ABSOLUTE VALUE INEQUALITIES

If a is a positive number, then $|x| < a$ is equivalent to $-a < x < a$.

Solve for y:

$$|y - 5| \leq 3$$

$$-3 \leq y - 5 \leq 3$$

$$-3 + 5 \leq y - 5 + 5 \leq 3 + 5 \quad \text{Add 5.}$$

$$2 \leq y \leq 8$$

The solution set is $[2, 8]$.

DEFINITIONS AND CONCEPTS	**EXAMPLES**

SECTION 3.5 ABSOLUTE VALUE INEQUALITIES (continued)

If a is a positive number, then $|x| > a$ is equivalent to $x < -a$ or $x > a$.

Solve for x:

$$\left|\frac{x}{2} - 3\right| > 7$$

$$\frac{x}{2} - 3 < -7 \quad \text{or} \quad \frac{x}{2} - 3 > 7$$

$$x - 6 < -14 \quad \text{or} \quad x - 6 > 14 \qquad \text{Multiply by 2.}$$

$$x < -8 \quad \text{or} \quad x > 20 \qquad \text{Add 6.}$$

The solution set is $(-\infty, -8) \cup (20, \infty)$.

SECTION 3.6 GRAPHING LINEAR INEQUALITIES IN TWO VARIABLES

If the equal sign in a linear equation in two variables is replaced with an inequality symbol, the result is a **linear inequality in two variables.**

To graph a linear inequality

1. Graph the boundary line by graphing the related equation. Draw the line solid if the inequality symbol is $\leq$ or $\geq$. Draw the line dashed if the inequality symbol is $<$ or $>$.

2. Choose a test point not on the line. Substitute its coordinates into the original inequality.

3. If the resulting inequality is true, shade the **half-plane** that contains the test point. If the inequality is not true, shade the half-plane that does not contain the test point.

Linear Inequalities in Two Variables

$$x \leq -5 \qquad y \geq 2$$
$$3x - 2y > 7 \qquad x < -5$$

Graph $2x - 4y > 4$.

1. Graph $2x - 4y = 4$. Draw a dashed line because the inequality symbol is $>$.

2. Check the test point $(0, 0)$ in the inequality $2x - 4y > 4$.

$$2 \cdot 0 - 4 \cdot 0 > 4 \qquad \text{Let } x = 0 \text{ and } y = 0.$$
$$0 > 4 \qquad \text{False}$$

3. The inequality is false, so we shade the half-plane that does not contain $(0, 0)$.

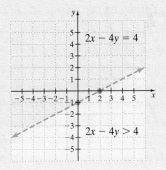

CHAPTER 3 REVIEW

(3.1) *Solve each equation algebraically and graphically.*

1. $4(x - 6) + 3 = 27$

2. $15(x + 2) - 6 = 18$

3. $5x + 15 = 3(x + 2) + 2(x - 3)$

4. $2x - 5 + 3(x - 4) = 5(x + 2) - 27$

5. $14 - 2(x + 3) = 3(x - 9) + 18$

6. $16 + 2(5 - x) = 19 - 3(x + 2)$

Solve each equation graphically. Round solutions to the nearest hundredth.

7. $0.4(x - 6) = \pi x + \sqrt{3}$

8. $1.7x + \sqrt{7} = -0.4x - \sqrt{6}$

(3.2) *Solve each linear inequality. Write your answers in interval notation.*

9. $3(x - 5) > -(x + 3)$

10. $-2(x + 7) \geq 3(x + 2)$

11. $4x - (5 + 2x) < 3x - 1$

12. $3(x - 8) < 7x + 2(5 - x)$

13. $24 \geq 6x - 2(3x - 5) + 2x$

14. $\dfrac{x}{3} + \dfrac{1}{2} > \dfrac{2}{3}$

15. $x + \dfrac{3}{4} < -\dfrac{x}{2} + \dfrac{9}{4}$

16. $\dfrac{x - 5}{2} \leq \dfrac{3}{8}(2x + 6)$

Solve.

17. George Boros can pay his housekeeper $15 per week to do his laundry, or he can have the laundromat do it at a cost of 50 cents per pound for the first 10 pounds and 40 cents for each additional pound. Use an inequality to find the weight at which it is more economical to use the housekeeper than the laundromat.

18. Ceramic firing temperatures usually range from 500° to 1000° Fahrenheit. Use a compound inequality to convert this range to the Celsius scale. Round to the nearest degree. $\left(\text{Use F} = \dfrac{9C + 160}{5}\right)$

19. In the Olympic gymnastics competition, Nana must average a score of 9.65 to win the silver medal. Seven of the eight judges have reported scores of 9.5, 9.7, 9.9, 9.7, 9.7, 9.6, and 9.5. Use an inequality to find the minimum score that Nana must receive from the last judge to win the silver medal.

20. Carol would like to pay cash for a car when she graduates from college and estimates that she can afford a car that costs between $4000 and $8000. She has saved $500 so far and plans to earn the rest of the money by working the next two summers. If Carol plans to save the same amount each summer, use a compound inequality to find the range of money she must save each summer to buy the car.

(3.3) Solve each inequality. Write your answers in interval notation.

21. $1 \leq 4x - 7 \leq 3$

22. $-2 \leq 8 + 5x < -1$

23. $-3 < 4(2x - 1) < 12$

24. $-6 < x - (3 - 4x) < -3$

25. $\dfrac{1}{6} < \dfrac{4x - 3}{3} \leq \dfrac{4}{5}$

26. $x \leq 2$ and $x > -5$

27. $3x - 5 > 6$ or $-x < -5$

(3.4) Solve each absolute value equation.

28. $|x - 7| = 9$

29. $|8 - x| = 3$

30. $|2x + 9| = 9$

31. $|-3x + 4| = 7$

32. $|3x - 2| + 6 = 10$

33. $5 + |6x + 1| = 5$

34. $-5 = |4x - 3|$

35. $|5 - 6x| + 8 = 3$

36. $-8 = |x - 3| - 10$

37. $\left|\dfrac{3x - 7}{4}\right| = 2$

38. $|6x + 1| = |15 + 4x|$

(3.5) Solve each absolute value inequality. Graph the solution set and write it in interval notation.

39. $|5x - 1| < 9$

40. $|6 + 4x| \geq 10$

41. $|3x| - 8 > 1$

42. $9 + |5x| < 24$

43. $|6x - 5| \leq -1$

44. $\left|3x + \dfrac{2}{5}\right| \geq 4$

45. $\left|\dfrac{x}{3} + 6\right| - 8 > -5$

46. $\left|\dfrac{4(x - 1)}{7}\right| + 10 < 2$

(3.6) Graph each linear inequality.

47. $\dfrac{1}{2}x - y < 2$ **48.** $3x + y > 4$

49. $3y \geq x$ **50.** $5x - 2y \leq 9$

51. $x > -2$ **52.** $y < 1$

53. Graph the intersection of $2x < 3y + 8$ and $y \geq -2$.

54. Graph the union of $y > 2x + 3$ or $x \leq -3$.

MIXED REVIEW

Solve.

55. $\dfrac{x - 2}{5} + \dfrac{x + 2}{2} = \dfrac{x + 4}{3}$

56. $\dfrac{2z - 3}{4} - \dfrac{4 - z}{2} = \dfrac{z + 1}{3}$

Solve. If an inequality, write your solutions in interval notation.

57. $\dfrac{3(x - 2)}{5} > \dfrac{-5(x - 2)}{3}$

58. $0 \leq \dfrac{2(3x + 4)}{5} \leq 3$

59. $x \leq 2$ or $x > -5$

60. $-2x \leq 6$ and $-2x + 3 < -7$

61. $|7x| - 26 = -5$

62. $\left|\dfrac{9 - 2x}{5}\right| = -3$

63. $|x - 3| = |7 + 2x|$

64. $|6x - 5| \geq -1$

65. $\left|\dfrac{4x - 3}{5}\right| < 1$

66. $48 + x \geq 5(2x + 4) - 2x$

CHAPTER 3 TEST

TEST PREP VIDEO Remember to use your Chapter Test Prep Video CD to help you study and view solutions to the test questions you need help with.

1. Solve $15x + 26 = -2(x + 1) - 1$ algebraically and graphically.

2. Solve $-3x - \sqrt{5} = \pi(x - 1)$ graphically. Round the solution to the nearest hundredth.

Solve. Write inequality solutions in interval notation.

3. $|6x - 5| - 3 = -2$

4. $|8 - 2t| = -6$

5. $3(2x - 7) - 4x > -(x + 6)$

6. $8 - \dfrac{x}{2} \geq 7$

7. $-3 < 2(x - 3) \leq 4$

8. $|3x + 1| > 5$

9. $x \leq -2$ and $x \leq -5$

10. $x \leq -2$ or $x \leq -5$

11. $-x > 1$ and $3x + 3 \geq x - 3$

12. $6x + 1 > 5x + 4$ or $1 - x > -4$

13. $|x - 5| - 4 < -2$

14. $\left|\dfrac{5x - 7}{2}\right| = 4$

15. $\left|17x - \dfrac{1}{5}\right| > -2$

16. $|x - 5| = |x + 2|$

Graph each inequality.

17. $x \leq -4$

18. $2x - y > 5$

19. The intersection of $2x + 4y < 6$ and $y \leq -4$

Use the given screen to solve each inequality. Write the solution in interval notation.

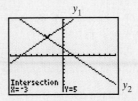

20. $y_1 < y_2$

21. $y_1 > y_2$

22. The company that makes Photoray sunglasses figures that the cost C to make x number of sunglasses weekly is given by $C = 3910 + 2.8x$, and the weekly revenue R is given by $R = 7.4x$. Use an inequality to find the number of sunglasses that must be made and sold to make a profit. (Revenue must exceed cost in order to make a profit.)

CHAPTER 3 CUMULATIVE REVIEW

List the elements in each set.

1. a. $\{x \mid x \text{ is a natural number greater than } 100\}$
   b. $\{x \mid x \text{ is a whole number between } 1 \text{ and } 6\}$

2. a. $\{x \mid x \text{ is an integer between } -3 \text{ and } 5\}$
   b. $\{x \mid x \text{ is a whole number between } 3 \text{ and } 5\}$

3. Find each value.
   a. $|3|$
   b. $\left|-\dfrac{1}{7}\right|$
   c. $-|2.7|$
   d. $-|-8|$
   e. $|0|$

4. Find the opposite of each number.
   a. $\dfrac{2}{3}$
   b. -9
   c. 1.5

5. Add.
   a. $-3 + (-11)$
   b. $3 + (-7)$
   c. $-10 + 15$
   d. $-8.3 + (-1.9)$
   e. $-\dfrac{2}{3} + \dfrac{3}{7}$

6. Subtract.
   a. $-2 - (-10)$
   b. $1.7 - 8.9$
   c. $-\dfrac{1}{2} - \dfrac{1}{4}$

7. Find the square roots.
   a. $\sqrt{9}$
   b. $\sqrt{25}$
   c. $\sqrt{\dfrac{1}{4}}$
   d. $-\sqrt{36}$
   e. $\sqrt{-36}$

8. Multiply or divide.
   a. $-3(-2)$
   b. $-\dfrac{3}{4}\left(-\dfrac{4}{7}\right)$
   c. $\dfrac{0}{-2}$
   d. $\dfrac{-20}{-2}$

9. Evaluate each algebraic expression when $x = 2$, $y = -1$, and $z = -3$.
   a. $z - y$
   b. $-2z^2$
   c. $\dfrac{2x + y}{z}$
   d. $-x^2 - 4x$

10. Find the roots.

   a. $\sqrt[4]{1}$ **b.** $\sqrt[3]{8}$

   c. $\sqrt[4]{81}$

11. Write each sentence using mathematical symbols.

   a. The sum of x and 5 is 20.

   b. Two times the sum of 3 and y amounts to 4.

   c. Subtract 8 from x, and the difference is the same as the product of 2 and x.

   d. The quotient of z and 9 amounts to 9 plus z.

12. Insert $<, >$, or $=$ between each pair of numbers to form a true statement.

   a. $-3 \quad -5$ **b.** $\dfrac{-12}{-4} \quad 3$

   c. $0 \quad -2$

13. Use the commutative property of addition to write an expression equivalent to $7x + 5$.

14. Use the associative property of multiplication to write an expression equivalent to $5 \cdot (7x)$. Then simplify the expression.

Solve for x.

15. $2x + 5 = 9$

16. $11.2 = 1.2 - 5x$

17. $6x - 4 = 2 + 6(x - 1)$

18. $2x + 1.5 = -0.2 + 1.6x$

19. Write the following as algebraic expressions. Then simplify.

   a. The sum of three consecutive integers, if x is the first consecutive integer.

   b. The perimeter of the triangle with sides of length x, $5x$, and $6x - 3$.

20. Write the following as algebraic expressions. Then simplify.

   a. The sum of three consecutive odd integers if x is the first consecutive integers.

   b. The perimeter of a square with side length $3x + 1$.

21. Find three numbers such that the second number is 3 more than twice the first number and the third number is four times the first number. The sum of the three numbers is 164.

22. Find two numbers such that the second number is 2 more than three times the first number and the difference of the two numbers is 24.

23. Solve $3y - 2x = 7$ for y.

24. Solve $7x - 4y = 10$ for x.

25. Solve $A = \dfrac{1}{2}(B + b)h$ for b.

26. Solve $P = 2l + 2w$ for l.

27. Write each in interval notation.

   a. $\{x \mid x \geq 2\}$ **b.** $\{x \mid x < -1\}$

   c. $\{x \mid 0.5 < x \leq 3\}$

28. Write each in interval notation.

   a. $\{x \mid x \leq -3\}$

   b. $\{x \mid -2 \leq x < 0.1\}$

Solve.

29. $-(x - 3) + 2 \leq 3(2x - 5) + x$

30. $2(7x - 1) - 5x > -(-7x) + 4$

31. **a.** $2(x + 3) > 2x + 1$

   b. $2(x + 3) < 2x + 1$

32. $4(x + 1) - 3 < 4x + 1$

33. If $A = \{x \mid x$ is an even number greater than 0 and less than 10$\}$ and $B = \{3, 4, 5, 6\}$, find $A \cap B$.

34. Find the union: $\{-2, 0, 2, 4\} \cup \{-1, 1, 3, 5\}$

35. Solve: $x - 7 < 2$ and $2x + 1 < 9$

36. Solve: $x + 3 \leq 1$ or $3x - 1 < 8$

37. If $A = \{x \mid x$ is an even number greater than 0 and less than 10$\}$ and $B = \{3, 4, 5, 6\}$, find $A \cup B$.

38. Find the intersection: $\{-2, 0, 2, 4\} \cap \{-1, 1, 3, 5\}$

39. Solve: $-2x - 5 < -3$ or $6x < 0$

40. Solve: $-2x - 5 < -3$ and $6x < 0$

Solve.

41. $|p| = 2$

42. $|x| = 5$

43. $\left| \dfrac{x}{2} - 1 \right| = 11$

44. $\left| \dfrac{y}{3} + 2 \right| = 10$

45. $|x - 3| = |5 - x|$

46. $|x + 3| = |7 - x|$

47. $|x| \leq 3$

48. $|x| > 1$

49. $|2x + 9| + 5 > 3$

50. $|3x + 1| + 9 < 1$

4

Systems of Linear Equations and Inequalities

MySpace and Facebook are both popular social networking Web sites offering interactive, user-submitted networks of friends, personal profiles, blogs, groups, music, videos, and photos.

MySpace is currently the world's sixth most popular Web site in any language and the third most popular Web site in the United States. Facebook was launched in early 2004 and is currently the seventh most visited Web site.

In Section 4.3, Exercise 53, page 300, we will form and use the functions graphed below to solve the system of equations.

In this chapter, two or more equations in two or more variables are solved simultaneously. Such a collection of equations is called a **system of equations.** Systems of equations are good mathematical models for many real-world problems because these problems may involve several related patterns.

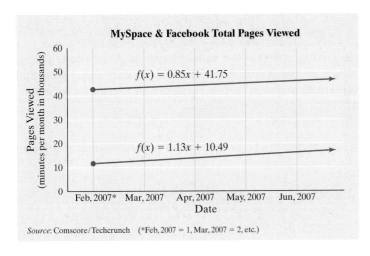

MySpace & Facebook Total Pages Viewed

$f(x) = 0.85x + 41.75$

$f(x) = 1.13x + 10.49$

Pages Viewed (minutes per month in thousands)

Feb, 2007* Mar, 2007 Apr, 2007 May, 2007 Jun, 2007

Date

Source: Comscore/Techcrunch   (*Feb, 2007 = 1, Mar, 2007 = 2, etc.)

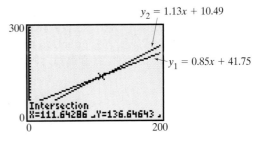

$y_2 = 1.13x + 10.49$

$y_1 = 0.85x + 41.75$

Intersection
X=111.64286 Y=136.64643

4.1 SOLVING SYSTEMS OF LINEAR EQUATIONS IN TWO VARIABLES

OBJECTIVES

1 Determine whether an ordered pair is a solution of a system of two linear equations.

2 Solve a system by graphing.

3 Solve a system by substitution.

4 Solve a system by elimination.

An important problem that often occurs in the fields of business and economics concerns the concepts of revenue and cost. For example, suppose that a small manufacturing company begins to manufacture and sell compact disc storage units. The revenue of a company is the company's income from selling these units, and the cost is the amount of money that a company spends to manufacture these units. The following coordinate system shows the graphs of revenue and cost for the storage units.

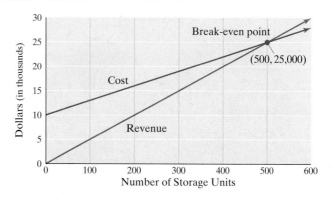

These lines intersect at the point (500, 25,000). This means that when 500 storage units are manufactured and sold, both cost and revenue are $25,000. In business, this point of intersection is called the **break-even point.** Notice that for x-values (units sold) less than 500, the cost graph is above the revenue graph, meaning that cost of manufacturing is greater than revenue, and so the company is losing money. For x-values (units sold) greater than 500, the revenue graph is above the cost graph, meaning that revenue is greater than cost, and so the company is making money.

Recall from Chapter 2 that each line is a graph of some linear equation in two variables. Both equations together form a **system of equations.** The common point of intersection is called the **solution of the system.** Some examples of systems of linear equations in two variables are

Systems of Linear Equations in Two Variables

$$\begin{cases} x - 2y = -7 \\ 3x + y = 0 \end{cases} \qquad \begin{cases} x = 5 \\ x + \dfrac{y}{2} = 9 \end{cases} \qquad \begin{cases} x - 3 = 2y + 6 \\ y = 1 \end{cases}$$

OBJECTIVE 1 ▶ Determining whether an ordered pair is a solution. Recall that a solution of an equation in two variables is an ordered pair (x, y) that makes the equation true. A **solution of a system** of two equations in two variables is an ordered pair (x, y) that makes both equations true.

EXAMPLE 1 Determine whether the given ordered pair is a solution of the system.

a. $\begin{cases} -x + y = 2 \\ 2x - y = -3 \end{cases} \quad (-1, 1)$ **b.** $\begin{cases} 5x + 3y = -1 \\ x - y = 1 \end{cases} \quad (-2, 3)$

Solution

a. We replace x with -1 and y with 1 in each equation.

$$\begin{array}{ll} -x + y = 2 & \text{First equation} \\ -(-1) + (1) \stackrel{?}{=} 2 & \text{Let } x = -1 \text{ and } y = 1. \\ 1 + 1 \stackrel{?}{=} 2 & \\ 2 = 2 & \text{True} \end{array}$$

$$\begin{array}{ll} 2x - y = -3 & \text{Second equation} \\ 2(-1) - (1) \stackrel{?}{=} -3 & \text{Let } x = -1 \text{ and } y = 1. \\ -2 - 1 \stackrel{?}{=} -3 & \\ -3 = -3 & \text{True} \end{array}$$

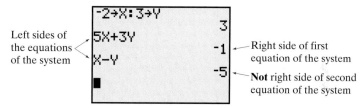

A calculator check for Example 1a.

Since $(-1, 1)$ makes *both* equations true, it is a solution. Using set notation, the solution set is $\{(-1, 1)\}$.

b. We replace x with -2 and y with 3 in each equation.

$5x + 3y = -1$ First equation $x - y = 1$ Second equation

$5(-2) + 3(3) \overset{?}{=} -1$ Let $x = -2$ and $y = 3$. $(-2) - (3) \overset{?}{=} 1$ Let $x = -2$ and $y = 3$.

$-10 + 9 \overset{?}{=} -1$ $-5 = 1$ False

$-1 = -1$ True

A calculator check for Example 1b.

Since the ordered pair $(-2, 3)$ does not make *both* equations true, it is not a solution of the system. $\square$

PRACTICE

1 Determine whether the given ordered pair is a solution of the system.

a. $\begin{cases} -x - 4y = 1 \\ 2x + y = 5 \end{cases}$ $(3, -1)$ **b.** $\begin{cases} 4x + y = -4 \\ -x + 3y = 8 \end{cases}$ $(-2, 4)$

Example 1 above shows how to determine that an ordered pair is a solution of a system of equations, but how do we find such a solution? Actually, there are various methods to find the solution. We will investigate several in this chapter: graphing, substitution, elimination, matrices, and determinants.

OBJECTIVE 2 ▶ **Solving a system by graphing.** To solve by graphing, we graph each equation in an appropriate window and find the coordinates of any points of intersection.

EXAMPLE 2 Solve the system by graphing.

$$\begin{cases} x + y = 2 \\ 3x - y = -2 \end{cases}$$

Solution Since the graph of a linear equation in two variables is a line, graphing two such equations yields two lines in a plane. To use a graphing utility, solve each equation for y.

$$\begin{cases} y = -x + 2 & \text{First equation} \\ y = 3x + 2 & \text{Second equation} \end{cases}$$

Graph $y_1 = -x + 2$ and $y_2 = 3x + 2$ and find the point of intersection.

When solving a system of equations by graphing, try the standard or integer window first and see if the intersection appears. If it does not appear in one of these windows, you may be able to see enough of the graph to estimate where the intersection will occur and adjust the window setting accordingly.

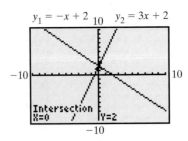

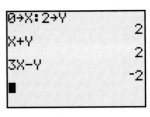

A check of the solution: $(0, 2)$.

Verify the ordered pair solution $(0, 2)$ by replacing x with 0 and y with 2 in both original equations and seeing that true statements result each time. The screen on the right above shows that the ordered pair $(0, 2)$ does satisfy both equations. We conclude therefore that $(0, 2)$ is the solution of the system. A system that has at least one solution, such as this one, is said to be **consistent.** ☐

PRACTICE

2 Solve each system by graphing. If the system has just one solution, find the solution.

a. $\begin{cases} 3x - 2y = 4 \\ -9x + 6y = -12 \end{cases}$ 　　**b.** $\begin{cases} y = 5x \\ 2x + y = 7 \end{cases}$ 　　**c.** $\begin{cases} y = \dfrac{3}{4}x + 1 \\ 3x - 4y = 12 \end{cases}$

DISCOVER THE CONCEPT

Use your graphing utility to solve the system

$$\begin{cases} x - 2y = 4 \\ x = 2y \end{cases}$$

In the discovery above, we see that solving each equation for y produces the following:

$x - 2y = 4$ 　First equation	$x = 2y$ 　Second equation
$-2y = -x + 4$ 　Subtract x from both sides.	$\dfrac{1}{2}x = y$ 　Divide both sides by 2.
$y = \dfrac{1}{2}x - 2$ 　Divide both sides by -2.	$y = \dfrac{1}{2}x$

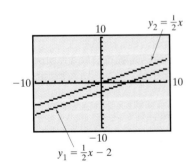

Notice that each equation is now in the form $y = mx + b$. From this form, we see that both lines have the same slope, $\dfrac{1}{2}$, but different y-intercepts, so they are parallel as shown to the left. Therefore, the system has no solution since the equations have no common solution (there are no intersection points). A system that has no solution is said to be **inconsistent.**

DISCOVER THE CONCEPT

Use your graphing utility to solve the system

$$\begin{cases} 2x + 4y = 10 \\ x + 2y = 5 \end{cases}$$

In the discovery above, we see that solving each equation for y produces the following:

$2x + 4y = 10$ 　First equation	$x + 2y = 5$ 　Second equation
$y = -\dfrac{1}{2}x + \dfrac{5}{2}$	$y = -\dfrac{1}{2}x + \dfrac{5}{2}$

> ▶ **Helpful Hint**
> - If a system of equations has *at least one solution,* the system is *consistent.*
> - If a system of equations has *no solution,* the system is *inconsistent.*

Notice that both lines have the same slope, $-\dfrac{1}{2}$, and the same y-intercept, $\dfrac{5}{2}$. This means that the graph of each equation is the same line.

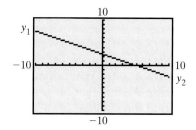

> ▶ **Helpful Hint**
> - If the graphs of two equations *differ,* they are *independent* equations.
> - If the graphs of two equations are the *same,* they *are dependent* equations.

To confirm this, notice that the entries for y_1 and y_2 are the same in the table shown on the right above. The equations have identical solutions and any ordered pair solution of one equation satisfies the other equation also. Thus, these equations are said to be **dependent equations.** The solution set of the system is $\{(x, y)\,|\,x + 2y = 5\}$ or, equivalently, $\{(x, y)\,|\,2x + 4y = 10\}$ since the equations describe identical ordered pairs. Written this way, the solution set is read "the set of all ordered pairs (x, y), such that $2x + 4y = 10$." There are therefore an infinite number of solutions to the system.

Concept Check ☑

The equations in the system are dependent and the system has an infinite number of solutions. Which ordered pairs below are solutions?

$$\begin{cases} -x + 3y = 4 \\ 2x + 8 = 6y \end{cases}$$

a. $(4, 0)$ **b.** $(-4, 0)$ **c.** $(-1, 1)$

We can summarize the information discovered in Example 2 as follows.

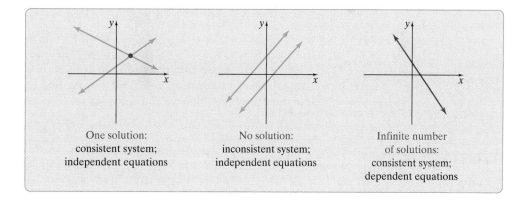

Concept Check ☑

How can you tell just by looking at the following system that it has no solution?

$$\begin{cases} y = 3x + 5 \\ y = 3x - 7 \end{cases}$$

How can you tell just by looking at the following system that it has infinitely many solutions?

$$\begin{cases} x + y = 5 \\ 2x + 2y = 10 \end{cases}$$

Answer to Concept Check: b, c; answers may vary

A graphing calculator is a very useful tool for approximating solutions to a system of equations in two variables. See the next example.

EXAMPLE 3 Solve the system by graphing. Approximate the solution to two decimal places.

$$\begin{cases} y + 2.6x = 5.6 \\ y - 4.3x = -4.9 \end{cases}$$

Solution First use a standard window and graph both equations on a single screen. The screen in the margin shows that the two lines intersect. To approximate the point of intersection, trace to the point of intersection and use an Intersect feature of the graphing calculator. We find that the approximate point of intersection is $(1.52, 1.64)$.

Because the solution is an approximation, notice that the numerical check with these approximations does not show equivalent expressions. For example, instead of $y + 2.6x = 5.6$, we have $y + 2.6x = 5.592$. The number 5.592 is close to 5.6, but not equal to 5.6. Keep this in mind when checking approximations. □

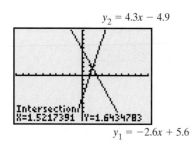

$y_2 = 4.3x - 4.9$

Intersection
X=1.5217391 Y=1.6434783

$y_1 = -2.6x + 5.6$

Solving graphically. x- and y-values are automatically stored to 14 decimal places.

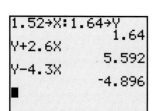

1.52→X:1.64→Y
                1.64
Y+2.6X
                5.592
Y-4.3X
                -4.896
■

The numerical check with decimal approximations.

PRACTICE
3 Solve by graphing. Approximate the solution to two decimal places.

$$\begin{cases} y - 0.25x = 1.6 \\ y + 1.03x = -5.1 \end{cases}$$

OBJECTIVE 3 ▶ Solving a system by substitution. Graphing the equations of a system by hand is often a good method of finding approximate solutions of a system, but it is not a reliable method of finding exact solutions of a system. We turn instead to two algebraic methods of solving systems. We use the first method, the **substitution method,** to solve the system

$$\begin{cases} 2x + 4y = -6 & \text{First equation} \\ x = 2y - 5 & \text{Second equation} \end{cases}$$

EXAMPLE 4 Use the substitution method to solve the system.

$$\begin{cases} 2x + 4y = -6 & \text{First equation} \\ x = 2y - 5 & \text{Second equation} \end{cases}$$

Solution In the second equation, we are told that x is equal to $2y - 5$. Since they are equal, we can _substitute_ $2y - 5$ for x in the first equation. This will give us an equation in one variable, which we can solve for y.

$$2x + 4y = -6 \qquad \text{First equation}$$
$$2(2y - 5) + 4y = -6 \qquad \text{Substitute } 2y - 5 \text{ for } x.$$
$$4y - 10 + 4y = -6$$
$$8y = 4$$
$$y = \frac{4}{8} = \frac{1}{2} \qquad \text{Solve for } y.$$

The y-coordinate of the solution is $\frac{1}{2}$. To find the x-coordinate, we replace y with $\frac{1}{2}$ in the second equation,

$$x = 2y - 5.$$
$$x = 2y - 5$$
$$x = 2\left(\frac{1}{2}\right) - 5 = 1 - 5 = -4$$

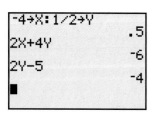

-4→X:1/2→Y
                .5
2X+4Y
                -6
2Y-5
                -4
■

A numeric check of the solution $\left(-4, \frac{1}{2}\right)$ for Example 4.

The ordered pair solution is $\left(-4, \frac{1}{2}\right)$. Check to see that $\left(-4, \frac{1}{2}\right)$ satisfies both equations of the system. □

PRACTICE
4   Use the substitution method to solve the system.

$$\begin{cases} y = 4x + 7 \\ 2x + y = 4 \end{cases}$$

The steps below summarize the substitution method.

Solving a System of Two Equations Using the Substitution Method

STEP 1. Solve one of the equations for one of its variables.

STEP 2. Substitute the expression for the variable found in Step 1 into the other equation.

STEP 3. Find the value of one variable by solving the equation from Step 2.

STEP 4. Find the value of the other variable by substituting the value found in Step 3 into the equation from Step 1.

STEP 5. Check the ordered pair solution in *both* original equations.

▶ **Helpful Hint**

If a system of equations contains equations with fractions, first clear the equations of fractions.

EXAMPLE 5   Use the substitution method to solve the system.

$$\begin{cases} -\dfrac{x}{6} + \dfrac{y}{2} = \dfrac{1}{2} \\ \dfrac{x}{3} - \dfrac{y}{6} = -\dfrac{3}{4} \end{cases}$$

Solution   First we multiply each equation by its least common denominator to clear the system of fractions. We multiply the first equation by 6 and the second equation by 12.

$$\begin{cases} 6\left(-\dfrac{x}{6} + \dfrac{y}{2}\right) = 6\left(\dfrac{1}{2}\right) \\ 12\left(\dfrac{x}{3} - \dfrac{y}{6}\right) = 12\left(-\dfrac{3}{4}\right) \end{cases}$$ simplifies to $$\begin{cases} -x + 3y = 3 & \text{First equation} \\ 4x - 2y = -9 & \text{Second equation} \end{cases}$$

▶ **Helpful Hint**

To avoid tedious fractions, solve for a variable whose coefficient is 1 or −1, if possible.

To use the substitution method, we now solve the first equation for x.

$$-x + 3y = 3 \quad \text{First equation}$$
$$3y - 3 = x \quad \text{Solve for } x.$$

Next we replace x with $3y - 3$ in the second equation.

$$4x - 2y = -9 \quad \text{Second equation}$$
$$4(3y - 3) - 2y = -9$$
$$12y - 12 - 2y = -9$$
$$10y = 3$$
$$y = \dfrac{3}{10} \quad \text{Solve for } y.$$

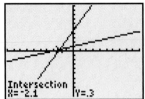

To find the corresponding x-coordinate, we replace y with $\frac{3}{10}$ in the equation $x = 3y - 3$. Then

$$x = 3\left(\frac{3}{10}\right) - 3 = \frac{9}{10} - 3 = \frac{9}{10} - \frac{30}{10} = -\frac{21}{10}$$

The ordered pair solution is $\left(-\frac{21}{10}, \frac{3}{10}\right)$ or equivalently $(-2.1, 0.3)$. We check this solution graphically to the left. □

PRACTICE
5 Use the substitution method to solve the system.

$$\begin{cases} -\dfrac{x}{3} + \dfrac{y}{4} = \dfrac{1}{2} \\ \dfrac{x}{4} - \dfrac{y}{2} = -\dfrac{1}{4} \end{cases}$$

OBJECTIVE 4 ▶ Solving a system by elimination. The **elimination method,** or **addition method,** is a second algebraic technique for solving systems of equations. For this method, we rely on a version of the addition property of equality, which states that "equals added to equals are equal."

> If $A = B$ and $C = D$ then $A + C = B + D$.

EXAMPLE 6 Use the elimination method to solve the system.

$$\begin{cases} x - 5y = -12 & \text{First equation} \\ -x + y = 4 & \text{Second equation} \end{cases}$$

Solution Since the left side of each equation is equal to the right side, we add equal quantities by adding the left sides of the equations and the right sides of the equations. This sum gives us an equation in one variable, y, which we can solve for y.

$$\begin{array}{ll} x - 5y = -12 & \text{First equation} \\ \underline{-x + y = 4} & \text{Second equation} \\ \qquad -4y = -8 & \text{Add.} \\ \qquad y = 2 & \text{Solve for } y. \end{array}$$

The y-coordinate of the solution is 2. To find the corresponding x-coordinate, we replace y with 2 in either original equation of the system. Let's use the second equation.

$$\begin{array}{ll} -x + y = 4 & \text{Second equation} \\ -x + 2 = 4 & \text{Let } y = 2. \\ -x = 2 \\ x = -2 \end{array}$$

The ordered pair solution is $(-2, 2)$. We check numerically (to the left) to see that $(-2, 2)$ satisfies both equations of the system. □

Satisfies
$x - 5y = -12$

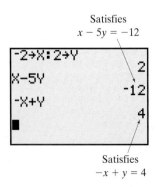

Satisfies
$-x + y = 4$

PRACTICE
6 Use the elimination method to solve the system.

$$\begin{cases} 3x - y = 5 \\ 5x + y = 11 \end{cases}$$

The steps below summarize the elimination method.

Solving a System of Two Linear Equations Using the Elimination Method

STEP 1. Rewrite each equation in standard form, $Ax + By = C$.

STEP 2. If necessary, multiply one or both equations by some nonzero number so that the coefficients of a variable are opposites of each other.

STEP 3. Add the equations.

STEP 4. Find the value of one variable by solving the equation from Step 3.

STEP 5. Find the value of the second variable by substituting the value found in Step 4 into either original equation.

STEP 6. Check the proposed ordered pair solution in *both* original equations.

EXAMPLE 7   Use the elimination method to solve the system.

$$\begin{cases} 3x - 2y = 10 \\ 4x - 3y = 15 \end{cases}$$

Solution   If we add the two equations, the sum will still be an equation in two variables. Notice, however, that we can eliminate y when the equations are added if we multiply both sides of the first equation by 3 and both sides of the second equation by -2. Then

$$\begin{cases} 3(3x - 2y) = 3(10) \\ -2(4x - 3y) = -2(15) \end{cases} \quad \text{simplifies to} \quad \begin{cases} 9x - 6y = 30 \\ -8x + 6y = -30 \end{cases}$$

Next we add the left sides and add the right sides.

$$\begin{array}{r} 9x - 6y = 30 \\ -8x + 6y = -30 \\ \hline x = 0 \end{array}$$

To find y, we let $x = 0$ in either equation of the system.

$$\begin{array}{ll} 3x - 2y = 10 & \text{First equation} \\ 3(0) - 2y = 10 & \text{Let } x = 0. \\ -2y = 10 & \\ y = -5 & \end{array}$$

The ordered pair solution is $(0, -5)$. Check to see that $(0, -5)$ satisfies both equations of the system.   □

PRACTICE

7   Use the elimination method to solve the system.

$$\begin{cases} 3x - 2y = -6 \\ 4x + 5y = -8 \end{cases}$$

EXAMPLE 8 Use the elimination method to solve the system.

$$\begin{cases} 3x + \dfrac{y}{2} = 2 \\ 6x + y = 5 \end{cases}$$

Solution If we multiply both sides of the first equation by -2, the coefficients of x in the two equations will be opposites. Then

$$\begin{cases} -2\left(3x + \dfrac{y}{2}\right) = -2(2) \\ 6x + y = 5 \end{cases} \quad \text{simplifies to} \quad \begin{cases} -6x - y = -4 \\ 6x + y = 5 \end{cases}$$

Now we can add the left sides and add the right sides.

$$\begin{array}{r} -6x - y = -4 \\ 6x + y = 5 \\ \hline 0 = 1 \quad \text{False} \end{array}$$

The resulting equation, $0 = 1$, is false for all values of y or x. Thus, the system has no solution. The solution set is $\{\ \}$ or $\varnothing$. This system is inconsistent, and the graphs of the equations are parallel lines. □

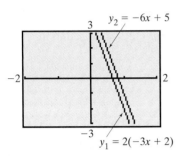

$y_2 = -6x + 5$

$y_1 = 2(-3x + 2)$

The two graphs appear to be parallel lines, supporting no solution to the system of Example 8.

PRACTICE
8 Use the elimination method to solve the system.

$$\begin{cases} 8x + y = 6 \\ 2x + \dfrac{y}{4} = -2 \end{cases}$$

EXAMPLE 9 Use the elimination method to solve the system.

$$\begin{cases} -5x - 3y = 9 \\ 10x + 6y = -18 \end{cases}$$

Solution To eliminate x when the equations are added, we multiply both sides of the first equation by 2. Then

$$\begin{cases} 2(-5x - 3y) = 2(9) \\ 10x + 6y = -18 \end{cases} \quad \text{simplifies to} \quad \begin{cases} -10x - 6y = 18 \\ 10x + 6y = -18 \end{cases}$$

Next we add the equations.

$$\begin{array}{r} -10x - 6y = 18 \\ 10x + 6y = -18 \\ \hline 0 = 0 \end{array}$$

The resulting equation, $0 = 0$, is true for all possible values of y or x. Notice in the original system that if both sides of the first equation are multiplied by -2, the result is the second equation. This means that the two equations are equivalent. They have the same solution set and there are an infinite number of solutions. Thus, the equations of this system are dependent, and the solution set of the system is

$$\{(x, y)\,|\,-5x - 3y = 9\} \quad \text{or, equivalently,} \quad \{(x, y)\,|\,10x + 6y = -18\}. \quad \square$$

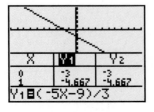

The graph (shown on a standard window) and table indicate that the graph of both equations is the same line. This supports the solution above for Example 9.

9 Use the elimination method to solve the system.

$$\begin{cases} -3x + 2y = -1 \\ 9x - 6y = 3 \end{cases}$$

> ▶ **Helpful Hint**
>
> Remember that not all ordered pairs are solutions of the system in Example 9, only the infinite number of ordered pairs that satisfy $-5x - 3y = 9$ or equivalently $10x + 6y = -18$.

EXAMPLE 10 **Finding the Break-Even Point**

A small manufacturing company manufactures and sells compact disc storage units. The revenue equation for these units is

$$y = 50x$$

where x is the number of units sold and y is the revenue, or income, in dollars for selling x units. The cost equation for these units is

$$y = 30x + 10,000$$

where x is the number of units manufactured and y is the total cost in dollars for manufacturing x units. Use these equations to find the number of units to be sold for the company to break even.

Solution The break-even point is found by solving the system

$$\begin{cases} y = 50x & \text{First equation} \\ y = 30x + 10,000 & \text{Second equation} \end{cases}$$

To solve the system, graph $y_1 = 50x$ and $y_2 = 30x + 10,000$ and find the point of intersection, the break-even point.

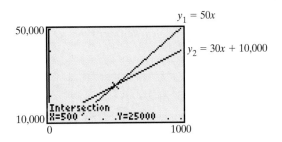

The ordered pair solution is $(500, 25,000)$. This means that the business must sell 500 compact disc storage units to break even. A hand-drawn graph of the equations in this system can be found at the beginning of this section. □

10 The revenue equation for a certain product is $y = 17x$, where x is the number of units sold and y is the revenue in dollars. The cost equation for the product is $y = 6x + 8030$, where x is the number of units manufactured and y is the cost in dollars for manufacturing x units. Find the number of units for the company to break even.

VOCABULARY & READINESS CHECK

Match each graph with the solution of the corresponding system.

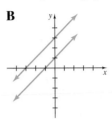 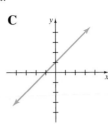 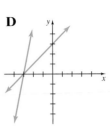

A   **B**   **C**   **D**

1. no solution   **2.** Infinite number of solutions   **3.** $(1, -2)$ **4.** $(-3, 0)$

4.1 EXERCISE SET

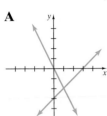

MyMathLab

PRACTICE   WATCH   DOWNLOAD   READ   REVIEW

Determine whether each given ordered pair is a solution of each system. See Example 1.

1. $\begin{cases} x - y = 3 \quad (2, -1) \\ 2x - 4y = 8 \end{cases}$

2. $\begin{cases} x - y = -4 \quad (-3, 1) \\ 2x + 10y = 4 \end{cases}$

3. $\begin{cases} 2x - 3y = -9 \quad (3, 5) \\ 4x + 2y = -2 \end{cases}$

4. $\begin{cases} 2x - 5y = -2 \quad (4, 2) \\ 3x + 4y = 4 \end{cases}$

5. $\begin{cases} 3x + 7y = -19 \\ -6x \quad = 5y + 8 \end{cases} \left(\frac{2}{3}, -3\right)$

6. $\begin{cases} 4x + 5y = -7 \\ -8x \quad = 3y - 1 \end{cases} \left(\frac{3}{4}, -2\right)$

A system of equations and the graph of each equation of the system is given below. Find the solution of the system and verify that it is the solution. See Example 1.

7. $\begin{cases} 2x + 5y = 8 \\ 6x + y = 10 \end{cases}$

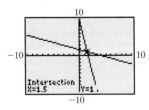

8. $\begin{cases} x + y = 1 \\ x - 2y = 4 \end{cases}$

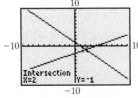

9. $\begin{cases} x - 4y = -5 \\ -3x - 8y = 0 \end{cases}$

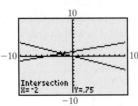

10. $\begin{cases} 2x - y = 8 \\ x - 3y = 11 \end{cases}$

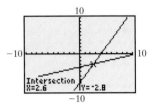

Solve each system by graphing. See Examples 2 and 3.

11. $\begin{cases} x + y = 1 \\ x - 2y = 4 \end{cases}$

12. $\begin{cases} 2x - y = 8 \\ x + 3y = 11 \end{cases}$

13. $\begin{cases} 2y - 4x = 0 \\ x + 2y = 5 \end{cases}$

14. $\begin{cases} 4x - y = 6 \\ x - y = 0 \end{cases}$

15. $\begin{cases} 3x - y = 4 \\ 6x - 2y = 4 \end{cases}$

16. $\begin{cases} -x + 3y = 6 \\ 3x - 9y = 9 \end{cases}$

17. Can a system consisting of two linear equations have exactly two solutions? Explain why or why not.

18. Suppose the graph of the equations in a system of two equations in two variables consists of a circle and a line. Discuss the possible number of solutions for this system.

Solve each system of equations by the substitution method. See Examples 4 and 5.

19. $\begin{cases} x + y = 10 \\ y = 4x \end{cases}$

20. $\begin{cases} 5x + 2y = -17 \\ x = 3y \end{cases}$

21. $\begin{cases} 4x - y = 9 \\ 2x + 3y = -27 \end{cases}$

22. $\begin{cases} 3x - y = 6 \\ -4x + 2y = -8 \end{cases}$

23. $\begin{cases} \dfrac{1}{2}x + \dfrac{3}{4}y = -\dfrac{1}{4} \\ \dfrac{3}{4}x - \dfrac{1}{4}y = 1 \end{cases}$

24. $\begin{cases} \dfrac{2}{5}x + \dfrac{1}{5}y = -1 \\ x + \dfrac{2}{5}y = -\dfrac{8}{5} \end{cases}$

25. $\begin{cases} \dfrac{x}{3} + y = \dfrac{4}{3} \\ -x + 2y = 11 \end{cases}$

26. $\begin{cases} \dfrac{x}{8} - \dfrac{y}{2} = 1 \\ \dfrac{x}{3} - y = 2 \end{cases}$

Solve each system of equations by the elimination method. See Examples 6 through 9.

27. $\begin{cases} -x + 2y = 0 \\ x + 2y = 5 \end{cases}$

28. $\begin{cases} -2x + 3y = 0 \\ 2x + 6y = 3 \end{cases}$

29. $\begin{cases} 5x + 2y = 1 \\ x - 3y = 7 \end{cases}$

30. $\begin{cases} 6x - y = -5 \\ 4x - 2y = 6 \end{cases}$

31. $\begin{cases} \dfrac{3}{4}x + \dfrac{5}{2}y = 11 \\ \dfrac{1}{16}x - \dfrac{3}{4}y = -1 \end{cases}$

32. $\begin{cases} \dfrac{2}{3}x + \dfrac{1}{4}y = -\dfrac{3}{2} \\ \dfrac{1}{2}x - \dfrac{1}{4}y = -2 \end{cases}$

33. $\begin{cases} 3x - 5y = 11 \\ 2x - 6y = 2 \end{cases}$

34. $\begin{cases} 6x - 3y = -3 \\ 4x + 5y = -9 \end{cases}$

35. $\begin{cases} x - 2y = 4 \\ 2x - 4y = 4 \end{cases}$

36. $\begin{cases} -x + 3y = 6 \\ 3x - 9y = 9 \end{cases}$

37. $\begin{cases} 3x + y = 1 \\ 2y = 2 - 6x \end{cases}$

38. $\begin{cases} y = 2x - 5 \\ 8x - 4y = 20 \end{cases}$

MIXED PRACTICE

Solve each system of equations.

39. $\begin{cases} 2x + 5y = 8 \\ 6x + y = 10 \end{cases}$

40. $\begin{cases} x - 4y = -5 \\ -3x - 8y = 0 \end{cases}$

41. $\begin{cases} x + y = 1 \\ x - 2y = 4 \end{cases}$

42. $\begin{cases} 2x - y = 8 \\ x + 3y = 11 \end{cases}$

43. $\begin{cases} \dfrac{1}{3}x + y = \dfrac{4}{3} \\ -\dfrac{1}{4}x - \dfrac{1}{2}y = -\dfrac{1}{4} \end{cases}$

44. $\begin{cases} \dfrac{3}{4}x - \dfrac{1}{2}y = -\dfrac{1}{2} \\ x + y = -\dfrac{3}{2} \end{cases}$

45. $\begin{cases} 2x + 6y = 8 \\ 3x + 9y = 12 \end{cases}$

46. $\begin{cases} x = 3y - 1 \\ 2x - 6y = -2 \end{cases}$

47. $\begin{cases} 4x + 2y = 5 \\ 2x + y = -1 \end{cases}$

48. $\begin{cases} 3x + 6y = 15 \\ 2x + 4y = 3 \end{cases}$

49. $\begin{cases} 10y - 2x = 1 \\ 5y = 4 - 6x \end{cases}$

50. $\begin{cases} 3x + 4y = 0 \\ 7x = 3y \end{cases}$

51. $\begin{cases} 5x - 2y = 27 \\ -3x + 5y = 18 \end{cases}$

52. $\begin{cases} 3x + 4y = 2 \\ 2x + 5y = -1 \end{cases}$

53. $\begin{cases} x = 3y + 2 \\ 5x - 15y = 10 \end{cases}$

54. $\begin{cases} y = \dfrac{1}{7}x + 3 \\ x - 7y = -21 \end{cases}$

55. $\begin{cases} 2x - y = -1 \\ y = -2x \end{cases}$

56. $\begin{cases} x = \dfrac{1}{5}y \\ x - y = -4 \end{cases}$

57. $\begin{cases} 2x = 6 \\ y = 5 - x \end{cases}$

58. $\begin{cases} x = 3y + 4 \\ -y = 5 \end{cases}$

59. $\begin{cases} \dfrac{x + 5}{2} = \dfrac{6 - 4y}{3} \\ \dfrac{3x}{5} = \dfrac{21 - 7y}{10} \end{cases}$

60. $\begin{cases} \dfrac{y}{5} = \dfrac{8 - x}{2} \\ x = \dfrac{2y - 8}{3} \end{cases}$

61. $\begin{cases} 4x - 7y = 7 \\ 12x - 21y = 24 \end{cases}$

62. $\begin{cases} 2x - 5y = 12 \\ -4x + 10y = 20 \end{cases}$

63. $\begin{cases} \dfrac{2}{3}x - \dfrac{3}{4}y = -1 \\ -\dfrac{1}{6}x + \dfrac{3}{8}y = 1 \end{cases}$

64. $\begin{cases} \dfrac{1}{2}x - \dfrac{1}{3}y = -3 \\ \dfrac{1}{8}x + \dfrac{1}{6}y = 0 \end{cases}$

65. $\begin{cases} 0.7x - 0.2y = -1.6 \\ 0.2x - y = -1.4 \end{cases}$

66. $\begin{cases} -0.7x + 0.6y = 1.3 \\ 0.5x - 0.3y = -0.8 \end{cases}$

67. $\begin{cases} 4x - 1.5y = 10.2 \\ 2x + 7.8y = -25.68 \end{cases}$

68. $\begin{cases} x - 3y = -5.3 \\ 6.3x + 6y = 3.96 \end{cases}$

REVIEW AND PREVIEW

Determine whether the given replacement values make each equation true or false. See Section 1.3.

69. $3x - 4y + 2z = 5; x = 1, y = 2,$ and $z = 5$

70. $x + 2y - z = 7; x = 2, y = -3,$ and $z = 3$

71. $-x - 5y + 3z = 15; x = 0, y = -1,$ and $z = 5$

72. $-4x + y - 8z = 4; x = 1, y = 0,$ and $z = -1$

Add the equations. See Section 4.1.

73. $3x + 2y - 5z = 10$
$-3x + 4y + z = 15$

74. $x + 4y - 5z = 20$
$2x - 4y - 2z = -17$

75. $10x + 5y + 6z = 14$
$-9x + 5y - 6z = -12$

76. $-9x - 8y - z = 31$
$9x + 4y - z = 12$

CONCEPT EXTENSIONS

The concept of supply and demand is used often in business. In general, as the unit price of a commodity increases, the demand for that commodity decreases. Also, as a commodity's unit price increases, the manufacturer normally increases the supply. The point where supply is equal to demand is called the equilibrium point. The following shows the graph of a demand equation and the graph of a supply equation for previously rented DVDs. The x-axis represents the number of DVDs in thousands, and the y-axis represents the cost of a DVD. Use this graph to answer Exercises 77 through 80. See Example 10.

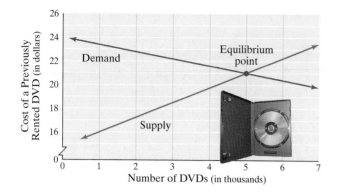

77. Find the number of DVDs and the price per DVD when supply equals demand.

78. When *x* is between 3 and 4, is supply greater than demand or is demand greater than supply?

79. When *x* is greater than 7, is supply greater than demand or is demand greater than supply?

80. For what *x*-values are the *y*-values corresponding to the supply equation greater than the *y*-values corresponding to the demand equation?

The revenue equation for a certain brand of toothpaste is y = 2.5x, where x is the number of tubes of toothpaste sold and y is the total income for selling x tubes. The cost equation is y = 0.9x + 3000, where x is the number of tubes of toothpaste manufactured and y is the cost of producing x tubes. The following set of axes shows the graph of the cost and revenue equations. Use this graph for Exercises 81 through 86. See Example 10.

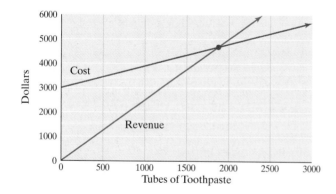

81. Find the coordinates of the point of intersection, or break-even point, by solving the system

$$\begin{cases} y = 2.5x \\ y = 0.9x + 3000 \end{cases}$$

82. Explain the meaning of the ordered pair point of intersection.

83. If the company sells 2000 tubes of toothpaste, does the company make money or lose money?

84. If the company sells 1000 tubes of toothpaste, does the company make money or lose money?

85. For what *x*-values will the company make a profit? (*Hint:* For what *x*-values is the revenue graph "higher" than the cost graph?)

86. For what *x*-values will the company lose money? (*Hint:* For what *x*-values is the revenue graph "lower" than the cost graph?)

87. Write a system of two linear equations in *x* and *y* that has the ordered pair solution (2, 5).

88. Which method would you use to solve the system?

$$\begin{cases} 5x - 2y = 6 \\ 2x + 3y = 5 \end{cases}$$

Explain your choice.

89. The amount *y* of red meat consumed per person in the United States (in pounds) in the year *x* can be modeled by the linear equation *y* = −0.3*x* + 113. The amount *y* of all poultry consumed per person in the United States (in pounds) in the year *x* can be modeled by the linear equation *y* = *x* + 68. In both models, *x* = 0 represents the year 2000. (*Source:* Based on data and forecasts from the Economic Research Service, U.S. Department of Agriculture)

   a. What does the slope of each equation tell you about the patterns of red meat and poultry consumption in the United States?

   b. Solve this system of equations. (Round your final results to the nearest whole numbers.)

   c. Explain the meaning of your answer to part **b.**

90. The number of books (in thousands) in the University of Texas libraries *y* for the years 2002 through 2005 can be modeled by the linear equation *y* = 230*x* + 8146. For the same time period, the number of books (in thousands) in the Columbia University libraries can be modeled by *y* = 611*x* + 7378, where *x* is the number of years since 2002. (*Source:* Association of Research Libraries)

   a. What does the slope of each equation tell you about the pattern of books in these two university libraries?

   b. Solve this system of equations. (Round your results to the nearest whole number.)

   c. Explain the meaning of your answer to part **b.**

Solve each system. To do so you may want to let $a = \dfrac{1}{x}$ (if x is in the denominator) and let $b = \dfrac{1}{y}$ (if y is in the denominator).

91.
$$\begin{cases} \dfrac{1}{x} + y = 12 \\ \dfrac{3}{x} - y = 4 \end{cases}$$

92.
$$\begin{cases} x + \dfrac{2}{y} = 7 \\ 3x + \dfrac{3}{y} = 6 \end{cases}$$

93. $\begin{cases} \dfrac{1}{x} + \dfrac{1}{y} = 5 \\ \dfrac{1}{x} - \dfrac{1}{y} = 1 \end{cases}$

94. $\begin{cases} \dfrac{2}{x} + \dfrac{3}{y} = 5 \\ \dfrac{5}{x} - \dfrac{3}{y} = 2 \end{cases}$

97. $\begin{cases} \dfrac{2}{x} - \dfrac{4}{y} = 5 \\ \dfrac{1}{x} - \dfrac{2}{y} = \dfrac{3}{2} \end{cases}$

95. $\begin{cases} \dfrac{2}{x} + \dfrac{3}{y} = -1 \\ \dfrac{3}{x} - \dfrac{2}{y} = 18 \end{cases}$

96. $\begin{cases} \dfrac{3}{x} - \dfrac{2}{y} = -18 \\ \dfrac{2}{x} + \dfrac{3}{y} = 1 \end{cases}$

98. $\begin{cases} \dfrac{5}{x} + \dfrac{7}{y} = 1 \\ -\dfrac{10}{x} - \dfrac{14}{y} = 0 \end{cases}$

📖 STUDY SKILLS BUILDER

Are You Familiar with Your Textbook Supplements?

There are many student supplements available for additional study. Below, I have listed some of these. See the preface of this text or your instructor for further information.

- *Chapter Test Prep Video CD.* This material is found in your textbook and is fully explained. The CD contains video clip solutions to the Chapter Test exercises in this text and are excellent help when studying for chapter tests.
- *Lecture Video CDs.* These video segments are keyed to each section of the text. The material is presented by me, Elayn Martin-Gay, and I have placed a video icon by each exercise in the text that I have worked on the video.
- *The Student Solutions Manual.* This contains worked-out solutions to odd-numbered exercises as well as every exercise in the Integrated Reviews, Chapter Reviews, Chapter Tests, and Cumulative Reviews.

- *Prentice Hall Tutor Center.* Mathematics questions may be phoned, faxed, or e-mailed to this center.
- *MyMathLab, MathXL, and Interact Math.* These are computer and Internet tutorials. This supplement may already be available to you somewhere on campus, for example at your local learning resource lab. Take a moment and find the name and location of any such lab on campus.

As usual, your instructor is your best source of information.

Let's see how you are doing with textbook supplements:

1. Name one way the Chapter Test Prep Video can help you prepare for a chapter test.
2. List any textbook supplements that you have found useful.
3. Have you located and visited a learning resource lab located on your campus?
4. List the textbook supplements that are currently housed in your campus' learning resource lab.

4.2 SOLVING SYSTEMS OF LINEAR EQUATIONS IN THREE VARIABLES

OBJECTIVE

1 Solve a system of three linear equations in three variables.

In this section, the algebraic methods of solving systems of two linear equations in two variables are extended to systems of three linear equations in three variables. We call the equation $3x - y + z = -15$, for example, a **linear equation in three variables** since there are three variables and each variable is raised only to the power 1. A solution of this equation is an **ordered triple (x, y, z)** that makes the equation a true statement. For example, the ordered triple $(2, 0, -21)$ is a solution of $3x - y + z = -15$ since replacing x with 2, y with 0, and z with -21 yields the true statement $3(2) - 0 + (-21) = -15$. The graph of this equation is a plane in three-dimensional space, just as the graph of a linear equation in two variables is a line in two-dimensional space.

Although we will not discuss the techniques for graphing equations in three variables, visualizing the possible patterns of intersecting planes gives us insight into the possible patterns of solutions of a system of three three-variable linear equations. There are four possible patterns.

1. Three planes have a single point in common. This point represents the single solution of the system. This system is **consistent.**

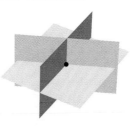

2. Three planes intersect at no point common to all three. This system has no solution. A few ways that this can occur are shown. This system is **inconsistent.**

3. Three planes intersect at all the points of a single line. The system has infinitely many solutions. This system is **consistent.**

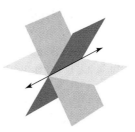

4. Three planes coincide at all points on the plane. The system is consistent, and the equations are **dependent.**

OBJECTIVE 1 ▶ Solving a system of three linear equations in three variables. Just as with systems of two equations in two variables, we can use the elimination or substitution method to solve a system of three equations in three variables. To use the elimination method, we eliminate a variable and obtain a system of two equations in two variables. Then we use the methods we learned in the previous section to solve the system of two equations.

EXAMPLE 1 Solve the system.

$$\begin{cases} 3x - y + z = -15 & \text{Equation (1)} \\ x + 2y - z = 1 & \text{Equation (2)} \\ 2x + 3y - 2z = 0 & \text{Equation (3)} \end{cases}$$

Solution Add equations (1) and (2) to eliminate z.

$$\begin{array}{r} 3x - y + z = -15 \\ \underline{x + 2y - z = 1} \\ 4x + y = -14 \quad \text{Equation (4)} \end{array}$$

> **Helpful Hint**
>
> Don't forget to add two other equations besides equations (1) and (2) *and* to **eliminate the same variable.**

Next, add two *other* equations and *eliminate z again*. To do so, multiply both sides of equation (1) by 2 and add this resulting equation to equation (3). Then

$$\begin{cases} 2(3x - y + z) = 2(-15) \\ 2x + 3y - 2z = 0 \end{cases} \quad \text{simplifies to} \quad \begin{cases} 6x - 2y + 2z = -30 \\ \underline{2x + 3y - 2z = 0} \\ 8x + y \quad\quad = -30 \quad \text{Equation (5)} \end{cases}$$

Satisfies
equation (1)

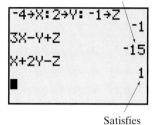

Now solve equations (4) and (5) for x and y. To solve by elimination, multiply both sides of equation (4) by -1 and add this resulting equation to equation (5). Then

$$\begin{cases} -1(4x + y) = -1(-14) \\ 8x + y = -30 \end{cases} \quad \text{simplifies to} \quad \begin{cases} -4x - y = 14 \\ \underline{8x + y = -30} \\ 4x \quad\quad = -16 \quad \text{Add the equations.} \\ x = -4 \quad \text{Solve for } x. \end{cases}$$

Satisfies
equation (2)

Satisfies
equation (3)

Replace x with -4 in equation (4) or (5).

$$4x + y = -14 \quad \text{Equation (4)}$$
$$4(-4) + y = -14 \quad \text{Let } x = -4.$$
$$y = 2 \quad \text{Solve for } y.$$

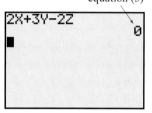

Finally, replace x with -4 and y with 2 in equation (1), (2), or (3).

$$x + 2y - z = 1 \quad \text{Equation (2)}$$
$$-4 + 2(2) - z = 1 \quad \text{Let } x = -4 \text{ and } y = 2.$$
$$-4 + 4 - z = 1$$
$$-z = 1$$
$$z = -1$$

A calculator check of Example 1.

The solution is $(-4, 2, -1)$. To check, let $x = -4$, $y = 2$, and $z = -1$ in all three original equations of the system. A paper and pencil check follows, and a calculator check is shown in the margin.

Equation (1)	**Equation (2)**	**Equation (3)**
$3x - y + z = -15$	$x + 2y - z = 1$	$2x + 3y - 2z = 0$
$3(-4) - 2 + (-1) \overset{?}{=} -15$	$-4 + 2(2) - (-1) \overset{?}{=} 1$	$2(-4) + 3(2) - 2(-1) \overset{?}{=} 0$
$-12 - 2 - 1 \overset{?}{=} -15$	$-4 + 4 + 1 \overset{?}{=} 1$	$-8 + 6 + 2 \overset{?}{=} 0$
$-15 = -15$	$1 = 1$	$0 = 0$
True	True	True

All three statements are true, so the solution is $(-4, 2, -1)$. ☐

PRACTICE
1 Solve the system. $\begin{cases} 3x + 2y - z = 0 \\ x - y + 5z = 2 \\ 2x + 3y + 3z = 7 \end{cases}$

EXAMPLE 2 Solve the system.

$$\begin{cases} 2x - 4y + 8z = 2 \quad (1) \\ -x - 3y + z = 11 \quad (2) \\ x - 2y + 4z = 0 \quad (3) \end{cases}$$

Solution   Add equations (2) and (3) to eliminate x, and the new equation is

$$-5y + 5z = 11 \quad (4)$$

To eliminate x again, multiply both sides of equation (2) by 2, and add the resulting equation to equation (1). Then

$$\begin{cases} 2x - 4y + 8z = 2 \\ 2(-x - 3y + z) = 2(11) \end{cases} \quad \begin{array}{l} \text{simplifies} \\ \text{to} \end{array} \quad \begin{cases} 2x - 4y + 8z = 2 \\ \underline{-2x - 6y + 2z = 22} \\ -10y + 10z = 24 \quad (5) \end{cases}$$

Next, solve for y and z using equations (4) and (5). Multiply both sides of equation (4) by -2, and add the resulting equation to equation (5).

$$\begin{cases} -2(-5y + 5z) = -2(11) \\ -10y + 10z = 24 \end{cases} \quad \begin{array}{l} \text{simplifies} \\ \text{to} \end{array} \quad \begin{cases} 10y - 10z = -22 \\ \underline{-10y + 10z = 24} \\ 0 = 2 \quad \text{False} \end{cases}$$

Since the statement is false, this system is inconsistent and has no solution. The solution set is the empty set $\{\ \}$ or $\varnothing$. $\square$

PRACTICE
2   Solve the system. $\begin{cases} 6x - 3y + 12z = 4 \\ -6x + 4y - 2z = 7 \\ -2x + y - 4z = 3 \end{cases}$

The elimination method is summarized next.

Solving a System of Three Linear Equations by the Elimination Method

STEP 1. Write each equation in standard form, $Ax + By + Cz = D$.

STEP 2. Choose a pair of equations and use the equations to eliminate a variable.

STEP 3. Choose any **other** pair of equations and eliminate the **same variable** as in Step 2.

STEP 4. Two equations in two variables should be obtained from Step 2 and Step 3. Use methods from Section 4.1 to solve this system for both variables.

STEP 5. To solve for the third variable, substitute the values of the variables found in Step 4 into any of the original equations containing the third variable.

STEP 6. Check the ordered triple solution in _all three_ original equations.

▶ **Helpful Hint**

Make sure you read closely and follow Step 3.

Concept Check ✓

In the system

$$\begin{cases} x + y + z = 6 & \text{Equation (1)} \\ 2x - y + z = 3 & \text{Equation (2)} \\ x + 2y + 3z = 14 & \text{Equation (3)} \end{cases}$$

equations (1) and (2) are used to eliminate y. Which action could be used to best finish solving? Why?

a. Use (1) and (2) to eliminate z.   **b.** Use (2) and (3) to eliminate y.

c. Use (1) and (3) to eliminate x.

Answer to Concept Check: b

EXAMPLE 3   Solve the system.

$$\begin{cases} 2x + 4y & = 1 & (1) \\ 4x & - 4z = -1 & (2) \\ & y - 4z = -3 & (3) \end{cases}$$

Solution   Notice that equation (2) has no term containing the variable y. Let us eliminate y using equations (1) and (3). Multiply both sides of equation (3) by -4, and add the resulting equation to equation (1). Then

$$\begin{cases} 2x + 4y & = 1 \\ -4(y - 4z) & = -4(-3) \end{cases} \quad \text{simplifies to} \quad \begin{cases} 2x + 4y & = 1 \\ -4y + 16z = 12 \\ \overline{2x \qquad + 16z = 13} & (4) \end{cases}$$

Next, solve for z using equations (4) and (2). Multiply both sides of equation (4) by -2 and add the resulting equation to equation (2).

$$\begin{cases} -2(2x + 16z) = -2(13) \\ 4x - 4z = -1 \end{cases} \quad \text{simplifies to} \quad \begin{cases} -4x - 32z = -26 \\ \underline{4x - 4z = -1} \\ \qquad -36z = -27 \\ \qquad z = \dfrac{3}{4} \end{cases}$$

Replace z with $\dfrac{3}{4}$ in equation (3) and solve for y.

$$y - 4\left(\dfrac{3}{4}\right) = -3 \quad \text{Let } z = \dfrac{3}{4} \text{ in equation (3).}$$
$$y - 3 = -3$$
$$y = 0$$

Replace y with 0 in equation (1) and solve for x.

$$2x + 4(0) = 1$$
$$2x = 1$$
$$x = \dfrac{1}{2}$$

A calculator check of Example 3.

The solution is $\left(\dfrac{1}{2}, 0, \dfrac{3}{4}\right)$. Check to see that this solution satisfies all three equations of the system. A calculator check is shown in the margin.   □

PRACTICE
3   Solve the system. $\begin{cases} 3x + 4y & = 0 \\ 9x & - 4z = 6 \\ -2y + 7z = 1 \end{cases}$

EXAMPLE 4   Solve the system.

$$\begin{cases} x - 5y - 2z = 6 & (1) \\ -2x + 10y + 4z = -12 & (2) \\ \dfrac{1}{2}x - \dfrac{5}{2}y - z = 3 & (3) \end{cases}$$

Solution   Multiply both sides of equation (3) by 2 to eliminate fractions, and multiply both sides of equation (2) by $-\dfrac{1}{2}$ so that the coefficient of x is 1. The resulting system is then

$$\begin{cases} x - 5y - 2z = 6 & (1) \\ x - 5y - 2z = 6 & \text{Multiply (2) by } -\dfrac{1}{2}. \\ x - 5y - 2z = 6 & \text{Multiply (3) by 2.} \end{cases}$$

All three equations are identical, and therefore equations (1), (2), and (3) are all equivalent. There are infinitely many solutions of this system. The equations are dependent. The solution set can be written as $\{(x, y, z) | x - 5y - 2z = 6\}$. □

PRACTICE
4 Solve the system. $\begin{cases} 2x + y - 3z = 6 \\ x + \dfrac{1}{2}y - \dfrac{3}{2}z = 3 \\ -4x - 2y + 6z = -12 \end{cases}$

As mentioned earlier, we can also use the substitution method to solve a system of linear equations in three variables.

EXAMPLE 5 Solve the system:

$$\begin{cases} x - 4y - 5z = 35 & (1) \\ x - 3y = 0 & (2) \\ -y + z = -55 & (3) \end{cases}$$

Solution Notice in equations (2) and (3) that a variable is missing. Also notice that both equations contain the variable y. Let's use the substitution method by solving equation (2) for x and equation (3) for z and substituting the results in equation (1).

$$\begin{aligned} x - 3y &= 0 & (2) \\ x &= 3y & \text{Solve equation (2) for } x. \\ -y + z &= -55 & (3) \\ z &= y - 55 & \text{Solve equation (3) for } z. \end{aligned}$$

Now substitute $3y$ for x and $y - 55$ for z in equation (1).

$$x - 4y - 5z = 35 \qquad (1)$$

▶ **Helpful Hint**
Do not forget to distribute.

$$\begin{aligned} 3y - 4y - 5(y - 55) &= 35 & \text{Let } x = 3y \text{ and } z = y - 55. \\ 3y - 4y - 5y + 275 &= 35 & \text{Use the distributive law and multiply.} \\ -6y + 275 &= 35 & \text{Combine like terms.} \\ -6y &= -240 & \text{Subtract 275 from both sides.} \\ y &= 40 & \text{Solve.} \end{aligned}$$

To find x, recall that $x = 3y$ and substitute 40 for y. Then $x = 3y$ becomes $x = 3 \cdot 40 = 120$. To find z, recall that $z = y - 55$ and substitute 40 for y, also. Then $z = y - 55$ becomes $z = 40 - 55 = -15$. The solution is $(120, 40, -15)$. □

PRACTICE
5 Solve the system. $\begin{cases} x + 2y + 4z = 16 \\ x + 2z = -4 \\ y - 3z = 30 \end{cases}$

4.2 EXERCISE SET

Solve.

1. Choose the equation(s) that has $(-1, 3, 1)$ as a solution.

   a. $x + y + z = 3$ **b.** $-x + y + z = 5$

   c. $-x + y + 2z = 0$ **d.** $x + 2y - 3z = 2$

2. Choose the equation(s) that has $(2, 1, -4)$ as a solution.

   a. $x + y + z = -1$ **b.** $x - y - z = -3$

   c. $2x - y + z = -1$ **d.** $-x - 3y - z = -1$

3. Use the result of Exercise 1 to determine whether $(-1, 3, 1)$ is a solution of the system below. Explain your answer.

$$\begin{cases} x + y + z = 3 \\ -x + y + z = 5 \\ x + 2y - 3z = 2 \end{cases}$$

4. Use the result of Exercise 2 to determine whether $(2, 1, -4)$ is a solution of the system below. Explain your answer.

$$\begin{cases} x + y + z = -1 \\ x - y - z = -3 \\ 2x - y + z = -1 \end{cases}$$

MIXED PRACTICE

Solve each system. See Examples 1 through 5.

5. $\begin{cases} x - y + z = -4 \\ 3x + 2y - z = 5 \\ -2x + 3y - z = 15 \end{cases}$

6. $\begin{cases} x + y - z = -1 \\ -4x - y + 2z = -7 \\ 2x - 2y - 5z = 7 \end{cases}$

7. $\begin{cases} x + y = 3 \\ 2y = 10 \\ 3x + 2y - 3z = 1 \end{cases}$

8. $\begin{cases} 5x = 5 \\ 2x + y = 4 \\ 3x + y - 4z = -15 \end{cases}$

9. $\begin{cases} 2x + 2y + z = 1 \\ -x + y + 2z = 3 \\ x + 2y + 4z = 0 \end{cases}$

10. $\begin{cases} 2x - 3y + z = 5 \\ x + y + z = 0 \\ 4x + 2y + 4z = 4 \end{cases}$

11. $\begin{cases} x - 2y + z = -5 \\ -3x + 6y - 3z = 15 \\ 2x - 4y + 2z = -10 \end{cases}$

12. $\begin{cases} 3x + y - 2z = 2 \\ -6x - 2y + 4z = -2 \\ 9x + 3y - 6z = 6 \end{cases}$

13. $\begin{cases} 4x - y + 2z = 5 \\ 2y + z = 4 \\ 4x + y + 3z = 10 \end{cases}$

14. $\begin{cases} 5y - 7z = 14 \\ 2x + y + 4z = 10 \\ 2x + 6y - 3z = 30 \end{cases}$

15. $\begin{cases} x + 5z = 0 \\ 5x + y = 0 \\ y - 3z = 0 \end{cases}$

16. $\begin{cases} x - 5y = 0 \\ x - z = 0 \\ -x + 5z = 0 \end{cases}$

17. $\begin{cases} 6x - 5z = 17 \\ 5x - y + 3z = -1 \\ 2x + y = -41 \end{cases}$

18. $\begin{cases} x + 2y = 6 \\ 7x + 3y + z = -33 \\ x - z = 16 \end{cases}$

19. $\begin{cases} x + y + z = 8 \\ 2x - y - z = 10 \\ x - 2y - 3z = 22 \end{cases}$

20. $\begin{cases} 5x + y + 3z = 1 \\ x - y + 3z = -7 \\ -x + y = 1 \end{cases}$

21. $\begin{cases} x + 2y - z = 5 \\ 6x + y + z = 7 \\ 2x + 4y - 2z = 5 \end{cases}$

22. $\begin{cases} 4x - y + 3z = 10 \\ x + y - z = 5 \\ 8x - 2y + 6z = 10 \end{cases}$

23. $\begin{cases} 2x - 3y + z = 2 \\ x - 5y + 5z = 3 \\ 3x + y - 3z = 5 \end{cases}$

24. $\begin{cases} 4x + y - z = 8 \\ x - y + 2z = 3 \\ 3x - y + z = 6 \end{cases}$

25. $\begin{cases} -2x - 4y + 6z = -8 \\ x + 2y - 3z = 4 \\ 4x + 8y - 12z = 16 \end{cases}$

26. $\begin{cases} -6x + 12y + 3z = -6 \\ 2x - 4y - z = 2 \\ -x + 2y + \dfrac{z}{2} = -1 \end{cases}$

27. $\begin{cases} 2x + 2y - 3z = 1 \\ y + 2z = -14 \\ 3x - 2y = -1 \end{cases}$

28. $\begin{cases} 7x + 4y = 10 \\ x - 4y + 2z = 6 \\ y - 2z = -1 \end{cases}$

29. $\begin{cases} x + 2y - z = 5 \\ -3x - 2y - 3z = 11 \\ 4x + 4y + 5z = -18 \end{cases}$

30. $\begin{cases} 3x - 3y + z = -1 \\ 3x - y - z = 3 \\ -6x + 2y + 2z = -6 \end{cases}$

31. $\begin{cases} \dfrac{3}{4}x - \dfrac{1}{3}y + \dfrac{1}{2}z = 9 \\ \dfrac{1}{6}x + \dfrac{1}{3}y - \dfrac{1}{2}z = 2 \\ \dfrac{1}{2}x - y + \dfrac{1}{2}z = 2 \end{cases}$

32. $\begin{cases} \dfrac{1}{3}x - \dfrac{1}{4}y + z = -9 \\ \dfrac{1}{2}x - \dfrac{1}{3}y - \dfrac{1}{4}z = -6 \\ x - \dfrac{1}{2}y - z = -8 \end{cases}$

REVIEW AND PREVIEW

Solve. See Section 1.6.

33. The sum of two numbers is 45 and one number is twice the other. Find the numbers.

34. The difference between two numbers is 5. Twice the smaller number added to five times the larger number is 53. Find the numbers.

Solve. See Section 1.5.

35. $2(x - 1) - 3x = x - 12$

36. $7(2x - 1) + 4 = 11(3x - 2)$

37. $-y - 5(y + 5) = 3y - 10$

38. $z - 3(z + 7) = 6(2z + 1)$

CONCEPT EXTENSIONS

39. Write a single linear equation in three variables that has $(-1, 2, -4)$ as a solution. (There are many possibilities.) Explain the process you used to write an equation.

40. Write a system of three linear equations in three variables that has $(2, 1, 5)$ as a solution. (There are many possibilities.) Explain the process you used to write an equation.

41. Write a system of linear equations in three variables that has the solution $(-1, 2, -4)$. Explain the process you used to write your system.

42. When solving a system of three equations in three unknowns, explain how to determine that a system has no solution.

43. The fraction $\frac{1}{24}$ can be written as the following sum:

$$\frac{1}{24} = \frac{x}{8} + \frac{y}{4} + \frac{z}{3}$$

where the numbers $x, y,$ and z are solutions of

$$\begin{cases} x + y + z = 1 \\ 2x - y + z = 0 \\ -x + 2y + 2z = -1 \end{cases}$$

Solve the system and see that the sum of the fractions is $\frac{1}{24}$.

44. The fraction $\frac{1}{18}$ can be written as the following sum:

$$\frac{1}{18} = \frac{x}{2} + \frac{y}{3} + \frac{z}{9}$$

where the numbers $x, y,$ and z are solutions of

$$\begin{cases} x + 3y + z = -3 \\ -x + y + 2z = -14 \\ 3x + 2y - z = 12 \end{cases}$$

Solve the system and see that the sum of the fractions is $\frac{1}{18}$.

Solving systems involving more than three variables can be accomplished with methods similar to those encountered in this section. Apply what you already know to solve each system of equations in four variables. Answers will be given in the form (x, y, z, w).

45. $\begin{cases} x + y - w = 0 \\ y + 2z + w = 3 \\ x - z = 1 \\ 2x - y - w = -1 \end{cases}$

46. $\begin{cases} 5x + 4y = 29 \\ y + z - w = -2 \\ 5x + z = 23 \\ y - z + w = 4 \end{cases}$

47. $\begin{cases} x + y + z + w = 5 \\ 2x + y + z + w = 6 \\ x + y + z = 2 \\ x + y = 0 \end{cases}$

48. $\begin{cases} 2x - z = -1 \\ y + z + w = 9 \\ y - 2w = -6 \\ x + y = 3 \end{cases}$

49. Write a system of three linear equations in three variables that are dependent equations.

50. What is the solution to the system in Exercise 49?

4.3 SYSTEMS OF LINEAR EQUATIONS AND PROBLEM SOLVING

OBJECTIVES

1 Solve problems that can be modeled by a system of two linear equations.

2 Solve problems with cost and revenue functions.

3 Solve problems that can be modeled by a system of three linear equations.

OBJECTIVE 1 ▶ Solving problems modeled by systems of two equations. Thus far, we have solved problems by writing one-variable equations and solving for the variable. Some of these problems can be solved, perhaps more easily, by writing a system of equations, as illustrated in this section.

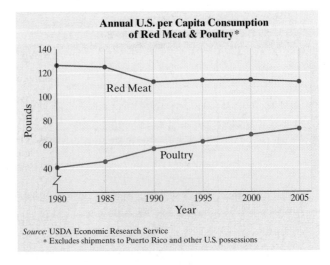

Annual U.S. per Capita Consumption of Red Meat & Poultry*

Source: USDA Economic Research Service
* Excludes shipments to Puerto Rico and other U.S. possessions

EXAMPLE 1 **Predicting Equal Consumption of Red Meat and Poultry**

America's consumption of red meat has decreased most years since 1980 while consumption of poultry has increased. The function $y = -0.59x + 124.6$ approximates the annual pounds of red meat consumed per capita, where x is the number of years since 1980. The function $y = 1.34x + 40.9$ approximates the annual pounds of poultry consumed per capita, where x is also the number of years since 1980. If this trend continues, determine the year in which the annual consumption of red meat and poultry is equal. (*Source:* Based on data from Economic Research Service, U.S. Dept. of Agriculture)

Solution

1. UNDERSTAND. Read and reread the problem and guess a year. Let's guess the year 2020. This year is 40 years since 1980, so $x = 40$. Now let $x = 40$ in each given function.

   Red meat: $y = -0.59x + 124.6 = -0.59(40) + 124.6 = 101$ pounds

   Poultry: $y = 1.34x + 40.9 = 1.34(40) + 40.9 = 94.5$ pounds

   Since the projected pounds in 2020 for red meat and poultry are not the same, we guessed incorrectly, but we do have a better understanding of the problem. We also know that the year will be later than 2020 since projected consumption of red meat is still greater than poultry that year.

2. TRANSLATE. We are already given the system of equations.

3. SOLVE. We want to know the year x in which pounds y are the same, so we solve the system:

   $$\begin{cases} y = -0.59x + 124.6 \\ y = 1.34x + 40.9 \end{cases}$$

   We solve by paper and pencil below, and a calculator check is shown in the margin. Since both equations are solved for y, one way to solve is to use the substitution method.

   $$y = -0.59x + 124.6 \quad \text{First equation}$$

   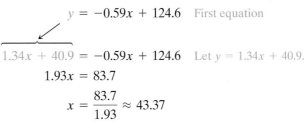

   $$1.34x + 40.9 = -0.59x + 124.6 \quad \text{Let } y = 1.34x + 40.9.$$

   $$1.93x = 83.7$$

   $$x = \frac{83.7}{1.93} \approx 43.37$$

4. INTERPRET. Since we are only asked to give the year, we need only solve for x.

Check: To check, see whether $x \approx 43.37$ gives approximately the same number of pounds of red meat and poultry.

   Red meat: $-0.59x + 124.6 = -0.59(43.37) + 124.6 = 99.01$ pounds

   Poultry: $1.34x + 40.9 = 1.34(43.37) + 40.9 = 99.02$ pounds

Since we rounded the number of years, the number of pounds do differ slightly. They differ only by 0.0041, so we can assume that we solved correctly.

State: The consumption of red meat and poultry will be the same about 43.37 years after 1980, or 2023.37. Thus, in the year 2023, we predict the consumption will be the same, about 99.01 pounds. □

PRACTICE

1 Read Example 1. If we use the years 1995, 2000, and 2005 only to write functions approximating the consumption of red meat and poultry, we have the following:

   Red meat: $y = -0.16x + 113.9$

   Poultry: $y = 1.06x + 62.3$

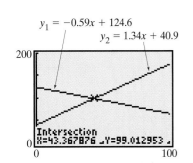

$y_1 = -0.59x + 124.6$

$y_2 = 1.34x + 40.9$

Intersection
X=43.367876 Y=99.012953

A calculator check for Example 1.

where x is the years since 1995 and y is pounds per year consumed.

a. Assuming this trend continues, predict the year in which the consumption of red meat and poultry will be the same.

b. Does your answer differ from the example? Why or why not?

Note: A similar exercise is found in Section 4.1, Exercise 89. In the example above, the data years used to generate the equations are 1980–2005. In Section 4.1, the data years used are 2000–2005. Note all the differing equations and answers.

EXAMPLE 2　Finding Unknown Numbers

A first number is 4 less than a second number. Four times the first number is 6 more than twice the second. Find the numbers.

Solution

1. UNDERSTAND. Read and reread the problem and guess a solution. If a first number is 10 and this is 4 less than a second number, the second number is 14. Four times the first number is 4(10), or 40. This is not equal to 6 more than twice the second number, which is 2(14) + 6 or 34. Although we guessed incorrectly, we now have a better understanding of the problem.

Since we are looking for two numbers, we will let

$$x = \text{first number}$$
$$y = \text{second number}$$

2. TRANSLATE. Since we have assigned two variables to this problem, we will translate the given facts into two equations. For the first statement we have

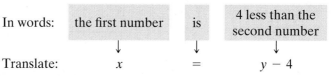

In words:	the first number	is	4 less than the second number
Translate:	x	$=$	$y - 4$

Next we translate the second statement into an equation.

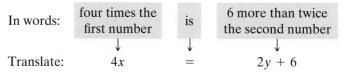

In words:	four times the first number	is	6 more than twice the second number
Translate:	$4x$	$=$	$2y + 6$

3. SOLVE. Here we solve the system

$$\begin{cases} x = y - 4 \\ 4x = 2y + 6 \end{cases}$$

Since the first equation expresses x in terms of y, we will use substitution. We substitute $y - 4$ for x in the second equation and solve for y.

$$4x = 2y + 6 \quad \text{Second equation}$$

$$4(y - 4) = 2y + 6$$
$$4y - 16 = 2y + 6 \quad \text{Let } x = y - 4.$$
$$2y = 22$$
$$y = 11$$

Now we replace y with 11 in the equation $x = y - 4$ and solve for x. Then $x = y - 4$ becomes $x = 11 - 4 = 7$. The ordered pair solution of the system is $(7, 11)$.

4. INTERPRET. Since the solution of the system is (7, 11), then the first number we are looking for is 7 and the second number is 11.

Check: Notice that 7 *is* 4 less than 11, and 4 times 7 *is* 6 more than twice 11. The proposed numbers, 7 and 11, are correct.

State: The numbers are 7 and 11. □

PRACTICE
2   A first number is 5 more than a second number. Twice the first number is 2 less than 3 times the second number. Find the numbers.

EXAMPLE 3   Finding the Rate of Speed

Two cars leave Indianapolis, one traveling east and the other west. After 3 hours they are 297 miles apart. If one car is traveling 5 mph faster than the other, what is the speed of each?

Solution

1. UNDERSTAND. Read and reread the problem. Let's guess a solution and use the formula $d = rt$ (distance = rate · time) to check. Suppose that one car is traveling at a rate of 55 miles per hour. This means that the other car is traveling at a rate of 50 miles per hour since we are told that one car is traveling 5 mph faster than the other. To find the distance apart after 3 hours, we will first find the distance traveled by each car. One car's distance is rate · time = 55(3) = 165 miles. The other car's distance is rate · time = 50(3) = 150 miles. Since one car is traveling east and the other west, their distance apart is the sum of their distances, or 165 miles + 150 miles = 315 miles. Although this distance apart is not the required distance of 297 miles, we now have a better understanding of the problem.

Let's model the problem with a system of equations. We will let

$$x = \text{speed of one car}$$
$$y = \text{speed of the other car}$$

We summarize the information on the following chart. Both cars have traveled 3 hours. Since distance = rate · time, their distances are $3x$ and $3y$ miles, respectively.

	Rate	•	Time	=	Distance
One Car	x		3		$3x$
Other Car	y		3		$3y$

2. TRANSLATE. We can now translate the stated conditions into two equations.

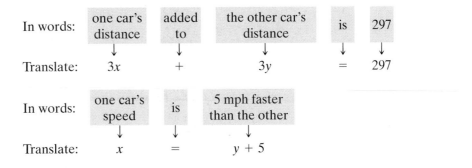

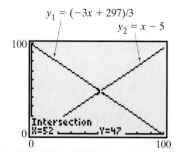

$y_1 = (-3x + 297)/3$

$y_2 = x - 5$

A graphical solution. We choose the window $[0, 100, 10]$ for the x and y windows, which indicates that the speed is between 0 and 100 mph.

3. SOLVE. Here we solve the system

$$\begin{cases} 3x + 3y = 297 \\ x = y + 5 \end{cases}$$

Again, the substitution method is appropriate. We replace x with $y + 5$ in the first equation and solve for y.

$$3x + 3y = 297 \quad \text{First equation}$$
$$3(y + 5) + 3y = 297 \quad \text{Let } x = y + 5.$$
$$3y + 15 + 3y = 297$$
$$6y = 282$$
$$y = 47$$

To find x, we replace y with 47 in the equation $x = y + 5$. Then $x = 47 + 5 = 52$. The ordered pair solution of the system is $(52, 47)$. A graphical solution is shown in the margin.

4. INTERPRET. The solution $(52, 47)$ means that the cars are traveling at 52 mph and 47 mph, respectively.

Check: Notice that one car is traveling 5 mph faster than the other. Also, if one car travels 52 mph for 3 hours, the distance is $3(52) = 156$ miles. The other car traveling for 3 hours at 47 mph travels a distance of $3(47) = 141$ miles. The sum of the distances $156 + 141$ is 297 miles, the required distance.

State: The cars are traveling at 52 mph and 47 mph. □

> ▶ **Helpful Hint**
>
> Don't forget to attach units, if appropriate.

PRACTICE

3 In 2007, the French train TGV V150 became the fastest conventional rail train in the world. It broke the 1990 record of the next fastest conventional rail train, the French TGV Atlantique. Assume the V150 and the Atlantique left the same station in Paris, with one heading west and one heading east. After 2 hours, they were 2150 kilometers apart. If the V150 is 75 kph faster than the Atlantique, what is the speed of each?

EXAMPLE 4 Mixing Solutions

Lynn Pike, a pharmacist, needs 70 liters of a 50% alcohol solution. She has available a 30% alcohol solution and an 80% alcohol solution. How many liters of each solution should she mix to obtain 70 liters of a 50% alcohol solution?

Solution

1. UNDERSTAND. Read and reread the problem. Next, guess the solution. Suppose that we need 20 liters of the 30% solution. Then we need $70 - 20 = 50$ liters of the 80% solution. To see if this gives us 70 liters of a 50% alcohol solution, let's find the amount of pure alcohol in each solution.

number of liters	×	alcohol strength	=	amount of pure alcohol
20 liters	×	0.30	=	6 liters
50 liters	×	0.80	=	40 liters
70 liters	×	0.50	=	35 liters

Since 6 liters + 40 liters = 46 liters and not 35 liters, our guess is incorrect, but we have gained some insight as to how to model and check this problem.

We will let

$$x = \text{amount of 30\% solution, in liters}$$
$$y = \text{amount of 80\% solution, in liters}$$

and use a table to organize the given data.

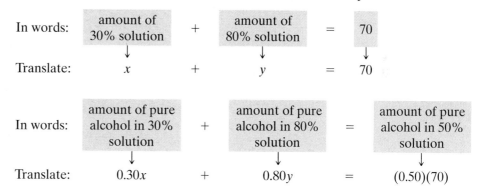

	Number of Liters	Alcohol Strength	Amount of Pure Alcohol
30% Solution	x	30%	$0.30x$
80% Solution	y	80%	$0.80y$
50% Solution Needed	70	50%	$(0.50)(70)$

2. TRANSLATE. We translate the stated conditions into two equations.

In words: | amount of 30% solution | $+$ | amount of 80% solution | $=$ | 70

Translate: x $+$ y $=$ 70

In words: | amount of pure alcohol in 30% solution | $+$ | amount of pure alcohol in 80% solution | $=$ | amount of pure alcohol in 50% solution

Translate: $0.30x$ $+$ $0.80y$ $=$ $(0.50)(70)$

3. SOLVE. Here we solve the system

$$\begin{cases} x + y = 70 \\ 0.30x + 0.80y = (0.50)(70) \end{cases}$$

To solve this system, we use the elimination method. We multiply both sides of the first equation by -3 and both sides of the second equation by 10. Then

$$\begin{cases} -3(x + y) = -3(70) \\ 10(0.30x + 0.80y) = 10(0.50)(70) \end{cases} \quad \begin{matrix} \text{simplifies} \\ \text{to} \end{matrix} \quad \begin{cases} -3x - 3y = -210 \\ \underline{3x + 8y = 350} \\ 5y = 140 \\ y = 28 \end{cases}$$

Now we replace y with 28 in the equation $x + y = 70$ and find that $x + 28 = 70$, or $x = 42$.

The ordered pair solution of the system is $(42, 28)$.

4. INTERPRET.

Check: Check the solution in the same way that we checked our guess.

State: The pharmacist needs to mix 42 liters of 30% solution and 28 liters of 80% solution to obtain 70 liters of 50% solution.   □

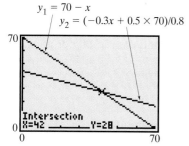

$y_1 = 70 - x$
$y_2 = (-0.3x + 0.5 \times 70)/0.8$

A graphical solution. The x and y windows are both $[0, 70, 10]$ since the total amount is 70 liters.

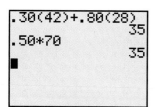

A calculator check for Example 4. (We already see by inspection that $42 + 28 = 70$, as needed.)

PRACTICE

4 Keith Robinson is a chemistry teacher who needs 1 liter of a solution of 5% hydrochloric acid to carry out an experiment. If he only has a stock solution of 99% hydrochloric acid, how much water (0% acid) and how much stock solution (99%) of HCL must he mix to get 1 liter of 5% solution? Round answers to the nearest hundredth of a liter.

Concept Check ☑

Suppose you mix an amount of 25% acid solution with an amount of 60% acid solution. You then calculate the acid strength of the resulting acid mixture. For which of the following results should you suspect an error in your calculation? Why?

a. 14%   **b.** 32%   **c.** 55%

OBJECTIVE 2 ▶ **Solving problems with cost and revenue functions.** Recall that businesses are often computing cost and revenue functions or equations to predict sales, to determine whether prices need to be adjusted, and to see whether the company is making or losing money. Recall also that the value at which revenue equals cost is called the break-even point. When revenue is less than cost, the company is losing money; when revenue is greater than cost, the company is making money.

EXAMPLE 5 **Finding a Break-Even Point**

A manufacturing company recently purchased $3000 worth of new equipment to offer new personalized stationery to its customers. The cost of producing a package of personalized stationery is $3.00, and it is sold for $5.50. Find the number of packages that must be sold for the company to break even.

Solution

1. **UNDERSTAND.** Read and reread the problem. Notice that the cost to the company will include a one-time cost of $3000 for the equipment and then $3.00 per package produced. The revenue will be $5.50 per package sold.

   To model this problem, we will let

   $$x = \text{number of packages of personalized stationery}$$
   $$C(x) = \text{total cost for producing } x \text{ packages of stationery}$$
   $$R(x) = \text{total revenue for selling } x \text{ packages of stationery}$$

2. **TRANSLATE.** The revenue equation is

In words:	revenue for selling x packages of stationery	=	price per package	·	number of packages
	↓		↓		↓
Translate:	$R(x)$	=	5.5	·	x

The cost equation is

In words:	cost for producing x packages of stationery	=	cost per package	·	number of packages	+	cost for equipment
	↓		↓		↓	↓	↓
Translate:	$C(x)$	=	3	·	x	+	3000

Since the break-even point is when $R(x) = C(x)$, we solve the equation

$$5.5x = 3x + 3000$$

3. SOLVE.

$$5.5x = 3x + 3000$$
$$2.5x = 3000 \qquad \text{Subtract } 3x \text{ from both sides.}$$
$$x = 1200 \qquad \text{Divide both sides by 2.5.}$$

4. INTERPRET.

Check: To see whether the break-even point occurs when 1200 packages are produced and sold, see if revenue equals cost when $x = 1200$. When $x = 1200$, $R(x) = 5.5x = 5.5(1200) = 6600$ and $C(x) = 3x + 3000 = 3(1200) + 3000 = 6600$. Since $R(1200) = C(1200) = 6600$, the break-even point is 1200.

State: The company must sell 1200 packages of stationery to break even. The graph of this system is shown.

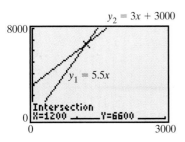

A calculator graph of Example 5.

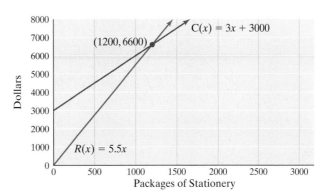

PRACTICE

5   An online-only electronics firm recently purchased $3000 worth of new equipment to create shock-proof packaging for its products. The cost of producing one shock-proof package is $2.50, and the firm charges the customer $4.50 for the packaging. Find the number of packages that must be sold for the company to break even.

OBJECTIVE 3 ▶ Solving problems modeled by systems of three equations. To introduce problem solving by writing a system of three linear equations in three variables, we solve a problem about triangles.

EXAMPLE 6   **Finding Angle Measures**

The measure of the largest angle of a triangle is 80° more than the measure of the smallest angle, and the measure of the remaining angle is 10° more than the measure of the smallest angle. Find the measure of each angle.

Solution

1. UNDERSTAND. Read and reread the problem. Recall that the sum of the measures of the angles of a triangle is 180°. Then guess a solution. If the smallest angle measures 20°, the measure of the largest angle is 80° more, or 20° + 80° = 100°. The measure of the remaining angle is 10° more than the measure of the smallest angle, or 20° + 10° = 30°. The sum of these three angles is 20° + 100° + 30° = 150°, not the required 180°. We now know that the measure of the smallest angle is greater than 20°.

   To model this problem we will let

$$x = \text{degree measure of the smallest angle}$$
$$y = \text{degree measure of the largest angle}$$
$$z = \text{degree measure of the remaining angle}$$

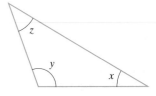

2. TRANSLATE. We translate the given information into three equations.

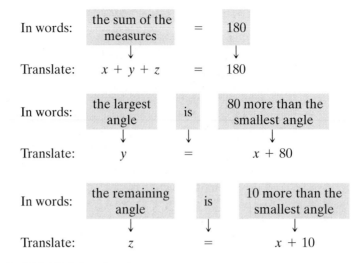

In words: the sum of the measures = 180

Translate: $x + y + z$ = 180

In words: the largest angle | is | 80 more than the smallest angle

Translate: y = $x + 80$

In words: the remaining angle | is | 10 more than the smallest angle

Translate: z = $x + 10$

3. SOLVE. We solve the system

$$\begin{cases} x + y + z = 180 \\ y = x + 80 \\ z = x + 10 \end{cases}$$

Since y and z are both expressed in terms of x, we will solve using the substitution method. We substitute $y = x + 80$ and $z = x + 10$ in the first equation. Then

$$x + y + z = 180 \quad \text{First equation}$$

$$x + (x + 80) + (x + 10) = 180 \quad \text{Let } y = x + 80 \text{ and } z = x + 10.$$

$$3x + 90 = 180$$

$$3x = 90$$

$$x = 30$$

Then $y = x + 80 = 30 + 80 = 110$, and $z = x + 10 = 30 + 10 = 40$. The ordered triple solution is $(30, 110, 40)$.

4. INTERPRET.

Check: Notice that $30° + 40° + 110° = 180°$. Also, the measure of the largest angle, $110°$, is $80°$ more than the measure of the smallest angle, $30°$. The measure of the remaining angle, $40°$, is $10°$ more than the measure of the smallest angle, $30°$. ☐

PRACTICE

6 The measure of the largest angle of a triangle is $40°$ more than the measure of the smallest angle, and the measure of the remaining angle is $20°$ more than the measure of the smallest angle. Find the measure of each angle.

4.3 EXERCISE SET

PRACTICE WATCH DOWNLOAD READ REVIEW

MIXED PRACTICE

Solve. See Examples 1 through 4.

1. One number is two more than a second number. Twice the first is 4 less than 3 times the second. Find the numbers.

2. Three times one number minus a second is 8, and the sum of the numbers is 12. Find the numbers.

3. The United States has the world's only "large deck" aircraft carriers which can hold up to 72 aircraft. The Enterprise class carrier is longest in length while the Nimitz class carrier is the second longest. The total length of these two carriers is 2193 feet while the difference of their lengths is only 9 feet. (*Source: U.S.A. Today*)

a. Find the length of each class carrier.

b. If a football field has a length of 100 yards, determine the length of the Enterprise class carrier in terms of number of football fields.

4. The rate of growth of participation (age 7 and older) in sports featured in the X-Games has slowed in recent years, but still surpasses that for some older sports such as football. The most popular X-Game sport is inline roller skating, followed by skateboarding. In 2005, the total number of participants in both sports was 25.1 million. If the number of participants in skateboarding was 14.2 million less than twice the number of participants in inline skating, find the number of participants in each sport. (*Source:* National Sporting Goods Association)

5. A B747 aircraft flew 6 hours with the wind. The return trip took 7 hours against the wind. If the speed of the plane in still air is 13 times the speed of the wind, find the wind speed and the speed of the plane in still air.

6. During a multi-day camping trip, Terry Watkins rowed 17 hours downstream. It took 26.5 hours rowing upstream to travel the same distance. If the speed of the current is 6.8 kilometers per hour less than his rowing speed in still water, find his rowing speed and the speed of the current.

7. Find how many quarts of 4% butterfat milk and 1% butterfat milk should be mixed to yield 60 quarts of 2% butterfat milk.

8. A pharmacist needs 500 milliliters of a 20% phenobarbital solution but has only 5% and 25% phenobarbital solutions available. Find how many milliliters of each he should mix to get the desired solution.

9. In 2005, the United Kingdom was the most popular host country in which U.S. students traveling abroad studied. Italy was the second most popular destination. A total of 56,929 students visited one of the two countries. If 7213 more U.S. students studied in the United Kingdom than in Italy, find how many students studied abroad in each country. (*Source:* Institute of International Education, *Open Doors 2006*)

10. Harvard University and Cornell University are each known for their excellent libraries, and each is participating with Google to put their collections into Google's searchable database. In 2005, Harvard libraries contained 266,791 more printed volumes than twice the number of printed volumes in the libraries of Cornell. Together, these two great libraries house 23,199,904 printed volumes. Find the number of printed volumes in each library. (*Source:* Association of Research Libraries)

11. Karen Karlin bought some large frames for $15 each and some small frames for $8 each at a closeout sale. If she bought 22 frames for $239, find how many of each type she bought.

12. Hilton University Drama Club sold 311 tickets for a play. Student tickets cost 50 cents each; nonstudent tickets cost $1.50. If total receipts were $385.50, find how many tickets of each type were sold.

13. One number is two less than a second number. Twice the first is 4 more than 3 times the second. Find the numbers.

14. Twice a first number plus a second number is 42, and the first number minus the second number is −6. Find the numbers.

15. In the United States, the percent of women using the Internet is increasing faster than the percent of men. The function $y = 5.3x + 39.5$ can be used to estimate the percent of females using the Internet, while the function $y = 4.5x + 45.5$ can be used to estimate the percent of males. For both functions, x is the number of years since 2000. Use these functions to give the year in which the percent of females using the Internet equals the percent of males. (*Source:* Pew Internet & American Life Project)

16. The percent of car vehicle sales has been decreasing over a ten-year period while the percent of light truck (pickups, sport-utility vans, and minivans) vehicles has been increasing. For the years 2000–2006, the function $y = -x + 54.2$ can be used to estimate the percent of new car vehicle sales in the United States, while the function $y = x + 45.8$ can be used to estimate the percent of light truck vehicle sales. For both functions, x is the number of years since 2000. (*Source:* USA Today, Environmental Protection Agency, "Light-Duty Automotive Technology and Fuel Economy Trends: 1975–2006")

   a. Calculate the year in which the percent of new car sales equaled the percent of light truck sales.

   b. Before the actual 2001 vehicle sales data was published, *USA Today* predicted that light truck sales would likely be greater than car sales in the year 2001. Does your finding in part **a** agree with this statement?

17. An office supply store in San Diego sells 7 writing tablets and 4 pens for $6.40. Also, 2 tablets and 19 pens cost $5.40. Find the price of each.

18. A Candy Barrel shop manager mixes M&M's worth $2.00 per pound with trail mix worth $1.50 per pound. Find how many pounds of each she should use to get 50 pounds of a party mix worth $1.80 per pound.

19. A Piper airplane and a B737 aircraft cross each other (at different altitudes) traveling in opposite directions. The B737 travels 5 times the speed of the Piper. If in 4 hours, they are 2160 miles apart, find the speed of each aircraft.

20. Two cyclists start at the same point and travel in opposite directions. One travels 4 mph faster than the other. In 4 hours they are 112 miles apart. Find how fast each is traveling.

21. While it is said that trains opened up the American West to settlement, U.S. railroad miles have been on the decline for decades. On the other hand, the miles of roads in the U.S. highway system have been increasing. The function $y = -1379.4x + 150,604$ represents the U.S. railroad miles, while the function $y = 478.4x + 157,838$ models the number of U.S. highway miles. For each function, x is the number of years after 1995. (*Source:* Association of American Railroads, Federal Highway Administration)

a. Explain how the decrease in railroad miles can be verified by their given function while the increase in highway miles can be verified by their given function.

b. Find the year in which it is estimated that the number of U.S. railroad miles and the number of U.S. highway miles were the same.

22. The annual U.S. per capita consumption of whole milk has decreased since 1980, while the per capita consumption of lower fat milk has increased. For the years 1980–2005, the function $y = -0.40x + 15.9$ approximates the annual U.S. per capita consumption of whole milk in gallons, and the function $y = 0.14x + 11.9$ approximates the annual U.S. per capita consumption of lower fat milk in gallons. Determine the year in which the per capita consumption of whole milk equaled the per capita consumption of lower fat milk. (*Source:* Economic Research Service: U.S.D.A.)

△ **23.** The perimeter of a triangle is 93 centimeters. If two sides are equally long and the third side is 9 centimeters longer than the others, find the lengths of the three sides.

24. Jack Reinholt, a car salesman, has a choice of two pay arrangements: a weekly salary of $200 plus 5% commission on sales, or a straight 15% commission. Find the amount of weekly sales for which Jack's earnings are the same regardless of the pay arrangement.

25. Hertz car rental agency charges $25 daily plus 10 cents per mile. Budget charges $20 daily plus 25 cents per mile. Find the daily mileage for which the Budget charge for the day is twice that of the Hertz charge for the day.

26. Carroll Blakemore, a drafting student, bought three templates and a pencil one day for $6.45. Another day he bought two pads of paper and four pencils for $7.50. If the price of a pad of paper is three times the price of a pencil, find the price of each type of item.

△ **27.** In the figure, line l and line m are parallel lines cut by transversal t. Find the values of x and y.

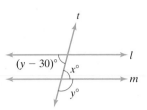

△ **28.** Find the values of x and y in the following isosceles triangle.

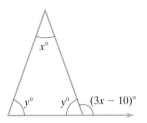

Given the cost function $C(x)$ and the revenue function $R(x)$, find the number of units x that must be sold to break even. See Example 5.

29. $C(x) = 30x + 10,000; R(x) = 46x$

30. $C(x) = 12x + 15,000; R(x) = 32x$

31. $C(x) = 1.2x + 1500; R(x) = 1.7x$

32. $C(x) = 0.8x + 900; R(x) = 2x$

33. $C(x) = 75x + 160,000; R(x) = 200x$

34. $C(x) = 105x + 70,000; R(x) = 245x$

35. The planning department of Abstract Office Supplies has been asked to determine whether the company should introduce a new computer desk next year. The department estimates that $6000 of new manufacturing equipment will need to be purchased and that the cost of constructing each desk will be $200. The department also estimates that the revenue from each desk will be $450.

a. Determine the revenue function $R(x)$ from the sale of x desks.

b. Determine the cost function $C(x)$ for manufacturing x desks.

c. Find the break-even point.

36. Baskets, Inc., is planning to introduce a new woven basket. The company estimates that $500 worth of new equipment will be needed to manufacture this new type of basket and that it will cost $15 per basket to manufacture. The company also estimates that the revenue from each basket will be $31.

a. Determine the revenue function $R(x)$ from the sale of x baskets.

b. Determine the cost function $C(x)$ for manufacturing x baskets.

c. Find the break-even point.

Solve. See Example 6.

37. Rabbits in a lab are to be kept on a strict daily diet that includes 30 grams of protein, 16 grams of fat, and 24 grams of carbohydrates. The scientist has only three food mixes available with the following grams of nutrients per unit.

	Protein	*Fat*	*Carbohydrate*
Mix A	4	6	3
Mix B	6	1	2
Mix C	4	1	12

Find how many units of each mix are needed daily to meet each rabbit's dietary need.

38. Gerry Gundersen mixes different solutions with concentrations of 25%, 40%, and 50% to get 200 liters of a 32% solution. If he uses twice as much of the 25% solution as of the 40% solution, find how many liters of each kind he uses.

△ 39. The perimeter of a quadrilateral (four-sided polygon) is 29 inches. The longest side is twice as long as the shortest side. The other two sides are equally long and are 2 inches longer than the shortest side. Find the length of all four sides.

△ 40. The measure of the largest angle of a triangle is 90° more than the measure of the smallest angle, and the measure of the remaining angle is 30° more than the measure of the smallest angle. Find the measure of each angle.

41. The sum of three numbers is 40. One number is five more than a second number. It is also twice the third. Find the numbers.

42. The sum of the digits of a three-digit number is 15. The tens-place digit is twice the hundreds-place digit, and the ones-place digit is 1 less than the hundreds-place digit. Find the three-digit number.

43. Diana Taurasi, of the Phoenix Mercury, was the WNBA's top scorer for the 2006 regular season, with a total of 860 points. The number of two-point field goals that Taurasi made was 65 less than double the number of three-point field goals she made. The number of free throws (each worth one point) she made was 34 less than the number of two-point field goals she made. Find how many free throws, two-point field goals, and three-point field goals Diana Taurasi made during the 2006 regular season. (*Source:* Women's National Basketball Association)

44. During the 2006 NBA playoffs, the top scoring player was Dwyane Wade of the Miami Heat. Wade scored a total of 654 points during the playoffs. The number of free throws (each worth one point) he made was three less than the number of two-point field goals he made. He also made 27 fewer three-point field goals than one-fifth the number of two-point

field goals. How many free throws, two-point field goals, and three-point field goals did Dwyane Wade make during the 2006 playoffs? (*Source:* National Basketball Association)

△ 45. Find the values of x, y, and z in the following triangle.

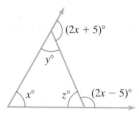

△ 46. The sum of the measures of the angles of a quadrilateral is 360°. Find the value of x, y, and z in the following quadrilateral.

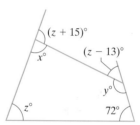

REVIEW AND PREVIEW

Multiply both sides of equation (1) by 2, and add the resulting equation to equation (2). See Section 4.2.

47. $3x - y + z = 2$ (1)
$-x + 2y + 3z = 6$ (2)

48. $2x + y + 3z = 7$ (1)
$-4x + y + 2z = 4$ (2)

Multiply both sides of equation (1) by -3, and add the resulting equation to equation (2). See Section 4.2.

49. $x + 2y - z = 0$ (1)
$3x + y - z = 2$ (2)

50. $2x - 3y + 2z = 5$ (1)
$x - 9y + z = -1$ (2)

CONCEPT EXTENSIONS

51. The number of personal bankruptcy petitions filed in the United States was constantly increasing from the early 1980s until the bankruptcy laws were changed in 2006. In 2006, the number of petitions filed was only 25,765 more than the number of petitions filed in 1996. The total number of personal bankruptcies filed in these two years was 2,144,653. Find the

number of personal bankruptcies filed in each year. (*Source: Based on data from the Administrative Office of the United States Courts*)

52. In 2006, the median weekly earnings for male postal service mail carriers in the United States was $126 more than the median weekly earnings for female mail postal service carriers. The median weekly earnings for female postal service mail carriers was 0.86 times that of their male counterparts. Also in 2006, the median weekly earnings for female lawyers in the United States was $540 less than the median weekly earnings for male lawyers. The median weekly earnings of male lawyers was 1.4 times that of their female counterparts. (*Source:* Based on data from the Bureau of Labor Statistics)

   a. Find the median weekly earnings for female postal service mail carriers in the United States in 2006.

   b. Find the median weekly earnings for female lawyers in the United States in 2006.

   c. Of the four groups of workers described in the problem, which group makes the greatest weekly earnings? Which group makes the least weekly earnings?

53. MySpace and Facebook are both popular social networking Web sites. The function $f(x) = 0.85x + 41.75$ represents the MySpace minutes (in thousands)/month while the function $f(x) = 1.13x + 10.49$ represents the Facebook minutes (in thousands)/month. In both of these functions $x = 1$ represents Feb. 2007, $x = 2$ represents March 2007, and so on.

   a. Solve the system formed by these functions. Round each coordinate to the nearest whole number.

   b. Use your answer to part **a** to predict the month and year in which the minutes (in thousands)/month for MySpace and Facebook are the same.

54. Find the values of a, b, and c such that the equation $y = ax^2 + bx + c$ has ordered pair solutions $(1, 6)$, $(-1, -2)$, and $(0, -1)$. To do so, substitute each ordered pair solution into the equation. Each time, the result is an equation in three unknowns: a, b, and c. Then solve the resulting system of three linear equations in three unknowns, a, b, and c.

55. Find the values of a, b, and c such that the equation $y = ax^2 + bx + c$ has ordered pair solutions $(1, 2)$, $(2, 3)$, and $(-1, 6)$. (*Hint:* See Exercise 53.)

56. Data (x, y) for the total number y (in thousands) of college-bound students who took the ACT assessment in the year x are $(0, 1065)$, $(1, 1070)$, and $(3, 1175)$, where $x = 0$ represents 2000 and $x = 1$ represents 2001. Find the values of a, b, and c such that the equation $y = ax^2 + bx + c$ models this data. According to your model, how many students will take the ACT in 2009? (*Source:* ACT, Inc.)

57. Monthly normal rainfall data (x, y) for Portland, Oregon, are $(4, 2.47)$, $(7, 0.6)$, $(8, 1.1)$, where x represents time in months (with $x = 1$ representing January) and y represents rainfall in inches. Find the values of a, b, and c rounded to 2 decimal places such that the equation $y = ax^2 + bx + c$ models this data. According to your model, how much rain should Portland expect during September? (*Source:* National Climatic Data Center)

INTEGRATED REVIEW SYSTEMS OF LINEAR EQUATIONS

Sections 4.1–4.3

The graphs of various systems of equations are shown. Match each graph with the solution of its corresponding system.

A 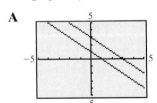 B 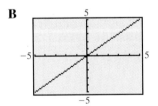 C 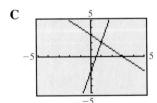 D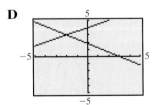

1. Solution: $(1, 2)$

2. Solution: $(-2, 3)$

3. No solution

4. Infinite number of solutions

Solve each system by elimination or substitution.

5. $\begin{cases} x + y = 4 \\ \quad\; y = 3x \end{cases}$

6. $\begin{cases} x - y = -4 \\ \quad\; y = 4x \end{cases}$

7. $\begin{cases} x + y = 1 \\ x - 2y = 4 \end{cases}$

8. $\begin{cases} 2x - y = 8 \\ x + 3y = 11 \end{cases}$

9. $\begin{cases} 2x + 5y = 8 \\ 6x + y = 10 \end{cases}$

10. $\begin{cases} \dfrac{1}{8}x - \dfrac{1}{2}y = -\dfrac{5}{8} \\ -3x - 8y = 0 \end{cases}$

11. $\begin{cases} 4x - 7y = 7 \\ 12x - 21y = 24 \end{cases}$

12. $\begin{cases} 2x - 5y = 3 \\ -4x + 10y = -6 \end{cases}$

13. $\begin{cases} y = \dfrac{1}{3}x \\ 5x - 3y = 4 \end{cases}$

14. $\begin{cases} y = \dfrac{1}{4}x \\ 2x - 4y = 3 \end{cases}$

15. $\begin{cases} x + y \phantom{{}+z} = 2 \\ -3y + z = -7 \\ 2x + y - z = -1 \end{cases}$

16. $\begin{cases} \phantom{x+{}}y + 2z = -3 \\ x - 2y \phantom{{}+2z} = 7 \\ 2x - y + z = 5 \end{cases}$

17. $\begin{cases} 2x + 4y - 6z = 3 \\ -x + y - z = 6 \\ x + 2y - 3z = 1 \end{cases}$

18. $\begin{cases} x - y + 3z = 2 \\ -2x + 2y - 6z = -4 \\ 3x - 3y + 9z = 6 \end{cases}$

19. $\begin{cases} x + y - 4z = 5 \\ x - y + 2z = -2 \\ 3x + 2y + 4z = 18 \end{cases}$

20. $\begin{cases} 2x - y + 3z = 2 \\ x + y - 6z = 0 \\ 3x + 4y - 3z = 6 \end{cases}$

21. A first number is 8 less than a second number. Twice the first number is 11 more than the second number. Find the numbers.

22. The sum of the measures of the angles of a quadrilateral is 360°. The two smallest angles of the quadrilateral have the same measure. The third angle measures 30° more than the measure of one of the smallest angles and the fourth angle measures 50° more than the measure of one of the smallest angles. Find the measure of each angle.

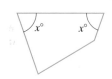

4.4 SOLVING SYSTEMS OF EQUATIONS BY MATRICES

OBJECTIVES

1 Use matrices to solve a system of two equations.

2 Use matrices to solve a system of three equations.

3 Use matrices and a graphing utility to solve a system.

By now, you may have noticed that the solution of a system of equations depends on the coefficients of the equations in the system and not on the variables. In this section, we introduce solving a system of equations by a **matrix.**

OBJECTIVE 1 ▶ Using matrices to solve a system of two equations. A matrix (plural: **matrices**) is a rectangular array of numbers. The following are examples of matrices.

$$\begin{bmatrix} 1 & 0 \\ 0 & 1 \end{bmatrix} \qquad \begin{bmatrix} 2 & 1 & 3 & -1 \\ 0 & -1 & 4 & 5 \\ -6 & 2 & 1 & 0 \end{bmatrix} \qquad \begin{bmatrix} a & b & c \\ d & e & f \end{bmatrix}$$

The numbers aligned horizontally in a matrix are in the same **row.** The numbers aligned vertically are in the same **column.**

$$\begin{matrix} \text{row 1} \rightarrow \\ \text{row 2} \rightarrow \end{matrix} \begin{bmatrix} 2 & 1 & 0 \\ -1 & 6 & 2 \end{bmatrix}$$

This matrix has 2 rows and 3 columns. It is called a 2 × 3 (read "two by three") matrix.

column 1
column 2
column 3

To see the relationship between systems of equations and matrices, study the example below.

▶ Helpful Hint

Before writing the corresponding matrix associated with a system of equations, make sure that the equations are written in standard form.

System of Equations
(in standard form)

$$\begin{cases} 2x - 3y = 6 & \text{Equation 1} \\ x + y = 0 & \text{Equation 2} \end{cases}$$

Corresponding Matrix

$$\begin{bmatrix} 2 & -3 & \vdots & 6 \\ 1 & 1 & \vdots & 0 \end{bmatrix} \quad \begin{matrix} \text{Row 1} \\ \text{Row 2} \end{matrix}$$

Notice that the rows of the matrix correspond to the equations in the system. The coefficients of each variable are placed to the left of a vertical dashed line. The constants are placed to the right. Each of these numbers in the matrix is called an **element.**

The method of solving systems by matrices is to write this matrix as an equivalent matrix from which we easily identify the solution. Two matrices are equivalent if they represent systems that have the same solution set. The following **row operations** can be performed on matrices, and the result is an equivalent matrix.

Elementary Row Operations

1. Any two rows in a matrix may be interchanged.

2. The elements of any row may be multiplied (or divided) by the same nonzero number.

3. The elements of any row may be multiplied (or divided) by a nonzero number and added to their corresponding elements in any other row.

▶ **Helpful Hint**

Notice that these *row* operations are the same operations that we can perform on *equations* in a system.

To solve a system of two equations in x and y by matrices, write the corresponding matrix associated with the system. Then use elementary row operations to write equivalent matrices until you have a matrix of the form

$$\begin{bmatrix} 1 & a & \vdots & b \\ 0 & 1 & \vdots & c \end{bmatrix},$$

where a, b, and c are constants. Why? If a matrix associated with a system of equations is in this form, we can easily solve for x and y. For example,

Matrix		*System of Equations*
$\begin{bmatrix} 1 & 2 & \vdots & -3 \\ 0 & 1 & \vdots & 5 \end{bmatrix}$	corresponds to	$\begin{cases} 1x + 2y = -3 \\ 0x + 1y = 5 \end{cases}$ or $\begin{cases} x + 2y = -3 \\ y = 5 \end{cases}$

In the second equation, we have $y = 5$. Substituting this in the first equation, we have $x + 2(5) = -3$ or $x = -13$. The solution of the system is the ordered pair $(-13, 5)$.

EXAMPLE 1 Use matrices to solve the system.

$$\begin{cases} x + 3y = 5 \\ 2x - y = -4 \end{cases}$$

Solution The corresponding matrix is $\begin{bmatrix} 1 & 3 & \vdots & 5 \\ 2 & -1 & \vdots & -4 \end{bmatrix}$. We use elementary row operations to write an equivalent matrix that looks like $\begin{bmatrix} 1 & a & \vdots & b \\ 0 & 1 & \vdots & c \end{bmatrix}$.

For the matrix given, the element in the first row, first column is already 1, as desired. Next we write an equivalent matrix with a 0 below the 1. To do this, we multiply row 1 by -2 and add to row 2. *We will change only row 2.*

$$\begin{bmatrix} 1 & 3 & \vdots & 5 \\ -2(1) + 2 & -2(3) + (-1) & \vdots & -2(5) + (-4) \end{bmatrix} \text{ simplifies to } \begin{bmatrix} 1 & 3 & \vdots & 5 \\ 0 & -7 & \vdots & -14 \end{bmatrix}$$

↑ ↑ ↑ ↑ ↑ ↑

row 1   row 2   row 1   row 2   row 1   row 2
element element element element element element

Now we change the -7 to a 1 by use of an elementary row operation. We divide row 2 by -7, then

$$\begin{bmatrix} 1 & 3 & | & 5 \\ \dfrac{0}{-7} & \dfrac{-7}{-7} & | & \dfrac{-14}{-7} \end{bmatrix} \quad \text{simplifies to} \quad \begin{bmatrix} 1 & 3 & | & 5 \\ 0 & 1 & | & 2 \end{bmatrix}$$

This last matrix corresponds to the system

$$\begin{cases} x + 3y = 5 \\ \quad\;\; y = 2 \end{cases}$$

To find x, we let $y = 2$ in the first equation, $x + 3y = 5$.

$$x + 3y = 5 \qquad \text{First equation}$$
$$x + 3(2) = 5 \qquad \text{Let } y = 2.$$
$$x = -1$$

The ordered pair solution is $(-1, 2)$. Check to see that this ordered pair satisfies both equations. □

PRACTICE
1   Use matrices to solve the system.

$$\begin{cases} x + 4y = -2 \\ 3x - \;\; y = 7 \end{cases}$$

EXAMPLE 2   Use matrices to solve the system.

$$\begin{cases} 2x - \;\; y = 3 \\ 4x - 2y = 5 \end{cases}$$

Solution   The corresponding matrix is $\begin{bmatrix} 2 & -1 & | & 3 \\ 4 & -2 & | & 5 \end{bmatrix}$. To get 1 in the row 1, column 1 position, we divide the elements of row 1 by 2.

$$\begin{bmatrix} \dfrac{2}{2} & -\dfrac{1}{2} & | & \dfrac{3}{2} \\ 4 & -2 & | & 5 \end{bmatrix} \quad \text{simplifies to} \quad \begin{bmatrix} 1 & -\dfrac{1}{2} & | & \dfrac{3}{2} \\ 4 & -2 & | & 5 \end{bmatrix}$$

To get 0 under the 1, we multiply the elements of row 1 by -4 and add the new elements to the elements of row 2.

$$\begin{bmatrix} 1 & -\dfrac{1}{2} & | & \dfrac{3}{2} \\ -4(1) + 4 & -4\left(-\dfrac{1}{2}\right) - 2 & | & -4\left(\dfrac{3}{2}\right) + 5 \end{bmatrix} \quad \text{simplifies to} \quad \begin{bmatrix} 1 & -\dfrac{1}{2} & | & \dfrac{3}{2} \\ 0 & 0 & | & -1 \end{bmatrix}$$

The corresponding system is $\begin{cases} x - \dfrac{1}{2}y = \dfrac{3}{2} \\ \qquad\quad 0 = -1 \end{cases}$. The equation $0 = -1$ is false for all y- or x-values; hence the system is inconsistent and has no solution. □

PRACTICE
2   Use matrices to solve the system.

$$\begin{cases} x - 3y = 3 \\ -2x + 6y = 4 \end{cases}$$

Concept Check ✓

Consider the system

$$\begin{cases} 2x - 3y = 8 \\ x + 5y = -3 \end{cases}$$

What is wrong with its corresponding matrix shown below?

$$\begin{bmatrix} 2 & -3 & \vdots & 8 \\ 0 & 5 & \vdots & -3 \end{bmatrix}$$

OBJECTIVE 2 ▶ Using matrices to solve a system of three equations. To solve a system of three equations in three variables using matrices, we will write the corresponding matrix in the form

$$\begin{bmatrix} 1 & a & b & \vdots & d \\ 0 & 1 & c & \vdots & e \\ 0 & 0 & 1 & \vdots & f \end{bmatrix}$$

EXAMPLE 3   Use matrices to solve the system.

$$\begin{cases} x + 2y + z = 2 \\ -2x - y + 2z = 5 \\ x + 3y - 2z = -8 \end{cases}$$

Solution   The corresponding matrix is $\begin{bmatrix} 1 & 2 & 1 & \vdots & 2 \\ -2 & -1 & 2 & \vdots & 5 \\ 1 & 3 & -2 & \vdots & -8 \end{bmatrix}$. Our goal is to

write an equivalent matrix with 1's along the diagonal (see the numbers in red) and 0's below the 1's. The element in row 1, column 1 is already 1. Next we get 0's for each element in the rest of column 1. To do this, first we multiply the elements of row 1 by 2 and add the new elements to row 2. Also, we multiply the elements of row 1 by −1 and add the new elements to the elements of row 3. We *do not change row 1.* Then

$$\begin{bmatrix} 1 & 2 & 1 & \vdots & 2 \\ 2(1) - 2 & 2(2) - 1 & 2(1) + 2 & \vdots & 2(2) + 5 \\ -1(1) + 1 & -1(2) + 3 & -1(1) - 2 & \vdots & -1(2) - 8 \end{bmatrix} \text{ simplifies to } \begin{bmatrix} 1 & 2 & 1 & \vdots & 2 \\ 0 & 3 & 4 & \vdots & 9 \\ 0 & 1 & -3 & \vdots & -10 \end{bmatrix}$$

We continue down the diagonal and use elementary row operations to get 1 where the element 3 is now. To do this, we interchange rows 2 and 3.

$$\begin{bmatrix} 1 & 2 & 1 & \vdots & 2 \\ 0 & 3 & 4 & \vdots & 9 \\ 0 & 1 & -3 & \vdots & -10 \end{bmatrix} \text{ is equivalent to } \begin{bmatrix} 1 & 2 & 1 & \vdots & 2 \\ 0 & 1 & -3 & \vdots & -10 \\ 0 & 3 & 4 & \vdots & 9 \end{bmatrix}$$

Next we want the new row 3, column 2 element to be 0. We multiply the elements of row 2 by −3 and add the result to the elements of row 3.

$$\begin{bmatrix} 1 & 2 & 1 & \vdots & 2 \\ 0 & 1 & -3 & \vdots & -10 \\ -3(0) + 0 & -3(1) + 3 & -3(-3) + 4 & \vdots & -3(-10) + 9 \end{bmatrix} \text{ simplifies to }$$

$$\begin{bmatrix} 1 & 2 & 1 & \vdots & 2 \\ 0 & 1 & -3 & \vdots & -10 \\ 0 & 0 & 13 & \vdots & 39 \end{bmatrix}$$

Answer to Concept Check:

matrix should be $\begin{bmatrix} 2 & -3 & \vdots & 8 \\ 1 & 5 & \vdots & -3 \end{bmatrix}$

Finally, we divide the elements of row 3 by 13 so that the final diagonal element is 1.

$$\begin{bmatrix} 1 & 2 & 1 & \vdots & 2 \\ 0 & 1 & -3 & \vdots & -10 \\ \dfrac{0}{13} & \dfrac{0}{13} & \dfrac{13}{13} & \vdots & \dfrac{39}{13} \end{bmatrix} \quad \text{simplifies to} \quad \begin{bmatrix} 1 & 2 & 1 & \vdots & 2 \\ 0 & 1 & -3 & \vdots & -10 \\ 0 & 0 & 1 & \vdots & 3 \end{bmatrix}$$

This matrix corresponds to the system

$$\begin{cases} x + 2y + z = 2 \\ \quad\quad y - 3z = -10 \\ \quad\quad\quad\quad z = 3 \end{cases}$$

We identify the z-coordinate of the solution as 3. Next we replace z with 3 in the second equation and solve for y.

$$\begin{aligned} y - 3z &= -10 \quad \text{Second equation} \\ y - 3(3) &= -10 \quad \text{Let } z = 3. \\ y &= -1 \end{aligned}$$

To find x, we let $z = 3$ and $y = -1$ in the first equation.

$$\begin{aligned} x + 2y + z &= 2 \quad \text{First equation} \\ x + 2(-1) + 3 &= 2 \quad \text{Let } z = 3 \text{ and } y = -1. \\ x &= 1 \end{aligned}$$

The ordered triple solution is $(1, -1, 3)$. Check to see that it satisfies all three equations in the original system. □

PRACTICE
3   Use matrices to solve the system.

$$\begin{cases} x + 3y - z = 0 \\ 2x + y + 3z = 5 \\ -x - 2y + 4z = 7 \end{cases}$$

OBJECTIVE 3 ▶ Using matrices and a graphing utility to solve systems. Thus far, we have solved systems of equations by matrices by writing the augmented matrix of the system as an equivalent matrix with 1's along the diagonal we'll call the main diagonal, and 0's below the 1's. Another way to solve a system of equations by matrices is to use the elementary row operations and write the augmented matrix of the system as an equivalent matrix with 1's along this main diagonal and 0's *above* and below these 1's. This form of a matrix is given a special name: the **reduced row echelon form.**

Although we will not practice writing matrices in reduced row echelon form using pencil and paper, we can easily see an advantage of this form, and we will use this form with a graphing utility.

Below is a system of equations, its corresponding augmented matrix, and an equivalent matrix in **reduced row echelon** form found by performing elementary row operations on the augmented matrix.

System of Equations	*Augmented Matrix of the System*	*Matrix in Reduced Row Echelon Form*
$3x + y + 2z = 5$		
$x - y + z = 0$	$\begin{bmatrix} 3 & 1 & 2 & \vdots & 5 \\ 1 & -1 & 1 & \vdots & 0 \\ -2 & 2 & -1 & \vdots & -1 \end{bmatrix}$	$\begin{bmatrix} 1 & 0 & 0 & \vdots & 2 \\ 0 & 1 & 0 & \vdots & 1 \\ 0 & 0 & 1 & \vdots & -1 \end{bmatrix}$
$-2x + 2y - z = -1$		

The corresponding system equivalent to the original system is

$$\begin{cases} 1x + 0y + 0z = 2 \\ 0x + 1y + 0z = 1 \\ 0x + 0y + 1z = -1 \end{cases} \quad \text{or} \quad \begin{cases} x = 2 \\ y = 1 \\ z = -1 \end{cases}$$

Notice that the solution of the system is $(2, 1, -1)$, which is found by reading the last column of the associated matrix written in reduced row echelon form.

Although using elementary row operations to write an augmented matrix in equivalent reduced row echelon form may be tedious for us, a calculator can do this quickly and accurately.

EXAMPLE 4 Solve the system of equations in Example 1 using matrices and a calculator.

$$\begin{cases} x + 3y = 5 \\ 2x - y = -4 \end{cases}$$

Solution This is the same system of equations solved in Example 1. Recall that the augmented matrix associated with this system is

$$\begin{bmatrix} 1 & 3 & | & 5 \\ 2 & -1 & | & -4 \end{bmatrix}$$

Using the matrix edit feature, put the corresponding entries in matrix A. Then use a feature of your calculator to find the equivalent reduced row echelon form.

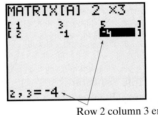

Row 2 column 3 entry, for example

Augmented matrix in reduced row echelon form

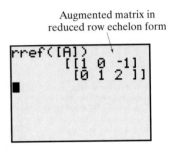

The augmented matrix in reduced row echelon form is $\begin{bmatrix} 1 & 0 & | & -1 \\ 0 & 1 & | & 2 \end{bmatrix}$, which corresponds to the system

$$\begin{cases} 1x + 0y = -1 \\ 0x + 1y = 2 \end{cases} \quad \text{or} \quad \begin{cases} x = -1 \\ y = 2 \end{cases}$$

The solution of the system is $(-1, 2)$, as confirmed in Example 1. □

PRACTICE
4 Solve the system of equations in Practice 1 using matrices and a calculator.

EXAMPLE 5 Solve the system in Example 3 using matrices and a calculator.

$$\begin{cases} x + 2y + z = 2 \\ -2x - y + 2z = 5 \\ x + 3y - 2z = -8 \end{cases}$$

Solution Enter the augmented matrix in matrix A and then use your calculator to find the reduced row echelon form of matrix A.

Dimensions of augmented matrix

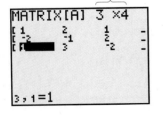

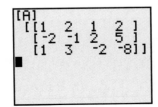

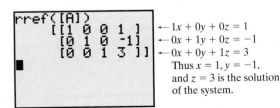

$\leftarrow 1x + 0y + 0z = 1$
$\leftarrow 0x + 1y + 0z = -1$
$\leftarrow 0x + 0y + 1z = 3$
Thus $x = 1, y = -1,$ and $z = 3$ is the solution of the system.

Augmented matrix in reduced row echelon form.

This confirms that the solution of the system of equations is $(1, -1, 3)$.   ☐

PRACTICE
5   Solve the system in Practice 3 using matrices and a calculator.

VOCABULARY & READINESS CHECK

Use the choices below to fill in each blank.

    column   element   row   matrix

1. A _____ is a rectangular array of numbers.

2. Each of the numbers in a matrix is called an _____.

3. The numbers aligned horizontally in a matrix are in the same _____.

4. The numbers aligned vertically in a matrix are in the same _____.

Answer true or false for each statement about operations within a matrix forming an equivalent matrix.

5. Any two columns may be interchanged. _____

6. Any two rows may be interchanged. _____

7. The elements in a row may be added to their corresponding elements in another row. _____

8. The elements of a column may be multiplied by any nonzero number. _____

4.4 | EXERCISE SET

MyMathLab   PRACTICE   WATCH   DOWNLOAD   READ   REVIEW

Solve each system of linear equations using matrices. See Example 1.

1. $\begin{cases} x + y = 1 \\ x - 2y = 4 \end{cases}$ **2.** $\begin{cases} 2x - y = 8 \\ x + 3y = 11 \end{cases}$

3. $\begin{cases} x + 3y = 2 \\ x + 2y = 0 \end{cases}$ **4.** $\begin{cases} 4x - y = 5 \\ 3x + 3y = 0 \end{cases}$

Solve each system of linear equations using matrices. See Example 2.

5. $\begin{cases} x - 2y = 4 \\ 2x - 4y = 4 \end{cases}$ **6.** $\begin{cases} -x + 3y = 6 \\ 3x - 9y = 9 \end{cases}$

7. $\begin{cases} 3x - 3y = 9 \\ 2x - 2y = 6 \end{cases}$ **8.** $\begin{cases} 9x - 3y = 6 \\ -18x + 6y = -12 \end{cases}$

Solve each system of linear equations using matrices. See Example 3.

9. $\begin{cases} x + y = 3 \\ 2y = 10 \\ 3x + 2y - 4z = 12 \end{cases}$

10. $\begin{cases} 5x = 5 \\ 2x + y = 4 \\ 3x + y - 5z = -15 \end{cases}$

11. $\begin{cases} 2y - z = -7 \\ x + 4y + z = -4 \\ 5x - y + 2z = 13 \end{cases}$

12. $\begin{cases} 4y + 3z = -2 \\ 5x - 4y = 1 \\ -5x + 4y + z = -3 \end{cases}$

MIXED PRACTICE

Solve each system of linear equations using matrices. See Examples 1 through 3.

13. $\begin{cases} x - 4 = 0 \\ x + y = 1 \end{cases}$

14. $\begin{cases} 3y = 6 \\ x + y = 7 \end{cases}$

15. $\begin{cases} x + y + z = 2 \\ 2x - z = 5 \\ 3y + z = 2 \end{cases}$

16. $\begin{cases} x + 2y + z = 5 \\ x - y - z = 3 \\ y + z = 2 \end{cases}$

17. $\begin{cases} 5x - 2y = 27 \\ -3x + 5y = 18 \end{cases}$

18. $\begin{cases} 4x - y = 9 \\ 2x + 3y = -27 \end{cases}$

19. $\begin{cases} 4x - 7y = 7 \\ 12x - 21y = 24 \end{cases}$

20. $\begin{cases} 2x - 5y = 12 \\ -4x + 10y = 20 \end{cases}$

21. $\begin{cases} 4x - y + 2z = 5 \\ 2y + z = 4 \\ 4x + y + 3z = 10 \end{cases}$

22. $\begin{cases} 5y - 7z = 14 \\ 2x + y + 4z = 10 \\ 2x + 6y - 3z = 30 \end{cases}$

23. $\begin{cases} 4x + y + z = 3 \\ -x + y - 2z = -11 \\ x + 2y + 2z = -1 \end{cases}$

24. $\begin{cases} x + y + z = 9 \\ 3x - y + z = -1 \\ -2x + 2y - 3z = -2 \end{cases}$

REVIEW AND PREVIEW

Determine whether each graph is the graph of a function. See Section 2.2.

25.

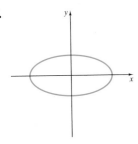

26.

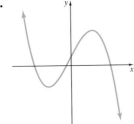

27.

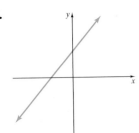

28.

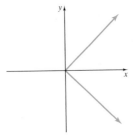

Evaluate. See Section 1.3.

29. $(-1)(-5) - (6)(3)$

30. $(2)(-8) - (-4)(1)$

31. $(4)(-10) - (2)(-2)$

32. $(-7)(3) - (-2)(-6)$

33. $(-3)(-3) - (-1)(-9)$

34. $(5)(6) - (10)(10)$

CONCEPT EXTENSIONS

Solve. See the Concept Check in the section.

35. For the system $\begin{cases} x + z = 7 \\ y + 2z = -6, \\ 3x - y = 0 \end{cases}$ which is the correct corresponding matrix?

a. $\begin{bmatrix} 1 & 1 & | & 7 \\ 1 & 2 & | & -6 \\ 3 & -1 & | & 0 \end{bmatrix}$ **b.** $\begin{bmatrix} 1 & 0 & 1 & | & 7 \\ 1 & 2 & 0 & | & -6 \\ 3 & -1 & 0 & | & 0 \end{bmatrix}$

c. $\begin{bmatrix} 1 & 0 & 1 & | & 7 \\ 0 & 1 & 2 & | & -6 \\ 3 & -1 & 0 & | & 0 \end{bmatrix}$

36. For the system $\begin{cases} x - 6 = 0 \\ 2x - 3y = 1 \end{cases}$, which is the correct corresponding matrix?

a. $\begin{bmatrix} 1 & -6 & | & 0 \\ 2 & -3 & | & 1 \end{bmatrix}$ **b.** $\begin{bmatrix} 1 & 0 & | & 6 \\ 2 & -3 & | & 1 \end{bmatrix}$ **c.** $\begin{bmatrix} 1 & 0 & | & -6 \\ 2 & -3 & | & 1 \end{bmatrix}$

37. The percent y of U.S. households that owned a black-and-white television set between the years 1980 and 1993 can be modeled by the linear equation $2.3x + y = 52$, where x represents the number of years after 1980. Similarly, the percent y of U.S. households that owned a microwave oven during this same period can be modeled by the linear equation $-5.4x + y = 14$. (*Source:* Based on data from the Energy Information Administration, U.S. Department of Energy)

a. The data used to form these two models was incomplete. It is impossible to tell from the data the year in which the percent of households owning black-and-white television sets was the same as the percent of households owning microwave ovens. Use matrix methods to estimate the year in which this occurred.

b. Did more households own black-and-white television sets or microwave ovens in 1980? In 1993? What trends do these models show? Does this seem to make sense? Why or why not?

c. According to the models, when will the percent of households owning black-and-white television sets reach 0%?

d. Do you think your answer to part **c** is accurate? Why or why not?

38. The most popular amusement park in the world currently (according to annual attendance) is Disney World's Magic Kingdom, whose annual attendance in thousands can be approximated by the equation $y = 455x + 14123$, where x is the number of years after 2001. This theme park stole the title from Tokyo Disneyland, which had been in first place for many years. The yearly attendance for Tokyo Disneyland, in thousands, can be represented by the equation $y = -776x + 15985$. Find the last year when attendance at Tokyo Disneyland was greater then attendance at Magic Kingdom. (*Source:* Amusement Business, and TEA Park World)

39. For the system $\begin{cases} 2x - 3y = 8 \\ x + 5y = -3 \end{cases}$, explain what is wrong with writing the corresponding matrix as $\begin{bmatrix} 2 & 3 & | & 8 \\ 0 & 5 & | & -3 \end{bmatrix}$.

STUDY SKILLS BUILDER

Are You Satisfied with Your Performance on a Particular Quiz or Exam?

If not, don't forget to analyze your quiz or exam and look for common errors. Were most of your errors a result of:

- *Carelessness?* Did you turn in your quiz or exam before the allotted time expired? If so, resolve next time to use the entire time allotted. Any extra time can be spent checking your work.

- *Running out of time?* If so, make a point to better manage your time on your next quiz or exam. Try completing any questions that you are unsure of last and delay checking your work until all questions have been answered.

- *Not understanding a concept?* If so, review that concept and correct your work. Try to understand how this happened so that you make sure it doesn't happen before the next quiz or exam.

- *Test conditions?* When studying for a quiz or exam, make sure you place yourself in conditions similar to test conditions. For example, before your next quiz or exam, use a few sheets of blank paper and take a sample test without the aid of your notes or text.

(See your instructor or use the Chapter Test at the end of each chapter.)

Exercises

1. Have you corrected all your previous quizzes and exams?

2. List any errors you have found common to two or more of your graded papers.

3. Is one of your common errors not understanding a concept? If so, are you making sure you understand all the concepts for the next quiz or exam?

4. Is one of your common errors making careless mistakes? If so, are you now taking all the time allotted to check over your work so that you can minimize the number of careless mistakes?

5. Are you satisfied with your grades thus far on quizzes and tests?

6. If your answer to Exercise 5 is no, are there any more suggestions you can make to your instructor or yourself to help? If so, list them here and share these with your instructor.

4.5 SYSTEMS OF LINEAR INEQUALITIES

OBJECTIVE

1 Graph a system of linear inequalities.

OBJECTIVE 1 ▶ Graphing systems of linear inequalities. In Section 3.7 we solved linear inequalities in two variables as well as their union and intersection. Just as two or more linear equations make a system of linear equations, two or more linear inequalities make a **system of linear inequalities.** Systems of inequalities are very important in a process called linear programming. Many businesses use linear programming to find the most profitable way to use limited resources such as employees, machines, or buildings.

A **solution of a system of linear inequalities** is an ordered pair that satisfies each inequality in the system. The set of all such ordered pairs is the solution set of the system. Graphing this set gives us a picture of the solution set. We can graph a system of inequalities by graphing each inequality in the system and identifying the region of overlap.

Graphing the Solutions of a System of Linear Inequalities

STEP 1. Graph each inequality in the system on the same set of axes.

STEP 2. The solutions of the system are the points common to the graphs of all the inequalities in the system.

EXAMPLE 1 Graph the solutions of the system: $\begin{cases} 3x \geq y \\ x + 2y \leq 8 \end{cases}$

Solution We begin by graphing each inequality on the *same* set of axes. The graph of the solutions of the system is the region contained in the graphs of both inequalities. In other words, it is their intersection.

First let's graph $3x \geq y$. The boundary line is the graph of $3x = y$. We sketch a solid boundary line since the inequality $3x \geq y$ means $3x > y$ or $3x = y$. The test point $(1, 0)$ satisfies the inequality, so we shade the half-plane that includes $(1, 0)$.

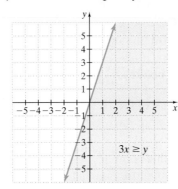

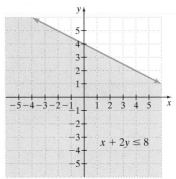

Next we sketch a solid boundary line $x + 2y = 8$ on the same set of axes. The test point $(0, 0)$ satisfies the inequality $x + 2y \leq 8$, so we shade the half-plane that includes $(0, 0)$. (For clarity, the graph of $x + 2y \leq 8$ is shown here on a separate set of axes.) An ordered pair solution of the system must satisfy both inequalities. These solutions are points that lie in both shaded regions. The solution of the system is the darkest shaded region. This solution includes parts of both boundary lines.

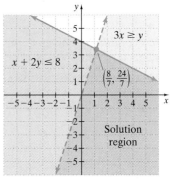

PRACTICE 1 Graph the solutions of the system: $\begin{cases} 4x \geq y \\ x + 3y \geq 6 \end{cases}$.

TECHNOLOGY NOTE

When solving systems of inequalities as in Example 1, some graphing utilities have APPS (installed or downloadable applications) that make it easier to see the shaded portion of the intersection of the graphs. One such App is Inequalz.

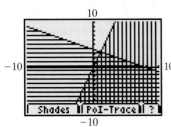

Example 1 using the Inequalz App.

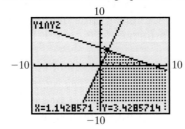

Finding the Point of Intersection (PoI).

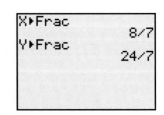

Point of Intersection Coordinates.

In linear programming, it is sometimes necessary to find the coordinates of the **corner point:** the point at which the two boundary lines intersect. To find the corner point for the system of Example 1, we solve the related linear system

$$\begin{cases} 3x = y \\ x + 2y = 8 \end{cases}$$

using either the substitution or the elimination method. The lines intersect at $\left(\dfrac{8}{7}, \dfrac{24}{7}\right)$, the corner point of the graph.

EXAMPLE 2 Graph the solutions of the system: $\begin{cases} x - y < 2 \\ x + 2y > -1 \\ y < 2 \end{cases}$

Solution First we graph all three inequalities on the same set of axes. All boundary lines are dashed lines since the inequality symbols are $<$ and $>$. The solution of the system is the region shown by the darkest shading. In this example, the boundary lines are *not* a part of the solution.

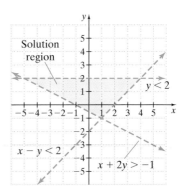

PRACTICE 2 Graph the solutions of the system: $\begin{cases} x - y < 1 \\ y < 4 \\ 3x + y > -3 \end{cases}$.

Concept Check ☑

Describe the solution of the system of inequalities: $\begin{cases} x \le 2 \\ x \ge 2 \end{cases}$

EXAMPLE 3 Graph the solutions of the system: $\begin{cases} -3x + 4y \le 12 \\ x \le 3 \\ x \ge 0 \\ y \ge 0 \end{cases}$

Solution We graph the inequalities on the same set of axes. The intersection of the inequalities is the solution region. It is the only region shaded in this graph and includes the portions of all four boundary lines that border the shaded region.

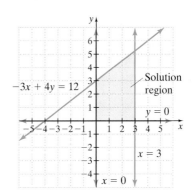

PRACTICE 3 Graph the solutions of the system: $\begin{cases} -2x + 5y \le 10 \\ x \le 4 \\ x \ge 0 \\ y \ge 0 \end{cases}$.

Answer to Concept Check:
the line $x = 2$

VOCABULARY & READINESS CHECK

Use the choices below to fill in each blank. Not all choices will be used.

solution   union   system
corner   intersection

1. Two or more linear inequalities form a _____ of linear inequalities.
2. An ordered pair that satisfies each inequality in a system is a _____ of the system.
3. The point where two boundary lines intersect is a _____ point.
4. The solution region of a system of inequalities consists of the _____ of the solution regions of the inequalities in the system.

4.5 | EXERCISE SET

MyMathLab   PRACTICE   WATCH   DOWNLOAD   READ   REVIEW

MIXED PRACTICE

Graph the solutions of each system of linear inequalities. See Examples 1 through 3.

1. $\begin{cases} y \geq x + 1 \\ y \geq 3 - x \end{cases}$

2. $\begin{cases} y \geq x - 3 \\ y \geq -1 - x \end{cases}$

3. $\begin{cases} y < 3x - 4 \\ y \leq x + 2 \end{cases}$

4. $\begin{cases} y \leq 2x + 1 \\ y > x + 2 \end{cases}$

5. $\begin{cases} y < -2x - 2 \\ y > x + 4 \end{cases}$

6. $\begin{cases} y \leq 2x + 4 \\ y \geq -x - 5 \end{cases}$

7. $\begin{cases} y \geq -x + 2 \\ y \leq 2x + 5 \end{cases}$

8. $\begin{cases} y \geq x - 5 \\ y \leq -3x + 3 \end{cases}$

9. $\begin{cases} x \geq 3y \\ x + 3y \leq 6 \end{cases}$

10. $\begin{cases} -2x < y \\ x + 2y < 3 \end{cases}$

11. $\begin{cases} x \leq 2 \\ y \geq -3 \end{cases}$

12. $\begin{cases} x \geq -3 \\ y \geq -2 \end{cases}$

13. $\begin{cases} y \geq 1 \\ x < -3 \end{cases}$

14. $\begin{cases} y > 2 \\ x \geq -1 \end{cases}$

15. $\begin{cases} y + 2x \geq 0 \\ 5x - 3y \leq 12 \\ y \leq 2 \end{cases}$

16. $\begin{cases} y + 2x \leq 0 \\ 5x + 3y \geq -2 \\ y \leq 4 \end{cases}$

17. $\begin{cases} 3x - 4y \geq -6 \\ 2x + y \leq 7 \\ y \geq -3 \end{cases}$

18. $\begin{cases} 4x - y \geq -2 \\ 2x + 3y \leq -8 \\ y \geq -5 \end{cases}$

19. $\begin{cases} 2x + y \leq 5 \\ x \leq 3 \\ x \geq 0 \\ y \geq 0 \end{cases}$

20. $\begin{cases} 3x + y \leq 4 \\ x \leq 4 \\ x \geq 0 \\ y \geq 0 \end{cases}$

Match each system of inequalities to the corresponding graph.

A
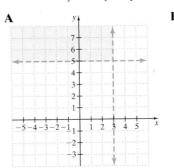

B
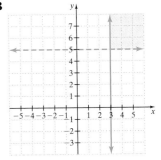

C
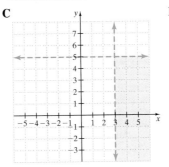

D
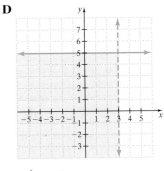

21. $\begin{cases} y < 5 \\ x > 3 \end{cases}$

22. $\begin{cases} y > 5 \\ x < 3 \end{cases}$

23. $\begin{cases} y \leq 5 \\ x < 3 \end{cases}$

24. $\begin{cases} y > 5 \\ x \geq 3 \end{cases}$

REVIEW

Evaluate each expression. See Section 1.3.

25. $(-3)^2$

26. $(-5)^3$

27. $\left(\dfrac{2}{3}\right)^2$

28. $\left(\dfrac{3}{4}\right)^3$

Perform each indicated operation. See Section 1.3.

29. $(-2)^2 - (-3) + 2(-1)$

30. $5^2 - 11 + 3(-5)$

31. $8^2 + (-13) - 4(-2)$

32. $(-12)^2 + (-1)(2) - 6$

CONCEPT EXTENSIONS

Solve. See the Concept Check in this section.

33. Describe the solution of the system: $\begin{cases} y \le 3 \\ y \ge 3 \end{cases}$.

34. Describe the solution of the system: $\begin{cases} x \le 5 \\ x \le 3 \end{cases}$.

35. Explain how to decide which region to shade to show the solution region of the following system.

$$\begin{cases} x \ge 3 \\ y \ge -2 \end{cases}$$

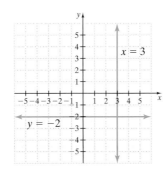

36. Tony Noellert budgets his time at work today. Part of the day he can write bills; the rest of the day he can use to write purchase orders. The total time available is at most 8 hours. Less than 3 hours is to be spent writing bills.

   a. Write a system of inequalities to describe the situation. (Let x = hours available for writing bills and y = hours available for writing purchase orders.)

   b. Graph the solutions of the system.

CHAPTER 4 GROUP ACTIVITY

Another Mathematical Model

Sometimes mathematical models other than linear models are appropriate for data. Suppose that an equation of the form $y = ax^2 + bx + c$ is an appropriate model for the ordered pairs $(x_1, y_1), (x_2, y_2)$, and (x_3, y_3). Then it is necessary to find the values of $a, b,$ and c such that the given ordered pairs are solutions of the equation $y = ax^2 + bx + c$. To do so, substitute each ordered pair into the equation. Each time, the result is an equation in three unknowns: $a, b,$ and c. Solving the resulting system of three linear equations in three unknowns will give the required values of $a, b,$ and c.

1. The table gives the total beef supply (in billions of pounds) in the United States in each of the years listed.

   a. Write the data as ordered pairs of the form (x, y), where y is the beef supply (in billions of pounds) in the year x ($x = 0$ represents 2000).

   b. Find the values of $a, b,$ and c such that the equation $y = ax^2 + bx + c$ models this data.

   c. Verify that the model you found in part **b** gives each of the ordered pair solutions from part **a.**

   d. According to the model, what was the U.S. beef supply in 2005?

Total U.S. Beef Supply	
Year	**Beef Supply (Billions of Pounds)**
2002	27
2004	24.6
2006	25

(*Source:* Economic Research Service, U.S. Department of Agriculture)

2. The table gives Toyota Hybrid sales figures for each of the years listed.

   a. Write the data as ordered pairs of the form (x, y), where y is sales in the year x ($x = 0$ represents 2000).

   b. Find the values of $a, b,$ and c such that the equation $y = ax^2 + bx + c$ models this data.

   c. According to the model, what were the total sales in 2005?

Total Toyota Hybrid Sales	
Year	**Sales in Thousands**
2002	41
2004	135
2006	313

(*Source:* Toyota Motor Corporation)

3. a. Make up an equation of the form $y = ax^2 + bx + c$.

   b. Find three ordered pair solutions of the equation.

   c. Without revealing your equation from part **a,** exchange lists of ordered pair solutions with another group.

   d. Use the method described above to find the values of $a, b,$ and c such that the equation $y = ax^2 + bx + c$ has the ordered pair solutions you received from the other group.

   e. Check with the other group to see if your equation from part **d** is the correct one.

CHAPTER 4 VOCABULARY CHECK

Fill in each blank with one of the words or phrases listed below.

matrix   consistent   system of equations
solution   inconsistent   square

1. Two or more linear equations in two variables form a _____.

2. A _____ of a system of two equations in two variables is an ordered pair that makes both equations true.

3. A(n) _____ system of equations has at least one solution.

4. If a matrix has the same number of rows and columns, it is called a _____ matrix.

5. A(n) _____ system of equations has no solution.

6. A _____ is a rectangular array of numbers.

> ▶ **Helpful Hint**
>
> Are you preparing for your test? Don't forget to take the Chapter 4 Test on page 321. Then check your answers at the back of the text and use the Chapter Test Prep Video CD to see the fully worked-out solutions to any of the exercises you want to review.

CHAPTER 4 HIGHLIGHTS

DEFINITIONS AND CONCEPTS	EXAMPLES

SECTION 4.1 SOLVING SYSTEMS OF LINEAR EQUATIONS IN TWO VARIABLES

A **system of linear equations** consists of two or more linear equations.

$$\begin{cases} x - 3y = 6 \\ y = \dfrac{1}{2}x \end{cases} \qquad \begin{cases} x + 2y - z = 1 \\ 3x - y + 4z = 0 \\ 5y + z = 6 \end{cases}$$

A **solution** of a system of two equations in two variables is an ordered pair (x, y) that makes both equations true.

Determine whether $(2, -5)$ is a solution of the system.

$$\begin{cases} x + y = -3 \\ 2x - 3y = 19 \end{cases}$$

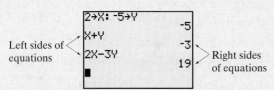

Left sides of equations

Right sides of equations

A numerical check that $(2, -5)$ is the solution of the system to the right.

Replace x with 2 and y with -5 in both equations.

$$x + y = -3 \qquad\qquad 2x - 3y = 19$$
$$2 + (-5) \stackrel{?}{=} -3 \qquad\qquad 2(2) - 3(-5) \stackrel{?}{=} 19$$
$$-3 = -3 \quad \text{True} \qquad\qquad 4 + 15 \stackrel{?}{=} 19$$
$$19 = 19 \quad \text{True}$$

$(2, -5)$ is a solution of the system.

Geometrically, a solution of a system in two variables is a point of intersection of the graphs of the equations.

Solve by graphing: $\begin{cases} y = 2x - 1 \\ x + 2y = 13 \end{cases}$

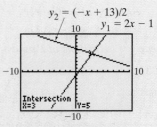

The solution is $(3, 5)$.

| **DEFINITIONS AND CONCEPTS** | **EXAMPLES** |

A system of equations with at least one solution is a **consistent system.** A system that has no solution is an **inconsistent system.**

If the graphs of two linear equations are identical, the equations are **dependent.**

If their graphs are different, the equations are **independent.**

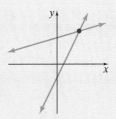

One solution:
Independent equations
Consistent system

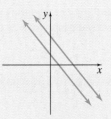

No solution:
Independent equations
Inconsistent system

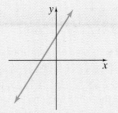

Infinite number of solutions:
Dependent equations
Consistent system

Solving a System of Linear Equations by the Substitution Method

Step 1. Solve one equation for a variable.

Step 2. Substitute the expression for the variable into the other equation.

Step 3. Solve the equation from Step 2 to find the value of one variable.

Step 4. Substitute the value from Step 3 in either original equation to find the value of the other variable.

Step 5. Check the solution in both equations.

Solve by substitution:

$$\begin{cases} y = x + 2 \\ 3x - 2y = -5 \end{cases}$$

Since the first equation is solved for y, substitute $x + 2$ for y in the second equation.

$$\begin{aligned} 3x - 2y &= -5 \quad \text{Second equation} \\ 3x - 2(x + 2) &= -5 \quad \text{Let } y = x + 2. \\ 3x - 2x - 4 &= -5 \\ x - 4 &= -5 \quad \text{Simplify.} \\ x &= -1 \quad \text{Add 4.} \end{aligned}$$

To find y, let $x = -1$ in $y = x + 2$, so $y = -1 + 2 = 1$. The solution $(-1, 1)$ checks.

Solving a System of Linear Equations by the Elimination Method

Step 1. Rewrite each equation in standard form, $Ax + By = C$.

Step 2. Multiply one or both equations by a nonzero number so that the coefficients of a variable are opposites.

Step 3. Add the equations.

Step 4. Find the value of the remaining variable by solving the resulting equation.

Step 5. Substitute the value from Step 4 into either original equation to find the value of the other variable.

Step 6. Check the solution in both equations.

Solve by elimination:

$$\begin{cases} x - 3y = -3 \\ -2x + y = 6 \end{cases}$$

Multiply both sides of the first equation by 2.

$$\begin{aligned} 2x - 6y &= -6 \\ \underline{-2x + y = 6} \\ -5y &= 0 \quad \text{Add.} \\ y &= 0 \quad \text{Divide by } -5. \end{aligned}$$

To find x, let $y = 0$ in an original equation.

$$\begin{aligned} x - 3y &= -3 \\ x - 3 \cdot 0 &= -3 \\ x &= -3 \end{aligned}$$

The solution $(-3, 0)$ checks.

| **DEFINITIONS AND CONCEPTS** | **EXAMPLES** |

A **solution** of an equation in three variables x, y, and z is an **ordered triple** (x, y, z) that makes the equation a true statement.

Verify that $(-2, 1, 3)$ is a solution of $2x + 3y - 2z = -7$. Replace x with -2, y with 1, and z with 3.

$$2(-2) + 3(1) - 2(3) \stackrel{?}{=} -7$$

$$-4 + 3 - 6 \stackrel{?}{=} -7$$

$$-7 = -7 \quad \text{True}$$

$(-2, 1, 3)$ is a solution.

Solving a System of Three Linear Equations by the Elimination Method

Step 1. Write each equation in standard form, $Ax + By + Cz = D$.

Step 2. Choose a pair of equations and use them to eliminate a variable.

Step 3. Choose any other pair of equations and eliminate the same variable.

Step 4. Solve the system of two equations in two variables from Steps 2 and 3.

Step 5. Solve for the third variable by substituting the values of the variables from Step 4 into any of the original equations.

Step 6. Check the solution in all three original equations.

Solve:

$$\begin{cases} 2x + y - z = 0 & (1) \\ x - y - 2z = -6 & (2) \\ -3x - 2y + 3z = -22 & (3) \end{cases}$$

1. Each equation is written in standard form.

2.
$$\begin{array}{ll} 2x + y - z = 0 & (1) \\ \underline{x - y - 2z = -6} & (2) \\ 3x \quad\quad - 3z = -6 & (4) \quad \text{Add.} \end{array}$$

3. Eliminate y from equations (1) and (3) also.

$$\begin{array}{ll} 4x + 2y - 2z = 0 & \text{Multiply equation} \\ \underline{-3x - 2y + 3z = -22} \;\; (3) & \text{(1) by 2.} \\ x \quad\quad + z = -22 & (5) \quad \text{Add.} \end{array}$$

4. Solve.

$$\begin{cases} 3x - 3z = -6 & (4) \\ x + z = -22 & (5) \end{cases}$$

$$\begin{array}{ll} x - z = -2 & \text{Divide equation (4) by 3.} \\ \underline{x + z = -22} & (5) \\ 2x \quad\quad = -24 \\ x \quad\quad = -12 \end{array}$$

To find z, use equation (5).

$$x + z = -22$$
$$-12 + z = -22$$
$$z = -10$$

5. To find y, use equation (1).

$$2x + y - z = 0$$
$$2(-12) + y - (-10) = 0$$
$$-24 + y + 10 = 0$$
$$y = 14$$

```
-12→X:14→Y: -10→Z
:2X+Y-Z
                    0
X-Y-2Z
                   -6
-3X-2Y+3Z
                  -22
■
```

A numerical check of the solution.

6. The solution $(-12, 14, -10)$ checks as shown to the left.

DEFINITIONS AND CONCEPTS	EXAMPLES

SECTION 4.3   SYSTEMS OF LINEAR EQUATIONS AND PROBLEM SOLVING

	Two numbers have a sum of 11. Twice one number is 3 less than 3 times the other. Find the numbers.
1. UNDERSTAND the problem.	**1.** Read and reread. x = one number y = other number
2. TRANSLATE.	**2.** In words: sum of numbers is 11 $\downarrow$ $\downarrow$ $\downarrow$ Translate: $x + y$ $=$ 11 In words: twice one number is 3 less than 3 times the other number $\downarrow$ $\downarrow$ $\downarrow$ Translate: $2x$ $=$ $3y - 3$
3. SOLVE.	**3.** Solve the system: $\begin{cases} x + y = 11 \\ 2x = 3y - 3 \end{cases}$ In the first equation, $x = 11 - y$. Substitute into the other equation. $$2x = 3y - 3$$ $$2(11 - y) = 3y - 3$$ $$22 - 2y = 3y - 3$$ $$-5y = -25$$ $$y = 5$$ Replace y with 5 in the equation $x = 11 - y$. Then $x = 11 - 5 = 6$. The solution is $(6, 5)$.
4. INTERPRET.	**4.** *Check:* See that $6 + 5 = 11$ is the required sum and that twice 6 is 3 times 5 less 3. *State:* The numbers are 6 and 5.

SECTION 4.4   SOLVING SYSTEMS OF EQUATIONS BY MATRICES

A **matrix** is a rectangular array of numbers.	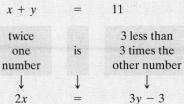
The **matrix** corresponding to a system is composed of the coefficients of the variables and the constants of the system.	The matrix corresponding to the system $\begin{cases} x - y = 1 \\ 2x + y = 11 \end{cases}$ is $\begin{bmatrix} 1 & -1 & \vdots & 1 \\ 2 & 1 & \vdots & 11 \end{bmatrix}$

(continued)

| **DEFINITIONS AND CONCEPTS** | **EXAMPLES** |

The following **row operations** can be performed on matrices, and the result is an equivalent matrix.

Elementary row operations:

1. Interchange any two rows.
2. Multiply (or divide) the elements of one row by the same nonzero number.
3. Multiply (or divide) the elements of one row by the same nonzero number and add them to their corresponding elements in any other row.

Use matrices to solve: $\begin{cases} x - y = 1 \\ 2x + y = 11 \end{cases}$.

The corresponding matrix is

$$\left[\begin{array}{cc:c} 1 & -1 & 1 \\ 2 & 1 & 11 \end{array}\right]$$

Use row operations to write an equivalent matrix with 1's along the diagonal and 0's below each 1 in the diagonal. Multiply row 1 by -2 and add to row 2. Change row 2 only.

$$\left[\begin{array}{cc:c} 1 & -1 & 1 \\ -2(1) + 2 & -2(-1) + 1 & -2(1) + 11 \end{array}\right]$$

simplifies to $\left[\begin{array}{cc:c} 1 & -1 & 1 \\ 0 & 3 & 9 \end{array}\right]$

Divide row 2 by 3.

$$\left[\begin{array}{cc:c} 1 & -1 & 1 \\ \dfrac{0}{3} & \dfrac{3}{3} & \dfrac{9}{3} \end{array}\right] \quad \text{simplifies to} \quad \left[\begin{array}{cc:c} 1 & -1 & 1 \\ 0 & 1 & 3 \end{array}\right]$$

This matrix corresponds to the system

$$\begin{cases} x - y = 1 \\ y = 3 \end{cases}$$

Let $y = 3$ in the first equation.

$$x - 3 = 1$$
$$x = 4$$

The ordered pair solution is $(4, 3)$.

Alternatively, you can use a calculator to enter and write the matrix in reduced now echelon form.

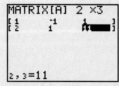
Enter matrix values.

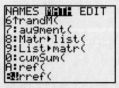
Under matrix math select reduced row echelon form (rref).

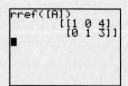
Finding the reduced row echelon menu.

DEFINITIONS AND CONCEPTS	EXAMPLES

SECTION 4.5   SYSTEMS OF LINEAR INEQUALITIES

A **system of linear inequalities** consists of two or more linear inequalities.

To graph a system of inequalities, graph each inequality in the system. The overlapping region is the solution of the system.

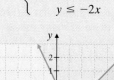

$$\begin{cases} x - y \geq 3 \\ \quad y \leq -2x \end{cases}$$

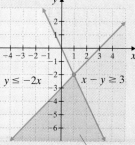

Solution region

CHAPTER 4 REVIEW

(4.1) *Solve each system of equations in two variables by each method: (a) graphing, (b) substitution, and (c) elimination.*

1. $\begin{cases} 3x + 10y = 1 \\ x + 2y = -1 \end{cases}$

2. $\begin{cases} y = \dfrac{1}{2}x + \dfrac{2}{3} \\ 4x + 6y = 4 \end{cases}$

3. $\begin{cases} 2x - 4y = 22 \\ 5x - 10y = 15 \end{cases}$

4. $\begin{cases} 3x - 6y = 12 \\ 2y = x - 4 \end{cases}$

5. $\begin{cases} \dfrac{1}{2}x - \dfrac{3}{4}y = -\dfrac{1}{2} \\ \dfrac{1}{8}x + \dfrac{3}{4}y = \dfrac{19}{8} \end{cases}$

6. The revenue equation for a certain style of backpack is $y = 32x$, where x is the number of backpacks sold and y is the income in dollars for selling x backpacks. The cost equation for these units is $y = 15x + 25{,}500$, where x is the number of backpacks manufactured and y is the cost in dollars for manufacturing x backpacks. Find the number of units to be sold for the company to break even. (*Hint:* Solve the system of equations formed by the two given equations.)

(4.2) *Solve each system of equations in three variables.*

7. $\begin{cases} x \quad\;\; + z = 4 \\ 2x - y \quad\;\; = 4 \\ x + y - z = 0 \end{cases}$

8. $\begin{cases} 2x + 5y \quad\;\; = 4 \\ x - 5y + z = -1 \\ 4x \quad\quad - z = 11 \end{cases}$

9. $\begin{cases} \quad\;\; 4y + 2z = 5 \\ 2x + 8y \quad\;\; = 5 \\ 6x + \quad\;\; 4z = 1 \end{cases}$

10. $\begin{cases} 5x + 7y \quad\quad = 9 \\ \quad\;\; 14y - z = 28 \\ 4x \quad\quad\;\; + 2z = -4 \end{cases}$

11. $\begin{cases} 3x - 2y + 2z = 5 \\ -x + 6y + z = 4 \\ 3x + 14y + 7z = 20 \end{cases}$

12. $\begin{cases} x + 2y + 3z = 11 \\ \quad\;\; y + 2z = 3 \\ 2x \quad\quad + 2z = 10 \end{cases}$

13. $\begin{cases} 7x - 3y + 2z = 0 \\ 4x - 4y - z = 2 \\ 5x + 2y + 3z = 1 \end{cases}$

14. $\begin{cases} x - 3y - 5z = -5 \\ 4x - 2y + 3z = 13 \\ 5x + 3y + 4z = 22 \end{cases}$

(4.3) *Use systems of equations to solve.*

15. The sum of three numbers is 98. The sum of the first and second is two more than the third number, and the second is four times the first. Find the numbers.

16. One number is three times a second number, and twice the sum of the numbers is 168. Find the numbers.

17. Two cars leave Chicago, one traveling east and the other west. After 4 hours they are 492 miles apart. If one car is traveling 7 mph faster than the other, find the speed of each.

18. The foundation for a rectangular Hardware Warehouse has a length three times the width and is 296 feet around. Find the dimensions of the building.

19. James Callahan has available a 10% alcohol solution and a 60% alcohol solution. Find how many liters of each solution he should mix to make 50 liters of a 40% alcohol solution.

20. An employee at See's Candy Store needs a special mixture of candy. She has creme-filled chocolates that sell for $3.00 per pound, chocolate-covered nuts that sell for $2.70 per pound, and chocolate-covered raisins that sell for $2.25 per pound. She wants to have twice as many raisins as nuts in the mixture. Find how many pounds of each she should use to make 45 pounds worth $2.80 per pound.

21. Chris Kringler has $2.77 in her coin jar—all in pennies, nickels, and dimes. If she has 53 coins in all and four more nickels than dimes, find how many of each type of coin she has.

22. If $10,000 and $4000 are invested such that $1250 in interest is earned in one year, and if the rate of interest on the larger investment is 2% more than that on the smaller investment, find the rates of interest.

23. The perimeter of an isosceles (two sides equal) triangle is 73 centimeters. If the unequal side is 7 centimeters longer than the two equal sides, find the lengths of the three sides.

24. The sum of three numbers is 295. One number is five more than a second and twice the third. Find the numbers.

(4.4) *Use matrices to solve each system.*

25. $\begin{cases} 3x + 10y = 1 \\ x + 2y = -1 \end{cases}$

26. $\begin{cases} 3x - 6y = 12 \\ 2y = x - 4 \end{cases}$

27. $\begin{cases} 3x - 2y = -8 \\ 6x + 5y = 11 \end{cases}$

28. $\begin{cases} 6x - 6y = -5 \\ 10x - 2y = 1 \end{cases}$

29. $\begin{cases} 3x - 6y = 0 \\ 2x + 4y = 5 \end{cases}$

30. $\begin{cases} 5x - 3y = 10 \\ -2x + y = -1 \end{cases}$

31. $\begin{cases} 0.2x - 0.3y = -0.7 \\ 0.5x + 0.3y = 1.4 \end{cases}$

32. $\begin{cases} 3x + 2y = 8 \\ 3x - y = 5 \end{cases}$

33. $\begin{cases} x + z = 4 \\ 2x - y = 0 \\ x + y - z = 0 \end{cases}$

34. $\begin{cases} 2x + 5y = 4 \\ x - 5y + z = -1 \\ 4x - z = 11 \end{cases}$

35. $\begin{cases} 3x - y = 11 \\ x + 2z = 13 \\ y - z = -7 \end{cases}$

36. $\begin{cases} 5x + 7y + 3z = 9 \\ 14y - z = 28 \\ 4x + 2z = -4 \end{cases}$

37. $\begin{cases} 7x - 3y + 2z = 0 \\ 4x - 4y - z = 2 \\ 5x + 2y + 3z = 1 \end{cases}$

38. $\begin{cases} x + 2y + 3z = 14 \\ y + 2z = 3 \\ 2x - 2z = 10 \end{cases}$

(4.5) *Graph the solution of each system of linear inequalities.*

39. $\begin{cases} y \ge 2x - 3 \\ y \le -2x + 1 \end{cases}$

40. $\begin{cases} y \le -3x - 3 \\ y \le 2x + 7 \end{cases}$

41. $\begin{cases} x + 2y > 0 \\ x - y \le 6 \end{cases}$

42. $\begin{cases} x - 2y \ge 7 \\ x + y \le -5 \end{cases}$

43. $\begin{cases} 3x - 2y \le 4 \\ 2x + y \ge 5 \\ y \le 4 \end{cases}$

44. $\begin{cases} 4x - y \le 0 \\ 3x - 2y \ge -5 \\ y \ge -4 \end{cases}$

45. $\begin{cases} x + 2y \le 5 \\ x \le 2 \\ x \ge 0 \\ y \ge 0 \end{cases}$

46. $\begin{cases} x + 3y \le 7 \\ y \le 5 \\ x \ge 0 \\ y \ge 0 \end{cases}$

MIXED REVIEW

Solve each system.

47. $\begin{cases} y = x - 5 \\ y = -2x + 2 \end{cases}$

48. $\begin{cases} \dfrac{2}{5}x + \dfrac{3}{4}y = 1 \\ x + 3y = -2 \end{cases}$

49. $\begin{cases} 5x - 2y = 10 \\ x = \dfrac{2}{5}y + 2 \end{cases}$

50. $\begin{cases} x - 4y = 4 \\ \dfrac{1}{8}x - \dfrac{1}{2}y = 3 \end{cases}$

51. $\begin{cases} x - 3y + 2z = 0 \\ 9y - z = 22 \\ 5x + 3z = 10 \end{cases}$

52. One number is five less than three times a second number. If the sum of the numbers is 127, find the numbers.

53. The perimeter of a triangle is 126 units. The length of one side is twice the length of the shortest side. The length of the third side is fourteen more than the length of the shortest side. Find the lengths of the sides of the triangles.

54. Graph the solution of the system: $\begin{cases} y \le 3x - \dfrac{1}{2} \\ 3x + 4y \ge 6 \end{cases}$.

55. In the United States, the consumer spending on VCR decks is decreasing while the spending on DVD players is increasing. For the years 1998–2003, the function $y = -443x + 2584$ estimates the millions of dollars spent on purchasing VCR decks while the function $y = 500x + 551$ estimates the millions of dollars spent on purchasing DVD players. For both functions, x is the number of years since 1998. Use these equations to determine the year in which the amount of money spent on VCR decks equals the amount of money spent on DVD players. (*Source:* Consumer Electronics Association)

CHAPTER 4 TEST

TEST PREP VIDEO — Remember to use the Chapter Test Prep Video CD to see the fully worked-out solutions to any of the exercises you want to review.

Solve each system of equations graphically and then solve by the elimination method or the substitution method.

1. $\begin{cases} 2x - y = -1 \\ 5x + 4y = 17 \end{cases}$

2. $\begin{cases} 7x - 14y = 5 \\ x = 2y \end{cases}$

Solve each system.

3. $\begin{cases} 4x - 7y = 29 \\ 2x + 5y = -11 \end{cases}$

4. $\begin{cases} 15x + 6y = 15 \\ 10x + 4y = 10 \end{cases}$

5. $\begin{cases} 2x - 3y = 4 \\ 3y + 2z = 2 \\ x - z = -5 \end{cases}$

6. $\begin{cases} 3x - 2y - z = -1 \\ 2x - 2y = 4 \\ 2x - 2z = -12 \end{cases}$

7. $\begin{cases} \dfrac{x}{2} + \dfrac{y}{4} = -\dfrac{3}{4} \\ x + \dfrac{3}{4}y = -4 \end{cases}$

Use matrices to solve each system.

8. $\begin{cases} x - y = -2 \\ 3x - 3y = -6 \end{cases}$

9. $\begin{cases} x + 2y = -1 \\ 2x + 5y = -5 \end{cases}$

10. $\begin{cases} x - y - z = 0 \\ 3x - y - 5z = -2 \\ 2x + 3y = -5 \end{cases}$

11. A motel in New Orleans charges $90 per day for double occupancy and $80 per day for single occupancy. If 80 rooms are occupied for a total of $6930, how many rooms of each kind are occupied?

12. The research department of a company that manufactures children's fruit drinks is experimenting with a new flavor. A 17.5% fructose solution is needed, but only 10% and 20% solutions are available. How many gallons of a 10% fructose solution should be mixed with a 20% fructose solution to obtain 20 gallons of a 17.5% fructose solution?

13. A company that manufactures boxes recently purchased $2000 worth of new equipment to offer gift boxes to its customers. The cost of producing a package of gift boxes is $1.50 and it is sold for $4.00. Find the number of packages that must be sold for the company to break even.

14. The measure of the largest angle of a triangle is 3 less than five times the measure of the smallest angle. The measure of the remaining angle is 1 less than twice the measure of the smallest angle. Find the measure of each angle.

Graph the solutions of the system of linear inequalities.

15. $\begin{cases} 2y - x \geq 1 \\ x + y \geq -4 \\ y \leq 2 \end{cases}$

CHAPTER 4 CUMULATIVE REVIEW

1. Determine whether each statement is true or false.

  a. $3 \in \{x \mid x \text{ is a natural number}\}$

  b. $7 \notin \{1, 2, 3\}$

2. Determine whether each statement is true or false.

  a. $\{0, 7\} \subseteq \{0, 2, 4, 6, 8\}$

  b. $\{1, 3, 5\} \subseteq \{1, 3, 5, 7\}$

3. Simplify the following expressions.

  a. $11 + 2 - 7$

  b. $-5 - 4 + 2$

4. Subtract.

  a. $-7 - (-2)$

  b. $14 - 38$

5. Write the additive inverse, or opposite, of each.

  a. 4

  b. $\dfrac{3}{7}$

  c. -11.2

6. Write the reciprocal of each.

  a. 5

  b. $-\dfrac{2}{3}$

7. Use the distributive property to multiply.

  a. $3(2x + y)$

  b. $-(3x - 1)$

  c. $0.7a(b - 2)$

8. Multiply.

  a. $7(3x - 2y + 4)$

  b. $-(-2s - 3t)$

9. Use the distributive property to simplify each expression.

  a. $3x - 5x + 4$

  b. $7yz + yz$

  c. $4z + 6.1$

10. Simplify.

  a. $5y^2 - 1 + 2(y^2 + 2)$

  b. $(7.8x - 1.2) - (5.6x - 2.4)$

Solve.

11. $-4x - 1 + 5x = 9x + 3 - 7x$

12. $8y - 14 = 6y - 14$

13. $0.3x + 0.1 = 0.27x - 0.02$

14. $2(m - 6) - m = 4(m - 3) - 3m$

15. A pennant in the shape of an isosceles triangle is to be constructed for the Slidell High School Athletic Club and sold at a fund-raiser. The company manufacturing the pennant charges according to perimeter, and the athletic club has determined that a perimeter of 149 centimeters should make a nice profit. If each equal side of the triangle is twice the length of the third side, increased by 12 centimeters, find the lengths of the sides of the triangular pennant.

16. A quadrilateral has 4 angles whose sum is $360°$. In a particular quadrilateral, two angles have the same measure. A third angle is $10°$ more than the measure of one of the equal angles, and the fourth angle is half the measure of one of the equal angles. Find the measures of the angles.

17. Solve: $3x + 4 \geq 2x - 6$. Graph the solution set.

18. Solve: $5(2x - 1) > -5$

19. Solve: $2 < 4 - x < 7$

20. Solve: $-1 < \dfrac{-2x - 1}{3} < 1$

21. Solve: $|2x| + 5 = 7$

22. Solve: $|x - 5| = 4$

23. Solve for m: $|m - 6| < 2$

24. $|2x + 1| > 5$

25. Plot each ordered pair on a Cartesian coordinate system and name the quadrant or axis in which the point is located.

   a. $(2, -1)$ **b.** $(0, 5)$ **c.** $(-3, 5)$

   d. $(-2, 0)$ **e.** $\left(-\dfrac{1}{2}, -4\right)$ **f.** $(1.5, 1.5)$

26. Name the quadrant or axis in which each point is located.

   a. $(-1, -5)$

   b. $(4, -2)$

   c. $(0, 2)$

27. Is the relation $x = y^2$ also a function?

28. Graph: $-2x + \dfrac{1}{2}y = -2$

29. If $f(x) = 7x^2 - 3x + 1, g(x) = 3x - 2$, and $h(x) = x^2$, find the following.

   a. $f(1)$ **b.** $g(3)$

   c. $h(-2)$

30. If $f(x) = 3x^2$, find the following.

   a. $f(5)$ **b.** $f(-2)$

31. Graph $g(x) = 2x + 1$. Compare this graph with the graph of $f(x) = 2x$.

32. Find the slope of the line containing $(-2, 6)$ and $(0, 9)$.

33. Find the slope and the y-intercept of the line $3x - 4y = 4$.

34. Find the slope and y-intercept of the line defined by $y = 2$.

35. Are the following pairs of lines parallel, perpendicular, or neither?

   a. $3x + 7y = 21$
        $6x + 14y = 7$

   b. $-x + 3y = 2$
        $2x + 6y = 5$

   c. $2x - 3y = 12$
        $6x + 4y = 16$

36. Find an equation of the line through $(0, -9)$ with slope $\dfrac{1}{5}$.

37. Find an equation of the line through points $(4, 0)$ and $(-4, -5)$. Write the equation using function notation.

38. Find an equation of the line through $(-2, 6)$ perpendicular to $f(x) = \dfrac{1}{2}x - \dfrac{1}{3}$.

39. Graph $3x \geq y$.

40. Graph: $x \geq 1$.

41. Determine whether the given ordered pair is a solution of the system.

   a. $\begin{cases} -x + y = 2 \\ 2x - y = -3 \end{cases}$ $(-1, 1)$

   b. $\begin{cases} 5x + 3y = -1 \\ x - y = 1 \end{cases}$ $(-2, 3)$

42. Solve the system:

$$\begin{cases} 5x + y = -2 \\ 4x - 2y = -10 \end{cases}$$

43. Solve the system.

$$\begin{cases} 3x - y + z = -15 \\ x + 2y - z = 1 \\ 2x + 3y - 2z = 0 \end{cases}$$

44. Solve the system:

$$\begin{cases} x - 2y + z = 0 \\ 3x - y - 2z = -15 \\ 2x - 3y + 3z = 7 \end{cases}$$

45. Use matrices to solve the system.

$$\begin{cases} x + 3y = 5 \\ 2x - y = -4 \end{cases}$$

46. Solve the system:

$$\begin{cases} -6x + 8y = 0 \\ 9x - 12y = 2 \end{cases}$$

5 Exponents, Polynomials, and Polynomial Functions

To remember the Great Lakes, remember the word "**HOMES.**"

H = Huron
O = Ontario
M = Michigan
E = Erie
S = Superior

The Great Lakes are a group of five lakes in North America on or near the United States-Canada border. They were formed at the end of the last ice age, about 10,000 years ago.

The Great Lakes are the largest freshwater lake group on this Earth. They contain about 84% of North America's surface fresh water and 21% of the world's supply. The total surface area of the Great Lakes is about the size of Wyoming. Lake Superior alone is larger than the state of South Carolina. In Section 5.2, Exercise 99, page 339, we will explore the mass of the water in Lake Superior.

Linear equations are important for solving problems. They are not sufficient, however, to solve all problems. Many real-world phenomena are modeled by polynomials. We begin this chapter by reviewing exponents. We will then study operations on polynomials and how polynomials can be used in problem solving.

The volume of water (in cubic feet) in Lake Superior, given in scientific notation.

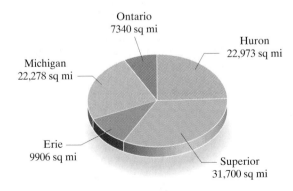

Surface Areas of Great Lakes

Ontario 7340 sq mi
Huron 22,973 sq mi
Michigan 22,278 sq mi
Erie 9906 sq mi
Superior 31,700 sq mi

5.1 EXPONENTS AND SCIENTIFIC NOTATION

OBJECTIVES

1 Use the product rule for exponents.

2 Evaluate expressions raised to the 0 power.

3 Use the quotient rule for exponents.

4 Evaluate expressions raised to the negative nth power.

5 Convert between scientific notation and standard notation.

OBJECTIVE 1 ▶ Using the product rule. Recall that exponents may be used to write repeated factors in a more compact form. As we have seen in the previous chapters, exponents can be used when the repeated factor is a number or a variable. For example,

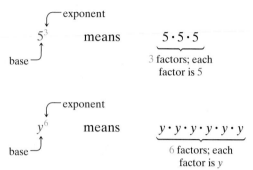

Expressions such as 5^3 and y^6 that contain exponents are called **exponential expressions.**

Exponential expressions can be multiplied, divided, added, subtracted, and themselves raised to powers. In this section, we review operations on exponential expressions.

DISCOVER THE CONCEPT

a. Use a calculator to evaluate each pair of expressions. Then compare the results.

$$2^5 \cdot 2^4 \text{ and } 2^9$$
$$(3.7)^2 \cdot (3.7)^3 \text{ and } (3.7)^5$$
$$(-5)^4 \cdot (-5)^2 \text{ and } (-5)^6$$

b. From the results of part **a**, write a pattern you observed for multiplying exponential expressions with the same base.

c. Use your findings to simplify the product

$$x^2 \cdot x^3$$

In part **a** in the box above, we see that the expressions in each pair are equal since they simplify to the same number, as shown in the screens below. It appears that the product of exponential expressions with the same base is the base raised to the sum of the exponents. From this, we reason that

$$x^2 \cdot x^3 = x^{2+3} = x^5$$

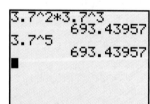
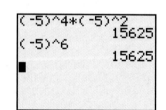

To check, use the definition of a^n:

$$x^2 \cdot x^3 = \underbrace{(x \cdot x)(x \cdot x \cdot x)}_{x \text{ is a factor } 5 \text{ times}} = x^5$$

This suggests the following.

Product Rule for Exponents

If m and n are positive integers and a is a real number, then

$$a^m \cdot a^n = a^{m+n}$$

In other words, the *product* of exponential expressions with a common base is the common base raised to a power equal to the *sum* of the exponents of the factors.

EXAMPLE 1 Use the product rule to simplify.

a. $2^2 \cdot 2^5$ **b.** $x^7 x^3$ **c.** $y \cdot y^2 \cdot y^4$

Solution

a. $2^2 \cdot 2^5 = 2^{2+5} = 2^7$

b. $x^7 x^3 = x^{7+3} = x^{10}$

c. $y \cdot y^2 \cdot y^4 = (y^1 \cdot y^2) \cdot y^4$
$$= y^3 \cdot y^4$$
$$= y^7$$

PRACTICE

1 Use the product rule to simplify.

a. $3^4 \cdot 3^2$ **b.** $x^5 \cdot x^2$ **c.** $y \cdot y^3 \cdot y^5$

EXAMPLE 2 Use the product rule to simplify.

a. $(3x^6)(5x)$ **b.** $(-2.4x^3 p^2)(4xp^{10})$

Solution Here, we use properties of multiplication to group together like bases.

a. $(3x^6)(5x) = 3(5)x^6 x^1 = 15x^7$

b. $(-2.4x^3 p^2)(4xp^{10}) = -2.4(4)x^3 x^1 p^2 p^{10} = -9.6x^4 p^{12}$

PRACTICE

2 Use the product rule to simplify.

a. $(5z^3)(7z)$ **b.** $(-4.1t^5 q^3)(5tq^5)$

OBJECTIVE 2 ▶ Evaluating expressions raised to the 0 power. The definition of a^n does not include the possibility that n might be 0. But if it did, then, by the product rule,

$$\underbrace{a^0 \cdot a^n}_{} = a^{0+n} = a^n = \underbrace{1 \cdot a^n}_{}$$

From this, we reasonably define that $a^0 = 1$, as long as a does not equal 0.

> **Zero Exponent**
> If a does not equal 0, then $a^0 = 1$.

EXAMPLE 3 Evaluate the following.

a. 7^0 **b.** -7^0 **c.** $(2x + 5)^0$ **d.** $2x^0$

Solution

a. $7^0 = 1$

b. Without parentheses, only 7 is raised to the 0 power.
$$-7^0 = -(7^0) = -(1) = -1$$

c. $(2x + 5)^0 = 1$

d. $2x^0 = 2(1) = 2$

PRACTICE

3 Evaluate the following.

a. 5^0 **b.** -5^0 **c.** $(3x - 8)^0$ **d.** $3x^0$

A calculator check of Example 3b, c, d.

OBJECTIVE 3 ▶ Using the quotient rule. Next, we discover a pattern that occurs when dividing exponential expressions with a common base.

DISCOVER THE CONCEPT

a. Use a calculator to evaluate each pair of expressions. Then compare the results.

$$\frac{7^5}{7^2} \text{ and } 7^3$$

$$\frac{(-3)^9}{(-3)^4} \text{ and } (-3)^5$$

b. From the results of part **a**, write a pattern you observed for dividing exponential expressions with the same base.

c. Use your findings to simplify the quotient $\frac{x^9}{x^2}$.

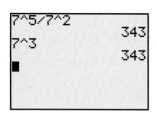 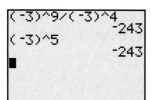

In part **a** in the box above, we see that each pair of expressions is equal since they simplify to the same number as shown in the screens to the left. It appears that the quotient of exponential expressions with the same base is the base raised to the difference of the exponents. From this, we reason that

$$\frac{x^9}{x^2} = x^{9-2} = x^7$$

To check, we begin with the definition of a^n to simplify $\frac{x^9}{x^2}$.

$$\frac{x^9}{x^2} = \frac{x \cdot x \cdot x \cdot x \cdot x \cdot x \cdot x \cdot x \cdot x}{x \cdot x} = x^7$$

(Assume for the next two sections that denominators containing variables are not 0.) Notice that the result is exactly the same if we subtract the exponents.

$$\frac{x^9}{x^2} = x^{9-2} = x^7$$

This suggests the following.

> **Quotient Rule for Exponents**
>
> If a is a nonzero real number and n and m are integers, then
>
> $$\frac{a^m}{a^n} = a^{m-n}$$

In other words, the *quotient* of exponential expressions with a common base is the common base raised to a power equal to the *difference* of the exponents.

EXAMPLE 4 Use the quotient rule to simplify.

a. $\dfrac{x^7}{x^4}$ **b.** $\dfrac{5^8}{5^2}$ **c.** $\dfrac{20x^6}{4x^5}$ **d.** $\dfrac{12y^{10}z^7}{14y^8z^7}$

Solution

a. $\dfrac{x^7}{x^4} = x^{7-4} = x^3$

b. $\dfrac{5^8}{5^2} = 5^{8-2} = 5^6$

c. $\dfrac{20x^6}{4x^5} = 5x^{6-5} = 5x^1,$ or $5x$

d. $\dfrac{12y^{10}z^7}{14y^8z^7} = \dfrac{6}{7}y^{10-8} \cdot z^{7-7} = \dfrac{6}{7}y^2z^0 = \dfrac{6}{7}y^2,$ or $\dfrac{6y^2}{7}$

PRACTICE
4 Use the quotient rule to simplify.

a. $\dfrac{z^8}{z^3}$ **b.** $\dfrac{3^9}{3^3}$ **c.** $\dfrac{45x^7}{5x^3}$ **d.** $\dfrac{24a^{14}b^6}{18a^7b^6}$

OBJECTIVE 4 ▶ Evaluating exponents raised to the negative *n*th power. When the exponent of the denominator is larger than the exponent of the numerator, applying the quotient rule yields a negative exponent. For example,

$$\frac{x^3}{x^5} = x^{3-5} = x^{-2}$$

Using the definition of a^n, though, gives us

$$\frac{x^3}{x^5} = \frac{x \cdot x \cdot x}{x \cdot x \cdot x \cdot x \cdot x} = \frac{1}{x^2}$$

From this, we reasonably define $x^{-2} = \dfrac{1}{x^2}$ or, in general, $a^{-n} = \dfrac{1}{a^n}$.

> **Negative Exponents**
> If a is a real number other than 0 and n is a positive integer, then
> $$a^{-n} = \frac{1}{a^n}$$

EXAMPLE 5 Simplify and write with positive exponents only.

a. 5^{-2} **b.** $(-4)^{-4}$ **c.** $2x^{-3}$ **d.** $(3x)^{-1}$

e. $\dfrac{m^5}{m^{15}}$ **f.** $\dfrac{3^3}{3^6}$ **g.** $2^{-1} + 3^{-2}$ **h.** $\dfrac{1}{t^{-5}}$

Solution

a. $5^{-2} = \dfrac{1}{5^2} = \dfrac{1}{25}$

b. $(-4)^{-4} = \dfrac{1}{(-4)^4} = \dfrac{1}{256}$

c. $2x^{-3} = 2 \cdot \dfrac{1}{x^3} = \dfrac{2}{x^3}$ Without parentheses, only x is raised to the -3 power.

d. $(3x)^{-1} = \dfrac{1}{(3x)^1} = \dfrac{1}{3x}$ With parentheses, both 3 and x are raised to the -1 power.

e. $\dfrac{m^5}{m^{15}} = m^{5-15} = m^{-10} = \dfrac{1}{m^{10}}$

f. $\dfrac{3^3}{3^6} = 3^{3-6} = 3^{-3} = \dfrac{1}{3^3} = \dfrac{1}{27}$

g. $2^{-1} + 3^{-2} = \dfrac{1}{2^1} + \dfrac{1}{3^2} = \dfrac{1}{2} + \dfrac{1}{9} = \dfrac{9}{18} + \dfrac{2}{18} = \dfrac{11}{18}$

h. $\dfrac{1}{t^{-5}} = \dfrac{1}{\dfrac{1}{t^5}} = 1 \div \dfrac{1}{t^5} = 1 \cdot \dfrac{t^5}{1} = t^5$

```
5^-2▸Frac
            1/25
2⁻1+3^-2▸Frac
            11/18
■
```
A calculator check for
Example 5a and g.

5 Simplify and write with positive exponents only.

a. 6^{-2} **b.** $(-2)^{-6}$ **c.** $3x^{-5}$ **d.** $(5y)^{-1}$ **e.** $\dfrac{k^4}{k^{11}}$

f. $\dfrac{5^3}{5^5}$ **g.** $5^{-1} + 2^{-2}$ **h.** $\dfrac{1}{z^{-8}}$

> **▶ Helpful Hint**
>
> Notice that when a factor containing an exponent is moved from the numerator to the denominator or from the denominator to the numerator, the sign of its exponent changes.
>
> $$x^{-3} = \frac{1}{x^3}, \qquad 5^{-2} = \frac{1}{5^2} = \frac{1}{25}$$
>
> $$\frac{1}{y^{-4}} = y^4, \qquad \frac{1}{2^{-3}} = 2^3 = 8$$

EXAMPLE 6   Simplify and write with positive exponents only.

a. $\dfrac{x^{-9}}{x^2}$ **b.** $\dfrac{5p^4}{p^{-3}}$ **c.** $\dfrac{2^{-3}}{2^{-1}}$ **d.** $\dfrac{2x^{-7}y^2}{10xy^{-5}}$ **e.** $\dfrac{(3x^{-3})(x^2)}{x^6}$

Solution

a. $\dfrac{x^{-9}}{x^2} = x^{-9-2} = x^{-11} = \dfrac{1}{x^{11}}$ **b.** $\dfrac{5p^4}{p^{-3}} = 5 \cdot p^{4-(-3)} = 5p^7$

c. $\dfrac{2^{-3}}{2^{-1}} = 2^{-3-(-1)} = 2^{-2} = \dfrac{1}{2^2} = \dfrac{1}{4}$ **d.** $\dfrac{2x^{-7}y^2}{10xy^{-5}} = \dfrac{x^{-7-1} \cdot y^{2-(-5)}}{5} = \dfrac{x^{-8}y^7}{5} = \dfrac{y^7}{5x^8}$

e. Simplify the numerator first.

$$\frac{(3x^{-3})(x^2)}{x^6} = \frac{3x^{-3+2}}{x^6} = \frac{3x^{-1}}{x^6} = 3x^{-1-6} = 3x^{-7} = \frac{3}{x^7}$$ □

6 Simplify and write with positive exponents only.

a. $\dfrac{z^{-8}}{z^3}$ **b.** $\dfrac{7t^3}{t^{-5}}$ **c.** $\dfrac{3^{-2}}{3^{-4}}$ **d.** $\dfrac{5a^{-5}b^3}{15a^2b^{-4}}$ **e.** $\dfrac{(2x^{-5})(x^6)}{x^5}$

Concept Check ☑

Find and correct the error in the following:

$$\frac{y^{-6}}{y^{-2}} = y^{-6-2} = y^{-8} = \frac{1}{y^8}$$

EXAMPLE 7   Simplify. Assume that a and t are nonzero integers and that x is not 0.

a. $x^{2a} \cdot x^3$ **b.** $\dfrac{x^{2t-1}}{x^{t-5}}$

Solution

a. $x^{2a} \cdot x^3 = x^{2a+3}$ Use the product rule.

Answer to Concept Check:

$\dfrac{y^{-6}}{y^{-2}} = y^{-6-(-2)} = y^{-4} = \dfrac{1}{y^4}$

b. $\dfrac{x^{2t-1}}{x^{t-5}} = x^{(2t-1)-(t-5)}$ Use the quotient rule.

$= x^{2t-1-t+5} = x^{t+4}$ □

PRACTICE
7   Simplify. Assume that a and t are nonzero integers and that x is not 0.

a. $x^{3a} \cdot x^4$ **b.** $\dfrac{x^{3t-2}}{x^{t-3}}$

— TECHNOLOGY NOTE —

To determine the scientific notation capabilities of your graphing utility, consult your owner's manual.

OBJECTIVE 5 ▶ Converting between scientific notation and standard notation. Very large and very small numbers occur frequently in nature. For example, the distance between the Earth and the Sun is approximately 150,000,000 kilometers. A helium atom has a diameter of 0.000 000 022 centimeters. It can be tedious to write these very large and very small numbers in standard notation like this. **Scientific notation** is a convenient shorthand notation for writing very large and very small numbers.

Helium atom

0.000 000 022
centimeters

Scientific Notation

A positive number is written in **scientific notation** if it is written as the product of a number a, where $1 \leq a < 10$, and an integer power r of 10:

$$a \times 10^r$$

The following are examples of numbers written in scientific notation.

diameter of helium atom: 2.2×10^{-8} cm; 1.5×10^8 km ← approximate distance between Earth and Sun

---

Writing a Number in Scientific Notation

STEP 1. Move the decimal point in the original number until the new number has a value between 1 and 10.

STEP 2. Count the number of decimal places the decimal point was moved in Step 1. If the original number is 10 or greater, the count is positive. If the original number is less than 1, the count is negative.

STEP 3. Write the product of the new number in Step 1 by 10 raised to an exponent equal to the count found in Step 2.

EXAMPLE 8   Write each number in scientific notation.

a. 730,000   **b.** 0.00000104

Solution

a. STEP 1. Move the decimal point until the number is between 1 and 10.

730,000.

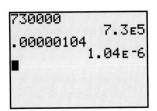

A calculator check for
Example 8.

STEP 2. The decimal point is moved 5 places and the original number is 10 or greater, so the count is positive 5.

STEP 3. $730,000 = 7.3 \times 10^5$.

b. STEP 1. Move the decimal point until the number is between 1 and 10.

$$0.00000104$$

STEP 2. The decimal point is moved 6 places and the original number is less than 1, so the count is -6.

STEP 3. $0.00000104 = 1.04 \times 10^{-6}$.

The screen to the left verifies the results of Example 8. This screen was generated with the calculator in scientific notation mode. ☐

PRACTICE
8 Write each number in scientific notation.

a. 65,000   **b.** 0.000038

To write a scientific notation number in standard form, we reverse the preceding steps.

> **Writing a Scientific Notation Number in Standard Notation**
> Move the decimal point in the number the same number of places as the exponent on 10. If the exponent is positive, move the decimal point to the right. If the exponent is negative, move the decimal point to the left.

EXAMPLE 9   Write each number in standard notation.

a. 7.7×10^8 **b.** 1.025×10^{-3}

Solution

a. $7.7 \times 10^8 = 770,000,000$ Since the exponent is positive, move the decimal point 8 places to the right. Add zeros as needed.

b. $1.025 \times 10^{-3} = 0.001025$ Since the exponent is negative, move the decimal point 3 places to the left. Add zeros as needed. ☐

PRACTICE
9 Write each number in standard notation.

a. 6.2×10^5 **b.** 3.109×10^{-2}

The calculator is in normal
mode for the first calculation and in scientific mode
for the second calculation.
See Example 9a.

Answers to Concept Check:

a, c, d

Concept Check ☑

Which of the following numbers have values that are less than 1?

a. 3.5×10^{-5} **b.** 3.5×10^5 **c.** -3.5×10^5 **d.** -3.5×10^{-5}

VOCABULARY & READINESS CHECK

State the base of the exponent 5 in each expression.

1. $9x^5$ **2.** yz^5 **3.** -3^5 **4.** $(-3)^5$ **5.** $(y^7)^5$ **6.** $9 \cdot 2^5$

Write each expression with positive exponents.

7. $5x^{-1}y^{-2}$ **8.** $7xy^{-4}$ **9.** $a^2b^{-1}c^{-5}$ **10.** $a^{-4}b^2c^{-6}$ **11.** $\dfrac{y^{-2}}{x^{-4}}$ **12.** $\dfrac{x^{-7}}{z^{-3}}$

5.1 | EXERCISE SET

Use the product rule to simplify each expression. See Examples 1 and 2.

1. $4^2 \cdot 4^3$

2. $3^3 \cdot 3^5$

3. $x^5 \cdot x^3$

4. $a^2 \cdot a^9$

5. $m \cdot m^7 \cdot m^6$

6. $n \cdot n^{10} \cdot n^{12}$

7. $(4xy)(-5x)$

8. $(-7xy)(7y)$

9. $(-4x^3p^2)(4y^3x^3)$

10. $(-6a^2b^3)(-3ab^3)$

Evaluate each expression. See Example 3.

11. -8^0

12. $(-9)^0$

13. $(4x + 5)^0$

14. $(3x - 1)^0$

15. $-x^0$

16. $-5x^0$

17. $4x^0 + 5$

18. $8x^0 + 1$

Use the quotient rule to simplify. See Example 4.

19. $\dfrac{a^5}{a^2}$

20. $\dfrac{x^9}{x^4}$

21. $-\dfrac{26z^{11}}{2z^7}$

22. $-\dfrac{16x^5}{8x}$

23. $\dfrac{x^9y^6}{x^8y^6}$

24. $\dfrac{a^{12}b^2}{a^9b}$

25. $\dfrac{12x^4y^7}{9xy^5}$

26. $\dfrac{24a^{10}b^{11}}{10ab^3}$

27. $\dfrac{-36a^5b^7c^{10}}{6ab^3c^4}$

28. $\dfrac{49a^3bc^{14}}{-7abc^8}$

Simplify and write using positive exponents only. See Examples 5 and 6.

29. 4^{-2}

30. 2^{-3}

31. $(-3)^{-3}$

32. $(-6)^{-2}$

33. $\dfrac{x^7}{x^{15}}$

34. $\dfrac{z}{z^3}$

35. $5a^{-4}$

36. $10b^{-1}$

37. $\dfrac{x^{-7}}{y^{-2}}$

38. $\dfrac{p^{-13}}{q^{-3}}$

39. $\dfrac{x^{-2}}{x^5}$

40. $\dfrac{z^{-12}}{z^{10}}$

41. $\dfrac{8r^4}{2r^{-4}}$

42. $\dfrac{3s^3}{15s^{-3}}$

43. $\dfrac{x^{-9}x^4}{x^{-5}}$

44. $\dfrac{y^{-7}y}{y^8}$

45. $\dfrac{2a^{-6}b^2}{18ab^{-5}}$

46. $\dfrac{18ab^{-6}}{3a^{-3}b^6}$

47. $\dfrac{(24x^8)(x)}{20x^{-7}}$

48. $\dfrac{(30z^2)(z^5)}{55z^{-4}}$

MIXED PRACTICE

Simplify and write using positive exponents only. See Examples 1 through 6.

49. $-7x^3 \cdot 20x^9$

50. $-3y \cdot -9y^4$

51. $x^7 \cdot x^8 \cdot x$

52. $y^6 \cdot y \cdot y^9$

53. $2x^3 \cdot 5x^7$

54. $-3z^4 \cdot 10z^7$

55. $(5x)^0 + 5x^0$

56. $4y^0 - (4y)^0$

57. $\dfrac{z^{12}}{z^{15}}$

58. $\dfrac{x^{11}}{x^{20}}$

59. $3^0 - 3t^0$

60. $4^0 + 4x^0$

61. $\dfrac{y^{-3}}{y^{-7}}$

62. $\dfrac{y^{-6}}{y^{-9}}$

63. $4^{-1} + 3^{-2}$

64. $1^{-3} - 4^{-2}$

65. $3x^{-1}$

66. $(4x)^{-1}$

67. $\dfrac{r^4}{r^{-4}}$

68. $\dfrac{x^{-5}}{x^3}$

69. $\dfrac{x^{-7}y^{-2}}{x^2y^2}$

70. $\dfrac{a^{-5}b^7}{a^{-2}b^{-3}}$

71. $(-4x^2y)(3x^4)(-2xy^5)$

72. $(-6a^4b)(2b^3)(-3ab^6)$

73. $2^{-4} \cdot x$

74. $5^{-2} \cdot y$

75. $\dfrac{5^{17}}{5^{13}}$

76. $\dfrac{10^{25}}{10^{23}}$

77. $\dfrac{8^{-7}}{8^{-6}}$

78. $\dfrac{13^{-10}}{13^{-9}}$

79. $\dfrac{9^{-5}a^4}{9^{-3}a^{-1}}$

80. $\dfrac{11^{-9}b^3}{11^{-7}b^{-4}}$

81. $\dfrac{14x^{-2}yz^{-4}}{2xyz}$

82. $\dfrac{30x^{-7}yz^{-14}}{3xyz}$

Simplify. Assume that variables in the exponents represent nonzero integers and that x, y, and z are not 0. See Example 7.

83. $x^5 \cdot x^{7a}$

84. $y^{2p} \cdot y^{9p}$

85. $\dfrac{x^{3t-1}}{x^t}$

86. $\dfrac{y^{4p-2}}{y^{3p}}$

87. $x^{4a} \cdot x^7$

88. $x^{9y} \cdot x^{-7y}$

89. $\dfrac{z^{6x}}{z^7}$

90. $\dfrac{y^6}{y^{4z}}$

91. $\dfrac{x^{3t} \cdot x^{4t-1}}{x^t}$

92. $\dfrac{z^{5x} \cdot z^{x-7}}{z^x}$

Write each number in scientific notation. See Example 8.

93. 31,250,000

94. 678,000

95. 0.016

96. 0.007613

97. 67,413

98. 36,800,000

99. 0.0125

100. 0.00084

101. 0.000053

102. 98,700,000,000

Write each number in scientific notation.

103. Total revenues for Wal-Mart in fiscal year ending January 2007 were $344,992,000,000. (*Source:* Wal-Mart Stores, Inc.)

104. The University of Texas system has more than 170,000 students statewide. (*Source:* University of Texas)

105. On a recent day, the Apple iTunes Store featured more than 3,500,000 songs to buy for $0.99 each.

106. In 2006, approximately 61,049,000 passengers passed through the Los Angeles International Airport. (*Source:* Los Angeles International Airport)

107. Lake Mead, created from the Colorado River by the Hoover Dam, has a capacity of 124,000,000,000 cubic feet of water. (*Source:* U.S. Bureau of Reclamation)

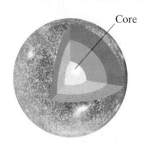

108. The temperature of the core of the sun is about 27,000,000°F.

Core

109. A pulsar is a rotating neutron star that gives off sharp, regular pulses of radio waves. For one particular pulsar, the rate of pulses is every 0.001 second.

△ **110.** To convert from cubic inches to cubic meters, multiply by 0.0000164.

Write each number in standard notation, without exponents. See Example 9.

111. 3.6×10^{-9}

112. 2.7×10^{-5}

113. 9.3×10^{7}

114. 6.378×10^{8}

115. 1.278×10^{6}

116. 7.6×10^{4}

117. 7.35×10^{12}

118. 1.66×10^{-5}

119. 4.03×10^{-7}

120. 8.007×10^{8}

Write each number in standard notation.

121. The estimated world population in 1 A.D. was 3.0×10^{8}. (*Source: World Almanac and Book of Facts*)

122. There are 3.949×10^{6} miles of highways, roads, and streets in the United States. (*Source:* Bureau of Transportation Statistics)

123. In 2006, teenagers had an estimated spending power of 1.53×10^{11} dollars. (*Source:* A. C. Neilsen Research)

124. Each day, an estimated 1.2×10^{9} beverages consumed throughout the world are Coca-Cola products. (*Source:* Coca-Cola)

REVIEW AND PREVIEW

Evaluate. See Sections 1.3 and 5.1.

125. $(5 \cdot 2)^{2}$

126. $5^{2} \cdot 2^{2}$

127. $\left(\dfrac{3}{4}\right)^{3}$

128. $\dfrac{3^{3}}{4^{3}}$

129. $(2^{3})^{2}$

130. $(2^{2})^{3}$

CONCEPT EXTENSIONS

131. Explain how to convert a number from standard notation to scientific notation.

132. Explain how to convert a number from scientific notation to standard notation.

133. Explain why $(-5)^{0}$ simplifies to 1 but -5^{0} simplifies to -1.

134. Explain why both $4x^{0} - 3y^{0}$ and $(4x - 3y)^{0}$ simplify to 1.

135. Simplify where possible.

   a. $x^{a} \cdot x^{a}$ **b.** $x^{a} + x^{a}$

   c. $\dfrac{x^{a}}{x^{b}}$ **d.** $x^{a} \cdot x^{b}$

   e. $x^{a} + x^{b}$

136. Which numbers are equal to 36,000? Of these, which is written in scientific notation?

   a. 36×10^{3} **b.** 360×10^{2}

   c. 0.36×10^{5} **d.** 3.6×10^{4}

Without calculating, determine which number is larger.

137. 7^{11} or 7^{13}

138. 5^{10} or 5^{9}

139. 7^{-11} or 7^{-13}

140. 5^{-10} or 5^{-9}

What to Do the Day of an Exam

Your first exam may be soon. On the day of an exam, don't forget to try the following:

- Allow yourself plenty of time to arrive.
- Read the directions on the test carefully.
- Read each problem carefully as you take your test. Make sure that you answer the question asked.
- Watch your time and pace yourself so that you may attempt each problem on your test.
- Check your work and answers.
- **Do not turn your test in early.** If you have extra time, spend it double-checking your work.

Good luck!

Answer the following questions based on your most recent mathematics exam, whenever that was.

1. How soon before class did you arrive?
2. Did you read the directions on the test carefully?
3. Did you make sure you answered the question asked for each problem on the exam?
4. Were you able to attempt each problem on your exam?
5. If your answer to question 4 is no, list reasons why.
6. Did you have extra time on your exam?
7. If your answer to question 6 is yes, describe how you spent that extra time.

5.2 MORE WORK WITH EXPONENTS AND SCIENTIFIC NOTATION

OBJECTIVES

1. Use the power rules for exponents.
2. Use exponent rules and definitions to simplify exponential expressions.
3. Compute, using scientific notation.

OBJECTIVE 1 ▶ Using the power rules. The volume of the cube shown whose side measures x^2 units is $(x^2)^3$ cubic units. To simplify an expression such as $(x^2)^3$, we use the definition of a^n. Then

$$(x^2)^3 = \underbrace{(x^2)(x^2)(x^2)}_{x^2 \text{ is a factor 3 times}} = x^{2+2+2} = x^6$$

Notice that the result is exactly the same if the exponents are multiplied.

$$(x^2)^3 = x^{2 \cdot 3} = x^6$$

x^2 units

This suggests that the power of an exponential expression raised to a power is the product of the exponents. To discover additional rules for exponents see below.

DISCOVER THE CONCEPT

a. Use a calculator to evaluate each pair of expressions. Then compare the results.

$$(-2 \cdot 3)^7 \quad \text{and} \quad (-2)^7 \cdot 3^7$$
$$(3 \cdot 5)^{-2} \quad \text{and} \quad 3^{-2} \cdot 5^{-2}$$
$$\left(\frac{6}{7}\right)^4 \quad \text{and} \quad \frac{6^4}{7^4}$$
$$\left(\frac{3}{4}\right)^{-5} \quad \text{and} \quad \frac{3^{-5}}{4^{-5}}$$

b. From the results of part **a**, write a pattern you observed for raising a product to a power and for raising a quotient to a power.

c. Use your findings to simplify $(xy)^3$ and $\left(\frac{r}{s}\right)^7$.

The results of part **a** are below.

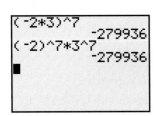 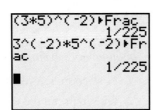 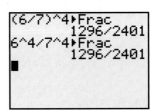 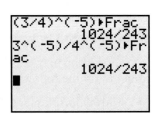

It appears that a product raised to a power is equal to the product of each factor raised to the power. Also, a quotient raised to a power is equal to the quotient of the numerator raised to the power and the denominator raised to the power. From this, we reason that

$$(xy)^3 = x^3y^3 \quad \text{and} \quad \left(\frac{r}{s}\right)^7 = \frac{r^7}{s^7}.$$

These power rules suggested above are given in the following box.

The Power Rule and Power of a Product or Quotient Rules for Exponents

If a and b are real numbers and m and n are integers, then

$$(a^m)^n = a^{m \cdot n} \qquad \text{Power rule}$$
$$(ab)^m = a^m b^m \qquad \text{Power of a product}$$
$$\left(\frac{a}{b}\right)^n = \frac{a^n}{b^n}(b \neq 0) \qquad \text{Power of a quotient}$$

EXAMPLE 1 Use the power rule to simplify the following expressions. Use positive exponents to write all results.

a. $(x^5)^7$ **b.** $(2^2)^3$ **c.** $(5^{-1})^2$ **d.** $(y^{-3})^{-4}$

Solution

a. $(x^5)^7 = x^{5 \cdot 7} = x^{35}$ **b.** $(2^2)^3 = 2^{2 \cdot 3} = 2^6 = 64$

c. $(5^{-1})^2 = 5^{-1 \cdot 2} = 5^{-2} = \dfrac{1}{5^2} = \dfrac{1}{25}$ **d.** $(y^{-3})^{-4} = y^{-3(-4)} = y^{12}$ □

PRACTICE

1 Use the power rule to simplify the following expressions. Use positive exponents to write all results.

a. $(z^3)^5$ **b.** $(5^2)^2$ **c.** $(3^{-1})^3$ **d.** $(x^{-4})^{-6}$

EXAMPLE 2 Use the power rules to simplify the following. Use positive exponents to write all results.

a. $(5x^2)^3$ **b.** $\left(\dfrac{2}{3}\right)^3$ **c.** $\left(\dfrac{3p^4}{q^5}\right)^2$ **d.** $\left(\dfrac{2^{-3}}{y}\right)^{-2}$ **e.** $(x^{-5}y^2z^{-1})^7$

Solution

a. $(5x^2)^3 = 5^3 \cdot (x^2)^3 = 5^3 \cdot x^{2 \cdot 3} = 125x^6$

b. $\left(\dfrac{2}{3}\right)^3 = \dfrac{2^3}{3^3} = \dfrac{8}{27}$

c. $\left(\dfrac{3p^4}{q^5}\right)^2 = \dfrac{(3p^4)^2}{(q^5)^2} = \dfrac{3^2 \cdot (p^4)^2}{(q^5)^2} = \dfrac{9p^8}{q^{10}}$

d. $\left(\dfrac{2^{-3}}{y}\right)^{-2} = \dfrac{(2^{-3})^{-2}}{y^{-2}}$

$\qquad\qquad = \dfrac{2^6}{y^{-2}} = 64y^2$ Use the negative exponent rule.

e. $(x^{-5}y^2z^{-1})^7 = (x^{-5})^7 \cdot (y^2)^7 \cdot (z^{-1})^7$

$\qquad\qquad = x^{-35}y^{14}z^{-7} = \dfrac{y^{14}}{x^{35}z^7}$ $\square$

PRACTICE

2   Use the power rules to simplify the following. Use positive exponents to write all results.

a. $(2x^3)^5$ **b.** $\left(\dfrac{3}{5}\right)^2$ **c.** $\left(\dfrac{2a^5}{b^7}\right)^4$ **d.** $\left(\dfrac{3^{-2}}{x}\right)^{-1}$ **e.** $(a^{-2}b^{-5}c^4)^{-2}$

OBJECTIVE 2 ▶ Using exponent rules and definitions to simplify. In the next few examples, we practice the use of several of the rules and definitions for exponents. The following is a summary of these rules and definitions.

Summary of Rules for Exponents

If a and b are real numbers and m and n are integers, then

Product rule	$a^m \cdot a^n = a^{m+n}$	
Zero exponent	$a^0 = 1$	$(a \neq 0)$
Negative exponent	$a^{-n} = \dfrac{1}{a^n}$	$(a \neq 0)$
Quotient rule	$\dfrac{a^m}{a^n} = a^{m-n}$	$(a \neq 0)$
Power rule	$(a^m)^n = a^{m \cdot n}$	
Power of a product	$(ab)^m = a^m \cdot b^m$	
Power of a quotient	$\left(\dfrac{a}{b}\right)^m = \dfrac{a^m}{b^m}$	$(b \neq 0)$

EXAMPLE 3   Simplify each expression. Use positive exponents to write the answers.

a. $(2x^0y^{-3})^{-2}$ **b.** $\left(\dfrac{x^{-5}}{x^{-2}}\right)^{-3}$ **c.** $\left(\dfrac{2}{7}\right)^{-2}$ **d.** $\dfrac{5^{-2}x^{-3}y^{11}}{x^2y^{-5}}$

Solution

a. $(2x^0y^{-3})^{-2} = 2^{-2}(x^0)^{-2}(y^{-3})^{-2}$

$\qquad\qquad = 2^{-2}x^0y^6$

$\qquad\qquad = \dfrac{1(y^6)}{2^2}$ Write x^0 as 1.

$\qquad\qquad = \dfrac{y^6}{4}$

b. $\left(\dfrac{x^{-5}}{x^{-2}}\right)^{-3} = \dfrac{(x^{-5})^{-3}}{(x^{-2})^{-3}} = \dfrac{x^{15}}{x^6} = x^{15-6} = x^9$

c. $\left(\dfrac{2}{7}\right)^{-2} = \dfrac{2^{-2}}{7^{-2}} = \dfrac{7^2}{2^2} = \dfrac{49}{4}$

d. $\dfrac{5^{-2}x^{-3}y^{11}}{x^2y^{-5}} = (5^{-2})\left(\dfrac{x^{-3}}{x^2}\right)\left(\dfrac{y^{11}}{y^{-5}}\right) = 5^{-2}x^{-3-2}y^{11-(-5)} = 5^{-2}x^{-5}y^{16}$

$$= \dfrac{y^{16}}{5^2 x^5} = \dfrac{y^{16}}{25x^5}$$

PRACTICE

3 Simplify each expression. Use positive exponents to write the answer.

a. $(3ab^{-5})^{-3}$ 　　　 **b.** $\left(\dfrac{y^{-7}}{y^{-4}}\right)^{-5}$ 　　　 **c.** $\left(\dfrac{3}{8}\right)^{-2}$ 　　　 **d.** $\dfrac{9^{-2}a^{-4}b^3}{a^2b^{-5}}$

EXAMPLE 4 Simplify each expression. Use positive exponents to write the answers.

a. $\left(\dfrac{3x^2y}{y^{-9}z}\right)^{-2}$ 　　　　　　 **b.** $\left(\dfrac{3a^2}{2x^{-1}}\right)^3\left(\dfrac{x^{-3}}{4a^{-2}}\right)^{-1}$

Solution There is often more than one way to simplify exponential expressions. Here, we will simplify inside the parentheses if possible before we apply the power rules for exponents.

a. $\left(\dfrac{3x^2y}{y^{-9}z}\right)^{-2} = \left(\dfrac{3x^2y^{10}}{z}\right)^{-2} = \dfrac{3^{-2}x^{-4}y^{-20}}{z^{-2}} = \dfrac{z^2}{3^2x^4y^{20}} = \dfrac{z^2}{9x^4y^{20}}$

b. $\left(\dfrac{3a^2}{2x^{-1}}\right)^3\left(\dfrac{x^{-3}}{4a^{-2}}\right)^{-1} = \dfrac{27a^6}{8x^{-3}} \cdot \dfrac{x^3}{4^{-1}a^2}$

$$= \dfrac{27 \cdot 4 \cdot a^6 x^3 x^3}{8 \cdot a^2} = \dfrac{27a^4x^6}{2}$$

PRACTICE

4 Simplify each expression. Use positive exponents to write the answers.

a. $\left(\dfrac{5a^4b}{a^{-8}c}\right)^{-3}$ 　　　　　　 **b.** $\left(\dfrac{2x^4}{5y^{-2}}\right)^3\left(\dfrac{x^{-4}}{10y^{-2}}\right)^{-1}$

EXAMPLE 5 Simplify each expression. Assume that a and b are integers and that x and y are not 0.

a. $x^{-b}(2x^b)^2$ 　　　　　 **b.** $\dfrac{(y^{3a})^2}{y^{a-6}}$

Solution

a. $x^{-b}(2x^b)^2 = x^{-b}2^2x^{2b} = 4x^{-b+2b} = 4x^b$

b. $\dfrac{(y^{3a})^2}{y^{a-6}} = \dfrac{y^{6a}}{y^{a-6}} = y^{6a-(a-6)} = y^{6a-a+6} = y^{5a+6}$

PRACTICE

5 Simplify each expression. Assume that a and b are integers and that x and y are not 0.

a. $x^{-2a}(3x^a)^3$

b. $\dfrac{(y^{3b})^3}{y^{4b-3}}$

OBJECTIVE 3 ▶ Computing, using scientific notation. To perform operations on numbers written in scientific notation, we use properties of exponents.

EXAMPLE 6 Perform the indicated operations. Write each result in scientific notation.

a. $(8.1 \times 10^5)(5 \times 10^{-7})$

b. $\dfrac{1.2 \times 10^4}{3 \times 10^{-2}}$

Solution

a. $(8.1 \times 10^5)(5 \times 10^{-7}) = 8.1 \times 5 \times 10^5 \times 10^{-7}$

$= 40.5 \times 10^{-2}$ Not in scientific notation because 40.5 is not between 1 and 10.

$= (4.05 \times 10^1) \times 10^{-2}$

$= 4.05 \times 10^{-1}$

b. $\dfrac{1.2 \times 10^4}{3 \times 10^{-2}} = \left(\dfrac{1.2}{3}\right)\left(\dfrac{10^4}{10^{-2}}\right) = 0.4 \times 10^{4-(-2)}$

$= 0.4 \times 10^6 = (4 \times 10^{-1}) \times 10^6 = 4 \times 10^5$ ☐

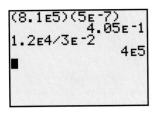

A calculator check for Example 6.

PRACTICE

6 Perform the indicated operations. Write each result in scientific notation.

a. $(3.4 \times 10^4)(5 \times 10^{-7})$

b. $\dfrac{5.6 \times 10^8}{4 \times 10^{-2}}$

EXAMPLE 7 Use scientific notation to simplify $\dfrac{2000 \times 0.000021}{700}$. Write the result in scientific notation.

Solution $\dfrac{2000 \times 0.000021}{700} = \dfrac{(2 \times 10^3)(2.1 \times 10^{-5})}{7 \times 10^2} = \dfrac{2(2.1)}{7} \cdot \dfrac{10^3 \cdot 10^{-5}}{10^2}$

$= 0.6 \times 10^{-4}$

$= (6 \times 10^{-1}) \times 10^{-4}$

$= 6 \times 10^{-5}$ ☐

```
(2000*.000021)/7
00
              6E-5
■
```

A calculator check for Example 7.

PRACTICE

7 Use scientific notation to simplify $\dfrac{2400 \times 0.0000014}{800}$. Write the result in scientific notation.

VOCABULARY & READINESS CHECK

Simplify. See Examples 1 through 4.

1. $(x^4)^5$ **2.** $(5^6)^2$ **3.** $x^4 \cdot x^5$ **4.** $x^7 \cdot x^8$ **5.** $(y^6)^7$

6. $(x^3)^4$ **7.** $(z^4)^9$ **8.** $(z^3)^7$ **9.** $(z^{-6})^{-3}$ **10.** $(y^{-4})^{-2}$

5.2 EXERCISE SET

MyMathLab | PRACTICE | WATCH | DOWNLOAD | READ | REVIEW

Simplify. Write each answer using positive exponents only. See Examples 1 and 2.

1. $(3^{-1})^2$

2. $(2^{-2})^2$

3. $(x^4)^{-9}$

4. $(y^7)^{-3}$

5. $(y)^{-5}$

6. $(z^{-1})^{10}$

7. $(3x^2y^3)^2$

8. $(4x^3yz)^2$

9. $\left(\dfrac{2x^5}{y^{-3}}\right)^4$

10. $\left(\dfrac{3a^{-4}}{b^7}\right)^3$

11. $(a^2bc^{-3})^{-6}$

12. $(6x^{-6}y^7z^0)^{-2}$

13. $\left(\dfrac{x^7y^{-3}}{z^{-4}}\right)^{-5}$

14. $\left(\dfrac{a^{-2}b^{-5}}{c^{-11}}\right)^{-6}$

15. $(5^{-1})^3$

Simplify. Write each answer using positive exponents only. See Examples 3 and 4.

16. $\left(\dfrac{a^{-4}}{a^{-5}}\right)^{-2}$

17. $\left(\dfrac{x^{-9}}{x^{-4}}\right)^{-3}$

18. $\left(\dfrac{2a^{-2}b^5}{4a^2b^7}\right)^{-2}$

19. $\left(\dfrac{5x^7y^4}{10x^3y^{-2}}\right)^{-3}$

20. $\dfrac{4^{-1}x^2yz}{x^{-2}yz^3}$

21. $\dfrac{8^{-2}x^{-3}y^{11}}{x^2y^{-5}}$

22. $\left(\dfrac{6p^6}{p^{12}}\right)^2$

23. $\left(\dfrac{4p^6}{p^9}\right)^3$

24. $(-8y^3xa^{-2})^{-3}$

25. $(-xy^0x^2a^3)^{-3}$

26. $\left(\dfrac{x^{-2}y^{-2}}{a^{-3}}\right)^{-7}$

27. $\left(\dfrac{x^{-1}y^{-2}}{5^{-3}}\right)^{-5}$

MIXED PRACTICE

Simplify. Write each answer using positive exponents.

28. $(8^2)^{-1}$

29. $(x^7)^{-9}$

30. $(y^{-4})^5$

31. $\left(\dfrac{7}{8}\right)^3$

32. $\left(\dfrac{4}{3}\right)^2$

33. $(4x^2)^2$

34. $(-8x^3)^2$

35. $(-2^{-2}y)^3$

36. $(-4^{-6}y^{-6})^{-4}$

37. $\left(\dfrac{4^{-4}}{y^3x}\right)^{-2}$

38. $\left(\dfrac{7^{-3}}{ab^2}\right)^{-2}$

39. $\left(\dfrac{1}{4}\right)^{-3}$

40. $\left(\dfrac{1}{8}\right)^{-2}$

41. $\left(\dfrac{3x^5}{6x^4}\right)^4$

42. $\left(\dfrac{8^{-3}}{y^2}\right)^{-2}$

43. $\dfrac{(y^3)^{-4}}{y^3}$

44. $\dfrac{2(y^3)^{-3}}{y^{-3}}$

45. $\left(\dfrac{2x^{-3}}{y^{-1}}\right)^{-3}$

46. $\left(\dfrac{n^5}{2m^{-2}}\right)^{-4}$

47. $\dfrac{3^{-2}a^{-5}b^6}{4^{-2}a^{-7}b^{-3}}$

48. $\dfrac{2^{-3}m^{-4}n^{-5}}{5^{-2}m^{-5}n}$

49. $(4x^6y^5)^{-2}(6x^4y^3)$

50. $(5xy)^3(z^{-2})^{-3}$

51. $x^6(x^6bc)^{-6}$

52. $2(y^2b)^{-4}$

53. $\dfrac{2^{-3}x^2y^{-5}}{5^{-2}x^7y^{-1}}$

54. $\dfrac{7^{-1}a^{-3}b^5}{a^2b^{-2}}$

55. $\left(\dfrac{2x^2}{y^4}\right)^3\left(\dfrac{2x^5}{y}\right)^{-2}$

56. $\left(\dfrac{3z^{-2}}{y}\right)^2\left(\dfrac{9y^{-4}}{z^{-3}}\right)^{-1}$

Simplify the following. Assume that variables in the exponents represent integers and that all other variables are not 0. See Example 5.

57. $(x^{3a+6})^3$

58. $(x^{2b+7})^2$

59. $\dfrac{x^{4a}(x^{4a})^3}{x^{4a-2}}$

60. $\dfrac{x^{-5y+2}x^{2y}}{x}$

61. $(b^{5x-2})^2$

62. $(c^{2a+3})^3$

63. $\dfrac{(y^{2a})^8}{y^{a-3}}$

64. $\dfrac{(y^{4a})^7}{y^{2a-1}}$

65. $\left(\dfrac{2x^{3t}}{x^{2t-1}}\right)^4$

66. $\left(\dfrac{3y^{5a}}{y^{-a+1}}\right)^2$

67. $\dfrac{25x^{2a+1}y^{a-1}}{5x^{3a+1}y^{2a-3}}$

68. $\dfrac{16x^{-5-3a}y^{-2a-b}}{2x^{-5+3b}y^{-2b-a}}$

Perform each indicated operation. Write each answer in scientific notation. See Examples 6 and 7.

69. $(5 \times 10^{11})(2.9 \times 10^{-3})$

70. $(3.6 \times 10^{-12})(6 \times 10^9)$

71. $(2 \times 10^5)^3$

72. $(3 \times 10^{-7})^3$

73. $\dfrac{3.6 \times 10^{-4}}{9 \times 10^2}$

74. $\dfrac{1.2 \times 10^9}{2 \times 10^{-5}}$

75. $\dfrac{0.0069}{0.023}$

76. $\dfrac{0.00048}{0.0016}$

77. $\dfrac{18,200 \times 100}{91,000}$

78. $\dfrac{0.0003 \times 0.0024}{0.0006 \times 20}$

79. $\dfrac{6000 \times 0.006}{0.009 \times 400}$

80. $\dfrac{0.00016 \times 300}{0.064 \times 100}$

81. $\dfrac{0.00064 \times 2000}{16,000}$

82. $\dfrac{0.00072 \times 0.003}{0.00024}$

83. $\dfrac{66,000 \times 0.001}{0.002 \times 0.003}$

84. $\dfrac{0.0007 \times 11,000}{0.001 \times 0.0001}$

85. $\dfrac{9.24 \times 10^{15}}{(2.2 \times 10^{-2})(1.2 \times 10^{-5})}$

86. $\dfrac{(2.6 \times 10^{-3})(4.8 \times 10^{-4})}{1.3 \times 10^{-12}}$

Solve.

87. A computer can add two numbers in about 10^{-8} second. Express in scientific notation how long it would take this computer to do this task 200,000 times.

△ **88.** To convert from square inches to square meters, multiply by 6.452×10^{-4}. The area of the following square is 4×10^{-2} square inches. Convert this area to square meters.

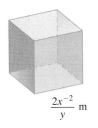

4×10^{-2} sq in.

△ **89.** To convert from cubic inches to cubic meters, multiply by 1.64×10^{-5}. A grain of salt is in the shape of a cube. If an average size of a grain of salt is 3.8×10^{-6} cubic inches, convert this volume to cubic meters.

REVIEW AND PREVIEW

Simplify each expression. See Section 1.4.

90. $-5y + 4y - 18 - y$

91. $12m - 14 - 15m - 1$

92. $-3x - (4x - 2)$

93. $-9y - (5 - 6y)$

94. $3(z - 4) - 2(3z + 1)$

95. $5(x - 3) - 4(2x - 5)$

CONCEPT EXTENSIONS

△ **96.** Each side of the cube shown is $\dfrac{2x^{-2}}{y}$ meters. Find its volume.

$\dfrac{2x^{-2}}{y}$ m

△ **97.** The lot shown is in the shape of a parallelogram with base $\dfrac{3x^{-1}}{y^{-3}}$ feet and height $5x^{-7}$ feet. Find its area.

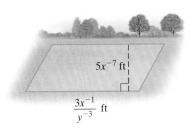

$5x^{-7}$ ft

$\dfrac{3x^{-1}}{y^{-3}}$ ft

98. The density D of an object is equivalent to the quotient of its mass M and volume V. Thus $D = \dfrac{M}{V}$. Express in scientific notation the density of an object whose mass is 500,000 pounds and whose volume is 250 cubic feet.

99. The density of ordinary water is 3.12×10^{-2} tons per cubic foot. The volume of water in the largest of the Great Lakes, Lake Superior, is 4.269×10^{14} cubic feet. Use the formula $D = \dfrac{M}{V}$ (see Exercise 98) to find the mass (in tons) of the water in Lake Superior. Express your answer in scientific notation. (*Source:* National Ocean Service)

100. Is there a number a such that $a^{-1} = a^{1}$? If so, give the value of a.

101. Is there a number a such that a^{-2} is a negative number? If so, give the value of a.

102. Explain whether 0.4×10^{-5} is written in scientific notation.

103. The estimated population of the United States in 2007 was 3.016×10^{8}. The land area of the United States is 3.536×10^{6} square miles. Find the population density (number of people per square mile) for the United States in 2007. Round to the nearest whole number. (*Source:* U.S. Census Bureau)

104. In 2006, the value of goods imported into the United States was $\$1.855 \times 10^{12}$. The estimated population of the United States in 2006 was 2.98×10^{8}. Find the average value of imports per person in the United States for 2006. Round to the nearest dollar. (*Sources:* U.S. Census Bureau, Bureau of Economic Analysis)

105. The largest subway system in the world (based on passenger volume) is in Tokyo, Japan, with an estimated 2.82×10^9 riders per year. The subway system in Toronto boasts an estimated 4.44×10^8 riders per year. How many times greater is the Tokyo subway volume than the Toronto subway volume? Round to the nearest tenth. (*Source:* Tokyo and Toronto Transit Authorities)

106. Explain whether 0.4×10^{-5} is written in scientific notation.

107. In 2006, the population of Beijing was approximately 15.38×10^6. To prepare for the 2008 Summer Olympic Games, the Chinese put forth a massive effort to increase the numbers of their population who speak a foreign language. By the end of 2006, about 4.87×10^6 people in Beijing could boast that they speak a foreign language. What percent of the residents of Beijing could speak a foreign language at the end of 2006? Round to the nearest tenth of a percent. (*Source: China Daily*)

108. In 2006, office space in downtown New York City was estimated to be 4.21×10^8 sq ft, while office space in downtown Dallas was 4.8×10^7 sq feet. How many times greater is the square footage of office space in New York City than Dallas? Round to the nearest tenth. (*Source:* National Center for Real Estate Research)

5.3 POLYNOMIALS AND POLYNOMIAL FUNCTIONS

OBJECTIVES

1 Identify term, constant, polynomial, monomial, binomial, trinomial, and the degree of a term and of a polynomial.

2 Define polynomial functions.

3 Review combining like terms.

4 Add polynomials.

5 Subtract polynomials.

6 Recognize the graph of a polynomial function from the degree of the polynomial.

OBJECTIVE 1 ▶ Identifing polynomial terms and degrees of terms and polynomials. A **term** is a number or the product of a number and one or more variables raised to powers. The **numerical coefficient,** or simply the **coefficient,** is the numerical factor of a term.

Term	Numerical Coefficient of Term
$-1.2x^5$	-1.2
x^3y	1
$-z$	-1
2	2
$\dfrac{x^9}{7}$ $\left(\text{or } \dfrac{1}{7}x^9\right)$	$\dfrac{1}{7}$

If a term contains only a number, it is called a **constant term,** or simply a **constant.**

A **polynomial** is a finite sum of terms in which all variables are raised to nonnegative integer powers and no variables appear in any denominator.

Polynomials	Not Polynomials	
$4x^5y + 7xz$	$5x^{-3} + 2x$	Negative integer exponent
$-5x^3 + 2x + \dfrac{2}{3}$	$\dfrac{6}{x^2} - 5x + 1$	Variable in denominator

A polynomial that contains only one variable is called a **polynomial in one variable.** For example, $3x^2 - 2x + 7$ is a **polynomial in x.** This polynomial in x is written in *descending order* since the terms are listed in descending order of the variable's exponents. (The term 7 can be thought of as $7x^0$.) The following examples are polynomials in one variable written in **descending order.**

$$4x^3 - 7x^2 + 5 \qquad y^2 - 4 \qquad 8a^4 - 7a^2 + 4a$$

A **monomial** is a polynomial consisting of one term. A **binomial** is a polynomial consisting of two terms. A **trinomial** is a polynomial consisting of three terms.

Monomials	Binomials	Trinomials
ax^2	$x + y$	$x^2 + 4xy + y^2$
$-3x$	$6y^2 - 2$	$-x^4 + 3x^3 + 1$
4	$\dfrac{5}{7}z^3 - 2z$	$8y^2 - 2y - 10$

By definition, all monomials, binomials, and trinomials are also polynomials.
Each term of a polynomial has a **degree.**

> **Degree of a Term**
> The **degree of a term** is the sum of the exponents on the *variables* contained in the term.

EXAMPLE 1   Find the degree of each term.

a. $3x^2$ **b.** -2^3x^5 **c.** y **d.** $12x^2yz^3$ **e.** 5.27

Solution

a. The exponent on x is 2, so the degree of the term is 2.
b. The exponent on x is 5, so the degree of the term is 5. (Recall that the degree is the sum of the exponents on only the *variables*.)
c. The degree of y, or y^1, is 1.
d. The degree is the sum of the exponents on the variables, or $2 + 1 + 3 = 6$.
e. The degree of 5.27, which can be written as $5.27x^0$, is 0. ☐

PRACTICE

1   Find the degree of each term.

a. $4x^5$ **b.** -4^3y^3 **c.** z **d.** $65a^3b^7c$ **e.** 36

From the preceding example, we can say that the degree of a constant is 0. Also, the term 0 has no degree.
Each polynomial also has a degree.

> **Degree of a Polynomial**
> The **degree of a polynomial** is the largest degree of all its terms.

EXAMPLE 2   Find the degree of each polynomial and indicate whether the polynomial is also a monomial, binomial, or trinomial.

	Polynomial	Degree	Classification
a.	$7x^3 - \dfrac{3}{4}x + 2$	3	Trinomial
b.	$-xyz$	$1 + 1 + 1 = 3$	Monomial
c.	$x^4 - 16.5$	4	Binomial

☐

PRACTICE

2   Find the degree of each polynomial and indicate whether the polynomial is also a monomial, binomial, or trinomial.

	Polynomial	Degree	Classification
a.	$3x^4 + 2x^2 - 3$		
b.	$9abc^3$		
c.	$8x^5 + 5x^3$		

EXAMPLE 3 Find the degree of the polynomial

$$3xy + x^2y^2 - 5x^2 - 6.7$$

Solution The degree of each term is

$$3xy + x^2y^2 - 5x^2 - 6.7$$
$$\quad\downarrow\qquad\downarrow\qquad\downarrow\qquad\downarrow$$
Degree: 2   4   2   0

The largest degree of any term is 4, so the degree of this polynomial is 4. ☐

PRACTICE

3 Find the degree of the polynomial $2x^3y - 3x^3y^2 - 9y^5 + 9.6$.

OBJECTIVE 2 ▶ Defining polynomial functions. At times, it is convenient to use function notation to represent polynomials. For example, we may write $P(x)$ to represent the polynomial $3x^2 - 2x - 5$. In symbols, this is

$$P(x) = 3x^2 - 2x - 5$$

This function is called a **polynomial function** because the expression $3x^2 - 2x - 5$ is a polynomial.

> **▶ Helpful Hint**
>
> Recall that the symbol $P(x)$ **does not mean** P times x. It is a special symbol used to denote a function.

If $P(x) = 3x^2 - 2x - 5$, let's evaluate $P(2)$. Below is a review of methods that we have used to evaluate functions at given values.

$$P(x) = 3(x)^2 - 2x - 5$$
$$P(2) = 3 \cdot 2^2 - 2 \cdot 2 - 5$$
$$P(2) = 3$$

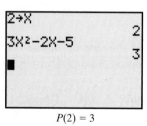

$P(2) = 3$

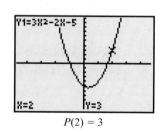

$P(2) = 3$

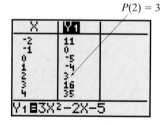

$P(2) = 3$

EXAMPLE 4 If $P(x) = 3x^2 - 2x - 5$, find the following. Use a graph to check part **a** and a table to check part **b**.

a. $P(1)$ **b.** $P(-2)$

Solution

a. Substitute 1 for x in $P(x) = 3x^2 - 2x - 5$ and simplify.

$$P(x) = 3x^2 - 2x - 5$$
$$P(1) = 3(1)^2 - 2(1) - 5 = -4$$

b. Substitute -2 for x in $P(x) = 3x^2 - 2x - 5$ and simplify.

$$P(x) = 3x^2 - 2x - 5$$
$$P(-2) = 3(-2)^2 - 2(-2) - 5 = 11$$

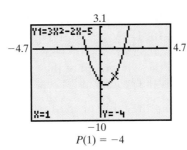

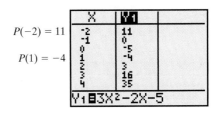

$P(-2) = 11$

$P(1) = -4$

$P(1) = -4$

Calculator checks for Example 4a and b.   □

$P(1) = -4$

PRACTICE
4   If $P(x) = -5x^2 + 2x - 8$, find the following.

a. $P(-1)$ **b.** $P(3)$

Many real-world phenomena are modeled by polynomial functions. If the polynomial function model is given, we can often find the solution of a problem by evaluating the function at a certain value.

EXAMPLE 5   **Finding the Height of an Object**

The world's highest bridge, the Millau Viaduct in France, is 1125 feet above the River Tarn. An object is dropped from the top of this bridge. Neglecting air resistance, the height of the object at time t seconds is given by the polynomial function $P(t) = -16t^2 + 1125$.

a. Find the height of the object when $t = 1$ second and when $t = 8$ seconds.

b. Approximate, to the nearest second, the time when the object hits the ground.

c. Approximate, to the nearest tenth of a second, the time when the object hits the ground.

Solution

a. To find the height of the object at 1 second, we find $P(1)$. Below on the left, we use paper and pencil. As an alternative, since we are asked to find height at 1 second and 8 seconds; we may set table start at 1 and table increment at 1. These screens are shown to the right.

$P(1) = 1109$

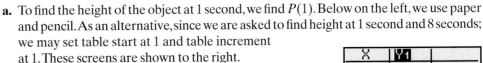

$$P(t) = -16t^2 + 1125$$
$$P(1) = -16(1)^2 + 1125$$
$$P(1) = 1109$$

When $t = 1$ second, the height of the object is 1109 feet.

To find the height of the object at 8 seconds, we find $P(8)$.

$$P(t) = -16t^2 + 1125$$
$$P(8) = -16(8)^2 + 1125$$
$$P(8) = 101$$

$P(8) = 101$

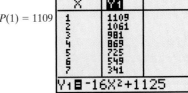

When $t = 8$ seconds, the height of the object is 101 feet. Notice that as time t increases, the height of the object decreases.

b. The object will hit the ground when $P(x)$ or $y = 0$. Since

$$P(8) = 101 \quad \text{and} \quad P(9) = -171$$

the object hits the ground between 8 and 9 seconds. Since 101 is closer to 0 than -171, the time to the nearest second is 8 seconds.

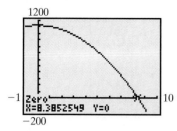

c. Given the table on the previous page, graph y_1 in an appropriate window. We graph y_1 in a $[-1, 10, 1]$ by $[-200, 1200, 100]$ window. Recall that the object hits the ground when $P(x)$ or $y = 0$. Thus, use a zero or root function of your graphing utility to find the x-intercept of the graph in the first quadrant. To the nearest tenth, the object hits the ground at 8.4 seconds.

Notice from the table or the first quadrant portion of the graph that as time x increases, the height of the object decreases. ☐

PRACTICE
5 The largest natural bridge is in the canyons at the base of Navajo Mountain, Utah. From the base to the top of the arch, it measures 290 feet. Neglecting air resistance, the height of an object dropped off the bridge is given by the polynomial function $P(t) = -16t^2 + 290$ at time t seconds. **a.** Find the height of the object at time $t = 0$ second and $t = 2$ seconds. **b.** Approximate, to the nearest second, the time when the object hits the ground. **c.** Approximate, to the nearest tenth of a second, the time when the object hits the ground.

OBJECTIVE 3 ▶ Combining like terms review. Before we add polynomials, recall that terms are considered to be **like terms** if they contain exactly the same variables raised to exactly the same powers.

Like Terms	Unlike Terms
$-5x^2, -x^2$	$4x^2, 3x$
$7xy^3z, -2xzy^3$	$12x^2y^3, -2xy^3$

To simplify a polynomial, **combine like terms** by using the distributive property. For example, by the distributive property,

$$5x + 7x = (5 + 7)x = 12x$$

EXAMPLE 6 Simplify by combining like terms.

a. $-12x^2 + 7x^2 - 6x$ **b.** $3xy - 2x + 5xy - x$

Solution By the distributive property,

a. $-12x^2 + 7x^2 - 6x = (-12 + 7)x^2 - 6x = -5x^2 - 6x$

b. Use the associative and commutative properties to group together like terms; then combine.

$$3xy - 2x + 5xy - x = 3xy + 5xy - 2x - x$$
$$= (3 + 5)xy + (-2 - 1)x$$
$$= 8xy - 3x$$ ☐

▶ **Helpful Hint**
These two terms are unlike terms. They cannot be combined.

PRACTICE
6 Simplify by combining like terms.

a. $8x^4 - 5x^4 - 5x$ **b.** $4ab - 5b + 3ab + 2b$

OBJECTIVE 4 ▶ Adding polynomials. Now we have reviewed the necessary skills to add polynomials.

Adding Polynomials
To add polynomials, combine all like terms.

EXAMPLE 7   Add.

a. $(7x^3y - xy^3 + 11) + (6x^3y - 4)$ **b.** $(3a^3 - b + 2a - 5) + (a + b + 5)$

Solution

a. To add, remove the parentheses and group like terms.

$$(7x^3y - xy^3 + 11) + (6x^3y - 4)$$
$$= 7x^3y - xy^3 + 11 + 6x^3y - 4$$
$$= 7x^3y + 6x^3y - xy^3 + 11 - 4 \quad \text{Group like terms.}$$
$$= 13x^3y - xy^3 + 7 \quad\quad\quad \text{Combine like terms.}$$

b.
$$(3a^3 - b + 2a - 5) + (a + b + 5)$$
$$= 3a^3 - b + 2a - 5 + a + b + 5$$
$$= 3a^3 - b + b + 2a + a - 5 + 5 \quad \text{Group like terms.}$$
$$= 3a^3 + 3a \quad\quad\quad\quad\quad \text{Combine like terms.} \quad \square$$

PRACTICE
7   Add.

a. $(3a^4b - 5ab^2 + 7) + (9ab^2 - 12)$ **b.** $(2x^5 - 3y + x - 6) + (4y - 2x - 3)$

EXAMPLE 8   Add $11x^3 - 12x^2 + x - 3$ and $x^3 - 10x + 5$.

Solution $(11x^3 - 12x^2 + x - 3) + (x^3 - 10x + 5)$
$$= 11x^3 + x^3 - 12x^2 + x - 10x - 3 + 5 \quad \text{Group like terms.}$$
$$= 12x^3 - 12x^2 - 9x + 2 \quad\quad\quad\quad\quad \text{Combine like terms.} \quad \square$$

PRACTICE
8   Add $5x^3 - 3x^2 - 9x - 8$ and $x^3 + 9x^2 + 2x$.

Sometimes it is more convenient to add polynomials vertically. To do this, line up like terms beneath one another and add like terms.

OBJECTIVE 5 ▶ Subtracting polynomials. The definition of subtraction of real numbers can be extended to apply to polynomials. To subtract a number, we add its opposite.

$$a - b = a + (-b)$$

Likewise, to subtract a polynomial, we add its opposite. In other words, if P and Q are polynomials, then

$$P - Q = P + (-Q)$$

The polynomial $-Q$ is the **opposite,** or **additive inverse,** of the polynomial Q. We can find $-Q$ by writing the opposite of each term of Q.

> **Subtracting Polynomials**
> To subtract a polynomial, add its opposite.

For example,

To subtract, change the signs; then add.

$$(3x^2 + 4x - 7) - (3x^2 - 2x - 5) = (3x^2 + 4x - 7) + (-3x^2 + 2x + 5)$$

$$= 3x^2 + 4x - 7 - 3x^2 + 2x + 5$$

$$= 6x - 2 \qquad \text{Combine like terms.}$$

Concept Check ☑

Which polynomial is the opposite of $16x^3 - 5x + 7$?

a. $-16x^3 - 5x + 7$ **b.** $-16x^3 + 5x - 7$

c. $16x^3 + 5x + 7$ **d.** $-16x^3 + 5x + 7$

EXAMPLE 9 Subtract: $(12z^5 - 12z^3 + z) - (-3z^4 + z^3 + 12z)$

Solution To subtract, add the opposite of the second polynomial to the first polynomial.

$$(12z^5 - 12z^3 + z) - (-3z^4 + z^3 + 12z)$$

$$= 12z^5 - 12z^3 + z + 3z^4 - z^3 - 12z) \qquad \text{Add the opposite of the polynomial being subtracted.}$$

$$= 12z^5 + 3z^4 - 12z^3 - z^3 + z - 12z \qquad \text{Group like terms.}$$

$$= 12z^5 + 3z^4 - 13z^3 - 11z \qquad \text{Combine like terms.} \qquad \square$$

PRACTICE

9 Subtract: $(13a^4 - 7a^3 - 9) - (-2a^4 + 8a^3 - 12)$.

Concept Check ☑

Why is the following subtraction incorrect?

$$(7z - 5) - (3z - 4)$$

$$= 7z - 5 - 3z - 4$$

$$= 4z - 9$$

EXAMPLE 10 Subtract $4x^3y^2 - 3x^2y^2 + 2y^2$ from $10x^3y^2 - 7x^2y^2$.

Solution If we subtract 2 from 8, the difference is $8 - 2 = 6$. Notice the order of the numbers, and then write "Subtract $4x^3y^2 - 3x^2y^2 + 2y^2$ from $10x^3y^2 - 7x^2y^2$" as a mathematical expression.

$$(10x^3y^2 - 7x^2y^2) - (4x^3y^2 - 3x^2y^2 + 2y^2)$$

$$= 10x^3y^2 - 7x^2y^2 - 4x^3y^2 + 3x^2y^2 - 2y^2 \quad \text{Remove parentheses.}$$

$$= 6x^3y^2 - 4x^2y^2 - 2y^2 \qquad\qquad\qquad \text{Combine like terms.} \qquad \square$$

PRACTICE

10 Subtract $5x^2y^2 - 3xy^2 + 5y^3$ from $11x^2y^2 - 7xy^2$.

Answers to Concept Check:

b;

With parentheses removed, the expression should be

$$7z - 5 - 3z + 4 = 4z - 1$$

To add or subtract polynomials vertically, just remember to line up like terms. For example, perform the subtraction $(10x^3y^2 - 7x^2y^2) - (4x^3y^2 - 3x^2y^2 + 2y^2)$ vertically.

Add the opposite of the second polynomial.

$$10x^3y^2 - 7x^2y^2$$
$$\underline{-(4x^3y^2 - 3x^2y^2 + 2y^2)} \quad \text{is equivalent to}$$

$$10x^3y^2 - 7x^2y^2$$
$$\underline{-4x^3y^2 + 3x^2y^2 - 2y^2}$$
$$6x^3y^2 - 4x^2y^2 - 2y^2$$

Polynomial functions, like polynomials, can be added, subtracted, multiplied, and divided. For example, if

$$P(x) = x^2 + x + 1$$

then

$$2P(x) = 2(x^2 + x + 1) = 2x^2 + 2x + 2 \quad \text{Use the distributive property.}$$

Also, if $Q(x) = 5x^2 - 1$, then $P(x) + Q(x) = (x^2 + x + 1) + (5x^2 - 1) = 6x^2 + x$.

A useful business and economics application of subtracting polynomial functions is finding the profit function $P(x)$ when given a revenue function $R(x)$ and a cost function $C(x)$. In business, it is true that

$$\text{profit} = \text{revenue} - \text{cost, or}$$
$$P(x) = R(x) - C(x)$$

For example, if the revenue function is $R(x) = 7x$ and the cost function is $C(x) = 2x + 5000$, then the profit function is

$$P(x) = R(x) - C(x)$$

or

$$P(x) = 7x - (2x + 5000) \quad \text{Substitute } R(x) = 7x$$
$$P(x) = 5x - 5000 \qquad\qquad \text{and } C(x) = 2x + 5000.$$

Problem-solving exercises involving profit are in the exercise set.

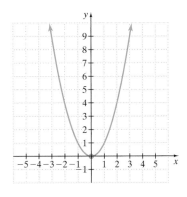

OBJECTIVE 6 ▶ Recognizing graphs of polynomial functions from their degree. In this section, we reviewed how to find the degree of a polynomial. Knowing the degree of a polynomial can help us recognize the graph of the related polynomial function. For example, we know from Section 3.1 that the graph of the polynomial function $f(x) = x^2$ is a parabola as shown to the left.

The polynomial x^2 has degree 2. The graphs of all polynomial functions of degree 2 will have this same general shape—opening upward, as shown, or downward. Graphs of polynomial functions of degree 2 or 3 will, in general, resemble one of the graphs shown next.

General Shapes of Graphs of Polynomial Functions

Degree 2

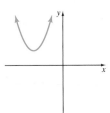

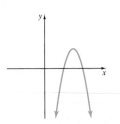

Coefficient of x^2
is a positive number.

Coefficient of x^2
is a negative number.

Degree 3

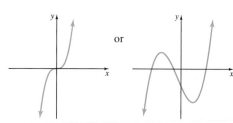

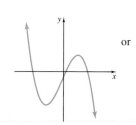

Coefficient of x^3
is a positive number.

Coefficient of x^3
is a negative number.

EXAMPLE 11 Determine which of the following graphs most closely resembles the graph of $f(x) = 5x^3 - 6x^2 + 2x + 3$

A

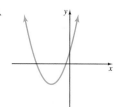

B

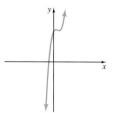

C

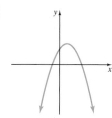

D

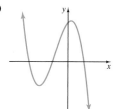

Solution The degree of $f(x)$ is 3, which means that its graph has the shape of B or D. The coefficient of x^3 is 5, a positive number, so the graph has the shape of B. ☐

PRACTICE

11 Determine which of the following graphs most closely resembles the graph of $f(x) = x^3 - 3$.

A

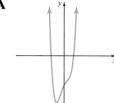

B

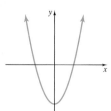

C

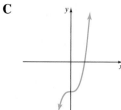

D

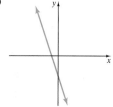

EXAMPLE 12 Match each graph with its equation and give the coordinates of the y-intercept of each graph.

a. $f(x) = x^2 + x - 5$ **b.** $g(x) = x^3 - x^2 + x + 2$

c. $h(x) = x^3 - 4x^2 + 2x - 3$ **d.** $F(x) = -x^3 + x - 6$

e. $H(x) = -x^2 + 5x + 1$

1.
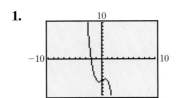

2.
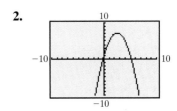

3.
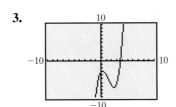

4.
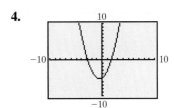

5.
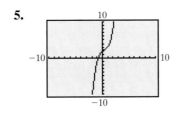

Solution

a. The polynomial function $f(x) = x^2 + x - 5$ has degree 2, so the shape of its graph is a parabola. Since the coefficient of x^2 is the positive number 1, the parabola opens upward and has the shape of option 4. Its y-intercept is $(0, -5)$.

b. The polynomial function $g(x) = x^3 - x^2 + x + 2$ has degree 3, so the shape of its graph is option 1, 3, or 5. Since the coefficient of x^3 is the positive number 1, it has the shape of option 3 or 5. The y-intercept of $g(x)$ is $(0, 2)$, so the graph resembles option 5.

c. The graph of $h(x)$ resembles option 3. The y-intercept of $h(x)$ is $(0, -3)$.

d. Since the coefficient of x^3 is -1, the graph of $F(x)$ resembles option 1. The y-intercept is $(0, -6)$.

e. The graph of $H(x)$ is a parabola opening downward, since the coefficient of x^2 is -1. Its graph is option 2. The y-intercept is $(0, 1)$.   □

PRACTICE

12 Give the coordinates of the y-intercept of the graph of each equation and describe the shape of its graph.

a. $f(x) = x^2 - x - 12$ **b.** $g(x) = -3x^2 + 5x + 6$

c. $h(x) = x^3 + 4$ **d.** $f(x) = -x^3 + 2x^2 - 5x - 1$

TECHNOLOGY NOTE

A graphing calculator may be used to visualize addition and subtraction of polynomials in one variable. For example, to visualize the following polynomial subtraction statement

$$(3x^2 - 6x + 9) - (x^2 - 5x + 6) = 2x^2 - x + 3$$

graph both

$$Y_1 = (3x^2 - 6x + 9) - (x^2 - 5x + 6) \quad \text{Left side of equation}$$

and

$$Y_2 = 2x^2 - x + 3 \quad \text{Right side of equation}$$

on the same screen and see that their graphs coincide. (*Note:* If the graphs do not coincide, we can be sure that a mistake has been made in combining polynomials or in calculator keystrokes. If the graphs appear to coincide, we cannot be sure that our work is correct. This is because it is possible for the graphs to differ so slightly that we do not notice it.)

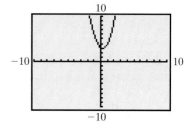

The graphs of Y_1 and Y_2 are shown. The graphs appear to coincide, so the subtraction statement

$$(3x^2 - 6x + 9) - (x^2 - 5x + 6) = 2x^2 - x + 3$$

appears to be correct. You may want to use this method to check the reasonableness of operations on polynomials in one variable.

VOCABULARY & READINESS CHECK

Use the choices below to fill in each blank. Not all choices will be used.

monomial	trinomial	like	degree	coefficient
binomial	polynomial	unlike	variables	term

1. The numerical factor of a term is the _____.

2. A _____ is a finite sum of terms in which all variables are raised to nonnegative integer powers and no variables appear in any denominator.

3. A _____ is a polynomial with 2 terms.

4. A _____ is a polynomial with 1 term.

5. A _____ is a polynomial with 3 terms.

6. The degree of a term is the sum of the exponents on the _____ in the term.

7. The _____ of a polynomial is the largest degree of all its terms.

8. _____ terms contain the same variables raised to the same powers.

Add or subtract, if possible.

9. $5x + x$ **10.** $5x - x$ **11.** $y + y$ **12.** $z^2 + z^2$ **13.** $7xy^2 - y^2$ **14.** $x^3 - 9x^3$

5.3 | EXERCISE SET

Find the degree of each term. See Example 1.

1. 4

2. 7

3. $5x^2$

4. $-z^3$

5. $-3xy^2$

6. $12x^3z$

7. -8^7y^3

8. $-9^{11}y^5$

9. $3.78ab^3c^5$

10. $9.11r^2st^{12}$

Find the degree of each polynomial and indicate whether the polynomial is a monomial, binomial, trinomial, or none of these. See Examples 2 and 3.

11. $6x + 0.3$

12. $7x - 0.8$

13. $3x^2 - 2x + 5$

14. $5x^2 - 3x - 2$

15. -3^4xy^2

16. -7^5abc

17. $x^2y - 4xy^2 + 5x + y^4$

18. $-2x^2y - 3y^2 + 4x + y^5$

If $P(x) = x^2 + x + 1$ and $Q(x) = 5x^2 - 1$, find the following. See Example 4.

19. $P(7)$

20. $Q(4)$

21. $Q(-10)$

22. $P(-4)$

23. $Q\left(\dfrac{1}{4}\right)$

24. $P\left(\dfrac{1}{2}\right)$

Refer to Example 5 for Exercises 25 through 28.

25. Find the height of the object at $t = 2$ seconds.

26. Find the height of the object at $t = 4$ seconds.

27. Find the height of the object at $t = 6$ seconds.

28. Approximate (to the nearest second) how long it takes before the object hits the ground. (*Hint:* The object hits the ground when $P(x) = 0$.)

Simplify by combining like terms. See Example 6.

29. $5y + y$

30. $-x + 3x$

31. $4x + 7x - 3$

32. $-8y + 9y + 4y^2$

33. $4xy + 2x - 3xy - 1$

34. $-8xy^2 + 4x - x + 2xy^2$

35. $7x^2 - 2xy + 5y^2 - x^2 + xy + 11y^2$

36. $-a^2 + 18ab - 2b^2 + 14a^2 - 12ab - b^2$

MIXED PRACTICE

Perform the indicated operations. See Examples 7 through 10.

37. $(9y^2 - 8) + (9y^2 - 9)$

38. $(x^2 + 4x - 7) + (8x^2 + 9x - 7)$

39. Add $(x^2 + xy - y^2)$ and $(2x^2 - 4xy + 7y^2)$.

40. Add $(4x^3 - 6x^2 + 5x + 7)$ and $(2x^2 + 6x - 3)$.

41. $\begin{aligned} x^2 - 6x + 3 \\ +\ \underline{(2x + 5)} \end{aligned}$

42. $\begin{aligned} -2x^2 + 3x - 9 \\ +\ \underline{(2x - 3)} \end{aligned}$

43. $(9y^2 - 7y + 5) - (8y^2 - 7y + 2)$

44. $(2x^2 + 3x + 12) - (5x - 7)$

45. Subtract $(6x^2 - 3x)$ from $(4x^2 + 2x)$.

46. Subtract $(xy + x - y)$ from $(xy + x - 3)$.

47. $\begin{aligned} 3x^2 - 4x + 8 \\ -\ \underline{(5x^2 - 7)} \end{aligned}$

48. $\begin{aligned} -3x^2 - 4x + 8 \\ -\ \underline{(5x + 12)} \end{aligned}$

49. $(5x - 11) + (-x - 2)$

50. $(3x^2 - 2x) + (5x^2 - 9x)$

51. $(7x^2 + x + 1) - (6x^2 + x - 1)$

52. $(4x - 4) - (-x - 4)$

53. $(7x^3 - 4x + 8) + (5x^3 + 4x + 8x)$

54. $(9xyz + 4x - y) + (-9xyz - 3x + y + 2)$

55. $(9x^3 - 2x^2 + 4x - 7) - (2x^3 - 6x^2 - 4x + 3)$

56. $(3x^2 + 6xy + 3y^2) - (8x^2 - 6xy - y^2)$

57. Add $(y^2 + 4yx + 7)$ and $(-19y^2 + 7yx + 7)$.

58. Subtract $(x - 4)$ from $(3x^2 - 4x + 5)$.

59. $(3x^3 - b + 2a - 6) + (-4x^3 + b + 6a - 6)$

60. $(5x^2 - 6) + (2x^2 - 4x + 8)$

61. $(4x^2 - 6x + 2) - (-x^2 + 3x + 5)$

62. $(5x^2 + x + 9) - (2x^2 - 9)$

63. $(-3x + 8) + (-3x^2 + 3x - 5)$

64. $(5y^2 - 2y + 4) + (3y + 7)$

65. $(-3 + 4x^2 + 7xy^2) + (2x^3 - x^2 + xy^2)$

66. $(-3x^2y + 4) - (-7x^2y - 8y)$

67. $\begin{aligned} 6y^2 - 6y + 4 \\ -\underline{(-y^2 - 6y + 7)} \end{aligned}$

68. $\begin{aligned} -4x^3 + 4x^2 - 4x \\ -\underline{(2x^3 - 2x^2 + 3x)} \end{aligned}$

69. $\begin{aligned} 3x^2 + 15x + 8 \\ +\underline{(2x^2 + 7x + 8)} \end{aligned}$

70. $\begin{aligned} 9x^2 + 9x - 4 \\ +\underline{(7x^2 - 3x - 4)} \end{aligned}$

71. $\left(\dfrac{1}{2}x^2 - \dfrac{1}{3}x^2y + 2y^3\right) + \left(\dfrac{1}{4}x^2 - \dfrac{8}{3}x^2y - \dfrac{1}{2}y^3\right)$

72. $\left(\frac{2}{5}a^2 - ab + \frac{4}{3}b^2\right) + \left(\frac{1}{5}a^2b - ab + \frac{5}{6}b^2\right)$

73. Find the sum of $(5q^4 - 2q^2 - 3q)$ and $(-6q^4 + 3q^2 + 5)$.

74. Find the sum of $(5y^4 - 7y^2 + x^2 - 3)$ and $(-3y^4 + 2y^2 + 4)$.

75. Subtract $(3x + 7)$ from the sum of $(7x^2 + 4x + 9)$ and $(8x^2 + 7x - 8)$.

76. Subtract $(9x + 8)$ from the sum of $(3x^2 - 2x - x^3 + 2)$ and $(5x^2 - 8x - x^3 + 4)$.

77. Find the sum of $(4x^4 - 7x^2 + 3)$ and $(2 - 3x^4)$.

78. Find the sum of $(8x^4 - 14x^2 + 6)$ and $(-12x^6 - 21x^4 - 9x^2)$.

79. $\left(\frac{2}{3}x^2 - \frac{1}{6}x + \frac{5}{6}\right) - \left(\frac{1}{3}x^2 + \frac{5}{6}x - \frac{1}{6}\right)$

80. $\left(\frac{3}{16}x^2 + \frac{5}{8}x - \frac{1}{4}\right) - \left(\frac{5}{16}x^2 - \frac{3}{8}x + \frac{3}{4}\right)$

Solve. See Example 5.

The surface area of a rectangular box is given by the polynomial function

$$f(x) = 2HL + 2LW + 2HW$$

and is measured in square units. In business, surface area is often calculated to help determine cost of materials.

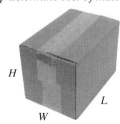

△ **81.** A rectangular box is to be constructed to hold a new camcorder. The box is to have dimensions 5 inches by 4 inches by 9 inches. Find the surface area of the box.

△ **82.** Suppose it has been determined that a box of dimensions 4 inches by 4 inches by 8.5 inches can be used to contain the camcorder in Exercise 81. Find the surface area of this box and calculate the square inches of material saved by using this box instead of the box in Exercise 81.

A projectile is fired upward from the ground with an initial velocity of 300 feet per second. Neglecting air resistance, the height of the projectile at any time t can be described by the polynomial function

$$P(t) = -16t^2 + 300t$$

Use the following table of values to answer Exercise 83.

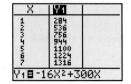

83. Find the height of the projectile at the given times.
  a. $t = 1$ second   **b.** $t = 2$ seconds
  c. $t = 3$ seconds   **d.** $t = 4$ seconds
  e. Generate the table above for positive integer x-values and explain why the height increases and then decreases as time passes.
  f. Use the table from part **e** and approximate (to the nearest second) how long before the object hits the ground.

84. An object is thrown upward with an initial velocity of 25 feet per second from the top of the 984-foot-high Eiffel Tower in Paris, France. The height of the object at any time t can be described by the polynomial function $P(t) = -16t^2 + 25t + 984$. Find the height of the projectile at each given time. (*Source:* Council on Tall Buildings and Urban Habitat, Lehigh University)
  a. $t = 1$ second   **b.** $t = 3$ seconds
  c. $t = 5$ seconds
  d. Approximate (to the nearest second) how long before the object hits the ground.

85. The polynomial function $P(x) = 45x - 100,000$ models the relationship between the number of computer briefcases x that a company sells and the profit the company makes, $P(x)$. Find $P(4000)$, the profit from selling 4000 computer briefcases.

86. The total cost (in dollars) for MCD, Inc., Manufacturing Company to produce x blank audiocassette tapes per week is given by the polynomial function $C(x) = 0.8x + 10,000$. Find the total cost of producing 20,000 tapes per week.

87. The total revenues (in dollars) for MCD, Inc., Manufacturing Company from selling x blank audiocasette tapes per week is given by the polynomial function $R(x) = 2x$. Find the total revenue from selling 20,000 tapes per week.

88. In business, profit equals revenue minus cost, or $P(x) = R(x) - C(x)$. Find the profit function for MCD, Inc., by subtracting the given functions in Exercises 86 and 87.

Match each equation with its graph. See Example 11.

89. $f(x) = 3x^2 - 2$

90. $h(x) = 5x^3 - 6x + 2$

91. $g(x) = -2x^3 - 3x^2 + 3x - 2$

92. $F(x) = -2x^2 - 6x + 2$

A   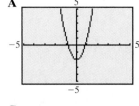   B

C   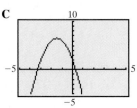   D

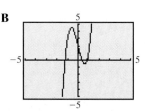

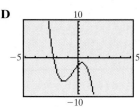

REVIEW AND PREVIEW

Multiply. See Section 1.4.

93. $5(3x - 2)$

94. $-7(2z - 6y)$

95. $-2(x^2 - 5x + 6)$

96. $5(-3y^2 - 2y + 7)$

CONCEPT EXTENSIONS

Solve. See the Concept Checks in this section.

97. Which polynomial(s) is the opposite of $8x - 6$?

   a. $-(8x - 6)$ **b.** $8x + 6$

   c. $-8x + 6$ **d.** $-8x - 6$

98. Which polynomial(s) is the opposite of $-y^5 + 10y^3 - 2.3$?

   a. $y^5 + 10y^3 + 2.3$ **b.** $-y^5 - 10y^3 - 2.3$

   c. $y^5 + 10y^3 - 2.3$ **d.** $y^5 - 10y^3 + 2.3$

99. Correct the subtraction.

$$(12x - 1.7) - (15x + 6.2) = 12x - 1.7 - 15x + 6.2$$
$$= -3x + 4.5$$

100. Correct the addition.

$$(12x - 1.7) + (15x + 6.2) = 12x - 1.7 + 15x + 6.2$$
$$= 27x + 7.9$$

101. Write a function, $P(x)$, so that $P(0) = 7$.

102. Write a function, $R(x)$, so that $R(1) = 2$.

103. In your own words, describe how to find the degree of a term.

104. In your own words, describe how to find the degree of a polynomial.

Perform the indicated operations.

105. $(4x^{2a} - 3x^a + 0.5) - (x^{2a} - 5x^a - 0.2)$

106. $(9y^{5a} - 4y^{3a} + 1.5y) - (6y^{5a} - y^{3a} + 4.7y)$

107. $(8x^{2y} - 7x^y + 3) + (-4x^{2y} + 9x^y - 14)$

108. $(14z^{5x} + 3z^{2x} + z) - (2z^{5x} - 10z^{2x} + 3z)$

Find each perimeter.

109.

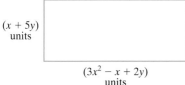

$(x + 5y)$ units

$(3x^2 - x + 2y)$ units

110.

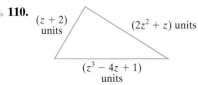

$(z + 2)$ units

$(2z^2 + z)$ units

$(z^3 - 4z + 1)$ units

If $P(x) = 3x + 3, Q(x) = 4x^2 - 6x + 3$, and $R(x) = 5x^2 - 7$, find the following.

111. $P(x) + Q(x)$

112. $R(x) + P(x)$

113. $Q(x) - R(x)$

114. $P(x) - Q(x)$

115. $2[Q(x)] - R(x)$

116. $-5[P(x)] - Q(x)$

117. $3[R(x)] + 4[P(x)]$

118. $2[Q(x)] + 7[R(x)]$

*If $P(x)$ is the polynomial given, find **a.** $P(a)$, **b.** $P(-x)$, and **c.** $P(x + h)$.*

119. $P(x) = 2x - 3$

120. $P(x) = 8x + 3$

121. $P(x) = 4x$

122. $P(x) = -4x$

123. $P(x) = 4x - 1$

124. $P(x) = 3x - 2$

125. The function $f(x) = 0.07x^2 - 0.8x + 3.6$ can be used to approximate the amazing growth of the number of Web logs (Blogs) appearing on the Internet from January 2004 to October 2006, where January 2004 = 1 for x, February 2004 = 2 for x, and so on, and y is the number of blogs (in millions). Round answers to the nearest tenth of a million. (*Note:* This is one company's tracking of the cumulative number of blogs. These numbers vary greatly according to source and activity of blog.) (*Source:* Technorati)

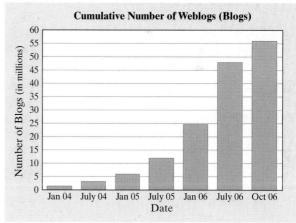

Cumulative Number of Weblogs (Blogs)

    a. Approximate the number of Web logs on the Internet in January 2004.

    b. Approximate the number of Web logs on the Internet in October 2006.

    c. Use this function to approximate the number of Web logs on the Internet in May 2009. (65 months)

    d. From parts **a**, **b**, and **c**, determine whether the number of Web logs on the Internet is increasing at a steady rate. Explain why or why not.

126. The function $f(x) = -61x^2 - 5530x + 585{,}753$ can be used to approximate the number of international students studying in the United States during the academic years 2002 through 2006, where x is the number of years after 2002 and $f(x)$ is the number of international students.

(*Source:* Institute of International Education: *Open Doors 2006*)

a. Approximate the number of international students studying in the United States in 2003.

b. Approximate the number of international students studying in the United States in 2005.

c. Use the function to predict the number of international students studying in the United States in 2010.

127. The function $f(x) = -0.39x^2 + 2.49x + 38.7$ can be used to approximate the number of Americans under age 65 without health insurance during the period 1999–2005, where x is the number of years after 1999 and $f(x)$ is the number in millions of Americans. Round answers to the nearest tenth of a million. (*Source:* National Center for Health Statistics)

a. Approximate the number of Americans under 65 without health insurance in 2003.

b. Use the function to predict the number of Americans under 65 without health insurance in 2008.

128. University libraries need to continue to grow to meet the educational needs of their students. The function $f(x) = -14x^2 + 269.2x + 7414.2$ approximates the number of written volumes (in thousands) to be found in the University of California, Los Angeles, libraries for the years from 2002 to 2005, where x is the number of years after 2002. (*Source:* Association of Research Libraries)

a. Approximate the number of volumes in 2004.

b. Use the function to estimate the number of volumes at UCLA libraries in 2008.

129. Sport utility vehicle (SUV) sales in the United States have increased since 1995. The function $f(x) = -0.005x^2 + 0.377x + 1.71$ can be used to approximate the number of SUV sales during the years 1995–2004, where x is the number of years after 1995 and $f(x)$ is the SUV sales (in millions). Round answers to the nearest tenth of a million. (*Source:* Bureau of Transportation Statistics)

a. Approximate the number of SUVs sold in 2003.

b. Use the function to predict the number of SUVs sold in 2010.

130. The function $f(x) = 873x^2 - 4104x + 40,263$ can be used to approximate the number of AIDS cases diagnosed in the United States from 2001 to 2005, where x is the number of years since 2001. (*Source:* Based on data from the U.S. Centers for Disease Control and Prevention)

a. Approximate the number of AIDS cases diagnosed in the United States in 2001.

b. Approximate the number of AIDS cases diagnosed in the United States in 2003.

c. Approximate the number of AIDS cases diagnosed in the United States in 2005.

d. Describe the trend in the number of AIDS cases diagnosed during the period covered by this model.

5.4 MULTIPLYING POLYNOMIALS

OBJECTIVES

1 Multiply two polynomials.

2 Multiply binomials.

3 Square binomials.

4 Multiply the sum and difference of two terms.

5 Multiply three or more polynomials.

6 Evaluate polynomial functions.

OBJECTIVE 1 ▶ Multiplying two polynomials. Properties of real numbers and exponents are used continually in the process of multiplying polynomials. To multiply monomials, for example, we apply the commutative and associative properties of real numbers and the product rule for *exponents*.

EXAMPLE 1 Multiply.

a. $(2x^3)(5x^6)$ b. $(7y^4z^4)(-xy^{11}z^5)$

Solution Group like bases and apply the product rule for exponents.

a. $(2x^3)(5x^6) = 2(5)(x^3)(x^6) = 10x^9$

b. $(7y^4z^4)(-xy^{11}z^5) = 7(-1)x(y^4y^{11})(z^4z^5) = -7xy^{15}z^9$ ☐

PRACTICE

1 Multiply.

a. $(3x^4)(2x^2)$ b. $(-5m^4np^3)(-8mnp^5)$

To multiply a monomial by a polynomial other than a monomial, we use an expanded form of the distributive property.

$$a(b + c + d + \cdots + z) = ab + ac + ad + \cdots + az$$

Notice that the monomial a is multiplied by each term of the polynomial.

▶ **Helpful Hint**

See Sections 5.1 and 5.2 to review exponential expressions further.

EXAMPLE 2   Multiply.

a. $2x(5x - 4)$ **b.** $-3x^2(4x^2 - 6x + 1)$ **c.** $-xy(7x^2y + 3xy - 11)$

Solution   Apply the distributive property.

a. $2x(5x - 4) = 2x(5x) + 2x(-4)$ Use the distributive property.
$\qquad\qquad\qquad = 10x^2 - 8x$ Multiply.

b. $-3x^2(4x^2 - 6x + 1) = -3x^2(4x^2) + (-3x^2)(-6x) + (-3x^2)(1)$
$\qquad\qquad\qquad\qquad\qquad = -12x^4 + 18x^3 - 3x^2$

c. $-xy(7x^2y + 3xy - 11) = -xy(7x^2y) + (-xy)(3xy) + (-xy)(-11)$
$\qquad\qquad\qquad\qquad\qquad = -7x^3y^2 - 3x^2y^2 + 11xy$ □

PRACTICE
2   Multiply.

a. $3x(7x - 1)$ **b.** $-5a^2(3a^2 - 6a + 5)$ **c.** $-mn^3(5m^2n^2 + 2mn - 5m)$

To multiply any two polynomials, we can use the following.

> **Multiplying Two Polynomials**
> To multiply any two polynomials, use the distributive property and multiply each term of one polynomial by each term of the other polynomial. Then combine any like terms.

Concept Check ☑

Find the error:

$$4x(x - 5) + 2x$$
$$= 4x(x) + 4x(-5) + 4x(2x)$$
$$= 4x^2 - 20x + 8x^2$$
$$= 12x^2 - 20x$$

EXAMPLE 3   Multiply and simplify the product if possible.

a. $(x + 3)(2x + 5)$ **b.** $(2x - 3)(5x^2 - 6x + 7)$

Solution

a. Multiply each term of $(x + 3)$ by $(2x + 5)$.

$(x + 3)(2x + 5) = x(2x + 5) + 3(2x + 5)$ Apply the distributive property.
$\qquad\qquad\qquad = 2x^2 + 5x + 6x + 15$ Apply the distributive property again.
$\qquad\qquad\qquad = 2x^2 + 11x + 15$ Combine like terms.

b. Multiply each term of $(2x - 3)$ by each term of $(5x^2 - 6x + 7)$.

$(2x - 3)(5x^2 - 6x + 7) = 2x(5x^2 - 6x + 7) + (-3)(5x^2 - 6x + 7)$
$\qquad\qquad\qquad\qquad\qquad = 10x^3 - 12x^2 + 14x - 15x^2 + 18x - 21$
$\qquad\qquad\qquad\qquad\qquad = 10x^3 - 27x^2 + 32x - 21$ Combine like terms.   □

Answer to Concept Check:
$4x(x - 5) + 2x$
$= 4x(x) + 4x(-5) + 2x$
$= 4x^2 - 20x + 2x$
$= 4x^2 - 18x$

PRACTICE
3   Multiply and simplify the product if possible.

a. $(x + 5)(2x + 3)$ **b.** $(3x - 1)(x^2 - 6x + 2)$

Sometimes polynomials are easier to multiply vertically, in the same way we multiply real numbers. When multiplying vertically, we line up like terms in the **partial products** vertically. This makes combining like terms easier.

EXAMPLE 4 Multiply vertically $(4x^2 + 7)(x^2 + 2x + 8)$.

Solution

$$
\begin{array}{r}
x^2 + 2x + 8 \\
4x^2 + 7 \\
\hline
7x^2 + 14x + 56 \qquad 7(x^2 + 2x + 8) \\
4x^4 + 8x^3 + 32x^2 \qquad\qquad 4x^2(x^2 + 2x + 8) \\
\hline
4x^4 + 8x^3 + 39x^2 + 14x + 56 \quad \text{Combine like terms.}
\end{array}
$$

□

PRACTICE
4 Multiply vertically: $(3x^2 + 2)(x^2 - 4x - 5)$.

OBJECTIVE 2 ▶ Multiplying binomials. When multiplying a binomial by a binomial, we can use a special order of multiplying terms, called the **FOIL** order. The letters of FOIL stand for "First-Outer-Inner-Last." To illustrate this method, let's multiply $(2x - 3)$ by $(3x + 1)$.

Multiply the **F**irst terms of each binomial. $(2x - 3)(3x + 1)$ **F** $2x(3x) = 6x^2$

Multiply the **O**uter terms of each binomial. $(2x - 3)(3x + 1)$ **O** $2x(1) = 2x$

Multiply the **I**nner terms of each binomial. $(2x - 3)(3x + 1)$ **I** $-3(3x) = -9x$

Multiply the **L**ast terms of each binomial. $(2x - 3)(3x + 1)$ **L** $-3(1) = -3$
Combine like terms.

$$6x^2 + 2x - 9x - 3 = 6x^2 - 7x - 3$$

TECHNOLOGY NOTE

A graphing utility may be used to visualize operations on polynomials. To visualize a multiplication statement such as

$$(2x - 3)(3x + 1) = 6x^2 - 7x - 3,$$

graph $y_1 = (2x - 3)(3x + 1)$ and $y_2 = 6x^2 - 7x - 3$ on the same set of axes and see that their graphs coincide.

Note: If the graphs do not coincide, we can be sure that a mistake has been made in multiplying the polynomials or in entering keystrokes. If the graphs appear to coincide, we cannot be sure that our work is correct. This is because it is possible for the graphs to differ so slightly that we do not notice it.

The graphs of y_1 and y_2 are shown to the left. The graphs *appear* to coincide so the multiplication statement

$$(2x - 3)(3x + 1) = 6x^2 - 7x - 3$$

appears to be correct.

A table of values, such as the one to the left showing y_1 and y_2, can also help confirm operations on polynomials.

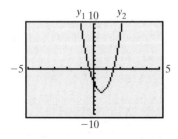

EXAMPLE 5   Use the FOIL order to multiply $(x - 1)(x + 2)$.

Solution

$$
\begin{array}{cccc}
\text{First} & \text{Outer} & \text{Inner} & \text{Last} \\
\downarrow & \downarrow & \downarrow & \downarrow
\end{array}
$$

$$(x - 1)(x + 2) = x \cdot x + 2 \cdot x + (-1)x + (-1)(2)$$
$$= x^2 + 2x - x - 2$$
$$= x^2 + x - 2 \quad \text{Combine like terms.} \qquad \square$$

PRACTICE
5   Use the FOIL order to multiply $(x - 5)(x + 3)$.

EXAMPLE 6   Multiply.

a. $(2x - 7)(3x - 4)$ **b.** $(3x^2 + y)(5x^2 - 2y)$

Solution

$$
\begin{array}{cccc}
\text{First} & \text{Outer} & \text{Inner} & \text{Last} \\
\downarrow & \downarrow & \downarrow & \downarrow
\end{array}
$$

a. $(2x - 7)(3x - 4) = 2x(3x) + 2x(-4) + (-7)(3x) + (-7)(-4)$
$$= 6x^2 - 8x - 21x + 28$$
$$= 6x^2 - 29x + 28$$

$$
\begin{array}{cccc}
\text{F} & \text{O} & \text{I} & \text{L} \\
\downarrow & \downarrow & \downarrow & \downarrow
\end{array}
$$

b. $(3x^2 + y)(5x^2 - 2y) = 15x^4 - 6x^2y + 5x^2y - 2y^2$
$$= 15x^4 - x^2y - 2y^2 \qquad \square$$

PRACTICE
6   Multiply.

a. $(3x - 5)(2x - 7)$ **b.** $(2x^2 - 3y)(4x^2 + y)$

OBJECTIVE 3 ▶ Squaring binomials. The **square of a binomial** is a special case of the product of two binomials. By the FOIL order for multiplying two binomials, we have

$$(a + b)^2 = (a + b)(a + b)$$

$$
\begin{array}{cccc}
\text{F} & \text{O} & \text{I} & \text{L} \\
\downarrow & \downarrow & \downarrow & \downarrow
\end{array}
$$

$$= a^2 + ab + ba + b^2$$
$$= a^2 + 2ab + b^2$$

This product can be visualized geometrically by analyzing areas.

Area of larger square: $(a + b)^2$
Sum of areas of smaller rectangles: $a^2 + 2ab + b^2$
Thus, $(a + b)^2 = a^2 + 2ab + b^2$

The same pattern occurs for the square of a difference. In general,

---

Square of a Binomial

$$(a + b)^2 = a^2 + 2ab + b^2 \qquad (a - b)^2 = a^2 - 2ab + b^2$$

In other words, a binomial squared is the sum of the first term squared, twice the product of both terms, and the second term squared.

EXAMPLE 7 Multiply.

a. $(x + 5)^2$ **b.** $(x - 9)^2$ **c.** $(3x + 2z)^2$ **d.** $(4m^2 - 3n)^2$

Solution

$(a + b)^2 = a^2 + 2 \cdot a \cdot b + b^2$

a. $\downarrow \quad \downarrow \quad \downarrow \quad \downarrow \downarrow \downarrow \quad \downarrow$

$(x + 5)^2 = x^2 + 2 \cdot x \cdot 5 + 5^2 = x^2 + 10x + 25$

b. $(x - 9)^2 = x^2 - 2 \cdot x \cdot 9 + 9^2 = x^2 - 18x + 81$

c. $(3x + 2z)^2 = (3x)^2 + 2(3x)(2z) + (2z)^2 = 9x^2 + 12xz + 4z^2$

d. $(4m^2 - 3n)^2 = (4m^2)^2 - 2(4m^2)(3n) + (3n)^2 = 16m^4 - 24m^2n + 9n^2$ ☐

PRACTICE

7 Multiply.

a. $(x + 6)^2$ **b.** $(x - 2)^2$ **c.** $(3x + 5y)^2$ **d.** $(3x^2 - 8b)^2$

── **TECHNOLOGY NOTE** ──

To verify that
$(x + 5)^2 \neq x^2 + 25$,
for example, graph
$y_1 = (x + 5)^2$ and
$y_2 = x^2 + 25$ using the
same standard window.
Since the two graphs do
not coincide, the expres-
sions are not equivalent.

▶ **Helpful Hint**

Note that $(a + b)^2 = a^2 + 2ab + b^2$, **not** $a^2 + b^2$. Also,

$(a - b)^2 = a^2 - 2ab + b^2$, **not** $a^2 - b^2$.

OBJECTIVE 4 ▶ Multiplying the sum and difference of two terms. Another special product applies to the sum and difference of the same two terms. Multiply $(a + b)(a - b)$ to see a pattern.

$$(a + b)(a - b) = a^2 - ab + ba - b^2$$
$$= a^2 - b^2$$

Product of the Sum and Difference of Two Terms

$$(a + b)(a - b) = a^2 - b^2$$

The product of the sum and difference of the same two terms is the difference of the first term squared and the second term squared.

EXAMPLE 8 Multiply.

a. $(x - 3)(x + 3)$ **b.** $(4y + 1)(4y - 1)$

c. $(x^2 + 2y)(x^2 - 2y)$ **d.** $\left(3m^2 - \dfrac{1}{2}\right)\left(3m^2 + \dfrac{1}{2}\right)$

Solution

$(a + b)(a - b) = a^2 - b^2$

a. $\downarrow \quad \downarrow \downarrow \quad \downarrow \quad \downarrow \quad \downarrow$

$(x + 3)(x - 3) = x^2 - 3^2 = x^2 - 9$

b. $(4y + 1)(4y - 1) = (4y)^2 - 1^2 = 16y^2 - 1$

c. $(x^2 + 2y)(x^2 - 2y) = (x^2)^2 - (2y)^2 = x^4 - 4y^2$

d. $\left(3m^2 - \dfrac{1}{2}\right)\left(3m^2 + \dfrac{1}{2}\right) = (3m^2)^2 - \left(\dfrac{1}{2}\right)^2 = 9m^4 - \dfrac{1}{4}$ ☐

PRACTICE

8 Multiply.

a. $(x - 7)(x + 7)$ **b.** $(2a + 5)(2a - 5)$

c. $\left(5x^2 + \dfrac{1}{4}\right)\left(5x^2 - \dfrac{1}{4}\right)$ **d.** $(a^3 - 4b^2)(a^3 + 4b^2)$

EXAMPLE 9   Multiply $[3 + (2a + b)]^2$.

Solution   Think of 3 as the first term and $(2a + b)$ as the second term, and apply the method for squaring a binomial.

$$[a \quad + \quad b]^2 = a^2 + 2(a) \cdot \quad b \quad + \quad b^2$$
$$[3 + \overbrace{(2a + b)}]^2 = 3^2 + 2(3)\overbrace{(2a + b)} + \overbrace{(2a + b)}^2$$
$$= 9 + 6(2a + b) + (2a + b)^2$$
$$= 9 + 12a + 6b + (2a)^2 + 2(2a)(b) + b^2 \quad \text{Square } (2a + b).$$
$$= 9 + 12a + 6b + 4a^2 + 4ab + b^2$$

PRACTICE
9   Multiply $[2 + (3x - y)]^2$.

EXAMPLE 10   Multiply $[(5x - 2y) - 1][(5x - 2y) + 1]$.

Solution   Think of $(5x - 2y)$ as the first term and 1 as the second term, and apply the method for the product of the sum and difference of two terms.

$$(a \quad - b) \quad (a \quad + b) = \quad a^2 \quad - b^2$$
$$[\overbrace{(5x - 2y)} - 1][\overbrace{(5x - 2y)} + 1] = \overbrace{(5x - 2y)}^2 - 1^2$$
$$= (5x)^2 - 2(5x)(2y) + (2y)^2 - 1 \quad \text{Square}$$
$$= 25x^2 - 20xy + 4y^2 - 1 \qquad (5x - 2y).$$

PRACTICE
10   Multiply $[(3x - y) - 5][(3x - y) + 5]$.

OBJECTIVE 5 ▶ Multiplying three or more polynomials. To multiply three or more polynomials, more than one method may be needed.

EXAMPLE 11   Multiply: $(x - 3)(x + 3)(x^2 - 9)$

Solution   We multiply the first two binomials, the sum and difference of two terms. Then we multiply the resulting two binomials, the square of a binomial.

$$(x - 3)(x + 3)(x^2 - 9) = (x^2 - 9)(x^2 - 9) \quad \text{Multiply } (x - 3)(x + 3).$$
$$= (x^2 - 9)^2$$
$$= x^4 - 18x^2 + 81 \quad \text{Square } (x^2 - 9).$$

PRACTICE
11   Multiply $(x + 4)(x - 4)(x^2 - 16)$.

OBJECTIVE 6 ▶ Evaluating polynomial functions. Our work in multiplying polynomials is often useful in evaluating polynomial functions.

EXAMPLE 12   If $f(x) = x^2 + 5x - 2$, find $f(a + 1)$.

Solution   To find $f(a + 1)$, replace x with the expression $a + 1$ in the polynomial function $f(x)$.

$$f(x) = x^2 + 5x - 2$$
$$f(a + 1) = (a + 1)^2 + 5(a + 1) - 2$$
$$= a^2 + 2a + 1 + 5a + 5 - 2$$
$$= a^2 + 7a + 4$$

PRACTICE
12   If $f(x) = x^2 - 3x + 5$, find $f(h + 1)$.

VOCABULARY & READINESS CHECK

Use the choices to fill in each blank.

1. $(6x^3)\left(\dfrac{1}{2}x^3\right) = $ _____

    a. $3x^3$ **b.** $3x^6$ **c.** $10x^6$ **d.** $\dfrac{13}{2}x^6$

2. $(x + 7)^2 = $ _____

    a. $x^2 + 49$ **b.** $x^2 - 49$ **c.** $x^2 + 14x + 49$ **d.** $x^2 + 7x + 49$

3. $(x + 7)(x - 7) = $ _____

    a. $x^2 + 49$ **b.** $x^2 - 49$ **c.** $x^2 + 14x - 49$ **d.** $x^2 + 7x - 49$

4. The product of $(3x - 1)(4x^2 - 2x + 1)$ is a polynomial of degree _____.

    a. 3   **b.** 12   **c.** $12x^3$ **d.** 2

5. If $f(x) = x^2 + 1$ then $f(a + 1) = $ _____

    a. $(a + 1)^2$ **b.** $a + 1$ **c.** $(a + 1)^2 + (a + 1)$ **d.** $(a + 1)^2 + 1$

6. $[x + (2y + 1)]^2 = $ _____

    a. $[x + (2y + 1)][x - (2y + 1)]$ **b.** $[x + (2y + 1)][x + (2y + 1)]$ **c.** $[x + (2y + 1)][x + (2y - 1)]$

5.4 | EXERCISE SET

Multiply. See Examples 1 through 4.

1. $(-4x^3)(3x^2)$

2. $(-6a)(4a)$

3. $3x(4x + 7)$

4. $5x(6x - 4)$

5. $-6xy(4x + y)$

6. $-8y(6xy + 4x)$

7. $-4ab(xa^2 + ya^2 - 3)$

8. $-6b^2z(z^2a + baz - 3b)$

9. $(x - 3)(2x + 4)$

10. $(y + 5)(3y - 2)$

11. $(2x + 3)(x^3 - x + 2)$

12. $(a + 2)(3a^2 - a + 5)$

13. $3x - 2$
    $\times\ 5x + 1$

14. $2z - 4$
    $\times\ 6z - 2$

15. $3m^2 + 2m - 1$
    $\times\ \qquad 5m + 2$

16. $2x^2 - 3x - 4$
    $\times\ \qquad x + 5$

Multiply the binomials. See Examples 5 and 6.

17. $(x - 3)(x + 4)$

18. $(c - 3)(c + 1)$

19. $(5x + 8y)(2x - y)$

20. $(2n - 9m)(n - 7m)$

21. $(3x - 1)(x + 3)$

22. $(5d - 3)(d + 6)$

23. $\left(3x + \dfrac{1}{2}\right)\left(3x - \dfrac{1}{2}\right)$

24. $\left(2x - \dfrac{1}{3}\right)\left(2x + \dfrac{1}{3}\right)$

25. $(5x^2 - 2y^2)(x^2 - 3y^2)$

26. $(4x^2 - 5y^2)(x^2 - 2y^2)$

Multiply, using special product methods. See Examples 7 and 8.

27. $(x + 4)^2$

28. $(x - 5)^2$

29. $(6y - 1)(6y + 1)$

30. $(7x - 9)(7x + 9)$

31. $(3x - y)^2$

32. $(4x - z)^2$

33. $(5b - 6y)(5b + 6y)$

34. $(2x - 4y)(2x + 4y)$

Multiply, using special product methods. See Examples 9 and 10.

35. $[3 + (4b + 1)]^2$

36. $[5 - (3b - 3)]^2$

37. $[(2s - 3) - 1][(2s - 3) + 1]$

38. $[(2y + 5) + 6][(2y + 5) - 6]$

39. $[(xy + 4) - 6]^2$

40. $[(2a^2 + 4a) + 1]^2$

41. Explain when the FOIL method can be used to multiply polynomials.

42. Explain why the product of $(a + b)$ and $(a - b)$ is not a trinomial.

Multiply. See Example 11.

43. $(x + y)(x - y)(x^2 - y^2)$

44. $(z - y)(z + y)(z^2 - y^2)$

45. $(x - 2)^4$

46. $(x - 1)^4$

47. $(x - 5)(x + 5)(x^2 + 25)$

48. $(x + 3)(x - 3)(x^2 + 9)$

MIXED PRACTICE

Multiply.

49. $(3x + 1)(3x + 5)$

50. $(4x - 5)(5x + 6)$

51. $(2x^3 + 5)(5x^2 + 4x + 1)$

52. $(3y^3 - 1)(3y^3 - 6y + 1)$

53. $(7x - 3)(7x + 3)$

54. $(4x + 1)(4x - 1)$

55. $\begin{array}{r} 3x^2 + 4x - 4 \\ \times \qquad 3x + 6 \end{array}$

56. $\begin{array}{r} 6x^2 + 2x - 1 \\ \times \qquad 3x - 6 \end{array}$

57. $\left(4x + \dfrac{1}{3}\right)\left(4x - \dfrac{1}{2}\right)$

58. $\left(4y - \dfrac{1}{3}\right)\left(3y - \dfrac{1}{8}\right)$

59. $(6x + 1)^2$

60. $(4x + 7)^2$

61. $(x^2 + 2y)(x^2 - 2y)$

62. $(3x + 2y)(3x - 2y)$

63. $-6a^2b^2[5a^2b^2 - 6a - 6b]$

64. $7x^2y^3(-3ax - 4xy + z)$

65. $(a - 4)(2a - 4)$

66. $(2x - 3)(x + 1)$

67. $(7ab + 3c)(7ab - 3c)$

68. $(3xy - 2b)(3xy + 2b)$

69. $(m - 4)^2$

70. $(x + 2)^2$

71. $(3x + 1)^2$

72. $(4x + 6)^2$

73. $(y - 4)(y - 3)$

74. $(c - 8)(c + 2)$

75. $(x + y)(2x - 1)(x + 1)$

76. $(z + 2)(z - 3)(2z + 1)$

77. $(3x^2 + 2x - 1)^2$

78. $(4x^2 + 4x - 4)^2$

79. $(3x + 1)(4x^2 - 2x + 5)$

80. $(2x - 1)(5x^2 - x - 2)$

If $f(x) = x^2 - 3x$, find the following. See Example 12.

81. $f(a)$

82. $f(c)$

83. $f(a + h)$

84. $f(a + 5)$

85. $f(b - 2)$

86. $f(a - b)$

REVIEW AND PREVIEW

Use the slope-intercept form of a line, $y = mx + b$, to find the slope of each line. See Section 2.4.

87. $y = -2x + 7$

88. $y = \dfrac{3}{2}x - 1$

89. $3x - 5y = 14$

90. $x + 7y = 2$

Use the vertical line test to determine which of the following are graphs of functions. See Section 2.2.

91.

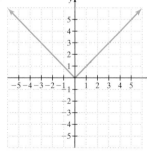

92.

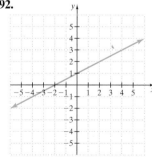

CONCEPT EXTENSIONS

Solve. See the Concept Check in this section.

93. Find the error: $7y(3z - 2) + 1$
$$= 21yz - 14y + 7y$$
$$= 21yz - 7y$$

94. Find the error: $2x + 3x(12 - x)$
$$= 5x(2 - x)$$
$$= 60x - 5x^2$$

95. Explain how to multiply a polynomial by a polynomial.

96. Explain why $(3x + 2)^2$ does not equal $9x^2 + 4$.

97. If $F(x) = x^2 + 3x + 2$, find
  a. $F(a + h)$
  b. $F(a)$
  c. $F(a + h) - F(a)$

98. If $g(x) = x^2 + 2x + 1$, find
  a. $g(a + h)$
  b. $g(a)$
  c. $g(a + h) - g(a)$

Multiply. Assume that variables represent positive integers.

99. $5x^2y^n(6y^{n+1} - 2)$
100. $-3yz^n(2y^3z^{2n} - 1)$
101. $(x^a + 5)(x^{2a} - 3)$
102. $(x^a + y^{2b})(x^a - y^{2b})$

For Exercises 103 through 106, write the result as a simplified polynomial.

△ **103.** Find the area of the circle. Do not approximate π.

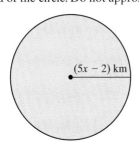
(5x − 2) km

△ **104.** Find the volume of the cylinder. Do not approximate π.

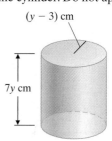
(y − 3) cm
7y cm

Find the area of each shaded region.

△ **105.**
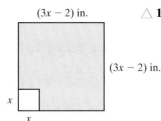
(3x − 2) in.
(3x − 2) in.
x
x

△ **106.**
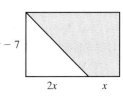
x − 7
2x
x

107. Perform each indicated operation. Explain the difference between the two problems.
  a. $(3x + 5) + (3x + 7)$
  b. $(3x + 5)(3x + 7)$

108. Explain when the FOIL method can be used to multiply polynomials.

If $R(x) = x + 5$, $Q(x) = x^2 - 2$, and $P(x) = 5x$, find the following.

109. $P(x) \cdot R(x)$
110. $P(x) \cdot Q(x)$
111. $[Q(x)]^2$
112. $[R(x)]^2$
113. $R(x) \cdot Q(x)$
114. $P(x) \cdot R(x) \cdot Q(x)$

 STUDY SKILLS BUILDER

Are You Getting All the Mathematics Help That You Need?

Remember that, in addition to your instructor, there are many places to get help with your mathematics course. For example,

- This text has an accompanying video lesson for every section and worked-out solutions to every Chapter Test exercise on video.
- The back of the book contains answers to odd-numbered exercises and selected solutions.
- A *student Solutions Manual* is available that contains worked-out solutions to odd-numbered exercises as well as solutions to every exercise in the Integrated Reviews, Chapter Reviews, Chapter Tests, and Cumulative Reviews.

- Don't forget to check with your instructor for other local resources available to you, such as a tutor center.

Exercises

1. List items you find helpful in the text and all student supplements to this text.

2. List all the campus help that is available to you for this course.

3. List any help (besides the textbook) from Exercises 1 and 2 above that you are using.

4. List any help (besides the textbook) that you feel you should try.

5. Write a goal for yourself that includes trying anything you listed in Exercise 4 during the next week.

5.5 THE GREATEST COMMON FACTOR AND FACTORING BY GROUPING

OBJECTIVE 1 ▶ Identifying the GCF. **Factoring** is the reverse process of multiplying. It is the process of writing a polynomial as a product.

$$6x^2 + 13x - 5 = (3x - 1)(2x + 5)$$

factoring →

← multiplying

In the next few sections, we review techniques for factoring polynomials. These techniques are used at the end of this chapter to solve polynomial equations.

To factor a polynomial, we first factor out the greatest common factor (GCF) of its terms, using the distributive property. The GCF of a list of terms or monomials is the product of the GCF of the numerical coefficients and each GCF of the powers of a common variable.

Finding the GCF of a List of Monomials

STEP 1. Find the GCF of the numerical coefficients.

STEP 2. Find the GCF of the variable factors.

STEP 3. The product of the factors found in Steps 1 and 2 is the GCF of the monomials.

EXAMPLE 1   Find the GCF of $20x^3y$, $10x^2y^2$, and $35x^3$.

Solution   The GCF of the numerical coefficients 20, 10, and 35 is 5, the largest integer that is a factor of each integer. The GCF of the variable factors x^3, x^2, and x^3 is x^2 because x^2 is the largest factor common to all three powers of x. The variable y is not a common factor because it does not appear in all three monomials. The GCF is thus

$$5 \cdot x^2, \quad \text{or} \quad 5x^2$$

To see this in factored form,

$$20x^3y = 2 \cdot 2 \cdot 5 \cdot x^2 \cdot x \cdot y$$
$$10x^2y^2 = 2 \cdot 5 \cdot x^2 \cdot y$$
$$35x^3 = 3 \cdot 5 \cdot x^2 \cdot x$$
$$\text{GCF} = 5 \cdot x^2$$

PRACTICE
1   Find the GCF of $32x^4y^2$, $48x^3y$, and $24y^2$.

OBJECTIVE 2 ▶ Factoring out the GCF of a polynomial's terms. A first step in factoring polynomials is to use the distributive property and write the polynomial as a product of the GCF of its monomial terms and a simpler polynomial. This is called **factoring out** the GCF.

EXAMPLE 2   Factor.

a. $8x^2 + 4$ **b.** $5y - 2z^4$ **c.** $6x^2 - 3x^3 + 12x^4$

Solution

a. The GCF of terms $8x^2$ and 4 is 4.

$$8x^2 + 4 = 4 \cdot 2x^2 + 4 \cdot 1 \quad \text{Factor out 4 from each term.}$$
$$= 4(2x^2 + 1) \quad \text{Apply the distributive property.}$$

The factored form of $8x^2 + 4$ is $4(2x^2 + 1)$. To check, multiply $4(2x^2 + 1)$ to see that the product is $8x^2 + 4$.

b. There is no common factor of the terms $5y$ and $-2z^4$ other than 1 (or -1).

c. The greatest common factor of $6x^2$, $-3x^3$, and $12x^4$ is $3x^2$. Thus,

$$6x^2 - 3x^3 + 12x^4 = 3x^2 \cdot 2 - 3x^2 \cdot x + 3x^2 \cdot 4x^2$$
$$= 3x^2(2 - x + 4x^2)$$ □

PRACTICE
2   Factor.

a. $6x^2 + 9 + 15x$ **b.** $3x - 8y^3$ **c.** $8a^4 - 2a^3$

> ▶ **Helpful Hint**
> To verify that the GCF has been factored out correctly, multiply the factors together and see that their product is the original polynomial.

EXAMPLE 3   Factor $17x^3y^2 - 34x^4y^2$.

Solution   The GCF of the two terms is $17x^3y^2$, which we factor out of each term.

$$17x^3y^2 - 34x^4y^2 = 17x^3y^2 \cdot 1 - 17x^3y^2 \cdot 2x$$
$$= 17x^3y^2(1 - 2x)$$ □

PRACTICE
3   Factor $64x^5y^2 - 8x^3y^2$.

> ▶ **Helpful Hint**
> If the GCF happens to be one of the terms in the polynomial, a factor of 1 will remain for this term when the GCF is factored out. For example, in the polynomial $21x^2 + 7x$, the GCF of $21x^2$ and $7x$ is $7x$, so
> $$21x^2 + 7x = 7x \cdot 3x + 7x \cdot 1 = 7x(3x + 1)$$

Concept Check ☑
Which factorization of $12x^2 + 9x - 3$ is correct?

a. $3(4x^2 + 3x + 1)$ **b.** $3(4x^2 + 3x - 1)$ **c.** $3(4x^2 + 3x - 3)$ **d.** $3(4x^2 + 3x)$

EXAMPLE 4   Factor $-3x^3y + 2x^2y - 5xy$.

Solution   Two possibilities are shown for factoring this polynomial. First, the common factor xy is factored out.

$$-3x^3y + 2x^2y - 5xy = xy(-3x^2 + 2x - 5)$$

Also, the common factor $-xy$ can be factored out as shown.

$$-3x^3y + 2x^2y - 5xy = -xy(3x^2) + (-xy)(-2x) + (-xy)(5)$$
$$= -xy(3x^2 - 2x + 5)$$

Both of these alternatives are correct. □

PRACTICE
4   Factor $-9x^4y^2 + 5x^2y^2 + 7xy^2$.

Answer to Concept Check: b

EXAMPLE 5   Factor $2(x - 5) + 3a(x - 5)$.

Solution   The greatest common factor is the binomial factor $(x - 5)$.

$$2(x - 5) + 3a(x - 5) = (x - 5)(2 + 3a)$$ $\square$

PRACTICE
5   Factor $3(x + 4) + 5b(x + 4)$.

EXAMPLE 6   Factor $7x(x^2 + 5y) - (x^2 + 5y)$.

Solution $7x(x^2 + 5y) - (x^2 + 5y) = 7x(x^2 + 5y) - 1(x^2 + 5y)$
$$= (x^2 + 5y)(7x - 1)$$ $\square$

> **Helpful Hint**
> Notice that we wrote $-(x^2 + 5y)$ as $-1(x^2 + 5y)$ to aid in factoring.

PRACTICE
6   Factor $8b(a^3 + 2y) - (a^3 + 2y)$.

OBJECTIVE 3 ▶ Factoring polynomials by grouping.   Sometimes it is possible to factor a polynomial by grouping the terms of the polynomial and looking for common factors in each group. This method of factoring is called **factoring by grouping.**

EXAMPLE 7   Factor $ab - 6a + 2b - 12$.

Solution   First look for the GCF of all four terms. The GCF of all four terms is 1. Next group the first two terms and the last two terms and factor out common factors from each group.

$$ab - 6a + 2b - 12 = (ab - 6a) + (2b - 12)$$

Factor a from the first group and 2 from the second group.

$$= a(b - 6) + 2(b - 6)$$

Now we see a GCF of $(b - 6)$. Factor out $(b - 6)$ to get

$$a(b - 6) + 2(b - 6) = (b - 6)(a + 2)$$

Check: To check, multiply $(b - 6)$ and $(a + 2)$ to see that the product is $ab - 6a + 2b - 12$. $\square$

PRACTICE
7   Factor $xy + 2y - 10 - 5x$.

> **Helpful Hint**
> Notice that the polynomial $a(b - 6) + 2(b - 6)$ is _not_ in factored form. It is a _sum_, not a _product_. The factored form is $(b - 6)(a + 2)$.

EXAMPLE 8   Factor $x^3 + 5x^2 + 3x + 15$.

Solution $x^3 + 5x^2 + 3x + 15 = (x^3 + 5x^2) + (3x + 15)$ Group pairs of terms.
$$= x^2(x + 5) + 3(x + 5)$$ Factor each binomial.
$$= (x + 5)(x^2 + 3)$$ Factor out the common factor, $(x + 5)$. $\square$

PRACTICE
8   Factor $a^3 + 2a^2 + 5a + 10$.

EXAMPLE 9 Factor $m^2n^2 + m^2 - 2n^2 - 2$.

Solution $m^2n^2 + m^2 - 2n^2 - 2 = (m^2n^2 + m^2) + (-2n^2 - 2)$ Group pairs of terms.

$$= m^2(n^2 + 1) - 2(n^2 + 1)$$ Factor each binomial.

$$= (n^2 + 1)(m^2 - 2)$$ Factor out the common factor, $(n^2 + 1)$. □

PRACTICE
9 Factor $x^2y^2 + 3y^2 - 5x^2 - 15$.

EXAMPLE 10 Factor $xy + 2x - y - 2$.

Solution $xy + 2x - y - 2 = (xy + 2x) + (-y - 2)$ Group pairs of terms.

$$= x(y + 2) - 1(y + 2)$$ Factor each binomial.

$$= (y + 2)(x - 1)$$ Factor out the common factor, $(y + 2)$. □

PRACTICE
10 Factor $pq + 3p - q - 3$.

VOCABULARY & READINESS CHECK

Use the choices below to fill in each blank. Some choices will be used more than once and some not at all.

| least | greatest | sum | product | factoring | x^3 | x^7 | true | false |

1. The reverse process of multiplying is _____.

2. The greatest common factor (GCF) of x^7, x^3, x^5 is ____.

3. In general, the GCF of a list of common variables raised to powers is the _____ exponent in the list.

4. Factoring means writing as a _____.

5. True or false: A factored form of $2xy^3 + 10xy$ is $2xy \cdot y^2 + 2xy \cdot 5$. _____

6. True or false: A factored form of $x^3 - 6x^2 + x$ is $x(x^2 - 6x)$. _____

7. True or false: A factored form of $5x - 5y + x^3 - x^2y$ is $5(x - y) + x^2(x - y)$. _____

8. True or false: A factored form of $5x - 5y + x^3 - x^2y$ is $(x - y)(5 + x^2)$. _____

Find the GCF of each list of monomials.

9. $6, 12$　　　　　**10.** $9, 27$　　　　　**11.** $15x, 10$　　　　　**12.** $9x, 12$

13. $13x, 2x$　　　　**14.** $4y, 5y$　　　　**15.** $7x, 14x$　　　　**16.** $8z, 4z$

5.5 EXERCISE SET

Find the GCF of each list of monomials. See Example 1.

1. a^8, a^5, a^3

2. b^9, b^2, b^5

3. $x^2y^3z^3, y^2z^3, xy^2z^2$

4. $xy^2z^3, x^2y^2z^2, x^2y^3$

5. $6x^3y, 9x^2y^2, 12x^2y$

6. $4xy^2, 16xy^3, 8x^2y^2$

7. $10x^3yz^3, 20x^2z^5, 45xz^3$

8. $12y^2z^4, 9xy^3z^4, 15x^2y^2z^3$

Factor out the GCF in each polynomial. See Examples 2 through 6.

9. $18x - 12$

10. $21x + 14$

11. $4y^2 - 16xy^3$

12. $3z - 21xz^4$

13. $6x^5 - 8x^4 + 2x^3$

14. $9x + 3x^2 - 6x^3$

15. $8a^3b^3 - 4a^2b^2 + 4ab + 16ab^2$

16. $12a^3b - 6ab + 18ab^2 - 18a^2b$

17. $6(x + 3) + 5a(x + 3)$

18. $2(x - 4) + 3y(x - 4)$

19. $2x(z + 7) + (z + 7)$

20. $x(y - 2) + (y - 2)$

21. $3x(x^2 + 5) - 2(x^2 + 5)$

22. $4x(2y + 3) - 5(2y + 3)$

23. When $3x^2 - 9x + 3$ is factored, the result is $3(x^2 - 3x + 1)$. Explain why it is necessary to include the term 1 in this factored form.

24. Construct a trinomial whose GCF is $5x^2y^3$.

Factor each polynomial by grouping. See Examples 7 through 10.

25. $ab + 3a + 2b + 6$

26. $ab + 2a + 5b + 10$

27. $ac + 4a - 2c - 8$

28. $bc + 8b - 3c - 24$

29. $2xy - 3x - 4y + 6$

30. $12xy - 18x - 10y + 15$

31. $12xy - 8x - 3y + 2$

32. $20xy - 15x - 4y + 3$

MIXED PRACTICE

Factor each polynomial

33. $6x^3 + 9$

34. $6x^2 - 8$

35. $x^3 + 3x^2$

36. $x^4 - 4x^3$

37. $8a^3 - 4a$

38. $12b^4 + 3b^2$

39. $-20x^2y + 16xy^3$

40. $-18xy^3 + 27x^4y$

41. $10a^2b^3 + 5ab^2 - 15ab^3$

42. $10ef - 20e^2f^3 + 30e^3f$

43. $9abc^2 + 6a^2bc - 6ab + 3bc$

44. $4a^2b^2c - 6ab^2c - 4ac + 8a$

45. $4x(y - 2) - 3(y - 2)$

46. $8y(z + 8) - 3(z + 8)$

47. $6xy + 10x + 9y + 15$

48. $15xy + 20x + 6y + 8$

49. $xy + 3y - 5x - 15$

50. $xy + 4y - 3x - 12$

51. $6ab - 2a - 9b + 3$

52. $16ab - 8a - 6b + 3$

53. $12xy + 18x + 2y + 3$

54. $20xy + 8x + 5y + 2$

55. $2m(n - 8) - (n - 8)$

56. $3a(b - 4) - (b - 4)$

57. $15x^3y^2 - 18x^2y^2$

58. $12x^4y^2 - 16x^3y^3$

59. $2x^2 + 3xy + 4x + 6y$

60. $3x^2 + 12x + 4xy + 16y$

61. $5x^2 + 5xy - 3x - 3y$

62. $4x^2 + 2xy - 10x - 5y$

63. $x^3 + 3x^2 + 4x + 12$

64. $x^3 + 4x^2 + 3x + 12$

65. $x^3 - x^2 - 2x + 2$

66. $x^3 - 2x^2 - 3x + 6$

REVIEW AND PREVIEW

Simplify the following. See Section 5.1.

67. $(5x^2)(11x^5)$

68. $(7y)(-2y^3)$

69. $(5x^2)^3$

70. $(-2y^3)^4$

Find each product by using the FOIL order of multiplying binomials. See Section 5.4.

71. $(x + 2)(x - 5)$

72. $(x - 7)(x - 1)$

73. $(x + 3)(x + 2)$

74. $(x - 4)(x + 2)$

75. $(y - 3)(y - 1)$

76. $(s + 8)(s + 10)$

CONCEPT EXTENSIONS

Solve. See the Concept Check in this section.

77. Which factorization of $10x^2 - 2x - 2$ is correct?

   a. $2(5x^2 - x + 1)$ **b.** $2(5x^2 - x)$

   c. $2(5x^2 - x - 2)$ **d.** $2(5x^2 - x - 1)$

78. Which factorization of $x^4 + 5x^3 - x^2$ is correct?

   a. $-1(x^4 + 5x^3 + x^2)$ **b.** $x^2(x^2 + 5x^3 - x^2)$

   c. $x^2(x^2 + 5x - 1)$ **d.** $5x^2(x^2 + 5x - 5)$

Solve.

79. The area of the material needed to manufacture a tin can is given by the polynomial $2\pi r^2 + 2\pi rh$, where the radius is r and height is h. Factor this expression.

80. To estimate the cost of a new product, one expression used by the production department is $4\pi r^2 + \dfrac{4}{3}\pi r^3$. Write an equivalent expression by factoring $4\pi r^2$ from both terms.

81. At the end of T years, the amount of money A in a savings account earning simple interest from an initial investment of \$5600 at rate r is given by the formula $A = 5600 + 5600rt$. Write an equivalent equation by factoring the expression $5600 + 5600rt$.

82. An open-topped box has a square base and a height of 10 inches. If each of the bottom edges of the box has length x inches, find the amount of material needed to construct the box. Write the answer in factored form.

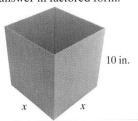

83. Explain why $9(5 - x) + y(5 - x)$ is not a factored form of $45 - 9x + 5y - xy$.

84. Construct a 4-term polynomial whose greatest common factor is $2a^3b^4$.

85. A factored polynomial can be in many forms. For example, a factored form of $xy - 3x - 2y + 6$ is $(x - 2)(y - 3)$. Which of the following is not a factored form of $xy - 3x - 2y + 6$?

  a. $(2 - x)(3 - y)$ **b.** $(-2 + x)(-3 + y)$

  c. $(y - 3)(x - 2)$ **d.** $(-x + 2)(-y + 3)$

86. Consider the following sequence of algebraic steps:

$$x^3 - 6x^2 + 2x - 10 = (x^3 - 6x^2) + (2x - 10)$$
$$= x^2(x - 6) + 2(x - 5)$$

Explain whether the final result is the factored form of the original polynomial.

87. Which factorization of $12x^2 + 9x + 3$ is correct?

  a. $3(4x^2 + 3x + 1)$ **b.** $3(4x^2 + 3x - 1)$

  c. $3(4x^2 + 3x - 3)$ **d.** $3(4x^2 + 3x)$

88. The amount E of voltage in an electrical circuit is given by the formula

$$IR_1 + IR_2 = E$$

Write an equivalent equation by factoring the expression $IR_1 + IR_2$.

89. At the end of T years, the amount of money A in a savings account earning simple interest from an initial investment of P dollars at rate R is given by the formula

$$A = P + PRT$$

Write an equivalent equation by factoring the expression $P + PRT$.

Factor out the greatest common factor. Assume that variables used as exponents represent positive integers.

90. $x^{3n} - 2x^{2n} + 5x^n$

91. $3y^n + 3y^{2n} + 5y^{8n}$

92. $6x^{8a} - 2x^{5a} - 4x^{3a}$

93. $3x^{5a} - 6x^{3a} + 9x^{2a}$

94. An object is thrown upward from the ground with an initial velocity of 64 feet per second. The height $h(t)$ in feet of the object after t seconds is given by the polynomial function

$$h(t) = -16t^2 + 64t$$

  a. Write an equivalent factored expression for the function $h(t)$ by factoring $-16t^2 + 64t$.

  b. Find $h(1)$ by using

$$h(t) = -16t^2 + 64t$$

  and then by using the factored form of $h(t)$.

  c. Explain why the values found in part **b** are the same.

95. An object is dropped from the gondola of a hot-air balloon at a height of 224 feet. The height $h(t)$ of the object after t seconds is given by the polynomial function

$$h(t) = -16t^2 + 224$$

224 ft

  a. Write an equivalent factored expression for the function $h(t)$ by factoring $-16t^2 + 224$.

  b. Find $h(2)$ by using $h(t) = -16t^2 + 224$ and then by using the factored form of the function.

  c. Explain why the values found in part **b** are the same.

5.6 FACTORING TRINOMIALS

OBJECTIVES

1 Factor trinomials of the form $x^2 + bx + c$.

2 Factor trinomials of the form $ax^2 + bx + c$.
  a. Method 1—Trial and Check
  b. Method 2—Grouping

3 Factor by substitution.

OBJECTIVE 1 ▶ Factoring trinomials of the form $x^2 + bx + c$. In the previous section, we used factoring by grouping to factor four-term polynomials. In this section, we present techniques for factoring trinomials. Since $(x - 2)(x + 5) = x^2 + 3x - 10$, we say that $(x - 2)(x + 5)$ is a factored form of $x^2 + 3x - 10$. Taking a close look at how $(x - 2)$ and $(x + 5)$ are multiplied suggests a pattern for factoring trinomials of the form

$$x^2 + bx + c$$

$$(x - 2)(x + 5) = x^2 + 3x - 10$$

$$-2 + 5$$
$$-2 \cdot 5$$

The pattern for factoring is summarized next.

> **Factoring a Trinomial of the Form $x^2 + bx + c$**
> Find two numbers whose product is c and whose sum is b. The factored form of $x^2 + bx + c$ is
>
> $$(x + \text{one number})(x + \text{other number})$$

EXAMPLE 1   Factor $x^2 + 10x + 16$.

Solution   We look for two integers whose product is 16 and whose sum is 10. Since our integers must have a positive product and a positive sum, we look at only positive factors of 16.

Positive Factors of 16	Sum of Factors
1, 16	$1 + 16 = 17$
4, 4	$4 + 4 = 8$
2, 8	$2 + 8 = 10$ Correct pair

The correct pair of numbers is 2 and 8 because their product is 16 and their sum is 10. Thus,

$$x^2 + 10x + 16 = (x + 2)(x + 8)$$

Check: To check, see that $(x + 2)(x + 8) = x^2 + 10x + 16$.   □

PRACTICE
1   Factor $x^2 + 5x + 6$.

TECHNOLOGY NOTE

Just as a graphing utility can be used to visualize the multiplication of polynomials, a graphing utility can also be used to visualize the factorization of a polynomial. For example, to visualize the factorization from Example 1, graph $y_1 = x^2 + 10x + 16$ and $y_2 = (x + 2)(x + 8)$ and see that the graphs coincide.

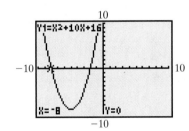

Notice that the graph and the table values are the same for y_1 and y_2.
     Also notice the x-value of the x-intercepts and the factors. Look for a pattern again in the next example.

EXAMPLE 2   Factor $x^2 - 12x + 35$.

Solution   We need to find two integers whose product is 35 and whose sum is -12. Since our integers must have a positive product and a negative sum, we consider only negative factors of 35.

Negative Factors of 35	Sum of Factors
$-1, -35$	$-1 + (-35) = -36$
$-5, -7$	$-5 + (-7) = -12$ Correct pair

The numbers are -5 and -7.

Notice the x-intercepts and look for a pattern that connects the x-value of the x-intercepts and the factors.

$$x^2 - 12x + 35 = [x + (-5)][x + (-7)]$$
$$= (x - 5)(x - 7)$$

Check: To check, see that $(x - 5)(x - 7) = x^2 - 12x + 35$. ☐

PRACTICE
2 Factor $x^2 - 11x + 24$.

DISCOVER THE CONCEPT

a. The following polynomials are factored. Graph each related polynomial function shown in a standard window. Find the x-intercepts of the graph and compare them with the factors of the polynomial.

Polynomial	*Factors*	*Related Polynomial Function*	*x-Intercepts*
$x^2 - x - 12$	$(x - 4)(x + 3)$	$y = x^2 - x - 12$	
$x^2 + 10x + 16$	$(x + 2)(x + 8)$	$y = x^2 + 10x + 16$	
$x^2 - 8x + 15$	$(x - 3)(x - 5)$	$y = x^2 - 8x + 15$	
$x^2 - 9x - 10$	$(x + 1)(x - 10)$	$y = x^2 - 9x - 10$	

b. Find a connection between the factors and the x-intercepts above. If you know the x-intercepts of the related graph, can it help you factor the polynomial?

c. If a function has x-intercepts of 2 and -8, what are two factors of the related polynomial? Graph a related function in factored form and see if the x-intercepts are 2 and -8.

In the above discovery, we find that if the graph of a polynomial function has an x-intercept at a, then the related polynomial has a factor of $(x - a)$. Also, if a polynomial has a factor of $(x - a)$, the graph of the related polynomial function has an x-intercept at a.

EXAMPLE 3 Factor $5x^3 - 30x^2 - 35x$.

Solution First we factor out the greatest common factor, $5x$.

$$5x^3 - 30x^2 - 35x = 5x(x^2 - 6x - 7)$$

Next we try to factor $x^2 - 6x - 7$ by finding two numbers whose product is -7 and whose sum is -6. The numbers are 1 and -7.

$$5x^3 - 30x^2 - 35x = 5x(x^2 - 6x - 7)$$
$$= 5x(x + 1)(x - 7)$$ ☐

PRACTICE
3 Factor $3x^3 - 9x^2 - 30x$.

▶ **Helpful Hint**
If the polynomial to be factored contains a common factor that is factored out, don't forget to include that common factor in the final factored form of the original polynomial.

EXAMPLE 4 Factor $2n^2 - 38n + 80$.

Solution The terms of this polynomial have a greatest common factor of 2, which we factor out first.

$$2n^2 - 38n + 80 = 2(n^2 - 19n + 40)$$

Next we factor $n^2 - 19n + 40$ by finding two numbers whose product is 40 and whose sum is -19. Both numbers must be negative since their sum is -19. Possibilities are

$$-1 \text{ and } -40, \qquad -2 \text{ and } -20, \qquad -4 \text{ and } -10, \qquad -5 \text{ and } -8$$

None of the pairs has a sum of -19, so no further factoring with integers is possible. The factored form of $2n^2 - 38n + 80$ is

$$2n^2 - 38n + 80 = 2(n^2 - 19n + 40) \qquad \qquad \square$$

PRACTICE
4   Factor $2b^2 - 18b - 22$.

We call a polynomial such as $n^2 - 19n + 40$ that cannot be factored further, a **prime polynomial.**

OBJECTIVE 2 ▶ Factoring trinomials of the form $ax^2 + bx + c$. Next, we factor trinomials of the form $ax^2 + bx + c$, where the coefficient a of x^2 is not 1. Don't forget that the first step in factoring any polynomial is to factor out the greatest common factor of its terms. We will review two methods here. The first method we'll call trial and check.

EXAMPLE 5   **Method 1—Trial and Check**

Factor $2x^2 + 11x + 15$.

Solution   Factors of $2x^2$ are $2x$ and x. Let's try these factors as first terms of the binomials.

$$2x^2 + 11x + 15 = (2x + \quad)(x + \quad)$$

Next we try combinations of factors of 15 until the correct middle term, $11x$, is obtained. We will try only positive factors of 15 since the coefficient of the middle term, 11, is positive. Positive factors of 15 are 1 and 15 and 3 and 5.

$$(2x + 1)(x + 15)$$
$$\begin{array}{c} 1x \\ \underline{30x} \\ 31x \end{array} \text{ Incorrect middle term}$$

$$(2x + 15)(x + 1)$$
$$\begin{array}{c} 15x \\ \underline{2x} \\ 17x \end{array} \text{ Incorrect middle term}$$

$$(2x + 3)(x + 5)$$
$$\begin{array}{c} 3x \\ \underline{10x} \\ 13x \end{array} \text{ Incorrect middle term}$$

$$(2x + 5)(x + 3)$$
$$\begin{array}{c} 5x \\ \underline{6x} \\ 11x \end{array} \text{ Correct middle term}$$

Thus, the factored form of $2x^2 + 11x + 15$ is $(2x + 5)(x + 3)$. $\qquad \square$

PRACTICE
5   Factor $2x^2 + 13x + 6$.

> **Factoring a Trinomial of the Form $ax^2 + bx + c$**
>
> **STEP 1.** Write all pairs of factors of ax^2.
>
> **STEP 2.** Write all pairs of factors of c, the constant term.
>
> **STEP 3.** Try various combinations of these factors until the correct middle term, bx, is found.
>
> **STEP 4.** If no combination exists, the polynomial is **prime.**

EXAMPLE 6 Factor $3x^2 - x - 4$.

Solution Factors of $3x^2$: $3x \cdot x$
Factors of -4: $-1 \cdot 4$, $1 \cdot -4$, $-2 \cdot 2$, $2 \cdot -2$

Let's try possible combinations of these factors.

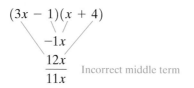

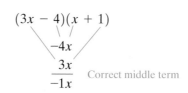

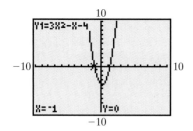

Notice that the x-intercept $(-1, 0)$ indicates that $(x + 1)$ is one of the factors.

Thus, $3x^2 - x - 4 = (3x - 4)(x + 1)$. ☐

PRACTICE

6 Factor $4x^2 + 5x - 6$.

> ▶ **Helpful Hint—Sign Patterns**
>
> A positive constant in a trinomial tells us to look for two numbers with the same sign. The sign of the coefficient of the middle term tells us whether the signs are both positive or both negative.
>
> both positive   same sign
> ↓   ↓
> $2x^2 + 7x + 3 = (2x + 1)(x + 3)$
>
> both negative   same sign
> ↓   ↓
> $2x^2 - 7x + 3 = (2x - 1)(x - 3)$
>
> A negative constant in a trinomial tells us to look for two numbers with opposite signs.
>
> opposite signs
> ↓
> $2x^2 - 5x - 3 = (2x + 1)(x - 3)$
>
> opposite signs
> ↓
> $2x^2 + 5x - 3 = (2x - 1)(x + 3)$

EXAMPLE 7 Factor $12x^3y - 22x^2y + 8xy$.

Solution First we factor out the greatest common factor of the terms of this trinomial, $2xy$.

$$12x^3y - 22x^2y + 8xy = 2xy(6x^2 - 11x + 4)$$

Now we try to factor the trinomial $6x^2 - 11x + 4$.
Factors of $6x^2$: $2x \cdot 3x$, $6x \cdot x$

Let's try $2x$ and $3x$.

$$2xy(6x^2 - 11x + 4) = 2xy(2x + \quad)(3x + \quad)$$

The constant term, 4, is positive and the coefficient of the middle term, -11, is negative, so we factor 4 into negative factors only.

Negative factors of 4: $-4(-1), \quad -2(-2)$

Let's try -4 and -1.

$$2xy(2x - 4)(3x - 1)$$

$$-12x$$
$$\underline{-2x}$$
$$-14x \qquad \text{Incorrect middle term}$$

This combination cannot be correct, because one of the factors, $(2x - 4)$, has a common factor of 2. This cannot happen if the polynomial $6x^2 - 11x + 4$ has no common factors.

Now let's try -1 and -4.

$$2xy(2x - 1)(3x - 4)$$

$$-3x$$
$$\underline{-8x}$$
$$-11x \qquad \text{Correct middle term}$$

Thus,

$$12x^3y - 22x^2y + 8xy = 2xy(2x - 1)(3x - 4)$$

If this combination had not worked, we would have tried -2 and -2 as factors of 4 and then $6x$ and x as factors of $6x^2$. $\qquad\square$

PRACTICE
7 Factor $18b^4 - 57b^3 + 30b^2$.

▶ **Helpful Hint**

If a trinomial has no common factor (other than 1), then none of its binomial factors will contain a common factor (other than 1).

EXAMPLE 8 Factor $16x^2 + 24xy + 9y^2$.

Solution No greatest common factor can be factored out of this trinomial.

Factors of $16x^2$: $16x \cdot x, \qquad 8x \cdot 2x, \qquad 4x \cdot 4x$

Factors of $9y^2$: $y \cdot 9y, \qquad 3y \cdot 3y$

We try possible combinations until the correct factorization is found.

$$16x^2 + 24xy + 9y^2 = (4x + 3y)(4x + 3y) \quad \text{or} \quad (4x + 3y)^2 \qquad\square$$

PRACTICE
8 Factor $25x^2 + 20xy + 4y^2$.

The trinomial $16x^2 + 24xy + 9y^2$ in Example 8 is an example of a **perfect square trinomial** since its factors are two identical binomials. In the next section, we examine a special method for factoring perfect square trinomials.

Method 2—Grouping

There is another method we can use when factoring trinomials of the form $ax^2 + bx + c$: Write the trinomial as a four-term polynomial, and then factor by grouping.

Factoring a Trinomial of the Form $ax^2 + bx + c$ by Grouping

STEP 1. Find two numbers whose product is $a \cdot c$ and whose sum is b.

STEP 2. Write the term bx as a sum by using the factors found in Step 1.

STEP 3. Factor by grouping.

EXAMPLE 9   Factor $6x^2 + 13x + 6$.

Solution   In this trinomial, $a = 6$, $b = 13$, and $c = 6$.

STEP 1. Find two numbers whose product is $a \cdot c$, or $6 \cdot 6 = 36$, and whose sum is b, 13. The two numbers are 4 and 9.

STEP 2. Write the middle term, $13x$, as the sum $4x + 9x$.

$$6x^2 + 13x + 6 = 6x^2 + 4x + 9x + 6$$

STEP 3. Factor $6x^2 + 4x + 9x + 6$ by grouping.

$$(6x^2 + 4x) + (9x + 6) = 2x(3x + 2) + 3(3x + 2)$$
$$= (3x + 2)(2x + 3) \qquad \square$$

PRACTICE
9   Factor $20x^2 + 23x + 6$.

Concept Check ☑

Name one way that a factorization can be checked.

EXAMPLE 10   Factor $18x^2 - 9x - 2$.

Solution   In this trinomial, $a = 18$, $b = -9$, and $c = -2$.

STEP 1. Find two numbers whose product is $a \cdot c$ or $18(-2) = -36$ and whose sum is b, -9. The two numbers are -12 and 3.

STEP 2. Write the middle term, $-9x$, as the sum $-12x + 3x$.

$$18x^2 - 9x - 2 = 18x^2 - 12x + 3x - 2$$

STEP 3. Factor by grouping.

$$(18x^2 - 12x) + (3x - 2) = 6x(3x - 2) + 1(3x - 2)$$
$$= (3x - 2)(6x + 1). \qquad \square$$

PRACTICE
10   Factor $15x^2 + 4x - 3$.

Answer to Concept Check:
Answers may vary. A sample is:
By multiplying the factors to see
that the product is the original
polynomial.

OBJECTIVE 3 ▶ Factoring by substitution. A complicated-looking polynomial may be a simpler trinomial "in disguise." Revealing the simpler trinomial is possible by substitution.

EXAMPLE 11   Factor $2(a + 3)^2 - 5(a + 3) - 7$.

Solution   The quantity $(a + 3)$ is in two of the terms of this polynomial. **Substitute** x for $(a + 3)$, and the result is the following simpler trinomial.

$$2(a + 3)^2 - 5(a + 3) - 7 \quad \text{Original trinomial.}$$

$$= 2(x)^2 - 5(x) - 7 \quad \text{Substitute } x \text{ for } (a + 3).$$

Now factor $2x^2 - 5x - 7$.

$$2x^2 - 5x - 7 = (2x - 7)(x + 1)$$

But the quantity in the original polynomial was $(a + 3)$, not x. Thus, we need to reverse the substitution and replace x with $(a + 3)$.

$$(2x - 7)(x + 1) \quad \text{Factored expression.}$$

$$= [2(a + 3) - 7][(a + 3) + 1] \quad \text{Substitute } (a + 3) \text{ for } x.$$

$$= (2a + 6 - 7)(a + 3 + 1) \quad \text{Remove inside parentheses.}$$

$$= (2a - 1)(a + 4) \quad \text{Simplify.}$$

Thus, $2(a + 3)^2 - 5(a + 3) - 7 = (2a - 1)(a + 4)$. □

PRACTICE
11   Factor $3(x + 1)^2 - 7(x + 1) - 20$.

EXAMPLE 12   Factor $5x^4 + 29x^2 - 42$.

Solution   Again, substitution may help us factor this polynomial more easily. Since this polynomial contains the variable x, we will choose a different substitution variable. Let $y = x^2$, so $y^2 = (x^2)^2$, or x^4. Then

$$5x^4 + 29x^2 - 42$$

becomes

$$5y^2 + 29y - 42$$

which factors as

$$5y^2 + 29y - 42 = (5y - 6)(y + 7)$$

Next, replace y with x^2 to get

$$(5x^2 - 6)(x^2 + 7)$$ □

PRACTICE
12   Factor $6x^4 - 11x^2 - 10$.

VOCABULARY & READINESS CHECK

1. Find two numbers whose product is 10 and whose sum is 7.
2. Find two numbers whose product is 12 and whose sum is 8.
3. Find two numbers whose product is 24 and whose sum is 11.
4. Find two numbers whose product is 30 and whose sum is 13.

5.6 | EXERCISE SET

 MyMathLab

PRACTICE   WATCH   DOWNLOAD   READ   REVIEW

Factor each trinomial. See Examples 1 through 4.

1. $x^2 + 9x + 18$

2. $x^2 + 9x + 20$

3. $x^2 - 12x + 32$

4. $x^2 - 12x + 27$

5. $x^2 + 10x - 24$

6. $x^2 + 3x - 54$

7. $x^2 - 2x - 24$

8. $x^2 - 9x - 36$

9. $3x^2 - 18x + 24$

10. $x^2y^2 + 4xy^2 + 3y^2$

11. $4x^2z + 28xz + 40z$

12. $5x^2 - 45x + 70$

A polynomial function is graphed on each screen. Use the graph to write factors of the related polynomial. Each tick mark on the x-axis is 1 unit, and it can be assumed that both x-intercepts are integer values.

13.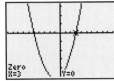

14.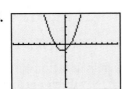

15.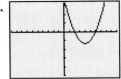

16.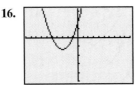

Factor each trinomial. See Examples 5 through 10.

17. $5x^2 + 16x + 3$

18. $3x^2 + 8x + 4$

19. $2x^2 - 11x + 12$

20. $3x^2 - 19x + 20$

21. $2x^2 + 25x - 20$

22. $6x^2 + 13x + 8$

23. $4x^2 - 12x + 9$

24. $25x^2 - 30x + 9$

25. $12x^2 + 10x - 50$

26. $12y^2 - 48y + 45$

27. $3y^4 - y^3 - 10y^2$

28. $2x^2z + 5xz - 12z$

29. $2x^2 - 5xy - 3y^2$

30. $6x^2 + 11xy + 4y^2$

31. $28y^2 + 22y + 4$

32. $24y^3 - 2y^2 - y$

33. $2x^2 + 15x - 27$

34. $3x^2 + 14x + 15$

Use substitution to factor each polynomial completely. See Examples 11 and 12.

35. $x^4 + x^2 - 6$

36. $x^4 - x^2 - 20$

37. $(5x + 1)^2 + 8(5x + 1) + 7$

38. $(3x - 1)^2 + 5(3x - 1) + 6$

39. $x^6 - 7x^3 + 12$

40. $x^6 - 4x^3 - 12$

41. $(a + 5)^2 - 5(a + 5) - 24$

42. $(3c + 6)^2 + 12(3c + 6) - 28$

MIXED PRACTICE

Factor each polynomial completely. See Examples 1 through 12.

43. $x^2 - 24x - 81$

44. $x^2 - 48x - 100$

45. $x^2 - 15x - 54$

46. $x^2 - 15x + 54$

47. $3x^2 - 6x + 3$

48. $8x^2 - 8x + 2$

49. $3x^2 - 5x - 2$

50. $5x^2 - 14x - 3$

51. $8x^2 - 26x + 15$

52. $12x^2 - 17x + 6$

53. $18x^4 + 21x^3 + 6x^2$

54. $20x^5 + 54x^4 + 10x^3$

55. $x^2 + 8xz + 7z^2$

56. $a^2 - 2ab - 15b^2$

57. $x^2 - x - 12$

58. $x^2 + 4x - 5$

59. $3a^2 + 12ab + 12b^2$

60. $2x^2 + 16xy + 32y^2$

61. $x^2 + 4x + 5$

62. $x^2 + 6x + 8$

63. $2(x + 4)^2 + 3(x + 4) - 5$

64. $3(x + 3)^2 + 2(x + 3) - 5$

65. $6x^2 - 49x + 30$

66. $4x^2 - 39x + 27$

67. $x^4 - 5x^2 - 6$

68. $x^4 - 5x^2 + 6$

69. $6x^3 - x^2 - x$

70. $12x^3 + x^2 - x$

71. $12a^2 - 29ab + 15b^2$

72. $16y^2 + 6yx - 27x^2$

73. $9x^2 + 30x + 25$

74. $4x^2 + 6x + 9$

75. $3x^2y - 11xy + 8y$

76. $5xy^2 - 9xy + 4x$

77. $2x^2 + 2x - 12$

78. $3x^2 + 6x - 45$

79. $(x - 4)^2 + 3(x - 4) - 18$

80. $(x - 3)^2 - 2(x - 3) - 8$

81. $2x^6 + 3x^3 - 9$

82. $3x^6 - 14x^3 + 8$

83. $72xy^4 - 24xy^2z + 2xz^2$

84. $36xy^2 - 48xyz^2 + 16xz^4$

85. $2x^3y + 2x^2y - 12xy$

86. $3x^2y^3 + 6x^2y^2 - 45x^2y$

87. $x^2 + 6xy + 5y^2$

88. $x^2 + 6xy + 8y^2$

REVIEW AND PREVIEW

Multiply. See Section 5.4.

89. $(x - 3)(x + 3)$

90. $(x - 4)(x + 4)$

91. $(2x + 1)^2$

92. $(3x + 5)^2$

93. $(x - 2)(x^2 + 2x + 4)$

94. $(y + 1)(y^2 - y + 1)$

CONCEPT EXTENSIONS

95. Find all positive and negative integers b such that $x^2 + bx + 6$ is factorable.

96. Find all positive and negative integers b such that $x^2 + bx - 10$ is factorable.

97. The volume $V(x)$ of a box in terms of its height x is given by the function $V(x) = x^3 + 2x^2 - 8x$. Factor this expression for $V(x)$.

98. Based on your results from Exercise 97, find the length and width of the box if the height is 5 inches and the dimensions of the box are whole numbers.

99. Suppose that a movie is being filmed in New York City. An action shot requires an object to be thrown upward with an initial velocity of 80 feet per second off the top of 1 Madison Square Plaza, a height of 576 feet. The height $h(t)$ in feet of the object after t seconds is given by the function $h(t) = -16t^2 + 80t + 576$. (*Source: The World Almanac*)

a. Find the height of the object at $t = 0$ seconds, $t = 2$ seconds, $t = 4$ seconds, and $t = 6$ seconds.

b. Explain why the height of the object increases and then decreases as time passes.

576 ft

c. Factor the polynomial $-16t^2 + 80t + 576$.

100. Suppose that an object is thrown upward with an initial velocity of 64 feet per second off the edge of a 960-foot cliff. The height $h(t)$ in feet of the object after t seconds is given by the function

$$h(t) = -16t^2 + 64t + 960$$

a. Find the height of the object at $t = 0$ seconds, $t = 3$ seconds, $t = 6$ seconds, and $t = 9$ seconds.

b. Explain why the height of the object increases and then decreases as time passes.

c. Factor the polynomial $-16t^2 + 64t + 960$.

Factor. Assume that variables used as exponents represent positive integers.

101. $x^{2n} + 10x^n + 16$

102. $x^{2n} - 7x^n + 12$

103. $x^{2n} - 3x^n - 18$

104. $x^{2n} + 7x^n - 18$

105. $2x^{2n} + 11x^n + 5$

106. $3x^{2n} - 8x^n + 4$

107. $4x^{2n} - 12x^n + 9$

108. $9x^{2n} + 24x^n + 16$

5.7 FACTORING BY SPECIAL PRODUCTS

OBJECTIVES

1. Factor a perfect square trinomial.

2. Factor the difference of two squares.

3. Factor the sum or difference of two cubes.

OBJECTIVE 1 ▶ Factoring a perfect square trinomial. In the previous section, we considered a variety of ways to factor trinomials of the form $ax^2 + bx + c$. In one particular example, we factored $16x^2 + 24xy + 9y^2$ as

$$16x^2 + 24xy + 9y^2 = (4x + 3y)^2$$

Recall that $16x^2 + 24xy + 9y^2$ is a perfect square trinomial because its factors are two identical binomials. A perfect square trinomial can be factored quickly if you recognize the trinomial as a perfect square.

A trinomial is a perfect square trinomial if it can be written so that its first term is the square of some quantity a, its last term is the square of some quantity b, and its middle term is twice the product of the quantities a and b. The following special formulas can be used to factor perfect square trinomials.

Perfect Square Trinomials

$$a^2 + 2ab + b^2 = (a + b)^2$$
$$a^2 - 2ab + b^2 = (a - b)^2$$

TECHNOLOGY NOTE

Below is the graph of $y = x^2 + 10x + 25$ or $y = (x + 5)^2$.

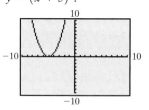

Notice that because the factors of the polynomial are the same, $(x + 5)$, only one x-intercept is generated, as shown by the graph.

Notice that these formulas above are the same special products from Section 5.4 for the square of a binomial.

From

we see that

$$a^2 + 2ab + b^2 = (a + b)^2,$$

$$16x^2 + 24xy + 9y^2 = (4x)^2 + 2(4x)(3y) + (3y)^2 = (4x + 3y)^2$$

EXAMPLE 1   Factor $m^2 + 10m + 25$.

Solution   Notice that the first term is a square: $m^2 = (m)^2$; the last term is a square: $25 = 5^2$; and $10m = 2 \cdot 5 \cdot m$.

Thus,

$$m^2 + 10m + 25 = m^2 + 2(m)(5) + 5^2 = (m + 5)^2$$ □

PRACTICE
1   Factor $b^2 + 16b + 64$.

EXAMPLE 2 Factor $3a^2x - 12abx + 12b^2x$.

Solution The terms of this trinomial have a GCF of $3x$, which we factor out first.

$$3a^2x - 12abx + 12b^2x = 3x(a^2 - 4ab + 4b^2)$$

Now, the polynomial $a^2 - 4ab + 4b^2$ is a perfect square trinomial. Notice that the first term is a square: $a^2 = (a)^2$; the last term is a square: $4b^2 = (2b)^2$; and $4ab = 2(a)(2b)$. The factoring can now be completed as

$$3x(a^2 - 4ab + 4b^2) = 3x(a - 2b)^2$$

□

PRACTICE

2 Factor $45x^2b - 30xb + 5b$.

▶ **Helpful Hint**

If you recognize a trinomial as a perfect square trinomial, use the special formulas to factor. However, methods for factoring trinomials in general from Section 5.6 will also result in the correct factored form.

OBJECTIVE 2 ▶ **Factoring the difference of two squares.** We now factor special types of binomials, beginning with the **difference of two squares.** The special product pattern presented in Section 5.4 for the product of a sum and a difference of two terms is used again here. However, the emphasis is now on factoring rather than on multiplying.

Difference of Two Squares

$$a^2 - b^2 = (a + b)(a - b)$$

Notice that a binomial is a difference of two squares when it is the difference of the square of some quantity a and the square of some quantity b.

EXAMPLE 3 Factor the following.

a. $x^2 - 9$ **b.** $16y^2 - 9$ **c.** $50 - 8y^2$ **d.** $x^2 - \dfrac{1}{4}$

Solution

a. $x^2 - 9 = x^2 - 3^2$

$ = (x + 3)(x - 3)$

b. $16y^2 - 9 = (4y)^2 - 3^2$

$ = (4y + 3)(4y - 3)$

c. First factor out the common factor of 2.

$50 - 8y^2 = 2(25 - 4y^2)$

$ = 2(5 + 2y)(5 - 2y)$

d. $x^2 - \dfrac{1}{4} = x^2 - \left(\dfrac{1}{2}\right)^2 = \left(x + \dfrac{1}{2}\right)\left(x - \dfrac{1}{2}\right)$

□

PRACTICE

3 Factor the following.

a. $x^2 - 16$

b. $25b^2 - 49$

c. $45 - 20x^2$

d. $y^2 - \dfrac{1}{81}$

The binomial $x^2 + 9$ is a **sum of two squares** and cannot be factored by using real numbers. **In general, except for factoring out a GCF, the sum of two squares usually cannot be factored by using real numbers.**

> ▶ **Helpful Hint**
>
> The sum of two squares whose GCF is 1 usually cannot be factored by using real numbers. For example, $x^2 + 9$ is called a prime polynomial.

EXAMPLE 4   Factor the following.

a. $p^4 - 16$ **b.** $(x + 3)^2 - 36$

Solution

a. $p^4 - 16 = (p^2)^2 - 4^2$
$= (p^2 + 4)(p^2 - 4)$

The binomial factor $p^2 + 4$ cannot be factored by using real numbers, but the binomial factor $p^2 - 4$ is a difference of squares.

$$(p^2 + 4)(p^2 - 4) = (p^2 + 4)(p + 2)(p - 2)$$

b. Factor $(x + 3)^2 - 36$ as the difference of squares.

$$\begin{aligned}
(x + 3)^2 - 36 &= (x + 3)^2 - 6^2 \\
&= [(x + 3) + 6][(x + 3) - 6] \quad \text{Factor.} \\
&= [x + 3 + 6][x + 3 - 6] \quad \text{Remove parentheses.} \\
&= (x + 9)(x - 3) \quad \text{Simplify.} \qquad \square
\end{aligned}$$

PRACTICE
4   Factor the following.

a. $x^4 - 10000$ **b.** $(x + 2)^2 - 49$

Concept Check ☑
Is $(x - 4)(y^2 - 9)$ completely factored? Why or why not?

EXAMPLE 5   Factor $x^2 + 4x + 4 - y^2$.

Solution   Factoring by grouping comes to mind since the sum of the first three terms of this polynomial is a perfect square trinomial.

$$\begin{aligned}
x^2 + 4x + 4 - y^2 &= (x^2 + 4x + 4) - y^2 \quad \text{Group the first three terms.} \\
&= (x + 2)^2 - y^2 \quad \text{Factor the perfect square trinomial.}
\end{aligned}$$

This is not factored yet since we have a *difference*, not a *product*. Since $(x + 2)^2 - y^2$ is a difference of squares, we have

$$\begin{aligned}
(x + 2)^2 - y^2 &= [(x + 2) + y][(x + 2) - y] \\
&= (x + 2 + y)(x + 2 - y) \qquad \square
\end{aligned}$$

PRACTICE
5   Factor $m^2 + 6m + 9 - n^2$.

Answer to Concept Check:
no; $(y^2 - 9)$ can be factored

OBJECTIVE 3 ▶ Factoring the sum or difference of two cubes. Although the sum of two squares usually cannot be factored, the sum of two cubes, as well as the difference of two cubes, can be factored as follows.

Sum and Difference of Two Cubes

$$a^3 + b^3 = (a + b)(a^2 - ab + b^2)$$
$$a^3 - b^3 = (a - b)(a^2 + ab + b^2)$$

To check the first pattern, let's find the product of $(a + b)$ and $(a^2 - ab + b^2)$.

$$(a + b)(a^2 - ab + b^2) = a(a^2 - ab + b^2) + b(a^2 - ab + b^2)$$
$$= a^3 - a^2b + ab^2 + a^2b - ab^2 + b^3$$
$$= a^3 + b^3$$

EXAMPLE 6   Factor $x^3 + 8$.

Solution   First we write the binomial in the form $a^3 + b^3$. Then we use the formula

$$a^3 + b^3 = (a + b)(a^2 - a \cdot b + b^2), \quad \text{where } a \text{ is } x \text{ and } b \text{ is } 2.$$

$$x^3 + 8 = x^3 + 2^3 = (x + 2)(x^2 - x \cdot 2 + 2^2)$$

Thus, $x^3 + 8 = (x + 2)(x^2 - 2x + 4)$ ☐

PRACTICE
6   Factor $x^3 + 64$.

EXAMPLE 7   Factor $p^3 + 27q^3$.

Solution
$$p^3 + 27q^3 = p^3 + (3q)^3$$
$$= (p + 3q)[p^2 - (p)(3q) + (3q)^2]$$
$$= (p + 3q)(p^2 - 3pq + 9q^2)$$ ☐

PRACTICE
7   Factor $a^3 + 8b^3$.

EXAMPLE 8   Factor $y^3 - 64$.

Solution   This is a difference of cubes since $y^3 - 64 = y^3 - 4^3$.

From

$$a^3 - b^3 = (a - b)(a^2 + a \cdot b + b^2)$$

$$y^3 - 4^3 = (y - 4)(y^2 + y \cdot 4 + 4^2)$$

$$= (y - 4)(y^2 + 4y + 16)$$ ☐

PRACTICE
8   Factor $27 - y^3$.

▶ **Helpful Hint**

When factoring sums or differences of cubes, be sure to notice the sign patterns.

Same sign

$$x^3 + y^3 = (x + y)(x^2 - xy + y^2)$$

Opposite sign

Always positive

Same sign

$$x^3 - y^3 = (x - y)(x^2 + xy + y^2)$$

Opposite sign

EXAMPLE 9   Factor $125q^2 - n^3q^2$.

Solution   First we factor out a common factor of q^2.

$$125q^2 - n^3q^2 = q^2(125 - n^3)$$
$$= q^2(5^3 - n^3)$$

Opposite sign   Positive

$$= q^2(5 - n)[5^2 + (5)(n) + (n^2)]$$
$$= q^2(5 - n)(25 + 5n + n^2)$$

Thus, $125q^2 - n^3q^2 = q^2(5 - n)(25 + 5n + n^2)$. The trinomial $25 + 5n + n^2$ cannot be factored further. □

PRACTICE
9   Factor $b^3x^2 - 8x^2$.

VOCABULARY & READINESS CHECK

Write each term as a square. For example, $25x^2$ as a square is $(5x)^2$.

1. $81y^2$ **2.** $4z^2$ **3.** $64x^6$ **4.** $49y^6$

Write each number or term as a cube.

5. 125   **6.** 216   **7.** $8x^3$ **8.** $27y^3$ **9.** $64x^6$ **10.** x^3y^6

5.7 | EXERCISE SET

Factor the following. See Examples 1 and 2.

1. $x^2 + 6x + 9$

2. $x^2 - 10x + 25$

3. $4x^2 - 12x + 9$

4. $25x^2 + 10x + 1$

5. $3x^2 - 24x + 48$

6. $x^3 + 14x^2 + 49x$

7. $9y^2x^2 + 12yx^2 + 4x^2$

8. $32x^2 - 16xy + 2y^2$

Factor the following. See Examples 3 through 5.

9. $x^2 - 25$

10. $y^2 - 100$

11. $9 - 4z^2$

12. $16x^2 - y^2$

13. $(y + 2)^2 - 49$

14. $(x - 1)^2 - z^2$

15. $64x^2 - 100$

16. $4x^2 - 36$

Factor the following. See Examples 6 through 9.

17. $x^3 + 27$

18. $y^3 + 1$

19. $z^3 - 1$

20. $x^3 - 8$

21. $m^3 + n^3$

22. $r^3 + 125$

23. $x^3y^2 - 27y^2$

24. $64 - p^3$

25. $a^3b + 8b^4$

26. $8ab^3 + 27a^4$

27. $125y^3 - 8x^3$

28. $54y^3 - 128$

Factor the following. See Example 5.

29. $x^2 + 6x + 9 - y^2$

30. $x^2 + 12x + 36 - y^2$

31. $x^2 - 10x + 25 - y^2$

32. $x^2 - 18x + 81 - y^2$

33. $4x^2 + 4x + 1 - z^2$

34. $9y^2 + 12y + 4 - x^2$

MIXED PRACTICE

Factor each polynomial completely.

35. $9x^2 - 49$

36. $25x^2 - 4$

37. $x^2 - 12x + 36$

38. $x^2 - 18x + 81$

39. $x^4 - 81$

40. $x^4 - 256$

41. $x^2 + 8x + 16 - 4y^2$

42. $x^2 + 14x + 49 - 9y^2$

43. $(x + 2y)^2 - 9$

44. $(3x + y)^2 - 25$

45. $x^3 - 216$

46. $8 - a^3$

47. $x^3 + 125$

48. $x^3 + 216$

49. $4x^2 + 25$

50. $16x^2 + 25$

51. $4a^2 + 12a + 9$

52. $9a^2 - 30a + 25$

53. $18x^2y - 2y$

54. $12xy^2 - 108x$

55. $8x^3 + y^3$

56. $27x^3 - y^3$

57. $x^6 - y^3$

58. $x^3 - y^6$

59. $x^2 + 16x + 64 - x^4$

60. $x^2 + 20x + 100 - x^4$

61. $3x^6y^2 + 81y^2$

62. $x^2y^9 + x^2y^3$

63. $(x + y)^3 + 125$

64. $(x + y)^3 + 27$

65. $(2x + 3)^3 - 64$

66. $(4x + 2)^3 - 125$

REVIEW AND PREVIEW

Solve the following equations. See Section 1.5.

67. $x - 5 = 0$

68. $x + 7 = 0$

69. $3x + 1 = 0$

70. $5x - 15 = 0$

71. $-2x = 0$

72. $3x = 0$

73. $-5x + 25 = 0$

74. $-4x - 16 = 0$

CONCEPT EXTENSIONS

Determine whether each polynomial is factored completely or not. See the Concept Check in this section.

75. $5x(x^2 - 4)$

76. $x^2y^2(x^3 - y^3)$

77. $7y(a^2 + a + 1)$

78. $9z(x^2 + 4)$

△ **79.** A manufacturer of metal washers needs to determine the cross-sectional area of each washer. If the outer radius of the washer is R and the radius of the hole is r, express the area of the washer as a polynomial. Factor this polynomial completely.

△ **80.** Express the area of the shaded region as a polynomial. Factor the polynomial completely.

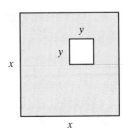

Express the volume of each solid as a polynomial. To do so, subtract the volume of the "hole" from the volume of the larger solid. Then factor the resulting polynomial.

△ **81.**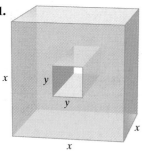

△ **82.**

Find the value of c that makes each trinomial a perfect square trinomial.

83. $x^2 + 6x + c$

84. $y^2 + 10y + c$

85. $m^2 - 14m + c$

86. $n^2 - 2n + c$

87. $x^2 + cx + 16$

88. $x^2 + cx + 36$

89. Factor $x^6 - 1$ completely, using the following methods from this chapter.

    a. Factor the expression by treating it as the difference of two squares, $(x^3)^2 - 1^2$.

    b. Factor the expression by treating it as the difference of two cubes, $(x^2)^3 - 1^3$.

    c. Are the answers to parts **a** and **b** the same? Why or why not?

Factor. Assume that variables used as exponents represent positive integers.

90. $x^{2n} - 25$

91. $x^{2n} - 36$

92. $36x^{2n} - 49$

93. $25x^{2n} - 81$

94. $x^{4n} - 16$

95. $x^{4n} - 625$

INTEGRATED REVIEW   OPERATIONS ON POLYNOMIALS AND FACTORING STRATEGIES

Sections 5.1–5.7

OPERATIONS ON POLYNOMIALS

Perform each indicated operation.

1. $(-y^2 + 6y - 1) + (3y^2 - 4y - 10)$

2. $(5z^4 - 6z^2 + z + 1) - (7z^4 - 2z + 1)$

3. Subtract $(x - 5)$ from $(x^2 - 6x + 2)$.

4. $(2x^2 + 6x - 5) + (5x^2 - 10x)$

5. $(5x - 3)^2$

6. $(5x^2 - 14x - 3) \div (5x + 1)$

7. $(2x^4 - 3x^2 + 5x - 2) \div (x + 2)$

8. $(4x - 1)(x^2 - 3x - 2)$

FACTORING STRATEGIES

The key to proficiency in factoring polynomials is to practice until you are comfortable with each technique. A strategy for factoring polynomials completely is given next.

> **Factoring a Polynomial**
>
> **STEP 1.** Are there any common factors? If so, factor out the greatest common factor.
>
> **STEP 2.** How many terms are in the polynomial?
>
>   **a.** If there are *two* terms, decide if one of the following formulas may be applied:
>
>   **i.** Difference of two squares: $a^2 - b^2 = (a - b)(a + b)$
>
>   **ii.** Difference of two cubes: $a^3 - b^3 = (a - b)(a^2 + ab + b^2)$
>
>   **iii.** Sum of two cubes: $a^3 + b^3 = (a + b)(a^2 - ab + b^2)$
>
>   **b.** If there are *three* terms, try one of the following:
>
>   **i.** Perfect square trinomial: $a^2 + 2ab + b^2 = (a + b)^2$
> $$a^2 - 2ab + b^2 = (a - b)^2$$
>
>   **ii.** If not a perfect square trinomial, factor by using the methods presented in Section 5.5.
>
>   **c.** If there are *four* or more terms, try factoring by grouping.
>
> **STEP 3.** See whether any factors in the factored polynomial can be factored further.

A few examples are worked for you below.

EXAMPLE 1 Factor each polynomial completely.

a. $8a^2b - 4ab$ **b.** $36x^2 - 9$ **c.** $2x^2 - 5x - 7$

d. $5p^2 + 5 + qp^2 + q$ **e.** $9x^2 + 24x + 16$ **f.** $y^2 + 25$

Solution

a. STEP 1. The terms have a common factor of $4ab$, which we factor out.

$$8a^2b - 4ab = 4ab(2a - 1)$$

STEP 2. There are two terms, but the binomial $2a - 1$ is not the difference of two squares or the sum or difference of two cubes.

STEP 3. The factor $2a - 1$ cannot be factored further.

b. STEP 1. Factor out a common factor of 9.

$$36x^2 - 9 = 9(4x^2 - 1)$$

STEP 2. The factor $4x^2 - 1$ has two terms, and it is the difference of two squares.

$$9(4x^2 - 1) = 9(2x + 1)(2x - 1)$$

STEP 3. No factor with more than one term can be factored further.

c. STEP 1. The terms of $2x^2 - 5x - 7$ contain no common factor other than 1 or -1.

STEP 2. There are three terms. The trinomial is not a perfect square, so we factor by methods from Section 5.6.

$$2x^2 - 5x - 7 = (2x - 7)(x + 1)$$

STEP 3. No factor with more than one term can be factored further.

d. STEP 1. There is no common factor of all terms of $5p^2 + 5 + qp^2 + q$.

STEP 2. The polynomial has four terms, so try factoring by grouping.

$$5p^2 + 5 + qp^2 + q = (5p^2 + 5) + (qp^2 + q) \quad \text{Group the terms.}$$
$$= 5(p^2 + 1) + q(p^2 + 1)$$
$$= (p^2 + 1)(5 + q)$$

STEP 3. No factor can be factored further.

e. STEP 1. The terms of $9x^2 + 24x + 16$ contain no common factor other than 1 or -1.

STEP 2. The trinomial $9x^2 + 24x + 16$ is a perfect square trinomial, and $9x^2 + 24x + 16 = (3x + 4)^2$.

STEP 3. No factor can be factored further.

f. STEP 1. There is no common factor of $y^2 + 25$ other than 1.

STEP 2. This binomial is the sum of two squares and is prime.

STEP 3. The binomial $y^2 + 25$ cannot be factored further. ☐

PRACTICE

1 Factor each polynomial completely.

a. $12x^2y - 3xy$ **b.** $49x^2 - 4$

c. $5x^2 + 2x - 3$ **d.** $3x^2 + 6 + x^3 + 2x$

e. $4x^2 + 20x + 25$ **f.** $b^2 + 100$

EXAMPLE 2 Factor each completely.

a. $27a^3 - b^3$ **b.** $3n^2m^4 - 48m^6$ **c.** $2x^2 - 12x + 18 - 2z^2$

d. $8x^4y^2 + 125xy^2$ **e.** $(x - 5)^2 - 49y^2$

Solution

a. This binomial is the difference of two cubes.

$$27a^3 - b^3 = (3a)^3 - b^3$$
$$= (3a - b)[(3a)^2 + (3a)(b) + b^2]$$
$$= (3a - b)(9a^2 + 3ab + b^2)$$

b. $3n^2m^4 - 48m^6 = 3m^4(n^2 - 16m^2)$ Factor out the GCF, $3m^4$.

$$= 3m^4(n + 4m)(n - 4m)$$ Factor the difference of squares.

c. $2x^2 - 12x + 18 - 2z^2 = 2(x^2 - 6x + 9 - z^2)$ The GCF is 2.

$$= 2[(x^2 - 6x + 9) - z^2]$$ Group the first three terms together.

$$= 2[(x - 3)^2 - z^2]$$ Factor the perfect square trinomial.

$$= 2[(x - 3) + z][(x - 3) - z]$$ Factor the difference of squares.

$$= 2(x - 3 + z)(x - 3 - z)$$

d. $8x^4y^2 + 125xy^2 = xy^2(8x^3 + 125)$ The GCF is xy^2.

$$= xy^2[(2x)^3 + 5^3]$$

$$= xy^2(2x + 5)[(2x)^2 - (2x)(5) + 5^2]$$ Factor the sum of cubes.

$$= xy^2(2x + 5)(4x^2 - 10x + 25)$$

e. This binomial is the difference of squares.

$$(x - 5)^2 - 49y^2 = (x - 5)^2 - (7y)^2$$

$$= [(x - 5) + 7y][(x - 5) - 7y]$$

$$= (x - 5 + 7y)(x - 5 - 7y)$$

PRACTICE

2 Factor each polynomial completely.

a. $64x^3 + y^3$

b. $7x^2y^2 - 63y^4$

c. $3x^2 + 12x + 12 - 3b^2$

d. $x^5y^4 + 27x^2y$

e. $(x + 7)^2 - 81y^2$

Factor completely.

9. $x^2 - 8x + 16 - y^2$ **10.** $12x^2 - 22x - 20$

11. $x^4 - x$ **12.** $(2x + 1)^2 - 3(2x + 1) + 2$

13. $14x^2y - 2xy$ **14.** $24ab^2 - 6ab$

15. $4x^2 - 16$ **16.** $9x^2 - 81$

17. $3x^2 - 8x - 11$

18. $5x^2 - 2x - 3$

19. $4x^2 + 8x - 12$

20. $6x^2 - 6x - 12$

21. $4x^2 + 36x + 81$

22. $25x^2 + 40x + 16$

23. $8x^3 + 125y^3$

24. $27x^3 - 64y^3$

25. $64x^2y^3 - 8x^2$

26. $27x^5y^4 - 216x^2y$

27. $(x + 5)^3 + y^3$

28. $(y - 1)^3 + 27x^3$

29. $(5a - 3)^2 - 6(5a - 3) + 9$

30. $(4r + 1)^2 + 8(4r + 1) + 16$

31. $7x^2 - 63x$

32. $20x^2 + 23x + 6$

33. $ab - 6a + 7b - 42$

34. $20x^2 - 220x + 600$

35. $x^4 - 1$

36. $15x^2 - 20x$

37. $10x^2 - 7x - 33$

38. $45m^3n^3 - 27m^2n^2$

39. $5a^3b^3 - 50a^3b$

40. $x^4 + x$

41. $16x^2 + 25$

42. $20x^3 + 20y^3$

43. $10x^3 - 210x^2 + 1100x$

44. $9y^2 - 42y + 49$

45. $64a^3b^4 - 27a^3b$

46. $y^4 - 16$

47. $2x^3 - 54$

48. $2sr + 10s - r - 5$

49. $3y^5 - 5y^4 + 6y - 10$

50. $64a^2 + b^2$

51. $100z^3 + 100$

52. $250x^4 - 16x$

53. $4b^2 - 36b + 81$

54. $2a^5 - a^4 + 6a - 3$

55. $(y - 6)^2 + 3(y - 6) + 2$

56. $(c + 2)^2 - 6(c + 2) + 5$

△ **57.** Express the area of the shaded region as a polynomial. Factor the polynomial completely.

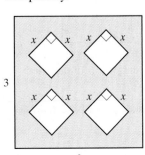

5.8 SOLVING EQUATIONS BY FACTORING AND PROBLEM SOLVING

OBJECTIVE 1 ▶ Solving polynomial equations by factoring. In this section, your efforts to learn factoring start to pay off. We use factoring to solve polynomial equations, which in turn helps us solve problems that can be modeled by polynomial equations and also helps us sketch the graph of polynomial functions.

A **polynomial equation** is the result of setting two polynomials equal to each other. Examples of polynomial equations are

$$3x^3 - 2x^2 = x^2 + 2x - 1 \qquad 2.6x + 7 = -1.3 \qquad -5x^2 - 5 = -9x^2 - 2x + 1$$

A polynomial equation is in **standard form** if one side of the equation is 0. In standard form the polynomial equations above are

$$3x^3 - 3x^2 - 2x + 1 = 0 \qquad 2.6x + 8.3 = 0 \qquad 4x^2 + 2x - 6 = 0$$

The degree of a simplified polynomial equation in standard form is the same as the highest degree of any of its terms. A polynomial equation of degree 2 is also called a **quadratic equation.**

A solution of a polynomial equation in one variable is a value of the variable that makes the equation true. The method presented in this section for solving polynomial equations is called the **factoring method.** This method is based on the **zero-factor property.**

Zero-Factor Property

If a and b are real numbers and $a \cdot b = 0$, then $a = 0$ or $b = 0$. This property is true for three or more factors also.

In other words, if the product of two or more real numbers is zero, then at least one number must be zero.

EXAMPLE 1 Solve: $(x + 2)(x - 6) = 0$. Check numerically and graphically.

Solution By the zero-factor property, $(x + 2)(x - 6) = 0$ only if $x + 2 = 0$ or $x - 6 = 0$.

$$x + 2 = 0 \quad \text{or} \quad x - 6 = 0 \quad \text{Apply the zero-factor property.}$$
$$x = -2 \quad \text{or} \quad x = 6 \quad \text{Solve each linear equation.}$$

To check numerically, let $x = -2$ and then let $x = 6$ in the original equation, as shown in the screen below to the left. To check graphically, see the screen below to the right.

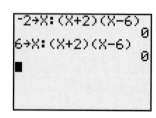
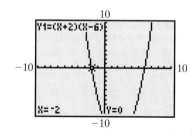

Notice that the *x*-intercepts of the graph are $(-2, 0)$ and $(6, 0)$.

Both -2 and 6 check, so they are both solutions. The solution set is $\{-2, 6\}$. □

PRACTICE

1 Solve: $(x + 8)(x - 5) = 0$.

EXAMPLE 2 Solve: $2x^2 + 9x - 5 = 0$. Check numerically and graphically.

Solution To use the zero-factor property, one side of the equation must be 0, and the other side must be in factored form.

$$2x^2 + 9x - 5 = 0$$
$$(2x - 1)(x + 5) = 0 \qquad \text{Factor.}$$
$$2x - 1 = 0 \quad \text{or} \quad x + 5 = 0 \qquad \text{Set each factor equal to zero.}$$
$$2x = 1$$
$$x = \frac{1}{2} \quad \text{or} \qquad x = -5 \qquad \text{Solve each linear equation.}$$

A numerical check and graphical check are shown.

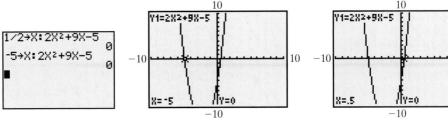

A numerical check   A graphical check

The solutions are -5 and $\dfrac{1}{2}$, or the solution set is $\left\{ -5, \dfrac{1}{2} \right\}$. ☐

PRACTICE
2 Solve: $3x^2 + 10x - 8 = 0$.

Solving Polynomial Equations by Factoring

STEP 1. Write the equation in standard form so that one side of the equation is 0.

STEP 2. Factor the polynomial completely.

STEP 3. Set each factor containing a variable equal to 0.

STEP 4. Solve the resulting equations.

STEP 5. Check each solution in the original equation.

Since it is not always possible to factor a polynomial, not all polynomial equations can be solved by factoring. Other methods of solving polynomial equations are presented in Chapter 8.

EXAMPLE 3 Solve: $x(2x - 7) = 4$.

Solution First, write the equation in standard form; then, factor.

$$x(2x - 7) = 4$$
$$2x^2 - 7x = 4 \qquad \text{Multiply.}$$
$$2x^2 - 7x - 4 = 0 \qquad \text{Write in standard form.}$$
$$(2x + 1)(x - 4) = 0 \qquad \text{Factor.}$$
$$2x + 1 = 0 \quad \text{or} \quad x - 4 = 0 \qquad \text{Set each factor equal to zero.}$$
$$2x = -1 \qquad \text{Solve.}$$
$$x = -\frac{1}{2} \quad \text{or} \qquad x = 4$$

The solutions are $-\dfrac{1}{2}$ and 4. Check both solutions in the original equation, as shown to the left. ☐

PRACTICE
3 Solve: $x(3x + 14) = -8$.

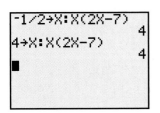

A numerical check of Example 3.

> ▶ **Helpful Hint**
>
> To apply the zero-factor property, one side of the equation must be 0, and the other side of the equation must be factored. To solve the equation $x(2x - 7) = 4$, for example, you may **not** set each factor equal to 4.

EXAMPLE 4   Solve: $3(x^2 + 4) + 5 = -6(x^2 + 2x) + 13$.

Solution   Rewrite the equation so that one side is 0.

$$3(x^2 + 4) + 5 = -6(x^2 + 2x) + 13$$
$$3x^2 + 12 + 5 = -6x^2 - 12x + 13 \qquad \text{Apply the distributive property.}$$
$$9x^2 + 12x + 4 = 0 \qquad \text{Rewrite the equation so that one side is 0.}$$
$$(3x + 2)(3x + 2) = 0 \qquad \text{Factor.}$$
$$3x + 2 = 0 \quad \text{or} \quad 3x + 2 = 0 \qquad \text{Set each factor equal to 0.}$$
$$3x = -2 \quad \text{or} \quad 3x = -2$$
$$x = -\frac{2}{3} \quad \text{or} \quad x = -\frac{2}{3} \qquad \text{Solve each equation.}$$

Notice that the identical factors, $(3x + 2)$, led to a single solution, $-\dfrac{2}{3}$. The solution is $-\dfrac{2}{3}$.

We choose to check this time by graphing and using the intersection-of-graphs method as shown in the screen to the left. The intersection is at $x \approx -0.66667$, which is an approximation for $-\dfrac{2}{3}$.

TECHNOLOGY NOTE

When checking a solution graphically, if the graphical solution is given in decimal form and your algebraic solution is in fractional form, recall that you can convert to fraction form using a fraction command.

$y_1 = 3(x^2 + 4) + 5$

$y_2 = -6(x^2 + 2x) + 13$

PRACTICE

4   Solve: $8(x^2 + 3) + 4 = -8x(x + 3) + 19$.

If the equation contains fractions, we clear the equation of fractions as a first step.

EXAMPLE 5   Solve: $2x^2 = \dfrac{17}{3}x + 1$.

Solution
$$2x^2 = \frac{17}{3}x + 1$$
$$3(2x^2) = 3\left(\frac{17}{3}x + 1\right) \qquad \text{Clear the equation of fractions.}$$
$$6x^2 = 17x + 3 \qquad \text{Apply the distributive property.}$$
$$6x^2 - 17x - 3 = 0 \qquad \text{Rewrite the equation in standard form.}$$
$$(6x + 1)(x - 3) = 0 \qquad \text{Factor.}$$
$$6x + 1 = 0 \quad \text{or} \quad x - 3 = 0 \qquad \text{Set each factor equal to zero.}$$
$$6x = -1$$
$$x = -\frac{1}{6} \quad \text{or} \quad x = 3 \qquad \text{Solve each equation.}$$

The solutions are $-\dfrac{1}{6}$ and 3. To check, graph $y_1 = 2x^2$ and $y_2 = \dfrac{17}{3}x + 1$. Notice that the graph of y_1 is a parabola and the graph of y_2 is a line.

$\left(-\frac{1}{6}, \frac{1}{18}\right)$ $(3, 18)$

The two graphs intersect twice.

Verify the solutions of $x = -\dfrac{1}{6}$ and $x = 3$.

PRACTICE

5   Solve: $4x^2 = \dfrac{15}{2}x + 1$.

EXAMPLE 6   Solve: $x^3 = 4x$.

Solution

$$x^3 = 4x$$

$$x^3 - 4x = 0 \qquad \text{Rewrite the equation so that one side is 0.}$$

$$x(x^2 - 4) = 0 \qquad \text{Factor out the GCF, } x.$$

$$x(x + 2)(x - 2) = 0 \qquad \text{Factor the difference of squares.}$$

$$x = 0 \quad \text{or} \quad x + 2 = 0 \quad \text{or} \quad x - 2 = 0 \qquad \text{Set each factor equal to 0.}$$

$$x = 0 \quad \text{or} \qquad x = -2 \quad \text{or} \qquad x = 2 \qquad \text{Solve each equation.}$$

The solutions are -2, 0, and 2. Check graphically by graphing $y_1 = x^3 - 4x$ and checking x-intercepts, as shown to the left.   □

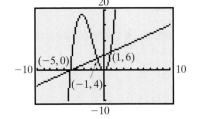

Example 6 check: The x-values of the x-intercepts are -2, 0, and 2. Notice that the third-degree equation has three solutions.

PRACTICE

6   Solve: $x^3 = 2x^2 + 3x$.

Notice that the *third*-degree equation of Example 6 yielded *three* solutions.

EXAMPLE 7   Solve: $x^3 + 5x^2 = x + 5$.

Solution   First, write the equation so that one side is 0.

$$x^3 + 5x^2 - x - 5 = 0$$

$$(x^3 - x) + (5x^2 - 5) = 0 \qquad \text{Factor by grouping.}$$

$$x(x^2 - 1) + 5(x^2 - 1) = 0$$

$$(x^2 - 1)(x + 5) = 0$$

$$(x + 1)(x - 1)(x + 5) = 0 \qquad \text{Factor the difference of squares.}$$

$$x + 1 = 0 \quad \text{or} \quad x - 1 = 0 \quad \text{or} \quad x + 5 = 0 \qquad \text{Set each factor equal to 0.}$$

$$x = -1 \quad \text{or} \qquad x = 1 \quad \text{or} \qquad x = -5 \qquad \text{Solve each equation.}$$

The solutions are -5, -1, and 1. The screen to the left verifies the solutions. Check in the original equation.   □

PRACTICE

7   Solve: $x^3 - 9x = 18 - 2x^2$.

Concept Check ☑

Which solution strategies are incorrect? Why?

a. Solve $(y - 2)(y + 2) = 4$ by setting each factor equal to 4.

b. Solve $(x + 1)(x + 3) = 0$ by setting each factor equal to 0.

c. Solve $z^2 + 5z + 6 = 0$ by factoring $z^2 + 5z + 6$ and setting each factor equal to 0.

d. Solve $x^2 + 6x + 8 = 10$ by factoring $x^2 + 6x + 8$ and setting each factor equal to 0.

OBJECTIVE 2 ▶ Solving problems modeled by polynomial equations. Some problems may be modeled by polynomial equations. To solve these problems, we use the same problem-solving steps that were introduced in Section 2.2. When solving these problems, keep in mind that a solution of an equation that models a problem is not always a solution to the problem. For example, a person's weight or the length of a side of a geometric figure is always a positive number. Discard solutions that do not make sense as solutions of the problem.

Answer to Concept Check:
a and d; the zero-factor property works only if one side of the equation is 0

EXAMPLE 8 **Finding the Return Time of a Rocket**

An Alpha III model rocket is launched from the ground with an A8–3 engine. Without a parachute, the height of the rocket h at time t seconds is approximated by the equation

$$h = -16t^2 + 144t$$

Find how long it takes the rocket to return to the ground.

Solution

1. UNDERSTAND. Read and reread the problem. The equation $h = -16t^2 + 144t$ models the height of the rocket. Familiarize yourself with this equation by finding a few values.

   When $t = 1$ second, the height of the rocket is

   $$h = -16(1)^2 + 144(1) = 128 \text{ feet}$$

   When $t = 2$ seconds, the height of the rocket is

   $$h = -16(2)^2 + 144(2) = 224 \text{ feet}$$

2. TRANSLATE. To find how long it takes the rocket to return to the ground, we want to know what value of t makes the height h equal to 0. That is, we want to solve $h = 0$.

   $$-16t^2 + 144t = 0$$

3. SOLVE the quadratic equation by factoring.

   $$-16t^2 + 144t = 0$$
   $$-16t(t - 9) = 0$$
   $$-16t = 0 \quad \text{or} \quad t - 9 = 0$$
   $$t = 0 \qquad\qquad t = 9$$

350

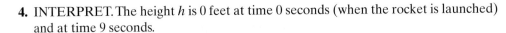

Zero
X=9   Y=0

−100

The rocket is at ground level at 0 seconds and 9 seconds.

4. INTERPRET. The height h is 0 feet at time 0 seconds (when the rocket is launched) and at time 9 seconds.

Check: See that the height of the rocket at 9 seconds equals 0.

$$h = -16(9)^2 + 144(9) = -1296 + 1296 = 0$$

State: The rocket returns to the ground 9 seconds after it is launched. □

PRACTICE

8 A model rocket is launched from the ground. Its height h in feet at time t seconds is approximated by the equation $h = -16t^2 + 96t$. Find how long it takes the rocket to return to the ground.

Some of the exercises at the end of this section make use of the **Pythagorean theorem.** Before we review this theorem, recall that a **right triangle** is a triangle that contains a 90° angle, or right angle. The **hypotenuse** of a right triangle is the side opposite the right angle and is the longest side of the triangle. The **legs** of a right triangle are the other sides of the triangle.

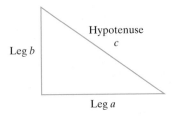

Hypotenuse
c

Leg b

Leg a

> **Pythagorean Theorem**
>
> In a right triangle, the sum of the squares of the lengths of the two legs is equal to the square of the length of the hypotenuse.
>
> $$(\text{leg})^2 + (\text{leg})^2 = (\text{hypotenuse})^2 \quad \text{or} \quad a^2 + b^2 = c^2$$

EXAMPLE 9   **Using the Pythagorean Theorem**

While framing an addition to an existing home, Kim Menzies, a carpenter, used the Pythagorean theorem to determine whether a wall was "square"—that is, whether the wall formed a right angle with the floor. He used a triangle whose sides are three consecutive integers. Find a right triangle whose sides are three consecutive integers.

Solution

1. UNDERSTAND. Read and reread the problem.

Let x, $x + 1$, and $x + 2$ be three consecutive integers. Since these integers represent lengths of the sides of a right triangle, we have

x = one leg
$x + 1$ = other leg
$x + 2$ = hypotenuse (longest side)

2. TRANSLATE. By the Pythagorean theorem, we have

In words: $(\text{leg})^2$ $+$ $(\text{leg})^2$ $=$ $(\text{hypotenuse})^2$

Translate: $(x)^2$ $+$ $(x + 1)^2$ $=$ $(x + 2)^2$

3. SOLVE the equation.

$$x^2 + (x + 1)^2 = (x + 2)^2$$
$$x^2 + x^2 + 2x + 1 = x^2 + 4x + 4 \qquad \text{Multiply.}$$
$$2x^2 + 2x + 1 = x^2 + 4x + 4$$
$$x^2 - 2x - 3 = 0 \qquad \text{Write in standard form.}$$
$$(x - 3)(x + 1) = 0$$
$$x - 3 = 0 \quad \text{or} \quad x + 1 = 0$$
$$x = 3 \qquad\qquad x = -1$$

4. INTERPRET. Discard $x = -1$ since length cannot be negative. If $x = 3$, then $x + 1 = 4$ and $x + 2 = 5$.

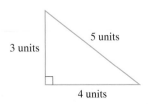

Check: To check, see that $(\text{leg})^2 + (\text{leg})^2 = (\text{hypotenuse})^2$

$$3^2 + 4^2 = 5^2$$
$$9 + 16 = 25 \quad \text{True}$$

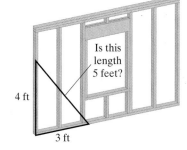

State: The lengths of the sides of the right triangle are 3, 4, and 5 units. Kim used this information, for example, by marking off lengths of 3 and 4 feet on the floor and framing, respectively. If the diagonal length between these marks was 5 feet, the wall was "square." If not, adjustments were made. ☐

PRACTICE

9 Find a right triangle whose sides are consecutive even integers.

OBJECTIVE 3 ▶ Finding the *x*-intercepts of polynomial functions. Recall that to find the *x*-intercepts of the graph of a function, let $f(x)$ or $y = 0$ and solve for x. This fact again gives us a visual interpretation of the results of this section.

From Example 1, we know that the solutions of the equation $(x + 2)(x - 6) = 0$ are -2 and 6. These solutions give us important information about the related polynomial function $p(x) = (x + 2)(x - 6)$. We know that when x is -2 or when x is 6, the value of $p(x)$ is 0.

$$p(x) = (x + 2)(x - 6)$$
$$p(-2) = (-2 + 2)(-2 - 6) = (0)(-8) = 0$$
$$p(6) = (6 + 2)(6 - 6) = (8)(0) = 0$$

Thus, we know that $(-2, 0)$ and $(6, 0)$ are the *x*-intercepts of the graph of $p(x)$.

We also know that the graph of $p(x)$ does not cross the *x*-axis at any other point. For this reason, and the fact that $p(x) = (x + 2)(x - 6) = x^2 - 4x - 12$ has degree 2, we conclude that the graph of p must look something like one of these two graphs:

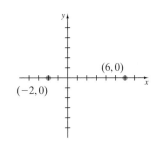

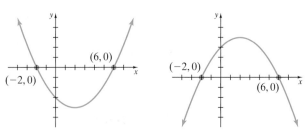

In a later chapter, we explore these graphs more fully. For the moment, know that the solutions of a polynomial equation are the *x*-intercepts of the graph of the related function and that the *x*-intercepts of the graph of a polynomial function are the solutions of the related polynomial equation. These values are also called **roots,** or **zeros,** of a polynomial function.

EXAMPLE 10 **Match Each Function with Its Graph**

$$f(x) = (x - 3)(x + 2) \quad g(x) = x(x + 2)(x - 2) \quad h(x) = (x - 2)(x + 2)(x - 1)$$

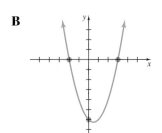
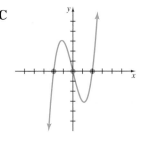

Solution The graph of the function $f(x) = (x - 3)(x + 2)$ has two x-intercepts, $(3, 0)$ and $(-2, 0)$, because the equation $0 = (x - 3)(x + 2)$ has two solutions, 3 and -2.

The graph of $f(x)$ is graph B.

The graph of the function $g(x) = x(x + 2)(x - 2)$ has three x-intercepts $(0, 0)$, $(-2, 0)$, and $(2, 0)$, because the equation $0 = x(x + 2)(x - 2)$ has three solutions, $0, -2$, and 2.

The graph of $g(x)$ is graph C.

The graph of the function $h(x) = (x - 2)(x + 2)(x - 1)$ has three x-intercepts, $(-2, 0)$, $(1, 0)$, and $(2, 0)$, because the equation $0 = (x - 2)(x + 2)(x - 1)$ has three solutions, $-2, 1$, and 2.

The graph of $h(x)$ is graph A. ☐

PRACTICE
10 Match each function with its graph.

$$f(x) = (x - 1)(x + 3) \quad g(x) = x(x + 3)(x - 2) \quad h(x) = (x - 3)(x + 2)(x - 2)$$

A 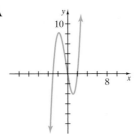 **B** 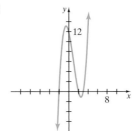 **C**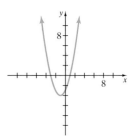

VOCABULARY & READINESS CHECK

Solve each equation for the variable. See Example 1.

1. $(x - 3)(x + 5) = 0$

2. $(y + 5)(y + 3) = 0$

3. $(z - 3)(z + 7) = 0$

4. $(c - 2)(c - 4) = 0$

5. $x(x - 9) = 0$

6. $w(w + 7) = 0$

5.8 EXERCISE SET

Solve each equation. See Example 1.

1. $(x + 3)(3x - 4) = 0$

2. $(5x + 1)(x - 2) = 0$

3. $3(2x - 5)(4x + 3) = 0$

4. $8(3x - 4)(2x - 7) = 0$

Solve each equation. See Examples 2 through 5.

5. $x^2 + 11x + 24 = 0$

6. $y^2 - 10y + 24 = 0$

7. $12x^2 + 5x - 2 = 0$

8. $3y^2 - y - 14 = 0$

9. $z^2 + 9 = 10z$

10. $n^2 + n = 72$

11. $x(5x + 2) = 3$

12. $n(2n - 3) = 2$

13. $x^2 - 6x = x(8 + x)$

14. $n(3 + n) = n^2 + 4n$

15. $\dfrac{z^2}{6} - \dfrac{z}{2} - 3 = 0$

16. $\dfrac{c^2}{20} - \dfrac{c}{4} + \dfrac{1}{5} = 0$

17. $\dfrac{x^2}{2} + \dfrac{x}{20} = \dfrac{1}{10}$

18. $\dfrac{y^2}{30} = \dfrac{y}{15} + \dfrac{1}{2}$

19. $\dfrac{4t^2}{5} = \dfrac{t}{5} + \dfrac{3}{10}$

20. $\dfrac{5x^2}{6} - \dfrac{7x}{2} + \dfrac{2}{3} = 0$

Solve each equation. See Examples 6 and 7.

21. $(x + 2)(x - 7)(3x - 8) = 0$

22. $(4x + 9)(x - 4)(x + 1) = 0$

23. $y^3 = 9y$

24. $n^3 = 16n$

25. $x^3 - x = 2x^2 - 2$

26. $m^3 = m^2 + 12m$

27. Explain how solving $2(x - 3)(x - 1) = 0$ differs from solving $2x(x - 3)(x - 1) = 0$.

28. Explain why the zero-factor property works for more than two numbers whose product is 0.

MIXED PRACTICE

Solve each equation.

29. $(2x + 7)(x - 10) = 0$

30. $(x + 4)(5x - 1) = 0$

31. $3x(x - 5) = 0$

32. $4x(2x + 3) = 0$

33. $x^2 - 2x - 15 = 0$

34. $x^2 + 6x - 7 = 0$

35. $12x^2 + 2x - 2 = 0$

36. $8x^2 + 13x + 5 = 0$

37. $w^2 - 5w = 36$

38. $x^2 + 32 = 12x$

39. $25x^2 - 40x + 16 = 0$

40. $9n^2 + 30n + 25 = 0$

41. $2r^3 + 6r^2 = 20r$

42. $-2t^3 = 108t - 30t^2$

43. $z(5z - 4)(z + 3) = 0$

44. $2r(r + 3)(5r - 4) = 0$

45. $2z(z + 6) = 2z^2 + 12z - 8$

46. $3c^2 - 8c + 2 = c(3c - 8)$

47. $(x - 1)(x + 4) = 24$

48. $(2x - 1)(x + 2) = -3$

49. $\dfrac{x^2}{4} - \dfrac{5}{2}x + 6 = 0$

50. $\dfrac{x^2}{18} + \dfrac{x}{2} + 1 = 0$

51. $y^2 + \dfrac{1}{4} = -y$

52. $\dfrac{x^2}{10} + \dfrac{5}{2} = x$

53. $y^3 + 4y^2 = 9y + 36$

54. $x^3 + 5x^2 = x + 5$

55. $2x^3 = 50x$

56. $m^5 = 36m^3$

57. $x^2 + (x + 1)^2 = 61$

58. $y^2 + (y + 2)^2 = 34$

59. $m^2(3m - 2) = m$

60. $x^2(5x + 3) = 26x$

61. $3x^2 = -x$

62. $y^2 = -5y$

63. $x(x - 3) = x^2 + 5x + 7$

64. $z^2 - 4z + 10 = z(z - 5)$

65. $3(t - 8) + 2t = 7 + t$

66. $7c - 2(3c + 1) = 5(4 - 2c)$

67. $-3(x - 4) + x = 5(3 - x)$

68. $-4(a + 1) - 3a = -7(2a - 3)$

69. Which solution strategies are incorrect? Why?

   a. Solve $(y - 2)(y + 2) = 4$ by setting each factor equal to 4.

   b. Solve $(x + 1)(x + 3) = 0$ by setting each factor equal to 0.

   c. Solve $z^2 + 5z + 6 = 0$ by factoring $z^2 + 5z + 6$ and setting each factor equal to 0.

   d. Solve $x^2 + 6x + 8 = 10$ by factoring $x^2 + 6x + 8$ and setting each factor equal to 0.

70. Describe two ways a linear equation differs from a quadratic equation.

Solve. See Examples 8 and 9.

71. One number exceeds another by five, and their product is 66. Find the numbers.

72. If the sum of two numbers is 4 and their product is $\dfrac{15}{4}$, find the numbers.

73. An electrician needs to run a cable from the top of a 60-foot tower to a transmitter box located 45 feet away from the base of the tower. Find how long he should make the cable.

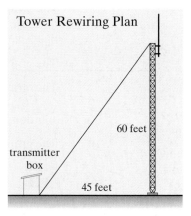

Tower Rewiring Plan

60 feet

transmitter box

45 feet

74. A stereo system installer needs to run speaker wire along the two diagonals of a rectangular room whose dimensions are 40 feet by 75 feet. Find how much speaker wire she needs.

75 ft

40 ft

75. If the cost, $C(x)$, for manufacturing x units of a certain product is given by $C(x) = x^2 - 15x + 50$, find the number of units manufactured at a cost of $9500.

76. Determine whether any three consecutive integers represent the lengths of the sides of a right triangle.

77. The shorter leg of a right triangle is 3 centimeters less than the other leg. Find the length of the two legs if the hypotenuse is 15 centimeters.

78. The longer leg of a right triangle is 4 feet longer than the other leg. Find the length of the two legs if the hypotenuse is 20 feet.

79. Marie Mulroney has a rectangular board 12 inches by 16 inches around which she wants to put a uniform border of shells. If she has enough shells for a border whose area is 128 square inches, determine the width of the border.

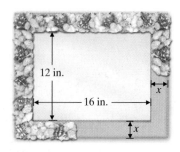

12 in.

16 in.

x

x

80. A gardener has a rose garden that measures 30 feet by 20 feet. He wants to put a uniform border of pine bark around the outside of the garden. Find how wide the border should be if he has enough pine bark to cover 336 square feet.

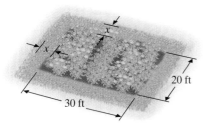

81. While hovering near the top of Ribbon Falls in Yosemite National Park at 1600 feet, a helicopter pilot accidentally drops his sunglasses. The height $h(t)$ of the sunglasses after t seconds is given by the polynomial function

$$h(t) = -16t^2 + 1600$$

When will the sunglasses hit the ground?

82. After t seconds, the height $h(t)$ of a model rocket launched from the ground into the air is given by the function
$$h(t) = -16t^2 + 80t$$
Find how long it takes the rocket to reach a height of 96 feet.

83. The floor of a shed has an area of 90 square feet. The floor is in the shape of a rectangle whose length is 3 feet less than twice the width. Find the length and the width of the floor of the shed.

84. A vegetable garden with an area of 200 square feet is to be fertilized. If the length of the garden is 1 foot less than three times the width, find the dimensions of the garden.

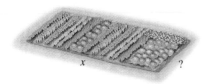

85. The function $W(x) = 0.5x^2$ gives the number of servings of wedding cake that can be obtained from a two-layer x-inch square wedding cake tier. What size square wedding cake tier is needed to serve 50 people? (*Source:* Based on data from the *Wilton 2000 Yearbook of Cake Decorating*)

86. Use the function in Exercise 85 to determine what size square wedding cake tier is needed to serve 200 people.

87. Suppose that a movie is being filmed in New York City. An action shot requires an object to be thrown upward with an initial velocity of 80 feet per second off the top of 1 Madison Square Plaza, a height of 576 feet. The height $h(t)$ in feet of the object after t seconds is given by the function

$$h(t) = -16t^2 + 80t + 576$$

Determine how long before the object strikes the ground. (See Exercise 99, Section 5.6.) (*Source: The World Almanac*)

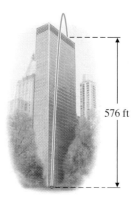

576 ft

88. Suppose that an object is thrown upward with an initial velocity of 64 feet per second off the edge of a 960-foot cliff. The height $h(t)$ in feet of the object after t seconds is given by the function

$$h(t) = -16t^2 + 64t + 960$$

Determine how long before the object strikes the ground. (See Exercise 100, Section 5.6.)

Match each polynomial function with its graph (A–F). See Example 10.

89. $f(x) = (x - 2)(x + 5)$

90. $g(x) = (x + 1)(x - 6)$

91. $h(x) = x(x + 3)(x - 3)$

92. $F(x) = (x + 1)(x - 2)(x + 5)$

93. $G(x) = 2x^2 + 9x + 4$

94. $H(x) = 2x^2 - 7x - 4$

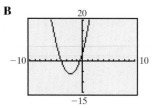

C

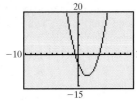

D

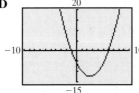

E

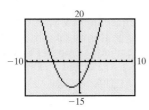

F

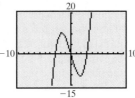

99. Draw a function with intercepts $(-3, 0)$, $(5, 0)$, and $(0, 4)$.

100. Draw a function with intercepts $(-7, 0)$, $\left(-\frac{1}{2}, 0\right)$, $(4, 0)$, and $(0, -1)$.

CONCEPT EXTENSIONS

Each exercise contains an error. Find and correct the error. See the Concept Check in the section.

101. $(x - 5)(x + 2) = 0$

$x - 5 = 0$ or $x + 2 = 0$

$x = -5$ or $x = -2$

102. $(4x - 5)(x + 7) = 0$

$4x - 5 = 0$ or $x + 7 = 0$

$x = \frac{4}{5}$ or $x = -7$

103. $y(y - 5) = -6$

$y = -6$ or $y - 5 = -5$

$y = -7$ or $y = 0$

104. $3x^2 - 19x = 14$

$-16x = 14$

$x = -\frac{14}{16}$

$x = -\frac{7}{8}$

Solve.

105. $(x^2 + x - 6)(3x^2 - 14x - 5) = 0$

106. $(x^2 - 9)(x^2 + 8x + 16) = 0$

107. Is the following step correct? Why or why not?

$$x(x - 3) = 5$$
$$x = 5 \quad \text{or} \quad x - 3 = 5$$

Write a quadratic equation that has the given numbers as solutions.

108. $5, 3$

109. $6, 7$

110. $-1, 2$

111. $4, -3$

REVIEW AND PREVIEW

Write the x- and y-intercepts for each graph and determine whether the graph is the graph of a function. See Sections 2.1 and 2.2.

95.

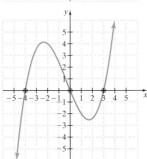

96.

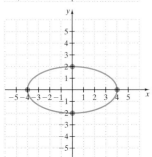

97.

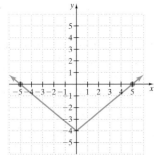

98.

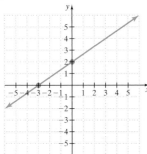

THE BIGGER PICTURE   SOLVING EQUATIONS AND INEQUALITIES

Continue the outline started in Section 1.5 and continued in Sections 3.2, 3.3, and 3.5. Write how to recognize and how to solve quadratic equations by factoring.

Solving Equations and Inequalities

I. **Equations**

    A. Linear equations (Sec. 1.5 and 3.1)

    B. Absolute value equations (Sec. 3.4)

    C. Quadratic equations: Equation can be written in the standard form $ax^2 + bx + c = 0$, with $a \neq 0$.

    Solve: $2x^2 - 7x = 9$

    $2x^2 - 7x - 9 = 0$ Write in standard form so that equation is equal to 0.

    $(2x - 9)(x + 1) = 0$ Factor.

    $2x - 9 = 0$ or $x + 1 = 0$ Set each factor equal to 0.

    $x = \dfrac{9}{2}$ or $x = -1$ Solve.

II. **Inequalities**

    A. Linear inequalities (Sec. 3.2)

    B. Compound inequalities (Sec. 3.3)

    C. Absolute value inequalities (Sec. 3.5)

Solve. If an inequality, write your answer in interval notation.

1. $|7x - 3| = |5x + 9|$

2. $\left|\dfrac{x + 2}{5}\right| < 1$

3. $3(x - 6) + 2 = 9 + 5(3x - 1)$

4. $(x - 6)(2x + 3) = 0$

5. $|-3x + 10| \geq -2$

6. $|-2x - 5| = 11$

7. $x(x - 7) = 30$

8. $8x - 4 \geq 15x - 4$

CHAPTER 5 GROUP ACTIVITY

Finding the Largest Area

This activity may be completed by working in groups or individually.

A picture framer has a piece of wood that measures 1 inch wide by 50 inches long. She would like to make a picture frame with the largest possible interior area. Complete the following activity to help her determine the dimensions of the frame that she should use to achieve her goal.

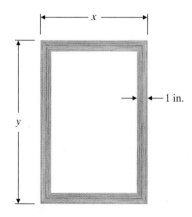

1. Use the situation shown in the figure to write an equation in x and y for the *outer* perimeter of the frame. (Remember that the outer perimeter will equal 50 inches.)

2. Use your equation from Question 1 to help you find the value of y for each value of x given in the table. Complete the y column of the table. (*Note:* The first two columns of the table give possible combinations for the outer dimensions of the frame.)

3. How is the interior width of the frame related to the exterior width of the frame? How is the interior height of the frame related to the exterior height of the frame? Use these relationships to complete the two columns of the table labeled "Interior Width" and "Interior Height."

4. Complete the last column of the table labeled "Interior Area" by using the columns of dimensions for the interior width and height.

5. From the table, what appears to be the largest interior area of the frame? Which exterior dimensions of the frame provide this area?

6. Use the patterns in the table to write an algebraic expression in terms of x for the interior width of the frame.

7. Use the patterns in the table to write an algebraic expression in terms of y for the interior height of the frame.

8. Use the perimeter equation from Question 1 to rewrite the algebraic expression for the interior height of the frame in terms of x.

Frame's Interior Dimensions				
x	y	*Interior Width*	*Interior Height*	*Interior Area*
2.0				
2.5				
3.0				
3.5				
4.0				
4.5				
5.0				
5.5				
6.0				
6.5				
7.0				
7.5				
8.0				
8.5				

Frame's Interior Dimensions				
x	y	*Interior Width*	*Interior Height*	*Interior Area*
9.0				
9.5				
10.0				
10.5				
11.0				
11.5				
12.0				
12.5				
13.0				
13.5				
14.0				
14.5				
15.0				

9. Find a function A that gives the interior area of the frame in terms of its exterior width x. (*Hint:* Study the patterns in the table. How could the expressions from Questions 6 and 8 be used to write this function?)

10. Graph the function A. Locate and label the point from the table that represents the maximum interior area. Describe the location of the point in relation to the rest of the graph.

CHAPTER 5 VOCABULARY CHECK

Fill in each blank with one of the words or phrases listed below.

| quadratic equation | scientific notation | polynomial | exponents | 1 | monomial |
| binomial | trinomial | degree of a polynomial | degree of a term | 0 | factoring |

1. A _____ is a finite sum of terms in which all variables are raised to nonnegative integer powers and no variables appear in any denominator.

2. _____ is the process of writing a polynomial as a product.

3. _____ are used to write repeated factors in a more compact form.

4. The _____ is the sum of the exponents on the variables contained in the term.

5. A _____ is a polynomial with one term.

6. If a is not 0, $a^0 =$ _____ .

7. A _____ is a polynomial with three terms.

8. A polynomial equation of degree 2 is also called a _____ .

9. A positive number is written in _____ if it is written as the product of a number a, such that $1 \le a < 10$, and a power of 10.

10. The _____ is the largest degree of all of its terms.

11. A _____ is a polynomial with two terms.

12. If a and b are real numbers and $a \cdot b =$ _____ , then $a = 0$ or $b = 0$.

▶ **Helpful Hint**

Are you preparing for your test? Don't forget to take the Chapter 5 Test on page 406. Then check your answers at the back of the text and use the Chapter Test Prep Video CD to see the fully worked-out solutions to any of the exercises you want to review.

CHAPTER 5 HIGHLIGHTS

| **DEFINITIONS AND CONCEPTS** | **EXAMPLES** |

SECTION 5.1 EXPONENTS AND SCIENTIFIC NOTATION

Product rule: $a^m \cdot a^n = a^{m+n}$

$x^2 \cdot x^3 = x^5$

Zero exponent: $a^0 = 1, a \neq 0$

$7^0 = 1, (-10)^0 = 1$

Quotient rule: $\dfrac{a^m}{a^n} = a^{m-n}, a \neq 0$

$\dfrac{y^{10}}{y^4} = y^{10-4} = y^6$

Negative exponent: $a^{-n} = \dfrac{1}{a^n}, a \neq 0$

$3^{-2} = \dfrac{1}{3^2} = \dfrac{1}{9}, \dfrac{x^{-5}}{x^{-7}} = x^{-5-(-7)} = x^2$

A positive number is written in **scientific notation** if it is written as the product of a number a, where $1 \leq a < 10$, and an integer power of 10: $a \times 10^r$.

Numbers written in scientific notation

$568{,}000 = 5.68 \times 10^5$

$0.0002117 = 2.117 \times 10^{-4}$

SECTION 5.2 MORE WORK WITH EXPONENTS AND SCIENTIFIC NOTATION

Power rules:

$$(a^m)^n = a^{m \cdot n}$$
$$(ab)^m = a^m b^m$$
$$\left(\dfrac{a}{b}\right)^n = \dfrac{a^n}{b^n}, b \neq 0$$

$$(7^8)^2 = 7^{16}$$
$$(2y)^3 = 2^3 y^3 = 8y^3$$
$$\left(\dfrac{5x^{-3}}{x^2}\right)^{-2} = \dfrac{5^{-2} x^6}{x^{-4}}$$
$$= 5^{-2} \cdot x^{6-(-4)}$$
$$= \dfrac{x^{10}}{5^2}, \text{ or } \dfrac{x^{10}}{25}$$

SECTION 5.3 POLYNOMIALS AND POLYNOMIAL FUNCTIONS

A **polynomial** is a finite sum of terms in which all variables have exponents raised to nonnegative integer powers and no variables appear in the denominator.

Polynomials

$1.3x^2$ (monomial)

$-\dfrac{1}{3}y + 5$ (binomial)

$6z^2 - 5z + 7$ (trinomial)

A function P is a **polynomial function** if $P(x)$ is a polynomial.

For the polynomial function

$$P(x) = -x^2 + 6x - 12, \text{ find } P(-2)$$
$$P(-2) = -(-2)^2 + 6(-2) - 12 = -28.$$

To add polynomials, combine all like terms.

Add

$$(3y^2x - 2yx + 11) + (-5y^2x - 7)$$
$$= -2y^2x - 2yx + 4$$

To subtract polynomials, change the signs of the terms of the polynomial being subtracted, then add.

Subtract

$$(-2z^3 - z + 1) - (3z^3 + z - 6)$$
$$= -2z^3 - z + 1 - 3z^3 - z + 6$$
$$= -5z^3 - 2z + 7$$

DEFINITIONS AND CONCEPTS	**EXAMPLES**

SECTION 5.4 MULTIPLYING POLYNOMIALS

To multiply two polynomials, use the distributive property and multiply each term of one polynomial by each term of the other polynomial; then combine like terms.

Special products

$$(a + b)^2 = a^2 + 2ab + b^2$$
$$(a - b)^2 = a^2 - 2ab + b^2$$
$$(a + b)(a - b) = a^2 - b^2$$

Multiply

$$(x^2 - 2x)(3x^2 - 5x + 1)$$
$$= 3x^4 - 5x^3 + x^2 - 6x^3 + 10x^2 - 2x$$
$$= 3x^4 - 11x^3 + 11x^2 - 2x$$
$$(3m + 2n)^2 = 9m^2 + 12mn + 4n^2$$
$$(z^2 - 5)^2 = z^4 - 10z^2 + 25$$
$$(7y + 1)(7y - 1) = 49y^2 - 1$$

The FOIL method may be used when multiplying two binomials.

Multiply

$$(x^2 + 5)(2x^2 - 9)$$

$$\quad\quad\text{F}\quad\quad\text{O}\quad\quad\text{I}\quad\quad\text{L}$$
$$= x^2(2x^2) + x^2(-9) + 5(2x^2) + 5(-9)$$
$$= 2x^4 - 9x^2 + 10x^2 - 45$$
$$= 2x^4 + x^2 - 45$$

SECTION 5.5 THE GREATEST COMMON FACTOR AND FACTORING BY GROUPING

The greatest common factor (GCF) of the terms of a polynomial is the product of the GCF of the numerical coefficients and the GCF of the variable factors.

Factor: $14xy^3 - 2xy^2 = 2 \cdot 7 \cdot x \cdot y^3 - 2 \cdot x \cdot y^2$.

The GCF is $2 \cdot x \cdot y^2$, or $2xy^2$.

$$14xy^3 - 2xy^2 = 2xy^2(7y - 1)$$

To factor a polynomial by grouping, group the terms so that each group has a common factor. Factor out these common factors. Then see if the new groups have a common factor.

Factor $x^4y - 5x^3 + 2xy - 10$.

$$x^4y - 5x^3 + 2xy - 10 = x^3(xy - 5) + 2(xy - 5)$$
$$= (xy - 5)(x^3 + 2)$$

SECTION 5.6 FACTORING TRINOMIALS

To factor $ax^2 + bx + c$, by trial and check,

Step 1. Write all pairs of factors of ax^2.

Step 2. Write all pairs of factors of c.

Step 3. Try combinations of these factors until the middle term bx is found.

Factor $28x^2 - 27x - 10$.

Factors of $28x^2$: $28x$ and x, $2x$ and $14x$, $4x$ and $7x$.

Factors of -10: -2 and 5, 2 and -5, -10 and 1, 10 and -1.

$$28x^2 - 27x - 10 = (7x + 2)(4x - 5)$$

To factor $ax^2 + bx + c$ by grouping,

Find two numbers whose product is $a \cdot c$ and whose sum is b, write the term bx using the two numbers found, and then factor by grouping.

$$2x^2 - 3x - 5 = 2x^2 - 5x + 2x - 5$$
$$= x(2x - 5) + 1(2x - 5)$$
$$= (2x - 5)(x + 1)$$

DEFINITIONS AND CONCEPTS	**EXAMPLES**

Perfect square trinomial

$$a^2 + 2ab + b^2 = (a + b)^2$$
$$a^2 - 2ab + b^2 = (a - b)^2$$

Difference of two squares

$$a^2 - b^2 = (a + b)(a - b)$$

Sum and difference of two cubes

$$a^3 + b^3 = (a + b)(a^2 - ab + b^2)$$
$$a^3 - b^3 = (a - b)(a^2 + ab + b^2)$$

To factor a polynomial

Step   1. Factor out the GCF.

Step   2. If the polynomial is a binomial, see if it is a difference of two squares or a sum or difference of two cubes. If it is a trinomial, see if it is a perfect square trinomial. If not, try factoring by methods of Section 5.6. If it is a polynomial with 4 or more terms, try factoring by grouping.

Step   3. See if any factors can be factored further.

Factor

$$25x^2 + 30x + 9 = (5x + 3)^2$$
$$49z^2 - 28z + 4 = (7z - 2)^2$$

$$36x^2 - y^2 = (6x + y)(6x - y)$$
$$8y^3 + 1 = (2y + 1)(4y^2 - 2y + 1)$$
$$27p^3 - 64q^3 = (3p - 4q)(9p^2 + 12pq + 16q^2)$$

Factor $10x^4y + 5x^2y - 15y$.

$$10x^4y + 5x^2y - 15y = 5y(2x^4 + x^2 - 3)$$
$$= 5y(2x^2 + 3)(x^2 - 1)$$
$$= 5y(2x^2 + 3)(x + 1)(x - 1)$$

To solve polynomial equations by factoring:

Step   1. Write the equation so that one side is 0.

Step   2. Factor the polynomial completely.

Step   3. Set each factor equal to 0.

Step   4. Solve the resulting equations.

Step   5. Check each solution.

Solve

$$2x^3 - 5x^2 = 3x$$
$$2x^3 - 5x^2 - 3x = 0$$
$$x(2x + 1)(x - 3) = 0$$
$$x = 0 \quad \text{or} \quad 2x + 1 = 0 \quad \text{or} \quad x - 3 = 0$$
$$x = 0 \quad \text{or} \quad x = -\frac{1}{2} \quad \text{or} \quad x = 3$$

The solutions are 0, $-\dfrac{1}{2}$, and 3.

CHAPTER 5 REVIEW

(5.1) Evaluate.

1. $(-2)^2$

2. $(-3)^4$

3. -2^2

4. -3^4

5. 8^0

6. -9^0

7. -4^{-2}

8. $(-4)^{-2}$

Simplify each expression. Use only positive exponents.

9. $-xy^2 \cdot y^3 \cdot xy^2z$

10. $(-4xy)(-3xy^2b)$

11. $a^{-14} \cdot a^5$

12. $\dfrac{a^{16}}{a^{17}}$

13. $\dfrac{x^{-7}}{x^4}$

14. $\dfrac{9a(a^{-3})}{18a^{15}}$

15. $\dfrac{y^{6p-3}}{y^{6p+2}}$

Write in scientific notation.

16. 36,890,000

17. -0.000362

Write each number without exponents.

18. 1.678×10^{-6}

19. 4.1×10^5

(5.2) Simplify. Use only positive exponents.

20. $(8^5)^3$

21. $\left(\dfrac{a}{4}\right)^2$

22. $(3x)^3$

23. $(-4x)^{-2}$

24. $\left(\dfrac{6x}{5}\right)^2$

25. $(8^6)^{-3}$

26. $\left(\dfrac{4}{3}\right)^{-2}$

27. $(-2x^3)^{-3}$

28. $\left(\dfrac{8p^6}{4p^4}\right)^{-2}$

29. $(-3x^{-2}y^2)^3$

30. $\left(\dfrac{x^{-5}y^{-3}}{z^3}\right)^{-5}$

31. $\dfrac{4^{-1}x^3yz}{x^{-2}yx^4}$

32. $(5xyz)^{-4}(x^{-2})^{-3}$

33. $\dfrac{2(3yz)^{-3}}{y^{-3}}$

Simplify each expression.

34. $x^{4a}(3x^{5a})^3$

35. $\dfrac{4y^{3x-3}}{2y^{2x+4}}$

Use scientific notation to find the quotient. Express each quotient in scientific notation.

36. $\dfrac{(0.00012)(144{,}000)}{0.0003}$

37. $\dfrac{(-0.00017)(0.00039)}{3000}$

Simplify. Use only positive exponents.

38. $\dfrac{27x^{-5}y^5}{18x^{-6}y^2} \cdot \dfrac{x^4y^{-2}}{x^{-2}y^3}$

39. $\dfrac{3x^5}{y^{-4}} \cdot \dfrac{(3xy^{-3})^{-2}}{(z^{-3})^{-4}}$

40. $\dfrac{(x^w)^2}{(x^{w-4})^{-2}}$

(5.3) Find the degree of each polynomial.

41. $x^2y - 3xy^3z + 5x + 7y$ **42.** $3x + 2$

Simplify by combining like terms.

43. $4x + 8x - 6x^2 - 6x^2y$

44. $-8xy^3 + 4xy^3 - 3x^3y$

Add or subtract as indicated.

45. $(3x + 7y) + (4x^2 - 3x + 7) + (y - 1)$

46. $(4x^2 - 6xy + 9y^2) - (8x^2 - 6xy - y^2)$

47. $(3x^2 - 4b + 28) + (9x^2 - 30) - (4x^2 - 6b + 20)$

48. Add $(9xy + 4x^2 + 18)$ and $(7xy - 4x^3 - 9x)$.

49. Subtract $(x - 7)$ from the sum of $(3x^2y - 7xy - 4)$ and $(9x^2y + x)$.

50. $\begin{array}{r} x^2 - 5x + 7 \\ -(x + 4) \\ \hline \end{array}$

51. $\begin{array}{r} x^3 \quad + 2xy^2 - y \\ + (x - 4xy^2 \; -7) \\ \hline \end{array}$

If $P(x) = 9x^2 - 7x + 8$, find the following.

52. $P(6)$ **53.** $P(-2)$

54. $P(-3)$

If $P(x) = 2x - 1$ and $Q(x) = x^2 + 2x - 5$, find the following.

55. $P(x) + Q(x)$

56. $2[P(x)] - Q(x)$

△ **57.** Find the perimeter of the rectangle.

$x^2y + 5$
cm

$2x^2y - 6x + 1$
cm

(5.4) Multiply.

58. $-6x(4x^2 - 6x + 1)$

59. $-4ab^2(3ab^3 + 7ab + 1)$

60. $(x - 4)(2x + 9)$

61. $(-3xa + 4b)^2$

62. $(9x^2 + 4x + 1)(4x - 3)$

63. $(5x - 9y)(3x + 9y)$

64. $\left(x - \dfrac{1}{3}\right)\left(x + \dfrac{2}{3}\right)$

65. $(x^2 + 9x + 1)^2$

Multiply, using special products.

66. $(3x - y)^2$

67. $(4x + 9)^2$

68. $(x + 3y)(x - 3y)$

69. $[4 + (3a - b)][4 - (3a - b)]$

70. If $P(x) = 2x - 1$ and $Q(x) = x^2 + 2x - 5$, find $P(x) \cdot Q(x)$.

△ **71.** Find the area of the rectangle.

$3y - 7z$
units

$3y + 7z$
units

Multiply. Assume that all variable exponents represent integers.

72. $4a^b(3a^{b+2} - 7)$

73. $(4xy^z - b)^2$

74. $(3x^a - 4)(3x^a + 4)$

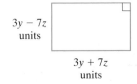

(5.5) *Factor out the greatest common factor.*

75. $16x^3 - 24x^2$

76. $36y - 24y^2$

77. $6ab^2 + 8ab - 4a^2b^2$

78. $14a^2b^2 - 21ab^2 + 7ab$

79. $6a(a + 3b) - 5(a + 3b)$

80. $4x(x - 2y) - 5(x - 2y)$

81. $xy - 6y + 3x - 18$

82. $ab - 8b + 4a - 32$

83. $pq - 3p - 5q + 15$

84. $x^3 - x^2 - 2x + 2$

△ **85.** A smaller square is cut from a larger rectangle. Write the area of the shaded region as a factored polynomial.

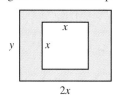

(5.6) *Completely factor each polynomial.*

86. $x^2 - 14x - 72$

87. $x^2 + 16x - 80$

88. $2x^2 - 18x + 28$

89. $3x^2 + 33x + 54$

90. $2x^3 - 7x^2 - 9x$

91. $3x^2 + 2x - 16$

92. $6x^2 + 17x + 10$

93. $15x^2 - 91x + 6$

94. $4x^2 + 2x - 12$

95. $9x^2 - 12x - 12$

96. $y^2(x + 6)^2 - 2y(x + 6)^2 - 3(x + 6)^2$

97. $(x + 5)^2 + 6(x + 5) + 8$

98. $x^4 - 6x^2 - 16$

99. $x^4 + 8x^2 - 20$

(5.7) *Factor each polynomial completely.*

100. $x^2 - 100$

101. $x^2 - 81$

102. $2x^2 - 32$

103. $6x^2 - 54$

104. $81 - x^4$

105. $16 - y^4$

106. $(y + 2)^2 - 25$

107. $(x - 3)^2 - 16$

108. $x^3 + 216$

109. $y^3 + 512$

110. $8 - 27y^3$

111. $1 - 64y^3$

112. $6x^4y + 48xy$

113. $2x^5 + 16x^2y^3$

114. $x^2 - 2x + 1 - y^2$

115. $x^2 - 6x + 9 - 4y^2$

116. $4x^2 + 12x + 9$

117. $16a^2 - 40ab + 25b^2$

△ **118.** The volume of the cylindrical shell is $\pi R^2 h - \pi r^2 h$ cubic units. Write this volume as a factored expression.

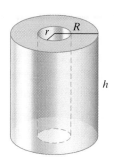

(5.8) *Solve each polynomial equation for the variable.*

119. $(3x - 1)(x + 7) = 0$

120. $3(x + 5)(8x - 3) = 0$

121. $5x(x - 4)(2x - 9) = 0$

122. $6(x + 3)(x - 4)(5x + 1) = 0$

123. $2x^2 = 12x$

124. $4x^3 - 36x = 0$

125. $(1 - x)(3x + 2) = -4x$

126. $2x(x - 12) = -40$

127. $3x^2 + 2x = 12 - 7x$

128. $2x^2 + 3x = 35$

129. $x^3 - 18x = 3x^2$

130. $19x^2 - 42x = -x^3$

131. $12x = 6x^3 + 6x^2$

132. $8x^3 + 10x^2 = 3x$

133. The sum of a number and twice its square is 105. Find the number.

△ **134.** The length of a rectangular piece of carpet is 5 meters less than twice its width. Find the dimensions of the carpet if its area is 33 square meters.

135. A scene from an adventure film calls for a stunt dummy to be dropped from above the second-story platform of the

Eiffel Tower, a distance of 400 feet. Its height $h(t)$ at time t seconds is given by

$$h(t) = -16t^2 + 400$$

Determine when the stunt dummy will reach the ground.

400 ft

MIXED REVIEW

136. The Royal Gorge suspension bridge in Colorado is 1053 feet above the Arkansas River. Neglecting air resistance, the height of an object dropped off the bridge is given by the polynomial function $P(t) = -16t^2 + 1053$ after time t seconds. Find the height of the object when $t = 1$ second and when $t = 8$ seconds.

Perform the indicated operation.

137. $(x + 5)(3x^2 - 2x + 1)$

138. $(3x^2 + 4x - 1.2) - (5x^2 - x + 5.7)$

139. $(3x^2 + 4x - 1.2) + (5x^2 - x + 5.7)$

140. $\left(7ab - \dfrac{1}{2}\right)^2$

If $P(x) = -x^2 + x - 4$, find

141. $P(5)$ **142.** $P(-2)$

Factor each polynomial completely.

143. $12y^5 - 6y^4$

144. $x^2y + 4x^2 - 3y - 12$

145. $6x^2 - 34x - 12$

146. $y^2(4x + 3)^2 - 19y(4x + 3)^2 - 20(4x + 3)^2$

147. $4z^7 - 49z^5$

148. $5x^4 + 4x^2 - 9$

Solve each equation.

149. $8x^2 = 24x$ **150.** $x(x - 11) = 26$

CHAPTER 5 TEST

TEST PREP VIDEO

Remember to use the Chapter Test Prep Video CD to see the fully worked-out solutions to any of the exercises you want to review.

Simplify. Use positive exponents to write the answers.

1. $(-9x)^{-2}$ **2.** $-3xy^{-2}(4xy^2)z$

3. $\dfrac{6^{-1}a^2b^{-3}}{3^{-2}a^{-5}b^2}$ **4.** $\left(\dfrac{-xy^{-5}z}{xy^3}\right)^{-5}$

Write Exercises 5 and 6 in scientific notation.

5. 630,000,000   **6.** 0.01200

7. Write 5×10^{-6} without exponents.

8. Use scientific notation to find the quotient.

$$\dfrac{(0.0024)(0.00012)}{0.00032}$$

Perform the indicated operations.

9. $(4x^3y - 3x - 4) - (9x^3y + 8x + 5)$

10. $-3xy(4x + y)$

11. $(3x + 4)(4x - 7)$

12. $(5a - 2b)(5a + 2b)$

13. $(6m + n)^2$

14. $(2x - 1)(x^2 - 6x + 4)$

Factor each polynomial completely.

15. $16x^3y - 12x^2y^4$

16. $x^2 - 13x - 30$

17. $4y^2 + 20y + 25$

18. $6x^2 - 15x - 9$

19. $4x^2 - 25$

20. $x^3 + 64$

21. $3x^2y - 27y^3$

22. $6x^2 + 24$

23. $16y^3 - 2$

24. $x^2y - 9y - 3x^2 + 27$

Solve the equation for the variable.

25. $3n(7n - 20) = 96$

26. $(x + 2)(x - 2) = 5(x + 4)$

27. $2x^3 + 5x^2 = 8x + 20$

△ **28.** Write the area of the shaded region as a factored polynomial.

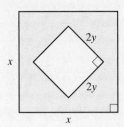

29. A pebble is hurled upward from the top of the Canada Trust Tower, which is 880 feet tall, with an initial velocity of 96 feet per second. Neglecting air resistance, the height $h(t)$ of the pebble after t seconds is given by the polynomial function

$$h(t) = -16t^2 + 96t + 880$$

   a. Find the height of the pebble when $t = 1$.

   b. Find the height of the pebble when $t = 5.1$.

   c. When will the pebble hit the ground?

CHAPTER 5 CUMULATIVE REVIEW

1. Find the roots.

   a. $\sqrt[3]{27}$

   b. $\sqrt[5]{1}$

   c. $\sqrt[4]{16}$

2. Find the roots.

   a. $\sqrt[3]{64}$

   b. $\sqrt[4]{81}$

   c. $\sqrt[5]{32}$

3. Solve: $2(x - 3) = 5x - 9$.

4. Solve: $0.3y + 2.4 = 0.1y + 4$

5. Karen Estes just received an inheritance of $10,000 and plans to place all the money in a savings account that pays 5% compounded quarterly to help her son go to college in 3 years. How much money will be in the account in 3 years?

6. A gallon of latex paint can cover 400 square feet. How many gallon containers of paint should be bought to paint two coats on each wall of a rectangular room whose dimensions are 14 feet by 18 feet? (Assume 8-foot ceilings.)

7. Solve and graph the solution set.

   a. $\dfrac{1}{4}x \le \dfrac{3}{8}$

   b. $-2.3x < 6.9$

8. Solve. Graph the solution set and write it in interval notation.

$x + 2 \le \dfrac{1}{4}(x - 7)$

Solve.

9. $-1 \le \dfrac{2x}{3} + 5 \le 2$

10. Solve: $-\dfrac{1}{3} < \dfrac{3x + 1}{6} \le \dfrac{1}{3}$

11. $|y| = 0$

12. Solve: $8 + |4c| = 24$

13. $\left|2x - \dfrac{1}{10}\right| < -13$

14. Solve: $|5x - 1| + 9 > 5$

15. Graph the linear equation $y = \dfrac{1}{3}x$.

16. Graph the linear equation $y = 3x$.

17. Evaluate $f(2), f(-6)$, and $f(0)$ for the function

$$f(x) = \begin{cases} 2x + 3 & \text{if } x \le 0 \\ -x - 1 & \text{if } x > 0 \end{cases}$$

Write your results in ordered pair form.

18. If $f(x) = 3x^2 + 2x + 3$, find $f(-3)$.

19. Graph $x = 2$.

20. Graph $y - 5 = 0$

21. Find the slope of the line $y = 2$.

22. Find the slope of the line $f(x) = -2x - 3$.

23. Find an equation of the horizontal line containing the point $(2, 3)$.

24. Find the equation of the vertical line containing the point $(-3, 2)$.

25. Graph the union of $x + \dfrac{1}{2}y \ge -4$ or $y \le -2$.

26. Find the equation of the line containing the point $(-2, 3)$ and slope of 0.

27. Use the substitution method to solve the system.
$$\begin{cases} 2x + 4y = -6 \\ x = 2y - 5 \end{cases}$$

28. Use the substitution method to solve the system.
$$\begin{cases} 4x - 2y = 8 \\ y = 3x - 6 \end{cases}$$

29. Solve the system. $\begin{cases} 2x + 4y = 1 \\ 4x - 4z = -1 \\ y - 4z = -3 \end{cases}$

30. Solve the system. $\begin{cases} x + y - \dfrac{3}{2}z = \dfrac{1}{2} \\ -y - 2z = 14 \\ x - \dfrac{2}{3}y = -\dfrac{1}{3} \end{cases}$

31. A first number is 4 less than a second number. Four times the first number is 6 more than twice the second. Find the numbers.

32. One solution contains 20% acid and a second solution contains 60% acid. How many ounces of each solution should be mixed in order to have 50 ounces of a 30% acid solution?

33. Use matrices to solve the system. $\begin{cases} 2x - y = 3 \\ 4x - 2y = 5 \end{cases}$

34. Use matrices to solve the system. $\begin{cases} 4y = 8 \\ x + y = 7 \end{cases}$

35. The measure of the largest angle of a triangle is 80° more than the measure of the smallest angle, and the measure of the remaining angle is 10° more than the measure of the smallest angle. Find the measure of each angle.

36. Find the equation of the line with slope $\frac{1}{2}$, through the point $(0, 5)$. Write the equation using function notation.

37. Write each number in scientific notation.
a. 730,000
b. 0.00000104

38. Write each number in scientific notation.
a. 8,250,000
b. 0.0000346

39. Simplify each expression. Use positive exponents to write the answers.
a. $(2x^0y^{-3})^{-2}$ **b.** $\left(\dfrac{x^{-5}}{x^{-2}}\right)^{-3}$
c. $\left(\dfrac{2}{7}\right)^{-2}$ **d.** $\dfrac{5^{-2}x^{-3}y^{11}}{x^2y^{-5}}$

40. Simplify each expression. Use positive exponents to write the answers.
a. $(4a^{-1}b^0)^{-3}$ **b.** $\left(\dfrac{a^{-6}}{a^{-8}}\right)^{-2}$
c. $\left(\dfrac{2}{3}\right)^{-3}$ **d.** $\dfrac{3^{-2}a^{-2}b^{12}}{a^4b^{-5}}$

41. Find the degree of the polynomial $3xy + x^2y^2 - 5x^2 - 6.7$.

42. Subtract $(5x^2 + 3x)$ from $(3x^2 - 2x)$.

43. Multiply.
a. $(2x^3)(5x^6)$
b. $(7y^4z^4)(-xy^{11}z^5)$

44. Multiply.
a. $(3y^6)(4y^2)$
b. $(6a^3b^2)(-a^2bc^4)$

Factor.

45. $17x^3y^2 - 34x^4y^2$

46. Factor completely $12x^3y - 3xy^3$.

47. $x^2 + 10x + 16$

48. Factor $5a^2 + 14a - 3$.

Solve.

49. Solve $2x^2 + 9x - 5 = 0$.

50. Solve $3x^2 - 10x - 8 = 0$.

6

Rational Expressions

Movies still attract more people (1,448 million annually) than either theme parks (341 million) or major professional league sports (137 million) combined. U.S. theater admissions grew 3.3% in 2006 and movie advertising is changing, with advertising on TV or in the newspaper decreasing and online advertising increasing.

The graph below shows yearly rating changes for top grossing films. In Section 6.6, Exercises 60 and 61, page 464, you will have the opportunity to solve applications about movie ratings and costs to make a movie.

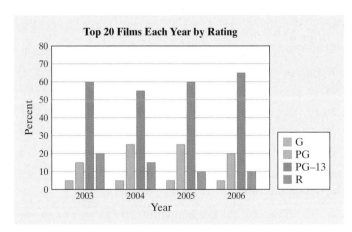

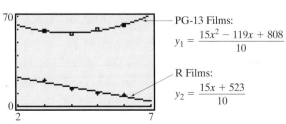

The calculator screen shows function models for the percent changes of R and PG-13 top grossing films for years 2003–2006.

PG-13 Films:
$$y_1 = \frac{15x^2 - 119x + 808}{10}$$

R Films:
$$y_2 = \frac{15x + 523}{10}$$

Polynomials are to algebra what integers are to arithmetic. We have added, subtracted, multiplied, and raised polynomials to powers, each operation yielding another polynomial, just as these operations on integers yield another integer. But when we divide one integer by another, the result may or may not be another integer. Likewise, when we divide one polynomial by another, we may or may not get a polynomial in return. The quotient $x \div (x + 1)$ is not a polynomial; it is a *rational expression* that can be written as
$$\frac{x}{x + 1}.$$

In this chapter, we study these new algebraic forms known as rational expressions and the *rational functions* they generate.

409

6.1 RATIONAL FUNCTIONS AND MULTIPLYING AND DIVIDING RATIONAL EXPRESSIONS

OBJECTIVES

1 Find the domain of a rational expression.

2 Simplify rational expressions.

3 Multiply rational expressions.

4 Divide rational expressions.

5 Use rational functions in applications.

Recall that a *rational number,* or *fraction,* is a number that can be written as the quotient $\frac{p}{q}$ of two integers p and q as long as q is not 0. A **rational expression** is an expression that can be written as the quotient $\frac{P}{Q}$ of two polynomials P and Q as long as Q is not 0.

Examples of Rational Expressions

$$\frac{3x + 7}{2} \qquad \frac{5x^2 - 3}{x - 1} \qquad \frac{7x - 2}{2x^2 + 7x + 6}$$

Rational expressions are sometimes used to describe functions. For example, we call the function $f(x) = \frac{x^2 + 2}{x - 3}$ a **rational function** since $\frac{x^2 + 2}{x - 3}$ is a rational expression.

OBJECTIVE 1 ▶ Finding the domain of a rational expression. As with fractions, a rational expression is **undefined** if the denominator is 0. If a variable in a rational expression is replaced with a number that makes the denominator 0, we say that the rational expression is **undefined** for this value of the variable. For example, the rational expression $\frac{x^2 + 2}{x - 3}$ is undefined when x is 3, because replacing x with 3 results in a denominator of 0. For this reason, we must exclude 3 from the domain of the function $f(x) = \frac{x^2 + 2}{x - 3}$.

The domain of f is then

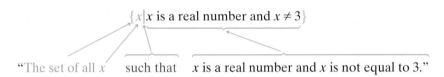

$$\{x \mid x \text{ is a real number and } x \neq 3\}$$

"The set of all x such that x is a real number and x is not equal to 3."

In this section, we will use this set builder notation to write domains. Unless told otherwise, we assume that the domain of a function described by an equation is the set of all real numbers for which the equation is defined.

EXAMPLE 1   Find the domain of each rational function.

a. $f(x) = \dfrac{8x^3 + 7x^2 + 20}{2}$ **b.** $g(x) = \dfrac{7x + 2}{x - 3}$ **c.** $f(x) = \dfrac{9x - 13}{x^2 - 2x - 15}$

Solution   The domain of each function will contain all real numbers except those values that make the denominator 0.

a. No matter what the value of x, the denominator of $f(x) = \dfrac{8x^3 + 7x^2 + 20}{2}$ is never 0, so the domain of f is $\{x \mid x \text{ is a real number}\}$.

b. To find the values of x that make the denominator of $g(x)$ equal to 0, we solve the equation "denominator = 0":

$$x - 3 = 0, \quad \text{or} \quad x = 3$$

The domain must exclude 3 since the rational expression is undefined when x is 3. The domain of g is $\{x \mid x \text{ is a real number and } x \neq 3\}$.

c. We find the domain by setting the denominator equal to 0.

$$x^2 - 2x - 15 = 0 \quad \text{Set the denominator equal to 0 and solve.}$$
$$(x - 5)(x + 3) = 0$$
$$x - 5 = 0 \quad \text{or} \quad x + 3 = 0$$
$$x = 5 \quad \text{or} \quad x = -3$$

If x is replaced with 5 or with -3, the rational expression is undefined.

The domain of f is $\{x \mid x \text{ is a real number and } x \neq 5, x \neq -3\}$.

PRACTICE
1 Find the domain of each rational function.

a. $f(x) = \dfrac{4x^5 - 3x^2 + 2}{-6}$ **b.** $g(x) = \dfrac{6x^2 + 1}{x + 3}$ **c.** $h(x) = \dfrac{8x - 3}{x^2 - 5x + 6}$

Let's use a graphing utility to confirm the domain of the function in Example 1b, $g(x) = \dfrac{7x + 2}{x - 3}$.

To confirm this domain, graph $y_1 = \dfrac{7x + 2}{x - 3}$. The domain of $g(x)$ does not include 3, so the graph of $g(x)$ does not exist at $x = 3$. If we graph $g(x)$ in dot mode, the graph shows that the function is undefined at $x = 3$ and no ordered pair solutions exist with an x-value of 3. See the graph in the margin.

The graph of $g(x) = \dfrac{7x + 2}{x - 3}$ confirms that the domain is $\{x \mid x \text{ is a real number}$ and $x \neq 3\}$.

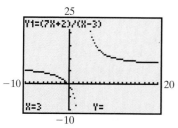

In dot mode the graphing utility plots only the ordered pairs that make the equation true. Notice y_1 is undefined when $x = 3$ and thus a point does not exist on the graph of the function.

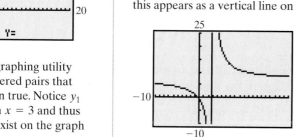
EXAMPLE 2 Find the domain of each rational function. Then graph each rational function and use the graph to confirm the domain.

a. $f(x) = \dfrac{x + 1}{x^2 - 4}$ **b.** $h(x) = \dfrac{x^2}{2x^2 + 7x - 4}$

Solution **a.** To find the domain of $f(x) = \dfrac{x + 1}{x^2 - 4}$, find the values that make the denominator 0.

$$x^2 - 4 = 0$$
$$(x - 2)(x + 2) = 0$$
$$x - 2 = 0 \text{ or } x + 2 = 0$$
$$x = 2 \text{ or } x = -2$$

This means that both 2 and −2 make $f(x)$ undefined. Thus the domain is $\{x \mid x$ is a real number and $x \neq \pm 2\}$. The graph below to the left shows that the function is undefined for $x = -2$ and $x = 2$.

b. To find the domain, solve "denominator = 0."

$$2x^2 + 7x - 4 = 0$$
$$(2x - 1)(x + 4) = 0$$
$$2x - 1 = 0 \text{ or } x + 4 = 0$$
$$x = \frac{1}{2} \text{ or } x = -4$$

Thus, the domain is $\{x \mid x$ is a real number and $x \neq \frac{1}{2}, x \neq -4\}$. The graph below to the right shows the function undefined at $x = -4$ and $x = \frac{1}{2}$.

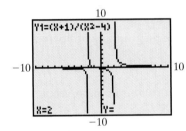
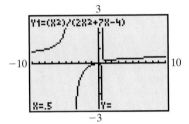

PRACTICE
2 Find the domain of each rational function. Then graph each rational function and use the graph to confirm the domain.

a. $f(x) = \dfrac{9x - 14}{x + 3}$
     b. $g(x) = \dfrac{5x}{3x^2 - 8x + 5}$

Concept Check ☑

For which of these values (if any) is the rational expression $\dfrac{x - 3}{x^2 + 2}$ undefined?

a. 2 **b.** 3 **c.** −2 **d.** 0 **e.** None of these

OBJECTIVE 2 ▶ Simplifying rational expressions. Recall that a fraction is in lowest terms or simplest form if the numerator and denominator have no common factors other than 1 (or −1). For example, $\dfrac{3}{13}$ is in lowest terms since 3 and 13 have no common factors other than 1 (or −1).

To **simplify** a rational expression, or to write it in lowest terms, we use a method similar to simplifying a fraction.

Recall that to simplify a fraction, we essentially "remove factors of 1." Our ability to do this comes from these facts:

- If $c \neq 0$, then $\dfrac{c}{c} = 1$. For example, $\dfrac{7}{7} = 1$ and $\dfrac{-8.65}{-8.65} = 1$.

- $n \cdot 1 = n$. For example, $-5 \cdot 1 = -5$, $126.8 \cdot 1 = 126.8$, and $\dfrac{a}{b} \cdot 1 = \dfrac{a}{b}, b \neq 0$.

In other words, we have the following:

$$\frac{a \cdot c}{b \cdot c} = \underbrace{\frac{a}{b} \cdot \frac{c}{c}}_{} = \frac{a}{b}$$

Since $\dfrac{a}{b} \cdot 1 = \dfrac{a}{b}$

Let's practice simplifying a fraction by simplifying $\dfrac{15}{65}$.

$$\frac{15}{65} = \frac{3 \cdot 5}{13 \cdot 5} = \frac{3}{13} \cdot \frac{5}{5} = \frac{3}{13} \cdot 1 = \frac{3}{13}$$

Let's use the same technique and simplify the rational expression $\dfrac{(x+2)^2}{x^2-4}$.

$$\frac{(x+2)^2}{x^2-4} = \frac{(x+2)(x+2)}{(x-2)(x+2)}$$

$$= \frac{(x+2)}{(x-2)} \cdot \frac{x+2}{x+2}$$

$$= \frac{x+2}{x-2} \cdot 1$$

$$= \frac{x+2}{x-2}$$

This means that the rational expression $\dfrac{(x+2)^2}{x^2-4}$ has the same value as the rational expression $\dfrac{x+2}{x-2}$ for all values of x except 2 and -2. (Remember that when x is 2, the denominators of both rational expressions are 0 and that when x is -2, the original rational expression has a denominator of 0.)

As we simplify rational expressions, we will assume that the simplified rational expression is equivalent to the original rational expression for all real numbers except those for which either denominator is 0.

Just as for numerical fractions, we can use a shortcut notation. Remember that as long as exact factors in both the numerator and denominator are divided out, we are "removing a factor of 1." We can use the following notation:

$$\frac{(x+2)^2}{x^2-4} = \frac{(x+2)\,(x+2)}{(x-2)\,(x+2)} \qquad \text{A factor of 1 is identified by the shading.}$$

$$= \frac{x+2}{x-2} \qquad \text{"Remove" the factor of 1.}$$

This "removing a factor of 1" is stated in the principle below:

In this table, $y_1 = \dfrac{(x+2)^2}{x^2-4}$ and $y_2 = \dfrac{x+2}{x-2}$.

Notice that both expressions are undefined at $x = 2$ (and an error message is given), and when $x = -2$, y_1 is not defined.

Fundamental Principle of Rational Expressions

For any rational expression $\dfrac{P}{Q}$ and any polynomial R, where $R \neq 0$,

$$\frac{PR}{QR} = \frac{P}{Q} \cdot \frac{R}{R} = \frac{P}{Q} \cdot 1 = \frac{P}{Q}$$

or, simply,

$$\frac{PR}{QR} = \frac{P}{Q}$$

In general, the following steps may be used to simplify rational expressions or to write a rational expression in lowest terms.

> **Simplifying or Writing a Rational Expression in Lowest Terms**
> **STEP 1.** Completely factor the numerator and denominator of the rational expression.
> **STEP 2.** Divide out factors common to the numerator and denominator. (This is the same as "removing a factor of 1.")

For now, we assume that variables in a rational expression do not represent values that make the denominator 0.

EXAMPLE 3 Simplify each rational expression.

a. $\dfrac{2x^2}{10x^3 - 2x^2}$ **b.** $\dfrac{9x^2 + 13x + 4}{8x^2 + x - 7}$

Solution

a. $\dfrac{2x^2}{10x^3 - 2x^2} = \dfrac{2x^2 \cdot 1}{2x^2 (5x - 1)} = 1 \cdot \dfrac{1}{5x - 1} = \dfrac{1}{5x - 1}$

b. $\dfrac{9x^2 + 13x + 4}{8x^2 + x - 7} = \dfrac{(9x + 4)\,(x + 1)}{(8x - 7)\,(x + 1)}$ Factor the numerator and denominator.

$\qquad = \dfrac{9x + 4}{8x - 7} \cdot 1$ Since $\dfrac{x + 1}{x + 1} = 1$

$\qquad = \dfrac{9x + 4}{8x - 7}$ Simplest form   □

PRACTICE
3 Simplify each rational expression.

a. $\dfrac{5z^4}{10z^5 - 5z^4}$ **b.** $\dfrac{5x^2 + 13x + 6}{6x^2 + 7x - 10}$

EXAMPLE 4 Simplify each rational expression.

a. $\dfrac{2 + x}{x + 2}$ **b.** $\dfrac{2 - x}{x - 2}$

Solution

a. $\dfrac{2 + x}{x + 2} = \dfrac{x + 2}{x + 2} = 1$ By the commutative property of addition, $2 + x = x + 2$.

b. $\dfrac{2 - x}{x - 2}$

The terms in the numerator of $\dfrac{2 - x}{x - 2}$ differ by sign from the terms of the denominator, so the polynomials are opposites of each other and the expression simplifies to -1. To see this, we factor out -1 from the numerator or the denominator. If -1 is factored from the numerator, then

$$\frac{2 - x}{x - 2} = \frac{-1(-2 + x)}{x - 2} = \frac{-1\,(x - 2)}{x - 2} = \frac{-1}{1} = -1$$

> ▶ **Helpful Hint**
>
> When the numerator and the denominator of a rational expression are opposites of each other, the expression simplifies to −1.

If −1 is factored from the denominator, the result is the same.

$$\frac{2 - x}{x - 2} = \frac{2 - x}{-1(-x + 2)} = \frac{2 - x}{-1\,(2 - x)} = \frac{1}{-1} = -1$$

PRACTICE

4   Simplify each rational expression.

   a. $\dfrac{x + 3}{3 + x}$ b. $\dfrac{3 - x}{x - 3}$

DISCOVER THE CONCEPT

a. Consider the expression $\dfrac{2x^2 - 18}{x^2 - 2x - 3}$. Graph the numerator as $y_1 = 2x^2 - 18$ and the denominator as $y_2 = x^2 - 2x - 3$ in the same window.

b. Completely factor the numerator and the denominator of the original expression.

c. What do you notice about the common factor in the expression and the point of intersection of the graphs of the numerator and denominator?

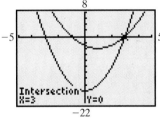

Graphs of $y_1 = 2x^2 - 18$ and $y_2 = x^2 - 2x - 3$ both intersect the x-axis at $x = 3$.

In the above discovery, we notice that both graphs intersect the x-axis at $x = 3$. In part **b**, we find that $(x - 3)$ is a common factor of the numerator and denominator.

$$\frac{2x^2 - 18}{x^2 - 2x - 3} = \frac{2(x^2 - 9)}{x^2 - 2x - 3} = \frac{2(x + 3)(x - 3)}{(x + 1)(x - 3)}$$

In general, if the graphs of the numerator and denominator of a rational expression share an x-intercept c, then $(x - c)$ is a common factor of the numerator and denominator of the rational expression.

EXAMPLE 5   Simplify $\dfrac{18 - 2x^2}{x^2 - 2x - 3}$.

Solution
$$\frac{18 - 2x^2}{x^2 - 2x - 3} = \frac{2(9 - x^2)}{(x + 1)(x - 3)} \qquad \text{Factor.}$$

$$= \frac{2(3 + x)(3 - x)}{(x + 1)(x - 3)} \qquad \text{Factor completely.}$$

$$= \frac{2(3 + x)\cdot -1\,(x - 3)}{(x + 1)\,(x - 3)} \qquad \begin{array}{l}\text{Notice the opposites } 3 - x \\ \text{and } x - 3. \text{ Write } 3 - x \text{ as} \\ -1(x - 3) \text{ and simplify.}\end{array}$$

$$= -\frac{2(3 + x)}{x + 1}$$

PRACTICE

5   Simplify $\dfrac{20 - 5x^2}{x^2 + x - 6}$.

> ▶ **Helpful Hint**
>
> Recall that for a fraction $\dfrac{a}{b}$,
>
> $$\frac{a}{-b} = \frac{-a}{b} = -\frac{a}{b}$$
>
> For example
>
> $$\frac{-(x+1)}{(x+2)} = \frac{(x+1)}{-(x+2)} = -\frac{x+1}{x+2}$$

Concept Check ✓

Which of the following expressions are equivalent to $\dfrac{x}{8-x}$?

a. $\dfrac{-x}{x-8}$ **b.** $\dfrac{-x}{8-x}$ **c.** $\dfrac{x}{x-8}$ **d.** $\dfrac{-x}{-8+x}$

EXAMPLE 6   Simplify each rational expression. Check graphically.

a. $\dfrac{x^3 + 8}{2 + x}$ **b.** $\dfrac{2y^2 + 2}{y^3 - 5y^2 + y - 5}$

Solution

a.
$$\frac{x^3 + 8}{2 + x} = \frac{(x+2)(x^2 - 2x + 4)}{x + 2}$$ Factor the sum of the two cubes.

$$= x^2 - 2x + 4$$ Divide out common factors.

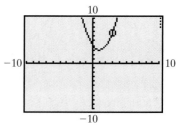

Notice in y_2 that the graph style was changed to visualize that y_1 and y_2 have the same graph.

b.
$$\frac{2y^2 + 2}{y^3 - 5y^2 + y - 5} = \frac{2(y^2 + 1)}{(y^3 - 5y^2) + (y - 5)}$$ Factor the numerator.

$$= \frac{2(y^2 + 1)}{y^2(y - 5) + 1(y - 5)}$$ Factor the denominator by grouping.

$$= \frac{2(y^2 + 1)}{(y - 5)(y^2 + 1)}$$

$$= \frac{2}{y - 5}$$ Divide out common factors.   □

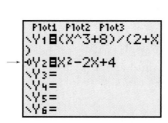

Here, $y_1 = \dfrac{2x^2 + 2}{x^3 - 5x^2 + x - 5}$ and $y_2 = \dfrac{2}{y - 5}$. Notice that both graphs are undefined when x is 5.

PRACTICE
6   Simplify each rational expression.

a. $\dfrac{x^3 + 64}{4 + x}$ **b.** $\dfrac{5z^2 + 10}{z^3 - 3z^2 + 2z - 6}$

Answers to Concept Check:
a and d
no; answers may vary

Concept Check ✓

Does $\dfrac{n}{n+2}$ simplify to $\dfrac{1}{2}$? Why or why not?

OBJECTIVE 3 ▶ Multiplying rational expressions. Arithmetic operations on rational expressions are performed in the same way as they are on rational numbers.

Multiplying Rational Expressions

The rule for multiplying rational expressions is

$$\frac{P}{Q} \cdot \frac{R}{S} = \frac{PR}{QS} \quad \text{as long as } Q \neq 0 \text{ and } S \neq 0.$$

To multiply rational expressions, you may use these steps:

STEP 1. Completely factor each numerator and denominator.

STEP 2. Use the rule above and multiply the numerators and the denominators.

STEP 3. Simplify the product by dividing the numerator and denominator by their common factors.

When we multiply rational expressions, notice that we factor each numerator and denominator first. This helps when we apply the fundamental principle to write the product in simplest form.

EXAMPLE 7   Multiply.

a. $\dfrac{1 + 3n}{2n} \cdot \dfrac{2n - 4}{3n^2 - 2n - 1}$

b. $\dfrac{x^3 - 1}{-3x + 3} \cdot \dfrac{15x^2}{x^2 + x + 1}$

Solution

a. $\dfrac{1 + 3n}{2n} \cdot \dfrac{2n - 4}{3n^2 - 2n - 1} = \dfrac{1 + 3n}{2n} \cdot \dfrac{2(n - 2)}{(3n + 1)(n - 1)}$ Factor.

$$= \frac{(1 + 3n) \cdot 2(n - 2)}{2n(3n + 1)(n - 1)}$$ Multiply.

$$= \frac{n - 2}{n(n - 1)}$$ Divide out common factors.

b. $\dfrac{x^3 - 1}{-3x + 3} \cdot \dfrac{15x^2}{x^2 + x + 1} = \dfrac{(x - 1)(x^2 + x + 1)}{-3(x - 1)} \cdot \dfrac{15x^2}{x^2 + x + 1}$ Factor.

$$= \frac{(x - 1)(x^2 + x + 1) \cdot 3 \cdot 5x^2}{-1 \cdot 3(x - 1)(x^2 + x + 1)}$$ Factor.

$$= \frac{5x^2}{-1} = -5x^2$$ Simplest form   □

PRACTICE
7   Multiply.

a. $\dfrac{2 + 5n}{3n} \cdot \dfrac{6n + 3}{5n^2 - 3n - 2}$

b. $\dfrac{x^3 - 8}{-6x + 12} \cdot \dfrac{6x^2}{x^2 + 2x + 4}$

OBJECTIVE 4 ▶ Dividing rational expressions. Recall that two numbers are reciprocals of each other if their product is 1. Similarly, if $\dfrac{P}{Q}$ is a rational expression, then $\dfrac{Q}{P}$ is its **reciprocal,** since

$$\frac{P}{Q} \cdot \frac{Q}{P} = \frac{P \cdot Q}{Q \cdot P} = 1$$

The following are examples of expressions and their reciprocals.

Expression	Reciprocal
$\dfrac{3}{x}$	$\dfrac{x}{3}$
$\dfrac{2 + x^2}{4x - 3}$	$\dfrac{4x - 3}{2 + x^2}$
x^3	$\dfrac{1}{x^3}$
0	no reciprocal

Dividing Rational Expressions

The rule for dividing rational expressions is

$$\frac{P}{Q} \div \frac{R}{S} = \frac{P}{Q} \cdot \frac{S}{R} = \frac{PS}{QR} \quad \text{as long as } Q \neq 0, S \neq 0, \text{ and } R \neq 0.$$

To divide by a rational expression, use the rule above and multiply by its reciprocal. Then simplify if possible.

Notice that division of rational expressions is the same as for rational numbers.

EXAMPLE 8   Divide.

a. $\dfrac{8m^2}{3m^2 - 12} \div \dfrac{40}{2 - m}$

b. $\dfrac{18y^2 + 9y - 2}{24y^2 - 10y + 1} \div \dfrac{3y^2 + 17y + 10}{8y^2 + 18y - 5}$

Solution

a. $\dfrac{8m^2}{3m^2 - 12} \div \dfrac{40}{2 - m} = \dfrac{8m^2}{3m^2 - 12} \cdot \dfrac{2 - m}{40}$ Multiply by the reciprocal of the divisor.

$= \dfrac{8m^2(2 - m)}{3(m + 2)(m - 2) \cdot 40}$ Factor and multiply.

$= \dfrac{8m^2 \cdot -1\,(m - 2)}{3(m + 2)\,(m - 2) \cdot 8 \cdot 5}$ Write $(2 - m)$ as $-1(m - 2)$.

$= -\dfrac{m^2}{15(m + 2)}$ Simplify.

b. $\dfrac{18y^2 + 9y - 2}{24y^2 - 10y + 1} \div \dfrac{3y^2 + 17y + 10}{8y^2 + 18y - 5}$

$= \dfrac{18y^2 + 9y - 2}{24y^2 - 10y + 1} \cdot \dfrac{8y^2 + 18y - 5}{3y^2 + 17y + 10}$ Multiply by the reciprocal.

$= \dfrac{(6y - 1)\,(3y + 2)}{(6y - 1)\,(4y - 1)} \cdot \dfrac{(4y - 1)\,(2y + 5)}{(3y + 2)\,(y + 5)}$ Factor.

$= \dfrac{2y + 5}{y + 5}$ Simplest form   ☐

PRACTICE
8   Divide.

a. $\dfrac{6y^3}{3y^2 - 27} \div \dfrac{42}{3 - y}$

b. $\dfrac{10x^2 + 23x - 5}{5x^2 - 51x + 10} \div \dfrac{2x^2 + 9x + 10}{7x^2 - 68x - 20}$

> ▶ **Helpful Hint**
> When dividing rational expressions, do not divide out common factors until the division problem is rewritten as a multiplication problem.

EXAMPLE 9   Perform each indicated operation.

$$\frac{x^2 - 25}{(x + 5)^2} \cdot \frac{3x + 15}{4x} \div \frac{x^2 - 3x - 10}{x}$$

Solution $\dfrac{x^2 - 25}{(x + 5)^2} \cdot \dfrac{3x + 15}{4x} \div \dfrac{x^2 - 3x - 10}{x}$

$$= \frac{x^2 - 25}{(x + 5)^2} \cdot \frac{3x + 15}{4x} \cdot \frac{x}{x^2 - 3x - 10} \qquad \text{To divide, multiply by the reciprocal}$$

$$= \frac{(x + 5)(x - 5)}{(x + 5)(x + 5)} \cdot \frac{3(x + 5)}{4x} \cdot \frac{x}{(x - 5)(x + 2)}$$

$$= \frac{3}{4(x + 2)} \qquad\qquad □$$

PRACTICE
9   Perform each indicated operation.

$$\frac{x^2 - 16}{(x - 4)^2} \cdot \frac{5x - 20}{3x} \div \frac{x^2 + x - 12}{x}$$

OBJECTIVE 5 ▶ Using rational functions in applications. Rational functions occur often in real-life situations.

EXAMPLE 10   **Cost for Pressing Compact Discs**

For the ICL Production Company, the rational function $C(x) = \dfrac{2.6x + 10,000}{x}$ describes the company's cost per disc of pressing x compact discs. Find the cost per disc for pressing:

a. 100 compact discs
b. 1000 compact discs

Solution

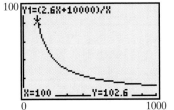

a. $C(100) = \dfrac{2.6(100) + 10,000}{100} = \dfrac{10,260}{100} = 102.6$

The cost per disc for pressing 100 compact discs is $102.60.

b. $C(1000) = \dfrac{2.6(1000) + 10,000}{1000} = \dfrac{12,600}{1000} = 12.6$

The cost per disc for pressing 1000 compact discs is $12.60. Notice that as more compact discs are produced, the cost per disc decreases.   □

PRACTICE
10   A company's cost per T-shirt for silk screening x T-shirts is given by the rational function $C(x) = \dfrac{3.2x + 400}{x}$. Find the cost per T-shirt for printing:

a. 100 T-shirts   **b.** 1000 T-shirts

VOCABULARY & READINESS CHECK

Use the choices below to fill in each blank. Some choices may not be used.

1	true	rational	simplified	$\dfrac{-a}{-b}$	$\dfrac{-a}{b}$	$\dfrac{a}{-b}$
−1	false	domain	0			

1. A _____ expression is an expression that can be written as the quotient $\dfrac{P}{Q}$ of two polynomials P and Q as long as $Q \neq 0$.

2. A rational expression is undefined if the denominator is _____.

3. The _____ of the rational function $f(x) = \dfrac{2}{x}$ is $\{x | x$ is a real number and $x \neq 0\}$.

4. A rational expression is _____ if the numerator and denominator have no common factors other than 1 or −1.

5. The expression $\dfrac{x^2 + 2}{2 + x^2}$ simplifies to _____.

6. The expression $\dfrac{y - z}{z - y}$ simplifies to _____.

7. For a rational expression, $-\dfrac{a}{b} =$ _____ = _____.

8. True or false: $\dfrac{a - 6}{a + 2} = \dfrac{-(a - 6)}{-(a + 2)} = \dfrac{-a + 6}{-a - 2}$. _____

Multiply.

9. $\dfrac{x}{5} \cdot \dfrac{y}{2}$ **10.** $\dfrac{y}{6} \cdot \dfrac{z}{5}$ **11.** $\dfrac{2}{x} \cdot \dfrac{y}{3}$ **12.** $\dfrac{a}{5} \cdot \dfrac{7}{b}$ **13.** $\dfrac{m}{6} \cdot \dfrac{m}{6}$ **14.** $\dfrac{9}{x} \cdot \dfrac{8}{x}$

6.1 EXERCISE SET

MyMathLab

PRACTICE   WATCH   DOWNLOAD   READ   REVIEW

Find the domain of each rational function, then graph the function and use the graph to confirm the domain. See Examples 1 and 2.

1. $f(x) = \dfrac{5x - 7}{4}$

2. $g(x) = \dfrac{4 - 3x}{2}$

3. $s(t) = \dfrac{t^2 + 1}{2t}$

4. $v(t) = -\dfrac{5t + t^2}{3t}$

5. $f(x) = \dfrac{3x}{7 - x}$

6. $f(x) = \dfrac{-4x}{-2 + x}$

7. $f(x) = \dfrac{x}{3x - 1}$

8. $g(x) = \dfrac{-2}{2x + 5}$

9. $R(x) = \dfrac{3 + 2x}{x^3 + x^2 - 2x}$

10. $h(x) = \dfrac{5 - 3x}{2x^2 - 14x + 20}$

11. $C(x) = \dfrac{x + 3}{x^2 - 4}$

12. $R(x) = \dfrac{5}{x^2 - 7x}$

Simplify each rational expression. See Examples 3 through 6.

13. $\dfrac{8x - 16x^2}{8x}$

14. $\dfrac{3x - 6x^2}{3x}$

15. $\dfrac{x^2 - 9}{3 + x}$

16. $\dfrac{x^2 - 25}{5 + x}$

17. $\dfrac{9y - 18}{7y - 14}$

18. $\dfrac{6y - 18}{2y - 6}$

19. $\dfrac{x^2 + 6x - 40}{x + 10}$

20. $\dfrac{x^2 - 8x + 16}{x - 4}$

21. $\dfrac{x - 9}{9 - x}$

22. $\dfrac{x - 4}{4 - x}$

23. $\dfrac{x^2 - 49}{7 - x}$

24. $\dfrac{x^2 - y^2}{y - x}$

25. $\dfrac{2x^2 - 7x - 4}{x^2 - 5x + 4}$

26. $\dfrac{3x^2 - 11x + 10}{x^2 - 7x + 10}$

27. $\dfrac{x^3 - 125}{2x - 10}$

28. $\dfrac{4x + 4}{x^3 + 1}$

29. $\dfrac{3x^2 - 5x - 2}{6x^3 + 2x^2 + 3x + 1}$

30. $\dfrac{2x^2 - x - 3}{2x^3 - 3x^2 + 2x - 3}$

31. $\dfrac{9x^2 - 15x + 25}{27x^3 + 125}$

32. $\dfrac{8x^3 - 27}{4x^2 + 6x + 9}$

Multiply and simplify. See Example 7.

33. $\dfrac{2x - 4}{15} \cdot \dfrac{6}{2 - x}$

34. $\dfrac{10 - 2x}{7} \cdot \dfrac{14}{5x - 25}$

35. $\dfrac{18a - 12a^2}{4a^2 + 4a + 1} \cdot \dfrac{4a^2 + 8a + 3}{4a^2 - 9}$

36. $\dfrac{a - 5b}{a^2 + ab} \cdot \dfrac{b^2 - a^2}{10b - 2a}$

37. $\dfrac{9x + 9}{4x + 8} \cdot \dfrac{2x + 4}{3x^2 - 3}$

38. $\dfrac{2x^2 - 2}{10x + 30} \cdot \dfrac{12x + 36}{3x - 3}$

39. $\dfrac{2x^3 - 16}{6x^2 + 6x - 36} \cdot \dfrac{9x + 18}{3x^2 + 6x + 12}$

40. $\dfrac{x^2 - 3x + 9}{5x^2 - 20x - 105} \cdot \dfrac{x^2 - 49}{x^3 + 27}$

41. $\dfrac{a^3 + a^2b + a + b}{5a^3 + 5a} \cdot \dfrac{6a^2}{2a^2 - 2b^2}$

42. $\dfrac{4a^2 - 8a}{ab - 2b + 3a - 6} \cdot \dfrac{8b + 24}{3a + 6}$

43. $\dfrac{x^2 - 6x - 16}{2x^2 - 128} \cdot \dfrac{x^2 + 16x + 64}{3x^2 + 30x + 48}$

44. $\dfrac{2x^2 + 12x - 32}{x^2 + 16x + 64} \cdot \dfrac{x^2 + 10x + 16}{x^2 - 3x - 10}$

Divide and simplify. See Example 8.

45. $\dfrac{2x}{5} \div \dfrac{6x + 12}{5x + 10}$

46. $\dfrac{7}{3x} \div \dfrac{14 - 7x}{18 - 9x}$

47. $\dfrac{a + b}{ab} \div \dfrac{a^2 - b^2}{4a^3b}$

48. $\dfrac{6a^2b^2}{a^2 - 4} \div \dfrac{3ab^2}{a - 2}$

49. $\dfrac{x^2 - 6x + 9}{x^2 - x - 6} \div \dfrac{x^2 - 9}{4}$

50. $\dfrac{x^2 - 4}{3x + 6} \div \dfrac{2x^2 - 8x + 8}{x^2 + 4x + 4}$

51. $\dfrac{x^2 - 6x - 16}{2x^2 - 128} \div \dfrac{x^2 + 10x + 16}{x^2 + 16x + 64}$

52. $\dfrac{a^2 - a - 6}{a^2 - 81} \div \dfrac{a^2 - 7a - 18}{4a + 36}$

53. $\dfrac{3x - x^2}{x^3 - 27} \div \dfrac{x}{x^2 + 3x + 9}$

54. $\dfrac{x^2 - 3x}{x^3 - 27} \div \dfrac{2x}{2x^2 + 6x + 18}$

55. $\dfrac{8b + 24}{3a + 6} \div \dfrac{ab - 2b + 3a - 6}{a^2 - 4a + 4}$

56. $\dfrac{2a^2 - 2b^2}{a^3 + a^2b + a + b} \div \dfrac{6a^2}{a^3 + a}$

MIXED PRACTICE

Perform each indicated operation. See Examples 3 through 9.

57. $\dfrac{x^2 - 9}{4} \cdot \dfrac{x^2 - x - 6}{x^2 - 6x + 9}$

58. $\dfrac{x^2 - 4}{9} \cdot \dfrac{x^2 - 6x + 9}{x^2 - 5x + 6}$

59. $\dfrac{2x^2 - 4x - 30}{5x^2 - 40x - 75} \div \dfrac{x^2 - 8x + 15}{x^2 - 6x + 9}$

60. $\dfrac{4a + 36}{a^2 - 7a - 18} \div \dfrac{a^2 - a - 6}{a^2 - 81}$

61. Simplify: $\dfrac{r^3 + s^3}{r + s}$

62. Simplify: $\dfrac{m^3 - n^3}{m - n}$

63. $\dfrac{4}{x} \div \dfrac{3xy}{x^2} \cdot \dfrac{6x^2}{x^4}$

64. $\dfrac{4}{x} \cdot \dfrac{3xy}{x^2} \div \dfrac{6x^2}{x^4}$

65. $\dfrac{3x^2 - 5x - 2}{y^2 + y - 2} \cdot \dfrac{y^2 + 4y - 5}{12x^2 + 7x + 1} \div \dfrac{5x^2 - 9x - 2}{8x^2 - 2x - 1}$

66. $\dfrac{x^2 + x - 2}{3y^2 - 5y - 2} \cdot \dfrac{12y^2 + y - 1}{x^2 + 4x - 5} \div \dfrac{8y^2 - 6y + 1}{5y^2 - 9y - 2}$

67. $\dfrac{5a^2 - 20}{3a^2 - 12a} \div \dfrac{a^3 + 2a^2}{2a^2 - 8a} \cdot \dfrac{9a^3 + 6a^2}{2a^2 - 4a}$

68. $\dfrac{5a^2 - 20}{3a^2 - 12a} \div \left(\dfrac{a^3 + 2a^2}{2a^2 - 8a} \cdot \dfrac{9a^3 + 6a^2}{2a^2 - 4a} \right)$

69. $\dfrac{5x^4 + 3x^2 - 2}{x - 1} \cdot \dfrac{x + 1}{x^4 - 1}$

70. $\dfrac{3x^4 - 10x^2 - 8}{x - 2} \cdot \dfrac{3x + 6}{15x^2 + 10}$

Find each function value. See Example 10.

71. If $f(x) = \dfrac{x + 8}{2x - 1}$, find $f(2)$, $f(0)$, and $f(-1)$.

72. If $f(x) = \dfrac{x - 2}{-5 + x}$, find $f(-5)$, $f(0)$, and $f(10)$.

73. If $g(x) = \dfrac{x^2 + 8}{x^3 - 25x}$, find $g(3)$, $g(-2)$, and $g(1)$.

74. If $s(t) = \dfrac{t^3 + 1}{t^2 + 1}$, find $s(-1)$, $s(1)$, and $s(2)$.

75. The total revenue from the sale of a popular book is approximated by the rational function $R(x) = \dfrac{1000x^2}{x^2 + 4}$, where x is the number of years since publication and $R(x)$ is the total revenue in millions of dollars.

   a. Find the total revenue at the end of the first year.

   b. Find the total revenue at the end of the second year.

   c. Find the revenue during the second year only.

   d. Find the domain of function R.

76. The function $f(x) = \dfrac{100,000x}{100 - x}$ models the cost in dollars for removing x percent of the pollutants from a bayou in which a nearby company dumped creosol.

   a. Find the cost of removing 20% of the pollutants from the bayou. [*Hint:* Find $f(20)$.]

   b. Find the cost of removing 60% of the pollutants and then 80% of the pollutants.

   c. Find $f(90)$, then $f(95)$, and then $f(99)$. What happens to the cost as x approaches 100%?

   d. Find the domain of function f.

REVIEW AND PREVIEW

Perform each indicated operation. See Section 1.3.

77. $\dfrac{4}{5} + \dfrac{3}{5}$

78. $\dfrac{4}{10} - \dfrac{7}{10}$

79. $\dfrac{5}{28} - \dfrac{2}{21}$

80. $\dfrac{5}{13} + \dfrac{2}{7}$

81. $\dfrac{3}{8} + \dfrac{1}{2} - \dfrac{3}{16}$

82. $\dfrac{2}{9} - \dfrac{1}{6} + \dfrac{2}{3}$

CONCEPT EXTENSIONS

Solve. For Exercises 83 and 84, see the first Concept Check in this section; for Exercises 85 and 86, see the second Concept Check.

83. Which of the expressions are equivalent to $\dfrac{x}{5 - x}$?

   a. $\dfrac{-x}{5 - x}$ **b.** $\dfrac{-x}{-5 + x}$

   c. $\dfrac{x}{x - 5}$ **d.** $\dfrac{-x}{x - 5}$

84. Which of the expressions are equivalent to $\dfrac{-2 + x}{x}$?

   a. $\dfrac{2 - x}{-x}$ **b.** $-\dfrac{2 - x}{x}$

   c. $\dfrac{x - 2}{x}$ **d.** $\dfrac{x - 2}{-x}$

85. Does $\dfrac{x}{x + 5}$ simplify to $\dfrac{1}{5}$? Why or why not?

86. Does $\dfrac{x + 7}{x}$ simplify to 7? Why or why not?

△ **87.** Find the area of the rectangle.

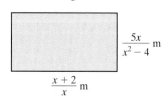

$\dfrac{5x}{x^2 - 4}$ m

$\dfrac{x + 2}{x}$ m

△ **88.** Find the area of the triangle.

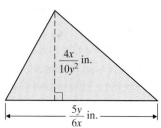

$\dfrac{4x}{10y^2}$ in.

$\dfrac{5y}{6x}$ in.

△ **89.** A parallelogram has an area of $\dfrac{x^2 + x - 2}{x^3}$ square feet and a height of $\dfrac{x^2}{x - 1}$ feet. Express the length of its base as a rational expression in x. (*Hint:* Since $A = b \cdot h$, then $b = \dfrac{A}{h}$ or $b = A \div h$.)

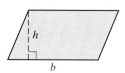

90. A lottery prize of $\dfrac{15x^3}{y^2}$ dollars is to be divided among $5x$ people. Express the amount of money each person is to receive as a rational expression in x and y.

91. In your own words explain how to simplify a rational expression.

92. In your own words, explain the difference between multiplying rational expressions and dividing rational expressions.

93. Decide whether each rational expression equals 1, -1, or neither.

   a. $\dfrac{x + 5}{5 + x}$ **b.** $\dfrac{x - 5}{5 - x}$

   c. $\dfrac{x + 5}{x - 5}$ **d.** $\dfrac{-x - 5}{x + 5}$

   e. $\dfrac{x - 5}{-x + 5}$ **f.** $\dfrac{-5 + x}{x - 5}$

94. In our definition of division for

$$\dfrac{P}{Q} \div \dfrac{R}{S}$$

we stated that $Q \neq 0$, $S \neq 0$, and $R \neq 0$. Explain why R cannot equal 0.

95. Find the polynomial in the second numerator such that the following statement is true.

$$\dfrac{x^2 - 4}{x^2 - 7x + 10} \cdot \dfrac{?}{2x^2 + 11x + 14} = 1$$

96. In your own words, explain how to find the domain of a rational function.

97. Graph a portion of the function $f(x) = \dfrac{20x}{100 - x}$. To do so, complete the given table, plot the points, and then connect the plotted points with a smooth curve.

x	0	10	30	50	70	90	95	99
y or $f(x)$								

98. The domain of the function $f(x) = \dfrac{1}{x}$ is all real numbers except 0. This means that the graph of this function will be in two pieces: one piece corresponding to x-values less than 0 and one piece corresponding to x-values greater than 0. Graph the function by completing the following tables, separately plotting the points, and connecting each set of plotted points with a smooth curve.

x	$\frac{1}{4}$	$\frac{1}{2}$	1	2	4
y or $f(x)$					

x	-4	-2	-1	$-\frac{1}{2}$	$-\frac{1}{4}$
y or $f(x)$					

Perform the indicated operation. Write all answers in lowest terms.

99. $\dfrac{x^{2n} - 4}{7x} \cdot \dfrac{14x^3}{x^n - 2}$

100. $\dfrac{x^{2n} + 4x^n + 4}{4x - 3} \cdot \dfrac{8x^2 - 6x}{x^n + 2}$

101. $\dfrac{y^{2n} + 9}{10y} \cdot \dfrac{y^n - 3}{y^{4n} - 81}$

102. $\dfrac{y^{4n} - 16}{y^{2n} + 4} \cdot \dfrac{6y}{y^n + 2}$

103. $\dfrac{y^{2n} - y^n - 2}{2y^n - 4} \div \dfrac{y^{2n} - 1}{1 + y^n}$

104. $\dfrac{y^{2n} + 7y^n + 10}{10} \div \dfrac{y^{2n} + 4y^n + 4}{5y^n + 25}$

📖 STUDY SKILLS BUILDER

Are You Satisfied with Your Performance in this Course thus Far?

To see if there is room for improvement, answer these questions:

1. Am I attending all classes and arriving on time?
2. Am I working and checking my homework assignments on time?
3. Am I getting help (from my instructor or a campus learning resource lab) when I need it?
4. In addition to my instructor, am I using the text supplements that might help me?
5. Am I satisfied with my performance on quizzes and exams?

If you answered no to any of these questions, read or reread Section 1.1 for suggestions in these areas. Also, you might want to contact your instructor for additional feedback.

6.2 ADDING AND SUBTRACTING RATIONAL EXPRESSIONS

OBJECTIVES

1. Add or subtract rational expressions with common denominators.

2. Identify the least common denominator of two or more rational expressions.

3. Add or subtract rational expressions with unlike denominators.

OBJECTIVE 1 ▶ Adding or subtracting rational expressions with common denominators. Rational expressions, like rational numbers, can be added or subtracted. We add or subtract rational expressions in the same way that we add or subtract rational numbers (fractions).

Adding or Subtracting Rational Expressions with Common Denominators

If $\dfrac{P}{Q}$ and $\dfrac{R}{Q}$ are rational expressions, then

$$\frac{P}{Q} + \frac{R}{Q} = \frac{P + R}{Q} \quad \text{and} \quad \frac{P}{Q} - \frac{R}{Q} = \frac{P - R}{Q}$$

To add or subtract rational expressions with common denominators, add or subtract the numerators and write the sum or difference over the common denominator.

EXAMPLE 1   Add or subtract.

a. $\dfrac{x}{4} + \dfrac{5x}{4}$ **b.** $\dfrac{5}{7z^2} + \dfrac{x}{7z^2}$ **c.** $\dfrac{x^2}{x+7} - \dfrac{49}{x+7}$ **d.** $\dfrac{x}{3y^2} - \dfrac{x+1}{3y^2}$

Solution   The rational expressions have common denominators, so add or subtract their numerators and place the sum or difference over their common denominator.

a. $\dfrac{x}{4} + \dfrac{5x}{4} = \dfrac{x+5x}{4} = \dfrac{6x}{4} = \dfrac{3x}{2}$ Add the numerators and write the result over the common denominator.

b. $\dfrac{5}{7z^2} + \dfrac{x}{7z^2} = \dfrac{5+x}{7z^2}$

c. $\dfrac{x^2}{x+7} - \dfrac{49}{x+7} = \dfrac{x^2 - 49}{x+7}$ Subtract the numerators and write the result over the common denominator.

$\qquad = \dfrac{(x+7)(x-7)}{x+7}$ Factor the numerator.

$\qquad = x - 7$ Simplify.

d. $\dfrac{x}{3y^2} - \dfrac{x+1}{3y^2} = \dfrac{x - (x+1)}{3y^2}$ Subtract the numerators.

$\qquad = \dfrac{x - x - 1}{3y^2}$ Use the distributive property.

$\qquad = -\dfrac{1}{3y^2}$ Simplify.   ☐

▶ **Helpful Hint**

Very Important: Be sure to insert parentheses here so that the entire numerator is subtracted.

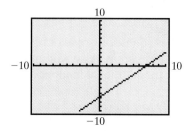

10

−10 ┤├ 10

−10

A graphing utility can be used to provide partial support of the result of operations on rational expressions. For example, the graphs of $y_1 = \dfrac{x^2}{x+7} - \dfrac{49}{x+7}$ and $y_2 = x - 7$ appear to coincide. This provides partial support for the algebraic solution in part c of Example 1. The trace feature can be used to verify that the two graphs coincide (except when x is -7).

PRACTICE

1   Add or subtract.

a. $\dfrac{9}{11z^2} + \dfrac{x}{11z^2}$ **b.** $\dfrac{x}{8} + \dfrac{5x}{8}$ **c.** $\dfrac{x^2}{x+4} - \dfrac{16}{x+4}$ **d.** $\dfrac{z}{2a^2} - \dfrac{z+3}{2a^2}$

Concept Check ✓

Find and correct the error.

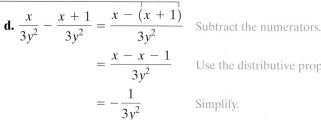

$$\dfrac{3+2y}{y^2-1} - \dfrac{y+3}{y^2-1} = \dfrac{3+2y-y+3}{y^2-1}$$

$$= \dfrac{y+6}{y^2-1}$$

OBJECTIVE 2 ▶ **Identifying the least common denominator of rational expressions.**
To add or subtract rational expressions with unlike denominators, first write the rational expressions as equivalent rational expressions with common denominators.

The **least common denominator (LCD)** is usually the easiest common denominator to work with. The LCD of a list of rational expressions is a polynomial of least degree whose factors include the denominator factors in the list.

Use the following steps to find the LCD.

Answer to Concept Check:

$\dfrac{3+2y}{y^2-1} - \dfrac{y+3}{y^2-1}$

$\qquad = \dfrac{3+2y-y-3}{y^2-1} = \dfrac{y}{y^2-1}$

Finding the Least Common Denominator (LCD)

STEP 1. Factor each denominator completely.

STEP 2. The LCD is the product of all unique factors each raised to a power equal to the greatest number of times that the factor appears in any factored denominator.

EXAMPLE 2 Find the LCD of the rational expressions in each list.

a. $\dfrac{2}{3x^5y^2}, \dfrac{3z}{5xy^3}$

b. $\dfrac{7}{z+1}, \dfrac{z}{z-1}$

c. $\dfrac{m-1}{m^2-25}, \dfrac{2m}{2m^2-9m-5}, \dfrac{7}{m^2-10m+25}$

d. $\dfrac{x}{x^2-4}, \dfrac{11}{6-3x}$

Solution

a. First we factor each denominator.

$$3x^5y^2 = 3 \cdot x^5 \cdot y^2$$
$$5xy^3 = 5 \cdot x \cdot y^3$$
$$\text{LCD} = 3 \cdot 5 \cdot x^5 \cdot y^3 = 15x^5y^3$$

> ▶ **Helpful Hint**
>
> The greatest power of x is 5, so we have a factor of x^5. The greatest power of y is 3, so we have a factor of y^3.

b. The denominators $z+1$ and $z-1$ do not factor further. Thus,

$$\text{LCD} = (z+1)(z-1)$$

c. We first factor each denominator.

$$m^2 - 25 = (m+5)(m-5)$$
$$2m^2 - 9m - 5 = (2m+1)(m-5)$$
$$m^2 - 10m + 25 = (m-5)(m-5)$$
$$\text{LCD} = (m+5)(2m+1)(m-5)^2$$

d. Factor each denominator.

$$x^2 - 4 = (x+2)(x-2)$$
$$6 - 3x = 3(2-x) = 3(-1)(x-2)$$
$$\text{LCD} = 3(-1)(x+2)(x-2)$$
$$= -3(x+2)(x-2)$$

> ▶ **Helpful Hint**
>
> $(x-2)$ and $(2-x)$ are opposite factors. Notice that -1 was factored from $(2-x)$ so that the factors are identical.

> ▶ **Helpful Hint**
>
> If opposite factors occur, do not use both in the LCD. Instead, factor -1 from one of the opposite factors so that the factors are then identical.

PRACTICE
2 Find the LCD of the rational expression in each list.

a. $\dfrac{7}{6x^3y^5}, \dfrac{2}{9x^2y^4}$

b. $\dfrac{11}{x-2}, \dfrac{x}{x+3}$

c. $\dfrac{b+2}{b^2-16}, \dfrac{8}{b^2-8b+16}, \dfrac{5b}{2b^2-5b-12}$

d. $\dfrac{y}{y^2-9}, \dfrac{3}{12-4y}$

OBJECTIVE 3 ▶ Adding or subtracting rational expressions with unlike denominators.
To add or subtract rational expressions with unlike denominators, we write each rational expression as an equivalent rational expression so that their denominators are alike.

> **Adding or Subtracting Rational Expressions with Unlike Denominators**
>
> **STEP 1.** Find the LCD of the rational expressions.
>
> **STEP 2.** Write each rational expression as an equivalent rational expression whose denominator is the LCD found in Step 1.
>
> **STEP 3.** Add or subtract numerators, and write the result over the common denominator.
>
> **STEP 4.** Simplify the resulting rational expression.

EXAMPLE 3 Perform the indicated operation.

a. $\dfrac{2}{x^2 y} + \dfrac{5}{3x^3 y}$ **b.** $\dfrac{3x}{x+2} + \dfrac{2x}{x-2}$ **c.** $\dfrac{2x-6}{x-1} - \dfrac{4}{1-x}$

Solution

a. The LCD is $3x^3 y$. Write each fraction as an equivalent fraction with denominator $3x^3 y$. To do this, we multiply both the numerator and denominator of each fraction by the factors needed to obtain the LCD as denominator.

The first fraction is multiplied by $\dfrac{3x}{3x}$ so that the new denominator is the LCD.

$$\dfrac{2}{x^2 y} + \dfrac{5}{3x^3 y} = \dfrac{2 \cdot 3x}{x^2 y \cdot 3x} + \dfrac{5}{3x^3 y} \qquad \text{The second expression already has a denominator of } 3x^3 y.$$

$$= \dfrac{6x}{3x^3 y} + \dfrac{5}{3x^3 y}$$

$$= \dfrac{6x + 5}{3x^3 y} \qquad \text{Add the numerators.}$$

b. The LCD is the product of the two denominators: $(x+2)(x-2)$.

$$\dfrac{3x}{x+2} + \dfrac{2x}{x-2} = \dfrac{3x \cdot (x-2)}{(x+2) \cdot (x-2)} + \dfrac{2x \cdot (x+2)}{(x-2) \cdot (x+2)} \qquad \text{Write equivalent rational expressions.}$$

$$= \dfrac{3x^2 - 6x}{(x+2)(x-2)} + \dfrac{2x^2 + 4x}{(x+2)(x-2)} \qquad \text{Multiply in the numerators.}$$

$$= \dfrac{3x^2 - 6x + 2x^2 + 4x}{(x+2)(x-2)} \qquad \text{Add the numerators.}$$

$$= \dfrac{5x^2 - 2x}{(x+2)(x-2)} \qquad \text{Simplify the numerator.}$$

X	Y1	Y2
-3	10.2	10.2
-2	ERROR	ERROR
-1	-2.333	-2.333
0	0	0
1	-1	-1
2	ERROR	ERROR
3	7.8	7.8

Y₁◼3X/(X+2)+2X/...

This table of
$y_1 = \dfrac{3x}{x+2} + \dfrac{2x}{x-2}$ and
$y_2 = \dfrac{5x^2 - 2x}{(x+2)(x-2)}$ provides
partial support for the result of part b in Example 3. Notice that both -2 and 2 give error messages indicating each would make the denominator zero.

c. The LCD is either $x - 1$ or $1 - x$. To get a common denominator of $x - 1$, we factor -1 from the denominator of the second rational expression.

$$\dfrac{2x-6}{x-1} - \dfrac{4}{1-x} = \dfrac{2x-6}{x-1} - \dfrac{4}{-1(x-1)} \qquad \text{Write } 1 - x \text{ as } -1(x-1).$$

$$= \dfrac{2x-6}{x-1} - \dfrac{-1 \cdot 4}{x-1} \qquad \text{Write } \dfrac{4}{-1(x-1)} \text{ as } \dfrac{-1 \cdot 4}{x-1}.$$

$$= \dfrac{2x - 6 - (-4)}{x-1} \qquad \text{Combine the numerators.}$$

$$= \dfrac{2x - 6 + 4}{x-1} \qquad \text{Simplify.}$$

$$= \dfrac{2x - 2}{x-1}$$

$$= \frac{2(x-1)}{x-1} \qquad \text{Factor.}$$

$$= 2 \qquad \text{Simplest form} \qquad \square$$

PRACTICE
3   Perform the indicated operation.

a. $\dfrac{4}{p^3q} + \dfrac{3}{5p^4q}$ **b.** $\dfrac{4}{y+3} + \dfrac{5y}{y-3}$ **c.** $\dfrac{3z-18}{z-5} - \dfrac{3}{5-z}$

EXAMPLE 4   Subtract $\dfrac{5k}{k^2-4} - \dfrac{2}{k^2+k-2}$.

Solution $\dfrac{5k}{k^2-4} - \dfrac{2}{k^2+k-2} = \dfrac{5k}{(k+2)(k-2)} - \dfrac{2}{(k+2)(k-1)}$ Factor each denominator to find the LCD.

The LCD is $(k+2)(k-2)(k-1)$. We write equivalent rational expressions with the LCD as denominators.

$$\frac{5k}{(k+2)(k-2)} - \frac{2}{(k+2)(k-1)}$$

$$= \frac{5k \cdot (k-1)}{(k+2)(k-2) \cdot (k-1)} - \frac{2 \cdot (k-2)}{(k+2)(k-1) \cdot (k-2)} \qquad \text{Write equivalent rational expressions.}$$

$$= \frac{5k^2 - 5k}{(k+2)(k-2)(k-1)} - \frac{2k-4}{(k+2)(k-2)(k-1)} \qquad \text{Multiply in the numerators.}$$

> ▶ **Helpful Hint**
> **Very Important:** Because we are subtracting; notice the sign change on 4.

$$= \frac{5k^2 - 5k - 2k + 4}{(k+2)(k-2)(k-1)} \qquad \text{Subtract the numerators.}$$

$$= \frac{5k^2 - 7k + 4}{(k+2)(k-2)(k-1)} \qquad \text{Simplify.} \qquad \square$$

PRACTICE
4   Subtract $\dfrac{t}{t^2-25} - \dfrac{3}{t^2-3t-10}$.

EXAMPLE 5   Add $\dfrac{2x-1}{2x^2-9x-5} + \dfrac{x+3}{6x^2-x-2}$.

Solution

$$\frac{2x-1}{2x^2-9x-5} + \frac{x+3}{6x^2-x-2} = \frac{2x-1}{(2x+1)(x-5)} + \frac{x+3}{(2x+1)(3x-2)} \qquad \text{Factor the denominators.}$$

The LCD is $(2x+1)(x-5)(3x-2)$.

$$= \frac{(2x-1) \cdot (3x-2)}{(2x+1)(x-5) \cdot (3x-2)} + \frac{(x+3) \cdot (x-5)}{(2x+1)(3x-2) \cdot (x-5)}$$

$$= \frac{6x^2 - 7x + 2}{(2x+1)(x-5)(3x-2)} + \frac{x^2 - 2x - 15}{(2x+1)(x-5)(3x-2)} \qquad \text{Multiply in the numerators.}$$

$$= \frac{6x^2 - 7x + 2 + x^2 - 2x - 15}{(2x+1)(x-5)(3x-2)} \qquad \text{Add the numerators.}$$

$$= \frac{7x^2 - 9x - 13}{(2x+1)(x-5)(3x-2)} \qquad \text{Simplify.} \qquad \square$$

PRACTICE
5 Add $\dfrac{2x + 3}{3x^2 - 5x - 2} + \dfrac{x - 6}{6x^2 - 13x - 5}$.

EXAMPLE 6 Perform each indicated operation.

$$\frac{7}{x - 1} + \frac{10x}{x^2 - 1} - \frac{5}{x + 1}$$

Solution $\dfrac{7}{x - 1} + \dfrac{10x}{x^2 - 1} - \dfrac{5}{x + 1} = \dfrac{7}{x - 1} + \dfrac{10x}{(x - 1)(x + 1)} - \dfrac{5}{x + 1}$ Factor the denominators.

The LCD is $(x - 1)(x + 1)$.

$$= \frac{7 \cdot (x + 1)}{(x - 1) \cdot (x + 1)} + \frac{10x}{(x - 1)(x + 1)} - \frac{5 \cdot (x - 1)}{(x + 1) \cdot (x - 1)}$$

$$= \frac{7x + 7}{(x - 1)(x + 1)} + \frac{10x}{(x - 1)(x + 1)} - \frac{5x - 5}{(x + 1)(x - 1)} \quad \text{Multiply in the numerators.}$$

$$= \frac{7x + 7 + 10x - 5x + 5}{(x - 1)(x + 1)} \quad \text{Add and subtract the numerators.}$$

$$= \frac{12x + 12}{(x - 1)(x + 1)} \quad \text{Simplify.}$$

$$= \frac{12\,(x + 1)}{(x - 1)\,(x + 1)} \quad \text{Factor the numerator.}$$

$$= \frac{12}{x - 1} \quad \text{Divide out common factors.}$$

PRACTICE
6 Perform each indicated operation.

$$\frac{2}{x - 2} + \frac{3x}{x^2 - x - 2} - \frac{1}{x + 1}$$

VOCABULARY & READINESS CHECK

Name the operation(s) below that make each statement true.

   a. Addition   **b.** Subtraction   **c.** Multiplication   **d.** Division

1. The denominators must be the same before performing the operation. _____
2. To perform this operation, you multiply the first rational expression by the reciprocal of the second rational expression. _____
3. Numerator times numerator all over denominator times denominator. _____
4. These operations are commutative (order doesn't matter.) _____

For the rational expressions $\dfrac{5}{y}$ and $\dfrac{7}{y}$, perform each operation mentally.

5. Addition   **6.** Subtraction   **7.** Multiplication   **8.** Division
    _____   _____   _____   _____

Be careful when subtracting! For example, $\dfrac{8}{x+1} - \dfrac{x+5}{x+1} = \dfrac{8-(x+5)}{x+1} = \dfrac{3-x}{x+1}$ or $\dfrac{-x+3}{x+1}$.

Use this example to help you perform the subtractions.

9. $\dfrac{5}{2x} - \dfrac{x+1}{2x} = $ _____

10. $\dfrac{9}{5x} - \dfrac{6-x}{5x} = $ _____

11. $\dfrac{y+11}{y-2} - \dfrac{y-5}{y-2} = $ _____

12. $\dfrac{z-1}{z+6} - \dfrac{z+4}{z+6} = $ _____

6.2 | EXERCISE SET

MyMathLab

PRACTICE · WATCH · DOWNLOAD · READ · REVIEW

Add or subtract as indicated. Simplify each answer. See Example 1.

1. $\dfrac{2}{xz^2} - \dfrac{5}{xz^2}$

2. $\dfrac{4}{x^2y} - \dfrac{2}{x^2y}$

3. $\dfrac{2}{x-2} + \dfrac{x}{x-2}$

4. $\dfrac{x}{5-x} + \dfrac{7}{5-x}$

5. $\dfrac{x^2}{x+2} - \dfrac{4}{x+2}$

6. $\dfrac{x^2}{x+6} - \dfrac{36}{x+6}$

7. $\dfrac{2x-6}{x^2+x-6} + \dfrac{3-3x}{x^2+x-6}$

8. $\dfrac{5x+2}{x^2+2x-8} + \dfrac{2-4x}{x^2+2x-8}$

9. $\dfrac{x-5}{2x} - \dfrac{x+5}{2x}$

10. $\dfrac{x+4}{4x} - \dfrac{x-4}{4x}$

Find the LCD of the rational expressions in each list. See Example 2.

11. $\dfrac{2}{7}, \dfrac{3}{5x}$

12. $\dfrac{4}{5y}, \dfrac{3}{4y^2}$

13. $\dfrac{3}{x}, \dfrac{2}{x+1}$

14. $\dfrac{5}{2x}, \dfrac{7}{2+x}$

15. $\dfrac{12}{x+7}, \dfrac{8}{x-7}$

16. $\dfrac{1}{2x-1}, \dfrac{8}{2x+1}$

17. $\dfrac{5}{3x+6}, \dfrac{2x}{2x-4}$

18. $\dfrac{2}{3a+9}, \dfrac{5}{5a-15}$

19. $\dfrac{2a}{a^2-b^2}, \dfrac{1}{a^2-2ab+b^2}$

20. $\dfrac{2a}{a^2+8a+16}, \dfrac{7a}{a^2+a-12}$

21. $\dfrac{x}{x^2-9}, \dfrac{5}{x}, \dfrac{7}{12-4x}$

22. $\dfrac{9}{x^2-25}, \dfrac{1}{50-10x}, \dfrac{6}{x}$

Add or subtract as indicated. Simplify each answer. See Examples 3a and 3b.

23. $\dfrac{4}{3x} + \dfrac{3}{2x}$

24. $\dfrac{10}{7x} + \dfrac{5}{2x}$

25. $\dfrac{5}{2y^2} - \dfrac{2}{7y}$

26. $\dfrac{4}{11x^4} - \dfrac{1}{4x^2}$

27. $\dfrac{x-3}{x+4} - \dfrac{x+2}{x-4}$

28. $\dfrac{x-1}{x-5} - \dfrac{x+2}{x+5}$

29. $\dfrac{1}{x-5} - \dfrac{19-2x}{(x-5)(x+4)}$

30. $\dfrac{4x-2}{(x-5)(x+4)} - \dfrac{2}{x+4}$

Perform the indicated operation. If possible, simplify your answer. See Example 3c.

31. $\dfrac{1}{a-b} + \dfrac{1}{b-a}$

32. $\dfrac{1}{a-3} - \dfrac{1}{3-a}$

33. $\dfrac{x+1}{1-x} + \dfrac{1}{x-1}$

34. $\dfrac{5}{1-x} - \dfrac{1}{x-1}$

35. $\dfrac{5}{x-2} + \dfrac{x+4}{2-x}$

36. $\dfrac{3}{5-x} + \dfrac{x+2}{x-5}$

Perform each indicated operation. If possible, simplify your answer. See Examples 4 through 6.

37. $\dfrac{y+1}{y^2-6y+8} - \dfrac{3}{y^2-16}$

38. $\dfrac{x+2}{x^2-36} - \dfrac{x}{x^2+9x+18}$

39. $\dfrac{x+4}{3x^2+11x+6} + \dfrac{x}{2x^2+x-15}$

40. $\dfrac{x+3}{5x^2+12x+4} + \dfrac{6}{x^2-x-6}$

41. $\dfrac{7}{x^2-x-2} - \dfrac{x-1}{x^2+4x+3}$

42. $\dfrac{a}{a^2+10a+25} - \dfrac{4-a}{a^2+6a+5}$

43. $\dfrac{x}{x^2-8x+7} - \dfrac{x+2}{2x^2-9x-35}$

44. $\dfrac{x}{x^2-7x+6} - \dfrac{x+4}{3x^2-2x-1}$

45. $\dfrac{2}{a^2+2a+1} + \dfrac{3}{a^2-1}$

46. $\dfrac{9x+2}{3x^2-2x-8} + \dfrac{7}{3x^2+x-4}$

MIXED PRACTICE

Add or subtract as indicated. If possible, simplify your answer. See Examples 1 through 6.

47. $\dfrac{4}{3x^2y^3} + \dfrac{5}{3x^2y^3}$

48. $\dfrac{7}{2xy^4} + \dfrac{1}{2xy^4}$

49. $\dfrac{13x - 5}{2x} - \dfrac{13x + 5}{2x}$

50. $\dfrac{17x + 4}{4x} - \dfrac{17x - 4}{4x}$

51. $\dfrac{3}{2x + 10} + \dfrac{8}{3x + 15}$

52. $\dfrac{10}{3x - 3} + \dfrac{1}{7x - 7}$

53. $\dfrac{-2}{x^2 - 3x} - \dfrac{1}{x^3 - 3x^2}$

54. $\dfrac{-3}{2a + 8} - \dfrac{8}{a^2 + 4a}$

55. $\dfrac{ab}{a^2 - b^2} + \dfrac{b}{a + b}$

56. $\dfrac{x}{25 - x^2} + \dfrac{2}{3x - 15}$

57. $\dfrac{5}{x^2 - 4} - \dfrac{3}{x^2 + 4x + 4}$

58. $\dfrac{3z}{z^2 - 9} - \dfrac{2}{3 - z}$

59. $\dfrac{3x}{2x^2 - 11x + 5} + \dfrac{7}{x^2 - 2x - 15}$

60. $\dfrac{2x}{3x^2 - 13x + 4} + \dfrac{5}{x^2 - 2x - 8}$

61. $\dfrac{2}{x + 1} - \dfrac{3x}{3x + 3} + \dfrac{1}{2x + 2}$

62. $\dfrac{5}{3x - 6} - \dfrac{x}{x - 2} + \dfrac{3 + 2x}{5x - 10}$

63. $\dfrac{3}{x + 3} + \dfrac{5}{x^2 + 6x + 9} - \dfrac{x}{x^2 - 9}$

64. $\dfrac{x + 2}{x^2 - 2x - 3} + \dfrac{x}{x - 3} - \dfrac{x}{x + 1}$

65. $\dfrac{x}{x^2 - 9} + \dfrac{3}{x^2 - 6x + 9} - \dfrac{1}{x + 3}$

66. $\dfrac{3}{x^2 - 9} - \dfrac{x}{x^2 - 6x + 9} + \dfrac{1}{x + 3}$

67. $\left(\dfrac{1}{x} + \dfrac{2}{3}\right) - \left(\dfrac{1}{x} - \dfrac{2}{3}\right)$

68. $\left(\dfrac{1}{2} + \dfrac{2}{x}\right) - \left(\dfrac{1}{2} - \dfrac{1}{x}\right)$

Perform the indicated operation. If possible, simplify your answer.

69. $\left(\dfrac{2}{3} - \dfrac{1}{x}\right) \cdot \left(\dfrac{3}{x} + \dfrac{1}{2}\right)$

70. $\left(\dfrac{2}{3} - \dfrac{1}{x}\right) \div \left(\dfrac{3}{x} + \dfrac{1}{2}\right)$

71. $\left(\dfrac{2a}{3}\right)^2 \div \left(\dfrac{a^2}{a + 1} - \dfrac{1}{a + 1}\right)$

72. $\left(\dfrac{x + 2}{2x} - \dfrac{x - 2}{2x}\right) \cdot \left(\dfrac{5x}{4}\right)^2$

73. $\left(\dfrac{2x}{3}\right)^2 \div \left(\dfrac{x}{3}\right)^2$

74. $\left(\dfrac{2x}{3}\right)^2 \cdot \left(\dfrac{3}{x}\right)^2$

75. $\left(\dfrac{x}{x + 1} - \dfrac{x}{x - 1}\right) \div \dfrac{x}{2x + 2}$

76. $\dfrac{x}{2x + 2} \div \left(\dfrac{x}{x + 1} + \dfrac{x}{x - 1}\right)$

77. $\dfrac{4}{x} \cdot \left(\dfrac{2}{x + 2} - \dfrac{2}{x - 2}\right)$

78. $\dfrac{1}{x + 1} \cdot \left(\dfrac{5}{x} + \dfrac{2}{x - 3}\right)$

REVIEW AND PREVIEW

Use the distributive property to multiply the following. See Section 1.4.

79. $12\left(\dfrac{2}{3} + \dfrac{1}{6}\right)$

80. $14\left(\dfrac{1}{7} + \dfrac{3}{14}\right)$

81. $x^2\left(\dfrac{4}{x^2} + 1\right)$

82. $5y^2\left(\dfrac{1}{y^2} - \dfrac{1}{5}\right)$

Find each root. See Section 1.3.

83. $\sqrt{100}$

84. $\sqrt{25}$

85. $\sqrt[3]{8}$

86. $\sqrt[3]{27}$

87. $\sqrt[4]{81}$

88. $\sqrt[4]{16}$

Use the Pythagorean theorem to find each unknown length of a right triangle. See Section 5.8.

△ **89.**

3 meters

4 meters

△ **90.**

7 feet

24 feet

CONCEPT EXTENSIONS

Find and correct each error. See the Concept Check in this section.

91. $\dfrac{2x - 3}{x^2 + 1} - \dfrac{x - 6}{x^2 + 1} = \dfrac{2x - 3 - x - 6}{x^2 + 1}$

$$= \dfrac{x - 9}{x^2 + 1}$$

92.
$$\frac{7}{x+7} - \frac{x+3}{x+7} = \frac{7-x-3}{(x+7)^2}$$
$$= \frac{-x+4}{(x+7)^2}$$

△ **93.** Find the perimeter and the area of the square.

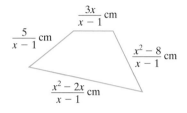
$\frac{x}{x+5}$ ft

△ **94.** Find the perimeter of the quadrilateral.

$\frac{3x}{x-1}$ cm

$\frac{5}{x-1}$ cm

$\frac{x^2-8}{x-1}$ cm

$\frac{x^2-2x}{x-1}$ cm

95. When is the LCD of two rational expressions equal to the product of their denominators? $\left(\textit{Hint:}\text{ What is the LCD of }\frac{1}{x}\text{ and }\frac{7}{x+5}?\right)$

96. When is the LCD of two rational expressions with different denominators equal to one of the denominators? $\left(\textit{Hint:}\right.$ What is the LCD of $\frac{3x}{x+2}$ and $\frac{7x+1}{(x+2)^3}?\left.\right)$

97. In your own words, explain how to add rational expressions with different denominators.

98. In your own words, explain how to multiply rational expressions.

99. In your own words, explain how to divide rational expressions.

100. In your own words, explain how to subtract rational expressions with different denominators.

Perform each indicated operation. (Hint: First write each expression with positive exponents.)

101. $x^{-1} + (2x)^{-1}$ **102.** $y^{-1} + (4y)^{-1}$

103. $4x^{-2} - 3x^{-1}$ **104.** $(4x)^{-2} - (3x)^{-1}$

Use a graphing calculator to support the results of each exercise.

105. Exercise 3   **106.** Exercise 4

6.3 | SIMPLIFYING COMPLEX FRACTIONS

OBJECTIVES

1 Simplify complex fractions by simplifying the numerator and denominator and then dividing.

2 Simplify complex fractions by multiplying by a common denominator.

3 Simplify expressions with negative exponents.

OBJECTIVE 1 ▶ Simplifying complex fractions: Method 1. A rational expression whose numerator, denominator, or both contain one or more rational expressions is called a **complex rational expression** or a **complex fraction.**

Complex Fractions

$$\frac{\frac{1}{a}}{\frac{b}{2}} \qquad \frac{\frac{x}{2y^2}}{\frac{6x-2}{9y}} \qquad \frac{x+\frac{1}{y}}{y+1}$$

The parts of a complex fraction are

$$\frac{\left.\dfrac{x}{y+2}\right\}}{\left.7+\dfrac{1}{y}\right\}}$$

← Numerator of complex fraction

← Main fraction bar

← Denominator of complex fraction

Our goal in this section is to simplify complex fractions. A complex fraction is simplified when it is in the form $\frac{P}{Q}$, where P and Q are polynomials that have no common factors. Two methods of simplifying complex fractions are introduced. The first method evolves from the definition of a fraction as a quotient.

> **Simplifying a Complex Fraction: Method 1**
>
> **STEP 1.** Simplify the numerator and the denominator of the complex fraction so that each is a single fraction.
>
> **STEP 2.** Perform the indicated division by multiplying the numerator of the complex fraction by the reciprocal of the denominator of the complex fraction.
>
> **STEP 3.** Simplify if possible.

EXAMPLE 1   Simplify each complex fraction.

a. $\dfrac{\dfrac{2x}{27y^2}}{\dfrac{6x^2}{9}}$ **b.** $\dfrac{\dfrac{5x}{x+2}}{\dfrac{10}{x-2}}$ **c.** $\dfrac{\dfrac{x}{y^2}+\dfrac{1}{y}}{\dfrac{y}{x^2}+\dfrac{1}{x}}$

Solution

a. The numerator of the complex fraction is already a single fraction, and so is the denominator. Perform the indicated division by multiplying the numerator, $\dfrac{2x}{27y^2}$, by the reciprocal of the denominator, $\dfrac{6x^2}{9}$. Then simplify.

$$\frac{\dfrac{2x}{27y^2}}{\dfrac{6x^2}{9}} = \frac{2x}{27y^2} \div \frac{6x^2}{9}$$

$$= \frac{2x}{27y^2} \cdot \frac{9}{6x^2} \qquad \text{Multiply by the reciprocal of } \frac{6x^2}{9}.$$

$$= \frac{2x \cdot 9}{27y^2 \cdot 6x^2}$$

$$= \frac{1}{9xy^2}$$

▶ **Helpful Hint**

Both the numerator and denominator are single fractions, so we perform the indicated division.

b. $\dfrac{\dfrac{5x}{x+2}}{\dfrac{10}{x-2}} = \frac{5x}{x+2} \div \frac{10}{x-2} = \frac{5x}{x+2} \cdot \frac{x-2}{10}$ Multiply by the reciprocal of $\frac{10}{x-2}$.

$$= \frac{5x(x-2)}{2\cdot 5(x+2)}$$

$$= \frac{x(x-2)}{2(x+2)} \qquad \text{Simplify.}$$

c. First simplify the numerator and the denominator of the complex fraction separately so that each is a single fraction. Then perform the indicated division.

$$\frac{\dfrac{x}{y^2} + \dfrac{1}{y}}{\dfrac{y}{x^2} + \dfrac{1}{x}} = \frac{\dfrac{x}{y^2} + \dfrac{1 \cdot y}{y \cdot y}}{\dfrac{y}{x^2} + \dfrac{1 \cdot x}{x \cdot x}}$$ Simplify the numerator. The LCD is y^2.

Simplify the denominator. The LCD is x^2.

$$= \frac{\dfrac{x + y}{y^2}}{\dfrac{y + x}{x^2}}$$ Add.

$$= \frac{x + y}{y^2} \cdot \frac{x^2}{y + x}$$ Multiply by the reciprocal of $\dfrac{y + x}{x^2}$.

$$= \frac{x^2 (x + y)}{y^2 (y + x)}$$

$$= \frac{x^2}{y^2}$$ Simplify.   □

PRACTICE
1   Simplify each complex fraction.

a. $\dfrac{\dfrac{5k}{36m}}{\dfrac{15k}{9}}$ **b.** $\dfrac{\dfrac{8x}{x - 4}}{\dfrac{3}{x + 4}}$ **c.** $\dfrac{\dfrac{5}{a} + \dfrac{b}{a^2}}{\dfrac{5a}{b^2} + \dfrac{1}{b}}$

Concept Check ☑

Which of the following are equivalent to $\dfrac{\dfrac{1}{x}}{\dfrac{3}{y}}$?

a. $\dfrac{1}{x} \div \dfrac{3}{y}$ **b.** $\dfrac{1}{x} \cdot \dfrac{y}{3}$ **c.** $\dfrac{1}{x} \div \dfrac{y}{3}$

OBJECTIVE 2 ▶ Simplifying complex fractions: Method 2. Next we look at another method of simplifying complex fractions. With this method we multiply the numerator and the denominator of the complex fraction by the LCD of all fractions in the complex fraction.

> **Simplifying a Complex Fraction: Method 2**
> **STEP 1.** Multiply the numerator and the denominator of the complex fraction by the LCD of the fractions in both the numerator and the denominator.
>
> **STEP 2.** Simplify.

Answer to Concept Check:
a and b

EXAMPLE 2 Simplify each complex fraction.

a. $\dfrac{\dfrac{5x}{x+2}}{\dfrac{10}{x-2}}$ **b.** $\dfrac{\dfrac{x}{y^2}+\dfrac{1}{y}}{\dfrac{y}{x^2}+\dfrac{1}{x}}$

Solution

a. The least common denominator of $\dfrac{5x}{x+2}$ and $\dfrac{10}{x-2}$ is $(x+2)(x-2)$. Multiply both the numerator, $\dfrac{5x}{x+2}$, and the denominator, $\dfrac{10}{x-2}$, by the LCD.

$$\dfrac{\dfrac{5x}{x+2}}{\dfrac{10}{x-2}} = \dfrac{\left(\dfrac{5x}{x+2}\right)\cdot(x+2)(x-2)}{\left(\dfrac{10}{x-2}\right)\cdot(x+2)(x-2)} \quad \text{Multiply numerator and denominator by the LCD.}$$

$$= \dfrac{5x\cdot(x-2)}{2\cdot5\cdot(x+2)} \quad \text{Simplify.}$$

$$= \dfrac{x(x-2)}{2(x+2)} \quad \text{Simplify.}$$

b. The least common denominator of $\dfrac{x}{y^2}, \dfrac{1}{y}, \dfrac{y}{x^2}$, and $\dfrac{1}{x}$ is x^2y^2.

$$\dfrac{\dfrac{x}{y^2}+\dfrac{1}{y}}{\dfrac{y}{x^2}+\dfrac{1}{x}} = \dfrac{\left(\dfrac{x}{y^2}+\dfrac{1}{y}\right)\cdot x^2y^2}{\left(\dfrac{y}{x^2}+\dfrac{1}{x}\right)\cdot x^2y^2} \quad \text{Multiply the numerator and denominator by the LCD.}$$

$$= \dfrac{\dfrac{x}{y^2}\cdot x^2y^2 + \dfrac{1}{y}\cdot x^2y^2}{\dfrac{y}{x^2}\cdot x^2y^2 + \dfrac{1}{x}\cdot x^2y^2} \quad \text{Use the distributive property.}$$

$$= \dfrac{x^3+x^2y}{y^3+xy^2} \quad \text{Simplify.}$$

$$= \dfrac{x^2(x+y)}{y^2(y+x)} \quad \text{Factor.}$$

$$= \dfrac{x^2}{y^2} \quad \text{Simplify.} \qquad \square$$

PRACTICE
2 Use Method 2 to simplify:

a. $\dfrac{\dfrac{8x}{x-4}}{\dfrac{3}{x+4}}$ **b.** $\dfrac{\dfrac{b}{a^2}+\dfrac{1}{a}}{\dfrac{a}{b^2}+\dfrac{1}{b}}$

OBJECTIVE 3 ▶ Simplifying expressions with negative exponents. If an expression contains negative exponents, write the expression as an equivalent expression with positive exponents.

EXAMPLE 3   Simplify.

$$\frac{x^{-1} + 2xy^{-1}}{x^{-2} - x^{-2}y^{-1}}$$

Solution   This fraction does not appear to be a complex fraction. If we write it by using only positive exponents, however, we see that it is a complex fraction.

$$\frac{x^{-1} + 2xy^{-1}}{x^{-2} - x^{-2}y^{-1}} = \frac{\dfrac{1}{x} + \dfrac{2x}{y}}{\dfrac{1}{x^2} - \dfrac{1}{x^2y}}$$

The LCD of $\dfrac{1}{x}, \dfrac{2x}{y}, \dfrac{1}{x^2}$, and $\dfrac{1}{x^2y}$ is x^2y. Multiply both the numerator and denominator by x^2y.

$$= \frac{\left(\dfrac{1}{x} + \dfrac{2x}{y}\right) \cdot x^2y}{\left(\dfrac{1}{x^2} - \dfrac{1}{x^2y}\right) \cdot x^2y}$$

$$= \frac{\dfrac{1}{x} \cdot x^2y + \dfrac{2x}{y} \cdot x^2y}{\dfrac{1}{x^2} \cdot x^2y - \dfrac{1}{x^2y} \cdot x^2y}$$ Apply the distributive property.

$$= \frac{xy + 2x^3}{y - 1} \quad \text{or} \quad \frac{x(y + 2x^2)}{y - 1}$$ Simplify.   □

PRACTICE
3   Simplify: $\dfrac{3x^{-1} + x^{-2}y^{-1}}{y^{-2} + xy^{-1}}$.

EXAMPLE 4   Simplify: $\dfrac{(2x)^{-1} + 1}{2x^{-1} - 1}$

Solution $\dfrac{(2x)^{-1} + 1}{2x^{-1} - 1} = \dfrac{\dfrac{1}{2x} + 1}{\dfrac{2}{x} - 1}$ Write using positive exponents.

▶ **Helpful Hint**

Don't forget that $(2x)^{-1} = \dfrac{1}{2x}$,

but $2x^{-1} = 2 \cdot \dfrac{1}{x} = \dfrac{2}{x}$.

$$= \frac{\left(\dfrac{1}{2x} + 1\right) \cdot 2x}{\left(\dfrac{2}{x} - 1\right) \cdot 2x}$$ The LCD of $\dfrac{1}{2x}$ and $\dfrac{2}{x}$ is $2x$.

$$= \frac{\dfrac{1}{2x} \cdot 2x + 1 \cdot 2x}{\dfrac{2}{x} \cdot 2x - 1 \cdot 2x}$$ Use the distributive property.

$$= \frac{1 + 2x}{4 - 2x} \quad \text{or} \quad \frac{1 + 2x}{2(2 - x)}$$ Simplify.   □

PRACTICE
4   Simplify: $\dfrac{(3x)^{-1} - 2}{5x^{-1} + 2}$.

VOCABULARY & READINESS CHECK

Complete the steps by writing the simplified complex fraction.

1. $\dfrac{\dfrac{7}{x}}{\dfrac{1}{x} + \dfrac{z}{x}} = \dfrac{x\left(\dfrac{7}{x}\right)}{x\left(\dfrac{1}{x}\right) + x\left(\dfrac{z}{x}\right)} = $ _____

2. $\dfrac{\dfrac{x}{4}}{\dfrac{x^2}{2} + \dfrac{1}{4}} = \dfrac{4\left(\dfrac{x}{4}\right)}{4\left(\dfrac{x^2}{2}\right) + 4\left(\dfrac{1}{4}\right)} = $ _____

Write each with positive exponents.

3. $x^{-2} = $ _____

4. $y^{-3} = $ _____

5. $2x^{-1} = $ _____

6. $(2x)^{-1} = $ _____

7. $(9y)^{-1} = $ _____

8. $9y^{-2} = $ _____

6.3 | EXERCISE SET

Simplify each complex fraction. See Examples 1 and 2.

1. $\dfrac{\dfrac{10}{3x}}{\dfrac{5}{6x}}$

2. $\dfrac{\dfrac{15}{2x}}{\dfrac{5}{6x}}$

3. $\dfrac{1 + \dfrac{2}{5}}{2 + \dfrac{3}{5}}$

4. $\dfrac{2 + \dfrac{1}{7}}{3 - \dfrac{4}{7}}$

5. $\dfrac{\dfrac{4}{x-1}}{\dfrac{x}{x-1}}$

6. $\dfrac{\dfrac{x}{x+2}}{\dfrac{2}{x+2}}$

7. $\dfrac{1 - \dfrac{2}{x}}{x + \dfrac{4}{9x}}$

8. $\dfrac{5 - \dfrac{3}{x}}{x + \dfrac{2}{3x}}$

9. $\dfrac{\dfrac{4x^2 - y^2}{xy}}{\dfrac{2}{y} - \dfrac{1}{x}}$

10. $\dfrac{\dfrac{x^2 - 9y^2}{xy}}{\dfrac{1}{y} - \dfrac{3}{x}}$

11. $\dfrac{\dfrac{x+1}{3}}{\dfrac{2x-1}{6}}$

12. $\dfrac{\dfrac{x+3}{12}}{\dfrac{4x-5}{15}}$

13. $\dfrac{\dfrac{2}{x} + \dfrac{3}{x^2}}{\dfrac{4}{x^2} - \dfrac{9}{x}}$

14. $\dfrac{\dfrac{2}{x^2} + \dfrac{1}{x}}{\dfrac{4}{x^2} - \dfrac{1}{x}}$

15. $\dfrac{\dfrac{1}{x} + \dfrac{2}{x^2}}{x + \dfrac{8}{x^2}}$

16. $\dfrac{\dfrac{1}{y} + \dfrac{3}{y^2}}{y + \dfrac{27}{y^2}}$

17. $\dfrac{\dfrac{4}{5-x} + \dfrac{5}{x-5}}{\dfrac{2}{x} + \dfrac{3}{x-5}}$

18. $\dfrac{\dfrac{3}{x-4} - \dfrac{2}{4-x}}{\dfrac{2}{x-4} - \dfrac{2}{x}}$

19. $\dfrac{\dfrac{x+2}{x} - \dfrac{2}{x-1}}{\dfrac{x+1}{x} + \dfrac{x+1}{x-1}}$

20. $\dfrac{\dfrac{5}{a+2} - \dfrac{1}{a-2}}{\dfrac{3}{2+a} + \dfrac{6}{2-a}}$

21. $\dfrac{\dfrac{2}{x} + 3}{\dfrac{4}{x^2} - 9}$

22. $\dfrac{2 + \dfrac{1}{x}}{4x - \dfrac{1}{x}}$

23. $\dfrac{1 - \dfrac{x}{y}}{\dfrac{x^2}{y^2} - 1}$

24. $\dfrac{1 - \dfrac{2}{x}}{x - \dfrac{4}{x}}$

25. $\dfrac{\dfrac{-2x}{x-y}}{\dfrac{y}{x^2}}$

26. $\dfrac{\dfrac{7y}{x^2 + xy}}{\dfrac{y^2}{x^2}}$

27. $\dfrac{\dfrac{2}{x} + \dfrac{1}{x^2}}{\dfrac{y}{x^2}}$

28. $\dfrac{\dfrac{5}{x^2} - \dfrac{2}{x}}{\dfrac{1}{x} + 2}$

29. $\dfrac{\dfrac{x}{9} - \dfrac{1}{x}}{1 + \dfrac{3}{x}}$

30. $\dfrac{\dfrac{x}{4} - \dfrac{4}{x}}{1 - \dfrac{4}{x}}$

31. $\dfrac{\dfrac{x-1}{x^2-4}}{1 + \dfrac{1}{x-2}}$

32. $\dfrac{\dfrac{x+3}{x^2-9}}{1 + \dfrac{1}{x-3}}$

33. $\dfrac{\dfrac{2}{x+5} + \dfrac{4}{x+3}}{\dfrac{3x+13}{x^2+8x+15}}$

34. $\dfrac{\dfrac{2}{x+2} + \dfrac{6}{x+7}}{\dfrac{4x+13}{x^2+9x+14}}$

Simplify. See Examples 3 and 4.

35. $\dfrac{x^{-1}}{x^{-2}+y^{-2}}$

36. $\dfrac{a^{-3}+b^{-1}}{a^{-2}}$

37. $\dfrac{2a^{-1}+3b^{-2}}{a^{-1}-b^{-1}}$

38. $\dfrac{x^{-1}+y^{-1}}{3x^{-2}+5y^{-2}}$

39. $\dfrac{1}{x-x^{-1}}$

40. $\dfrac{x^{-2}}{x+3x^{-1}}$

41. $\dfrac{a^{-1}+1}{a^{-1}-1}$

42. $\dfrac{a^{-1}-4}{4+a^{-1}}$

43. $\dfrac{3x^{-1}+(2y)^{-1}}{x^{-2}}$

44. $\dfrac{5x^{-2}-3y^{-1}}{x^{-1}+y^{-1}}$

45. $\dfrac{2a^{-1}+(2a)^{-1}}{a^{-1}+2a^{-2}}$

46. $\dfrac{a^{-1}+2a^{-2}}{2a^{-1}+(2a)^{-1}}$

47. $\dfrac{5x^{-1}+2y^{-1}}{x^{-2}y^{-2}}$

48. $\dfrac{x^{-2}y^{-2}}{5x^{-1}+2y^{-1}}$

49. $\dfrac{5x^{-1}-2y^{-1}}{25x^{-2}-4y^{-2}}$

50. $\dfrac{3x^{-1}+3y^{-1}}{4x^{-2}-9y^{-2}}$

REVIEW AND PREVIEW

Simplify. See Sections 5.1 and 5.2.

51. $\dfrac{3x^3y^2}{12x}$

52. $\dfrac{-36xb^3}{9xb^2}$

53. $\dfrac{144x^5y^5}{-16x^2y}$

54. $\dfrac{48x^3y^2}{-4xy}$

Solve the following. See Section 3.4.

55. $|x-5| = 9$

56. $|2y+1| = 1$

CONCEPT EXTENSIONS

Solve. See the Concept Check in this section.

57. Which of the following are equivalent to $\dfrac{\dfrac{x+1}{9}}{\dfrac{y-2}{5}}$?

a. $\dfrac{x+1}{9} \div \dfrac{y-2}{5}$　**b.** $\dfrac{x+1}{9} \cdot \dfrac{y-2}{5}$　**c.** $\dfrac{x+1}{9} \cdot \dfrac{5}{y-2}$

58. Which of the following are equivalent to $\dfrac{\dfrac{a}{7}}{\dfrac{b}{13}}$?

a. $\dfrac{a}{7} \cdot \dfrac{b}{13}$　**b.** $\dfrac{a}{7} \div \dfrac{b}{13}$　**c.** $\dfrac{a}{7} \div \dfrac{13}{b}$　**d.** $\dfrac{a}{7} \cdot \dfrac{13}{b}$

59. When the source of a sound is traveling toward a listener, the pitch that the listener hears due to the Doppler effect is given by the complex rational compression $\dfrac{a}{1-\dfrac{s}{770}}$, where a is the

actual pitch of the sound and s is the speed of the sound source. Simplify this expression.

60. In baseball, the earned run average (ERA) statistic gives the average number of earned runs scored on a pitcher per game. It is computed with the following expression: $\dfrac{E}{\dfrac{I}{9}}$, where E is the number of earned runs scored on a pitcher and I is the total number of innings pitched by the pitcher. Simplify this expression.

61. Which of the following are equivalent to $\dfrac{\dfrac{1}{x}}{\dfrac{3}{y}}$?

a. $\dfrac{1}{x} \div \dfrac{3}{y}$　**b.** $\dfrac{1}{x} \cdot \dfrac{y}{3}$　**c.** $\dfrac{1}{x} \div \dfrac{y}{3}$

62. In your own words, explain one method for simplifying a complex fraction.

Simplify.

63. $\dfrac{1}{1+(1+x)^{-1}}$

64. $\dfrac{(x+2)^{-1}+(x-2)^{-1}}{(x^2-4)^{-1}}$

65. $\dfrac{x}{1-\dfrac{1}{1+\dfrac{1}{x}}}$

66. $\dfrac{x}{1-\dfrac{1}{1-\dfrac{1}{x}}}$

67. $\dfrac{\dfrac{2}{y^2} - \dfrac{5}{xy} - \dfrac{3}{x^2}}{\dfrac{2}{y^2} + \dfrac{7}{xy} + \dfrac{3}{x^2}}$

68. $\dfrac{\dfrac{2}{x^2} - \dfrac{1}{xy} - \dfrac{1}{y^2}}{\dfrac{1}{x^2} - \dfrac{3}{xy} + \dfrac{2}{y^2}}$

69. $\dfrac{3(a + 1)^{-1} + 4a^{-2}}{(a^3 + a^2)^{-1}}$

70. $\dfrac{9x^{-1} - 5(x - y)^{-1}}{4(x - y)^{-1}}$

In the study of calculus, the difference quotient $\dfrac{f(a + h) - f(a)}{h}$ *is often found and simplified. Find and simplify this quotient for each function f(x) by following steps **a** through **d**.*

a. *Find* $(a + h)$.

b. *Find f(a).*

c. *Use steps **a** and **b** to find* $\dfrac{f(a + h) - f(a)}{h}$.

d. *Simplify the result of step **c**.*

71. $f(x) = \dfrac{1}{x}$

72. $f(x) = \dfrac{5}{x}$

73. $\dfrac{3}{x + 1}$

74. $\dfrac{2}{x^2}$

📖 STUDY SKILLS BUILDER

How Are You Doing?

If you haven't done so yet, take a few moments and think about how you are doing in this course. Are you working toward your goal of successfully completing this course? Is your performance on homework, quizzes, and tests satisfactory? If not, you might want to see your instructor to see if he/she has any suggestions on how you can improve your performance. Reread Section 1.1 for ideas on places to get help with your mathematics course.

Answer the following.

1. List any textbook supplements you are using to help you through this course.

2. List any campus resources you are using to help you through this course.

3. Write a short paragraph describing how you are doing in your mathematics course.

4. If improvement is needed, list ways that you can work toward improving your situation as described in Exercise 3.

6.4 DIVIDING POLYNOMIALS: LONG DIVISION AND SYNTHETIC DIVISION

OBJECTIVES

1 Divide a polynomial by a monomial.

2 Divide by a polynomial.

3 Use synthetic division to divide a polynomial by a binomial.

4 Use the remainder theorem to evaluate polynomials.

OBJECTIVE 1 ▶ Dividing a polynomial by a monomial. Recall that a rational expression is a quotient of polynomials. An equivalent form of a rational expression can be obtained by performing the indicated division. For example, the rational expression $\dfrac{10x^3 - 5x^2 + 20x}{5x}$ can be thought of as the polynomial $10x^3 - 5x^2 + 20x$ divided by the monomial $5x$. To perform this division of a polynomial by a monomial (which we do below), recall the following addition fact for fractions with a common denominator.

$$\frac{a}{c} + \frac{b}{c} = \frac{a + b}{c}$$

If a, b, and c are monomials, we might read this equation from right to left and gain insight into dividing a polynomial by a monomial.

Dividing a Polynomial by a Monomial
Divide each term in the polynomial by the monomial.

$$\frac{a + b}{c} = \frac{a}{c} + \frac{b}{c}, \quad \text{where } c \neq 0$$

EXAMPLE 1 Divide $10x^3 - 5x^2 + 20x$ by $5x$.

Solution We divide each term of $10x^3 - 5x^2 + 20x$ by $5x$ and simplify.

$$\frac{10x^3 - 5x^2 + 20x}{5x} = \frac{10x^3}{5x} - \frac{5x^2}{5x} + \frac{20x}{5x} = 2x^2 - x + 4$$

Check: To check, see that (quotient) (divisor) = dividend, or

$$(2x^2 - x + 4)(5x) = 10x^3 - 5x^2 + 20x. \qquad \square$$

PRACTICE
1 Divide $18a^3 - 12a^2 + 30a$ by $6a$.

EXAMPLE 2 Divide: $\dfrac{3x^5y^2 - 15x^3y - x^2y - 6x}{x^2y}$.

Solution We divide each term in the numerator by x^2y.

$$\frac{3x^5y^2 - 15x^3y - x^2y - 6x}{x^2y} = \frac{3x^5y^2}{x^2y} - \frac{15x^3y}{x^2y} - \frac{x^2y}{x^2y} - \frac{6x}{x^2y}$$

$$= 3x^3y - 15x - 1 - \frac{6}{xy} \qquad \square$$

PRACTICE
2 Divide: $\dfrac{5a^3b^4 - 8a^2b^3 + ab^2 - 8b}{ab^2}$.

OBJECTIVE 2 ▶ Dividing by a polynomial. To divide a polynomial by a polynomial other than a monomial, we use **long division.** Polynomial long division is similar to long division of real numbers. We review long division of real numbers by dividing 7 into 296.

$$
\begin{array}{r}
42 \\
7\overline{)296} \\
\underline{-28} \qquad 4(7) = 28. \\
16 \qquad \text{Subtract and bring down the next digit in the dividend.} \\
\underline{-14} \qquad 2(7) = 14. \\
2 \qquad \text{Subtract. The remainder is 2.}
\end{array}
$$

Divisor:

The quotient is $42\dfrac{2\ (\text{remainder})}{7\ (\text{divisor})}$.

Check: To check, notice that

$$42(7) + 2 = 296, \text{ the dividend.}$$

This same division process can be applied to polynomials, as shown next.

EXAMPLE 3 Divide $2x^2 - x - 10$ by $x + 2$.

Solution $2x^2 - x - 10$ is the dividend, and $x + 2$ is the divisor.

STEP 1. Divide $2x^2$ by x.

$$
\begin{array}{r}
2x \\
x + 2\overline{)2x^2 - x - 10}
\end{array}
\qquad \frac{2x^2}{x} = 2x, \text{ so } 2x \text{ is the first term of the quotient.}
$$

STEP 2. Multiply $2x(x + 2)$.

$$\begin{array}{r} 2x \\ x + 2\overline{)2x^2 - x - 10} \\ 2x^2 + 4x \end{array}$$

$2x(x + 2)$
Like terms are lined up vertically.

STEP 3. Subtract $(2x^2 + 4x)$ from $(2x^2 - x - 10)$ by changing the signs of $(2x^2 + 4x)$ and adding.

$$\begin{array}{r} 2x \\ x + 2\overline{)\ 2x^2 - x - 10} \\ \underline{\not\mp 2x^2 \not\mp 4x} \\ -5x \end{array}$$

STEP 4. Bring down the next term, -10, and start the process over.

$$\begin{array}{r} 2x \\ x + 2\overline{)\ 2x^2 - x - 10} \\ \underline{\not\mp 2x^2 \not\mp 4x}\quad \downarrow \\ -5x - 10 \end{array}$$

STEP 5. Divide $-5x$ by x.

$$\begin{array}{r} 2x - 5 \\ x + 2\overline{)\ 2x^2 - x - 10} \\ \underline{\not\mp 2x^2 \not\mp 4x} \\ -5x - 10 \end{array}$$

$\dfrac{-5x}{x} = -5$, so -5 is the second term of the quotient.

STEP 6. Multiply $-5(x + 2)$.

$$\begin{array}{r} 2x - 5 \\ x + 2\overline{)\ 2x^2 - x - 10} \\ \underline{\not\mp 2x^2 \not\mp 4x} \\ -5x - 10 \\ \underline{-5x - 10} \end{array}$$

Multiply: $-5(x + 2)$. Like terms are lined up vertically.

STEP 7. Subtract by changing signs of $-5x - 10$ and adding.

$$\begin{array}{r} 2x - 5 \\ x + 2\overline{)\ 2x^2 - x - 10} \\ \underline{\not\mp 2x^2 \not\mp 4x} \\ -5x - 10 \\ \underline{\not\pm 5x \not\pm 10} \\ 0 \end{array}$$

Subtract.
Remainder

By graphing $y_1 = 2x^2 - x - 10$ and $y_2 = x + 2$ in the same window, we see that the graphs share the x-intercept $x = -2$. We know this means that $x + 2$ is a common factor of $2x^2 - x - 10$ and $x + 2$. In other words, $x + 2$ divides into $2x^2 - x - 10$ evenly without a remainder.

Then $\dfrac{2x^2 - x - 10}{x + 2} = 2x - 5$. There is no remainder.

Check: Check this result by multiplying $2x - 5$ by $x + 2$. Their product is $(2x - 5)(x + 2) = 2x^2 - x - 10$, the dividend. □

PRACTICE 3 Divide $3x^2 + 7x - 6$ by $x + 3$.

EXAMPLE 4   Divide: $(6x^2 - 19x + 12) \div (3x - 5)$.

Solution

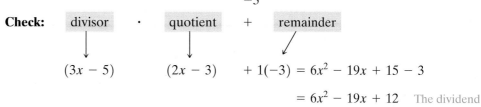

$$\begin{array}{r} 2x \\ 3x - 5 \overline{)6x^2 - 19x + 12} \\ \underline{6x^2 - 10x} \quad \downarrow \\ -9x + 12 \end{array}$$

Divide $\dfrac{6x^2}{3x} = 2x$.

Multiply $2x(3x - 5)$.
Subtract by adding the opposite.
Bring down the next term, $+12$.

$$\begin{array}{r} 2x - 3 \\ 3x - 5 \overline{)6x^2 - 19x + 12} \\ \underline{6x^2 - 10x} \\ -9x + 12 \\ \underline{-9x - 15} \\ -3 \end{array}$$

Divide $\dfrac{-9x}{3x} = -3$.

Multiply $-3(3x - 5)$.
Subtract by adding the opposite.

Check: divisor   ·   quotient   +   remainder

$(3x - 5)$ $(2x - 3)$ $+ 1(-3) = 6x^2 - 19x + 15 - 3$

$= 6x^2 - 19x + 12$ The dividend

The division checks, so

$$\frac{6x^2 - 19x + 12}{3x - 5} = 2x - 3 + \frac{-3}{3x - 5}$$

$$\text{or} \quad 2x - 3 - \frac{3}{3x - 5}$$

▶ **Helpful Hint**
This fraction is the remainder over the divisor.

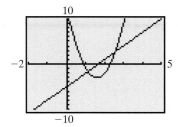

Notice that the graphs of $y_1 = 6x^2 - 19x + 12$ and $y_2 = 3x - 5$ do not have a common x-intercept. Therefore, $3x - 5$ does not divide into $6x^2 - 19x + 12$ evenly, and there will be a remainder.

PRACTICE
4   Divide $(6x^2 - 7x + 2)$ by $(2x - 1)$.

EXAMPLE 5   Divide: $(7x^3 + 16x^2 + 2x - 1) \div (x + 4)$.

Solution

$$\begin{array}{r} 7x^2 - 12x + 50 \\ x + 4 \overline{)7x^3 + 16x^2 + 2x - 1} \\ \underline{7x^3 - 28x^2} \\ -12x^2 + 2x \\ \underline{-12x^2 - 48x} \\ 50x - 1 \\ \underline{50x - 200} \\ -201 \end{array}$$

Divide $\dfrac{7x^3}{x} = 7x^2$.

$7x^2(x + 4)$

Subtract. Bring down $2x$.

$\dfrac{-12x^2}{x} = -12x$, a term of the quotient.

$-12x(x + 4)$ Subtract. Bring down -1.

$\dfrac{50x}{x} = 50$, a term of the quotient.

$50(x + 4)$.   Subtract.

Thus, $\dfrac{7x^3 + 16x^2 + 2x - 1}{x + 4} = 7x^2 - 12x + 50 + \dfrac{-201}{x + 4}$ or

$$7x^2 - 12x + 50 - \frac{201}{x + 4}.$$

PRACTICE
5   Divide $(5x^3 + 9x^2 - 10x + 30) \div (x + 3)$.

EXAMPLE 6 Divide $3x^4 + 2x^3 - 8x + 6$ by $x^2 - 1$.

Solution Before dividing, we represent any "missing powers" by the product of 0 and the variable raised to the missing power. There is no x^2 term in the dividend, so we include $0x^2$ to represent the missing term. Also, there is no x term in the divisor, so we include $0x$ in the divisor.

$$
\begin{array}{r}
3x^2 + 2x + 3 \\
x^2 + 0x - 1 \overline{)3x^4 + 2x^3 + 0x^2 - 8x + 6} \\
\underline{3x^4 + 0x^3 - 3x^2} \\
2x^3 + 3x^2 - 8x \\
\underline{2x^3 + 0x^2 - 2x} \\
3x^2 - 6x + 6 \\
\underline{3x^2 + 0x - 3} \\
-6x + 9
\end{array}
$$

$\dfrac{3x^4}{x^2} = 3x^2$

$3x^2(x^2 + 0x - 1)$

Subtract. Bring down $-8x$.

$\dfrac{2x^3}{x^2} = 2x$, a term of the quotient.

$2x(x^2 + 0x - 1)$

Subtract. Bring down 6.

$\dfrac{3x^2}{x^2} = 3$, a term of the quotient.

$3(x^2 + 0x - 1)$

Subtract.

The division process is finished when the degree of the remainder polynomial is less than the degree of the divisor. Thus,

$$
\frac{3x^4 + 2x^3 - 8x + 6}{x^2 - 1} = 3x^2 + 2x + 3 + \frac{-6x + 9}{x^2 - 1}
$$

PRACTICE
6 Divide $2x^4 + 3x^3 - 5x + 2$ by $x^2 + 1$.

EXAMPLE 7 Divide $27x^3 + 8$ by $3x + 2$.

Solution We replace the missing terms in the dividend with $0x^2$ and $0x$.

$$
\begin{array}{r}
9x^2 - 6x + 4 \\
3x + 2 \overline{)27x^3 + 0x^2 + 0x + 8} \\
\underline{27x^3 + 18x^2} \\
-18x^2 + 0x \\
\underline{-18x^2 - 12x} \\
12x + 8 \\
\underline{12x + 8}
\end{array}
$$

$9x^2(3x + 2)$

Subtract. Bring down $0x$.

$-6x(3x + 2)$

Subtract. Bring down 8.

$4(3x + 2)$

Thus, $\dfrac{27x^3 + 8}{3x + 2} = 9x^2 - 6x + 4$.

PRACTICE
7 Divide $64x^3 - 125$ by $4x - 5$.

Concept Check ☑

In a division problem, the divisor is $4x^3 - 5$. The division process can be stopped when which of these possible remainder polynomials is reached?

a. $2x^4 + x^2 - 3$　　**b.** $x^3 - 5^2$　　**c.** $4x^2 + 25$

OBJECTIVE 3 ▶ Using synthetic division to divide a polynomial by a binomial. When a polynomial is to be divided by a binomial of the form $x - c$, a shortcut process called **synthetic division** may be used. On the left is an example of long division, and on the right, the same example showing the coefficients of the variables only.

$$
\begin{array}{r}
2x^2 + 5x + 2 \\
x - 3\overline{)2x^3 - x^2 - 13x + 1} \\
\underline{2x^3 - 6x^2} \\
5x^2 - 13x \\
\underline{5x^2 - 15x} \\
2x + 1 \\
\underline{2x - 6} \\
7
\end{array}
\qquad
\begin{array}{r}
2 \quad 5 \quad 2 \\
1 - 3\overline{)2 - 1 - 13 + 1} \\
\underline{2 - 6} \\
5 - 13 \\
\underline{5 - 15} \\
2 + 1 \\
\underline{2 - 6} \\
7
\end{array}
$$

Notice that as long as we keep coefficients of powers of x in the same column, we can perform division of polynomials by performing algebraic operations on the coefficients only. This shortcut process of dividing with coefficients only in a special format is called synthetic division. To find $(2x^3 - x^2 - 13x + 1) \div (x - 3)$ by synthetic division, follow the next example.

EXAMPLE 8   Use synthetic division to divide $2x^3 - x^2 - 13x + 1$ by $x - 3$.

Solution   To use synthetic division, the divisor must be in the form $x - c$. Since we are dividing by $x - 3$, c is 3. Write down 3 and the coefficients of the dividend.

$$
\begin{array}{c|cccc}
c & & & & \\
3 & 2 & -1 & -13 & 1 \\
& & & & \\
\hline
& 2 & & &
\end{array}
$$

Next, draw a line and bring down the first coefficient of the dividend.

$$
\begin{array}{c|cccc}
3 & 2 & -1 & -13 & 1 \\
& & 6 & & \\
\hline
& 2 & & &
\end{array}
$$

Multiply $3 \cdot 2$ and write down the product, 6.

$$
\begin{array}{c|cccc}
3 & 2 & -1 & -13 & 1 \\
& & 6 & & \\
\hline
& 2 & 5 & &
\end{array}
$$

Add $-1 + 6$. Write down the sum, 5.

$$
\begin{array}{c|cccc}
3 & 2 & -1 & -13 & 1 \\
& & 6 & 15 & \\
\hline
& 2 & 5 & 2 &
\end{array}
$$

$3 \cdot 5 = 15$.

$-13 + 15 = 2$.

$$
\begin{array}{c|cccc}
3 & 2 & -1 & -13 & 1 \\
& & 6 & 15 & 6 \\
\hline
& 2 & 5 & 2 & 7
\end{array}
$$

$3 \cdot 2 = 6$.

$1 + 6 = 7$.

The quotient is found in the bottom row. The numbers 2, 5, and 2 are the coefficients of the quotient polynomial, and the number 7 is the remainder. The degree of the quotient polynomial is one less than the degree of the dividend. In our example, the degree of the dividend is 3, so the degree of the quotient polynomial is 2. As we found when we performed the long division, the quotient is

$$2x^2 + 5x + 2, \quad \text{remainder 7}$$

or

$$2x^2 + 5x + 2 + \frac{7}{x - 3} \qquad \square$$

PRACTICE

8   Use synthetic division to divide $4x^3 - 3x^2 + 6x + 5$ by $x - 1$.

EXAMPLE 9 Use synthetic division to divide $x^4 - 2x^3 - 11x^2 + 5x + 34$ by $x + 2$.

Solution The divisor is $x + 2$, which we write in the form $x - c$ as $x - (-2)$. Thus, c is -2. The dividend coefficients are $1, -2, -11, 5$, and 34.

$$
\begin{array}{r|rrrrr}
-2 & 1 & -2 & -11 & 5 & 34 \\
& & -2 & 8 & 6 & -22 \\
\hline
& 1 & -4 & -3 & 11 & 12
\end{array}
$$

The dividend is a fourth-degree polynomial, so the quotient polynomial is a third-degree polynomial. The quotient is $x^3 - 4x^2 - 3x + 11$ with a remainder of 12. Thus,

$$\frac{x^4 - 2x^3 - 11x^2 + 5x + 34}{x + 2} = x^3 - 4x^2 - 3x + 11 + \frac{12}{x + 2}$$

PRACTICE

9 Use synthetic division to divide $x^4 + 3x^3 - 5x^2 + 6x + 12$ by $x + 3$.

Concept Check ✓

Which division problems are candidates for the synthetic division process?

a. $(3x^2 + 5) \div (x + 4)$ **b.** $(x^3 - x^2 + 2) \div (3x^3 - 2)$

c. $(y^4 + y - 3) \div (x^2 + 1)$ **d.** $x^5 \div (x - 5)$

▶ **Helpful Hint**

Before dividing by synthetic division, write the dividend in descending order of variable exponents. Any "missing powers" of the variable should be represented by 0 times the variable raised to the missing power.

DISCOVER THE CONCEPT

Let $P(x) = 2x^3 - 4x^2 + 5$.

a. Find $P(2)$ by substitution.
b. Use synthetic division to find the remainder when $P(x)$ is divided by $x - 2$.
c. Compare the results of parts **a** and **b**.
d. Now read Example 10 below.

EXAMPLE 10 If $P(x) = 2x^3 - 4x^2 + 5$,

a. Find $P(2)$ by substitution.
b. Use synthetic division to find the remainder when $P(x)$ is divided by $x - 2$.

Solution

a. $P(x) = 2x^3 - 4x^2 + 5$
$P(2) = 2(2)^3 - 4(2)^2 + 5$
$\quad = 2(8) - 4(4) + 5 = 16 - 16 + 5 = 5$

Thus, $P(2) = 5$.

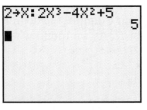

A calculator check for Example 10 showing that $P(2) = 5$.

b. The coefficients of $P(x)$ are $2, -4, 0$, and 5. The number 0 is a coefficient of the missing power of x^1. The divisor is $x - 2$, so c is 2.

The remainder when $P(x)$ is divided by $x - 2$ is 5.

PRACTICE
10   If $P(x) = x^3 - 5x - 2$,

a. Find $P(2)$ by substitution.

b. Use synthetic division to find the remainder when $P(x)$ is divided by $x - 2$.

OBJECTIVE 4 ▶ Using the remainder theorem to evaluate polynomials. Notice in the preceding example that $P(2) = 5$ and that the remainder when $P(x)$ is divided by $x - 2$ is 5. This is no accident. This illustrates the **remainder theorem.**

> **Remainder Theorem**
> If a polynomial $P(x)$ is divided by $x - c$, then the remainder is $P(c)$.

EXAMPLE 11   Use the remainder theorem and synthetic division to find $P(4)$ if
$$P(x) = 4x^6 - 25x^5 + 35x^4 + 17x^2.$$

Solution   To find $P(4)$ by the remainder theorem, we divide $P(x)$ by $x - 4$. The coefficients of $P(x)$ are 4, −25, 35, 0, 17, 0, and 0. Also, c is 4.

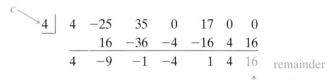

Thus, $P(4) = 16$, the remainder.   □

PRACTICE
11   Use the remainder theorem and synthetic division to find $P(3)$ if $P(x) = 2x^5 - 18x^4 + 90x^2 + 59x$.

6.4 | EXERCISE SET

Divide. See Examples 1 and 2.

1. $4a^2 + 8a$ by $2a$

2. $6x^4 - 3x^3$ by $3x^2$

3. $\dfrac{12a^5b^2 + 16a^4b}{4a^4b}$

4. $\dfrac{4x^3y + 12x^2y^2 - 4xy^3}{4xy}$

5. $\dfrac{4x^2y^2 + 6xy^2 - 4y^2}{2x^2y}$

6. $\dfrac{6x^5y + 75x^4y - 24x^3y^2}{3x^4y}$

Divide. See Examples 3 through 7.

7. $(x^2 + 3x + 2) \div (x + 2)$

8. $(y^2 + 7y + 10) \div (y + 5)$

9. $(2x^2 - 6x - 8) \div (x + 1)$

10. $(3x^2 + 19x + 20) \div (x + 5)$

11. $2x^2 + 3x - 2$ by $2x + 4$

12. $6x^2 - 17x - 3$ by $3x - 9$

13. $(4x^3 + 7x^2 + 8x + 20) \div (2x + 4)$

14. $(8x^3 + 18x^2 + 16x + 24) \div (4x + 8)$

15. $(2x^2 + 6x^3 - 18x - 6) \div (3x + 1)$

16. $(4x - 15x^2 + 10x^3 - 6) \div (2x - 3)$

17. $(3x^5 - x^3 + 4x^2 - 12x - 8) \div (x^2 - 2)$

18. $(2x^5 - 6x^4 + x^3 - 4x + 3) \div (x^2 - 3)$

19. $\left(2x^4 + \dfrac{1}{2}x^3 + x^2 + x\right) \div (x - 2)$

20. $\left(x^4 - \dfrac{2}{3}x^3 + x\right) \div (x - 3)$

Use synthetic division to divide. See Examples 8 and 9.

21. $\dfrac{x^2 + 3x - 40}{x - 5}$

22. $\dfrac{x^2 - 14x + 24}{x - 2}$

23. $\dfrac{x^2 + 5x - 6}{x + 6}$

24. $\dfrac{x^2 + 12x + 32}{x + 4}$

25. $\dfrac{x^3 - 7x^2 - 13x + 5}{x - 2}$

26. $\dfrac{x^3 + 6x^2 + 4x - 7}{x + 5}$

27. $\dfrac{4x^2 - 9}{x - 2}$

28. $\dfrac{3x^2 - 4}{x - 1}$

MIXED PRACTICE

Divide. See Examples 1 through 9.

29. $\dfrac{4x^7y^4 + 8xy^2 + 4xy^3}{4xy^3}$

30. $\dfrac{15x^3y - 5x^2y + 10xy^2}{5x^2y}$

31. $(10x^3 - 5x^2 - 12x + 1) \div (2x - 1)$

32. $(20x^3 - 8x^2 + 5x - 5) \div (5x - 2)$

33. $(2x^3 - 6x^2 - 4) \div (x - 4)$

34. $(3x^3 + 4x - 10) \div (x + 2)$

35. $\dfrac{2x^4 - 13x^3 + 16x^2 - 9x + 20}{x - 5}$

36. $\dfrac{3x^4 + 5x^3 - x^2 + x - 2}{x + 2}$

37. $\dfrac{7x^2 - 4x + 12 + 3x^3}{x + 1}$

38. $\dfrac{4x^3 + x^4 - x^2 - 16x - 4}{x - 2}$

39. $\dfrac{3x^3 + 2x^2 - 4x + 1}{x - \dfrac{1}{3}}$

40. $\dfrac{9y^3 + 9y^2 - y + 2}{y + \dfrac{2}{3}}$

41. $\dfrac{x^3 - 1}{x - 1}$ **42.** $\dfrac{y^3 - 8}{y - 2}$

43. $(25xy^2 + 75xyz + 125x^2yz) \div (-5x^2y)$

44. $(x^6y^6 - x^3y^3z + 7x^3y) \div (-7yz^2)$

45. $(9x^5 + 6x^4 - 6x^2 - 4x) \div (3x + 2)$

46. $(5x^4 - 5x^2 + 10x^3 - 10x) \div (5x + 10)$

For the given polynomial $P(x)$ and the given c, use the remainder theorem to find $P(c)$. See Examples 10 and 11.

47. $P(x) = x^3 + 3x^2 - 7x + 4; 1$

48. $P(x) = x^3 + 5x^2 - 4x - 6; 2$

49. $P(x) = 3x^3 - 7x^2 - 2x + 5; -3$

50. $P(x) = 4x^3 + 5x^2 - 6x - 4; -2$

51. $P(x) = 4x^4 + x^2 - 2; -1$

52. $P(x) = x^4 - 3x^2 - 2x + 5; -2$

53. $P(x) = 2x^4 - 3x^2 - 2; \dfrac{1}{3}$

54. $P(x) = 4x^4 - 2x^3 + x^2 - x - 4; \dfrac{1}{2}$

55. $P(x) = x^5 + x^4 - x^3 + 3; \dfrac{1}{2}$

56. $P(x) = x^5 - 2x^3 + 4x^2 - 5x + 6; \dfrac{2}{3}$

REVIEW AND PREVIEW

Solve each equation for x. See Sections 1.5 and 5.8.

57. $7x + 2 = x - 3$ **58.** $4 - 2x = 17 - 5x$

59. $x^2 = 4x - 4$ **60.** $5x^2 + 10x = 15$

61. $\dfrac{x}{3} - 5 = 13$ **62.** $\dfrac{2x}{9} + 1 = \dfrac{7}{9}$

Factor the following. See Sections 5.5 and 5.7.

63. $x^3 - 1$

64. $8y^3 + 1$

65. $125z^3 + 8$

66. $a^3 - 27$

67. $xy + 2x + 3y + 6$

68. $x^2 - x + xy - y$

69. $x^3 - 9x$

70. $2x^3 - 32x$

CONCEPT EXTENSIONS

Which division problems are candidates for the synthetic division process? See the Concept Checks in this section.

71. $(5x^2 - 3x + 2) \div (x + 2)$

72. $(x^4 - 6) \div (x^3 + 3x - 1)$

73. $(x^7 - 2) \div (x^5 + 1)$

74. $(3x^2 + 7x - 1) \div \left(x - \dfrac{1}{3}\right)$

75. In a long division exercise, if the divisor is $9x^3 - 2x$, then the division process can be stopped when the degree of the remainder is
   a. 1   **b.** 3   **c.** 9   **d.** 2

76. In a division exercise, if the divisor is $x - 3$, then the division process can be stopped when the degree of the remainder is
   a. 1   **b.** 0   **c.** 2   **d.** 3

△ **77.** A board of length $(3x^4 + 6x^2 - 18)$ meters is to be cut into three pieces of the same length. Find the length of each piece.

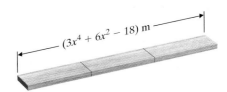

△ **78.** The perimeter of a regular hexagon is given to be $(12x^5 - 48x^3 + 3)$ miles. Find the length of each side.

△ **79.** If the area of the rectangle is $(15x^2 - 29x - 14)$ square inches, and its length is $(5x + 2)$ inches, find its width.

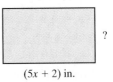

$(5x + 2)$ in.

△ **80.** If the area of a parallelogram is $(2x^2 - 17x + 35)$ square centimeters and its base is $(2x - 7)$ centimeters, find its height.

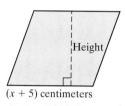

(2x − 7) cm

△ **81.** If the area of a parallelogram is $(x^4 - 23x^2 + 9x - 5)$ square centimeters and its base is $(x + 5)$ centimeters, find its height.

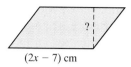

Height

(x + 5) centimeters

△ **82.** If the volume of a box is $(x^4 + 6x^3 - 7x^2)$ cubic meters, its height is x^2 meters, and its length is $(x + 7)$ meters, find its width.

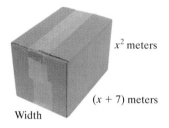

x^2 meters

(x + 7) meters

Width

Divide.

83. $\left(x^4 + \dfrac{2}{3}x^3 + x\right) \div (x - 1)$

84. $\left(2x^3 + \dfrac{9}{2}x^2 - 4x - 10\right) \div (x + 2)$

85. $\left(3x^4 - x - x^3 + \dfrac{1}{2}\right) \div (2x - 1)$

86. $\left(2x^4 + \dfrac{1}{2}x^3 - \dfrac{1}{4}x^2 + x\right) \div (2x + 1)$

87. $(5x^4 - 2x^2 + 10x^3 - 4x) \div (5x + 10)$

88. $(9x^5 + 6x^4 - 6x^2 - 4x) \div (3x + 2)$

For each given f(x) and g(x), find $\dfrac{f(x)}{g(x)}$. Also find any x-values that are not in the domain of $\dfrac{f(x)}{g(x)}$. (Note: Since g(x) is in the denominator, g(x) cannot be 0.)

89. $f(x) = 25x^2 - 5x + 30; g(x) = 5x$

90. $f(x) = 12x^4 - 9x^3 + 3x - 1; g(x) = 3x$

91. $f(x) = 7x^4 - 3x^2 + 2; g(x) = x - 2$

92. $f(x) = 2x^3 - 4x^2 + 1; g(x) = x + 3$

93. Try performing the following division without changing the order of the terms. Describe why this makes the process more complicated. Then perform the division again after putting the terms in the dividend in descending order of exponents.

$$\frac{4x^2 - 12x - 12 + 3x^3}{x - 2}$$

94. Explain how to check polynomial long division.

95. Explain an advantage of using the remainder theorem instead of direct substitution.

96. Explain an advantage of using synthetic division instead of long division.

We say that 2 is a factor of 8 because 2 divides 8 evenly, or with a remainder of 0. In the same manner, the polynomial $x - 2$ is a factor of the polynomial $x^3 - 14x^2 + 24x$ because the remainder is 0 when $x^3 - 14x^2 + 24x$ is divided by $x - 2$. Use this information for Exercises 97 through 99.

97. Use synthetic division to show that $x + 3$ is a factor of $x^3 + 3x^2 + 4x + 12$.

98. Use synthetic division to show that $x - 2$ is a factor of $x^3 - 2x^2 - 3x + 6$.

99. From the remainder theorem, the polynomial $x - c$ is a factor of a polynomial function $P(x)$ if $P(c)$ is what value?

100. If a polynomial is divided by $x - 5$, the quotient is $2x^2 + 5x - 6$ and the remainder is 3. Find the original polynomial.

101. If a polynomial is divided by $x + 3$, the quotient is $x^2 - x + 10$ and the remainder is -2. Find the original polynomial.

102. eBay is the leading online auction house. eBay's annual net profit can be modeled by the polynomial function $P(x) = -7x^3 + 94x^2 - 76x + 59$, where $P(x)$ is net profit in millions of dollars and x is the year after 2000. eBay's annual revenue can be modeled by the function $R(x) = 939x - 194$, where $R(x)$ is revenue in millions of dollars and x is years since 2000. (*Source: eBay, Inc.*)

a. Given that

$$\text{Net profit margin} = \frac{\text{net profit}}{\text{revenue}},$$

write a function, $m(x)$, that models eBay's net profit margin.

b. Use part **a** to predict eBay's profit margin in 2010. Round to the nearest hundredth.

6.5 SOLVING EQUATIONS CONTAINING RATIONAL EXPRESSIONS

OBJECTIVE

1 Solve equations containing rational expressions.

> ▶ **Helpful Hint**
>
> The method described here is for equations only. It may *not* be used for performing operations on expressions.

OBJECTIVE 1 ▶ Solving equations containing rational expressions. In this section, we solve equations containing rational expressions. Before beginning this section, make sure that you understand the difference between an *equation* and an *expression*. An **equation** contains an equal sign and an **expression** does not.

Equation	*Expression*
$\dfrac{x}{2} + \dfrac{x}{6} = \dfrac{2}{3}$	$\dfrac{x}{2} + \dfrac{x}{6}$

Solving Equations Containing Rational Expressions

To solve *equations* containing rational expressions, first clear the equation of fractions by multiplying both sides of the equation by the LCD of all rational expressions. Then solve as usual.

Concept Check ☑

True or false? Clearing fractions is valid when solving an equation and when simplifying rational expressions. Explain.

EXAMPLE 1 Solve $\dfrac{4x}{5} + \dfrac{3}{2} = \dfrac{3x}{10}$ algebraically and check graphically.

Solution The LCD of $\dfrac{4x}{5}, \dfrac{3}{2}$, and $\dfrac{3x}{10}$ is 10. We multiply both sides of the equation by 10.

$$\frac{4x}{5} + \frac{3}{2} = \frac{3x}{10}$$

$$10\left(\frac{4x}{5} + \frac{3}{2}\right) = 10\left(\frac{3x}{10}\right) \quad \text{Multiply both sides by the LCD.}$$

$$10 \cdot \frac{4x}{5} + 10 \cdot \frac{3}{2} = 10 \cdot \frac{3x}{10} \quad \text{Use the distributive property.}$$

$$8x + 15 = 3x \quad \text{Simplify.}$$

$$15 = -5x \quad \text{Subtract } 8x \text{ from both sides.}$$

$$-3 = x \quad \text{Solve.}$$

Check: To check this solution graphically, we use the intersection-of-graphs method shown below.

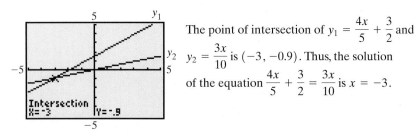

The point of intersection of $y_1 = \dfrac{4x}{5} + \dfrac{3}{2}$ and $y_2 = \dfrac{3x}{10}$ is $(-3, -0.9)$. Thus, the solution of the equation $\dfrac{4x}{5} + \dfrac{3}{2} = \dfrac{3x}{10}$ is $x = -3$.

The solution is -3 or the solution set is $\{-3\}$.

PRACTICE
1 Solve: $\dfrac{5x}{4} - \dfrac{3}{2} = \dfrac{7x}{8}$.

Answer to Concept Check:
false; answers may vary

The important difference of the equations in this section is that the denominator of a rational expression may contain a variable. Recall that a rational expression is undefined for values of the variable that make the denominator 0. If a proposed solution makes the denominator 0, then it must be rejected as a solution of the original equation. Such proposed solutions are called **extraneous solutions.**

EXAMPLE 2   Solve $\dfrac{3}{x} - \dfrac{x+21}{3x} = \dfrac{5}{3}$ algebraically and check graphically.

Solution   The LCD of the denominators x, $3x$, and 3 is $3x$. We multiply both sides by $3x$.

$$\frac{3}{x} - \frac{x+21}{3x} = \frac{5}{3}$$

$$3x\left(\frac{3}{x} - \frac{x+21}{3x}\right) = 3x\left(\frac{5}{3}\right) \qquad \text{Multiply both sides by the LCD.}$$

$$3x \cdot \frac{3}{x} - 3x \cdot \frac{x+21}{3x} = 3x \cdot \frac{5}{3} \qquad \text{Use the distributive property.}$$

$$9 - (x + 21) = 5x \qquad \text{Simplify.}$$

$$9 - x - 21 = 5x$$

$$-12 = 6x$$

$$-2 = x \qquad \text{Solve.}$$

The proposed solution is -2.

Check: A check using a graphing utility is shown below.

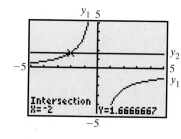

The intersection of $y_1 = \dfrac{3}{x} - \dfrac{x+21}{3x}$ and $y_2 = \dfrac{5}{3}$ verifies that the solution of the equation $\dfrac{3}{x} - \dfrac{x+21}{3x} = \dfrac{5}{3}$ is $x = -2$.

The solution is -2 or the solution set is $\{-2\}$. □

PRACTICE
2   Solve: $\dfrac{6}{x} - \dfrac{x+9}{5x} = \dfrac{2}{5}$.

The following steps may be used to solve equations containing rational expressions.

Solving an Equation Containing Rational Expressions

STEP 1. Multiply both sides of the equation by the LCD of all rational expressions in the equation.

STEP 2. Simplify both sides.

STEP 3. Determine whether the equation is linear, quadratic, or higher degree and solve accordingly.

STEP 4. Check the solution in the original equation.

EXAMPLE 3 Solve: $\dfrac{x+6}{x-2} = \dfrac{2(x+2)}{x-2}$.

Solution First we multiply both sides of the equation by the LCD, $x-2$. (Remember, we can only do this if $x \neq 2$ so that we are not multiplying by 0.)

$$\frac{x+6}{x-2} = \frac{2(x+2)}{x-2}$$

$$(x-2) \cdot \frac{x+6}{x-2} = (x-2) \cdot \frac{2(x+2)}{x-2} \qquad \text{Multiply both sides by } x-2.$$

$$x + 6 = 2(x+2) \qquad \text{Simplify.}$$

$$x + 6 = 2x + 4 \qquad \text{Use the distributive property.}$$

$$2 = x \qquad \text{Solve.}$$

TECHNOLOGY NOTE

Recall that when replacing the variable with a number that results in a zero denominator, a graphing utility will give an error message. In Example 3, store 2 in x, evaluate $\dfrac{(x+6)}{(x-2)}$, and see the results.

From above, we assumed that $x \neq 2$, so this equation has no solution. This will also show as we attempt to check this proposed solution.

Check: The proposed solution is 2. Notice that 2 makes a denominator 0 in the original equation. This can also be seen in a check. Check the proposed solution 2 in the original equation.

$$\frac{x+6}{x-2} = \frac{2(x+2)}{x-2}$$

$$\frac{2+6}{2-2} = \frac{2(2+2)}{2-2}$$

$$\frac{8}{0} = \frac{2(4)}{0}$$

The denominators are 0, so 2 is not a solution of the original equation. The solution is $\{\ \}$ or $\varnothing$.

PRACTICE
3 Solve: $\dfrac{x-5}{x+3} = \dfrac{2(x-1)}{x+3}$.

EXAMPLE 4 Solve: $\dfrac{2x}{2x-1} + \dfrac{1}{x} = \dfrac{1}{2x-1}$.

Solution The LCD is $x(2x-1)$. Multiply both sides by $x(2x-1)$. By the distributive property, this is the same as multiplying each term by $x(2x-1)$.

$$x(2x-1) \cdot \frac{2x}{2x-1} + x(2x-1) \cdot \frac{1}{x} = x(2x-1) \cdot \frac{1}{2x-1}$$

$$x(2x) + (2x-1) = x \qquad \text{Simplify.}$$

$$2x^2 + 2x - 1 - x = 0$$

$$2x^2 + x - 1 = 0$$

$$(x+1)(2x-1) = 0$$

$$x + 1 = 0 \quad \text{or} \quad 2x - 1 = 0$$

$$x = -1 \qquad\qquad x = \frac{1}{2}$$

The number $\dfrac{1}{2}$ makes the denominator $2x-1$ equal 0, so it is not a solution. The solution is -1.

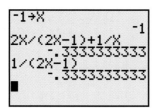

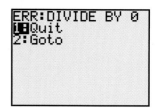

A numerical check that -1 is the solution to the equation in Example 4.

The error message that results when $\frac{1}{2}$ is checked as a proposed solution in Example 4.   □

PRACTICE 4   Solve: $\dfrac{5x}{5x - 1} + \dfrac{1}{x} = \dfrac{1}{5x - 1}$.

EXAMPLE 5   Solve: $\dfrac{2x}{x - 3} + \dfrac{6 - 2x}{x^2 - 9} = \dfrac{x}{x + 3}$.

Solution   We factor the second denominator to find that the LCD is $(x + 3)(x - 3)$. We multiply both sides of the equation by $(x + 3)(x - 3)$. By the distributive property, this is the same as multiplying each term by $(x + 3)(x - 3)$.

$$\frac{2x}{x - 3} + \frac{6 - 2x}{x^2 - 9} = \frac{x}{x + 3}$$

$$(x + 3)(x - 3) \cdot \frac{2x}{x - 3} + (x + 3)(x - 3) \cdot \frac{6 - 2x}{(x + 3)(x - 3)}$$

$$= (x + 3)(x - 3)\left(\frac{x}{x + 3}\right)$$

$$2x(x + 3) + (6 - 2x) = x(x - 3) \quad \text{Simplify.}$$

$$2x^2 + 6x + 6 - 2x = x^2 - 3x \quad \text{Use the distributive property.}$$

Next we solve this quadratic equation by the factoring method. To do so, we first write the equation so that one side is 0.

$$x^2 + 7x + 6 = 0$$

$$(x + 6)(x + 1) = 0 \quad \text{Factor.}$$

$$x = -6 \text{ or } x = -1 \quad \text{Set each factor equal to 0.}$$

Neither -6 nor -1 makes any denominator 0 so they are both solutions. The solutions are -6 and -1.   □

PRACTICE 5   Solve: $\dfrac{2}{x - 2} - \dfrac{5 + 2x}{x^2 - 4} = \dfrac{x}{x + 2}$.

EXAMPLE 6   Solve: $\dfrac{z}{2z^2 + 3z - 2} - \dfrac{1}{2z} = \dfrac{3}{z^2 + 2z}$.

Solution   Factor the denominators to find that the LCD is $2z(z + 2)(2z - 1)$. Multiply both sides by the LCD. Remember, by using the distributive property, this is the same as multiplying each term by $2z(z + 2)(2z - 1)$.

$$\frac{z}{2z^2 + 3z - 2} - \frac{1}{2z} = \frac{3}{z^2 + 2z}$$

$$\frac{z}{(2z - 1)(z + 2)} - \frac{1}{2z} = \frac{3}{z(z + 2)}$$

$$2z(z + 2)(2z - 1) \cdot \frac{z}{(2z - 1)(z + 2)} - 2z(z + 2)(2z - 1) \cdot \frac{1}{2z}$$

$$= 2z(z + 2)(2z - 1) \cdot \frac{3}{z(z + 2)} \quad \text{Apply the distributive property.}$$

$$2z(z) - (z + 2)(2z - 1) = 3 \cdot 2(2z - 1) \quad \text{Simplify.}$$

$$2z^2 - (2z^2 + 3z - 2) = 12z - 6$$

$$2z^2 - 2z^2 - 3z + 2 = 12z - 6$$

$$-3z + 2 = 12z - 6$$

$$-15z = -8$$

$$z = \frac{8}{15} \quad \text{Solve.}$$

```
8/15→Z:Z/(2Z²+3Z
-2)-1/((2Z)
        2.220394737
3/(Z²+2Z)
        2.220394737
■
```

A numerical check for Example 6. The solution is $z = \dfrac{8}{15}$.

The proposed solution $\dfrac{8}{15}$ does not make any denominator 0; the solution is $\dfrac{8}{15}$. ☐

PRACTICE

6 Solve: $\dfrac{z}{2z^2 - z - 6} - \dfrac{1}{3z} = \dfrac{2}{z^2 - 2z}$.

VOCABULARY & READINESS CHECK

Choose the least common denominator (LCD) for the rational expressions in each equation. Do not solve.

1. $\dfrac{x}{7} - \dfrac{x}{2} = \dfrac{1}{2}$; LCD = _____

 a. 7 **b.** 2 **c.** 14 **d.** 28

2. $\dfrac{9}{x + 1} + \dfrac{5}{(x + 1)^2} = \dfrac{x}{x + 1}$; LCD = _____

 a. $x + 1$ **b.** $(x + 1)^2$ **c.** $(x + 1)^3$

3. $\dfrac{7}{x - 4} = \dfrac{x}{x^2 - 16} + \dfrac{1}{x + 4}$; LCD = _____

 a. $(x + 4)(x - 4)$ **b.** $x - 4$ **c.** $x + 4$ **d.** $(x^2 - 16)(x - 4)(x + 4)$

4. $3 = \dfrac{1}{x - 5} - \dfrac{2}{x^2 - 5x}$; LCD = _____

 a. $x - 5$ **b.** $3(x - 5)$ **c.** $3x(x - 5)$ **d.** $x(x - 5)$

6.5 EXERCISE SET

MyMathLab Powered by CourseCompass™ and MathXL® PRACTICE WATCH DOWNLOAD READ REVIEW

Solve each equation. See Examples 1 and 2.

1. $\dfrac{x}{2} - \dfrac{x}{3} = 12$

2. $x = \dfrac{x}{2} - 4$

3. $\dfrac{x}{3} = \dfrac{1}{6} + \dfrac{x}{4}$

4. $\dfrac{x}{2} = \dfrac{21}{10} - \dfrac{x}{5}$

5. $\dfrac{2}{x} + \dfrac{1}{2} = \dfrac{5}{x}$

6. $\dfrac{5}{3x} + 1 = \dfrac{7}{6}$

7. $\dfrac{x^2 + 1}{x} = \dfrac{5}{x}$

8. $\dfrac{x^2 - 14}{2x} = -\dfrac{5}{2x}$

Solve each equation. See Examples 3 through 6.

9. $\dfrac{x + 5}{x + 3} = \dfrac{2}{x + 3}$

10. $\dfrac{x - 7}{x - 1} = \dfrac{11}{x - 1}$

11. $\dfrac{5}{x - 2} - \dfrac{2}{x + 4} = -\dfrac{4}{x^2 + 2x - 8}$

12. $\dfrac{1}{x - 1} + \dfrac{1}{x + 1} = \dfrac{2}{x^2 - 1}$

13. $\dfrac{1}{x - 1} = \dfrac{2}{x + 1}$

14. $\dfrac{6}{x + 3} = \dfrac{4}{x - 3}$

15. $\dfrac{x^2 - 23}{2x^2 - 5x - 3} + \dfrac{2}{x - 3} = \dfrac{-1}{2x + 1}$

16. $\dfrac{4x^2 - 24x}{3x^2 - x - 2} + \dfrac{3}{3x + 2} = \dfrac{-4}{x - 1}$

17. $\dfrac{1}{x - 4} - \dfrac{3x}{x^2 - 16} = \dfrac{2}{x + 4}$

18. $\dfrac{3}{2x + 3} - \dfrac{1}{2x - 3} = \dfrac{4}{4x^2 - 9}$

19. $\dfrac{1}{x - 4} = \dfrac{8}{x^2 - 16}$

20. $\dfrac{2}{x^2 - 4} = \dfrac{1}{2x - 4}$

21. $\dfrac{1}{x - 2} - \dfrac{2}{x^2 - 2x} = 1$

22. $\dfrac{12}{3x^2 + 12x} = 1 - \dfrac{1}{x + 4}$

MIXED PRACTICE

Solve each equation. See Examples 1 through 6.

23. $\dfrac{5}{x} = \dfrac{20}{12}$

24. $\dfrac{2}{x} = \dfrac{10}{5}$

25. $1 - \dfrac{4}{a} = 5$

26. $7 + \dfrac{6}{a} = 5$

27. $\dfrac{x^2 + 5}{x} - 1 = \dfrac{5(x + 1)}{x}$

28. $\dfrac{x^2 + 6}{x} + 5 = \dfrac{2(x + 3)}{x}$

29. $\dfrac{1}{2x} - \dfrac{1}{x + 1} = \dfrac{1}{3x^2 + 3x}$

30. $\dfrac{2}{x - 5} + \dfrac{1}{2x} = \dfrac{5}{3x^2 - 15x}$

31. $\dfrac{1}{x} - \dfrac{x}{25} = 0$

32. $\dfrac{x}{4} + \dfrac{5}{x} = 3$

33. $5 - \dfrac{2}{2y - 5} = \dfrac{3}{2y - 5}$

34. $1 - \dfrac{5}{y + 7} = \dfrac{4}{y + 7}$

35. $\dfrac{x - 1}{x + 2} = \dfrac{2}{3}$

36. $\dfrac{6x + 7}{2x + 9} = \dfrac{5}{3}$

37. $\dfrac{x + 3}{x + 2} = \dfrac{1}{x + 2}$

38. $\dfrac{2x + 1}{4 - x} = \dfrac{9}{4 - x}$

39. $\dfrac{1}{a - 3} + \dfrac{2}{a + 3} = \dfrac{1}{a^2 - 9}$

40. $\dfrac{12}{9 - a^2} + \dfrac{3}{3 + a} = \dfrac{2}{3 - a}$

41. $\dfrac{64}{x^2 - 16} + 1 = \dfrac{2x}{x - 4}$

42. $2 + \dfrac{3}{x} = \dfrac{2x}{x + 3}$

43. $\dfrac{-15}{4y + 1} + 4 = y$

44. $\dfrac{36}{x^2 - 9} + 1 = \dfrac{2x}{x + 3}$

45. $\dfrac{28}{x^2 - 9} + \dfrac{2x}{x - 3} + \dfrac{6}{x + 3} = 0$

46. $\dfrac{x^2 - 20}{x^2 - 7x + 12} = \dfrac{3}{x - 3} + \dfrac{5}{x - 4}$

47. $\dfrac{x + 2}{x^2 + 7x + 10} = \dfrac{1}{3x + 6} - \dfrac{1}{x + 5}$

48. $\dfrac{3}{2x - 5} + \dfrac{2}{2x + 3} = 0$

REVIEW AND PREVIEW

Write each sentence as an equation and solve. See Section 1.6.

49. Four more than 3 times a number is 19.

50. The sum of two consecutive integers is 147.

51. The length of a rectangle is 5 inches more than the width. Its perimeter is 50 inches. Find the length and width.

52. The sum of a number and its reciprocal is $\dfrac{5}{2}$.

The following graph is from statistics gathered for the National Health and Nutrition Examination Survey. Use this histogram to answer Exercises 53 through 57. (Source: Economic Research Service: USDA) See Section 1.6.

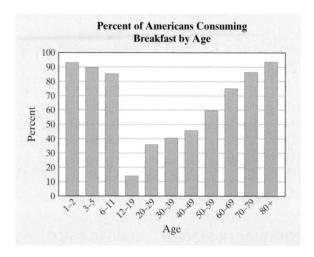

53. What percent of Americans ages 20–29 eat breakfast regularly?

54. What percent of Americans over age 80 eat breakfast regularly?

55. What age category shows the smallest percentage of Americans who eat breakfast regularly?

56. What percent of Americans ages 40–49 eat breakfast regularly?

57. According to the New York City Department of Education, there were about 284,000 high schools students at the end of 2006. Approximately how many of these students would you expect to have eaten breakfast regularly? Round to the nearest ten thousand.

CONCEPT EXTENSIONS

58. In your own words, explain the differences between equations and expressions.

59. In your own words, explain why it is necessary to check solutions to equations containing rational expressions.

60. The average cost of producing x game disks for a computer is given by the function $f(x) = 3.3 + \dfrac{5400}{x}$. Find the number of game disks that must be produced for the average cost to be $5.10.

61. The average cost of producing x electric pencil sharpeners is given by the function $f(x) = 20 + \dfrac{4000}{x}$. Find the number of electric pencil sharpeners that must be produced for the average cost to be \$25.

Solve each equation. Begin by writing each equation with positive exponents only.

62. $x^{-2} - 19x^{-1} + 48 = 0$

63. $x^{-2} - 5x^{-1} - 36 = 0$

64. $p^{-2} + 4p^{-1} - 5 = 0$

65. $6p^{-2} - 5p^{-1} + 1 = 0$

Solve each equation. Round solutions to two decimal places.

66. $\dfrac{1.4}{x - 2.6} = \dfrac{-3.5}{x + 7.1}$

67. $\dfrac{-8.5}{x + 1.9} = \dfrac{5.7}{x - 3.6}$

68. $\dfrac{10.6}{y} - 14.7 = \dfrac{9.92}{3.2} + 7.6$

69. $\dfrac{12.2}{x} + 17.3 = \dfrac{9.6}{x} - 14.7$

Solve each equation by substitution.

For example, to solve Exercise 70, first let $u = x - 1$. After substituting, we have $u^2 + 3u + 2 = 0$. Solve for u and then substitute back to solve for x.

70. $(x - 1)^2 + 3(x - 1) + 2 = 0$

71. $(4 - x)^2 - 5(4 - x) + 6 = 0$

72. $\left(\dfrac{3}{x - 1}\right)^2 + 2\left(\dfrac{3}{x - 1}\right) + 1 = 0$

73. $\left(\dfrac{5}{2 + x}\right)^2 + \left(\dfrac{5}{2 + x}\right) - 20 = 0$

Use a graphing calculator to verify the solution of each given exercise.

74. Exercise 23   **75.** Exercise 24

76. Exercise 35   **77.** Exercise 36

THE BIGGER PICTURE   SOLVING EQUATIONS AND INEQUALITIES

Continue the outline from Sections 1.5, 3.2, 3.3, 3.5, and 5.8. Write how to recognize and how to solve equations with rational expressions in your own words. For example:

Solving Equations and Inequalities

I. Equations

   A. Linear equations (Sec. 1.5 and 3.1)

   B. Absolute value equations (Sec. 3.4)

   C. Quadratic equations (Sec. 5.8)

   D. Equations with rational expressions: Equation contains rational expressions.

$$\frac{7}{x - 1} + \frac{3}{x + 1} = \frac{x + 3}{x^2 - 1} \qquad \text{Equation with rational expressions.}$$

$$(x - 1)(x + 1) \cdot \frac{7}{x - 1} + (x - 1)(x + 1) \cdot \frac{3}{x + 1} = (x - 1)(x + 1) \cdot \frac{x + 3}{(x - 1)(x + 1)} \qquad \text{Multiply by the LCD.}$$

$$7(x + 1) + 3(x - 1) = x + 3 \qquad \text{Simplify.}$$

$$7x + 7 + 3x - 3 = x + 3$$

$$9x = -1$$

$$x = -\frac{1}{9} \qquad \text{Solve.}$$

II. Inequalities

    A. Linear inequalities (Sec. 3.2)

    B. Compound inequalities (Sec. 3.3)

    C. Absolute value inequalities (Sec. 3.5)

Solve. Write solutions to inequalities using interval notation.

1. $|-7x + 1| < 15$

2. $|-7x + 1| = 15$

3. $x^2 - 121 = 0$

4. $\dfrac{8}{x + 2} - \dfrac{3}{x - 1} = \dfrac{x + 6}{x^2 + x - 2}$

5. $9x + 6 = 4x - 2$

6. $3x \leq 6$ or $-x \geq 5$

7. $3x \leq 6$ and $-x \geq 5$

8. $-9 \leq -3x + 21 < 0$

9. $\left|\dfrac{2x - 1}{5}\right| > 7$

10. $15x^3 - 16x^2 = 7x$

INTEGRATED REVIEW EXPRESSIONS AND EQUATIONS CONTAINING RATIONAL EXPRESSIONS

Sections 6.1–6.5

It is very important that you understand the difference between an expression and an equation containing rational expressions. An equation contains an equal sign; an expression does not.

Expression to be Simplified

$$\frac{x}{2} + \frac{x}{6}$$

Write both rational expressions with the LCD, 6, as the denominator.

$$\frac{x}{2} + \frac{x}{6} = \frac{x \cdot 3}{2 \cdot 3} + \frac{x}{6}$$

$$= \frac{3x}{6} + \frac{x}{6}$$

$$= \frac{4x}{6} = \frac{2x}{3}$$

Equation to be Solved

$$\frac{x}{2} + \frac{x}{6} = \frac{2}{3}$$

Multiply both sides by the LCD, 6.

$$6\left(\frac{1}{2} + \frac{x}{6}\right) = 6\left(\frac{2}{3}\right)$$

$$3 + x = 4$$

$$x = 1$$

Check to see that the solution set is 1.

> ▶ **Helpful Hint**
>
> Remember: Equations can be cleared of fractions; expressions cannot.

Perform each indicated operation and simplify, or solve the equation for the variable.

1. $\dfrac{x}{2} = \dfrac{1}{8} + \dfrac{x}{4}$

2. $\dfrac{x}{4} = \dfrac{3}{2} + \dfrac{x}{10}$

3. $\dfrac{1}{8} + \dfrac{x}{4}$

4. $\dfrac{3}{2} + \dfrac{x}{10}$

5. $\dfrac{4}{x + 2} - \dfrac{2}{x - 1}$

6. $\dfrac{5}{x - 2} - \dfrac{10}{x + 4}$

7. $\dfrac{4}{x + 2} = \dfrac{2}{x - 1}$

8. $\dfrac{5}{x - 2} = \dfrac{10}{x + 4}$

9. $\dfrac{2}{x^2 - 4} = \dfrac{1}{x + 2} - \dfrac{3}{x - 2}$

10. $\dfrac{3}{x^2 - 25} = \dfrac{1}{x + 5} + \dfrac{2}{x - 5}$

11. $\dfrac{5}{x^2 - 3x} + \dfrac{4}{2x - 6}$

12. $\dfrac{5}{x^2 - 3x} \div \dfrac{4}{2x - 6}$

13. $\dfrac{x - 1}{x + 1} + \dfrac{x + 7}{x - 1} = \dfrac{4}{x^2 - 1}$

14. $\left(1 - \dfrac{y}{x}\right) \div \left(1 - \dfrac{x}{y}\right)$

15. $\dfrac{a^2 - 9}{a - 6} \cdot \dfrac{a^2 - 5a - 6}{a^2 - a - 6}$

16. $\dfrac{2}{a-6} + \dfrac{3a}{a^2-5a-6} - \dfrac{a}{5a+5}$

17. $\dfrac{2x+3}{3x-2} = \dfrac{4x+1}{6x+1}$

18. $\dfrac{5x-3}{2x} = \dfrac{10x+3}{4x+1}$

19. $\dfrac{a}{9a^2-1} + \dfrac{2}{6a-2}$

20. $\dfrac{3}{4a-8} - \dfrac{a+2}{a^2-2a}$

21. $-\dfrac{3}{x^2} - \dfrac{1}{x} + 2 = 0$

22. $\dfrac{x}{2x+6} + \dfrac{5}{x^2-9}$

23. $\dfrac{x-8}{x^2-x-2} + \dfrac{2}{x-2}$

24. $\dfrac{x-8}{x^2-x-2} + \dfrac{2}{x-2} = \dfrac{3}{x+1}$

25. $\dfrac{3}{a} - 5 = \dfrac{7}{a} - 1$

26. $\dfrac{7}{3z-9} + \dfrac{5}{z}$

Use $\dfrac{x}{5} - \dfrac{x}{4} = \dfrac{1}{10}$ *and* $\dfrac{x}{5} - \dfrac{x}{4} + \dfrac{1}{10}$ *for Exercises 27 and 28.*

27. a. Which one above is an expression?
   b. Describe the first step to simplify this expression.
   c. Simplify the expression.

28. a. Which one above is an equation?
   b. Describe the first step to solve this equation.
   c. Solve the equation.

For each exercise, choose the correct statement. * *Each figure represents a real number and no denominators are 0.*

29. a. $\dfrac{\triangle + \square}{\triangle} = \square$ **b.** $\dfrac{\triangle + \square}{\triangle} = 1 + \dfrac{\square}{\triangle}$ **c.** $\dfrac{\triangle + \square}{\triangle} = \dfrac{\square}{\triangle}$ **d.** $\dfrac{\triangle + \square}{\triangle} = 1 + \square$ **e.** $\dfrac{\triangle + \square}{\triangle - \square} = -1$

30. a. $\dfrac{\triangle}{\square} + \dfrac{\square}{\triangle} = \dfrac{\triangle + \square}{\square + \triangle} = 1$ **b.** $\dfrac{\triangle}{\square} + \dfrac{\square}{\triangle} = \dfrac{\triangle + \square}{\triangle \square}$ **c.** $\dfrac{\triangle}{\square} + \dfrac{\square}{\triangle} = \triangle\triangle + \square\square$

   d. $\dfrac{\triangle}{\square} + \dfrac{\square}{\triangle} = \dfrac{\triangle\triangle + \square\square}{\square\triangle}$ **e.** $\dfrac{\triangle}{\square} + \dfrac{\square}{\triangle} = \dfrac{\triangle\square}{\square\triangle} = 1$

31. a. $\dfrac{\triangle}{\square} \cdot \dfrac{\bigcirc}{\square} = \dfrac{\triangle\bigcirc}{\square}$ **b.** $\dfrac{\triangle}{\square} \cdot \dfrac{\bigcirc}{\square} = \triangle\bigcirc$ **c.** $\dfrac{\triangle}{\square} \cdot \dfrac{\bigcirc}{\square} = \dfrac{\triangle + \bigcirc}{\square + \square}$ **d.** $\dfrac{\triangle}{\square} \cdot \dfrac{\bigcirc}{\square} = \dfrac{\triangle\bigcirc}{\square\square}$

32. a. $\dfrac{\triangle}{\square} \div \dfrac{\bigcirc}{\triangle} = \dfrac{\triangle\triangle}{\square\bigcirc}$ **b.** $\dfrac{\triangle}{\square} \div \dfrac{\bigcirc}{\triangle} = \dfrac{\bigcirc\square}{\triangle\triangle}$ **c.** $\dfrac{\triangle}{\square} \div \dfrac{\bigcirc}{\triangle} = \dfrac{\bigcirc}{\square}$ **d.** $\dfrac{\triangle}{\square} \div \dfrac{\bigcirc}{\triangle} = \dfrac{\triangle + \triangle}{\square + \bigcirc}$

33. a. $\dfrac{\dfrac{\triangle + \square}{\bigcirc}}{\dfrac{\triangle}{\bigcirc}} = \square$ **b.** $\dfrac{\dfrac{\triangle + \square}{\bigcirc}}{\dfrac{\triangle}{\bigcirc}} = \dfrac{\triangle\triangle + \triangle\square}{\bigcirc\bigcirc}$ **c.** $\dfrac{\dfrac{\triangle + \square}{\bigcirc}}{\dfrac{\triangle}{\bigcirc}} = 1 + \square$ **d.** $\dfrac{\dfrac{\triangle + \square}{\bigcirc}}{\dfrac{\triangle}{\bigcirc}} = \dfrac{\triangle + \square}{\triangle}$

My thanks to Kelly Champagne for permission to use her Exercises for 29 through 33.

6.6 RATIONAL EQUATIONS AND PROBLEM SOLVING

OBJECTIVES

1 Solve an equation containing rational expressions for a specified variable.

2 Solve problems by writing equations containing rational expressions.

OBJECTIVE 1 ▶ Solving equations with rational expressions for a specified variable.
In Section 1.8 we solved equations for a specified variable. In this section, we continue practicing this skill by solving equations containing rational expressions for a specified variable. The steps given in Section 1.8 for solving equations for a specified variable are repeated here.

Solving Equations for a Specified Variable

STEP 1. Clear the equation of fractions or rational expressions by multiplying each side of the equation by the least common denominator (LCD) of all denominators in the equation.

STEP 2. Use the distributive property to remove grouping symbols such as parentheses.

STEP 3. Combine like terms on each side of the equation.

STEP 4. Use the addition property of equality to rewrite the equation as an equivalent equation with terms containing the specified variable on one side and all other terms on the other side.

STEP 5. Use the distributive property and the multiplication property of equality to get the specified variable alone.

EXAMPLE 1   Solve: $\dfrac{1}{x} + \dfrac{1}{y} = \dfrac{1}{z}$ for x.

Solution   To clear this equation of fractions, we multiply both sides of the equation by xyz, the LCD of $\dfrac{1}{x}, \dfrac{1}{y}$, and $\dfrac{1}{z}$.

$$\frac{1}{x} + \frac{1}{y} = \frac{1}{z}$$

$$xyz\left(\frac{1}{x} + \frac{1}{y}\right) = xyz\left(\frac{1}{z}\right) \quad \text{Multiply both sides by } xyz.$$

$$xyz\left(\frac{1}{x}\right) + xyz\left(\frac{1}{y}\right) = xyz\left(\frac{1}{z}\right) \quad \text{Use the distributive property.}$$

$$yz + xz = xy \quad \text{Simplify.}$$

Notice the two terms that contain the specified variable x.

Next, we subtract xz from both sides so that all terms containing the specified variable x are on one side of the equation and all other terms are on the other side.

$$yz = xy - xz$$

Now we use the distributive property to factor x from $xy - xz$ and then the multiplication property of equality to solve for x.

$$yz = x(y - z)$$

$$\frac{yz}{y - z} = x \quad \text{or} \quad x = \frac{yz}{y - z} \quad \text{Divide both sides by } y - z.$$ □

PRACTICE
1   Solve: $\dfrac{1}{a} - \dfrac{1}{b} = \dfrac{1}{c}$ for a.

OBJECTIVE 2 ▶ Solving problems modeled by equations with rational expressions.
Problem solving sometimes involves modeling a described situation with an equation containing rational expressions. In Examples 2 through 5, we practice solving such problems and use the problem-solving steps first introduced in Section 2.2.

EXAMPLE 2   Finding an Unknown Number

If a certain number is subtracted from the numerator and added to the denominator of $\frac{9}{19}$, the new fraction is equivalent to $\frac{1}{3}$. Find the number.

Solution

1. UNDERSTAND the problem. Read and reread the problem and try guessing the solution. For example, if the unknown number is 3, we have

$$\frac{9 - 3}{19 + 3} = \frac{1}{3}$$

To see if this is a true statement, we simplify the fraction on the left side.

$$\frac{6}{22} = \frac{1}{3} \quad \text{or} \quad \frac{3}{11} = \frac{1}{3} \quad \text{False}$$

Since this is not a true statement, 3 is not the correct number. Remember that the purpose of this step is not to guess the correct solution but to gain an understanding of the problem posed.

   We will let n = the number to be subtracted from the numerator and added to the denominator.

2. TRANSLATE the problem.

In words:	when the number is subtracted from the numerator and added to the denominator of the fraction $\frac{9}{19}$	this is equivalent to	$\frac{1}{3}$
	↓	↓	↓
Translate:	$\dfrac{9 - n}{19 + n}$	$=$	$\dfrac{1}{3}$

3. SOLVE the equation for n.

$$\frac{9 - n}{19 + n} = \frac{1}{3}$$

To solve for n, we begin by multiplying both sides by the LCD of $3(19 + n)$.

$$3(19 + n) \cdot \frac{9 - n}{19 + n} = 3(19 + n) \cdot \frac{1}{3} \quad \text{Multiply both sides by the LCD.}$$
$$3(9 - n) = 19 + n \quad \text{Simplify.}$$
$$27 - 3n = 19 + n$$
$$8 = 4n$$
$$2 = n \quad \text{Solve.}$$

4. INTERPRET the results.

Check: If we subtract 2 from the numerator and add 2 to the denominator of $\frac{9}{19}$, we have $\frac{9 - 2}{19 + 2} = \frac{7}{21} = \frac{1}{3}$, and the problem checks.

State: The unknown number is 2. ☐

PRACTICE

2   Find a number that when added to the numerator and subtracted from the denominator of $\frac{3}{11}$ results in a fraction equivalent to $\frac{5}{2}$.

A **ratio** is the quotient of two number or two quantities. Since rational expressions are quotients of quantities, rational expressions are ratios, also. A **proportion** is a mathematical statement that two ratios are equal.

EXAMPLE 3   Calculating Homes Heated by Electricity

In the United States, 8 out of every 25 homes are heated by electricity. At this rate, how many homes in a community of 36,000 homes would you predict are heated by electricity? (*Source: 2005 American Housing Survey for the United States*)

Solution

1. UNDERSTAND. Read and reread the problem. Try to estimate a reasonable solution. For example, since 8 is less than $\frac{1}{3}$ of 25, we might reason that the solution would be less than $\frac{1}{3}$ of 36,000 or 12,000.

   Let's let x = number of homes in the community heated by electricity.

2. TRANSLATE.

   $$\begin{array}{r} \text{homes heated by electricity} \rightarrow \\ \text{total homes} \rightarrow \end{array} \quad \frac{8}{25} = \frac{x}{36{,}000} \quad \begin{array}{l} \leftarrow \text{homes heated by electricity} \\ \leftarrow \text{total homes} \end{array}$$

3. SOLVE. To solve this proportion we can multiply both sides by the LCD, 36,000, or we can set cross products equal. We will set cross products equal.

   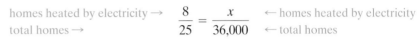

   $$\frac{8}{25} = \frac{x}{36{,}000}$$

   $$25x = 8 \cdot 36{,}000$$

   $$x = \frac{288{,}000}{25}$$

   $$x = 11{,}520$$

4. INTERPRET.

Check: To check, replace x with 11,520 in the proportion and see that a true statement results. Notice that our answer is reasonable since it is less than 12,000 as we stated above.

State: We predict that 11,520 homes are heated by electricity.   □

PRACTICE

3   In the United States, 1 out of 12 homes is heated by fuel oil. At this rate, how many homes in a community of 36,000 homes are heated by fuel oil? (*Source: 2005 American Housing Survey for the United States*)

The following work example leads to an equation containing rational expressions.

EXAMPLE 4   Calculating Work Hours

Melissa Scarlatti can clean the house in 4 hours, whereas her husband, Zack, can do the same job in 5 hours. They have agreed to clean together so that they can finish in time to watch a movie on TV that starts in 2 hours. How long will it take them to clean the house together? Can they finish before the movie starts?

Solution

1. UNDERSTAND. Read and reread the problem. The key idea here is the relationship between the *time* (in hours) it takes to complete the job and the *part of the job* completed in 1 unit of time (1 hour). For example, if the *time* it takes Melissa to complete the job is 4 hours, the part of the job she can complete in 1 hour is $\frac{1}{4}$. Similarly, Zack can complete $\frac{1}{5}$ of the job in 1 hour.

We will let $t = $ *the time* in hours it takes Melissa and Zack to clean the house together. Then $\frac{1}{t}$ represents the *part of the job* they complete in 1 hour. We summarize the given information in a chart.

	Hours to Complete the Job	*Part of Job Completed in 1 Hour*
MELISSA ALONE	4	$\frac{1}{4}$
ZACK ALONE	5	$\frac{1}{5}$
TOGETHER	t	$\frac{1}{t}$

2. TRANSLATE.

In words:	part of job Melissa can complete in 1 hour	added to	part of job Zack can complete in 1 hour	is equal to	part of job they can complete together in 1 hour
	↓	↓	↓	↓	↓
Translate:	$\frac{1}{4}$	$+$	$\frac{1}{5}$	$=$	$\frac{1}{t}$

3. SOLVE.

$$\frac{1}{4} + \frac{1}{5} = \frac{1}{t}$$

$$20t\left(\frac{1}{4} + \frac{1}{5}\right) = 20t\left(\frac{1}{t}\right) \qquad \text{Multiply both sides by the LCD, } 20t.$$

$$5t + 4t = 20$$

$$9t = 20$$

$$t = \frac{20}{9} \quad \text{or} \quad 2\frac{2}{9} \quad \text{Solve.}$$

4. INTERPRET.

Check: The proposed solution is $2\frac{2}{9}$. That is, Melissa and Zack would take $2\frac{2}{9}$ hours to clean the house together. This proposed solution is reasonable since $2\frac{2}{9}$ hours is more than half of Melissa's time and less than half of Zack's time. Check this solution in the originally stated problem.

State: Melissa and Zack can clean the house together in $2\frac{2}{9}$ hours. They cannot complete the job before the movie starts. □

PRACTICE

4 Elissa Juarez can clean the animal cages at the animal shelter where she volunteers in 3 hours. Bill Stiles can do the same job in 2 hours. How long would it take them to clean the cages if they work together?

EXAMPLE 5 **Finding the Speed of a Current**

Steve Deitmer takes $1\frac{1}{2}$ times as long to go 72 miles upstream in his boat as he does to return. If the boat cruises at 30 mph in still water, what is the speed of the current?

Solution

1. UNDERSTAND. Read and reread the problem. Guess a solution. Suppose that the current is 4 mph. The speed of the boat upstream is slowed down by the current: $30 - 4$, or 26 mph, and the speed of the boat downstream is speeded up by the current: $30 + 4$, or 34 mph. Next let's find out how long it takes to travel 72 miles upstream and 72 miles downstream. To do so, we use the formula $d = rt$, or $\dfrac{d}{r} = t$.

<table>
<tr><td align="center">***Upstream***</td><td align="center">***Downstream***</td></tr>
<tr><td align="center">$\dfrac{d}{r} = t$</td><td align="center">$\dfrac{d}{r} = t$</td></tr>
<tr><td align="center">$\dfrac{72}{26} = t$</td><td align="center">$\dfrac{72}{34} = t$</td></tr>
<tr><td align="center">$2\dfrac{10}{13} = t$</td><td align="center">$2\dfrac{2}{17} = t$</td></tr>
</table>

Since the time upstream $\left(2\dfrac{10}{13}\text{ hours}\right)$ is not $1\dfrac{1}{2}$ times the time downstream $\left(2\dfrac{2}{17}\text{ hours}\right)$, our guess is not correct. We do, however, have a better understanding of the problem.

We will let

$$x = \text{the speed of the current}$$
$$30 + x = \text{the speed of the boat downstream}$$
$$30 - x = \text{the speed of the boat upstream}$$

This information is summarized in the following chart, where we use the formula $\dfrac{d}{r} = t$.

	Distance	**Rate**	**Time** $\left(\dfrac{d}{r}\right)$
UPSTREAM	72	$30 - x$	$\dfrac{72}{30 - x}$
DOWNSTREAM	72	$30 + x$	$\dfrac{72}{30 + x}$

2. TRANSLATE. Since the time spent traveling upstream is $1\dfrac{1}{2}$ times the time spent traveling downstream, we have

In words:	time upstream	is	$1\dfrac{1}{2}$	times	time downstream
	$\downarrow$	$\downarrow$	$\downarrow$	$\downarrow$	$\downarrow$
Translate:	$\dfrac{72}{30 - x}$	$=$	$\dfrac{3}{2}$	$\cdot$	$\dfrac{72}{30 + x}$

3. SOLVE. $\dfrac{72}{30 - x} = \dfrac{3}{2} \cdot \dfrac{72}{30 + x}$

First we multiply both sides by the LCD, $2(30 + x)(30 - x)$.

$$2(30 + x)(30 - x) \cdot \dfrac{72}{30 - x} = 2(30 + x)(30 - x)\left(\dfrac{3}{2} \cdot \dfrac{72}{30 + x}\right)$$

$$72 \cdot 2(30 + x) = 3 \cdot 72 \cdot (30 - x) \quad \text{Simplify.}$$

$$2(30 + x) = 3(30 - x) \quad \text{Divide both sides by 72.}$$

$$60 + 2x = 90 - 3x \quad \text{Use the distributive property.}$$

$$5x = 30$$

$$x = 6 \quad \text{Solve.}$$

4. INTERPRET.

Check: Check the proposed solution of 6 mph in the originally stated problem.

State: The current's speed is 6 mph.

PRACTICE
5 An airplane flying from Los Angeles to Boston at a speed of 450 mph had a tail-wind assisting its flight. At the same time, there was another flight doing the same speed, going from Boston to Los Angeles. This second flight encountered a headwind. It took the pilot heading west $1\frac{1}{4}$ times as long to travel from Boston to Los Angeles as it took the pilot flying east from Los Angeles to Boston. What was the speed of the wind?

6.6 EXERCISE SET

PRACTICE   WATCH   DOWNLOAD   READ   REVIEW

Solve each equation for the specified variable. See Example 1.

1. $F = \frac{9}{5}C + 32$ for C (Meteorology)

2. $V = \frac{1}{3}\pi r^2 h$ for h (Volume)

3. $Q = \frac{A - I}{L}$ for I (Finance)

4. $P = 1 - \frac{C}{S}$ for S (Finance)

5. $\frac{1}{R} = \frac{1}{R_1} + \frac{1}{R_2}$ for R (Electronics)

6. $\frac{1}{R} = \frac{1}{R_1} + \frac{1}{R_2}$ for R_1 (Electronics)

7. $S = \frac{n(a + L)}{2}$ for n (Sequences)

8. $S = \frac{n(a + L)}{2}$ for a (Sequences)

9. $A = \frac{h(a + b)}{2}$ for b (Geometry)

10. $A = \frac{h(a + b)}{2}$ for h (Geometry)

11. $\frac{P_1 V_1}{T_1} = \frac{P_2 V_2}{T_2}$ for T_2 (Chemistry)

12. $H = \frac{kA(T_1 - T_2)}{L}$ for T_2 (Physics)

13. $f = \frac{f_1 f_2}{f_1 + f_2}$ for f_2 (Optics)

14. $I = \frac{E}{R + r}$ for r (Electronics)

15. $\lambda = \frac{2L}{n}$ for L (Physics)

16. $S = \frac{a_1 - a_n r}{1 - r}$ for a_1 (Sequences)

17. $\frac{\theta}{\omega} = \frac{2L}{c}$ for c

18. $F = \frac{-GMm}{r^2}$ for M (Physics)

Solve. See Example 2.

19. The sum of a number and 5 times its reciprocal is 6. Find the number(s).

20. The quotient of a number and 9 times its reciprocal is 1. Find the number(s).

21. If a number is added to the numerator of $\frac{12}{41}$ and twice the number is added to the denominator of $\frac{12}{41}$, the resulting fraction is equivalent to $\frac{1}{3}$. Find the number.

22. If a number is subtracted from the numerator of $\frac{13}{8}$ and added to the denominator of $\frac{13}{8}$, the resulting fraction is equivalent to $\frac{2}{5}$. Find the number.

Solve. See Example 3.

23. An Arabian camel can drink 15 gallons of water in 10 minutes. At this rate, how much water can the camel drink in 3 minutes? (*Source:* Grolier, Inc.)

24. An Arabian camel can travel 20 miles in 8 hours, carrying a 300-pound load on its back. At this rate, how far can the camel travel in 10 hours? (*Source:* Grolier, Inc.)

25. In 2005, 5.5 out of every 50 Coast Guard personnel were women. If there are 40,639 total Coast Guard personnel on active duty, estimate the number of women. Round to the nearest whole. (*Source: The World Almanac,* 2007)

26. In 2005, 42.8 out of every 50 Navy personnel were men. If there are 353,496 total Navy personnel on active duty, estimate the number of men. Round to the nearest whole. (*Source: The World Almanac,* 2007)

Solve. See Example 4.

27. An experienced roofer can roof a house in 26 hours. A beginning roofer needs 39 hours to complete the same job. Find how long it takes for the two to do the job together.

28. Alan Cantrell can word process a research paper in 6 hours. With Steve Isaac's help, the paper can be processed in 4 hours. Find how long it takes Steve to word process the paper alone.

29. Three postal workers can sort a stack of mail in 20 minutes, 30 minutes, and 60 minutes, respectively. Find how long it takes them to sort the mail if all three work together.

30. A new printing press can print newspapers twice as fast as the old one can. The old one can print the afternoon edition in 4 hours. Find how long it takes to print the afternoon edition if both printers are operating.

Solve. See Example 5.

31. Mattie Evans drove 150 miles in the same amount of time that it took a turbopropeller plane to travel 600 miles. The speed of the plane was 150 mph faster than the speed of the car. Find the speed of the plane.

32. An F-100 plane and a Toyota truck leave the same town at sunrise and head for a town 450 miles away. The speed of the plane is three times the speed of the truck, and the plane arrives 6 hours ahead of the truck. Find the speed of the truck.

33. The speed of Lazy River's current is 5 mph. If a boat travels 20 miles downstream in the same time that it takes to travel 10 miles upstream, find the speed of the boat in still water.

34. The speed of a boat in still water is 24 mph. If the boat travels 54 miles upstream in the same time that it takes to travel 90 miles downstream, find the speed of the current.

MIXED PRACTICE

Solve.

35. The sum of the reciprocals of two consecutive integers is $-\dfrac{15}{56}$. Find the two integers.

36. The sum of the reciprocals of two consecutive odd integers is $\dfrac{20}{99}$. Find the two integers.

37. One hose can fill a goldfish pond in 45 minutes, and two hoses can fill the same pond in 20 minutes. Find how long it takes the second hose alone to fill the pond.

38. If Sarah Clark can do a job in 5 hours and Dick Belli and Sarah working together can do the same job in 2 hours, find how long it takes Dick to do the job alone.

39. Two trains going in opposite directions leave at the same time. One train travels 15 mph faster than the other. In 6 hours the trains are 630 miles apart. Find the speed of each.

40. The speed of a bicyclist is 10 mph faster than the speed of a walker. If the bicyclist travels 26 miles in the same amount of time that the walker travels 6 miles, find the speed of the bicyclist.

41. A giant tortoise can travel 0.17 miles in 1 hour. At this rate, how long would it take the tortoise to travel 1 mile? Round to the nearest tenth of an hour. (*Source: The World Almanac*)

42. A black mamba snake can travel 88 feet in 3 seconds. At this rate, how long does it take to travel 300 feet (the length of a football field)? Round to the nearest tenth of a second. (*Source: The World Almanac*)

43. A local dairy has three machines to fill half-gallon milk cartons. The machines can fill the daily quota in 5 hours, 6 hours, and 7.5 hours, respectively. Find how long it takes to fill the daily quota if all three machines are running.

44. The inlet pipe of an oil tank can fill the tank in 1 hour, 30 minutes. The outlet pipe can empty the tank in 1 hour. Find how long it takes to empty a full tank if both pipes are open.

45. A plane flies 465 miles with the wind and 345 miles against the wind in the same length of time. If the speed of the wind is 20 mph, find the speed of the plane in still air.

46. Two rockets are launched. The first travels at 9000 mph. Fifteen minutes later the second is launched at 10,000 mph. Find the distance at which both rockets are an equal distance from Earth.

47. Two joggers, one averaging 8 mph and one averaging 6 mph, start from a designated initial point. The slower jogger arrives at the end of the run a half-hour after the other jogger. Find the distance of the run.

48. A semi truck travels 300 miles through the flatland in the same amount of time that it travels 180 miles through the Great Smoky Mountains. The rate of the truck is 20 miles per hour slower in the mountains than in the flatland. Find both the flatland rate and mountain rate.

49. The denominator of a fraction is 1 more than the numerator. If both the numerator and the denominator are decreased by 3, the resulting fraction is equivalent to $\dfrac{4}{5}$. Find the fraction.

50. The numerator of a fraction is 4 less than the denominator. If both the numerator and the denominator are increased by 2, the resulting fraction is equivalent to $\dfrac{2}{3}$. Find the fraction.

51. In 2 minutes, a conveyor belt can move 300 pounds of recyclable aluminum from the delivery truck to a storage area. A smaller belt can move the same quantity of cans the same distance in 6 minutes. If both belts are used, find how long it takes to move the cans to the storage area.

52. Gary Marcus and Tony Alva work at Lombardo's Pipe and Concrete. Mr. Lombardo is preparing an estimate for a customer. He knows that Gary can lay a slab of concrete in 6 hours. Tony can lay the same size slab in 4 hours. If both work on the job and the cost of labor is $45.00 per hour, determine what the labor estimate should be.

53. Smith Engineering is in the process of reviewing the salaries of their surveyors. During this review, the company found that an experienced surveyor can survey a roadbed in 4 hours. An apprentice surveyor needs 5 hours to survey the same stretch of road. If the two work together, find how long it takes them to complete the job.

54. Mr. Dodson can paint his house by himself in four days. His son will need an additional day to complete the job if he works by himself. If they work together, find how long it takes to paint the house.

55. Cyclist Lance Armstrong of the United States won the Tours de France a record seven times. This inspired an amateur cyclist to train for a local road race. He rode the first 20-mile portion of his workout at a constant rate. For the 16-mile cool-down portion of his workout, he reduced his speed by 2 miles per hour. Each portion of the workout took equal time. Find the cyclist's rate during the first portion and his rate during the cool-down portion.

56. The world record for the largest white bass caught is held by Ronald Sprouse of Virginia. The bass weighed 6 pounds 13 ounces. If Ronald rows to his favorite fishing spot 9 miles downstream in the same amount of time that he rows 3 miles upstream and if the current is 6 mph, find how long it takes him to cover the 12 miles.

57. An experienced bricklayer can construct a small wall in 3 hours. An apprentice can complete the job in 6 hours. Find how long it takes if they work together.

58. Scanner A can scan a document in 3 hours. Scanner B takes 5 hours to do the same job. If both scanners are used, how long will it take for the document to be scanned?

59. A marketing manager travels 1080 miles in a corporate jet and then an additional 240 miles by car. If the car ride takes 1 hour longer, and if the rate of the jet is 6 times the rate of the car, find the time the manager travels by jet and find the time she travels by car.

60. In a recent year, 13 out of 20 top grossing movies were rated PG-13. At this rate, how many movies in a year with 599 new releases would you predict to be rated PG-13? Round to the nearest whole movie. (*Source:* Motion Picture Association)

61. In a recent year, 5 out of 7 movies cost between $50 and $99 million to make. At this rate, how many movies in a year with 599 new releases would you predict to cost between $50 and $99 million to make? Round to the nearest whole movie. (*Source:* Motion Picture Association)

REVIEW AND PREVIEW

Solve each equation for x. See Section 1.5.

62. $\dfrac{x}{5} = \dfrac{x+2}{3}$

63. $\dfrac{x}{4} = \dfrac{x+3}{6}$

64. $\dfrac{x-3}{2} = \dfrac{x-5}{6}$

65. $\dfrac{x-6}{4} = \dfrac{x-2}{5}$

CONCEPT EXTENSIONS

Calculating body-mass index (BMI) is a way to gauge whether a person should lose weight. Doctors recommend that body-mass index values fall between 19 and 25. The formula for body-mass index B is $B = \dfrac{705w}{h^2}$, where w is weight in pounds and h is height in inches. Use this formula to answer Exercises 66 and 67.

66. A patient is 5 ft 8 in. tall. What should his or her weight be to have a body-mass index of 25? Round to the nearest whole pound.

67. A doctor recorded a body-mass index of 47 on a patient's chart. Later, a nurse notices that the doctor recorded the patient's weight as 240 pounds but neglected to record the patient's height. Explain how the nurse can use the information from the chart to find the patient's height. Then find the height.

In physics, when the source of a sound is traveling toward an observer, the relationship between the actual pitch a of the sound and the pitch h that the observer hears due to the Doppler effect is described by the formula $h = \dfrac{a}{1 - \dfrac{s}{770}}$, where s is the speed of the sound source in miles per hour. Use this formula to answer Exercise 68.

68. An emergency vehicle has a single-tone siren with the pitch of the musical note E. As it approaches an observer standing by the road, the vehicle is traveling 50 mph. Is the pitch that the observer hears due to the Doppler effect lower or higher than the actual pitch? To which musical note is the pitch that the observer hears closest?

Pitch of an Octave of Musical Notes in Hertz (Hz)	
Note	*Pitch*
Middle C	261.63
D	293.66
E	329.63
F	349.23
G	392.00
A	440.00
B	493.88

Note: Greater numbers indicate higher pitches (acoustically).
(*Source:* American Standards Association)

In electronics, the relationship among the resistances R_1 and R_2 of two resistors wired in a parallel circuit and their combined resistance R is described by the formula $\dfrac{1}{R} = \dfrac{1}{R_1} + \dfrac{1}{R_2}$. Use this formula to solve Exercises 69 through 71.

69. If the combined resistance is 2 ohms and one of the two resistances is 3 ohms, find the other resistance.

70. Find the combined resistance of two resistors of 12 ohms each when they are wired in a parallel circuit.

71. The relationship among resistance of two resistors wired in a parallel circuit and their combined resistance may be extended to three resistors of resistances R_1, R_2, and R_3. Write an equation you believe may describe the relationship, and use it to find the combined resistance if R_1 is 5, R_2 is 6, and R_3 is 2.

6.7 VARIATION AND PROBLEM SOLVING

OBJECTIVE 1 ▶ Solving problems involving direct variation. A very familiar example of direct variation is the relationship of the circumference C of a circle to its radius r. The formula $C = 2\pi r$ expresses that the circumference is always 2π times the radius. In other words, C is always a constant multiple (2π) of r. Because it is, we say that C **varies directly as r,** that C **varies directly with r,** or that C **is directly proportional to r.**

> **Direct Variation**
>
> **y varies directly as x,** or **y is directly proportional to x,** if there is a nonzero constant k such that
>
> $$y = kx$$
>
> The number k is called the **constant of variation** or the **constant of proportionality.**

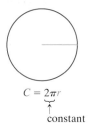

$C = 2\pi r$

$\underbrace{}$
constant

In the above definition, the relationship described between x and y is a linear one. In other words, the graph of $y = kx$ is a line. The slope of the line is k, and the line passes through the origin.

For example, the graph of the direct variation equation $C = 2\pi r$ is shown. The horizontal axis represents the radius r, and the vertical axis is the circumference C. From the graph we can read that when the radius is 6 units, the circumference is approximately 38 units. Also, when the circumference is 45 units, the radius is between 7 and 8 units. Notice that as the radius increases, the circumference increases.

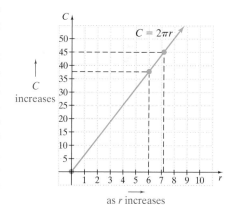

EXAMPLE 1   Suppose that y varies directly as x. If y is 5 when x is 30, find the constant of variation and the direct variation equation.

Solution   Since y varies directly as x, we write $y = kx$. If $y = 5$ when $x = 30$, we have that

$$y = kx$$
$$5 = k(30) \quad \text{Replace } y \text{ with 5 and } x \text{ with 30.}$$
$$\frac{1}{6} = k \quad\quad \text{Solve for } k.$$

The constant of variation is $\dfrac{1}{6}$.

After finding the constant of variation k, the direct variation equation can be written as $y = \dfrac{1}{6}x$.   □

PRACTICE

1   Suppose that y varies directly as x. If y is 20 when x is 15, find the constant of variation and the direct variation equation.

EXAMPLE 2   Using Direct Variation and Hooke's Law

Hooke's law states that the distance a spring stretches is directly proportional to the weight attached to the spring. If a 40-pound weight attached to a spring stretches the spring 5 inches, find the distance that a 65-pound weight attached to a spring stretches the spring.

Solution

1. UNDERSTAND. Read and reread the problem. Notice that we are given that the distance a spring stretches is **directly proportional** to the weight attached. We let

    d = the distance stretched

    w = the weight attached

    The constant of variation is represented by k.

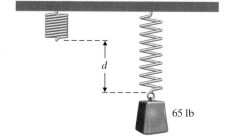

2. TRANSLATE. Because d is directly proportional to w, we write

    $$d = kw$$

3. SOLVE. When a weight of 40 pounds is attached, the spring stretches 5 inches. That is, when $w = 40$, $d = 5$.

    $$d = kw$$
    $$5 = k(40) \quad \text{Replace } d \text{ with 5 and } w \text{ with 40.}$$
    $$\frac{1}{8} = k \quad\quad \text{Solve for } k.$$

    Now when we replace k with $\frac{1}{8}$ in the equation

    $$d = kw, \text{ we have}$$
    $$d = \frac{1}{8}w$$

    To find the stretch when a weight of 65 pounds is attached, we replace w with 65 to find d.

    $$d = \frac{1}{8}(65)$$
    $$= \frac{65}{8} = 8\frac{1}{8} \quad \text{or} \quad 8.125$$

4. INTERPRET.

Check: Check the proposed solution of 8.125 inches in the original problem.

State: The spring stetches 8.125 inches when a 65-pound weight is attached.   □

PRACTICE

2   Use Hooke's law as stated in Example 2. If a 36-pound weight attached to a spring stretches the spring 9 inches, find the distance that a 75-pound weight attached to the spring stretches the spring.

OBJECTIVE 2 ▶ Solving problems involving inverse variation. When y is proportional to the **reciprocal** of another variable x, we say that **y varies inversely as x,** or that **y is inversely proportional to x.** An example of the inverse variation relationship is the relationship between the pressure that a gas exerts and the volume of its container. As the volume of a container decreases, the pressure of the gas it contains increases.

Inverse Variation

y **varies inversely as** *x,* or *y* **is inversely proportional to** *x,* if there is a nonzero constant *k* such that

$$y = \frac{k}{x}$$

The number *k* is called the **constant of variation** or the **constant of proportionality.**

Notice that $y = \frac{k}{x}$ is a rational equation. Its graph for $k > 0$ and $x > 0$ is shown. From the graph, we can see that as *x* increases, *y* decreases.

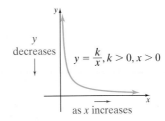

$$y = \frac{k}{x}, k > 0, x > 0$$

y decreases

as *x* increases

EXAMPLE 3   Suppose that *u* varies inversely as *w*. If *u* is 3 when *w* is 5, find the constant of variation and the inverse variation equation.

Solution   Since *u* varies inversely as *w*, we have $u = \frac{k}{w}$. We let $u = 3$ and $w = 5$, and we solve for *k*.

$$u = \frac{k}{w}$$

$$3 = \frac{k}{5} \quad \text{Let } u = 3 \text{ and } w = 5.$$

$$15 = k \quad \text{Multiply both sides by 5.}$$

The constant of variation *k* is 15. This gives the inverse variation equation

$$u = \frac{15}{w}$$

PRACTICE
3   Suppose that *b* varies inversely as *a*. If *b* is 5 when *a* is 9, find the constant of variation and the inverse variation equation.

EXAMPLE 4   **Using Inverse Variation and Boyle's Law**

Boyle's law says that if the temperature stays the same, the pressure *P* of a gas is inversely proportional to the volume *V*. If a cylinder in a steam engine has a pressure of 960 kilopascals when the volume is 1.4 cubic meters, find the pressure when the volume increases to 2.5 cubic meters.

Solution

1. UNDERSTAND. Read and reread the problem. Notice that we are given that the pressure of a gas is *inversely proportional* to the volume. We will let *P* = the pressure and *V* = the volume. The constant of variation is represented by *k*.
2. TRANSLATE. Because *P* is inversely proportional to *V*, we write

$$P = \frac{k}{V}$$

When $P = 960$ kilopascals, the volume $V = 1.4$ cubic meters. We use this information to find k.

$$960 = \frac{k}{1.4} \quad \text{Let } P = 960 \text{ and } V = 1.4.$$

$$1344 = k \quad \text{Multiply both sides by 1.4.}$$

Thus, the value of k is 1344. Replacing k with 1344 in the variation equation, we have

$$P = \frac{1344}{V}$$

Next we find P when V is 2.5 cubic meters.

3. SOLVE.

$$P = \frac{1344}{2.5} \quad \text{Let } V = 2.5.$$

$$= 537.6$$

4. INTERPRET.

Check: Check the proposed solution in the original problem.

State: When the volume is 2.5 cubic meters, the pressure is 537.6 kilopascals. ☐

PRACTICE
4 Use Boyle's law as stated in Example 4. When $P = 350$ kilopascals and $V = 2.8$ cubic meters, find the pressure when the volume decreases to 1.5 cubic meters.

OBJECTIVE 3 ▶ Solving problems involving joint variation. Sometimes the ratio of a variable to the product of many other variables is constant. For example, the ratio of distance traveled to the product of speed and time traveled is always 1.

$$\frac{d}{rt} = 1 \qquad \text{or} \qquad d = rt$$

Such a relationship is called **joint variation.**

> **Joint Variation**
>
> If the ratio of a variable y to the product of two or more variables is constant, then **y varies jointly as,** or **is jointly proportional to,** the other variables. If
>
> $$y = kxz$$
>
> then the number k is the **constant of variation** or the **constant of proportionality.**

Concept Check ☑

Which type of variation is represented by the equation $xy = 8$? Explain.

a. Direct variation   **b.** Inverse variation   **c.** Joint variation

⚠ **EXAMPLE 5**   **Expressing Surface Area**

The lateral surface area of a cylinder varies jointly as its radius and height. Express this surface area S in terms of radius r and height h.

Solution   Because the surface area varies jointly as the radius r and the height h, we equate S to a constant multiple of r and h.

$$S = krh$$

In the equation $S = krh$, it can be determined that the constant k is 2π, and we then have the formula $S = 2\pi rh$. (The lateral surface area formula does not include the areas of the two circular bases.)   □

PRACTICE
5   The area of a regular polygon varies jointly as its apothem and its perimeter. Express the area in terms of the apothem a and the perimeter p.

OBJECTIVE 4 ▶ Solving problems involving combined variation. Some examples of variation involve combinations of direct, inverse, and joint variation. We will call these variations **combined variation.**

EXAMPLE 6   Suppose that y varies directly as the square of x. If y is 24 when x is 2, find the constant of variation and the variation equation.

Solution   Since y varies directly as the square of x, we have

$$y = kx^2$$

Now let $y = 24$ and $x = 2$ and solve for k.

$$y = kx^2$$
$$24 = k \cdot 2^2$$
$$24 = 4k$$
$$6 = k$$

The constant of variation is 6, so the variation equation is

$$y = 6x^2$$ □

PRACTICE
6   Suppose that y varies inversely as the cube of x. If y is $\dfrac{1}{2}$ when x is 2, find the constant of variation and the variation equation.

△ **EXAMPLE 7**   **Finding Column Weight**

The maximum weight that a circular column can support is directly proportional to the fourth power of its diameter and is inversely proportional to the square of its height. A 2-meter-diameter column that is 8 meters in height can support 1 ton. Find the weight that a 1-meter-diameter column that is 4 meters in height can support.

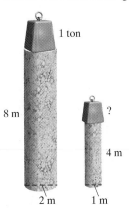

Solution

1. **UNDERSTAND.** Read and reread the problem. Let w = weight, d = diameter, h = height, and k = the constant of variation.

2. **TRANSLATE.** Since w is directly proportional to d^4 and inversely proportional to h^2, we have

$$w = \frac{kd^4}{h^2}$$

3. **SOLVE.** To find k, we are given that a 2-meter-diameter column that is 8 meters in height can support 1 ton. That is, $w = 1$ when $d = 2$ and $h = 8$, or

$$1 = \frac{k \cdot 2^4}{8^2} \quad \text{Let } w = 1, d = 2, \text{ and } h = 8.$$

$$1 = \frac{k \cdot 16}{64}$$

$$4 = k \quad \text{Solve for } k.$$

Now replace k with 4 in the equation $w = \dfrac{kd^4}{h^2}$ and we have

$$w = \frac{4d^4}{h^2}$$

To find weight w for a 1-meter-diameter column that is 4 meters in height, let $d = 1$ and $h = 4$.

$$w = \frac{4 \cdot 1^4}{4^2}$$

$$w = \frac{4}{16} = \frac{1}{4}$$

4. **INTERPRET.**

Check: Check the proposed solution in the original problem.

State: The 1-meter-diameter column that is 4 meters in height can hold $\dfrac{1}{4}$ ton of weight. □

PRACTICE
7 Suppose that y varies directly as z and inversely as the cube of x. If y is 15 when $z = 5$ and $x = 3$, find the constant of variation and the variation equation.

VOCABULARY & READINESS CHECK

State whether each equation represents direct, inverse, or joint variation.

1. $y = 5x$

2. $y = \dfrac{700}{x}$

3. $y = 5xz$

4. $y = \dfrac{1}{2}abc$

5. $y = \dfrac{9.1}{x}$

6. $y = 2.3x$

7. $y = \dfrac{2}{3}x$

8. $y = 3.1\,st$

6.7 EXERCISE SET

MyMathLab PRACTICE WATCH DOWNLOAD READ REVIEW

If y varies directly as x, find the constant of variation and the direct variation equation for each situation. See Example 1.

1. $y = 4$ when $x = 20$

2. $y = 5$ when $x = 30$

3. $y = 6$ when $x = 4$

4. $y = 12$ when $x = 8$

5. $y = 7$ when $x = \dfrac{1}{2}$

6. $y = 11$ when $x = \dfrac{1}{3}$

7. $y = 0.2$ when $x = 0.8$

8. $y = 0.4$ when $x = 2.5$

Solve. See Example 2.

9. The weight of a synthetic ball varies directly with the cube of its radius. A ball with a radius of 2 inches weighs 1.20 pounds. Find the weight of a ball of the same material with a 3-inch radius.

10. At sea, the distance to the horizon is directly proportional to the square root of the elevation of the observer. If a person who is 36 feet above the water can see 7.4 miles, find how far a person 64 feet above the water can see. Round to the nearest tenth of a mile.

11. The amount P of pollution varies directly with the population N of people. Kansas City has a population of 442,000 and produces 260,000 tons of pollutants. Find how many tons of pollution we should expect St. Louis to produce, if we know that its population is 348,000. Round to the nearest whole ton. (*Population Source: The World Almanac, 2005*)

12. Charles's law states that if the pressure P stays the same, the volume V of a gas is directly proportional to its temperature T. If a balloon is filled with 20 cubic meters of a gas at a temperature of 300 K, find the new volume if the temperature rises 360 K while the pressure stays the same.

If y varies inversely as x, find the constant of variation and the inverse variation equation for each situation. See Example 3.

13. $y = 6$ when $x = 5$

14. $y = 20$ when $x = 9$

15. $y = 100$ when $x = 7$

16. $y = 63$ when $x = 3$

17. $y = \dfrac{1}{8}$ when $x = 16$

18. $y = \dfrac{1}{10}$ when $x = 40$

19. $y = 0.2$ when $x = 0.7$

20. $y = 0.6$ when $x = 0.3$

Solve. See Example 4.

21. Pairs of markings a set distance apart are made on highways so that police can detect drivers exceeding the speed limit. Over a fixed distance, the speed R varies inversely with the time T. In one particular pair of markings, R is 45 mph when T is 6 seconds. Find the speed of a car that travels the given distance in 5 seconds.

22. The weight of an object on or above the surface of Earth varies inversely as the square of the distance between the object and Earth's center. If a person weighs 160 pounds on

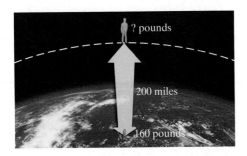

Earth's surface, find the individual's weight if he moves 200 miles above Earth. Round to the nearest whole pound. (Assume that Earth's radius is 4000 miles.)

23. If the voltage V in an electric circuit is held constant, the current I is inversely proportional to the resistance R. If the current is 40 amperes when the resistance is 270 ohms, find the current when the resistance is 150 ohms.

24. Because it is more efficient to produce larger numbers of items, the cost of producing Dysan computer disks is inversely proportional to the number produced. If 4000 can be produced at a cost of $1.20 each, find the cost per disk when 6000 are produced.

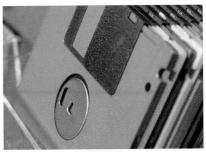

25. The intensity I of light varies inversely as the square of the distance d from the light source. If the distance from the light source is doubled (see the figure), determine what happens to the intensity of light at the new location.

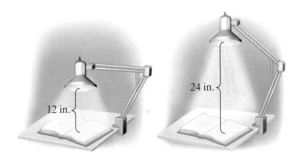

26. The maximum weight that a circular column can hold is inversely proportional to the square of its height. If an 8-foot column can hold 2 tons, find how much weight a 10-foot column can hold.

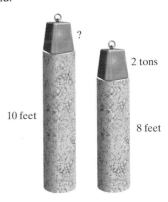

MIXED PRACTICE

Write each statement as an equation. Use k as the constant of variation. See Example 5.

27. x varies jointly as y and z.

28. P varies jointly as R and the square of S.

29. r varies jointly as s and the cube of t.

30. a varies jointly as b and c.

For each statement, find the constant of variation and the variation equation. See Examples 5 and 6.

31. y varies directly as the cube of x; $y = 9$ when $x = 3$

32. y varies directly as the cube of x; $y = 32$ when $x = 4$

33. y varies directly as the square root of x; $y = 0.4$ when $x = 4$

34. y varies directly as the square root of x; $y = 2.1$ when $x = 9$

35. y varies inversely as the square of x; $y = 0.052$ when $x = 5$

36. y varies inversely as the square of x; $y = 0.011$ when $x = 10$

37. y varies jointly as x and the cube of z; $y = 120$ when $x = 5$ and $z = 2$

38. y varies jointly as x and the square of z; $y = 360$ when $x = 4$ and $z = 3$

Solve. See Example 7.

39. The maximum weight that a rectangular beam can support varies jointly as its width and the square of its height and inversely as its length. If a beam $\frac{1}{2}$ foot wide, $\frac{1}{3}$ foot high, and 10 feet long can support 12 tons, find how much a similar beam can support if the beam is $\frac{2}{3}$ foot wide, $\frac{1}{2}$ foot high, and 16 feet long.

40. The number of cars manufactured on an assembly line at a General Motors plant varies jointly as the number of workers and the time they work. If 200 workers can produce 60 cars in 2 hours, find how many cars 240 workers should be able to make in 3 hours.

41. The volume of a cone varies jointly as its height and the square of its radius. If the volume of a cone is 32π cubic inches when the radius is 4 inches and the height is 6 inches, find the volume of a cone when the radius is 3 inches and the height is 5 inches.

42. When a wind blows perpendicularly against a flat surface, its force is jointly proportional to the surface area and the speed of the wind. A sail whose surface area is 12 square feet experiences a 20-pound force when the wind speed is 10 miles per hour. Find the force on an 8-square-foot sail if the wind speed is 12 miles per hour.

43. The intensity of light (in foot-candles) varies inversely as the square of x, the distance in feet from the light source. The intensity of light 2 feet from the source is 80 foot-candles. How far away is the source if the intensity of light is 5 foot-candles?

44. The horsepower that can be safely transmitted to a shaft varies jointly as the shaft's angular speed of rotation (in revolutions per minute) and the cube of its diameter. A 2-inch shaft making 120 revolutions per minute safely transmits 40 horsepower. Find how much horsepower can be safely transmitted by a 3-inch shaft making 80 revolutions per minute.

MIXED PRACTICE

Write an equation to describe each variation. Use k for the constant of proportionality. See Examples 1 through 7.

45. y varies directly as x

46. p varies directly as q

47. a varies inversely as b

48. y varies inversely as x

49. y varies jointly as x and z

50. y varies jointly as q, r, and t

51. y varies inversely as x^3

52. y varies inversely as a^4

53. y varies directly as x and inversely as p^2

54. y varies directly as a^5 and inversely as b

REVIEW AND PREVIEW

Find the exact circumference and area of each circle. See the inside cover for a list of geometric formulas.

55.

56.

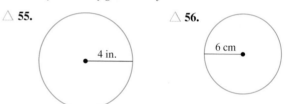

57.

58.

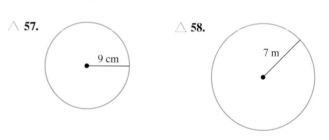

Find each square root. See Section 1.3.

59. $\sqrt{81}$

60. $\sqrt{36}$

61. $\sqrt{1}$

62. $\sqrt{4}$

63. $\sqrt{\dfrac{1}{4}}$

64. $\sqrt{\dfrac{1}{25}}$

65. $\sqrt{\dfrac{4}{9}}$

66. $\sqrt{\dfrac{25}{121}}$

CONCEPT EXTENSIONS

Solve. See the Concept Check in this section. Choose the type of variation that each equation represents. **a.** *Direct variation* **b.** *Inverse variation* **c.** *Joint variation*

67. $y = \dfrac{2}{3}x$

68. $y = \dfrac{0.6}{x}$

69. $y = 9ab$

70. $xy = \dfrac{2}{11}$

71. The horsepower to drive a boat varies directly as the cube of the speed of the boat. If the speed of the boat is to double, determine the corresponding increase in horsepower required.

72. The volume of a cylinder varies jointly as the height and the square of the radius. If the height is halved and the radius is doubled, determine what happens to the volume.

73. Suppose that y varies directly as x. If x is doubled, what is the effect on y?

74. Suppose that y varies directly as x^2. If x is doubled, what is the effect on y?

Complete the following table for the inverse variation $y = \dfrac{k}{x}$ over each given value of k. Plot the points on a rectangular coordinate system.

x	$\dfrac{1}{4}$	$\dfrac{1}{2}$	1	2	4
$y = \dfrac{k}{x}$					

75. $k = 3$　**76.** $k = 1$　**77.** $k = \dfrac{1}{2}$　**78.** $k = 5$

CHAPTER 6 GROUP ACTIVITY

Fastest-Growing Occupations

We reviewed fastest-growing occupations originally in Chapter 2. In this chapter, let's study this important data in terms of percents.

According to U.S. Bureau of Labor Statistics projections, the careers listed in the following table will be among the top twenty-five fastest-growing jobs into the next decade, according to the percent increase in the number of jobs.

Employment (in thousands)				
Occupation	2004	2014	% Change	Rank
Medical assistants	387	589		
Preschool teachers	431	573		
Computer software engineers	800	1168		
Personal and home care aides	701	988		
Physician assistants	62	93		
Network administrators	278	385		
Postsecondary teachers	1628	2153		
Dental hygienists	158	226		
Network systems and data communications analysts	231	357		
Home health aides	624	974		

What do all these fast-growing occupations have in common? They all require knowledge of math! For some careers, such as management analysts, registered nurses, and computer software engineers, the ways math is used on the job may be obvious. For other occupations, the use of math may not be quite as apparent. However, tasks common to many jobs, like filling in a time sheet, writing up an expense or mileage report, planning a budget, figuring a bill, ordering supplies, and even making a work schedule, all require math.

Group Activity

1. Find the percent change in the number of jobs available from 2004 to 2014 for each occupation in the list.

2. Rank these top-ten occupations according to percent growth, from greatest to least.

3. Which occupation will be the fastest growing during this period?

4. How many occupations will have 50% or more positions in 2014 than in 2004?

5. Which of the listed occupations will be the slowest growing during this period?

📖 STUDY SKILLS BUILDER

Are You Preparing for a Test on Chapter 6?

Below I have listed some common trouble areas for students in Chapter 6. After studying for your test—but before taking your test—read these.

- Make sure you know the difference in the following:

Simplify: $\dfrac{\dfrac{3}{x}}{\dfrac{1}{x} - \dfrac{5}{y}}$

Solve: $\dfrac{5x}{6} - \dfrac{1}{2} = \dfrac{5x}{12}$

Subtract: $\dfrac{1}{2x} - \dfrac{7}{x-3}$

Multiply numerator and denominator by the LCD.

$$\dfrac{\dfrac{3}{x} \cdot xy}{\dfrac{1}{x} \cdot xy - \dfrac{5}{y} \cdot xy}$$

$$= \dfrac{3y}{y - 5x}$$

Multiply both sides by the LCD.

$$12 \cdot \dfrac{5x}{6} - 12 \cdot \dfrac{1}{2} = 12 \cdot \dfrac{5x}{12}$$

$$2 \cdot 5x - 6 = 5x$$

$$10x - 6 = 5x$$

$$5x = 6$$

$$x = \dfrac{6}{5}$$

Write each expression as an equivalent expression with the LCD.

$$\dfrac{1 \cdot (x - 3)}{2x \cdot (x - 3)} - \dfrac{7 \cdot 2x}{(x - 3) \cdot 2x}$$

$$= \dfrac{x - 3}{2x(x - 3)} - \dfrac{14x}{2x(x - 3)}$$

$$= \dfrac{-13x - 3}{2x(x - 3)}$$

Remember: This is simply a checklist of common trouble areas. For a review of Chapter 6, see the Highlights and Chapter Review at the end of this chapter.

CHAPTER 6 VOCABULARY CHECK

Fill in each blank with one of the words or phrases listed below.

rational expression   equation   complex fraction   opposites   synthetic division

least common denominator   expression   long division   jointly   directly   inversely

1. A rational expression whose numerator, denominator, or both contain one or more rational expressions is called a
   _____ .

2. To divide a polynomial by a polynomial other than a monomial, we use _____ .

3. In the equation $y = kx$, y varies _____ as x.

4. In the equation $y = \dfrac{k}{x}$, y varies _____ as x.

5. The _____ of a list of rational expressions is a polynomial of least degree whose factors include the denominator factors in the list.

6. When a polynomial is to be divided by a binomial of the form $x - c$, a shortcut process called _____ may be used.

7. In the equation $y = kxz$, y varies _____ as x and z.

8. The expressions $(x - 5)$ and $(5 - x)$ are called _____ .

9. A _____ is an expression that can be written as the quotient $\dfrac{P}{Q}$ of two polynomials P and Q as long as Q is not 0.

10. Which is an expression and which is an equation? An example of an _____ is $\dfrac{2}{x} + \dfrac{2}{x^2} = 7$ and an example of an _____ is $\dfrac{2}{x} + \dfrac{5}{x^2}$.

▶ **Helpful Hint**

Are you preparing for your test? Don't forget to take the Chapter 6 Test on page 481. Then check your answers at the back of the text and use the Chapter Test Prep Video CD to see the fully worked-out solutions to any of the exercises you want to review.

CHAPTER 6 HIGHLIGHTS

DEFINITIONS AND CONCEPTS	EXAMPLES

SECTION 6.1   RATIONAL FUNCTIONS AND MULTIPLYING AND DIVIDING RATIONAL EXPRESSIONS

A **rational expression** is the quotient $\dfrac{P}{Q}$ of two polynomials P and Q, as long as Q is not 0.

$$\frac{2x - 6}{7}, \quad \frac{t^2 - 3t + 5}{t - 1}$$

To Simplify a Rational Expression

Step 1. Completely factor the numerator and the denominator.

Step 2. Apply the fundamental principle of rational expressions.

Simplify.

$$\frac{2x^2 + 9x - 5}{x^2 - 25} = \frac{(2x - 1)(x + 5)}{(x - 5)(x + 5)}$$
$$= \frac{2x - 1}{x - 5}$$

To Multiply Rational Expressions

Step 1. Completely factor numerators and denominators.

Step 2. Multiply the numerators and multiply the denominators.

Step 3. Apply the fundamental principle of rational expressions.

Multiply $\dfrac{x^3 + 8}{12x - 18} \cdot \dfrac{14x^2 - 21x}{x^2 + 2x}$.

$$= \frac{(x + 2)(x^2 - 2x + 4)}{6(2x - 3)} \cdot \frac{7x(2x - 3)}{x(x + 2)}$$
$$= \frac{7(x^2 - 2x + 4)}{6}$$

To Divide Rational Expressions

Multiply the first rational expression by the reciprocal of the second rational expression.

Divide $\dfrac{x^2 + 6x + 9}{5xy - 5y} \div \dfrac{x + 3}{10y}$.

$$= \frac{(x + 3)(x + 3)}{5y(x - 1)} \cdot \frac{2 \cdot 5y}{x + 3}$$
$$= \frac{2(x + 3)}{x - 1}$$

A **rational function** is a function described by a rational expression.

$$f(x) = \frac{2x - 6}{7}, \quad h(t) = \frac{t^2 - 3t + 5}{t - 1}$$

SECTION 6.2   ADDING AND SUBTRACTING RATIONAL EXPRESSIONS

To Add or Subtract Rational Expressions

Step 1. Find the LCD.

Step 2. Write each rational expression as an equivalent rational expression whose denominator is the LCD.

Step 3. Add or subtract numerators and write the result over the common denominator.

Step 4. Simplify the resulting rational expression.

Subtract $\dfrac{3}{x + 2} - \dfrac{x + 1}{x - 3}$.

$$= \frac{3 \cdot (x - 3)}{(x + 2) \cdot (x - 3)} - \frac{(x + 1) \cdot (x + 2)}{(x - 3) \cdot (x + 2)}$$
$$= \frac{3(x - 3) - (x + 1)(x + 2)}{(x + 2)(x - 3)}$$
$$= \frac{3x - 9 - (x^2 + 3x + 2)}{(x + 2)(x - 3)}$$
$$= \frac{3x - 9 - x^2 - 3x - 2}{(x + 2)(x - 3)}$$
$$= \frac{-x^2 - 11}{(x + 2)(x - 3)}$$

DEFINITIONS AND CONCEPTS	**EXAMPLES**

SECTION 6.3 SIMPLIFYING COMPLEX FRACTIONS

Method 1: Simplify the numerator and the denominator so that each is a single fraction. Then perform the indicated division and simplify if possible.

Simplify $\dfrac{\dfrac{x+2}{x}}{x-\dfrac{4}{x}}$.

Method 1: $\dfrac{\dfrac{x+2}{x}}{\dfrac{x\cdot x}{1\cdot x}-\dfrac{4}{x}}=\dfrac{\dfrac{x+2}{x}}{\dfrac{x^2-4}{x}}$

$=\dfrac{x+2}{x}\cdot\dfrac{x}{(x+2)(x-2)}=\dfrac{1}{x-2}$

Method 2: Multiply the numerator and the denominator of the complex fraction by the LCD of the fractions in both the numerator and the denominator. Then simplify if possible.

Method 2: $\dfrac{\left(\dfrac{x+2}{x}\right)\cdot x}{\left(x-\dfrac{4}{x}\right)\cdot x}=\dfrac{x+2}{x\cdot x-\dfrac{4}{x}\cdot x}$

$=\dfrac{x+2}{x^2-4}=\dfrac{x+2}{(x+2)(x-2)}=\dfrac{1}{x-2}$

SECTION 6.4 DIVIDING POLYNOMIALS: LONG DIVISION AND SYNTHETIC DIVISION

To divide a polynomial by a monomial: Divide each term in the polynomial by the monomial.

Divide $\dfrac{12a^5b^3-6a^2b^2+ab}{6a^2b^2}$

$=\dfrac{12a^5b^3}{6a^2b^2}-\dfrac{6a^2b^2}{6a^2b^2}+\dfrac{ab}{6a^2b^2}$

$=2a^3b-1+\dfrac{1}{6ab}$

To divide a polynomial by a polynomial other than a monomial:

Use long division.

Divide $2x^3-x^2-8x-1$ by $x-2$.

$$
\begin{array}{r}
2x^2+3x-2 \\
x-2\,\overline{\smash{\big)}\,2x^3-\ x^2-8x-1} \\
\underline{2x^3-4x^2} \\
3x^2-8x \\
\underline{3x^2-6x} \\
-2x-1 \\
\underline{-2x+4} \\
-5
\end{array}
$$

The quotient is $2x^2+3x-2-\dfrac{5}{x-2}$.

A shortcut method called **synthetic division** may be used to divide a polynomial by a binomial of the form $x-c$.

Use synthetic division to divide $2x^3-x^2-8x-1$ by $x-2$.

$$
\begin{array}{r|rrrr}
2 & 2 & -1 & -8 & -1 \\
  & \downarrow & 4 & 6 & -4 \\
\hline
  & 2 & 3 & -2 & -5
\end{array}
$$

The quotient is $2x^2+3x-2-\dfrac{5}{x-2}$.

DEFINITIONS AND CONCEPTS

EXAMPLES

SECTION 6.5   SOLVING EQUATIONS CONTAINING RATIONAL EXPRESSIONS

To solve an equation containing rational expressions: Multiply both sides of the equation by the LCD of all rational expressions. Then apply the distributive property and simplify. Solve the resulting equation and then check each proposed solution to see whether it makes the denominator 0. If so, it is an **extraneous solution.**

Solve $x - \dfrac{3}{x} = \dfrac{1}{2}$.

$$2x\left(x - \frac{3}{x}\right) = 2x\left(\frac{1}{2}\right) \quad \text{The LCD is } 2x.$$

$$2x \cdot x - 2x\left(\frac{3}{x}\right) = 2x\left(\frac{1}{2}\right) \quad \text{Distribute.}$$

$$2x^2 - 6 = x$$
$$2x^2 - x - 6 = 0 \quad \text{Subtract } x.$$
$$(2x + 3)(x - 2) = 0 \quad \text{Factor.}$$

$$x = -\frac{3}{2} \quad \text{or} \quad x = 2 \quad \text{Solve.}$$

Both $-\dfrac{3}{2}$ and 2 check. The solutions are 2 and $-\dfrac{3}{2}$.

SECTION 6.6   RATIONAL EQUATIONS AND PROBLEM SOLVING

Solving an Equation for a Specified Variable

Treat the specified variable as the only variable of the equation and solve as usual.

Problem-Solving Steps to Follow

Solve for x.

$$A = \frac{2x + 3y}{5}$$
$$5A = 2x + 3y \quad \text{Multiply both sides by 5.}$$
$$5A - 3y = 2x \quad \text{Subtract } 3y \text{ from both sides.}$$
$$\frac{5A - 3y}{2} = x \quad \text{Divide both sides by 2.}$$

Jeanee and David Dillon volunteer every year to clean a strip of Lake Ponchartrain Beach. Jeanee can clean all the trash in this area of beach in 6 hours; David takes 5 hours. Find how long it will take them to clean the area of beach together.

1. UNDERSTAND.

1. Read and reread the problem.

Let x = time in hours that it takes Jeanee and David to clean the beach together.

	Hours to Complete	**Part Completed in 1 Hour**
Jeanee Alone	6	$\dfrac{1}{6}$
David Alone	5	$\dfrac{1}{5}$
Together	x	$\dfrac{1}{x}$

(continued)

DEFINITIONS AND CONCEPTS	EXAMPLES

2. TRANSLATE.

2. In words:

part Jeanee can complete in 1 hour	+	part David can complete in 1 hour	=	part they can complete together in 1 hour
$\downarrow$		$\downarrow$		$\downarrow$

Translate:

$$\frac{1}{6} \quad + \quad \frac{1}{5} \quad = \quad \frac{1}{x}$$

3. SOLVE.

3. $\dfrac{1}{6} + \dfrac{1}{5} = \dfrac{1}{x}$ Multiply both sides by $30x$.

$$5x + 6x = 30$$
$$11x = 30$$
$$x = \frac{30}{11} \quad \text{or} \quad 2\frac{8}{11}$$

4. INTERPRET.

4. *Check* and then *state*. Together, they can clean the beach in $2\dfrac{8}{11}$ hours.

y **varies directly as** x, or y is **directly proportional to** x, if there is a nonzero constant k such that

$$y = kx$$

y **varies inversely as** x, or y is **inversely proportional to** x, if there is a nonzero constant k such that

$$y = \frac{k}{x}$$

y **varies jointly as** x and z, or y is **jointly proportional to** x and z, if there is a nonzero constant k such that

$$y = kxz$$

The circumference of a circle C varies directly as its radius r.

$$C = \underset{k}{2\pi} r$$

Pressure P varies inversely with volume V.

$$P = \frac{k}{V}$$

The lateral surface area S of a cylinder varies jointly as its radius r and height h.

$$S = \underset{k}{2\pi} rh$$

CHAPTER 6 REVIEW

(6.1) *Find the domain for each rational function.*

1. $f(x) = \dfrac{3 - 5x}{7}$

2. $g(x) = \dfrac{2x + 4}{11}$

3. $F(x) = \dfrac{-3x^2}{x - 5}$

4. $h(x) = \dfrac{4x}{3x - 12}$

5. $f(x) = \dfrac{x^3 + 2}{x^2 + 8x}$

6. $G(x) = \dfrac{20}{3x^2 - 48}$

Write each rational expression in lowest terms.

7. $\dfrac{x - 12}{12 - x}$

8. $\dfrac{5x - 15}{25x - 75}$

9. $\dfrac{2x}{2x^2 - 2x}$

10. $\dfrac{x + 7}{x^2 - 49}$

11. $\dfrac{2x^2 + 4x - 30}{x^2 + x - 20}$

12. The average cost (per bookcase) of manufacturing x bookcases is given by the rational function.

$$C(x) = \frac{35x + 4200}{x}$$

  a. Find the average cost per bookcase of manufacturing 50 bookcases.

  b. Find the average cost per bookcase of manufacturing 100 bookcases.

  c. As the number of bookcases increases, does the average cost per bookcase increase or decrease? (See parts **a** and **b**.)

Perform each indicated operation. Write your answers in lowest terms.

13. $\dfrac{4 - x}{5} \cdot \dfrac{15}{2x - 8}$

14. $\dfrac{x^2 - 6x + 9}{2x^2 - 18} \cdot \dfrac{4x + 12}{5x - 15}$

15. $\dfrac{a - 4b}{a^2 + ab} \cdot \dfrac{b^2 - a^2}{8b - 2a}$

16. $\dfrac{x^2 - x - 12}{2x^2 - 32} \cdot \dfrac{x^2 + 8x + 16}{3x^2 + 21x + 36}$

17. $\dfrac{4x + 8y}{3} \div \dfrac{5x + 10y}{9}$

18. $\dfrac{x^2 - 25}{3} \div \dfrac{x^2 - 10x + 25}{x^2 - x - 20}$

19. $\dfrac{a - 4b}{a^2 + ab} \div \dfrac{20b - 5a}{b^2 - a^2}$

20. $\dfrac{3x + 3}{x - 1} \div \dfrac{x^2 - 6x - 7}{x^2 - 1}$

21. $\dfrac{2x - x^2}{x^3 - 8} \div \dfrac{x^2}{x^2 + 2x + 4}$

22. $\dfrac{5x - 15}{3 - x} \cdot \dfrac{x + 2}{10x + 20} \cdot \dfrac{x^2 - 9}{x^2 - x - 6}$

(6.2) *Find the LCD of the rational expressions in each list.*

23. $\dfrac{5}{4x^2y^5}, \dfrac{3}{10x^2y^4}, \dfrac{x}{6y^4}$

24. $\dfrac{5}{2x}, \dfrac{7}{x - 2}$

25. $\dfrac{3}{5x}, \dfrac{2}{x - 5}$

26. $\dfrac{1}{5x^3}, \dfrac{4}{x^2 + 3x - 28}, \dfrac{11}{10x^2 - 30x}$

Perform each indicated operation. Write your answers in lowest terms.

27. $\dfrac{4}{x - 4} + \dfrac{x}{x - 4}$

28. $\dfrac{4}{3x^2} + \dfrac{2}{3x^2}$

29. $\dfrac{1}{x - 2} - \dfrac{1}{4 - 2x}$

30. $\dfrac{1}{10 - x} + \dfrac{x - 1}{x - 10}$

31. $\dfrac{x}{9 - x^2} - \dfrac{2}{5x - 15}$

32. $2x + 1 - \dfrac{1}{x - 3}$

33. $\dfrac{2}{a^2 - 2a + 1} + \dfrac{3}{a^2 - 1}$

34. $\dfrac{x}{9x^2 + 12x + 16} - \dfrac{3x + 4}{27x^3 - 64}$

Perform each indicated operation. Write your answers in lowest terms.

35. $\dfrac{2}{x - 1} - \dfrac{3x}{3x - 3} + \dfrac{1}{2x - 2}$

△ **36.** Find the perimeter of the heptagon (a polygon with seven sides).

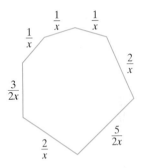

(6.3) *Simplify each complex fraction.*

37. $\dfrac{1 - \dfrac{3x}{4}}{2 + \dfrac{x}{4}}$

38. $\dfrac{\dfrac{x^2}{15}}{\dfrac{x + 1}{5x}}$

39. $\dfrac{2 - \dfrac{3}{2x}}{x - \dfrac{2}{5x}}$

40. $\dfrac{1 + \dfrac{x}{y}}{\dfrac{x^2}{y^2} - 1}$

41. $\dfrac{\dfrac{5}{x} + \dfrac{1}{xy}}{\dfrac{3}{x^2}}$

42. $\dfrac{\dfrac{x}{3} - \dfrac{3}{x}}{1 + \dfrac{3}{x}}$

43. $\dfrac{\dfrac{1}{x - 1} + 1}{\dfrac{1}{x + 1} - 1}$

44. $\dfrac{\dfrac{x - 3}{x + 3} + \dfrac{x + 3}{x - 3}}{\dfrac{x - 3}{x + 3} - \dfrac{x + 3}{x - 3}}$

If $f(x) = \dfrac{3}{x}, x \neq 0$, find each of the following.

45. $f(a + h)$ **46.** $f(a)$

47. Use Exercises 45 and 46 to find $\dfrac{f(a + h) - f(a)}{h}$.

48. Simplify the results of Exercise 47.

(6.4)

49. $(4xy + 2x^2 - 9) \div 4xy$

50. Divide $12xb^2 + 16xb^4$ by $4xb^3$.

51. $(3x^4 - 25x^2 - 20) \div (x - 3)$

52. $(-x^2 + 2x^4 + 5x - 12) \div (x + 2)$

53. $(2x^3 + 3x^2 - 2x + 2) \div (2x + 3)$

54. $(3x^4 + 5x^3 + 7x^2 + 3x - 2) \div (x^2 + x + 2)$

Use synthetic division to find each quotient.

55. $(3x^3 + 12x - 4) \div (x - 2)$

56. $(x^5 - 1) \div (x + 1)$

57. $(x^3 - 81) \div (x - 3)$

58. $(3x^4 - 2x^2 + 10) \div (x + 2)$

If $P(x) = 3x^5 - 9x + 7$, use the remainder theorem to find the following.

59. $P(4)$

60. $P(-5)$

△ **61.** $P\left(-\dfrac{1}{2}\right)$

62. If the area of the rectangle is $(x^4 - x^3 - 6x^2 - 6x + 18)$ square miles and its width is $(x - 3)$ miles, find the length.

$$\boxed{\begin{array}{c} x^4 - x^3 - 6x^2 - 6x + 18 \\ \text{square miles} \end{array}} \quad \begin{array}{c} x - 3 \\ \text{miles} \end{array}$$

(6.5) Solve each equation.

63. $\dfrac{3}{x} + \dfrac{1}{3} = \dfrac{5}{x}$

64. $\dfrac{2x + 3}{5x - 9} = \dfrac{3}{2}$

65. $\dfrac{1}{x - 2} - \dfrac{3x}{x^2 - 4} = \dfrac{2}{x + 2}$

66. $\dfrac{7}{x} - \dfrac{x}{7} = 0$

Solve each equation or perform each indicated operation. Simplify.

67. $\dfrac{5}{x^2 - 7x} + \dfrac{4}{2x - 14}$

68. $3 - \dfrac{5}{x} - \dfrac{2}{x^2} = 0$

69. $\dfrac{4}{3 - x} - \dfrac{7}{2x - 6} + \dfrac{5}{x}$

(6.6) Solve each equation for the specified variable.

△ **70.** $A = \dfrac{h(a + b)}{2}$ for a

71. $\dfrac{1}{R} = \dfrac{1}{R_1} + \dfrac{1}{R_2}$ for R_2

72. $I = \dfrac{E}{R + r}$ for R

73. $A = P + Prt$ for r

74. $H = \dfrac{kA(T_1 - T_2)}{L}$ for A

Solve.

75. The sum of a number and twice its reciprocal is 3. Find the number(s).

76. If a number is added to the numerator of $\dfrac{3}{7}$, and twice that number is added to the denominator of $\dfrac{3}{7}$, the result is equivalent to $\dfrac{10}{21}$. Find the number.

77. Three boys can paint a fence in 4 hours, 5 hours, and 6 hours, respectively. Find how long it will take all three boys to paint the fence.

78. If Sue Katz can type a certain number of mailing labels in 6 hours and Tom Neilson and Sue working together can type the same number of mailing labels in 4 hours, find how long it takes Tom alone to type the mailing labels.

79. The speed of a Ranger boat in still water is 32 mph. If the boat travels 72 miles upstream in the same time that it takes to travel 120 miles downstream, find the current of the stream.

80. The speed of a jogger is 3 mph faster than the speed of a walker. If the jogger travels 14 miles in the same amount of time that the walker travels 8 miles, find the speed of the walker.

(6.7) Solve each variation problem.

81. A is directly proportional to B. If $A = 6$ when $B = 14$, find A when $B = 21$.

82. According to Boyle's law, the pressure exerted by a gas is inversely proportional to the volume, as long as the temperature stays the same. If a gas exerts a pressure of 1250 kilopascals when the volume is 2 cubic meters, find the volume when the pressure is 800 kilopascals.

MIXED REVIEW

For expressions, perform the indicated operation and/or simplify. For equations, solve the equation for the unknown variable.

83. $\dfrac{22x + 8}{11x + 4}$

84. $\dfrac{xy - 3x + 2y - 6}{x^2 + 4x + 4}$

85. $\dfrac{2}{5x} \div \dfrac{4 - 18x}{6 - 27x}$

86. $\dfrac{7x + 28}{2x + 4} \div \dfrac{x^2 + 2x - 8}{x^2 - 2x - 8}$

87. $\dfrac{5a^2 - 20}{a^3 + 2a^2 + a + 2} \div \dfrac{7a}{a^3 + a}$

88. $\dfrac{4a + 8}{5a^2 - 20} \cdot \dfrac{3a^2 - 6a}{a + 3} \div \dfrac{2a^2}{5a + 15}$

89. $\dfrac{7}{2x} + \dfrac{5}{6x}$

90. $\dfrac{x - 2}{x + 1} - \dfrac{x - 3}{x - 1}$

91. $\dfrac{2x + 1}{x^2 + x - 6} + \dfrac{2 - x}{x^2 + x - 6}$

92. $\dfrac{2}{x^2 - 16} - \dfrac{3x}{x^2 + 8x + 16} + \dfrac{3}{x + 4}$

93. $\dfrac{\dfrac{1}{x} - \dfrac{2}{3x}}{\dfrac{5}{2x} - \dfrac{1}{3}}$

94. $\dfrac{2}{1 - \dfrac{2}{x}}$

95. $\dfrac{\dfrac{x^2 + 5x - 6}{4x + 3}}{\dfrac{(x + 6)^2}{8x + 6}}$

96. $\dfrac{\dfrac{3}{x - 1} - \dfrac{2}{1 - x}}{\dfrac{2}{x - 1} - \dfrac{2}{x}}$

97. $4 + \dfrac{8}{x} = 8$

98. $\dfrac{x - 2}{x^2 - 7x + 10} = \dfrac{1}{5x - 10} - \dfrac{1}{x - 5}$

99. The denominator of a fraction is 2 more than the numerator. If the numerator is decreased by 3 and the denominator is increased by 5, the resulting fraction is equivalent to $\dfrac{2}{3}$. Find the fraction.

100. The sum of the reciprocals of two consecutive even integers is $-\dfrac{9}{40}$. Find the two integers.

101. The inlet pipe of a water tank can fill the tank in 2 hours and 30 minutes. The outlet pipe can empty the tank in 2 hours. Find how long it takes to empty a full tank if both pipes are open.

102. Timmy Garnica drove 210 miles in the same amount of time that it took a DC-10 jet to travel 1715 miles. The speed of the jet was 430 mph faster than the speed of the car. Find the speed of the jet.

103. Two Amtrak trains traveling on parallel tracks leave Tucson at the same time. In 6 hours the faster train is 382 miles from Tucson and the trains are 112 miles apart. Find how fast each train is traveling.

104. C is inversely proportional to D. If $C = 12$ when $D = 8$, find C when $D = 24$.

105. The surface area of a sphere varies directly as the square of its radius. If the surface area is 36π square inches when the radius is 3 inches, find the surface area when the radius is 4 inches.

106. Divide $(x^3 - x^2 + 3x^4 - 2)$ by $(x - 4)$.

CHAPTER 6 TEST TEST PREP

Remember to use the Chapter Test Prep Video CD to see the fully worked-out solutions to any of the exercises you want to review.

Find the domain of each rational function.

1. $f(x) = \dfrac{5x^2}{1 - x}$

2. $g(x) = \dfrac{9x^2 - 9}{x^2 + 4x + 3}$

Write each rational expression in lowest terms.

3. $\dfrac{7x - 21}{24 - 8x}$

4. $\dfrac{x^2 - 4x}{x^2 + 5x - 36}$

5. $\dfrac{x^3 - 8}{x - 2}$

Perform the indicated operation. If possible, simplify your answer.

6. $\dfrac{2x^3 + 16}{6x^2 + 12x} \cdot \dfrac{5}{x^2 - 2x + 4}$

7. $\dfrac{5}{4x^3} + \dfrac{7}{4x^3}$

8. $\dfrac{3x^2 - 12}{x^2 + 2x - 8} \div \dfrac{6x + 18}{x + 4}$

9. $\dfrac{4x - 12}{2x - 9} \div \dfrac{3 - x}{4x^2 - 81} \cdot \dfrac{x + 3}{5x + 15}$

10. $\dfrac{3 + 2x}{10 - x} + \dfrac{13 + x}{x - 10}$

11. $\dfrac{2x^2 + 7}{2x^4 - 18x^2} - \dfrac{6x + 7}{2x^4 - 18x^2}$

12. $\dfrac{3}{x^2 - x - 6} + \dfrac{2}{x^2 - 5x + 6}$

13. $\dfrac{5}{x - 7} - \dfrac{2x}{3x - 21} + \dfrac{x}{2x - 14}$

14. $\dfrac{3x}{5} \cdot \left(\dfrac{5}{x} - \dfrac{5}{2x}\right)$

Simplify each complex fraction.

15. $\dfrac{\dfrac{5}{x} - \dfrac{7}{3x}}{\dfrac{9}{8x} - \dfrac{1}{x}}$

16. $\dfrac{\dfrac{x^2 - 5x + 6}{x + 3}}{\dfrac{x^2 - 4x + 4}{x^2 - 9}}$

Divide.

17. $(4x^2y + 9x + 3xz) \div 3xz$

18. $(4x^3 - 5x) \div (2x + 1)$

19. Use synthetic division to divide $(4x^4 - 3x^3 - x - 1)$ by $(x + 3)$.

20. If $P(x) = 4x^4 + 7x^2 - 2x - 5$, use the remainder theorem to find $P(-2)$.

Solve each equation for x.

21. $\dfrac{x}{x-4} = 3 - \dfrac{4}{x-4}$

22. $\dfrac{3}{x+2} - \dfrac{1}{5x} = \dfrac{2}{5x^2 + 10x}$

23. $\dfrac{x^2 + 8}{x} - 1 = \dfrac{2(x+4)}{x}$

24. Solve for x: $\dfrac{x+b}{a} = \dfrac{4x - 7a}{b}$

25. The product of one more than a number and twice the reciprocal of the number is $\dfrac{12}{5}$. Find the number.

26. If Jan can weed the garden in 2 hours and her husband can weed it in 1 hour and 30 minutes, find how long it takes them to weed the garden together.

27. Suppose that W is inversely proportional to V. If $W = 20$ when $V = 12$, find W when $V = 15$.

28. Suppose that Q is jointly proportional to R and the square of S. If $Q = 24$ when $R = 3$ and $S = 4$, find Q when $R = 2$ and $S = 3$.

29. When an anvil is dropped into a gorge, the speed with which it strikes the ground is directly proportional to the square root of the distance it falls. An anvil that falls 400 feet hits the ground at a speed of 160 feet per second. Find the height of a cliff over the gorge if a dropped anvil hits the ground at a speed of 128 feet per second.

CHAPTER 6 CUMULATIVE REVIEW

1. Translate each phrase to an algebraic expression. Use the variable x to represent each unknown number.

  a. Eight times a number

  b. Three more than eight times a number

  c. The quotient of a number and -7

  d. One and six-tenths subtracted from twice a number

  e. Six less than a number

  f. Twice the sum of four and a number

2. Translate each phrase to an algebraic expression. Use the variable x to represent each unknown number.

  a. One third subtracted from a number

  b. Six less than five times a number

  c. Three more than eight times a number

  d. The quotient of seven and the difference of two and a number.

3. Solve for y: $\dfrac{y}{3} - \dfrac{y}{4} = \dfrac{1}{6}$

4. Solve $\dfrac{x}{7} + \dfrac{x}{5} = \dfrac{12}{5}$

5. The formula $C = \dfrac{5}{9}(F - 32)$ converts degrees Fahrenheit to degrees Celsius. Use this formula and the table feature of your calculator to complete the table. If necessary, round values to the nearest hundredth.

Fahrenheit	-4	10	32	70	100
Celsius					

6. Olivia has scores of 78, 65, 82, and 79 on her algebra tests. Use an inequality to find the minimum score she can make on her final exam to pass the course with a 78 average or higher, given that the final exam counts as two tests.

7. Solve: $\left| \dfrac{3x+1}{2} \right| = -2$

8. Solve: $\left| \dfrac{2x-1}{3} \right| + 6 = 3$

9. Solve for x: $\left| \dfrac{2(x+1)}{3} \right| \leq 0$

10. Solve for x: $\left| \dfrac{3(x-1)}{4} \right| \geq 2$

11. Graph the equation $y = -2x + 3$.

12. Graph the equation $y = -x + 3$.

13. Which of the following relations are also functions?

  a. $\{(-2, 5), (2, 7), (-3, 5), (9, 9)\}$

  b.

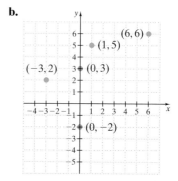

  c.

Input	Correspondence	Output
People in a certain city	Each person's age	The set of nonnegative integers

14. If $f(x) = -x^2 + 3x - 2$, find

    a. f(0)   **b.** $f(-3)$ **c.** $f\left(\dfrac{1}{3}\right)$

15. Graph $x - 3y = 6$ by plotting intercept points.

16. Graph $3x - y = 6$ by plotting x- and y-intercepts.

17. Find an equation of the line with slope -3 containing the point $(1, -5)$. Write the equation in slope-intercept form $y = mx + b$.

18. Find an equation of the line with slope $\dfrac{1}{2}$ containing the point $(-1, 3)$. Use function notation to write the equation.

19. Graph the intersection of $x \geq 1$ and $y \geq 2x - 1$.

20. Graph the union of $2x + y \leq 4$ or $y > 2$.

21. Use the elimination method to solve the system.
$$\begin{cases} 3x - 2y = 10 \\ 4x - 3y = 15 \end{cases}$$

22. Use the substitution method to solve the system.
$$\begin{cases} -2x + 3y = 6 \\ 3x - y = 5 \end{cases}$$

23. Solve the system. $\begin{cases} 2x - 4y + 8z = 2 \\ -x - 3y + z = 11 \\ x - 2y + 4z = 0 \end{cases}$

24. Solve the system. $\begin{cases} 2x - 2y + 4z = 6 \\ -4x - y + z = -8 \\ 3x - y + z = 6 \end{cases}$

25. The measure of the largest angle of a triangle is $80°$ more than the measure of the smallest angle, and the measure of the remaining angle is $10°$ more than the measure of the smallest angle. Find the measure of each angle.

26. Kernersville office supply sold three reams of paper and two boxes of manila folders for $21.90. Also, five reams of paper and one box of manila folders cost $24.25. Find the price of a ream of paper and a box of manila folders.

27. Use matrices to solve the system. $\begin{cases} x + 2y + z = 2 \\ -2x - y + 2z = 5 \\ x + 3y - 2z = -8 \end{cases}$

28. Use matrices to solve the system. $\begin{cases} x + y + z = 9 \\ 2x - 2y + 3z = 2 \\ -3x + y - z = 1 \end{cases}$

29. Evaluate the following.

    a. 7^0 **b.** -7^0

    c. $(2x + 5)^0$ **d.** $2x^0$

30. Simplify the following. Write answers with positive exponents.

    a. $2^{-2} + 3^{-1}$ **b.** $-6a^0$ **c.** $\dfrac{x^{-5}}{x^{-2}}$

31. Simplify each. Assume that a and b are integers and that x and y are not 0.

    a. $x^{-b}(2x^b)^2$ **b.** $\dfrac{(y^{3a})^2}{y^{a-6}}$

32. Simplify each. Assume that a and b are integers and that x and y are not 0.

    a. $3x^{4a}(4x^{-a})^2$ **b.** $\dfrac{(y^{4b})^3}{y^{2b-3}}$

33. Find the degree of each term.

    a. $3x^2$ **b.** -2^3x^5 **c.** y

    d. $12x^2yz^3$ **e.** 5.27

34. Subtract $(2x - 7)$ from $2x^2 + 8x - 3$.

35. Multiply $[3 + (2a + b)]^2$.

36. Multiply $[4 + (3x - y)]^2$.

37. Factor $ab - 6a + 2b - 12$.

38. Factor $xy + 2x - 5y - 10$.

39. Factor $2n^2 - 38n + 80$.

40. Factor $6x^2 - x - 35$.

41. Factor $x^2 + 4x + 4 - y^2$.

42. Factor $4x^2 - 4x + 1 - 9y^2$.

43. Solve $(x + 2)(x - 6) = 0$.

44. Solve $2x(3x + 1)(x - 3) = 0$.

45. Simplify:

    a. $\dfrac{2x^2}{10x^3 - 2x^2}$

    b. $\dfrac{9x^2 + 13x + 4}{8x^2 + x - 7}$

46. For the graph of $f(x)$, answer the following:

    a. Find the domain and range.

    b. List the x- and y-intercepts.

    c. Find the coordinates of the point with the greatest y-value.

    d. Find the coordinates of the point with the least y-value.

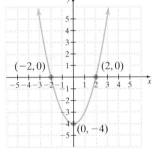

    e. List the x-values whose y-values are equal to 0.

    f. List the x-values whose y-values are less than 0.

    g. Find the solutions of $f(x) = 0$.

47. Subtract $\dfrac{5k}{k^2 - 4} - \dfrac{2}{k^2 + k - 2}$.

48. Subtract $\dfrac{5a}{a^2 - 4} - \dfrac{3}{2 - a}$.

49. Solve: $\dfrac{3}{x} - \dfrac{x + 21}{3x} = \dfrac{5}{3}$.

50. Solve: $\dfrac{3x - 4}{2x} = -\dfrac{8}{x}$.

7 Rational Exponents, Radicals, and Complex Numbers

In this chapter, radical notation is reviewed, and then rational exponents are introduced. As the name implies, rational exponents are exponents that are rational numbers. We present an interpretation of rational exponents that is consistent with the meaning and rules already established for integer exponents, and we present two forms of notation for roots: radical and exponent. We conclude this chapter with complex numbers, a natural extension of the real number system.

What is a zorb? Simply put, a zorb is a large inflated ball within a ball, and zorbing is a recreational activity which may involve rolling down a hill while strapped in a zorb. Zorbing started in New Zealand (as well as bungee jumping) and was invented by Andrew Akers and Dwane van der Sluis. The first site was set up in New Zealand's North Island. This downhill course has a length of about 490 feet, and you can reach speeds of up to 20 mph.

An example of a course is shown in the diagram below, and you can see the mathematics involved. In Section 7.3, Exercise 115, page 508, you will calculate the outer radius of a zorb, which would certainly be closely associated with the cost of production.

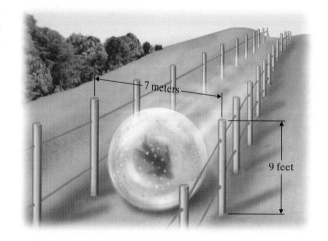

7 meters

9 feet

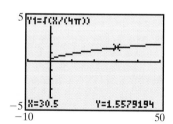

Surface area (A) vs. radius (r) of a sphere

The radius of a sphere with surface area A is calculated using the formula $r = \sqrt{\dfrac{A}{4\pi}}$. In this chapter, we learn that this can also be written as $r = \dfrac{1}{\sqrt{4\pi}} \cdot \sqrt{A}$. This means that r, radius, varies directly as the square root of A. The graph illustrates that as the surface area increases, so does the radius of a sphere.

7.1 RADICALS AND RADICAL FUNCTIONS

OBJECTIVES

1 Find square roots.

2 Approximate roots.

3 Find cube roots.

4 Find nth roots.

5 Find $\sqrt[n]{a^n}$ where a is a real number.

6 Graph square and cube root functions.

OBJECTIVE 1 ▶ Finding square roots. Recall from Section 1.3 that to find a **square root** of a number a, we find a number that was squared to get a.

Thus, because

$$5^2 = 25 \quad \text{and} \quad (-5)^2 = 25, \text{ then}$$

both 5 and -5 are square roots of 25.

Recall that we denote the **nonnegative**, or **principal**, **square root** with the **radical sign.**

$$\sqrt{25} = 5$$

We denote the **negative square root** with the **negative radical sign.**

$$-\sqrt{25} = -5$$

An expression containing a radical sign is called a **radical expression.** An expression within, or "under," a radical sign is called a **radicand.**

$$\text{radical expression}: \quad \overset{\text{radical sign}}{\sqrt{\underset{\text{radicand}}{a}}}$$

> **Principal and Negative Square Roots**
>
> If a is a nonnegative number, then
>
> $\sqrt{a}$ is the **principal,** or **nonnegative, square root** of a
>
> $-\sqrt{a}$ is the **negative square root** of a

 EXAMPLE 1   Simplify. Assume that all variables represent positive numbers.

a. $\sqrt{36}$ **b.** $\sqrt{0}$ **c.** $\sqrt{\dfrac{4}{49}}$ **d.** $\sqrt{0.25}$

e. $\sqrt{x^6}$ **f.** $\sqrt{9x^{12}}$ **g.** $-\sqrt{81}$ **h.** $\sqrt{-81}$

Solution

a. $\sqrt{36} = 6$ because $6^2 = 36$ and 6 is not negative.

b. $\sqrt{0} = 0$ because $0^2 = 0$ and 0 is not negative.

c. $\sqrt{\dfrac{4}{49}} = \dfrac{2}{7}$ because $\left(\dfrac{2}{7}\right)^2 = \dfrac{4}{49}$ and $\dfrac{2}{7}$ is not negative.

d. $\sqrt{0.25} = 0.5$ because $(0.5)^2 = 0.25$.

e. $\sqrt{x^6} = x^3$ because $(x^3)^2 = x^6$.

f. $\sqrt{9x^{12}} = 3x^6$ because $(3x^6)^2 = 9x^{12}$.

g. $-\sqrt{81} = -9$. The negative in front of the radical indicates the negative square root of 81.

h. $\sqrt{-81}$ is not a real number.   □

PRACTICE

1   Simplify. Assume that all variables represent positive numbers.

a. $\sqrt{49}$ **b.** $\sqrt{\dfrac{0}{1}}$ **c.** $\sqrt{\dfrac{16}{81}}$ **d.** $\sqrt{0.64}$

e. $\sqrt{z^8}$ **f.** $\sqrt{16b^4}$ **g.** $-\sqrt{36}$ **h.** $\sqrt{-36}$

TECHNOLOGY NOTE

Most graphing utilities will give a message such as the one below when taking the square root of a negative number when you are in real mode.

```
ERR:NONREAL ANS
1▪Quit
2:Goto
```

Recall from Section 1.3 our discussion of the square root of a negative number. For example, can we simplify $\sqrt{-4}$? That is, can we find a real number whose square is -4? No, there is no real number whose square is -4, and we say that $\sqrt{-4}$ is not a real number. In general:

The square root of a negative number is not a real number. (See the first Technology Note in this section.)

> ▶ **Helpful Hint**
> - Remember: $\sqrt{0} = 0$
> - Don't forget, the square root of a negative number, such as $\sqrt{-9}$, is not a real number. In Section 7.7, we will see what kind of a number $\sqrt{-9}$ is.

OBJECTIVE 2 ▶ **Approximating roots.** Recall that numbers such as 1, 4, 9, and 25 are called **perfect squares,** since $1 = 1^2, 4 = 2^2, 9 = 3^2$, and $25 = 5^2$. Square roots of perfect square radicands simplify to rational numbers. What happens when we try to simplify a root such as $\sqrt{3}$? Since there is no rational number whose square is 3, then $\sqrt{3}$ is not a rational number. It is called an **irrational number,** and we can find a decimal **approximation** of it. To find decimal approximations, use a calculator. For example, an approximation for $\sqrt{3}$ is

$$\sqrt{3} \approx 1.732$$
$$\uparrow$$
$$\text{approximation symbol}$$

To see if the approximation is reasonable, notice that since

$$1 < 3 < 4, \text{ then}$$
$$\sqrt{1} < \sqrt{3} < \sqrt{4}, \text{ or}$$
$$1 < \sqrt{3} < 2.$$

We found $\sqrt{3} \approx 1.732$, a number between 1 and 2, so our result is reasonable.

EXAMPLE 2 Use a calculator to approximate $\sqrt{20}$. Round the approximation to 3 decimal places and check to see that your approximation is reasonable.

$$\sqrt{20} \approx 4.472 \quad \text{A calculator screen is shown in the margin.}$$

Solution Is this reasonable? Since $16 < 20 < 25$, then $\sqrt{16} < \sqrt{20} < \sqrt{25}$, or $4 < \sqrt{20} < 5$. The approximation is between 4 and 5 and thus is reasonable. ☐

PRACTICE
2 Use a calculator to approximate $\sqrt{45}$. Round the approximation to three decimal places and check to see that your approximation is reasonable.

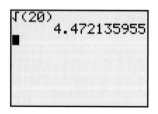

A calculator solution for Example 2.

OBJECTIVE 3 ▶ **Finding cube roots.** Finding roots can be extended to other roots such as cube roots. For example, since $2^3 = 8$, we call 2 the **cube root** of 8. In symbols, we write

$$\sqrt[3]{8} = 2$$

> **Cube Root**
> The **cube root** of a real number a is written as $\sqrt[3]{a}$, and
> $$\sqrt[3]{a} = b \text{ only if } b^3 = a$$

From this definition, we have

$$\sqrt[3]{64} = 4 \text{ since } 4^3 = 64$$
$$\sqrt[3]{-27} = -3 \text{ since } (-3)^3 = -27$$
$$\sqrt[3]{x^3} = x \text{ since } x^3 = x^3$$

Notice that, unlike with square roots, *it is possible to have a negative radicand when finding a cube root*. This is so because the *cube* of a negative number is a negative number. Therefore, the *cube root* of a negative number is a negative number.

EXAMPLE 3   Find the cube roots.

a. $\sqrt[3]{1}$ **b.** $\sqrt[3]{-64}$ **c.** $\sqrt[3]{\dfrac{8}{125}}$ **d.** $\sqrt[3]{x^6}$ **e.** $\sqrt[3]{-27x^9}$

Solution

a. $\sqrt[3]{1} = 1$ because $1^3 = 1$.
b. $\sqrt[3]{-64} = -4$ because $(-4)^3 = -64$.
c. $\sqrt[3]{\dfrac{8}{125}} = \dfrac{2}{5}$ because $\left(\dfrac{2}{5}\right)^3 = \dfrac{8}{125}$.
d. $\sqrt[3]{x^6} = x^2$ because $(x^2)^3 = x^6$.
e. $\sqrt[3]{-27x^9} = -3x^3$ because $(-3x^3)^3 = -27x^9$.

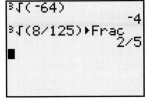

A calculator check for Example 3b and c.

PRACTICE
3   Find the cube roots.

a. $\sqrt[3]{-1}$ **b.** $\sqrt[3]{27}$ **c.** $\sqrt[3]{\dfrac{27}{64}}$ **d.** $\sqrt[3]{x^{12}}$ **e.** $\sqrt[3]{-8x^3}$

OBJECTIVE 4 ▶ Finding *n*th roots. Just as we can raise a real number to powers other than 2 or 3, we can find roots other than square roots and cube roots. In fact, we can find the **nth root** of a number, where *n* is any natural number. In symbols, the *n*th root of *a* is written as $\sqrt[n]{a}$, where *n* is called the **index.** The index 2 is usually omitted for square roots.

> ▶ **Helpful Hint**
> If the index is even, such as $\sqrt{\ }$, $\sqrt[4]{\ }$, $\sqrt[6]{\ }$, and so on, the radicand must be nonnegative for the root to be a real number. For example,
> $$\sqrt[4]{16} = 2, \text{ but } \sqrt[4]{-16} \text{ is not a real number.}$$
> $$\sqrt[6]{64} = 2, \text{ but } \sqrt[6]{-64} \text{ is not a real number.}$$
> If the index is odd, such as $\sqrt[3]{\ }$, $\sqrt[5]{\ }$, and so on, the radicand may be any real number. For example,
> $$\sqrt[3]{64} = 4 \quad \text{and} \quad \sqrt[3]{-64} = -4$$
> $$\sqrt[5]{32} = 2 \quad \text{and} \quad \sqrt[5]{-32} = -2$$

Concept Check ☑

Which one is not a real number?

a. $\sqrt[3]{-15}$ **b.** $\sqrt[4]{-15}$ **c.** $\sqrt[5]{-15}$ **d.** $\sqrt{(-15)^2}$

EXAMPLE 4   Simplify the following expressions.

a. $\sqrt[4]{81}$ **b.** $\sqrt[5]{-243}$ **c.** $-\sqrt{25}$ **d.** $\sqrt[4]{-81}$ **e.** $\sqrt[3]{64x^3}$

Solution

a. $\sqrt[4]{81} = 3$ because $3^4 = 81$ and 3 is positive.
b. $\sqrt[5]{-243} = -3$ because $(-3)^5 = -243$.
c. $-\sqrt{25} = -5$ because -5 is the opposite of $\sqrt{25}$.

A calculator check for Example 4a, b, c.

Answer to Concept Check: b

d. $\sqrt[4]{-81}$ is not a real number. There is no real number that, when raised to the fourth power, is -81.

e. $\sqrt[3]{64x^3} = 4x$ because $(4x)^3 = 64x^3$. ☐

PRACTICE

4 Simplify the following expressions.

a. $\sqrt[4]{10000}$ **b.** $\sqrt[5]{-1}$ **c.** $-\sqrt{81}$ **d.** $\sqrt[4]{-625}$ **e.** $\sqrt[3]{27x^9}$

OBJECTIVE 5 ▶ Finding $\sqrt[n]{a^n}$ where a is a real number. Recall that the notation $\sqrt{a^2}$ indicates the positive square root of a^2 only. For example,

$$\sqrt{(-5)^2} = \sqrt{25} = 5$$

When variables are present in the radicand and it is unclear whether the variable represents a positive number or a negative number, absolute value bars are sometimes needed to ensure that the result is a positive number. For example,

$$\sqrt{x^2} = |x|$$

This ensures that the result is positive. This same situation may occur when the index is any *even* positive integer. When the index is any *odd* positive integer, absolute value bars are not necessary.

Finding $\sqrt[n]{a^n}$

If n is an *even* positive integer, then $\sqrt[n]{a^n} = |a|$.

If n is an *odd* positive integer, then $\sqrt[n]{a^n} = a$.

EXAMPLE 5 Simplify.

a. $\sqrt{(-3)^2}$ **b.** $\sqrt{x^2}$ **c.** $\sqrt[4]{(x-2)^4}$ **d.** $\sqrt[3]{(-5)^3}$

e. $\sqrt[5]{(2x-7)^5}$ **f.** $\sqrt{25x^2}$ **g.** $\sqrt{x^2 + 2x + 1}$

Solution

a. $\sqrt{(-3)^2} = |-3| = 3$ When the index is even, the absolute value bars ensure us that our result is not negative.

b. $\sqrt{x^2} = |x|$

c. $\sqrt[4]{(x-2)^4} = |x-2|$

d. $\sqrt[3]{(-5)^3} = -5$

e. $\sqrt[5]{(2x-7)^5} = 2x - 7$ Absolute value bars are not needed when the index is odd.

f. $\sqrt{25x^2} = 5|x|$

g. $\sqrt{x^2 + 2x + 1} = \sqrt{(x+1)^2} = |x+1|$ ☐

PRACTICE

5 Simplify.

a. $\sqrt{(-4)^2}$ **b.** $\sqrt{x^{14}}$ **c.** $\sqrt[4]{(x+7)^4}$ **d.** $\sqrt[3]{(-7)^3}$

e. $\sqrt[5]{(3x-5)^5}$ **f.** $\sqrt{49x^2}$ **g.** $\sqrt{x^2 + 4x + 4}$

OBJECTIVE 6 ▶ Graphing square and cube root functions. Recall that an equation in x and y describes a function if each x-value is paired with exactly one y-value. With this in mind, does the equation

$$y = \sqrt{x}$$

describe a function? First, notice that replacement values for x must be nonnegative real numbers, since $\sqrt{x}$ is not a real number if $x < 0$. The notation $\sqrt{x}$ denotes the principal square root of x, so for every nonnegative number x, there is exactly one number, $\sqrt{x}$. Therefore, $y = \sqrt{x}$ describes a function, and we may write it as

$$f(x) = \sqrt{x}$$

In general, radical functions are functions of the form

$$f(x) = \sqrt[n]{x}.$$

Recall that the domain of a function in x is the set of all possible replacement values of x. This means that if n is even, the domain is the set of all nonnegative numbers, $\{x \mid x \geq 0\}$ or $[0, \infty)$ in interval notation. If n is odd, the domain is the set of all real numbers or $(-\infty, \infty)$. Keep this in mind as we find function values.

EXAMPLE 6 If $f(x) = \sqrt{x - 4}$ and $g(x) = \sqrt[3]{x + 2}$, find each function value.

a. $f(8)$ **b.** $f(6)$ **c.** $g(-1)$ **d.** $g(1)$

Solution

a. $f(8) = \sqrt{8 - 4} = \sqrt{4} = 2$ **b.** $f(6) = \sqrt{6 - 4} = \sqrt{2}$

c. $g(-1) = \sqrt[3]{-1 + 2} = \sqrt[3]{1} = 1$ **d.** $g(1) = \sqrt[3]{1 + 2} = \sqrt[3]{3}$

□

PRACTICE

6 If $f(x) = \sqrt{x + 5}$ and $g(x) = \sqrt[3]{x - 3}$, find each function value.

a. $f(11)$ **b.** $f(-1)$ **c.** $g(11)$ **d.** $g(-5)$

▶ **Helpful Hint**

Notice that for the function $f(x) = \sqrt{x - 4}$, the domain includes all real numbers that make the radicand ≥ 0. To see what numbers these are, solve $x - 4 \geq 0$ and find that $x \geq 4$. The domain is $\{x \mid x \geq 4\}$ or $[4, \infty)$ in interval notation.

The domain of the cube root function $g(x) = \sqrt[3]{x + 2}$ is the set of real numbers, or $(-\infty, \infty)$.

EXAMPLE 7 Identify the domain of the square root function $f(x) = \sqrt{x}$ and then graph the function.

Solution Recall that we graphed this function in Section 2.7. To become familiar with this function again, we find function values for $f(x)$. For example,

$$f(0) = \sqrt{0} = 0$$
$$f(1) = \sqrt{1} = 1$$
$$f(4) = \sqrt{4} = 2$$
$$f(9) = \sqrt{9} = 3$$

Choosing perfect squares for x ensures us that $f(x)$ is a rational number, but it is important to stress that $f(x) = \sqrt{x}$ is defined for all nonnegative real numbers. For example,

$$f(3) = \sqrt{3} \approx 1.732$$

The domain of this function is the set of all nonnegative numbers, $\{x \mid x \geq 0\}$ or $[0, \infty)$. On the next page, we graph $y_1 = \sqrt{x}$ in a $[-3, 10, 1]$ by $[-2, 10, 1]$ window. The table on the next page shows that a negative x-value gives an error message since $\sqrt{x}$ is not a real number when x is negative.

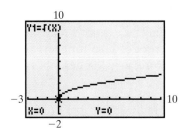

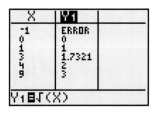

Notice that the graph of this function passes the vertical line test, as expected. ☐

PRACTICE
7 Graph the square root function $h(x) = \sqrt{x + 2}$.

Recall from Sections 2.1 and 2.2 that the equation $f(x) = \sqrt[3]{x}$ also describes a function. Here x may be any real number, so the domain of this function is the set of all real numbers, or $(-\infty, \infty)$. A few function values are given next.

$$f(0) = \sqrt[3]{0} = 0$$
$$f(1) = \sqrt[3]{1} = 1$$
$$f(-1) = \sqrt[3]{-1} = -1$$
$$\left.\begin{array}{l} f(6) = \sqrt[3]{6} \\ f(-6) = \sqrt[3]{-6} \end{array}\right\}$$ Here, there is no rational number whose cube is 6. Thus, the radicals do not simplify to rational numbers.
$$f(8) = \sqrt[3]{8} = 2$$
$$f(-8) = \sqrt[3]{-8} = -2$$

EXAMPLE 8 Identify the domain of the function $f(x) = \sqrt[3]{x}$ and then graph the function.

Solution The domain of this function is the set of all real numbers. Below, we graph $y_1 = \sqrt[3]{x}$ in a $[-10, 10, 1]$ by $[-5, 5, 1]$ window.

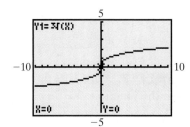

The graph of this function passes the vertical line test, as expected. ☐

PRACTICE
8 Graph the function $f(x) = \sqrt[3]{x} - 4$.

VOCABULARY & READINESS CHECK

Use the choices below to fill in each blank. Not all choices will be used.

is	cubes	$-\sqrt{a}$	radical sign	index
is not	squares	$\sqrt{-a}$	radicand	

1. In the expression $\sqrt[n]{a}$, the n is called the _____, the $\sqrt{}$ is called the _____, and a is called the _____.
2. If $\sqrt{a}$ is the positive square root of a, $a \neq 0$, then _____ is the negative square root of a.
3. The square root of a negative number _____ a real number.
4. Numbers such as 1, 4, 9, and 25 are called perfect _____ whereas numbers such as 1, 8, 27, and 125 are called perfect _____.

Fill in the blank.

5. The domain of the function $f(x) = \sqrt{x}$ is _____.

6. The domain of the function $f(x) = \sqrt[3]{x}$ is _____.

7. If $f(16) = 4$, the corresponding ordered pair is _____.

8. If $g(-8) = -2$, the corresponding ordered pair is _____.

Choose the correct letter or letters. No pencil is needed, just think your way through these.

9. Which radical is not a real number?

   a. $\sqrt{3}$ **b.** $-\sqrt{11}$ **c.** $\sqrt[3]{-10}$ **d.** $\sqrt{-10}$

10. Which radical(s) simplify to 3?

   a. $\sqrt{9}$ **b.** $\sqrt{-9}$ **c.** $\sqrt[3]{27}$ **d.** $\sqrt[3]{-27}$

11. Which radical(s) simplify to -3?

   a. $\sqrt{9}$ **b.** $\sqrt{-9}$ **c.** $\sqrt[3]{27}$ **d.** $\sqrt[3]{-27}$

12. Which radical does not simplify to a whole number?

   a. $\sqrt{64}$ **b.** $\sqrt[3]{64}$ **c.** $\sqrt{8}$ **d.** $\sqrt[3]{8}$

7.1 | EXERCISE SET

MyMathLab®
              PRACTICE   WATCH   DOWNLOAD   READ   REVIEW

Simplify. Assume that variables represent positive real numbers. See Example 1.

1. $\sqrt{100}$ **2.** $\sqrt{400}$

3. $\sqrt{\dfrac{1}{4}}$ **4.** $\sqrt{\dfrac{9}{25}}$

5. $\sqrt{0.0001}$ **6.** $\sqrt{0.04}$

7. $-\sqrt{36}$ **8.** $-\sqrt{9}$

9. $\sqrt{x^{10}}$ **10.** $\sqrt{x^{16}}$

11. $\sqrt{16y^6}$ **12.** $\sqrt{64y^{20}}$

Use a calculator to approximate each square root to 3 decimal places. Check to see that each approximation is reasonable. See Example 2.

13. $\sqrt{7}$ **14.** $\sqrt{11}$

15. $\sqrt{38}$ **16.** $\sqrt{56}$

17. $\sqrt{200}$ **18.** $\sqrt{300}$

Find each cube root. See Example 3.

19. $\sqrt[3]{64}$ **20.** $\sqrt[3]{27}$

21. $\sqrt[3]{\dfrac{1}{8}}$ **22.** $\sqrt[3]{\dfrac{27}{64}}$

23. $\sqrt[3]{-1}$ **24.** $\sqrt[3]{-125}$

25. $\sqrt[3]{x^{12}}$ **26.** $\sqrt[3]{x^{15}}$

27. $\sqrt[3]{-27x^9}$ **28.** $\sqrt[3]{-64x^6}$

Find each root. Assume that all variables represent nonnegative real numbers. See Example 4.

29. $-\sqrt[4]{16}$ **30.** $\sqrt[5]{-243}$

31. $\sqrt[4]{-16}$ **32.** $\sqrt{-16}$

33. $\sqrt[5]{-32}$ **34.** $\sqrt[5]{-1}$

35. $\sqrt[5]{x^{20}}$ **36.** $\sqrt[4]{x^{20}}$

37. $\sqrt[6]{64x^{12}}$ **38.** $\sqrt[5]{-32x^{15}}$

39. $\sqrt{81x^4}$ **40.** $\sqrt[4]{81x^4}$

41. $\sqrt[4]{256x^8}$ **42.** $\sqrt{256x^8}$

Simplify. Assume that the variables represent any real number. See Example 5.

43. $\sqrt{(-8)^2}$ **44.** $\sqrt{(-7)^2}$

45. $\sqrt[3]{(-8)^3}$ **46.** $\sqrt[5]{(-7)^5}$

47. $\sqrt{4x^2}$ **48.** $\sqrt[4]{16x^4}$

49. $\sqrt[3]{x^3}$ **50.** $\sqrt[5]{x^5}$

51. $\sqrt{(x-5)^2}$ **52.** $\sqrt{(y-6)^2}$

53. $\sqrt{x^2 + 4x + 4}$

   (*Hint:* Factor the polynomial first.)

54. $\sqrt{x^2 - 8x + 16}$

   (*Hint:* Factor the polynomial first.)

MIXED PRACTICE

Simplify each radical. Assume that all variables represent positive real numbers.

55. $-\sqrt{121}$ **56.** $-\sqrt[3]{125}$

57. $\sqrt[3]{8x^3}$ **58.** $\sqrt{16x^8}$

59. $\sqrt{y^{12}}$ **60.** $\sqrt[3]{y^{12}}$

61. $\sqrt{25a^2b^{20}}$ **62.** $\sqrt{9x^4y^6}$

63. $\sqrt[3]{-27x^{12}y^9}$ **64.** $\sqrt[3]{-8a^{21}b^6}$

65. $\sqrt[4]{a^{16}b^4}$ **66.** $\sqrt[4]{x^8y^{12}}$

67. $\sqrt[5]{-32x^{10}y^5}$ **68.** $\sqrt[5]{-243z^{15}}$

69. $\sqrt{\dfrac{25}{49}}$

70. $\sqrt{\dfrac{4}{81}}$

71. $\sqrt{\dfrac{x^2}{4y^2}}$

72. $\sqrt{\dfrac{y^{10}}{9x^6}}$

73. $-\sqrt[3]{\dfrac{z^{21}}{27x^3}}$

74. $-\sqrt[3]{\dfrac{64a^3}{b^9}}$

75. $\sqrt[4]{\dfrac{x^4}{16}}$

76. $\sqrt[4]{\dfrac{y^4}{81x^4}}$

If $f(x) = \sqrt{2x + 3}$ and $g(x) = \sqrt[3]{x - 8}$, find the following function values. See Example 6.

77. $f(0)$

78. $g(0)$

79. $g(7)$

80. $f(-1)$

81. $g(-19)$

82. $f(3)$

83. $f(2)$

84. $g(1)$

For Exercises 85 through 88, match the graph with its equation. All graphs are in a $[-10, 10, 1]$ by $[-5, 5, 5]$ window. Also give the domain of each equation in interval notation. See Example 7.

A.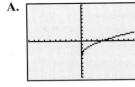

B.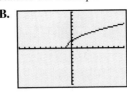

C.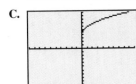

D.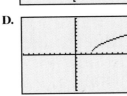

85. $f(x) = \sqrt{x} + 2$

86. $f(x) = \sqrt{x} - 2$

87. $f(x) = \sqrt{x} - 3$

88. $f(x) = \sqrt{x} + 1$

For Exercises 89 through 92, match the graph with its equation. All graphs are in a $[-4.7, 4.7, 1]$ by $[-5, 5, 1]$ window. Also give the domain of each function in interval notation. See Example 8.

A.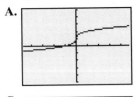

B.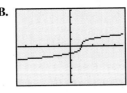

C.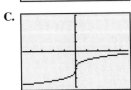

D.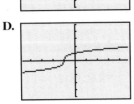

89. $f(x) = \sqrt[3]{x} + 1$

90. $f(x) = \sqrt[3]{x} - 2$

91. $g(x) = \sqrt[3]{x - 1}$

92. $g(x) = \sqrt[3]{x + 1}$

REVIEW AND PREVIEW

Simplify each exponential expression. See Sections 5.1 and 5.2.

93. $(-2x^3y^2)^5$

94. $(4y^6z^7)^3$

95. $(-3x^2y^3z^5)(20x^5y^7)$

96. $(-14a^5bc^2)(2abc^4)$

97. $\dfrac{7x^{-1}y}{14(x^5y^2)^{-2}}$

98. $\dfrac{(2a^{-1}b^2)^3}{(8a^2b)^{-2}}$

CONCEPT EXTENSIONS

Which of the following are not real numbers? See the Concept Check in this section.

99. $\sqrt{-17}$

100. $\sqrt[3]{-17}$

101. $\sqrt[10]{-17}$

102. $\sqrt[15]{-17}$

103. Explain why $\sqrt{-64}$ is not a real number.

104. Explain why $\sqrt[3]{-64}$ is a real number.

For Exercises 105 through 108, do not use a calculator.

105. $\sqrt{160}$ is closest to

   a. 10   **b.** 13   **c.** 20   **d.** 40

106. $\sqrt{1000}$ is closest to

   a. 10   **b.** 30   **c.** 100   **d.** 500

107. The perimeter of the triangle is closest to

   a. 12   **b.** 18

   c. 66   **d.** 132

$\sqrt{30}$ $\sqrt{10}$

$\sqrt{90}$

108. The length of the bent wire is closest to

   a. 5   **b.** $\sqrt{28}$

   c. 7   **d.** 14

$\sqrt{8}$

$\sqrt{20}$

The Mosteller formula for calculating adult body surface area is $B = \sqrt{\dfrac{hw}{3131}}$, where B is an individual's body surface area in square meters, h is the individual's height in inches, and w is the individual's weight in pounds. Use this information to answer Exercises 109 and 110. Round answers to 2 decimal places.

109. Find the body surface area of an individual who is 66 inches tall and who weighs 135 pounds.

110. Find the body surface area of an individual who is 74 inches tall and who weighs 225 pounds.

111. Suppose that a friend tells you that $\sqrt{13} \approx 5.7$. Without a calculator, how can you convince your friend that he or she must have made an error?

112. Escape velocity is the minimum speed that an object must reach to escape a planet's pull of gravity. Escape velocity v is given by the equation $v = \sqrt{\dfrac{2Gm}{r}}$, where m is the mass of the planet, r is its radius, and G is the universal gravitational constant, which has a value of $G = 6.67 \times 10^{-11}$ m^3/kg · sec^2. The mass of Earth is 5.97×10^{24} kg and its radius is 6.37×10^6 m. Use this information to find the escape velocity for Earth. Round to the nearest whole number. (*Source: National Space Science Data Center*)

 STUDY SKILLS BUILDER

How Are Your Homework Assignments Going?

Remember that it is important to keep up with homework. Why? Many concepts in mathematics build on each other. Often, your understanding of a day's lecture depends on an understanding of the previous day's material.

   To complete a homework assignment, remember these four things:

- Attempt all of it.
- Check it.
- Correct it.
- If needed, ask questions about it.

Take a moment and review your completed homework assignments. Answer the exercises below based on this review.

1. Approximate the fraction of your homework you have attempted.

2. Approximate the fraction of your homework you have checked (if possible).

3. If you are able to check your homework, have you corrected it when errors have been found?

4. What do you do if you do not understand a concept while working on homework?

7.2 RATIONAL EXPONENTS

OBJECTIVES

1 Understand the meaning of $a^{1/n}$.

2 Understand the meaning of $a^{m/n}$.

3 Understand the meaning of $a^{-m/n}$.

4 Use rules for exponents to simplify expressions that contain rational exponents.

5 Use rational exponents to simplify radical expressions.

OBJECTIVE 1 ▶ Understanding the meaning of $a^{1/n}$. So far in this text, we have not defined expressions with rational exponents such as $3^{1/2}$, $x^{2/3}$, and $-9^{-1/4}$. We will define these expressions so that the rules for exponents will apply to these rational exponents as well.

   Suppose that $x = 5^{1/3}$. Then

$$x^3 = \left(5^{1/3}\right)^3 = 5^{1/3 \cdot 3} = 5^1 \text{ or } 5$$

$$\underset{\text{for exponents}}{\underbrace{\text{using rules}}}$$

Since $x^3 = 5$, then x is the number whose cube is 5, or $x = \sqrt[3]{5}$. Notice that we also know that $x = 5^{1/3}$. This means

$$5^{1/3} = \sqrt[3]{5}$$

Definition of $a^{1/n}$

If n is a positive integer greater than 1 and $\sqrt[n]{a}$ is a real number, then

$$a^{1/n} = \sqrt[n]{a}$$

Notice that the denominator of the rational exponent corresponds to the index of the radical.

EXAMPLE 1   Use radical notation to write the following. Simplify if possible.

a. $4^{1/2}$ **b.** $64^{1/3}$ **c.** $x^{1/4}$ **d.** $0^{1/6}$ **e.** $-9^{1/2}$ **f.** $(81x^8)^{1/4}$ **g.** $(5y)^{1/3}$

Solution

a. $4^{1/2} = \sqrt{4} = 2$

b. $64^{1/3} = \sqrt[3]{64} = 4$

c. $x^{1/4} = \sqrt[4]{x}$

d. $0^{1/6} = \sqrt[6]{0} = 0$

e. $-9^{1/2} = -\sqrt{9} = -3$

f. $(81x^8)^{1/4} = \sqrt[4]{81x^8} = 3x^2$

g. $(5y)^{1/3} = \sqrt[3]{5y}$

□

PRACTICE
1 Use radical notation to write the following. Simplify if possible.

a. $36^{1/2}$ **b.** $1000^{1/3}$ **c.** $x^{1/5}$ **d.** $1^{1/4}$ **e.** $-64^{1/2}$
f. $(125x^9)^{1/3}$ **g.** $(3x)^{1/4}$

OBJECTIVE 2 ▶ Understanding the meaning of $a^{m/n}$. As we expand our use of exponents to include $\dfrac{m}{n}$, we define their meaning so that rules for exponents still hold true. For example, by properties of exponents,

$$8^{2/3} = (8^{1/3})^2 = (\sqrt[3]{8})^2 \quad \text{or}$$
$$8^{2/3} = (8^2)^{1/3} = \sqrt[3]{8^2}$$

> **Definition of $a^{m/n}$**
> If m and n are positive integers greater than 1 with $\dfrac{m}{n}$ in lowest terms, then
> $$a^{m/n} = \sqrt[n]{a^m} = (\sqrt[n]{a})^m$$
> as long as $\sqrt[n]{a}$ is a real number.

Notice that the denominator n of the rational exponent corresponds to the index of the radical. The numerator m of the rational exponent indicates that the base is to be raised to the mth power. This means

$$8^{2/3} = \sqrt[3]{8^2} = \sqrt[3]{64} = 4 \quad \text{or}$$
$$8^{2/3} = (\sqrt[3]{8})^2 = 2^2 = 4$$

From simplifying $8^{2/3}$, can you see that it doesn't matter whether you raise to a power first and then take the nth root or you take the nth root first and then raise to a power?

> ▶ **Helpful Hint**
> Most of the time, $(\sqrt[n]{a})^m$ will be easier to calculate than $\sqrt[n]{a^m}$.

EXAMPLE 2 Use radical notation to write the following. Then simplify if possible.

a. $4^{3/2}$ **b.** $-16^{3/4}$ **c.** $(-27)^{2/3}$
d. $\left(\dfrac{1}{9}\right)^{3/2}$ **e.** $(4x-1)^{3/5}$

Solution

a. $4^{3/2} = (\sqrt{4})^3 = 2^3 = 8$ **b.** $-16^{3/4} = -(\sqrt[4]{16})^3 = -(2)^3 = -8$

c. $(-27)^{2/3} = (\sqrt[3]{-27})^2 = (-3)^2 = 9$ **d.** $\left(\dfrac{1}{9}\right)^{3/2} = \left(\sqrt{\dfrac{1}{9}}\right)^3 = \left(\dfrac{1}{3}\right)^3 = \dfrac{1}{27}$

e. $(4x-1)^{3/5} = \sqrt[5]{(4x-1)^3}$

PRACTICE
2 Use radical notation to write the following. Simplify if possible.

a. $16^{3/2}$ **b.** $-1^{3/5}$ **c.** $-(81)^{3/4}$
d. $\left(\dfrac{1}{25}\right)^{3/2}$ **e.** $(3x+2)^{5/9}$

> ▶ **Helpful Hint**
> The *denominator* of a rational exponent is the index of the corresponding radical. For example, $x^{1/5} = \sqrt[5]{x}$ and $z^{2/3} = \sqrt[3]{z^2}$, or $z^{2/3} = \left(\sqrt[3]{z}\right)^2$.

OBJECTIVE 3 ▶ Understanding the meaning of $a^{-m/n}$. The rational exponents we have given meaning to exclude negative rational numbers. To complete the set of definitions, we define $a^{-m/n}$.

> **Definition of $a^{-m/n}$**
>
> $$a^{-m/n} = \frac{1}{a^{m/n}}$$
>
> as long as $a^{m/n}$ is a nonzero real number.

EXAMPLE 3 Write each expression with a positive exponent, and then simplify.

a. $16^{-3/4}$ **b.** $(-27)^{-2/3}$

Solution

a. $16^{-3/4} = \dfrac{1}{16^{3/4}} = \dfrac{1}{\left(\sqrt[4]{16}\right)^3} = \dfrac{1}{2^3} = \dfrac{1}{8}$

b. $(-27)^{-2/3} = \dfrac{1}{(-27)^{2/3}} = \dfrac{1}{\left(\sqrt[3]{-27}\right)^2} = \dfrac{1}{(-3)^2} = \dfrac{1}{9}$

PRACTICE
3 Write each expression with a positive exponent; then simplify.

a. $9^{-3/2}$ **b.** $(-64)^{-2/3}$

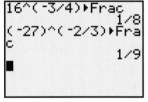

A calculator check for Example 3.

```
16^(-3/4)▶Frac
             1/8
(-27)^(-2/3)▶Fra
c
             1/9
■
```

```
9^(-3/2)▶Frac
            1/27
9^(3/2)
              27
(-27)^(-1/3)▶Fra
c
            -1/3
■
```

A calculator check for the Helpful Hint.

> ▶ **Helpful Hint**
> If an expression contains a negative rational exponent, such as $9^{-3/2}$, you may want to first write the expression with a positive exponent and then interpret the rational exponent. Notice that the sign of the base is not affected by the sign of its exponent. For example,
>
> $$9^{-3/2} = \frac{1}{9^{3/2}} = \frac{1}{\left(\sqrt{9}\right)^3} = \frac{1}{27}$$
>
> Also,
>
> $$(-27)^{-1/3} = \frac{1}{(-27)^{1/3}} = -\frac{1}{3}$$

Concept Check ✓

Which one is correct?

a. $-8^{2/3} = \dfrac{1}{4}$ **b.** $8^{-2/3} = -\dfrac{1}{4}$ **c.** $8^{-2/3} = -4$ **d.** $-8^{-2/3} = -\dfrac{1}{4}$

OBJECTIVE 4 ▶ Using rules for exponents to simplify expressions. It can be shown that the properties of integer exponents hold for rational exponents. By using these properties and definitions, we can now simplify expressions that contain rational exponents.

These rules are repeated here for review.

Note: For the remainder of this chapter, we will assume that variables represent positive real numbers. Since this is so, we need not insert absolute value bars when we simplify even roots.

Answer to Concept Check: d

Summary of Exponent Rules

If m and n are rational numbers, and a, b, and c are numbers for which the expressions below exist, then

Product rule for exponents:	$a^m \cdot a^n = a^{m+n}$
Power rule for exponents:	$(a^m)^n = a^{m \cdot n}$
Power rules for products and quotients:	$(ab)^n = a^n b^n$ and
	$\left(\dfrac{a}{c}\right)^n = \dfrac{a^n}{c^n}, c \neq 0$
Quotient rule for exponents:	$\dfrac{a^m}{a^n} = a^{m-n}, a \neq 0$
Zero exponent:	$a^0 = 1, a \neq 0$
Negative exponent:	$a^{-n} = \dfrac{1}{a^n}, a \neq 0$

EXAMPLE 4   Use properties of exponents to simplify. Write results with only positive exponents.

a. $b^{1/3} \cdot b^{5/3}$ **b.** $x^{1/2} x^{1/3}$ **c.** $\dfrac{7^{1/3}}{7^{4/3}}$

d. $y^{-4/7} \cdot y^{6/7}$ **e.** $\dfrac{(2x^{2/5} y^{-1/3})^5}{x^2 y}$

Solution

a. $b^{1/3} \cdot b^{5/3} = b^{(1/3 + 5/3)} = b^{6/3} = b^2$

b. $x^{1/2} x^{1/3} = x^{(1/2 + 1/3)} = x^{3/6 + 2/6} = x^{5/6}$ Use the product rule.

c. $\dfrac{7^{1/3}}{7^{4/3}} = 7^{1/3 - 4/3} = 7^{-3/3} = 7^{-1} = \dfrac{1}{7}$ Use the quotient rule.

d. $y^{-4/7} \cdot y^{6/7} = y^{-4/7 + 6/7} = y^{2/7}$ Use the product rule.

e. We begin by using the power rule $(ab)^m = a^m b^m$ to simplify the numerator.

$$\dfrac{(2x^{2/5} y^{-1/3})^5}{x^2 y} = \dfrac{2^5 (x^{2/5})^5 (y^{-1/3})^5}{x^2 y} = \dfrac{32 x^2 y^{-5/3}}{x^2 y} \quad \text{Use the power rule and simplify.}$$

$$= 32 x^{2-2} y^{-5/3 - 3/3} \quad \text{Apply the quotient rule.}$$

$$= 32 x^0 y^{-8/3}$$

$$= \dfrac{32}{y^{8/3}}$$

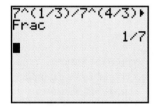

A calculator check for Example 4c.

PRACTICE
4   Use properties of exponents to simplify.

a. $y^{2/3} \cdot y^{8/3}$ **b.** $x^{3/5} \cdot x^{1/4}$ **c.** $\dfrac{9^{2/7}}{9^{9/7}}$

d. $b^{4/9} \cdot b^{-2/9}$ **e.** $\dfrac{\left(3x^{1/4} y^{-2/3}\right)^4}{x^4 y}$

EXAMPLE 5   Multiply.

a. $z^{2/3}(z^{1/3} - z^5)$ **b.** $(x^{1/3} - 5)(x^{1/3} + 2)$

Solution

a. $z^{2/3}(z^{1/3} - z^5) = z^{2/3}z^{1/3} - z^{2/3}z^5$ Apply the distributive property.

$= z^{(2/3+1/3)} - z^{(2/3+5)}$ Use the product rule.

$= z^{3/3} - z^{(2/3+15/3)}$

$= z - z^{17/3}$

b. $(x^{1/3} - 5)(x^{1/3} + 2) = x^{2/3} + 2x^{1/3} - 5x^{1/3} - 10$ Think of $(x^{1/3} - 5)$ and $(x^{1/3} + 2)$

$= x^{2/3} - 3x^{1/3} - 10$ as 2 binomials, and FOIL.

PRACTICE

5 Multiply.

a. $x^{3/5}(x^{1/3} - x^2)$ **b.** $(x^{1/2} + 6)(x^{1/2} - 2)$

EXAMPLE 6 Factor $x^{-1/2}$ from the expression $3x^{-1/2} - 7x^{5/2}$. Assume that all variables represent positive numbers.

Solution

$$3x^{-1/2} - 7x^{5/2} = (x^{-1/2})(3) - (x^{-1/2})(7x^{6/2})$$
$$= x^{-1/2}(3 - 7x^3)$$

To check, multiply $x^{-1/2}(3 - 7x^3)$ to see that the product is $3x^{-1/2} - 7x^{5/2}$.

PRACTICE

6 Factor $x^{-1/5}$ from the expression $2x^{-1/5} - 7x^{4/5}$.

OBJECTIVE 5 ▶ Using rational exponents to simplify radical expressions. Some radical expressions are easier to simplify when we first write them with rational exponents. We can simplify some radical expressions by first writing the expression with rational exponents. Use properties of exponents to simplify, and then convert back to radical notation.

EXAMPLE 7 Use rational exponents to simplify. Assume that variables represent positive numbers.

a. $\sqrt[8]{x^4}$ **b.** $\sqrt[6]{25}$ **c.** $\sqrt[4]{r^2s^6}$

Solution

a. $\sqrt[8]{x^4} = x^{4/8} = x^{1/2} = \sqrt{x}$

b. $\sqrt[6]{25} = 25^{1/6} = (5^2)^{1/6} = 5^{2/6} = 5^{1/3} = \sqrt[3]{5}$

c. $\sqrt[4]{r^2s^6} = (r^2s^6)^{1/4} = r^{2/4}s^{6/4} = r^{1/2}s^{3/2} = (rs^3)^{1/2} = \sqrt{rs^3}$

```
6 ˣ√25
          1.709975947
³√(5)
          1.709975947
■
```

A calculator check for Example 7b.

PRACTICE

7 Use rational exponents to simplify. Assume that the variables represent positive numbers.

a. $\sqrt[9]{x^3}$ **b.** $\sqrt[4]{36}$ **c.** $\sqrt[8]{a^4b^2}$

EXAMPLE 8 Use rational exponents to write as a single radical.

a. $\sqrt{x} \cdot \sqrt[4]{x}$ **b.** $\dfrac{\sqrt{x}}{\sqrt[3]{x}}$ **c.** $\sqrt[3]{3} \cdot \sqrt{2}$

Solution

a. $\sqrt{x} \cdot \sqrt[4]{x} = x^{1/2} \cdot x^{1/4} = x^{1/2+1/4}$

$\qquad = x^{3/4} = \sqrt[4]{x^3}$

b. $\dfrac{\sqrt{x}}{\sqrt[3]{x}} = \dfrac{x^{1/2}}{x^{1/3}} = x^{1/2-1/3} = x^{3/6-2/6}$

$\qquad = x^{1/6} = \sqrt[6]{x}$

c. $\sqrt[3]{3} \cdot \sqrt{2} = 3^{1/3} \cdot 2^{1/2}$ Write with rational exponents.

$\qquad = 3^{2/6} \cdot 2^{3/6}$ Write the exponents so that they have the same denominator.

$\qquad = (3^2 \cdot 2^3)^{1/6}$ Use $a^n b^n = (ab)^n$.

$\qquad = \sqrt[6]{3^2 \cdot 2^3}$ Write with radical notation.

$\qquad = \sqrt[6]{72}$ Multiply $3^2 \cdot 2^3$.

```
³√(3)*√(2)
        2.039648903
6 ×√72
        2.039648903
■
```

A calculator check for
Example 8c.

PRACTICE

8   Use rational expressions to write each of the following as a single radical.

a. $\sqrt[3]{x} \cdot \sqrt[4]{x}$ **b.** $\dfrac{\sqrt[3]{y}}{\sqrt[5]{y}}$ **c.** $\sqrt[3]{5} \cdot \sqrt{3}$

VOCABULARY & READINESS CHECK

Answer each true or false.

1. $9^{-1/2}$ is a positive number. _____

2. $9^{-1/2}$ is a whole number. _____

3. $\dfrac{1}{a^{-m/n}} = a^{m/n}$ (where $a^{m/n}$ is a nonzero real number). _____

Fill in the blank with the correct choice.

4. To simplify $x^{2/3} \cdot x^{1/5}$, _____ the exponents.

    a. add   **b.** subtract   **c.** multiply   **d.** divide

5. To simplify $(x^{2/3})^{1/5}$, _____ the exponents.

    a. add   **b.** subtract   **c.** multiply   **d.** divide

6. To simplify $\dfrac{x^{2/3}}{x^{1/5}}$, _____ the exponents.

    a. add   **b.** subtract   **c.** multiply   **d.** divide

Choose the correct letter for each exercise. Letters will be used more than once. No pencil is needed. Just think about the meaning of each expression.

A = 2, B = −2, C = not a real number

7. $4^{1/2}$ _____   **8.** $-4^{1/2}$ _____   **9.** $(-4)^{1/2}$ _____   **10.** $8^{1/3}$ _____   **11.** $-8^{1/3}$ _____   **12.** $(-8)^{1/3}$ _____

7.2 | EXERCISE SET

Use radical notation to write each expression. Simplify if possible. See Example 1.

1. $49^{1/2}$

2. $64^{1/3}$

3. $27^{1/3}$

4. $8^{1/3}$

5. $\left(\dfrac{1}{16}\right)^{1/4}$

6. $\left(\dfrac{1}{64}\right)^{1/2}$

7. $169^{1/2}$

8. $81^{1/4}$

9. $2m^{1/3}$

10. $(2m)^{1/3}$

11. $(9x^4)^{1/2}$

12. $(16x^8)^{1/2}$

13. $(-27)^{1/3}$

14. $-64^{1/2}$

15. $-16^{1/4}$

16. $(-32)^{1/5}$

Use radical notation to write each expression. Simplify if possible. See Example 2.

17. $16^{3/4}$

18. $4^{5/2}$

19. $(-64)^{2/3}$

20. $(-8)^{4/3}$

21. $(-16)^{3/4}$

22. $(-9)^{3/2}$

23. $(2x)^{3/5}$

24. $2x^{3/5}$

25. $(7x + 2)^{2/3}$

26. $(x - 4)^{3/4}$

27. $\left(\dfrac{16}{9}\right)^{3/2}$

28. $\left(\dfrac{49}{25}\right)^{3/2}$

Write with positive exponents. Simplify if possible. See Example 3.

29. $8^{-4/3}$

30. $64^{-2/3}$

31. $(-64)^{-2/3}$

32. $(-8)^{-4/3}$

33. $(-4)^{-3/2}$

34. $(-16)^{-5/4}$

35. $x^{-1/4}$

36. $y^{-1/6}$

37. $\dfrac{1}{a^{-2/3}}$

38. $\dfrac{1}{n^{-8/9}}$

39. $\dfrac{5}{7x^{-3/4}}$

40. $\dfrac{2}{3y^{-5/7}}$

Use the properties of exponents to simplify each expression. Write with positive exponents. See Example 4.

41. $a^{2/3}a^{5/3}$

42. $b^{9/5}b^{8/5}$

43. $x^{-2/5} \cdot x^{7/5}$

44. $y^{4/3} \cdot y^{-1/3}$

45. $3^{1/4} \cdot 3^{3/8}$

46. $5^{1/2} \cdot 5^{1/6}$

47. $\dfrac{y^{1/3}}{y^{1/6}}$

48. $\dfrac{x^{3/4}}{x^{1/8}}$

49. $(4u^2)^{3/2}$

50. $(32^{1/5}x^{2/3})^3$

51. $\dfrac{b^{1/2}b^{3/4}}{-b^{1/4}}$

52. $\dfrac{a^{1/4}a^{-1/2}}{a^{2/3}}$

53. $\dfrac{(x^3)^{1/2}}{x^{7/2}}$

54. $\dfrac{y^{11/3}}{(y^5)^{1/3}}$

55. $\dfrac{(3x^{1/4})^3}{x^{1/12}}$

56. $\dfrac{(2x^{1/5})^4}{x^{3/10}}$

57. $\dfrac{(y^3z)^{1/6}}{y^{-1/2}z^{1/3}}$

58. $\dfrac{(m^2n)^{1/4}}{m^{-1/2}n^{5/8}}$

59. $\dfrac{(x^3y^2)^{1/4}}{(x^{-5}y^{-1})^{-1/2}}$

60. $\dfrac{(a^{-2}b^3)^{1/8}}{(a^{-3}b)^{-1/4}}$

Multiply. See Example 5.

61. $y^{1/2}(y^{1/2} - y^{2/3})$

62. $x^{1/2}(x^{1/2} + x^{3/2})$

63. $x^{2/3}(x - 2)$

64. $3x^{1/2}(x + y)$

65. $(2x^{1/3} + 3)(2x^{1/3} - 3)$

66. $(y^{1/2} + 5)(y^{1/2} + 5)$

Factor the common factor from the given expression. See Example 6.

67. $x^{8/3}; x^{8/3} + x^{10/3}$

68. $x^{3/2}; x^{5/2} - x^{3/2}$

69. $x^{1/5}; x^{2/5} - 3x^{1/5}$

70. $x^{2/7}; x^{3/7} - 2x^{2/7}$

71. $x^{-1/3}; 5x^{-1/3} + x^{2/3}$

72. $x^{-3/4}; x^{-3/4} + 3x^{1/4}$

Use rational exponents to simplify each radical. Assume that all variables represent positive numbers. See Example 7.

73. $\sqrt[6]{x^3}$

74. $\sqrt[9]{a^3}$

75. $\sqrt[6]{4}$

76. $\sqrt[4]{36}$

77. $\sqrt[4]{16x^2}$

78. $\sqrt[8]{4y^2}$

79. $\sqrt[8]{x^4y^4}$

80. $\sqrt[9]{y^6z^3}$

81. $\sqrt[12]{a^8b^4}$

82. $\sqrt[10]{a^5b^5}$

83. $\sqrt[4]{(x + 3)^2}$

84. $\sqrt[8]{(y + 1)^4}$

Use rational expressions to write as a single radical expression. See Example 8.

85. $\sqrt[3]{y} \cdot \sqrt[5]{y^2}$

86. $\sqrt[3]{y^2} \cdot \sqrt[6]{y}$

87. $\dfrac{\sqrt[3]{b^2}}{\sqrt[4]{b}}$

88. $\dfrac{\sqrt[4]{a}}{\sqrt[5]{a}}$

89. $\sqrt[3]{x} \cdot \sqrt[4]{x} \cdot \sqrt[8]{x^3}$

90. $\sqrt[6]{y} \cdot \sqrt[3]{y} \cdot \sqrt[5]{y^2}$

91. $\dfrac{\sqrt[3]{a^2}}{\sqrt[6]{a}}$

92. $\dfrac{\sqrt[5]{b^2}}{\sqrt[10]{b^3}}$

93. $\sqrt{3} \cdot \sqrt[3]{4}$

94. $\sqrt[3]{5} \cdot \sqrt{2}$

95. $\sqrt[5]{7} \cdot \sqrt[3]{y}$

96. $\sqrt[4]{5} \cdot \sqrt[3]{x}$

97. $\sqrt{5r} \cdot \sqrt[3]{s}$

98. $\sqrt[3]{b} \cdot \sqrt[5]{4a}$

REVIEW AND PREVIEW

Write each integer as a product of two integers such that one of the factors is a perfect square. For example, write 18 as $9 \cdot 2$, because 9 is a perfect square.

99. 75

100. 20

101. 48

102. 45

Write each integer as a product of two integers such that one of the factors is a perfect cube. For example, write 24 as $8 \cdot 3$, because 8 is a perfect cube.

103. 16

104. 56

105. 54

106. 80

CONCEPT EXTENSIONS

Basal metabolic rate (BMR) is the number of calories per day a person needs to maintain life. A person's basal metabolic rate $B(w)$ in calories per day can be estimated with the function $B(w) = 70w^{3/4}$, where w is the person's weight in kilograms. Use this information to answer Exercises 107 and 108.

107. Estimate the BMR for a person who weighs 60 kilograms. Round to the nearest calorie. (*Note:* 60 kilograms is approximately 132 pounds.)

108. Estimate the BMR for a person who weighs 90 kilograms. Round to the nearest calorie. (*Note:* 90 kilograms is approximately 198 pounds.)

The number of cellular telephone subscriptions in the United States from 1996 through 2006 can be modeled by the function $f(x) = 33.3x^{4/5}$, where y is the number of cellular telephone subscriptions in millions, x years after 1996. (Source: Based on data from the Cellular Telecommunications & Internet Association, 1994–2000). Use this information to answer Exercises 109 and 110.

109. Use this model to estimate the number of cellular telephone subscriptions in the United States in 2006. Round to the nearest tenth of a million.

110. Predict the number of cellular telephone subscriptions in the United States in 2010. Round to the nearest tenth of a million.

Fill in each box with the correct expression.

111. $\Box \cdot a^{2/3} = a^{3/3}$, or a

112. $\Box \cdot x^{1/8} = x^{4/8}$, or $x^{1/2}$

113. $\dfrac{\Box}{x^{-2/5}} = x^{3/5}$

114. $\dfrac{\Box}{y^{-3/4}} = y^{4/4}$, or y

Use a calculator to write a four-decimal-place approximation of each number.

115. $8^{1/4}$

116. $20^{1/5}$

117. $18^{3/5}$

118. $76^{5/7}$

119. In physics, the speed of a wave traveling over a stretched string with tension t and density u is given by the expression $\dfrac{\sqrt{t}}{\sqrt{u}}$. Write this expression with rational exponents.

120. In electronics, the angular frequency of oscillations in a certain type of circuit is given by the expression $(LC)^{-1/2}$. Use radical notation to write this expression.

7.3 SIMPLIFYING RADICAL EXPRESSIONS

OBJECTIVES

1. Use the product rule for radicals.
2. Use the quotient rule for radicals.
3. Simplify radicals.
4. Use the distance and midpoint formulas.

OBJECTIVE 1 ▶ Using the product rule. It is possible to simplify some radicals that do not evaluate to rational numbers. To do so, we use a product rule and a quotient rule for radicals. To discover the product rule, notice the following pattern.

$$\sqrt{9} \cdot \sqrt{4} = 3 \cdot 2 = 6$$
$$\sqrt{9 \cdot 4} = \sqrt{36} = 6$$

Since both expressions simplify to 6, it is true that

$$\sqrt{9} \cdot \sqrt{4} = \sqrt{9 \cdot 4}$$

This pattern suggests the following product rule for radicals.

Product Rule for Radicals

If $\sqrt[n]{a}$ and $\sqrt[n]{b}$ are real numbers, then

$$\sqrt[n]{a} \cdot \sqrt[n]{b} = \sqrt[n]{ab}$$

Notice that the product rule is the relationship $a^{1/n} \cdot b^{1/n} = (ab)^{1/n}$ stated in radical notation.

EXAMPLE 1   Multiply.

a. $\sqrt{3} \cdot \sqrt{5}$ **b.** $\sqrt{21} \cdot \sqrt{x}$ **c.** $\sqrt[3]{4} \cdot \sqrt[3]{2}$

d. $\sqrt[4]{5y^2} \cdot \sqrt[4]{2x^3}$ **e.** $\sqrt{\dfrac{2}{a}} \cdot \sqrt{\dfrac{b}{3}}$

Solution

a. $\sqrt{3} \cdot \sqrt{5} = \sqrt{3 \cdot 5} = \sqrt{15}$

b. $\sqrt{21} \cdot \sqrt{x} = \sqrt{21x}$

c. $\sqrt[3]{4} \cdot \sqrt[3]{2} = \sqrt[3]{4 \cdot 2} = \sqrt[3]{8} = 2$

d. $\sqrt[4]{5y^2} \cdot \sqrt[4]{2x^3} = \sqrt[4]{5y^2 \cdot 2x^3} = \sqrt[4]{10y^2x^3}$

e. $\sqrt{\dfrac{2}{a}} \cdot \sqrt{\dfrac{b}{3}} = \sqrt{\dfrac{2}{a} \cdot \dfrac{b}{3}} = \sqrt{\dfrac{2b}{3a}}$ □

PRACTICE

1   Multiply.

a. $\sqrt{5} \cdot \sqrt{7}$ **b.** $\sqrt{13} \cdot \sqrt{z}$ **c.** $\sqrt[4]{125} \cdot \sqrt[4]{5}$

d. $\sqrt[3]{5y} \cdot \sqrt[3]{3x^2}$ **e.** $\sqrt{\dfrac{5}{m}} \cdot \sqrt{\dfrac{t}{2}}$

OBJECTIVE 2 ▶ Using the quotient rule.   To discover a quotient rule for radicals, notice the following pattern.

$$\sqrt{\dfrac{4}{9}} = \dfrac{2}{3}$$

$$\dfrac{\sqrt{4}}{\sqrt{9}} = \dfrac{2}{3}$$

Since both expressions simplify to $\dfrac{2}{3}$, it is true that

$$\sqrt{\dfrac{4}{9}} = \dfrac{\sqrt{4}}{\sqrt{9}}$$

This pattern suggests the following quotient rule for radicals.

Quotient Rule for Radicals

If $\sqrt[n]{a}$ and $\sqrt[n]{b}$ are real numbers and $\sqrt[n]{b}$ is not zero, then

$$\sqrt[n]{\dfrac{a}{b}} = \dfrac{\sqrt[n]{a}}{\sqrt[n]{b}}$$

Notice that the quotient rule is the relationship $\left(\dfrac{a}{b}\right)^{1/n} = \dfrac{a^{1/n}}{b^{1/n}}$ stated in radical notation. We can use the quotient rule to simplify radical expressions by reading the rule from left to right, or to divide radicals by reading the rule from right to left.

For example,

$$\sqrt{\frac{x}{16}} = \frac{\sqrt{x}}{\sqrt{16}} = \frac{\sqrt{x}}{4} \qquad \text{Using } \sqrt[n]{\frac{a}{b}} = \frac{\sqrt[n]{a}}{\sqrt[n]{b}}$$

$$\frac{\sqrt{75}}{\sqrt{3}} = \sqrt{\frac{75}{3}} = \sqrt{25} = 5 \qquad \text{Using } \frac{\sqrt[n]{a}}{\sqrt[n]{b}} = \sqrt[n]{\frac{a}{b}}$$

Note: *Recall that from Section 7.2 on, we assume that variables represent positive real numbers. Since this is so, we need not insert absolute value bars when we simplify even roots.*

EXAMPLE 2   Use the quotient rule to simplify.

a. $\sqrt{\frac{25}{49}}$ **b.** $\sqrt{\frac{x}{9}}$ **c.** $\sqrt[3]{\frac{8}{27}}$ **d.** $\sqrt[4]{\frac{3}{16y^4}}$

Solution

a. $\sqrt{\frac{25}{49}} = \frac{\sqrt{25}}{\sqrt{49}} = \frac{5}{7}$

b. $\sqrt{\frac{x}{9}} = \frac{\sqrt{x}}{\sqrt{9}} = \frac{\sqrt{x}}{3}$

c. $\sqrt[3]{\frac{8}{27}} = \frac{\sqrt[3]{8}}{\sqrt[3]{27}} = \frac{2}{3}$

d. $\sqrt[4]{\frac{3}{16y^4}} = \frac{\sqrt[4]{3}}{\sqrt[4]{16y^4}} = \frac{\sqrt[4]{3}}{2y}$

```
√(25)/√(49)►Frac
                5/7
³√(8/27)►Frac
                2/3
■
```

A calculator check for Example 2a and c.

PRACTICE
2   Use the quotient rule to simplify.

a. $\sqrt{\frac{36}{49}}$ **b.** $\sqrt{\frac{z}{16}}$ **c.** $\sqrt[3]{\frac{125}{8}}$ **d.** $\sqrt[4]{\frac{5}{81x^8}}$

OBJECTIVE 3 ▶ Simplifying radicals. Both the product and quotient rules can be used to simplify a radical. If the product rule is read from right to left, we have that

$$\sqrt[n]{ab} = \sqrt[n]{a} \cdot \sqrt[n]{b}.$$

This is used to simplify the following radicals.

EXAMPLE 3   Simplify the following.

a. $\sqrt{50}$ **b.** $\sqrt[3]{24}$ **c.** $\sqrt{26}$ **d.** $\sqrt[4]{32}$

Solution

a. Factor 50 such that one factor is the largest perfect square that divides 50. The largest perfect square factor of 50 is 25, so we write 50 as $25 \cdot 2$ and use the product rule for radicals to simplify.

$$\sqrt{50} = \sqrt{25 \cdot 2} = \sqrt{25} \cdot \sqrt{2} = 5\sqrt{2}$$
$$\underset{\text{The largest perfect square}}{\uparrow}$$
factor of 50

▶ **Helpful Hint**
Don't forget that, for example, $5\sqrt{2}$ means $5 \cdot \sqrt{2}$.

b. $\sqrt[3]{24} = \sqrt[3]{8 \cdot 3} = \sqrt[3]{8} \cdot \sqrt[3]{3} = 2\sqrt[3]{3}$
$$\underset{\text{The largest perfect cube factor of 24}}{\uparrow}$$

```
√(50)
        7.071067812
5√(2)
        7.071067812
■
```

A calculator check for Example 3a.

c. $\sqrt{26}$ The largest perfect square factor of 26 is 1, so $\sqrt{26}$ cannot be simplified further.

d. $\sqrt[4]{32} = \sqrt[4]{16 \cdot 2} = \sqrt[4]{16} \cdot \sqrt[4]{2} = 2\sqrt[4]{2}$

$\qquad\qquad$ ↑＿ The largest fourth power factor of 32 $\qquad\qquad\qquad\qquad\qquad\qquad$ □

PRACTICE
3 Simplify the following.

a. $\sqrt{98}$ $\qquad\qquad$ **b.** $\sqrt[3]{54}$ $\qquad\qquad$ **c.** $\sqrt{35}$ $\qquad\qquad$ **d.** $\sqrt[4]{243}$

After simplifying a radical such as a square root, always check the radicand to see that it contains no other perfect square factors. It may, if the largest perfect square factor of the radicand was not originally recognized. For example,

$$\sqrt{200} = \sqrt{4 \cdot 50} = \sqrt{4} \cdot \sqrt{50} = 2\sqrt{50}$$

Notice that the radicand 50 still contains the perfect square factor 25. This is because 4 is not the largest perfect square factor of 200. We continue as follows.

$$2\sqrt{50} = 2\sqrt{25 \cdot 2} = 2 \cdot \sqrt{25} \cdot \sqrt{2} = 2 \cdot 5 \cdot \sqrt{2} = 10\sqrt{2}$$

The radical is now simplified since 2 contains no perfect square factors (other than 1).

▶ **Helpful Hint**

To help you recognize largest perfect power factors of a radicand, it will help if you are familiar with some perfect powers. A few are listed below.

Perfect Squares	1,	4,	9,	16,	25,	36,	49,	64,	81,	100,	121,	144
	1^2	2^2	3^2	4^2	5^2	6^2	7^2	8^2	9^2	10^2	11^2	12^2

Perfect Cubes	1,	8,	27,	64,	125
	1^3	2^3	3^3	4^3	5^3

Perfect Fourth Powers	1,	16,	81,	256
	1^4	2^4	3^4	4^4

In general, we say that a radicand of the form $\sqrt[n]{a}$ is simplified when the radicand a contains no factors that are perfect nth powers (other than 1 or −1).

EXAMPLE 4 Use the product rule to simplify.

a. $\sqrt{25x^3}$ $\qquad\qquad$ **b.** $\sqrt[3]{54x^6y^8}$ $\qquad\qquad$ **c.** $\sqrt[4]{81z^{11}}$

Solution

a. $\sqrt{25x^3} = \sqrt{25x^2 \cdot x}$ $\qquad\qquad$ Find the largest perfect square factor.

$\qquad\quad = \sqrt{25x^2} \cdot \sqrt{x}$ $\qquad\qquad$ Apply the product rule.

$\qquad\quad = 5x\sqrt{x}$ $\qquad\qquad\qquad\quad$ Simplify.

b. $\sqrt[3]{54x^6y^8} = \sqrt[3]{27 \cdot 2 \cdot x^6 \cdot y^6 \cdot y^2}$ $\quad$ Factor the radicand and identify perfect cube factors.

$\qquad\qquad = \sqrt[3]{27x^6y^6 \cdot 2y^2}$

$\qquad\qquad = \sqrt[3]{27x^6y^6} \cdot \sqrt[3]{2y^2}$ $\qquad$ Apply the product rule.

$\qquad\qquad = 3x^2y^2\sqrt[3]{2y^2}$ $\qquad\qquad$ Simplify.

c. $\sqrt[4]{81z^{11}} = \sqrt[4]{81 \cdot z^8 \cdot z^3}$ $\qquad\qquad$ Factor the radicand and identify perfect fourth power factors.

$\qquad\qquad = \sqrt[4]{81z^8} \cdot \sqrt[4]{z^3}$ $\qquad\quad$ Apply the product rule.

$\qquad\qquad = 3z^2\sqrt[4]{z^3}$ $\qquad\qquad\qquad$ Simplify. $\qquad\qquad$ □

PRACTICE
4 Use the product rule to simplify.

a. $\sqrt{36z^7}$ $\qquad\qquad$ **b.** $\sqrt[3]{32p^4q^7}$ $\qquad\qquad$ **c.** $\sqrt[4]{16x^{15}}$

EXAMPLE 5   Use the quotient rule to divide, and simplify if possible.

a. $\dfrac{\sqrt{20}}{\sqrt{5}}$ **b.** $\dfrac{\sqrt{50x}}{2\sqrt{2}}$ **c.** $\dfrac{7\sqrt[3]{48x^4y^8}}{\sqrt[3]{6y^2}}$ **d.** $\dfrac{2\sqrt[4]{32a^8b^6}}{\sqrt[4]{a^{-1}b^2}}$

Solution

a. $\dfrac{\sqrt{20}}{\sqrt{5}} = \sqrt{\dfrac{20}{5}}$ Apply the quotient rule.

$\qquad\quad = \sqrt{4}$ Simplify.

$\qquad\quad = 2$ Simplify.

b. $\dfrac{\sqrt{50x}}{2\sqrt{2}} = \dfrac{1}{2}\cdot\sqrt{\dfrac{50x}{2}}$ Apply the quotient rule.

$\qquad\quad = \dfrac{1}{2}\cdot\sqrt{25x}$ Simplify.

$\qquad\quad = \dfrac{1}{2}\cdot\sqrt{25}\cdot\sqrt{x}$ Factor 25x.

$\qquad\quad = \dfrac{1}{2}\cdot 5\cdot\sqrt{x}$ Simplify.

$\qquad\quad = \dfrac{5}{2}\sqrt{x}$

c. $\dfrac{7\sqrt[3]{48x^4y^8}}{\sqrt[3]{6y^2}} = 7\cdot\sqrt[3]{\dfrac{48x^4y^8}{6y^2}}$ Apply the quotient rule.

$\qquad\quad = 7\cdot\sqrt[3]{8x^4y^6}$ Simplify.

$\qquad\quad = 7\sqrt[3]{8x^3y^6\cdot x}$ Factor.

$\qquad\quad = 7\cdot\sqrt[3]{8x^3y^6}\cdot\sqrt[3]{x}$ Apply the product rule.

$\qquad\quad = 7\cdot 2xy^2\cdot\sqrt[3]{x}$ Simplify.

$\qquad\quad = 14xy^2\sqrt[3]{x}$

d. $\dfrac{2\sqrt[4]{32a^8b^6}}{\sqrt[4]{a^{-1}b^2}} = 2\sqrt[4]{\dfrac{32a^8b^6}{a^{-1}b^2}} = 2\sqrt[4]{32a^9b^4} = 2\sqrt[4]{16\cdot a^8\cdot b^4\cdot 2\cdot a}$

$\qquad\qquad = 2\sqrt[4]{16a^8b^4}\cdot\sqrt[4]{2a} = 2\cdot 2a^2b\cdot\sqrt[4]{2a} = 4a^2b\sqrt[4]{2a}$ □

PRACTICE
5   Use the quotient rule to divide and simplify.

a. $\dfrac{\sqrt{80}}{\sqrt{5}}$ **b.** $\dfrac{\sqrt{98z}}{3\sqrt{2}}$ **c.** $\dfrac{5\sqrt[3]{40x^5y^7}}{\sqrt[3]{5y}}$ **d.** $\dfrac{3\sqrt[5]{64x^9y^8}}{\sqrt[5]{x^{-1}y^2}}$

Concept Check ✓
Find and correct the error:

$$\dfrac{\sqrt[3]{27}}{\sqrt{9}} = \sqrt[3]{\dfrac{27}{9}} = \sqrt[3]{3}$$

OBJECTIVE 4 ▶ Using the distance and midpoint formulas. Now that we know how to simplify radicals, we can derive and use the distance formula. The midpoint formula is often confused with the distance formula, so to clarify both, we will also review the midpoint formula.

Answer to Concept Check:
$\dfrac{\sqrt[3]{27}}{\sqrt{9}} = \dfrac{3}{3} = 1$

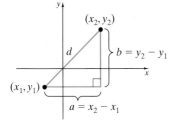

The Cartesian coordinate system helps us visualize a distance between points. To find the distance between two points, we use the distance formula, which is derived from the Pythagorean theorem.

To find the distance d between two points (x_1, y_1) and (x_2, y_2) as shown to the left, notice that the length of leg a is $x_2 - x_1$ and that the length of leg b is $y_2 - y_1$.

Thus, the Pythagorean theorem tells us that

$$d^2 = a^2 + b^2$$

or

$$d^2 = (x_2 - x_1)^2 + (y_2 - y_1)^2$$

or

$$d = \sqrt{(x_2 - x_1)^2 + (y_2 - y_1)^2}$$

This formula gives us the distance between any two points on the real plane.

> **Distance Formula**
> The distance d between two points (x_1, y_1) and (x_2, y_2) is given by
> $$d = \sqrt{(x_2 - x_1)^2 + (y_2 - y_1)^2}$$

EXAMPLE 6   Find the distance between $(2, -5)$ and $(1, -4)$. Give an exact distance and a three-decimal-place approximation.

Solution   To use the distance formula, it makes no difference which point we call (x_1, y_1) and which point we call (x_2, y_2). We will let $(x_1, y_1) = (2, -5)$ and $(x_2, y_2) = (1, -4)$.

$$\begin{aligned}
d &= \sqrt{(x_2 - x_1)^2 + (y_2 - y_1)^2} \\
&= \sqrt{(1 - 2)^2 + [-4 - (-5)]^2} \\
&= \sqrt{(-1)^2 + (1)^2} \\
&= \sqrt{1 + 1} \\
&= \sqrt{2} \approx 1.414
\end{aligned}$$

The distance between the two points is exactly $\sqrt{2}$ units, or approximately 1.414 units.

PRACTICE

6   Find the distance between $P(-3, 7)$ and $Q(-2, 3)$. Give an exact distance and a three-decimal-place approximation.

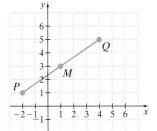

The **midpoint** of a line segment is the **point** located exactly halfway between the two end points of the line segment. On the graph to the left, the point M is the midpoint of line segment PQ. Thus, the distance between M and P equals the distance between M and Q.

Note: We usually need no knowledge of roots to calculate the midpoint of a line segment. We review midpoint here only because it is often confused with the distance between two points.

The x-coordinate of M is at half the distance between the x-coordinates of P and Q, and the y-coordinate of M is at half the distance between the y-coordinates of P and Q. That is, the x-coordinate of M is the average of the x-coordinates of P and Q; the y-coordinate of M is the average of the y-coordinates of P and Q.

> **Midpoint Formula**
> The midpoint of the line segment whose end points are (x_1, y_1) and (x_2, y_2) is the point with coordinates
>
> $$\left(\frac{x_1 + x_2}{2}, \frac{y_1 + y_2}{2} \right)$$

EXAMPLE 7 Find the midpoint of the line segment that joins points $P(-3, 3)$ and $Q(1, 0)$.

Solution Use the midpoint formula. It makes no difference which point we call (x_1, y_1) or which point we call (x_2, y_2). Let $(x_1, y_1) = (-3, 3)$ and $(x_2, y_2) = (1, 0)$.

$$\text{midpoint} = \left(\frac{x_1 + x_2}{2}, \frac{y_1 + y_2}{2} \right)$$

$$= \left(\frac{-3 + 1}{2}, \frac{3 + 0}{2} \right)$$

$$= \left(\frac{-2}{2}, \frac{3}{2} \right)$$

$$= \left(-1, \frac{3}{2} \right)$$

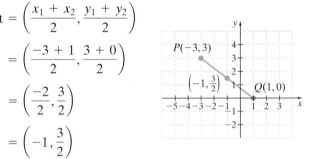

The midpoint of the segment is $\left(-1, \dfrac{3}{2} \right)$.

PRACTICE

7 Find the midpoint of the line segment that joins points $P(5, -2)$ and $Q(8, -6)$.

> ▶ **Helpful Hint**
> The distance between two points is a distance. The midpoint of a line segment is the point halfway between the end points of the segment.
>
> distance—measured in units
>
> midpoint—it is a point

VOCABULARY & READINESS CHECK

Use the choices below to fill in each blank. Some choices may be used more than once.

distance   midpoint   point

1. The _____ of a line segment is a _____ exactly halfway between the two end points of the line segment.

2. The _____ formula is $d = \sqrt{(x_2 - x_1)^2 + (y_2 - y_1)^2}$.

3. The _____ formula is $\left(\dfrac{x_1 + x_2}{2}, \dfrac{y_1 + y_2}{2} \right)$.

Answer true or false. Assume all radicals represent nonzero real numbers.

4. $\sqrt[n]{a} \cdot \sqrt[n]{b} = \sqrt[n]{ab}$ _____

5. $\sqrt[3]{7} \cdot \sqrt[3]{11} = \sqrt[3]{18}$ _____

6. $\sqrt[3]{7} \cdot \sqrt{11} = \sqrt{77}$ _____

7. $\sqrt{x^7 y^8} = \sqrt{x^7} \cdot \sqrt{y^8}$ _____

8. $\dfrac{\sqrt[n]{a}}{\sqrt[n]{b}} = \sqrt[n]{\dfrac{a}{b}}$ _____

9. $\dfrac{\sqrt[3]{12}}{\sqrt[3]{4}} = \sqrt[3]{8}$ _____

10. $\dfrac{\sqrt[n]{x^7}}{\sqrt[n]{x}} = \sqrt[n]{x^6}$ _____

7.3 | EXERCISE SET

MyMathLab Powered by CourseCompass™ and MathXL™

 PRACTICE   WATCH   DOWNLOAD   READ   REVIEW

Use the product rule to multiply. See Example 1.

1. $\sqrt{7} \cdot \sqrt{2}$

2. $\sqrt{11} \cdot \sqrt{10}$

3. $\sqrt[4]{8} \cdot \sqrt[4]{2}$

4. $\sqrt[4]{27} \cdot \sqrt[4]{3}$

5. $\sqrt[3]{4} \cdot \sqrt[3]{9}$

6. $\sqrt[3]{10} \cdot \sqrt[3]{5}$

7. $\sqrt{2} \cdot \sqrt{3x}$

8. $\sqrt{3y} \cdot \sqrt{5x}$

9. $\sqrt{\dfrac{7}{x}} \cdot \sqrt{\dfrac{2}{y}}$

10. $\sqrt{\dfrac{6}{m}} \cdot \sqrt{\dfrac{n}{5}}$

11. $\sqrt[4]{4x^3} \cdot \sqrt[4]{5}$

12. $\sqrt[4]{ab^2} \cdot \sqrt[4]{27ab}$

Use the quotient rule to simplify. See Examples 2 and 3.

13. $\sqrt{\dfrac{6}{49}}$

14. $\sqrt{\dfrac{8}{81}}$

15. $\sqrt{\dfrac{2}{49}}$

16. $\sqrt{\dfrac{5}{121}}$

17. $\sqrt[4]{\dfrac{x^3}{16}}$

18. $\sqrt[4]{\dfrac{y}{81x^4}}$

19. $\sqrt[3]{\dfrac{4}{27}}$

20. $\sqrt[3]{\dfrac{3}{64}}$

21. $\sqrt[4]{\dfrac{8}{x^8}}$

22. $\sqrt[4]{\dfrac{a^3}{81}}$

23. $\sqrt[3]{\dfrac{2x}{81y^{12}}}$

24. $\sqrt[3]{\dfrac{3}{8x^6}}$

25. $\sqrt{\dfrac{x^2 y}{100}}$

26. $\sqrt{\dfrac{y^2 z}{36}}$

27. $\sqrt{\dfrac{5x^2}{4y^2}}$

28. $\sqrt{\dfrac{y^{10}}{9x^6}}$

29. $-\sqrt[3]{\dfrac{z^7}{27x^3}}$

30. $-\sqrt[3]{\dfrac{64a}{b^9}}$

Simplify. See Examples 3 and 4.

31. $\sqrt{32}$

32. $\sqrt{27}$

33. $\sqrt[3]{192}$

34. $\sqrt[3]{108}$

35. $5\sqrt{75}$

36. $3\sqrt{8}$

37. $\sqrt{24}$

38. $\sqrt{20}$

39. $\sqrt{100x^5}$

40. $\sqrt{64y^9}$

41. $\sqrt[3]{16y^7}$

42. $\sqrt[3]{64y^9}$

43. $\sqrt[4]{a^8 b^7}$

44. $\sqrt[5]{32z^{12}}$

45. $\sqrt{y^5}$

46. $\sqrt[3]{y^5}$

47. $\sqrt{25a^2 b^3}$

48. $\sqrt{9x^5 y^7}$

49. $\sqrt[5]{-32x^{10}y}$

50. $\sqrt[5]{-243z^9}$

51. $\sqrt[3]{50x^{14}}$

52. $\sqrt[3]{40y^{10}}$

53. $-\sqrt{32a^8 b^7}$

54. $-\sqrt{20ab^6}$

55. $\sqrt{9x^7 y^9}$

56. $\sqrt{12r^9 s^{12}}$

57. $\sqrt[3]{125r^9 s^{12}}$

58. $\sqrt[3]{8a^6 b^9}$

Use the quotient rule to divide. Then simplify if possible. See Example 5.

59. $\dfrac{\sqrt{14}}{\sqrt{7}}$

60. $\dfrac{\sqrt{45}}{\sqrt{9}}$

61. $\dfrac{\sqrt[3]{24}}{\sqrt[3]{3}}$

62. $\dfrac{\sqrt[3]{10}}{\sqrt[3]{2}}$

63. $\dfrac{5\sqrt[4]{48}}{\sqrt[4]{3}}$

64. $\dfrac{7\sqrt[4]{162}}{\sqrt[4]{2}}$

65. $\dfrac{\sqrt{x^5 y^3}}{\sqrt{xy}}$

66. $\dfrac{\sqrt{a^7 b^6}}{\sqrt{a^3 b^2}}$

67. $\dfrac{8\sqrt[3]{54m^7}}{\sqrt[3]{2m}}$

68. $\dfrac{\sqrt[3]{128x^3}}{-3\sqrt[3]{2x}}$

69. $\dfrac{3\sqrt{100x^2}}{2\sqrt{2x^{-1}}}$

70. $\dfrac{\sqrt{270y^2}}{5\sqrt{3y^{-4}}}$

71. $\dfrac{\sqrt[4]{96a^{10}b^3}}{\sqrt[4]{3a^2 b^3}}$

72. $\dfrac{\sqrt[5]{64x^{10}y^3}}{\sqrt[5]{2x^3 y^{-7}}}$

Find the distance between each pair of points. Give an exact distance and a three-decimal-place approximation. See Example 6.

73. $(5, 1)$ and $(8, 5)$

74. $(2, 3)$ and $(14, 8)$

75. $(-3, 2)$ and $(1, -3)$

76. $(3, -2)$ and $(-4, 1)$

77. $(-9, 4)$ and $(-8, 1)$

78. $(-5, -2)$ and $(-6, -6)$

79. $\left(0, -\sqrt{2}\right)$ and $\left(\sqrt{3}, 0\right)$

80. $\left(-\sqrt{5}, 0\right)$ and $\left(0, \sqrt{7}\right)$

81. $(1.7, -3.6)$ and $(-8.6, 5.7)$

82. $(9.6, 2.5)$ and $(-1.9, -3.7)$

Find the midpoint of the line segment whose end points are given. See Example 7.

83. $(6, -8), (2, 4)$

84. $(3, 9), (7, 11)$

85. $(-2, -1), (-8, 6)$

86. $(-3, -4), (6, -8)$

87. $(7, 3), (-1, -3)$

88. $(-2, 5), (-1, 6)$

89. $\left(\dfrac{1}{2}, \dfrac{3}{8}\right), \left(-\dfrac{3}{2}, \dfrac{5}{8}\right)$

90. $\left(-\dfrac{2}{5}, \dfrac{7}{15}\right), \left(-\dfrac{2}{5}, -\dfrac{4}{15}\right)$

91. $\left(\sqrt{2}, 3\sqrt{5}\right), \left(\sqrt{2}, -2\sqrt{5}\right)$

92. $\left(\sqrt{8}, -\sqrt{12}\right), \left(3\sqrt{2}, 7\sqrt{3}\right)$

93. $(4.6, -3.5), (7.8, -9.8)$

94. $(-4.6, 2.1), (-6.7, 1.9)$

REVIEW AND PREVIEW

Perform each indicated operation. See Sections 1.4 and 5.4.

95. $6x + 8x$

96. $(6x)(8x)$

97. $(2x + 3)(x - 5)$

98. $(2x + 3) + (x - 5)$

99. $9y^2 - 8y^2$

100. $(9y^2)(-8y^2)$

101. $-3(x + 5)$

102. $-3 + x + 5$

103. $(x - 4)^2$

104. $(2x + 1)^2$

CONCEPT EXTENSIONS

Find and correct the error. See the Concept Check in this section.

105. $\dfrac{\sqrt[3]{64}}{\sqrt{64}} = \sqrt[3]{\dfrac{64}{64}} = \sqrt[3]{1} = 1$

106. $\dfrac{\sqrt[4]{16}}{\sqrt{4}} = \sqrt[4]{\dfrac{16}{4}} = \sqrt[4]{4}$

Simplify. See the Concept Check in this section. Assume variables represent positive numbers.

107. $\sqrt[5]{x^{35}}$

108. $\sqrt[6]{y^{48}}$

109. $\sqrt[4]{a^{12}b^4c^{20}}$

110. $\sqrt[3]{a^9b^{21}c^3}$

111. $\sqrt[3]{z^{32}}$

112. $\sqrt[5]{x^{49}}$

113. $\sqrt[7]{q^{17}r^{40}s^7}$

114. $\sqrt[4]{p^{11}q^4r^{45}}$

115. The formula for the radius r of a sphere with surface area A is given by $r = \sqrt{\dfrac{A}{4\pi}}$. Calculate the radius of a standard zorb whose outside surface area is 32.17 sq m. Round to the nearest tenth. (See the chapter opener, page 484. Source: Zorb, Ltd.)

116. The formula for the surface area A of a cone with height h and radius r is given by
$$A = \pi r \sqrt{r^2 + h^2}$$
   a. Find the surface area of a cone whose height is 3 centimeters and whose radius is 4 centimeters.
   b. Approximate to two decimal places the surface area of a cone whose height is 7.2 feet and whose radius is 6.8 feet.

117. The owner of Knightime Video has determined that the demand equation for renting older releases is given by the equation $F(x) = 0.6\sqrt{49 - x^2}$, where x is the price in dollars per two-day rental and $F(x)$ is the number of times the video is demanded per week.
   a. Approximate to one decimal place the demand per week of an older release if the rental price is $3 per two-day rental.
   b. Approximate to one decimal place the demand per week of an older release if the rental price is $5 per two-day rental.
   c. Explain how the owner of the video store can use this equation to predict the number of copies of each tape that should be in stock.

118. Before Mount Vesuvius, a volcano in Italy, erupted violently in 79 A.D., its height was 4190 feet. Vesuvius was roughly cone-shaped, and its base had a radius of approximately 25,200 feet. Use the formula for the surface area of a cone, given in Exercise 116, to approximate the surface area this volcano had before it erupted. (*Source:* Global Volcanism Network)

4190 ft

25,200 ft

7.4 ADDING, SUBTRACTING, AND MULTIPLYING RADICAL EXPRESSIONS

OBJECTIVES

1 Add or subtract radical expressions.

2 Multiply radical expressions.

OBJECTIVE 1 ▶ Adding or subtracting radical expressions. We have learned that sums or differences of like terms can be simplified. To simplify these sums or differences, we use the distributive property. For example,

$$2x + 3x = (2 + 3)x = 5x \quad \text{and} \quad 7x^2y - 4x^2y = (7 - 4)x^2y = 3x^2y$$

The distributive property can also be used to add **like radicals.**

> **Like Radicals**
> Radicals with the same index and the same radicand are like radicals.

For example, $2\sqrt{7} + 3\sqrt{7} = (2 + 3)\sqrt{7} = 5\sqrt{7}$. Also,

Like radicals

$$5\sqrt{3x} - 7\sqrt{3x} = (5 - 7)\sqrt{3x} = -2\sqrt{3x}$$

The expression $2\sqrt{7} + 2\sqrt[3]{7}$ cannot be simplified further since $2\sqrt{7}$ and $2\sqrt[3]{7}$ are not like radicals.

Unlike radicals

EXAMPLE 1 Add or subtract as indicated. Assume all variables represent positive real numbers.

a. $4\sqrt{11} + 8\sqrt{11}$ **b.** $5\sqrt[3]{3x} - 7\sqrt[3]{3x}$ **c.** $2\sqrt{7} + 2\sqrt[3]{7}$

Solution

a. $4\sqrt{11} + 8\sqrt{11} = (4 + 8)\sqrt{11} = 12\sqrt{11}$

b. $5\sqrt[3]{3x} - 7\sqrt[3]{3x} = (5 - 7)\sqrt[3]{3x} = -2\sqrt[3]{3x}$

c. $2\sqrt{7} + 2\sqrt[3]{7}$

This expression cannot be simplified since $2\sqrt{7}$ and $2\sqrt[3]{7}$ do not contain like radicals.

☐

PRACTICE

1 Add or subtract as indicated.

a. $3\sqrt{17} + 5\sqrt{17}$ **b.** $7\sqrt[3]{5z} - 12\sqrt[3]{5z}$ **c.** $3\sqrt{2} + 5\sqrt[3]{2}$

When adding or subtracting radicals, always check first to see whether any radicals can be simplified.

Concept Check ☑

True or false?

$$\sqrt{a} + \sqrt{b} = \sqrt{a + b}$$

Explain.

Answer to Concept Check:
false; answers may vary

EXAMPLE 2 Add or subtract. Assume that variables represent positive real numbers.

a. $\sqrt{20} + 2\sqrt{45}$ **b.** $\sqrt[3]{54} - 5\sqrt[3]{16} + \sqrt[3]{2}$ **c.** $\sqrt{27x} - 2\sqrt{9x} + \sqrt{72x}$
d. $\sqrt[3]{98} + \sqrt{98}$ **e.** $\sqrt[3]{48y^4} + \sqrt[3]{6y^4}$

Solution   First, simplify each radical. Then add or subtract any like radicals.

a.
$$\begin{aligned}
\sqrt{20} + 2\sqrt{45} &= \sqrt{4\cdot 5} + 2\sqrt{9\cdot 5} && \text{Factor 20 and 45.}\\
&= \sqrt{4}\cdot\sqrt{5} + 2\cdot\sqrt{9}\cdot\sqrt{5} && \text{Use the product rule.}\\
&= 2\cdot\sqrt{5} + 2\cdot 3\cdot\sqrt{5} && \text{Simplify } \sqrt{4} \text{ and } \sqrt{9}.\\
&= 2\sqrt{5} + 6\sqrt{5} && \text{Add like radicals.}\\
&= 8\sqrt{5}
\end{aligned}$$

A calculator check for
Example 2b.

> ▶ **Helpful Hint**
> None of these terms contain like radicals. We can simplify no further.

b.
$$\begin{aligned}
&\sqrt[3]{54} - 5\sqrt[3]{16} + \sqrt[3]{2}\\
&= \sqrt[3]{27}\cdot\sqrt[3]{2} - 5\cdot\sqrt[3]{8}\cdot\sqrt[3]{2} + \sqrt[3]{2} && \text{Factor and use the product rule.}\\
&= 3\cdot\sqrt[3]{2} - 5\cdot 2\cdot\sqrt[3]{2} + \sqrt[3]{2} && \text{Simplify } \sqrt[3]{27} \text{ and } \sqrt[3]{8}.\\
&= 3\sqrt[3]{2} - 10\sqrt[3]{2} + \sqrt[3]{2} && \text{Write } 5\cdot 2 \text{ as 10.}\\
&= -6\sqrt[3]{2} && \text{Combine like radicals.}
\end{aligned}$$

c.
$$\begin{aligned}
&\sqrt{27x} - 2\sqrt{9x} + \sqrt{72x}\\
&= \sqrt{9}\cdot\sqrt{3x} - 2\cdot\sqrt{9}\cdot\sqrt{x} + \sqrt{36}\cdot\sqrt{2x} && \text{Factor and use the product rule.}\\
&= 3\cdot\sqrt{3x} - 2\cdot 3\cdot\sqrt{x} + 6\cdot\sqrt{2x} && \text{Simplify } \sqrt{9} \text{ and } \sqrt{36}.\\
&= 3\sqrt{3x} - 6\sqrt{x} + 6\sqrt{2x} && \text{Write } 2\cdot 3 \text{ as 6.}
\end{aligned}$$

d.
$$\begin{aligned}
\sqrt[3]{98} + \sqrt{98} &= \sqrt[3]{98} + \sqrt{49}\cdot\sqrt{2} && \text{Factor and use the product rule.}\\
&= \sqrt[3]{98} + 7\sqrt{2} && \text{No further simplification is possible.}
\end{aligned}$$

e.
$$\begin{aligned}
\sqrt[3]{48y^4} + \sqrt[3]{6y^4} &= \sqrt[3]{8y^3}\cdot\sqrt[3]{6y} + \sqrt[3]{y^3}\cdot\sqrt[3]{6y} && \text{Factor and use the product rule.}\\
&= 2y\sqrt[3]{6y} + y\sqrt[3]{6y} && \text{Simplify } \sqrt[3]{8y^3} \text{ and } \sqrt[3]{y^3}.\\
&= 3y\sqrt[3]{6y} && \text{Combine like radicals.} \quad\square
\end{aligned}$$

PRACTICE
2   Add or subtract.

a. $\sqrt{24} + 3\sqrt{54}$ **b.** $\sqrt[3]{24} - 4\sqrt[3]{81} + \sqrt[3]{3}$ **c.** $\sqrt{75x} - 3\sqrt{27x} + \sqrt{12x}$
d. $\sqrt{40} + \sqrt[3]{40}$ **e.** $\sqrt[3]{81x^4} + \sqrt[3]{3x^4}$

Let's continue to assume that variables represent positive real numbers.

EXAMPLE 3 Add or subtract as indicated.

a. $\dfrac{\sqrt{45}}{4} - \dfrac{\sqrt{5}}{3}$ **b.** $\sqrt[3]{\dfrac{7x}{8}} + 2\sqrt[3]{7x}$

Solution

a.
$$\begin{aligned}
\frac{\sqrt{45}}{4} - \frac{\sqrt{5}}{3} &= \frac{3\sqrt{5}}{4} - \frac{\sqrt{5}}{3} && \text{To subtract, notice that the LCD is 12.}\\[2mm]
&= \frac{3\sqrt{5}\cdot 3}{4\cdot 3} - \frac{\sqrt{5}\cdot 4}{3\cdot 4} && \text{Write each expression as an equivalent expression with a denominator of 12.}\\[2mm]
&= \frac{9\sqrt{5}}{12} - \frac{4\sqrt{5}}{12} && \text{Multiply factors in the numerator and the denominator.}\\[2mm]
&= \frac{5\sqrt{5}}{12} && \text{Subtract.}
\end{aligned}$$

b. $\sqrt[3]{\dfrac{7x}{8}} + 2\sqrt[3]{7x} = \dfrac{\sqrt[3]{7x}}{\sqrt[3]{8}} + 2\sqrt[3]{7x}$ Apply the quotient rule for radicals.

$= \dfrac{\sqrt[3]{7x}}{2} + 2\sqrt[3]{7x}$ Simplify.

$= \dfrac{\sqrt[3]{7x}}{2} + \dfrac{2\sqrt[3]{7x} \cdot 2}{2}$ Write each expression as an equivalent expression with a denominator of 2.

$= \dfrac{\sqrt[3]{7x}}{2} + \dfrac{4\sqrt[3]{7x}}{2}$

$= \dfrac{5\sqrt[3]{7x}}{2}$ Add.   □

PRACTICE
3   Add or subtract as indicated.

a. $\dfrac{\sqrt{28}}{3} - \dfrac{\sqrt{7}}{4}$ **b.** $\sqrt[3]{\dfrac{6y}{64}} + 3\sqrt[3]{6y}$

OBJECTIVE 2 ▶ Multiplying radical expressions. We can multiply radical expressions by using many of the same properties used to multiply polynomial expressions. For instance, to multiply $\sqrt{2}(\sqrt{6} - 3\sqrt{2})$, we use the distributive property and multiply $\sqrt{2}$ by each term inside the parentheses.

$\sqrt{2}(\sqrt{6} - 3\sqrt{2}) = \sqrt{2}(\sqrt{6}) - \sqrt{2}(3\sqrt{2})$ Use the distributive property.

$= \sqrt{2 \cdot 6} - 3\sqrt{2 \cdot 2}$

$= \sqrt{2 \cdot 2 \cdot 3} - 3 \cdot 2$ Use the product rule for radicals.

$= 2\sqrt{3} - 6$

EXAMPLE 4   Multiply.

a. $\sqrt{3}(5 + \sqrt{30})$ **b.** $(\sqrt{5} - \sqrt{6})(\sqrt{7} + 1)$ **c.** $(7\sqrt{x} + 5)(3\sqrt{x} - \sqrt{5})$

d. $(4\sqrt{3} - 1)^2$ **e.** $(\sqrt{2x} - 5)(\sqrt{2x} + 5)$ **f.** $(\sqrt{x-3} + 5)^2$

Solution

a. $\sqrt{3}(5 + \sqrt{30}) = \sqrt{3}(5) + \sqrt{3}(\sqrt{30})$

$= 5\sqrt{3} + \sqrt{3 \cdot 30}$

$= 5\sqrt{3} + \sqrt{3 \cdot 3 \cdot 10}$

$= 5\sqrt{3} + 3\sqrt{10}$

b. To multiply, we can use the FOIL method.

$$\overset{\text{First}\quad\text{Outer}\quad\text{Inner}\quad\text{Last}}{(\sqrt{5} - \sqrt{6})(\sqrt{7} + 1) = \sqrt{5} \cdot \sqrt{7} + \sqrt{5} \cdot 1 - \sqrt{6} \cdot \sqrt{7} - \sqrt{6} \cdot 1}$$

$= \sqrt{35} + \sqrt{5} - \sqrt{42} - \sqrt{6}$

c. $(7\sqrt{x} + 5)(3\sqrt{x} - \sqrt{5}) = 7\sqrt{x}(3\sqrt{x}) - 7\sqrt{x}(\sqrt{5}) + 5(3\sqrt{x}) - 5(\sqrt{5})$

$= 21x - 7\sqrt{5x} + 15\sqrt{x} - 5\sqrt{5}$

d. $(4\sqrt{3} - 1)^2 = (4\sqrt{3} - 1)(4\sqrt{3} - 1)$

$= 4\sqrt{3}(4\sqrt{3}) - 4\sqrt{3}(1) - 1(4\sqrt{3}) - 1(-1)$

$= 16 \cdot 3 - 4\sqrt{3} - 4\sqrt{3} + 1$

$= 48 - 8\sqrt{3} + 1$

$= 49 - 8\sqrt{3}$

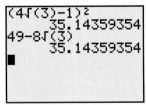

A calculator check for Example 4d.

e. $\left(\sqrt{2x} - 5\right)\left(\sqrt{2x} + 5\right) = \sqrt{2x} \cdot \sqrt{2x} + 5\sqrt{2x} - 5\sqrt{2x} - 5 \cdot 5$

$$= 2x - 25$$

f. $\left(\underbrace{\sqrt{x-3}}_{a} + \underbrace{5}_{b}\right)^2 = \underbrace{\left(\sqrt{x-3}\right)^2}_{a^2} + \underbrace{2 \cdot}_{+2\cdot} \underbrace{\sqrt{x-3}}_{a} \cdot \underbrace{5}_{\cdot b} + \underbrace{5^2}_{+b^2}$

$$= x - 3 + 10\sqrt{x-3} + 25 \qquad \text{Simplify.}$$

$$= x + 22 + 10\sqrt{x-3} \qquad \text{Combine like terms.} \qquad \square$$

PRACTICE
4 Multiply.

a. $\sqrt{5}\left(2 + \sqrt{15}\right)$

b. $\left(\sqrt{2} - \sqrt{5}\right)\left(\sqrt{6} + 2\right)$

c. $\left(3\sqrt{z} - 4\right)\left(2\sqrt{z} + 3\right)$

d. $\left(\sqrt{6} - 3\right)^2$

e. $\left(\sqrt{5x} + 3\right)\left(\sqrt{5x} - 3\right)$

f. $\left(\sqrt{x+2} + 3\right)^2$

VOCABULARY & READINESS CHECK

Complete the table with "Like" or "Unlike."

Terms	Like or Unlike Radical Terms?
1. $\sqrt{7}, \sqrt[3]{7}$	
2. $\sqrt[3]{x^2 y}, \sqrt[3]{yx^2}$	
3. $\sqrt[3]{abc}, \sqrt[3]{cba}$	
4. $2x\sqrt{5}, 2x\sqrt{10}$	

Simplify. Assume that all variables represent positive real numbers.

5. $2\sqrt{3} + 4\sqrt{3} =$ _____

6. $5\sqrt{7} + 3\sqrt{7} =$ _____

7. $8\sqrt{x} - \sqrt{x} =$ _____

8. $3\sqrt{y} - \sqrt{y} =$ _____

9. $7\sqrt[3]{x} + \sqrt[3]{x} =$ _____

10. $8\sqrt[3]{z} + \sqrt[3]{z} =$ _____

Add or subtract if possible.

11. $\sqrt{11} + \sqrt[3]{11} =$ _____

12. $9\sqrt{13} - \sqrt[4]{13} =$ _____

13. $8\sqrt[3]{2x} + 3\sqrt[3]{2x} - \sqrt[3]{2x} =$ _____

14. $8\sqrt[3]{2x} + 3\sqrt[3]{2x^2} - \sqrt[3]{2x} =$ _____

7.4 | EXERCISE SET

MyMathLab
PRACTICE   WATCH   DOWNLOAD   READ   REVIEW

Add or subtract. See Examples 1 through 3.

1. $\sqrt{8} - \sqrt{32}$

2. $\sqrt{27} - \sqrt{75}$

3. $2\sqrt{2x^3} + 4x\sqrt{8x}$

4. $3\sqrt{45x^3} + x\sqrt{5x}$

5. $2\sqrt{50} - 3\sqrt{125} + \sqrt{98}$

6. $4\sqrt{32} - \sqrt{18} + 2\sqrt{128}$

7. $\sqrt[3]{16x} - \sqrt[3]{54x}$

8. $2\sqrt[3]{3a^4} - 3a\sqrt[3]{81a}$

9. $\sqrt{9b^3} - \sqrt{25b^3} + \sqrt{49b^3}$

10. $\sqrt{4x^7} + 9x^2\sqrt{x^3} - 5x\sqrt{x^5}$

11. $\dfrac{5\sqrt{2}}{3} + \dfrac{2\sqrt{2}}{5}$

12. $\dfrac{\sqrt{3}}{2} + \dfrac{4\sqrt{3}}{3}$

13. $\sqrt[3]{\dfrac{11}{8}} - \dfrac{\sqrt[3]{11}}{6}$

14. $\dfrac{2\sqrt[3]{4}}{7} - \dfrac{\sqrt[3]{4}}{14}$

15. $\dfrac{\sqrt{20x}}{9} + \sqrt{\dfrac{5x}{9}}$

16. $\dfrac{3x\sqrt{7}}{5} + \sqrt{\dfrac{7x^2}{100}}$

17. $7\sqrt{9} - 7 + \sqrt{3}$

18. $\sqrt{16} - 5\sqrt{10} + 7$

19. $2 + 3\sqrt{y^2} - 6\sqrt{y^2} + 5$

20. $3\sqrt{7} - \sqrt[3]{x} + 4\sqrt{7} - 3\sqrt[3]{x}$

21. $3\sqrt{108} - 2\sqrt{18} - 3\sqrt{48}$

22. $-\sqrt{75} + \sqrt{12} - 3\sqrt{3}$

23. $-5\sqrt[3]{625} + \sqrt[3]{40}$

24. $-2\sqrt[3]{108} - \sqrt[3]{32}$

25. $\sqrt{9b^3} - \sqrt{25b^3} + \sqrt{16b^3}$

26. $\sqrt{4x^7y^5} + 9x^2\sqrt{x^3y^5} - 5xy\sqrt{x^5y^3}$

27. $5y\sqrt{8y} + 2\sqrt{50y^3}$

28. $3\sqrt{8x^2y^3} - 2x\sqrt{32y^3}$

29. $\sqrt[3]{54xy^3} - 5\sqrt[3]{2xy^3} + y\sqrt[3]{128x}$

30. $2\sqrt[3]{24x^3y^4} + 4x\sqrt[3]{81y^4}$

31. $6\sqrt[3]{11} + 8\sqrt{11} - 12\sqrt{11}$

32. $3\sqrt[3]{5} + 4\sqrt{5}$

33. $-2\sqrt[4]{x^7} + 3\sqrt[4]{16x^7}$

34. $6\sqrt[3]{24x^3} - 2\sqrt[3]{81x^3} - x\sqrt[3]{3}$

35. $\dfrac{4\sqrt{3}}{3} - \dfrac{\sqrt{12}}{3}$

36. $\dfrac{\sqrt{45}}{10} + \dfrac{7\sqrt{5}}{10}$

37. $\dfrac{\sqrt[3]{8x^4}}{7} + \dfrac{3x\sqrt[3]{x}}{7}$

38. $\dfrac{\sqrt[4]{48}}{5x} - \dfrac{2\sqrt[4]{3}}{10x}$

39. $\sqrt{\dfrac{28}{x^2}} + \sqrt{\dfrac{7}{4x^2}}$

40. $\dfrac{\sqrt{99}}{5x} - \sqrt{\dfrac{44}{x^2}}$

41. $\sqrt[3]{\dfrac{16}{27}} - \dfrac{\sqrt[3]{54}}{6}$

42. $\dfrac{\sqrt[3]{3}}{10} + \sqrt[3]{\dfrac{24}{125}}$

43. $-\dfrac{\sqrt[3]{2x^4}}{9} + \sqrt[3]{\dfrac{250x^4}{27}}$

44. $\dfrac{\sqrt[3]{y^5}}{8} + \dfrac{5y\sqrt[3]{y^2}}{4}$

△ **45.** Find the perimeter of the trapezoid.

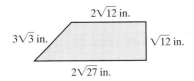

$2\sqrt{12}$ in.

$3\sqrt{3}$ in. $\sqrt{12}$ in.

$2\sqrt{27}$ in.

△ **46.** Find the perimeter of the triangle.

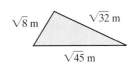

$\sqrt{8}$ m $\sqrt{32}$ m

$\sqrt{45}$ m

Multiply, and then simplify if possible. See Example 4.

47. $\sqrt{7}\left(\sqrt{5} + \sqrt{3}\right)$

48. $\sqrt{5}\left(\sqrt{15} - \sqrt{35}\right)$

49. $\left(\sqrt{5} - \sqrt{2}\right)^2$

50. $\left(3x - \sqrt{2}\right)\left(3x - \sqrt{2}\right)$

51. $\sqrt{3x}\left(\sqrt{3} - \sqrt{x}\right)$

52. $\sqrt{5y}\left(\sqrt{y} + \sqrt{5}\right)$

53. $\left(2\sqrt{x} - 5\right)\left(3\sqrt{x} + 1\right)$

54. $\left(8\sqrt{y} + z\right)\left(4\sqrt{y} - 1\right)$

55. $\left(\sqrt[3]{a} - 4\right)\left(\sqrt[3]{a} + 5\right)$

56. $\left(\sqrt[3]{a} + 2\right)\left(\sqrt[3]{a} + 7\right)$

57. $6\left(\sqrt{2} - 2\right)$

58. $\sqrt{5}\left(6 - \sqrt{5}\right)$

59. $\sqrt{2}\left(\sqrt{2} + x\sqrt{6}\right)$

60. $\sqrt{3}\left(\sqrt{3} - 2\sqrt{5x}\right)$

61. $\left(2\sqrt{7} + 3\sqrt{5}\right)\left(\sqrt{7} - 2\sqrt{5}\right)$

62. $\left(\sqrt{6} - 4\sqrt{2}\right)\left(3\sqrt{6} + \sqrt{2}\right)$

63. $\left(\sqrt{x} - y\right)\left(\sqrt{x} + y\right)$

64. $\left(\sqrt{3x} + 2\right)\left(\sqrt{3x} - 2\right)$

65. $\left(\sqrt{3} + x\right)^2$

66. $\left(\sqrt{y} - 3x\right)^2$

67. $\left(\sqrt{5x} - 2\sqrt{3x}\right)\left(\sqrt{5x} - 3\sqrt{3x}\right)$

68. $\left(5\sqrt{7x} - \sqrt{2x}\right)\left(4\sqrt{7x} + 6\sqrt{2x}\right)$

69. $\left(\sqrt[3]{4} + 2\right)\left(\sqrt[3]{2} - 1\right)$

70. $\left(\sqrt[3]{3} + \sqrt[3]{2}\right)\left(\sqrt[3]{9} - \sqrt[3]{4}\right)$

71. $\left(\sqrt[3]{x} + 1\right)\left(\sqrt[3]{x^2} - \sqrt[3]{x} + 1\right)$

72. $\left(\sqrt[3]{3x} + 2\right)\left(\sqrt[3]{9x^2} - 2\sqrt[3]{3x} + 4\right)$

73. $\left(\sqrt{x - 1} + 5\right)^2$

74. $\left(\sqrt{3x + 1} + 2\right)^2$

75. $\left(\sqrt{2x + 5} - 1\right)^2$

76. $\left(\sqrt{x - 6} - 7\right)^2$

REVIEW AND PREVIEW

Factor each numerator and denominator. Then simplify if possible. See Section 6.1.

77. $\dfrac{2x - 14}{2}$

78. $\dfrac{8x - 24y}{4}$

79. $\dfrac{7x - 7y}{x^2 - y^2}$

80. $\dfrac{x^3 - 8}{4x - 8}$

81. $\dfrac{6a^2b - 9ab}{3ab}$

82. $\dfrac{14r - 28r^2s^2}{7rs}$

83. $\dfrac{-4 + 2\sqrt{3}}{6}$

84. $\dfrac{-5 + 10\sqrt{7}}{5}$

CONCEPT EXTENSIONS

△ **85.** Find the perimeter and area of the rectangle.

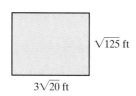

$3\sqrt{20}$ ft

$\sqrt{125}$ ft

△ **86.** Find the area and perimeter of the trapezoid. (*Hint:* The area of a trapezoid is the product of half the height $6\sqrt{3}$ meters and the sum of the bases $2\sqrt{63}$ and $7\sqrt{7}$ meters.)

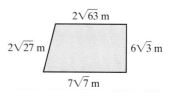

$2\sqrt{63}$ m

$2\sqrt{27}$ m

$6\sqrt{3}$ m

$7\sqrt{7}$ m

87. a. Add: $\sqrt{3} + \sqrt{3}$.

    b. Multiply: $\sqrt{3} \cdot \sqrt{3}$.

    c. Describe the differences in parts **a** and **b**.

88. Multiply: $\left(\sqrt{2} + \sqrt{3} - 1\right)^2$.

89. Explain how simplifying $2x + 3x$ is similar to simplifying $2\sqrt{x} + 3\sqrt{x}$.

90. Explain how multiplying $(x - 2)(x + 3)$ is similar to multiplying $\left(\sqrt{x} - \sqrt{2}\right)\left(\sqrt{x} + 3\right)$.

📖 STUDY SKILLS BUILDER

Have You Decided to Successfully Complete This Course?

Hopefully by now, one of your current goals is to successfully complete this course.

If it is not a goal of yours, ask yourself why? One common reason is fear of failure. Amazingly enough, fear of failure alone can be strong enough to keep many of us from doing our best in any endeavor. Another common reason is that you simply haven't taken the time to make successfully completing this course one of your goals.

Anytime you are registered for a course, successfully completing that course should probably be a goal. How do you do this? Start by writing this goal in your mathematics notebook. Then list steps you will take to ensure success. A great first step is to read or reread Section 1.1 and make a commitment to try the suggestions in this section.

Good luck, and don't forget that a positive attitude will make a big difference.

Let's see how you are doing.

1. Have you made the decision to make "successfully completing this course" a goal of yours? If not, please list reasons that this has not happened. Study your list and talk to your instructor about this.

2. If your answer to Exercise 1 is yes, take a moment and list, in your notebook, further specific goals that will help you achieve this major goal of successfully completing this course. (For example, "My goal this semester is not to miss any of my mathematics classes.")

3. Rate your commitment to this course with a number between 1 and 5. Use the diagram below to help.

High Commitment		Average Commitment		Not Committed at All
5	4	3	2	1

4. If you have rated your personal commitment level (from the exercise above) as a 1, 2, or 3, list the reasons why this is so. Then determine whether it is possible to increase your commitment level to a 4 or 5.

7.5 RATIONALIZING DENOMINATORS AND NUMERATORS OF RADICAL EXPRESSIONS

OBJECTIVES

1 Rationalize denominators.

2 Rationalize denominators having two terms.

3 Rationalize numerators.

OBJECTIVE 1 ▶ Rationalizing denominators of radical expressions. Often in mathematics, it is helpful to write a radical expression such as $\dfrac{\sqrt{3}}{\sqrt{2}}$ either without a radical in the denominator or without a radical in the numerator. The process of writing this expression as an equivalent expression but without a radical in the denominator is called **rationalizing the denominator.** To rationalize the denominator of $\dfrac{\sqrt{3}}{\sqrt{2}}$, we use the fundamental principle of fractions and multiply the numerator and the denominator by $\sqrt{2}$. Recall that this is the same as multiplying by $\dfrac{\sqrt{2}}{\sqrt{2}}$, which simplifies to 1.

$$\frac{\sqrt{3}}{\sqrt{2}} = \frac{\sqrt{3} \cdot \sqrt{2}}{\sqrt{2} \cdot \sqrt{2}} = \frac{\sqrt{6}}{\sqrt{4}} = \frac{\sqrt{6}}{2}$$

In this section, we continue to assume that variables represent positive real numbers.

EXAMPLE 1   Rationalize the denominator of each expression.

a. $\dfrac{2}{\sqrt{5}}$ **b.** $\dfrac{2\sqrt{16}}{\sqrt{9x}}$ **c.** $\sqrt[3]{\dfrac{1}{2}}$

Solution

a. To rationalize the denominator, we multiply the numerator and denominator by a factor that makes the radicand in the denominator a perfect square.

$$\frac{2}{\sqrt{5}} = \frac{2 \cdot \sqrt{5}}{\sqrt{5} \cdot \sqrt{5}} = \frac{2\sqrt{5}}{5} \qquad \text{The denominator is now rationalized.}$$

b. First, we simplify the radicals and then rationalize the denominator.

$$\frac{2\sqrt{16}}{\sqrt{9x}} = \frac{2(4)}{3\sqrt{x}} = \frac{8}{3\sqrt{x}}$$

To rationalize the denominator, multiply the numerator and denominator by $\sqrt{x}$. Then

$$\frac{8}{3\sqrt{x}} = \frac{8 \cdot \sqrt{x}}{3\sqrt{x} \cdot \sqrt{x}} = \frac{8\sqrt{x}}{3x}$$

c. $\sqrt[3]{\dfrac{1}{2}} = \dfrac{\sqrt[3]{1}}{\sqrt[3]{2}} = \dfrac{1}{\sqrt[3]{2}}$. Now we rationalize the denominator. Since $\sqrt[3]{2}$ is a cube root, we want to multiply by a value that will make the radicand 2 a perfect cube. If we multiply $\sqrt[3]{2}$ by $\sqrt[3]{2^2}$, we get $\sqrt[3]{2^3} = \sqrt[3]{8} = 2$.

$$\frac{1 \cdot \sqrt[3]{2^2}}{\sqrt[3]{2} \cdot \sqrt[3]{2^2}} = \frac{\sqrt[3]{4}}{\sqrt[3]{2^3}} = \frac{\sqrt[3]{4}}{2} \qquad \text{Multiply the numerator and denominator by } \sqrt[3]{2^2} \text{ and then simplify.}$$

```
2√(5)/5
         .894427191
2/√(5)
         .894427191
■
```

A calculator check for Example 1a.

PRACTICE

1   Rationalize the denominator of each expression.

a. $\dfrac{5}{\sqrt{3}}$ **b.** $\dfrac{3\sqrt{25}}{\sqrt{4x}}$ **c.** $\sqrt[3]{\dfrac{2}{9}}$

Concept Check ✓

Determine by which number both the numerator and denominator can be multiplied to rationalize the denominator of the radical expression.

a. $\dfrac{1}{\sqrt[3]{7}}$　　　　**b.** $\dfrac{1}{\sqrt[4]{8}}$

EXAMPLE 2 Rationalize the denominator of $\sqrt{\dfrac{7x}{3y}}$.

Solution　$\sqrt{\dfrac{7x}{3y}} = \dfrac{\sqrt{7x}}{\sqrt{3y}}$　Use the quotient rule. No radical may be simplified further.

$= \dfrac{\sqrt{7x} \cdot \sqrt{3y}}{\sqrt{3y} \cdot \sqrt{3y}}$　Multiply numerator and denominator by $\sqrt{3y}$ so that the radicand in the denominator is a perfect square.

$= \dfrac{\sqrt{21xy}}{3y}$　Use the product rule in the numerator and denominator. Remember that $\sqrt{3y} \cdot \sqrt{3y} = 3y$. □

PRACTICE
2　Rationalize the denominator of $\sqrt{\dfrac{3z}{5y}}$.

EXAMPLE 3 Rationalize the denominator of $\dfrac{\sqrt[4]{x}}{\sqrt[4]{81y^5}}$.

Solution　First, simplify each radical if possible.

$\dfrac{\sqrt[4]{x}}{\sqrt[4]{81y^5}} = \dfrac{\sqrt[4]{x}}{\sqrt[4]{81y^4} \cdot \sqrt[4]{y}}$　Use the product rule in the denominator.

$= \dfrac{\sqrt[4]{x}}{3y\sqrt[4]{y}}$　Write $\sqrt[4]{81y^4}$ as $3y$.

$= \dfrac{\sqrt[4]{x} \cdot \sqrt[4]{y^3}}{3y\sqrt[4]{y} \cdot \sqrt[4]{y^3}}$　Multiply numerator and denominator by $\sqrt[4]{y^3}$ so that the radicand in the denominator is a perfect fourth power.

$= \dfrac{\sqrt[4]{xy^3}}{3y\sqrt[4]{y^4}}$　Use the product rule in the numerator and denominator.

$= \dfrac{\sqrt[4]{xy^3}}{3y^2}$　In the denominator, $\sqrt[4]{y^4} = y$ and $3y \cdot y = 3y^2$. □

PRACTICE
3　Rationalize the denominator of $\dfrac{\sqrt[3]{z^2}}{\sqrt[3]{27x^4}}$.

OBJECTIVE 2 ▶ Rationalizing denominators having two terms. Remember the product of the sum and difference of two terms?

$$(a + b)(a - b) = a^2 - b^2$$

These two expressions are called **conjugates** of each other.

To rationalize a numerator or denominator that is a sum or difference of two terms, we use conjugates. To see how and why this works, let's rationalize the denominator of the expression $\dfrac{5}{\sqrt{3} - 2}$. To do so, we multiply both the numerator and the denominator by $\sqrt{3} + 2$, the **conjugate** of the denominator $\sqrt{3} - 2$, and see what happens.

$$\frac{5}{\sqrt{3} - 2} = \frac{5\left(\sqrt{3} + 2\right)}{\left(\sqrt{3} - 2\right)\left(\sqrt{3} + 2\right)}$$

$$= \frac{5\left(\sqrt{3} + 2\right)}{\left(\sqrt{3}\right)^2 - 2^2} \qquad \text{Multiply the sum and difference of two terms: } (a + b)(a - b) = a^2 - b^2.$$

$$= \frac{5\left(\sqrt{3} + 2\right)}{3 - 4}$$

$$= \frac{5\left(\sqrt{3} + 2\right)}{-1}$$

$$= -5\left(\sqrt{3} + 2\right) \quad \text{or} \quad -5\sqrt{3} - 10$$

Notice in the denominator that the product of $(\sqrt{3} - 2)$ and its conjugate, $(\sqrt{3} + 2)$, is -1. In general, the product of an expression and its conjugate will contain no radical terms. This is why, when rationalizing a denominator or a numerator containing two terms, we multiply by its conjugate. Examples of conjugates are

$$\sqrt{a} - \sqrt{b} \quad \text{and} \quad \sqrt{a} + \sqrt{b}$$
$$x + \sqrt{y} \quad \text{and} \quad x - \sqrt{y}$$

EXAMPLE 4   Rationalize each denominator.

a. $\dfrac{2}{3\sqrt{2} + 4}$ **b.** $\dfrac{\sqrt{6} + 2}{\sqrt{5} - \sqrt{3}}$ **c.** $\dfrac{2\sqrt{m}}{3\sqrt{x} + \sqrt{m}}$

Solution

a. Multiply the numerator and denominator by the conjugate of the denominator, $3\sqrt{2} + 4$.

$$\frac{2}{3\sqrt{2} + 4} = \frac{2\left(3\sqrt{2} - 4\right)}{\left(3\sqrt{2} + 4\right)\left(3\sqrt{2} - 4\right)}$$

$$= \frac{2\left(3\sqrt{2} - 4\right)}{\left(3\sqrt{2}\right)^2 - 4^2}$$

$$= \frac{2\left(3\sqrt{2} - 4\right)}{18 - 16}$$

$$= \frac{2\left(3\sqrt{2} - 4\right)}{2}, \quad \text{or} \quad 3\sqrt{2} - 4$$

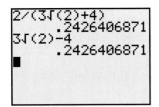

```
2/(3√(2)+4)
       .2426406871
3√(2)-4
       .2426406871
■
```

A calculator check for Example 4a. Notice the use of parentheses in the first expression.

It is often useful to leave a numerator in factored form to help determine whether the expression can be simplified.

A calculator check for Example 4b. Notice the use of parentheses in the numerators and the denominator containing more than one term.

b. Multiply the numerator and denominator by the conjugate of $\sqrt{5} - \sqrt{3}$.

$$\frac{\sqrt{6} + 2}{\sqrt{5} - \sqrt{3}} = \frac{(\sqrt{6} + 2)(\sqrt{5} + \sqrt{3})}{(\sqrt{5} - \sqrt{3})(\sqrt{5} + \sqrt{3})}$$

$$= \frac{\sqrt{6}\sqrt{5} + \sqrt{6}\sqrt{3} + 2\sqrt{5} + 2\sqrt{3}}{(\sqrt{5})^2 - (\sqrt{3})^2}$$

$$= \frac{\sqrt{30} + \sqrt{18} + 2\sqrt{5} + 2\sqrt{3}}{5 - 3}$$

$$= \frac{\sqrt{30} + 3\sqrt{2} + 2\sqrt{5} + 2\sqrt{3}}{2}$$

c. Multiply by the conjugate of $3\sqrt{x} + \sqrt{m}$ to eliminate the radicals from the denominator.

$$\frac{2\sqrt{m}}{3\sqrt{x} + \sqrt{m}} = \frac{2\sqrt{m}(3\sqrt{x} - \sqrt{m})}{(3\sqrt{x} + \sqrt{m})(3\sqrt{x} - \sqrt{m})} = \frac{6\sqrt{mx} - 2m}{(3\sqrt{x})^2 - (\sqrt{m})^2}$$

$$= \frac{6\sqrt{mx} - 2m}{9x - m}$$

PRACTICE
4   Rationalize the denominator.

a. $\dfrac{5}{3\sqrt{5} + 2}$ **b.** $\dfrac{\sqrt{2} + 5}{\sqrt{3} - \sqrt{5}}$ **c.** $\dfrac{3\sqrt{x}}{2\sqrt{x} + \sqrt{y}}$

OBJECTIVE 3 ▶ Rationalizing numerators. As mentioned earlier, it is also often helpful to write an expression such as $\dfrac{\sqrt{3}}{\sqrt{2}}$ as an equivalent expression without a radical in the numerator. This process is called **rationalizing the numerator.** To rationalize the numerator of $\dfrac{\sqrt{3}}{\sqrt{2}}$, we multiply the numerator and the denominator by $\sqrt{3}$.

$$\frac{\sqrt{3}}{\sqrt{2}} = \frac{\sqrt{3} \cdot \sqrt{3}}{\sqrt{2} \cdot \sqrt{3}} = \frac{\sqrt{9}}{\sqrt{6}} = \frac{3}{\sqrt{6}}$$

EXAMPLE 5   Rationalize the numerator of $\dfrac{\sqrt{7}}{\sqrt{45}}$.

Solution   First we simplify $\sqrt{45}$.

$$\frac{\sqrt{7}}{\sqrt{45}} = \frac{\sqrt{7}}{\sqrt{9 \cdot 5}} = \frac{\sqrt{7}}{3\sqrt{5}}$$

Next we rationalize the numerator by multiplying the numerator and the denominator by $\sqrt{7}$.

$$\frac{\sqrt{7}}{3\sqrt{5}} = \frac{\sqrt{7} \cdot \sqrt{7}}{3\sqrt{5} \cdot \sqrt{7}} = \frac{7}{3\sqrt{5} \cdot 7} = \frac{7}{3\sqrt{35}}$$

PRACTICE
5   Rationalize the numerator of $\dfrac{\sqrt{32}}{\sqrt{80}}$.

EXAMPLE 6   Rationalize the numerator of $\dfrac{\sqrt[3]{2x^2}}{\sqrt[3]{5y}}$.

Solution   The numerator and the denominator of this expression are already simplified. To rationalize the numerator, $\sqrt[3]{2x^2}$, we multiply the numerator and denominator by a factor that will make the radicand a perfect cube. If we multiply $\sqrt[3]{2x^2}$ by $\sqrt[3]{4x}$, we get $\sqrt[3]{8x^3} = 2x$.

$$\frac{\sqrt[3]{2x^2}}{\sqrt[3]{5y}} = \frac{\sqrt[3]{2x^2} \cdot \sqrt[3]{4x}}{\sqrt[3]{5y} \cdot \sqrt[3]{4x}} = \frac{\sqrt[3]{8x^3}}{\sqrt[3]{20xy}} = \frac{2x}{\sqrt[3]{20xy}}$$

☐

PRACTICE
6   Rationalize the numerator of $\dfrac{\sqrt[3]{5b}}{\sqrt[3]{2a}}$.

EXAMPLE 7   Rationalize the numerator of $\dfrac{\sqrt{x}+2}{5}$.

Solution   We multiply the numerator and the denominator by the conjugate of the numerator, $\sqrt{x}+2$.

$$\frac{\sqrt{x}+2}{5} = \frac{\left(\sqrt{x}+2\right)\left(\sqrt{x}-2\right)}{5\left(\sqrt{x}-2\right)} \qquad \text{Multiply by } \sqrt{x}-2, \text{ the conjugate of } \sqrt{x}+2.$$

$$= \frac{\left(\sqrt{x}\right)^2 - 2^2}{5\left(\sqrt{x}-2\right)} \qquad (a+b)(a-b) = a^2 - b^2$$

$$= \frac{x-4}{5\left(\sqrt{x}-2\right)}$$

☐

PRACTICE
7   Rationalize the numerator of $\dfrac{\sqrt{x}-3}{4}$.

VOCABULARY & READINESS CHECK

Use the choices below to fill in each blank. Not all choices will be used.

rationalizing the numerator	conjugate	$\dfrac{\sqrt{3}}{\sqrt{3}}$
rationalizing the denominator	$\dfrac{5}{5}$	

1. The _____ of $a + b$ is $a - b$.

2. The process of writing an equivalent expression, but without a radical in the denominator, is called _____.

3. The process of writing an equivalent expression, but without a radical in the numerator, is called _____.

4. To rationalize the denominator of $\dfrac{5}{\sqrt{3}}$, we multiply by _____.

Find the conjugate of each expression.

5. $\sqrt{2} + x$ **6.** $\sqrt{3} + y$ **7.** $5 - \sqrt{a}$ **8.** $6 - \sqrt{b}$
9. $-7\sqrt{5} + 8\sqrt{x}$ **10.** $-9\sqrt{2} - 6\sqrt{y}$

7.5 EXERCISE SET

Rationalize each denominator. See Examples 1 through 3.

1. $\dfrac{\sqrt{2}}{\sqrt{7}}$

2. $\dfrac{\sqrt{3}}{\sqrt{2}}$

3. $\sqrt{\dfrac{1}{5}}$

4. $\sqrt{\dfrac{1}{2}}$

5. $\sqrt{\dfrac{4}{x}}$

6. $\sqrt{\dfrac{25}{y}}$

7. $\dfrac{4}{\sqrt[3]{3}}$

8. $\dfrac{6}{\sqrt[3]{9}}$

9. $\dfrac{3}{\sqrt{8x}}$

10. $\dfrac{5}{\sqrt{27a}}$

11. $\dfrac{3}{\sqrt[3]{4x^2}}$

12. $\dfrac{5}{\sqrt[3]{3y}}$

13. $\dfrac{9}{\sqrt{3a}}$

14. $\dfrac{x}{\sqrt{5}}$

15. $\dfrac{3}{\sqrt[3]{2}}$

16. $\dfrac{5}{\sqrt[3]{9}}$

17. $\dfrac{2\sqrt{3}}{\sqrt{7}}$

18. $\dfrac{-5\sqrt{2}}{\sqrt{11}}$

19. $\sqrt{\dfrac{2x}{5y}}$

20. $\sqrt{\dfrac{13a}{2b}}$

21. $\sqrt[3]{\dfrac{3}{5}}$

22. $\sqrt[3]{\dfrac{7}{10}}$

23. $\sqrt{\dfrac{3x}{50}}$

24. $\sqrt{\dfrac{11y}{45}}$

25. $\dfrac{1}{\sqrt{12z}}$

26. $\dfrac{1}{\sqrt{32x}}$

27. $\dfrac{\sqrt[3]{2y^2}}{\sqrt[3]{9x^2}}$

28. $\dfrac{\sqrt[3]{3x}}{\sqrt[3]{4y^4}}$

29. $\sqrt[4]{\dfrac{81}{8}}$

30. $\sqrt[4]{\dfrac{1}{9}}$

31. $\sqrt[4]{\dfrac{16}{9x^7}}$

32. $\sqrt[5]{\dfrac{32}{m^6 n^{13}}}$

33. $\dfrac{5a}{\sqrt[5]{8a^9 b^{11}}}$

34. $\dfrac{9y}{\sqrt[4]{4y^9}}$

Rationalize each denominator. See Example 4.

35. $\dfrac{6}{2 - \sqrt{7}}$

36. $\dfrac{3}{\sqrt{7} - 4}$

37. $\dfrac{-7}{\sqrt{x} - 3}$

38. $\dfrac{-8}{\sqrt{y} + 4}$

39. $\dfrac{\sqrt{2} - \sqrt{3}}{\sqrt{2} + \sqrt{3}}$

40. $\dfrac{\sqrt{3} + \sqrt{4}}{\sqrt{2} - \sqrt{3}}$

41. $\dfrac{\sqrt{a} + 1}{2\sqrt{a} - \sqrt{b}}$

42. $\dfrac{2\sqrt{a} - 3}{2\sqrt{a} + \sqrt{b}}$

43. $\dfrac{8}{1 + \sqrt{10}}$

44. $\dfrac{-3}{\sqrt{6} - 2}$

45. $\dfrac{\sqrt{x}}{\sqrt{x} + \sqrt{y}}$

46. $\dfrac{2\sqrt{a}}{2\sqrt{x} - \sqrt{y}}$

47. $\dfrac{2\sqrt{3} + \sqrt{6}}{4\sqrt{3} - \sqrt{6}}$

48. $\dfrac{4\sqrt{5} + \sqrt{2}}{2\sqrt{5} - \sqrt{2}}$

Rationalize each numerator. See Examples 5 and 6.

49. $\sqrt{\dfrac{5}{3}}$

50. $\sqrt{\dfrac{3}{2}}$

51. $\sqrt{\dfrac{18}{5}}$

52. $\sqrt{\dfrac{12}{7}}$

53. $\dfrac{\sqrt{4x}}{7}$

54. $\dfrac{\sqrt{3x^5}}{6}$

55. $\dfrac{\sqrt[3]{5y^2}}{\sqrt[3]{4x}}$

56. $\dfrac{\sqrt[3]{4x}}{\sqrt[3]{z^4}}$

57. $\sqrt{\dfrac{2}{5}}$

58. $\sqrt{\dfrac{3}{7}}$

59. $\dfrac{\sqrt{2x}}{11}$

60. $\dfrac{\sqrt{y}}{7}$

61. $\sqrt[3]{\dfrac{7}{8}}$

62. $\sqrt[3]{\dfrac{25}{2}}$

63. $\dfrac{\sqrt[3]{3x^5}}{10}$

64. $\sqrt[3]{\dfrac{9y}{7}}$

65. $\sqrt{\dfrac{18x^4 y^6}{3z}}$

66. $\sqrt{\dfrac{8x^5 y}{2z}}$

67. When rationalizing the denominator of $\dfrac{\sqrt{5}}{\sqrt{7}}$, explain why both the numerator and the denominator must be multiplied by $\sqrt{7}$.

68. When rationalizing the numerator of $\dfrac{\sqrt{5}}{\sqrt{7}}$, explain why both the numerator and the denominator must be multiplied by $\sqrt{5}$.

Rationalize each numerator. See Example 7.

69. $\dfrac{2 - \sqrt{11}}{6}$

70. $\dfrac{\sqrt{15} + 1}{2}$

71. $\dfrac{2 - \sqrt{7}}{-5}$

72. $\dfrac{\sqrt{5} + 2}{\sqrt{2}}$

73. $\dfrac{\sqrt{x} + 3}{\sqrt{x}}$

74. $\dfrac{5 + \sqrt{2}}{\sqrt{2x}}$

75. $\dfrac{\sqrt{2} - 1}{\sqrt{2} + 1}$

76. $\dfrac{\sqrt{8} - \sqrt{3}}{\sqrt{2} + \sqrt{3}}$

77. $\dfrac{\sqrt{x} + 1}{\sqrt{x} - 1}$

78. $\dfrac{\sqrt{x} + \sqrt{y}}{\sqrt{x} - \sqrt{y}}$

REVIEW AND PREVIEW

Solve each equation. See Sections 1.5 and 5.8.

79. $2x - 7 = 3(x - 4)$

80. $9x - 4 = 7(x - 2)$

81. $(x - 6)(2x + 1) = 0$

82. $(y + 2)(5y + 4) = 0$

83. $x^2 - 8x = -12$

84. $x^3 = x$

CONCEPTS EXTENSIONS

Determine the smallest number both the numerator and denominator should be multiplied by to rationalize the denominator of the radical expression. See the Concept Check in this section.

85. $\dfrac{9}{\sqrt[3]{5}}$

86. $\dfrac{5}{\sqrt{27}}$

△ **87.** The formula of the radius r of a sphere with surface area A is

$$r = \sqrt{\dfrac{A}{4\pi}}$$

Rationalize the denominator of the radical expression in this formula.

△ **88.** The formula for the radius r of a cone with height 7 centimeters and volume V is

$$r = \sqrt{\dfrac{3V}{7\pi}}$$

Rationalize the numerator of the radical expression in this formula.

7 cm

r

89. Explain why rationalizing the denominator does not change the value of the original expression.

90. Explain why rationalizing the numerator does not change the value of the original expression.

INTEGRATED REVIEW RADICALS AND RATIONAL EXPONENTS

Sections 7.1–7.5

Find each root. Throughout this review, assume that all variables represent positive real numbers.

1. $\sqrt{81}$

2. $\sqrt[3]{-8}$

3. $\sqrt[4]{\dfrac{1}{16}}$

4. $\sqrt{x^6}$

5. $\sqrt[3]{y^9}$

6. $\sqrt{4y^{10}}$

7. $\sqrt[5]{-32y^5}$

8. $\sqrt[4]{81b^{12}}$

Use radical notation to rewrite each expression. Simplify if possible.

9. $36^{1/2}$

10. $(3y)^{1/4}$

11. $64^{-2/3}$

12. $(x + 1)^{3/5}$

Use the properties of exponents to simplify each expression. Write with positive exponents.

13. $y^{-1/6} \cdot y^{7/6}$

14. $\dfrac{(2x^{1/3})^4}{x^{5/6}}$

15. $\dfrac{x^{1/4}x^{3/4}}{x^{-1/4}}$

16. $4^{1/3} \cdot 4^{2/5}$

Use rational exponents to simplify each radical.

17. $\sqrt[3]{8x^6}$

18. $\sqrt[12]{a^9b^6}$

Use rational exponents to write each as a single radical expression.

19. $\sqrt[4]{x} \cdot \sqrt{x}$

20. $\sqrt{5} \cdot \sqrt[3]{2}$

Simplify.

21. $\sqrt{40}$

22. $\sqrt[4]{16x^7y^{10}}$

23. $\sqrt[3]{54x^4}$

24. $\sqrt[5]{-64b^{10}}$

Multiply or divide. Then simplify if possible.

25. $\sqrt{5} \cdot \sqrt{x}$ **26.** $\sqrt[3]{8x} \cdot \sqrt[3]{8x^2}$ **27.** $\dfrac{\sqrt{98y^6}}{\sqrt{2y}}$ **28.** $\dfrac{\sqrt[4]{48a^9b^3}}{\sqrt[4]{ab^3}}$

Perform each indicated operation.

29. $\sqrt{20} - \sqrt{75} + 5\sqrt{7}$ **30.** $\sqrt[3]{54y^4} - y\sqrt[3]{16y}$ **31.** $\sqrt{3}\left(\sqrt{5} - \sqrt{2}\right)$

32. $\left(\sqrt{7} + \sqrt{3}\right)^2$ **33.** $\left(2x - \sqrt{5}\right)\left(2x + \sqrt{5}\right)$ **34.** $\left(\sqrt{x + 1} - 1\right)^2$

Rationalize each denominator.

35. $\sqrt{\dfrac{7}{3}}$ **36.** $\dfrac{5}{\sqrt[3]{2x^2}}$ **37.** $\dfrac{\sqrt{3} - \sqrt{7}}{2\sqrt{3} + \sqrt{7}}$

Rationalize each numerator.

38. $\sqrt{\dfrac{7}{3}}$ **39.** $\sqrt[3]{\dfrac{9y}{11}}$ **40.** $\dfrac{\sqrt{x} - 2}{\sqrt{x}}$

7.6 RADICAL EQUATIONS AND PROBLEM SOLVING

OBJECTIVES

1 Solve equations that contain radical expressions.

2 Use the Pythagorean theorem to model problems.

OBJECTIVE 1 ▶ Solving equations that contain radical expressions. In this section, we present techniques to solve equations containing radical expressions such as

$$\sqrt{2x - 3} = 9$$

We use the power rule to help us solve these radical equations.

> **Power Rule**
>
> If both sides of an equation are raised to the same power, **all** solutions of the original equation are **among** the solutions of the new equation.

This property *does not* say that raising both sides of an equation to a power yields an equivalent equation. A solution of the new equation *may or may not* be a solution of the original equation. For example, $(-2)^2 = 2^2$, but $-2 \neq 2$. Thus, *each solution of the new equation must be checked* to make sure it is a solution of the original equation. Recall that a proposed solution that is not a solution of the original equation is called an **extraneous solution.**

EXAMPLE 1   Solve: $\sqrt{2x - 3} = 9$. Use a graphing utility to check.

Solution   We use the power rule to square both sides of the equation to eliminate the radical.

$$\sqrt{2x - 3} = 9$$
$$\left(\sqrt{2x - 3}\right)^2 = 9^2$$
$$2x - 3 = 81$$
$$2x = 84$$
$$x = 42$$

Now check the solution using a graphing utility. To check using the intersection-of-graphs method, graph $y_1 = \sqrt{2x - 3}$ and $y_2 = 9$ in a $[0, 95, 10]$ by $[-5, 15, 1]$ window. The intersection has x-value 42, the solution, as shown.

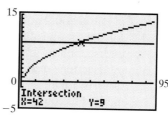

The solution checks, so we conclude that the solution is 42 or the solution set is $\{42\}$. □

PRACTICE
1 Solve: $\sqrt{3x - 5} = 7$.

In the next example, we choose to solve the equation graphically, using the x-intercept method, since the equation is written so that one side is 0.

EXAMPLE 2 Solve: $\sqrt{-10x - 1} + 3x = 0$.

Solution Graph $y_1 = \sqrt{-10x - 1} + 3x$ in a decimal window.

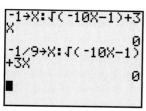

A numerical check showing -1 and $-\dfrac{1}{9}$ are both solutions of the equation in Example 2.

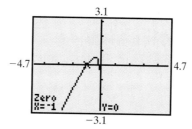 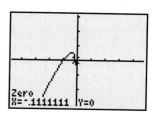

The equation has two solutions since there are two x-intercepts. The x-intercepts are $x = -1$ and $x = -0.11111\ldots$, a repeating decimal. Use the fraction command to write $-0.\overline{1}$ as the equivalent fraction $-\dfrac{1}{9}$.

Check the solutions numerically as in the screen to the left.

Both solutions check. The solutions are $-\dfrac{1}{9}$ and -1 or the solution set is $\left\{ -\dfrac{1}{9}, -1 \right\}$. □

PRACTICE
2 Solve: $\sqrt{3 - 2x} - 4x = 0$.

The following steps may be used to solve a radical equation.

Solving a Radical Equation Algebraically

STEP 1. Isolate one radical on one side of the equation.

STEP 2. Raise each side of the equation to a power equal to the index of the radical and simplify.

STEP 3. If the equation still contains a radical term, repeat Steps 1 and 2. If not, solve the equation.

STEP 4. Check all proposed solutions in the original equation.

Helpful Hint

To solve a radical equation graphically, use the intersection-of-graphs method or the x-intercept method.

When do we solve a radical equation algebraically and when do we solve it graphically? If given a choice, it depends on the complexity of the equation itself. An equation that contains only one radical may be a good candidate for the algebraic process. Recall that by raising each side to a power, you are introducing the possibility of extraneous roots or solutions. For this reason, a check is mandatory. However, when solving graphically, you can visualize the numbers of solutions immediately. For the intersection-of-graphs method, the number of real solutions is equal to the number of intersections of the two graphs. For the x-intercept method, the number of real solutions is equal to the number of x-intercepts. We check after solving by a graphing method to confirm that the solutions are exact.

In Example 3, we choose to solve graphically by the intersection-of-graphs method and we confirm the solution algebraically.

EXAMPLE 3   Solve $\sqrt[3]{x + 1} + 5 = 3$ graphically and check algebraically.

Solution   Graph $y_1 = \sqrt[3]{x + 1} + 5$ and $y_2 = 3$. The intersection of the two graphs has an x-value of -9, the solution of the equation.

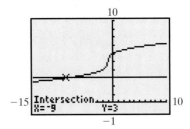

To check algebraically, first, isolate the radical by subtracting 5 from both sides of the equation.

$$\sqrt[3]{x + 1} + 5 = 3$$
$$\sqrt[3]{x + 1} = -2$$

Next we raise both sides of the equation to the third power to eliminate the radical.

$$\left(\sqrt[3]{x + 1}\right)^3 = (-2)^3$$
$$x + 1 = -8$$
$$x = -9$$

The solution checks in the original equation, so the solution is -9. $\square$

PRACTICE
3   Solve: $\sqrt[3]{x - 2} + 1 = 3$.

In Example 4, we solve algebraically and check graphically. Notice that in solving algebraically, an extraneous solution is introduced. However, graphically we see that there is only one solution because there is only one point of intersection for the two graphs.

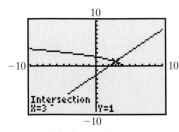

To check the results of Example 4, graph $y_1 = \sqrt{4 - x}$ and $y_2 = x - 2$. The x-value of the point of intersection is 3, as expected.

EXAMPLE 4   Solve algebraically: $\sqrt{4 - x} = x - 2$. Then check algebraically and graphically.

Solution

$$\sqrt{4 - x} = x - 2$$
$$\left(\sqrt{4 - x}\right)^2 = (x - 2)^2$$
$$4 - x = x^2 - 4x + 4$$
$$x^2 - 3x = 0 \qquad \text{Write the quadratic equation in standard form.}$$
$$x(x - 3) = 0 \qquad \text{Factor.}$$
$$x = 0 \quad \text{or} \quad x - 3 = 0 \qquad \text{Set each factor equal to 0.}$$
$$x = 3$$

Check: $\sqrt{4 - x} = x - 2$ $\sqrt{4 - x} = x - 2$

$\sqrt{4 - 0} \overset{?}{=} 0 - 2$ Let $x = 0$. $\sqrt{4 - 3} \overset{?}{=} 3 - 2$ Let $x = 3$.

$2 = -2$ False $1 = 1$ True

The proposed solution 3 checks, but 0 does not. Since 0 is an extraneous solution, the only solution is 3. A graphical check is shown in the margin of the previous page.   □

PRACTICE

4   Solve: $\sqrt{16 + x} = x - 4$.

> ▶ **Helpful Hint**
>
> In Example 4, notice that $(x - 2)^2 = x^2 - 4x + 4$. Make sure binomials are squared correctly.

Concept Check ☑

How can you immediately tell that the equation $\sqrt{2y + 3} = -4$ has no real solution?

In Example 5 we solve graphically. Use the intersection of graphs method and confirm the solution algebraically.

EXAMPLE 5   Solve: $\sqrt{2x + 5} + \sqrt{2x} = 3$.

Solution   Graph $y_1 = \sqrt{2x + 5} + \sqrt{2x}$ and $y_2 = 3$ in a $[-10, 10, 1]$ by $[-1, 10, 1]$ window. The intersection of the two graphs is $x = 0.22222\ldots$, a repeating decimal. Convert the value of x to fraction form using the fraction command. The solution is $\frac{2}{9}$.

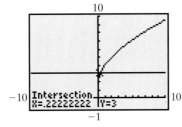 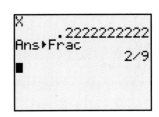

To check, we confirm the solution algebraically.

We get one radical alone by subtracting $\sqrt{2x}$ from both sides.

$$\sqrt{2x + 5} + \sqrt{2x} = 3$$
$$\sqrt{2x + 5} = 3 - \sqrt{2x}$$

Now we use the power rule to begin eliminating the radicals. First we square both sides.

$$\left(\sqrt{2x + 5}\right)^2 = \left(3 - \sqrt{2x}\right)^2$$
$$2x + 5 = 9 - 6\sqrt{2x} + 2x \quad \text{Multiply } \left(3 - \sqrt{2x}\right)\left(3 - \sqrt{2x}\right).$$

There is still a radical in the equation, so we get a radical alone again. Then we square both sides.

$$2x + 5 = 9 - 6\sqrt{2x} + 2x \quad \text{Get the radical alone.}$$
$$6\sqrt{2x} = 4$$
$$36(2x) = 16 \qquad\qquad \text{Square both sides of the equation to eliminate the radical.}$$
$$72x = 16 \qquad\qquad \text{Multiply.}$$
$$x = \frac{16}{72} \qquad\qquad \text{Solve.}$$
$$x = \frac{2}{9} \qquad\qquad \text{Simplify.}$$

The proposed solution, $\frac{2}{9}$, checks in the original equation. The solution is $\frac{2}{9}$.   □

PRACTICE

5   Solve: $\sqrt{8x + 1} + \sqrt{3x} = 2$.

▶ **Helpful Hint**

Make sure expressions are squared correctly. In Example 5, we squared $\left(3 - \sqrt{2x}\right)$ as

$$\left(3 - \sqrt{2x}\right)^2 = \left(3 - \sqrt{2x}\right)\left(3 - \sqrt{2x}\right)$$
$$= 3 \cdot 3 - 3\sqrt{2x} - 3\sqrt{2x} + \sqrt{2x} \cdot \sqrt{2x}$$
$$= 9 - 6\sqrt{2x} + 2x$$

Concept Check ✓

What is wrong with the following solution?

$$\sqrt{2x + 5} + \sqrt{4 - x} = 8$$
$$\left(\sqrt{2x + 5} + \sqrt{4 - x}\right)^2 = 8^2$$
$$(2x + 5) + (4 - x) = 64$$
$$x + 9 = 64$$
$$x = 55$$

OBJECTIVE 2 ▶ Using the Pythagorean theorem. Recall that the Pythagorean theorem states that in a right triangle, the length of the hypotenuse squared equals the sum of the lengths of each of the legs squared.

Pythagorean Theorem

If a and b are the lengths of the legs of a right triangle and c is the length of the hypotenuse, then $a^2 + b^2 = c^2$.

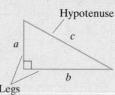

△ **EXAMPLE 6** Find the length of the unknown leg of the right triangle.

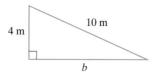

Solution In the formula $a^2 + b^2 = c^2$, c is the hypotenuse. Here, $c = 10$, the length of the hypotenuse, and $a = 4$. We solve for b. Then $a^2 + b^2 = c^2$ becomes

$$4^2 + b^2 = 10^2$$
$$16 + b^2 = 100$$
$$b^2 = 84 \quad \text{Subtract 16 from both sides.}$$
$$b = \pm\sqrt{84} = \pm\sqrt{4 \cdot 21} = \pm 2\sqrt{21}$$

Since b is a length and thus is positive, we will use the positive value only.

The unknown leg of the triangle is $2\sqrt{21}$ meters long. ☐

PRACTICE

6 Find the length of the unknown leg of the right triangle.

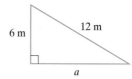

Answers to Concept Check:
$\left(\sqrt{2x + 5} + \sqrt{4 - x}\right)^2$ is not
$(2x + 5) + (4 - x)$.

⚠ **EXAMPLE 7**   **Calculating Placement of a Wire**

A 50-foot supporting wire is to be attached to a 75-foot antenna. Because of surrounding buildings, sidewalks, and roadways, the wire must be anchored exactly 20 feet from the base of the antenna.

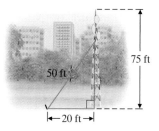

a. How high from the base of the antenna is the wire attached?

b. Local regulations require that a supporting wire be attached at a height no less than $\frac{3}{5}$ of the total height of the antenna. From part **a,** have local regulations been met?

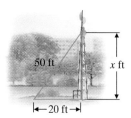

Solution

1. UNDERSTAND. Read and reread the problem. From the diagram we notice that a right triangle is formed with hypotenuse 50 feet and one leg 20 feet. Let x be the height from the base of the antenna to the attached wire.

2. TRANSLATE. Use the Pythagorean theorem.

$$a^2 + b^2 = c^2$$
$$20^2 + x^2 = 50^2 \quad a = 20, c = 50$$

3. SOLVE.

$$20^2 + x^2 = 50^2$$
$$400 + x^2 = 2500$$
$$x^2 = 2100 \qquad \text{Subtract 400 from both sides.}$$
$$x = \pm\sqrt{2100}$$
$$= \pm 10\sqrt{21}$$

4. INTERPRET. *Check* the work and *state* the solution.

Check: We will use only the positive value, $x = 10\sqrt{21}$, because x represents length. The wire is attached exactly $10\sqrt{21}$ feet from the base of the pole, or approximately 45.8 feet.

State: The supporting wire must be attached at a height no less than $\frac{3}{5}$ of the total height of the antenna. This height is $\frac{3}{5}$(75 feet), or 45 feet. Since we know from part **a** that the wire is to be attached at a height of approximately 45.8 feet, local regulations have been met.   □

PRACTICE

7   Keith Robinson bought two Siamese fighting fish, but when he got home he found he only had one rectangular tank that was 12 in. long, 7 in. wide, and 5 in. deep. Since the fish must be kept separated, he needs to insert a plastic divider in the diagonal of the tank. He already has a piece that is 5 in. in one dimension, but how long must it be to fit corner to corner in the tank?

VOCABULARY & READINESS CHECK

Use the choices below to fill in each blank. Not all choices will be used.

hypotenuse	right	$x^2 + 25$	$16 - 8\sqrt{7x} + 7x$
extraneous solution	legs	$x^2 - 10x + 25$	$16 + 7x$

1. A proposed solution that is not a solution of the original equation is called an _____.

2. The Pythagorean theorem states that $a^2 + b^2 = c^2$ where a and b are the lengths of the _____ of a _____ triangle and c is the length of the _____.

3. The square of $x - 5$, or $(x - 5)^2 = $ _____.

4. The square of $4 - \sqrt{7x}$, or $(4 - \sqrt{7x})^2 = $ _____.

7.6 EXERCISE SET

Solve. See Examples 1 and 2.

1. $\sqrt{2x} = 4$
2. $\sqrt{3x} = 3$
3. $\sqrt{x - 3} = 2$
4. $\sqrt{x + 1} = 5$
5. $\sqrt{2x} = -4$
6. $\sqrt{5x} = -5$
7. $\sqrt{4x - 3} - 5 = 0$
8. $\sqrt{x - 3} - 1 = 0$
9. $\sqrt{2x - 3} - 2 = 1$
10. $\sqrt{3x + 3} - 4 = 8$

Solve. See Example 3.

11. $\sqrt[3]{6x} = -3$
12. $\sqrt[3]{4x} = -2$
13. $\sqrt[3]{x - 2} - 3 = 0$
14. $\sqrt[3]{2x - 6} - 4 = 0$

Solve. See Examples 4 and 5.

15. $\sqrt{13 - x} = x - 1$
16. $\sqrt{2x - 3} = 3 - x$
17. $x - \sqrt{4 - 3x} = -8$
18. $2x + \sqrt{x + 1} = 8$
19. $\sqrt{y + 5} = 2 - \sqrt{y - 4}$
20. $\sqrt{x + 3} + \sqrt{x - 5} = 3$
21. $\sqrt{x - 3} + \sqrt{x + 2} = 5$
22. $\sqrt{2x - 4} - \sqrt{3x + 4} = -2$

MIXED PRACTICE

Solve. See Examples 1 through 5.

23. $\sqrt{3x - 2} = 5$
24. $\sqrt{5x - 4} = 9$
25. $-\sqrt{2x} + 4 = -6$
26. $-\sqrt{3x + 9} = -12$
27. $\sqrt{3x + 1} + 2 = 0$
28. $\sqrt{3x + 1} - 2 = 0$
29. $\sqrt[4]{4x + 1} - 2 = 0$
30. $\sqrt[4]{2x - 9} - 3 = 0$
31. $\sqrt{4x - 3} = 7$
32. $\sqrt{3x + 9} = 6$
33. $\sqrt[3]{6x - 3} - 3 = 0$
34. $\sqrt[3]{3x + 4} = 7$
35. $\sqrt[3]{2x - 3} - 2 = -5$
36. $\sqrt[3]{x - 4} - 5 = -7$
37. $\sqrt{x + 4} = \sqrt{2x - 5}$
38. $\sqrt[3]{3y + 6} = \sqrt[3]{7y - 6}$
39. $x - \sqrt{1 - x} = -5$
40. $x - \sqrt{x - 2} = 4$
41. $\sqrt[3]{-6x - 1} = \sqrt[3]{-2x - 5}$
42. $\sqrt[3]{-4x - 3} = \sqrt[3]{-x - 15}$

43. $\sqrt{5x - 1} - \sqrt{x + 2} = 3$
44. $\sqrt{2x - 1} - 4 = -\sqrt{x - 4}$
45. $\sqrt{2x - 1} = \sqrt{1 - 2x}$
46. $\sqrt{7x - 4} = \sqrt{4 - 7x}$
47. $\sqrt{3x + 4} - 1 = \sqrt{2x + 1}$
48. $\sqrt{x - 2} + 3 = \sqrt{4x + 1}$
49. $\sqrt{y + 3} - \sqrt{y - 3} = 1$
50. $\sqrt{x + 1} - \sqrt{x - 1} = 2$

Find the length of the unknown side of each triangle. See Example 6.

51.
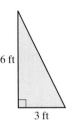

52. 7 in.
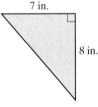

53.
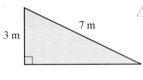

54. 4 cm
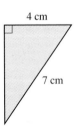

Find the length of the unknown side of each triangle. Give the exact length and a one-decimal-place approximation. See Example 6.

55.
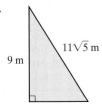

56.
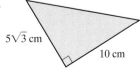

△ 57.

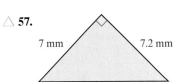

7 mm   7.2 mm

△ 58.

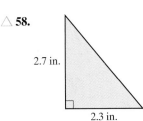

2.7 in.

2.3 in.

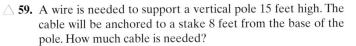

Solve. See Example 7. Give exact answers and two-decimal-place approximations where appropriate.

△ 59. A wire is needed to support a vertical pole 15 feet high. The cable will be anchored to a stake 8 feet from the base of the pole. How much cable is needed?

15 ft

8 ft

△ 60. The tallest structure in the United States is a TV tower in Blanchard, North Dakota. Its height is 2063 feet. A 2382-foot length of wire is to be used as a guy wire attached to the top of the tower. Approximate to the nearest foot how far from the base of the tower the guy wire must be anchored. (*Source:* U.S. Geological Survey)

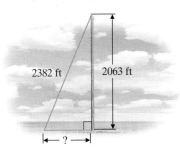

2382 ft   2063 ft

?

△ 61. A spotlight is mounted on the eaves of a house 12 feet above the ground. A flower bed runs between the house and the sidewalk, so the closest the ladder can be placed to the house is 5 feet. How long a ladder is needed so that an electrician can reach the place where the light is mounted?

12 ft

5 ft

△ 62. A wire is to be attached to support a telephone pole. Because of surrounding buildings, sidewalks, and roadways, the wire must be anchored exactly 15 feet from the base of the pole. Telephone company workers have only 30 feet of cable, and 2 feet of that must be used to attach the cable to the pole and

to the stake on the ground. How high from the base of the pole can the wire be attached?

15 ft

△ 63. The radius of the Moon is 1080 miles. Use the formula for the radius r of a sphere given its surface area A,

$$r = \sqrt{\frac{A}{4\pi}}$$

to find the surface area of the Moon. Round to the nearest square mile. (*Source:* National Space Science Data Center)

64. Police departments find it very useful to be able to approximate the speed of a car when they are given the distance that the car skidded before it came to a stop. If the road surface is wet concrete, the function $S(x) = \sqrt{10.5x}$ is used, where $S(x)$ is the speed of the car in miles per hour and x is the distance skidded in feet. Find how fast a car was moving if it skidded 280 feet on wet concrete.

65. The formula $v = \sqrt{2gh}$ gives the velocity v, in feet per second, of an object when it falls h feet accelerated by gravity g, in feet per second squared. If g is approximately 32 feet per second squared, find how far an object has fallen if its velocity is 80 feet per second.

66. Two tractors are pulling a tree stump from a field. If two forces A and B pull at right angles (90°) to each other, the size of the resulting force R is given by the formula $R = \sqrt{A^2 + B^2}$. If tractor A is exerting 600 pounds of force and the resulting force is 850 pounds, find how much force tractor B is exerting.

600 lb   ?

In psychology, it has been suggested that the number S of nonsense syllables that a person can repeat consecutively depends on his or her IQ score I according to the equation $S = 2\sqrt{I} - 9$.

67. Use this relationship to estimate the IQ of a person who can repeat 11 nonsense syllables consecutively.

68. Use this relationship to estimate the IQ of a person who can repeat 15 nonsense syllables consecutively.

*The **period** of a pendulum is the time it takes for the pendulum to make one full back-and-forth swing. The period of a pendulum depends on the length of the pendulum. The formula for the period P, in seconds, is $P = 2\pi\sqrt{\dfrac{l}{32}}$, where l is the length of the pendulum in feet. Use this formula for Exercises 69 through 74.*

69. Find the period of a pendulum whose length is 2 feet. Give an exact answer and a two-decimal-place approximation.

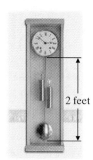

2 feet

70. Klockit sells a 43-inch lyre pendulum. Find the period of this pendulum. Round your answer to 2 decimal places. (*Hint:* First convert inches to feet.)

71. Find the length of a pendulum whose period is 4 seconds. Round your answer to 2 decimal places.

72. Find the length of a pendulum whose period is 3 seconds. Round your answer to 2 decimal places.

73. Study the relationship between period and pendulum length in Exercises 69 through 72 and make a conjecture about this relationship.

74. Galileo experimented with pendulums. He supposedly made conjectures about pendulums of equal length with different bob weights. Try this experiment. Make two pendulums 3 feet long. Attach a heavy weight (lead) to one and a light weight (a cork) to the other. Pull both pendulums back the same angle measure and release. Make a conjecture from your observations. (There is more about pendulums in the Chapter 7 Group Activity.)

If the three lengths of the sides of a triangle are known, Heron's formula can be used to find its area. If a, b, and c are the three lengths of the sides, Heron's formula for area is

$$A = \sqrt{s(s-a)(s-b)(s-c)}$$

where s is half the perimeter of the triangle, or $s = \dfrac{1}{2}(a + b + c)$.

Use this formula to find the area of each triangle. Give an exact answer and then a two-decimal place approximation.

△ **75.**

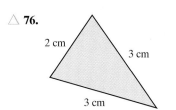

6 mi   10 mi

14 mi

△ **76.**

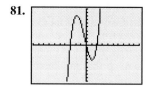

2 cm

3 cm

3 cm

77. Describe when Heron's formula might be useful.

78. In your own words, explain why you think s in Heron's formula is called the *semiperimeter*.

The maximum distance D(h) in kilometers that a person can see from a height h kilometers above the ground is given by the function $D(h) = 111.7\sqrt{h}$. Use this function for Exercises 79 and 80. Round your answers to two decimal places.

79. Find the height that would allow a person to see 80 kilometers.

80. Find the height that would allow a person to see 40 kilometers.

REVIEW AND PREVIEW

Use the vertical line test to determine whether each graph represents the graph of a function. See Section 2.2.

81.

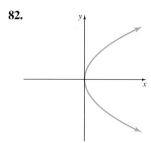

82.

83.

84.

85.

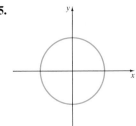

86.

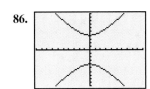

Simplify. See Section 6.3.

87. $\dfrac{\dfrac{x}{6}}{\dfrac{2x}{3} + \dfrac{1}{2}}$

88. $\dfrac{\dfrac{1}{y} + \dfrac{4}{5}}{\dfrac{-3}{20}}$

89. $\dfrac{\dfrac{z}{5} + \dfrac{1}{10}}{\dfrac{z}{20} - \dfrac{z}{5}}$

90. $\dfrac{\dfrac{1}{y} + \dfrac{1}{x}}{\dfrac{1}{y} - \dfrac{1}{x}}$

CONCEPT EXTENSIONS

91. Find the error in the following solution and correct. See the second Concept Check in this section.

$$\sqrt{5x - 1} + 4 = 7$$
$$(\sqrt{5x - 1} + 4)^2 = 7^2$$
$$5x - 1 + 16 = 49$$
$$5x = 34$$
$$x = \frac{34}{5}$$

92. Explain why proposed solutions of radical equations must be checked.

93. Solve: $\sqrt{\sqrt{x + 3} + \sqrt{x}} = \sqrt{3}$

94. The cost $C(x)$ in dollars per day to operate a small delivery service is given by $C(x) = 80\sqrt[3]{x} + 500$, where x is the number of deliveries per day. In July, the manager decides that it is necessary to keep delivery costs below $1620.00. Find the greatest number of deliveries this company can make per day and still keep overhead below $1620.00.

95. Consider the equations $\sqrt{2x} = 4$ and $\sqrt[3]{2x} = 4$.

   a. Explain the difference in solving these equations.

   b. Explain the similarity in solving these equations.

Example

For Exercises 96 through 99, see the example below.
Solve $(t^2 - 3t) - 2\sqrt{t^2 - 3t} = 0$.

Solution

Substitution can be used to make this problem somewhat simpler. Since $t^2 - 3t$ occurs more than once, let $x = t^2 - 3t$.

$$(t^2 - 3t) - 2\sqrt{t^2 - 3t} = 0$$
$$x - 2\sqrt{x} = 0$$
$$x = 2\sqrt{x}$$
$$x^2 = (2\sqrt{x})^2$$
$$x^2 = 4x$$
$$x^2 - 4x = 0$$
$$x(x - 4) = 0$$
$$x = 0 \quad \text{or} \quad x - 4 = 0$$
$$x = 4$$

Now we "undo" the substitution.
$x = 0$ Replace x with $t^2 - 3t$.

$$t^2 - 3t = 0$$
$$t(t - 3) = 0$$
$$t = 0 \quad \text{or} \quad t - 3 = 0$$
$$t = 3$$

$x = 4$ Replace x with $t^2 - 3t$.

$$t^2 - 3t = 4$$
$$t^2 - 3t - 4 = 0$$
$$(t - 4)(t + 1) = 0$$
$$t - 4 = 0 \quad \text{or} \quad t + 1 = 0$$
$$t = 4 \qquad\qquad t = -1$$

In this problem, we have four possible solutions: 0, 3, 4, and -1. All four solutions check in the original equation, so the solutions are $-1, 0, 3, 4$.

Solve. See the preceding example.

96. $3\sqrt{x^2 - 8x} = x^2 - 8x$

97. $\sqrt{(x^2 - x) + 7} = 2(x^2 - x) - 1$

98. $7 - (x^2 - 3x) = \sqrt{(x^2 - 3x) + 5}$

99. $x^2 + 6x = 4\sqrt{x^2 + 6x}$

THE BIGGER PICTURE   SOLVING EQUATIONS AND INEQUALITIES

Continue your outline from Sections 1.5, 3.2, 3.3, 3.5, 5.8, and 6.5. Write how to recognize and how to solve equations with radicals in your own words. For example:

Solving Equations and Inequalities

 I. Equations
   A. Linear equations (Sec. 1.5 and 3.1)
   B. Absolute value equations (Sec. 3.4)

 C. Quadratic and higher degree equations (Sec. 5.8)
 D. Equations with rational expressions (Sec. 6.5)
 E. Equations with radicals: Equation contains at least one root of a variable expression.

$$\sqrt{5x + 10} - 2 = x \qquad\qquad \text{Radical equation}$$
$$\sqrt{5x + 10} = x + 2 \qquad\qquad \text{Isolate the radical.}$$
$$\left(\sqrt{5x + 10}\right)^2 = (x + 2)^2 \qquad \text{Square both sides.}$$

$$5x + 10 = x^2 + 4x + 4 \qquad \text{Simplify.}$$
$$0 = x^2 - x - 6 \qquad \text{Write in standard form.}$$
$$0 = (x - 3)(x + 2) \qquad \text{Factor.}$$
$$x - 3 = 0 \quad \text{or} \quad x + 2 = 0 \qquad \text{Set each factor equal to 0.}$$
$$x = 3 \quad \text{or} \quad x = -2 \qquad \text{Solve.}$$

Both solutions check.

II. Inequalities
  A. Linear inequalities (Sec. 3.2)
  B. Compound inequalities (Sec. 3.3)
  C. Absolute value inequalities (Sec. 3.5)

Solve. Write inequality solutions in interval notation.

1. $\dfrac{x}{4} + \dfrac{x + 18}{20} = \dfrac{x - 5}{5}$

2. $|3x - 5| = 10$

3. $2x^2 - x = 45$

4. $-6 \le -5x - 1 \le 10$

5. $4(x - 1) + 3x > 1 + 2(x - 6)$

6. $\sqrt{x} + 14 = x - 6$

7. $x \ge 10$ or $-x < 5$

8. $\sqrt{3x - 1} + 4 = 1$

9. $|x - 2| > 15$

10. $5x - 4[x - 2(3x + 1)] = 25$

7.7 COMPLEX NUMBERS

OBJECTIVES

1 Write square roots of negative numbers in the form bi.

2 Add or subtract complex numbers.

3 Multiply complex numbers.

4 Divide complex numbers.

5 Raise i to powers.

OBJECTIVE 1 ▶ Writing numbers in the form *bi*. Our work with radical expressions has excluded expressions such as $\sqrt{-16}$ because $\sqrt{-16}$ is not a real number; there is no real number whose square is -16. In this section, we discuss a number system that includes roots of negative numbers. This number system is the **complex number system,** and it includes the set of real numbers as a subset. The complex number system allows us to solve equations such as $x^2 + 1 = 0$ that have no real number solutions. The set of complex numbers includes the **imaginary unit.**

> **Imaginary Unit**
> The imaginary unit, written i, is the number whose square is -1. That is,
> $$i^2 = -1 \quad \text{and} \quad i = \sqrt{-1}$$

To write the square root of a negative number in terms of i, use the property that if a is a positive number, then

$$\sqrt{-a} = \sqrt{-1} \cdot \sqrt{a}$$
$$= i \cdot \sqrt{a}$$

Using i, we can write $\sqrt{-16}$ as

$$\sqrt{-16} = \sqrt{-1 \cdot 16} = \sqrt{-1} \cdot \sqrt{16} = i \cdot 4, \text{ or } 4i$$

TECHNOLOGY NOTE

Many graphing utilities have a complex mode. Check your manual, and if yours has it, set your graphing utility to this mode for this chapter. Recall that a real number is a complex number, so in complex mode the graphing utility will display real numbers and imaginary numbers.

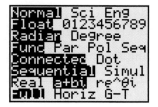

EXAMPLE 1 Write with i notation.

a. $\sqrt{-36}$ **b.** $\sqrt{-5}$ **c.** $-\sqrt{-20}$

Solution

> ▶ **Helpful Hint**
> Since $\sqrt{5}i$ can easily be confused with $\sqrt{5i}$, we write $\sqrt{5}i$ as $i\sqrt{5}$.

a. $\sqrt{-36} = \sqrt{-1 \cdot 36} = \sqrt{-1} \cdot \sqrt{36} = i \cdot 6$, or $6i$

b. $\sqrt{-5} = \sqrt{-1(5)} = \sqrt{-1} \cdot \sqrt{5} = i\sqrt{5}$.

c. $-\sqrt{-20} = -\sqrt{-1 \cdot 20} = -\sqrt{-1} \cdot \sqrt{4 \cdot 5} = -i \cdot 2\sqrt{5} = -2i\sqrt{5}$ □

This is an approximate answer, whereas $i\sqrt{5}$ is an exact answer.

```
√(-36)
              6i
√(-5)
     2.236067977i
■
```

A calculator check of Example 1a and b.

PRACTICE
1   Write with i notation.

a. $\sqrt{-4}$ **b.** $\sqrt{-7}$ **c.** $-\sqrt{-18}$

The product rule for radicals does not necessarily hold true for imaginary numbers. *To multiply square roots of negative numbers, first we write each number in terms of the imaginary unit i.* For example, to multiply $\sqrt{-4}$ and $\sqrt{-9}$, we first write each number in the form bi.

$$\sqrt{-4}\sqrt{-9} = 2i(3i) = 6i^2 = 6(-1) = -6 \quad \text{Correct}$$

We will also use this method to simplify quotients of square roots of negative numbers. Why? The product rule does not work for this example. In other words,

$$\sqrt{-4} \cdot \sqrt{-9} = \sqrt{(-4)(-9)} = \sqrt{36} = 6 \quad \text{Incorrect}$$

EXAMPLE 2 Multiply or divide as indicated.

a. $\sqrt{-3} \cdot \sqrt{-5}$ **b.** $\sqrt{-36} \cdot \sqrt{-1}$ **c.** $\sqrt{8} \cdot \sqrt{-2}$ **d.** $\dfrac{\sqrt{-125}}{\sqrt{5}}$

Solution

a. $\sqrt{-3} \cdot \sqrt{-5} = i\sqrt{3}\left(i\sqrt{5}\right) = i^2\sqrt{15} = -1\sqrt{15} = -\sqrt{15}$

b. $\sqrt{-36} \cdot \sqrt{-1} = 6i(i) = 6i^2 = 6(-1) = -6$

c. $\sqrt{8} \cdot \sqrt{-2} = 2\sqrt{2}\left(i\sqrt{2}\right) = 2i\left(\sqrt{2}\sqrt{2}\right) = 2i(2) = 4i$

d. $\dfrac{\sqrt{-125}}{\sqrt{5}} = \dfrac{i\sqrt{125}}{\sqrt{5}} = i\sqrt{25} = 5i$ □

PRACTICE
2   Multiply or divide as indicated.

a. $\sqrt{-5} \cdot \sqrt{-6}$ **b.** $\sqrt{-9} \cdot \sqrt{-1}$ **c.** $\sqrt{125} \cdot \sqrt{-5}$ **d.** $\dfrac{\sqrt{-27}}{\sqrt{3}}$

Now that we have practiced working with the imaginary unit, we define complex numbers.

> **Complex Numbers**
> A **complex number** is a number that can be written in the form $a + bi$, where a and b are real numbers.

Notice that the set of real numbers is a subset of the complex numbers since any real number can be written in the form of a complex number. For example,

$$16 = 16 + 0i$$

In general, a complex number $a + bi$ is a real number if $b = 0$. Also, a complex number is called a **pure imaginary number** if $a = 0$ and $b \neq 0$. For example,

$$3i = 0 + 3i \quad \text{and} \quad i\sqrt{7} = 0 + i\sqrt{7}$$

are pure imaginary numbers.

The following diagram shows the relationship between complex numbers and their subsets.

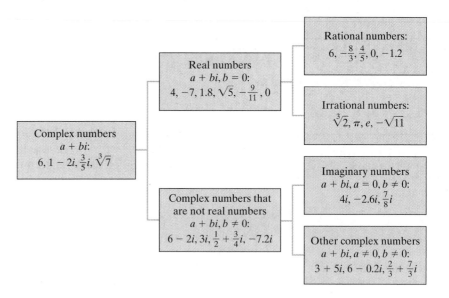

Concept Check ✓

True or false? Every complex number is also a real number.

OBJECTIVE 2 ▶ Adding or subtracting complex numbers. Two complex numbers $a + bi$ and $c + di$ are equal if and only if $a = c$ and $b = d$. Complex numbers can be added or subtracted by adding or subtracting their real parts and then adding or subtracting their imaginary parts.

Sum or Difference of Complex Numbers

If $a + bi$ and $c + di$ are complex numbers, then their sum is

$$(a + bi) + (c + di) = (a + c) + (b + d)i$$

Their difference is

$$(a + bi) - (c + di) = a + bi - c - di = (a - c) + (b - d)i$$

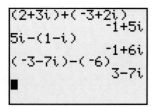

A calculator check for Example 3.

EXAMPLE 3 Add or subtract the complex numbers. Write the sum or difference in the form $a + bi$.

a. $(2 + 3i) + (-3 + 2i)$ **b.** $5i - (1 - i)$ **c.** $(-3 - 7i) - (-6)$

Solution

a. $(2 + 3i) + (-3 + 2i) = (2 - 3) + (3 + 2)i = -1 + 5i$

b. $5i - (1 - i) = 5i - 1 + i$
$$= -1 + (5 + 1)i$$
$$= -1 + 6i$$

c. $(-3 - 7i) - (-6) = -3 - 7i + 6$
$$= (-3 + 6) - 7i$$
$$= 3 - 7i \qquad \square$$

PRACTICE
3 Add or subtract the complex numbers. Write the sum or difference in the form $a + bi$.

a. $(3 - 5i) + (-4 + i)$ **b.** $4i - (3 - i)$ **c.** $(-5 - 2i) - (-8)$

OBJECTIVE 3 ▶ Multiplying complex numbers. To multiply two complex numbers of the form $a + bi$, we multiply as though they are binomials. Then we use the relationship $i^2 = -1$ to simplify.

EXAMPLE 4 Multiply the complex numbers. Write the product in the form $a + bi$.

a. $-7i \cdot 3i$ **b.** $3i(2 - i)$ **c.** $(2 - 5i)(4 + i)$
d. $(2 - i)^2$ **e.** $(7 + 3i)(7 - 3i)$

Solution

a. $-7i \cdot 3i = -21i^2$
$$= -21(-1) \qquad \text{Replace } i^2 \text{ with } -1.$$
$$= 21$$

b. $3i(2 - i) = 3i \cdot 2 - 3i \cdot i$ Use the distributive property.
$$= 6i - 3i^2 \qquad \text{Multiply.}$$
$$= 6i - 3(-1) \qquad \text{Replace } i^2 \text{ with } -1.$$
$$= 6i + 3$$
$$= 3 + 6i$$

Use the FOIL order below. (First, Outer, Inner, Last)

c. $(2 - 5i)(4 + i) = 2(4) + 2(i) - 5i(4) - 5i(i)$
$$ \text{F} \quad \text{O} \quad \text{I} \quad \text{L}$$
$$= 8 + 2i - 20i - 5i^2$$
$$= 8 - 18i - 5(-1) \qquad i^2 = -1$$
$$= 8 - 18i + 5$$
$$= 13 - 18i$$

d. $(2 - i)^2 = (2 - i)(2 - i)$
$$= 2(2) - 2(i) - 2(i) + i^2$$
$$= 4 - 4i + (-1) \qquad i^2 = -1$$
$$= 3 - 4i$$

e. $(7 + 3i)(7 - 3i) = 7(7) - 7(3i) + 3i(7) - 3i(3i)$
$$= 49 - 21i + 21i - 9i^2$$
$$= 49 - 9(-1) \qquad i^2 = -1$$
$$= 49 + 9$$
$$= 58 \qquad \square$$

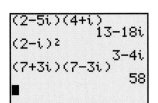

A calculator check for Example 4c, d, and e.

PRACTICE
4 Multiply the complex numbers. Write the product in the form $a + bi$.

a. $-4i \cdot 5i$ **b.** $5i(2 + i)$ **c.** $(2 + 3i)(6 - i)$
d. $(3 - i)^2$ **e.** $(9 + 2i)(9 - 2i)$

Notice that if you add, subtract, or multiply two complex numbers, just like real numbers, the result is a complex number.

OBJECTIVE 4 ▶ Dividing complex numbers. From Example 4e, notice that the product of $7 + 3i$ and $7 - 3i$ is a real number. These two complex numbers are called **complex conjugates** of one another. In general, we have the following definition.

Complex Conjugates

The complex numbers $(a + bi)$ and $(a - bi)$ are called **complex conjugates** of each other, and $(a + bi)(a - bi) = a^2 + b^2$.

To see that the product of a complex number $a + bi$ and its conjugate $a - bi$ is the real number $a^2 + b^2$, we multiply.

$$(a + bi)(a - bi) = a^2 - abi + abi - b^2i^2$$
$$= a^2 - b^2(-1)$$
$$= a^2 + b^2$$

We use complex conjugates to divide by a complex number.

EXAMPLE 5 Divide. Write in the form $a + bi$.

a. $\dfrac{2 + i}{1 - i}$ **b.** $\dfrac{7}{3i}$

Solution

a. Multiply the numerator and denominator by the complex conjugate of $1 - i$ to eliminate the imaginary number in the denominator.

$$\frac{2 + i}{1 - i} = \frac{(2 + i)(1 + i)}{(1 - i)(1 + i)}$$

$$= \frac{2(1) + 2(i) + 1(i) + i^2}{1^2 - i^2}$$

$$= \frac{2 + 3i - 1}{1 + 1} \qquad \text{Here, } i^2 = -1.$$

$$= \frac{1 + 3i}{2} \quad \text{or} \quad \frac{1}{2} + \frac{3}{2}i$$

(2+i)/(1-i)▶Frac
          1/2+3/2i
7/(3i)▶Frac
              -7/3i
■

A check for Example 5. Notice the need for parentheses in the denominator when simplifying part b.

b. Multiply the numerator and denominator by the conjugate of $3i$. Note that $3i = 0 + 3i$, so its conjugate is $0 - 3i$ or $-3i$.

$$\frac{7}{3i} = \frac{7(-3i)}{(3i)(-3i)} = \frac{-21i}{-9i^2} = \frac{-21i}{-9(-1)} = \frac{-21i}{9} = \frac{-7i}{3} \quad \text{or} \quad 0 - \frac{7}{3}i \qquad \square$$

PRACTICE

5 Divide. Write in the form $a + bi$.

a. $\dfrac{4 - i}{3 + i}$ **b.** $\dfrac{5}{2i}$

▶ **Helpful Hint**

Recall that division can be checked by multiplication.

To check that $\dfrac{2 + i}{1 - i} = \dfrac{1}{2} + \dfrac{3}{2}i$, in Example 5a, multiply $\left(\dfrac{1}{2} + \dfrac{3}{2}i\right)(1 - i)$ to verify that the product is $2 + i$.

OBJECTIVE 5 ▶ Finding powers of *i*. We can use the fact that $i^2 = -1$ to find higher powers of *i*. To find i^3, we rewrite it as the product of i^2 and *i*.

$$i^3 = i^2 \cdot i = (-1)i = -i$$
$$i^4 = i^2 \cdot i^2 = (-1) \cdot (-1) = 1$$

We continue this process and use the fact that $i^4 = 1$ and $i^2 = -1$ to simplify i^5 and i^6.

$$i^5 = i^4 \cdot i = 1 \cdot i = i$$
$$i^6 = i^4 \cdot i^2 = 1 \cdot (-1) = -1$$

If we continue finding powers of *i*, we generate the following pattern. Notice that the values i, -1, $-i$, and 1 repeat as *i* is raised to higher and higher powers.

$i^1 = i$	$i^5 = i$	$i^9 = i$
$i^2 = -1$	$i^6 = -1$	$i^{10} = -1$
$i^3 = -i$	$i^7 = -i$	$i^{11} = -i$
$i^4 = 1$	$i^8 = 1$	$i^{12} = 1$

This pattern allows us to find other powers of *i*. To do so, we will use the fact that $i^4 = 1$ and rewrite a power of *i* in terms of i^4. For example,

$$i^{22} = i^{20} \cdot i^2 = (i^4)^5 \cdot i^2 = 1^5 \cdot (-1) = 1 \cdot (-1) = -1.$$

EXAMPLE 6   Find the following powers of *i*.

a. i^7 **b.** i^{20} **c.** i^{46} **d.** i^{-12}

Solution

a. $i^7 = i^4 \cdot i^3 = 1(-i) = -i$

b. $i^{20} = (i^4)^5 = 1^5 = 1$

c. $i^{46} = i^{44} \cdot i^2 = (i^4)^{11} \cdot i^2 = 1^{11}(-1) = -1$

d. $i^{-12} = \dfrac{1}{i^{12}} = \dfrac{1}{(i^4)^3} = \dfrac{1}{(1)^3} = \dfrac{1}{1} = 1$

PRACTICE
6   Find the following powers of *i*.

a. i^9 **b.** i^{16} **c.** i^{34} **d.** i^{-24}

VOCABULARY & READINESS CHECK

Use the choices below to fill in each blank. Not all choices will be used.

-1	$\sqrt{-1}$	real	imaginary unit
1	$\sqrt{1}$	complex	pure imaginary

1. A _____ number is one that can be written in the form $a + bi$, where *a* and *b* are real numbers.

2. In the complex number system, *i* denotes the _____.

3. $i^2 = $ _____

4. $i = $ _____

5. A complex number, $a + bi$, is a _____ number if $b = 0$.

6. A complex number, $a + bi$, is a _____ number if $a = 0$ and $b \neq 0$.

Simplify. See Example 1.

7. $\sqrt{-81}$ **8.** $\sqrt{-49}$ **9.** $\sqrt{-7}$ **10.** $\sqrt{-3}$

11. $-\sqrt{16}$ **12.** $-\sqrt{4}$ **13.** $\sqrt{-64}$ **14.** $\sqrt{-100}$

7.7 | EXERCISE SET

Write in terms of i. See Example 1.

1. $\sqrt{-24}$

2. $\sqrt{-32}$

3. $-\sqrt{-36}$

4. $-\sqrt{-121}$

5. $8\sqrt{-63}$

6. $4\sqrt{-20}$

7. $-\sqrt{54}$

8. $\sqrt{-63}$

Multiply or divide. See Example 2.

9. $\sqrt{-2}\cdot\sqrt{-7}$

10. $\sqrt{-11}\cdot\sqrt{-3}$

11. $\sqrt{-5}\cdot\sqrt{-10}$

12. $\sqrt{-2}\cdot\sqrt{-6}$

13. $\sqrt{16}\cdot\sqrt{-1}$

14. $\sqrt{3}\cdot\sqrt{-27}$

15. $\dfrac{\sqrt{-9}}{\sqrt{3}}$

16. $\dfrac{\sqrt{49}}{\sqrt{-10}}$

17. $\dfrac{\sqrt{-80}}{\sqrt{-10}}$

18. $\dfrac{\sqrt{-40}}{\sqrt{-8}}$

Add or subtract. Write the sum or difference in the form a + bi. See Example 3.

19. $(4-7i)+(2+3i)$

20. $(2-4i)-(2-i)$

21. $(6+5i)-(8-i)$

22. $(8-3i)+(-8+3i)$

23. $6-(8+4i)$

24. $(9-4i)-9$

Multiply. Write the product in the form a + bi. See Example 4.

25. $-10i\cdot-4i$

26. $-2i\cdot-11i$

27. $6i(2-3i)$

28. $5i(4-7i)$

29. $\left(\sqrt{3}+2i\right)\left(\sqrt{3}-2i\right)$

30. $\left(\sqrt{5}-5i\right)\left(\sqrt{5}+5i\right)$

31. $(4-2i)^2$

32. $(6-3i)^2$

Write each quotient in the form a + bi. See Example 5.

33. $\dfrac{4}{i}$

34. $\dfrac{5}{6i}$

35. $\dfrac{7}{4+3i}$

36. $\dfrac{9}{1-2i}$

37. $\dfrac{3+5i}{1+i}$

38. $\dfrac{6+2i}{4-3i}$

39. $\dfrac{5-i}{3-2i}$

40. $\dfrac{6-i}{2+i}$

MIXED PRACTICE

Perform each indicated operation. Write the result in the form a + bi.

41. $(7i)(-9i)$

42. $(-6i)(-4i)$

43. $(6-3i)-(4-2i)$

44. $(-2-4i)-(6-8i)$

45. $-3i(-1+9i)$

46. $-5i(-2+i)$

47. $\dfrac{4-5i}{2i}$

48. $\dfrac{6+8i}{3i}$

49. $(4+i)(5+2i)$

50. $(3+i)(2+4i)$

51. $(6-2i)(3+i)$

52. $(2-4i)(2-i)$

53. $(8-3i)+(2+3i)$

54. $(7+4i)+(4-4i)$

55. $(1-i)(1+i)$

56. $(6+2i)(6-2i)$

57. $\dfrac{16+15i}{-3i}$

58. $\dfrac{2-3i}{-7i}$

59. $(9+8i)^2$

60. $(4-7i)^2$

61. $\dfrac{2}{3+i}$

62. $\dfrac{5}{3-2i}$

63. $(5-6i)-4i$

64. $(6-2i)+7i$

65. $\dfrac{2-3i}{2+i}$

66. $\dfrac{6+5i}{6-5i}$

67. $(2+4i)+(6-5i)$

68. $(5-3i)+(7-8i)$

69. $(\sqrt{3}+2i)(\sqrt{3}-2i)$

70. $(\sqrt{5}-5i)(\sqrt{5}+5i)$

71. $(4-2i)^2$

72. $(6-3i)^2$

Find each power of i. See Example 6.

73. i^8

74. i^{10}

75. i^{21}

76. i^{15}

77. i^{11}

78. i^{40}

79. i^{-6}

80. i^{-9}

81. $(2i)^6$

82. $(5i)^4$

83. $(-3i)^5$

84. $(-2i)^7$

REVIEW AND PREVIEW

Recall that the sum of the measures of the angles of a triangle is 180°. Find the unknown angle in each triangle.

△ 85.

△ 86.

Use synthetic division to divide the following. See Section 6.4.

87. $(x^3-6x^2+3x-4)\div(x-1)$

88. $(5x^4-3x^2+2)\div(x+2)$

Thirty people were recently polled about their average monthly balance in their checking accounts. The results of this poll are shown in the following histogram. Use this graph to answer Exercises 89 through 94. See Section 1.6.

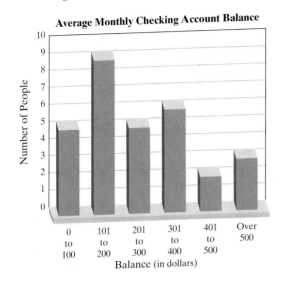

Average Monthly Checking Account Balance

89. How many people polled reported an average checking balance of $201 to $300?

90. How many people polled reported an average checking balance of $0 to $100?

91. How many people polled reported an average checking balance of $200 or less?

92. How many people polled reported an average checking balance of $301 or more?

93. What percent of people polled reported an average checking balance of $201 to $300?

94. What percent of people polled reported an average checking balance of $0 to $100?

CONCEPT EXTENSIONS

Write in the form $a + bi$.

95. $i^3 - i^4$

96. $i^8 - i^7$

97. $i^6 + i^8$

98. $i^4 + i^{12}$

99. $2 + \sqrt{-9}$

100. $5 - \sqrt{-16}$

101. $\dfrac{6 + \sqrt{-18}}{3}$

102. $\dfrac{4 - \sqrt{-8}}{2}$

103. $\dfrac{5 - \sqrt{-75}}{10}$

104. Describe how to find the conjugate of a complex number.

105. Explain why the product of a complex number and its complex conjugate is a real number.

Simplify.

106. $\left(8 - \sqrt{-3}\right) - \left(2 + \sqrt{-12}\right)$

107. $\left(8 - \sqrt{-4}\right) - \left(2 + \sqrt{-16}\right)$

108. Determine whether $2i$ is a solution of $x^2 + 4 = 0$.

109. Determine whether $-1 + i$ is a solution of $x^2 + 2x = -2$.

CHAPTER 7 GROUP ACTIVITY

Heron of Alexandria

Heron (also Hero) was a Greek mathematician and engineer. He lived and worked in Alexandria, Egypt, around 75 A.D. During his prolific work life, Heron developed a rotary steam engine called an aeolipile, a surveying tool called a dioptra, as well as a wind organ and a fire engine. As an engineer, he must have had the need to approximate square roots because he described an iterative method for doing so in his work *Metrica*. Heron's method for approximating a square root can be summarized as follows:

Suppose that x is not a perfect square and a^2 is the nearest perfect square to x. For a rough estimate of the value of $\sqrt{x}$, find the value of $y_1 = \frac{1}{2}\left(a + \frac{x}{a}\right)$. This estimate can be improved by calculating a second estimate using the first estimate y_1 in place of a: $y_2 = \frac{1}{2}\left(y_1 + \frac{x}{y_1}\right)$.

Repeating this process several times will give more and more accurate estimates of $\sqrt{x}$.

Critical Thinking

1. a. Which perfect square is closest to 80?

b. Use Heron's method for approximating square roots to calculate the first estimate of the square root of 80. Give an exact decimal answer.

c. Use the first estimate of the square root of 80 to find a more refined second estimate. Round this second estimate to 6 decimal places.

d. Use a calculator to find the actual value of the square root of 80. List all digits shown on your calculator's display.

e. Compare the actual value from part **d** to the values of the first and second estimates. What do you notice?

f. How many iterations of this process are necessary to get an estimate that differs no more than one digit from the actual value recorded in part **d**?

2. Repeat Question 1 for finding an estimate of the square root of 30.

3. Repeat Question 1 for finding an estimate of the square root of 4572.

4. Why would this iterative method have been important to people of Heron's era? Would you say that this method is as important today? Why or why not?

📖 STUDY SKILLS BUILDER

Are You Prepared for a Test on Chapter 7?

Below I have listed some common trouble areas for students in Chapter 7. After studying for your test, but before taking your test, read these.

- Remember how to convert an expression with rational expressions to one with radicals and one with radicals to one with rational expressions.

$$7^{2/3} = \sqrt[3]{7^2} \text{ or } (\sqrt[3]{7})^2$$
$$\sqrt[5]{4^3} = 4^{3/5}$$

- Remember the difference between $\sqrt{x} + \sqrt{x}$ and $\sqrt{x} \cdot \sqrt{x}, x > 0.$

$$\sqrt{x} + \sqrt{x} = 2\sqrt{x}$$
$$\sqrt{x} \cdot \sqrt{x} = x$$

- Don't forget the difference between rationalizing the denominator of $\sqrt{\dfrac{2}{x}}$ and rationalizing the denominator of $\dfrac{\sqrt{2}}{\sqrt{x} + 1}, x > 0.$

$$\sqrt{\frac{2}{x}} = \frac{\sqrt{2}}{\sqrt{x}} = \frac{\sqrt{2} \cdot \sqrt{x}}{\sqrt{x} \cdot \sqrt{x}} = \frac{\sqrt{2x}}{x}$$
$$\frac{\sqrt{2}}{\sqrt{x} + 1} = \frac{\sqrt{2}(\sqrt{x} - 1)}{(\sqrt{x} + 1)(\sqrt{x} - 1)} = \frac{\sqrt{2}(\sqrt{x} - 1)}{x - 1}$$

- Remember that the midpoint of a segment is a *point*. The x-coordinate is the average of the x-coordinates of the end points of the segment and the y-coordinate is the average of the y-coordinates of the end points of the segment.

The midpoint of the segment joining $(-1, 5)$ and $(3, 4)$ is $\left(\dfrac{-1 + 3}{2}, \dfrac{5 + 4}{2}\right)$ or $\left(1, \dfrac{9}{2}\right).$

- Remember that the distance formula gives the *distance* between two points. The distance between $(-1, 5)$ and $(3, 4)$ is

$$\sqrt{(3 - (-1))^2 + (4 - 5)^2} = \sqrt{4^2 + (-1)^2}$$
$$= \sqrt{16 + 1} = \sqrt{17} \text{ units}$$

Remember: This is simply a checklist of common trouble areas. For a review of Chapter 7, see the Highlights and Chapter Review at the end of this chapter.

CHAPTER 7 VOCABULARY CHECK

Fill in each blank with one of the words or phrases listed below.

| index | rationalizing | conjugate | principal square root | cube root | midpoint |
| complex number | like radicals | radicand | imaginary unit | distance | |

1. The _____ of $\sqrt{3} + 2$ is $\sqrt{3} - 2.$
2. The _____ of a nonnegative number a is written as $\sqrt{a}.$
3. The process of writing a radical expression as an equivalent expression but without a radical in the denominator is called _____ the denominator.
4. The _____, written i, is the number whose square is $-1.$
5. The _____ of a number is written as $\sqrt[3]{a}.$
6. In the notation $\sqrt[n]{a}$, n is called the _____ and a is called the _____.
7. Radicals with the same index and the same radicand are called _____.
8. A _____ is a number that can be written in the form $a + bi$, where a and b are real numbers.
9. The _____ formula is $d = \sqrt{(x_2 - x_1)^2 + (y_2 - y_1)^2}.$
10. The _____ formula is $\left(\dfrac{x_1 + x_2}{2}, \dfrac{y_1 + y_2}{2}\right).$

▶ **Helpful Hint**

Are you preparing for your test? Don't forget to take the Chapter 7 Test on page 547. Then check your answers at the back of the text and use the Chapter Test Prep Video CD to see the fully worked-out solutions to any of the exercises you want to review.

CHAPTER 7 HIGHLIGHTS

DEFINITIONS AND CONCEPTS	EXAMPLES

SECTION 7.1 RADICALS AND RADICAL FUNCTIONS

The **positive**, or **principal**, **square root** of a nonnegative number a is written as $\sqrt{a}$.

$$\sqrt{a} = b \text{ only if } b^2 = a \text{ and } b \geq 0$$

The **negative square root** of a is written as $-\sqrt{a}$.

The **cube root** of a real number a is written as $\sqrt[3]{a}$.

$$\sqrt[3]{a} = b \text{ only if } b^3 = a$$

If n is an even positive integer, then $\sqrt[n]{a^n} = |a|$.

If n is an odd positive integer, then $\sqrt[n]{a^n} = a$.

A **radical function** in x is a function defined by an expression containing a root of x.

$$\sqrt{36} = 6 \qquad \sqrt{\frac{9}{100}} = \frac{3}{10}$$

$$-\sqrt{36} = -6 \qquad \sqrt{0.04} = 0.2$$

$$\sqrt[3]{27} = 3 \qquad \sqrt[3]{-\frac{1}{8}} = -\frac{1}{2}$$

$$\sqrt[3]{y^6} = y^2 \qquad \sqrt[3]{64x^9} = 4x^3$$

$$\sqrt{(-3)^2} = |-3| = 3$$

$$\sqrt[3]{(-7)^3} = -7$$

If $f(x) = \sqrt{x} + 2$,

$$f(1) = \sqrt{(1)} + 2 = 1 + 2 = 3$$

$$f(3) = \sqrt{(3)} + 2 \approx 3.73$$

SECTION 7.2 RATIONAL EXPONENTS

$a^{1/n} = \sqrt[n]{a}$ if $\sqrt[n]{a}$ is a real number.

If m and n are positive integers greater than 1 with $\dfrac{m}{n}$ in lowest terms and $\sqrt[n]{a}$ is a real number, then

$$a^{m/n} = (a^{1/n})^m = \left(\sqrt[n]{a}\right)^m$$

$a^{-m/n} = \dfrac{1}{a^{m/n}}$ as long as $a^{m/n}$ is a nonzero number.

Exponent rules are true for rational exponents.

$$81^{1/2} = \sqrt{81} = 9$$

$$(-8x^3)^{1/3} = \sqrt[3]{-8x^3} = -2x$$

$$4^{5/2} = \left(\sqrt{4}\right)^5 = 2^5 = 32$$

$$27^{2/3} = \left(\sqrt[3]{27}\right)^2 = 3^2 = 9$$

$$16^{-3/4} = \frac{1}{16^{3/4}} = \frac{1}{\left(\sqrt[4]{16}\right)^3} = \frac{1}{2^3} = \frac{1}{8}$$

$$x^{2/3} \cdot x^{-5/6} = x^{2/3 - 5/6} = x^{-1/6} = \frac{1}{x^{1/6}}$$

$$(8^4)^{1/2} = 8^2 = 64$$

$$\frac{a^{4/5}}{a^{-2/5}} = a^{4/5 - (-2/5)} = a^{6/5}$$

(continued)

| **DEFINITIONS AND CONCEPTS** | **EXAMPLES** |

SECTION 7.3 SIMPLIFYING RADICAL EXPRESSIONS

Product and Quotient Rules

If $\sqrt[n]{a}$ and $\sqrt[n]{b}$ are real numbers,

$$\sqrt[n]{a} \cdot \sqrt[n]{b} = \sqrt[n]{a \cdot b}$$

$$\frac{\sqrt[n]{a}}{\sqrt[n]{b}} = \sqrt[n]{\frac{a}{b}}, \text{ provided } \sqrt[n]{b} \neq 0$$

A radical of the form $\sqrt[n]{a}$ is **simplified** when a contains no factors that are perfect nth powers.

Multiply or divide as indicated:

$$\sqrt{11} \cdot \sqrt{3} = \sqrt{33}$$

$$\frac{\sqrt[3]{40x}}{\sqrt[3]{5x}} = \sqrt[3]{8} = 2$$

$$\sqrt{40} = \sqrt{4 \cdot 10} = 2\sqrt{10}$$

$$\sqrt{36x^5} = \sqrt{36x^4 \cdot x} = 6x^2\sqrt{x}$$

$$\sqrt[3]{24x^7y^3} = \sqrt[3]{8x^6y^3 \cdot 3x} = 2x^2y\sqrt[3]{3x}$$

$$\sqrt{36x^4 \cdot x} = 6x^2\sqrt{x}$$

Distance Formula

The distance d between two points (x_1, y_1) and (x_2, y_2) is given by

$$d = \sqrt{(x_2 - x_1)^2 + (y_2 - y_1)^2}$$

Find the distance between points $(-1, 6)$ and $(-2, -4)$. Let $(x_1, y_1) = (-1, 6)$ and $(x_2, y_2) = (-2, -4)$.

$$d = \sqrt{(x_2 - x_1)^2 + (y_2 - y_1)^2}$$

$$= \sqrt{(-2 - (-1))^2 + (-4 - 6)^2}$$

$$= \sqrt{1 + 100} = \sqrt{101}$$

Midpoint Formula

The midpoint of the line segment whose end points are (x_1, y_1) and (x_2, y_2) is the point with coordinates

$$\left(\frac{x_1 + x_2}{2}, \frac{y_1 + y_2}{2} \right)$$

Find the midpoint of the line segment whose end points are $(-1, 6)$ and $(-2, -4)$.

$$\left(\frac{-1 + (-2)}{2}, \frac{6 + (-4)}{2} \right)$$

The midpoint is $\left(-\frac{3}{2}, 1 \right)$.

SECTION 7.4 ADDING, SUBTRACTING, AND MULTIPLYING RADICAL EXPRESSIONS

Radicals with the same index and the same radicand are **like radicals.**

The distributive property can be used to add like radicals.

$$5\sqrt{6} + 2\sqrt{6} = (5 + 2)\sqrt{6} = 7\sqrt{6}$$

$$= \sqrt[3]{3x} - 10\sqrt[3]{3x} + 3\sqrt[3]{10x}$$

$$= (-1 - 10)\sqrt[3]{3x} + 3\sqrt[3]{10x}$$

$$= -11\sqrt[3]{3x} + 3\sqrt[3]{10x}$$

Radical expressions are multiplied by using many of the same properties used to multiply polynomials.

Multiply:

$$(\sqrt{5} - \sqrt{2x})(\sqrt{2} + \sqrt{2x})$$

$$= \sqrt{10} + \sqrt{10x} - \sqrt{4x} - 2x$$

$$= \sqrt{10} + \sqrt{10x} - 2\sqrt{x} - 2x$$

$$(2\sqrt{3} - \sqrt{8x})(2\sqrt{3} + \sqrt{8x})$$

$$= 4(3) - 8x = 12 - 8x$$

DEFINITIONS AND CONCEPTS	**EXAMPLES**

SECTION 7.5 RATIONALIZING DENOMINATORS AND NUMERATORS OF RADICAL EXPRESSIONS

The **conjugate** of $a + b$ is $a - b$.

The conjugate of $\sqrt{7} + \sqrt{3}$ is $\sqrt{7} - \sqrt{3}$.

The process of writing the denominator of a radical expression without a radical is called **rationalizing the denominator.**

Rationalize each denominator.

$$\frac{\sqrt{5}}{\sqrt{3}} = \frac{\sqrt{5} \cdot \sqrt{3}}{\sqrt{3} \cdot \sqrt{3}} = \frac{\sqrt{15}}{3}$$

$$\frac{6}{\sqrt{7} + \sqrt{3}} = \frac{6(\sqrt{7} - \sqrt{3})}{(\sqrt{7} + \sqrt{3})(\sqrt{7} - \sqrt{3})}$$

$$= \frac{6(\sqrt{7} - \sqrt{3})}{7 - 3}$$

$$= \frac{6(\sqrt{7} - \sqrt{3})}{4} = \frac{3(\sqrt{7} - \sqrt{3})}{2}$$

The process of writing the numerator of a radical expression without a radical is called **rationalizing the numerator.**

Rationalize each numerator:

$$\frac{\sqrt[3]{9}}{\sqrt[3]{5}} = \frac{\sqrt[3]{9} \cdot \sqrt[3]{3}}{\sqrt[3]{5} \cdot \sqrt[3]{3}} = \frac{\sqrt[3]{27}}{\sqrt[3]{15}} = \frac{3}{\sqrt[3]{15}}$$

$$\frac{\sqrt{9} + \sqrt{3x}}{12} = \frac{(\sqrt{9} + \sqrt{3x})(\sqrt{9} - \sqrt{3x})}{12(\sqrt{9} - \sqrt{3x})}$$

$$= \frac{9 - 3x}{12(\sqrt{9} - \sqrt{3x})}$$

$$= \frac{3(3 - x)}{3 \cdot 4(3 - \sqrt{3x})} = \frac{3 - x}{4(3 - \sqrt{3x})}$$

SECTION 7.6 RADICAL EQUATIONS AND PROBLEM SOLVING

To Solve a Radical Equation

Step 1. Write the equation so that one radical is by itself on one side of the equation.

Step 2. Raise each side of the equation to a power equal to the index of the radical and simplify.

Step 3. If the equation still contains a radical, repeat Steps 1 and 2. If not, solve the equation.

Step 4. Check all proposed solutions in the original equation.

Solve: $x = \sqrt{4x + 9} + 3$.

1. $x - 3 = \sqrt{4x + 9}$

2. $(x - 3)^2 = (\sqrt{4x + 9})^2$
$x^2 - 6x + 9 = 4x + 9$

3. $x^2 - 10x = 0$
$x(x - 10) = 0$
$x = 0$ or $x = 10$

4. The proposed solution 10 checks, but 0 does not. The solution is 10.

(continued)

DEFINITIONS AND CONCEPTS	**EXAMPLES**

SECTION 7.6 RADICAL EQUATIONS AND PROBLEM SOLVING (continued)

To Solve a Radical Equation Graphically

Use the intersection-of-graphs method or the x-intercept method.

Solve $x = \sqrt{4x + 9} + 3$ using the intersection-of-graphs method.

Graph $y_1 = x$ and $y_2 = \sqrt{4x + 9} + 3$ in a $[-5, 20, 5]$ by $[-10, 15, 5]$ window.

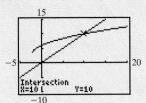

The intersection of the two graphs is at $x = 10$. The solution of the equation is 10.

See the algebraic solution for a check.

SECTION 7.7 COMPLEX NUMBERS

$$i^2 = -1 \text{ and } i = \sqrt{-1}$$

A **complex number** is a number that can be written in the form $a + bi$, where a and b are real numbers.

Simplify: $\sqrt{-9}$.

$$\sqrt{-9} = \sqrt{-1 \cdot 9} = \sqrt{-1} \cdot \sqrt{9} = i \cdot 3 \text{ or } 3i$$

Complex Numbers	**Written in Form $a + bi$**
12	$12 + 0i$
$-5i$	$0 + (-5)i$
$-2 - 3i$	$-2 + (-3)i$

Multiply,

$$\sqrt{-3} \cdot \sqrt{-7} = i\sqrt{3} \cdot i\sqrt{7}$$
$$= i^2\sqrt{21}$$
$$= -\sqrt{21}$$

To add or subtract complex numbers, add or subtract their real parts and then add or subtract their imaginary parts.

To multiply complex numbers, multiply as though they are binomials.

Perform each indicated operation.

$$(-3 + 2i) - (7 - 4i) = -3 + 2i - 7 + 4i$$
$$= -10 + 6i$$

$$(-7 - 2i)(6 + i) = -42 - 7i - 12i - 2i^2$$
$$= -42 - 19i - 2(-1)$$
$$= -42 - 19i + 2$$
$$= -40 - 19i$$

The complex numbers $(a + bi)$ and $(a - bi)$ are called **complex conjugates.**

The complex conjugate of

$$(3 + 6i) \text{ is } (3 - 6i).$$

Their product is a real number:

$$(3 - 6i)(3 + 6i) = 9 - 36i^2$$
$$= 9 - 36(-1) = 9 + 36 = 45$$

To divide complex numbers, multiply the numerator and the denominator by the conjugate of the denominator.

Divide.

$$\frac{4}{2 - i} = \frac{4(2 + i)}{(2 - i)(2 + i)}$$
$$= \frac{4(2 + i)}{4 - i^2}$$
$$= \frac{4(2 + i)}{5}$$
$$= \frac{8 + 4i}{5} = \frac{8}{5} + \frac{4}{5}i$$

CHAPTER 7 REVIEW

(7.1) *Find the root. Assume that all variables represent positive numbers.*

1. $\sqrt{81}$

2. $\sqrt[4]{81}$

3. $\sqrt[3]{-8}$

4. $\sqrt[4]{-16}$

5. $-\sqrt{\dfrac{1}{49}}$

6. $\sqrt{x^{64}}$

7. $-\sqrt{36}$

8. $\sqrt[3]{64}$

9. $\sqrt[3]{-a^6b^9}$

10. $\sqrt{16a^4b^{12}}$

11. $\sqrt[5]{32a^5b^{10}}$

12. $\sqrt[5]{-32x^{15}y^{20}}$

13. $\sqrt{\dfrac{x^{12}}{36y^2}}$

14. $\sqrt[3]{\dfrac{27y^3}{z^{12}}}$

Simplify. Use absolute value bars when necessary.

15. $\sqrt{(-x)^2}$

16. $\sqrt[4]{(x^2-4)^4}$

17. $\sqrt[3]{(-27)^3}$

18. $\sqrt[5]{(-5)^5}$

19. $-\sqrt[5]{x^5}$

20. $\sqrt[4]{16(2y+z)^{12}}$

21. $\sqrt{25(x-y)^{10}}$

22. $\sqrt[5]{-y^5}$

23. $\sqrt[9]{-x^9}$

Identify the domain and then graph each function.

24. $f(x) = \sqrt{x} + 3$

25. $g(x) = \sqrt[3]{x} - 3$; use the accompanying table.

x	-5	2	3	4	11
$g(x)$					

(7.2) *Evaluate the following.*

26. $\left(\dfrac{1}{81}\right)^{1/4}$

27. $\left(-\dfrac{1}{27}\right)^{1/3}$

28. $(-27)^{-1/3}$

29. $(-64)^{-1/3}$

30. $-9^{3/2}$

31. $64^{-1/3}$

32. $(-25)^{5/2}$

33. $\left(\dfrac{25}{49}\right)^{-3/2}$

34. $\left(\dfrac{8}{27}\right)^{-2/3}$

35. $\left(-\dfrac{1}{36}\right)^{-1/4}$

Write with rational exponents.

36. $\sqrt[3]{x^2}$

37. $\sqrt[5]{5x^2y^3}$

Write with radical notation.

38. $y^{4/5}$

39. $5(xy^2z^5)^{1/3}$

40. $(x+2y)^{-1/2}$

Simplify each expression. Assume that all variables represent positive numbers. Write with only positive exponents.

41. $a^{1/3}a^{4/3}a^{1/2}$

42. $\dfrac{b^{1/3}}{b^{4/3}}$

43. $(a^{1/2}a^{-2})^3$

44. $(x^{-3}y^6)^{1/3}$

45. $\left(\dfrac{b^{3/4}}{a^{-1/2}}\right)^8$

46. $\dfrac{x^{1/4}x^{-1/2}}{x^{2/3}}$

47. $\left(\dfrac{49c^{5/3}}{a^{-1/4}b^{5/6}}\right)^{-1}$

48. $a^{-1/4}(a^{5/4} - a^{9/4})$

Use a calculator and write a three-decimal-place approximation.

49. $\sqrt{20}$

50. $\sqrt[3]{-39}$

51. $\sqrt[4]{726}$

52. $56^{1/3}$

53. $-78^{3/4}$

54. $105^{-2/3}$

Use rational exponents to write each radical with the same index. Then multiply.

55. $\sqrt[3]{2} \cdot \sqrt{7}$

56. $\sqrt[3]{3} \cdot \sqrt[4]{x}$

(7.3) *Perform the indicated operations and then simplify if possible. For the remainder of this review, assume that variables represent positive numbers only.*

57. $\sqrt{3} \cdot \sqrt{8}$

58. $\sqrt[3]{7y} \cdot \sqrt[3]{x^2z}$

59. $\dfrac{\sqrt{44x^3}}{\sqrt{11x}}$

60. $\dfrac{\sqrt[4]{a^6b^{13}}}{\sqrt[4]{a^2b}}$

Simplify.

61. $\sqrt{60}$

62. $-\sqrt{75}$

63. $\sqrt[3]{162}$

64. $\sqrt[3]{-32}$

65. $\sqrt{36x^7}$

66. $\sqrt[3]{24a^5b^7}$

67. $\sqrt{\dfrac{p^{17}}{121}}$

68. $\sqrt[3]{\dfrac{y^5}{27x^6}}$

69. $\sqrt[4]{\dfrac{xy^6}{81}}$

70. $\sqrt{\dfrac{2x^3}{49y^4}}$

△ **71.** The formula for the radius r of a circle of area A is

$$r = \sqrt{\dfrac{A}{\pi}}$$

a. Find the exact radius of a circle whose area is 25 square meters.

b. Approximate to two decimal places the radius of a circle whose area is 104 square inches.

Find the distance between each pair of points. Give an exact value and a three-decimal-place approximation.

72. $(-6, 3)$ and $(8, 4)$

73. $(-4, -6)$ and $(-1, 5)$

74. $(-1, 5)$ and $(2, -3)$

75. $(-\sqrt{2}, 0)$ and $(0, -4\sqrt{6})$

76. $(-\sqrt{5}, -\sqrt{11})$ and $(-\sqrt{5}, -3\sqrt{11})$

77. $(7.4, -8.6)$ and $(-1.2, 5.6)$

Find the midpoint of each line segment whose end points are given.

78. $(2, 6); (-12, 4)$

79. $(-6, -5); (-9, 7)$

80. $(4, -6); (-15, 2)$

81. $\left(0, -\dfrac{3}{8}\right); \left(\dfrac{1}{10}, 0\right)$

82. $\left(\dfrac{3}{4}, -\dfrac{1}{7}\right); \left(-\dfrac{1}{4}, -\dfrac{3}{7}\right)$

83. $(\sqrt{3}, -2\sqrt{6})$ and $(\sqrt{3}, -4\sqrt{6})$

(7.4) *Perform the indicated operation.*

84. $2\sqrt{50} - 3\sqrt{125} + \sqrt{98}$

85. $x\sqrt{75xy} - \sqrt{27x^3y}$

86. $\sqrt[3]{128} + \sqrt[3]{250}$

87. $3\sqrt[4]{32a^5} - a\sqrt[4]{162a}$

88. $\dfrac{5}{\sqrt{4}} + \dfrac{\sqrt{3}}{3}$

89. $\sqrt{\dfrac{8}{x^2}} - \sqrt{\dfrac{50}{16x^2}}$

90. $2\sqrt{32x^2y^3} - xy\sqrt{98y}$

91. $2a\sqrt[4]{32b^5} - 3b\sqrt[4]{162a^4b} + \sqrt[4]{2a^4b^5}$

Multiply and then simplify if possible.

92. $\sqrt{3}\left(\sqrt{27} - \sqrt{3}\right)$

93. $\left(\sqrt{x} - 3\right)^2$

94. $\left(\sqrt{5} - 5\right)\left(2\sqrt{5} + 2\right)$

95. $\left(2\sqrt{x} - 3\sqrt{y}\right)\left(2\sqrt{x} + 3\sqrt{y}\right)$

96. $\left(\sqrt{a} + 3\right)\left(\sqrt{a} - 3\right)$

97. $\left(\sqrt[3]{a} + 2\right)^2$

98. $\left(\sqrt[3]{5x} + 9\right)\left(\sqrt[3]{5x} - 9\right)$

99. $\left(\sqrt[3]{a} + 4\right)\left(\sqrt[3]{a^2} - 4\sqrt[3]{a} + 16\right)$

(7.5) *Rationalize each denominator.*

100. $\dfrac{3}{\sqrt{7}}$

101. $\sqrt{\dfrac{x}{12}}$

102. $\dfrac{5}{\sqrt[3]{4}}$

103. $\sqrt{\dfrac{24x^5}{3y^2}}$

104. $\sqrt[3]{\dfrac{15x^6y^7}{z^2}}$

105. $\dfrac{5}{2 - \sqrt{7}}$

106. $\dfrac{3}{\sqrt{y} - 2}$

107. $\dfrac{\sqrt{2} - \sqrt{3}}{\sqrt{2} + \sqrt{3}}$

Rationalize each numerator.

108. $\dfrac{\sqrt{11}}{3}$

109. $\sqrt{\dfrac{18}{y}}$

110. $\dfrac{\sqrt[3]{9}}{7}$

111. $\sqrt{\dfrac{24x^5}{3y^2}}$

112. $\sqrt[3]{\dfrac{xy^2}{10z}}$

113. $\dfrac{\sqrt{x} + 5}{-3}$

(7.6) *Solve each equation for the variable.*

114. $\sqrt{y - 7} = 5$

115. $\sqrt{2x} + 10 = 4$

116. $\sqrt[3]{2x - 6} = 4$

117. $\sqrt{x + 6} = \sqrt{x + 2}$

118. $2x - 5\sqrt{x} = 3$

119. $\sqrt{x + 9} = 2 + \sqrt{x - 7}$

Find each unknown length.

△ **120.**

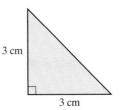

△ **121.**

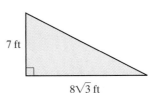

122. Beverly Hillis wants to determine the distance x across a pond on her property. She is able to measure the distances shown on the following diagram. Find how wide the lake is at the crossing point, indicated by the triangle, to the nearest tenth of a foot.

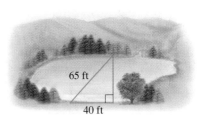

△ **123.** A pipe fitter needs to connect two underground pipelines that are offset by 3 feet, as pictured in the diagram. Neglecting the joints needed to join the pipes, find the length of the shortest possible connecting pipe rounded to the nearest hundredth of a foot.

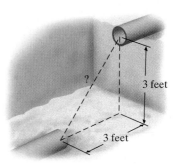

(7.7) Perform the indicated operation and simplify. Write the result in the form $a + bi$.

124. $\sqrt{-8}$

125. $-\sqrt{-6}$

126. $\sqrt{-4} + \sqrt{-16}$

127. $\sqrt{-2} \cdot \sqrt{-5}$

128. $(12 - 6i) + (3 + 2i)$

129. $(-8 - 7i) - (5 - 4i)$

130. $(2i)^6$

131. $-3i(6 - 4i)$

132. $(3 + 2i)(1 + i)$

133. $(2 - 3i)^2$

134. $\left(\sqrt{6} - 9i\right)\left(\sqrt{6} + 9i\right)$

135. $\dfrac{2 + 3i}{2i}$

136. $\dfrac{1 + i}{-3i}$

141. $\sqrt[4]{\dfrac{y^{20}}{16x^{12}}}$

142. $9^{1/2}$

143. $64^{-1/2}$

144. $\left(\dfrac{27}{64}\right)^{-2/3}$

145. $\dfrac{(x^{2/3}x^{-3})^3}{x^{-1/2}}$

146. $\sqrt{200x^9}$

147. $\sqrt{\dfrac{3n^3}{121m^{10}}}$

148. $3\sqrt{20} - 7x\sqrt[3]{40} + 3\sqrt[3]{5x^3}$

149. $(2\sqrt{x} - 5)^2$

150. Find the distance between $(-3, 5)$ and $(-8, 9)$.

151. Find the midpoint of the line segment joining $(-3, 8)$ and $(11, 24)$.

MIXED REVIEW

Simplify. Use absolute value bars when necessary.

137. $\sqrt[3]{x^3}$

138. $\sqrt{(x + 2)^2}$

Simplify. Assume that all variables represent positive real numbers. If necessary, write answers with positive exponents only.

139. $-\sqrt{100}$

140. $\sqrt[3]{-x^{12}y^3}$

Rationalize each denominator.

152. $\dfrac{7}{\sqrt{13}}$

153. $\dfrac{2}{\sqrt{x} + 3}$

Solve.

154. $\sqrt{x + 2} = x$

CHAPTER 7 TEST
TEST PREP **VIDEO** Remember to use the Chapter Test Prep Video CD to see the fully worked-out solutions to any of the exercises you want to review.

Raise to the power or find the root. Assume that all variables represent positive numbers. Write with only positive exponents.

1. $\sqrt{216}$

2. $-\sqrt[4]{x^{64}}$

3. $\left(\dfrac{1}{125}\right)^{1/3}$

4. $\left(\dfrac{1}{125}\right)^{-1/3}$

5. $\left(\dfrac{8x^3}{27}\right)^{2/3}$

6. $\sqrt[3]{-a^{18}b^9}$

7. $\left(\dfrac{64c^{4/3}}{a^{-2/3}b^{5/6}}\right)^{1/2}$

8. $a^{-2/3}(a^{5/4} - a^3)$

Find the root. Use absolute value bars when necessary.

9. $\sqrt[4]{(4xy)^4}$

10. $\sqrt[3]{(-27)^3}$

Rationalize the denominator. Assume that all variables represent positive numbers.

11. $\sqrt{\dfrac{9}{y}}$

12. $\dfrac{4 - \sqrt{x}}{4 + 2\sqrt{x}}$

13. $\dfrac{\sqrt[3]{ab}}{\sqrt[3]{ab^2}}$

14. Rationalize the numerator of $\dfrac{\sqrt{6} + x}{8}$ and simplify.

Perform the indicated operations. Assume that all variables represent positive numbers.

15. $\sqrt{125x^3} - 3\sqrt{20x^3}$

16. $\sqrt{3}\left(\sqrt{16} - \sqrt{2}\right)$

17. $\left(\sqrt{x} + 1\right)^2$

18. $\left(\sqrt{2} - 4\right)\left(\sqrt{3} + 1\right)$

19. $\left(\sqrt{5} + 5\right)\left(\sqrt{5} - 5\right)$

Use a calculator to approximate each to three decimal places.

20. $\sqrt{561}$

21. $386^{-2/3}$

Solve.

22. $x = \sqrt{x - 2} + 2$

23. $\sqrt{x^2 - 7} + 3 = 0$

24. $\sqrt[3]{x + 5} = \sqrt[3]{2x - 1}$

Perform the indicated operation and simplify. Write the result in the form $a + bi$.

25. $\sqrt{-2}$

26. $-\sqrt{-8}$

27. $(12 - 6i) - (12 - 3i)$

28. $(6 - 2i)(6 + 2i)$

29. $(4 + 3i)^2$

30. $\dfrac{1 + 4i}{1 - i}$

△ **31.** Find x.

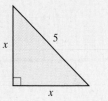

32. Identify the domain of $g(x)$. Then complete the accompanying table and graph $g(x)$.

$$g(x) = \sqrt{x + 2}$$

x	-2	-1	2	7
$g(x)$				

33. Find the distance between the points $(-6, 3)$ and $(-8, -7)$.

34. Find the distance between the points $(-2\sqrt{5}, \sqrt{10})$ and $(-\sqrt{5}, 4\sqrt{10})$.

35. Find the midpoint of the line segment whose end points are $(-2, -5)$ and $(-6, 12)$.

36. Find the midpoint of the line segment whose end points are $\left(-\dfrac{2}{3}, -\dfrac{1}{5}\right)$ and $\left(-\dfrac{1}{3}, \dfrac{4}{5}\right)$.

Solve.

37. The function $V(r) = \sqrt{2.5r}$ can be used to estimate the maximum safe velocity V in miles per hour at which a car can travel if it is driven along a curved road with a *radius of curvature* r in feet. To the nearest whole number, find the maximum safe speed if a cloverleaf exit on an expressway has a radius of curvature of 300 feet.

38. Use the formula from Exercise 37 to find the radius of curvature if the safe velocity is 30 mph.

CHAPTER 7 CUMULATIVE REVIEW

1. Simplify each expression.

 a. $3xy - 2xy + 5 - 7 + xy$

 b. $7x^2 + 3 - 5(x^2 - 4)$

 c. $(2.1x - 5.6) - (-x - 5.3)$

 d. $\dfrac{1}{2}(4a - 6b) - \dfrac{1}{3}(9a + 12b - 1) + \dfrac{1}{4}$

2. Simplify each expression.

 a. $2(x - 3) + (5x + 3)$

 b. $4(3x + 2) - 3(5x - 1)$

 c. $7x + 2(x - 7) - 3x$

3. Solve for x: $\dfrac{x + 5}{2} + \dfrac{1}{2} = 2x - \dfrac{x - 3}{8}$

4. Solve: $\dfrac{a - 1}{2} + a = 2 - \dfrac{2a + 7}{8}$

5. A part-time salesperson earns $600 per month plus a commission of 20% of sales. Find the minimum amount of sales needed to receive a total income of at least $1500 per month.

6. The Smith family owns a lake house 121.5 miles from home. If it takes them $4\dfrac{1}{2}$ hours round-trip to drive from their house to their lake house, find their average speed.

7. Solve: $2|x| + 25 = 23$

8. Solve: $|3x - 2| + 5 = 5$

9. Solve: $\left|\dfrac{x}{3} - 1\right| - 2 \ge 0$

10. Solve: $\left|\dfrac{x}{2} - 1\right| \le 0$

11. Graph the equation $y = |x|$.

12. Graph $y = |x - 2|$.

13. Determine the domain and range of each relation.

 a. $\{(2, 3), (2, 4), (0, -1), (3, -1)\}$

 b.

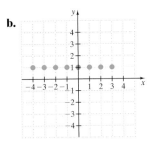

 c.

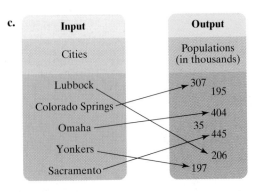

14. Find the domain and the range of each relation. Use the vertical line test to determine whether each graph is the graph of a function.

a.

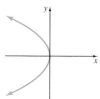

b.

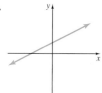

c.

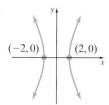

$(-2,0)$ $(2,0)$

15. Graph $y = -3$.

16. Graph $f(x) = -2$.

17. Find the slope of the line $x = -5$.

18. Find the slope of $y = -3$.

19. Use the substitution method to solve the system.

$$\begin{cases} -\dfrac{x}{6} + \dfrac{y}{2} = \dfrac{1}{2} \\ \dfrac{x}{3} - \dfrac{y}{6} = -\dfrac{3}{4} \end{cases}$$

20. Use the substitution method to solve the system.

$$\begin{cases} \dfrac{x}{6} - \dfrac{y}{2} = 1 \\ \dfrac{x}{3} - \dfrac{y}{4} = 2 \end{cases}$$

21. Use the product rule to simplify.

   a. $2^2 \cdot 2^5$

   b. $x^7 x^3$

   c. $y \cdot y^2 \cdot y^4$

22. At a seasonal clearance sale, Nana Long spent \$33.75. She paid \$3.50 for T-shirts and \$4.25 for shorts. If she bought 9 items, how many of each item did she buy?

23. Use scientific notation to simplify $\dfrac{2000 \times 0.000021}{700}$.

24. Use scientific notation to simplify and write the answer in scientific notation $\dfrac{0.0000035 \times 4000}{0.28}$.

25. If $P(x) = 3x^2 - 2x - 5$, find the following.

   a. $P(1)$

   b. $P(-2)$

26. Subtract $(2x - 5)$ from the sum of $(5x^2 - 3x + 6)$ and $(4x^2 + 5x - 3)$.

27. Multiply and simplify the product if possible.

   a. $(x + 3)(2x + 5)$

   b. $(2x - 3)(5x^2 - 6x + 7)$

28. Multiply and simplify the product if possible.

   a. $(y - 2)(3y + 4)$

   b. $(3y - 1)(2y^2 + 3y - 1)$

29. Find the GCF of $20x^3y$, $10x^2y^2$, and $35x^3$.

30. Factor $x^3 - x^2 + 4x - 4$.

31. Simplify each rational expression.

   a. $\dfrac{x^3 + 8}{2 + x}$

   b. $\dfrac{2y^2 + 2}{y^3 - 5y^2 + y - 5}$

32. Simplify each rational expression.

   a. $\dfrac{a^3 - 8}{2 - a}$

   b. $\dfrac{3a^2 - 3}{a^3 + 5a^2 - a - 5}$

33. Perform the indicated operation.

   a. $\dfrac{2}{x^2y} + \dfrac{5}{3x^3y}$

   b. $\dfrac{3x}{x + 2} + \dfrac{2x}{x - 2}$

   c. $\dfrac{2x - 6}{x - 1} - \dfrac{4}{1 - x}$

34. Perform the indicated operations.

   a. $\dfrac{3}{xy^2} - \dfrac{2}{3x^2y}$

   b. $\dfrac{5x}{x + 3} - \dfrac{2x}{x - 3}$

   c. $\dfrac{x}{x - 2} - \dfrac{5}{2 - x}$

35. Simplify each complex fraction.

   a. $\dfrac{\dfrac{5x}{x + 2}}{\dfrac{10}{x - 2}}$

   b. $\dfrac{\dfrac{x}{y^2} + \dfrac{1}{y}}{\dfrac{y}{x^2} + \dfrac{1}{x}}$

36. Simplify each complex fraction.

   a. $\dfrac{\dfrac{y - 2}{16}}{\dfrac{2y + 3}{12}}$

   b. $\dfrac{\dfrac{x}{16} - \dfrac{1}{x}}{1 - \dfrac{4}{x}}$

37. Divide $10x^3 - 5x^2 + 20x$ by $5x$.

38. Divide $x^3 - 2x^2 + 3x - 6$ by $x - 2$.

39. Use synthetic division to divide $2x^3 - x^2 - 13x + 1$ by $x - 3$.

40. Use synthetic division to divide $4y^3 - 12y^2 - y + 12$ by $y - 3$.

41. Solve: $\dfrac{x + 6}{x - 2} = \dfrac{2(x + 2)}{x - 2}$.

42. Solve: $\dfrac{28}{9 - a^2} = \dfrac{2a}{a - 3} + \dfrac{6}{a + 3}$.

43. Solve: $\dfrac{1}{x} + \dfrac{1}{y} = \dfrac{1}{z}$ for x.

44. Solve: $A = \dfrac{h(a + b)}{2}$ for a.

45. Suppose that u varies inversely as w. If u is 3 when w is 5, find the constant of variation and the inverse variation equation.

46. Suppose that y varies directly as x. If $y = 0.51$ when $x = 3$, find the constant of variation and the direct variation equation.

47. Write each expression with a positive exponent, and then simplify.

   a. $16^{-3/4}$

   b. $(-27)^{-2/3}$

48. Write each expression with a positive exponent, and then simplify.

   a. $(81)^{-3/4}$

   b. $(-125)^{-2/3}$

49. Rationalize the numerator of $\dfrac{\sqrt{x} + 2}{5}$.

50. Add or subtract.

   a. $\sqrt{36a^3} - \sqrt{144a^3} + \sqrt{4a^3}$

   b. $\sqrt[3]{128ab^3} - 3\sqrt[3]{2ab^3} + b\sqrt[3]{16a}$

   c. $\dfrac{\sqrt[3]{81}}{10} + \sqrt[3]{\dfrac{192}{125}}$

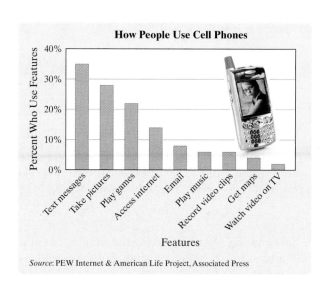

How People Use Cell Phones

Source: PEW Internet & American Life Project, Associated Press

The growth of cell phones, shown below, can be approximated by a quadratic function. More interesting information is probably given on the graph above. As shown, the cell phone is certainly no longer just a phone.

On page 597, Section 8.5, Exercises 79 and 80, you will have the opportunity to use the quadratic function below and discuss its limitations.

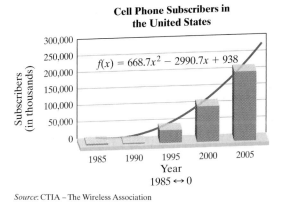

Cell Phone Subscribers in the United States

$f(x) = 668.7x^2 - 2990.7x + 938$

Year
1985 ↔ 0

Source: CTIA – The Wireless Association

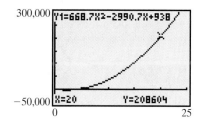

An important part of the study of algebra is learning to model and solve problems. Often, the model of a problem is a quadratic equation or a function containing a second-degree polynomial. In this chapter, we continue the work begun in Chapter 5, when we solved polynomial equations in one variable by factoring. Two additional methods of solving quadratic equations are analyzed, as well as methods of solving nonlinear inequalities in one variable.

8.1 SOLVING QUADRATIC EQUATIONS BY COMPLETING THE SQUARE

OBJECTIVES

1 Use the square root property to solve quadratic equations.

2 Solve quadratic equations by completing the square.

3 Use quadratic equations to solve problems.

TECHNOLOGY NOTE

Many graphing utilities have a complex mode. If yours does, set it to this mode for this chapter. Recall that a real number is a complex number, so in complex mode the graphing utility will display real numbers as well as complex numbers that are not real numbers.

OBJECTIVE 1 ▶ Using the square root property. In Chapter 5, we solved quadratic equations by factoring. Recall that a **quadratic**, or **second-degree, equation** is an equation that can be written in the form $ax^2 + bx + c = 0$, where a, b, and c are real numbers and a is not 0. To solve a quadratic equation such as $x^2 = 9$ by factoring, we use the zero-factor theorem. To use the zero-factor theorem, the equation must first be written in standard form, $ax^2 + bx + c = 0$.

$$x^2 = 9$$
$$x^2 - 9 = 0 \qquad \text{Subtract 9 from both sides.}$$
$$(x + 3)(x - 3) = 0 \qquad \text{Factor.}$$
$$x + 3 = 0 \quad \text{or} \quad x - 3 = 0 \quad \text{Set each factor equal to 0.}$$
$$x = -3 \qquad\qquad x = 3 \quad \text{Solve.}$$

The solution set is $\{-3, 3\}$, the positive and negative square roots of 9. Not all quadratic equations can be solved by factoring, so we need to explore other methods. Notice that the solutions of the equation $x^2 = 9$ are two numbers whose square is 9.

$$3^2 = 9 \qquad \text{and} \qquad (-3)^2 = 9$$

Thus, we can solve the equation $x^2 = 9$ by taking the square root of both sides. Be sure to include both $\sqrt{9}$ and $-\sqrt{9}$ as solutions since both $\sqrt{9}$ and $-\sqrt{9}$ are numbers whose square is 9.

$$x^2 = 9$$
$$\sqrt{x^2} = \pm\sqrt{9} \qquad \text{The notation } \pm\sqrt{9} \text{ (read as "plus or minus } \sqrt{9}\text{")}$$
$$x = \pm 3 \qquad\quad \text{indicates the pair of numbers } +\sqrt{9} \text{ and } -\sqrt{9}.$$

This illustrates the square root property.

> ▶ **Helpful Hint**
>
> The notation ± 3, for example, is read as "plus or minus 3." It is a shorthand notation for the pair of numbers $+3$ and -3.

Square Root Property

If b is a real number and if $a^2 = b$, then $a = \pm\sqrt{b}$.

EXAMPLE 1 Use the square root property to solve $x^2 = 50$.

Solution
$$x^2 = 50$$
$$x = \pm\sqrt{50} \quad \text{Use the square root property.}$$
$$x = \pm 5\sqrt{2} \quad \text{Simplify the radical.}$$

Check: Let $x = 5\sqrt{2}$.

$$x^2 = 50$$
$$\left(5\sqrt{2}\right)^2 \stackrel{?}{=} 50$$
$$25 \cdot 2 \stackrel{?}{=} 50$$
$$50 = 50 \quad \text{True}$$

Let $x = -5\sqrt{2}$.

$$x^2 = 50$$
$$\left(-5\sqrt{2}\right)^2 \stackrel{?}{=} 50$$
$$25 \cdot 2 \stackrel{?}{=} 50$$
$$50 = 50 \quad \text{True}$$

The solutions are $5\sqrt{2}$ and $-5\sqrt{2}$, or the solution set is $\{-5\sqrt{2}, 5\sqrt{2}\}$. ☐

```
5√(2)→X:X²
                50
-5√(2)→X:X²
                50
■
```

A calculator check of both solutions for Example 1.

PRACTICE
1   Use the square root property to solve $x^2 = 18$.

EXAMPLE 2 Use the square root property to solve $2x^2 - 14 = 0$.

Solution First we get the squared variable alone on one side of the equation.

$$2x^2 - 14 = 0$$
$$2x^2 = 14 \qquad \text{Add 14 to both sides.}$$
$$x^2 = 7 \qquad \text{Divide both sides by 2.}$$
$$x = \pm\sqrt{7} \qquad \text{Use the square root property.}$$

Check to see that the solutions are $\sqrt{7}$ and $-\sqrt{7}$, or the solution set is $\{-\sqrt{7}, \sqrt{7}\}$. ☐

PRACTICE
2 Use the square root property to solve $3x^2 - 30 = 0$.

EXAMPLE 3 Use the square root property to solve $(x + 1)^2 = 12$. Use a graphing utility to check.

Solution
$$(x + 1)^2 = 12$$
$$x + 1 = \pm\sqrt{12} \qquad \text{Use the square root property.}$$
$$x + 1 = \pm 2\sqrt{3} \qquad \text{Simplify the radical.}$$
$$x = -1 \pm 2\sqrt{3} \qquad \text{Subtract 1 from both sides.}$$

Check: Below is a check for $-1 + 2\sqrt{3}$. The check for $-1 - 2\sqrt{3}$ is almost the same and is left for you to do on your own.

$$(x + 1)^2 = 12$$
$$\left(-1 + 2\sqrt{3} + 1\right)^2 \stackrel{?}{=} 12$$
$$\left(2\sqrt{3}\right)^2 \stackrel{?}{=} 12$$
$$4 \cdot 3 \stackrel{?}{=} 12$$
$$12 = 12 \quad \text{True}$$

The solutions are $-1 + 2\sqrt{3}$ and $-1 - 2\sqrt{3}$.

To check graphically, we use the intersection-of-graphs method and graph

$$y_1 = (x + 1)^2 \quad \text{and} \quad y_2 = 12$$

The approximate points of intersection are shown below in the first two screens.

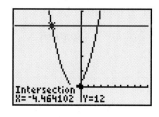 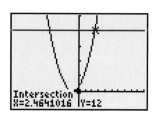 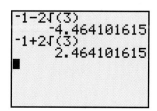

Notice that this graphical check gives approximate solutions only since $-1 + 2\sqrt{3}$ and $-1 - 2\sqrt{3}$ are irrational. The approximate solutions are $x \approx -4.464$ and $x \approx 2.464$. The exact solutions are $x = -1 + 2\sqrt{3}$ and $x = -1 - 2\sqrt{3}$.

Next, see that the x-values of the points of intersection approximate the exact solutions. The screen above and to the right confirms this. ☐

PRACTICE
3 Use the square root property to solve $(x + 3)^2 = 20$.

EXAMPLE 4   Use the square root property to solve $(2x - 5)^2 = -16$.

Solution

$$(2x - 5)^2 = -16$$

$$2x - 5 = \pm\sqrt{-16} \quad \text{Use the square root property.}$$

$$2x - 5 = \pm 4i \quad \text{Simplify the radical.}$$

$$2x = 5 \pm 4i \quad \text{Add 5 to both sides.}$$

$$x = \frac{5 \pm 4i}{2} \quad \text{Divide both sides by 2.}$$

The solutions are $\dfrac{5 + 4i}{2}$ and $\dfrac{5 - 4i}{2}$.

PRACTICE

4   Use the square root property to solve $(5x - 2)^2 = -9$.

Concept Check ☑

How do you know just by looking that $(x - 2)^2 = -4$ has complex, but not real, solutions?

DISCOVER THE CONCEPT

From the previous Example 4, we know that the equation $(2x - 5)^2 = -16$ has two complex, but not real, solutions. Use the intersection-of-graphs method to check.

a. Graph $y_1 = (2x - 5)^2$ and $y_2 = -16$.

b. Locate any points of intersection of the graphs.

c. Summarize the results of parts **a** and **b**, and what you think occurred.

```
(5+4i)/2→X:(2X-5
)²
               -16
(5-4i)/2→X:(2X-5
)²
               -16
■
```

The x-axis and the y-axis of the rectangular coordinate system include real numbers only. Thus, coordinates of points of the associated plane are real numbers only. This means that the intersection-of-graphs method on the rectangular coordinate system gives real number solutions of a related equation only. Since the graphs above do not intersect, the equation has no real number solutions.

To check Example 4 numerically, see the screen to the left.

OBJECTIVE 2 ▶ Solving by completing the square. Notice from Examples 3 and 4 that, if we write a quadratic equation so that one side is the square of a binomial, we can solve by using the square root property. To write the square of a binomial, we write perfect square trinomials. Recall that a perfect square trinomial is a trinomial that can be factored into two identical binomial factors.

Perfect Square Trinomials	*Factored Form*
$x^2 + 8x + 16$	$(x + 4)^2$
$x^2 - 6x + 9$	$(x - 3)^2$
$x^2 + 3x + \dfrac{9}{4}$	$\left(x + \dfrac{3}{2}\right)^2$

Notice that for each perfect square trinomial, **the constant term of the trinomial is the square of half the coefficient of the x-term.** For example,

$$x^2 + 8x + 16 \qquad\qquad x^2 - 6x + 9$$

$$\frac{1}{2}(8) = 4 \text{ and } 4^2 = 16 \qquad \frac{1}{2}(-6) = -3 \text{ and } (-3)^2 = 9$$

The process of writing a quadratic equation so that one side is a perfect square trinomial is called **completing the square.**

Answer to Concept Check:
answers may vary

EXAMPLE 5 Solve $p^2 + 2p = 4$ by completing the square.

Solution First, add the square of half the coefficient of p to both sides so that the resulting trinomial will be a perfect square trinomial. The coefficient of p is 2.

$$\frac{1}{2}(2) = 1 \quad \text{and} \quad 1^2 = 1$$

Add 1 to both sides of the original equation.

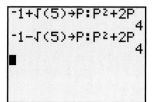

$$p^2 + 2p = 4$$
$$p^2 + 2p + 1 = 4 + 1 \quad \text{Add 1 to both sides.}$$
$$(p + 1)^2 = 5 \quad \text{Factor the trinomial; simplify the right side.}$$

We may now use the square root property and solve for p.

$$p + 1 = \pm\sqrt{5} \quad \text{Use the square root property.}$$
$$p = -1 \pm \sqrt{5} \quad \text{Subtract 1 from both sides.}$$

Notice that there are two solutions: $-1 + \sqrt{5}$ and $-1 - \sqrt{5}$. A numerical check is shown to the left. ☐

PRACTICE
5 Solve $b^2 + 4b = 3$ by completing the square.

EXAMPLE 6 Solve $m^2 - 7m - 1 = 0$ for m.

Algebraic Solution:

To solve by completing the square, we first add 1 to both sides of the equation so that the left side has no constant term.

$$m^2 - 7m - 1 = 0$$
$$m^2 - 7m = 1$$

Now find the constant term that makes the left side a perfect square trinomial by squaring half the coefficient of m. Add this constant to both sides of the equation.

$$\frac{1}{2}(-7) = -\frac{7}{2}$$

and

$$\left(-\frac{7}{2}\right)^2 = \frac{49}{4}$$

$$m^2 - 7m + \frac{49}{4} = 1 + \frac{49}{4} \quad \text{Add } \frac{49}{4} \text{ to both sides of the equation.}$$

$$\left(m - \frac{7}{2}\right)^2 = \frac{53}{4} \quad \text{Factor the perfect square trinomial and simplify the right side.}$$

$$m - \frac{7}{2} = \pm\sqrt{\frac{53}{4}} \quad \text{Apply the square root property.}$$

$$m = \frac{7}{2} \pm \frac{\sqrt{53}}{2} \quad \text{Add } \frac{7}{2} \text{ to both sides and simplify } \sqrt{\frac{53}{4}}.$$

$$m = \frac{7 \pm \sqrt{53}}{2} \quad \text{Simplify.}$$

Graphical Solution:

Enter $y_1 = x^2 - 7x - 1$ and find the x-intercepts of the graph, or the zeros by using the zero option on the Calc menu.

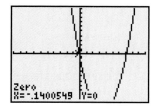
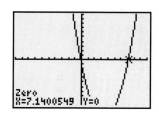

Rounded to the nearest hundredth we have x (or m) ≈ -0.14 and x (or m) ≈ 7.14. These are approximations of the exact solutions as shown below.

The solutions are $\dfrac{7 + \sqrt{53}}{2}$ and $\dfrac{7 - \sqrt{53}}{2}$. ☐

PRACTICE
6 Solve $p^2 - 3p + 1 = 0$ by completing the square.

EXAMPLE 7 Solve: $2x^2 - 8x + 3 = 0$.

Solution Our procedure for finding the constant term to complete the square works only if the coefficient of the squared variable term is 1. Therefore, to solve this equation, the first step is to divide both sides by 2, the coefficient of x^2.

$$2x^2 - 8x + 3 = 0$$

$$x^2 - 4x + \frac{3}{2} = 0 \qquad \text{Divide both sides by 2.}$$

$$x^2 - 4x = -\frac{3}{2} \quad \text{Subtract } \frac{3}{2} \text{ from both sides.}$$

Next find the square of half of -4.

$$\frac{1}{2}(-4) = -2 \quad \text{and} \quad (-2)^2 = 4$$

Add 4 to both sides of the equation to complete the square.

$$x^2 - 4x + 4 = -\frac{3}{2} + 4$$

$$(x - 2)^2 = \frac{5}{2} \qquad \text{Factor the perfect square and simplify the right side.}$$

$$x - 2 = \pm\sqrt{\frac{5}{2}} \qquad \text{Apply the square root property.}$$

$$x - 2 = \pm\frac{\sqrt{10}}{2} \qquad \text{Rationalize the denominator.}$$

$$x = 2 \pm \frac{\sqrt{10}}{2} \qquad \text{Add 2 to both sides.}$$

$$= \frac{4}{2} \pm \frac{\sqrt{10}}{2} \qquad \text{Find the common denominator.}$$

$$= \frac{4 \pm \sqrt{10}}{2} \qquad \text{Simplify.}$$

The solutions are $\dfrac{4 + \sqrt{10}}{2}$ and $\dfrac{4 - \sqrt{10}}{2}$. □

PRACTICE
7 Solve: $3x^2 - 12x + 1 = 0$.

The following steps may be used to solve a quadratic equation such as $ax^2 + bx + c = 0$ by completing the square. This method may be used whether or not the polynomial $ax^2 + bx + c$ is factorable.

Solving a Quadratic Equation in x by Completing the Square

STEP 1. If the coefficient of x^2 is 1, go to Step 2. Otherwise, divide both sides of the equation by the coefficient of x^2.

STEP 2. Isolate all variable terms on one side of the equation.

STEP 3. Complete the square for the resulting binomial by adding the square of half of the coefficient of x to both sides of the equation.

STEP 4. Factor the resulting perfect square trinomial and write it as the square of a binomial.

STEP 5. Use the square root property to solve for x.

EXAMPLE 8   Solve $3x^2 - 9x + 8 = 0$ by completing the square.

Solution $3x^2 - 9x + 8 = 0$

STEP 1. $x^2 - 3x + \dfrac{8}{3} = 0$ Divide both sides of the equation by 3.

STEP 2. $x^2 - 3x = -\dfrac{8}{3}$ Subtract $\dfrac{8}{3}$ from both sides.

Since $\dfrac{1}{2}(-3) = -\dfrac{3}{2}$ and $\left(-\dfrac{3}{2}\right)^2 = \dfrac{9}{4}$, we add $\dfrac{9}{4}$ to both sides of the equation.

STEP 3. $x^2 - 3x + \dfrac{9}{4} = -\dfrac{8}{3} + \dfrac{9}{4}$

STEP 4. $\left(x - \dfrac{3}{2}\right)^2 = -\dfrac{5}{12}$ Factor the perfect square trinomial.

STEP 5. $x - \dfrac{3}{2} = \pm\sqrt{-\dfrac{5}{12}}$ Apply the square root property.

$x - \dfrac{3}{2} = \pm\dfrac{i\sqrt{5}}{2\sqrt{3}}$ Simplify the radical.

$x - \dfrac{3}{2} = \pm\dfrac{i\sqrt{15}}{6}$ Rationalize the denominator.

$x = \dfrac{3}{2} \pm \dfrac{i\sqrt{15}}{6}$ Add $\dfrac{3}{2}$ to both sides.

$= \dfrac{9}{6} \pm \dfrac{i\sqrt{15}}{6}$ Find a common denominator.

$= \dfrac{9 \pm i\sqrt{15}}{6}$ Simplify.

The solutions are $\dfrac{9 + i\sqrt{15}}{6}$ and $\dfrac{9 - i\sqrt{15}}{6}$.   ☐

PRACTICE

8   Solve $2x^2 - 5x + 7 = 0$ by completing the square.

OBJECTIVE 3 ▶ Solving problems modeled by quadratic equations. Recall the **simple interest** formula $I = Prt$, where I is the interest earned, P is the principal, r is the rate of interest, and t is time in years. If \$100 is invested at a simple interest rate of 5% annually, at the end of 3 years the total interest I earned is

$$I = P \cdot r \cdot t$$

or

$$I = 100 \cdot 0.05 \cdot 3 = \$15$$

and the new principal is

$$\$100 + \$15 = \$115$$

Most of the time, the interest computed on money borrowed or money deposited is **compound interest.** Compound interest, unlike simple interest, is computed on original principal *and* on interest already earned. To see the difference between simple interest and compound interest, suppose that \$100 is invested at a rate of 5% compounded annually. To find the total amount of money at the end of 3 years, we calculate as follows.

$$I = P \cdot r \cdot t$$

First year: Interest $= \$100 \cdot 0.05 \cdot 1 = \$5.00$
New principal $= \$100.00 + \$5.00 = \$105.00$

Second year: Interest = $105.00 · 0.05 · 1 = $5.25
New principal = $105.00 + $5.25 = $110.25

Third year: Interest = $110.25 · 0.05 · 1 ≈ $5.51
New principal = $110.25 + $5.51 = $115.76

At the end of the third year, the total compound interest earned is $15.76, whereas the total simple interest earned is $15.

It is tedious to calculate compound interest as we did above, so we use a compound interest formula. The formula for calculating the total amount of money when interest is compounded annually is

$$A = P(1 + r)^t$$

where P is the original investment, r is the interest rate per compounding period, and t is the number of periods. For example, the amount of money A at the end of 3 years if $100 is invested at 5% compounded annually is

$$A = \$100(1 + 0.05)^3 \approx \$100(1.1576) = \$115.76$$

as we previously calculated.

EXAMPLE 9   Finding Interest Rates

Use the formula $A = P(1 + r)^t$ to find the interest rate r if $2000 compounded annually grows to $2420 in 2 years.

Solution

1. UNDERSTAND the problem. Since the $2000 is compounded annually, we use the compound interest formula. For this example, make sure that you understand the formula for compounding interest annually.

2. TRANSLATE. We substitute the given values into the formula.

$$A = P(1 + r)^t$$
$$2420 = 2000(1 + r)^2 \qquad \text{Let } A = 2420, P = 2000, \text{ and } t = 2.$$

3. SOLVE. Solve the equation for r. We will solve algebraically and graphically.

Algebraic Solution:

$$2420 = 2000(1 + r)^2$$

$$\frac{2420}{2000} = (1 + r)^2 \qquad \text{Divide both sides by 2000.}$$

$$\frac{121}{100} = (1 + r)^2 \qquad \text{Simplify the fraction.}$$

$$\pm\sqrt{\frac{121}{100}} = 1 + r \qquad \text{Use the square root property.}$$

$$\pm\frac{11}{10} = 1 + r \qquad \text{Simplify.}$$

$$-1 \pm \frac{11}{10} = r$$

$$-\frac{10}{10} \pm \frac{11}{10} = r$$

$$\frac{1}{10} = r \quad \text{or} \quad -\frac{21}{10} = r$$

Graphical Solution:

Enter $y_1 = 2000(1 + x)^2$ and $y_2 = 2420$. You may need to experiment a little before you find an appropriate window, but keep the following in mind. Since $y_2 = 2420$, use a window for the y-axis that includes this number. Since x represents the rate, use a much smaller window for the x-axis. The viewing window for the graph below is $[-2.5, 1, 0.1]$ for x and $[2000, 3000, 200]$ for y.

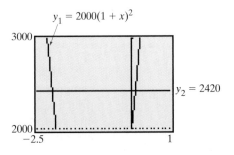

The two points of intersection have x-coordinates of -2.1 and 0.1.

4. INTERPRET. The rate cannot be negative, so we reject $-\dfrac{21}{10}$.

Check: $\dfrac{1}{10} = 0.10 = 10\%$ per year. If we invest \$2000 at 10% compounded annually, in 2 years the amount in the account would be $2000(1 + 0.10)^2 = 2420$ dollars, the desired amount.

State: The interest rate is 10% compounded annually.

PRACTICE
9   Use the formula from Example 9 to find the interest rate r if \$5000 compounded annually grows to \$5618 in 2 years.

VOCABULARY & READINESS CHECK

Use the choices below to fill in each blank. Not all choices will be used.

| binomial | $\sqrt{b}$ | $\pm\sqrt{b}$ | b^2 | 9 | 25 | completing the square |
| quadratic | $-\sqrt{b}$ | $\dfrac{b}{2}$ | $\left(\dfrac{b}{2}\right)^2$ | 3 | 5 | |

1. By the square root property, if b is a real number, and $a^2 = b$, then $a = $ _____.

2. A _____ equation can be written in the form $ax^2 + bx + c = 0, a \neq 0$.

3. The process of writing a quadratic equation so that one side is a perfect square trinomial is called _____.

4. A perfect square trinomial is one that can be factored as a _____ squared.

5. To solve $x^2 + 6x = 10$ by completing the square, add _____ to both sides.

6. To solve $x^2 + bx = c$ by completing the square, add _____ to both sides.

Fill in the blank with the number needed to make the expression a perfect square trinomial.

7. $m^2 + 2m + $ _____
8. $m^2 - 2m + $ _____
9. $y^2 - 14y + $ _____
10. $z^2 + z + $ _____

8.1 EXERCISE SET

MyMathLab   PRACTICE   WATCH   DOWNLOAD   READ   REVIEW

Use the square root property to solve each equation. These equations have real number solutions. See Examples 1 through 3.

1. $x^2 = 16$
2. $x^2 = 49$

3. $x^2 - 7 = 0$
4. $x^2 - 11 = 0$

5. $x^2 = 18$
6. $y^2 = 20$

7. $3z^2 - 30 = 0$
8. $2x^2 - 4 = 0$

9. $(x + 5)^2 = 9$
10. $(y - 3)^2 = 4$

11. $(z - 6)^2 = 18$
12. $(y + 4)^2 = 27$

13. $(2x - 3)^2 = 8$
14. $(4x + 9)^2 = 6$

Use the square root property to solve each equation. See Examples 1 through 4.

15. $x^2 + 9 = 0$
16. $x^2 + 4 = 0$

17. $x^2 - 6 = 0$
18. $y^2 - 10 = 0$

19. $2z^2 + 16 = 0$

20. $3p^2 + 36 = 0$

21. $(x - 1)^2 = -16$

22. $(y + 2)^2 = -25$

23. $(z + 7)^2 = 5$

24. $(x + 10)^2 = 11$

25. $(x + 3)^2 = -8$

26. $(y - 4)^2 = -18$

Add the proper constant to each binomial so that the resulting trinomial is a perfect square trinomial. Then factor the trinomial.

27. $x^2 + 16x + $ _____
28. $y^2 + 2y + $ _____

29. $z^2 - 12z + $ _____
30. $x^2 - 8x + $ _____

31. $p^2 + 9p +$ _____

32. $n^2 + 5n +$ _____

33. $x^2 + x +$ _____

34. $y^2 - y +$ _____

MIXED PRACTICE

Solve each equation by completing the square. These equations have real number solutions. See Examples 5 through 7.

35. $x^2 + 8x = -15$

36. $y^2 + 6y = -8$

37. $x^2 + 6x + 2 = 0$

38. $x^2 - 2x - 2 = 0$

39. $x^2 + x - 1 = 0$

40. $x^2 + 3x - 2 = 0$

41. $x^2 + 2x - 5 = 0$

42. $y^2 + y - 7 = 0$

43. $3p^2 - 12p + 2 = 0$

44. $2x^2 + 14x - 1 = 0$

45. $4y^2 - 12y - 2 = 0$

46. $6x^2 - 3 = 6x$

47. $2x^2 + 7x = 4$

48. $3x^2 - 4x = 4$

49. $x^2 - 4x - 5 = 0$

50. $y^2 + 6y - 8 = 0$

51. $x^2 + 8x + 1 = 0$

52. $x^2 - 10x + 2 = 0$

53. $3y^2 + 6y - 4 = 0$

54. $2y^2 + 12y + 3 = 0$

55. $2x^2 - 3x - 5 = 0$

56. $5x^2 + 3x - 2 = 0$

Solve each equation by completing the square. See Examples 5 through 8.

57. $y^2 + 2y + 2 = 0$

58. $x^2 + 4x + 6 = 0$

59. $x^2 - 6x + 3 = 0$

60. $x^2 - 7x - 1 = 0$

61. $2a^2 + 8a = -12$

62. $3x^2 + 12x = -14$

63. $x^2 + 10x + 28 = 0$

64. $y^2 + 8y + 18 = 0$

65. $z^2 + 3z - 4 = 0$

66. $y^2 + y - 2 = 0$

67. $2x^2 - 4x = -3$

68. $9x^2 - 36x = -40$

69. $3x^2 + 3x = 5$

70. $5y^2 - 15y = 1$

Use the graph to determine how many real number solutions exist for each equation.

71. $2x^2 - 3x - 5 = 0$
$y = 2x^2 - 3x - 5$

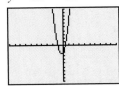

72. $5x^2 + 3x - 2 = 0$
$y = 5x^2 + 3x - 2$

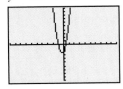

73. $x^2 + 2x + 2 = 0$
$y = x^2 + 2x + 2$

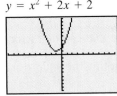

74. $x^2 + 4x + 6 = 0$
$y = x^2 + 4x + 6$

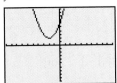

Use the formula $A = P(1 + r)^t$ to solve Exercises 75 through 78. See Example 9.

75. Find the rate r at which \$3000 compounded annually grows to \$4320 in 2 years.

76. Find the rate r at which \$800 compounded annually grows to \$882 in 2 years.

77. Find the rate at which \$15,000 compounded annually grows to \$16,224 in 2 years.

78. Find the rate at which \$2000 compounded annually grows to \$2880 in 2 years.

79. In your own words, what is the difference between simple interest and compound interest?

Neglecting air resistance, the distance $s(t)$ in feet traveled by a freely falling object is given by the function $s(t) = 16t^2$, where t is time in seconds. Use this formula to solve Exercises 80 through 83. Round answers to two decimal places.

80. The Petronas Towers in Kuala Lumpur, built in 1997, are the tallest buildings in Malaysia. Each tower is 1483 feet tall. How long would it take an object to fall to the ground from the top of one of the towers? (*Source:* Council on Tall Buildings and Urban Habitat, Lehigh University)

81. The height of the Chicago Beach Tower Hotel, built in 1998 in Dubai, United Arab Emirates, is 1053 feet. How long would it take an object to fall to the ground from the top of the building? (*Source:* Council on Tall Buildings and Urban Habitat, Lehigh University)

82. The height of the Nurek Dam in Tajikistan (part of the former USSR that borders Afghanistan) is 984 feet. How long would it take an object to fall from the top to the base of the dam? (*Source:* U.S. Committee on Large Dams of the International Commission on Large Dams)

83. The Hoover Dam, located on the Colorado River on the border of Nevada and Arizona near Las Vegas, is 725 feet tall. How long would it take an object to fall from the top to the base of the dam? (*Source:* U.S. Committee on Large Dams of the International Commission on Large Dams)

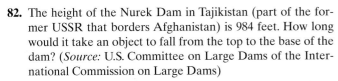

84. If you are depositing money in an account that pays 4%, would you prefer the interest to be simple or compound? Explain why.

85. If you are borrowing money at a rate of 10%, would you prefer the interest to be simple or compound? Explain why.

REVIEW AND PREVIEW

Simplify each expression. See Section 7.1.

86. $\dfrac{3}{4} - \sqrt{\dfrac{25}{16}}$

87. $\dfrac{3}{5} + \sqrt{\dfrac{16}{25}}$

88. $\dfrac{1}{2} - \sqrt{\dfrac{9}{4}}$

89. $\dfrac{9}{10} - \sqrt{\dfrac{49}{100}}$

Simplify each expression. See Section 6.1.

90. $\dfrac{6 + 4\sqrt{5}}{2}$

91. $\dfrac{10 - 20\sqrt{3}}{2}$

92. $\dfrac{3 - 9\sqrt{2}}{6}$

93. $\dfrac{12 - 8\sqrt{7}}{16}$

Evaluate $\sqrt{b^2 - 4ac}$ for each set of values. See Section 7.1.

94. $a = 2, b = 4, c = -1$

95. $a = 1, b = 6, c = 2$

96. $a = 3, b = -1, c = -2$

97. $a = 1, b = -3, c = -1$

CONCEPT EXTENSIONS

Without solving, determine whether the solutions of each equation are real numbers or complex, but not real, numbers. See the Concept Check in this section.

98. $(x + 1)^2 = -1$

99. $(y - 5)^2 = -9$

100. $3z^2 = 10$

101. $4x^2 = 17$

102. $(2y - 5)^2 + 7 = 3$

103. $(3m + 2)^2 + 4 = 1$

Find two possible missing terms so that each is a perfect square trinomial.

104. $x^2 + \underline{\quad} + 16$

105. $y^2 + \underline{\quad} + 9$

106. $z^2 + \underline{\quad} + \dfrac{25}{4}$

107. $x^2 + \underline{\quad} + \dfrac{1}{4}$

Solve.

△ **108.** The area of a square room is 225 square feet. Find the dimensions of the room.

△ **109.** The area of a circle is 36π square inches. Find the radius of the circle.

△ **110.** An isosceles right triangle has legs of equal length. If the hypotenuse is 20 centimeters long, find the length of each leg.

△ **111.** A 27-inch TV is advertised in the *Daily Sentry* newspaper. If 27 inches is the measure of the diagonal of the picture tube, find the measure of each side of the picture tube.

A common equation used in business is a demand equation. It expresses the relationship between the unit price of some commodity and the quantity demanded. For Exercises 112 and 113, p represents the unit price and x represents the quantity demanded in thousands.

112. A manufacturing company has found that the demand equation for a certain type of scissors is given by the equation $p = -x^2 + 47$. Find the demand for the scissors if the price is $11 per pair.

113. Acme, Inc., sells desk lamps and has found that the demand equation for a certain style of desk lamp is given by the equation $p = -x^2 + 15$. Find the demand for the desk lamp if the price is $7 per lamp.

8.2 SOLVING QUADRATIC EQUATIONS BY THE QUADRATIC FORMULA

OBJECTIVES

1 Solve quadratic equations by using the quadratic formula.

2 Determine the number and type of solutions of a quadratic equation by using the discriminant.

3 Solve geometric problems modeled by quadratic equations.

OBJECTIVE 1 ▶ Solving quadratic equations by using the quadratic formula. Any quadratic equation can be solved by completing the square. Since the same sequence of steps is repeated each time we complete the square, let's complete the square for a general quadratic equation, $ax^2 + bx + c = 0$, $a \neq 0$. By doing so, we find a pattern for the solutions of a quadratic equation known as the **quadratic formula.**

Recall that to complete the square for an equation such as $ax^2 + bx + c = 0$, we first divide both sides by the coefficient of x^2.

$$ax^2 + bx + c = 0$$

$$x^2 + \frac{b}{a}x + \frac{c}{a} = 0 \qquad \text{Divide both sides by } a \text{, the coefficient of } x^2.$$

$$x^2 + \frac{b}{a}x = -\frac{c}{a} \qquad \text{Subtract the constant } \frac{c}{a} \text{ from both sides.}$$

Next, find the square of half $\frac{b}{a}$, the coefficient of x.

$$\frac{1}{2}\left(\frac{b}{a}\right) = \frac{b}{2a} \quad \text{and} \quad \left(\frac{b}{2a}\right)^2 = \frac{b^2}{4a^2}$$

Add this result to both sides of the equation.

$$x^2 + \frac{b}{a}x + \frac{b^2}{4a^2} = -\frac{c}{a} + \frac{b^2}{4a^2} \qquad \text{Add } \frac{b^2}{4a^2} \text{ to both sides.}$$

$$x^2 + \frac{b}{a}x + \frac{b^2}{4a^2} = \frac{-c \cdot 4a}{a \cdot 4a} + \frac{b^2}{4a^2} \qquad \text{Find a common denominator on the right side.}$$

$$x^2 + \frac{b}{a}x + \frac{b^2}{4a^2} = \frac{b^2 - 4ac}{4a^2} \qquad \text{Simplify the right side.}$$

$$\left(x + \frac{b}{2a}\right)^2 = \frac{b^2 - 4ac}{4a^2} \qquad \text{Factor the perfect square trinomial on the left side.}$$

$$x + \frac{b}{2a} = \pm\sqrt{\frac{b^2 - 4ac}{4a^2}} \qquad \text{Apply the square root property.}$$

$$x + \frac{b}{2a} = \pm\frac{\sqrt{b^2 - 4ac}}{2a} \qquad \text{Simplify the radical.}$$

$$x = -\frac{b}{2a} \pm \frac{\sqrt{b^2 - 4ac}}{2a} \qquad \text{Subtract } \frac{b}{2a} \text{ from both sides.}$$

$$x = \frac{-b \pm \sqrt{b^2 - 4ac}}{2a} \qquad \text{Simplify.}$$

This equation identifies the solutions of the general quadratic equation in standard form and is called the quadratic formula. It can be used to solve any equation written in standard form, $ax^2 + bx + c = 0$, as long as a is not 0.

> **Quadratic Formula**
>
> A quadratic equation written in the form $ax^2 + bx + c = 0$ has the solutions
>
> $$x = \frac{-b \pm \sqrt{b^2 - 4ac}}{2a}$$

TECHNOLOGY NOTE

When evaluating the quadratic formula using a calculator, it is sometimes more convenient to evaluate the radicand separately. This prevents incorrect placement of parentheses.

EXAMPLE 1   Solve $3x^2 + 16x + 5 = 0$ for x.

Solution   This equation is in standard form, so $a = 3$, $b = 16$, and $c = 5$. Substitute these values into the quadratic formula.

$$x = \frac{-b \pm \sqrt{b^2 - 4ac}}{2a} \qquad \text{Quadratic formula}$$

$$= \frac{-16 \pm \sqrt{16^2 - 4(3)(5)}}{2 \cdot 3} \qquad \text{Use } a = 3, b = 16, \text{ and } c = 5.$$

$$= \frac{-16 \pm \sqrt{256 - 60}}{6}$$

$$= \frac{-16 \pm \sqrt{196}}{6} = \frac{-16 \pm 14}{6}$$

$$x = \frac{-16 + 14}{6} = -\frac{1}{3} \quad \text{or} \quad x = \frac{-16 - 14}{6} = -\frac{30}{6} = -5$$

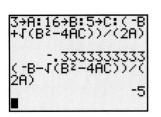

```
3→A:16→B:5→C:(-B
+√(B²-4AC))/(2A)
         -.3333333333
(-B-√(B²-4AC))/(
2A)
              -5
■
```

Checking Example 1 numerically by evaluating the formula using a calculator.

The solutions are $-\dfrac{1}{3}$ and -5, or the solution set is $\left\{-\dfrac{1}{3}, -5\right\}$. A numerical check is shown to the left. Notice the correct placement of parentheses.   □

PRACTICE
1   Solve $3x^2 - 5x - 2 = 0$ for x.

As usual, another way to check the solution of an equation is to use a graphical method such as the intersection-of-graphs method. Remember that this graphical method shows real number solutions only.

▶ **Helpful Hint**

To replace a, b, and c correctly in the quadratic formula, write the quadratic equation in standard form, $ax^2 + bx + c = 0$.

EXAMPLE 2   Solve: $2x^2 - 4x = 3$.

Solution   First write the equation in standard form by subtracting 3 from both sides.

$$2x^2 - 4x - 3 = 0$$

Now $a = 2$, $b = -4$, and $c = -3$. Substitute these values into the quadratic formula.

$$x = \frac{-b \pm \sqrt{b^2 - 4ac}}{2a}$$

$$= \frac{-(-4) \pm \sqrt{(-4)^2 - 4(2)(-3)}}{2 \cdot 2}$$

$$= \frac{4 \pm \sqrt{16 + 24}}{4}$$

$$= \frac{4 \pm \sqrt{40}}{4} = \frac{4 \pm 2\sqrt{10}}{4}$$

$$= \frac{2\left(2 \pm \sqrt{10}\right)}{2 \cdot 2} = \frac{2 \pm \sqrt{10}}{2}$$

The solutions are $\dfrac{2 + \sqrt{10}}{2}$ and $\dfrac{2 - \sqrt{10}}{2}$, or the solution set is $\left\{\dfrac{2 - \sqrt{10}}{2}, \dfrac{2 + \sqrt{10}}{2}\right\}$.

Since the solutions are real numbers, let's check using the intersection-of-graphs method. To do so, graph $y_1 = 2x^2 - 4x$ and $y_2 = 3$. Find the approximate points of intersection as shown next, and see that the exact solutions have the same approximations.

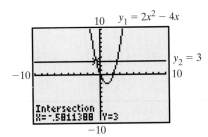
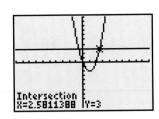
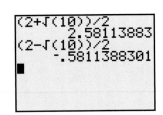

The screen to the right above shows that the exact solutions using the quadratic formula have the same approximations as those found by the intersection-of-graphs method. ☐

PRACTICE
2 Solve: $3x^2 - 8x = 2$.

▶ **Helpful Hint**

To simplify the expression $\dfrac{4 \pm 2\sqrt{10}}{4}$ in the preceding example, note that 2 is factored out of both terms of the numerator *before* simplifying.

$$\frac{4 \pm 2\sqrt{10}}{4} = \frac{2(2 \pm \sqrt{10})}{2 \cdot 2} = \frac{2 \pm \sqrt{10}}{2}$$

Concept Check ☑

For the quadratic equation $x^2 = 7$, which substitution is correct?

a. $a = 1, b = 0$, and $c = -7$ **b.** $a = 1, b = 0$, and $c = 7$

c. $a = 0, b = 0$, and $c = 7$ **d.** $a = 1, b = 1$, and $c = -7$

EXAMPLE 3 Solve: $\dfrac{1}{4} m^2 - m + \dfrac{1}{2} = 0$.

Solution We could use the quadratic formula with $a = \dfrac{1}{4}, b = -1$, and $c = \dfrac{1}{2}$. Instead, we find a simpler, equivalent standard form equation whose coefficients are not fractions. Multiply both sides of the equation by the LCD, 4, to clear fractions.

$$4\left(\frac{1}{4} m^2 - m + \frac{1}{2}\right) = 4 \cdot 0$$

$$m^2 - 4m + 2 = 0 \qquad \text{Simplify.}$$

Substitute $a = 1, b = -4$, and $c = 2$ into the quadratic formula and simplify.

$$m = \frac{-(-4) \pm \sqrt{(-4)^2 - 4(1)(2)}}{2 \cdot 1} = \frac{4 \pm \sqrt{16 - 8}}{2}$$

$$= \frac{4 \pm \sqrt{8}}{2} = \frac{4 \pm 2\sqrt{2}}{2} = \frac{2(2 \pm \sqrt{2})}{2}$$

$$= 2 \pm \sqrt{2}$$

The solutions are $2 + \sqrt{2}$ and $2 - \sqrt{2}$. ☐

PRACTICE
3 Solve: $\dfrac{1}{8}x^2 - \dfrac{1}{4}x - 2 = 0$.

Answer to Concept Check: a

EXAMPLE 4   Solve: $x = -3x^2 - 3$.

Solution   The equation in standard form is $3x^2 + x + 3 = 0$. Thus, let $a = 3$, $b = 1$, and $c = 3$ in the quadratic formula.

$$x = \frac{-1 \pm \sqrt{1^2 - 4(3)(3)}}{2 \cdot 3} = \frac{-1 \pm \sqrt{1 - 36}}{6} = \frac{-1 \pm \sqrt{-35}}{6} = \frac{-1 \pm i\sqrt{35}}{6}$$

The solutions are $\dfrac{-1 + i\sqrt{35}}{6}$ and $\dfrac{-1 - i\sqrt{35}}{6}$.

Since the solutions are not real numbers, we check numerically as shown below.

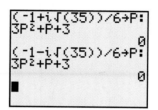

PRACTICE
4   Solve: $x = -2x^2 - 2$.

Concept Check ☑

What is the first step in solving $-3x^2 = 5x - 4$ using the quadratic formula?

In Example 1, the equation $3x^2 + 16x + 5 = 0$ had 2 real roots, $-\dfrac{1}{3}$ and -5. In Example 4, the equation $3x^2 + x + 3 = 0$ (written in standard form) had no real roots. How do their related graphs compare? Recall that the x-intercepts of $f(x) = 3x^2 + 16x + 5$ occur where $f(x) = 0$ or where $3x^2 + 16x + 5 = 0$. Since this equation has 2 real roots, the graph has 2 x-intercepts. Similarly, since the equation $3x^2 + x + 3 = 0$ has no real roots, the graph of $f(x) = 3x^2 + x + 3$ has no x-intercepts.

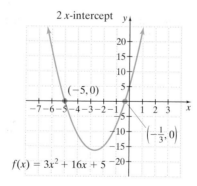

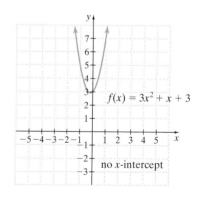

OBJECTIVE 2 ▶ Using the discriminant.   In the quadratic formula, $x = \dfrac{-b \pm \sqrt{b^2 - 4ac}}{2a}$, the radicand $b^2 - 4ac$ is called the **discriminant** because, by knowing its value, we can **discriminate** among the possible number and type of solutions of a quadratic equation. Possible values of the discriminant and their meanings are summarized next.

Answer to Concept Check:
Write the equation in standard form.

Discriminant

The following table corresponds the discriminant $b^2 - 4ac$ of a quadratic equation of the form $ax^2 + bx + c = 0$ with the number and type of solutions of the equation.

$b^2 - 4ac$	*Number and Type of Solutions*
Positive	Two real solutions
Zero	One real solution
Negative	Two complex but not real solutions

To see the results of the discriminant graphically, study the screens below showing examples of graphs of $y = ax^2 + bx + c$.

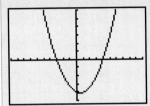 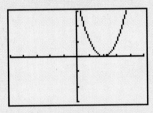 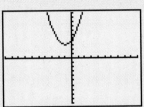

$b^2 - 4ac$ is positive $b^2 - 4ac = 0$ $b^2 - 4ac$ is negative
Two x-intercepts   One x-intercept   No x-intercept
Two real solutions   One real solution   Two complex, but not
    (double root)   real, solutions

EXAMPLE 5   Use the discriminant to determine the number and type of solutions of each quadratic equation.

a. $x^2 + 2x + 1 = 0$ **b.** $3x^2 + 2 = 0$ **c.** $2x^2 - 7x - 4 = 0$

Solution

a. In $x^2 + 2x + 1 = 0$, $a = 1$, $b = 2$, and $c = 1$. Thus,

$$b^2 - 4ac = 2^2 - 4(1)(1) = 0$$

Since $b^2 - 4ac = 0$, this quadratic equation has one real solution.

b. In this equation, $a = 3$, $b = 0$, and $c = 2$. Then $b^2 - 4ac = 0 - 4(3)(2) = -24$. Since $b^2 - 4ac$ is negative, the quadratic equation has two complex but not real solutions.

c. In this equation, $a = 2$, $b = -7$, and $c = -4$. Then

$$b^2 - 4ac = (-7)^2 - 4(2)(-4) = 81$$

Since $b^2 - 4ac$ is positive, the quadratic equation has two real solutions. To confirm graphically, see the results below.

a.

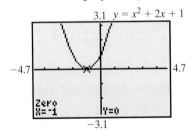

One x-intercept indicates one real solution.

b.

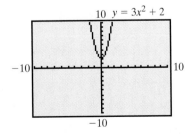

No x-intercepts indicates no real solutions, but two complex, not real, solutions.

c.

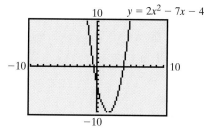

$y = 2x^2 - 7x - 4$

Two x-intercepts indicate two real
solutions.

□

PRACTICE

5   Use the discriminant to determine the number and type of solutions of each
quadratic equation.

a. $x^2 - 6x + 9 = 0$ **b.** $x^2 - 3x - 1 = 0$ **c.** $7x^2 + 11 = 0$

The discriminant helps us determine the number and type of solutions of a quadratic
equation, $ax^2 + bx + c = 0$. Recall that the solutions of this equation are the same as
the x-intercepts of its related graph, $f(x) = ax^2 + bx + c$. This means that the discriminant of $ax^2 + bx + c = 0$ also tells us the number of x-intercepts for the graph of
$f(x) = ax^2 + bx + c$, or equivalently $y = ax^2 + bx + c$.

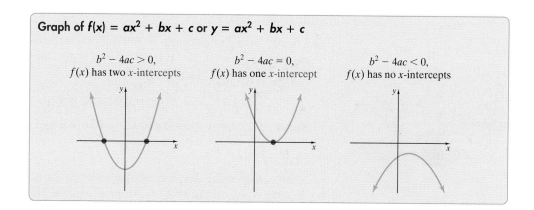

Graph of $f(x) = ax^2 + bx + c$ or $y = ax^2 + bx + c$

$b^2 - 4ac > 0$,
$f(x)$ has two x-intercepts

$b^2 - 4ac = 0$,
$f(x)$ has one x-intercept

$b^2 - 4ac < 0$,
$f(x)$ has no x-intercepts

OBJECTIVE 3 ▶ Solving problems modeled by quadratic equations. The quadratic
formula is useful in solving problems that are modeled by quadratic equations.

△   **EXAMPLE 6**   **Calculating Distance Saved**

At a local university, students often leave
the sidewalk and cut across the lawn to
save walking distance. Given the diagram of a favorite place to cut across the
lawn, approximate how many feet of
walking distance a student saves by cutting across the lawn instead of walking
on the sidewalk.

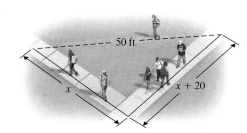

50 ft

x $x + 20$

Solution

1. UNDERSTAND. Read and reread the problem. In the diagram, notice that a triangle is formed. Since the corner of the block forms a right angle, we use the
   Pythagorean theorem for right triangles. You may want to review this theorem.

2. TRANSLATE. By the Pythagorean theorem, we have

In words: $(\text{leg})^2 + (\text{leg})^2 = (\text{hypotenuse})^2$

Translate: $x^2 + (x + 20)^2 = 50^2$

3. SOLVE.

Algebraic Solution:

$$x^2 + x^2 + 40x + 400 = 2500 \quad \text{Square } (x + 20) \text{ and } 50.$$

$$2x^2 + 40x - 2100 = 0 \quad \text{Set the equation equal to } 0.$$

$$x^2 + 20x - 1050 = 0 \quad \text{Divide by } 2.$$

Here, $a = 1, b = 20, c = -1050$. By the quadratic formula,

$$x = \frac{-20 \pm \sqrt{20^2 - 4(1)(-1050)}}{2 \cdot 1}$$

$$= \frac{-20 \pm \sqrt{400 + 4200}}{2} = \frac{-20 \pm \sqrt{4600}}{2}$$

$$= \frac{-20 \pm \sqrt{100 \cdot 46}}{2} = \frac{-20 \pm 10\sqrt{46}}{2}$$

$$= -10 \pm 5\sqrt{46} \quad \text{Simplify.}$$

Graphical Solution:

Let $y_1 = x^2 + (x + 20)^2$ and $y_2 = 50^2$. Since x represents feet, we are interested in positive solutions only. We also estimate the length labeled x to be less than 50, so we choose the x-window to be $[0, 50, 10]$. The y-window is chosen to be $[2000, 3000, 100]$ so it contains 2500. Find $x \approx -44$ and $x \approx 24$.

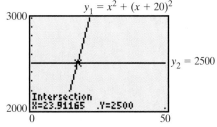

The positive solution is approximately 24.

---

4. INTERPRET.

Check: Your calculations in the quadratic formula. The length of a side of a triangle can't be negative, so we reject $-10 - 5\sqrt{46}$. Since $-10 + 5\sqrt{46} \approx 24$ feet, the walking distance along the sidewalk is

$$x + (x + 20) \approx 24 + (24 + 20) = 68 \text{ feet.}$$

State: A student saves about $68 - 50$ or 18 feet of walking distance by cutting across the lawn. ☐

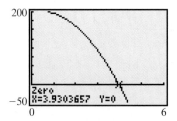

PRACTICE

6 Given the diagram, approximate to the nearest foot how many feet of walking distance a person can save by cutting across the lawn instead of walking on the sidewalk.

EXAMPLE 7 Calculating Landing Time

An object is thrown upward from the top of a 200-foot cliff with a velocity of 12 feet per second. The height h in feet of the object after t seconds is

$$h = -16t^2 + 12t + 200$$

How long after the object is thrown will it strike the ground? Round to the nearest tenth of a second.

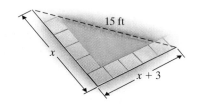

A graphical solution or check for Example 7. The graph of $y_1 = -16x^2 + 12x + 200$ is shown. The solution is the x-value of the x-intercept, as shown.

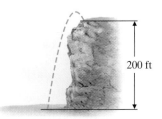

Solution

1. UNDERSTAND. Read and reread the problem.

2. TRANSLATE. Since we want to know when the object strikes the ground, we want to know when the height $h = 0$, or

$$0 = -16t^2 + 12t + 200$$

3. SOLVE. First we divide both sides of the equation by -4.

$$0 = 4t^2 - 3t - 50 \quad \text{Divide both sides by } -4.$$

Here, $a = 4$, $b = -3$, and $c = -50$. By the quadratic formula,

$$t = \frac{-(-3) \pm \sqrt{(-3)^2 - 4(4)(-50)}}{2 \cdot 4}$$

$$= \frac{3 \pm \sqrt{9 + 800}}{8}$$

$$= \frac{3 \pm \sqrt{809}}{8}$$

4. INTERPRET.

Check: We check our calculations from the quadratic formula. Since the time won't be negative, we reject the proposed solution

$$\frac{3 - \sqrt{809}}{8}.$$

State: The time it takes for the object to strike the ground is exactly

$$\frac{3 + \sqrt{809}}{8} \text{ seconds} \approx 3.9 \text{ seconds.} \qquad \square$$

PRACTICE

7 A toy rocket is shot upward at the edge of a building, 45 feet high, with an initial velocity of 20 feet per second. The height h in feet of the rocket after t seconds is

$$h = -16t^2 + 20t + 45$$

How long after the rocket is launched will it strike the ground? Round to the nearest tenth of a second.

VOCABULARY & READINESS CHECK

Fill in each blank.

1. The quadratic formula is _____ .

2. For $2x^2 + x + 1 = 0$, if $a = 2$, then $b =$ _____ and $c =$ _____ .

3. For $5x^2 - 5x - 7 = 0$, if $a = 5$, then $b =$ _____ and $c =$ _____ .

4. For $7x^2 - 4 = 0$, if $a = 7$, then $b =$ _____ and $c =$ _____ .

5. For $x^2 + 9 = 0$, if $c = 9$, then $a =$ _____ and $b =$ _____ .

6. The correct simplified form of $\dfrac{5 \pm 10\sqrt{2}}{5}$ is _____ .

   a. $1 \pm 10\sqrt{2}$ **b.** $2\sqrt{2}$ **c.** $1 \pm 2\sqrt{2}$ **d.** $\pm 5\sqrt{2}$

8.2 | EXERCISE SET

Use the quadratic formula to solve each equation. These equations have real number solutions only. See Examples 1 through 3.

1. $m^2 + 5m - 6 = 0$

2. $p^2 + 11p - 12 = 0$

3. $2y = 5y^2 - 3$

4. $5x^2 - 3 = 14x$

5. $x^2 - 6x + 9 = 0$

6. $y^2 + 10y + 25 = 0$

7. $x^2 + 7x + 4 = 0$

8. $y^2 + 5y + 3 = 0$

9. $8m^2 - 2m = 7$

10. $11n^2 - 9n = 1$

11. $3m^2 - 7m = 3$

12. $x^2 - 13 = 5x$

13. $\frac{1}{2}x^2 - x - 1 = 0$

14. $\frac{1}{6}x^2 + x + \frac{1}{3} = 0$

15. $\frac{2}{5}y^2 + \frac{1}{5}y = \frac{3}{5}$

16. $\frac{1}{8}x^2 + x = \frac{5}{2}$

17. $\frac{1}{3}y^2 = y + \frac{1}{6}$

18. $\frac{1}{2}y^2 = y + \frac{1}{2}$

19. $x^2 + 5x = -2$

20. $y^2 - 8 = 4y$

21. $(m + 2)(2m - 6) = 5(m - 1) - 12$

22. $7p(p - 2) + 2(p + 4) = 3$

MIXED PRACTICE

Use the quadratic formula to solve each equation. These equations have real solutions and complex, but not real, solutions. See Examples 1 through 4.

23. $x^2 + 6x + 13 = 0$

24. $x^2 + 2x + 2 = 0$

25. $(x + 5)(x - 1) = 2$

26. $x(x + 6) = 2$

27. $6 = -4x^2 + 3x$

28. $2 = -9x^2 - x$

29. $\frac{x^2}{3} - x = \frac{5}{3}$

30. $\frac{x^2}{2} - 3 = -\frac{9}{2}x$

31. $10y^2 + 10y + 3 = 0$

32. $3y^2 + 6y + 5 = 0$

33. $x(6x + 2) = 3$

34. $x(7x + 1) = 2$

35. $\frac{2}{5}y^2 + \frac{1}{5}y + \frac{3}{5} = 0$

36. $\frac{1}{8}x^2 + x + \frac{5}{2} = 0$

37. $\frac{1}{2}y^2 = y - \frac{1}{2}$

38. $\frac{2}{3}x^2 - \frac{20}{3}x = -\frac{100}{6}$

39. $(n - 2)^2 = 2n$

40. $\left(p - \frac{1}{2}\right)^2 = \frac{p}{2}$

Use the discriminant to determine the number and types of solutions of each equation. See Example 5.

41. $x^2 - 5 = 0$

42. $x^2 - 7 = 0$

43. $4x^2 + 12x = -9$

44. $9x^2 + 1 = 6x$

45. $3x = -2x^2 + 7$

46. $3x^2 = 5 - 7x$

47. $6 = 4x - 5x^2$

48. $5 - 4x + 12x^2 = 0$

Given the following graphs of quadratic equations of the form $y = f(x)$, find the number of real number solutions of the related equation $f(x) = 0$.

49. a.    **b.**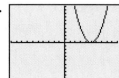

50. a. 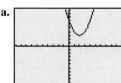   **b.**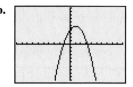

Solve. See Examples 6 and 7.

51. Nancy, Thelma, and John Varner live on a corner lot. Often, neighborhood children cut across their lot to save walking distance. Given the diagram below, approximate to the nearest foot how many feet of walking distance is saved by cutting across their property instead of walking around the lot.

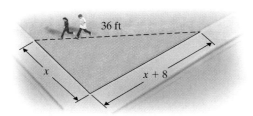

36 ft

x $x + 8$

52. Given the diagram below, approximate to the nearest foot how many feet of walking distance a person saves by cutting across the lawn instead of walking on the sidewalk.

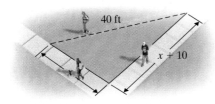

40 ft

$x + 10$

x

53. The hypotenuse of an isosceles right triangle is 2 centimeters longer than either of its legs. Find the exact length of each side. (*Hint:* An isosceles right triangle is a right triangle whose legs are the same length.)

54. The hypotenuse of an isosceles right triangle is one meter longer than either of its legs. Find the length of each side.

55. Bailey's rectangular dog pen for his Irish setter must have an area of 400 square feet. Also, the length must be 10 feet longer than the width. Find the dimensions of the pen.

?   ?

56. An entry in the Peach Festival Poster Contest must be rectangular and have an area of 1200 square inches. Furthermore, its length must be 20 inches longer than its width. Find the dimensions each entry must have.

57. A holding pen for cattle must be square and have a diagonal length of 100 meters.

  a. Find the length of a side of the pen.

  b. Find the area of the pen.

58. A rectangle is three times longer than it is wide. It has a diagonal of length 50 centimeters.

  a. Find the dimensions of the rectangle.

  b. Find the perimeter of the rectangle.

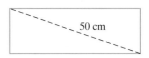

50 cm

59. The heaviest reported door in the world is the 708.6 ton radiation shield door in the National Institute for Fusion Science at Toki, Japan. If the height of the door is 1.1 feet longer than its width, and its front area (neglecting depth) is 1439.9 square feet, find its width and height [Interesting note: the door is 6.6 feet thick.] (*Source: Guinness World Records*)

60. Christi and Robbie Wegmann are constructing a rectangular stained glass window whose length is 7.3 inches longer than its width. If the area of the window is 569.9 square inches, find its width and length.

61. The base of a triangle is four more than twice its height. If the area of the triangle is 42 square centimeters, find its base and height.

x

62. If a point B divides a line segment such that the smaller portion is to the larger portion as the larger is to the whole, the whole is the length of the *golden ratio*.

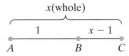

x(whole)

1      $x - 1$

A B C

The golden ratio was thought by the Greeks to be the most pleasing to the eye, and many of their buildings contained numerous examples of the golden ratio. The value of the golden ratio is the positive solution of

(smaller) $\dfrac{x - 1}{1} = \dfrac{1}{x}$ (larger)

(larger)   (whole)

Find this value.

The Wollomombi Falls in Australia have a height of 1100 feet. A pebble is thrown upward from the top of the falls with an initial velocity of 20 feet per second. The height of the pebble h after t seconds is given by the equation $h = -16t^2 + 20t + 1100$. Use this equation for Exercises 63 and 64.

63. How long after the pebble is thrown will it hit the ground? Round to the nearest tenth of a second.

64. How long after the pebble is thrown will it be 550 feet from the ground? Round to the nearest tenth of a second.

A ball is thrown downward from the top of a 180-foot building with an initial velocity of 20 feet per second. The height of the ball h after t seconds is given by the equation $h = -16t^2 - 20t + 180$. Use this equation to answer Exercises 65 and 66.

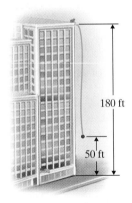

180 ft

50 ft

65. How long after the ball is thrown will it strike the ground? Round the result to the nearest tenth of a second.

66. How long after the ball is thrown will it be 50 feet from the ground? Round the result to the nearest tenth of a second.

REVIEW AND PREVIEW

Solve each equation. See Sections 6.5 and 7.6.

67. $\sqrt{5x - 2} = 3$

68. $\sqrt{y + 2} + 7 = 12$

69. $\dfrac{1}{x} + \dfrac{2}{5} = \dfrac{7}{x}$

70. $\dfrac{10}{z} = \dfrac{5}{z} - \dfrac{1}{3}$

Factor. See Section 5.7.

71. $x^4 + x^2 - 20$

72. $2y^4 + 11y^2 - 6$

73. $z^4 - 13z^2 + 36$

74. $x^4 - 1$

CONCEPT EXTENSIONS

For each quadratic equation, choose the correct substitution for a, b, and c in the standard form $ax^2 + bx + c = 0$.

75. $x^2 = -10$
  a. $a = 1, b = 0, c = -10$
  b. $a = 1, b = 0, c = 10$
  c. $a = 0, b = 1, c = -10$
  d. $a = 1, b = 1, c = 10$

76. $x^2 + 5 = -x$
  a. $a = 1, b = 5, c = -1$
  b. $a = 1, b = -1, c = 5$
  c. $a = 1, b = 5, c = 1$
  d. $a = 1, b = 1, c = 5$

77. Solve Exercise 1 by factoring. Explain the result.

78. Solve Exercise 2 by factoring. Explain the result.

Use the quadratic formula and a calculator to approximate each solution to the nearest tenth.

79. $2x^2 - 6x + 3 = 0$

80. $3.6x^2 + 1.8x - 4.3 = 0$

The accompanying graph shows the daily low temperatures for one week in New Orleans, Louisiana.

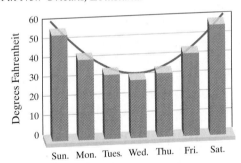

81. Which day of the week shows the greatest decrease in low temperature?

82. Which day of the week shows the greatest increase in low temperature?

83. Which day of the week had the lowest low temperature?

84. Use the graph to estimate the low temperature on Thursday.

Notice that the shape of the temperature graph is similar to the curve drawn. In fact, this graph can be modeled by the quadratic function $f(x) = 3x^2 - 18x + 56$, where $f(x)$ is the temperature in degrees Fahrenheit and x is the number of days from Sunday. (This graph is shown in blue.) Use this function to answer Exercises 85 and 86.

85. Use the quadratic function given to approximate the temperature on Thursday. Does your answer agree with the graph?

86. Use the function given and the quadratic formula to find when the temperature was 35° F. [*Hint:* Let $f(x) = 35$ and solve for x.] Round your answer to one decimal place and interpret your result. Does your answer agree with the graph?

87. The number of Starbucks stores can be modeled by the quadratic function $f(x) = 115x^2 + 711x + 3946$, where $f(x)$ is the number of Starbucks and x is the number of years after 2000. (*Source: Starbuck's Annual Report 2006*)

   a. Find the number of Starbucks in 2004.

   b. If the trend described by the model continues, predict the year after 2000 in which the number of Starbucks will be 25,000. Round to the nearest whole year.

88. The number of visitors to U.S. theme parks can be modeled by the quadratic equation $v(x) = 0.25x^2 + 2.6x + 315.6$, where $v(x)$ is the number of visitors (in millions) and x is the number of years after 2000. (*Source:* Price Waterhouse Coopers)

   a. Find the number of visitors to U.S. theme parks in 2005. Round to the nearest million.

   b. Find the projected number of visitors to U.S. theme parks in 2010. Round to the nearest million.

The solutions of the quadratic equation $ax^2 + bx + c = 0$ are
$$\frac{-b + \sqrt{b^2 - 4ac}}{2a} \text{ and } \frac{-b - \sqrt{b^2 - 4ac}}{2a}.$$

89. Show that the sum of these solutions is $\dfrac{-b}{a}$.

90. Show that the product of these solutions is $\dfrac{c}{a}$.

Use the quadratic formula to solve each quadratic equation.

91. $3x^2 - \sqrt{12}x + 1 = 0$
   (*Hint:* $a = 3, b = -\sqrt{12}, c = 1$)

92. $5x^2 + \sqrt{20}x + 1 = 0$

93. $x^2 + \sqrt{2}x + 1 = 0$

94. $x^2 - \sqrt{2}x + 1 = 0$

95. $2x^2 - \sqrt{3}x - 1 = 0$

96. $7x^2 + \sqrt{7}x - 2 = 0$

📖 STUDY SKILLS BUILDER

How Well Do You Know Your Textbook?

Let's check to see whether you are familiar with your textbook yet. For help, see Section 1.1 in this text.

1. What does the 🖥 icon mean?

2. What does the ⬌ icon mean?

3. What does the △ icon mean?

4. Where can you find a review for each chapter? What answers to this review can be found in the back of your text?

5. Each chapter contains an overview of the chapter along with examples. What is this feature called?

6. Each chapter contains a review of vocabulary. What is this feature called?

7. There are free CDs in your text. What content is contained on these CDs?

8. What is the location of the section that is entirely devoted to study skills?

9. There are Practice Problems that are contained in the margin of the text. What are they and how can they be used?

8.3 SOLVING EQUATIONS BY USING QUADRATIC METHODS

OBJECTIVES

1 Solve various equations that are quadratic in form.

2 Solve problems that lead to quadratic equations.

OBJECTIVE 1 ▶ Solving equations that are quadratic in form. In this section, we discuss various types of equations that can be solved in part by using the methods for solving quadratic equations.

   Once each equation is simplified, you may want to use these steps when deciding what method to use to solve the quadratic equation.

> **Solving a Quadratic Equation**
>
> **STEP 1.** If the equation is in the form $(ax + b)^2 = c$, use the square root property and solve. If not, go to Step 2.
>
> **STEP 2.** Write the equation in standard form: $ax^2 + bx + c = 0$.
>
> **STEP 3.** Try to solve the equation by the factoring method. If not possible, go to Step 4.
>
> **STEP 4.** Solve the equation by the quadratic formula.

The first example is a radical equation that becomes a quadratic equation once we square both sides.

EXAMPLE 1 Solve: $x - \sqrt{x} - 6 = 0$.

Solution Recall that to solve a radical equation, first get the radical alone on one side of the equation. Then square both sides.

$$x - 6 = \sqrt{x} \qquad \text{Add } \sqrt{x} \text{ to both sides.}$$
$$(x - 6)^2 = \left(\sqrt{x}\right)^2 \qquad \text{Square both sides.}$$
$$x^2 - 12x + 36 = x$$
$$x^2 - 13x + 36 = 0 \qquad \text{Set the equation equal to 0.}$$
$$(x - 9)(x - 4) = 0$$
$$x - 9 = 0 \quad \text{or} \quad x - 4 = 0$$
$$x = 9 \qquad\qquad x = 4$$

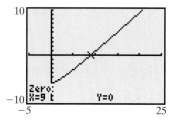

A graphical check for Example 1 shows that there is only one solution to the equation.

Check:

Let $x = 9$	Let $x = 4$
$x - \sqrt{x} - 6 = 0$	$x - \sqrt{x} - 6 = 0$
$9 - \sqrt{9} - 6 \stackrel{?}{=} 0$	$4 - \sqrt{4} - 6 \stackrel{?}{=} 0$
$9 - 3 - 6 \stackrel{?}{=} 0$	$4 - 2 - 6 \stackrel{?}{=} 0$
$0 = 0$ True	$-4 = 0$ False

The solution is 9 or the solution set is {9}. □

PRACTICE
1 Solve: $x - \sqrt{x + 1} - 5 = 0$.

EXAMPLE 2 Solve: $\dfrac{3x}{x - 2} - \dfrac{x + 1}{x} = \dfrac{6}{x(x - 2)}$.

Solution In this equation, x cannot be either 2 or 0, because these values cause denominators to equal zero. To solve for x, we first multiply both sides of the equation by $x(x - 2)$ to clear the fractions. By the distributive property, this means that we multiply each term by $x(x - 2)$.

$$x(x - 2)\left(\frac{3x}{x - 2}\right) - x(x - 2)\left(\frac{x + 1}{x}\right) = x(x - 2)\left[\frac{6}{x(x - 2)}\right]$$
$$3x^2 - (x - 2)(x + 1) = 6 \qquad \text{Simplify.}$$
$$3x^2 - (x^2 - x - 2) = 6 \qquad \text{Multiply.}$$
$$3x^2 - x^2 + x + 2 = 6$$
$$2x^2 + x - 4 = 0 \qquad \text{Simplify.}$$

This equation cannot be factored using integers, so we solve by the quadratic formula.

$$x = \frac{-1 \pm \sqrt{1^2 - 4(2)(-4)}}{2 \cdot 2} \qquad \begin{array}{l}\text{Use } a = 2, b = 1, \text{ and } c = -4 \\ \text{in the quadratic formula.}\end{array}$$
$$= \frac{-1 \pm \sqrt{1 + 32}}{4} \qquad \text{Simplify.}$$
$$= \frac{-1 \pm \sqrt{33}}{4}$$

Neither proposed solution will make the denominators 0.

The solutions are $\dfrac{-1 + \sqrt{33}}{4}$ and $\dfrac{-1 - \sqrt{33}}{4}$ or the solution set is $\left\{\dfrac{-1 + \sqrt{33}}{4}, \dfrac{-1 - \sqrt{33}}{4}\right\}$.

To check, we evaluate the left and right sides of the equation with the proposed solutions and see that a true statement results. Careful placement of parentheses is important, as shown below.

PRACTICE
2 Solve: $\dfrac{5x}{x + 1} - \dfrac{x + 4}{x} = \dfrac{3}{x(x + 1)}$.

EXAMPLE 3   Solve: $p^4 - 3p^2 - 4 = 0$.

Solution   First we factor the trinomial.

$$p^4 - 3p^2 - 4 = 0$$
$$(p^2 - 4)(p^2 + 1) = 0 \qquad \text{Factor.}$$
$$(p - 2)(p + 2)(p^2 + 1) = 0 \qquad \text{Factor further.}$$
$$p - 2 = 0 \quad \text{or} \quad p + 2 = 0 \quad \text{or} \quad p^2 + 1 = 0 \qquad \text{Set each factor equal to 0 and solve.}$$
$$p = 2 \qquad\qquad p = -2 \qquad\qquad p^2 = -1$$
$$p = \pm\sqrt{-1} = \pm i$$

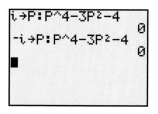

Example 3 numerical check for the imaginary solutions.

The solutions are $2, -2, i,$ and $-i$.

PRACTICE
3   Solve: $p^4 - 7p^2 - 144 = 0$.

> ▶ **Helpful Hint**
>
> Example 3 can be solved using substitution also. Think of $p^4 - 3p^2 - 4 = 0$ as
>
> $$(p^2)^2 - 3p^2 - 4 = 0 \qquad \text{Then let } x = p^2, \text{ and solve and substitute back.}$$
> $$\qquad\qquad\qquad\qquad \text{The solutions will be the same.}$$
> $$x^2 - 3x - 4 = 0$$

Concept Check ☑

a. True or false? The maximum number of solutions that a quadratic equation can have is 2.

b. True or false? The maximum number of solutions that an equation in quadratic form can have is 2.

EXAMPLE 4   Solve: $(x - 3)^2 - 3(x - 3) - 4 = 0$.

Solution   Notice that the quantity $(x - 3)$ is repeated in this equation. Sometimes it is helpful to substitute a variable (in this case other than x) for the repeated quantity. We will let $y = x - 3$. Then

$$(x - 3)^2 - 3(x - 3) - 4 = 0$$

becomes

$$y^2 - 3y - 4 = 0 \qquad \text{Let } x - 3 = y.$$
$$(y - 4)(y + 1) = 0 \qquad \text{Factor.}$$

To solve, we use the zero-factor property.

$$y - 4 = 0 \quad \text{or} \quad y + 1 = 0 \qquad \text{Set each factor equal to 0.}$$
$$y = 4 \qquad\qquad y = -1 \qquad \text{Solve.}$$

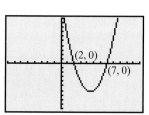

A graphical check for Example 4. The solutions are found from the x-intercepts of the graph of $y_1 = (x - 3)^2 - 3(x - 3) - 4$.

(2, 0)
(7, 0)

Answer to Concept Check:
a. true   **b.** false

> ▶ **Helpful Hint**
>
> When using substitution, don't forget to substitute back to the original variable.

To find values of x, we substitute back. That is, we substitute $x - 3$ for y.

$$x - 3 = 4 \quad \text{or} \quad x - 3 = -1$$
$$x = 7 \qquad\qquad x = 2$$

Both 2 and 7 check. The solutions are 2 and 7. ☐

PRACTICE
4   Solve: $(x + 2)^2 - 2(x + 2) - 3 = 0$.

EXAMPLE 5   Solve: $x^{2/3} - 5x^{1/3} + 6 = 0$.

Solution   The key to solving this equation is recognizing that $x^{2/3} = (x^{1/3})^2$. We replace $x^{1/3}$ with m so that

$$(x^{1/3})^2 - 5x^{1/3} + 6 = 0$$

becomes

$$m^2 - 5m + 6 = 0$$

Now we solve by factoring.

$$m^2 - 5m + 6 = 0$$
$$(m - 3)(m - 2) = 0 \qquad\qquad \text{Factor.}$$
$$m - 3 = 0 \quad \text{or} \quad m - 2 = 0 \quad \text{Set each factor equal to 0.}$$
$$m = 3 \qquad\qquad m = 2$$

Since $m = x^{1/3}$, we have

$$x^{1/3} = 3 \qquad \text{or} \quad x^{1/3} = 2$$
$$x = 3^3 = 27 \quad \text{or} \qquad x = 2^3 = 8$$

Both 8 and 27 check. The solutions are 8 and 27.

To visualize these solutions, graph $y_1 = x^{2/3} - 5x^{1/3} + 6$ and find the x-intercepts of the graph.

The x-intercepts are $(8, 0)$ and $(27, 0)$, so the solutions 8 and 27 check. ☐

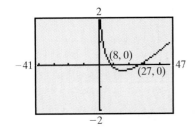

PRACTICE
5   Solve: $x^{2/3} - 5x^{1/3} + 4 = 0$.

OBJECTIVE 2 ▶ Solving problems that lead to quadratic equations. The next example is a work problem. This problem is modeled by a rational equation that simplifies to a quadratic equation.

EXAMPLE 6   **Finding Work Time**

Together, an experienced word processor and an apprentice word processor can create a word document in 6 hours. Alone, the experienced word processor can create the document 2 hours faster than the apprentice word processor can. Find the time in which each person can create the word document alone.

Solution

1. UNDERSTAND. Read and reread the problem. The key idea here is the relationship between the *time* (hours) it takes to complete the job and the *part of the job* completed in one unit of time (hour). For example, because they can complete the job together in 6 hours, the *part of the job* they can complete in 1 hour is $\frac{1}{6}$.

   Let

   x = the *time* in hours it takes the apprentice word processor to complete the job alone

   $x - 2$ = the *time* in hours it takes the experienced word processor to complete the job alone

We can summarize in a chart the information discussed

	Total Hours to Complete Job	Part of Job Completed in 1 Hour
Apprentice Word Processor	x	$\dfrac{1}{x}$
Experienced Word Processor	$x - 2$	$\dfrac{1}{x - 2}$
Together	6	$\dfrac{1}{6}$

2. TRANSLATE.

In words:	part of job completed by apprentice word processor in 1 hour	added to	part of job completed by experienced word processor in 1 hour	is equal to	part of job completed together in 1 hour
	↓	↓	↓	↓	↓
Translate:	$\dfrac{1}{x}$	$+$	$\dfrac{1}{x - 2}$	$=$	$\dfrac{1}{6}$

3. SOLVE.

Algebraic Solution:

$$\frac{1}{x} + \frac{1}{x - 2} = \frac{1}{6}$$

$$6x(x - 2)\left(\frac{1}{x} + \frac{1}{x - 2}\right) = 6x(x - 2) \cdot \frac{1}{6} \qquad \text{Multiply both sides by the LCD } 6x(x - 2).$$

$$6x(x - 2) \cdot \frac{1}{x} + 6x(x - 2) \cdot \frac{1}{x - 2} = 6x(x - 2) \cdot \frac{1}{6} \qquad \text{Use the distributive property.}$$

$$6(x - 2) + 6x = x(x - 2)$$
$$6x - 12 + 6x = x^2 - 2x$$
$$0 = x^2 - 14x + 12$$

Now we can substitute $a = 1$, $b = -14$, and $c = 12$ into the quadratic formula and simplify.

$$x = \frac{-(-14) \pm \sqrt{(-14)^2 - 4(1)(12)}}{2 \cdot 1} = \frac{14 \pm \sqrt{148}}{2}$$

$$x \approx \frac{14 + 12.2}{2} = 13.1 \quad \text{or} \quad x \approx \frac{14 - 12.2}{2} = 0.9$$

Graphical Solution:

Graph $y_1 = \dfrac{1}{x} + \dfrac{1}{x - 2}$ and $y_2 = \dfrac{1}{6}$ in a 0, 20, 5] by $[-1, 1, 1]$ window. The approximate intersections are $x \approx 0.9$ and $x \approx 13.1$. A hand sketch is shown below followed by graphing utility screens.

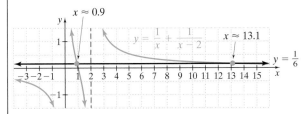

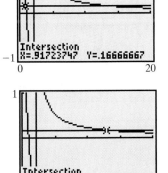

Since x represents time, we show the first-quadrant portion of the graph only.

4. INTERPRET.

Check: If the apprentice word processor completes the job alone in 0.9 hours, the experienced word processor completes the job alone in $x - 2 = 0.9 - 2 = -1.1$ hours. Since this is not possible, we reject the solution of 0.9. The approximate solution thus is 13.1 hours.

State: The apprentice word processor can complete the job alone in approximately 13.1 hours, and the experienced word processor can complete the job alone in approximately

$$x - 2 = 13.1 - 2 = 11.1 \text{ hours.} \qquad \Box$$

PRACTICE

6 Together, Katy and Steve can groom all the dogs at the Barkin' Doggie Day Care in 4 hours. Alone, Katy can groom the dogs 1 hour faster than Steve can groom the dogs alone. Find the time in which each of them can groom the dogs alone.

EXAMPLE 7 **Finding Driving Speeds**

Beach and Fargo are about 400 miles apart. A salesperson travels from Fargo to Beach one day at a certain speed. She returns to Fargo the next day and drives 10 mph faster. Her total travel time was $14\dfrac{2}{3}$ hours. Find her speed to Beach and the return speed to Fargo.

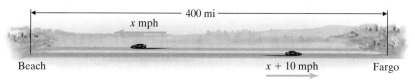

Solution

1. UNDERSTAND. Read and reread the problem. Let

$$x = \text{the speed to Beach, so}$$
$$x + 10 = \text{the return speed to Fargo.}$$

Then organize the given information in a table.

▶ **Helpful Hint**

Since $d = rt$, then $t = \dfrac{d}{r}$. The time column was completed using $\dfrac{d}{r}$.

	distance	=	rate	·	time	
To Beach	400		x		$\dfrac{400}{x}$	← distance
						← rate
Return to Fargo	400		$x + 10$		$\dfrac{400}{x + 10}$	← distance
						← rate

2. TRANSLATE.

In words: | time to Beach | + | return time to Fargo | = | $14\dfrac{2}{3}$ hours |

Translate: $\dfrac{400}{x}$ + $\dfrac{400}{x + 10}$ = $\dfrac{44}{3}$

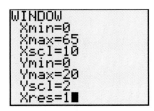

Below is a graphical solution for Example 7. Since x represents mph, we choose $[0, 65, 10]$. Since the y-window should include $\frac{44}{3}$, we choose $[0, 20, 2]$.

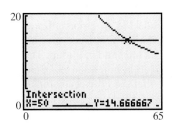

Here $x = 50$ indicates the speed.

3. SOLVE.

$$\frac{400}{x} + \frac{400}{x + 10} = \frac{44}{3}$$

$$\frac{100}{x} + \frac{100}{x + 10} = \frac{11}{3}$$ Divide both sides by 4.

$$3x(x + 10)\left(\frac{100}{x} + \frac{100}{x + 10}\right) = 3x(x + 10) \cdot \frac{11}{3}$$ Multiply both sides by the LCD, $3x(x + 10)$.

$$3x(x + 10) \cdot \frac{100}{x} + 3x(x + 10) \cdot \frac{100}{x + 10} = 3x(x + 10) \cdot \frac{11}{3}$$ Use the distributive property.

$$3(x + 10) \cdot 100 + 3x \cdot 100 = x(x + 10) \cdot 11$$

$$300x + 3000 + 300x = 11x^2 + 110x$$

$$0 = 11x^2 - 490x - 3000$$ Set equation equal to 0.

$$0 = (11x + 60)(x - 50)$$ Factor.

$$11x + 60 = 0 \quad \text{or} \quad x - 50 = 0$$ Set each factor equal to 0.

$$x = -\frac{60}{11} \text{ or } -5\frac{5}{11}; \quad x = 50$$

4. INTERPRET.

Check: The speed is not negative, so it's not $-5\frac{5}{11}$. The number 50 does check.

State: The speed to Beach was 50 mph and her return speed to Fargo was 60 mph.   □

PRACTICE

7   The 36-km S-shaped Hangzhou Bay Bridge is the longest cross-sea bridge in the world, linking Ningbo and Shanghai, China. A merchant drives over the bridge one morning from Ningbo to Shanghai in very heavy traffic and returns home that night driving 50 km per hour faster. The total travel time was 1.3 hours. Find the speed to Shanghai and the return speed to Ningbo.

8.3 | EXERCISE SET

Solve. See Example 1.

1. $2x = \sqrt{10 + 3x}$

2. $3x = \sqrt{8x + 1}$

3. $x - 2\sqrt{x} = 8$

4. $x - \sqrt{2x} = 4$

5. $\sqrt{9x} = x + 2$

6. $\sqrt{16x} = x + 3$

Solve. See Example 2.

7. $\frac{2}{x} + \frac{3}{x - 1} = 1$

8. $\frac{6}{x^2} = \frac{3}{x + 1}$

9. $\frac{3}{x} + \frac{4}{x + 2} = 2$

10. $\frac{5}{x - 2} + \frac{4}{x + 2} = 1$

11. $\frac{7}{x^2 - 5x + 6} = \frac{2x}{x - 3} - \frac{x}{x - 2}$

12. $\frac{11}{2x^2 + x - 15} = \frac{5}{2x - 5} - \frac{x}{x + 3}$

Solve. See Example 3.

13. $p^4 - 16 = 0$

14. $x^4 + 2x^2 - 3 = 0$

15. $4x^4 + 11x^2 = 3$

16. $z^4 = 81$

17. $z^4 - 13z^2 + 36 = 0$

18. $9x^4 + 5x^2 - 4 = 0$

Solve. See Examples 4 and 5.

19. $x^{2/3} - 3x^{1/3} - 10 = 0$

20. $x^{2/3} + 2x^{1/3} + 1 = 0$

21. $(5n + 1)^2 + 2(5n + 1) - 3 = 0$

22. $(m - 6)^2 + 5(m - 6) + 4 = 0$

23. $2x^{2/3} - 5x^{1/3} = 3$

24. $3x^{2/3} + 11x^{1/3} = 4$

25. $1 + \dfrac{2}{3t - 2} = \dfrac{8}{(3t - 2)^2}$

26. $2 - \dfrac{7}{x + 6} = \dfrac{15}{(x + 6)^2}$

27. $20x^{2/3} - 6x^{1/3} - 2 = 0$

28. $4x^{2/3} + 16x^{1/3} = -15$

MIXED PRACTICE

Solve. See Examples 1 through 5.

29. $a^4 - 5a^2 + 6 = 0$

30. $x^4 - 12x^2 + 11 = 0$

31. $\dfrac{2x}{x - 2} + \dfrac{x}{x + 3} = -\dfrac{5}{x + 3}$

32. $\dfrac{5}{x - 3} + \dfrac{x}{x + 3} = \dfrac{19}{x^2 - 9}$

33. $(p + 2)^2 = 9(p + 2) - 20$

34. $2(4m - 3)^2 - 9(4m - 3) = 5$

35. $2x = \sqrt{11x + 3}$

36. $4x = \sqrt{2x + 3}$

37. $x^{2/3} - 8x^{1/3} + 15 = 0$

38. $x^{2/3} - 2x^{1/3} - 8 = 0$

39. $y^3 + 9y - y^2 - 9 = 0$

40. $x^3 + x - 3x^2 - 3 = 0$

41. $2x^{2/3} + 3x^{1/3} - 2 = 0$

42. $6x^{2/3} - 25x^{1/3} - 25 = 0$

43. $x^{-2} - x^{-1} - 6 = 0$

44. $y^{-2} - 8y^{-1} + 7 = 0$

45. $x - \sqrt{x} = 2$

46. $x - \sqrt{3x} = 6$

47. $\dfrac{x}{x - 1} + \dfrac{1}{x + 1} = \dfrac{2}{x^2 - 1}$

48. $\dfrac{x}{x - 5} + \dfrac{5}{x + 5} = -\dfrac{1}{x^2 - 25}$

49. $p^4 - p^2 - 20 = 0$

50. $x^4 - 10x^2 + 9 = 0$

51. $(x + 3)(x^2 - 3x + 9) = 0$

52. $(x - 6)(x^2 + 6x + 36) = 0$

53. $1 = \dfrac{4}{x - 7} + \dfrac{5}{(x - 7)^2}$

54. $3 + \dfrac{1}{2p + 4} = \dfrac{10}{(2p + 4)^2}$

55. $27y^4 + 15y^2 = 2$

56. $8z^4 + 14z^2 = -5$

Solve. See Examples 6 and 7.

57. A jogger ran 3 miles, decreased her speed by 1 mile per hour, and then ran another 4 miles. If her total time jogging was $1\frac{3}{5}$ hours, find her speed for each part of her run.

58. Mark Keaton's workout consists of jogging for 3 miles, and then riding his bike for 5 miles at a speed 4 miles per hour faster than he jogs. If his total workout time is 1 hour, find his jogging speed and his biking speed.

59. A Chinese restaurant in Mandeville, Louisiana, has a large goldfish pond around the restaurant. Suppose that an inlet pipe and a hose together can fill the pond in 8 hours. The inlet pipe alone can complete the job in one hour less time than the hose alone. Find the time that the hose can complete the job alone and the time that the inlet pipe can complete the job alone. Round each to the nearest tenth of an hour.

60. A water tank on a farm in Flatonia, Texas, can be filled with a large inlet pipe and a small inlet pipe in 3 hours. The large inlet pipe alone can fill the tank in 2 hours less time than the small inlet pipe alone. Find the time to the nearest tenth of an hour each pipe can fill the tank alone.

61. Roma Sherry drove 330 miles from her hometown to Tucson. During her return trip, she was able to increase her speed by 11 mph. If her return trip took 1 hour less time, find her original speed and her speed returning home.

62. A salesperson drove to Portland, a distance of 300 miles. During the last 80 miles of his trip, heavy rainfall forced him to decrease his speed by 15 mph. If his total driving time was 6 hours, find his original speed and his speed during the rainfall.

63. Bill Shaughnessy and his son Billy can clean the house together in 4 hours. When the son works alone, it takes him an hour longer to clean than it takes his dad alone. Find how long to the nearest tenth of an hour it takes the son to clean alone.

64. Together, Noodles and Freckles eat a 50-pound bag of dog food in 30 days. Noodles by himself eats a 50-pound bag in 2 weeks less time than Freckles does by himself. How many days to the nearest whole day would a 50-pound bag of dog food last Freckles?

65. The product of a number and 4 less than the number is 96. Find the number.

66. A whole number increased by its square is two more than twice itself. Find the number.

△ **67.** Suppose that an open box is to be made from a square sheet of cardboard by cutting out squares from each corner as shown and then folding along the dotted lines. If the box is to have a volume of 300 cubic centimeters, find the original dimensions of the sheet of cardboard.

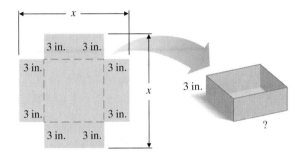

a. The ? in the drawing above will be the length (and also the width) of the box as shown. Represent this length in terms of x.

b. Use the formula for volume of a box, $V = l \cdot w \cdot h$, to write an equation in x.

c. Solve the equation for x and give the dimensions of the sheet of cardboard. Check your solution.

△ **68.** Suppose that an open box is to be made from a square sheet of cardboard by cutting out squares from each corner as shown and then folding along the dotted lines. If the box is to have a volume of 128 cubic inches, find the original dimensions of the sheet of cardboard.

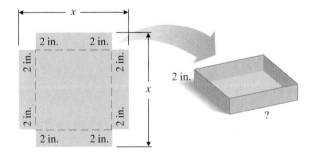

a. The ? in the drawing above will be the length (and also the width) of the box as shown. Represent this length in terms of x.

b. Use the formula for volume of a box, $V = l \cdot w \cdot h$, to write an equation in x.

c. Solve the equation for x and give the dimensions of the sheet of cardboard. Check your solution.

△ **69.** A sprinkler that sprays water in a circular motion is to be used to water a square garden. If the area of the garden is 920 square feet, find the smallest whole number *radius* that the sprinkler can be adjusted to so that the entire garden is watered.

△ **70.** Suppose that a square field has an area of 6270 square feet. See Exercise 69 and find a new sprinkler radius.

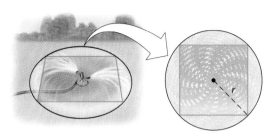

REVIEW AND PREVIEW

Solve each inequality. See Section 3.2.

71. $\dfrac{5x}{3} + 2 \le 7$

72. $\dfrac{2x}{3} + \dfrac{1}{6} \ge 2$

73. $\dfrac{y - 1}{15} > -\dfrac{2}{5}$

74. $\dfrac{z - 2}{12} < \dfrac{1}{4}$

Find the domain and range of each graphed relation. Decide which relations are also functions. See Section 2.2.

75.

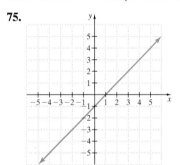

76.

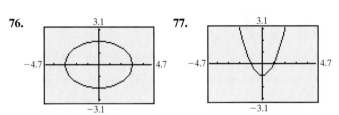

77.

78.

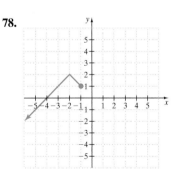

CONCEPT EXTENSIONS

Solve.

79. $y^3 + 9y - y^2 - 9 = 0$

80. $x^3 + x - 3x^2 - 3 = 0$

81. $x^{-2} - x^{-1} - 6 = 0$

82. $y^{-2} - 8y^{-1} + 7 = 0$

83. $2x^3 = -54$

84. $y^3 - 216 = 0$

85. Write a polynomial equation that has three solutions: 2, 5, and −7.

86. Write a polynomial equation that has three solutions: 0, $2i$, and −2i.

87. At the 2007 Grand Prix of Long Beach auto race, Simon Pagenaud posted the fastest lap speed, but Sebastian Bourdais won the race. One lap through the streets of Long Beach is 10,391 feet (1.968 miles) long. Pagenaud's fastest lap speed was 0.55 foot per second faster than Bourdais's fastest lap speed.

Traveling at these fastest speeds, Bourdais would have taken 0.25 second longer than Pagenaud to complete a lap. (*Source:* Championship Auto Racing Teams, Inc.)

  a. Find Sebastian Bourdais's fastest lap speed during the race. Round to two decimal places.

  b. Find Simon Pagenaud's fastest lap speed during the race. Round to two decimal places.

  c. Convert each speed to miles per hour. Round to one decimal place.

88. Use a graphing calculator to solve Exercise 29. Compare the solution with the solution from Exercise 29. Explain any differences.

INTEGRATED REVIEW   SUMMARY ON SOLVING QUADRATIC EQUATIONS

Sections 8.1–8.3

Use the square root property to solve each equation.

1. $x^2 - 10 = 0$

2. $x^2 - 14 = 0$

3. $(x - 1)^2 = 8$

4. $(x + 5)^2 = 12$

Solve each equation by completing the square.

5. $x^2 + 2x - 12 = 0$

6. $x^2 - 12x + 11 = 0$

7. $3x^2 + 3x = 5$

8. $16y^2 + 16y = 1$

Use the quadratic formula to solve each equation.

9. $2x^2 - 4x + 1 = 0$

10. $\frac{1}{2}x^2 + 3x + 2 = 0$

11. $x^2 + 4x = -7$

12. $x^2 + x = -3$

Solve each equation. Use a method of your choice.

13. $x^2 + 3x + 6 = 0$

14. $2x^2 + 18 = 0$

15. $x^2 + 17x = 0$

16. $4x^2 - 2x - 3 = 0$

17. $(x - 2)^2 = 27$

18. $\frac{1}{2}x^2 - 2x + \frac{1}{2} = 0$

19. $3x^2 + 2x = 8$

20. $2x^2 = -5x - 1$

21. $x(x - 2) = 5$

22. $x^2 - 31 = 0$

23. $5x^2 - 55 = 0$

24. $5x^2 + 55 = 0$

25. $x(x + 5) = 66$

26. $5x^2 + 6x - 2 = 0$

27. $2x^2 + 3x = 1$

28. The diagonal of a square room measures 20 feet. Find the exact length of a side of the room. Then approximate the length to the nearest tenth of a foot.

29. Together, Jack and Lucy Hoag can prepare a crawfish boil for a large party in 4 hours. Lucy alone can complete the job in 2 hours less time than Jack alone. Find the time that each person can prepare the crawfish boil alone. Round each time to the nearest tenth of an hour.

30. Diane Gray exercises at Total Body Gym. On the treadmill, she runs 5 miles, then increases her speed by 1 mile per hour and runs an additional 2 miles. If her total time on the tread mill is $1\frac{1}{3}$ hours, find her speed during each part of her run.

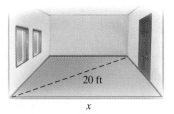

20 ft

x

8.4 NONLINEAR INEQUALITIES IN ONE VARIABLE

OBJECTIVES

1 Solve polynomial inequalities of degree 2 or greater.

2 Solve inequalities that contain rational expressions with variables in the denominator.

OBJECTIVE 1 ▶ Solving polynomial inequalities. Just as we can solve linear inequalities in one variable, so can we also solve quadratic inequalities in one variable. A **quadratic inequality** is an inequality that can be written so that one side is a quadratic expression and the other side is 0. Here are examples of quadratic inequalities in one variable. Each is written in **standard form.**

$$x^2 - 10x + 7 \le 0 \qquad 3x^2 + 2x - 6 > 0$$
$$2x^2 + 9x - 2 < 0 \qquad x^2 - 3x + 11 \ge 0$$

A solution of a quadratic inequality in one variable is a value of the variable that makes the inequality a true statement.

DISCOVER THE CONCEPT

Graph the quadratic function $y = x^2 - 3x - 10$.

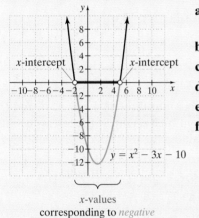

x-values corresponding to *negative* *y*-values

a. Find the values of x for which $y = 0$. (Recall that these are the *x*-intercepts of the graph.)

b. Find the *x*-values for which $y < 0$.

c. Find the *x*-values for which $y > 0$.

d. What is the solution of $y = 0$ or $x^2 - 3x - 10 = 0$?

e. What is the solution of $y < 0$ or $x^2 - 3x - 10 < 0$?

f. What is the solution of $y > 0$ or $x^2 - 3x - 10 > 0$?

Notice that the *x*-values for which y is positive are separated from the *x*-values for which y is negative by the *x*-intercepts. (Recall that the *x*-intercepts correspond to values of x for which $y = 0$.) Thus, the solution set of $x^2 - 3x - 10 < 0$ consists of all real numbers from -2 to 5, or in interval notation, $(-2, 5)$.

It is not necessary to graph $y = x^2 - 3x - 10$ to solve the related inequality $x^2 - 3x - 10 < 0$. Instead, we can draw a number line representing the *x*-axis and keep the following in mind: *A region on the number line for which the value of $x^2 - 3x - 10$ is positive is separated from a region on the number line for which the value of $x^2 - 3x - 10$ is negative by a value for which the expression is 0.*

Let's find these values for which the expression is 0 by solving the related equation:

$$x^2 - 3x - 10 = 0$$
$$(x - 5)(x + 2) = 0 \qquad \text{Factor.}$$
$$x - 5 = 0 \quad \text{or} \quad x + 2 = 0 \qquad \text{Set each factor equal to 0.}$$
$$x = 5 \quad x = -2 \qquad \text{Solve.}$$

These two numbers, -2 and 5, divide the number line into three regions. We will call the regions *A*, *B*, and *C*. These regions are important because, if the value of $x^2 - 3x - 10$ is negative when a number from a region is substituted for x, then $x^2 - 3x - 10$ is negative when any number in that region is substituted for x. The same is true if the value of $x^2 - 3x - 10$ is positive for a particular value of x in a region.

To see whether the inequality $x^2 - 3x - 10 < 0$ is true or false in each region, we choose a test point from each region and substitute its value for x in the inequality

$x^2 - 3x - 10 < 0$. If the resulting inequality is true, the region containing the test point is a solution region.

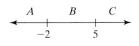

Region	Test Point Value	$(x - 5)(x + 2) < 0$	Result
A	-3	$(-8)(-1) < 0$	False
B	0	$(-5)(2) < 0$	True
C	6	$(1)(8) < 0$	False

The values in region B satisfy the inequality. The numbers -2 and 5 are not included in the solution set since the inequality symbol is $<$. The solution set is $(-2, 5)$, and its graph is shown.

EXAMPLE 1   Solve: $(x + 3)(x - 3) > 0$.

Solution   First we solve the related equation, $(x + 3)(x - 3) = 0$.

$$(x + 3)(x - 3) = 0$$
$$x + 3 = 0 \quad \text{or} \quad x - 3 = 0$$
$$x = -3 \qquad\qquad x = 3$$

The two numbers -3 and 3 separate the number line into three regions, A, B, and C.

Now we substitute the value of a test point from each region. If the test value satisfies the inequality, every value in the region containing the test value is a solution.

Region	Test Point Value	$(x + 3)(x - 3) > 0$	Result
A	-4	$(-1)(-7) > 0$	True
B	0	$(3)(-3) > 0$	False
C	4	$(7)(1) > 0$	True

The points in regions A and C satisfy the inequality. The numbers -3 and 3 are not included in the solution since the inequality symbol is $>$. The solution set is $(-\infty, -3) \cup (3, \infty)$, and its graph is shown.

PRACTICE
1   Solve: $(x - 4)(x + 3) > 0$.

The following steps may be used to solve a polynomial inequality.

Solving a Polynomial Inequality Algebraically

STEP 1. Write the inequality in standard form and then solve the related equation.

STEP 2. Separate the number line into regions with the solutions from Step 1.

STEP 3. For each region, choose a test point and determine whether its value satisfies the *original inequality*.

STEP 4. The solution set includes the regions whose test point value is a solution. If the inequality symbol is $\leq$ or $\geq$, the values from Step 1 are solutions; if $<$ or $>$, they are not.

Concept Check ☑

When choosing a test point in Step 4, why would the solutions from Step 2 not make good choices for test points?

EXAMPLE 2   Solve: $x^2 - 4x \leq 0$.

Solution   First we solve the related equation, $x^2 - 4x = 0$.

$$x^2 - 4x = 0$$
$$x(x - 4) = 0$$
$$x = 0 \quad \text{or} \quad x = 4$$

The numbers 0 and 4 separate the number line into three regions, A, B, and C.

$$\begin{array}{ccc} A & B & C \\ \hline & 0 & 4 \end{array}$$

We check a test value in each region in the original inequality. Values in region B satisfy the inequality. The numbers 0 and 4 are included in the solution since the inequality symbol is $\leq$. The solution set is $[0, 4]$, and its graph is shown.

$$\begin{array}{ccc} A & B & C \\ \hline F \ \ 0 & \text{T} & 4 \ \ F \end{array}$$

□

PRACTICE

2   Solve: $x^2 - 8x \leq 0$.

Next, we solve a quadratic inequality similar to Example 2 graphically.

EXAMPLE 3   Use a graphical approach to solve $x^2 - 4x \geq 0$.

Solution   The x-intercepts of the graph of $y = x^2 - 4x$ are found by solving $x^2 - 4x = 0$. We found these intercepts earlier to be $x = 0$ and $x = 4$. Next, graph $y = x^2 - 4x$. Solutions of the quadratic inequality are x-values where $y \geq 0$, or where the graph is on or above the x-axis.

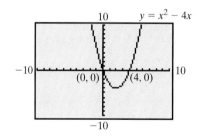

The x-values where the graph is on or above the x-axis are the x-intercepts of 0 and 4 and the x-values to the left of the x-intercept 0 and to the right of the x-intercept 4. In interval notation, the solution set is $(-\infty, 0] \cup [4, \infty)$.

□

PRACTICE

3   Use a graphical approach to solve $x^2 - 8x \geq 0$.

EXAMPLE 4   Use a graphical approach to solve $(x + 2)(x - 1)(x - 5) < 0$.

Solution   The x-intercepts of the graph are found by solving $(x + 2)(x - 1)(x - 5) = 0$. These x-intercepts are $-2, 1$, and 5. Graph $y = (x + 2)(x - 1)(x - 5)$. To solve $y < 0$, find x-values where the graph lies below the x-axis.

Again, the solution set of $y < 0$ contains x-values where the graph of y is below the x-axis. In interval notation, the solution set is $(-\infty, -2) \cup (1, 5)$.

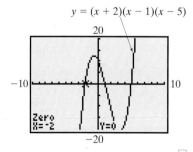

□

PRACTICE

4   Use a graphical approach to solve $(x + 3)(x - 2)(x + 1) \leq 0$.

EXAMPLE 5 Use a graphical approach to solve $x^2 + 2x > -4$.

Solution In standard form, the quadratic inequality is $x^2 + 2x + 4 > 0$. The graph of $y = x^2 + 2x + 4$ is shown below. Notice that the entire graph lies above the x-axis, so $y > 0$ for all values of x. The solution set in interval notation is $(-\infty, \infty)$.

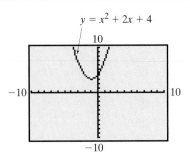

$y = x^2 + 2x + 4$

PRACTICE
5 Use a graphical approach to solve $x^2 + 3x < -7$.

OBJECTIVE 2 ▶ Solving rational inequalities. Inequalities containing rational expressions with variables in the denominator are solved by using a similar procedure.

EXAMPLE 6 Solve: $\dfrac{x + 2}{x - 3} \le 0$.

Solution First we find all values that make the denominator equal to 0. To do this, we solve $x - 3 = 0$ and find that $x = 3$.

Next, we solve the related equation $\dfrac{x + 2}{x - 3} = 0$.

$$\frac{x + 2}{x - 3} = 0$$

$$x + 2 = 0 \qquad \text{Multiply both sides by the LCD, } x - 3.$$

$$x = -2$$

Now we place these numbers on a number line and proceed as before, checking test point values in the original inequality.

$$\begin{array}{c c c}
A & B & C \\
\hline
\quad -2 \quad & \quad 3 \quad
\end{array}$$

Choose -3 *from region A.*   *Choose* 0 *from region B.*   *Choose* 4 *from region C.*

$$\frac{x + 2}{x - 3} \le 0 \qquad\qquad \frac{x + 2}{x - 3} \le 0 \qquad\qquad \frac{x + 2}{x - 3} \le 0$$

$$\frac{-3 + 2}{-3 - 3} \le 0 \qquad\qquad \frac{0 + 2}{0 - 3} \le 0 \qquad\qquad \frac{4 + 2}{4 - 3} \le 0$$

$$\frac{-1}{-6} \le 0 \qquad\qquad -\frac{2}{3} \le 0 \quad \text{True} \qquad\qquad 6 \le 0 \quad \text{False}$$

$$\frac{1}{6} \le 0 \quad \text{False}$$

The solution set is $[-2, 3)$. This interval includes -2 because -2 satisfies the original inequality. This interval does not include 3, because 3 would make the denominator 0.

$$\begin{array}{c c c}
A & B & C \\
\hline
\text{F } -2 \quad & \text{T} \quad 3 & \text{F}
\end{array}$$

To visualize this graphically, graph

$$y = \frac{x + 2}{x - 3}$$

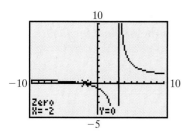

and find x-values where the graph is on or below the x-axis. The graph shown to the left is on the x-axis at $x = -2$. The graph is below the x-axis from -2 to the undefined value $x = 3$. Thus, we verify the solution set as $[-2, 3)$. ☐

PRACTICE
6 Solve: $\dfrac{x - 5}{x + 4} \le 0$.

The following steps may be used to solve a rational inequality with variables in the denominator.

> **Solving a Rational Inequality Algebraically**
> **STEP 1.** Solve for values that make all denominators 0.
>
> **STEP 2.** Solve the related equation.
>
> **STEP 3.** Separate the number line into regions with the solutions from Steps 1 and 2.
>
> **STEP 4.** For each region, choose a test point and determine whether its value satisfies the *original inequality.*
>
> **STEP 5.** The solution set includes the regions whose test point value is a solution. Check whether to include values from Step 2. Be sure *not* to include values that make any denominator 0.

To graphically solve a rational inequality with variables in the denominator, use the steps below.

> **Solving a Rational Inequality Graphically**
> **STEP 1.** Write the related equation so that the right side is 0.
>
> **STEP 2.** Enter the left side expression from Step 1 into the Y= editor and graph.
>
> **STEP 3.** If $<$ or $\le$, determine the intervals where the graph is below the x-axis. If $>$ or $\ge$, determine the intervals where the graph is above the x-axis.
>
> **STEP 4.** Write down the solution intervals from Step 3. If $\le$ or $\ge$, include the end points of the intervals as solutions by using brackets. Remember not to include values that make any expression undefined.

EXAMPLE 7 Solve: $\dfrac{5}{x + 1} < -2$.

Solution First we find values for x that make the denominator equal to 0.

$$x + 1 = 0$$
$$x = -1$$

Next we solve $\dfrac{5}{x + 1} = -2$.

$$(x + 1) \cdot \frac{5}{x + 1} = (x + 1) \cdot -2 \quad \text{Multiply both sides by the LCD, } x + 1.$$
$$5 = -2x - 2 \quad \text{Simplify.}$$
$$7 = -2x$$
$$-\frac{7}{2} = x$$

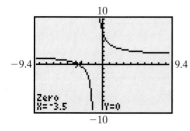

The graph of $y = \dfrac{5}{x + 1} + 2$.

The solution is found by the portion of the graph below the x-axis. That is, from the x-intercept $x = -3.5$ to where the expression is undefined at $x = -1$.

We use these two solutions to divide a number line into three regions and choose test points. Only a test point value from region B satisfies the *original inequality*. The solution set is $\left(-\dfrac{7}{2}, -1\right)$, and its graph is shown.

PRACTICE

7 Solve: $\dfrac{7}{x+3} < 5$.

VOCABULARY & READINESS CHECK

Write the graphed solution set in interval notation.

1.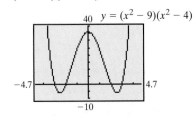
$-7 \qquad 3$

2.
$-1 \qquad 5$

3.
0

4.
-8

5.
$-12 \qquad -10$

6.
$-3 \qquad 4$

8.4 | EXERCISE SET

MyMathLab Powered by CourseCompass™ and MathXL®

MathXL PRACTICE   WATCH   DOWNLOAD   READ   REVIEW

Solve each quadratic inequality. Write the solution set in interval notation. See Examples 1 through 5.

1. $(x + 1)(x + 5) > 0$

2. $(x + 1)(x + 5) \le 0$

3. $(x - 3)(x + 4) \le 0$

4. $(x + 4)(x - 1) > 0$

5. $x^2 - 7x + 10 \le 0$

6. $x^2 + 8x + 15 \ge 0$

7. $3x^2 + 16x < -5$

8. $2x^2 - 5x < 7$

9. $(x - 6)(x - 4)(x - 2) > 0$

10. $(x - 6)(x - 4)(x - 2) \le 0$

11. $x(x - 1)(x + 4) \le 0$

12. $x(x - 6)(x + 2) > 0$

Use the related graph to solve each inequality. Write the solution set in interval notation. See Examples 3 through 6.

13. $(x^2 - 9)(x^2 - 4) \le 0$

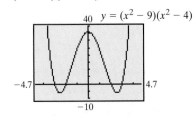
$y = (x^2 - 9)(x^2 - 4)$
40
$-4.7 \qquad 4.7$
-10

14. $(x^2 - 16)(x^2 - 1) \le 0$

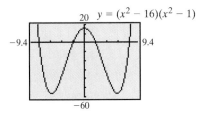
$y = (x^2 - 16)(x^2 - 1)$
20
$-9.4 \qquad 9.4$
-60

Solve each inequality. Write the solution set in interval notation. See Example 6.

15. $\dfrac{x + 7}{x - 2} < 0$

16. $\dfrac{x - 5}{x - 6} > 0$

17. $\dfrac{5}{x + 1} > 0$

18. $\dfrac{3}{y - 5} < 0$

Use the related graph to solve each inequality. Write the solution set in interval notation. See Example 7.

19. $\dfrac{x + 1}{x - 4} \ge 0$

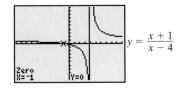
$y = \dfrac{x + 1}{x - 4}$
Zero
X=-1   Y=0

20. $\dfrac{x + 1}{x - 4} \leq 0$

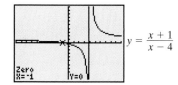

$y = \dfrac{x + 1}{x - 4}$

Zero
X=-1   Y=0

Solve each inequality. Write the solution set in interval notation. See Example 7.

21. $\dfrac{3}{x - 2} < 4$

22. $\dfrac{-2}{y + 3} > 2$

23. $\dfrac{x^2 + 6}{5x} \geq 1$

24. $\dfrac{y^2 + 15}{8y} \leq 1$

MIXED PRACTICE

Solve each inequality. Write the solution set in interval notation.

25. $(x - 8)(x + 7) > 0$

26. $(x - 5)(x + 1) < 0$

27. $(2x - 3)(4x + 5) \leq 0$

28. $(6x + 7)(7x - 12) > 0$

29. $x^2 > x$

30. $x^2 < 25$

31. $(2x - 8)(x + 4)(x - 6) \leq 0$

32. $(3x - 12)(x + 5)(2x - 3) \geq 0$

33. $6x^2 - 5x \geq 6$

34. $12x^2 + 11x \leq 15$

35. $4x^3 + 16x^2 - 9x - 36 > 0$

36. $x^3 + 2x^2 - 4x - 8 < 0$

37. $x^4 - 26x^2 + 25 \geq 0$

38. $16x^4 - 40x^2 + 9 \leq 0$

39. $(2x - 7)(3x + 5) > 0$

40. $(4x - 9)(2x + 5) < 0$

41. $\dfrac{x}{x - 10} < 0$

42. $\dfrac{x + 10}{x - 10} > 0$

43. $\dfrac{x - 5}{x + 4} \geq 0$

44. $\dfrac{x - 3}{x + 2} \leq 0$

45. $\dfrac{x(x + 6)}{(x - 7)(x + 1)} \geq 0$

46. $\dfrac{(x - 2)(x + 2)}{(x + 1)(x - 4)} \leq 0$

47. $\dfrac{-1}{x - 1} > -1$

48. $\dfrac{4}{y + 2} < -2$

49. $\dfrac{x}{x + 4} \leq 2$

50. $\dfrac{4x}{x - 3} \geq 5$

51. $\dfrac{z}{z - 5} \geq 2z$

52. $\dfrac{p}{p + 4} \leq 3p$

53. $\dfrac{(x + 1)^2}{5x} > 0$

54. $\dfrac{(2x - 3)^2}{x} < 0$

REVIEW AND PREVIEW

Recall that the graph of $f(x) + K$ is the same as the graph of $f(x)$ shifted K units upward if $K > 0$ and $|K|$ units downward if $K < 0$. Use the graph of $f(x) = |x|$ below to sketch the graph of each function. See Section 2.7.

55. $g(x) = |x| + 2$ **56.** $H(x) = |x| - 2$

57. $F(x) = |x| - 1$ **58.** $h(x) = |x| + 5$

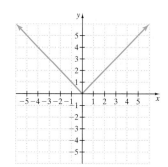

Use the graph of $f(x) = x^2$ below to sketch the graph of each function.

59. $F(x) = x^2 - 3$ **60.** $h(x) = x^2 - 4$

61. $H(x) = x^2 + 1$ **62.** $g(x) = x^2 + 3$

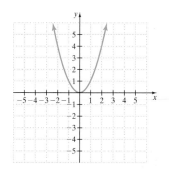

CONCEPT EXTENSIONS

63. Explain why $\dfrac{x + 2}{x - 3} > 0$ and $(x + 2)(x - 3) > 0$ have the same solutions.

64. Explain why $\dfrac{x + 2}{x - 3} \geq 0$ and $(x + 2)(x - 3) \geq 0$ do not have the same solutions.

Find all numbers that satisfy each of the following.

65. A number minus its reciprocal is less than zero. Find the numbers.

66. Twice a number added to its reciprocal is nonnegative. Find the numbers.

67. The total profit function $P(x)$ for a company producing x thousand units is given by

$$P(x) = -2x^2 + 26x - 44$$

Find the values of x for which the company makes a profit. [*Hint:* The company makes a profit when $P(x) > 0$.]

68. A projectile is fired straight up from the ground with an initial velocity of 80 feet per second. Its height $s(t)$ in feet at any time t is given by the function

$$s(t) = -16t^2 + 80t$$

Find the interval of time for which the height of the projectile is greater than 96 feet.

Use a graphing calculator to check each exercise.

69. Exercise 25

70. Exercise 26

71. Exercise 37

72. Exercise 38

THE BIGGER PICTURE   SOLVING EQUATIONS AND INEQUALITIES

Continue your outline from Sections 1.5, 3.2, 3.3, 3.5, 5.8, 6.5, and 7.6. Write how to recognize and how to solve quadratic equations and nonlinear inequalities in your own words. For example:

Solving Equations and Inequalities

I. Equations

   A. Linear equations (Sec. 1.5 and 3.1)

   B. Absolute value equations (Sec. 3.4)

   C. Quadratic and higher degree equations (Sec. 5.8, 8.1, 8.2, 8.3)

Solving by Factoring:

$$2x^2 - 7x = 9$$
$$2x^2 - 7x - 9 = 0$$
$$(2x - 9)(x + 1) = 0$$
$$2x - 9 = 0$$
or $\qquad x + 1 = 0$
$$x = \frac{9}{2} \quad \text{or} \quad x = -1$$

Solving by the Quadratic Formula:

$$2x^2 + x - 2 = 0$$
$$a = 2, b = 1, c = -2$$
$$x = \frac{-1 \pm \sqrt{(1)^2 - 4(2)(-2)}}{2 \cdot 2}$$
$$x = \frac{-1 \pm \sqrt{17}}{4}$$

   D. Equations with rational expressions (Sec. 6.5)

   E. Equations with radicals (Sec. 7.6)

II. Inequalities

   A. Linear inequalities (Sec. 3.2)

   B. Compound inequalities (Sec. 3.3)

   C. Absolute value inequalities (Sec. 3.5)

   D. Nonlinear inequalities

Polynomial Inequality	Rational Inequality with variable in denominator
$x^2 - x < 6$ $x^2 - x - 6 < 0$ $(x - 3)(x + 2) < 0$	$\dfrac{x - 5}{x + 1} \geq 0$

Polynomial: ✗ ✓ ✗ at -2, 3 → $(-2, 3)$

Rational: ✓ ✗ ✓ at -1, 5 → $(-\infty, -1) \cup [5, \infty)$

Solve. Write solutions to inequalities in interval notation.

1. $|x - 8| = |2x + 1|$

2. $0 < -x + 7 < 3$

3. $\sqrt{3x - 11} + 3 = x$

4. $x(3x + 1) = 1$

5. $\dfrac{x + 2}{x - 7} \leq 0$

6. $x(x - 6) + 4 = x^2 - 2(3 - x)$

7. $x(5x - 36) = -7$

8. $2x^2 - 4 \geq 7x$

9. $\left|\dfrac{x - 7}{3}\right| > 5$

10. $2(x - 5) + 4 < 1 + 7(x - 5) - x$

8.5 QUADRATIC FUNCTIONS AND THEIR GRAPHS

OBJECTIVE 1 ▶ Graphing $f(x) = x^2 + k$. We first graphed the quadratic equation $y = x^2$ in Section 2.1. In Section 2.2, we learned that this graph defines a function, and we wrote $y = x^2$ as $f(x) = x^2$. In these sections, we discovered that the graph of a quadratic function is a parabola opening upward or downward. In this section, we continue our study of quadratic functions and their graphs. (Much of the contents of this section is a review of shifting and reflecting techniques from Section 2.7.)

First, let's recall the definition of a quadratic function.

> **Quadratic Function**
>
> A quadratic function is a function that can be written in the form $f(x) = ax^2 + bx + c$, where a, b, and c are real numbers and $a \neq 0$.

Notice that equations of the form $y = ax^2 + bx + c$, where $a \neq 0$, define quadratic functions, since y is a function of x or $y = f(x)$.

Recall that if $a > 0$, the parabola opens upward and that if $a < 0$, the parabola opens downward. Also, the vertex of a parabola is the lowest point if the parabola opens upward and the highest point if the parabola opens downward. The axis of symmetry is the vertical line that passes through the vertex.

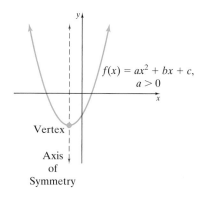

$f(x) = ax^2 + bx + c,$
$a > 0$

Vertex

Axis of Symmetry

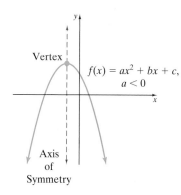

Vertex

$f(x) = ax^2 + bx + c,$
$a < 0$

Axis of Symmetry

DISCOVER THE CONCEPT

> Graph $f(x) = x^2$, $g(x) = x^2 + 3$, and $h(x) = x^2 - 5$. Compare the graphs of $g(x)$ and $h(x)$ to the graph of $f(x)$.

The graphs of $y_1 = x^2$, $y_2 = x^2 + 3$, and $y_3 = x^2 - 5$ and corresponding tables are shown below.

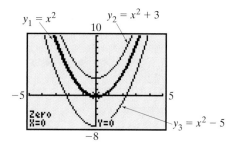

$y_1 = x^2$ $y_2 = x^2 + 3$

$y_3 = x^2 - 5$

Each y-value is increased by 3

Each y-value is decreased by 5

Notice that the graph of $y_2 = x^2 + 3$ is the same as the graph of $y_1 = x^2$, but shifted *upward* 3 units. Also, the graph of $y_3 = x^2 - 5$ is the same as the graph of $y_1 = x^2$, but shifted *downward* 5 units. The axis of symmetry is the same for all graphs: $x = 0$.

In general, we have the following properties.

> **Graphing the Parabola Defined by $f(x) = x^2 + k$**
>
> If k is positive, the graph of $f(x) = x^2 + k$ is the graph of $y = x^2$ shifted upward k units.
>
> If k is negative, the graph of $f(x) = x^2 + k$ is the graph of $y = x^2$ shifted downward $|k|$ units.
>
> The vertex is $(0, k)$, and the axis of symmetry is the y-axis.

EXAMPLE 1 Graph each function.

a. $F(x) = x^2 + 2$ **b.** $g(x) = x^2 - 3$

Solution

a. $F(x) = x^2 + 2$

The graph of $F(x) = x^2 + 2$ is obtained by shifting the graph of $y = x^2$ upward 2 units.

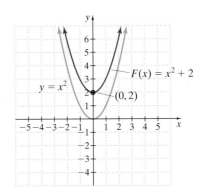

b. $g(x) = x^2 - 3$

The graph of $g(x) = x^2 - 3$ is obtained by shifting the graph of $y = x^2$ downward 3 units.

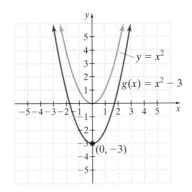

PRACTICE

1 Graph each function.

a. $f(x) = x^2 - 5$ **b.** $g(x) = x^2 + 3$

OBJECTIVE 2 ▶ Graphing $f(x) = (x - h)^2$. Now we will graph functions of the form $f(x) = (x - h)^2$.

DISCOVER THE CONCEPT

Graph $f(x) = x^2$, $g(x) = (x - 2)^2$, and $h(x) = (x + 4)^2$. Compare the graphs of $g(x)$ and $h(x)$ to the graph of $f(x)$.

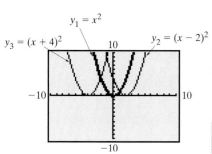

The graphs of $y_1 = x^2$, $y_2 = (x - 2)^2$, and $y_3 = (x + 4)^2$ are shown to the left. The graph of $y_2 = (x - 2)^2$ is the same as the graph of $y_1 = x^2$ *shifted to the right* 2 units. Also, the graph of $y_3 = (x + 4)^2$ is the same as the graph of $y_1 = x^2$ *shifted to the left* 4 units. Notice that the axes of symmetry have shifted also.

In general, we have the following properties.

> **Graphing the Parabola Defined by $f(x) = (x - h)^2$**
>
> If h is positive, the graph of $f(x) = (x - h)^2$ is the graph of $y = x^2$ shifted to the right h units.
> If h is negative, the graph of $f(x) = (x - h)^2$ is the graph of $y = x^2$ shifted to the left $|h|$ units.
> The vertex is $(h, 0)$, and the axis of symmetry is the vertical line $x = h$.

EXAMPLE 2 Graph each function.

a. $G(x) = (x - 3)^2$ **b.** $F(x) = (x + 1)^2$

Solution

a. The graph of $G(x) = (x - 3)^2$ is obtained by shifting the graph of $y = x^2$ to the right 3 units. The graph of $G(x)$ is below on the left.

b. The equation $F(x) = (x + 1)^2$ can be written as $F(x) = [x - (-1)]^2$. The graph of $F(x) = [x - (-1)]^2$ is obtained by shifting the graph of $y = x^2$ to the left 1 unit. The graph of $F(x)$ is below on the right.

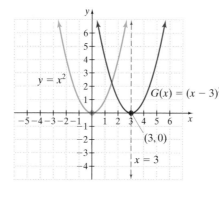 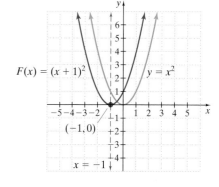

PRACTICE

2 Graph each function.

a. $G(x) = (x + 4)^2$ **b.** $H(x) = (x - 7)^2$

OBJECTIVE 3 ▶ **Graphing $f(x) = (x - h)^2 + k$.** As we will see in graphing functions of the form $f(x) = (x - h)^2 + k$, it is possible to combine vertical and horizontal shifts.

> **Graphing the Parabola Defined by $f(x) = (x - h)^2 + k$**
>
> The parabola has the same shape as $y = x^2$.
> The vertex is (h, k), and the axis of symmetry is the vertical line $x = h$.

EXAMPLE 3 Graph $F(x) = (x - 3)^2 + 1$.

Solution The graph of $F(x) = (x - 3)^2 + 1$ is the graph of $y = x^2$ shifted 3 units to the right and 1 unit up. The vertex is then $(3, 1)$, and the axis of symmetry is $x = 3$. A few ordered pair solutions are plotted to aid in graphing.

x	$F(x) = (x - 3)^2 + 1$
1	5
2	2
4	2
5	5

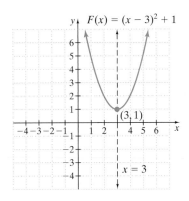

PRACTICE
3 Graph $f(x) = (x + 2)^2 + 2$.

OBJECTIVE 4 ▶ Graphing $f(x) = ax^2$. Next, we discover the change in the shape of the graph when the coefficient of x^2 is not 1.

DISCOVER THE CONCEPT

Graph $f(x) = ax^2$ for $a = 1, 3$, and $\dfrac{1}{2}$. Compare the other graphs to the graph of $f(x) = 1x^2$ or $f(x) = x^2$.

The graphs of $y_1 = x^2$, $y_2 = 3x^2$, and $y_3 = \dfrac{1}{2}x^2$ and corresponding tables are shown below.

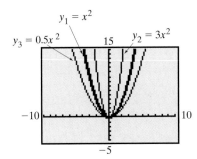

Compare the table of values. We see that for each x-value, the corresponding value of y_2 is triple the corresponding value of y_1. Similarly, the value of y_3 is half the value of y_1. The result is that the graph of $y_2 = 3x^2$ is narrower than the graph of $f(x) = x^2$ and the graph of $y_3 = \dfrac{1}{2}x^2$ is wider. The vertex for each graph is $(0, 0)$, and the axis of symmetry is the y-axis.

> **Graphing the Parabola Defined by $f(x) = ax^2$**
> If a is positive, the parabola opens upward, and if a is negative, the parabola opens downward.
> If $|a| > 1$, the graph of the parabola is narrower than the graph of $y = x^2$.
> If $|a| < 1$, the graph of the parabola is wider than the graph of $y = x^2$.

EXAMPLE 4 Graph $f(x) = -2x^2$.

Solution Because $a = -2$, a negative value, this parabola opens downward. Since $|-2| = 2$ and $2 > 1$, the parabola is narrower than the graph of $y = x^2$. The vertex is $(0, 0)$, and the axis of symmetry is the y-axis. We verify this by plotting a few points.

x	$f(x) = -2x^2$
-2	-8
-1	-2
0	0
1	-2
2	-8

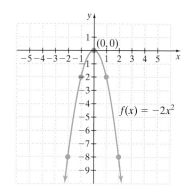

PRACTICE

4   Graph $f(x) = -\dfrac{1}{2}x^2$.

OBJECTIVE 5 ▶ **Graphing $f(x) = a(x - h)^2 + k$.** Now we will see the shape of the graph of a quadratic function of the form $f(x) = a(x - h)^2 + k$.

EXAMPLE 5   Find the vertex and the axis of symmetry of the graph of $g(x) = \dfrac{1}{2}(x + 2)^2 + 5$. Use a graphing utility to verify the results.

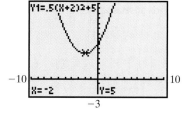

Solution   The function $g(x) = \dfrac{1}{2}(x + 2)^2 + 5$ may be written as $g(x) = \dfrac{1}{2}[x - (-2)]^2 + 5$. Thus, this graph is the same shape as the graph of $y = x^2$ shifted 2 units to the left and 5 units up, and it is wider because a is $\dfrac{1}{2}$. The vertex is $(-2, 5)$, and the axis of symmetry is $x = -2$. The screen to the left shows the graph with the vertex indicated.

PRACTICE

5   Graph $h(x) = \dfrac{1}{3}(x - 4)^2 - 3$.

In general, the following holds.

Graph of a Quadratic Function

The graph of a quadratic function written in the form $f(x) = a(x - h)^2 + k$ is a parabola with vertex (h, k). If $a > 0$, the parabola opens upward, and if $a < 0$, the parabola opens downward. The axis of symmetry is the line whose equation is $x = h$.

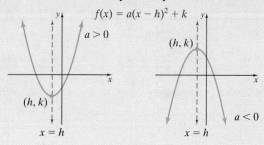

Concept Check ☑

Which description of the graph of $f(x) = -0.35(x + 3)^2 - 4$ is correct?

a. The graph opens downward and has its vertex at $(-3, 4)$.

b. The graph opens upward and has its vertex at $(-3, 4)$.

c. The graph opens downward and has its vertex at $(-3, -4)$.

d. The graph is narrower than the graph of $y = x^2$.

Answer to Concept Check: c

VOCABULARY & READINESS CHECK

Use the choices below to fill in each blank. Some choices will be used more than once.

upward   highest   parabola   downward   lowest   quadratic

1. A _____ function is one that can be written in the form $f(x) = ax^2 + bx + c, a \neq 0$.

2. The graph of a quadratic function is a _____ opening _____ or _____.

3. If $a > 0$, the graph of the quadratic function opens _____.

4. If $a < 0$, the graph of the quadratic function opens _____.

5. The vertex of a parabola is the _____ point if $a > 0$.

6. The vertex of a parabola is the _____ point if $a < 0$.

State the vertex of the graph of each quadratic function.

7. $f(x) = x^2$

8. $f(x) = -5x^2$

9. $g(x) = (x - 2)^2$

10. $g(x) = (x + 5)^2$

11. $f(x) = 2x^2 + 3$

12. $h(x) = x^2 - 1$

13. $g(x) = (x + 1)^2 + 5$

14. $h(x) = (x - 10)^2 - 7$

8.5 EXERCISE SET

MyMathLab   PRACTICE   WATCH   DOWNLOAD   READ   REVIEW

MIXED PRACTICE

Sketch the graph of each quadratic function. Label the vertex, and sketch and label the axis of symmetry. See Examples 1 through 3.

1. $f(x) = x^2 - 1$

2. $g(x) = x^2 + 3$

3. $h(x) = x^2 + 5$

4. $h(x) = x^2 - 4$

5. $g(x) = x^2 + 7$

6. $f(x) = x^2 - 2$

7. $f(x) = (x - 5)^2$

8. $g(x) = (x + 5)^2$

9. $h(x) = (x + 2)^2$

10. $H(x) = (x - 1)^2$

11. $G(x) = (x + 3)^2$

12. $f(x) = (x - 6)^2$

13. $f(x) = (x - 2)^2 + 5$

14. $g(x) = (x - 6)^2 + 1$

15. $h(x) = (x + 1)^2 + 4$

16. $G(x) = (x + 3)^2 + 3$

17. $g(x) = (x + 2)^2 - 5$

18. $h(x) = (x + 4)^2 - 6$

Sketch the graph of each quadratic function. Label the vertex, and sketch and label the axis of symmetry. See Example 4.

19. $g(x) = -x^2$

20. $f(x) = 5x^2$

21. $h(x) = \dfrac{1}{3}x^2$

22. $f(x) = -\dfrac{1}{4}x^2$

23. $H(x) = 2x^2$

24. $g(x) = -3x^2$

Sketch the graph of each quadratic function. Label the vertex, and sketch and label the axis of symmetry. See Example 5.

25. $f(x) = 2(x - 1)^2 + 3$

26. $g(x) = 4(x - 4)^2 + 2$

27. $h(x) = -3(x + 3)^2 + 1$

28. $f(x) = -(x - 2)^2 - 6$

29. $H(x) = \dfrac{1}{2}(x - 6)^2 - 3$

30. $G(x) = \dfrac{1}{5}(x + 4)^2 + 3$

MIXED PRACTICE

Sketch the graph of each quadratic function. Label the vertex, and sketch and label the axis of symmetry.

31. $f(x) = -(x - 2)^2$

32. $g(x) = -(x + 6)^2$

33. $F(x) = -x^2 + 4$

34. $H(x) = -x^2 + 10$

35. $F(x) = 2x^2 - 5$

36. $g(x) = \dfrac{1}{2}x^2 - 2$

37. $h(x) = (x - 6)^2 + 4$

38. $f(x) = (x - 5)^2 + 2$

39. $F(x) = \left(x + \dfrac{1}{2}\right)^2 - 2$

40. $H(x) = \left(x + \dfrac{1}{2}\right)^2 - 3$

41. $F(x) = \dfrac{3}{2}(x + 7)^2 + 1$

42. $g(x) = -\dfrac{3}{2}(x - 1)^2 - 5$

43. $f(x) = \dfrac{1}{4}x^2 - 9$

44. $H(x) = \dfrac{3}{4}x^2 - 2$

45. $G(x) = 5\left(x + \dfrac{1}{2}\right)^2$

46. $F(x) = 3\left(x - \dfrac{3}{2}\right)^2$

47. $h(x) = -(x - 1)^2 - 1$

48. $f(x) = -3(x + 2)^2 + 2$

49. $g(x) = \sqrt{3}(x + 5)^2 + \dfrac{3}{4}$

50. $G(x) = \sqrt{5}(x - 7)^2 - \dfrac{1}{2}$

51. $h(x) = 10(x + 4)^2 - 6$

52. $h(x) = 8(x + 1)^2 + 9$

53. $f(x) = -2(x - 4)^2 + 5$

54. $G(x) = -4(x + 9)^2 - 1$

REVIEW AND PREVIEW

Add the proper constant to each binomial so that the resulting trinomial is a perfect square trinomial. See Section 8.1.

55. $x^2 + 8x$

56. $y^2 + 4y$

57. $z^2 - 16z$

58. $x^2 - 10x$

59. $y^2 + y$

60. $z^2 - 3z$

Solve by completing the square. See Section 8.1.

61. $x^2 + 4x = 12$

62. $y^2 + 6y = -5$

63. $z^2 + 10z - 1 = 0$

64. $x^2 + 14x + 20 = 0$

65. $z^2 - 8z = 2$

66. $y^2 - 10y = 3$

CONCEPT EXTENSIONS

Solve. See the Concept Check in this section.

67. Which description of $f(x) = -213(x - 0.1)^2 + 3.6$ is correct?

Graph Opens	Vertex
a. upward	$(0.1, 3.6)$
b. upward	$(-213, 3.6)$
c. downward	$(0.1, 3.6)$
d. downward	$(-0.1, 3.6)$

68. Which description of $f(x) = 5\left(x + \dfrac{1}{2}\right)^2 + \dfrac{1}{2}$ is correct?

Graph Opens	Vertex
a. upward	$\left(\dfrac{1}{2}, \dfrac{1}{2}\right)$
b. upward	$\left(-\dfrac{1}{2}, \dfrac{1}{2}\right)$
c. downward	$\left(\dfrac{1}{2}, -\dfrac{1}{2}\right)$
d. downward	$\left(-\dfrac{1}{2}, -\dfrac{1}{2}\right)$

Write the equation of the parabola that has the same shape as $f(x) = 5x^2$ but with the following vertex.

69. $(2, 3)$

70. $(1, 6)$

71. $(-3, 6)$

72. $(4, -1)$

The shifting properties covered in this section apply to the graphs of all functions. Given the accompanying graph of $y = f(x)$, sketch the graph of each of the following.

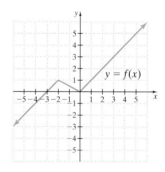

73. $y = f(x) + 1$

74. $y = f(x) - 2$

75. $y = f(x - 3)$

76. $y = f(x + 3)$

77. $y = f(x + 2) + 2$

78. $y = f(x - 1) + 1$

79. The quadratic function $f(x) = 668.8x^2 - 2990.1x + 939$ approximates the U.S. growth of cell phone subscribers between 1985 and 2006, where x is the number of years past 1985 and $f(x)$ is the number of subscribers in thousands. (This function differs slightly from the Chapter 8 opener as the data here extends to 2006.)

    a. Use this function to approximate the number of subscribers in 2004.

    b. Use this function to approximate the number of subscribers in 2006.

80. Use the function in Exercise 79.

    a. Predict the number of cell phone subscribers in 2012.

    b. Look up the current population of the United States.

    c. Based on your answers for parts **a.** and **b.**, discuss some limitations of using this quadratic function to predict data.

8.6 FURTHER GRAPHING OF QUADRATIC FUNCTIONS

OBJECTIVE 1 ▶ Writing quadratic functions in the form $y = a(x - h)^2 + k$. We know that the graph of a quadratic function is a parabola. If a quadratic function is written in the form

$$f(x) = a(x - h)^2 + k$$

we can easily find the vertex (h, k) and graph the parabola. To write a quadratic function in this form, complete the square. (See Section 8.1 for a review of completing the square.)

EXAMPLE 1   Graph $f(x) = x^2 - 4x - 12$. Find the vertex and any intercepts.

Solution   The graph of this quadratic function is a parabola. To find the vertex of the parabola, we will write the function in the form $y = (x - h)^2 + k$. To do this, we complete the square on the binomial $x^2 - 4x$. To simplify our work, we let $f(x) = y$.

$$y = x^2 - 4x - 12 \quad \text{Let } f(x) = y.$$

$$y + 12 = x^2 - 4x \qquad \begin{array}{l}\text{Add 12 to both sides to get}\\ \text{the } x\text{-variable terms alone.}\end{array}$$

Now we add the square of half of -4 to both sides.

$$\frac{1}{2}(-4) = -2 \quad \text{and} \quad (-2)^2 = 4$$

$$y + 12 + 4 = x^2 - 4x + 4 \qquad \text{Add 4 to both sides.}$$

$$y + 16 = (x - 2)^2 \qquad \text{Factor the trinomial.}$$

$$y = (x - 2)^2 - 16 \quad \text{Subtract 16 from both sides.}$$

$$f(x) = (x - 2)^2 - 16 \quad \text{Replace } y \text{ with } f(x).$$

From this equation, we can see that the vertex of the parabola is $(2, -16)$, a point in quadrant IV, and the axis of symmetry is the line $x = 2$.

Notice that $a = 1$. Since $a > 0$, the parabola opens upward. This parabola opening upward with vertex $(2, -16)$ will have two x-intercepts and one y-intercept. (See the Helpful Hint after this example.)

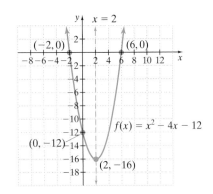

x-intercepts: let y or $f(x) = 0$	y-intercept: let $x = 0$
$f(x) = x^2 - 4x - 12$	$f(x) = x^2 - 4x - 12$
$0 = x^2 - 4x - 12$	$f(0) = 0^2 - 4 \cdot 0 - 12$
$0 = (x - 6)(x + 2)$	$= -12$
$0 = x - 6 \quad \text{or} \quad 0 = x + 2$	
$6 = x \qquad \qquad -2 = x$	

The two x-intercepts are $(6, 0)$ and $(-2, 0)$. The y-intercept is $(0, -12)$. The sketch of $f(x) = x^2 - 4x - 12$ is shown in the margin.

Notice that the axis of symmetry is always halfway between the x-intercepts. For the example above, halfway between -2 and 6 is $\dfrac{-2 + 6}{2} = 2$, and the axis of symmetry is $x = 2$. □

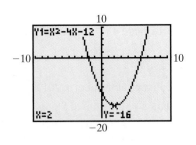

A check for Example 1.

PRACTICE

1   Graph $g(x) = x^2 - 2x - 3$. Find the vertex and any intercepts.

▶ **Helpful Hint**

Parabola Opens Upward	Parabola Opens Downward
Vertex in I or II: no x-intercept	Vertex in I or II: 2 x-intercepts
Vertex in III or IV: 2 x-intercepts	Vertex in III or IV: no x-intercept.

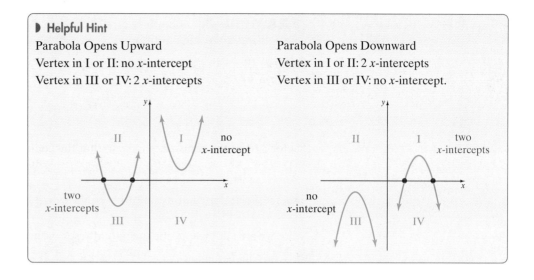

From the Helpful Hint above, you can see that the vertex of an upward parabola is its minimum point and the vertex of a downward parabola is its maximum point. We can find these points using the Calc menu and the maximum and minimum feature on a graphing utility.

EXAMPLE 2   Use a graphing utility to find the vertex and the axis of symmetry of the graph of $y = -2x^2 + 5x - 1$.

Solution   Graph $y_1 = -2x^2 + 5x - 1$ in a decimal window. Since the coefficent of x^2 is negative, the parabola opens downward and the function has a maximum value. Press maximum on the Calc menu. The calculator will find the maximum (or minimum) over a specified interval only, so we must first enter left and right end points or bounds for an interval. Next, we move the cursor as close as possible to what appears to be the vertex for a Guess.

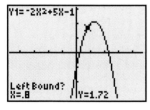

Left Bound

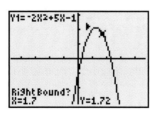

Right Bound

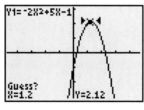

Guess

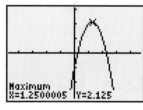

Maximum

The maximum value of the function is 2.125. The vertex is then $(1.25, 2.125)$. Since it is a downward parabola, the axis of symmetry is a vertical line through the vertex with the equation $x = 1.25$.

Note: If the parabola is an upward parabola, then use the minimum feature on the Calc menu.

□

PRACTICE

2   Find the vertex and axis of symmetry of the graph of $y = x^2 - x - 6$.

EXAMPLE 3 Graph $f(x) = 3x^2 + 3x + 1$. Find the vertex and any intercepts. Check using a graphing utility.

Solution Replace $f(x)$ with y and complete the square on x to write the equation in the form $y = a(x - h)^2 + k$.

$$y = 3x^2 + 3x + 1 \quad \text{Replace } f(x) \text{ with } y.$$
$$y - 1 = 3x^2 + 3x \quad \text{Isolate } x\text{-variable terms.}$$

Factor 3 from the terms $3x^2 + 3x$ so that the coefficient of x^2 is 1.

$$y - 1 = 3(x^2 + x) \quad \text{Factor out 3.}$$

The coefficient of x in the parentheses above is 1. Then $\dfrac{1}{2}(1) = \dfrac{1}{2}$ and $\left(\dfrac{1}{2}\right)^2 = \dfrac{1}{4}$.

Since we are adding $\dfrac{1}{4}$ inside the parentheses, we are really adding $3\left(\dfrac{1}{4}\right)$, so we *must* add $3\left(\dfrac{1}{4}\right)$ to the left side.

$$y - 1 + 3\left(\frac{1}{4}\right) = 3\left(x^2 + x + \frac{1}{4}\right)$$

$$y - \frac{1}{4} = 3\left(x + \frac{1}{2}\right)^2 \quad \text{Simplify the left side and factor the right side.}$$

$$y = 3\left(x + \frac{1}{2}\right)^2 + \frac{1}{4} \quad \text{Add } \frac{1}{4} \text{ to both sides.}$$

$$f(x) = 3\left(x + \frac{1}{2}\right)^2 + \frac{1}{4} \quad \text{Replace } y \text{ with } f(x).$$

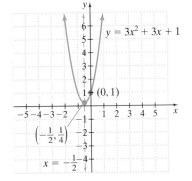

$y = 3x^2 + 3x + 1$

$(0, 1)$

$\left(-\frac{1}{2}, \frac{1}{4}\right)$

$x = -\frac{1}{2}$

Then $a = 3$, $h = -\dfrac{1}{2}$, and $k = \dfrac{1}{4}$. This means that the parabola opens upward with vertex $\left(-\dfrac{1}{2}, \dfrac{1}{4}\right)$ and that the axis of symmetry is the line $x = -\dfrac{1}{2}$.

To find the y-intercept, let $x = 0$. Then

$$f(0) = 3(0)^2 + 3(0) + 1 = 1$$

Thus the y-intercept is $(0, 1)$.

This parabola has no x-intercepts since the vertex is in the second quadrant and opens upward. Use the vertex, axis of symmetry, and y-intercept to sketch the parabola. □

PRACTICE

3 Graph $g(x) = 4x^2 + 4x + 3$. Find the vertex and any intercepts.

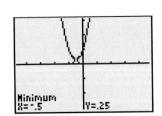

Minimum
X=-.5   Y=.25

A check for Example 3. This parabola opens up, so we used the minimum feature to calculate the vertex.

OBJECTIVE 2 ▶ Deriving a formula for finding the vertex. There is also a formula that may be used to find the vertex of a parabola. Now that we have practiced completing the square, we will show that the x-coordinate of the vertex of the graph of $f(x)$ or $y = ax^2 + bx + c$ can be found by the formula $x = \dfrac{-b}{2a}$. To do so, we complete the square on x and write the equation in the form $y = a(x - h)^2 + k$.

First, isolate the x-variable terms by subtracting c from both sides.

$$y = ax^2 + bx + c$$
$$y - c = ax^2 + bx$$

Next, factor a from the terms $ax^2 + bx$.

$$y - c = a\left(x^2 + \frac{b}{a}x\right)$$

Next, add the square of half of $\dfrac{b}{a}$, or $\left(\dfrac{b}{2a}\right)^2 = \dfrac{b^2}{4a^2}$, to the right side inside the parentheses. Because of the factor a, what we really added was $a\left(\dfrac{b^2}{4a^2}\right)$ and this must be added to the left side.

$$y - c + a\left(\frac{b^2}{4a^2}\right) = a\left(x^2 + \frac{b}{a}x + \frac{b^2}{4a^2}\right)$$

$$y - c + \frac{b^2}{4a} = a\left(x + \frac{b}{2a}\right)^2 \qquad \text{Simplify the left side and factor the right side.}$$

$$y = a\left(x + \frac{b}{2a}\right)^2 + c - \frac{b^2}{4a} \qquad \text{Add } c \text{ to both sides and subtract } \frac{b^2}{4a} \text{ from both sides.}$$

Compare this form with $f(x)$ or $y = a(x - h)^2 + k$ and see that h is $\dfrac{-b}{2a}$, which means that the x-coordinate of the vertex of the graph of $f(x) = ax^2 + bx + c$ is $\dfrac{-b}{2a}$.

> **Vertex Formula**
> The graph of $f(x) = ax^2 + bx + c$, when $a \neq 0$, is a parabola with vertex
> $$\left(\frac{-b}{2a}, f\left(\frac{-b}{2a}\right)\right)$$

Let's use this formula to find the vertex of the parabola we graphed in Example 1.

EXAMPLE 4   Find the vertex of the graph of $f(x) = x^2 - 4x - 12$.

Solution   In the quadratic function $f(x) = x^2 - 4x - 12$, notice that $a = 1, b = -4,$ and $c = -12$. Then

$$\frac{-b}{2a} = \frac{-(-4)}{2(1)} = 2$$

The x-value of the vertex is 2. To find the corresponding $f(x)$ or y-value, find $f(2)$. Then

$$f(2) = 2^2 - 4(2) - 12 = 4 - 8 - 12 = -16$$

The vertex is $(2, -16)$. These results agree with our findings in Example 1.   □

PRACTICE
4   Find the vertex of the graph of $g(x) = x^2 - 2x - 3$.

A calculator check of Example 4.

```
1→A: -4→B: -B/(2A)
                    2
2²-4(2)-12
                  -16
■
```

EXAMPLE 5   Use a graphing utility to graph $f(x) = 3 - 2x - x^2$. Find the vertex and any intercepts algebraically.

Solution   For $f(x) = 3 - 2x - x^2$, we have $a = -1, b = -2,$ and $c = 3$. Since $a < 0$, the parabola opens downward. We use the vertex formula to find the vertex.

$$x = \frac{-b}{2a} = \frac{-(-2)}{2(-1)} = -1$$

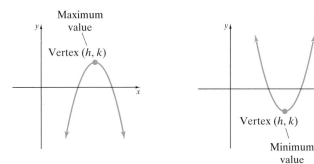

Also, $f(-1) = 4$, so the vertex has coordinates $(-1, 4)$. The vertex is in the second quadrant and the parabola opens downward, so the parabola has two x-intercepts and a y-intercept, as shown to the left.

To find the x-intercepts, let y or $f(x) = 0$ and solve for x.

$$f(x) = 3 - 2x - x^2$$
$$0 = 3 - 2x - x^2$$
$$0 = (3 + x)(1 - x)$$
$$3 + x = 0 \quad \text{or} \quad 1 - x = 0$$
$$x = -3 \quad \text{or} \quad 1 = x$$

The x-intercepts are $(-3, 0)$ and $(1, 0)$. To find the y-intercept, let $x = 0$.

$$f(0) = 3 - 2 \cdot 0 - 0^2 = 3$$

The y-intercept is $(0, 3)$.   □

PRACTICE

5   Use a graphing utility to graph $g(x) = -x^2 + 5x + 6$. Find the vertex and any intercepts algebraically.

OBJECTIVE 3 ▶ Finding minimum and maximum values. The vertex of a parabola gives us some important information about its corresponding quadratic function. The quadratic function whose graph is a parabola that opens upward has a minimum value, and the quadratic function whose graph is a parabola that opens downward has a maximum value. The $f(x)$ or y-value of the vertex is the minimum or maximum value of the function.

Maximum
value
Vertex (h, k)

Vertex (h, k)
Minimum
value

Concept Check ☑

Without making any calculations, tell whether the graph of $f(x) = 7 - x - 0.3x^2$ has a maximum value or a minimum value. Explain your reasoning.

EXAMPLE 6   **Finding Maximum Height**

A rock is thrown upward from the ground. Its height in feet above ground after t seconds is given by the function $f(t) = -16t^2 + 20t$. Find the maximum height of the rock and the number of seconds it took for the rock to reach its maximum height.

Solution

1. UNDERSTAND. The maximum height of the rock is the largest value of $f(t)$. Since the function $f(t) = -16t^2 + 20t$ is a quadratic function, its graph is a parabola. It opens downward since $-16 < 0$. Thus, the maximum value of $f(t)$ is the $f(t)$ or y-value of the vertex of its graph.

Answer to Concept Check:
$f(x)$ has a maximum value since it opens downward.

2. TRANSLATE. To find the vertex (h, k), notice that for $f(t) = -16t^2 + 20t$, $a = -16, b = 20$, and $c = 0$. We will use these values and the vertex formula

$$\left(\frac{-b}{2a}, f\left(\frac{-b}{2a} \right) \right)$$

3. SOLVE.

$$h = \frac{-b}{2a} = \frac{-20}{-32} = \frac{5}{8}$$

$$f\left(\frac{5}{8} \right) = -16\left(\frac{5}{8} \right)^2 + 20\left(\frac{5}{8} \right)$$

$$= -16\left(\frac{25}{64} \right) + \frac{25}{2}$$

$$= -\frac{25}{4} + \frac{50}{4} = \frac{25}{4}$$

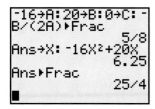

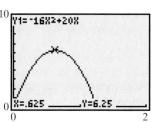

4. INTERPRET. Check: The graph of $f(t)$ is a parabola opening downward with vertex $\left(\frac{5}{8}, \frac{25}{4} \right)$. To *check*, graph $y_1 = -16x^2 + 20x$ in a $[0, 2, 1]$ by $[0, 10, 1]$ window. The maximum height occurs when $x = 0.625$ or $\frac{5}{8}$ seconds and is $6\frac{1}{4}$ feet.

State: This means that the rock's maximum height is $\frac{25}{4}$ feet, or $6\frac{1}{4}$ feet, which was reached in $\frac{5}{8}$ second.   □

PRACTICE

6   A ball is tossed upward from the ground. Its height in feet above ground after t seconds is given by the function $h(t) = -16t^2 + 24t$. Find the maximum height of the ball and the number of seconds it took for the ball to reach the maximum height.

VOCABULARY & READINESS CHECK

Fill in each blank.

1. If a quadratic function is in the form $f(x) = a(x - h)^2 + k$, the vertex of its graph is _____ .

2. The graph of $f(x) = ax^2 + bx + c, a \neq 0$, is a parabola whose vertex has x-value of _____ .

	Parabola Opens	*Vertex Location*	*Number of x-intercept(s)*	*Number of y-intercept(s)*
3.	up	Q I		
4.	up	Q III		
5.	down	Q II		
6.	down	Q IV		
7.	up	*x*-axis		
8.	down	*x*-axis		
9.		Q III	0	
10.		Q I	2	
11.		Q IV	2	
12.		Q II	0	

8.6 | EXERCISE SET

Find the vertex of the graph of each quadratic function. See Examples 1 through 5.

1. $f(x) = x^2 + 8x + 7$

2. $f(x) = x^2 + 6x + 5$

3. $f(x) = -x^2 + 10x + 5$

4. $f(x) = -x^2 - 8x + 2$

5. $f(x) = 5x^2 - 10x + 3$

6. $f(x) = -3x^2 + 6x + 4$

7. $f(x) = -x^2 + x + 1$

8. $f(x) = x^2 - 9x + 8$

Match each function with its graph. See Examples 1 through 5.

A
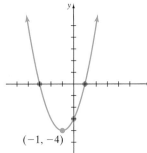
$(-1, -4)$

B
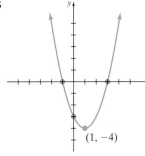
$(1, -4)$

C
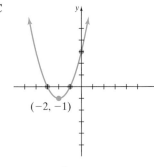
$(-2, -1)$

D
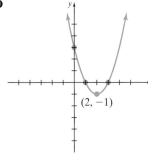
$(2, -1)$

9. $f(x) = x^2 - 4x + 3$

10. $f(x) = x^2 + 2x - 3$

11. $f(x) = x^2 - 2x - 3$

12. $f(x) = x^2 + 4x + 3$

MIXED PRACTICE

Find the vertex of the graph of each quadratic function. Determine whether the graph opens upward or downward, find any intercepts, and sketch the graph. See Examples 1 through 5.

13. $f(x) = x^2 + 4x - 5$

14. $f(x) = x^2 + 2x - 3$

 15. $f(x) = -x^2 + 2x - 1$

16. $f(x) = -x^2 + 4x - 4$

17. $f(x) = x^2 - 4$

18. $f(x) = x^2 - 1$

19. $f(x) = 4x^2 + 4x - 3$

20. $f(x) = 2x^2 - x - 3$

21. $f(x) = x^2 + 8x + 15$

22. $f(x) = x^2 + 10x + 9$

23. $f(x) = x^2 - 6x + 5$

24. $f(x) = x^2 - 4x + 3$

25. $f(x) = x^2 - 4x + 5$

26. $f(x) = x^2 - 6x + 11$

27. $f(x) = 2x^2 + 4x + 5$

28. $f(x) = 3x^2 + 12x + 16$

29. $f(x) = -2x^2 + 12x$

30. $f(x) = -4x^2 + 8x$

31. $f(x) = x^2 + 1$

32. $f(x) = x^2 + 4$

33. $f(x) = x^2 - 2x - 15$

34. $f(x) = x^2 - x - 12$

35. $f(x) = -5x^2 + 5x$

36. $f(x) = 3x^2 - 12x$

37. $f(x) = -x^2 + 2x - 12$

38. $f(x) = -x^2 + 8x - 17$

39. $f(x) = 3x^2 - 12x + 15$

40. $f(x) = 2x^2 - 8x + 11$

41. $f(x) = x^2 + x - 6$

42. $f(x) = x^2 + 3x - 18$

43. $f(x) = -2x^2 - 3x + 35$

44. $f(x) = 3x^2 - 13x - 10$

Solve. See Example 6.

45. If a projectile is fired straight upward from the ground with an initial speed of 96 feet per second, then its height h in feet after t seconds is given by the equation

$$h(t) = -16t^2 + 96t$$

Find the maximum height of the projectile.

46. If Rheam Gaspar throws a ball upward with an initial speed of 32 feet per second, then its height h in feet after t seconds is given by the equation

$$h(t) = -16t^2 + 32t$$

Find the maximum height of the ball.

47. The cost C in dollars of manufacturing x bicycles at Holladay's Production Plant is given by the function

$$C(x) = 2x^2 - 800x + 92,000.$$

a. Find the number of bicycles that must be manufactured to minimize the cost.

b. Find the minimum cost.

48. The Utah Ski Club sells calendars to raise money. The profit P, in cents, from selling x calendars is given by the equation $P(x) = 360x - x^2$.

a. Find how many calendars must be sold to maximize profit.

b. Find the maximum profit.

49. Find two numbers whose sum is 60 and whose product is as large as possible. [*Hint:* Let x and $60 - x$ be the two positive numbers. Their product can be described by the function $f(x) = x(60 - x)$.]

50. Find two numbers whose sum is 11 and whose product is as large as possible. (Use the hint for Exercise 49.)

51. Find two numbers whose difference is 10 and whose product is as small as possible. (Use the hint for Exercise 49.)

52. Find two numbers whose difference is 8 and whose product is as small as possible.

△ **53.** The length and width of a rectangle must have a sum of 40. Find the dimensions of the rectangle that will have the maximum area. (Use the hint for Exercise 49.)

△ **54.** The length and width of a rectangle must have a sum of 50. Find the dimensions of the rectangle that will have maximum area.

REVIEW AND PREVIEW

Sketch the graph of each function. See Section 8.5.

55. $f(x) = x^2 + 2$

56. $f(x) = (x - 3)^2$

57. $g(x) = x + 2$

58. $h(x) = x - 3$

59. $f(x) = (x + 5)^2 + 2$

60. $f(x) = 2(x - 3)^2 + 2$

61. $f(x) = 3(x - 4)^2 + 1$

62. $f(x) = (x + 1)^2 + 4$

63. $f(x) = -(x - 4)^2 + \dfrac{3}{2}$

64. $f(x) = -2(x + 7)^2 + \dfrac{1}{2}$

CONCEPT EXTENSIONS

Without calculating, tell whether each graph has a minimum value or a maximum value. See the Concept Check in the section.

65. $f(x) = 2x^2 - 5$

66. $g(x) = -7x^2 + x + 1$

67. $F(x) = 3 - \dfrac{1}{2}x^2$

68. $G(x) = 3 - \dfrac{1}{2}x + 0.8x^2$

Find the vertex of the graph of each quadratic function. Determine whether the graph opens upward or downward, find the y-intercept, approximate the x-intercepts to one decimal place, and sketch the graph.

69. $f(x) = x^2 + 10x + 15$

70. $f(x) = x^2 - 6x + 4$

71. $f(x) = 3x^2 - 6x + 7$

72. $f(x) = 2x^2 + 4x - 1$

Find the maximum or minimum value of each function. Approximate to two decimal places.

73. $f(x) = 2.3x^2 - 6.1x + 3.2$

74. $f(x) = 7.6x^2 + 9.8x - 2.1$

75. $f(x) = -1.9x^2 + 5.6x - 2.7$

76. $f(x) = -5.2x^2 - 3.8x + 5.1$

77. The number of McDonald's restaurants worldwide can be modeled by the quadratic equation $f(x) = -96x^2 + 1018x + 28{,}824$, where $f(x)$ is the number of McDonald's restaurants and x is the number of years after 2000. (*Source:* Based on data from McDonald's Corporation)

  a. Will this function have a maximum or minimum? How can you tell?

  b. According to this model, in what year was the number of McDonald's restaurants at its maximum/minimum?

  c. What is the maximum/minimum number of McDonald's restaurants predicted?

78. Methane is a gas produced by landfills, natural gas systems, and coal mining that contributes to the greenhouse effect and global warming. Projected methane emissions in the United States can be modeled by the quadratic function

$$f(x) = -0.072x^2 + 1.93x + 173.9$$

where $f(x)$ is the amount of methane produced in million metric tons and x is the number of years after 2000. (*Source:* Based on data from the U.S. Environmental Protection Agency, 2000–2020)

  a. According to this model, what will U.S. emissions of methane be in 2009? (Round to 2 decimal places.)

  b. Will this function have a maximum or a minimum? How can you tell?

  c. In what year will methane emissions in the United States be at their maximum/minimum? Round to the nearest whole year.

  d. What is the level of methane emissions for that year? (Use your rounded answer from part **c**.) (Round this answer to 2 decimals places.)

8.7 INTERPRETING DATA: LINEAR AND QUADRATIC MODELS

OBJECTIVE

1 Plot data points and calculate linear and quadratic regression models.

As we have seen thus far in our text, many situations arise that involve two related quantities. We saw that we can investigate and record certain facts about these situations, and that these are called **data.** To *model* the data means to find an equation that describes the relationship between the given data quantities. Visually, the graph of the *best fit equation* should have a majority of the plotted ordered pair data points on the graph or close to it. Recall from Section 2.6 that a technique called **regression** is used to determine an equation that best fits two-quantity data. In this text, we concentrate on modeling ordered pair data points with an equation that visually appears to best fit the data points. In this section, we will also check the r^2 or R^2 numbers, called coefficients of determination. The best fit occurs when r^2 or R^2 is close to 1.

Recall that an equation that best fits a set of ordered pair data points is called a regression equation. Keep in mind that no model, or equation, may fit the data exactly, but a model should fit the data closely enough so that useful predictions may be made using it. Most importantly, know that when we use an equation to make predictions in the future, we are assuming that future data is also modeled by the equation.

An important caution should be made about regression equations and using such equations for predictions.

Just because an r^2 or R^2 coefficient of determination is closer to 1, it absolutely does *not* mean that this equation will be a better predictor of future trends. Often, companies model past and present data using many regression equations. Predictions are then made based on R^2 values, the economy, and other restrictions and data.

OBJECTIVE 1 ▶ Plotting data points and using a linear or quadratic model. In this section, we concentrate on two models, linear and quadratic. These models are shown below for your review.

Linear Model
$y = mx + b$
slope: m
y-intercept: $(0, b)$

Quadratic Model
$y = ax^2 + bx + c$
$a > 0$, *parabola opens upward*
$a < 0$, *parabola opens downward*
vertex: (h, k) *where* $h = \dfrac{-b}{2a}$

EXAMPLE 1 Determining the Better Model for the Cost of a U.S. First-Class Postage Stamp

The cost of U.S. first-class postage stamps has risen over the years between 1917 and 2008.

Years since 1900	17	19	32	58	63	68	71	74	75	78	81
Cost	3	2	3	4	5	6	8	10	13	18	20

Years since 1900	85	88	91	101	102	106	107	108
Cost	22	25	29	34	37	39	41	42

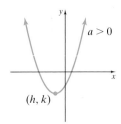

a. Graph the data in the table. Let $x = 0$ represent the year 1900.

b. Find a linear regression equation that models the data.

c. Find a quadratic equation that models the data.

d. Find which regression equation is the better fit by looking at the R^2 values and the graphs of each equation. Use that equation to predict the cost of a first-class stamp in the year 2020 if the trend continues.

Solution

a. Enter the number of years since 1900 in List 1 and the cost of the stamp in List 2. Draw a scatter plot as shown to the left below.

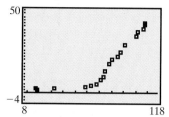
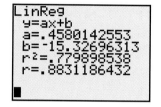

b. The linear regression equation is $y = 0.458x - 15.327$ as shown to the right above. (This equation was stored in Y_1.)

c.

The quadratic regression equation is $y = 0.008x^2 - 0.561x + 10.285$, as shown above. (This equation was stored in Y_2.)

d. From looking at the both regression equations' R^2 values, we see that the quadratic equation appears to be the better fit, since its R^2 value is closer to 1. The graph of the two equations with the data confirms that the quadratic equation $y = 0.008x^2 - 0.561x + 10.285$ is the better fit.

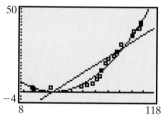

If the trend continues to rise at the same rate, we can predict the cost of a U.S. first-class postage stamp in the year 2020 to be approximately 58 cents.   ☐

PRACTICE

1 Use the Example 1 data from the years 1971 and later. Answer, parts **b**, **c**, and **d** from Example 1. For part **d**, use the regression equation whose r^2 or R^2 value is closer to 1.

EXAMPLE 2   Models for Fuel Consumption

The table below shows the average miles per gallon (MPG) for U.S. cars in the years shown. (*Source:* U.S. Federal Highway Administration)

Year	2000	2001	2002	2003	2004	2005	2006	2007
Average MPG	22.9	23.0	23.1	23.2	23.1	23.5	23.3	23.4

a. Graph the data points. Let x represent the number of years since 2000.

b. Find and graph a linear regression equation that models the data.

c. Use the equation to predict the average MPG for U.S. cars in the year 2013.

Solution

a. List the years in L_1 and the average MPG in L_2. (For L_1, enter the number of years past 2000.) Then draw a scatter plot as shown below.

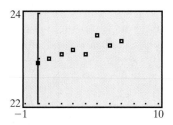

Notice that for the most part, the graph resembles the graph of a line.

b. To find and graph the linear regression equation that models this data, select LinReg $(ax + b)$ from the Stat Calc menu. Enter L_1, L_2, Y_1 so that the screen reads LinReg $(ax + b)$ L_1, L_2, Y_1. This instructs the calculator to use L_1 and L_2 as x and y lists and enters the regression equation in the Y= editor as Y_1.

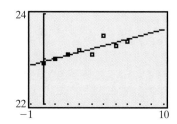

TECHNOLOGY NOTE

When predicting values of x that lie outside of the chosen window, it is easier to look at a table or evaluate the function $Y_1(x)$.

The linear regression equation is $y = 0.073x + 22.933$, where a and b values are rounded to three decimal places.

c. To predict the average MPG for the year 2013, we evaluate the linear regression equation we have stored in Y_1 for 13 ($2013 - 2000 = 13$). The predicted MPG for cars in the year 2013 is 23.9 MPG rounded to the nearest tenth.

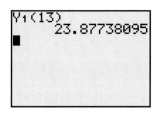

PRACTICE

2   Use the data in Example 2 for the years 2000, 2001, 2002, 2003, 2004, 2006, 2007. (Eliminate the 2005 data only.)

a. Find a linear regression equation that models the data.

b. Find a quadratic regression equation that models the data.

c. Use both equations to predict the average MPG for the year 2013.

d. Discuss the R^2 value for part **a** and why it is now closer to 1 as compared to Example 2.

For the next example, we study the increase in sport utility vehicle (SUV) sales.

EXAMPLE 3   **Models for SUV Sales**

The table below represents the retail sales in millions of sport utility vehicles (SUVs) for selected years 1996 through 2006.

Year	SUV Sales (in millions)
1996	2
1998	2.9
2000	3.5
2002	4.2
2004	4.7
2006	4.5

(*Source:* American Automobile Manufacturer's Assoc.)

a. Let x = the number of years since 1990 and graph the data points. Use the graph to visually identify the type of function that best fits the data.

b. Find the corresponding regression equation for the data and then graph the results.

c. If the sales trend continues in the same manner, predict the number of SUVs that we would expect to sell in the year 2011.

Solution

a. List the years in L_1 and the SUV sales in L_2. (For L_1, enter the number of years past 1990.) Then draw a scatter plot as shown below.

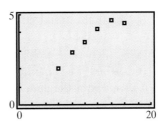

Since a quadratic function appears to best fit the data, we will model the data with a quadratic regression equation.

b. Select QuadReg from the Stat Calc menu. Enter L_1, L_2, Y_1, so that the screen reads QuadReg L_1, L_2, Y_1. This instructs the calculator to use L_1 and L_2 as x and y lists and enters the regression equation in the Y= editor as Y_1.

Quadratic regression equation.

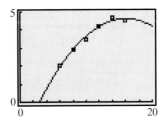

Graph of quadratic regression equation.

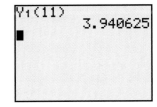

Projected SUV sales sales in 2011.

The quadratic regression equation is

$$y = -0.026x^2 + 0.845x - 2.169$$

with a, b, and c values rounded to 3 decimal places.

c. To predict SUV sales for the year 2011, we evaluate the quadratic regression equation we have stored in Y_1 for 2011. The predicted SUV sales for the year 2011 is 3.9 million SUVs, rounded to the nearest tenth of a million.   □

PRACTICE
3
The table shows yearly truck sales for a local dealership.

Year	1996	1999	2002	2005	2008
Trucks Sold	362	390	407	433	457

Answer Example 3 questions **a** and **b**. For part **c**, predict the truck sales in 2015.

In Example 4, we see a need for research institutions to forecast the projected amount of funding available for use in the future.

EXAMPLE 4   **Alzheimer's Research**

The funding for Alzheimer's disease research at the National Institutes of Health has doubled since 1994. The table below represents selected funding history.

Year	1998	2000	2002	2004	2006	2008
Funding (millions)	358	462	595	633	643	644

a. Let x = the number of years since 1990. Plot the data points and use the graph to identify the type of function that appears to best fit the data.

b. Find an appropriate regression equation for the data and then graph the results.

c. If the trend continues in the same pattern, find the amount of funding expected in 2012.

Solution

a. Enter the data in L_1 (years since 1990 and L_2 (funding in millions) and draw a scatter plot.
    The graph resembles a part of a parabola so we will model the data with a quadratic regression equation.

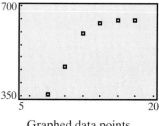

b. The regression results and the graph of the equation are shown on the screens below. The equation is $y = -4.496x^2 + 145.613x - 524.936$ with $a, b,$ and c rounded to three decimal places.

Graphed data points.

c. To forecast the amount of funding expected in the year 2012, we evaluate the quadratic regression equation we stored in Y_1 for $x = 22$. The expected funding for the year 2012 is approximately $503 million rounded to the nearest million dollars.

Quadratic regression equation.

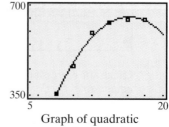

Graph of quadratic regression equation.

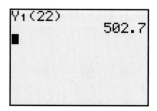

Predicted funding for 2012 in millions of dollars.   □

PRACTICE
4
Use the data in Example 4 for the years 2000, 2002, 2004, 2006, and 2008. Let x represent the number of years since the year 2000 and answer Example 4, parts **b** and **c**.

EXAMPLE 5   In the Chapter 2 opener (page 98), we saw a graph representing world diamond production. Use the table below, which gives the value in billions of dollars for the corresponding year. Let $x = 0$ be the number of years since the year 2000.

Year	2004	2005	2006	2007	2008
Value (in billions of dollars)	12	12.6	13	13.44	14

a. Find the curve of best fit.

b. Use the regression equation to predict the value in billions of dollars in the year 2012.

Solution

a. Graph a scatter plot of the data in a [4, 9, 1] by [11, 15, 1] window. We see that the graph appears to be linear and from the data, we can see that the value is increasing and we expect it to continue to increase over time.

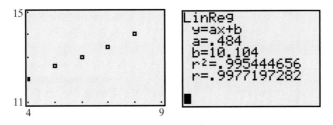

The linear regression equation is $y = 0.484x + 10.104$.

b. Using the equation we can see that we should expect $15.912 billion in the year 2012.

Notice that this equation is slightly different from the one given in the Chapter 2 opener, as here we use five pieces of data and the graph in the opener shows eight pieces of data. ☐

PRACTICE
5   The data below shows the yearly U.S. independent films released.

Year	2001	2002	2003	2004	2005	2006	2007
Number of films released	274	229	265	275	345	396	411

Source: Motion Picture Association of America

a. Let x represent the number of years since 2000. Find a linear or quadratic equation of best fit.

b. Use the results of part **a** to predict the number of independent films released in 2011.

In this section, we concentrated on linear and quadratic regression equations only. There are many other types of regression equations that can be used as models and there are advantages and disadvantages of each. Proceed with regression equations carefully and know that we have only looked at the tip of the regression analysis iceberg.

8.7 EXERCISE SET

MathXL PRACTICE | WATCH | DOWNLOAD | READ | REVIEW

Solve. Predictions using regression equations may differ slightly depending on the rounding of coefficients in the equation to different place values. Given the following graphs of data points, tell whether the graph is likely to be best represented by a linear model, a quadratic model, or neither. See Examples 1 through 5.

1. 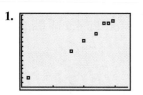   **2.**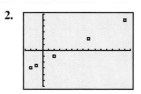

3. 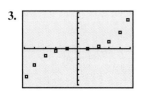   **4.**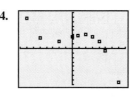

5. 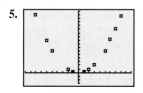   **6.**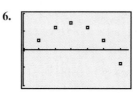

7.

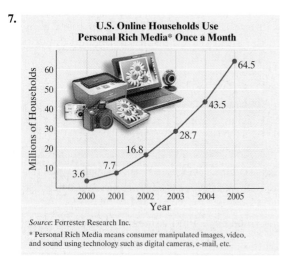

U.S. Online Households Use Personal Rich Media* Once a Month

Millions of Households

3.6, 7.7, 16.8, 28.7, 43.5, 64.5

Years: 2000, 2001, 2002, 2003, 2004, 2005

Year

Source: Forrester Research Inc.

* Personal Rich Media means consumer manipulated images, video, and sound using technology such as digital cameras, e-mail, etc.

8.

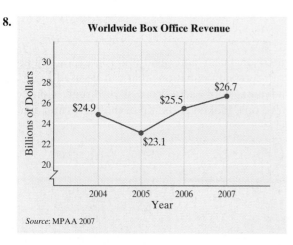

Worldwide Box Office Revenue

Billions of Dollars

$24.9, $23.1, $25.5, $26.7

Years: 2004, 2005, 2006, 2007

Year

Source: MPAA 2007

MIXED PRACTICE

For each exercise tell whether the graph is likely to be best represented by a linear model, a quadratic model, or neither. If linear or quadratic, find a linear or quadratic regression equation that models the given data. Round the coefficients to three decimal places. Then use the equation to answer the question. For Exercises 10, 11, and 12, let x = the number of years since 1990; otherwise, let x = the number of years since 2000. If the graphed data do not appear linear or quadratic, write "not linear or quadratic" and you are through with that exercise. See Examples 1 through 5.

9. The table below shows the number of U.S. wildfires in the years 2000–2006. If there is a trend, predict the number of U.S. wildfires in 2010.

Year	2000	2001	2002	2003	2004	2005	2006
No. of Wildfires (in thousands)	122.8	84.1	88.4	57.6	77.5	66.6	96.3

10. The table below shows the retail sales, in thousands, of passenger cars for the years 1997 through 2006. If there is a trend, predict the number of cars sold in the year 2011.

Year	Car Sales (in thousands)
1997	8272
1998	8142
1999	8677
2000	8852
2001	8422
2002	8103
2003	7610
2004	7506
2005	7667
2006	7780

11. The table below shows the median weekly earnings of full-time wage and salary workers 25 years and older with some years of college. If this trend continues, predict the median weekly earnings in 2012. For this exercise, find a linear and a quadratic regression equation and use both to predict the earnings in 2012.

Year	1990	1994	1996	1997	2000	2004
Weekly Earnings (in dollars)	476	499	518	528	574	580

12. In order to see the value of attaining a bachelor's degree, study the following data from the U.S. Bureau of Labor Statistics. This table shows mean weekly earnings of full-time wage and salary workers 25 years and older with a bachelor's degree. If there is a trend, predict the mean weekly earnings in 2012.

Year	1990	1994	1996	1997	2000	2004
Weekly Earnings (in dollars)	639	733	758	779	841	992

13. Use the data in Exercise 8 to write a quadratic regression equation. Then predict the worldwide box office revenue in 2011. Let x = the number of years since 2000.

14. Use the data in Exercise 7 to write a regression equation. Then predict the number of U.S. online households to use personal rich media once a year in the year 2012. Let $x =$ the number of years since 2000.

15. Suppose your regression equation predicts a population of 500,000,000 U.S. citizens in 2010. Discuss the limitations of such a regression equation.

16. Research and find the current number of U.S. households. Now compare this number with your predicted media-rich households in 2012. (See Exercise 14.) Discuss any limitations you may foresee with predictions and regression equations.

17. The table shows the number of global Internet users.

Year	2000	2001	2002	2003	2004	2005	2006	2007
Global Internet Users (in millions)	1125	1163	1175	1188	1200	1213	1225	1238

Let $x =$ the number of years since 2000. Round regression coefficients to 3 decimal places.

a. Write a linear regression equation.

b. Write a quadratic regression equation.

c. Use each regression equation to predict the number of global Internet users in the year 2012. Round to the nearest whole million.

d. Choose a prediction from part **c** that you think is the better prediction and explain why.

18. The table shows the number of North American broadband access lines.

Year	2000	2001	2002	2003	2004	2005	2006
North American broadband access lines (in millions)	15	27	41	58	87	93	99

Let $x =$ the number of years since 2000. Round regression coefficients to 3 decimal places.

a. Write a linear regression equation.

b. Write a quadratic regression equation.

c. Use each regression equation to predict the number of North American broadband access lines in 2013. Round to the nearest whole million.

d. Choose a prediction from part **c** that you think is the better prediction and explain why.

19. A tennis shoe manufacturer is studying the suggested selling price of a certain brand of tennis shoe. The suggested selling price cannot be too low or the cost of manufacturing the shoes will not be covered, and it cannot be too high or people will not purchase the shoes. Below are data collected for selling tennis shoes at various prices and the corresponding profit.

Price (in dollars)	39	49	65	75	85	95	105
Profit (in dollars)	9500	16,750	19,600	19,500	15,475	7500	1500

a. Plot the data points and use the graph to determine whether a linear or quadratic regression equation would best represent the data.

b. Find a regression equation for this data, where x is price of the tennis shoes and y is profit.

c. Using the model, find the projected profit if the shoes are sold for $80 and for $60.

d. Use the table to find the profit if the shoes are sold for $105. Is this a good selling price? Why or why not?

e. Using the model, find the selling price, to the nearest dollar, that will yield the maximum profit.

20. Below are data collected for selling men's and women's flip flop shoes.

Price (in dollars)	5	15	25	35	45	55
Profit (in dollars)	2750	5816	6225	5611	4527	3199

Answer Exercise 19, **a., b., e.**

Given the following data for four houses sold in comparable neighborhoods and their corresponding number of square feet, draw a scatter plot and find a linear regression equation that best fits the relationship between the number of square feet and the selling price of the house.

21.

Square Feet	1754	2151	2587	2671
Selling Price	$90,900	$130,000	$155,000	$172,500

22.

Square Feet	1519	2593	3005	3016
Selling Price	$91,238	$220,000	$254,000	$269,000

23. Use the equation found in Exercise 21 to predict the selling price of a house with 2400 square feet.

24. Use the equation found in Exercise 22 to predict the selling price of a house with 2200 square feet.

25. A quadratic equation that models the average miles per gallon (MPG) for U.S. cars only is given by:
$$y = -0.007x^2 + 0.127x + 22.86$$
where x is the year since 2000. Use this model to fill in the predicted MPG below and compare with the actual given data. (Round predicted MPG to one decimal place.)

x (year)	Actual MPG	Predicted MPG	Difference between Actual and Predicted MPG
2002	23.1		
2003	23.2		
2004	23.1		
2005	23.5		
2006	23.3		
2007	23.4		

26. Find a linear equation that models the annual consumption of cigarettes per person, where x is the year and y is the number of cigarettes. Use this model to fill in the predicted consumption of cigarettes below and compare with the actual given data.

x (year)	Actual Consumption (per person)	Predicted Consumption (per person)	Difference between Actual and Predicted
2000	1637		
2001	1626		
2002	1609		
2003	1545		

Plot the following points and then find a regression equation that best models the data. Let $x = 3$ represent March, $x = 4$ represent April, and so on. For each exercise, explain the limitation of the regression equation that you found.

27. The table below shows the average temperature in Wellington, New Zealand, for the months of March through October.

March	April	May	June	July	August	Sept	Oct
60	57	52	49	47	48	51	54

28. The table below shows the average temperature in Louisville, Kentucky, for the months of March through October.

March	April	May	June	July	August	Sept	Oct
45	57	65	74	78	76	70	58

State whether the relationship described can best be modeled by a linear function or a quadratic function.

29. The area of a square *and* the length of one side.

30. A salary of $300 per week plus 5% commission on sales *and* weekly sales.

31. A number of calculators sold each week at a store *and* their cost, plus tax.

32. The area of a circle *and* the radius of the circle

33. The average temperature of Cleveland, Ohio, *and* the months of March through October.

34. An oven is turned on and set to reach 400 degrees. After it reaches the desired temperature, it is shut off. Temperatures are recorded every 2 minutes, from the time the oven is turned on until it cools to room temperature.

35. The number of bacteria present in a person with respect to time when he is coming down with the flu, has the flu, and then recovers.

36. The distance traveled during a recent trip with respect to the amount of time elapsed.

REVIEW AND PREVIEW

Solve each of the following equations. See Sections 1.5 and 5.8.

37. $(x + 3)(x - 5) = 0$

38. $(x - 1)(x + 19) = 0$

39. $2x^2 - 7x - 15 = 0$

40. $6x^2 + 13x - 5 = 0$

41. $3(x - 4) + 2 = 5(x - 6)$

42. $4(2x + 7) = 2(x - 4)$

Find the y-intercept of the graph of each function. See Section 2.3.

43. $f(x) = x^3 + 3x^2 - 5x - 8$

44. $f(x) = 2x^3 + x^2 - 7x + 12$

45. $g(x) = x^2 - 3x + 5$

46. $g(x) = 3x^2 - 5x - 10$

CHAPTER 8 GROUP ACTIVITY

Fitting a Quadratic Model to Data

Throughout the twentieth century, the eating habits of Americans changed noticeably. Americans started consuming less whole milk and butter, and started consuming more skim and low-fat milk and margarine. We also started eating more poultry and fish. In this project, you will have the opportunity to investigate trends in per capita consumption of poultry during the twentieth century. This project may be completed by working in groups or individually.

We will start by finding a quadratic model, $y = ax^2 + bx + c$, that has ordered pair solutions that correspond to the data for U.S. per capita consumption of poultry given in the table. To do so, substitute each data pair into the equation. Each time, the result is an equation in three unknowns: a, b, and c. Because there are three pairs of data, we can form a system of three linear

equations in three unknowns. Solving for the values of a, b, and c gives a quadratic model that represents the given data.

U.S. per Capita Consumption of Poultry (in Pounds)		
Year	x	Poultry Consumption, y (in pounds)
1909	9	11
1957	57	22
2005	105	66

(*Source:* Economic Research Service, U.S. Department of Agriculture)

1. Write the system of equations that must be solved to find the values of a, b, and c needed for a quadratic model of the given data.

2. Solve the system of equations for a, b, and c. Recall the various methods of solving linear systems used in Chapter 4. You might consider using matrices or a graphing calculator to do so. Round to the nearest thousandth.

3. Write the quadratic model for the data. Note that the variable x represents the number of years after 1900.

4. In 1939, the actual U.S. per capita consumption of poultry was 12 pounds per person. Based on this information, how accurate do you think this model is for years other than those given in the table?

5. Use your model to estimate the per capita consumption of poultry in 1950.

6. According to the model, in what year was per capita consumption of poultry 50 pounds per person?

7. In what year was the per capita consumption of poultry at its lowest level? What was that level?

8. Who might be interested in a model like this and how would it be helpful?

 STUDY SKILLS BUILDER

Are You Preparing for a Test on Chapter 8?

Below I have listed some common trouble areas for students in Chapter 8. After studying for your test—but before taking your test—read these.

- Don't forget that to solve a quadratic equation such as $x^2 + 6x = 1$ by completing the square, add the square of half of 6 to both sides.

$$x^2 + 6x = 1$$
$$x^2 + 6x + 9 = 1 + 9 \qquad \text{Add 9 to both sides } \left(\frac{1}{2}(6) = 3 \text{ and } 3^2 = 9\right).$$
$$(x + 3)^2 = 10$$
$$x + 3 = \pm\sqrt{10}$$
$$x = -3 \pm \sqrt{10}$$

- Remember to write a quadratic equation in standard form, $(ax^2 + bx + c = 0)$, before using the quadratic formula to solve.

$$x(4x - 1) = 1$$
$$4x^2 - x - 1 = 0 \qquad \text{Write in standard form.}$$
$$x = \frac{-(-1) \pm \sqrt{(-1)^2 - 4(4)(-1)}}{2 \cdot 4} \qquad \text{Use the quadratic formula with } a = 4, b = -1, \text{ and } c = -1.$$
$$x = \frac{1 \pm \sqrt{17}}{8} \qquad \text{Simplify.}$$

- Review the steps for solving a quadratic equation in general on page 573.
- Don't forget how to graph a quadratic function in the form $f(x) = a(x - h)^2 + k$.

The graph of $f(x) = -2(x - 3)^2 - 1$

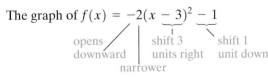

opens downward narrower | shift 3 units right | shift 1 unit down

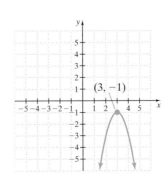

Remember: This is simply a checklist of common trouble areas. For a review of Chapter 8, see the Highlights and Chapter Review at the end of this chapter.

CHAPTER 8 VOCABULARY CHECK

Fill in each blank with one of the words or phrases listed below.

quadratic formula	quadratic	discriminant	$\pm\sqrt{b}$	$(h, 0)$
completing the square	quadratic inequality	(h, k)	$(0, k)$	$\dfrac{-b}{2a}$

1. The _____ helps us find the number and type of solutions of a quadratic equation.

2. If $a^2 = b$, then $a =$ _____ .

3. The graph of $f(x) = ax^2 + bx + c$, where a is not 0, is a parabola whose vertex has an x-value of _____ .

4. A(n) _____ is an inequality that can be written so that one side is a quadratic expression and the other side is 0.

5. The process of writing a quadratic equation so that one side is a perfect square trinomial is called _____ .

6. The graph of $f(x) = x^2 + k$ has vertex _____ .

7. The graph of $f(x) = (x - h)^2$ has vertex _____ .

8. The graph of $f(x) = (x - h)^2 + k$ has vertex _____ .

9. The formula $x = \dfrac{-b \pm \sqrt{b^2 - 4ac}}{2a}$ is called the _____ .

10. A _____ equation is one that can be written in the form $ax^2 + bx + c = 0$, where $a, b,$ and c are real numbers and a is not 0.

> ▶ **Helpful Hint**
>
> Are you preparing for your test? Don't forget to take the Chapter 8 Test on page 621. Then check your answers at the back of the text and use the Chapter Test Prep Video CD to see the fully worked-out solutions to any of the exercises you want to review.

CHAPTER 8 HIGHLIGHTS

DEFINITIONS AND CONCEPTS	EXAMPLES

SECTION 8.1 SOLVING QUADRATIC EQUATIONS BY COMPLETING THE SQUARE

DEFINITIONS AND CONCEPTS	EXAMPLES
Square root property If b is a real number and if $a^2 = b$, then $a = \pm\sqrt{b}$.	Solve: $(x + 3)^2 = 14$. $x + 3 = \pm\sqrt{14}$ $x = -3 \pm \sqrt{14}$
To solve a quadratic equation in x by completing the square	Solve: $3x^2 - 12x - 18 = 0$.
Step 1. If the coefficient of x^2 is not 1, divide both sides of the equation by the coefficient of x^2.	**1.** $x^2 - 4x - 6 = 0$
Step 2. Isolate the variable terms.	**2.** $x^2 - 4x = 6$
Step 3. Complete the square by adding the square of half of the coefficient of x to both sides.	**3.** $\dfrac{1}{2}(-4) = -2$ and $(-2)^2 = 4$ $x^2 - 4x + 4 = 6 + 4$
Step 4. Write the resulting trinomial as the square of a binomial.	**4.** $(x - 2)^2 = 10$
Step 5. Apply the square root property and solve for x.	**5.** $x - 2 = \pm\sqrt{10}$ $x = 2 \pm \sqrt{10}$

SECTION 8.2 SOLVING QUADRATIC EQUATIONS BY THE QUADRATIC FORMULA

DEFINITIONS AND CONCEPTS	EXAMPLES
A quadratic equation written in the form $ax^2 + bx + c = 0$ has solutions $x = \dfrac{-b \pm \sqrt{b^2 - 4ac}}{2a}$	Solve: $x^2 - x - 3 = 0$. $a = 1, b = -1, c = -3$ $x = \dfrac{-(-1) \pm \sqrt{(-1)^2 - 4(1)(-3)}}{2 \cdot 1}$ $x = \dfrac{1 \pm \sqrt{13}}{2}$

DEFINITIONS AND CONCEPTS	EXAMPLES

SECTION 8.3 SOLVING EQUATIONS BY USING QUADRATIC METHODS

Substitution is often helpful in solving an equation that contains a repeated variable expression.

Solve: $(2x + 1)^2 - 5(2x + 1) + 6 = 0$.

Let $m = 2x + 1$. Then

$m^2 - 5m + 6 = 0$ Let $m = 2x + 1$.

$(m - 3)(m - 2) = 0$

$m = 3$ or $m = 2$

$2x + 1 = 3$ or $2x + 1 = 2$ Substitute back.

$x = 1$ or $x = \dfrac{1}{2}$

SECTION 8.4 NONLINEAR INEQUALITIES IN ONE VARIABLE

To solve a polynomial inequality

Step 1. Write the inequality in standard form.

Step 2. Solve the related equation.

Step 3. Use solutions from Step 2 to separate the number line into regions.

Step 4. Use test points to determine whether values in each region satisfy the original inequality.

Step 5. Write the solution set as the union of regions whose test point value is a solution.

Solve: $x^2 \geq 6x$.

1. $x^2 - 6x \geq 0$

2. $x^2 - 6x = 0$

    $x(x - 6) = 0$

          $x = 0$ or $x = 6$

3.

4.

Region	Test Point Value	$x^2 \geq 6x$	Result
A	−2	$(-2)^2 \geq 6(-2)$	True
B	1	$1^2 \geq 6(1)$	False
C	7	$7^2 \geq 6(7)$	True

5.

The solution set is $(-\infty, 0] \cup [6, \infty)$.

To solve a rational inequality

Step 1. Solve for values that make all denominators 0.

Step 2. Solve the related equation.

Step 3. Use solutions from Steps 1 and 2 to separate the number line into regions.

Step 4. Use test points to determine whether values in each region satisfy the original inequality.

Step 5. Write the solution set as the union of regions whose test point value is a solution.

Solve: $\dfrac{6}{x - 1} < -2$.

1. $x - 1 = 0$ Set denominator equal to 0.

    $x = 1$

2. $\dfrac{6}{x - 1} = -2$

        $6 = -2(x - 1)$ Multiply by $(x - 1)$.

        $6 = -2x + 2$

        $4 = -2x$

    $-2 = x$

3.

4. Only a test value from region B satisfies the original inequality.

5.

The solution set is $(-2, 1)$.

DEFINITIONS AND CONCEPTS	**EXAMPLES**

Graph of a quadratic function

The graph of a quadratic function written in the form $f(x) = a(x - h)^2 + k$ is a parabola with vertex (h, k). If $a > 0$, the parabola opens upward; if $a < 0$, the parabola opens downward. The axis of symmetry is the line whose equation is $x = h$.

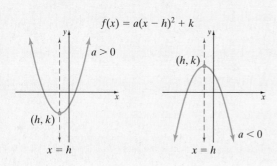

Graph $g(x) = 3(x - 1)^2 + 4$.

The graph is a parabola with vertex $(1, 4)$ and axis of symmetry $x = 1$. Since $a = 3$ is positive, the graph opens upward.

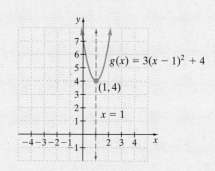

The graph of $f(x) = ax^2 + bx + c$, where $a \neq 0$, is a parabola with vertex

$$\left(\frac{-b}{2a}, f\left(\frac{-b}{2a} \right) \right)$$

Graph $f(x) = x^2 - 2x - 8$. Find the vertex and x- and y-intercepts.

$$\frac{-b}{2a} = \frac{-(-2)}{2 \cdot 1} = 1$$

$$f(1) = 1^2 - 2(1) - 8 = -9$$

The vertex is $(1, -9)$.

$$0 = x^2 - 2x - 8$$
$$0 = (x - 4)(x + 2)$$
$$x = 4 \quad \text{or} \quad x = -2$$

The x-intercepts are $(4, 0)$ and $(-2, 0)$.

$$f(0) = 0^2 - 2 \cdot 0 - 8 = -8$$

The y-intercept is $(0, -8)$.

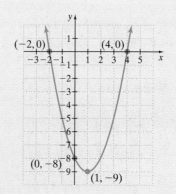

DEFINITIONS AND CONCEPTS	**EXAMPLES**

SECTION 8.7 INTERPRETING DATA: LINEAR AND QUADRATIC MODELS

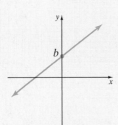

Linear Model
$y = mx + b$
slope: m
y-intercept: $(0, b)$

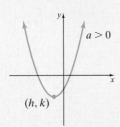

Quadratic Model
$y = ax^2 + bx + c$
$a > 0$, parabola opens up
$a < 0$, parabola opens down
vertex; (h, k) where $h = \dfrac{-b}{2a}$

Given the following graphs of data points, tell whether the graph can best be modeled by a linear model or a quadratic model

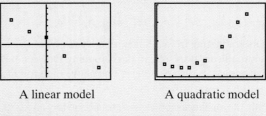

A linear model   A quadratic model

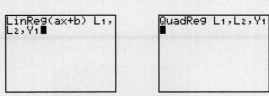

Indicates appropriate regression, x-list, y-list, y = position to store equation.

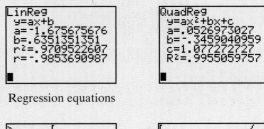

Regression equations

Graphs of regression equations.

CHAPTER 8 REVIEW

(8.1) *Solve by factoring.*

1. $x^2 - 15x + 14 = 0$ **2.** $7a^2 = 29a + 30$

Solve by using the square root property.

3. $4m^2 = 196$ **4.** $(5x - 2)^2 = 2$

Solve by completing the square.

5. $z^2 + 3z + 1 = 0$

6. $(2x + 1)^2 = x$

7. If P dollars are originally invested, the formula $A = P(1 + r)^2$ gives the amount A in an account paying interest rate r compounded annually after 2 years. Find the interest rate r such

that \$2500 increases to \$2717 in 2 years. Round the result to the nearest hundredth of a percent.

△ **8.** Two ships leave a port at the same time and travel at the same speed. One ship is traveling due north and the other due east. In a few hours, the ships are 150 miles apart. How many miles has each ship traveled? Give an exact answer and a one-decimal-place approximation.

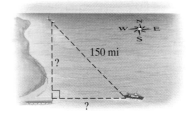

150 mi

(8.2) If the discriminant of a quadratic equation has the given value, determine the number and type of solutions of the equation.

9. -8

10. 48

11. 100

12. 0

Solve by using the quadratic formula.

13. $x^2 - 16x + 64 = 0$

14. $x^2 + 5x = 0$

15. $2x^2 + 3x = 5$

16. $9a^2 + 4 = 2a$

17. $6x^2 + 7 = 5x$

18. $(2x - 3)^2 = x$

19. Cadets graduating from military school usually toss their hats high into the air at the end of the ceremony. One cadet threw his hat so that its distance $d(t)$ in feet above the ground t seconds after it was thrown was $d(t) = -16t^2 + 30t + 6$.

    a. Find the distance above the ground of the hat 1 second after it was thrown.

    b. Find the time it took the hat to hit the ground. Give an exact time and a one-decimal-place approximation.

△ **20.** The hypotenuse of an isosceles right triangle is 6 centimeters longer than either of the legs. Find the length of the legs.

(8.3) Solve each equation for the variable.

21. $x^3 = 27$

22. $y^3 = -64$

23. $\dfrac{5}{x} + \dfrac{6}{x - 2} = 3$

24. $x^4 - 21x^2 - 100 = 0$

25. $x^{2/3} - 6x^{1/3} + 5 = 0$

26. $5(x + 3)^2 - 19(x + 3) = 4$

27. $a^6 - a^2 = a^4 - 1$

28. $y^{-2} + y^{-1} = 20$

29. Two postal workers, Jerome Grant and Tim Bozik, can sort a stack of mail in 5 hours. Working alone, Tim can sort the mail in 1 hour less time than Jerome can. Find the time that each postal worker can sort the mail alone. Round the result to one decimal place.

30. A negative number decreased by its reciprocal is $-\dfrac{24}{5}$. Find the number.

(8.4) Solve each inequality for x. Write each solution set in interval notation.

31. $2x^2 - 50 \leq 0$

32. $\dfrac{1}{4}x^2 < \dfrac{1}{16}$

33. $\dfrac{x - 5}{x - 6} < 0$

34. $(x^2 - 16)(x^2 - 1) > 0$

35. $\dfrac{(4x + 3)(x - 5)}{x(x + 6)} > 0$

36. $(x + 5)(x - 6)(x + 2) \leq 0$

37. $x^3 + 3x^2 - 25x - 75 > 0$

38. $\dfrac{x^2 + 4}{3x} \leq 1$

39. $\dfrac{(5x + 6)(x - 3)}{x(6x - 5)} < 0$

40. $\dfrac{3}{x - 2} > 2$

(8.5) Sketch the graph of each function. Label the vertex and the axis of symmetry.

41. $f(x) = x^2 - 4$

42. $g(x) = x^2 + 7$

43. $H(x) = 2x^2$

44. $h(x) = -\dfrac{1}{3}x^2$

45. $F(x) = (x - 1)^2$

46. $G(x) = (x + 5)^2$

47. $f(x) = (x - 4)^2 - 2$

48. $f(x) = -3(x - 1)^2 + 1$

(8.6) Sketch the graph of each function. Find the vertex and the intercepts.

49. $f(x) = x^2 + 10x + 25$

50. $f(x) = -x^2 + 6x - 9$

51. $f(x) = 4x^2 - 1$

52. $f(x) = -5x^2 + 5$

53. Find the vertex of the graph of $f(x) = -3x^2 - 5x + 4$. Determine whether the graph opens upward or downward, find the y-intercept, approximate the x-intercepts to one decimal place, and sketch the graph.

54. The function $h(t) = -16t^2 + 120t + 300$ gives the height in feet of a projectile fired from the top of a building in t seconds.

    a. When will the object reach a height of 350 feet? Round your answer to one decimal place.

    b. Explain why part **a** has two answers.

55. Find two numbers whose product is as large as possible, given that their sum is 420.

56. Write an equation of a quadratic function whose graph is a parabola that has vertex $(-3, 7)$ and that passes through the origin.

(8.7) *For each exercise,*

a. *Find a linear regression equation that models the data and use this equation to answer the question.*

b. *Find a quadratic regression equation that models the data and use this equation to answer the question.*

Round the coefficients to three decimal places.

57. According to the U.S. Bureau of the Census, Statistical Abstract of the United States, the U.S. population in millions is as given in the table below. If this trend continues, predict the population in the year 2015. Let x = the number of years since 1970.

Year	1970	1980	1990	2000	2008
Population (in millions)	203	227	249	275	303

58. The sales of downloaded digital music online (in billions of dollars) is shown in the graph below. If the trend continues to increase in this manner, predict the amount of sales for 2012. Let x = the number of years since 2000.

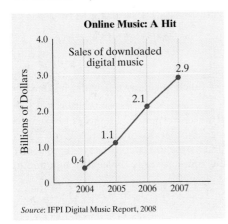

Source: IFPI Digital Music Report, 2008

MIXED REVIEW

Solve each equation.

59. $x^2 - x - 30 = 0$

60. $(9n + 1)^2 = 9$

61. $x^2 + x + 7 = 0$

62. $(3x - 4)^2 = 10x$

63. $x^2 + 11 = 0$

64. $(5a - 2)^2 - a = 0$

65. $\dfrac{7}{8} = \dfrac{8}{x^2}$

66. $x^{2/3} - 6x^{1/3} = -8$

67. $(2x - 3)(4x + 5) \geq 0$

68. $\dfrac{x(x + 5)}{4x - 3} \geq 0$

69. $\dfrac{3}{x - 2} > 2$

 70. The total amount of passenger traffic at Phoenix Sky Harbor International Airport in Phoenix, Arizona, during the period 1980 through 2005 can be modeled by the equation $y = 6.46x^2 + 1236.5x + 7289$, where y is the number of passengers enplaned and deplaned in thousands and x is the number of years after 1980. (*Source*: Based on data from The City of Phoenix Aviation Department, 1980–2005)

a. Estimate the passenger traffic at Phoenix Sky Harbor International Airport in 2000.

b. According to this model, in what year will passenger traffic at Phoenix Sky Harbor International Airport reach 60,000,000 passengers?

CHAPTER 8 TEST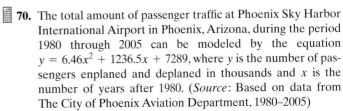

Remember to use the Chapter Test Prep Video CD to see the fully worked-out solutions to any of the exercises you want to review.

Solve each equation for the variable.

1. $5x^2 - 2x = 7$

2. $(x + 1)^2 = 10$

3. $m^2 - m + 8 = 0$

4. $u^2 - 6u + 2 = 0$

5. $7x^2 + 8x + 1 = 0$

6. $y^2 - 3y = 5$

7. $\dfrac{4}{x + 2} + \dfrac{2x}{x - 2} = \dfrac{6}{x^2 - 4}$

8. $x^5 + 3x^4 = x + 3$

9. $x^6 + 1 = x^4 + x^2$

10. $(x + 1)^2 - 15(x + 1) + 56 = 0$

Solve the equation for the variable by completing the square.

11. $x^2 - 6x = -2$

12. $2a^2 + 5 = 4a$

Solve each inequality for x. Write the solution set in interval notation.

13. $2x^2 - 7x > 15$

14. $(x^2 - 16)(x^2 - 25) \geq 0$

15. $\dfrac{5}{x+3} < 1$

16. $\dfrac{7x-14}{x^2-9} \le 0$

Graph each function. Label the vertex.

17. $f(x) = 3x^2$

18. $G(x) = -2(x-1)^2 + 5$

Graph each function. Find and label the vertex, y-intercept, and x-intercepts (if any).

19. $h(x) = x^2 - 4x + 4$

20. $F(x) = 2x^2 - 8x + 9$

21. Dave and Sandy Hartranft can paint a room together in 4 hours. Working alone, Dave can paint the room in 2 hours less time than Sandy can. Find how long it takes Sandy to paint the room alone.

22. A stone is thrown upward from a bridge. The stone's height in feet, $s(t)$, above the water t seconds after the stone is thrown is a function given by the equation $s(t) = -16t^2 + 32t + 256$.

   a. Find the maximum height of the stone.

   b. Find the time it takes the stone to hit the water. Round the answer to two decimal places.

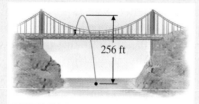

256 ft

23. Given the diagram shown, approximate to the nearest foot how many feet of walking distance a person saves by cutting across the lawn instead of walking on the sidewalk.

20 ft

$x + 8$

x

24. After steadily falling since 1990, the nation's birth rate started rising according to the chart below. (Birth rate per 1000 population)

Years	Birth rate
1990	16.7
1993	15.4
1996	14.4
1999	14.2
2002	13.9
2005	14.0

   a. Graph the data points and use the graph to identify the type of function (linear or quadratic) that best fits the data.

   b. Find the corresponding regression equation for the data and use it to predict the birth rate for 2012. (Round regression coefficients to three decimal places. Round your answer to the nearest hundredth.)

CHAPTER 8 CUMULATIVE REVIEW

1. Write each sentence using mathematical symbols.

   a. The sum of 5 and y is greater than or equal to 7.

   b. 11 is not equal to z.

   c. 20 is less than the difference of 5 and twice x.

2. Solve $|3x - 2| = -5$.

3. Find the slope of the line containing the points $(0, 3)$ and $(2, 5)$. Graph the line.

4. Use the elimination method to solve the system.
$$\begin{cases} -6x + y = 5 \\ 4x - 2y = 6 \end{cases}$$

5. Use the elimination method to solve the system.
$$\begin{cases} x - 5y = -12 \\ -x + y = 4 \end{cases}$$

6. Simplify. Use positive exponents to write each answer.

   a. $(a^{-2}bc^3)^{-3}$

   b. $\left(\dfrac{a^{-4}b^2}{c^3}\right)^{-2}$

   c. $\left(\dfrac{3a^8b^2}{12a^5b^5}\right)^{-2}$

7. Multiply.

   a. $(2x - 7)(3x - 4)$

   b. $(3x^2 + y)(5x^2 - 2y)$

8. Multiply.

   a. $(4a - 3)(7a - 2)$

   b. $(2a + b)(3a - 5b)$

9. Factor.

   a. $8x^2 + 4$

   b. $5y - 2z^4$

   c. $6x^2 - 3x^3 + 12x^4$

10. Factor.

   a. $9x^3 + 27x^2 - 15x$

   b. $2x(3y - 2) - 5(3y - 2)$

   c. $2xy + 6x - y - 3$

Factor the polynomials in Exercises 11 through 14.

11. $x^2 - 12x + 35$

12. $x^2 - 2x - 48$

13. $3a^2x - 12abx + 12b^2x$

14. Factor. $2ax^2 - 12axy + 18ay^2$

15. Solve $3(x^2 + 4) + 5 = -6(x^2 + 2x) + 13$.

16. Solve $2(a^2 + 2) - 8 = -2a(a - 2) - 5$.

17. Solve $x^3 = 4x$.

18. Find the vertex and any intercepts of $f(x) = x^2 + x - 12$.

19. Simplify $\dfrac{2x^2}{10x^3 - 2x^2}$.

20. Simplify $\dfrac{x^2 - 4x + 4}{2 - x}$.

21. Add $\dfrac{2x - 1}{2x^2 - 9x - 5} + \dfrac{x + 3}{6x^2 - x - 2}$.

22. Subtract $\dfrac{a + 1}{a^2 - 6a + 8} - \dfrac{3}{16 - a^2}$.

23. Simplify $\dfrac{x^{-1} + 2xy^{-1}}{x^{-2} - x^{-2}y^{-1}}$.

24. Simplify $\dfrac{(2a)^{-1} + b^{-1}}{a^{-1} + (2b)^{-1}}$.

25. Divide $\dfrac{3x^5y^2 - 15x^3y - x^2y - 6x}{x^2y}$.

26. Divide $x^3 - 3x^2 - 10x + 24$ by $x + 3$.

27. If $P(x) = 2x^3 - 4x^2 + 5$,

   a. Find $P(2)$ by substitution.

   b. Use synthetic division to find the remainder when $P(x)$ is divided by $x - 2$.

28. If $P(x) = 4x^3 - 2x^2 + 3$,

   a. Find $P(-2)$ by substitution.

   b. Use synthetic division to find the remainder when $P(x)$ is divided by $x + 2$.

29. Solve $\dfrac{4x}{5} + \dfrac{3}{2} = \dfrac{3x}{10}$.

30. Solve $\dfrac{x + 3}{x^2 + 5x + 6} = \dfrac{3}{2x + 4} - \dfrac{1}{x + 3}$.

31. If a certain number is subtracted from the numerator and added to the denominator of $\dfrac{9}{19}$, the new fraction is equivalent to $\dfrac{1}{3}$. Find the number.

32. Mr. Briley can roof his house in 24 hours. His son can roof the same house in 40 hours. If they work together, how long will it take to roof the house?

33. Suppose that y varies directly as x. If y is 5 when x is 30, find the constant of variation and the direct variation equation.

34. Suppose that y varies inversely as x. If y is 8 when x is 14, find the constant of variation and the inverse variation equation.

35. Simplify.

   a. $\sqrt{(-3)^2}$ **b.** $\sqrt{x^2}$

   c. $\sqrt[4]{(x - 2)^4}$ **d.** $\sqrt[3]{(-5)^3}$

   e. $\sqrt[5]{(2x - 7)^5}$ **f.** $\sqrt{25x^2}$

   g. $\sqrt{x^2 + 2x + 1}$

36. Simplify. Assume that the variables represent any real number.

   a. $\sqrt{(-2)^2}$ **b.** $\sqrt{y^2}$

   c. $\sqrt[4]{(a - 3)^4}$ **d.** $\sqrt[3]{(-6)^3}$

   e. $\sqrt[5]{(3x - 1)^5}$

37. Use rational exponents to simplify. Assume that variables represent positive numbers.

   a. $\sqrt[8]{x^4}$ **b.** $\sqrt[6]{25}$

   c. $\sqrt[4]{r^2 s^6}$

38. Use rational exponents to simplify. Assume that variables represent positive numbers.

   a. $\sqrt[4]{5^2}$ **b.** $\sqrt[12]{x^3}$

   c. $\sqrt[6]{x^2 y^4}$

39. Use the product rule to simplify.

   a. $\sqrt{25x^3}$ **b.** $\sqrt[3]{54x^6 y^8}$

   c. $\sqrt[4]{81z^{11}}$

40. Use the product rule to simplify. Assume that variables represent positive numbers.

   a. $\sqrt{64a^5}$ **b.** $\sqrt[3]{24a^7 b^9}$

   c. $\sqrt[4]{48x^9}$

41. Rationalize the denominator of each expression.

   a. $\dfrac{2}{\sqrt{5}}$ **b.** $\dfrac{2\sqrt{16}}{\sqrt{9x}}$

   c. $\sqrt[3]{\dfrac{1}{2}}$

42. Multiply. Simplify if possible.

   a. $\left(\sqrt{3} - 4\right)\left(2\sqrt{3} + 2\right)$

   b. $\left(\sqrt{5} - x\right)^2$

   c. $\left(\sqrt{a} + b\right)\left(\sqrt{a} - b\right)$

43. Solve $\sqrt{2x + 5} + \sqrt{2x} = 3$.

44. Solve $\sqrt{x - 2} = \sqrt{4x + 1} - 3$.

45. Divide. Write in the form $a + bi$.

   a. $\dfrac{2 + i}{1 - i}$ **b.** $\dfrac{7}{3i}$

46. Write each product in the form of $a + bi$.

   a. $3i(5 - 2i)$

   b. $(6 - 5i)^2$

   c. $\left(\sqrt{3} + 2i\right)\left(\sqrt{3} - 2i\right)$

47. Use the square root property to solve $(x + 1)^2 = 12$.

48. Use the square root property to solve $(y - 1)^2 = 24$.

49. Solve $x - \sqrt{x} - 6 = 0$.

50. Use the quadratic formula to solve $m^2 = 4m + 8$.

9

Exponential and Logarithmic Functions

In this chapter, we discuss two closely related functions: exponential and logarithmic functions. These functions are vital to applications in economics, finance, engineering, the sciences, education, and other fields. Models of tumor growth and learning curves are two examples of the uses of exponential and logarithmic functions.

pH (Potential of Hydrogen) is a measure of the acidity or alkalinity of a solution. Solutions with a pH less than 7 are considered acidic, those with a pH greater than 7 are considered basic (alkaline), and those equal to 7 are defined as "neutral." The pH scale is logarithmic and some examples are in the table below. Since pH is dependent on ionic activity, it can't easily be measured. One of the oldest ways to measure the pH of a solution is litmus paper. Litmus is a water-soluble mixture of different dyes extracted from lichens.

In Section 9.4, Exercise 106, page 657, we will calculate the pH for lemonade.

Representative pH values

Substance	pH
Hydrochloric Acid, 10M	−1.0
Lead-acid battery	0.5
Gastric acid	1.5 − 2.0
Lemon juice	2.4
Cola	2.5
Vinegar	2.9
Orange or apple juice	3.5
Tomato Juice	4.0
Beer	4.5
Acid Rain	<5.0
Coffee	5.0
Tea or healthy skin	5.5
Milk	6.5
Pure Water	7.0
Healthy human saliva	6.5 − 7.4
Blood	7.34 − 7.45
Seawater	7.7 − 8.3
Hand soap	9.0 − 10.0
Household ammonia	11.5
Bleach	12.5
Household lye	13.5

9.1 THE ALGEBRA OF FUNCTIONS; COMPOSITE FUNCTIONS

OBJECTIVE 1 ▶ Adding, subtracting, multiplying, and dividing functions. As we have seen in earlier chapters, it is possible to add, subtract, multiply, and divide functions. Although we have not stated it as such, the sums, differences, products, and quotients of functions are themselves functions. For example, if $f(x) = 3x$ and $g(x) = x + 1$, their product, $f(x) \cdot g(x) = 3x(x + 1) = 3x^2 + 3x$, is a new function. We can use the notation $(f \cdot g)(x)$ to denote this new function. Finding the sum, difference, product, and quotient of functions to generate new functions is called the **algebra of functions.**

Algebra of Functions

Let f and g be functions. New functions from f and g are defined as follows.

Sum	$(f + g)(x) = f(x) + g(x)$
Difference	$(f - g)(x) = f(x) - g(x)$
Product	$(f \cdot g)(x) = f(x) \cdot g(x)$
Quotient	$\left(\dfrac{f}{g}\right)(x) = \dfrac{f(x)}{g(x)}, \quad g(x) \neq 0$

EXAMPLE 1   If $f(x) = x - 1$ and $g(x) = 2x - 3$, find

a. $(f + g)(x)$ **b.** $(f - g)(x)$ **c.** $(f \cdot g)(x)$ **d.** $\left(\dfrac{f}{g}\right)(x)$

Solution   Use the algebra of functions and replace $f(x)$ by $x - 1$ and $g(x)$ by $2x - 3$. Then we simplify.

a. $(f + g)(x) = f(x) + g(x)$
$$= (x - 1) + (2x - 3)$$
$$= 3x - 4$$

b. $(f - g)(x) = f(x) - g(x)$
$$= (x - 1) - (2x - 3)$$
$$= x - 1 - 2x + 3$$
$$= -x + 2$$

c. $(f \cdot g)(x) = f(x) \cdot g(x)$
$$= (x - 1)(2x - 3)$$
$$= 2x^2 - 5x + 3$$

d. $\left(\dfrac{f}{g}\right)(x) = \dfrac{f(x)}{g(x)} = \dfrac{x - 1}{2x - 3}$, where $x \neq \dfrac{3}{2}$

PRACTICE
1   If $f(x) = x + 2$ and $g(x) = 3x + 5$, find

a. $(f + g)(x)$ **b.** $(f - g)(x)$ **c.** $(f \cdot g)(x)$ **d.** $\left(\dfrac{f}{g}\right)(x)$

There is an interesting but not surprising relationship between the graphs of functions and the graphs of their sum, difference, product, and quotient. For example, the graph of $(f + g)(x)$ can be found by adding the graph of $f(x)$ to the graph of $g(x)$. We add two graphs by adding y-values of corresponding x-values.

If $f(x) = \dfrac{1}{2}x + 2$ and $g(x) = \dfrac{1}{3}x^2 + 4$, then

$$(f + g)(x) = f(x) + g(x)$$
$$= \left(\frac{1}{2}x + 2\right) + \left(\frac{1}{3}x^2 + 4\right)$$
$$= \frac{1}{3}x^2 + \frac{1}{2}x + 6.$$

To visualize this addition of functions with a graphing calculator, graph

$$Y_1 = \frac{1}{2}x + 2, \qquad Y_2 = \frac{1}{3}x^2 + 4, \qquad Y_3 = \frac{1}{3}x^2 + \frac{1}{2}x + 6$$

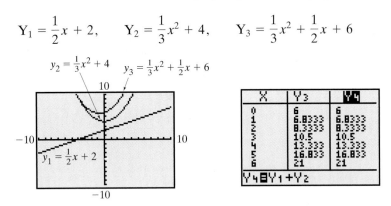

Use a TABLE feature to verify that for a given x-value, $Y_1 + Y_2 = Y_3$. For example, verify that when $x = 0$, $Y_1 = 2$, $Y_2 = 4$, and $Y_3 = 2 + 4 = 6$.

OBJECTIVE 2 ▶ Constructing composite functions. Another way to combine functions is called **function composition.** To understand this new way of combining functions, study the diagrams below. The left diagram below shows degrees Celsius $f(x)$ as a function of degrees Fahrenheit x. The right diagram shows Kelvins $g(x)$ as a function of degrees Celsius x. (The Kelvin scale is a temperature scale devised by Lord Kelvin in 1848.) The function represented by the first diagram we will call f, and the second function we will call g.

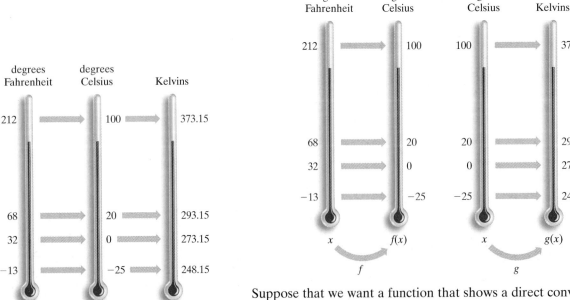

Suppose that we want a function that shows a direct conversion from degrees Fahrenheit to Kelvins. In other words, suppose that a function is needed that shows Kelvins as a function of degrees Fahrenheit. This can easily be done because the output of the first function $f(x)$ is the same as the input of the second function. If we use $f(x)$ to represent this, then we get the diagram in the margin.

Since the output of the first function is used as the input of the second function, we write the new function as $g(f(x))$. The new function is formed from the composition of the other two functions. The mathematical symbol for this composition is $(g \circ f)(x)$. Thus, $(g \circ f)(x) = g(f(x))$. For example, $g(f(-13)) = 248.15$, and so on.

It is possible to find an equation for the composition of the two functions f and g. In other words, we can find a function that converts degrees Fahrenheit directly to Kelvins. The function $f(x) = \dfrac{5}{9}(x - 32)$ converts degrees Fahrenheit to degrees Celsius, and the function $g(x) = x + 273.15$ converts degrees Celsius to Kelvins. Thus,

$$(g \circ f)(x) = g(f(x)) = g\left(\frac{5}{9}(x - 32)\right) = \frac{5}{9}(x - 32) + 273.15$$

In general, the notation $\mathbf{g(f(x))}$ means "g composed with f" and can be written as $\mathbf{(g \circ f)(x)}$. Also $f(g(x))$, or $(f \circ g)(x)$, means "f composed with g."

Composition of Functions

The composition of functions f and g is

$$(f \circ g)(x) = f(g(x))$$

▶ **Helpful Hint**

$(f \circ g)(x)$ does not mean the same as $(f \cdot g)(x)$.

$$(f \circ g)(x) = f(g(x)) \text{ while } (f \cdot g)(x) = f(x) \cdot g(x)$$

Composition of functions   Multiplication of functions

EXAMPLE 2   If $f(x) = x^2$ and $g(x) = x + 3$, find each composition.

a. $(f \circ g)(2)$ and $(g \circ f)(2)$ **b.** $(f \circ g)(x)$ and $(g \circ f)(x)$

Solution

a. $(f \circ g)(2) = f(g(2))$

$\qquad\qquad = f(5)$ Replace $g(2)$ with 5. [Since $g(x) = x + 3$, then

$\qquad\qquad = 5^2 = 25$ $g(2) = 2 + 3 = 5$.]

$(g \circ f)(2) = g(f(2))$

$\qquad\qquad = g(4)$ Since $f(x) = x^2$, then $f(2) = 2^2 = 4$.

$\qquad\qquad = 4 + 3 = 7$

b. $(f \circ g)(x) = f(g(x))$

$\qquad\qquad = f(x + 3)$ Replace $g(x)$ with $x + 3$.

$\qquad\qquad = (x + 3)^2$ $f(x + 3) = (x + 3)^2$

$\qquad\qquad = x^2 + 6x + 9$ Square $(x + 3)$.

$(g \circ f)(x) = g(f(x))$

$\qquad\qquad = g(x^2)$ Replace $f(x)$ with x^2.

$\qquad\qquad = x^2 + 3$ $g(x^2) = x^2 + 3$ □

PRACTICE

2   If $f(x) = x^2 + 1$ and $g(x) = 3x - 5$, find

a. $(f \circ g)(4)$ **b.** $(f \circ g)(x)$

$\quad (g \circ f)(4)$ $(g \circ f)(x)$

Here are two different ways that composite functions can be evaluated using a graphing calculator. Below, we find $(f \circ g)(2)$ from Example 2a.

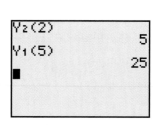

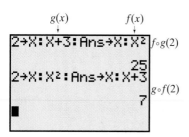

Enter $y_1 = x^2$ and $y_2 = x + 3$. Evaluate $(f \circ g)(2)$ by first evaluating $Y_2(2)$ and then, since $Y_2(2) = 5$, evaluating $Y_1(5)$.

Alternatively, first replace the variable in $g(x)$ with 2 and evaluate the expression. We then take that result and evaluate the $f(x)$ function there.

EXAMPLE 3   If $f(x) = |x|$ and $g(x) = x - 2$, find each composition.

a. $(f \circ g)(x)$ **b.** $(g \circ f)(x)$

Solution

a. $(f \circ g)(x) = f(g(x)) = f(x - 2) = |x - 2|$
b. $(g \circ f)(x) = g(f(x)) = g(|x|) = |x| - 2$ □

> ▶ **Helpful Hint**
>
> In Examples 2 and 3, notice that $(g \circ f)(x) \neq (f \circ g)(x)$. In general, $(g \circ f)(x)$ *may* or *may not* equal $(f \circ g)(x)$.

PRACTICE
3   If $f(x) = x^2 + 5$ and $g(x) = x + 3$, find each composition.

a. $(f \circ g)(x)$ **b.** $(g \circ f)(x)$

EXAMPLE 4   If $f(x) = 5x, g(x) = x - 2$, and $h(x) = \sqrt{x}$, write each function as a composition using two of the given functions.

a. $F(x) = \sqrt{x - 2}$ **b.** $G(x) = 5x - 2$

Solution

a. Notice the order in which the function F operates on an input value x. First, 2 is subtracted from x. This is the function $g(x) = x - 2$. Then the square root *of that result* is taken. The square root function is $h(x) = \sqrt{x}$. This means that $F = h \circ g$. To check, we find $h \circ g$.

$$F(x) = (h \circ g)(x) = h(g(x)) = h(x - 2) = \sqrt{x - 2}$$

b. Notice the order in which the function G operates on an input value x. First, x is multiplied by 5, and then 2 is subtracted from the result. This means that $G = g \circ f$. To check, we find $g \circ f$.

$$G(x) = (g \circ f)(x) = g(f(x)) = g(5x) = 5x - 2$$ □

PRACTICE
4   If $f(x) = 3x, g(x) = x - 4$, and $h(x) = |x|$, write each function as a composition using two of the given functions.

a. $F(x) = |x - 4|$ **b.** $G(x) = 3x - 4$

VOCABULARY & READINESS CHECK

Match each function with its definition.

1. $(f \circ g)(x)$ **4.** $(g \circ f)(x)$ **A.** $g(f(x))$ **D.** $\dfrac{f(x)}{g(x)}, g(x) \neq 0$

2. $(f \cdot g)(x)$ **5.** $\left(\dfrac{f}{g}\right)(x)$ **B.** $f(x) + g(x)$ **E.** $f(x) \cdot g(x)$

3. $(f - g)(x)$ **6.** $(f + g)(x)$ **C.** $f(g(x))$ **F.** $f(x) - g(x)$

9.1 | EXERCISE SET

MyMathLab   PRACTICE   WATCH   DOWNLOAD   READ   REVIEW

For the functions f and g, find ***a.*** $(f + g)(x)$, ***b.*** $(f - g)(x)$,
c. $(f \cdot g)(x)$, *and* ***d.*** $\left(\dfrac{f}{g}\right)(x)$. *See Example 1.*

1. $f(x) = x - 7, g(x) = 2x + 1$

2. $f(x) = x + 4, g(x) = 5x - 2$

3. $f(x) = x^2 + 1, g(x) = 5x$

4. $f(x) = x^2 - 2, g(x) = 3x$

5. $f(x) = \sqrt{x}, g(x) = x + 5$

6. $f(x) = \sqrt[3]{x}, g(x) = x - 3$

7. $f(x) = -3x, g(x) = 5x^2$

8. $f(x) = 4x^3, g(x) = -6x$

If $f(x) = x^2 - 6x + 2, g(x) = -2x,$ *and* $h(x) = \sqrt{x},$ *find each composition. See Example 2.*

9. $(f \circ g)(2)$ **10.** $(h \circ f)(-2)$

11. $(g \circ f)(-1)$ **12.** $(f \circ h)(1)$

13. $(g \circ h)(0)$ **14.** $(h \circ g)(0)$

Find $(f \circ g)(x)$ *and* $(g \circ f)(x)$. *See Examples 2 and 3.*

15. $f(x) = x^2 + 1, g(x) = 5x$

16. $f(x) = x - 3, g(x) = x^2$

17. $f(x) = 2x - 3, g(x) = x + 7$

18. $f(x) = x + 10, g(x) = 3x + 1$

19. $f(x) = x^3 + x - 2, g(x) = -2x$

20. $f(x) = -4x, g(x) = x^3 + x^2 - 6$

21. $f(x) = |x|, g(x) = 10x - 3$

22. $f(x) = |x|, g(x) = 14x - 8$

23. $f(x) = \sqrt{x}, g(x) = -5x + 2$

24. $f(x) = 7x - 1, g(x) = \sqrt[3]{x}$

If $f(x) = 3x, g(x) = \sqrt{x},$ *and* $h(x) = x^2 + 2,$ *write each function as a composition using two of the given functions. See Example 4.*

25. $H(x) = \sqrt{x^2 + 2}$

26. $G(x) = \sqrt{3x}$

27. $F(x) = 9x^2 + 2$

28. $H(x) = 3x^2 + 6$

29. $G(x) = 3\sqrt{x}$

30. $F(x) = x + 2$

Find f(x) and g(x) so that the given function $h(x) = (f \circ g)(x)$.

31. $h(x) = (x + 2)^2$

32. $h(x) = |x - 1|$

33. $h(x) = \sqrt{x + 5} + 2$

34. $h(x) = (3x + 4)^2 + 3$

35. $h(x) = \dfrac{1}{2x - 3}$

36. $h(x) = \dfrac{1}{x + 10}$

REVIEW AND PREVIEW

Solve each equation for y. See Section 1.8.

37. $x = y + 2$

38. $x = y - 5$

39. $x = 3y$

40. $x = -6y$

41. $x = -2y - 7$

42. $x = 4y + 7$

CONCEPT EXTENSIONS

Given that $f(-1) = 4 \quad g(-1) = -4$

$\qquad f(0) = 5 \qquad g(0) = -3$

$\qquad f(2) = 7 \qquad g(2) = -1$

$\qquad f(7) = 1 \qquad g(7) = 4$

Find each function value.

43. $(f + g)(2)$

44. $(f - g)(7)$

45. $(f \circ g)(2)$

46. $(g \circ f)(2)$

47. $(f \cdot g)(7)$

48. $(f \cdot g)(0)$

49. $\left(\dfrac{f}{g}\right)(-1)$

50. $\left(\dfrac{g}{f}\right)(-1)$

51. If you are given $f(x)$ and $g(x)$, explain in your own words how to find $(f \circ g)(x)$, and then how to find $(g \circ f)(x)$.

52. Given $f(x)$ and $g(x)$, describe in your own words the difference between $(f \circ g)(x)$ and $(f \cdot g)(x)$.

Solve.

53. Business people are concerned with cost functions, revenue functions, and profit functions. Recall that the profit $P(x)$ obtained from x units of a product is equal to the revenue $R(x)$ from selling the x units minus the cost $C(x)$ of manufacturing the x units. Write an equation expressing this relationship among $C(x), R(x)$, and $P(x)$.

54. Suppose the revenue $R(x)$ for x units of a product can be described by $R(x) = 25x$, and the cost $C(x)$ can be described by $C(x) = 50 + x^2 + 4x$. Find the profit $P(x)$ for x units. (See Exercise 53.)

📖 STUDY SKILLS BUILDER

Tips for Studying for an Exam

To prepare for an exam, try the following study techniques.

- Start the study process days before your exam.
- Make sure that you are up-to-date on your assignments.
- If there is a topic that you are unsure of, use one of the many resources that are available to you. For example,

  See your instructor.

  Visit a learning resource center on campus.

  Read the textbook material and examples on the topic.

  View a video on the topic.

- Reread your notes and carefully review the Chapter Highlights at the end of any chapter.
- Work the review exercises at the end of the chapter. Check your answers and correct any mistakes. If you have trouble, use a resource listed above.
- Find a quiet place to take the Chapter Test found at the end of the chapter. Do not use any resources when taking this sample test. This way, you will have a clear indication of how prepared you are for your exam.

Check your answers and make sure that you correct any missed exercises.

- Get lots of rest the night before the exam. It's hard to show how well you know the material if your brain is foggy from lack of sleep.

Good luck and keep a positive attitude.

Let's see how you did on your last exam.

1. How many days before your last exam did you start studying?
2. Were you up-to-date on your assignments at that time or did you need to catch up on assignments?
3. List the most helpful text supplement (if you used one).
4. List the most helpful campus supplement (if you used one).
5. List your process for preparing for a mathematics test.
6. Was this process helpful? In other words, were you satisfied with your performance on your exam?
7. If not, what changes can you make in your process that will make it more helpful to you?

9.2 INVERSE FUNCTIONS

OBJECTIVES

1 Determine whether a function is a one-to-one function.

2 Use the horizontal line test to decide whether a function is a one-to-one function.

3 Find the inverse of a function.

4 Find the equation of the inverse of a function.

5 Graph functions and their inverses.

6 Determine whether two functions are inverses of each other.

OBJECTIVE 1 ▶ Determining whether a function is one-to-one. In the next section, we begin a study of two new functions: exponential and logarithmic functions. As we learn more about these functions, we will discover that they share a special relation to each other: They are inverses of each other.

Before we study these functions, we need to learn about inverses. We begin by defining one-to-one functions.

Study the following diagram.

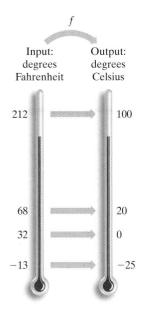

Recall that since each degrees Fahrenheit (input) corresponds to exactly one degrees Celsius (output), this pairing of inputs and outputs does describe a function. Also notice that each output corresponds to exactly one input. This type of function is given a special name—a one-to-one function.

Does the set $f = \{(0, 1), (2, 2), (-3, 5), (7, 6)\}$ describe a one-to-one function? It is a function since each x-value corresponds to a unique y-value. For this particular function f, each y-value also corresponds to a unique x-value. Thus, this function is also a **one-to-one function.**

One-to-One Function

For a **one-to-one function,** each x-value (input) corresponds to only one y-value (output), and each y-value (output) corresponds to only one x-value (input).

EXAMPLE 1   Determine whether each function described is one-to-one.

a. $f = \{(6, 2), (5, 4), (-1, 0), (7, 3)\}$

b. $g = \{(3, 9), (-4, 2), (-3, 9), (0, 0)\}$

c. $h = \{(1, 1), (2, 2), (10, 10), (-5, -5)\}$

d.

Mineral (Input)	Talc	Gypsum	Diamond	Topaz	Stibnite
Hardness on the Mohs Scale (Output)	1	2	10	8	2

e.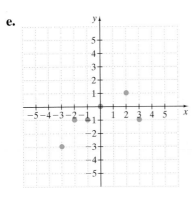

f.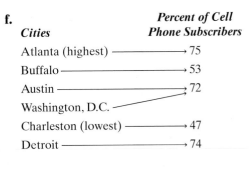

Cities	Percent of Cell Phone Subscribers
Atlanta (highest)	75
Buffalo	53
Austin	72
Washington, D.C.	
Charleston (lowest)	47
Detroit	74

Solution

a. *f* is one-to-one since each *y*-value corresponds to only one *x*-value.

b. *g* is not one-to-one because the *y*-value 9 in (3, 9) and (−3, 9) corresponds to two different *x*-values.

c. *h* is a one-to-one function since each *y*-value corresponds to only one *x*-value.

d. This table does not describe a one-to-one function since the output 2 corresponds to two different inputs, gypsum and stibnite.

e. This graph does not describe a one-to-one function since the *y*-value −1 corresponds to three different *x*-values, −2, −1, and 3.

f. The mapping is not one-to-one since 72% corresponds to Austin and Washington, D.C. □

PRACTICE
1   Determine whether each function described is one-to-one.

a. $f = \{(4, -3), (3, -4), (2, 7), (5, 0)\}$

b. $g = \{(8, 4), (-2, 0), (6, 4), (2, 6)\}$

c. $h = \{(2, 4), (1, 3), (4, 6), (-2, 4)\}$

d.

Year	1950	1963	1968	1975	1997	2002
Federal Minimum Wage	$0.75	$1.25	$1.60	$2.10	$5.15	$5.15

e.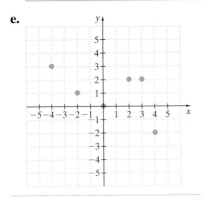

f.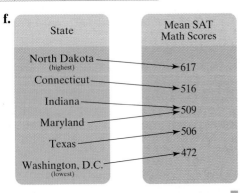

State	Mean SAT Math Scores
North Dakota (highest)	617
Connecticut	516
Indiana	509
Maryland	506
Texas	472
Washington, D.C. (lowest)	

OBJECTIVE 2 ▶ Using the horizontal line test. Recall that we recognize the graph of a function when it passes the vertical line test. Since every *x*-value of the function corresponds to exactly one *y*-value, each vertical line intersects the function's graph at most once. The graph shown (left), for instance, is the graph of a function.

    Is this function a *one-to-one* function? The answer is no. To see why not, notice that the *y*-value of the ordered pair (−3, 3), for example, is the same as the *y*-value of the ordered pair (3, 3). In other words, the *y*-value 3 corresponds to two *x*-values, −3 and 3. This function is therefore not one-to-one.

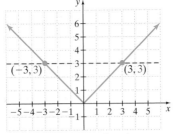

Not a one-to-one function.

To test whether a graph is the graph of a one-to-one function, apply the vertical line test to see if it is a function, and then apply a similar **horizontal line test** to see if it is a one-to-one function.

> **Horizontal Line Test**
>
> If every horizontal line intersects the graph of a function at most once, then the function is a one-to-one function.

EXAMPLE 2 Determine whether each graph is the graph of a one-to-one function.

a.

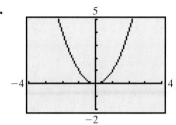

b.

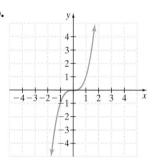

c.

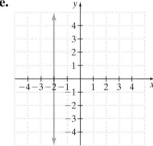

d.

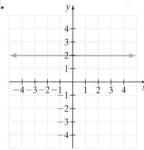

e.

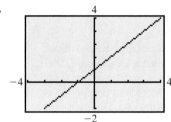

Solution Graphs **a, b, c,** and **d** all pass the vertical line test, so only these graphs are graphs of functions. But, of these, only **b** and **c** pass the horizontal line test, so only **b** and **c** are graphs of one-to-one functions. □

PRACTICE
2 Determine whether each graph is the graph of a one-to-one function.

a.

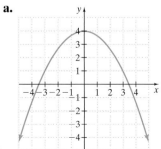

b.

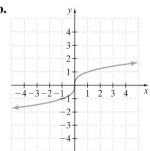

c.

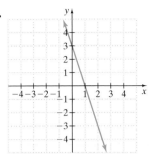

d.

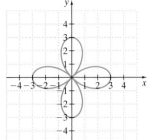

e.

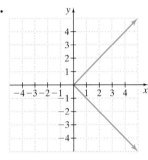

> ▶ **Helpful Hint**
>
> All linear equations are one-to-one functions except those whose graphs are horizontal or vertical lines. A vertical line does not pass the vertical line test and hence is not the graph of a function. A horizontal line is the graph of a function but does not pass the horizontal line test and hence is not the graph of a one-to-one function.
>
>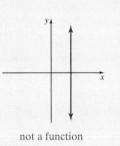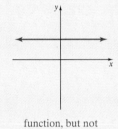
>
> not a function   function, but not one-to-one

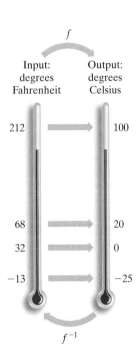

OBJECTIVE 3 ▶ Finding the inverse of a function. One-to-one functions are special in that their graphs pass both the vertical and horizontal line tests. They are special, too, in another sense: For each one-to-one function, we can find its **inverse function** by switching the coordinates of the ordered pairs of the function, or the inputs and the outputs. For example,

the inverse of the one-to-one function $f = \{(2, -3), (5, 10), (9, 1)\}$ is $\{(-3, 2), (10, 5), (1, 9)\}$.

For a function f, we use the notation f^{-1}, read "f inverse," to denote its inverse function. Notice that since the coordinates of each ordered pair have been switched, the domain (set of inputs) of f is the range (set of outputs) of f^{-1}, and the range of f is the domain of f^{-1}.

     The diagram to the left shows that the inverse of the one-to-one function f with ordered pairs of the form (degrees Fahrenheit, degrees Celsuis) is the function f^{-1} with ordered pairs of the form (degrees Celsuis, degrees Fahrenheit). Notice that the ordered pair $(-13, -25)$ of the function, for example, becomes the ordered pair $(-25, -13)$ of its inverse.

> **Inverse Function**
>
> The inverse of a one-to-one function f is the one-to-one function f^{-1} that consists of the set of all ordered pairs (y, x) where (x, y) belongs to f.

▶ **Helpful Hint**

If a function is not one-to-one, it does not have an inverse function.

EXAMPLE 3   Find the inverse of the one-to-one function.

$$f = \{(0, 1), (-2, 7), (3, -6), (4, 4)\}$$

Solution $f^{-1} = \{(1, 0), (7, -2), (-6, 3), (4, 4)\}$

Switch coordinates of each ordered pair.   ☐

PRACTICE
3   Find the inverse of the one-to-one function.

$$f(x) = \{(3, 4), (-2, 0), (2, 8), (6, 6)\}$$

▶ **Helpful Hint**

The symbol f^{-1} is the single symbol used to denote the inverse of the function f.

It is read as "f inverse." This symbol *does not mean* $\dfrac{1}{f}$.

Concept Check ✓

Suppose that f is a one-to-one function and that $f(1) = 5$.

a. Write the corresponding ordered pair.

b. Write one point that we know must belong to the inverse function f^{-1}.

OBJECTIVE 4 ▶ Finding the equation of the inverse of a function. If a one-to-one function f is defined as a set of ordered pairs, we can find f^{-1} by interchanging the x- and y-coordinates of the ordered pairs. If a one-to-one function f is given in the form of an equation, we can find f^{-1} by using a similar procedure.

Finding the Inverse of a One-to-One Function $f(x)$

STEP 1.   Replace $f(x)$ with y.

STEP 2.   Interchange x and y.

STEP 3.   Solve the equation for y.

STEP 4.   Replace y with the notation $f^{-1}(x)$.

EXAMPLE 4   Find an equation of the inverse of $f(x) = x + 3$.

Solution $f(x) = x + 3$

STEP 1. $y = x + 3$ Replace $f(x)$ with y.

STEP 2. $x = y + 3$ Interchange x and y.

STEP 3. $x - 3 = y$ Solve for y.

STEP 4. $f^{-1}(x) = x - 3$ Replace y with $f^{-1}(x)$.

Answer to Concept Check:
a. $(1, 5)$ **b.** $(5, 1)$

The inverse of $f(x) = x + 3$ is $f^{-1}(x) = x - 3$. Notice that, for example,

$$f(1) = 1 + 3 = 4 \quad \text{and} \quad f^{-1}(4) = 4 - 3 = 1$$

Ordered pair: $(1, 4)$ Ordered pair: $(4, 1)$

The coordinates are
switched, as expected.

☐

PRACTICE

4   Find the equation of the inverse of $f(x) = 6 - x$.

EXAMPLE 5   Find the equation of the inverse of $f(x) = 3x - 5$.

Solution $f(x) = 3x - 5$

STEP 1. $y = 3x - 5$ Replace $f(x)$ with y.

STEP 2. $x = 3y - 5$ Interchange x and y.

STEP 3. $3y = x + 5$ Solve for y.

$$y = \frac{x + 5}{3}$$

STEP 4. $f^{-1}(x) = \dfrac{x + 5}{3}$ Replace y with $f^{-1}(x)$.

The inverse of $f(x) = 3x - 5$ is $f^{-1}(x) = \dfrac{x + 5}{3}$.

DISCOVER THE CONCEPT

a. Use your graphing utility to graph both $f(x) = 3x - 5$ as $y_1 = 3x - 5$ and $f^{-1}(x) = \dfrac{x + 5}{3}$ as $y_2 = \dfrac{(x + 5)}{3}$ using a square window.

b. Trace to the point $(1, -2)$ on the graph of y_1. Is there a point $(-2, 1)$ on y_2? Try this for a point on y_2. If (a, b) is a point of y_2, is (b, a) a point of y_1?

c. Next include the graph of $y_3 = x$ in the display. How are the graphs of $f(x) = 3x - 5$ and $f^{-1}(x) = \dfrac{x + 5}{3}$ related to the graph of $y = x$?

A hand-drawn graph of $f(x)$ and $f^{-1}(x)$ is below on the same set of axes along with selected ordered pair solutions.

$f(x) = 3x - 5$	
x	$y = f(x)$
1	-2
0	-5
$\dfrac{5}{3}$	0

$f^{-1}(x) = \dfrac{x + 5}{3}$	
x	$y = f^{-1}(x)$
-2	1
-5	0
0	$\dfrac{5}{3}$

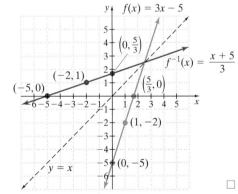

☐

PRACTICE

5   Find the equation of the inverse of $f(x) = 5x + 2$. Graph f and f^{-1} on the same set of axes.

OBJECTIVE 5 ▶ Graphing inverse functions. Notice in the Discover the Concept that the graphs of f and f^{-1} in Example 5 are mirror images of each other, and the "mirror" is the dashed line $y = x$. This is true for every function and its inverse. For this reason, we say that *the graphs of f and f^{-1} are symmetric about the line $y = x$.*

To see why this happens, study the graph of a few ordered pairs and their switched coordinates in the diagram to the right.

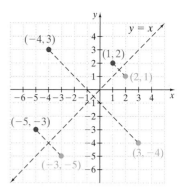

EXAMPLE 6   Graph the inverse of each function.

Solution   The function is graphed in blue and the inverse is graphed in red.

a.

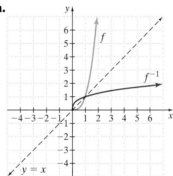

b.

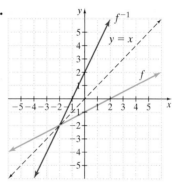

PRACTICE
6   Graph the inverse of each function.

a.

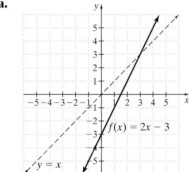

b.

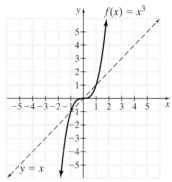

TECHNOLOGY NOTE

The inverse of a function in Y_1 can be drawn on a graph using the Draw menu and the Draw Inv Y_1 option. Notice you cannot trace on the graph of the inverse, as it has not been plotted.

OBJECTIVE 6 ▶ Determining whether functions are inverses of each other. Study the table of values below from Example 5.

X	Y1
0	-5
1	-2
2	1

Y1█3X-5

X	Y2
-5	0
-2	1
1	2

Y2█(X+5)/3

Notice in the table of values above that $f(0) = -5$ and $f^{-1}(-5) = 0$, as expected. Also, for example, $f(1) = -2$ and $f^{-1}(-2) = 1$. In words, we say that for some input x, the function f^{-1} takes the output of x, called $f(x)$, back to x.

$$x \rightarrow f(x) \quad \text{and} \quad f^{-1}(f(x)) \rightarrow x$$
$$\downarrow \quad \downarrow \quad\quad\quad \downarrow \quad \downarrow$$
$$f(0) = -5 \quad \text{and} \quad f^{-1}(-5) = 0$$
$$f(1) = -2 \quad \text{and} \quad f^{-1}(-2) = 1$$

In general,

If f is a one-to-one function, then the inverse of f is the function f^{-1} such that

$$(f^{-1} \circ f)(x) = x \quad \text{and} \quad (f \circ f^{-1})(x) = x$$

EXAMPLE 7 Show that if $f(x) = 3x + 2$, then $f^{-1}(x) = \dfrac{x - 2}{3}$.

Solution See that $(f^{-1} \circ f)(x) = x$ and $(f \circ f^{-1})(x) = x$.

$$(f^{-1} \circ f)(x) = f^{-1}(f(x))$$
$$= f^{-1}(3x + 2) \qquad \text{Replace } f(x) \text{ with } 3x + 2.$$
$$= \frac{3x + 2 - 2}{3}$$
$$= \frac{3x}{3}$$
$$= x$$

$$(f \circ f^{-1})(x) = f(f^{-1}(x))$$
$$= f\left(\frac{x - 2}{3}\right) \qquad \text{Replace } f^{-1}(x) \text{ with } \frac{x - 2}{3}.$$
$$= 3\left(\frac{x - 2}{3}\right) + 2$$
$$= x - 2 + 2$$
$$= x$$

To visually see these results, graph $y_1 = 3x + 2$, $y_2 = \dfrac{x - 2}{3}$, and $y_3 = x$ using a square window, as shown below. Graphically, we see that the graphs of f and f^{-1} are mirror images of each other across the line $y = x$.

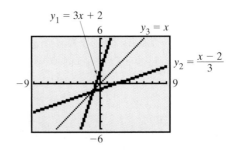

We can also make a table of values for f and f^{-1} and see that if $f(a) = b$ [the ordered pair (a, b)], then $f^{-1}(b) = a$ [the ordered pair (b, a)].

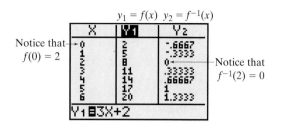

Notice that
$f(0) = 2$

Notice that
$f^{-1}(2) = 0$

$y_1 = f(x)$ $y_2 = f^{-1}(x)$

PRACTICE
7 Show that if $f(x) = 4x - 1$, then $f^{-1}(x) = \dfrac{x+1}{4}$.

VOCABULARY & READINESS CHECK

Use the choices below to fill in each blank. Some choices will not be used and some will be used more than once.

vertical $(3, 7)$ $(11, 2)$ $y = x$ x

horizontal $(7, 3)$ $(2, 11)$ $\dfrac{1}{f}$ the inverse of f

1. If $f(2) = 11$, the corresponding ordered pair is _____.
2. The symbol f^{-1} means _____.
3. If $(7, 3)$ is an ordered pair solution of $f(x)$, and $f(x)$ has an inverse, then an ordered pair solution of $f^{-1}(x)$ is _____.
4. To tell whether a graph is the graph of a function, use the _____ line test.
5. To tell whether the graph of a function is also a one-to-one function, use the _____ line test.
6. The graphs of f and f^{-1} are symmetric about the _____ line.
7. Two functions are inverses of each other if $(f \circ f^{-1})(x) = $ _____ and $(f^{-1} \circ f)(x) = $ ____.

9.2 | EXERCISE SET

Determine whether each function is a one-to-one function. If it is one-to-one, list the inverse function by switching coordinates, or inputs and outputs. See Examples 1 and 3.

1. $f = \{(-1, -1), (1, 1), (0, 2), (2, 0)\}$

2. $g = \{(8, 6), (9, 6), (3, 4), (-4, 4)\}$

3. $h = \{(10, 10)\}$

4. $r = \{(1, 2), (3, 4), (5, 6), (6, 7)\}$

5. $f = \{(11, 12), (4, 3), (3, 4), (6, 6)\}$

6. $g = \{(0, 3), (3, 7), (6, 7), (-2, -2)\}$

7.

Month of 2007 (Input)	January	February	March	April
Unemployment Rate in Percent	4.6	4.6	4.4	4.4

(*Source:* Bureau of Labor Statistics, U.S. Department of Housing and Urban Development)

8.

State (Input)	Wisconsin	Ohio	Georgia	Colorado	California	Arizona
Electoral Votes (Output)	10	20	15	9	55	10

9.

State (Input)	California	Maryland	Nevada	Florida	North Dakota
Rank in Population (Output)	1	19	35	4	48

(*Source:* U.S. Bureau of the Census)

△ **10.**

Shape (Input)	Triangle	Pentagon	Quadrilateral	Hexagon	Decagon
Number of Sides (Output)	3	5	4	6	10

Given the one-to-one function $f(x) = x^3 + 2$, *find the following.*
[*Hint: You do not need to find the equation for* $f^{-1}(x)$.]

11. a. $f(1)$
 b. $f^{-1}(3)$

12. a. $f(0)$
 b. $f^{-1}(2)$

13. a. $f(-1)$
 b. $f^{-1}(1)$

14. a. $f(-2)$
 b. $f^{-1}(-6)$

Determine whether the graph of each function is the graph of a one-to-one function. See Example 2.

15.

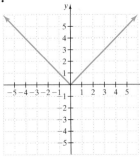

16.

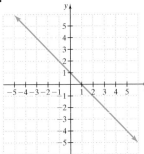

17.

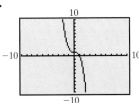

18.

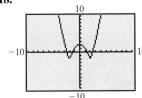

19.

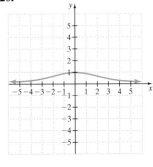

20.

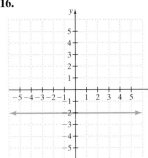

21.

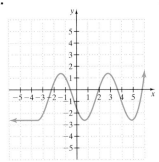

22.

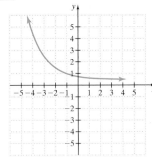

MIXED PRACTICE

Each of the following functions is one-to-one. Find the inverse of each function and graph the function and its inverse on the same set of axes. See Examples 4 and 5.

23. $f(x) = x + 4$

24. $f(x) = x - 5$

25. $f(x) = 2x - 3$

26. $f(x) = 4x + 9$

27. $f(x) = \dfrac{1}{2}x - 1$

28. $f(x) = -\dfrac{1}{2}x + 2$

29. $f(x) = x^3$

30. $f(x) = x^3 - 1$

Find the inverse of each one-to-one function. See Examples 4 and 5.

31. $f(x) = 5x + 2$

32. $f(x) = 6x - 1$

33. $f(x) = \dfrac{x - 2}{5}$

34. $f(x) = \dfrac{4x - 3}{2}$

35. $f(x) = \sqrt[3]{x}$

36. $f(x) = \sqrt[3]{x + 1}$

37. $f(x) = \dfrac{5}{3x + 1}$

38. $f(x) = \dfrac{7}{2x + 4}$

39. $f(x) = (x + 2)^3$

40. $f(x) = (x - 5)^3$

Graph the inverse of each function on the same set of axes. See Example 6.

41.
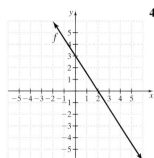

42.
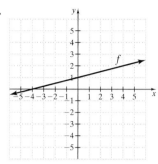

43.
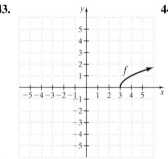

44.
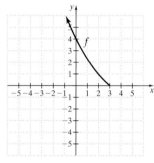

45.
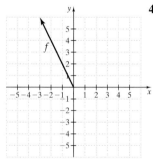

46.
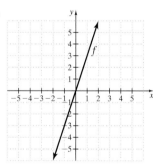

Solve. See Example 7.

47. If $f(x) = 2x + 1$, show that $f^{-1}(x) = \dfrac{x - 1}{2}$.

48. If $f(x) = 3x - 10$, show that $f^{-1}(x) = \dfrac{x + 10}{3}$.

49. If $f(x) = x^3 + 6$, show that $f^{-1}(x) = \sqrt[3]{x - 6}$.

50. If $f(x) = x^3 - 5$, show that $f^{-1}(x) = \sqrt[3]{x + 5}$.

REVIEW AND PREVIEW

Evaluate each of the following. See Section 7.2.

51. $25^{1/2}$

52. $49^{1/2}$

53. $16^{3/4}$

54. $27^{2/3}$

55. $9^{-3/2}$

56. $81^{-3/4}$

If $f(x) = 3^x$, find the following. In Exercises 59 and 60, give an exact answer and a two-decimal-place approximation. See Sections 2.2, 5.1, and 7.2.

57. $f(2)$

58. $f(0)$

59. $f\left(\dfrac{1}{2}\right)$

60. $f\left(\dfrac{2}{3}\right)$

CONCEPT EXTENSIONS

Solve. See the Concept Check in this section.

61. Suppose that f is a one-to-one function and that $f(2) = 9$.

    a. Write the corresponding ordered pair.

    b. Name one ordered pair that we know is a solution of the inverse of f, or f^{-1}.

62. Suppose that F is a one-to-one function and that $F\left(\dfrac{1}{2}\right) = -0.7$.

    a. Write the corresponding ordered pair.

    b. Name one ordered pair that we know is a solution of the inverse of F, or F^{-1}.

For Exercises 63 and 64,

    a. Write the ordered pairs for $f(x)$ whose points are highlighted. (Include the points whose coordinates are given.)

    b. Write the corresponding ordered pairs for the inverse of f, f^{-1}.

    c. Graph the ordered pairs for f^{-1} found in part **b.**

    d. Graph $f^{-1}(x)$ by drawing a smooth curve through the plotted points.

63.
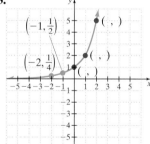

64.
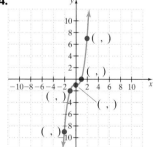

65. If you are given the graph of a function, describe how you can tell from the graph whether the function has an inverse.

66. Describe the appearance of the graphs of a function and its inverse.

Find the inverse of each given one-to-one function. Then use a graphing calculator to graph the function and its inverse on a square window.

67. $f(x) = 3x + 1$

68. $f(x) = -2x - 6$

69. $f(x) = \sqrt[3]{x + 1}$

70. $f(x) = x^3 - 3$

9.3 EXPONENTIAL FUNCTIONS

OBJECTIVES

1 Graph exponential functions.

2 Solve equations of the form $b^x = b^y$.

3 Solve problems modeled by exponential equations.

OBJECTIVE 1 ▶ Graphing exponential functions. In earlier chapters, we gave meaning to exponential expressions such as 2^x, where x is a rational number. For example,

$$2^3 = 2 \cdot 2 \cdot 2 \qquad \text{Three factors; each factor is 2}$$
$$2^{3/2} = (2^{1/2})^3 = \sqrt{2} \cdot \sqrt{2} \cdot \sqrt{2} \qquad \text{Three factors; each factor is } \sqrt{2}$$

When x is an irrational number (for example, $\sqrt{3}$), what meaning can we give to $2^{\sqrt{3}}$?

It is beyond the scope of this book to give precise meaning to 2^x if x is irrational. We can confirm your intuition and say that $2^{\sqrt{3}}$ is a real number, and since $1 < \sqrt{3} < 2$, then $2^1 < 2^{\sqrt{3}} < 2^2$. We can also use a calculator and approximate $2^{\sqrt{3}}$: $2^{\sqrt{3}} \approx 3.321997$. In fact, as long as the base b is positive, b^x is a real number for all real numbers x. Finally, the rules of exponents apply whether x is rational or irrational, as long as b is positive. In this section, we are interested in functions of the form $f(x) = b^x$, where $b > 0$. A function of this form is called an **exponential function.**

Exponential Function

A function of the form

$$f(x) = b^x$$

is called an **exponential function** if $b > 0$, b is not 1, and x is a real number.

Next, we practice graphing exponential functions.

EXAMPLE 1 Graph the exponential functions defined by $f(x) = 2^x$ and $g(x) = 3^x$ on the same set of axes.

Solution Graph each function by plotting points. Set up a table of values for each of the two functions.

If each set of points is plotted and connected with a smooth curve, the following graphs result.

$f(x) = 2^x$	x	0	1	2	3	-1	-2
	$f(x)$	1	2	4	8	$\frac{1}{2}$	$\frac{1}{4}$

$g(x) = 3^x$	x	0	1	2	3	-1	-2
	$g(x)$	1	3	9	27	$\frac{1}{3}$	$\frac{1}{9}$

PRACTICE

1 Graph the exponential functions defined by $f(x) = 2^x$ and $g(x) = 7^x$ on the same set of axes.

A number of things should be noted about the two graphs of exponential functions in Example 1. First, the graphs show that $f(x) = 2^x$ and $g(x) = 3^x$ are one-to-one functions since each graph passes the vertical and horizontal line tests. The y-intercept of each graph is $(0, 1)$, but neither graph has an x-intercept. From the graph, we can also see that the domain of each function is all real numbers and that the range is $(0, \infty)$. We can also see that as x-values are increasing, y-values are increasing also.

EXAMPLE 2 Graph the exponential functions $y = \left(\dfrac{1}{2}\right)^x$ and $y = \left(\dfrac{1}{3}\right)^x$ on the same set of axes.

Solution As before, plot points and connect them with a smooth curve.

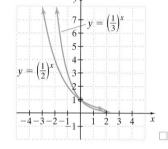

$y = \left(\dfrac{1}{2}\right)^x$	x	0	1	2	3	-1	-2
	y	1	$\dfrac{1}{2}$	$\dfrac{1}{4}$	$\dfrac{1}{8}$	2	4

$y = \left(\dfrac{1}{3}\right)^x$	x	0	1	2	3	-1	-2
	y	1	$\dfrac{1}{3}$	$\dfrac{1}{9}$	$\dfrac{1}{27}$	3	9

PRACTICE
2 Graph the exponential functions $f(x) = \left(\dfrac{1}{3}\right)^x$ and $g(x) = \left(\dfrac{1}{5}\right)^x$ on the same set of axes.

Each function in Example 2 again is a one-to-one function. The y-intercept of both is $(0, 1)$. The domain is the set of all real numbers, and the range is $(0, \infty)$.

Notice the difference between the graphs of Example 1 and the graphs of Example 2. An exponential function is always increasing if the base is greater than 1.

When the base is between 0 and 1, the graph is always decreasing. The following figures summarize these characteristics of exponential functions.

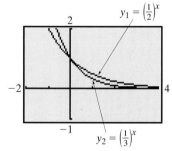

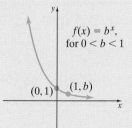

To check, graph $y_1 = \left(\dfrac{1}{2}\right)^x$ and $y_2 = \left(\dfrac{1}{3}\right)^x$ using the same window. Compare the screens to the calculations done by hand.

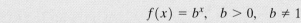

$$f(x) = b^x, \quad b > 0, \quad b \neq 1$$

- one-to-one function
- y-intercept $(0, 1)$
- no x-intercept
- domain: $(-\infty, \infty)$
- range: $(0, \infty)$

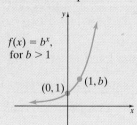

$f(x) = b^x$, for $b > 1$

$(0, 1)$ $(1, b)$

$f(x) = b^x$, for $0 < b < 1$

$(0, 1)$ $(1, b)$

EXAMPLE 3   Graph the exponential function $f(x) = 3^{x+2}$.

Solution   As before, we find and plot a few ordered pair solutions. Then we connect the points with a smooth curve.

$y = 3^{x+2}$	
x	y
0	9
-1	3
-2	1
-3	$\dfrac{1}{3}$
-4	$\dfrac{1}{9}$

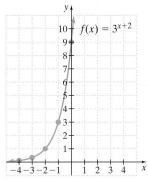

PRACTICE
3   Graph the exponential function $f(x) = 2^{x+3}$.

Concept Check ☑

Which functions are exponential functions?

a. $f(x) = x^3$ **b.** $g(x) = \left(\dfrac{2}{3}\right)^x$ **c.** $h(x) = 5^{x-2}$ **d.** $w(x) = (2x)^2$

OBJECTIVE 2 ▶ Solving equations of the form $b^x = b^y$. We have seen that an exponential function $y = b^x$ is a one-to-one function. Another way of stating this fact is a property that we can use to solve exponential equations.

> **Uniqueness of b^x**
> Let $b > 0$ and $b \neq 1$. Then $b^x = b^y$ is equivalent to $x = y$.

EXAMPLE 4   Solve each equation for x.

a. $2^x = 16$ **b.** $9^x = 27$ **c.** $4^{x+3} = 8^x$ **d.** $5^x = 10$

Solution

a. We write 16 as a power of 2 and then use the uniqueness of b^x to solve.

$$2^x = 16$$
$$2^x = 2^4$$

Since the bases are the same and are nonnegative, by the uniqueness of b^x, we then have that the exponents are equal. Thus,

$$x = 4$$

The solution is 4, or the solution set is $\{4\}$.

Algebraic Solution:

b. Notice that both 9 and 27 are powers of 3.

$$9^x = 27$$
$$(3^2)^x = 3^3 \quad \text{Write 9 and 27 as powers of 3.}$$
$$3^{2x} = 3^3$$
$$2x = 3 \quad \text{Apply the uniqueness of } b^x.$$
$$x = \frac{3}{2} \quad \text{Divide by 2.}$$

Graphical Solution:

Graph $y_1 = 9^x$ and $y_2 = 27$.

The x-value of the point of intersection is 1.5.

Answer to Concept Check:
b, c

To check, replace x with $\dfrac{3}{2}$ in the original expression, $9^x = 27$. The solution is $\dfrac{3}{2}$.

Algebraic Solution:

c. Write both 4 and 8 as powers of 2.

$$4^{x+3} = 8^x$$
$$(2^2)^{x+3} = (2^3)^x$$
$$2^{2x+6} = 2^{3x}$$
$$2x + 6 = 3x \qquad \text{Apply the uniqueness of } b^x.$$
$$6 = x \qquad \text{Subtract } 2x \text{ from both sides.}$$

Graphical Solution:

Graph $y_1 = 4^{x+3}$ and $y_2 = 8^x$.

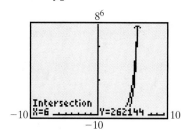

The x-value of the point of intersection is 6.

The solution is 6.

d. For $5^x = 10$, notice that 5 and 10 cannot be easily written as powers of a common base. To solve this equation, we will approximate the solution graphically. Graph $y_1 = 5^x$ and $y_2 = 10$.

The point of intersection is approximately $(1.431, 10)$, which indicates that the solution of the equation is $x \approx 1.431$. To check, replace x with 1.431 in the original equation.

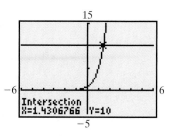

$$5^x = 10$$
$$5^{1.431} = 10 \qquad \text{Let } x = 1.431.$$
$$10.00520695 \approx 10 \qquad \text{Approximately true}$$

The solution is approximately 1.431. □

A calculator check for Example 4d.

PRACTICE

4 Solve each equation for x.

a. $3^x = 9$ **b.** $8^x = 16$ **c.** $125^x = 25^{x-2}$ **d.** $7^x = 21$

As we see in Example 4d, often the two sides of an equation cannot easily be written as powers of a common base. We explore how to find the exact solution to an equation such as $5^x = 10$ with the help of **logarithms** later.

OBJECTIVE 3 ▶ Solving problems modeled by exponential equations. The bar graph to the left shows the increase in the number of cellular phone users. Notice that the graph of the exponential function $y = 110.6(1.132)^x$ approximates the heights of the bars.

Note: Compare this graph with the Chapter 8 opener graph, page 551, on the bottom of the page. Notice how using different data years leads to different models—one quadratic and one exponential.

The graph to the left shows just one example of how the world abounds with patterns that can be modeled by exponential functions. Another application of an exponential function has to do with interest rates on loans.

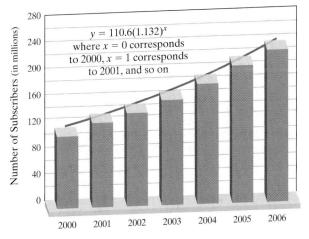

Cellular Phone Subscribers

$y = 110.6(1.132)^x$ where $x = 0$ corresponds to 2000, $x = 1$ corresponds to 2001, and so on

Source: Cellular Telecommunications & Internet Association

The exponential function defined by $A = P\left(1 + \dfrac{r}{n}\right)^{nt}$ models the dollars A accrued (or owed) after P dollars are invested (or loaned) at an annual rate of interest r compounded n times each year for t years. This function is known as the compound interest formula.

EXAMPLE 5　Using the Compound Interest Formula

Find the amount owed at the end of 5 years if $1600 is loaned at a rate of 9% compounded monthly.

Solution　We use the formula $A = P\left(1 + \dfrac{r}{n}\right)^{nt}$, with the following values.

$P = \$1600$ (the amount of the loan)

$r = 9\% = 0.09$ (the annual rate of interest)

$n = 12$ (the number of times interest is compounded each year)

$t = 5$ (the duration of the loan, in years)

$$A = P\left(1 + \frac{r}{n}\right)^{nt} \qquad \text{Compound interest formula}$$

$$= 1600\left(1 + \frac{0.09}{12}\right)^{12(5)} \qquad \text{Substitute known values.}$$

$$= 1600(1.0075)^{60}$$

Use your calculator to approximate A. Two ways to do this are shown below.

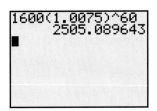

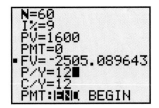

Home Screen　　　　　　　　　Finance Menu

Thus, the amount A owed is approximately $2505.09.　　　　　　□

PRACTICE
5　Find the amount owed at the end of 4 years if $3000 is loaned at a rate of 7% compounded semiannually (twice a year).

EXAMPLE 6　Estimating Percent of Radioactive Material

As a result of the Chernobyl nuclear accident, radioactive debris was carried through the atmosphere. One immediate concern was the impact that the debris had on the milk supply. The percent y of radioactive material in raw milk after t days is estimated by $y = 100(2.7)^{-0.1t}$. Estimate the expected percent of radioactive material in the milk after 30 days.

Solution　Replace t with 30 in the given equation.

$$y = 100(2.7)^{-0.1t}$$

$$= 100(2.7)^{-0.1(30)} \quad \text{Let } t = 30.$$

$$= 100(2.7)^{-3}$$

$$\approx 5.0805$$

Thus, approximately 5% of the radioactive material still remained in the milk supply after 30 days.

We can use a graphing calculator and its TRACE feature to solve Example 6 graphically.

To estimate the expected percent of radioactive material in the milk after 30 days, enter $Y_1 = 100(2.7)^{-0.1x}$. (The variable t in Example 6 is changed to x here to better accommodate our work on the graphing calculator.) The graph does not appear on a standard viewing window, so we need to determine an appropriate viewing window. Because it doesn't make sense to look at radioactivity *before* the Chernobyl nuclear accident, we use Xmin = 0. We are interested in finding the percent of radioactive material in the milk when $x = 30$, so we choose Xmax = 35 to leave enough space to see the graph at $x = 30$. Because the values of y are percents, it seems appropriate that $0 \le y \le 100$. (We also use Xscl = 5 and Yscl = 10.) Now we graph the function.

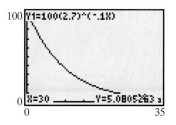

Notice from the graph that as the number of days, x, increases, the percent of radioactive material, y, decreases.

We can use the TRACE feature to obtain an approximation of the expected percent of radioactive material in the milk when $x = 30$. (A TABLE feature may also be used to approximate the percent.)

X	Y1
0	100
10	37.037
20	13.717
30	5.0805
40	1.8817
50	.69692
60	.25812

$Y_1 = 100(2.7)^{\wedge}(-....$

The percent of radioactive material in the milk 30 days after the Chernobyl accident was 5.08%, accurate to two decimal places.   □

PRACTICE

6   If a single sheet of glass prevents 5% of the incoming light from passing through it, then the percent p of light that passes through n successive sheets of glass is given approximately by the function $p(n) = 100(2.7)^{-0.05n}$. Estimate the expected percent of light that will pass through 10 sheets of glass. Round to the nearest hundredth of a percent.

EXAMPLE 7   Use a graphing calculator to find each percent. Approximate your solutions so that they are accurate to two decimal places. Estimate the expected percent of radioactive material in the milk from the Chernobyl nuclear accident after:

a. 2 days   **b.** 10 days   **c.** 15 days   **d.** 25 days

Solution   Use a table similar to the one above with $y_1 = 100(2.7)^{-0.1x}$. In Ask mode, let $x = 2, 10, 15$, and 25. The results are

a. 81.98%   **b.** 37.04%   **c.** 22.54%   **d.** 8.35%   □

PRACTICE

7   Estimate the expected percent of radioactive material in the milk from the Chernobyl accident after (round answers to 2 decimal places):

a. 4 days   **b.** 35 days

VOCABULARY & READINESS CHECK

Use the choices to fill in each blank.

1. A function such as $f(x) = 2^x$ is a(n) _____ function.

  A. linear **B.** quadratic **C.** exponential

2. If $7^x = 7^y$, then _____.

  A. $x = 7^y$ **B.** $x = y$ **C.** $y = 7^x$ **D.** $7 = 7^y$

Answer the questions about the graph of $y = 2^x$, shown to the right.

3. Is this a one-to-one function? _____

4. Is there an x-intercept? _____ If so, name the coordinates. _____

5. Is there a y-intercept? _____ If so, name the coordinates. _____

6. The domain of this function, in interval notation, is _____ .

7. The range of this function, in interval notation, is _____ .

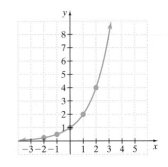

9.3 EXERCISE SET

 MyMathLab PRACTICE WATCH DOWNLOAD READ REVIEW

Graph each exponential function. See Examples 1 through 3.

1. $y = 4^x$

2. $y = 5^x$

3. $y = 2^x + 1$

4. $y = 3^x - 1$

5. $y = \left(\dfrac{1}{4}\right)^x$

6. $y = \left(\dfrac{1}{5}\right)^x$

7. $y = \left(\dfrac{1}{2}\right)^x - 2$

8. $y = \left(\dfrac{1}{3}\right)^x + 2$

9. $y = -2^x$

10. $y = -3^x$

11. $y = -\left(\dfrac{1}{4}\right)^x$

12. $y = -\left(\dfrac{1}{5}\right)^x$

13. $f(x) = 2^{x+1}$

14. $f(x) = 3^{x-1}$

15. $f(x) = 4^{x-2}$

16. $f(x) = 2^{x+3}$

Match each exponential equation with its graph below or in the next column. See Examples 1 through 3.

17. $f(x) = \left(\dfrac{1}{2}\right)^x$

18. $f(x) = \left(\dfrac{1}{4}\right)^x$

19. $f(x) = 2^x$

20. $f(x) = 3^x$

A.

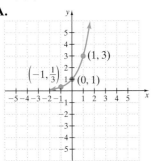

B.

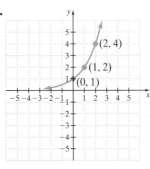

C.

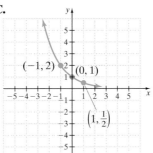

D.

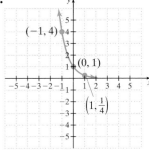

Solve each equation for x. See Example 4.

21. $3^x = 27$

22. $6^x = 36$

23. $16^x = 8$

24. $64^x = 16$

25. $32^{2x-3} = 2$

26. $9^{2x+1} = 81$

27. $\dfrac{1}{4} = 2^{3x}$

28. $\dfrac{1}{27} = 3^{2x}$

29. $5^x = 625$

30. $2^x = 64$

31. $4^x = 8$

32. $32^x = 4$

33. $27^{x+1} = 9$

34. $125^{x-2} = 25$

35. $81^{x-1} = 27^{2x}$

36. $4^{3x-7} = 32^{2x}$

Solve. Unless otherwise indicated, round results to one decimal place. See Examples 6 and 7.

37. One type of uranium has a daily radioactive decay rate of 0.4%. If 30 pounds of this uranium is available today, find how much will still remain after 50 days. Use $y = 30(2.7)^{-0.004t}$, and let t be 50.

38. The nuclear waste from an atomic energy plant decays at a rate of 3% each century. If 150 pounds of nuclear waste is disposed of, find how much of it will still remain after 10 centuries. Use $y = 150(2.7)^{-0.03t}$, and let t be 10.

39. National Park Service personnel are trying to increase the size of the bison population of Theodore Roosevelt National Park. If 260 bison currently live in the park, and if the population's rate of growth is 2.5% annually, find how many bison (rounded to the nearest whole) there should be in 10 years. Use $y = 260(2.7)^{0.025t}$.

40. The size of the rat population of a wharf area grows at a rate of 8% monthly. If there are 200 rats in January, find how many rats (rounded to the nearest whole) should be expected by next January. Use $y = 200(2.7)^{0.08t}$.

41. A rare isotope of a nuclear material is very unstable, decaying at a rate of 15% each second. Find how much isotope remains 10 seconds after 5 grams of the isotope is created. Use $y = 5(2.7)^{-0.15t}$.

42. An accidental spill of 75 grams of radioactive material in a local stream has led to the presence of radioactive debris decaying at a rate of 4% each day. Find how much debris still remains after 14 days. Use $y = 75(2.7)^{-0.04t}$.

43. The atmospheric pressure p, in Pascals, on a weather balloon decreases with increasing height. This pressure is related to the number of kilometers h above sea level by the function $p(h) = 760(2.7)^{-0.145h}$. Round to the nearest tenth of a Pascal.

  a. Find the atmospheric pressure at a height of 1 kilometer.

  b. Find the atmospheric pressure at a height of 10 kilometers.

44. An unusually wet spring has caused the size of the Cape Cod mosquito population to increase by 8% each day. If an estimated 200,000 mosquitoes are on Cape Cod on May 12, find how many mosquitoes will inhabit the Cape on May 25. Use $y = 200,000(2.7)^{0.08t}$. Round to the nearest thousand.

45. The equation $y = 84,949(1.096)^x$ models the number of American college students who studied abroad each year from 1995 through 2006. In the equation, y is the number of American students studying abroad and x represents the number of years after 1995. Round answers to the nearest whole. (*Source:* Based on data from the Institute of International Education, Open Doors 2006)

  a. Estimate the number of American students studying abroad in 2000.

  b. Assuming this equation continues to be valid in the future, use this equation to predict the number of American students studying abroad in 2020.

46. Carbon dioxide (CO_2) is a greenhouse gas that contributes to global warming. Partially due to the combustion of fossil fuels, the amount of CO_2 in Earth's atmosphere has been increasing by 0.4% annually over the past century. In 2000, the concentration of CO_2 in the atmosphere was 369.4 parts per million by volume. To make the following predictions, use $y = 369.4(1.004)^t$, where y is the concentration of CO_2 in parts per million and t is the number of years after 2000. Round answers to the nearest tenth. (*Sources:* Based on data from the United Nations Environment Programme and the Carbon Dioxide Information Analysis Center)

  a. Predict the concentration of CO_2 in the atmosphere in the year 2015.

  b. Predict the concentration of CO_2 in the atmosphere in the year 2030.

Solve. Use $A = P\left(1 + \dfrac{r}{n}\right)^{nt}$. Round answers to two decimal places. See Example 5.

47. Find the amount Erica owes at the end of 3 years if $6000 is loaned to her at a rate of 8% compounded monthly.

48. Find the amount owed at the end of 5 years if $3000 is loaned at a rate of 10% compounded quarterly.

49. Find the total amount Janina has in a college savings account if $2000 was invested and earned 6% compounded semiannually for 12 years.

50. Find the amount accrued if $500 is invested and earns 7% compounded monthly for 4 years.

The formula $y = 18(1.24)^x$ gives the number of cellular phone users y (in millions) in the United States for the years 1994 through 2006. In this formula, $x = 0$ corresponds to 1994, $x = 1$ corresponds to 1995, and so on. Use this formula to solve Exercises 51 and 52. Round results to the nearest whole million.

51. Use this model to predict the number of cellular phone users in the year 2010.

52. Use this model to predict the number of cellular phone users in the year 2014.

REVIEW AND PREVIEW

Solve each equation. See Sections 1.5 and 5.8.

53. $5x - 2 = 18$

54. $3x - 7 = 11$

55. $3x - 4 = 3(x + 1)$

56. $2 - 6x = 6(1 - x)$

57. $x^2 + 6 = 5x$

58. $18 = 11x - x^2$

By inspection, find the value for x that makes each statement true. See Section 5.1.

59. $2^x = 8$

60. $3^x = 9$

61. $5^x = \dfrac{1}{5}$

62. $4^x = 1$

CONCEPT EXTENSIONS

63. Explain why the graph of an exponential function $y = b^x$ contains the point $(1, b)$.

64. Explain why an exponential function $y = b^x$ has a y-intercept of $(0, 1)$.

Graph.

65. $y = |3^x|$

66. $y = \left|\left(\dfrac{1}{3}\right)^x\right|$

67. $y = 3^{|x|}$

68. $y = \left(\dfrac{1}{3}\right)^{|x|}$

69. Graph $y = 2^x$ and $y = \left(\dfrac{1}{2}\right)^{-x}$ on the same set of axes. Describe what you see and why.

70. Graph $y = 2^x$ and $x = 2^y$ on the same set of axes. Describe what you see.

Use a graphing calculator to solve. Estimate each result to two decimal places.

71. From Exercise 37, estimate the number of pounds of uranium that will be available after 120 days.

72. From Exercise 37, estimate the number of pounds of uranium that will be available after 100 days.

73. From Exercise 42, estimate the amount of debris that remains after 10 days.

74. From Exercise 42, estimate the amount of debris that remains after 20 days.

9.4 LOGARITHMIC FUNCTIONS

OBJECTIVES

1. Write exponential equations with logarithmic notation and write logarithmic equations with exponential notation.

2. Solve logarithmic equations by using exponential notation.

3. Identify and graph logarithmic functions.

OBJECTIVE 1 ▶ Using logarithmic notation. Since the exponential function $f(x) = 2^x$ is a one-to-one function, it has an inverse.

We can create a table of values for f^{-1} by switching the coordinates in the accompanying table of values for $f(x) = 2^x$.

$$f(x) = 2^x$$

x	$y = f(x)$
-3	$\dfrac{1}{8}$
-2	$\dfrac{1}{4}$
-1	$\dfrac{1}{2}$
0	1
1	2
2	4
3	8

x	$y = f^{-1}(x)$
$\dfrac{1}{8}$	-3
$\dfrac{1}{4}$	-2
$\dfrac{1}{2}$	-1
1	0
2	1
4	2
8	3

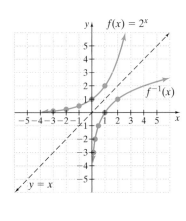

The graphs of $f(x)$ and its inverse are shown above. Notice that the graphs of f and f^{-1} are symmetric about the line $y = x$, as expected.

Now we would like to be able to write an equation for f^{-1}. To do so, we follow the steps for finding an inverse.

$$f(x) = 2^x$$

STEP 1. Replace $f(x)$ by y. $\qquad\qquad y = 2^x$

STEP 2. Interchange x and y. $\qquad\qquad x = 2^y$

STEP 3. Solve for y.

At this point, we are stuck. To solve this equation for y, a new notation, the **logarithmic notation,** is needed. The symbol $\log_b x$ means "the power to which b is raised in order to produce a result of x."

$$\log_b x = y \quad \text{means} \quad b^y = x$$

We say that $\log_b x$ is "the logarithm of x to the base b" or "the log of x to the base b."

Logarithmic Definition

If $b > 0$ and $b \neq 1$, then

$$y = \log_b x \text{ means } x = b^y$$

for every $x > 0$ and every real number y.

Before returning to the function $x = 2^y$ and solving it for y in terms of x, let's practice using the new notation $\log_b x$.

It is important to be able to write exponential equations from logarithmic notation, and vice versa. The following table shows examples of both forms.

> ▶ **Helpful Hint**
>
> Notice that a *logarithm* is an *exponent*. In other words, $\log_3 9$ is the *power* that we raise 3 to in order to get 9.

Logarithmic Equation	Corresponding Exponential Equation
$\log_3 9 = 2$	$3^2 = 9$
$\log_6 1 = 0$	$6^0 = 1$
$\log_2 8 = 3$	$2^3 = 8$
$\log_4 \dfrac{1}{16} = -2$	$4^{-2} = \dfrac{1}{16}$
$\log_8 2 = \dfrac{1}{3}$	$8^{1/3} = 2$

EXAMPLE 1   Write as an exponential equation.

a. $\log_5 25 = 2$ **b.** $\log_6 \dfrac{1}{6} = -1$ **c.** $\log_2 \sqrt{2} = \dfrac{1}{2}$ **d.** $\log_7 x = 5$

Solution

a. $\log_5 25 = 2$ means $5^2 = 25$

b. $\log_6 \dfrac{1}{6} = -1$ means $6^{-1} = \dfrac{1}{6}$

c. $\log_2 \sqrt{2} = \dfrac{1}{2}$ means $2^{1/2} = \sqrt{2}$

d. $\log_7 x = 5$ means $7^5 = x$

PRACTICE
1   Write as an exponential equation.

a. $\log_3 81 = 4$ **b.** $\log_5 \dfrac{1}{5} = -1$ **c.** $\log_7 \sqrt{7} = \dfrac{1}{2}$ **d.** $\log_{13} y = 4$

EXAMPLE 2   Write as a logarithmic equation.

a. $9^3 = 729$ **b.** $6^{-2} = \dfrac{1}{36}$ **c.** $5^{1/3} = \sqrt[3]{5}$ **d.** $\pi^4 = x$

Solution

a. $9^3 = 729$ means $\log_9 729 = 3$

b. $6^{-2} = \dfrac{1}{36}$ means $\log_6 \dfrac{1}{36} = -2$

c. $5^{1/3} = \sqrt[3]{5}$ means $\log_5 \sqrt[3]{5} = \dfrac{1}{3}$

d. $\pi^4 = x$ means $\log_\pi x = 4$ □

PRACTICE

2 Write as a logarithmic equation.

a. $4^3 = 64$ **b.** $6^{1/3} = \sqrt[3]{6}$ **c.** $5^{-3} = \dfrac{1}{125}$ **d.** $\pi^7 = z$

EXAMPLE 3 Find the value of each logarithmic expression.

a. $\log_4 16$ **b.** $\log_{10} \dfrac{1}{10}$ **c.** $\log_9 3$

Solution

a. $\log_4 16 = 2$ because $4^2 = 16$

b. $\log_{10} \dfrac{1}{10} = -1$ because $10^{-1} = \dfrac{1}{10}$

c. $\log_9 3 = \dfrac{1}{2}$ because $9^{1/2} = \sqrt{9} = 3$ □

PRACTICE

3 Find the value of each logarithmic expression.

a. $\log_3 9$ **b.** $\log_2 \dfrac{1}{8}$ **c.** $\log_{49} 7$

TECHNOLOGY NOTE

A change of base formula, such as

$$\log_a b = \frac{\log b}{\log a}$$

can be used to evaluate logarithms on a calculator. (See Section 9.6, where we also find that log means $\log_{10}$.) For Example 3a,

$$\log_4 16 = \frac{\log 16}{\log 4}.$$

```
log(16)/log(4)
                2
log(1/10)/log(10
)
               -1
log(3)/log(9)
               .5
■
```

A calculator check for Example 3.

> ▶ **Helpful Hint**
>
> Another method for evaluating logarithms such as those in Example 3 is to set the expressions equal to x and then write them in exponential form to find x. For example:
>
> **a.** $\log_4 16 = x$ means $4^x = 16$. Since $4^2 = 16$, $x = 2$ or $\log_4 16 = 2$.
>
> **b.** $\log_{10} \dfrac{1}{10} = x$ means $10^x = \dfrac{1}{10}$. Since $10^{-1} = \dfrac{1}{10}$, $x = -1$ or $\log_{10} \dfrac{1}{10} = -1$.
>
> **c.** $\log_9 3 = x$ means $9^x = 3$. Since $9^{1/2} = 3$, $x = \dfrac{1}{2}$ or $\log_9 3 = \dfrac{1}{2}$.

OBJECTIVE 2 ▶ Solving logarithmic equations. The ability to interchange the logarithmic and exponential forms of a statement is often the key to solving logarithmic equations.

EXAMPLE 4 Solve each equation for x.

a. $\log_4 \dfrac{1}{4} = x$ **b.** $\log_5 x = 3$ **c.** $\log_x 25 = 2$ **d.** $\log_3 1 = x$ **e.** $\log_b 1 = x$

Solution

a. $\log_4 \dfrac{1}{4} = x$ means $4^x = \dfrac{1}{4}$. Solve $4^x = \dfrac{1}{4}$ for x.

$$4^x = \frac{1}{4}$$

$$4^x = 4^{-1}$$

Since the bases are the same, by the uniqueness of b^x, we have that

$$x = -1$$

The solution is -1 or the solution set is $\{-1\}$. To check, see that $\log_4 \frac{1}{4} = -1$, since $4^{-1} = \frac{1}{4}$.

b. $\log_5 x = 3$

$\qquad 5^3 = x$ Write as an exponential equation.

$\qquad 125 = x$

The solution is 125.

c. $\log_x 25 = 2$

$\qquad x^2 = 25$ Write as an exponential equation. Here, $x > 0$, $x \neq 1$.

$\qquad x = 5$

Even though $(-5)^2 = 25$, the base b of a logarithm must be positive. The solution is 5.

d. $\log_3 1 = x$

$\qquad 3^x = 1$ Write as an exponential equation.

$\qquad 3^x = 3^0$ Write 1 as 3^0.

$\qquad x = 0$ Use the uniqueness of b^x.

The solution is 0.

e. $\log_b 1 = x$

$\qquad b^x = 1$ Write as an exponential equation. Here, $b > 0$ and $b \neq 1$.

$\qquad b^x = b^0$ Write 1 as b^0.

$\qquad x = 0$ Apply the uniqueness of b^x.

The solution is 0.   □

PRACTICE

4   Solve each equation for x.

a. $\log_5 \frac{1}{25} = x$ **b.** $\log_x 8 = 3$ **c.** $\log_6 x = 2$

d. $\log_{13} 1 = x$ **e.** $\log_h 1 = x$

In Example 4e we proved an important property of logarithms. That is, $\log_b 1$ is always 0. This property as well as two important others are given next.

Properties of Logarithms

If b is a real number, $b > 0$, and $b \neq 1$, then

1. $\log_b 1 = 0$

2. $\log_b b^x = x$

3. $b^{\log_b x} = x$

To see Property **2.**, $\log_b b^x = x$, change the logarithmic form to exponential form. Then, $\log_b b^x = x$ means $b^x = b^x$. In exponential form, the statement is true, so in logarithmic form, the statement is also true. To understand Property **3.**, $b^{\log_b x} = x$, write this exponential equation as an equivalent logarithm.

EXAMPLE 5   Simplify.

a. $\log_3 3^2$ **b.** $\log_7 7^{-1}$ **c.** $5^{\log_5 3}$ **d.** $2^{\log_2 6}$

Solution

a. From Property 2, $\log_3 3^2 = 2$.   **b.** From Property 2, $\log_7 7^{-1} = -1$.

c. From Property 3, $5^{\log_5 3} = 3$.   **d.** From Property 3, $2^{\log_2 6} = 6$.   □

5 Simplify.

a. $\log_5 5^4$ **b.** $\log_9 9^{-2}$ **c.** $6^{\log_6 5}$ **d.** $7^{\log_7 4}$

OBJECTIVE 3 ▶ Graphing logarithmic functions. Let us now return to the function $f(x) = 2^x$ and write an equation for its inverse, $f^{-1}(x)$. Recall our earlier work.

$$f(x) = 2^x$$

STEP 1. Replace $f(x)$ by y. $\qquad y = 2^x$

STEP 2. Interchange x and y. $\qquad x = 2^y$

Having gained proficiency with the notation $\log_b x$, we can now complete the steps for writing the inverse equation by writing $x = 2^y$ as an equivalent logarithm.

STEP 1. Solve for y. $\qquad\qquad y = \log_2 x$

STEP 2. Replace y with $f^{-1}(x)$. $\qquad f^{-1}(x) = \log_2 x$

Thus, $f^{-1}(x) = \log_2 x$ defines a function that is the inverse function of the function $f(x) = 2^x$. The function $f^{-1}(x)$ or $y = \log_2 x$ is called a **logarithmic function.**

Logarithmic Function

If x is a positive real number, b is a constant positive real number, and b is not 1, then a **logarithmic function** is a function that can be defined by

$$f(x) = \log_b x$$

The domain of f is the set of positive real numbers, and the range of f is the set of real numbers.

Concept Check ☑

Let $f(x) = \log_3 x$ and $g(x) = 3^x$. These two functions are inverses of each other. Since $(2, 9)$ is an ordered pair solution of $g(x)$ or $g(2) = 9$, what ordered pair do we know to be a solution of $f(x)$? Also, find $f(9)$. Explain why.

We can explore logarithmic functions by graphing them.

EXAMPLE 6 Graph the logarithmic function $y = \log_2 x$.

Solution First we write the equation with exponential notation as $2^y = x$. Then we find some ordered pair solutions that satisfy this equation. Finally, we plot the points and connect them with a smooth curve. The domain of this function is $(0, \infty)$, and the range is all real numbers.

Since $x = 2^y$ is solved for x, we choose y-values and compute corresponding x-values.

If $y = 0, x = 2^0 = 1$.
If $y = 1, x = 2^1 = 2$.
If $y = 2, x = 2^2 = 4$.
If $y = -1, x = 2^{-1} = \dfrac{1}{2}$.

$x = 2^y$	y
1	0
2	1
4	2
$\dfrac{1}{2}$	-1

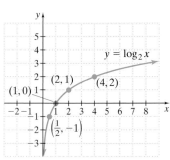

Notice that the x-intercept is $(1, 0)$ and there is no y-intercept.

To check using the graphing utility, we need to use the change of base formula. (See the Technology Note on page 652.) Graph $y_1 = \dfrac{\log(x)}{\log(2)}$ in a decimal window.

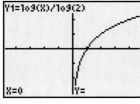

Notice there is no y-intercept.

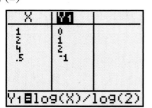

A table of values using the calculator.

PRACTICE
6   Graph the logarithmic function $y = \log_7 x$.

EXAMPLE 7   Graph the logarithmic function $f(x) = \log_{1/3} x$.

Solution   Replace $f(x)$ with y, and write the result with exponential notation.

$$f(x) = \log_{1/3} x$$

$$y = \log_{1/3} x \quad \text{Replace } f(x) \text{ with } y.$$

$$\left(\frac{1}{3}\right)^y = x \quad \text{Write in exponential form.}$$

Now we can find ordered pair solutions that satisfy $\left(\dfrac{1}{3}\right)^y = x$, plot these points, and connect them with a smooth curve.

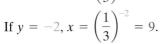

A calculator check of Example 7 using the change of base formula. A table may also be used to help check.

If $y = 0$, $x = \left(\dfrac{1}{3}\right)^0 = 1$.

If $y = 1$, $x = \left(\dfrac{1}{3}\right)^1 = \dfrac{1}{3}$.

If $y = -1$, $x = \left(\dfrac{1}{3}\right)^{-1} = 3$.

If $y = -2$, $x = \left(\dfrac{1}{3}\right)^{-2} = 9$.

$x = \left(\dfrac{1}{3}\right)^y$	y
1	0
$\dfrac{1}{3}$	1
3	-1
9	-2

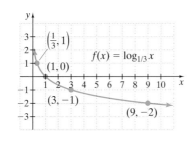

The domain of this function is $(0, \infty)$, and the range is the set of all real numbers. The x-intercept is $(1, 0)$ and there is no y-intercept.

PRACTICE
7   Graph the logarithmic function $y = \log_{1/4} x$.

The following figures summarize characteristics of logarithmic functions.

$$f(x) = \log_b x, b > 0, b \neq 1$$

- one-to-one function
- x-intercept $(1, 0)$
- no y-intercept

- domain: $(0, \infty)$
- range: $(-\infty, \infty)$

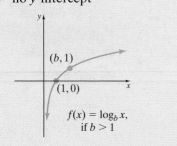

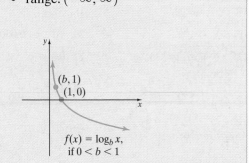

VOCABULARY & READINESS CHECK

Use the choices to fill in each blank.

1. A function, such as $y = \log_2 x$, is a(n) _____ function.

   A. linear   **B.** logarithmic   **C.** quadratic   **D.** exponential

2. If $y = \log_2 x$, then _____.

   A. $x = y$ **B.** $2^x = y$ **C.** $2^y = x$ **D.** $2y = x$

Answer the questions about the graph of $y = \log_2 x$, shown to the left.

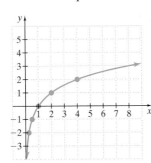

3. Is this a one-to-one function? _____

4. Is there an x-intercept? _____ If so, name the coordinates. _____

5. Is there a y-intercept? _____ If so, name the coordinates. _____

6. The domain of this function, in interval notation, is _____ .

7. The range of this function, in interval notation, is _____ .

9.4 | EXERCISE SET

Write each as an exponential equation. See Example 1.

 1. $\log_6 36 = 2$ **2.** $\log_2 32 = 5$

3. $\log_3 \dfrac{1}{27} = -3$ **4.** $\log_5 \dfrac{1}{25} = -2$

5. $\log_{10} 1000 = 3$ **6.** $\log_{10} 10 = 1$

7. $\log_9 x = 4$ **8.** $\log_8 y = 7$

9. $\log_\pi \dfrac{1}{\pi^2} = -2$ **10.** $\log_e \dfrac{1}{e} = -1$

 11. $\log_7 \sqrt{7} = \dfrac{1}{2}$ **12.** $\log_{11} \sqrt[4]{11} = \dfrac{1}{4}$

13. $\log_{0.7} 0.343 = 3$ **14.** $\log_{1.2} 1.44 = 2$

15. $\log_3 \dfrac{1}{81} = -4$ **16.** $\log_{1/4} 16 = -2$

Write each as a logarithmic equation. See Example 2.

 17. $2^4 = 16$ **18.** $5^3 = 125$

19. $10^2 = 100$ **20.** $10^4 = 10{,}000$

21. $\pi^3 = x$ **22.** $\pi^5 = y$

 23. $10^{-1} = \dfrac{1}{10}$ **24.** $10^{-2} = \dfrac{1}{100}$

25. $4^{-2} = \dfrac{1}{16}$ **26.** $3^{-4} = \dfrac{1}{81}$

27. $5^{1/2} = \sqrt{5}$ **28.** $4^{1/3} = \sqrt[3]{4}$

Find the value of each logarithmic expression. See Examples 3 and 5.

 29. $\log_2 8$ **30.** $\log_3 9$

31. $\log_3 \dfrac{1}{9}$ **32.** $\log_2 \dfrac{1}{32}$

33. $\log_{25} 5$ **34.** $\log_8 \dfrac{1}{2}$

 35. $\log_{1/2} 2$ **36.** $\log_{2/3} \dfrac{4}{9}$

 37. $\log_6 1$ **38.** $\log_9 9$

39. $\log_{10} 100$ **40.** $\log_{10} \dfrac{1}{10}$

41. $\log_3 81$ **42.** $\log_2 16$

43. $\log_4 \dfrac{1}{64}$ **44.** $\log_3 \dfrac{1}{9}$

Solve. See Example 4.

45. $\log_3 9 = x$ **46.** $\log_2 8 = x$

47. $\log_3 x = 4$ **48.** $\log_2 x = 3$

49. $\log_x 49 = 2$ **50.** $\log_x 8 = 3$

51. $\log_2 \dfrac{1}{8} = x$ **52.** $\log_3 \dfrac{1}{81} = x$

53. $\log_3 \dfrac{1}{27} = x$

54. $\log_5 \dfrac{1}{125} = x$

55. $\log_8 x = \dfrac{1}{3}$

56. $\log_9 x = \dfrac{1}{2}$

57. $\log_4 16 = x$

58. $\log_2 16 = x$

59. $\log_{3/4} x = 3$

60. $\log_{2/3} x = 2$

61. $\log_x 100 = 2$

62. $\log_x 27 = 3$

63. $\log_2 2^4 = x$

64. $\log_6 6^{-2} = x$

65. $3^{\log_3 5} = x$

66. $5^{\log_5 7} = x$

67. $\log_x \dfrac{1}{7} = \dfrac{1}{2}$

68. $\log_x 2 = -\dfrac{1}{3}$

Simplify. See Example 5.

69. $\log_5 5^3$

70. $\log_6 6^2$

71. $2^{\log_2 3}$

72. $7^{\log_7 4}$

73. $\log_9 9$

74. $\log_8 (8)^{-1}$

Graph each logarithmic function. Label any intercepts. See Examples 6 and 7.

75. $y = \log_3 x$

76. $y = \log_8 x$

77. $f(x) = \log_{1/4} x$

78. $f(x) = \log_{1/2} x$

79. $f(x) = \log_5 x$

80. $f(x) = \log_6 x$

81. $f(x) = \log_{1/6} x$

82. $f(x) = \log_{1/5} x$

REVIEW AND PREVIEW

Simplify each rational expression. See Section 6.1.

83. $\dfrac{x + 3}{3 + x}$

84. $\dfrac{x - 5}{5 - x}$

85. $\dfrac{x^2 - 8x + 16}{2x - 8}$

86. $\dfrac{x^2 - 3x - 10}{2 + x}$

Add or subtract as indicated. See Section 6.2.

87. $\dfrac{2}{x} + \dfrac{3}{x^2}$

88. $\dfrac{3x}{x + 3} + \dfrac{9}{x + 3}$

89. $\dfrac{m^2}{m + 1} - \dfrac{1}{m + 1}$

90. $\dfrac{5}{y + 1} - \dfrac{4}{y - 1}$

CONCEPT EXTENSIONS

Solve. See the Concept Check in this section.

91. Let $f(x) = \log_5 x$. Then $g(x) = 5^x$ is the inverse of $f(x)$. The ordered pair $(2, 25)$ is a solution of the function $g(x)$.

    a. Write this solution using function notation.

    b. Write an ordered pair that we know to be a solution of $f(x)$.

    c. Use the answer to part **b** and write the solution using function notation.

92. Let $f(x) = \log_{0.3} x$. Then $g(x) = 0.3^x$ is the inverse of $f(x)$. The ordered pair $(3, 0.027)$ is a solution of the function $g(x)$.

    a. Write this solution using function notation.

    b. Write an ordered pair that we know to be a solution of $f(x)$.

    c. Use the answer to part **b** and write the solution using function notation.

93. Explain why negative numbers are not included as logarithmic bases.

94. Explain why 1 is not included as a logarithmic base.

Solve by first writing as an exponential.

95. $\log_7 (5x - 2) = 1$

96. $\log_3 (2x + 4) = 2$

97. Simplify: $\log_3(\log_5 125)$

98. Simplify: $\log_7(\log_4(\log_2 16))$

Graph each function and its inverse function on the same set of axes. Label any intercepts.

99. $y = 4^x$; $y = \log_4 x$

100. $y = 3^x$; $y = \log_3 x$

101. $y = \left(\dfrac{1}{3}\right)^x$; $y = \log_{1/3} x$

102. $y = \left(\dfrac{1}{2}\right)^x$; $y = \log_{1/2} x$

103. Explain why the graph of the function $y = \log_b x$ contains the point $(1, 0)$ no matter what b is.

104. $\log_3 10$ is between which two integers? Explain your answer.

105. The formula $\log_{10}(1 - k) = \dfrac{-0.3}{H}$ models the relationship between the half-life H of a radioactive material and its rate of decay k. Find the rate of decay of the iodine isotope I-131 if its half-life is 8 days. Round to four decimal places.

106. The formula $\text{pH} = -\log_{10}(\text{H}^+)$ provides the pH for a liquid, where H^+ stands for the concentration of hydronium ions. Find the pH of lemonade, whose concentration of hydronium ions is 0.0050 moles/liter.

9.5 PROPERTIES OF LOGARITHMS

OBJECTIVES

1 Use the product property of logarithms.

2 Use the quotient property of logarithms.

3 Use the power property of logarithms.

4 Use the properties of logarithms together.

In the previous section we explored some basic properties of logarithms. We now introduce and explore additional properties. Because a logarithm is an exponent, logarithmic properties are just restatements of exponential properties.

OBJECTIVE 1 ▶ Using the product property. The first of these properties is called the **product property of logarithms,** because it deals with the logarithm of a product.

Product Property of Logarithms

If x, y, and b are positive real numbers and $b \neq 1$, then

$$\log_b xy = \log_b x + \log_b y$$

To prove this, let $\log_b x = M$ and $\log_b y = N$. Now write each logarithm with exponential notation.

$$\log_b x = M \qquad \text{is equivalent to} \qquad b^M = x$$
$$\log_b y = N \qquad \text{is equivalent to} \qquad b^N = y$$

Multiply the left sides and the right sides of the exponential equations, and we have that

$$xy = (b^M)(b^N) = b^{M+N}$$

If we write the equation $xy = b^{M+N}$ in equivalent logarithmic form, we have

$$\log_b xy = M + N$$

But since $M = \log_b x$ and $N = \log_b y$, we can write

$$\log_b xy = \log_b x + \log_b y \quad \text{Let } M = \log_b x \text{ and } N = \log_b y.$$

In other words, the logarithm of a product is the sum of the logarithms of the factors. This property is sometimes used to simplify logarithmic expressions.

In the examples that follow, assume that variables represent positive numbers.

EXAMPLE 1 Write each sum as a single logarithm.

a. $\log_{11} 10 + \log_{11} 3$ **b.** $\log_3 \dfrac{1}{2} + \log_3 12$ **c.** $\log_2(x + 2) + \log_2 x$

Solution

In each case, both terms have a common logarithmic base.

a. $\log_{11} 10 + \log_{11} 3 = \log_{11}(10 \cdot 3)$ Apply the product property.

$$= \log_{11} 30$$

b. $\log_3 \dfrac{1}{2} + \log_3 12 = \log_3 \left(\dfrac{1}{2} \cdot 12 \right) = \log_3 6$

c. $\log_2(x + 2) + \log_2 x = \log_2[(x + 2) \cdot x] = \log_2(x^2 + 2x)$

▶ **Helpful Hint**

Check your logarithm properties. Make sure you understand that $\log_2(x + 2)$ *is not* $\log_2 x + \log_2 2$.

PRACTICE

1 Write each sum as a single logarithm.

a. $\log_8 5 + \log_8 3$

b. $\log_2 \dfrac{1}{3} + \log_2 18$

c. $\log_5(x - 1) + \log_5(x + 1)$

OBJECTIVE 2 ▶ **Using the quotient property.** The second property is the **quotient property of logarithms.**

> **Quotient Property of Logarithms**
> If x, y, and b are positive real numbers and $b \neq 1$, then
> $$\log_b \frac{x}{y} = \log_b x - \log_b y$$

The proof of the quotient property of logarithms is similar to the proof of the product property. Notice that the quotient property says that the logarithm of a quotient is the difference of the logarithms of the dividend and divisor.

Concept Check ☑

Which of the following is the correct way to rewrite $\log_5 \frac{7}{2}$?

a. $\log_5 7 - \log_5 2$ **b.** $\log_5(7 - 2)$ **c.** $\dfrac{\log_5 7}{\log_5 2}$ **d.** $\log_5 14$

EXAMPLE 2   Write each difference as a single logarithm.

a. $\log_{10} 27 - \log_{10} 3$ **b.** $\log_5 8 - \log_5 x$ **c.** $\log_3(x^2 + 5) - \log_3(x^2 + 1)$

Solution   All terms have a common logarithmic base.

a. $\log_{10} 27 - \log_{10} 3 = \log_{10} \dfrac{27}{3} = \log_{10} 9$

b. $\log_5 8 - \log_5 x = \log_5 \dfrac{8}{x}$

c. $\log_3(x^2 + 5) - \log_3(x^2 + 1) = \log_3 \dfrac{x^2 + 5}{x^2 + 1}$ Apply the quotient property.   □

PRACTICE
2   Write each difference as a single logarithm.

a. $\log_5 18 - \log_5 6$ **b.** $\log_6 x - \log_6 3$ **c.** $\log_4(x^2 + 1) - \log_4(x^2 + 3)$

OBJECTIVE 3 ▶ **Using the power property.** The third and final property we introduce is the **power property of logarithms.**

> **Power Property of Logarithms**
> If x and b are positive real numbers, $b \neq 1$, and r is a real number, then
> $$\log_b x^r = r \log_b x$$

EXAMPLE 3   Use the power property to rewrite each expression.

a. $\log_5 x^3$ **b.** $\log_4 \sqrt{2}$

Solution

a. $\log_5 x^3 = 3 \log_5 x$ **b.** $\log_4 \sqrt{2} = \log_4 2^{1/2} = \dfrac{1}{2} \log_4 2$ □

PRACTICE
3   Use the power property to rewrite each expression.

a. $\log_7 x^8$ **b.** $\log_5 \sqrt[4]{7}$

Answer to Concept Check: a

OBJECTIVE 4 ▶ Using the properties together. Many times we must use more than one property of logarithms to simplify a logarithmic expression.

EXAMPLE 4 Write as a single logarithm.

a. $2 \log_5 3 + 3 \log_5 2$ **b.** $3 \log_9 x - \log_9(x + 1)$ **c.** $\log_4 25 + \log_4 3 - \log_4 5$

Solution   In each case, all terms have a common logarithmic base.

a. $2 \log_5 3 + 3 \log_5 2 = \log_5 3^2 + \log_5 2^3$ Apply the power property.

$$= \log_5 9 + \log_5 8$$

$$= \log_5(9 \cdot 8)$$ Apply the product property.

$$= \log_5 72$$

b. $3 \log_9 x - \log_9(x + 1) = \log_9 x^3 - \log_9(x + 1)$ Apply the power property.

$$= \log_9 \frac{x^3}{x + 1}$$ Apply the quotient property.

c. Use both the product and quotient properties.

$$\log_4 25 + \log_4 3 - \log_4 5 = \log_4(25 \cdot 3) - \log_4 5 \quad \text{Apply the product property.}$$

$$= \log_4 75 - \log_4 5 \quad \text{Simplify.}$$

$$= \log_4 \frac{75}{5} \quad \text{Apply the quotient property.}$$

$$= \log_4 15 \quad \text{Simplify.}$$

**PRACTICE
4**   Write as a single logarithm.

a. $2 \log_5 4 + 5 \log_5 2$ **b.** $2 \log_8 x - \log_8(x + 3)$ **c.** $\log_7 12 + \log_7 5 - \log_7 4$

EXAMPLE 5 Write each expression as sums or differences of multiples of logarithms.

a. $\log_3 \dfrac{5 \cdot 7}{4}$ **b.** $\log_2 \dfrac{x^5}{y^2}$

Solution

a. $\log_3 \dfrac{5 \cdot 7}{4} = \log_3(5 \cdot 7) - \log_3 4$ Apply the quotient property.

$$= \log_3 5 + \log_3 7 - \log_3 4 \quad \text{Apply the product property.}$$

b. $\log_2 \dfrac{x^5}{y^2} = \log_2(x^5) - \log_2(y^2)$ Apply the quotient property.

$$= 5 \log_2 x - 2 \log_2 y \quad \text{Apply the power property.}$$

**PRACTICE
5**   Write each expression as sums or differences of multiples of logarithms.

a. $\log_5 \dfrac{4 \cdot 3}{7}$ **b.** $\log_4 \dfrac{a^2}{b^5}$

▶ Helpful Hint

Notice that we are not able to simplify further a logarithmic expression such as $\log_5(2x - 1)$. None of the basic properties gives a way to write the logarithm of a difference in some equivalent form.

Concept Check ☑

What is wrong with the following?

$$\log_{10}(x^2 + 5) = \log_{10} x^2 + \log_{10} 5$$
$$= 2 \log_{10} x + \log_{10} 5$$

Use a numerical example to demonstrate that the result is incorrect.

EXAMPLE 6   If $\log_b 2 = 0.43$ and $\log_b 3 = 0.68$, use the properties of logarithms to evaluate.

a. $\log_b 6$ **b.** $\log_b 9$ **c.** $\log_b \sqrt{2}$

Solution

a. $\log_b 6 = \log_b(2 \cdot 3)$ Write 6 as $2 \cdot 3$.

$\qquad = \log_b 2 + \log_b 3$ Apply the product property.

$\qquad = 0.43 + 0.68$ Substitute given values.

$\qquad = 1.11$ Simplify.

b. $\log_b 9 = \log_b 3^2$ Write 9 as 3^2.

$\qquad = 2 \log_b 3$

$\qquad = 2(0.68)$ Substitute 0.68 for $\log_b 3$.

$\qquad = 1.36$ Simplify.

c. First, recall that $\sqrt{2} = 2^{1/2}$. Then

$\log_b \sqrt{2} = \log_b 2^{1/2}$ Write $\sqrt{2}$ as $2^{1/2}$.

$\qquad = \dfrac{1}{2} \log_b 2$ Apply the power property.

$\qquad = \dfrac{1}{2}(0.43)$ Substitute the given value.

$\qquad = 0.215$ Simplify.   ☐

PRACTICE

6   If $\log_b 5 = 0.83$ and $\log_b 3 = 0.56$, use the properties of logarithms to evaluate.

a. $\log_b 15$ **b.** $\log_b 25$ **c.** $\log_b \sqrt{3}$

A summary of the basic properties of logarithms that we have developed so far is given next.

Properties of Logarithms

If x, y, and b are positive real numbers, $b \neq 1$, and r is a real number, then

1. $\log_b 1 = 0$

2. $\log_b b^x = x$

3. $b^{\log_b x} = x$

4. $\log_b xy = \log_b x + \log_b y$ Product property.

5. $\log_b \dfrac{x}{y} = \log_b x - \log_b y$ Quotient property.

6. $\log_b x^r = r \log_b x$ Power property.

Answer to Concept Check:
The properties do not give any way to simplify the logarithm of a sum; answers may vary.

VOCABULARY & READINESS CHECK

Select the correct choice.

1. $\log_b 12 + \log_b 3 = \log_b$ ____

a. 36 **b.** 15 **c.** 4 **d.** 9

2. $\log_b 12 - \log_b 3 = \log_b$ ___

a. 36 **b.** 15 **c.** 4 **d.** 9

3. $7 \log_b 2 =$ _____

a. $\log_b 14$ **b.** $\log_b 2^7$ **c.** $\log_b 7^2$ **d.** $(\log_b 2)^7$

4. $\log_b 1 =$ ___

a. b **b.** 1 **c.** 0 **d.** no answer

5. $b^{\log_b x} =$ ___

a. x **b.** b **c.** 1 **d.** 0

6. $\log_5 5^2 =$ ___

a. 25 **b.** 2 **c.** 5^5 **d.** 32

9.5 | EXERCISE SET

Write each sum as a single logarithm. Assume that variables represent positive numbers. See Example 1.

1. $\log_5 2 + \log_5 7$

2. $\log_3 8 + \log_3 4$

3. $\log_4 9 + \log_4 x$

4. $\log_2 x + \log_2 y$

5. $\log_6 x + \log_6 (x + 1)$

6. $\log_5 y^3 + \log_5 (y - 7)$

7. $\log_{10} 5 + \log_{10} 2 + \log_{10} (x^2 + 2)$

8. $\log_6 3 + \log_6 (x + 4) + \log_6 5$

Write each difference as a single logarithm. Assume that variables represent positive numbers. See Examples 2 and 4.

9. $\log_5 12 - \log_5 4$

10. $\log_7 20 - \log_7 4$

11. $\log_3 8 - \log_3 2$

12. $\log_5 12 - \log_5 3$

13. $\log_2 x - \log_2 y$

14. $\log_3 12 - \log_3 z$

15. $\log_2 (x^2 + 6) - \log_2 (x^2 + 1)$

16. $\log_7 (x + 9) - \log_7 (x^2 + 10)$

Use the power property to rewrite each expression. See Example 3.

17. $\log_3 x^2$

18. $\log_2 x^5$

19. $\log_4 5^{-1}$

20. $\log_6 7^{-2}$

21. $\log_5 \sqrt{y}$

22. $\log_5 \sqrt[3]{x}$

MIXED PRACTICE

Write each as a single logarithm. Assume that variables represent positive numbers. See Example 4.

23. $\log_2 5 + \log_2 x^3$

24. $\log_5 2 + \log_5 y^2$

25. $3\log_4 2 + \log_4 6$

26. $2\log_3 5 + \log_3 2$

27. $3\log_5 x + 6\log_5 z$

28. $2\log_7 y + 6\log_7 z$

29. $\log_4 2 + \log_4 10 - \log_4 5$

30. $\log_6 18 + \log_6 2 - \log_6 9$

31. $\log_7 6 + \log_7 3 - \log_7 4$

32. $\log_8 5 + \log_8 15 - \log_8 20$

33. $\log_{10} x - \log_{10} (x + 1) + \log_{10} (x^2 - 2)$

34. $\log_9 (4x) - \log_9 (x - 3) + \log_9 (x^3 + 1)$

35. $3\log_2 x + \dfrac{1}{2}\log_2 x - 2\log_2 (x + 1)$

36. $2\log_5 x + \dfrac{1}{3}\log_5 x - 3\log_5 (x + 5)$

37. $2\log_8 x - \dfrac{2}{3}\log_8 x + 4\log_8 x$

38. $5\log_6 x - \dfrac{3}{4}\log_6 x + 3\log_6 x$

Write each expression as a sum or difference of logarithms. Assume that variables represent positive numbers. See Example 5.

39. $\log_3 \dfrac{4y}{5}$

40. $\log_7 \dfrac{5x}{4}$

41. $\log_4 \dfrac{2}{9z}$

42. $\log_9 \dfrac{7}{8y}$

43. $\log_2 \dfrac{x^3}{y}$

44. $\log_5 \dfrac{x}{y^4}$

45. $\log_b \sqrt{7x}$

46. $\log_b \sqrt{\dfrac{3}{y}}$

47. $\log_6 x^4 y^5$

48. $\log_2 y^3 z$

49. $\log_5 x^3 (x + 1)$

50. $\log_3 x^2 (x - 9)$

51. $\log_6 \dfrac{x^2}{x + 3}$

52. $\log_3 \dfrac{(x + 5)^2}{x}$

If $\log_b 3 = 0.5$ and $\log_b 5 = 0.7$, evaluate each expression. See Example 6.

53. $\log_b 15$

54. $\log_b 25$

55. $\log_b \dfrac{5}{3}$

56. $\log_b \dfrac{3}{5}$

57. $\log_b \sqrt{5}$

58. $\log_b \sqrt[4]{3}$

If $\log_b 2 = 0.43$ and $\log_b 3 = 0.68$, evaluate each expression. See Example 6.

59. $\log_b 8$

60. $\log_b 81$

61. $\log_b \dfrac{3}{9}$

62. $\log_b \dfrac{4}{32}$

63. $\log_b \sqrt{\dfrac{2}{3}}$

64. $\log_b \sqrt{\dfrac{3}{2}}$

REVIEW AND PREVIEW

65. Graph the functions $y = 10^x$ and $y = \log_{10} x$ on the same set of axes. See Section 9.4.

Evaluate each expression. See Section 9.4.

66. $\log_{10} 100$

67. $\log_{10} \dfrac{1}{10}$

68. $\log_7 7^2$

69. $\log_7 \sqrt{7}$

CONCEPT EXTENSIONS

Solve. See the Concept Checks in this section.

70. Which of the following is the correct way to rewrite $\log_3 \dfrac{14}{11}$?

   a. $\dfrac{\log_3 14}{\log_3 11}$ **b.** $\log_3 14 - \log_3 11$

   c. $\log_3 (14 - 11)$ **d.** $\log_3 154$

71. Which of the following is the correct way to rewrite $\log_9 \dfrac{21}{3}$?

   a. $\log_9 7$ **b.** $\log_9 (21 - 3)$

   c. $\dfrac{\log_9 21}{\log_9 3}$ **d.** $\log_9 21 - \log_9 3$

Answer the following true or false. Study your logarithm properties carefully before answering.

72. $\log_2 x^3 = 3\log_2 x$

73. $\log_3(x + y) = \log_3 x + \log_3 y$

74. $\dfrac{\log_7 10}{\log_7 5} = \log_7 2$

75. $\log_7 \dfrac{14}{8} = \log_7 14 - \log_7 8$

76. $\dfrac{\log_7 x}{\log_7 y} = (\log_7 x) - (\log_7 y)$

77. $(\log_3 6) \cdot (\log_3 4) = \log_3 24$

78. It is true that $\log 8 = \log(8 \cdot 1) = \log 8 + \log 1$. Explain how $\log 8$ can equal $\log 8 + \log 1$.

INTEGRATED REVIEW FUNCTIONS AND PROPERTIES OF LOGARITHMS

Sections 9.1–9.5

If $f(x) = x - 6$ and $g(x) = x^2 + 1$, find each value.

1. $(f + g)(x)$

2. $(f - g)(x)$

3. $(f \cdot g)(x)$

4. $\left(\dfrac{f}{g}\right)(x)$

If $f(x) = \sqrt{x}$ and $g(x) = 3x - 1$, find each function.

5. $(f \circ g)(x)$

6. $(g \circ f)(x)$

Determine whether each is a one-to-one function. If it is, find its inverse.

7. $f = \{(-2, 6), (4, 8), (2, -6), (3, 3)\}$

8. $g = \{(4, 2), (-1, 3), (5, 3), (7, 1)\}$

Determine whether the graph of each function is one-to-one.

9.
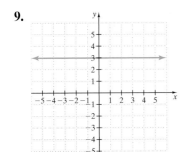

10.
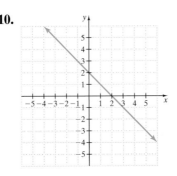

11.
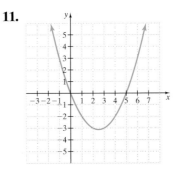

Each function listed is one-to-one. Find the inverse of each function.

12. $f(x) = 3x$

13. $f(x) = x + 4$

14. $f(x) = 5x - 1$

15. $f(x) = 3x + 2$

Graph each function.

16. $y = \left(\dfrac{1}{2}\right)^x$

17. $y = 2^x + 1$

18. $y = \log_3 x$

19. $y = \log_{1/3} x$

Solve.

20. $2^x = 8$

21. $9 = 3^{x-5}$

22. $4^{x-1} = 8^{x+2}$

23. $25^x = 125^{x-1}$

24. $\log_4 16 = x$

25. $\log_{49} 7 = x$

26. $\log_2 x = 5$

27. $\log_x 64 = 3$

28. $\log_x \dfrac{1}{125} = -3$

29. $\log_3 x = -2$

Write each as a single logarithm.

30. $5 \log_2 x$

31. $x \log_2 5$

32. $3 \log_5 x - 5 \log_5 y$

33. $9 \log_5 x + 3 \log_5 y$

34. $\log_2 x + \log_2(x - 3) - \log_2(x^2 + 4)$

35. $\log_3 y - \log_3(y + 2) + \log_3(y^3 + 11)$

Write each expression as sums or differences of multiples of logarithms.

36. $\log_7 \dfrac{9x^2}{y}$

37. $\log_6 \dfrac{5y}{z^2}$

9.6 COMMON LOGARITHMS, NATURAL LOGARITHMS, AND CHANGE OF BASE

OBJECTIVES

1 Identify common logarithms and approximate them by calculator.

2 Evaluate common logarithms of powers of 10.

3 Identify natural logarithms and approximate them by calculator.

4 Evaluate natural logarithms of powers of e.

5 Use the change of base formula.

In this section we look closely at two particular logarithmic bases. These two logarithmic bases are used so frequently that logarithms to their bases are given special names. **Common logarithms** are logarithms to base 10. **Natural logarithms** are logarithms to base *e*, which we introduce in this section. The work in this section is based on the use of the calculator, which has both the common "log" $\boxed{\text{LOG}}$ and the natural "log" $\boxed{\text{LN}}$ keys.

OBJECTIVE 1 ▶ Approximating common logarithms. Logarithms to base 10, common logarithms, are used frequently because our number system is a base 10 decimal system. The notation log *x* means the same as $\log_{10} x$.

> **Common Logarithms**
>
> $$\log x \text{ means } \log_{10} x$$

EXAMPLE 1

a. Use a calculator to approximate log 7 to four decimal places.

b. Use a graphing utility to graph $f(x) = \log_{10} x$ in an appropriate window. From the graph, complete the ordered pair (7,).

Solution

a. See the screen to the left. To four decimal places, $\log 7 \approx 0.8451$.

b. We know from Section 9.4 that the domain of *f* is the positive real numbers. Define the window to be $[-1, 9, 1]$ by $[-3, 3, 1]$.

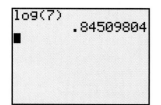

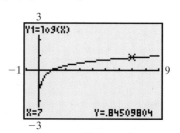

The approximate ordered pair $(7, 0.8451)$ is on the graph of $f(x) = \log_{10} x$.

Notice that parts **a** and **b** show two methods for approximating log 7. ☐

PRACTICE

1 Use a calculator to approximate log 15 to four decimal places.

OBJECTIVE 2 ▶ Evaluating common logarithms of powers of 10. To evaluate the common log of a power of 10, a calculator is not needed. According to the property of logarithms,

$$\log_b b^x = x$$

It follows that if b is replaced with 10, we have

$$\log 10^x = x$$

▶ **Helpful Hint**
Remember that $\log 10^x$ means $\log_{10} 10^x = x$.

EXAMPLE 2   Find the exact value of each logarithm.

a. $\log 10$ **b.** $\log 1000$ **c.** $\log \dfrac{1}{10}$ **d.** $\log \sqrt{10}$

Solution

a. $\log 10 = \log 10^1 = 1$ **b.** $\log 1000 = \log 10^3 = 3$

c. $\log \dfrac{1}{10} = \log 10^{-1} = -1$ **d.** $\log \sqrt{10} = \log 10^{1/2} = \dfrac{1}{2}$ □

PRACTICE
2   Find the exact value of each logarithm.

a. $\log \dfrac{1}{100}$ **b.** $\log 100{,}000$ **c.** $\log \sqrt[5]{10}$ **d.** $\log 0.001$

As we will soon see, equations containing common logs are useful models of many natural phenomena.

EXAMPLE 3   Solve $\log x = 1.2$ for x. Give an exact solution, and then approximate the solution to four decimal places.

Solution   Remember that the base of a common log is understood to be 10.

▶ **Helpful Hint**
The understood base is 10.

$$\log x = 1.2$$
$$10^{1.2} = x \qquad \text{Write with exponential notation.}$$

The exact solution is $10^{1.2}$. To four decimal places, $x \approx 15.8489$.

We can check this solution with a graphing utility by graphing the left side and the right side of the original equation and finding the point of intersection.

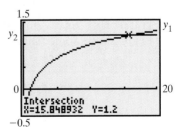

The point of intersection of $y_1 = \log x$ and $y_2 = 1.2$ is approximately $(15.8489, 1.2)$. Thus, the solution of the equation $\log x = 1.2$ is $x \approx 15.8489$.   □

PRACTICE
3   Solve $\log x = 3.4$ for x. Give an exact solution, and then approximate the solution to four decimal places.

The Richter scale measures the intensity, or magnitude, of an earthquake. The formula for the magnitude R of an earthquake is $R = \log\left(\dfrac{a}{T}\right) + B$, where a is the amplitude in micrometers of the vertical motion of the ground at the recording station, T is the number of seconds between successive seismic waves, and B is an adjustment factor that takes into account the weakening of the seismic wave as the distance increases from the epicenter of the earthquake.

EXAMPLE 4 Finding the Magnitude of an Earthquake

Find an earthquake's magnitude on the Richter scale if a recording station measures an amplitude of 300 micrometers and 2.5 seconds between waves. Assume that B is 4.2. Approximate the solution to the nearest tenth.

Solution Substitute the known values into the formula for earthquake intensity.

$$R = \log\left(\frac{a}{T}\right) + B \qquad \text{Richter scale formula}$$

$$= \log\left(\frac{300}{2.5}\right) + 4.2 \quad \text{Let } a = 300, T = 2.5, \text{ and } B = 4.2.$$

$$= \log(120) + 4.2$$

$$\approx 2.1 + 4.2 \qquad \text{Approximate log 120 by 2.1.}$$

$$= 6.3$$

This earthquake had a magnitude of 6.3 on the Richter scale. □

PRACTICE

4 Find an earthquake's magnitude on the Richter scale if a recording station measures an amplitude of 450 micrometers and 4.2 seconds between waves with $B = 3.6$. Approximate the solution to the nearest tenth.

OBJECTIVE 3 ▶ Approximating natural logarithms. **Natural logarithms** are also frequently used, especially to describe natural events; hence the label "natural logarithm." Natural logarithms are logarithms to the base e, which is a constant approximately equal to 2.7183. The number e is an irrational number, as is π. The notation $\log_e x$ is usually abbreviated to $\ln x$. (The abbreviation ln is read "el en.")

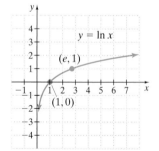

Natural Logarithms

$$\ln x \text{ means } \log_e x$$

The graph of $y = \ln x$ is shown to the left.
The table below shows some values of the natural logarithm function.

X	Y1
0	ERROR
1	0
2	.69315
3	1.0986
4	1.3863
5	1.6094
6	1.7918

Y1◻ln(X)

EXAMPLE 5   Use a calculator to approximate ln 8 to four decimal places.

Solution   See the screen below.

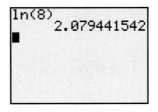

To four decimal places,

$$\ln 8 \approx 2.0794$$

PRACTICE
5   Use a calculator to approximate ln 13 to four decimal places.

OBJECTIVE 4 ▶ Evaluating natural logarithms of powers of e.   As a result of the property $\log_b b^x = x$, we know that $\log_e e^x = x$, or **ln e^x = x.**

Since $\ln e^x = x$, then $\ln e^5 = 5$, $\ln e^{22} = 22$, and so on. Also,

$$\ln e^1 = 1 \text{ or simply } \ln e = 1.$$

That is why the graph of $y = \ln x$ shown on the previous page passes through $(e, 1)$.
If $x = e$, then $y = \ln e = 1$, thus the ordered pair is $(e, 1)$.

EXAMPLE 6   Find the exact value of each natural logarithm.

a. $\ln e^3$ **b.** $\ln \sqrt[5]{e}$

Solution

a. $\ln e^3 = 3$ **b.** $\ln \sqrt[5]{e} = \ln e^{1/5} = \dfrac{1}{5}$

PRACTICE
6   Find the exact value of each natural logarithm.

a. $\ln e^4$ **b.** $\ln \sqrt[3]{e}$

EXAMPLE 7   Solve $\ln 3x = 5$. Give an exact solution, and then approximate the solution to four decimal places.

Solution   Remember that the base of a natural logarithm is understood to be e.

$$\ln 3x = 5$$
$$e^5 = 3x \quad \text{Write with exponential notation.}$$
$$\frac{e^5}{3} = x \quad \text{Solve for } x.$$

▶ **Helpful Hint**
The understood base is e.

The exact solution is $\dfrac{e^5}{3}$. To four decimal places,

$$x \approx 49.4711.$$

PRACTICE
7   Solve $\ln 5x = 8$. Give an exact solution, and then approximate the solution to four decimal places.

Recall from Section 9.3 the formula $A = P\left(1 + \dfrac{r}{n}\right)^{nt}$ for compound interest, where n represents the number of compoundings per year. When interest is compounded continuously, the formula $A = Pe^{rt}$ is used, where r is the annual interest rate and interest is compounded continuously for t years.

EXAMPLE 8 **Finding Final Loan Payment**

Find the amount owed at the end of 5 years if \$1600 is loaned at a rate of 9% compounded continuously.

Solution Use the formula $A = Pe^{rt}$, where

$$P = \$1600 \text{ (the size of the loan)}$$
$$r = 9\% = 0.09 \text{ (the rate of interest)}$$
$$t = 5 \text{ (the 5-year duration of the loan)}$$
$$A = Pe^{rt}$$
$$= 1600e^{0.09(5)} \quad \text{Substitute in known values.}$$
$$= 1600e^{0.45}$$

Now we can use a calculator to approximate the solution.

$$A \approx 2509.30$$

The total amount of money owed is \$2509.30. ☐

PRACTICE

8 Find the amount owed at the end of 4 years if \$2400 is borrowed at a rate of 6% compounded continuously.

OBJECTIVE 5 ▶ Using the change of base formula. Calculators are handy tools for approximating natural and common logarithms. Unfortunately, some calculators cannot be used to approximate logarithms to bases other than e or 10—at least not directly. In such cases, we use the change of base formula.

Change of Base

If a, b, and c are positive real numbers and neither b nor c is 1, then

$$\log_b a = \frac{\log_c a}{\log_c b}$$

EXAMPLE 9 Approximate $\log_5 3$ to four decimal places.

Solution Use the change of base property to write $\log_5 3$ as a quotient of logarithms to base 10.

$$\log_5 3 = \frac{\log 3}{\log 5} \quad \text{Use the change of base property. In the change of base property, we let } a = 3, b = 5, \text{ and } c = 10.$$

$$\approx \frac{0.4771213}{0.69897} \quad \text{Approximate logarithms by calculator.}$$

$$\approx 0.6826062 \quad \text{Simplify by calculator.}$$

To four decimal places, $\log_5 3 \approx 0.6826$. ☐

```
log(3)/log(5)
         .6826061945
■
```

PRACTICE

9 Approximate $\log_8 5$ to four decimal places.

Answer to Concept Check:

$$f(x) = \frac{\log x}{\log 5}$$

Concept Check ✓

If a graphing calculator cannot directly evaluate logarithms to base 5, describe how you could use the graphing calculator to graph the function $f(x) = \log_5 x$.

VOCABULARY & READINESS CHECK

Use the choices to fill in each blank.

1. The base of log 7 is ____.
   a. e **b.** 7   **c.** 10   **d.** no answer

2. The base of ln 7 is ___.
   a. e **b.** 7   **c.** 10   **d.** no answer

3. $\log_{10} 10^7 =$ ___.
   a. e **b.** 7   **c.** 10   **d.** no answer

4. $\log_7 1 =$ ___.
   a. e **b.** 7   **c.** 10   **d.** 0

5. $\log_e e^5 =$ ___.
   a. e **b.** 5   **c.** 0   **d.** 1

6. Study exercise 5 to the left. Then answer: $\ln e^5 =$ ___.
   a. e **b.** 5   **c.** 0   **d.** 1

7. $\log_2 7 =$ _____   (There may be more than one answer.)

   a. $\dfrac{\log 7}{\log 2}$ **b.** $\dfrac{\ln 7}{\ln 2}$ **c.** $\dfrac{\log 2}{\log 7}$ **d.** $\log \dfrac{7}{2}$

9.6 | EXERCISE SET

MyMathLab®   *Powered by CourseCompass™ and MathXL®*   Math XL PRACTICE   WATCH   DOWNLOAD   READ   REVIEW

MIXED PRACTICE

Use a calculator to approximate each logarithm to four decimal places. See Examples 1 and 5.

1. log 8

2. log 6

3. log 2.31

4. log 4.86

5. ln 2

6. ln 3

7. ln 0.0716

8. ln 0.0032

9. log 12.6

10. log 25.9

11. ln 5

12. ln 7

13. log 41.5

14. ln 41.5

15. Use a calculator and try to approximate log 0. Describe what happens and explain why.

16. Use a calculator and try to approximate ln 0. Describe what happens and explain why.

Find the exact value. See Examples 2 and 6.

17. log 100

18. log 10,000

19. $\log\left(\dfrac{1}{1000}\right)$

20. $\log\left(\dfrac{1}{100}\right)$

21. $\ln e^2$

22. $\ln e^4$

23. $\ln \sqrt[4]{e}$

24. $\ln \sqrt[5]{e}$

25. $\log 10^3$

26. $\log 10^7$

27. $\ln e^{-7}$

28. $\ln e^{-5}$

29. log 0.0001

30. log 0.001

31. $\ln \sqrt{e}$

32. $\log \sqrt{10}$

Solve each equation for x. Give an exact solution and a four-decimal-place approximation. See Examples 3 and 7.

33. $\ln 2x = 7$

34. $\ln 5x = 9$

35. $\log x = 1.3$

36. $\log x = 2.1$

37. $\log 2x = 1.1$

38. $\log 3x = 1.3$

39. $\ln x = 1.4$

40. $\ln x = 2.1$

41. $\ln(3x - 4) = 2.3$

42. $\ln(2x + 5) = 3.4$

43. $\log x = 2.3$

44. $\log x = 3.1$

45. $\ln x = -2.3$

46. $\ln x = -3.7$

47. $\log(2x + 1) = -0.5$

48. $\log(3x - 2) = -0.8$

49. $\ln 4x = 0.18$

50. $\ln 3x = 0.76$

Approximate each logarithm to four decimal places. See Example 9.

51. $\log_2 3$

52. $\log_3 2$

53. $\log_{1/2} 5$

54. $\log_{1/3} 2$

55. $\log_4 9$

56. $\log_9 4$

57. $\log_3 \dfrac{1}{6}$

58. $\log_6 \dfrac{2}{3}$

59. $\log_8 6$

60. $\log_6 8$

Use the formula $R = \log\left(\dfrac{a}{T}\right) + B$ to find the intensity R on the Richter scale of the earthquakes that fit the descriptions given. Round answers to one decimal place. See Example 4.

61. Amplitude a is 200 micrometers, time T between waves is 1.6 seconds, and B is 2.1.

62. Amplitude a is 150 micrometers, time T between waves is 3.6 seconds, and B is 1.9.

63. Amplitude a is 400 micrometers, time T between waves is 2.6 seconds, and B is 3.1.

64. Amplitude a is 450 micrometers, time T between waves is 4.2 seconds, and B is 2.7.

Use the formula $A = Pe^{rt}$ to solve. See Example 8.

65. Find how much money Dana Jones has after 12 years if $1400 is invested at 8% interest compounded continuously.

66. Determine the size of an account in which $3500 earns 6% interest compounded continuously for 1 year.

67. Find the amount of money Barbara Mack owes at the end of 4 years if 6% interest is compounded continuously on her $2000 debt.

68. Find the amount of money for which a $2500 certificate of deposit is redeemable if it has been paying 10% interest compounded continuously for 3 years.

REVIEW AND PREVIEW

Solve each equation for x. See Sections 1.5, 1.8, and 5.8.

69. $6x - 3(2 - 5x) = 6$

70. $2x + 3 = 5 - 2(3x - 1)$

71. $2x + 3y = 6x$

72. $4x - 8y = 10x$

73. $x^2 + 7x = -6$

74. $x^2 + 4x = 12$

Solve each system of equations. See Section 4.1.

75. $\begin{cases} x + 2y = -4 \\ 3x - y = 9 \end{cases}$

76. $\begin{cases} 5x + y = 5 \\ -3x - 2y = -10 \end{cases}$

CONCEPT EXTENSIONS

77. Without using a calculator, explain which of $\log 50$ or $\ln 50$ must be larger and why.

78. Without using a calculator, explain which of $\log 50^{-1}$ or $\ln 50^{-1}$ must be larger and why.

Graph each function by finding ordered pair solutions, plotting the solutions, and then drawing a smooth curve through the plotted points.

79. $f(x) = e^x$

80. $f(x) = e^{2x}$

81. $f(x) = e^{-3x}$

82. $f(x) = e^{-x}$

83. $f(x) = e^x + 2$

84. $f(x) = e^x - 3$

85. $f(x) = e^{x-1}$

86. $f(x) = e^{x+4}$

87. $f(x) = 3e^x$

88. $f(x) = -2e^x$

89. $f(x) = \ln x$

90. $f(x) = \log x$

91. $f(x) = -2 \log x$

92. $f(x) = 3 \ln x$

93. $f(x) = \log(x + 2)$

94. $f(x) = \log(x - 2)$

95. $f(x) = \ln x - 3$

96. $f(x) = \ln x + 3$

97. Graph $f(x) = e^x$ (Exercise 79), $f(x) = e^x + 2$ (Exercise 83), and $f(x) = e^x - 3$ (Exercise 84) on the same screen. Discuss any trends shown on the graphs.

98. Graph $f(x) = \ln x$ (Exercise 89), $f(x) = \ln x - 3$ (Exercise 95), and $f(x) = \ln x + 3$ (Exercise 96) on the same screen. Discuss any trends shown on the graphs.

📖 STUDY SKILLS BUILDER

What to Do the Day of an Exam

On the day of an exam, don't forget to try the following:

- Allow yourself plenty of time to arrive.
- Read the directions on the test carefully.
- Read each problem carefully as you take your test. Make sure that you answer the question asked.
- Watch your time and pace yourself so that you may attempt each problem on your test.
- Check your work and answers.
- **Do not turn your test in early.** If you have extra time, spend it double-checking your work.

Good luck!

Answer the following questions based on your most recent mathematics exam, whenever that was.

1. How soon before class did you arrive?

2. Did you read the directions on the test carefully?

3. Did you make sure you answered the question asked for each problem on the exam?

4. Were you able to attempt each problem on your exam?

5. If your answer to Question 4 is no, list reasons why.

6. Did you have extra time on your exam?

7. If your answer to Question 6 is yes, describe how you spent that extra time.

9.7 EXPONENTIAL AND LOGARITHMIC EQUATIONS AND APPLICATIONS

OBJECTIVES

1 Solve exponential equations.

2 Solve logarithmic equations.

3 Solve problems that can be modeled by exponential and logarithmic equations.

OBJECTIVE 1 ▶ Solving exponential equations. In Section 9.3 we solved exponential equations such as $2^x = 16$ by writing 16 as a power of 2 and applying the uniqueness of b^x.

$$2^x = 16$$
$$2^x = 2^4 \quad \text{Write 16 as } 2^4.$$
$$x = 4 \quad \text{Use the uniqueness of } b^x.$$

Solving the equation in this manner is possible since 16 is a power of 2. If solving an equation such as $2^x = a\ number$, where the number is not a power of 2, we use logarithms. For example, to solve an equation such as $3^x = 7$, we use the fact that $f(x) = \log_b x$ is a one-to-one function. Another way of stating this fact is as a property of equality.

> **Logarithm Property of Equality**
>
> Let a, b, and c be real numbers such that $\log_b a$ and $\log_b c$ are real numbers and b is not 1. Then
>
> $$\log_b a = \log_b c \text{ is equivalent to } a = c$$

EXAMPLE 1   Solve: $3^x = 7$.

Solution   To solve, we use the logarithm property of equality and take the logarithm of both sides. For this example, we use the common logarithm.

$$3^x = 7$$
$$\log 3^x = \log 7 \quad \text{Take the common log of both sides.}$$
$$x \log 3 = \log 7 \quad \text{Apply the power property of logarithms.}$$
$$x = \frac{\log 7}{\log 3} \quad \text{Divide both sides by } \log 3.$$

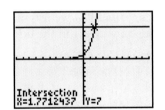

To verify the solution of the equation $3^x = 7$, graph $y_1 = 3^x$ and $y_2 = 7$. See that $x \approx 1.7712$.

The exact solution is $\dfrac{\log 7}{\log 3}$. If a decimal approximation is preferred,

$$\frac{\log 7}{\log 3} \approx \frac{0.845098}{0.4771213} \approx 1.7712 \text{ to four decimal places.}$$

The solution is $\dfrac{\log 7}{\log 3}$, or *approximately* 1.7712.

PRACTICE

1   Solve: $5^x = 9$.

OBJECTIVE 2 ▶ Solving logarithmic equations. By applying the appropriate properties of logarithms, we can solve a broad variety of logarithmic equations.

EXAMPLE 2   Solve: $\log_4(x - 2) = 2$.

Solution   Notice that $x - 2$ must be positive, so x must be greater than 2. With this in mind, we first write the equation with exponential notation.

$$\log_4(x - 2) = 2$$
$$4^2 = x - 2$$
$$16 = x - 2$$
$$18 = x \quad \text{Add 2 to both sides.}$$

Check: To check, numerically, we replace x with 18 in the original equation.

$$\log_4(x - 2) = 2$$
$$\log_4(18 - 2) \overset{?}{=} 2 \quad \text{Let } x = 18.$$
$$\log_4 16 \overset{?}{=} 2$$
$$4^2 = 16 \quad \text{True}$$

The solution is 18.

To check graphically,

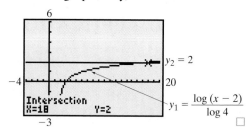

$$y_2 = 2$$
$$y_1 = \frac{\log(x - 2)}{\log 4}$$

PRACTICE

2 Solve: $\log_2(x - 1) = 5$.

EXAMPLE 3 Solve: $\log_2 x + \log_2(x - 1) = 1$.

Solution Notice that $x - 1$ must be positive, so x must be greater than 1. We use the product property on the left side of the equation.

$$\log_2 x + \log_2(x - 1) = 1$$
$$\log_2 x(x - 1) = 1 \quad \text{Apply the product property.}$$
$$\log_2(x^2 - x) = 1$$

Next we write the equation with exponential notation and solve for x.

$$2^1 = x^2 - x$$
$$0 = x^2 - x - 2 \quad \text{Subtract 2 from both sides.}$$
$$0 = (x - 2)(x + 1) \quad \text{Factor.}$$
$$0 = x - 2 \quad \text{or} \quad 0 = x + 1 \quad \text{Set each factor equal to 0.}$$
$$2 = x \qquad\qquad -1 = x$$

Recall that -1 cannot be a solution because x must be greater than 1. If we forgot this, we would still reject -1 after checking. To see this, we replace x with -1 in the original equation.

$$\log_2 x + \log_2(x - 1) = 1$$
$$\log_2(-1) + \log_2(-1 - 1) \overset{?}{=} 1 \quad \text{Let } x = -1.$$

Because the logarithm of a negative number is undefined, -1 is rejected. Check to see that the solution is 2.

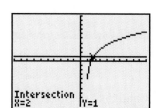

A graphic check for Example 3. Use the change of base formula to enter $y_1 = \log x/\log 2 + \log(x - 1)/\log 2$ and $y_2 = 1$.

PRACTICE

3 Solve: $\log_5 x + \log_5(x + 4) = 1$.

EXAMPLE 4 Solve: $\log(x + 2) - \log x = 2$.

We use the quotient property of logarithms on the left side of the equation.

Solution $\log(x + 2) - \log x = 2$

$$\log \frac{x + 2}{x} = 2 \qquad \text{Apply the quotient property.}$$
$$10^2 = \frac{x + 2}{x} \qquad \text{Write using exponential notation.}$$
$$100 = \frac{x + 2}{x} \qquad \text{Simplify.}$$
$$100x = x + 2 \qquad \text{Multiply both sides by } x.$$
$$99x = 2 \qquad \text{Subtract } x \text{ from both sides.}$$
$$x = \frac{2}{99} \qquad \text{Divide both sides by 99.}$$

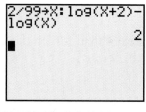

A numerical check for Example 4.

Verify that the solution is $\dfrac{2}{99}$.

PRACTICE
4   Solve: $\log(x + 3) - \log x = 1$.

OBJECTIVE 3 ▶ Solving problems modeled by exponential and logarithmic equations.
Logarithmic and exponential functions are used in a variety of scientific, technical, and business settings. A few examples follow.

EXAMPLE 5   **Estimating Population Size**

The population size y of a community of lemmings varies according to the relationship $y = y_0 e^{0.15t}$. In this formula, t is time in months, and y_0 is the initial population at time 0. Estimate the population after 6 months if there were originally 5000 lemmings.

Solution   We substitute 5000 for y_0 and 6 for t.

$$y = y_0 e^{0.15t}$$
$$= 5000 e^{0.15(6)} \quad \text{Let } t = 6 \text{ and } y_0 = 5000.$$
$$= 5000 e^{0.9} \quad \text{Multiply.}$$

Using a calculator, we find that $y \approx 12{,}298.016$. In 6 months the population will be approximately 12,300 lemmings.   □

PRACTICE
5   The population size y of a group of rabbits varies according to the relationship $y = y_0 e^{0.916t}$. In this formula, t is time in years and y_0 is the initial population at time $t = 0$. Estimate the population in three years if there were originally 60 rabbits.

EXAMPLE 6   **Doubling an Investment**

How long does it take an investment of $2000 to double if it is invested at 5% interest compounded quarterly? The necessary formula is $A = P\left(1 + \dfrac{r}{n}\right)^{nt}$, where A is the accrued (or owed) amount, P is the principal invested, r is the annual rate of interest, n is the number of compounding periods per year, and t is the number of years.

Solution   We are given that $P = \$2000$ and $r = 5\% = 0.05$. Compounding quarterly means 4 times a year, so $n = 4$. The investment is to double, so A must be $4000. Substitute these values and solve for t.

$$A = P\left(1 + \frac{r}{n}\right)^{nt}$$
$$4000 = 2000\left(1 + \frac{0.05}{4}\right)^{4t} \quad \text{Substitute in known values.}$$
$$4000 = 2000(1.0125)^{4t} \quad \text{Simplify } 1 + \frac{0.05}{4}.$$
$$2 = (1.0125)^{4t} \quad \text{Divide both sides by 2000.}$$
$$\log 2 = \log 1.0125^{4t} \quad \text{Take the logarithm of both sides.}$$
$$\log 2 = 4t(\log 1.0125) \quad \text{Apply the power property.}$$
$$\frac{\log 2}{4 \log 1.0125} = t \quad \text{Divide both sides by } 4 \log 1.0125.$$
$$13.949408 \approx t \quad \text{Approximate by calculator.}$$

Thus, it takes nearly 14 years for the money to double in value.   □

PRACTICE
6   How long does it take for an investment of $3000 to double if it is invested at 7% interest compounded monthly? Round to the nearest whole year.

EXAMPLE 7   **Tracking an Investment**

Suppose that you invest $1500 at an annual rate of 8% compounded monthly.

a. Make a table giving the value of the investment at the end of each year for the next 14 years.

b. How long does it take the investment to triple?

Solution

a. First, let $P = \$1500$, $r = 0.08$, and $n = 12$ (for monthly compounding) in the formula $A = P\left(1 + \dfrac{r}{n}\right)^{nt}$. Using the function $A = 1500\left(1 + \dfrac{0.08}{12}\right)^{12x}$, we can use a graphing utility to make the following table, shown below in two screens:

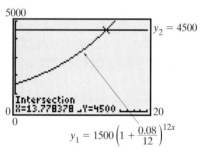

> **▶ Helpful Hint**
>
> Looking at a table helps to determine the size of the window when working with exponential functions.

X	Y1
1	1624.5
2	1759.3
3	1905.4
4	2063.5
5	2234.8
6	2420.3
7	2621.1

Y1⊟1500(1+.08/1…

X	Y1
8	2838.7
9	3074.3
10	3329.5
11	3605.8
12	3905.1
13	4229.2
14	4580.2

Y1⊟1500(1+.08/1…

b. Notice that when the investment has tripled, the accrued amount A is $4500. Use a graphing utility to approximate the value of x that gives $A = 4500$. We can see from the table in part a that the value of the investment is $4500 between $x = 13$ years and $x = 14$ years. Using this information, we can determine an appropriate viewing window, such as $[0, 20, 1]$ by $[0, 5000, 1000]$. Graph $y_1 = 1500(1 + 0.08/12)^{12x}$ and $y_2 = 4500$ in the same window. As usual, the point of intersection of the two curves is the solution.

The screen in the margin shows that it takes $x \approx 13.78$ years, or 13 years and 9 months (rounded to the nearest month), for the original investment to triple in value to $4500. ☐

(margin graph) 5000 … $y_2 = 4500$ … Intersection X=13.778378 Y=4500 … 20 … 0 … 0 … $y_1 = 1500\left(1 + \dfrac{0.08}{12}\right)^{12x}$

PRACTICE

7   Find how long it takes an investment of $5000 to grow to $6000 if it is invested at 5% interest compounded quarterly.

9.7 EXERCISE SET

 MyMathLab Math XL PRACTICE   WATCH   DOWNLOAD   READ   REVIEW

Solve each equation. Give an exact solution, and also approximate the solution to four decimal places. See Example 1.

1. $3^x = 6$

2. $4^x = 7$

3. $3^{2x} = 3.8$

4. $5^{3x} = 5.6$

5. $2^{x-3} = 5$

6. $8^{x-2} = 12$

7. $9^x = 5$

8. $3^x = 11$

9. $4^{x+7} = 3$

10. $6^{x+3} = 2$

MIXED PRACTICE

Solve each equation. See Examples 1 through 4.

11. $7^{3x-4} = 11$

12. $5^{2x-6} = 12$

13. $e^{6x} = 5$

14. $e^{2x} = 8$

15. $\log_2(x + 5) = 4$

16. $\log_6(x^2 - x) = 1$

17. $\log_3 x^2 = 4$

18. $\log_2 x^2 = 6$

19. $\log_4 2 + \log_4 x = 0$

20. $\log_3 5 + \log_3 x = 1$

21. $\log_2 6 - \log_2 x = 3$

22. $\log_4 10 - \log_4 x = 2$

23. $\log_4 x + \log_4(x + 6) = 2$

24. $\log_3 x + \log_3(x + 6) = 3$

25. $\log_5(x + 3) - \log_5 x = 2$

26. $\log_6(x + 2) - \log_6 x = 2$

27. $\log_3(x - 2) = 2$

28. $\log_2(x - 5) = 3$

29. $\log_4(x^2 - 3x) = 1$

30. $\log_8(x^2 - 2x) = 1$

31. $\ln 5 + \ln x = 0$

32. $\ln 3 + \ln (x - 1) = 0$

33. $3 \log x - \log x^2 = 2$

34. $2 \log x - \log x = 3$

35. $\log_2 x + \log_2(x + 5) = 1$

36. $\log_4 x + \log_4(x + 7) = 1$

37. $\log_4 x - \log_4(2x - 3) = 3$

38. $\log_2 x - \log_2(3x + 5) = 4$

39. $\log_2 x + \log_2(3x + 1) = 1$

40. $\log_3 x + \log_3(x - 8) = 2$

Solve. See Example 5.

41. The size of the wolf population at Isle Royale National Park increases at a rate of 4.3% per year. If the size of the current population is 83 wolves, find how many there should be in 5 years. Use $y = y_0 e^{0.043t}$ and round to the nearest whole.

42. The number of victims of a flu epidemic is increasing at a rate of 7.5% per week. If 20,000 persons are currently infected, find in how many days we can expect 45,000 to have the flu. Use $y = y_0 e^{0.075t}$ and round to the nearest whole. (*Hint:* Don't forget to convert your answer to days.)

43. The size of the population of Belize is increasing at a rate of 2.3% per year. If 294,380 people lived in Belize in 2007, find how many inhabitants there will be by 2015. Round to the nearest thousand. Use $y = y_0 e^{0.023t}$. (*Source: CIA 2007 World Factbook*)

44. In 2007, 1730 million people were citizens of India. Find how long it will take India's population to reach a size of 2000 million (that is, 2 billion) if the population size is growing at a rate of 1.6% per year. Use $y = y_0 e^{0.016t}$ and round to the nearest tenth. (*Source:* U.S. Bureau of the Census, International Data Base)

45. In 2007, Germany had a population of 82,400 thousand. At that time, Germany's population was declining at a rate of 0.033% per year. If this continues, how long will it take Germany's population to reach 82,000 thousand? Use $y = y_0 e^{-0.00033t}$ and round to the nearest tenth. (*Source: CIA 2007 World Factbook*)

46. The population of the United States has been increasing at a rate of 0.894% per year. If there were 301,140,000 people living in the United States in 2007, how many inhabitants will there be by 2020? Use $y = y_0 e^{0.00894t}$ and round to the nearest ten-thousand. (*Source: CIA 2007 World Factbook*)

Use the formula $A = P\left(1 + \dfrac{r}{n}\right)^{nt}$ *to solve these compound interest problems. Round to the nearest tenth. See Examples 6 and 7.*

47. Find how long it takes $600 to double if it is invested at 7% interest compounded monthly.

48. Find how long it takes $600 to double if it is invested at 12% interest compounded monthly.

49. Find how long it takes a $1200 investment to earn $200 interest if it is invested at 9% interest compounded quarterly.

50. Find how long it takes a $1500 investment to earn $200 interest if it is invested at 10% compounded semiannually.

51. Find how long it takes $1000 to double if it is invested at 8% interest compounded semiannually.

52. Find how long it takes $1000 to double if it is invested at 8% interest compounded monthly.

The formula $w = 0.00185h^{2.67}$ *is used to estimate the normal weight* w *of a boy* h *inches tall. Use this formula to solve the height-weight problems. Round to the nearest tenth.*

53. Find the expected weight of a boy who is 35 inches tall.

54. Find the expected weight of a boy who is 43 inches tall.

55. Find the expected height of a boy who weighs 85 pounds.

56. Find the expected height of a boy who weighs 140 pounds.

The formula $P = 14.7e^{-0.21x}$ *gives the average atmospheric pressure* P, *in pounds per square inch, at an altitude* x, *in miles above sea level. Use this formula to solve these pressure problems. Round answers to the nearest tenth.*

57. Find the average atmospheric pressure of Denver, which is 1 mile above sea level.

58. Find the average atmospheric pressure of Pikes Peak, which is 2.7 miles above sea level.

59. Find the elevation of a Delta jet if the atmospheric pressure outside the jet is 7.5 lb/in.^{2}.

60. Find the elevation of a remote Himalayan peak if the atmospheric pressure atop the peak is 6.5 lb/in.^{2}.

Psychologists call the graph of the formula $t = \dfrac{1}{c}\ln\left(\dfrac{A}{A - N}\right)$ *the learning curve, since the formula relates time* t *passed, in weeks, to a measure* N *of learning achieved, to a measure* A *of maximum learning possible, and to a measure* c *of an individual's learning style. Round to the nearest week.*

61. Norman is learning to type. If he wants to type at a rate of 50 words per minute (N is 50) and his expected maximum rate is 75 words per minute (A is 75), find how many weeks it should take him to achieve his goal. Assume that c is 0.09.

62. An experiment with teaching chimpanzees sign language shows that a typical chimp can master a maximum of 65 signs. Find how many weeks it should take a chimpanzee to master 30 signs if c is 0.03.

63. Janine is working on her dictation skills. She wants to take dictation at a rate of 150 words per minute and believes that the maximum rate she can hope for is 210 words per minute. Find how many weeks it should take her to achieve the 150 words per minute level if c is 0.07.

64. A psychologist is measuring human capability to memorize nonsense syllables. Find how many weeks it should take a subject to learn 15 nonsense syllables if the maximum possible to learn is 24 syllables and c is 0.47.

REVIEW AND PREVIEW

If $x = -2$, $y = 0$, *and* $z = 3$, *find the value of each expression. See Section 1.3.*

65. $\dfrac{x^2 - y + 2z}{3x}$

66. $\dfrac{x^3 - 2y + z}{2z}$

67. $\dfrac{3z - 4x + y}{x + 2z}$

68. $\dfrac{4y - 3x + z}{2x + y}$

Find the inverse function of each one-to-one function. See Section 9.2.

69. $f(x) = 5x + 2$

70. $f(x) = \dfrac{x - 3}{4}$

CONCEPT EXTENSIONS

The formula $y = y_0 e^{kt}$ *gives the population size y of a population that experiences an annual rate of population growth k (given as a decimal). In this formula, t is time in years and y_0 is the initial population at time 0. Use this formula to solve Exercises 71 and 72.*

71. In 2000, the population of Arizona was 5,130,632. By 2006, the population had grown to 6,123,106. Find the annual rate of population growth over this period. Round your answer to the nearest tenth of a percent. (*Source:* State of Arizona)

72. In 2000, the population of Nevada was 2,018,456. By 2006, the population had grown to 2,495,529. Find the annual rate of population growth over this period. Round your answer to the nearest tenth of a percent. (*Source:* State of Nevada)

73. When solving a logarithmic equation, explain why you must check possible solutions in the original equation.

74. Solve $5^x = 9$ by taking the common logarithm of both sides of the equation. Next, solve this equation by taking the natural logarithm of both sides. Compare your solutions. Are they the same? Why or why not?

Solve using exponential regression features on a graphing calculator.

75. The amount of online retail spending in the United States has risen sharply since 2000.

    a. Use the data in the table below to find and graph the exponential regression equation for the data with y representing the amount of online spending in the United States in the billions of dollars. (Let $x = 0$ represent the year 2000.)

Year (0 represents 2000)	U.S. Online Spending (in billions of dollars)
0	24.1
3	53.9
4	67.3
5	80.9
6	95.3

    b. Find the approximate amount of spending for the year 2001.

    c. Predict the amount of online spending in the year 2011.

76. Using the exponential regression equation from Exercise 75, find the amount of online spending in the year 2002. Predict the amount of online spending to be made in the year 2009.

77. The revenue for sales of digital cameras in the United States is shown in the table below.

Year (0 represents 2000)	U.S. Revenue (in millions of U.S. dollars)
0	1825
1	1972
2	2794
3	3921
4	4739
5	7468

    a. Use the data in this table to find and graph the exponential regression equation for the data with y representing the revenue of sales of digital cameras in the United States in millions of U.S. dollars. (Let $x = 0$ represent the year 2000.)

    b. Find the approximate revenue for the year 1999.

    c. Predict the approximate revenue in the year 2009.

78. Using the exponential regression equation from Exercise 77, find the revenue in the year 1998. Predict the revenue in the year 2010.

79. U.S. health spending is shown in the table below.

    a. Use the data in this table to find and graph the exponential regression equation for the data with y representing the health spending in billions of U.S. dollars. (Let $x = 0$ represent the year 1960.)

Year (0 represents 1960)	U.S. Health Spending (in billions of dollars)
0	27.6
10	75.1
20	254.9
30	717.3
35	1020.4
40	1358.5
42	1607.9
43	1740.6
44	1877.6

    b. Find the approximate health spending for the year 2001.

    c. Predict the health spending in the year 2010.

80. Using the exponential regression equation from Exercise 79, find the health spending in the year 1987. Predict the health spending in the year 2011.

Use a graphing utility to solve each equation. Round all solutions to two decimal places.

81. $e^{0.3x} = 8$

82. $10^{0.5x} = 7$

83. $2 \log(-5.6x + 1.3) + x + 1 = 0$

84. $\ln(1.3x - 2.1) + 3.5x - 5 = 0$

THE BIGGER PICTURE   SOLVING EQUATIONS AND INEQUALITIES

Continue your outline from Sections 1.5, 3.2, 3.3, 3.5, 5.8, 6.5, 7.6, and 8.4. Write how to recognize and how to solve exponential and logarithmic equations in your own words. For example:

Solving Equations and Inequalities

I. Equations

    A. Linear equations (Sec. 1.5 and 3.1)

    B. Absolute value equations (Sec. 3.4)

    C. Quadratic and higher degree equations (Sec. 5.8 and Chapter 8)

    D. Equations with rational expressions (Sec. 6.5)

    E. Equations with radicals (Sec. 7.6)

    F. Exponential Equations—equations with variables in the exponent.

1. If we can write both expressions with the same base, then set the exponents equal to each other and solve.

$$9^x = 27^{x+1}$$
$$(3^2)^x = (3^3)^{x+1}$$
$$3^{2x} = 3^{3x+3}$$
$$2x = 3x + 3$$
$$-3 = x$$

2. If we can't write both expressions with the same base, then solve using logarithms.

$$5^x = 7$$
$$\log 5^x = \log 7$$
$$x \log 5 = \log 7$$
$$x = \frac{\log 7}{\log 5}$$
$$x \approx 1.2091$$

    G. Logarithmic Equations—equations with logarithms of variable expressions

$$\log 7 + \log(x + 3) = 2 \quad \text{Write equation with a single logarithm on}$$
$$\log 7(x + 3) = 2 \quad \text{one side and a constant on the other side.}$$
$$10^2 = 7(x + 3) \quad \text{Use definition of logarithm.}$$
$$100 = 7x + 21 \quad \text{Multiply.}$$
$$79 = 7x$$
$$\frac{79}{7} = x \quad \text{Solve.}$$

II. Inequalities

    A. Linear inequalities (Sec. 3.2)

    B. Compound inequalities (Sec. 3.3)

    C. Absolute value inequalities (Sec. 3.5)

    D. Nonlinear inequalities (Sec. 8.4)

        1. Polynomial inequalities

        2. Rational inequalities

Solve. Write solutions to inequalities in interval notation.

1. $8^x = 2^{x-3}$

2. $11^x = 5$

3. $-7x + 3 \le -5x + 13$

4. $-7 \le 3x + 6 \le 0$

5. $|5y + 3| < 3$

6. $(x - 6)(5x + 1) = 0$

7. $\log_{13} 8 + \log_{13}(x - 1) = 1$

8. $\left| \dfrac{3x - 1}{4} \right| = 2$

9. $|7x + 1| > -2$

10. $x^2 = 4$

11. $(x + 5)^2 = 3$

12. $\log_7(4x^2 - 27x) = 1$

CHAPTER 9 GROUP ACTIVITY

Sound Intensity

The decibel (dB) measures sound intensity, or the relative loudness or strength of a sound. One decibel is the smallest difference in sound levels that is detectable by humans. The decibel is a logarithmic unit. This means that for approximately every 3-decibel increase in sound intensity, the relative loudness of the sound is doubled. For example, a 35 dB sound is twice as loud as a 32 dB sound.

In the modern world, noise pollution has increasingly become a concern. Sustained exposure to high sound intensities can lead to hearing loss. Regular exposure to 90 dB sounds can eventually lead to loss of hearing. Sounds of 130 dB and more can cause permanent loss of hearing instantaneously.

The relative loudness of a sound D in decibels is given by the equation

$$D = 10\log_{10}\frac{I}{10^{-16}}$$

where I is the intensity of a sound given in watts per square centimeter. Some sound intensities of common noises are listed in the table in order of increasing sound intensity.

Group Activity

1. Work together to create a table of the relative loudness (in decibels) of the sounds listed in the table.

2. Research the loudness of other common noises. Add these sounds and their decibel levels to your table. Be sure to list the sounds in order of increasing sound intensity.

Some Sound Intensities of Common Noises		
Noise	*Intensity* (watts/cm²)	*Decibels*
Whispering	10^{-15}	
Rustling leaves	$10^{-14.2}$	
Normal conversation	10^{-13}	
Background noise in a quiet residence	$10^{-12.2}$	
Typewriter	10^{-11}	
Air conditioning	10^{-10}	
Freight train at 50 feet	$10^{-8.5}$	
Vacuum cleaner	10^{-8}	
Nearby thunder	10^{-7}	
Air hammer	$10^{-6.5}$	
Jet plane at takeoff	10^{-6}	
Threshold of pain	10^{-4}	

 STUDY SKILLS BUILDER

Are You Prepared for a Test on Chapter 9?

Below I have listed some common trouble areas for students in Chapter 9. After studying for your test—but before taking your test—read these.

- Don't forget how to find the composition of two functions.

 If $f(x) = x^2 + 5$ and $g(x) = 3x$, then

$$(f \circ g)(x) = f[g(x)] = f(3x)$$
$$= (3x)^2 + 5 = 9x^2 + 5$$
$$(g \circ f)(x) = g[f(x)] = g(x^2 + 5)$$
$$= 3(x^2 + 5) = 3x^2 + 15$$

- Don't forget that f^{-1} is a special notation used to denote the inverse of a function.

 Let's find the inverse of the one-to-one function $f(x) = 3x - 5$.

$$\begin{aligned} f(x) &= 3x - 5 \\ y &= 3x - 5 \quad \text{Replace } f(x) \text{ with } y. \\ x &= 3y - 5 \quad \text{Interchange } x \text{ and } y. \\ x + 5 &= 3y \\ \frac{x + 5}{3} &= y \qquad \text{Solve for } y. \\ f^{-1}(x) &= \frac{x + 5}{3} \quad \text{Replace } y \text{ with } f^{-1}(x). \end{aligned}$$

- Don't forget that $y = \log_b x$ means $b^y = x$.
 Thus, $3 = \log_5 125$ means $5^3 = 125$.
- Remember rules for logarithms.
 $$\log_b 3x = \log_b 3 + \log_b x$$
 $$\log_b(3 + x)\text{ cannot be simplified in the same manner.}$$

Remember: This is simply a checklist of common trouble areas. For a review of Chapter 9, see the Highlights and Chapter Review at the end of this chapter.

CHAPTER 9 VOCABULARY CHECK

Fill in each blank with one of the words or phrases listed below.

inverse   common   composition   symmetric   exponential

vertical   logarithmic   natural   horizontal

1. For each one-to-one function, we can find its _____ function by switching the coordinates of the ordered pairs of the function.

2. The _____ of functions f and g is $(f \circ g)(x) = f(g(x))$.

3. A function of the form $f(x) = b^x$ is called an _____ function if $b > 0, b$ is not 1, and x is a real number.

4. The graphs of f and f^{-1} are _____ about the line $y = x$.

5. _____ logarithms are logarithms to base e.

6. _____ logarithms are logarithms to base 10.

7. To see whether a graph is the graph of a one-to-one function, apply the _____ line test to see if it is a function, and then apply the _____ line test to see if it is a one-to-one function.

8. A _____ function is a function that can be defined by $f(x) = \log_b x$ where x is a positive real number, b is a constant positive real number, and b is not 1.

> ▶ **Helpful Hint**
>
> Are you preparing for your test? Don't forget to take the Chapter 9 Test on page 685. Then check your answers at the back of the text and use the Chapter Test Prep Video CD to see the fully worked-out solutions to any of the exercises you want to review.

CHAPTER 9 HIGHLIGHTS

DEFINITIONS AND CONCEPTS	EXAMPLES

SECTION 9.1 THE ALGEBRA OF FUNCTIONS; COMPOSITE FUNCTIONS

Algebra of Functions

Sum $\quad (f + g)(x) = f(x) + g(x)$

Difference $\quad (f - g)(x) = f(x) - g(x)$

Product $\quad (f \cdot g)(x) = f(x) \cdot g(x)$

Quotient $\quad \left(\dfrac{f}{g}\right)(x) = \dfrac{f(x)}{g(x)}, g(x) \neq 0$

Composite Functions

The notation $(f \circ g)(x)$ means "f composed with g."

$$(f \circ g)(x) = f(g(x))$$
$$(g \circ f)(x) = g(f(x))$$

If $f(x) = 7x$ and $g(x) = x^2 + 1$,

$$(f + g)(x) = f(x) + g(x) = 7x + x^2 + 1$$
$$(f - g)(x) = f(x) - g(x) = 7x - (x^2 + 1)$$
$$= 7x - x^2 - 1$$
$$(f \cdot g)(x) = f(x) \cdot g(x) = 7x(x^2 + 1)$$
$$= 7x^3 + 7x$$
$$\left(\frac{f}{g}\right)(x) = \frac{f(x)}{g(x)} = \frac{7x}{x^2 + 1}$$

If $f(x) = x^2 + 1$ and $g(x) = x - 5$, find $(f \circ g)(x)$.

$$(f \circ g)(x) = f(g(x))$$
$$= f(x - 5)$$
$$= (x - 5)^2 + 1$$
$$= x^2 - 10x + 26$$

SECTION 9.2 INVERSE FUNCTIONS

If f is a function, then f is a **one-to-one function** only if each y-value (output) corresponds to only one x-value (input).

Horizontal Line Test

If every horizontal line intersects the graph of a function at most once, then the function is a one-to-one function.

Determine whether each graph is a one-to-one function.

A **B**

C

Graphs **A** and **C** pass the vertical line test, so only these are graphs of functions. Of graphs **A** and **C**, only graph **A** passes the horizontal line test, so only graph **A** is the graph of a one-to-one function.

(continued)

DEFINITIONS AND CONCEPTS	**EXAMPLES**

SECTION 9.2 INVERSE FUNCTIONS (continued)

The **inverse** of a one-to-one function f is the one-to-one function f^{-1} that is the set of all ordered pairs (b, a) such that (a, b) belongs to f.

To Find the Inverse of a One-to-One Function f(x)

Step 1. Replace $f(x)$ with y.
Step 2. Interchange x and y.
Step 3. Solve for y.
Step 4. Replace y with $f^{-1}(x)$.

Find the inverse of $f(x) = 2x + 7$.

$y = 2x + 7$ Replace $f(x)$ with y.

$x = 2y + 7$ Interchange x and y.

$2y = x - 7$ Solve for y.

$y = \dfrac{x - 7}{2}$

$f^{-1}(x) = \dfrac{x - 7}{2}$ Replace y with $f^{-1}(x)$.

The inverse of $f(x) = 2x + 7$ is $f^{-1}(x) = \dfrac{x - 7}{2}$.

SECTION 9.3 EXPONENTIAL FUNCTIONS

A function of the form $f(x) = b^x$ is an **exponential function,** where $b > 0$, $b \neq 1$, and x is a real number.

Graph the exponential function $y = 4^x$.

x	y
-2	$\dfrac{1}{16}$
-1	$\dfrac{1}{4}$
0	1
1	4
2	16

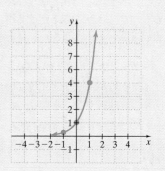

Uniqueness of b^x

If $b > 0$ and $b \neq 1$, then $b^x = b^y$ is equivalent to $x = y$.

Solve $2^{x+5} = 8$.

$2^{x+5} = 2^3$ Write 8 as 2^3.

$x + 5 = 3$ Use the uniqueness of b^x.

$x = -2$ Subtract 5 from both sides.

SECTION 9.4 LOGARITHMIC FUNCTIONS

Logarithmic Definition

If $b > 0$ and $b \neq 1$, then

$y = \log_b x$ means $x = b^y$

for any positive number x and real number y.

Logarithmic Form	*Corresponding Exponential Statement*
$\log_5 25 = 2$	$5^2 = 25$
$\log_9 3 = \dfrac{1}{2}$	$9^{1/2} = 3$

Properties of Logarithms

If b is a real number, $b > 0$, and $b \neq 1$, then

$\log_b 1 = 0, \quad \log_b b^x = x, \quad b^{\log_b x} = x$

$\log_5 1 = 0, \quad \log_7 7^2 = 2, \quad 3^{\log_3 6} = 6$

DEFINITIONS AND CONCEPTS	**EXAMPLES**

SECTION 9.4   LOGARITHMIC FUNCTIONS (continued)

Logarithmic Function

If $b > 0$ and $b \neq 1$, then a **logarithmic function** is a function that can be defined as

$$f(x) = \log_b x$$

The domain of f is the set of positive real numbers, and the range of f is the set of real numbers.

Graph $y = \log_3 x$.

Write $y = \log_3 x$ as $3^y = x$. Plot the ordered pair solutions listed in the table, and connect them with a smooth curve.

x	y
3	1
1	0
$\dfrac{1}{3}$	-1
$\dfrac{1}{9}$	-2

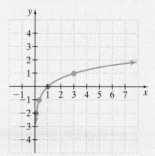

SECTION 9.5   PROPERTIES OF LOGARITHMS

Let x, y, and b be positive numbers and $b \neq 1$.

Product Property

$$\log_b xy = \log_b x + \log_b y$$

Quotient Property

$$\log_b \frac{x}{y} = \log_b x - \log_b y$$

Power Property

$$\log_b x^r = r \log_b x$$

Write as a single logarithm.

$2 \log_5 6 + \log_5 x - \log_5(y + 2)$

$\quad = \log_5 6^2 + \log_5 x - \log_5(y + 2)$ Power property

$\quad = \log_5 36 \cdot x - \log_5(y + 2)$ Product property

$\quad = \log_5 \dfrac{36x}{y + 2}$ Quotient property

SECTION 9.6   COMMON LOGARITHMS, NATURAL LOGARITHMS, AND CHANGE OF BASE

Common Logarithms

$$\log x \quad \text{means} \quad \log_{10} x$$

Natural Logarithms

$$\ln x \quad \text{means} \quad \log_e x$$

Continuously Compounded Interest Formula

$$A = Pe^{rt}$$

where r is the annual interest rate for P dollars invested for t years.

$\log 5 = \log_{10} 5 \approx 0.69897$

$\ln 7 = \log_e 7 \approx 1.94591$

Find the amount in an account at the end of 3 years if $1000 is invested at an interest rate of 4% compounded continuously.

Here, $t = 3$ years, $P = \$1000$, and $r = 0.04$.

$$A = Pe^{rt}$$

$$= 1000e^{0.04(3)}$$

$$\approx \$1127.50$$

SECTION 9.7   EXPONENTIAL AND LOGARITHMIC EQUATIONS AND APPLICATIONS

Logarithm Property of Equality

Let $\log_b a$ and $\log_b c$ be real numbers and $b \neq 1$. Then

$$\log_b a = \log_b c \text{ is equivalent to } a = c$$

Solve $2^x = 5$.

$\log 2^x = \log 5$ Log property of equality

$x \log 2 = \log 5$ Power property

$x = \dfrac{\log 5}{\log 2}$ Divide both sides by log 2.

$x \approx 2.3219$ Use a calculator.

CHAPTER 9   REVIEW

(9.1) *If* $f(x) = x - 5$ *and* $g(x) = 2x + 1$, *find*

1. $(f + g)(x)$

2. $(f - g)(x)$

3. $(f \cdot g)(x)$

4. $\left(\dfrac{g}{f}\right)(x)$

If $f(x) = x^2 - 2$, $g(x) = x + 1$, *and* $h(x) = x^3 - x^2$, *find each composition.*

5. $(f \circ g)(x)$

6. $(g \circ f)(x)$

7. $(h \circ g)(2)$

8. $(f \circ f)(x)$

9. $(f \circ g)(-1)$

10. $(h \circ h)(2)$

(9.2) *Determine whether each function is a one-to-one function. If it is one-to-one, list the elements of its inverse.*

11. $h = \{(-9, 14), (6, 8), (-11, 12), (15, 15)\}$

12. $f = \{(-5, 5), (0, 4), (13, 5), (11, -6)\}$

13.

U.S. Region (Input)	West	Midwest	South	Northeast
Rank in Automobile Thefts (Output)	2	4	1	3

 14.

Shape (Input)	Square	Triangle	Parallelogram	Rectangle
Number of Sides (Output)	4	3	4	4

Given that $f(x) = \sqrt{x + 2}$ *is a one-to-one function, find the following.*

15. a. $f(7)$

   b. $f^{-1}(3)$

16. a. $f(-1)$

   b. $f^{-1}(1)$

Determine whether each function is a one-to-one function.

17.

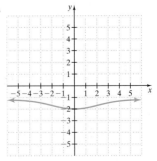

18.

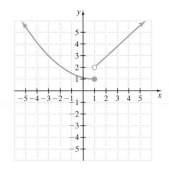

19.

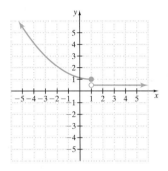

20.

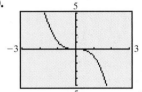

Find an equation defining the inverse function of the given one-to-one function.

21. $f(x) = x - 9$

22. $f(x) = x + 8$

23. $f(x) = 6x + 11$

24. $f(x) = 12x$

25. $f(x) = x^3 - 5$

26. $f(x) = \sqrt[3]{x + 2}$

27. $g(x) = \dfrac{12x - 7}{6}$

28. $r(x) = \dfrac{13}{2}x - 4$

On the same set of axes, graph the given one-to-one function and its inverse.

29. $g(x) = \sqrt{x}$

30. $h(x) = 5x - 5$

31. Find the inverse of the one-to-one function $f(x) = 2x - 3$. Then graph both $f(x)$ and $f^{-1}(x)$ with a square window.

(9.3) *Solve each equation for x.*

32. $4^x = 64$

33. $3^x = \dfrac{1}{9}$

34. $2^{3x} = \dfrac{1}{16}$

35. $5^{2x} = 125$

36. $9^{x+1} = 243$

37. $8^{3x-2} = 4$

Graph each exponential function.

38. $y = 3^x$

39. $y = \left(\dfrac{1}{3}\right)^x$

40. $y = 4 \cdot 2^x$

41. $y = 2^x + 4$

Use the formula $A = P\left(1 + \dfrac{r}{n}\right)^{nt}$ to solve the interest problems. In this formula,

A = amount accrued (or owed)

P = principal invested (or loaned)

r = rate of interest

n = number of compounding periods per year

t = time in years

42. Find the amount accrued if $1600 is invested at 9% interest compounded semiannually for 7 years.

43. A total of $800 is invested in a 7% certificate of deposit for which interest is compounded quarterly. Find the value that this certificate will have at the end of 5 years.

44. Use a graphing calculator to verify the results of Exercise 40.

(9.4) Write each equation with logarithmic notation.

45. $49 = 7^2$

46. $2^{-4} = \dfrac{1}{16}$

Write each logarithmic equation with exponential notation.

47. $\log_{1/2} 16 = -4$

48. $\log_{0.4} 0.064 = 3$

Solve for x.

49. $\log_4 x = -3$

50. $\log_3 x = 2$

51. $\log_3 1 = x$

52. $\log_4 64 = x$

53. $\log_x 64 = 2$

54. $\log_x 81 = 4$

55. $\log_4 4^5 = x$

56. $\log_7 7^{-2} = x$

57. $5^{\log_5 4} = x$

58. $2^{\log_2 9} = x$

59. $\log_2(3x - 1) = 4$

60. $\log_3(2x + 5) = 2$

61. $\log_4(x^2 - 3x) = 1$

62. $\log_8(x^2 + 7x) = 1$

Graph each pair of equations on the same coordinate system.

63. $y = 2^x$ and $y = \log_2 x$

64. $y = \left(\dfrac{1}{2}\right)^x$ and $y = \log_{1/2} x$

(9.5) Write each of the following as single logarithms.

65. $\log_3 8 + \log_3 4$

66. $\log_2 6 + \log_2 3$

67. $\log_7 15 - \log_7 20$

68. $\log 18 - \log 12$

69. $\log_{11} 8 + \log_{11} 3 - \log_{11} 6$

70. $\log_5 14 + \log_5 3 - \log_5 21$

71. $2 \log_5 x - 2 \log_5(x + 1) + \log_5 x$

72. $4 \log_3 x - \log_3 x + \log_3(x + 2)$

Use properties of logarithms to write each expression as a sum or difference of multiples of logarithms.

73. $\log_3 \dfrac{x^3}{x + 2}$

74. $\log_4 \dfrac{x + 5}{x^2}$

75. $\log_2 \dfrac{3x^2 y}{z}$

76. $\log_7 \dfrac{yz^3}{x}$

If $\log_b 2 = 0.36$ and $\log_b 5 = 0.83$, find the following.

77. $\log_b 50$

78. $\log_b \dfrac{4}{5}$

(9.6) Use a calculator to approximate the logarithm to four decimal places.

79. $\log 3.6$

80. $\log 0.15$

81. $\ln 1.25$

82. $\ln 4.63$

Find the exact value.

83. $\log 1000$

84. $\log \dfrac{1}{10}$

85. $\ln \dfrac{1}{e}$

86. $\ln e^4$

Solve each equation for x.

87. $\ln(2x) = 2$

88. $\ln(3x) = 1.6$

89. $\ln(2x - 3) = -1$

90. $\ln(3x + 1) = 2$

Use the formula $\ln \dfrac{I}{I_0} = -kx$ to solve radiation problems. In this formula,

x = depth in millimeters

I = intensity of radiation

I_0 = initial intensity

k = a constant measure dependent on the material

Round answers to two decimal places.

91. Find the depth at which the intensity of the radiation passing through a lead shield is reduced to 3% of the original intensity if the value of k is 2.1.

92. If k is 3.2, find the depth at which 2% of the original radiation will penetrate.

Approximate the logarithm to four decimal places.

93. $\log_5 1.6$

94. $\log_3 4$

Use the formula $A = Pe^{rt}$ to solve the interest problems in which interest is compounded continuously. In this formula,

A = amount accrued (or owed)

P = principal invested (or loaned)

r = rate of interest

t = time in years

95. Bank of New York offers a 5-year, 6% continuously compounded investment option. Find the amount accrued if $1450 is invested.

96. Find the amount to which a $940 investment grows if it is invested at 11% compounded continuously for 3 years.

(9.7) *Solve each exponential equation for x. Give an exact solution and also approximate the solution to four decimal places.*

97. $3^{2x} = 7$

98. $6^{3x} = 5$

99. $3^{2x+1} = 6$

100. $4^{3x+2} = 9$

101. $5^{3x-5} = 4$

102. $8^{4x-2} = 3$

103. $2 \cdot 5^{x-1} = 1$

104. $3 \cdot 4^{x+5} = 2$

Solve the equation for x.

105. $\log_5 2 + \log_5 x = 2$

106. $\log_3 x + \log_3 10 = 2$

107. $\log(5x) - \log(x + 1) = 4$

108. $\ln(3x) - \ln(x - 3) = 2$

109. $\log_2 x + \log_2 2x - 3 = 1$

110. $-\log_6(4x + 7) + \log_6 x = 1$

Use the formula $y = y_0 e^{kt}$ to solve the population growth problems. In this formula,

y = size of population

y_0 = initial count of population

k = rate of growth written as a decimal

t = time

Round each answer to the nearest whole.

111. The population of mallard ducks in Nova Scotia is expected to grow at a rate of 6% per week during the spring migration. If 155,000 ducks are already in Nova Scotia, find how many are expected by the end of 4 weeks.

112. The population of Armenia is declining at a rate of 0.129% per year. If the population in 2007 was 2,971,650, find the expected population by the year 2015. (*Source:* U.S. Bureau of the Census, International Data Base)

113. China is experiencing an annual growth rate of 0.606%. In 2007, the population of China was 1,321,851,888. How long will it take for the population to be 1,500,000,000? Round to the nearest tenth. (*Source: CIA 2007 World Factbook*)

114. In 2007, Canada had a population of 33,390,141. How long will it take for Canada to double its population if the growth rate is 0.9% annually? Round to the nearest tenth. (*Source: CIA 2007 World Factbook*)

115. Malaysia's population is increasing at a rate of 1.8% per year. How long will it take the 2007 population of 24,821,286 to double in size? Round to the nearest tenth. (*Source: CIA 2007 World Factbook*)

Use the compound interest equation $A = P\left(1 + \dfrac{r}{n}\right)^{nt}$ to solve the following. (See the directions for Exercises 42 and 43 for an explanation of this formula. Round answers to the nearest tenth.)

116. Find how long it will take a $5000 investment to grow to $10,000 if it is invested at 8% interest compounded quarterly.

117. An investment of $6000 has grown to $10,000 while the money was invested at 6% interest compounded monthly. Find how long it was invested.

Use a graphing utility to solve each equation. Round all solutions to two decimal places.

118. $e^x = 2$ **119.** $10^{0.3x} = 7$

Solve using exponential regression features on a graphing calculator.

120. Classroom costs have risen sharply since 1940. The chart below lists the costs for public elementary and secondary schools in billions of dollars. Use the chart to find the exponential regression equation and predict the amount of cost for the year 2010. (Let $x = 40$ represent the year 1940.)

Year (40 represents 1940)	*Cost $ (in billions)*
40	3
50	10
60	25
70	48
80	100
90	200
100	275

MIXED REVIEW

Solve each equation.

121. $7^{4x} = 49$ **122.** $3^x = \dfrac{1}{81}$

123. $\log_4 4 = x$ **124.** $8^{3x-2} = 32$

125. $\log_5(x^2 - 4x) = 1$ **126.** $\log_3 x = 4$

127. $\ln x = -3.2$ **128.** $\log_4(3x - 1) = 2$

129. $\ln x - \ln 2 = 1$ **130.** $\log_5 x + \log_5 10 = 2$

131. $\log_6 x - \log_6(4x + 7) = 1$

CHAPTER 9 TEST

If $f(x) = x$ and $g(x) = 2x - 3$, find the following.

1. $(f \cdot g)(x)$

2. $(f - g)(x)$

If $f(x) = x$, $g(x) = x - 7$, and $h(x) = x^2 - 6x + 5$, find the following.

3. $(f \circ h)(0)$

4. $(g \circ f)(x)$

5. $(g \circ h)(x)$

On the same set of axes, graph the given one-to-one function and its inverse.

6. $f(x) = 7x - 14$

Determine whether the given graph is the graph of a one-to-one function.

7.

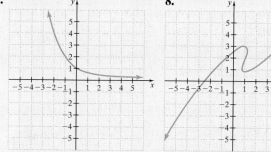

8.

Determine whether each function is one-to-one. If it is one-to-one, find an equation or a set of ordered pairs that defines the inverse function of the given function.

9. $f(x) = 6 - 2x$

10. $f = \{(0, 0), (2, 3), (-1, 5)\}$

11.

Word (Input)	Dog	Cat	House	Desk	Circle
First Letter of Word (Output)	d	c	h	d	c

Use the properties of logarithms to write each expression as a single logarithm.

12. $\log_3 6 + \log_3 4$

13. $\log_5 x + 3 \log_5 x - \log_5 (x + 1)$

14. Write the expression $\log_6 \dfrac{2x}{y^3}$ as the sum or difference of multiples of logarithms.

15. If $\log_b 3 = 0.79$ and $\log_b 5 = 1.16$, find the value of $\log_b \dfrac{3}{25}$.

16. Approximate $\log_7 8$ to four decimal places.

17. Solve $8^{x-1} = \dfrac{1}{64}$ for x. Give an exact solution.

18. Solve $3^{2x+5} = 4$ for x. Give an exact solution, and also approximate the solution to four decimal places.

Solve each logarithmic equation for x. Give an exact solution.

19. $\log_3 x = -2$

20. $\ln \sqrt{e} = x$

21. $\log_8 (3x - 2) = 2$

22. $\log_5 x + \log_5 3 = 2$

23. $\log_4 (x + 1) - \log_4 (x - 2) = 3$

24. Solve $\ln (3x + 7) = 1.31$ accurate to four decimal places.

25. Graph $y = \left(\dfrac{1}{2}\right)^x + 1$.

26. Graph the functions $y = 3^x$ and $y = \log_3 x$ on the same coordinate system.

Use the formula $A = P\left(1 + \dfrac{r}{n}\right)^{nt}$ to solve Exercises 27 and 28.

27. Find the amount in the account if \$4000 is invested for 3 years at 9% interest compounded monthly.

28. Find how long it will take \$2000 to grow to \$3000 if the money is invested at 7% interest compounded semiannually. Round to the nearest whole.

Use the population growth formula $y = y_0 e^{kt}$ to solve Exercises 29 and 30.

29. The prairie dog population of the Grand Rapids area now stands at 57,000 animals. If the population is growing at a rate of 2.6% annually, find how many prairie dogs there will be in that area 5 years from now.

30. In an attempt to save an endangered species of wood duck, naturalists would like to increase the wood duck population from 400 to 1000 ducks. If the annual population growth rate is 6.2%, find how long it will take the naturalists to reach their goal. Round to the nearest whole year.

31. The formula $\log(1 + k) = \dfrac{0.3}{D}$ relates the doubling time D, in days, and the growth rate k for a population of mice. Find the rate at which the population is increasing if the doubling time is 56 days. Round to the nearest tenth of a percent.

32. Use a graphing calculator to approximate the solution of
$$e^{0.2x} = e^{-0.4x} + 2$$
to two decimal places.

CHAPTER 9 CUMULATIVE REVIEW

1. Multiply.

   a. $(-8)(-1)$ **b.** $(-2)\dfrac{1}{6}$

   c. $-1.2(0.3)$ **d.** $0(-11)$

   e. $\left(\dfrac{1}{5}\right)\left(-\dfrac{10}{11}\right)$ **f.** $(7)(1)(-2)(-3)$

   g. $8(-2)(0)$

2. Solve: $\dfrac{1}{3}(x-2) = \dfrac{1}{4}(x+1)$

3. Graph $y = x^2$.

4. Find the equation of a line through $(-2, 6)$ and perpendicular to $f(x) = -3x + 4$. Write the equation using function notation.

5. Solve the system.

$$\begin{cases} x - 5y - 2z = 6 \\ -2x + 10y + 4z = -12 \\ \dfrac{1}{2}x - \dfrac{5}{2}y - z = 3 \end{cases}$$

6. Line l and line m are parallel lines cut by transversal t. Find the values of x and y.

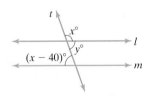

7. Use the quotient rule to simplify.

   a. $\dfrac{x^7}{x^4}$ **b.** $\dfrac{5^8}{5^2}$

   c. $\dfrac{20x^6}{4x^5}$ **d.** $\dfrac{12y^{10}z^7}{14y^8z^7}$

8. Use the power rules to simplify the following. Use positive exponents to write all results.

   a. $(4a^3)^2$ **b.** $\left(-\dfrac{2}{3}\right)^3$

   c. $\left(\dfrac{4a^5}{b^3}\right)^3$ **d.** $\left(\dfrac{3^{-2}}{x}\right)^{-3}$

   e. $(a^{-2}b^3c^{-4})^{-2}$

9. For the ICL Production Company, the rational function $C(x) = \dfrac{2.6x + 10{,}000}{x}$ describes the company's cost per disc of pressing x compact discs. Find the cost per disc for pressing:

   a. 100 compact discs

   b. 1000 compact discs

10. Multiply.

   a. $(3x - 1)^2$

   b. $\left(\dfrac{1}{2}x + 3\right)\left(\dfrac{1}{2}x - 3\right)$

   c. $(2x - 5)(6x + 7)$

11. Add or subtract.

   a. $\dfrac{x}{4} + \dfrac{5x}{4}$ **b.** $\dfrac{5}{7z^2} + \dfrac{x}{7z^2}$

   c. $\dfrac{x^2}{x+7} - \dfrac{49}{x+7}$ **d.** $\dfrac{x}{3y^2} - \dfrac{x+1}{3y^2}$

12. Perform the indicated operations and simplify if possible.
$$\dfrac{5}{x-2} + \dfrac{3}{x^2+4x+4} - \dfrac{6}{x+2}$$

13. Divide $3x^4 + 2x^3 - 8x + 6$ by $x^2 - 1$.

14. Simplify each complex fraction.

   a. $\dfrac{\dfrac{a}{5}}{\dfrac{a-1}{10}}$ **b.** $\dfrac{\dfrac{3}{2+a} + \dfrac{6}{2-a}}{\dfrac{5}{a+2} - \dfrac{1}{a-2}}$

   c. $\dfrac{x^{-1} + y^{-1}}{xy}$

15. Solve: $\dfrac{2x}{2x-1} + \dfrac{1}{x} = \dfrac{1}{2x-1}$

16. Divide $x^3 - 8$ by $x - 2$.

17. Steve Deitmer takes $1\dfrac{1}{2}$ times as long to go 72 miles upstream in his boat as he does to return. If the boat cruises at 30 mph in still water, what is the speed of the current?

18. Use synthetic division to divide: $(8x^2 - 12x - 7) \div (x - 2)$

19. Simplify the following expressions.

   a. $\sqrt[4]{81}$ **b.** $\sqrt[5]{-243}$

   c. $-\sqrt{25}$ **d.** $\sqrt[4]{-81}$

   e. $\sqrt[3]{64x^3}$

20. Solve $\dfrac{1}{a+5} = \dfrac{1}{3a+6} - \dfrac{a+2}{a^2+7x+10}$

21. Use rational exponents to write as a single radical.

   a. $\sqrt{x} \cdot \sqrt[4]{x}$ **b.** $\dfrac{\sqrt{x}}{\sqrt[3]{x}}$ **c.** $\sqrt[3]{3} \cdot \sqrt{2}$

22. Suppose that y varies directly as x. If $y = \dfrac{1}{2}$ when $x = 12$, find the constant of variation and the direct variation equation.

23. Multiply.

    a. $\sqrt{3}\left(5 + \sqrt{30}\right)$

    b. $\left(\sqrt{5} - \sqrt{6}\right)\left(\sqrt{7} + 1\right)$

    c. $\left(7\sqrt{x} + 5\right)\left(3\sqrt{x} - \sqrt{5}\right)$

    d. $\left(4\sqrt{3} - 1\right)^2$

    e. $\left(\sqrt{2x} - 5\right)\left(\sqrt{2x} + 5\right)$

    f. $\left(\sqrt{x - 3} + 5\right)^2$

24. Find each root. Assume that all variables represent non-negative real numbers.

    a. $\sqrt[4]{81}$ **b.** $\sqrt[3]{-27}$ **c.** $\sqrt{\dfrac{9}{64}}$

    d. $\sqrt[4]{x^{12}}$ **e.** $\sqrt[3]{-125y^6}$

25. Rationalize the denominator of $\dfrac{\sqrt[4]{x}}{\sqrt[4]{81y^5}}$.

26. Multiply.

    a. $a^{1/4}\left(a^{3/4} - a^{7/4}\right)$

    b. $\left(x^{1/2} - 3\right)\left(x^{1/2} + 5\right)$

27. Solve $\sqrt{4 - x} = x - 2$.

28. Use the quotient rule to divide and simplify if possible.

    a. $\dfrac{\sqrt{54}}{\sqrt{6}}$ **b.** $\dfrac{\sqrt{108a^2}}{3\sqrt{3}}$

    c. $\dfrac{3\sqrt[3]{81a^5b^{10}}}{\sqrt[3]{3b^4}}$

29. Solve $3x^2 - 9x + 8 = 0$ by completing the square.

30. Add or subtract as indicated.

    a. $\dfrac{\sqrt{20}}{3} + \dfrac{\sqrt{5}}{4}$

    b. $\sqrt[3]{\dfrac{24x}{27}} - \dfrac{\sqrt[3]{3x}}{2}$

31. Solve $\dfrac{3x}{x - 2} - \dfrac{x + 1}{x} = \dfrac{6}{x(x - 2)}$.

32. Rationalize the denominator. $\sqrt[3]{\dfrac{27}{m^4n^8}}$

33. Solve $x^2 - 4x \le 0$.

34. Find the length of the unknown side of the triangle.

35. Graph $F(x) = (x - 3)^2 + 1$.

36. Find the following powers of i.

    a. i^8 **b.** i^{21}

    c. i^{42} **d.** i^{-13}

37. If $f(x) = x - 1$ and $g(x) = 2x - 3$, find

    a. $(f + g)(x)$ **b.** $(f - g)(x)$

    c. $(f \cdot g)(x)$ **d.** $\left(\dfrac{f}{g}\right)(x)$

38. Solve $4x^2 + 8x - 1 = 0$ by completing the square.

39. Find an equation of the inverse of $f(x) = x + 3$.

40. Solve by using the quadratic formula. $\left(x - \dfrac{1}{2}\right)^2 = \dfrac{x}{2}$

41. Find the value of each logarithmic expression.

    a. $\log_4 16$ **b.** $\log_{10}\dfrac{1}{10}$ **c.** $\log_9 3$

42. Graph $f(x) = -(x + 1)^2 + 1$. Find the vertex and the axis of symmetry.

10 Conic Sections

In Chapter 8, we analyzed some of the important connections between a parabola and its equation. Parabolas are interesting in their own right but are more interesting still because they are part of a collection of curves known as conic sections. This chapter is devoted to quadratic equations in two variables and their conic section graphs: the parabola, circle, ellipse, and hyperbola.

When the sun rises above the horizon on June 21 or 22, thousands of people gather to witness and celebrate summer solstice at Stonehenge. Stonehenge is a megalithic ruin located on the Salisbury Plain in Wiltshire, England. It is a series of earth, timber and stone structures that were constructed, revised, and reconstructed over a period of 1400 years or so. Although no one can say for certain what its true purpose was, there have been multiple theories. The best-known theory is the one from eighteenth-century British antiquarian William Stukeley. He believed that Stonehenge was a temple, possibly an ancient cult center for the Druids. Despite the fact that we don't know its purpose for certain, Stonehenge acts as a prehistoric timepiece, allowing us to theorize what it would have been like during the Neolithic Period, and who could have built this megalithic wonder. In Section 10.1, Exercise 72, on page 697, we will explore the dimensions of the outer stone circle, known as the Sarsen Circle, or Stonehenge. (*Source:* The Discovery Channel)

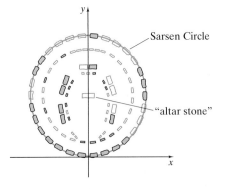

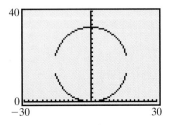

A graphing calculator illustration of the Sarsen Circle.

10.1 THE PARABOLA AND THE CIRCLE

OBJECTIVES

1 Graph parabolas of the form $x = a(y - k)^2 + h$ and $y = a(x - h)^2 + k$.

2 Graph circles of the form $(x - h)^2 + (y - k)^2 = r^2$.

3 Write the equation of a circle, given its center and radius.

4 Find the center and the radius of a circle, given its equation.

Conic sections derive their name because each conic section is the intersection of a right circular cone and a plane. The circle, parabola, ellipse, and hyperbola are the conic sections.

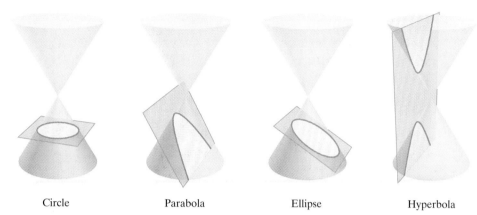

Circle   Parabola   Ellipse   Hyperbola

OBJECTIVE 1 ▶ Graphing parabolas in standard form. Thus far, we have seen that $f(x)$ or $y = a(x - h)^2 + k$ is the equation of a parabola that opens upward if $a > 0$ or downward if $a < 0$. Parabolas can also open left or right, or even on a slant. Equations of these parabolas are not functions of x, of course, since a parabola opening any way other than upward or downward fails the vertical line test. In this section, we introduce parabolas that open to the left and to the right. Parabolas opening on a slant will not be developed in this book.

Just as $y = a(x - h)^2 + k$ is the equation of a parabola that opens upward or downward, $x = a(y - k)^2 + h$ is the equation of a parabola that opens to the right or to the left. The parabola opens to the right if $a > 0$ and to the left if $a < 0$. The parabola has vertex (h, k), and its axis of symmetry is the line $y = k$.

Parabolas

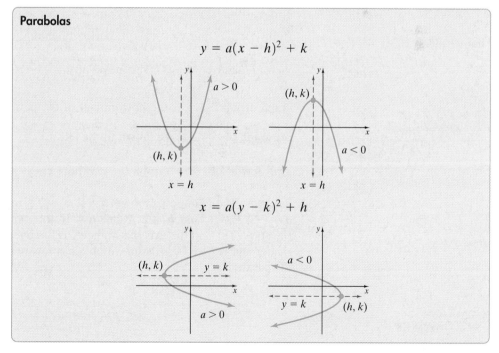

The equations $y = a(x - h)^2 + k$ and $x = a(y - k)^2 + h$ are called **standard forms.**

Concept Check ✓

Does the graph of the parabola given by the equation $x = -3y^2$ open to the left, to the right, upward, or downward?

EXAMPLE 1 Graph the parabola $x = 2y^2$.

Solution Written in standard form, the equation $x = 2y^2$ is $x = 2(y - 0)^2 + 0$, with $a = 2, h = 0$, and $k = 0$. Its graph is a parabola with vertex $(0, 0)$, and its axis of symmetry is the line $y = 0$. Since $a > 0$, this parabola opens to the right. The table shows a few more ordered pair solutions of $x = 2y^2$. Its graph is also shown.

x	y
8	-2
2	-1
0	0
2	1
8	2

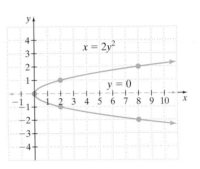

PRACTICE
1 Graph the parabola $x = \dfrac{1}{2}y^2$.

Notice that the parabola in Example 1 does not pass the vertical line test and is not the graph of a function. This means that we cannot enter a single function into the Y= editor to graph $x = 2y^2$. Instead, we solve for y and enter two functions separately, as shown below.

To graph $x = 2y^2$ using a graphing utility, solve the equation for y.

$$x = 2y^2$$
$$\frac{x}{2} = y^2$$
$$\pm\sqrt{\frac{x}{2}} = y$$

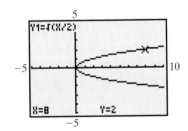

The graph of $y_1 = \sqrt{\frac{x}{2}}$ and $y_2 = -y_1$.

Both $y = \sqrt{\dfrac{x}{2}}$ and $y = -\sqrt{\dfrac{x}{2}}$ describe functions. Graph $y_1 = \sqrt{\dfrac{x}{2}}$ and $y_2 = -\sqrt{\dfrac{x}{2}}$ (or $y_1 = \sqrt{\dfrac{x}{2}}$ and $y_2 = -y_1$). The graph of $y = \sqrt{\dfrac{x}{2}}$ is the top half of the parabola and the graph of $y = -\sqrt{\dfrac{x}{2}}$ is the bottom half of the parabola. The two graphs together form the graph of $x = 2y^2$, as shown to the left.

EXAMPLE 2 Graph the parabola $x = -3(y - 1)^2 + 2$.

Solution The equation $x = -3(y - 1)^2 + 2$ is in the form $x = a(y - k)^2 + h$, with $a = -3, k = 1$, and $h = 2$. Since $a < 0$, the parabola opens to the left. The vertex (h, k) is $(2, 1)$, and the axis of symmetry is the line $y = 1$. When $y = 0, x = -1$, so the x-intercept is $(-1, 0)$. Again, we obtain a few ordered pair solutions and then graph the parabola.

TECHNOLOGY NOTE

There is a Conics App to check the graph of a conic section that is not a function.

---

x	y
2	1
-1	0
-1	2
-10	3
-10	-1

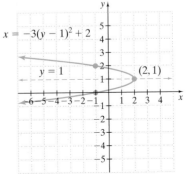

2   Graph the parabola $x = -2(y + 4)^2 - 1$.

EXAMPLE 3   Graph $y = -x^2 - 2x + 15$.

Solution   Complete the square on x to write the equation in standard form.

$$y - 15 = -x^2 - 2x \qquad \text{Subtract 15 from both sides.}$$

$$y - 15 = -1(x^2 + 2x) \qquad \text{Factor } -1 \text{ from the terms } -x^2 - 2x.$$

The coefficient of x is 2. Find the square of half of 2.

$$\frac{1}{2}(2) = 1 \quad \text{and} \quad 1^2 = 1$$

$$y - 15 - 1(1) = -1(x^2 + 2x + 1) \qquad \text{Add } -1(1) \text{ to both sides.}$$

$$y - 16 = -1(x + 1)^2 \qquad \begin{array}{l}\text{Simplify the left side and} \\ \text{factor the right side.}\end{array}$$

$$y = -(x + 1)^2 + 16 \qquad \text{Add 16 to both sides.}$$

The equation is now in standard form $y = a(x - h)^2 + k$, with $a = -1, h = -1$, and $k = 16$.

The vertex is then (h, k), or $(-1, 16)$.

A second method for finding the vertex is by using the formula $\dfrac{-b}{2a}$.

$$x = \frac{-(-2)}{2(-1)} = \frac{2}{-2} = -1$$

$$y = -(-1)^2 - 2(-1) + 15 = -1 + 2 + 15 = 16$$

Again, we see that the vertex is $(-1, 16)$, and the axis of symmetry is the vertical line $x = -1$. The y-intercept is $(0, 15)$. Now we can use a few more ordered pair solutions to graph the parabola.

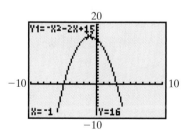

A calculator check of Example 3.

x	y
-1	16
0	15
-2	15
1	12
-3	12
3	0
-5	0

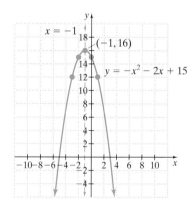

3   Graph $y = -x^2 + 4x + 6$.

EXAMPLE 4   Graph $x = 2y^2 + 4y + 5$.

Solution   Notice that this equation is quadratic in y, so its graph is a parabola that opens to the left or the right. We can complete the square on y or we can use the formula $\dfrac{-b}{2a}$ to find the vertex.

Since the equation is quadratic in y, the formula gives us the y-value of the vertex.

$$y = \frac{-4}{2 \cdot 2} = \frac{-4}{4} = -1$$

$$x = 2(-1)^2 + 4(-1) + 5 = 2 \cdot 1 - 4 + 5 = 3$$

The vertex is $(3, -1)$, and the axis of symmetry is the line $y = -1$. The parabola opens to the right since $a > 0$. The x-intercept is $(5, 0)$.

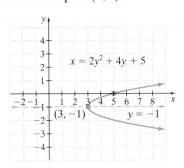

PRACTICE

4 Graph $x = 3y^2 + 6y + 4$.

EXAMPLE 5 Use a graphing utility to graph $x = -y^2 - 4y - 1$.

Solution Solve the equation for y. To do so, write the equation so that one side is 0 and then use the quadratic formula.

$$x = -y^2 - 4y - 1$$

$$y^2 + 4y + (x + 1) = 0$$

Since we are solving for y, $a = 1$, $b = 4$, and $c = x + 1$. By the quadratic formula,

$$y = \frac{-b \pm \sqrt{b^2 - 4ac}}{2a}$$

or

$$y = \frac{-4 \pm \sqrt{4^2 - 4(1)(x + 1)}}{2(1)}.$$

To enter these expressions on a graphing utility, it often is convenient to store the expression under the radical, $4^2 - 4(1)(x + 1)$, in y_1. Then deselect y_1, but use it to form y_2 and y_3 as shown below.

$$y_2 = \frac{-4 + \sqrt{y_1}}{2}$$

$$y_3 = \frac{-4 - \sqrt{y_1}}{2}$$

Enter y_2 and y_3 and graph these two equations in the standard window as shown below. (Don't forget to deselect y_1 so that its graph is not shown. The purpose of entering y_1 is to simplify the expressions for y_2 and y_3.)

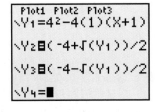
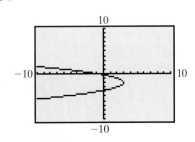

PRACTICE

5 Use a graphing utility to graph $x = y^2 + 4y - 1$.

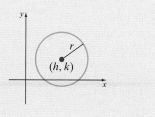

OBJECTIVE 2 ▶ Graphing circles in standard form. Another conic section is the **circle.** A circle is the set of all points in a plane that are the same distance from a fixed point called the **center.** The distance is called the **radius** of the circle. To find a standard equation for a circle, let (h, k) represent the center of the circle, and let (x, y) represent any point on the circle. The distance between (h, k) and (x, y) is defined to be the circle's radius, r units. We can find this distance r by using the distance formula.

$$r = \sqrt{(x - h)^2 + (y - k)^2}$$

$$r^2 = (x - h)^2 + (y - k)^2 \qquad \text{Square both sides.}$$

Circle

The graph of $(x - h)^2 + (y - k)^2 = r^2$ is a circle with center (h, k) and radius r.

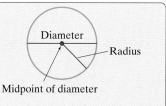

The equation $(x - h)^2 + (y - k)^2 = r^2$ is called **standard form.**

If an equation can be written in the standard form

$$(x - h)^2 + (y - k)^2 = r^2$$

then its graph is a circle, which we can draw by graphing the center (h, k) and using the radius r.

▶ **Helpful Hint**

Notice that the radius is the *distance* from the center of the circle to any point of the circle. Also notice that the *midpoint* of a diameter of a circle is the center of the circle.

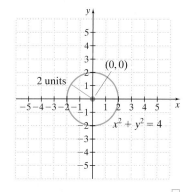

EXAMPLE 6   Graph $x^2 + y^2 = 4$.

Solution   The equation can be written in standard form as

$$(x - 0)^2 + (y - 0)^2 = 2^2$$

The center of the circle is $(0, 0)$, and the radius is 2. Its graph is shown.

PRACTICE
6   Graph $x^2 + y^2 = 25$.

▶ **Helpful Hint**

Notice the difference between the equation of a circle and the equation of a parabola. The equation of a circle contains both x^2 and y^2 terms on the same side of the equation with equal coefficients. The equation of a parabola has either an x^2 term or a y^2 term but not both.

EXAMPLE 7   Graph $(x + 1)^2 + y^2 = 8$. Use a graphing utility to check.

Algebraic Solution

The equation can be written as $(x + 1)^2 + (y - 0)^2 = 8$ with $h = -1, k = 0$, and $r = \sqrt{8}$. The center is $(-1, 0)$, and the radius is $\sqrt{8} = 2\sqrt{2} \approx 2.8$.

> **TECHNOLOGY NOTE**
>
> For the graph to appear circular, a square window must be used. Recall that a square window is one whose tick marks are equally spaced on the x- and y-axes.

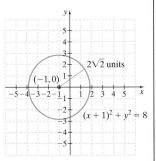

Graphical Solution

Solve the equation for y.

$$(x + 1)^2 + y^2 = 8$$
$$y^2 = 8 - (x + 1)^2$$
$$y = \pm\sqrt{8 - (x + 1)^2}$$

Graph $y_1 = \sqrt{8 - (x + 1)^2}$ and $y_2 = -y_1$ in a decimal window as shown below.

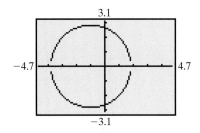

PRACTICE
7   Graph $(x - 3)^2 + (y + 2)^2 = 4$. Use a graphing utility to check.

Concept Check ☑

In the graph of the equation $(x - 3)^2 + (y - 2)^2 = 5$, what is the distance between the center of the circle and any point on the circle?

EXAMPLE 8   Use a graphing utility to graph $x^2 + y^2 = 25$.

Solution   To graph an equation such as $x^2 + y^2 = 25$ with a graphing calculator, we first solve the equation for y.

$$x^2 + y^2 = 25$$
$$y^2 = 25 - x^2$$
$$y = \pm\sqrt{25 - x^2}$$

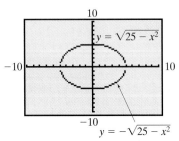

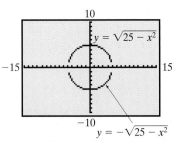

The graph of $y = \sqrt{25 - x^2}$ will be the top half of the circle, and the graph of $y = -\sqrt{25 - x^2}$ will be the bottom half of the circle.

To graph, press $\boxed{\text{Y=}}$ and enter $Y_1 = \sqrt{25 - x^2}$ and $Y_2 = -\sqrt{25 - x^2}$. Insert parentheses around $25 - x^2$ so that $\sqrt{25 - x^2}$ and not $\sqrt{25} - x^2$ is graphed.

The top graph to the left does not appear to be a circle because we are currently using a standard window and the screen is rectangular. This causes the tick marks on the x-axis to be farther apart than the tick marks on the y-axis and, thus, creates the distorted circle. If we want the graph to appear circular, we must define a square window by using a feature of the graphing calculator or by redefining the window to show the x-axis from -15 to 15 and the y-axis from -10 to 10. Using a square window, the graph appears as shown on the bottom to the left.

PRACTICE
8   Use a graphing calculator to graph $x^2 + y^2 = 20$.

Answer to Concept Check:

$\sqrt{5}$ units

OBJECTIVE 3 ▶ Writing equations of circles. Since a circle is determined entirely by its center and radius, this information is all we need to write the equation of a circle.

EXAMPLE 9 Find an equation of the circle with center $(-7, 3)$ and radius 10.

Solution Using the given values $h = -7$, $k = 3$, and $r = 10$, we write the equation

$$(x - h)^2 + (y - k)^2 = r^2$$

or

$$[x - (-7)]^2 + (y - 3)^2 = 10^2 \quad \text{Substitute the given values.}$$

or

$$(x + 7)^2 + (y - 3)^2 = 100 \qquad \square$$

PRACTICE
9 Find the equation of a circle with center $(-2, -5)$ and radius 9.

OBJECTIVE 4 ▶ Finding the center and the radius of a circle. To find the center and the radius of a circle from its equation, write the equation in standard form. To write the equation of a circle in standard form, we complete the square on both x and y.

EXAMPLE 10 Graph $x^2 + y^2 + 4x - 8y = 16$.

Solution Since this equation contains x^2 and y^2 terms on the same side of the equation with equal coefficients, its graph is a circle. To write the equation in standard form, group the terms involving x and the terms involving y, and then complete the square on each variable.

$$(x^2 + 4x) + (y^2 - 8y) = 16$$

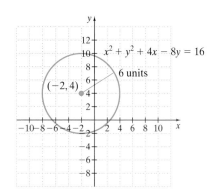

Thus, $\frac{1}{2}(4) = 2$ and $2^2 = 4$. Also, $\frac{1}{2}(-8) = -4$ and $(-4)^2 = 16$. Add 4 and then 16 to both sides.

$$(x^2 + 4x + 4) + (y^2 - 8y + 16) = 16 + 4 + 16$$
$$(x + 2)^2 + (y - 4)^2 = 36 \qquad \text{Factor.}$$

This circle has the center $(-2, 4)$ and radius 6, as shown. $\qquad \square$

PRACTICE
10 Graph $x^2 + y^2 + 6x - 2y = 6$.

VOCABULARY & READINESS CHECK

Use the choices below to fill in each blank. Some choices may be used more than once.

radius	center	vertex
diameter	circle	conic sections

1. The circle, parabola, ellipse, and hyperbola are called the _____.

2. For a parabola that opens upward, the lowest point is the _____.

3. A _____ is the set of all points in a plane that are the same distance from a fixed point. The fixed point is called the _____.

4. The midpoint of a diameter of a circle is the _____.

5. The distance from the center of a circle to any point of the circle is called the _____.

6. Twice a circle's radius is its _____.

The graph of each equation is a parabola. Determine whether the parabola opens upward, downward, to the left, or to the right.

7. $y = x^2 - 7x + 5$

8. $y = -x^2 + 16$

9. $x = -y^2 - y + 2$

10. $x = 3y^2 + 2y - 5$

11. $y = -x^2 + 2x + 1$

12. $x = -y^2 + 2y - 6$

10.1 | EXERCISE SET

MyMathLab *Powered by CourseCompass™ and MathXL®*

 PRACTICE   WATCH   DOWNLOAD   READ   REVIEW

The graph of each equation is a parabola. Find the vertex of the parabola and sketch its graph. See Examples 1 through 5.

1. $x = 3y^2$

2. $x = -2y^2$

3. $x = (y - 2)^2 + 3$

4. $x = (y - 4)^2 - 1$

5. $y = 3(x - 1)^2 + 5$

6. $x = -4(y - 2)^2 + 2$

7. $x = y^2 + 6y + 8$

8. $x = y^2 - 6y + 6$

9. $y = x^2 + 10x + 20$

10. $y = x^2 + 4x - 5$

11. $x = -2y^2 + 4y + 6$

12. $x = 3y^2 + 6y + 7$

The graph of each equation is a circle. Find the center and the radius, and then sketch. See Examples 6, 7, and 10.

13. $x^2 + y^2 = 9$

14. $x^2 + y^2 = 100$

15. $x^2 + (y - 2)^2 = 1$

16. $(x - 3)^2 + y^2 = 9$

17. $(x - 5)^2 + (y + 2)^2 = 1$

18. $(x + 3)^2 + (y + 3)^2 = 4$

19. $x^2 + y^2 + 6y = 0$

20. $x^2 + 10x + y^2 = 0$

21. $x^2 + y^2 + 2x - 4y = 4$

22. $x^2 + 6x - 4y + y^2 = 3$

23. $x^2 + y^2 - 4x - 8y - 2 = 0$ 24. $x^2 + y^2 - 2x - 6y - 5 = 0$

Write an equation of the circle with the given center and radius. See Example 9.

25. $(2, 3); 6$

26. $(-7, 6); 2$

27. $(0, 0); \sqrt{3}$

28. $(0, -6); \sqrt{2}$

29. $(-5, 4); 3\sqrt{5}$

30. the origin; $4\sqrt{7}$

31. Explain the error in the statement: The graph of $x^2 + (y + 3)^2 = 10$ is a circle with center $(0, -3)$ and radius 5.

MIXED PRACTICE

Sketch the graph of each equation. If the graph is a parabola, find its vertex. If the graph is a circle, find its center and radius.

32. $x = y^2 + 2$

33. $x = y^2 - 3$

34. $y = (x + 3)^2 + 3$

35. $y = (x - 2)^2 - 2$

36. $x^2 + y^2 = 49$

37. $x^2 + y^2 = 1$

38. $x = (y - 1)^2 + 4$

39. $x = (y + 3)^2 - 1$

40. $(x + 3)^2 + (y - 1)^2 = 9$ 41. $(x - 2)^2 + (y - 2)^2 = 16$

42. $x = -2(y + 5)^2$

43. $x = -(y - 1)^2$

44. $x^2 + (y + 5)^2 = 5$

45. $(x - 4)^2 + y^2 = 7$

46. $y = 3(x - 4)^2 + 2$

47. $y = 5(x + 5)^2 + 3$

48. $2x^2 + 2y^2 = \dfrac{1}{2}$

49. $\dfrac{x^2}{8} + \dfrac{y^2}{8} = 2$

50. $y = x^2 - 2x - 15$

51. $y = x^2 + 7x + 6$

52. $x^2 + y^2 + 6x + 10y - 2 = 0$

53. $x^2 + y^2 + 2x + 12y - 12 = 0$

54. $x = y^2 + 6y + 2$

55. $x = y^2 + 8y - 4$

56. $x^2 + y^2 - 8y + 5 = 0$

57. $x^2 - 10y + y^2 + 4 = 0$

58. $x = -2y^2 - 4y$

59. $x = -3y^2 + 30y$

60. $\dfrac{x^2}{3} + \dfrac{y^2}{3} = 2$

61. $5x^2 + 5y^2 = 25$

62. $y = 4x^2 - 40x + 105$

63. $y = 5x^2 - 20x + 16$

REVIEW AND PREVIEW

Graph each equation. See Section 2.3.

64. $y = 2x + 5$

65. $y = -3x + 3$

66. $y = 3$

67. $x = -2$

Rationalize each denominator and simplify, if possible. See Section 7.5.

68. $\dfrac{1}{\sqrt{3}}$ **69.** $\dfrac{\sqrt{5}}{\sqrt{8}}$ **70.** $\dfrac{4\sqrt{7}}{\sqrt{6}}$ **71.** $\dfrac{10}{\sqrt{5}}$

CONCEPT EXTENSIONS

72. The Sarsen Circle: The the first image that comes to mind when one thinks of Stonehenge is the very large sandstone blocks with sandstone lintels across the top. The Sarsen Circle of Stonehenge is the outer circle of the sandstone blocks, each of which weighs up to 50 tons. There were originally 30 of these monolithic blocks, but only 17 remain upright to this day. The "altar stone" lies at the center of this circle, which has a diameter of 33 meters.

   a. What is the radius of the Sarsen circle?

   b. What is the circumference of the Sarsen circle? Round your result to 2 decimal places.

   c. Since there were originally 30 Sarsen stones located on the circumference, how far apart would the centers of the stones have been? Round to the nearest tenth of a meter.

   d. Using the axes in the drawing, what are the coordinates of the center of the circle?

   e. Use parts **a** and **d** to write the equation of the Sarsen circle.

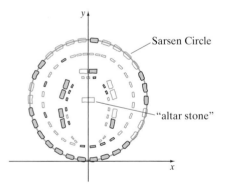

73. Opened in 2000 to honor the millennium, the British Airways London Eye is the world's biggest observation wheel. Each of the 32 enclosed capsules, which each hold 25 passengers, completes a full rotation every 30 minutes. Its diameter is 135 meters, and it is constructed on London's South Bank, to allow passengers to enter the Eye at ground level. (*Source: Guinness Book of World Records*)

   a. What is the radius of the London Eye?

   b. How close is the wheel to the ground?

   c. How high is the center of the wheel from the ground?

   d. If the wheel is drawn on a rectangular coordinate system so that the base (bottom) of the wheel lies at $(0, 0)$, what are the coordinates of the center of the wheel?

   e. Use parts **a** and **d** to write the equation of the Eye.

74. In 1893, Pittsburgh bridge builder George Ferris designed and built a gigantic revolving steel wheel whose height was 264 feet and diameter was 250 feet. This Ferris wheel opened at the 1893 exposition in Chicago. It had 36 wooden cars, each capable of holding 60 passengers. (*Source: The Handy Science Answer Book*)

   a. What was the radius of this Ferris wheel?

   b. How close was the wheel to the ground?

   c. How high was the center of the wheel from the ground?

   d. Using the axes in the drawing, what are the coordinates of the center of the wheel?

   e. Use parts **a** and **d** to write the equation of the wheel.

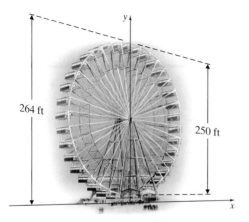

75. As of this writing, the world's largest-diameter Ferris wheel currently in operation is the Star of Nanchung in Jiangxi Province, China. It has 60 compartments, each of which carries eight people. It is 160 meters tall, and the diameter of the wheel is 153 meters. (*Source:* China News Agency)

   a. What is the radius of this Ferris wheel?

   b. How close is the wheel to the ground?

   c. How high is the center of the wheel from the ground?

   d. Using the axes in the drawing, what are the coordinates of the center of the wheel?

   e. Use parts **a** and **d** to write the equation of the wheel.

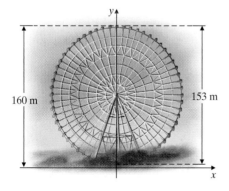

76. If you are given a list of equations of circles and parabolas and none are in standard form, explain how you would determine which is an equation of a circle and which is an equation of a parabola. Explain also how you would distinguish the upward or downward parabolas from the left-opening or right-opening parabolas.

Solve.

77. Cindy Brown, an architect, is drawing plans on grid paper for a circular pool with a fountain in the middle. The paper is marked off in centimeters, and each centimeter represents 1 foot. On the paper, the diameter of the "pool" is 20 centimeters, and "fountain" is the point $(0, 0)$.

   a. Sketch the architect's drawing. Be sure to label the axes.

   b. Write an equation that describes the circular pool.

   c. Cindy plans to place a circle of lights around the fountain such that each light is 5 feet from the fountain. Write an equation for the circle of lights and sketch the circle on your drawing.

78. A bridge constructed over a bayou has a supporting arch in the shape of a parabola. Find an equation of the parabolic arch if the length of the road over the arch is 100 meters and the maximum height of the arch is 40 meters.

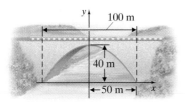

Use a graphing calculator to verify each exercise. Use a square viewing window. See Example 8.

79. Exercise 61   **80.** Exercise 60

81. Exercise 63   **82.** Exercise 62

10.2 THE ELLIPSE AND THE HYPERBOLA

OBJECTIVES

1 Define and graph an ellipse.

2 Define and graph a hyperbola.

OBJECTIVE 1 ▶ Graphing ellipses. An **ellipse** can be thought of as the set of points in a plane such that the sum of the distances of those points from two fixed points is constant. Each of the two fixed points is called a **focus.** (The plural of focus is **foci.**) The point midway between the foci is called the **center.**

    An ellipse may be drawn by hand by using two thumbtacks, a piece of string, and a pencil. Secure the two thumbtacks in a piece of cardboard, for example, and tie each end of the string to a tack. Use your pencil to pull the string tight and draw the ellipse. The two thumbtacks are the foci of the drawn ellipse.

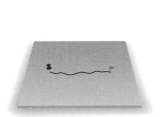 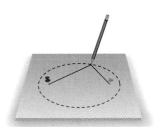 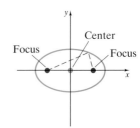

Ellipse with Center (0, 0)

The graph of an equation of the form $\dfrac{x^2}{a^2} + \dfrac{y^2}{b^2} = 1$ is an ellipse with center $(0, 0)$.

The x-intercepts are $(a, 0)$ and $(-a, 0)$, and the y-intercepts are $(0, b)$, and $(0, -b)$.

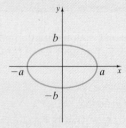

The **standard form** of an ellipse with center $(0, 0)$ is $\dfrac{x^2}{a^2} + \dfrac{y^2}{b^2} = 1$.

EXAMPLE 1   Graph $\dfrac{x^2}{9} + \dfrac{y^2}{16} = 1$.

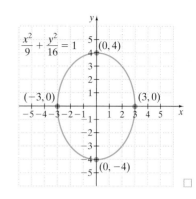

Solution   The equation is of the form $\dfrac{x^2}{a^2} + \dfrac{y^2}{b^2} = 1$, with $a = 3$ and $b = 4$, so its graph is an ellipse with center $(0, 0)$, x-intercepts $(3, 0)$ and $(-3, 0)$, and y-intercepts $(0, 4)$ and $(0, -4)$.

PRACTICE
1   Graph $\dfrac{x^2}{25} + \dfrac{y^2}{4} = 1$.

EXAMPLE 2   Graph $4x^2 + 16y^2 = 64$. Use a graphing utility to check.

Solution   Although this equation contains a sum of squared terms in x and y on the same side of an equation, this is not the equation of a circle since the coefficients of x^2 and y^2 are not the same. The graph of this equation is an ellipse. Since the standard form of the equation of an ellipse has 1 on one side, divide both sides of this equation by 64.

$$4x^2 + 16y^2 = 64$$

$$\frac{4x^2}{64} + \frac{16y^2}{64} = \frac{64}{64} \quad \text{Divide both sides by 64.}$$

$$\frac{x^2}{16} + \frac{y^2}{4} = 1 \quad \text{Simplify.}$$

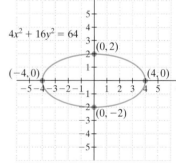

We now recognize the equation of an ellipse with $a = 4$ and $b = 2$. This ellipse has center $(0, 0)$, x-intercepts $(4, 0)$ and $(-4, 0)$, and y-intercepts $(0, 2)$ and $(0, -2)$.

To check using a graphing utility, use the same procedure as for graphing a circle. First, solve the equation for y.

$$4x^2 + 16y^2 = 64$$

$$16y^2 = 64 - 4x^2 \quad \text{Subtract } 4x^2 \text{ from both sides.}$$

$$y^2 = \frac{64 - 4x^2}{16} \quad \text{Divide both sides by 16.}$$

$$y = \pm\sqrt{\frac{64 - 4x^2}{16}} \quad \text{Solve for } y.$$

$$y = \pm\frac{\sqrt{64 - 4x^2}}{4} \quad \text{Simplify.}$$

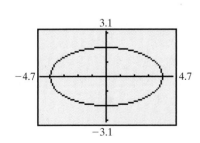

Graph $y_1 = \dfrac{\sqrt{64 - 4x^2}}{4}$ and $y_2 = -y_1$. (Insert two sets of parentheses in the radicand as $\sqrt{\left(\dfrac{(64 - 4x^2)}{4}\right)}$ so that the desired graph is obtained.) The graph is shown to the left.

PRACTICE
2   Graph $9x^2 + 4y^2 = 36$.

The center of an ellipse is not always $(0, 0)$, as shown in the next example.

EXAMPLE 3   Graph $\dfrac{(x+3)^2}{25} + \dfrac{(y-2)^2}{36} = 1$.

Solution   The center of this ellipse is found in a way that is similar to finding the center of a circle. This ellipse has center $(-3, 2)$. Notice that $a = 5$ and $b = 6$. To find four points on the graph of the ellipse, first graph the center, $(-3, 2)$. Since $a = 5$, count 5 units right and then 5 units left of the point with coordinates $(-3, 2)$. Next, since $b = 6$, start at $(-3, 2)$ and count 6 units up and then 6 units down to find two more points on the ellipse.

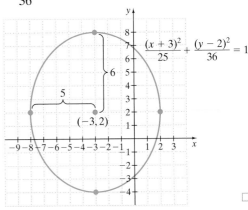

PRACTICE
3   Graph $\dfrac{(x-4)^2}{49} + \dfrac{(y+1)^2}{81} = 1$.

Concept Check ☑️

In the graph of the equation $\dfrac{x^2}{64} + \dfrac{y^2}{36} = 1$, which distance is longer: the distance between the x-intercepts or the distance between the y-intercepts? How much longer? Explain.

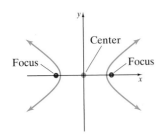

OBJECTIVE 2 ▶ Graphing hyperbolas. The final conic section is the **hyperbola.** A hyperbola is the set of points in a plane such that the absolute value of the difference of the distances from two fixed points is constant. Each of the two fixed points is called a **focus.** The point midway between the foci is called the **center.**

Using the distance formula, we can show that the graph of $\dfrac{x^2}{a^2} - \dfrac{y^2}{b^2} = 1$ is a hyperbola with center $(0, 0)$ and x-intercepts $(a, 0)$ and $(-a, 0)$. Also, the graph of $\dfrac{y^2}{b^2} - \dfrac{x^2}{a^2} = 1$ is a hyperbola with center $(0, 0)$ and y-intercepts $(0, b)$ and $(0, -b)$.

Hyperbola with Center (0, 0)

The graph of an equation of the form $\dfrac{x^2}{a^2} - \dfrac{y^2}{b^2} = 1$ is a hyperbola with center $(0, 0)$ and x-intercepts $(a, 0)$ and $(-a, 0)$.

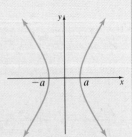

The graph of an equation of the form $\dfrac{y^2}{b^2} - \dfrac{x^2}{a^2} = 1$ is a hyperbola with center $(0, 0)$ and y-intercepts $(0, b)$ and $(0, -b)$.

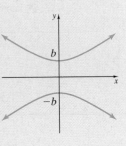

The equations $\dfrac{x^2}{a^2} - \dfrac{y^2}{b^2} = 1$ and $\dfrac{y^2}{b^2} - \dfrac{x^2}{a^2} = 1$ are the **standard forms** for the equation of a hyperbola.

> ▶ **Helpful Hint**
>
> Notice the difference between the equation of an ellipse and a hyperbola. The equation of the ellipse contains x^2 and y^2 terms on the same side of the equation with same-sign coefficients. For a hyperbola, the coefficients on the same side of the equation have different signs.

Graphing a hyperbola such as $\dfrac{y^2}{b^2} - \dfrac{x^2}{a^2} = 1$ is made easier by recognizing one of its important characteristics. Examining the figure below, notice how the sides of the branches of the hyperbola extend indefinitely and seem to approach the dashed lines in the figure. These dashed lines are called the **asymptotes** of the hyperbola.

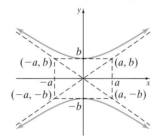

To sketch these lines, or asymptotes, draw a rectangle with vertices $(a, b), (-a, b),$ $(a, -b),$ and $(-a, -b)$. The asymptotes of the hyperbola are the extended diagonals of this rectangle.

EXAMPLE 4   Graph $\dfrac{x^2}{16} - \dfrac{y^2}{25} = 1$.

Solution   This equation has the form $\dfrac{x^2}{a^2} - \dfrac{y^2}{b^2} = 1$, with $a = 4$ and $b = 5$. Thus, its graph is a hyperbola that opens to the left and right. It has center $(0, 0)$ and x-intercepts $(4, 0)$ and $(-4, 0)$. To aid in graphing the hyperbola, we first sketch its asymptotes. The extended diagonals of the rectangle with corners $(4, 5), (4, -5), (-4, 5),$ and $(-4, -5)$ are the asymptotes of the hyperbola. Then we use the asymptotes to aid in sketching the hyperbola.

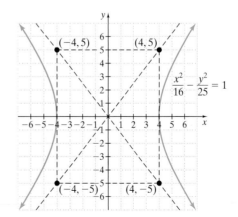

PRACTICE
4   Graph $\dfrac{x^2}{9} - \dfrac{y^2}{16} = 1$.

EXAMPLE 5   Graph $4y^2 - 9x^2 = 36$. Use a graphing utility to check.

Solution   Since this is a difference of squared terms in x and y on the same side of the equation, its graph is a hyperbola, as opposed to an ellipse or a circle. The standard form of the equation of a hyperbola has a 1 on one side, so divide both sides of the equation by 36.

$$4y^2 - 9x^2 = 36$$

$$\frac{4y^2}{36} - \frac{9x^2}{36} = \frac{36}{36} \quad \text{Divide both sides by 36.}$$

$$\frac{y^2}{9} - \frac{x^2}{4} = 1 \quad \text{Simplify.}$$

The equation is of the form $\frac{y^2}{b^2} - \frac{x^2}{a^2} = 1$, with $a = 2$ and $b = 3$, so the hyperbola is centered at $(0, 0)$ with y-intercepts $(0, 3)$ and $(0, -3)$. The sketch of the hyperbola is shown.

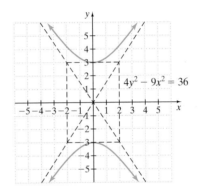

To check, solve the equation for y.

$$4y^2 - 9x^2 = 36$$

$$4y^2 = 36 + 9x^2 \quad \text{Add } 9x^2 \text{ to both sides.}$$

$$y^2 = \frac{36 + 9x^2}{4} \quad \text{Divide both sides by 4.}$$

$$y = \pm\sqrt{\frac{36 + 9x^2}{4}} \quad \text{Solve for } y.$$

$$y = \pm\frac{\sqrt{36 + 9x^2}}{2} \quad \text{Simplify.}$$

Graph $y_1 = \frac{\sqrt{36 + 9x^2}}{2}$ and $y_2 = -y_1$. The graph is shown below.

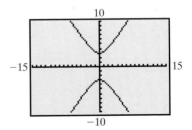

PRACTICE
5   Graph $9y^2 - 25x^2 = 225$.

VOCABULARY & READINESS CHECK

Use the choices below to fill in each blank. Some choices will be used more than once and some not at all.

ellipse	$(0, 0)$	focus	$(a, 0)$ and $(-a, 0)$	$(0, a)$ and $(0, -a)$
hyperbola	center	x	$(b, 0)$ and $(-b, 0)$	$(0, b)$ and $(0, -b)$
		y		

1. A(n) _____ is the set of points in a plane such that the absolute value of the differences of their distances from two fixed points is constant.

2. A(n) _____ is the set of points in a plane such that the sum of their distances from two fixed points is constant.

For exercises 1 and 2 above,

3. The two fixed points are each called a _____.

4. The point midway between the foci is called the _____.

5. The graph of $\dfrac{x^2}{a^2} - \dfrac{y^2}{b^2} = 1$ is a(n) _____ with center _____ and _____-intercepts of _____.

6. The graph of $\dfrac{x^2}{b^2} + \dfrac{y^2}{a^2} = 1$ is a(n) _____ with center _____ and x-intercepts of _____.

Identify the graph of each equation as an ellipse or a hyperbola.

7. $\dfrac{x^2}{16} + \dfrac{y^2}{4} = 1$

8. $\dfrac{x^2}{16} - \dfrac{y^2}{4} = 1$

9. $x^2 - 5y^2 = 3$

10. $-x^2 + 5y^2 = 3$

11. $-\dfrac{y^2}{25} + \dfrac{x^2}{36} = 1$

12. $\dfrac{y^2}{25} + \dfrac{x^2}{36} = 1$

10.2 EXERCISE SET

Sketch the graph of each equation. See Examples 1 and 2.

1. $\dfrac{x^2}{4} + \dfrac{y^2}{25} = 1$

2. $\dfrac{x^2}{16} + \dfrac{y^2}{9} = 1$

3. $\dfrac{x^2}{9} + y^2 = 1$

4. $x^2 + \dfrac{y^2}{4} = 1$

5. $9x^2 + y^2 = 36$

6. $x^2 + 4y^2 = 16$

7. $4x^2 + 25y^2 = 100$

8. $36x^2 + y^2 = 36$

Sketch the graph of each equation. See Example 3.

9. $\dfrac{(x + 1)^2}{36} + \dfrac{(y - 2)^2}{49} = 1$

10. $\dfrac{(x - 3)^2}{9} + \dfrac{(y + 3)^2}{16} = 1$

11. $\dfrac{(x - 1)^2}{4} + \dfrac{(y - 1)^2}{25} = 1$

12. $\dfrac{(x + 3)^2}{16} + \dfrac{(y + 2)^2}{4} = 1$

Sketch the graph of each equation. See Examples 4 and 5.

13. $\dfrac{x^2}{4} - \dfrac{y^2}{9} = 1$

14. $\dfrac{x^2}{36} - \dfrac{y^2}{36} = 1$

15. $\dfrac{y^2}{25} - \dfrac{x^2}{16} = 1$

16. $\dfrac{y^2}{25} - \dfrac{x^2}{49} = 1$

17. $x^2 - 4y^2 = 16$

18. $4x^2 - y^2 = 36$

19. $16y^2 - x^2 = 16$

20. $4y^2 - 25x^2 = 100$

21. If you are given a list of equations of circles, parabolas, ellipses, and hyperbolas, explain how you could distinguish the different conic sections from their equations.

MIXED PRACTICE

Identify whether each equation, when graphed, will be a parabola, circle, ellipse, or hyperbola. Sketch the graph of each equation.

22. $(x - 7)^2 + (y - 2)^2 = 4$

23. $y = x^2 + 4$

24. $y = x^2 + 12x + 36$

25. $\dfrac{x^2}{4} + \dfrac{y^2}{9} = 1$

26. $\dfrac{y^2}{9} - \dfrac{x^2}{9} = 1$

27. $\dfrac{x^2}{16} - \dfrac{y^2}{4} = 1$

28. $\dfrac{x^2}{16} + \dfrac{y^2}{4} = 1$

29. $x^2 + y^2 = 16$

30. $x = y^2 + 4y - 1$

31. $x = -y^2 + 6y$

32. $9x^2 - 4y^2 = 36$

33. $9x^2 + 4y^2 = 36$

34. $\dfrac{(x - 1)^2}{49} + \dfrac{(y + 2)^2}{25} = 1$

35. $y^2 = x^2 + 16$

36. $\left(x + \dfrac{1}{2}\right)^2 + \left(y - \dfrac{1}{2}\right)^2 = 1$

37. $y = -2x^2 + 4x - 3$

REVIEW AND PREVIEW

Solve each inequality. See Section 3.3.

38. $x < 5$ and $x < 1$ **39.** $x < 5$ or $x < 1$

40. $2x - 1 \geq 7$ or $-3x \leq -6$ **41.** $2x - 1 \geq 7$ and $-3x \leq -6$

Perform the indicated operations. See Sections 5.1 and 5.3.

42. $(2x^3)(-4x^2)$ **43.** $2x^3 - 4x^3$

44. $-5x^2 + x^2$ **45.** $(-5x^2)(x^2)$

CONCEPT EXTENSIONS

The graph of each equation is an ellipse. Determine which distance is longer, the distance between the x-intercepts or the distance between the y-intercepts. How much longer? See the Concept Check in this section.

46. $\dfrac{x^2}{16} + \dfrac{y^2}{25} = 1$ **47.** $\dfrac{x^2}{100} + \dfrac{y^2}{49} = 1$

48. $4x^2 + y^2 = 16$ **49.** $x^2 + 4y^2 = 36$

50. We know that $x^2 + y^2 = 25$ is the equation of a circle. Rewrite the equation so that the right side is equal to 1. Which type of conic section does this equation form resemble? In fact, the circle is a special case of this type of conic section. Describe the conditions under which this type of conic section is a circle.

The orbits of stars, planets, comets, asteroids, and satellites all have the shape of one of the conic sections. Astronomers use a measure called eccentricity to describe the shape and elongation of an orbital path. For the circle and ellipse, eccentricity e is calculated with the formula $e = \dfrac{c}{d}$, where $c^2 = |a^2 - b^2|$ and d is the larger value of a or b. For a hyperbola, eccentricity e is calculated with the formula $e = \dfrac{c}{d}$, where $c^2 = a^2 + b^2$ and the value of d is equal to a if the hyperbola has x-intercepts or equal to b if the hyperbola has y-intercepts. Use equations A–H to answer Exercises 51–60.

A. $\dfrac{x^2}{36} - \dfrac{y^2}{13} = 1$ **B.** $\dfrac{x^2}{4} + \dfrac{y^2}{4} = 1$ **C.** $\dfrac{x^2}{25} + \dfrac{y^2}{16} = 1$

D. $\dfrac{y^2}{25} - \dfrac{x^2}{39} = 1$ **E.** $\dfrac{x^2}{17} + \dfrac{y^2}{81} = 1$ **F.** $\dfrac{x^2}{36} + \dfrac{y^2}{36} = 1$

G. $\dfrac{x^2}{16} - \dfrac{y^2}{65} = 1$ **H.** $\dfrac{x^2}{144} + \dfrac{y^2}{140} = 1$

51. Identify the type of conic section represented by each of equations A–H.

52. For each of equations A–H, identify the values of a^2 and b^2.

53. For each of equations A–H, calculate the value of c^2 and c.

54. For each of equations A–H, find the value of d.

55. For each of equations A–H, calculate the eccentricity e.

56. What do you notice about the values of e for the equations you identified as ellipses?

57. What do you notice about the values of e for the equations you identified as circles?

58. What do you notice about the values of e for the equations you identified as hyperbolas?

59. The eccentricity of a parabola is exactly 1. Use this information and the observations you made in Exercises 31, 32, and 33 to describe a way that could be used to identify the type of conic section based on its eccentricity value.

60. Graph each of the conic sections given in equations A–H. What do you notice about the shape of the ellipses for increasing values of eccentricity? Which is the most elliptical? Which is the least elliptical, that is, the most circular?

61. A planet's orbit about the Sun can be described as an ellipse. Consider the Sun as the origin of a rectangular coordinate system. Suppose that the x-intercepts of the elliptical path of the planet are $\pm 130{,}000{,}000$ and that the y-intercepts are $\pm 125{,}000{,}000$. Write the equation of the elliptical path of the planet.

62. Comets orbit the Sun in elongated ellipses. Consider the Sun as the origin of a rectangular coordinate system. Suppose that the equation of the path of the comet is

$$\frac{(x - 1{,}782{,}000{,}000)^2}{3.42 \cdot 10^{23}} + \frac{(y - 356{,}400{,}000)^2}{1.368 \cdot 10^{22}} = 1$$

Find the center of the path of the comet.

63. Use a graphing calculator to verify Exercise 33.

64. Use a graphing calculator to verify Exercise 6.

For Exercises 65 through 70, see the example below.

Example

Sketch the graph of $\dfrac{(x - 2)^2}{25} - \dfrac{(y - 1)^2}{9} = 1$.

Solution

This hyperbola has center $(2, 1)$. Notice that $a = 5$ and $b = 3$.

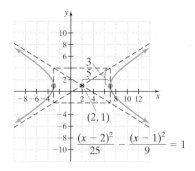

Sketch the graph of each equation.

65. $\dfrac{(x - 1)^2}{4} - \dfrac{(y + 1)^2}{25} = 1$ **66.** $\dfrac{(x + 2)^2}{9} - \dfrac{(y - 1)^2}{4} = 1$

67. $\dfrac{y^2}{16} - \dfrac{(x + 3)^2}{9} = 1$ **68.** $\dfrac{(y + 4)^2}{4} - \dfrac{x^2}{25} = 1$

69. $\dfrac{(x + 5)^2}{16} - \dfrac{(y + 2)^2}{25} = 1$ **70.** $\dfrac{(x - 3)^2}{9} - \dfrac{(y - 2)^2}{4} = 1$

INTEGRATED REVIEW GRAPHING CONIC SECTIONS

Following is a summary of conic sections.

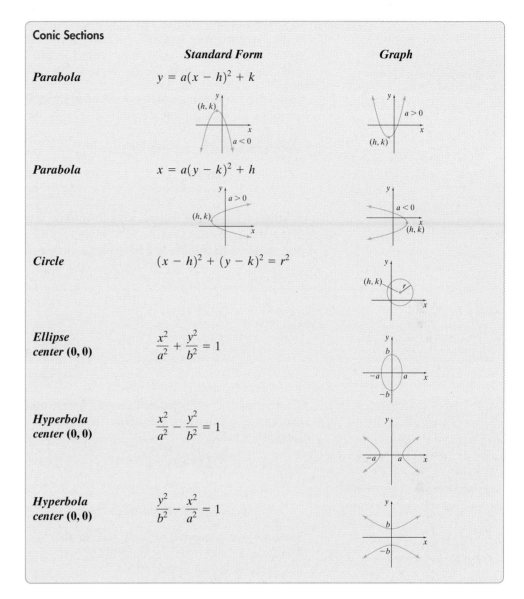

Conic Sections

	Standard Form	Graph
Parabola	$y = a(x - h)^2 + k$	
Parabola	$x = a(y - k)^2 + h$	
Circle	$(x - h)^2 + (y - k)^2 = r^2$	
Ellipse center $(0, 0)$	$\dfrac{x^2}{a^2} + \dfrac{y^2}{b^2} = 1$	
Hyperbola center $(0, 0)$	$\dfrac{x^2}{a^2} - \dfrac{y^2}{b^2} = 1$	
Hyperbola center $(0, 0)$	$\dfrac{y^2}{b^2} - \dfrac{x^2}{a^2} = 1$	

Identify whether each equation, when graphed, will be a parabola, circle, ellipse, or hyperbola. Then graph each equation.

1. $(x - 7)^2 + (y - 2)^2 = 4$

2. $y = x^2 + 4$

3. $y = x^2 + 12x + 36$

4. $\dfrac{x^2}{4} + \dfrac{y^2}{9} = 1$

5. $\dfrac{y^2}{9} - \dfrac{x^2}{9} = 1$

6. $\dfrac{x^2}{16} - \dfrac{y^2}{4} = 1$

7. $\dfrac{x^2}{16} + \dfrac{y^2}{4} = 1$

8. $x^2 + y^2 = 16$

9. $x = y^2 + 4y - 1$

10. $x = -y^2 + 6y$

11. $9x^2 - 4y^2 = 36$

12. $9x^2 + 4y^2 = 36$

13. $\dfrac{(x - 1)^2}{49} + \dfrac{(y + 2)^2}{25} = 1$

14. $y^2 = x^2 + 16$

15. $\left(x + \dfrac{1}{2}\right)^2 + \left(y - \dfrac{1}{2}\right)^2 = 1$

10.3 SOLVING NONLINEAR SYSTEMS OF EQUATIONS

OBJECTIVES

1 Solve a nonlinear system by substitution.

2 Solve a nonlinear system by elimination.

In Section 4.1, we used graphing, substitution, and elimination methods to find solutions of systems of linear equations in two variables. We now apply these same methods to nonlinear systems of equations in two variables. A **nonlinear system of equations** is a system of equations at least one of which is not linear. Since we will be graphing the equations in each system, we are interested in real number solutions only.

OBJECTIVE 1 ▶ Solving nonlinear systems by substitution. First, nonlinear systems are solved by the substitution method.

EXAMPLE 1 Solve the system

$$\begin{cases} y = \sqrt{x} \\ x^2 + y^2 = 6 \end{cases}$$

Solution This system is ideal for substitution since y is expressed in terms of x in the first equation. Notice that if $y = \sqrt{x}$, then both x and y must be nonnegative if they are real numbers. Substitute $\sqrt{x}$ for y in the second equation, and solve for x.

$$x^2 + y^2 = 6$$
$$x^2 + \left(\sqrt{x}\right)^2 = 6 \quad \text{Let } y = \sqrt{x}.$$
$$x^2 + x = 6$$
$$x^2 + x - 6 = 0$$
$$(x + 3)(x - 2) = 0$$
$$x = -3 \quad \text{or} \quad x = 2$$

The solution -3 is discarded because we have noted that x must be nonnegative. To see this, let $x = -3$ in the first equation. Then let $x = 2$ in the first equation to find a corresponding y-value.

Let $x = -3$.   Let $x = 2$.

$$y = \sqrt{x} \qquad\qquad\qquad y = \sqrt{x}$$
$$y = \sqrt{-3} \quad \text{Not a real number} \qquad y = \sqrt{2}$$

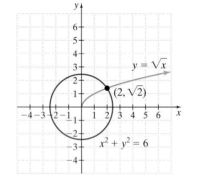

Since we are interested only in real number solutions, the only solution is $\left(2, \sqrt{2}\right)$. Check to see that this solution satisfies both equations. The graph of each equation in the system is shown to the left.

PRACTICE
1 Solve the system $\begin{cases} y = -\sqrt{x} \\ x^2 + y^2 = 20 \end{cases}$.

EXAMPLE 2 Solve the system. Use a graphical approach to check.

$$\begin{cases} x^2 - 3y = 1 \\ x - y = 1 \end{cases}$$

Solution We can solve this system by substitution if we solve one equation for one of the variables. Solving the first equation for x is not the best choice since doing so introduces a radical. Also, solving for y in the first equation introduces a fraction. We solve the second equation for y.

$$x - y = 1 \quad \text{Second equation}$$
$$x - 1 = y \quad \text{Solve for } y.$$

Replace y with $x - 1$ in the first equation, and then solve for x.

$$x^2 - 3y = 1 \quad \text{First equation}$$

$$x^2 - 3(x - 1) = 1 \quad \text{Replace } y \text{ with } x - 1.$$

$$x^2 - 3x + 3 = 1$$

$$x^2 - 3x + 2 = 0$$

$$(x - 2)(x - 1) = 0$$

$$x = 2 \quad \text{or} \quad x = 1$$

Let $x = 2$ and then let $x = 1$ in the equation $y = x - 1$ to find corresponding y-values.

Let $x = 2$.	Let $x = 1$.
$y = x - 1$	$y = x - 1$
$y = 2 - 1 = 1$	$y = 1 - 1 = 0$

The solutions are $(2, 1)$ and $(1, 0)$ or the solution set is $\{(2, 1), (1, 0)\}$. To check these solutions, graph each equation of the system. Since we are using a graphing utility, we first solve each equation for y.

$$x^2 - 3y = 1 \qquad\qquad\qquad x - y = 1$$

$$x^2 - 1 = 3y$$

$$y = \frac{x^2 - 1}{3} \quad \text{Solve for } y. \qquad\qquad y = x - 1 \quad \text{Solve for } y.$$

The graph of $y_1 = \dfrac{x^2 - 1}{3}$ and $y_2 = x - 1$ is shown below with the points of intersection noted. Since the coordinates of the points of intersection are the same as the ordered pair solutions found above, the solutions check.

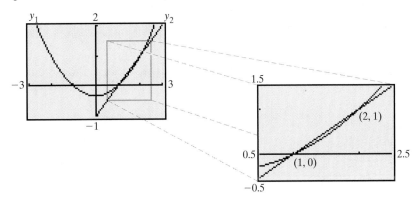

PRACTICE

2 Solve the system $\begin{cases} x^2 - 4y = 4 \\ x + y = -1 \end{cases}$.

EXAMPLE 3 Solve the system

$$\begin{cases} x^2 + y^2 = 4 \\ x + y = 3 \end{cases}$$

Solution We use the substitution method and solve the second equation for x.

$$x + y = 3 \qquad\qquad \text{Second equation}$$

$$x = 3 - y$$

Now we let $x = 3 - y$ in the first equation.

$$x^2 + y^2 = 4 \quad \text{First equation}$$

$$(3 - y)^2 + y^2 = 4 \quad \text{Let } x = 3 - y.$$
$$9 - 6y + y^2 + y^2 = 4$$
$$2y^2 - 6y + 5 = 0$$

By the quadratic formula, where $a = 2$, $b = -6$, and $c = 5$, we have

$$y = \frac{6 \pm \sqrt{(-6)^2 - 4 \cdot 2 \cdot 5}}{2 \cdot 2} = \frac{6 \pm \sqrt{-4}}{4}$$

Since $\sqrt{-4}$ is not a real number, there is no real solution, or $\varnothing$.

To check graphically, notice that $x^2 + y^2 = 4$ solved for y is $y = \pm\sqrt{4 - x^2}$, and $x + y = 3$ solved for y is $y = 3 - x$. Graph $y_1 = \sqrt{4 - x^2}$, $y_2 = -y_1$, and $y_3 = 3 - x$ in a decimal window.

The graph of the circle and the line do not intersect, as expected. The system has no solution.

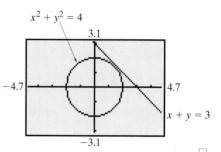

PRACTICE
3 Solve the system $\begin{cases} x^2 + y^2 = 9 \\ x - y = 5 \end{cases}$.

Concept Check ✓

Without solving, how can you tell that $x^2 + y^2 = 9$ and $x^2 + y^2 = 16$ do not have any points of intersection?

OBJECTIVE 2 ▶ Solving nonlinear systems by elimination. Some nonlinear systems may be solved by the elimination method.

EXAMPLE 4 Solve the system

$$\begin{cases} x^2 + 2y^2 = 10 \\ x^2 - y^2 = 1 \end{cases}$$

Solution We will use the elimination, or addition, method to solve this system. To eliminate x^2 when we add the two equations, multiply both sides of the second equation by -1. Then

$$\begin{cases} x^2 + 2y^2 = 10 \\ (-1)(x^2 - y^2) = -1 \cdot 1 \end{cases} \text{ is equivalent to } \begin{cases} x^2 + 2y^2 = 10 \\ \underline{-x^2 + y^2 = -1} \end{cases}$$

$$\begin{aligned} 3y^2 &= 9 \quad \text{Add.} \\ y^2 &= 3 \quad \text{Divide both sides by 3.} \\ y &= \pm\sqrt{3} \end{aligned}$$

To find the corresponding x-values, we let $y = \sqrt{3}$ and $y = -\sqrt{3}$ in either original equation. We choose the second equation.

Let $y = \sqrt{3}$.
$$x^2 - y^2 = 1$$
$$x^2 - (\sqrt{3})^2 = 1$$
$$x^2 - 3 = 1$$
$$x^2 = 4$$
$$x = \pm\sqrt{4} = \pm 2$$

Let $y = -\sqrt{3}$.
$$x^2 - y^2 = 1$$
$$x^2 - (-\sqrt{3})^2 = 1$$
$$x^2 - 3 = 1$$
$$x^2 = 4$$
$$x = \pm\sqrt{4} = \pm 2$$

By graphing the equations in the system first, we can see that there are four intersections.

Answer to Concept Check:

$x^2 + y^2 = 9$ is a circle inside the circle $x^2 + y^2 = 16$, therefore they do not have any points of intersection.

The solutions are $(2, \sqrt{3})$, $(-2, \sqrt{3})$, $(2, -\sqrt{3})$, and $(-2, -\sqrt{3})$. Check all four ordered pairs in both equations of the system. The graph of each equation in this system is shown.

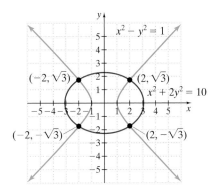

PRACTICE
4 Solve the system $\begin{cases} x^2 + 4y^2 = 16 \\ x^2 - y^2 = 1 \end{cases}$.

10.3 | EXERCISE SET

MyMathLab *Powered by CourseCompass and MathXL*

| MathXL PRACTICE | WATCH | DOWNLOAD | READ | REVIEW |

MIXED PRACTICE

Solve each nonlinear system of equations for real solutions. See Examples 1 through 4.

1. $\begin{cases} x^2 + y^2 = 25 \\ 4x + 3y = 0 \end{cases}$

2. $\begin{cases} x^2 + y^2 = 25 \\ 3x + 4y = 0 \end{cases}$

3. $\begin{cases} x^2 + 4y^2 = 10 \\ y = x \end{cases}$

4. $\begin{cases} 4x^2 + y^2 = 10 \\ y = x \end{cases}$

 5. $\begin{cases} y^2 = 4 - x \\ x - 2y = 4 \end{cases}$

6. $\begin{cases} x^2 + y^2 = 4 \\ x + y = -2 \end{cases}$

7. $\begin{cases} x^2 + y^2 = 9 \\ 16x^2 - 4y^2 = 64 \end{cases}$

8. $\begin{cases} 4x^2 + 3y^2 = 35 \\ 5x^2 + 2y^2 = 42 \end{cases}$

9. $\begin{cases} x^2 + 2y^2 = 2 \\ x - y = 2 \end{cases}$

10. $\begin{cases} x^2 + 2y^2 = 2 \\ x^2 - 2y^2 = 6 \end{cases}$

11. $\begin{cases} y = x^2 - 3 \\ 4x - y = 6 \end{cases}$

12. $\begin{cases} y = x + 1 \\ x^2 - y^2 = 1 \end{cases}$

13. $\begin{cases} y = x^2 \\ 3x + y = 10 \end{cases}$

14. $\begin{cases} 6x - y = 5 \\ xy = 1 \end{cases}$

15. $\begin{cases} y = 2x^2 + 1 \\ x + y = -1 \end{cases}$

16. $\begin{cases} x^2 + y^2 = 9 \\ x + y = 5 \end{cases}$

17. $\begin{cases} y = x^2 - 4 \\ y = x^2 - 4x \end{cases}$

18. $\begin{cases} x = y^2 - 3 \\ x = y^2 - 3y \end{cases}$

 19. $\begin{cases} 2x^2 + 3y^2 = 14 \\ -x^2 + y^2 = 3 \end{cases}$

20. $\begin{cases} 4x^2 - 2y^2 = 2 \\ -x^2 + y^2 = 2 \end{cases}$

21. $\begin{cases} x^2 + y^2 = 1 \\ x^2 + (y + 3)^2 = 4 \end{cases}$

22. $\begin{cases} x^2 + 2y^2 = 4 \\ x^2 - y^2 = 4 \end{cases}$

23. $\begin{cases} y = x^2 + 2 \\ y = -x^2 + 4 \end{cases}$

24. $\begin{cases} x = -y^2 - 3 \\ x = y^2 - 5 \end{cases}$

25. $\begin{cases} 3x^2 + y^2 = 9 \\ 3x^2 - y^2 = 9 \end{cases}$

26. $\begin{cases} x^2 + y^2 = 25 \\ \quad\;\; x = y^2 - 5 \end{cases}$

27. $\begin{cases} x^2 + 3y^2 = 6 \\ x^2 - 3y^2 = 10 \end{cases}$

28. $\begin{cases} x^2 + y^2 = 1 \\ \quad\;\; y = x^2 - 9 \end{cases}$

29. $\begin{cases} x^2 + y^2 = 36 \\ \quad\;\; y = \dfrac{1}{6}x^2 - 6 \end{cases}$

30. $\begin{cases} x^2 + y^2 = 16 \\ \quad\;\; y = -\dfrac{1}{4}x^2 + 4 \end{cases}$

REVIEW AND PREVIEW

Graph each inequality in two variables. See Section 3.6.

31. $x > -3$

32. $y \le 1$

33. $y < 2x - 1$

34. $3x - y \le 4$

Find the perimeter of each geometric figure. See Section 5.3.

△ **35.**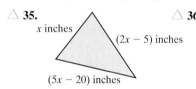
x inches
$(2x - 5)$ inches
$(5x - 20)$ inches

△ **36.**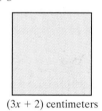
$(3x + 2)$ centimeters

△ **37.** $(x^2 + 3x + 1)$ meters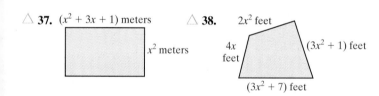
x^2 meters

△ **38.** $2x^2$ feet
$4x$ feet
$(3x^2 + 1)$ feet
$(3x^2 + 7)$ feet

CONCEPT EXTENSIONS

For the exercises below, see the Concept Check in this section.

39. Without graphing, how can you tell that the graphs of $x^2 + y^2 = 1$ and $x^2 + y^2 = 4$ do not have any points of intersection?

40. Without solving, how can you tell that the graphs of $y = 2x + 3$ and $y = 2x + 7$ do not have any points of intersection?

41. How many real solutions are possible for a system of equations whose graphs are a circle and a parabola? Draw diagrams to illustrate each possibility.

42. How many real solutions are possible for a system of equations whose graphs are an ellipse and a line? Draw diagrams to illustrate each possibility.

Solve.

43. The sum of the squares of two numbers is 130. The difference of the squares of the two numbers is 32. Find the two numbers.

44. The sum of the squares of two numbers is 20. Their product is 8. Find the two numbers.

△ **45.** During the development stage of a new rectangular keypad for a security system, it was decided that the area of the rectangle should be 285 square centimeters and the perimeter should be 68 centimeters. Find the dimensions of the keypad.

△ **46.** A rectangular holding pen for cattle is to be designed so that its perimeter is 92 feet and its area is 525 feet. Find the dimensions of the holding pen.

*Recall that in business, a demand function expresses the quantity of a commodity demanded as a function of the commodity's unit price. A supply function expresses the quantity of a commodity supplied as a function of the commodity's unit price. When the quantity produced and supplied is equal to the quantity demanded, then we have what is called **market equilibrium.***

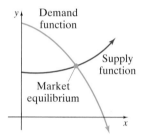

47. The demand function for a certain compact disc is given by the function

$$p = -0.01x^2 - 0.2x + 9$$

and the corresponding supply function is given by

$$p = 0.01x^2 - 0.1x + 3$$

where p is in dollars and x is in thousands of units. Find the equilibrium quantity and the corresponding price by solving the system consisting of the two given equations.

48. The demand function for a certain style of picture frame is given by the function

$$p = -2x^2 + 90$$

and the corresponding supply function is given by

$$p = 9x + 34$$

where p is in dollars and x is in thousands of units. Find the equilibrium quantity and the corresponding price by solving the system consisting of the two given equations.

Use a graphing calculator to verify the results of each exercise.

49. Exercise 3

50. Exercise 4

51. Exercise 23

52. Exercise 24

STUDY SKILLS BUILDER

Are You Prepared for Your Final Exam?

To prepare for your final exam, try the following study techniques:

- Review the material that you will be responsible for on your exam. This includes material from your textbook, your notebook, and any handouts from your instructor.
- Review any formulas that you may need to memorize.
- Check to see if your instructor or mathematics department will be conducting a final exam review.
- Check with your instructor to see whether final exams from previous semesters/quarters are available to students for review.

- Use your previously taken exams as a practice final exam. To do so, rewrite the test questions in mixed order on blank sheets of paper. This will help you prepare for exam conditions.
- If you are unsure of a few concepts, see your instructor or visit a learning lab for assistance. Also, view the video segment of any troublesome sections.
- If you need further exercises to work, try the Cumulative Reviews at the end of the chapters.

Once again, good luck! I hope you have enjoyed this textbook and your mathematics course.

10.4 NONLINEAR INEQUALITIES AND SYSTEMS OF INEQUALITIES

OBJECTIVES

1 Graph a nonlinear inequality.

2 Graph a system of nonlinear inequalities.

OBJECTIVE 1 ▶ Graphing nonlinear inequalities. From the Discover the Concept below, we see that the circle divides the plane into two regions. All points in the interior of the circle correspond to ordered pair solutions of $x^2 + y^2 < 100$. All points in the exterior of the circle correspond to ordered pair solutions of $x^2 + y^2 > 100$. Examples are shown on the screens below. Don't forget that all points of the circle correspond to ordered pair solutions of $x^2 + y^2 = 100$.

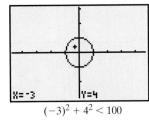

$$(-3)^2 + 4^2 < 100$$

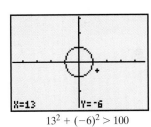

$$13^2 + (-6)^2 > 100$$

DISCOVER THE CONCEPT

Graph the circle defined by $x^2 + y^2 = 100$ in an integer window by graphing $y_1 = \sqrt{100 - x^2}$ and $y_2 = -\sqrt{100 - x^2}$ or $y_2 = -y_1$.

a. Move the cursor to several points in the interior of the circle. For each ordered pair of numbers displayed, compare the value of the expression $x^2 + y^2$ to 100.

b. Move the cursor to several points in the exterior of the circle. For each ordered pair of numbers displayed, compare the value of the expression $x^2 + y^2$ to 100.

c. What region do you think corresponds to ordered pair solutions of $x^2 + y^2 < 100$, and what region do you think corresponds to ordered pair solutions of $x^2 + y^2 > 100$?

In general, we can graph a nonlinear inequality in two variables in a way similar to the way we graphed a linear inequality in two variables in Section 3.6. First, we graph the related equation. The graph of this equation is our boundary. Then, using test points, we determine and shade the region whose points satisfy the inequality.

EXAMPLE 1 Graph $\dfrac{x^2}{9} + \dfrac{y^2}{16} \leq 1$.

Solution First, graph the equation $\dfrac{x^2}{9} + \dfrac{y^2}{16} = 1$. Sketch a solid curve since the graph of

$\dfrac{x^2}{9} + \dfrac{y^2}{16} \leq 1$ includes the graph of $\dfrac{x^2}{9} + \dfrac{y^2}{16} = 1$. The graph is an ellipse, and it divides the plane into two regions, the "inside" and the "outside" of the ellipse. To determine which region contains the solutions, select a test point in either region and determine whether the coordinates of the point satisfy the inequality. We choose $(0, 0)$ as the test point.

$$\dfrac{x^2}{9} + \dfrac{y^2}{16} \leq 1$$

$$\dfrac{0^2}{9} + \dfrac{0^2}{16} \leq 1 \quad \text{Let } x = 0 \text{ and } y = 0.$$

$$0 \leq 1 \quad \text{True}$$

Since this statement is true, the solution set is the region containing $(0, 0)$. The graph of the solution set includes the points on and inside the ellipse, as shaded in the figure.

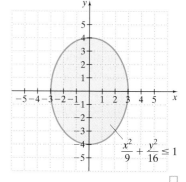

$\dfrac{x^2}{9} + \dfrac{y^2}{16} \leq 1$

PRACTICE
1   Graph $\dfrac{x^2}{36} + \dfrac{y^2}{16} \geq 1$.

EXAMPLE 2   Use a graphing utility to graph $y \geq x^2 - 3$.

Solution   Graph the related equation $y_1 = x^2 - 3$ and shade the region above the boundary curve since the inequality symbol is $\geq$.

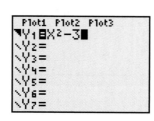

On some graphing calculators, the graph style can be changed in the Y= editor.

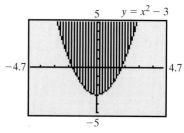

The solution region of the inequality $y \geq x^2 - 3$ is the region consisting of all points on the parabola itself and above the parabola.

PRACTICE
2   Use a graphing utility to graph $y \leq x^2 + 2$

EXAMPLE 3   Graph $4y^2 > x^2 + 16$.

Solution   The related equation is $4y^2 = x^2 + 16$. Subtract x^2 from both sides and divide both sides by 16, and we have $\dfrac{y^2}{4} - \dfrac{x^2}{16} = 1$, which is a hyperbola. Graph the hyperbola as a dashed curve since the graph of $4y^2 > x^2 + 16$ does *not* include the graph of $4y^2 = x^2 + 16$. The hyperbola divides the plane into three regions. Select a test point in each region—not on a boundary line—to determine whether that region contains solutions of the inequality.

***Test Region A* with (0, 4)**	***Test Region B* with (0, 0)**	***Test Region C* with (0, −4)**
$4y^2 > x^2 + 16$	$4y^2 > x^2 + 16$	$4y^2 > x^2 + 16$
$4(4)^2 > 0^2 + 16$	$4(0)^2 > 0^2 + 16$	$4(-4)^2 > 0^2 + 16$
$64 > 16$ True	$0 > 16$ False	$64 > 16$ True

The graph of the solution set includes the shaded regions A and C only, not the boundary.

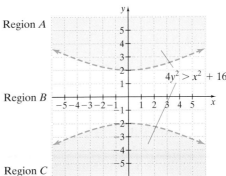

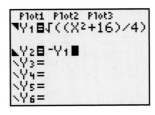

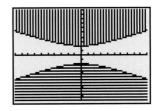

The solution to Example 3 using a graphing calculator.

PRACTICE
3   Graph $16y^2 > 9x^2 + 144$.

OBJECTIVE 2 ▶ Graphing systems of nonlinear inequalities. In Section 3.6 we graphed systems of linear inequalities. Recall that the graph of a system of inequalities is the intersection of the graphs of the inequalities.

EXAMPLE 4   Graph the system
$$\begin{cases} x \le 1 - 2y \\ y \le x^2 \end{cases}$$

Solution   We graph each inequality on the same set of axes. The intersection is shown in the first graph on the following page. It is the darkest shaded (appears purple) region along with its boundary lines. The coordinates of the points of intersection can be found by solving the related system.

$$\begin{cases} x = 1 - 2y \\ y = x^2 \end{cases}$$

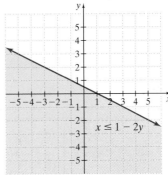

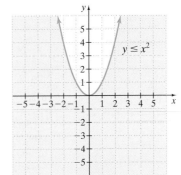

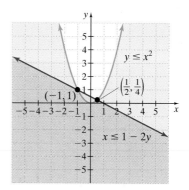

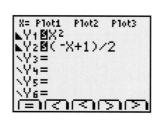 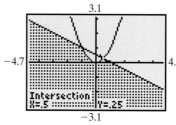 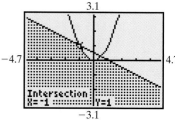

The solution to Example 4 using the Inequalz App. (See the Technology Note on page 310.) □

PRACTICE
4 Graph the system $\begin{cases} y \geq x^2 \\ y \leq -3x + 2 \end{cases}$.

EXAMPLE 5 Graph the system

$$\begin{cases} x^2 + y^2 < 25 \\ \dfrac{x^2}{9} - \dfrac{y^2}{25} < 1 \\ y < x + 3 \end{cases}$$

Solution We graph each inequality. The graph of $x^2 + y^2 < 25$ contains points "inside" the circle that has center $(0, 0)$ and radius 5. The graph of $\dfrac{x^2}{9} - \dfrac{y^2}{25} < 1$ is the region between the two branches of the hyperbola with x-intercepts -3 and 3 and center $(0, 0)$. The graph of $y < x + 3$ is the region "below" the line with slope 1 and y-intercept $(0, 3)$. The graph of the solution set of the system is the intersection of all the graphs, the darkest shaded region shown. The boundary of this region is not part of the solution.

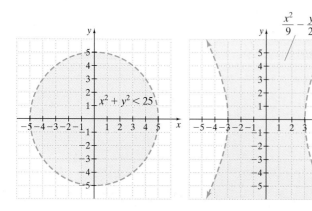 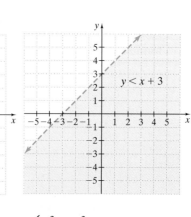 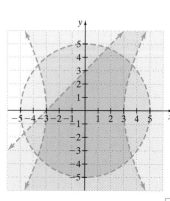

□

PRACTICE
5 Graph the system $\begin{cases} x^2 + y^2 < 16 \\ \dfrac{x^2}{4} - \dfrac{y^2}{9} < 1 \\ y < x + 3 \end{cases}$.

10.4 | EXERCISE SET

Graph each inequality. See Examples 1 through 3.

1. $y < x^2$

2. $y < -x^2$

3. $x^2 + y^2 \geq 16$

4. $x^2 + y^2 < 36$

5. $\dfrac{x^2}{4} - y^2 < 1$

6. $x^2 - \dfrac{y^2}{9} \geq 1$

7. $y > (x - 1)^2 - 3$

8. $y > (x + 3)^2 + 2$

9. $x^2 + y^2 \leq 9$

10. $x^2 + y^2 > 4$

11. $y > -x^2 + 5$

12. $y < -x^2 + 5$

13. $\dfrac{x^2}{4} + \dfrac{y^2}{9} \leq 1$

14. $\dfrac{x^2}{25} + \dfrac{y^2}{4} \geq 1$

15. $\dfrac{y^2}{4} - x^2 \leq 1$

16. $\dfrac{y^2}{16} - \dfrac{x^2}{9} > 1$

17. $y < (x - 2)^2 + 1$

18. $y > (x - 2)^2 + 1$

19. $y \leq x^2 + x - 2$

20. $y > x^2 + x - 2$

Graph each system. See Examples 4 and 5.

21. $\begin{cases} 4x + 3y \geq 12 \\ x^2 + y^2 < 16 \end{cases}$

22. $\begin{cases} 3x - 4y \leq 12 \\ x^2 + y^2 < 16 \end{cases}$

23. $\begin{cases} x^2 + y^2 \leq 9 \\ x^2 + y^2 \geq 1 \end{cases}$

24. $\begin{cases} x^2 + y^2 \geq 9 \\ x^2 + y^2 \geq 16 \end{cases}$

25. $\begin{cases} y > x^2 \\ y \geq 2x + 1 \end{cases}$

26. $\begin{cases} y \leq -x^2 + 3 \\ y \leq 2x - 1 \end{cases}$

27. $\begin{cases} x^2 + y^2 > 9 \\ y > x^2 \end{cases}$

28. $\begin{cases} x^2 + y^2 \leq 9 \\ y < x^2 \end{cases}$

29. $\begin{cases} \dfrac{x^2}{4} + \dfrac{y^2}{9} \geq 1 \\ x^2 + y^2 \geq 4 \end{cases}$

30. $\begin{cases} x^2 + (y - 2)^2 \geq 9 \\ \dfrac{x^2}{4} + \dfrac{y^2}{25} < 1 \end{cases}$

31. $\begin{cases} x^2 - y^2 \geq 1 \\ y \geq 0 \end{cases}$

32. $\begin{cases} x^2 - y^2 \geq 1 \\ x \geq 0 \end{cases}$

33. $\begin{cases} x + y \geq 1 \\ 2x + 3y < 1 \\ x > -3 \end{cases}$

34. $\begin{cases} x - y < -1 \\ 4x - 3y > 0 \\ y > 0 \end{cases}$

35. $\begin{cases} x^2 - y^2 < 1 \\ \dfrac{x^2}{16} + y^2 \leq 1 \\ x \geq -2 \end{cases}$

36. $\begin{cases} x^2 - y^2 \geq 1 \\ \dfrac{x^2}{16} + \dfrac{y^2}{4} \leq 1 \\ y \geq 1 \end{cases}$

REVIEW AND PREVIEW

Determine which graph is the graph of a function. See Section 2.2.

37.

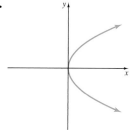

38.

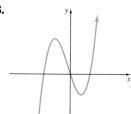

39.

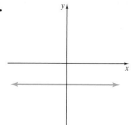

40.

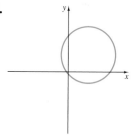

Find each function value if $f(x) = 3x^2 - 2$. See Section 2.2.

41. $f(-1)$

42. $f(-3)$

43. $f(a)$

44. $f(b)$

CONCEPT EXTENSIONS

45. Discuss how graphing a linear inequality such as $x + y < 9$ is similar to graphing a nonlinear inequality such as $x^2 + y^2 < 9$.

46. Discuss how graphing a linear inequality such as $x + y < 9$ is different from graphing a nonlinear inequality such as $x^2 + y^2 < 9$.

47. Graph the system $\begin{cases} y \leq x^2 \\ y \geq x + 2 \\ x \geq 0 \\ y \geq 0 \end{cases}$.

CHAPTER 10 GROUP ACTIVITY

Modeling Conic Sections

In this project, you will have the opportunity to construct and investigate a model of an ellipse. You will need two thumbtacks or nails, graph paper, cardboard, tape, string, a pencil, and a ruler. This project may be completed by working in groups or individually.

Follow these steps, answering any questions as you go.

1. Draw an *x*-axis and a *y*-axis on the graph paper as shown in Figure 1.

2. Place the graph paper on the cardboard and attach it with tape.

3. Locate two points on the *x*-axis, each about $1\frac{1}{2}$ inches from the origin and on opposite sides of the origin (see Figure 1). Insert thumbtacks (or nails) at each of these locations.

4. Fasten a 9-inch piece of string to the thumbtacks as shown in Figure 2. Use your pencil to draw and keep the string taut while you carefully move the pencil in a path all around the thumbtacks.

5. Using the grid of the graph paper as a guide, find an approximate equation of the ellipse you drew.

6. Experiment by moving the tacks closer together or farther apart and drawing new ellipses. What do you observe?

7. Write a paragraph explaining why the figure drawn by the pencil is an ellipse. How might you use the same materials to draw a circle?

8. (Optional) Choose one of the ellipses you drew with the string and pencil. Use a ruler to draw any six tangent lines to the ellipse. (A line is tangent to the ellipse if it intersects, or just touches, the ellipse at only one point. See Figure 3.) Extend the tangent lines to yield six points of intersection among the tangents. Use a straightedge to draw a line connecting each pair of opposite points of intersection. What do you observe? Repeat with a different ellipse. Can you make a conjecture about the relationship among the lines that connect opposite points of intersection?

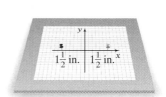

Figure 1

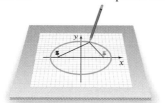

Figure 2

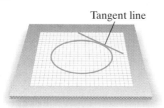

Figure 3

📖 **STUDY SKILLS BUILDER**

Are You Prepared for a Test on Chapter 10?

Below I have listed some common trouble areas for students in Chapter 10. After studying for your test—but before taking your test—read these.

- Don't forget to review all the standard forms for the conic sections.

- Don't forget that both methods, substitution and elimination, are available for solving nonlinear systems of equations.

$$\begin{cases} x^2 + y^2 = 7 \\ 2x^2 - 3y^2 = 4 \end{cases} \text{ is equivalent to }$$

$$\begin{cases} 3x^2 + 3y^2 = 21 \\ \underline{2x^2 - 3y^2 = 4} \\ 5x^2 \qquad\quad = 25 \\ \quad x^2 = 5 \\ \quad x = \pm\sqrt{5} \end{cases}$$

Letting $x = \pm\sqrt{5}$ in either original equation, and $y = \pm\sqrt{2}$, the solution set is $\{(\sqrt{5}, \sqrt{2}), (-\sqrt{5}, \sqrt{2}), (\sqrt{5}, -\sqrt{2}), (-\sqrt{5}, -\sqrt{2})\}$.

Remember: This is simply a checklist of common trouble areas. For a review of Chapter 10, see the Highlights and Chapter Review at the end of this chapter.

CHAPTER 10 VOCABULARY CHECK

Fill in each blank with one of the words or phrases listed below.

circle   radius   center

ellipse   hyperbola   nonlinear system of equations

1. A(n) _____ is the set of all points in a plane that are the same distance from a fixed point, called the _____ .

2. A _____ is a system of equations at least one of which is not linear.

3. A(n) _____ is the set of points on a plane such that the sum of the distances of those points from two fixed points is a constant.

4. In a circle, the distance from the center to a point of the circle is called its _____ .

5. A(n) _____ is the set of points in a plane such that the absolute value of the difference of the distance from two fixed points is constant.

▶ **Helpful Hint**

Are you preparing for your test? Don't forget to take the Chapter 10 Test on page 720. Then check your answers at the back of the text and use the Chapter Test Prep Video CD to see the fully worked-out solutions to any of the exercises you want to review.

CHAPTER 10 HIGHLIGHTS

DEFINITIONS AND CONCEPTS	**EXAMPLES**

SECTION 10.1   THE PARABOLA AND THE CIRCLE

Parabolas

$$y = a(x - h)^2 + k$$

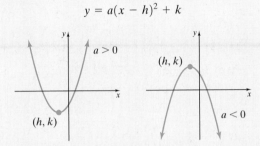

$$x = a(y - k)^2 + h$$

To use a graphing utility to graph an equation, we solve the equation for y.

Graph

$$x = 3y^2 - 12y + 13.$$

$$x - 13 = 3y^2 - 12y$$

$$x - 13 + 3(4) = 3(y^2 - 4y + 4) \quad \text{Add } 3(4) \text{ to both sides.}$$

$$x = 3(y - 2)^2 + 1$$

Since $a = 3$, this parabola opens to the right with vertex $(1, 2)$. Its axis of symmetry is $y = 2$. The x-intercept is $(13, 0)$.

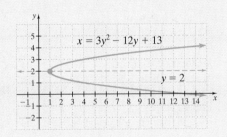

Graph $x = 3y^2 - 12y + 13$ using a graphing utility.

$$0 = 3y^2 - 12y + (13 - x)$$

$$a = 3, \quad b = -12, \quad c = 13 - x$$

Substitute these values in the quadratic formula.

$$y = \frac{12 \pm \sqrt{144 - 4(3)(13 - x)}}{2(3)}$$

Let y_1 be the expression under the radical. Then *deselect* y_1 and graph y_2 and y_3 as shown.

$$y_1 = 144 - 4(3)(13 - x)$$

$$y_2 = \frac{12 + \sqrt{y_1}}{6}, \quad y_3 = \frac{12 - \sqrt{y_1}}{6}$$

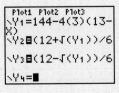

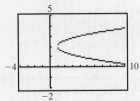

(continued)

DEFINITIONS AND CONCEPTS	**EXAMPLES**

Circle

The graph of $(x - h)^2 + (y - k)^2 = r^2$ is a circle with center (h, k) and radius r.

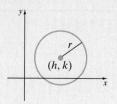

Graph $x^2 + (y + 3)^2 = 5$.

This equation can be written as

$$(x - 0)^2 + (y + 3)^2 = 5 \text{ with } h = 0,$$
$$k = -3, \text{ and } r = \sqrt{5}.$$

The center of this circle is $(0, -3)$, and the radius is $\sqrt{5}$.

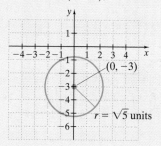

Ellipse with Center **(0, 0)**

The graph of an equation of the form $\dfrac{x^2}{a^2} + \dfrac{y^2}{b^2} = 1$ is an ellipse with center $(0, 0)$. The x-intercepts are $(a, 0)$ and $(-a, 0)$, and the y-intercepts are $(0, b)$ and $(0, -b)$.

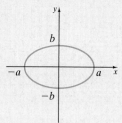

Graph $4x^2 + 9y^2 = 36$.

$$\frac{x^2}{9} + \frac{y^2}{4} = 1 \quad \text{Divide by 36.}$$

$$\frac{x^2}{3^2} + \frac{y^2}{2^2} = 1$$

The ellipse has center $(0, 0)$, x-intercepts $(3, 0)$ and $(-3, 0)$, and y-intercepts $(0, 2)$ and $(0, -2)$.

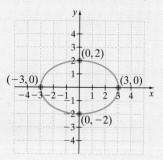

Hyperbola with Center **(0, 0)**

The graph of an equation of the form $\dfrac{x^2}{a^2} - \dfrac{y^2}{b^2} = 1$ is a hyperbola with center $(0, 0)$ and x-intercepts $(a, 0)$ and $(-a, 0)$.

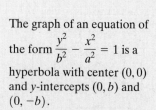

Graph $\dfrac{x^2}{9} - \dfrac{y^2}{4} = 1$. Here, $a = 3$ and $b = 2$.

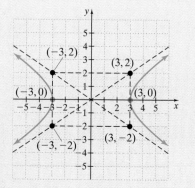

The graph of an equation of the form $\dfrac{y^2}{b^2} - \dfrac{x^2}{a^2} = 1$ is a hyperbola with center $(0, 0)$ and y-intercepts $(0, b)$ and $(0, -b)$.

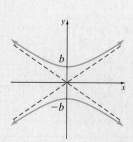

DEFINITIONS AND CONCEPTS	EXAMPLES

SECTION 10.3 SOLVING NONLINEAR SYSTEMS OF EQUATIONS

A **nonlinear system of equations** is a system of equations at least one of which is not linear. Both the substitution method and the elimination method may be used to solve a nonlinear system of equations.

Solve the nonlinear system $\begin{cases} y = x + 2 \\ 2x^2 + y^2 = 3 \end{cases}$.

Substitute $x + 2$ for y in the second equation.

$$2x^2 + y^2 = 3$$
$$2x^2 + (x + 2)^2 = 3$$
$$2x^2 + x^2 + 4x + 4 = 3$$
$$3x^2 + 4x + 1 = 0$$
$$(3x + 1)(x + 1) = 0$$
$$x = -\frac{1}{3}, x = -1$$

If $x = -\frac{1}{3}$, $y = x + 2 = -\frac{1}{3} + 2 = \frac{5}{3}$.
If $x = -1$, $y = x + 2 = -1 + 2 = 1$.

The solutions are $\left(-\frac{1}{3}, \frac{5}{3}\right)$ and $(-1, 1)$.

SECTION 10.4 NONLINEAR INEQUALITIES AND SYSTEMS OF INEQUALITIES

The graph of a system of inequalities is the intersection of the graphs of the inequalities.

Graph the system $\begin{cases} x \geq y^2 \\ x + y \leq 4 \end{cases}$.

The graph of the system is the pink shaded region along with its boundary lines.

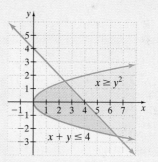

CHAPTER 10 REVIEW

(10.1) *Write an equation of the circle with the given center and radius.*

1. center $(-4, 4)$; radius 3

2. center $(5, 0)$; radius 5

3. center $(-7, -9)$; radius $\sqrt{11}$

4. center $(0, 0)$; radius $\frac{7}{2}$

Sketch the graph of the equation. If the graph is a circle, find its center. If the graph is a parabola, find its vertex.

5. $x^2 + y^2 = 7$

6. $x = 2(y - 5)^2 + 4$

7. $x = -(y + 2)^2 + 3$

8. $(x - 1)^2 + (y - 2)^2 = 4$

9. $y = -x^2 + 4x + 10$

10. $x = -y^2 - 4y + 6$

11. $x = \frac{1}{2}y^2 + 2y + 1$

12. $y = -3x^2 + \frac{1}{2}x + 4$

13. $x^2 + y^2 + 2x + y = \dfrac{3}{4}$ **14.** $x^2 + y^2 - 3y = \dfrac{7}{4}$

15. $4x^2 + 4y^2 + 16x + 8y = 1$

(10.2) Sketch the graph of each equation.

16. $x^2 + \dfrac{y^2}{4} = 1$ **17.** $x^2 - \dfrac{y^2}{4} = 1$

18. $\dfrac{x^2}{5} + \dfrac{y^2}{5} = 1$ **19.** $\dfrac{x^2}{5} - \dfrac{y^2}{5} = 1$

20. $-5x^2 + 25y^2 = 125$ **21.** $4y^2 + 9x^2 = 36$

22. $x^2 - y^2 = 1$ **23.** $\dfrac{(x+3)^2}{9} + \dfrac{(y-4)^2}{25} = 1$

24. $y^2 = x^2 + 9$ **25.** $x^2 = 4y^2 - 16$

26. $100 - 25x^2 = 4y^2$

(10.3) Solve each system of equations.

27. $\begin{cases} y = 2x - 4 \\ y^2 = 4x \end{cases}$

28. $\begin{cases} x^2 + y^2 = 4 \\ x - y = 4 \end{cases}$

29. $\begin{cases} y = x + 2 \\ y = x^2 \end{cases}$

30. $\begin{cases} x^2 + 4y^2 = 16 \\ x^2 + y^2 = 4 \end{cases}$

31. $\begin{cases} 4x - y^2 = 0 \\ 2x^2 + y^2 = 16 \end{cases}$

32. $\begin{cases} x^2 + 2y = 9 \\ 5x - 2y = 5 \end{cases}$

33. $\begin{cases} y = 3x^2 + 5x - 4 \\ y = 3x^2 - x + 2 \end{cases}$

34. $\begin{cases} x^2 - 3y^2 = 1 \\ 4x^2 + 5y^2 = 21 \end{cases}$

△ **35.** Find the length and the width of a room whose area is 150 square feet and whose perimeter is 50 feet.

36. What is the greatest number of real solutions possible for a system of two equations whose graphs are an ellipse and a hyperbola?

(10.4) Graph the inequality or system of inequalities.

37. $y \le -x^2 + 3$ **38.** $x^2 + y^2 < 9$

39. $\begin{cases} 2x \le 4 \\ x + y \ge 1 \end{cases}$ **40.** $\dfrac{x^2}{4} + \dfrac{y^2}{9} \ge 1$

41. $\begin{cases} x^2 + y^2 < 4 \\ x^2 - y^2 \le 1 \end{cases}$ **42.** $\begin{cases} x^2 + y^2 \le 16 \\ x^2 + y^2 \ge 4 \end{cases}$

MIXED REVIEW

43. Write an equation of the circle with center $(-7, 8)$ and radius 5.

Graph each equation.

44. $3x^2 + 6x + 3y^2 = 9$ **45.** $y = x^2 + 6x + 9$

46. $x = y^2 + 6y + 9$ **47.** $\dfrac{y^2}{4} - \dfrac{x^2}{16} = 1$

48. $\dfrac{y^2}{4} + \dfrac{x^2}{16} = 1$ **49.** $\dfrac{(x-2)^2}{4} + (y-1)^2 = 1$

50. $y^2 = x^2 + 6$ **51.** $y^2 + x^2 = 4x + 6$

52. $x^2 + y^2 - 8y = 0$ **53.** $6(x-2)^2 + 9(y+5)^2 = 36$

54. $\dfrac{x^2}{16} - \dfrac{y^2}{25} = 1$

Solve each system of equations.

55. $\begin{cases} y = x^2 - 5x + 1 \\ y = -x + 6 \end{cases}$

56. $\begin{cases} x^2 + y^2 = 10 \\ 9x^2 + y^2 = 18 \end{cases}$

Graph each inequality or system of inequalities.

57. $x^2 - y^2 < 1$ **58.** $\begin{cases} y > x^2 \\ x + y \ge 3 \end{cases}$

CHAPTER 10 TEST

Remember to use the Chapter Test Prep Video CD to see the fully worked-out solutions to any of the exercises you want to review.

Sketch the graph of each equation.

1. $x^2 + y^2 = 36$ **2.** $x^2 - y^2 = 36$

3. $16x^2 + 9y^2 = 144$ **4.** $y = x^2 - 8x + 16$

5. $x^2 + y^2 + 6x = 16$ **6.** $x = y^2 + 8y - 3$

7. $\dfrac{(x-4)^2}{16} + \dfrac{(y-3)^2}{9} = 1$ **8.** $y^2 - x^2 = 1$

Solve each system.

9. $\begin{cases} x^2 + y^2 = 26 \\ x^2 - 2y^2 = 23 \end{cases}$ **10.** $\begin{cases} y = x^2 - 5x + 6 \\ y = 2x \end{cases}$

Graph the solution of each system.

11. $\begin{cases} 2x + 5y \geq 10 \\ \quad\quad y \geq x^2 + 1 \end{cases}$

12. $\begin{cases} \dfrac{x^2}{4} + y^2 \leq 1 \\ \quad x + y > 1 \end{cases}$

13. $\begin{cases} x^2 + y^2 \geq 4 \\ x^2 + y^2 < 16 \\ \quad\quad y \geq 0 \end{cases}$

14. A bridge has an arch in the shape of a half-ellipse. If the equation of the ellipse, measured in feet, is $100x^2 + 225y^2 = 22,500$, find the height of the arch from the road and the width of the arch.

CHAPTER 10 CUMULATIVE REVIEW

1. Use the associative property of multiplication to write an expression equivalent to $4 \cdot (9y)$. Then simplify the equivalent expression.

2. Solve $3x + 4 > 1$ and $2x - 5 \leq 9$. Write the solution in interval notation.

3. Graph $x = -2y$ by plotting intercepts.

4. Find the slope of the line that goes through $(3, 2)$ and $(1, -4)$.

5. Use the elimination method to solve the system:

$$\begin{cases} 3x + \dfrac{y}{2} = 2 \\ 6x + y = 5 \end{cases}$$

6. Two planes leave Greensboro, one traveling north and the other south. After 2 hours they are 650 miles apart. If one plane is flying 25 mph faster than the other, what is the speed of each?

7. Use the power rules to simplify the following. Use positive exponents to write all results.

a. $(5x^2)^3$

b. $\left(\dfrac{2}{3}\right)^3$

c. $\left(\dfrac{3p^4}{q^5}\right)^2$

d. $\left(\dfrac{2^{-3}}{y}\right)^{-2}$

e. $(x^{-5}y^2z^{-1})^7$

8. Use the quotient rule to simplify.

a. $\dfrac{4^8}{4^3}$

b. $\dfrac{y^{11}}{y^5}$

c. $\dfrac{32x^7}{4x^6}$

d. $\dfrac{18a^{12}b^6}{12a^8b^6}$

9. Solve $2x^2 = \dfrac{17}{3}x + 1$.

10. Factor.

a. $3y^2 + 14y + 15$

b. $20a^5 + 54a^4 + 10a^3$

c. $(y - 3)^2 - 2(y - 3) - 8$

11. Perform each indicated operation. $\dfrac{7}{x - 1} + \dfrac{10x}{x^2 - 1} - \dfrac{5}{x + 1}$

12. Perform the indicated operation and simplify if possible.

$$\dfrac{2}{3a - 15} - \dfrac{a}{25 - a^2}$$

13. Simplify each complex fraction.

a. $\dfrac{\dfrac{2x}{27y^2}}{\dfrac{6x^2}{9}}$

b. $\dfrac{\dfrac{5x}{x + 2}}{\dfrac{10}{x - 2}}$

c. $\dfrac{\dfrac{x}{y^2} + \dfrac{1}{y}}{\dfrac{y}{x^2} + \dfrac{1}{x}}$

14. Simplify each complex fraction.

a. $(a^{-1} - b^{-1})^{-1}$

b. $\dfrac{2 - \dfrac{1}{x}}{4x - \dfrac{1}{x}}$

15. Divide $2x^2 - x - 10$ by $x + 2$.

16. Solve $\dfrac{2}{x + 3} = \dfrac{1}{x^2 - 9} - \dfrac{1}{x - 3}$.

17. Use the remainder theorem and synthetic division to find $P(4)$ if

$$P(x) = 4x^6 - 25x^5 + 35x^4 + 17x^2.$$

18. Suppose that y varies inversely as x. If $y = 3$ when $x = \dfrac{2}{3}$, find the constant of variation and the direct variation equation.

19. Solve: $\dfrac{2x}{x - 3} + \dfrac{6 - 2x}{x^2 - 9} = \dfrac{x}{x + 3}$.

20. Simplify the following expressions. Assume that all variables represent nonnegative real numbers.

a. $\sqrt[5]{-32}$

b. $\sqrt[4]{625}$

c. $-\sqrt{36}$

d. $-\sqrt[3]{-27x^3}$

e. $\sqrt{144y^2}$

21. Melissa Scarlatti can clean the house in 4 hours, whereas her husband, Zack, can do the same job in 5 hours. They have agreed to clean together so that they can finish in time to watch a movie on TV that starts in 2 hours. How long will it take them to clean the house together? Can they finish before the movie starts?

22. Use the quotient rule to simplify.

a. $\dfrac{\sqrt{32}}{\sqrt{4}}$

b. $\dfrac{\sqrt[3]{240y^2}}{5\sqrt[3]{3y^{-4}}}$

c. $\dfrac{\sqrt[5]{64x^9y^2}}{\sqrt[5]{2x^2y^{-8}}}$

23. Find the cube roots.

a. $\sqrt[3]{1}$

b. $\sqrt[3]{-64}$

c. $\sqrt[3]{\dfrac{8}{125}}$

d. $\sqrt[3]{x^6}$

e. $\sqrt[3]{-27x^9}$

24. Multiply and simplify if possible.

a. $\sqrt{5}\left(2 + \sqrt{15}\right)$

b. $\left(\sqrt{3} - \sqrt{5}\right)\left(\sqrt{7} - 1\right)$

c. $\left(2\sqrt{5} - 1\right)^2$

d. $\left(3\sqrt{2} + 5\right)\left(3\sqrt{2} - 5\right)$

25. Multiply.

a. $z^{2/3}\left(z^{1/3} - z^5\right)$

b. $\left(x^{1/3} - 5\right)\left(x^{1/3} + 2\right)$

26. Rationalize the denominator. $\dfrac{-2}{\sqrt{3} + 3}$

27. Use the quotient rule to divide, and simplify if possible.

a. $\dfrac{\sqrt{20}}{\sqrt{5}}$

b. $\dfrac{\sqrt{50x}}{2\sqrt{2}}$

c. $\dfrac{7\sqrt[3]{48x^4y^8}}{\sqrt[3]{6y^2}}$

d. $\dfrac{2\sqrt[4]{32a^8b^6}}{\sqrt[4]{a^{-1}b^2}}$

28. Solve: $\sqrt{2x - 3} = x - 3$.

29. Add or subtract as indicated.

a. $\dfrac{\sqrt{45}}{4} - \dfrac{\sqrt{5}}{3}$

b. $\sqrt[3]{\dfrac{7x}{8}} + 2\sqrt[3]{7x}$

30. Use the discriminant to determine the number and type of solutions for $9x^2 - 6x = -4$.

31. Rationalize the denominator of $\sqrt{\dfrac{7x}{3y}}$.

32. Solve: $\dfrac{4}{x - 2} - \dfrac{x}{x + 2} = \dfrac{16}{x^2 - 4}$.

33. Solve: $\sqrt{2x - 3} = 9$.

34. Solve: $x^3 + 2x^2 - 4x \geq 8$.

35. Find the following powers of i.

a. i^7

b. i^{20}

c. i^{46}

d. i^{-12}

36. Graph $f(x) = (x + 2)^2 - 1$.

37. Solve $p^2 + 2p = 4$ by completing the square.

38. Find the maximum value of $f(x) = -x^2 - 6x + 4$.

39. Solve: $\dfrac{1}{4}m^2 - m + \dfrac{1}{2} = 0$.

40. Find the inverse of $f(x) = \dfrac{x + 1}{2}$.

41. Solve: $p^4 - 3p^2 - 4 = 0$.

42. If $f(x) = x^2 - 3x + 2$ and $g(x) = -3x + 5$, find

a. $(f \circ g)(x)$

b. $(f \circ g)(-2)$

c. $(g \circ f)(x)$

d. $(g \circ f)(5)$

43. Solve: $\dfrac{x + 2}{x - 3} \leq 0$.

44. Graph $4x^2 + 9y^2 = 36$.

45. Graph $g(x) = \dfrac{1}{2}(x + 2)^2 + 5$. Find the vertex and the axis of symmetry.

46. Solve each equation for x.

a. $64^x = 4$

b. $125^{x-3} = 25$

c. $\dfrac{1}{81} = 3^{2x}$

47. Find the vertex of the graph of $f(x) = x^2 - 4x - 12$.

48. Graph the system: $\begin{cases} x + 2y < 8 \\ \quad\ y \geq x^2 \end{cases}$

49. Find the distance between $(2, -5)$ and $(1, -4)$. Give an exact distance and a three-decimal-place approximation.

50. Solve the system $\begin{cases} x^2 + y^2 = 36 \\ \quad\ y = x + 6 \end{cases}$

CHAPTER
11
Sequences, Series, and the Binomial Theorem

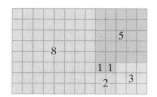

A tiling with squares whose sides are successive Fibonacci numbers in length

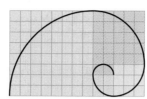

A Fibonacci spiral, created by drawing arcs connecting the opposite corners of squares in the Fibonacci tiling

The Fibonacci Sequence is a special sequence in which the first two terms are 1 and each term thereafter is the sum of the two previous terms:

$$1, 1, 2, 3, 5, 8, 13, 21, \ldots$$

The Fibonacci numbers are named after Leonardo of Pisa, known as Fibonacci, although there is some evidence that these numbers had been described earlier in India.

There are numerous interesting facts about this sequence, and some are shown on the diagrams on this page. In Section 11.1, Exercise 46, page 727, you will have the opportunity to check a formula for this sequence.

Having explored in some depth the concept of function, we turn now in this final chapter to *sequences*. In one sense, a sequence is simply an ordered list of numbers. In another sense, a sequence is itself a function. Phenomena modeled by such functions are everywhere around us. The starting place for all mathematics is the sequence of natural numbers: 1, 2, 3, 4, and so on.

Sequences lead us to *series,* which are a sum of ordered numbers. Through series we gain new insight, for example about the expansion of a binomial $(a + b)^n$, the concluding topic of this book.

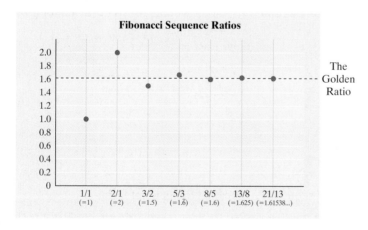

Fibonacci Sequence Ratios

The Golden Ratio

| 1/1 | 2/1 | 3/2 | 5/3 | 8/5 | 13/8 | 21/13 |
| (=1) | (=2) | (=1.5) | (=1.$\overline{6}$) | (=1.6) | (=1.625) | (=1.61538...) |

The Fibonacci sequence created by the formula

$$u_n = \frac{1}{\sqrt{5}}\left[\left(\frac{1 + \sqrt{5}}{2}\right)^n - \left(\frac{1 - \sqrt{5}}{2}\right)^n\right]$$

The ratio of successive numbers in the Fibonacci Sequence approaches a number called the golden ratio or golden number, which is approximately 1.618034.

11.1 SEQUENCES

OBJECTIVES

1 Write the terms of a sequence given its general term.

2 Find the general term of a sequence.

3 Solve applications that involve sequences.

Suppose that a town's present population of 100,000 is growing by 5% each year. After the first year, the town's population will be

$$100{,}000 + 0.05(100{,}000) = 105{,}000$$

After the second year, the town's population will be

$$105{,}000 + 0.05(105{,}000) = 110{,}250$$

After the third year, the town's population will be

$$110{,}250 + 0.05(110{,}250) \approx 115{,}763$$

If we continue to calculate, the town's yearly population can be written as the **infinite sequence** of numbers

$$105{,}000, 110{,}250, 115{,}763, \ldots$$

If we decide to stop calculating after a certain year (say, the fourth year), we obtain the **finite sequence**

$$105{,}000, \quad 110{,}250, \quad 115{,}763, \quad 121{,}551$$

TECHNOLOGY NOTE

Some graphing calculators have a designated sequence mode such as the one below. This mode allows the user to list terms of a sequence, investigate tables, and graph sequences in terms of n, where n is a natural number. Consult your owner's manual for specific instructions on how to use this feature.

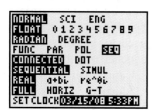

Sequences

An infinite sequence is a function whose domain is the set of natural numbers $\{1, 2, 3, 4, \ldots\}$.

A finite sequence is a function whose domain is the set of natural numbers $\{1, 2, 3, 4, \ldots, n\}$, where n is some natural number.

OBJECTIVE 1 ▶ Writing the terms of a sequence. Given the sequence $2, 4, 8, 16, \ldots$, we say that each number is a **term** of the sequence. Because a sequence is a function, we could describe it by writing $f(n) = 2^n$, where n is a natural number. Instead, we use the notation

$$a_n = 2^n$$

Some function values are

$$
\begin{aligned}
a_1 &= 2^1 = 2 && \text{First term of the sequence} \\
a_2 &= 2^2 = 4 && \text{Second term} \\
a_3 &= 2^3 = 8 && \text{Third term} \\
a_4 &= 2^4 = 16 && \text{Fourth term} \\
a_{10} &= 2^{10} = 1024 && \text{Tenth term}
\end{aligned}
$$

The nth term of the sequence a_n is called the **general term.**

▶ **Helpful Hint**

If it helps, think of a sequence as simply a list of values in which a position is assigned. For the sequence directly above,

Value: 2,   4,   8,   16, ..., 1024
         ↑   ↑   ↑   ↑   ↑
Position 1st 2nd 3rd 4th   10th

EXAMPLE 1 Write the first five terms of the sequence whose general term is given by

$$a_n = n^2 - 1$$

Solution Evaluate a_n, where n is $1, 2, 3, 4,$ and 5.

$$
\begin{aligned}
a_n &= n^2 - 1 \\
a_1 &= 1^2 - 1 = 0 && \text{Replace } n \text{ with 1.}
\end{aligned}
$$

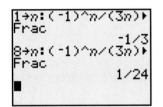

Here we use a sequence mode and list the terms of the sequence in a table.

$$a_2 = 2^2 - 1 = 3 \quad \text{Replace } n \text{ with 2.}$$
$$a_3 = 3^2 - 1 = 8 \quad \text{Replace } n \text{ with 3.}$$
$$a_4 = 4^2 - 1 = 15 \quad \text{Replace } n \text{ with 4.}$$
$$a_5 = 5^2 - 1 = 24 \quad \text{Replace } n \text{ with 5.}$$

Thus, the first five terms of the sequence $a_n = n^2 - 1$ are 0, 3, 8, 15, and 24. □

PRACTICE
1 Write the first five terms of the sequence whose general term is given by $a_n = 5 + n^2$.

EXAMPLE 2 If the general term of a sequence is given by $a_n = \dfrac{(-1)^n}{3n}$, find

a. the first term of the sequence

b. a_8

c. the one-hundredth term of the sequence

d. a_{15}

Solution

a. $a_1 = \dfrac{(-1)^1}{3(1)} = -\dfrac{1}{3}$ Replace n with 1.

b. $a_8 = \dfrac{(-1)^8}{3(8)} = \dfrac{1}{24}$ Replace n with 8.

c. $a_{100} = \dfrac{(-1)^{100}}{3(100)} = \dfrac{1}{300}$ Replace n with 100.

d. $a_{15} = \dfrac{(-1)^{15}}{3(15)} = -\dfrac{1}{45}$ Replace n with 15. □

Single terms of a sequence may be found by evaluating the general term expression at given values.

PRACTICE
2 If the general term of a sequence is given by $a_n = \dfrac{(-1)^n}{5n}$, find

a. the first term of the sequence

b. a_4

c. the thirtieth term of the sequence

d. a_{19}

OBJECTIVE 2 ▶ Finding the general term of a sequence. Suppose we know the first few terms of a sequence and want to find a general term that fits the pattern of the first few terms.

EXAMPLE 3 Find a general term a_n of the sequence whose first few terms are given.

a. $1, 4, 9, 16, \ldots$

b. $\dfrac{1}{1}, \dfrac{1}{2}, \dfrac{1}{3}, \dfrac{1}{4}, \dfrac{1}{5}, \ldots$

c. $-3, -6, -9, -12, \ldots$

d. $\dfrac{1}{2}, \dfrac{1}{4}, \dfrac{1}{8}, \dfrac{1}{16}, \ldots$

Solution

a. These numbers are the squares of the first four natural numbers, so a general term might be $a_n = n^2$.

b. These numbers are the reciprocals of the first five natural numbers, so a general term might be $a_n = \dfrac{1}{n}$.

c. These numbers are the product of -3 and the first four natural numbers, so a general term might be $a_n = -3n$.

— TECHNOLOGY NOTE —

Another command that may be available on your calculator for listing the terms of a sequence is the sequence command. An advantage of this command is that it allows the terms to be written as fractions.

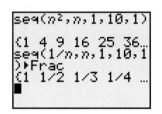

d. Notice that the denominators double each time.

$$\frac{1}{2}, \quad \frac{1}{2 \cdot 2}, \quad \frac{1}{2(2 \cdot 2)}, \quad \frac{1}{2(2 \cdot 2 \cdot 2)}$$

or

$$\frac{1}{2^1}, \quad \frac{1}{2^2}, \quad \frac{1}{2^3}, \quad \frac{1}{2^4}$$

We might then suppose that the general term is $a_n = \dfrac{1}{2^n}$. ☐

PRACTICE
3   Find the general term a_n of the sequence whose first few terms are given.

a. $1, 3, 5, 7, \ldots$

b. $3, 9, 27, 81, \ldots$

c. $\dfrac{1}{2}, \dfrac{2}{3}, \dfrac{3}{4}, \dfrac{4}{5}, \ldots$

d. $-\dfrac{1}{2}, -\dfrac{1}{3}, -\dfrac{1}{4}, -\dfrac{1}{5}, \ldots$

OBJECTIVE 3 ▶ Solving applications modeled by sequences. Sequences model many phenomena of the physical world, as illustrated by the following example.

EXAMPLE 4   **Finding a Puppy's Weight Gain**

The amount of weight, in pounds, a puppy gains in each month of its first year is modeled by a sequence whose general term is $a_n = n + 4$, where n is the number of the month. Write the first five terms of the sequence, and find how much weight the puppy should gain in its fifth month.

Solution   Evaluate $a_n = n + 4$ when n is $1, 2, 3, 4,$ and 5.

$$a_1 = 1 + 4 = 5$$
$$a_2 = 2 + 4 = 6$$
$$a_3 = 3 + 4 = 7$$
$$a_4 = 4 + 4 = 8$$
$$a_5 = 5 + 4 = 9$$

The puppy should gain 9 pounds in its fifth month. ☐

PRACTICE
4   The value v, in dollars, of an office copier depreciates according to the sequence $v_n = 3950(0.8)^n$, where n is the time in years. Find the value of the copier after three years.

VOCABULARY & READINESS CHECK

Use the choices below to fill in each blank.

     infinite   finite   general

1. The nth term of the sequence a_n is called the _____ term.

2. A(n) _____ sequence is a function whose domain is $\{1, 2, 3, 4, \ldots, n\}$, where n is some natural number.

3. A(n) _____ sequence is a function whose domain is $\{1, 2, 3, 4, \ldots\}$.

Write the first term of each sequence.

4. $a_n = 7^n; a_1 = $ _____.

5. $a_n = \dfrac{(-1)^n}{n}; a_1 = $ _____.

6. $a_n = (-1)^n \cdot n^4; a_1 = $ _____.

11.1 EXERCISE SET

Write the first five terms of each sequence whose general term is given. See Example 1.

1. $a_n = n + 4$

2. $a_n = 5 - n$

3. $a_n = (-1)^n$

4. $a_n = (-2)^n$

5. $a_n = \dfrac{1}{n + 3}$

6. $a_n = \dfrac{1}{7 - n}$

7. $a_n = 2n$

8. $a_n = -6n$

9. $a_n = -n^2$

10. $a_n = n^2 + 2$

11. $a_n = 2^n$

12. $a_n = 3^{n-2}$

13. $a_n = 2n + 5$

14. $a_n = 1 - 3n$

15. $a_n = (-1)^n n^2$

16. $a_n = (-1)^{n+1}(n - 1)$

Find the indicated term for each sequence whose general term is given. See Example 2.

17. $a_n = 3n^2; a_5$

18. $a_n = -n^2; a_{15}$

19. $a_n = 6n - 2; a_{20}$

20. $a_n = 100 - 7n; a_{50}$

21. $a_n = \dfrac{n + 3}{n}; a_{15}$

22. $a_n = \dfrac{n}{n + 4}; a_{24}$

23. $a_n = (-3)^n; a_6$

24. $a_n = 5^{n+1}; a_3$

25. $a_n = \dfrac{n - 2}{n + 1}; a_6$

26. $a_n = \dfrac{n + 3}{n + 4}; a_8$

27. $a_n = \dfrac{(-1)^n}{n}; a_8$

28. $a_n = \dfrac{(-1)^n}{2n}; a_{100}$

29. $a_n = -n^2 + 5; a_{10}$

30. $a_n = 8 - n^2; a_{20}$

31. $a_n = \dfrac{(-1)^n}{n + 6}; a_{19}$

32. $a_n = \dfrac{n - 4}{(-2)^n}; a_6$

Find a general term a_n for each sequence whose first four terms are given. See Example 3.

33. $3, 7, 11, 15$

34. $2, 7, 12, 17$

35. $-2, -4, -8, -16$

36. $-4, 16, -64, 256$

37. $\dfrac{1}{3}, \dfrac{1}{9}, \dfrac{1}{27}, \dfrac{1}{81}$

38. $\dfrac{2}{5}, \dfrac{2}{25}, \dfrac{2}{125}, \dfrac{2}{625}$

Solve. See Example 4.

39. The distance, in feet, that a Thermos dropped from a cliff falls in each consecutive second is modeled by a sequence whose general term is $a_n = 32n - 16$, where n is the number of seconds. Find the distance the Thermos falls in the second, third, and fourth seconds.

40. The population size of a culture of bacteria triples every hour such that its size is modeled by the sequence $a_n = 50(3)^{n-1}$, where n is the number of the hour just beginning. Find the size of the culture at the beginning of the fourth hour and the size of the culture at the beginning of the first hour.

41. Mrs. Laser agrees to give her son Mark an allowance of $0.10 on the first day of his 14-day vacation, $0.20 on the second day, $0.40 on the third day, and so on. Write an equation of a sequence whose terms correspond to Mark's allowance. Find the allowance Mark will receive on the last day of his vacation.

42. A small theater has 10 rows with 12 seats in the first row, 15 seats in the second row, 18 seats in the third row, and so on. Write an equation of a sequence whose terms correspond to the seats in each row. Find the number of seats in the eighth row.

43. The number of cases of a new infectious disease is doubling every year such that the number of cases is modeled by a sequence whose general term is $a_n = 75(2)^{n-1}$, where n is the number of the year just beginning. Find how many cases there will be at the beginning of the sixth year. Find how many cases there were at the beginning of the first year.

44. A new college had an initial enrollment of 2700 students in 2000, and each year the enrollment increases by 150 students. Find the enrollment for each of 5 years, beginning with 2000.

45. An endangered species of sparrow had an estimated population of 800 in 2000, and scientists predict that its population will decrease by half each year. Estimate the population in 2004. Estimate the year the sparrow will be extinct.

46. A **Fibonacci sequence** is a special type of sequence in which the first two terms are 1, and each term thereafter is the sum of the two previous terms: 1, 1, 2, 3, 5, 8, etc. The formula for the nth Fibonacci term is $a_n = \dfrac{1}{\sqrt{5}}\left[\left(\dfrac{1 + \sqrt{5}}{2}\right)^n - \left(\dfrac{1 - \sqrt{5}}{2}\right)^n\right]$.
Verify that the first two terms of the Fibonacci sequence are each 1.

REVIEW AND PREVIEW

Sketch the graph of each quadratic function. See Section 8.5.

47. $f(x) = (x - 1)^2 + 3$

48. $f(x) = (x - 2)^2 + 1$

49. $f(x) = 2(x + 4)^2 + 2$

50. $f(x) = 3(x - 3)^2 + 4$

Find the distance between each pair of points. See Section 7.3.

51. $(-4, -1)$ and $(-7, -3)$

52. $(-2, -1)$ and $(-1, 5)$

53. $(2, -7)$ and $(-3, -3)$

54. $(10, -14)$ and $(5, -11)$

CONCEPT EXTENSIONS

Use a graphing utility to find the first five terms of each sequence. Round each term after the first to four decimal places.

55. $a_n = \dfrac{1}{\sqrt{n}}$

56. $\dfrac{\sqrt{n}}{\sqrt{n} + 1}$

57. $a_n = \left(1 + \dfrac{1}{n}\right)^n$

58. $a_n = \left(1 + \dfrac{0.05}{n}\right)^n$

11.2 ARITHMETIC AND GEOMETRIC SEQUENCES

OBJECTIVES

1 Identify arithmetic sequences and their common differences.

2 Identify geometric sequences and their common ratios.

OBJECTIVE 1 ▶ Identifying arithmetic sequences. Find the first four terms of the sequence whose general term is $a_n = 5 + (n - 1)3$.

$$a_1 = 5 + (1 - 1)3 = 5 \qquad \text{Replace } n \text{ with 1.}$$
$$a_2 = 5 + (2 - 1)3 = 8 \qquad \text{Replace } n \text{ with 2.}$$
$$a_3 = 5 + (3 - 1)3 = 11 \qquad \text{Replace } n \text{ with 3.}$$
$$a_4 = 5 + (4 - 1)3 = 14 \qquad \text{Replace } n \text{ with 4.}$$

The first four terms are $5, 8, 11$, and 14. Notice that the difference of any two successive terms is 3.

$$8 - 5 = 3$$
$$11 - 8 = 3$$
$$14 - 11 = 3$$
$$\vdots$$
$$a_n - a_{n-1} = 3$$

$\uparrow$ $\uparrow$
nth   previous
term   term

Because the difference of any two successive terms is a constant, we call the sequence an **arithmetic sequence,** or an **arithmetic progression.** The constant difference d in successive terms is called the **common difference.** In this example, d is 3.

TECHNOLOGY NOTE

On most graphing utilities, the n indicating the nth term of a sequence is different from alpha N.

n	$u(n)$
1	5
2	6
3	7
4	8
5	9
6	10
7	11

$u(n) = n + 4$

> **Arithmetic Sequence and Common Difference**
> An **arithmetic sequence** is a sequence in which each term (after the first) differs from the preceding term by a constant amount d. The constant d is called the **common difference** of the sequence.

The sequence $2, 6, 10, 14, 18, \ldots$ is an arithmetic sequence. Its common difference is 4. Given the first term a_1 and the common difference d of an arithmetic sequence, we can find any term of the sequence.

EXAMPLE 1 Write the first five terms of the arithmetic sequence whose first term is 7 and whose common difference is 2.

Solution

$$a_1 = 7$$
$$a_2 = 7 + 2 = 9$$
$$a_3 = 9 + 2 = 11$$
$$a_4 = 11 + 2 = 13$$
$$a_5 = 13 + 2 = 15$$

The first five terms are $7, 9, 11, 13, 15$. □

PRACTICE

1 Write the first five terms of the arithmetic sequence whose first term is 4 and whose common difference is 5.

Notice the general pattern of the terms in Example 1.

$$a_1 = 7$$
$$a_2 = 7 + 2 = 9 \qquad \text{or} \quad a_2 = a_1 + d$$
$$a_3 = 9 + 2 = 11 \qquad \text{or} \quad a_3 = a_2 + d = (a_1 + d) + d = a_1 + 2d$$
$$a_4 = 11 + 2 = 13 \qquad \text{or} \quad a_4 = a_3 + d = (a_1 + 2d) + d = a_1 + 3d$$
$$a_5 = 13 + 2 = 15 \qquad \text{or} \quad a_5 = a_4 + d = (a_1 + 3d) + d = a_1 + 4d$$

$\longrightarrow$ (subscript $-$ 1) is multiplier $\longrightarrow$

The pattern on the right suggests that the general term a_n of an arithmetic sequence is given by

$$a_n = a_1 + (n-1)d$$

> **General Term of an Arithmetic Sequence**
> The general term a_n of an arithmetic sequence is given by
> $$a_n = a_1 + (n-1)d$$
> where a_1 is the first term and d is the common difference.

EXAMPLE 2   Consider the arithmetic sequence whose first term is 3 and common difference is -5.

a. Write an expression for the general term a_n.

b. Find the twentieth term of this sequence.

Solution

a. Since this is an arithmetic sequence, the general term a_n is given by $a_n = a_1 + (n-1)d$. Here, $a_1 = 3$ and $d = -5$, so

$$
\begin{aligned}
a_n &= 3 + (n-1)(-5) && \text{Let } a_1 = 3 \text{ and } d = -5.\\
&= 3 - 5n + 5 && \text{Multiply.}\\
&= 8 - 5n && \text{Simplify.}
\end{aligned}
$$

b. $a_n = 8 - 5n$

$$
\begin{aligned}
a_{20} &= 8 - 5 \cdot 20 && \text{Let } n = 20.\\
&= 8 - 100 = -92
\end{aligned}
$$

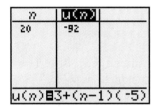

The general term of the sequence is $u(n) = 3 + (n-1)(-5)$ and the 20th term is -92 using the Table in Ask mode.

PRACTICE

2   Consider the arithmetic sequence whose first term is 2 and whose common difference is -3.

a. Write an expression for the general term a_n.

b. Find the twelfth term of the sequence.

EXAMPLE 3   Find the eleventh term of the arithmetic sequence whose first three terms are 2, 9, and 16.

Solution   Since the sequence is arithmetic, the eleventh term is

$$a_{11} = a_1 + (11-1)d = a_1 + 10d$$

We know a_1 is the first term of the sequence, so $a_1 = 2$. Also, d is the constant difference of terms, so $d = a_2 - a_1 = 9 - 2 = 7$. Thus,

$$
\begin{aligned}
a_{11} &= a_1 + 10d\\
&= 2 + 10 \cdot 7 && \text{Let } a_1 = 2 \text{ and } d = 7.\\
&= 72
\end{aligned}
$$

The check for Example 3 shows the first 3 terms are 2, 9, and 16.

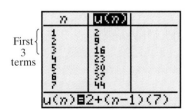

The 11th term is 72.

PRACTICE

3   Find the ninth term of the arithmetic sequence whose first three terms are 3, 9, and 15.

EXAMPLE 4   If the third term of an arithmetic sequence is 12 and the eighth term is 27, find the fifth term.

Solution   We need to find a_1 and d to write the general term, which then enables us to find a_5, the fifth term. The given facts about terms a_3 and a_8 lead to a system of linear equations.

$$\begin{cases} a_3 = a_1 + (3-1)d \\ a_8 = a_1 + (8-1)d \end{cases} \text{ or } \begin{cases} 12 = a_1 + 2d \\ 27 = a_1 + 7d \end{cases}$$

3rd term ⟶
5th term ⟶
8th term ⟶

$u(n)\blacksquare 6+(n-1)(3)$

The table shows that the 3rd term of the sequence is 12, the 8th term is 27, and the 5th term is 18.

Next, we solve the system $\begin{cases} 12 = a_1 + 2d \\ 27 = a_1 + 7d \end{cases}$ by elimination. Multiply both sides of the second equation by -1 so that

$$\begin{cases} 12 = a_1 + 2d \\ -1(27) = -1(a_1 + 7d) \end{cases} \begin{array}{c} \text{simplifies} \\ \text{to} \end{array} \begin{cases} 12 = \quad a_1 + 2d \\ \underline{-27 = -a_1 - 7d} \end{cases}$$

$$\begin{array}{ll} -15 = \qquad -5d & \text{Add the equations.} \\ 3 = d & \text{Divide both sides by } -5. \end{array}$$

To find a_1, let $d = 3$ in $12 = a_1 + 2d$. Then

$$12 = a_1 + 2(3)$$
$$12 = a_1 + 6$$
$$6 = a_1$$

Thus, $a_1 = 6$ and $d = 3$, so

$$a_n = 6 + (n-1)(3)$$
$$= 6 + 3n - 3$$
$$= 3 + 3n$$

and

$$a_5 = 3 + 3 \cdot 5 = 18 \qquad \square$$

PRACTICE
4   If the third term of an arithmetic sequence is 23 and the eighth term is 63, find the sixth term.

EXAMPLE 5   **Finding Salary**

Donna Theime has an offer for a job starting at $40,000 per year and guaranteeing her a raise of $1600 per year for the next 5 years. Write the general term for the arithmetic sequence that models Donna's potential annual salaries, and find her salary for the fourth year.

Solution   The first term, a_1, is 40,000, and d is 1600. So

$$a_n = 40,000 + (n-1)(1600) = 38,400 + 1600n$$
$$a_4 = 38,400 + 1600 \cdot 4 = 44,800$$

Her salary for the fourth year will be $44,800. $\qquad \square$

PRACTICE
5   A starting salary for a consulting company is $57,000 per year, with guaranteed annual increases of $2200 for the next 4 years. Write the general term for the arithmetic sequence that models the potential annual salaries, and find the salary for the third year.

$u(n)\blacksquare 40000+(n-1\ldots$

Verify that her salary for the fourth year is $44,800.

OBJECTIVE 2 ▶ Identifying geometric sequences. We now investigate a **geometric sequence**, also called a **geometric progression**. In the sequence 5, 15, 45, 135, ... , each term after the first is the *product* of 3 and the preceding term. This pattern of multiplying by a constant to get the next term defines a geometric sequence. The constant is called the **common ratio** because it is the ratio of any term (after the first) to its preceding term.

$$\frac{15}{5} = 3$$

$$\frac{45}{15} = 3$$

$$\frac{135}{45} = 3$$

$$\vdots$$

nth term $\longrightarrow$ $\dfrac{a_n}{a_{n-1}} = 3$ $\longleftarrow$ previous term

---

Geometric Sequence and Common Ratio

A **geometric sequence** is a sequence in which each term (after the first) is obtained by multiplying the preceding term by a constant r. The constant r is called the **common ratio** of the sequence.

The sequence $12, 6, 3, \dfrac{3}{2}, \ldots$ is geometric since each term after the first is the product of the previous term and $\dfrac{1}{2}$.

EXAMPLE 6   Write the first five terms of a geometric sequence whose first term is 7 and whose common ratio is 2.

Solution

$$a_1 = 7$$
$$a_2 = 7(2) = 14$$
$$a_3 = 14(2) = 28$$
$$a_4 = 28(2) = 56$$
$$a_5 = 56(2) = 112$$

The first five terms are $7, 14, 28, 56$, and 112. □

PRACTICE

6   Write the first four terms of a geometric sequence whose first term is 8 and whose common ratio is -3.

Notice the general pattern of the terms in Example 6.

$$a_1 = 7$$
$$a_2 = 7(2) = 14 \quad \text{or} \quad a_2 = a_1(r)$$
$$a_3 = 14(2) = 28 \quad \text{or} \quad a_3 = a_2(r) = (a_1 \cdot r) \cdot r = a_1 r^2$$
$$a_4 = 28(2) = 56 \quad \text{or} \quad a_4 = a_3(r) = (a_1 \cdot r^2) \cdot r = a_1 r^3$$
$$a_5 = 56(2) = 112 \quad \text{or} \quad a_5 = a_4(r) = (a_1 \cdot r^3) \cdot r = a_1 r^4$$

$\longrightarrow$ (subscript -1) is power $\longrightarrow$

The pattern on the right above suggests that the general term of a geometric sequence is given by $a_n = a_1 r^{n-1}$.

General Term of a Geometric Sequence

The general term a_n of a geometric sequence is given by

$$a_n = a_1 r^{n-1}$$

where a_1 is the first term and r is the common ratio.

EXAMPLE 7   Find the eighth term of the geometric sequence whose first term is 12 and whose common ratio is $\frac{1}{2}$.

<u>Solution</u>   Since this is a geometric sequence, the general term a_n is given by

$$a_n = a_1 r^{n-1}$$

Here $a_1 = 12$ and $r = \frac{1}{2}$, so $a_n = 12\left(\frac{1}{2}\right)^{n-1}$. Evaluate a_n for $n = 8$.

$$a_8 = 12\left(\frac{1}{2}\right)^{8-1} = 12\left(\frac{1}{2}\right)^7 = 12\left(\frac{1}{128}\right) = \frac{3}{32}$$

The seq command is used in Example 7 so that the terms are displayed in fractions. The input is seq(expression, variable, start, end, increment).

PRACTICE
7   Find the seventh term of the geometric sequence whose first term is 64 and whose common ratio is $\frac{1}{4}$.

EXAMPLE 8   Find the fifth term of the geometric sequence whose first three terms are 2, -6, and 18.

<u>Solution</u>   Since the sequence is geometric and $a_1 = 2$, the fifth term must be $a_1 r^{5-1}$, or $2r^4$. We know that r is the common ratio of terms, so r must be $\frac{-6}{2}$, or -3. Thus,

$$a_5 = 2r^4$$
$$a_5 = 2(-3)^4 = 162$$

A calculator check for Example 8.

PRACTICE
8   Find the seventh term of the geometric sequence whose first three terms are -3, 6, and -12.

EXAMPLE 9   If the second term of a geometric sequence is $\frac{5}{4}$ and the third term is $\frac{5}{16}$, find the first term and the common ratio.

<u>Solution</u>   Notice that $\frac{5}{16} \div \frac{5}{4} = \frac{1}{4}$, so $r = \frac{1}{4}$. Then

$$a_2 = a_1\left(\frac{1}{4}\right)^{2-1}$$
$$\frac{5}{4} = a_1\left(\frac{1}{4}\right)^1, \quad \text{or} \quad a_1 = 5 \quad \text{Replace } a_2 \text{ with } \frac{5}{4}.$$

The first term is 5.

A calculator check for Example 9. If the common ratio is a fraction, use the seq command.

PRACTICE
9   If the second term of a geometric sequence is $\frac{9}{2}$ and the third term is $\frac{27}{4}$, find the first term and the common ratio.

EXAMPLE 10   **Predicting Population of a Bacterial Culture**

The population size of a bacterial culture growing under controlled conditions is doubling each day. Predict how large the culture will be at the beginning of day 7 if it measures 10 units at the beginning of day 1.

<u>Solution</u>   Since the culture doubles in size each day, the population sizes are modeled by a geometric

The bacterial culture measures 640 units at the beginning of day 7.

sequence. Here $a_1 = 10$ and $r = 2$. Thus,

$$a_n = a_1 r^{n-1} = 10(2)^{n-1} \quad \text{and} \quad a_7 = 10(2)^{7-1} = 640$$

The bacterial culture should measure 640 units at the beginning of day 7. ☐

PRACTICE

10 After applying a test antibiotic, the population of a bacterial culture is reduced by one-half every day. Predict how large the culture will be at the start of day 7 if it measures 4800 units at the beginning of day 1.

VOCABULARY & READINESS CHECK

Use the choices below to fill in each blank. Some choices may be used more than once and some not at all.

first	arithmetic	difference
last	geometric	ratio

1. A(n) _____ sequence is one in which each term (after the first) is obtained by multiplying the preceding term by a constant r. The constant r is called the common _____.

2. A(n) _____ sequence is one in which each term (after the first) differs from the preceding term by a constant amount d. The constant d is called the common _____.

3. The general term of an arithmetic sequence is $a_n = a_1 + (n - 1)d$, where a_1 is the _____ term and d is the common _____.

4. The general term of a geometric sequence is $a_n = a_1 r^{n-1}$, where a_1 is the _____ term and r is the common _____.

11.2 | EXERCISE SET

Write the first five terms of the arithmetic or geometric sequence whose first term, a_1, and common difference, d, or common ratio, r, are given. See Examples 1 and 6.

1. $a_1 = 4; d = 2$

2. $a_1 = 3; d = 10$

3. $a_1 = 6; d = -2$

4. $a_1 = -20; d = 3$

5. $a_1 = 1; r = 3$

6. $a_1 = -2; r = 2$

7. $a_1 = 48; r = \dfrac{1}{2}$

8. $a_1 = 1; r = \dfrac{1}{3}$

Find the indicated term of each sequence. See Examples 2 and 7.

9. The eighth term of the arithmetic sequence whose first term is 12 and whose common difference is 3

10. The twelfth term of the arithmetic sequence whose first term is 32 and whose common difference is -4

11. The fourth term of the geometric sequence whose first term is 7 and whose common ratio is -5

12. The fifth term of the geometric sequence whose first term is 3 and whose common ratio is 3

13. The fifteenth term of the arithmetic sequence whose first term is -4 and whose common difference is -4

14. The sixth term of the geometric sequence whose first term is 5 and whose common ratio is -4

Find the indicated term of each sequence. See Examples 3 and 8.

15. The ninth term of the arithmetic sequence $0, 12, 24, \ldots$

16. The thirteenth term of the arithmetic sequence $-3, 0, 3, \ldots$

17. The twenty-fifth term of the arithmetic sequence $20, 18, 16, \ldots$

18. The ninth term of the geometric sequence $5, 10, 20, \ldots$

19. The fifth term of the geometric sequence $2, -10, 50, \ldots$

20. The sixth term of the geometric sequence $\dfrac{1}{2}, \dfrac{3}{2}, \dfrac{9}{2}, \ldots$

Find the indicated term of each sequence. See Examples 4 and 9.

21. The eighth term of the arithmetic sequence whose fourth term is 19 and whose fifteenth term is 52

22. If the second term of an arithmetic sequence is 6 and the tenth term is 30, find the twenty-fifth term.

23. If the second term of an arithmetic progression is -1 and the fourth term is 5, find the ninth term.

24. If the second term of a geometric progression is 15 and the third term is 3, find a_1 and r.

25. If the second term of a geometric progression is $-\dfrac{4}{3}$ and the third term is $\dfrac{8}{3}$, find a_1 and r.

26. If the third term of a geometric sequence is 4 and the fourth term is -12, find a_1 and r.

27. Explain why 14, 10, and 6 may be the first three terms of an arithmetic sequence when it appears we are subtracting instead of adding to get the next term.

28. Explain why 80, 20, and 5 may be the first three terms of a geometric sequence when it appears we are dividing instead of multiplying to get the next term.

MIXED PRACTICE

Given are the first three terms of a sequence that is either arithmetic or geometric. If the sequence is arithmetic, find a_1 and d. If a sequence is geometric, find a_1 and r.

29. 2, 4, 6

30. 8, 16, 24

31. 5, 10, 20

32. 2, 6, 18

33. $\dfrac{1}{2}, \dfrac{1}{10}, \dfrac{1}{50}$

34. $\dfrac{2}{3}, \dfrac{4}{3}, 2$

35. $x, 5x, 25x$

36. $y, -3y, 9y$

37. $p, p + 4, p + 8$

38. $t, t - 1, t - 2$

Find the indicated term of each sequence.

39. The twenty-first term of the arithmetic sequence whose first term is 14 and whose common difference is $\dfrac{1}{4}$

40. The fifth term of the geometric sequence whose first term is 8 and whose common ratio is -3

41. The fourth term of the geometric sequence whose first term is 3 and whose common ratio is $-\dfrac{2}{3}$

42. The fourth term of the arithmetic sequence whose first term is 9 and whose common difference is 5

43. The fifteenth term of the arithmetic sequence $\dfrac{3}{2}, 2, \dfrac{5}{2}, \ldots$

44. The eleventh term of the arithmetic sequence $2, \dfrac{5}{3}, \dfrac{4}{3}, \ldots$

45. The sixth term of the geometric sequence $24, 8, \dfrac{8}{3}, \ldots$

46. The eighteenth term of the arithmetic sequence $5, 2, -1, \ldots$

47. If the third term of an arithmetic sequence is 2 and the seventeenth term is -40, find the tenth term.

48. If the third term of a geometric sequence is -28 and the fourth term is -56, find a_1 and r.

Solve. See Examples 5 and 10.

49. An auditorium has 54 seats in the first row, 58 seats in the second row, 62 seats in the third row, and so on. Find the general term of this arithmetic sequence and the number of seats in the twentieth row.

50. A triangular display of cans in a grocery store has 20 cans in the first row, 17 cans in the next row, and so on, in an arithmetic sequence. Find the general term and the number of cans in the fifth row. Find how many rows there are in the display and how many cans are in the top row.

51. The initial size of a virus culture is 6 units, and it triples its size every day. Find the general term of the geometric sequence that models the culture's size.

52. A real estate investment broker predicts that a certain property will increase in value 15% each year. Thus, the yearly property values can be modeled by a geometric sequence whose common ratio r is 1.15. If the initial property value was $500,000, write the first four terms of the sequence and predict the value at the end of the third year.

53. A rubber ball is dropped from a height of 486 feet, and it continues to bounce one-third the height from which it last fell. Write out the first five terms of this geometric sequence and find the general term. Find how many bounces it takes for the ball to rebound less than 1 foot.

54. On the first swing, the length of the arc through which a pendulum swings is 50 inches. The length of each successive swing is 80% of the preceding swing. Determine whether this sequence is arithmetic or geometric. Find the length of the fourth swing.

55. Jose takes a job that offers a monthly starting salary of $4000 and guarantees him a monthly raise of $125 during his first year of training. Find the general term of this arithmetic sequence and his monthly salary at the end of his training.

56. At the beginning of Claudia Schaffer's exercise program, she rides 15 minutes on the Lifecycle. Each week she increases her riding time by 5 minutes. Write the general term of this arithmetic sequence, and find her riding time after 7 weeks. Find how many weeks it takes her to reach a riding time of 1 hour.

57. If a radioactive element has a half-life of 3 hours, then x grams of the element dwindles to $\dfrac{x}{2}$ grams after 3 hours. If a nuclear reactor has 400 grams of that radioactive element, find the amount of radioactive material after 12 hours.

REVIEW AND PREVIEW

Evaluate. See Section 1.3.

58. $5(1) + 5(2) + 5(3) + 5(4)$

59. $\dfrac{1}{3(1)} + \dfrac{1}{3(2)} + \dfrac{1}{3(3)}$

60. $2(2 - 4) + 3(3 - 4) + 4(4 - 4)$

61. $3^0 + 3^1 + 3^2 + 3^3$

62. $\dfrac{1}{4(1)} + \dfrac{1}{4(2)} + \dfrac{1}{4(3)}$

63. $\dfrac{8 - 1}{8 + 1} + \dfrac{8 - 2}{8 + 2} + \dfrac{8 - 3}{8 + 3}$

CONCEPT EXTENSIONS

Use a graphing utility to write the first four terms of the arithmetic or geometric sequence whose first term, a_1, and common difference, d, or common ratio, r, are given.

64. $a_1 = \$3720; d = -\$268.50$

65. $a_1 = \$11{,}782.40; r = 0.5$

66. $a_1 = 26.8; r = 2.5$

67. $a_1 = 19.652; d = -0.034$

68. Describe a situation in your life that can be modeled by a geometric sequence. Write an equation for the sequence.

69. Describe a situation in your life that can be modeled by an arithmetic sequence. Write an equation for the sequence.

11.3 SERIES

OBJECTIVES

1 Identify finite and infinite series and use summation notation.

2 Find partial sums.

OBJECTIVE 1 ▶ Identifying finite and infinite series and using summation notation. A person who conscientiously saves money by saving first $100 and then saving $10 more each month than he saved the preceding month is saving money according to the arithmetic sequence

$$a_n = 100 + 10(n - 1)$$

Following this sequence, he can predict how much money he should save for any particular month. But if he also wants to know how much money *in total* he has saved, say, by the fifth month, he must find the *sum* of an infinite first five terms of the sequence

$$\underbrace{100}_{a_1} + \underbrace{100 + 10}_{a_2} + \underbrace{100 + 20}_{a_3} + \underbrace{100 + 30}_{a_4} + \underbrace{100 + 40}_{a_5}$$

A sum of the terms of a sequence is called a **series** (the plural is also "series"). As our example here suggests, series are frequently used to model financial and natural phenomena.

A series is a **finite series** if it is the sum of a finite number of terms. A series is an **infinite series** if it is the sum of all the terms of an infinite sequence. For example,

Sequence	*Series*	
$5, 9, 13$	$5 + 9 + 13$	Finite; sum of 3 terms
$5, 9, 13, \ldots$	$5 + 9 + 13 + \cdots$	Infinite
$4, -2, 1, -\dfrac{1}{2}, \dfrac{1}{4}$	$4 + (-2) + 1 + \left(-\dfrac{1}{2}\right) + \left(\dfrac{1}{4}\right)$	Finite; sum of 5 terms
$4, -2, 1, \ldots$	$4 + (-2) + 1 + \cdots$	Infinite
$3, 6, \ldots, 99$	$3 + 6 + \cdots + 99$	Finite; sum of 33 terms

A shorthand notation for denoting a series when the general term of the sequence is known is called **summation notation.** The Greek uppercase letter **sigma,** Σ, is used to mean "sum." The expression $\displaystyle\sum_{n=1}^{5}(3n + 1)$ is read "the sum of $3n + 1$ as n goes from 1 to 5"; this expression means the sum of the first five terms of the sequence whose general term is $a_n = 3n + 1$. Often, the variable i is used instead of n in summation notation: $\displaystyle\sum_{i=1}^{5}(3i + 1)$. Whether we use n, i, k, or some other variable, the variable is called the **index of summation.** The notation $i = 1$ below the symbol Σ indicates the beginning value of i, and the number 5 above the symbol Σ indicates the ending value of i. Thus, the terms of the sequence are found by successively replacing i with the natural numbers 1, 2, 3, 4, 5. To find the sum, we write out the terms and then add.

$$\begin{aligned}\sum_{i=1}^{5}(3i + 1) &= (3 \cdot 1 + 1) + (3 \cdot 2 + 1) + (3 \cdot 3 + 1) \\ &\quad + (3 \cdot 4 + 1) + (3 \cdot 5 + 1) \\ &= 4 + 7 + 10 + 13 + 16 = 50\end{aligned}$$

TECHNOLOGY NOTE

You can use a graphing utility to evaluate summations. Use a sum command to add terms of a specified sequence.

```
sum(seq(3n+1,n,1
,5,1))
              50
■
```

EXAMPLE 1 Evaluate.

a. $\displaystyle\sum_{i=0}^{6} \frac{i-2}{2}$ **b.** $\displaystyle\sum_{i=3}^{5} 2^i$

Solution

a. $\displaystyle\sum_{i=0}^{6} \frac{i-2}{2} = \frac{0-2}{2} + \frac{1-2}{2} + \frac{2-2}{2} + \frac{3-2}{2} + \frac{4-2}{2} + \frac{5-2}{2} + \frac{6-2}{2}$

$$= (-1) + \left(-\frac{1}{2}\right) + 0 + \frac{1}{2} + 1 + \frac{3}{2} + 2$$

$$= \frac{7}{2}, \text{ or } 3\frac{1}{2}$$

b. $\displaystyle\sum_{i=3}^{5} 2^i = 2^3 + 2^4 + 2^5$

$$= 8 + 16 + 32$$

$$= 56$$

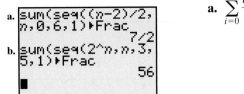

a. sum(seq((n-2)/2, n,0,6,1)▶Frac
7/2
b. sum(seq(2^n,n,3, 5,1)▶Frac
56

A calculator check for Example 1, part **a** and part **b**.

PRACTICE
1 Evaluate.

a. $\displaystyle\sum_{i=0}^{4} \frac{i-3}{4}$ **b.** $\displaystyle\sum_{i=2}^{5} 3^i$

EXAMPLE 2 Write each series with summation notation.

a. $3 + 6 + 9 + 12 + 15$ **b.** $\dfrac{1}{2} + \dfrac{1}{4} + \dfrac{1}{8} + \dfrac{1}{16}$

Solution

a. Since the *difference* of each term and the preceding term is 3, the terms correspond to the first five terms of the arithmetic sequence $a_n = a_1 + (n-1)d$ with $a_1 = 3$ and $d = 3$. So $a_n = 3 + (n-1)3 = 3n$, when simplified. Thus, in summation notation,

$$3 + 6 + 9 + 12 + 15 = \sum_{i=1}^{5} 3i.$$

b. Since each term is the *product* of the preceding term and $\dfrac{1}{2}$, these terms correspond to the first four terms of the geometric sequence $a_n = a_1 r^{n-1}$. Here $a_1 = \dfrac{1}{2}$ and $r = \dfrac{1}{2}$, so $a_n = \left(\dfrac{1}{2}\right)\left(\dfrac{1}{2}\right)^{n-1} = \left(\dfrac{1}{2}\right)^{1+(n-1)} = \left(\dfrac{1}{2}\right)^n$. In summation notation,

$$\frac{1}{2} + \frac{1}{4} + \frac{1}{8} + \frac{1}{16} = \sum_{i=1}^{4} \left(\frac{1}{2}\right)^i$$

PRACTICE
2 Write each series with summation notation.

a. $5 + 10 + 15 + 20 + 25 + 30$ **b.** $\dfrac{1}{5} + \dfrac{1}{25} + \dfrac{1}{125} + \dfrac{1}{625}$

OBJECTIVE 2 ▶ Finding partial sums. The sum of the first n terms of a sequence is a finite series known as a **partial sum,** S_n. Thus, for the sequence $a_1, a_2, \ldots, a_n$, the first three partial sums are

$$S_1 = a_1$$
$$S_2 = a_1 + a_2$$
$$S_3 = a_1 + a_2 + a_3$$

In general, S_n is the sum of the first n terms of a sequence.

$$S_n = \sum_{i=1}^{n} a_n$$

EXAMPLE 3 Find the sum of the first three terms of the sequence whose general term is $a_n = \dfrac{n+3}{2n}$.

Solution

$$S_3 = \sum_{i=1}^{3} \frac{i+3}{2i} = \frac{1+3}{2 \cdot 1} + \frac{2+3}{2 \cdot 2} + \frac{3+3}{2 \cdot 3}$$

$$= 2 + \frac{5}{4} + 1 = 4\frac{1}{4}$$

PRACTICE
3 Find the sum of the first four terms of the sequence whose general term is $a_n = \dfrac{2+3n}{n^2}$.

The next example illustrates how these sums model real-life phenomena.

EXAMPLE 4 **Number of Baby Gorillas Born**

The number of baby gorillas born at the San Diego Zoo is a sequence defined by $a_n = n(n-1)$, where n is the number of years the zoo has owned gorillas. Find the *total* number of baby gorillas born in the *first 4 years*.

Solution To solve, find the sum

$$S_4 = \sum_{i=1}^{4} i(i-1)$$

$$= 1(1-1) + 2(2-1) + 3(3-1) + 4(4-1)$$

$$= 0 + 2 + 6 + 12 = 20$$

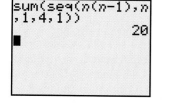

There were 20 gorillas born in the first 4 years.

PRACTICE
4 The number of strawberry plants growing in a garden is a sequence defined by $a_n = n(2n-1)$, where n is the number of years after planting a strawberry plant. Find the total number of strawberry plants after 5 years.

A calculator check for Example 4.

VOCABULARY & READINESS CHECK

Use the choices below to fill in each blank. Not all choices may be used.

index of summation   infinite   sigma   1   7

partial sum   finite   summation   5

1. A series is a(n) _____ series if it is the sum of all the terms of the sequence.

2. A series is a(n) _____ series if it is the sum of a finite number of terms.

3. A shorthand notation for denoting a series when the general term of the sequence is known is called _____ notation.

4. In the notation $\sum_{i=1}^{7}(5i - 2)$, the Σ is the Greek uppercase letter _____ and the i is called the _____.

5. The sum of the first n terms of a sequence is a finite series known as a _____.

6. For the notation in Exercise 4 above, the beginning value of i is _____ and the ending value of i is _____.

11.3 | EXERCISE SET

MyMathLab PRACTICE WATCH DOWNLOAD READ REVIEW

Evaluate. See Example 1.

1. $\sum_{i=1}^{4}(i-3)$

2. $\sum_{i=1}^{5}(i+6)$

3. $\sum_{i=4}^{7}(2i+4)$

4. $\sum_{i=2}^{3}(5i-1)$

5. $\sum_{i=2}^{4}(i^2-3)$

6. $\sum_{i=3}^{5}i^3$

7. $\sum_{i=1}^{3}\left(\frac{1}{i+5}\right)$

8. $\sum_{i=2}^{4}\left(\frac{2}{i+3}\right)$

9. $\sum_{i=1}^{3}\frac{1}{6i}$

10. $\sum_{i=1}^{3}\frac{1}{3i}$

11. $\sum_{i=2}^{6}3i$

12. $\sum_{i=3}^{6}-4i$

13. $\sum_{i=3}^{5}i(i+2)$

14. $\sum_{i=2}^{4}i(i-3)$

15. $\sum_{i=1}^{5}2^i$

16. $\sum_{i=1}^{4}3^{i-1}$

17. $\sum_{i=1}^{4}\frac{4i}{i+3}$

18. $\sum_{i=2}^{5}\frac{6-i}{6+i}$

Write each series with summation notation. See Example 2.

19. $1+3+5+7+9$

20. $4+7+10+13$

21. $4+12+36+108$

22. $5+10+20+40+80+160$

23. $12+9+6+3+0+(-3)$

24. $5+1+(-3)+(-7)$

25. $12+4+\frac{4}{3}+\frac{4}{9}$

26. $80+20+5+\frac{5}{4}+\frac{5}{16}$

27. $1+4+9+16+25+36+49$

28. $1+(-4)+9+(-16)$

Find each partial sum. See Example 3.

29. Find the sum of the first two terms of the sequence whose general term is $a_n=(n+2)(n-5)$.

30. Find the sum of the first two terms of the sequence whose general term is $a_n=n(n-6)$.

31. Find the sum of the first six terms of the sequence whose general term is $a_n=(-1)^n$.

32. Find the sum of the first seven terms of the sequence whose general term is $a_n=(-1)^{n-1}$.

33. Find the sum of the first four terms of the sequence whose general term is $a_n=(n+3)(n+1)$.

34. Find the sum of the first five terms of the sequence whose general term is $a_n=\dfrac{(-1)^n}{2n}$.

35. Find the sum of the first four terms of the sequence whose general term is $a_n=-2n$.

36. Find the sum of the first five terms of the sequence whose general term is $a_n=(n-1)^2$.

37. Find the sum of the first three terms of the sequence whose general term is $a_n=-\dfrac{n}{3}$.

38. Find the sum of the first three terms of the sequence whose general term is $a_n=(n+4)^2$.

Solve. See Example 4.

39. A gardener is making a triangular planting with 1 tree in the first row, 2 trees in the second row, 3 trees in the third row, and so on, for 10 rows. Write the sequence that describes the number of trees in each row. Find the total number of trees planted.

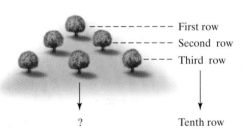

First row
Second row
Third row
? Tenth row

40. Some surfers at the beach form a human pyramid with 2 surfers in the top row, 3 surfers in the second row, 4 surfers in the third row, and so on. If there are 6 rows in the pyramid, write the sequence that describes the number of surfers in each row of the pyramid. Find the total number of surfers.

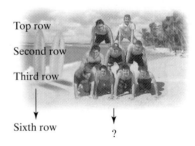

Top row
Second row
Third row
Sixth row ?

41. A culture of fungus starts with 6 units and doubles every day. Write the general term of the sequence that describes the growth of this fungus. Find the number of fungus units there will be at the beginning of the fifth day.

42. A bacterial colony begins with 100 bacteria and doubles every 6 hours. Write the general term of the sequence describing the growth of the bacteria. Find the number of bacteria there will be after 24 hours.

43. A bacterial colony begins with 50 bacteria and doubles every 12 hours. Write the sequence that describes the growth of the bacteria. Find the number of bacteria there will be after 48 hours.

44. The number of otters born each year in a new aquarium forms a sequence whose general term is $a_n = (n - 1)(n + 3)$. Find the number of otters born in the third year, and find the total number of otters born in the first three years.

45. The number of opossums killed each month on a new highway forms the sequence whose general term is $a_n = (n + 1)(n + 2)$, where n is the number of the months. Find the number of opossums killed in the fourth month, and find the total number killed in the first four months.

46. In 2007, the population of the Northern Spotted Owl continued to decline, and the owl remained on the endangered species list, as old-growth Northwest forests were logged. The size of the decrease in the population in a given year can be estimated by $200 - 6n$ pairs of birds. Find the decrease in population in 2010 if year 1 is 2007. Find the estimated total decrease in the spotted owl population for the years 2007 through 2010. (*Source:* United States Forest Service)

47. The amount of decay in pounds of a radioactive isotope each year is given by the sequence whose general term is $a_n = 100(0.5)^n$, where n is the number of the year. Find the amount of decay in the fourth year, and find the total amount of decay in the first four years.

48. Susan has a choice between two job offers. Job A has an annual starting salary of \$20,000 with guaranteed annual raises of \$1200 for the next four years, whereas job B has an annual starting salary of \$18,000 with guaranteed annual raises of \$2500 for the next four years. Compare the fifth partial sums for each sequence to determine which job would pay Susan more money over the next 5 years.

49. A pendulum swings a length of 40 inches on its first swing. Each successive swing is $\dfrac{4}{5}$ of the preceding swing. Find the length of the fifth swing and the total length swung during the first five swings. (Round to the nearest tenth of an inch.)

50. Explain the difference between a sequence and a series.

REVIEW AND PREVIEW

Evaluate. See Section 1.3.

51. $\dfrac{5}{1 - \dfrac{1}{2}}$

52. $\dfrac{-3}{1 - \dfrac{1}{7}}$

53. $\dfrac{\dfrac{1}{3}}{1 - \dfrac{1}{10}}$

54. $\dfrac{\dfrac{6}{11}}{1 - \dfrac{1}{10}}$

55. $\dfrac{3(1 - 2^4)}{1 - 2}$

56. $\dfrac{2(1 - 5^3)}{1 - 5}$

57. $\dfrac{10}{2}(3 + 15)$

58. $\dfrac{12}{2}(2 + 19)$

CONCEPT EXTENSIONS

59. a. Write the sum $\displaystyle\sum_{i=1}^{7}(i + i^2)$ without summation notation.

   b. Write the sum $\displaystyle\sum_{i=1}^{7}i + \sum_{i=1}^{7}i^2$ without summation notation.

   c. Compare the results of parts **a** and **b**.

   d. Do you think the following is true or false? Explain your answer.

$$\sum_{i=1}^{n}(a_n + b_n) = \sum_{i=1}^{n}a_n + \sum_{i=1}^{n}b_n$$

60. a. Write the sum $\displaystyle\sum_{i=1}^{6}5i^3$ without summation notation.

   b. Write the expression $5 \cdot \displaystyle\sum_{i=1}^{6}i^3$ without summation notation.

   c. Compare the results of parts **a** and **b**.

   d. Do you think the following is true or false? Explain your answer.

$$\sum_{i=1}^{n}c \cdot a_n = c \cdot \sum_{i=1}^{n}a_n, \text{ where } c \text{ is a constant}$$

INTEGRATED REVIEW   SEQUENCES AND SERIES

Write the first five terms of each sequence whose general term is given.

1. $a_n = n - 3$

2. $a_n = \dfrac{7}{1 + n}$

3. $a_n = 3^{n-1}$

4. $a_n = n^2 - 5$

Find the indicated term for each sequence.

5. $(-2)^n$; a_6

6. $-n^2 + 2$; a_4

7. $\dfrac{(-1)^n}{n}$; a_{40}

8. $\dfrac{(-1)^n}{2n}$; a_{41}

Write the first five terms of the arithmetic or geometric sequence whose first term is a_1 and whose common difference, d, or common ratio, r, are given.

9. $a_1 = 7; d = -3$

10. $a_1 = -3; r = 5$

11. $a_1 = 45; r = \dfrac{1}{3}$

12. $a_1 = -12; d = 10$

Find the indicated term of each sequence.

13. The tenth term of the arithmetic sequence whose first term is 20 and whose common difference is 9

14. The sixth term of the geometric sequence whose first term is 64 and whose common ratio is $\dfrac{3}{4}$

15. The seventh term of the geometric sequence $6, -12, 24, \ldots$

16. The twentieth term of the arithmetic sequence $-100, -85, -70, \ldots$

17. The fifth term of the arithmetic sequence whose fourth term is -5 and whose tenth term is -35

18. The fifth term of the geometric sequence whose fourth term is 1 and whose seventh term is $\dfrac{1}{125}$

Evaluate.

19. $\displaystyle\sum_{i=1}^{4} 5i$

20. $\displaystyle\sum_{i=1}^{7} (3i + 2)$

21. $\displaystyle\sum_{i=3}^{7} 2^{i-4}$

22. $\displaystyle\sum_{i=2}^{5} \dfrac{i}{i+1}$

Find each partial sum.

23. Find the sum of the first three terms of the sequence whose general term is $a_n = n(n - 4)$.

24. Find the sum of the first ten terms of the sequence whose general term is $a_n = (-1)^n(n + 1)$.

11.4 PARTIAL SUMS OF ARITHMETIC AND GEOMETRIC SEQUENCES

OBJECTIVES

1 Find the partial sum of an arithmetic sequence.

2 Find the partial sum of a geometric sequence.

3 Find the sum of the terms of an infinite geometric sequence.

OBJECTIVE 1 ▶ **Finding partial sums of arithmetic sequences.** Partial sums S_n are relatively easy to find when n is small—that is, when the number of terms to add is small. But when n is large, finding S_n can be tedious. For a large n, S_n is still relatively easy to find if the addends are terms of an arithmetic sequence or a geometric sequence.

For an arithmetic sequence, $a_n = a_1 + (n - 1)d$ for some first term a_1 and some common difference d. So S_n, the sum of the first n terms, is

$$S_n = a_1 + (a_1 + d) + (a_1 + 2d) + \cdots + (a_1 + (n - 1)d)$$

We might also find S_n by "working backward" from the nth term a_n, finding the preceding term a_{n-1}, by subtracting d each time.

$$S_n = a_n + (a_n - d) + (a_n - 2d) + \cdots + (a_n - (n - 1)d)$$

Now add the left sides of these two equations and add the right sides.

$$2S_n = (a_1 + a_n) + (a_1 + a_n) + (a_1 + a_n) + \cdots + (a_1 + a_n)$$

The d terms subtract out, leaving n sums of the first term, a_1, and last term, a_n. Thus, we write

$$2S_n = n(a_1 + a_n)$$

or

$$S_n = \frac{n}{2}(a_1 + a_n)$$

> **Partial Sum S_n of an Arithmetic Sequence**
> The partial sum S_n of the first n terms of an arithmetic sequence is given by
>
> $$S_n = \frac{n}{2}(a_1 + a_n)$$
>
> where a_1 is the first term of the sequence and a_n is the nth term.

EXAMPLE 1 Use the partial sum formula to find the sum of the first six terms of the arithmetic sequence $2, 5, 8, 11, 14, 17, \ldots$.

Solution Use the formula for S_n of an arithmetic sequence, replacing n with 6, a_1 with 2, and a_n with 17.

$$S_n = \frac{n}{2}(a_1 + a_n)$$

$$S_6 = \frac{6}{2}(2 + 17) = 3(19) = 57 \qquad \square$$

PRACTICE
1 Use the partial sum formula to find the sum of the first five terms of the arithmetic sequence $2, 9, 16, 23, 30$.

EXAMPLE 2 Find the sum of the first 30 positive integers.

Solution Because $1, 2, 3, \ldots, 30$ is an arithmetic sequence, use the formula for S_n with $n = 30$, $a_1 = 1$, and $a_n = 30$. Thus,

$$S_n = \frac{n}{2}(a_1 + a_n)$$

$$S_{30} = \frac{30}{2}(1 + 30) = 15(31) = 465 \qquad \square$$

PRACTICE
2 Find the sum of the first 50 positive integers.

EXAMPLE 3 **Stacking Rolls of Carpet**

Rolls of carpet are stacked in 20 rows with 3 rolls in the top row, 4 rolls in the next row, and so on, forming an arithmetic sequence. Find the total number of carpet rolls if there are 22 rolls in the bottom row.

— 3 rolls
— 4 rolls
— 5 rolls

Solution The list $3, 4, 5, \ldots, 22$ is the first 20 terms of an arithmetic sequence. Use the formula for S_n with $a_1 = 3$, $a_n = 22$, and $n = 20$ terms. Thus,

$$S_{20} = \frac{20}{2}(3 + 22) = 10(25) = 250$$

There are a total of 250 rolls of carpet. $\qquad \square$

PRACTICE
3 An ice sculptor created a gigantic castle-facade ice sculpture for First Night festivities in Boston. To get the volume of ice necessary, large blocks of ice were stacked atop each other. The topmost row was comprised of 6 blocks of ice, the next row of 7 blocks of ice, and so on, forming an arithmetic sequence. Find the total number of ice blocks needed if there were 15 blocks in the bottom row.

OBJECTIVE 2 ▶ Finding partial sums of geometric sequences. We can also derive a formula for the partial sum S_n of the first n terms of a geometric series. If $a_n = a_1 r^{n-1}$, then

$$S_n = a_1 + a_1 r + a_1 r^2 + \cdots + a_1 r^{n-1}$$

$$\begin{array}{cccc} \uparrow & \uparrow & \uparrow & \uparrow \\ \text{1st} & \text{2nd} & \text{3rd} & n\text{th} \\ \text{term} & \text{term} & \text{term} & \text{term} \end{array}$$

Multiply each side of the equation by $-r$.

$$-rS_n = -a_1 r - a_1 r^2 - a_1 r^3 - \cdots - a_1 r^n$$

Add the two equations.

$$S_n - rS_n = a_1 + (a_1 r - a_1 r) + (a_1 r^2 - a_1 r^2) + (a_1 r^3 - a_1 r^3) + \cdots - a_1 r^n$$
$$S_n - rS_n = a_1 - a_1 r^n$$

Now factor each side.

$$S_n(1 - r) = a_1(1 - r^n)$$

Solve for S_n by dividing both sides by $1 - r$. Thus,

$$S_n = \frac{a_1(1 - r^n)}{1 - r}$$

as long as r is not 1.

Partial Sum S_n of a Geometric Sequence

The partial sum S_n of the first n terms of a geometric sequence is given by

$$S_n = \frac{a_1(1 - r^n)}{1 - r}$$

where a_1 is the first term of the sequence, r is the common ratio, and $r \neq 1$.

EXAMPLE 4   Find the sum of the first six terms of the geometric sequence 5, 10, 20, 40, 80, 160.

Solution   Use the formula for the partial sum S_n of the terms of a geometric sequence. Here, $n = 6$, the first term $a_1 = 5$, and the common ratio $r = 2$.

$$S_n = \frac{a_1(1 - r^n)}{1 - r}$$

$$S_6 = \frac{5(1 - 2^6)}{1 - 2} = \frac{5(-63)}{-1} = 315 \qquad \square$$

PRACTICE
4   Find the sum of the first five terms of the geometric sequence $32, 8, 2, \dfrac{1}{2}, \dfrac{1}{8}$.

EXAMPLE 5   **Finding Amount of Donation**

A grant from an alumnus to a university specified that the university was to receive $800,000 during the first year and 75% of the preceding year's donation during each of the following 5 years. Find the total amount donated during the 6 years.

Solution   The donations are modeled by the first six terms of a geometric sequence. Evaluate S_n when $n = 6$, $a_1 = 800{,}000$, and $r = 0.75$.

$$S_6 = \frac{800{,}000[1 - (0.75)^6]}{1 - 0.75}$$

$$= \$2{,}630{,}468.75$$

The total amount donated during the 6 years is \$2,630,468.75. ☐

PRACTICE

5   A new youth center is being established in a downtown urban area. A philanthropic charity has agreed to help it get off the ground. The charity has pledged to donate \$250,000 in the first year, with 80% of the preceding year's donation for each of the following 6 years. Find the total amount donated during the 7 years.

OBJECTIVE 3 ▶ Finding sums of terms of infinite geometric sequences.   Is it possible to find the sum of all the terms of an infinite sequence? Examine the partial sums of the geometric sequence $\frac{1}{2}, \frac{1}{4}, \frac{1}{8}, \ldots$.

$$S_1 = \frac{1}{2}$$

$$S_2 = \frac{1}{2} + \frac{1}{4} = \frac{3}{4}$$

$$S_3 = \frac{1}{2} + \frac{1}{4} + \frac{1}{8} = \frac{7}{8}$$

$$S_4 = \frac{1}{2} + \frac{1}{4} + \frac{1}{8} + \frac{1}{16} = \frac{15}{16}$$

$$S_5 = \frac{1}{2} + \frac{1}{4} + \frac{1}{8} + \frac{1}{16} + \frac{1}{32} = \frac{31}{32}$$

$$\vdots$$

$$S_{10} = \frac{1}{2} + \frac{1}{4} + \frac{1}{8} + \cdots + \frac{1}{2^{10}} = \frac{1023}{1024}$$

Even though each partial sum is larger than the preceding partial sum, we see that each partial sum is closer to 1 than the preceding partial sum. If n gets larger and larger, then S_n gets closer and closer to 1. We say that 1 is the **limit** of S_n and also that 1 is the sum of the terms of this infinite sequence. In general, if $|r| < 1$, the following formula gives the sum of the terms of an infinite geometric sequence.

Sum of the Terms of an Infinite Geometric Sequence

The sum S_∞ of the terms of an infinite geometric sequence is given by

$$S_\infty = \frac{a_1}{1 - r}$$

where a_1 is the first term of the sequence, r is the common ratio, and $|r| < 1$. If $|r| \geq 1$, S_∞ does not exist.

What happens for other values of r? For example, in the following geometric sequence, $r = 3$.

$$6, 18, 54, 162, \ldots$$

Here, as n increases, the sum S_n increases also. This time, though, S_n does not get closer and closer to a fixed number but instead increases without bound.

EXAMPLE 6 Find the sum of the terms of the geometric sequence $2, \dfrac{2}{3}, \dfrac{2}{9}, \dfrac{2}{27}, \ldots$.

Solution For this geometric sequence, $r = \dfrac{1}{3}$. Since $|r| < 1$, we may use the formula for S_∞ of a geometric sequence, with $a_1 = 2$ and $r = \dfrac{1}{3}$.

$$S_\infty = \frac{a_1}{1 - r} = \frac{2}{1 - \dfrac{1}{3}} = \frac{2}{\dfrac{2}{3}} = 3 \qquad \square$$

The formula for the sum of the terms of an infinite geometric sequence can be used to write a repeating decimal as a fraction. For example,

$$0.33\overline{3} = \frac{3}{10} + \frac{3}{100} + \frac{3}{1000} + \cdots$$

This sum is the sum of the terms of an infinite geometric sequence whose first term a_1 is $\dfrac{3}{10}$ and whose common ratio r is $\dfrac{1}{10}$. Using the formula for S_∞,

$$S_\infty = \frac{a_1}{1 - r} = \frac{\dfrac{3}{10}}{1 - \dfrac{1}{10}} = \frac{1}{3}$$

So, $0.33\overline{3} = \dfrac{1}{3}$.

PRACTICE
6 Find the sum of the terms of the geometric sequence $7, \dfrac{7}{4}, \dfrac{7}{16}, \dfrac{7}{64}, \ldots$.

EXAMPLE 7 **Distance Traveled by a Pendulum**

On its first pass, a pendulum swings through an arc whose length is 24 inches. On each pass thereafter, the arc length is 75% of the arc length on the preceding pass. Find the total distance the pendulum travels before it comes to rest.

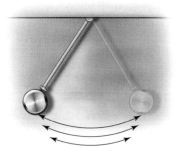

Solution We must find the sum of the terms of an infinite geometric sequence whose first term, a_1, is 24 and whose common ratio, r, is 0.75. Since $|r| < 1$, we may use the formula for S_∞.

$$S_\infty = \frac{a_1}{1 - r} = \frac{24}{1 - 0.75} = \frac{24}{0.25} = 96$$

The pendulum travels a total distance of 96 inches before it comes to rest. $\qquad \square$

PRACTICE
7 The manufacturers of the "perpetual bouncing ball" claim that the ball rises to 96% of its dropped height on each bounce of the ball. Find the total distance the ball travels before it comes to rest, if it is dropped from a height of 36 inches.

VOCABULARY & READINESS CHECK

Decide whether each sequence is geometric or arithmetic.

1. $5, 10, 15, 20, 25, \ldots;$ _____

2. $5, 10, 20, 40, 80, \ldots;$ _____

3. $-1, 3, -9, 27, -81 \ldots;$ _____

4. $-1, 1, 3, 5, 7, \ldots;$ _____

5. $-7, 0, 7, 14, 21, \ldots;$ _____

6. $-7, 7, -7, 7, -7, \ldots;$ _____

11.4 | EXERCISE SET

 MyMathLab

 Math XL
PRACTICE

 WATCH

 DOWNLOAD

 READ

REVIEW

Use the partial sum formula to find the partial sum of the given arithmetic or geometric sequence. See Examples 1 and 4.

1. Find the sum of the first six terms of the arithmetic sequence $1, 3, 5, 7, \ldots.$

2. Find the sum of the first seven terms of the arithmetic sequence $-7, -11, -15, \ldots.$

3. Find the sum of the first five terms of the geometric sequence $4, 12, 36, \ldots.$

4. Find the sum of the first eight terms of the geometric sequence $-1, 2, -4, \ldots.$

5. Find the sum of the first six terms of the arithmetic sequence $3, 6, 9, \ldots.$

6. Find the sum of the first four terms of the arithmetic sequence $-4, -8, -12, \ldots.$

7. Find the sum of the first four terms of the geometric sequence $2, \dfrac{2}{5}, \dfrac{2}{25}, \ldots.$

8. Find the sum of the first five terms of the geometric sequence $\dfrac{1}{3}, -\dfrac{2}{3}, \dfrac{4}{3}, \ldots.$

Solve. See Example 2.

9. Find the sum of the first ten positive integers.

10. Find the sum of the first eight negative integers.

11. Find the sum of the first four positive odd integers.

12. Find the sum of the first five negative odd integers.

Find the sum of the terms of each infinite geometric sequence. See Example 6.

13. $12, 6, 3, \ldots$

14. $45, 15, 5, \ldots$

15. $\dfrac{1}{10}, \dfrac{1}{100}, \dfrac{1}{1000}, \ldots$

16. $\dfrac{3}{5}, \dfrac{3}{20}, \dfrac{3}{80}, \ldots$

17. $-10, -5, -\dfrac{5}{2}, \ldots$

18. $-16, -4, -1, \ldots$

19. $2, -\dfrac{1}{4}, \dfrac{1}{32}, \ldots$

20. $-3, \dfrac{3}{5}, -\dfrac{3}{25}, \ldots$

21. $\dfrac{2}{3}, -\dfrac{1}{3}, \dfrac{1}{6}, \ldots$

22. $6, -4, \dfrac{8}{3}, \ldots$

MIXED PRACTICE

Solve.

23. Find the sum of the first ten terms of the sequence $-4, 1, 6, \ldots, 41$, where 41 is the tenth term.

24. Find the sum of the first twelve terms of the sequence $-3, -13, -23, \ldots, -113$, where -113 is the twelfth term.

25. Find the sum of the first seven terms of the sequence $3, \dfrac{3}{2}, \dfrac{3}{4}, \ldots.$

26. Find the sum of the first five terms of the sequence $-2, -6, -18, \ldots.$

27. Find the sum of the first five terms of the sequence $-12, 6, -3, \ldots.$

28. Find the sum of the first four terms of the sequence $-\dfrac{1}{4}, -\dfrac{3}{4}, -\dfrac{9}{4}, \ldots.$

29. Find the sum of the first twenty terms of the sequence $\dfrac{1}{2}, \dfrac{1}{4}, 0, \ldots, -\dfrac{17}{4}$, where $-\dfrac{17}{4}$ is the twentieth term.

30. Find the sum of the first fifteen terms of the sequence $-5, -9, -13, \ldots, -61$, where -61 is the fifteenth term.

31. If a_1 is 8 and r is $-\dfrac{2}{3}$, find S_3.

32. If a_1 is 10, a_{18} is $\dfrac{3}{2}$, and d is $-\dfrac{1}{2}$, find S_{18}.

Solve. See Example 3.

33. Modern Car Company has come out with a new car model. Market analysts predict that 4000 cars will be sold in the first month and that sales will drop by 50 cars per month after that during the first year. Write out the first five terms of the sequence, and find the number of sold cars predicted for the twelfth month. Find the total predicted number of sold cars for the first year.

34. A company that sends faxes charges $3 for the first page sent and $0.10 less than the preceding page for each additional page sent. The cost per page forms an arithmetic sequence. Write the first five terms of this sequence, and use a partial sum to find the cost of sending a nine-page document.

35. Sal has two job offers: Firm *A* starts at $22,000 per year and guarantees raises of $1000 per year, whereas Firm *B* starts at $20,000 and guarantees raises of $1200 per year. Over a 10-year period, determine the more profitable offer.

36. The game of pool uses 15 balls numbered 1 to 15. In the variety called rotation, a player who sinks a ball receives as many points as the number on the ball. Use an arithmetic series to find the score of a player who sinks all 15 balls.

Solve. See Example 5.

37. A woman made $30,000 during the first year she owned her business and made an additional 10% over the previous year in each subsequent year. Find how much she made during her fourth year of business. Find her total earnings during the first four years.

38. In free fall, a parachutist falls 16 feet during the first second, 48 feet during the second second, 80 feet during the third second, and so on. Find how far she falls during the eighth second. Find the total distance she falls during the first 8 seconds.

39. A trainee in a computer company takes 0.9 times as long to assemble each computer as he took to assemble the preceding computer. If it took him 30 minutes to assemble the first computer, find how long it takes him to assemble the fifth computer. Find the total time he takes to assemble the first five computers (round to the nearest minute).

40. On a gambling trip to Reno, Carol doubled her bet each time she lost. If her first losing bet was $5 and she lost six consecutive bets, find how much she lost on the sixth bet. Find the total amount lost on these six bets.

Solve. See Example 7.

41. A ball is dropped from a height of 20 feet and repeatedly rebounds to a height that is $\frac{4}{5}$ of its previous height. Find the total distance the ball covers before it comes to rest.

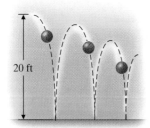

42. A rotating flywheel coming to rest makes 300 revolutions in the first minute and in each minute thereafter makes $\frac{2}{5}$ as many revolutions as in the preceding minute. Find how many revolutions the wheel makes before it comes to rest.

MIXED PRACTICE

Solve.

43. In the pool game of rotation, player *A* sinks balls numbered 1 to 9, and player *B* sinks the rest of the balls. Use an arithmetic series to find each player's score (see Exercise 36).

44. A godfather deposited $250 in a savings account on the day his godchild was born. On each subsequent birthday he deposited $50 more than he'd deposited the previous year. Find how much money he deposited on his godchild's twenty-first birthday. Find the total amount deposited over the 21 years.

45. During the holiday rush a business can rent a computer system for $200 the first day, with the rental fee decreasing $5 for each additional day. Find the fee paid for 20 days during the holiday rush.

46. The spraying of a field with insecticide killed 6400 weevils the first day, 1600 the second day, 400 the third day, and so on. Find the total number of weevils killed during the first 5 days.

47. A college student humorously asks his parents to charge him room and board according to this geometric sequence: $0.01 for the first day of the month, $0.02 for the second day, $0.04 for the third day, and so on. Find the total room and board he would pay for 30 days.

48. Following its television advertising campaign, a bank attracted 80 new customers the first day, 120 the second day, 160 the third day, and so on, in an arithmetic sequence. Find how many new customers were attracted during the first 5 days following its television campaign.

REVIEW AND PREVIEW

Evaluate. See Section 1.3.

49. $6 \cdot 5 \cdot 4 \cdot 3 \cdot 2 \cdot 1$

50. $8 \cdot 7 \cdot 6 \cdot 5 \cdot 4 \cdot 3 \cdot 2 \cdot 1$

51. $\dfrac{3 \cdot 2 \cdot 1}{2 \cdot 1}$

52. $\dfrac{5 \cdot 4 \cdot 3 \cdot 2 \cdot 1}{3 \cdot 2 \cdot 1}$

Multiply. See Section 5.4.

53. $(x + 5)^2$

54. $(x - 2)^2$

55. $(2x - 1)^3$

56. $(3x + 2)^3$

CONCEPT EXTENSIONS

57. Write $0.88\overline{8}$ as an infinite geometric series and use the formula for S_∞ to write it as a rational number.

58. Write $0.54\overline{54}$ as an infinite geometric series and use the formula S_∞ to write it as a rational number.

59. Explain whether the sequence $5, 5, 5, \ldots$ is arithmetic, geometric, neither, or both.

60. Describe a situation in everyday life that can be modeled by an infinite geometric series.

11.5 THE BINOMIAL THEOREM

OBJECTIVES

1 Use Pascal's triangle to expand binomials.

2 Evaluate factorials.

3 Use the binomial theorem to expand binomials.

4 Find the nth term in the expansion of a binomial raised to a positive power.

In this section, we learn how to **expand** binomials of the form $(a + b)^n$ easily. Expanding a binomial such as $(a + b)^n$ means to write the factored form as a sum. First, we review the patterns in the expansions of $(a + b)^n$.

$(a + b)^0 = 1$ 1 term

$(a + b)^1 = a + b$ 2 terms

$(a + b)^2 = a^2 + 2ab + b^2$ 3 terms

$(a + b)^3 = a^3 + 3a^2b + 3ab^2 + b^3$ 4 terms

$(a + b)^4 = a^4 + 4a^3b + 6a^2b^2 + 4ab^3 + b^4$ 5 terms

$(a + b)^5 = a^5 + 5a^4b + 10a^3b^2 + 10a^2b^3 + 5ab^4 + b^5$ 6 terms

Notice the following patterns.

1. The expansion of $(a + b)^n$ contains $n + 1$ terms. For example, for $(a + b)^3$, $n = 3$, and the expansion contains $3 + 1$ terms, or 4 terms.

2. The first term of the expansion of $(a + b)^n$ is a^n, and the last term is b^n.

3. The powers of a decrease by 1 for each term, whereas the powers of b increase by 1 for each term.

4. For each term of the expansion of $(a + b)^n$, the sum of the exponents of a and b is n. (For example, the sum of the exponents of $5a^4b$ is $4 + 1$, or 5, and the sum of the exponents of $10a^3b^2$ is $3 + 2$, or 5.)

OBJECTIVE 1 ▶ Using Pascal's triangle. There are patterns in the coefficients of the terms as well. Written in a triangular array, the coefficients are called **Pascal's triangle.**

$(a + b)^0$:					1					$n = 0$
$(a + b)^1$:				1		1				$n = 1$
$(a + b)^2$:			1		2		1			$n = 2$
$(a + b)^3$:		1		3		3		1		$n = 3$
$(a + b)^4$:	1		4		6		4		1	$n = 4$
$(a + b)^5$:	1	5		10		10		5	1	$n = 5$

Each row in Pascal's triangle begins and ends with 1. Any other number in a row is the sum of the two closest numbers above it. Using this pattern, we can write the next row, for $n = 6$, by first writing the number 1. Then we can add the consecutive numbers in the row for $n = 5$ and write each sum "between and below" the pair. We complete the row by writing a 1.

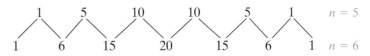

We can use Pascal's triangle and the patterns noted to expand $(a + b)^n$ without actually multiplying any terms.

EXAMPLE 1   Expand $(a + b)^6$.

Solution   Using the $n = 6$ row of Pascal's triangle as the coefficients and following the patterns noted, $(a + b)^6$ can be expanded as

$$a^6 + 6a^5b + 15a^4b^2 + 20a^3b^3 + 15a^2b^4 + 6ab^5 + b^6 \qquad \square$$

PRACTICE

1   Expand $(p + r)^7$.

OBJECTIVE 2 ▶ Evaluating factorials. For a large n, the use of Pascal's triangle to find coefficients for $(a + b)^n$ can be tedious. An alternative method for determining these coefficients is based on the concept of a **factorial.**

The **factorial of n,** written $n!$ (read "n factorial"), is the product of the first n consecutive natural numbers.

Factorial of n: $n!$

If n is a natural number, then $n! = n(n - 1)(n - 2)(n - 3) \cdots \cdot 3 \cdot 2 \cdot 1$. The factorial of 0, written $0!$, is defined to be 1.

For example, $3! = 3 \cdot 2 \cdot 1 = 6$, $5! = 5 \cdot 4 \cdot 3 \cdot 2 \cdot 1 = 120$, and $0! = 1$.

EXAMPLE 2 Evaluate each expression.

a. $\dfrac{5!}{6!}$ **b.** $\dfrac{10!}{7!3!}$ **c.** $\dfrac{3!}{2!1!}$ **d.** $\dfrac{7!}{7!0!}$

Solution

a. $\dfrac{5!}{6!} = \dfrac{5 \cdot 4 \cdot 3 \cdot 2 \cdot 1}{6 \cdot 5 \cdot 4 \cdot 3 \cdot 2 \cdot 1} = \dfrac{1}{6}$

b. $\dfrac{10!}{7!3!} = \dfrac{10 \cdot 9 \cdot 8 \cdot 7!}{7! \cdot 3 \cdot 2 \cdot 1} = \dfrac{10 \cdot 9 \cdot 8}{3 \cdot 2 \cdot 1} = 10 \cdot 3 \cdot 4 = 120$

c. $\dfrac{3!}{2!1!} = \dfrac{3 \cdot 2 \cdot 1}{2 \cdot 1 \cdot 1} = 3$

d. $\dfrac{7!}{7!0!} = \dfrac{7!}{7! \cdot 1} = 1$

PRACTICE
2 Evaluate each expression.

a. $\dfrac{6!}{7!}$ **b.** $\dfrac{8!}{4!2!}$ **c.** $\dfrac{5!}{4!1!}$ **d.** $\dfrac{9!}{9!0!}$

TECHNOLOGY NOTE

When evaluating factorials using a graphing utility, notice the need for parentheses grouping the denominators.

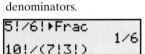

```
5!/6!▶Frac
            1/6
10!/(7!3!)
            120
3!/(2!1!)
             3
■
```

A check for Example 2, parts a, b, and c.

▶ Helpful Hint
We can use a calculator with a factorial key to evaluate a factorial. A calculator uses scientific notation for large results.

OBJECTIVE 3 ▶ Using the binomial theorem. It can be proved, although we won't do so here, that the coefficients of terms in the expansion of $(a + b)^n$ can be expressed in terms of factorials. Following patterns 1 through 4 given earlier and using the factorial expressions of the coefficients, we have what is known as the **binomial theorem.**

Binomial Theorem

If n is a positive integer, then

$$(a + b)^n = a^n + \frac{n}{1!}a^{n-1}b^1 + \frac{n(n-1)}{2!}a^{n-2}b^2$$
$$+ \frac{n(n-1)(n-2)}{3!}a^{n-3}b^3 + \cdots + b^n$$

We call the formula for $(a + b)^n$ given by the binomial theorem the **binomial formula.**

EXAMPLE 3   Use the binomial theorem to expand $(x + y)^{10}$.

Solution   Let $a = x$, $b = y$, and $n = 10$ in the binomial formula.

$$(x + y)^{10} = x^{10} + \frac{10}{1!}x^9 y + \frac{10 \cdot 9}{2!}x^8 y^2 + \frac{10 \cdot 9 \cdot 8}{3!}x^7 y^3 + \frac{10 \cdot 9 \cdot 8 \cdot 7}{4!}x^6 y^4$$

$$+ \frac{10 \cdot 9 \cdot 8 \cdot 7 \cdot 6}{5!}x^5 y^5 + \frac{10 \cdot 9 \cdot 8 \cdot 7 \cdot 6 \cdot 5}{6!}x^4 y^6$$

$$+ \frac{10 \cdot 9 \cdot 8 \cdot 7 \cdot 6 \cdot 5 \cdot 4}{7!}x^3 y^7$$

$$+ \frac{10 \cdot 9 \cdot 8 \cdot 7 \cdot 6 \cdot 5 \cdot 4 \cdot 3}{8!}x^2 y^8$$

$$+ \frac{10 \cdot 9 \cdot 8 \cdot 7 \cdot 6 \cdot 5 \cdot 4 \cdot 3 \cdot 2}{9!}xy^9 + y^{10}$$

$$= x^{10} + 10x^9 y + 45x^8 y^2 + 120x^7 y^3 + 210x^6 y^4 + 252x^5 y^5 + 210x^4 y^6$$

$$+ 120x^3 y^7 + 45x^2 y^8 + 10xy^9 + y^{10}$$

PRACTICE
3   Use the binomial theorem to expand $(a + b)^9$.

EXAMPLE 4   Use the binomial theorem to expand $(x + 2)^5$.

Solution   Let $a = x$ and $b = 2$ in the binomial formula.

$$(x + 2)^5 = x^5 + \frac{5}{1!}x^4(2) + \frac{5 \cdot 4}{2!}x^3(2)^2 + \frac{5 \cdot 4 \cdot 3}{3!}x^2(2)^3$$

$$+ \frac{5 \cdot 4 \cdot 3 \cdot 2}{4!}x(2)^4 + (2)^5$$

$$= x^5 + 10x^4 + 40x^3 + 80x^2 + 80x + 32$$

PRACTICE
4   Use the binomial theorem to expand $(a + 5)^3$.

EXAMPLE 5   Use the binomial theorem to expand $(3m - n)^4$.

Solution   Let $a = 3m$ and $b = -n$ in the binomial formula.

$$(3m - n)^4 = (3m)^4 + \frac{4}{1!}(3m)^3(-n) + \frac{4 \cdot 3}{2!}(3m)^2(-n)^2$$

$$+ \frac{4 \cdot 3 \cdot 2}{3!}(3m)(-n)^3 + (-n)^4$$

$$= 81m^4 - 108m^3 n + 54m^2 n^2 - 12mn^3 + n^4$$

PRACTICE
5   Use the binomial theorem to expand $(3x - 2y)^3$.

OBJECTIVE 4 ▶ Finding the _n_th term of a binomial expansion. Sometimes it is convenient to find a specific term of a binomial expansion without writing out the entire expansion. By studying the expansion of binomials, a pattern forms for each term. This pattern is most easily stated for the $(r + 1)$st term.

> **$(r + 1)$st Term in a Binomial Expansion**
>
> The $(r + 1)$st term of the expansion of $(a + b)^n$ is $\dfrac{n!}{r!(n - r)!}a^{n-r}b^r$.

TECHNOLOGY NOTE

To visually check Example 4, graph $y_1 = (x + 2)^5$ and $y_2 = x^5 + 10x^4 + 40x^3 + 80x^2 + 80x + 32$ in the same viewing window. If the graphs do not coincide, you should double-check your binomial expansion. In this case, the graphs appear to coincide, reinforcing the conclusion of Example 4.

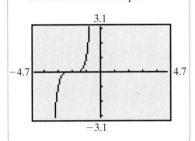

EXAMPLE 6 Find the eighth term in the expansion of $(2x - y)^{10}$.

Solution Use the formula, with $n = 10$, $a = 2x$, $b = -y$, and $r + 1 = 8$. Notice that, since $r + 1 = 8$, $r = 7$.

$$\frac{n!}{r!(n - r)!}a^{n-r}b^r = \frac{10!}{7!3!}(2x)^3(-y)^7$$

$$= 120(8x^3)(-y^7)$$

$$= -960x^3y^7$$

PRACTICE

6 Find the seventh term in the expansion of $(x - 4y)^{11}$.

VOCABULARY & READINESS CHECK

Fill in each blank.

1. $0! = $ _____ **2.** $1! = $ _____ **3.** $4! = $ _____ **4.** $2! = $ _____ **5.** $3!0! = $ _____ **6.** $0!2! = $ _____

11.5 EXERCISE SET

Use Pascal's triangle to expand the binomial. See Example 1.

1. $(m + n)^3$

2. $(x + y)^4$

3. $(c + d)^5$

4. $(a + b)^6$

5. $(y - x)^5$

6. $(q - r)^7$

7. Explain how to generate a row of Pascal's triangle.

8. Write the $n = 8$ row of Pascal's triangle.

Evaluate each expression. See Example 2.

9. $\dfrac{8!}{7!}$ **10.** $\dfrac{6!}{0!}$ **11.** $\dfrac{7!}{5!}$

12. $\dfrac{8!}{5!}$ **13.** $\dfrac{10!}{7!2!}$ **14.** $\dfrac{9!}{5!3!}$

15. $\dfrac{8!}{6!0!}$ **16.** $\dfrac{10!}{4!6!}$

MIXED PRACTICE

Use the binomial formula to expand each binomial. See Examples 3 through 5.

17. $(a + b)^7$

18. $(x + y)^8$

19. $(a + 2b)^5$

20. $(x + 3y)^6$

21. $(q + r)^9$

22. $(b + c)^6$

23. $(4a + b)^5$

24. $(3m + n)^4$

25. $(5a - 2b)^4$

26. $(m - 4)^6$

27. $(2a + 3b)^3$

28. $(4 - 3x)^5$

29. $(x + 2)^5$

30. $(3 + 2a)^4$

Find the indicated term. See Example 6.

31. The fifth term of the expansion of $(c - d)^5$

32. The fourth term of the expansion of $(x - y)^6$

33. The eighth term of the expansion of $(2c + d)^7$

34. The tenth term of the expansion of $(5x - y)^9$

35. The fourth term of the expansion of $(2r - s)^5$

36. The first term of the expansion of $(3q - 7r)^6$

37. The third term of the expansion of $(x + y)^4$

38. The fourth term of the expansion of $(a + b)^8$

39. The second term of the expansion of $(a + 3b)^{10}$

40. The third term of the expansion of $(m + 5n)^7$

REVIEW AND PREVIEW

Sketch the graph of each function. Decide whether each function is one-to-one. See Sections 2.2, 2.7, and 9.2.

41. $f(x) = |x|$

42. $g(x) = 3(x - 1)^2$

43. $H(x) = 2x + 3$

44. $F(x) = -2$

45. $f(x) = x^2 + 3$

46. $h(x) = -(x + 1)^2 - 4$

CONCEPT EXTENSIONS

47. Expand the expression $\left(\sqrt{x} + \sqrt{3}\right)^5$.

48. Find the term containing x^2 in the expansion of $\left(\sqrt{x} - \sqrt{5}\right)^6$.

Evaluate the following.

The notation $\binom{n}{r}$ means $\dfrac{n!}{r!(n - r)!}$. For example,

$$\binom{5}{3} = \frac{5!}{3!(5 - 3)!} = \frac{5!}{3!2!} = \frac{5 \cdot 4 \cdot 3 \cdot 2 \cdot 1}{(3 \cdot 2 \cdot 1) \cdot (2 \cdot 1)} = 10.$$

49. $\binom{9}{5}$

50. $\binom{4}{3}$

51. $\binom{8}{2}$

52. $\binom{12}{11}$

53. Show that $\binom{n}{n} = 1$ for any whole number n.

 STUDY SKILLS BUILDER

Are You Prepared for Your Final Exam?

To prepare for your final exam, try the following study techniques:

- Review the material that you will be responsible for on your exam. This includes material from your textbook, your notebook, and any handouts from your instructor.
- Review any formulas that you may need to memorize.
- Check to see if your instructor or mathematics department will be conducting a final exam review.
- Check with your instructor to see whether final exams from previous semesters/quarters are available to students for review.

- Use your previously taken exams as a practice final exam. To do so, rewrite the test questions in mixed order on blank sheets of paper. This will help you prepare for exam conditions.
- If you are unsure of a few concepts, see your instructor or visit a learning lab for assistance. Also, view the video segment of any troublesome sections.
- If you need further exercises to work, try the Cumulative Reviews at the end of the chapters.

Once again, good luck! I hope you have enjoyed this textbook and your mathematics course.

CHAPTER 11 GROUP ACTIVITY

Modeling College Tuition

Annual college tuition has steadily increased since 1970. According to the College Board, by the 2004–2005 academic year, the average annual tuition at a public 4-year university had increased to $5132. Similarly, average annual tuition at private 4-year universities had grown to $20,082.

Over the past few years, annual tuition at 4-year public universities has been increasing at an average rate of 7.9% per year. Over the same time period, annual tuition at 4-year private universities has been increasing at an average rate of $1068 per year. In this project, you will have the opportunity to model and investigate the trend in increasing tuition at public and private universities. This project may be completed by working in groups or individually.

1. Using the information given in the introductory paragraphs, decide whether the sequence of public university tuitions is arithmetic or geometric.

2. Using the information given in the introductory paragraphs, decide whether the sequence of private university tuitions is arithmetic or geometric.

3. Find the general term of the sequence that describes the pattern of average annual tuition for 4-year public universities. Let $n = 1$ represent the 2004–2005 academic year.

4. Find the general term of the sequence that describes the pattern of average annual tuition for private 4-year universities. Let $n = 1$ represent the 2004–2005 academic year.

Academic Year	n
2004–2005	1
2005–2006	2
2006–2007	3
2007–2008	4
2008–2009	5
2009–2010	6
2010–2011	7

5. Assuming that the rate of tuition increase remains the same, use the general term equation from Question 3 to find the average annual tuition at a 4-year public university for the 2007–2008 academic year.

6. Assuming that the rate of tuition increase remains the same, use the general term equation from Question 4 to find the

average annual tuition at a 4-year private university for the 2007–2008 academic year.

7. Use partial sums to find the average cost of a 4-year college education at a public university for a student who started college in the 2004–2005 academic year.

8. Use partial sums to find the average cost of a 4-year college education at a private university for a student who started college in the 2007–2008 academic year. (*Hint:* One way to do this is to find S_7 and subtract S_3 from it. If you use this method, explain why this gives the desired sum.)

9. (Optional) Use newspapers or news magazines to find a situation that can be modeled by a sequence. Briefly describe the situation, and decide whether it is an arithmetic or geometric sequence. Find an equation of the general term of the sequence.

CHAPTER 11 VOCABULARY CHECK

Fill in each blank with one of the words or phrases listed below.

general term   common difference   finite sequence   common ratio   Pascal's triangle

infinite sequence   factorial of n arithmetic sequence   geometric sequence   series

1. A(n) _____ is a function whose domain is the set of natural numbers $\{1, 2, 3, \ldots, n\}$, where n is some natural number.

2. The _____, written $n!$, is the product of the first n consecutive natural numbers.

3. A(n) _____ is a function whose domain is the set of natural numbers.

4. A(n) _____ is a sequence in which each term (after the first) is obtained by multiplying the preceding term by a constant amount r. The constant r is called the _____ of the sequence.

5. The sum of the terms of a sequence is called a _____.

6. The nth term of the sequence a_n is called the _____.

7. A(n) _____ is a sequence in which each term (after the first) differs from the preceding term by a constant amount d. The constant d is called the _____ of the sequence.

8. A(n) triangle array of the coefficients of the terms of the expansions of $(a + b)^n$ is called _____.

> ▶ **Helpful Hint**
>
> Are you preparing for your test? Don't forget to take the Chapter 11 Test on page 756. Then check your answers at the back of the text and use the Chapter Test Prep Video CD to see the fully worked-out solutions to any of the exercises you want to review.

CHAPTER 11 HIGHLIGHTS

DEFINITIONS AND CONCEPTS	EXAMPLES

SECTION 11.1 SEQUENCES

An **infinite sequence** is a function whose domain is the set of natural numbers $\{1, 2, 3, 4, \ldots\}$.

A **finite sequence** is a function whose domain is the set of natural numbers $\{1, 2, 3, 4, \ldots, n\}$, where n is some natural number.

The notation a_n, where n is a natural number, is used to denote a sequence.

Infinite Sequence
$$2, 4, 6, 8, 10, \ldots$$

Finite Sequence
$$1, -2, 3, -4, 5, -6$$

Write the first four terms of the sequence whose general term is $a_n = n^2 + 1$.

$$a_1 = 1^2 + 1 = 2$$
$$a_2 = 2^2 + 1 = 5$$
$$a_3 = 3^2 + 1 = 10$$
$$a_4 = 4^2 + 1 = 17$$

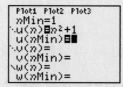

} First 4 terms of the sequence

A calculator check for writing the terms of the sequence $a_n = n^2 + 1$.

DEFINITIONS AND CONCEPTS	**EXAMPLES**

SECTION 11.2 ARITHMETIC AND GEOMETRIC SEQUENCES

An **arithmetic sequence** is a sequence in which each term differs from the preceding term by a constant amount d, called the **common difference.**

Arithmetic Sequence

$$5, 8, 11, 14, 17, 20, \ldots$$

Here, $a_1 = 5$ and $d = 3$.

The general term is

$$a_n = a_1 + (n - 1)d \text{ or}$$
$$a_n = 5 + (n - 1)3$$

The **general term** a_n of an arithmetic sequence is given by

$$a_n = a_1 + (n - 1)d$$

where a_1 is the first term and d is the common difference.

A **geometric sequence** is a sequence in which each term is obtained by multiplying the preceding term by a constant r, called the **common ratio.**

Geometric Sequence

$$12, -6, 3, -\frac{3}{2}, \ldots$$

Here $a_1 = 12$ and $r = -\frac{1}{2}$.

The general term is

$$a_n = a_1 r^{n-1} \text{ or}$$
$$a_n = 12\left(-\frac{1}{2}\right)^{n-1}$$

The **general term** a_n of a geometric sequence is given by

$$a_n = a_1 r^{n-1}$$

where a_1 is the first term and r is the common ratio.

SECTION 11.3 SERIES

A sum of the terms of a sequence is called a **series.**

A shorthand notation for denoting a series is called **summation notation:**

index of summation $\rightarrow \displaystyle\sum_{i=1}^{4} \rightarrow$ Greek letter sigma used to mean "sum"

Sequence	*Series*	
$3, 7, 11, 15$	$3 + 7 + 11 + 15$	finite
$3, 7, 11, 15, \ldots$	$3 + 7 + 11 + 15 + \cdots$	infinite

$$\sum_{i=1}^{4} 3^i = 3^1 + 3^2 + 3^3 + 3^4$$
$$= 3 + 9 + 27 + 81$$
$$= 120$$

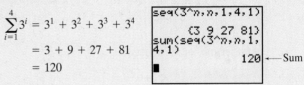

A calculator check of the sum.

SECTION 11.4 PARTIAL SUMS OF ARITHMETIC AND GEOMETRIC SEQUENCES

Partial sum, S_n, of the first n terms of an arithmetic sequence:

$$s_n = \frac{n}{2}(a_1 + a_n)$$

where a_1 is the first term and a_n is the nth term.

The sum of the first five terms of the arithmetic sequence

$$12, 24, 36, 48, 60, \ldots \text{ is}$$

$$S_5 = \frac{5}{2}(12 + 60) = 180$$

Partial sum, S_n, of the first n terms of a geometric sequence:

$$S_n = \frac{a_1(1 - r^n)}{1 - r}$$

where a_1 is the first term, r is the common ratio, and $r \neq 1$.

The sum of the first five terms of the geometric sequence

$$15, 30, 60, 120, 240, \ldots \text{ is}$$

$$S_5 = \frac{15(1 - 2^5)}{1 - 2} = 465$$

Sum of the terms of an infinite geometric sequence:

$$S_\infty = \frac{a_1}{1 - r}$$

where a_1 is the first term, r is the common ratio, and $|r| < 1$. (If $|r| \geq 1$, S_∞ does not exist.)

The sum of the terms of the infinite geometric sequence

$$1, \frac{1}{3}, \frac{1}{9}, \frac{1}{27}, \ldots \text{ is}$$

$$S_\infty = \frac{1}{1 - \frac{1}{3}} = \frac{3}{2}$$

DEFINITIONS AND CONCEPTS	**EXAMPLES**

SECTION 11.5   THE BINOMIAL THEOREM

The **factorial of n,** written $n!$, is the product of the first n consecutive natural numbers.

Binomial Theorem

If n is a positive integer, then

$$(a + b)^n = a^n + \frac{n}{1!}a^{n-1}b^1 + \frac{n(n-1)}{2!}a^{n-2}b^2$$
$$+ \frac{n(n-1)(n-2)}{3!}a^{n-3}b^3 + \cdots + b^n$$

$$5! = 5 \cdot 4 \cdot 3 \cdot 2 \cdot 1 = 120$$

Expand $(3x + y)^4$.

$$(3x + y)^4 = (3x)^4 + \frac{4}{1!}(3x)^3(y)^1$$
$$+ \frac{4 \cdot 3}{2!}(3x)^2(y)^2 + \frac{4 \cdot 3 \cdot 2}{3!}(3x)^1 y^3 + y^4$$
$$= 81x^4 + 108x^3y + 54x^2y^2 + 12xy^3 + y^4$$

CHAPTER 11   REVIEW

(11.1) *Find the indicated term(s) of the given sequence.*

1. The first five terms of the sequence $a_n = -3n^2$

2. The first five terms of the sequence $a_n = n^2 + 2n$

3. The one-hundredth term of the sequence $a_n = \dfrac{(-1)^n}{100}$

4. The fiftieth term of the sequence $a_n = \dfrac{2n}{(-1)^2}$

5. The general term a_n of the sequence $\dfrac{1}{6}, \dfrac{1}{12}, \dfrac{1}{18}, \dots$

6. The general term a_n of the sequence $-1, 4, -9, 16, \dots$

Solve the following applications.

7. The distance in feet that an olive falling from rest in a vacuum will travel during each second is given by an arithmetic sequence whose general term is $a_n = 32n - 16$, where n is the number of the second. Find the distance the olive will fall during the fifth, sixth, and seventh seconds.

8. A culture of yeast doubles every day in a geometric progression whose general term is $a_n = 100(2)^{n-1}$, where n is the number of the day just ending. Find how many days it takes the yeast culture to measure at least 10,000. Find the original measure of the yeast culture.

9. The Colorado Forest Service reported that western pine beetle infestation, which kills trees, affected approximately 660,000 acres of lodgepole forests in Colorado in 2006. The forest service predicts that during the next 5 years, the beetles will infest twice the number of acres per year as the year before. Write out the first 5 terms of this geometric sequence, and predict the number of acres of infested trees there will be in 2010.

10. The first row of an amphitheater contains 50 seats, and each row thereafter contains 8 additional seats. Write the first ten terms of this arithmetic progression, and find the number of seats in the tenth row.

(11.2)

11. Find the first five terms of the geometric sequence whose first term is -2 and whose common ratio is $\dfrac{2}{3}$.

12. Find the first five terms of the arithmetic sequence whose first term is 12 and whose common difference is -1.5.

13. Find the thirtieth term of the arithmetic sequence whose first term is -5 and whose common difference is 4.

14. Find the eleventh term of the arithmetic sequence whose first term is 2 and whose common difference is $\dfrac{3}{4}$.

15. Find the twentieth term of the arithmetic sequence whose first three terms are 12, 7, and 2.

16. Find the sixth term of the geometric sequence whose first three terms are 4, 6, and 9.

17. If the fourth term of an arithmetic sequence is 18 and the twentieth term is 98, find the first term and the common difference.

18. If the third term of a geometric sequence is -48 and the fourth term is 192, find the first term and the common ratio.

19. Find the general term of the sequence $\dfrac{3}{10}, \dfrac{3}{100}, \dfrac{3}{1000}, \dots$

20. Find a general term that satisfies the terms shown for the sequence 50, 58, 66, $\dots$

Determine whether each of the following sequences is arithmetic, geometric, or neither. If a sequence is arithmetic, find a_1 and d. If a sequence is geometric, find a_1 and r.

21. $\dfrac{8}{3}, 4, 6, \ldots$

22. $-10.5, -6.1, -1.7$

23. $7x, -14x, 28x$

24. $3x^2, 9x^4, 81x^8, \ldots$

Solve the following applications.

25. To test the bounce of a racquetball, the ball is dropped from a height of 8 feet. The ball is judged "good" if it rebounds at least 75% of its previous height with each bounce. Write out the first six terms of this geometric sequence (round to the nearest tenth). Determine if a ball is "good" that rebounds to a height of 2.5 feet after the fifth bounce.

26. A display of oil cans in an auto parts store has 25 cans in the bottom row, 21 cans in the next row, and so on, in an arithmetic progression. Find the general term and the number of cans in the top row.

27. Suppose that you save $1 the first day of a month, $2 the second day, $4 the third day, continuing to double your savings each day. Write the general term of this geometric sequence and find the amount you will save on the tenth day. Estimate the amount you will save on the thirtieth day of the month, and check your estimate with a calculator.

28. On the first swing, the length of an arc through which a pendulum swings is 30 inches. The length of the arc for each successive swing is 70% of the preceding swing. Find the length of the arc for the fifth swing.

29. Rosa takes a job that has a monthly starting salary of $900 and guarantees her a monthly raise of $150 during her 6-month training period. Find the general term of this sequence and her salary at the end of her training.

30. A sheet of paper is $\dfrac{1}{512}$-inch thick. By folding the sheet in half, the total thickness will be $\dfrac{1}{256}$ inch. A second fold produces a total thickness of $\dfrac{1}{128}$ inch. Estimate the thickness of the stack after 15 folds, and then check your estimate with a calculator.

(11.3) Write out the terms and find the sum for each of the following.

31. $\displaystyle\sum_{i=1}^{5}(2i - 1)$

32. $\displaystyle\sum_{i=1}^{5}i(i + 2)$

33. $\displaystyle\sum_{i=2}^{4}\dfrac{(-1)^i}{2i}$

34. $\displaystyle\sum_{i=3}^{5}5(-1)^{i-1}$

Find the partial sum of the given sequence.

35. S_4 of the sequence $a_n = (n - 3)(n + 2)$

36. S_6 of the sequence $a_n = n^2$

37. S_5 of the sequence $a_n = -8 + (n - 1)3$

38. S_3 of the sequence $a_n = 5(4)^{n-1}$

Write the sum with Σ notation.

39. $1 + 3 + 9 + 27 + 81 + 243$

40. $6 + 2 + (-2) + (-6) + (-10) + (-14) + (-18)$

41. $\dfrac{1}{4} + \dfrac{1}{16} + \dfrac{1}{64} + \dfrac{1}{256}$

42. $1 + \left(-\dfrac{3}{2}\right) + \dfrac{9}{4}$

Solve.

43. A yeast colony begins with 20 yeast and doubles every 8 hours. Write the sequence that describes the growth of the yeast, and find the total yeast after 48 hours.

44. The number of cranes born each year in a new aviary forms a sequence whose general term is $a_n = n^2 + 2n - 1$. Find the number of cranes born in the fourth year and the total number of cranes born in the first four years.

45. Harold has a choice between two job offers. Job A has an annual starting salary of $39,500 with guaranteed annual raises of $2200 for the next four years, whereas job B has an annual starting salary of $41,000 with guaranteed annual raises of $1400 for the next four years. Compare the salaries for the fifth year under each job offer.

46. A sample of radioactive waste is decaying such that the amount decaying in kilograms during year n is $a_n = 200(0.5)^n$. Find the amount of decay in the third year, and the total amount of decay in the first three years.

(11.4) Find the partial sum of the given sequence.

47. The sixth partial sum of the sequence $15, 19, 23, \ldots$

48. The ninth partial sum of the sequence $5, -10, 20, \ldots$

49. The sum of the first 30 odd positive integers

50. The sum of the first 20 positive multiples of 7

51. The sum of the first 20 terms of the sequence $8, 5, 2, \ldots$

52. The sum of the first eight terms of the sequence $\dfrac{3}{4}, \dfrac{9}{4}, \dfrac{27}{4}, \ldots$

53. S_4 if $a_1 = 6$ and $r = 5$

54. S_{100} if $a_1 = -3$ and $d = -6$

Find the sum of each infinite geometric sequence.

55. $5, \dfrac{5}{2}, \dfrac{5}{4}, \ldots$

56. $18, -2, \dfrac{2}{9}, \ldots$

57. $-20, -4, -\dfrac{4}{5}, \ldots$

58. $0.2, 0.02, 0.002, \ldots$

Solve.

59. A frozen yogurt store owner cleared \$20,000 the first year he owned his business and made an additional 15% over the previous year in each subsequent year. Find how much he made during his fourth year of business. Find his total earnings during the first 4 years (round to the nearest dollar).

60. On his first morning in a television assembly factory, a trainee takes 0.8 times as long to assemble each television as he took to assemble the one before. If it took him 40 minutes to assemble the first television, find how long it takes him to assemble the fourth television. Find the total time he takes to assemble the first four televisions (round to the nearest minute).

61. During the harvest season a farmer can rent a combine machine for \$100 the first day, with the rental fee decreasing \$7 for each additional day. Find how much the farmer pays for the rental on the seventh day. Find how much total rent the farmer pays for 7 days.

62. A rubber ball is dropped from a height of 15 feet and rebounds 80% of its previous height after each bounce. Find the total distance the ball travels before it comes to rest.

63. After a pond was sprayed once with insecticide, 1800 mosquitoes were killed the first day, 600 the second day, 200 the third day, and so on. Find the total number of mosquitoes killed during the first 6 days after the spraying (round to the nearest mosquito).

64. See Exercise 63. Find the day on which the insecticide is no longer effective, and find the total number of mosquitoes killed (round to the nearest mosquito).

65. Use the formula S_∞ to write $0.55\overline{5}$ as a fraction.

66. A movie theater has 27 seats in the first row, 30 seats in the second row, 33 seats in the third row, and so on. Find the total number of seats in the theater if there are 20 rows.

(11.5) Use Pascal's triangle to expand each binomial.

67. $(x + z)^5$

68. $(y - r)^6$

69. $(2x + y)^4$

70. $(3y - z)^4$

Use the binomial formula to expand the following.

71. $(b + c)^8$

72. $(x - w)^7$

73. $(4m - n)^4$

74. $(p - 2r)^5$

Find the indicated term.

75. The fourth term of the expansion of $(a + b)^7$

76. The eleventh term of the expansion of $(y + 2z)^{10}$

MIXED REVIEW

77. Evaluate: $\sum_{i=1}^{4} i^2(i + 1)$

78. Find the fifteenth term of the arithmetic sequence whose first three terms are 14, 8, and 2.

79. Find the sum of the infinite geometric sequence $27, 9, 3, 1, \ldots$

80. Expand: $(2x - 3)^4$

CHAPTER 11 TEST TEST PREP VIDEO

Remember to use the Chapter Test Prep Video CD to see the fully worked-out solutions to any of the exercises you want to review.

Find the indicated term(s) of the given sequence.

1. The first five terms of the sequence $a_n = \dfrac{(-1)^n}{n + 4}$

2. The eightieth term of the sequence $a_n = 10 + 3(n - 1)$

3. The general term of the sequence $\dfrac{2}{5}, \dfrac{2}{25}, \dfrac{2}{125}, \ldots$

4. The general term of the sequence $-9, 18, -27, 36, \ldots$

Find the partial sum of the given sequence.

5. S_5 of the sequence $a_n = 5(2)^{n-1}$

6. S_{30} of the sequence $a_n = 18 + (n - 1)(-2)$

7. S_∞ of the sequence $a_1 = 24$ and $r = \dfrac{1}{6}$

8. S_∞ of the sequence $\dfrac{3}{2}, -\dfrac{3}{4}, \dfrac{3}{8}, \ldots$

9. $\sum_{i=1}^{4} i(i - 2)$

10. $\sum_{i=2}^{4} 5(2)^i(-1)^{i-1}$

Expand each binomial.

11. $(a - b)^6$

12. $(2x + y)^5$

Solve the following applications.

13. The population of a small town is growing yearly according to the sequence defined by $a_n = 250 + 75(n - 1)$, where n is the number of the year just beginning. Predict the population at the beginning of the tenth year. Find the town's initial population.

14. A gardener is making a triangular planting with one shrub in the first row, three shrubs in the second row, five shrubs in the third row, and so on, for eight rows. Write the finite series of this sequence, and find the total number of shrubs planted.

15. A pendulum swings through an arc of length 80 centimeters on its first swing. On each successive swing, the length of the arc is $\dfrac{3}{4}$ the length of the arc on the preceding swing.

Find the length of the arc on the fourth swing, and find the total arc length for the first four swings.

16. See Exercise 15. Find the total arc length before the pendulum comes to rest.

17. A parachutist in free fall falls 16 feet during the first second, 48 feet during the second second, 80 feet during the third second, and so on. Find how far he falls during the tenth second. Find the total distance he falls during the first 10 seconds.

18. Use the formula S_∞ to write $0.42\overline{42}$ as a fraction.

CHAPTER 11 CUMULATIVE REVIEW

1. Divide.

  a. $\dfrac{20}{-4}$ **b.** $\dfrac{-9}{-3}$

  c. $-\dfrac{3}{8} \div 3$ **d.** $\dfrac{-40}{10}$

  e. $\dfrac{-1}{10} \div \dfrac{-2}{5}$ **f.** $\dfrac{8}{0}$

2. Simplify each expression.

  a. $3a - (4a + 3)$

  b. $(5x - 3) + (2x + 6)$

  c. $4(2x - 5) - 3(5x + 1)$

3. Suppose that a computer store just announced an 8% decrease in the price of a particular computer model. If this computer sells for $2162 after the decrease, find the original price of this computer.

4. Sara bought a digital camera for $344.50 including tax. If the tax rate is 6%, what was the price of the camera before taxes?

5. Solve: $3y - 2x = 7$ for y.

6. Find an equation of a line through $(3, -2)$ and parallel to $3x - 2y = 6$. Write the equation using function notation.

7. Use the product rule to multiply.

  a. $(3x^6)(5x)$

  b. $(-2.4x^3p^2)(4xp^{10})$

8. Solve: $y^3 + 5y^2 - y = 5$

9. Use synthetic division to divide $(x^4 - 2x^3 - 11x^2 + 5x + 34)$ by $(x + 2)$.

10. Perform the indicated operation and simplify, if possible.
$$\frac{5}{3a - 6} - \frac{a}{a - 2} + \frac{3 + 2a}{5a - 10}$$

11. Simplify the following.

  a. $\sqrt{50}$ **b.** $\sqrt[3]{24}$

  c. $\sqrt{26}$ **d.** $\sqrt[4]{32}$

12. Solve: $\sqrt{3x + 6} - \sqrt{7x - 6} = 0$

13. Use the formula $A = P(1 + r)^t$ to find the interest rate r if $2000 compounded annually grows to $2420 in 2 years.

14. Rationalize each denominator.

  a. $\sqrt[3]{\dfrac{4}{3x}}$ **b.** $\dfrac{\sqrt{2} + 1}{\sqrt{2} - 1}$

15. Solve: $(x - 3)^2 - 3(x - 3) - 4 = 0$.

16. Solve: $\dfrac{10}{(2x + 4)^2} - \dfrac{1}{2x + 4} = 3$.

17. Solve: $\dfrac{5}{x + 1} < -2$.

18. Graph $f(x) = (x + 2)^2 - 6$. Find the vertex and axis of symmetry.

19. A rock is thrown upward from the ground. Its height in feet above ground after t seconds is given by the function $f(t) = -16t^2 + 20t$. Find the maximum height of the rock and the number of seconds it took for the rock to reach its maximum height.

20. Find the vertex of $f(x) = x^2 + 3x - 18$.

21. If $f(x) = x^2$ and $g(x) = x + 3$, find each composition.

  a. $(f \circ g)(2)$ and $(g \circ f)(2)$

  b. $(f \circ g)(x)$ and $(g \circ f)(x)$

22. Find the inverse of $f(x) = -2x + 3$.

23. Find the inverse of the one-to-one function $f = \{(0, 1), (-2, 7), (3, -6), (4, 4)\}$.

24. If $f(x) = x^2 - 2$ and $g(x) = x + 1$, find each composition.

  a. $(f \circ g)(2)$ and $(g \circ f)(2)$

  b. $(f \circ g)(x)$ and $(g \circ f)(x)$

25. Solve each equation for x.

  a. $2^x = 16$ **b.** $9^x = 27$

  c. $4^{x+3} = 8^x$ **d.** $5^x = 10$

26. Solve each equation.

   a. $\log_2 32 = x$ **b.** $\log_4 \frac{1}{64} = x$

   c. $\log_{\frac{1}{2}} x = 5$

27. Simplify.

   a. $\log_3 3^2$ **b.** $\log_7 7^{-1}$

   c. $5^{\log_5 3}$ **d.** $2^{\log_2 6}$

28. Solve each equation for x.

   a. $4^x = 64$ **b.** $8^x = 32$

   c. $9^{x+4} = 243^x$

29. Write each sum as a single logarithm.

   a. $\log_{11} 10 + \log_{11} 3$

   b. $\log_3 \frac{1}{2} + \log_3 12$

   c. $\log_2(x + 2) + \log_2 x$

30. Find the exact value.

   a. $\log 100{,}000$ **b.** $\log 10^{-3}$

   c. $\ln \sqrt[5]{e}$ **d.** $\ln e^4$

31. Find the amount owed at the end of 5 years if $1600 is loaned at a rate of 9% compounded continuously.

32. Write each expression as a single logarithm.

   a. $\log_6 5 + \log_6 4$

   b. $\log_8 12 - \log_8 4$

   c. $2 \log_2 x + 3 \log_2 x - 2 \log_2(x - 1)$

33. Solve: $3^x = 7$.

34. Using $A = P\left(1 + \dfrac{r}{n}\right)^{nt}$, find how long it takes $5000 to double if it is invested at 2% interest compounded quarterly. Round to the nearest tenth.

35. Solve: $\log_4(x - 2) = 2$.

36. Solve: $\log_4 10 - \log_4 x = 2$.

37. Graph $\dfrac{x^2}{16} - \dfrac{y^2}{25} = 1$.

38. Find the distance between $(8, 5)$ and $(-2, 4)$.

39. Solve the system $\begin{cases} y = \sqrt{x} \\ x^2 + y^2 = 6 \end{cases}$.

40. Solve the system $\begin{cases} x^2 + y^2 = 36 \\ x - y = 6 \end{cases}$.

41. Graph $\dfrac{x^2}{9} + \dfrac{y^2}{16} \le 1$.

42. Graph $\begin{cases} y \ge x^2 \\ y \le 4 \end{cases}$.

43. Write the first five terms of the sequence whose general term is given by $a_n = n^2 - 1$.

44. If the general term of a sequence is $a_n = \dfrac{n}{n + 4}$, find a_8.

45. Find the eleventh term of the arithmetic sequence whose first three terms are 2, 9, and 16.

46. Find the sixth term of the geometric sequence 2, 10, 50, ...

47. Evaluate.

   a. $\displaystyle\sum_{i=0}^{6} \frac{i - 2}{2}$ **b.** $\displaystyle\sum_{i=3}^{5} 2^i$

48. Evaluate.

   a. $\displaystyle\sum_{i=0}^{4} i(i + 1)$ **b.** $\displaystyle\sum_{i=0}^{3} 2^i$

49. Find the sum of the first 30 positive integers.

50. Find the third term of the expansion of $(x - y)^6$.

Appendix A

The Bigger Picture/Practice Final Exam

A.1 THE BIGGER PICTURE—SOLVING EQUATIONS AND INEQUALITIES

I. Equations

A. Linear Equations (Secs. 1.5 and 3.1)

$$5(x - 2) = \frac{4(2x + 1)}{3}$$

$$3 \cdot 5(x - 2) = \cancel{3} \cdot \frac{4(2x + 1)}{\cancel{3}}$$

$$15x - 30 = 8x + 4$$

$$7x = 34$$

$$x = \frac{34}{7}$$

B. Absolute Value Equations (Sec. 3.4)

$$|3x - 1| = 8$$

$3x - 1 = 8$ or $3x - 1 = -8$

$3x = 9$ or $3x = -7$

$x = 3$ or $x = -\dfrac{7}{3}$

$$|x - 5| = |x + 1|$$

$x - 5 = x + 1$ or $x - 5 = -(x + 1)$

$\underline{-5 = 1}$ or $x - 5 = -x - 1$

No solution or $2x = 4$

or $x = 2$

C. Quadratic and Higher-Degree Equations (Secs. 5.8, 8.1, 8.2, 8.3)

$$2x^2 - 7x = 9$$
$$2x^2 - 7x - 9 = 0$$
$$(2x - 9)(x + 1) = 0$$
$$2x - 9 = 0 \quad \text{or} \quad x + 1 = 0$$
$$x = \frac{9}{2} \quad \text{or} \quad x = -1$$

$$2x^2 + x - 2 = 0$$
$$a = 2, \ b = 1, \ c = -2$$
$$x = \frac{-1 \pm \sqrt{1^2 - 4(2)(-2)}}{2 \cdot 2}$$
$$x = \frac{-1 \pm \sqrt{17}}{4}$$

D. Equations with Rational Expressions (Sec. 6.5)

$$\frac{7}{x - 1} + \frac{3}{x + 1} = \frac{x + 3}{x^2 - 1}$$

$$\cancel{(x - 1)}(x + 1) \cdot \frac{7}{\cancel{x - 1}} + (x - 1)\cancel{(x + 1)} \cdot \frac{3}{\cancel{x + 1}}$$

$$= \cancel{(x - 1)}\cancel{(x + 1)} \cdot \frac{x + 3}{\cancel{(x - 1)}\cancel{(x + 1)}}$$

$$7(x + 1) + 3(x - 1) = x + 3$$
$$7x + 7 + 3x - 3 = x + 3$$
$$9x = -1$$
$$x = -\frac{1}{9}$$

759

E. Equations with Radicals (Sec. 7.6)

$$\sqrt{5x + 10} - 2 = x$$
$$\sqrt{5x + 10} = x + 2$$
$$(\sqrt{5x + 10})^2 = (x + 2)^2$$
$$5x + 10 = x^2 + 4x + 4$$
$$0 = x^2 - x - 6$$
$$0 = (x - 3)(x + 2)$$
$$x - 3 = 0 \quad \text{or} \quad x + 2 = 0$$
$$x = 3 \quad \text{or} \quad x = -2$$

Both solutions check.

F. Exponential Equations (Secs. 9.3, 9.7)

$$9^x = 27^{x+1} \qquad\qquad 5^x = 7$$
$$(3^2)^x = (3^3)^{x+1} \qquad \log 5^x = \log 7$$
$$3^{2x} = 3^{3x+3} \qquad\quad x \log 5 = \log 7$$
$$2x = 3x + 3 \qquad\qquad x = \dfrac{\log 7}{\log 5}$$
$$-3 = x$$

G. Logarithmic Equations (Sec. 9.7)

$$\log 7 + \log(x + 3) = \log 5$$
$$\log 7(x + 3) = \log 5$$
$$7(x + 3) = 5$$
$$7x + 21 = 5$$
$$7x = -16$$
$$x = \dfrac{-16}{7}$$

II. Inequalities

A. Linear Inequalities (Sec. 3.2)

$$-3(x + 2) \geq 6$$
$$-3x - 6 \geq 6$$
$$-3x \geq 12$$
$$\dfrac{-3x}{-3} \leq \dfrac{12}{-3}$$
$$x \leq -4 \quad \text{or} \quad (-\infty, -4]$$

B. Compound Inequalities (Sec. 3.3)

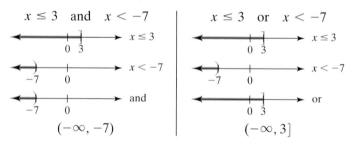

C. Absolute Value Inequalities (Sec. 3.5)

$$|x - 5| - 8 < -2 \qquad\qquad |2x + 1| \geq 17$$
$$|x - 5| < 6 \qquad 2x + 1 \geq 17 \quad \text{or} \quad 2x + 1 \leq -17$$
$$-6 < x - 5 < 6 \qquad\quad 2x \geq 16 \quad \text{or} \qquad 2x \leq -18$$
$$-1 < x < 11 \qquad\qquad x \geq 8 \quad \text{or} \qquad x \leq -9$$
$$(-1, 11) \qquad\qquad\quad (-\infty, -9] \cup [8, \infty)$$

D. Nonlinear Inequalities (Sec. 8.4)

$$x^2 - x < 6$$
$$x^2 - x - 6 < 0$$
$$(x - 3)(x + 2) < 0$$

$$(-2, 3)$$

$$\frac{x - 5}{x + 1} \geq 0$$

$$(-\infty, -1) \cup [5, \infty)$$

A.2 PRACTICE FINAL EXAM

Simplify. If needed, write answers with positive exponents only.

1. $\sqrt{216}$

2. $\dfrac{(4 - \sqrt{16}) - (-7 - 20)}{-2(1 - 4)^2}$

3. $\left(\dfrac{1}{125}\right)^{-1/3}$

4. $(-9x)^{-2}$

5. $\dfrac{\dfrac{5}{x} - \dfrac{7}{3x}}{\dfrac{9}{8x} - \dfrac{1}{x}}$

6. $\left(\dfrac{64c^{4/3}}{a^{-2/3}b^{5/6}}\right)^{1/2}$

7. Ann-Margaret Tober is deciding whether to accept a part-time sales position at Campo Electronics. She is offered a gross monthly pay of $1500 plus 5% commission on her sales.

   a. Complete the gross monthly pay table for the given amounts of sales.

Sales (Dollars)	8000	9000	10,000	11,000	12,000
Gross Monthly Pay (Dollars)					

   b. If she has sales of $11,000 per month, what is her gross annual pay?

   c. If Ann-Margaret decides she needs a gross monthly pay of $2200, how much must she sell every month?

Factor completely.

8. $3x^2y - 27y^3$

9. $16y^3 - 2$

10. $x^2y - 9y - 3x^2 + 27$

Perform the indicated operations and simplify if possible.

11. $(4x^3y - 3x - 4) - (9x^3y + 8x + 5)$

12. $(6m + n)^2$

13. $(2x - 1)(x^2 - 6x + 4)$

14. $\dfrac{3x^2 - 12}{x^2 + 2x - 8} \div \dfrac{6x + 18}{x + 4}$

15. $\dfrac{2x^2 + 7}{2x^4 - 18x^2} - \dfrac{6x + 7}{2x^4 - 18x^2}$

16. $\dfrac{3}{x^2 - x - 6} + \dfrac{2}{x^2 - 5x + 6}$

17. $\sqrt{125x^3} - 3\sqrt{20x^3}$

18. $(\sqrt{5} + 5)(\sqrt{5} - 5)$

19. $(4x^3 - 5x) \div (2x + 1)$ [Use long division.]

20. Solve $15x + 26 = -2(x + 1) - 1$ algebraically and graphically.

Solve each equation or inequality. Write inequality solutions using interval notation.

21. $|6x - 5| - 3 = -2$

22. $3n(7n - 20) = 96$

23. $-3 < 2(x - 3) \leq 4$

24. $|3x + 1| > 5$

25. $\dfrac{x^2 + 8}{x} - 1 = \dfrac{2(x + 4)}{x}$

26. $y^2 - 3y = 5$

27. $x = \sqrt{x - 2} + 2$

28. $2x^2 - 7x > 15$

29. Solve the system: $\begin{cases} \dfrac{x}{2} + \dfrac{y}{4} = -\dfrac{3}{4} \\ x + \dfrac{3}{4}y = -4 \end{cases}$

Graph the following.

30. $4x + 6y = 7$

31. $2x - y > 5$

32. $y = -3$

33. $g(x) = -|x + 2| - 1$. Also, find the domain and range of this function.

34. $h(x) = x^2 - 4x + 4$. Label the vertex and any intercepts.

35. $f(x) = \begin{cases} -\dfrac{1}{2}x & \text{if } x \leq 0 \\ 2x - 3 & \text{if } x > 0 \end{cases}$. Also, find the domain and range of this function.

Write equations of the following lines. Write each equation using function notation.

36. through $(4, -2)$ and $(6, -3)$

37. through $(-1, 2)$ and perpendicular to $3x - y = 4$

Find the distance or midpoint.

38. Find the distance between the points $(-6, 3)$ and $(-8, -7)$.

39. Find the midpoint of the line segment whose endpoints are $(-2, -5)$ and $(-6, 12)$.

Rationalize each denominator. Assume that variables represent positive numbers.

40. $\sqrt{\dfrac{9}{y}}$

41. $\dfrac{4 - \sqrt{x}}{4 + 2\sqrt{x}}$

Use the given screen to solve the inequality. Write the solution in interval notation.

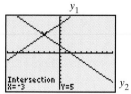

42. $y_1 < y_2$

Solve.

43. The most populous city in the United States is New York, although it is only the third most populous city in the world. Tokyo is the most populous city in the world. Second place is held by Seoul, Korea. Seoul's population is 1.3 million more than New York's and Tokyo's is 10.2 million less than twice the population of New York. If the sum of the populations of these three cities is 78.3 million, find the population of each city.

44. The product of one more than a number and twice the reciprocal of the number is $\dfrac{12}{5}$. Find the number.

45. After steadily falling since 1990, the nation's birth rate started rising according to the chart below. (Birth rate per 1000 population)

Years	Birth rate
1990	16.7
1993	15.4
1996	14.4
1999	14.2
2002	13.9
2005	14

a. Graph the data points and use the graph to identify the type of function (linear or quadratic) that best fits the data.
b. Find the corresponding regression equation for the data and use it to predict the birth rate for 2012. (Round regression coefficients to three decimal places. Round your answer to the nearest hundredth.)

46. Given the diagram shown, approximate to the nearest foot how many feet of walking distance a person saves by cutting across the lawn instead of walking on the sidewalk.

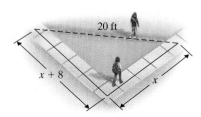

47. A stone is thrown upward from a bridge. The stone's height in feet, $s(t)$, above the water t seconds after the stone is thrown is a function given by the equation

$$s(t) = -16t^2 + 32t + 256$$

a. Find the maximum height of the stone.

b. Find the time it takes the stone to hit the water. Round the answer to two decimal places.

48. The research department of a company that manufactures children's fruit drinks is experimenting with a new flavor. A 17.5% fructose solution is needed, but only 10% and 20% solutions are available. How many gallons of a 10% fructose solution should be mixed with a 20% fructose solution to obtain 20 gallons of a 17.5% fructose solution?

Complex Numbers: Chapter 7

Perform the indicated operation and simplify. Write the result in the form a + bi.

49. $-\sqrt{-8}$

50. $(12 - 6i) - (12 - 3i)$

51. $(4 + 3i)^2$

52. $\dfrac{1 + 4i}{1 - i}$

Inverse, Exponential, and Logarithmic Functions: Chapter 9

53. If $g(x) = x - 7$ and $h(x) = x^2 - 6x + 5$, find $(g \circ h)(x)$.

54. Decide whether $f(x) = 6 - 2x$ is a one-to-one function. If it is, find its inverse.

55. Use properties of logarithms to write the expression as a single logarithm.

$$\log_5 x + 3\log_5 x - \log_5(x + 1)$$

Solve. Give exact solutions.

56. $8^{x-1} = \dfrac{1}{64}$

57. $3^{2x+5} = 4$. Give an exact solution and a 4-decimal place approximation.

58. $\log_8(3x - 2) = 2$

59. $\log_4(x + 1) - \log_4(x - 2) = 3$

60. Use a graphing calculator to approximate the solution of $e^{0.2x} = e^{-0.4x} + 2$ to two decimal places.

61. Graph $y = \left(\dfrac{1}{2}\right)^x + 1$

62. The prairie dog population of the Grand Rapids area now stands at 57,000 animals. If the population is growing at a rate of 2.6% annually, use the formula $y = y_0 e^{kt}$ to find how many prairie dogs there will be in that area 5 years from now.

Conic Sections: Chapter 10

Sketch the graph of each equation.

63. $x^2 - y^2 = 36$

64. $16x^2 + 9y^2 = 144$

65. $x^2 + y^2 + 6x = 16$

66. Solve the system:
$$\begin{cases} x^2 + y^2 = 26 \\ x^2 - 2y^2 = 23 \end{cases}$$

Sequences, Series, and the Binomial Theorem: Chapter 11

67. Find the first five terms of the sequence $a_n = \dfrac{(-1)^n}{n + 4}$.

68. Find the partial sum, S_5, of the sequence $a_n = 5(2)^{n-1}$.

69. Find S_∞ of the sequence $\dfrac{3}{2}, -\dfrac{3}{4}, \dfrac{3}{8}, \ldots$

70. Find $\displaystyle\sum_{i=1}^{4} i(i - 2)$

71. Expand: $(2x + y)^5$

Appendix B

Geometry

B.1 GEOMETRIC FORMULAS

Rectangle

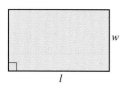

Perimeter: $P = 2l + 2w$
Area: $A = lw$

Square

Perimeter: $P = 4s$
Area: $A = s^2$

Triangle

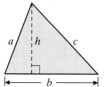

Perimeter: $P = a + b + c$
Area: $A = \frac{1}{2}bh$

Sum of Angles of Triangle

$A + B + C = 180°$
The sum of the measures of the three angles is 180°.

Right Triangles

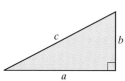

Perimeter: $P = a + b + c$
Area: $A = \frac{1}{2}ab$
One 90° (right) angle

Pythagorean Theorem (for right triangles)

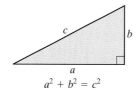

$a^2 + b^2 = c^2$

Isosceles Triangle

Triangle has:
two equal sides and
two equal angles.

Equilateral Triangle

Triangle has:
three equal sides and
three equal angles.
Measure of each angle is 60°.

Trapezoid

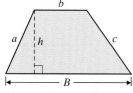

Perimeter: $P = a + b + c + B$
Area: $A = \frac{1}{2}h(B + b)$

Parallelogram

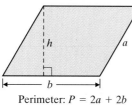

Perimeter: $P = 2a + 2b$
Area: $A = bh$

Circle

Circumference: $C = \pi d$
$C = 2\pi r$
Area: $A = \pi r^2$

Rectangular Solid

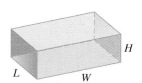

Volume: $V = LWH$
Surface Area:
$S = 2LW + 2HL + 2HW$

Cube

Volume: $V = s^3$
Surface Area: $S = 6s^2$

Cone

Volume: $V = \frac{1}{3}\pi r^2 h$
Lateral Surface Area:
$S = \pi r\sqrt{r^2 + h^2}$

Right Circular Cylinder

Volume: $V = \pi r^2 h$
Surface Area: $S = 2\pi r^2 + 2\pi rh$

Sphere

Volume: $V = \frac{4}{3}\pi r^3$
Surface Area: $S = 4\pi r^2$

Other Formulas
Distance: $d = rt$ (r = rate, t = time)
Percent: $p = br$ (p = percentage, b = base, r = rate)

Compound Interest: $A = P\left(1 + \dfrac{r}{n}\right)^{nt}$

(P = principal, r = annual interest rate, t = time in years, n = number of compoundings per year)

Temperature: $F = \dfrac{9}{5}C + 32 \quad C = \dfrac{5}{9}(F - 32)$

Simple Interest: $I = Prt$
(P = principal, r = annual interest rate, t = time in years)

B.2 REVIEW OF GEOMETRIC FIGURES

Plane figures have length and width but no thickness or depth

Name	Description	Figure
Polygon	Union of three or more coplanar line segments that intersect with each other only at each end point, with each end point shared by two segments.	
Triangle	Polygon with three sides (sum of measures of three angles is 180°).	
Scalene Triangle	Triangle with no sides of equal length.	
Isosceles Triangle	Triangle with two sides of equal length.	
Equilateral Triangle	Triangle with all sides of equal length.	
Right Triangle	Triangle that contains a right angle.	hypotenuse, leg, leg
Quadrilateral	Polygon with four sides (sum of measures of four angles is 360°).	
Trapezoid	Quadrilateral with exactly one pair of opposite sides parallel.	base, leg, parallel sides, leg, base
Isosceles Trapezoid	Trapezoid with legs of equal length.	
Parallelogram	Quadrilateral with both pairs of opposite sides parallel and equal in length.	

(continued)

Plane figures have length and width but no thickness or depth (continued)

Name	Description	Figure
Rhombus	Parallelogram with all sides of equal length.	
Rectangle	Parallelogram with four right angles.	
Square	Rectangle with all sides of equal length.	
Circle	All points in a plane the same distance from a fixed point called the **center.**	

Solids have length, width, and depth

Name	Description	Figure
Rectangular Solid	A solid with six sides, all of which are rectangles.	
Cube	A rectangular solid whose six sides are squares.	
Sphere	All points the same distance from a fixed point, called the center.	
Right Circular Cylinder	A cylinder with two circular bases that are perpendicular to its altitude.	
Right Circular Cone	A cone with a circular base that is perpendicular to its altitude.	

B.3 REVIEW OF VOLUME AND SURFACE AREA

A **convex solid** is a set of points, S, not all in one plane, such that for any two points A and B in S, all points between A and B are also in S. In this appendix, we will find the volume and surface area of special types of solids called polyhedrons. A solid formed by the intersection of a finite number of planes is called a **polyhedron.** The box below is an example of a polyhedron.

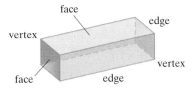

Each of the plane regions of the polyhedron is called a **face** of the polyhedron. If the intersection of two faces is a line segment, this line segment is an **edge** of the polyhedron. The intersections of the edges are the **vertices** of the polyhedron.

Volume is a measure of the space of a solid. The volume of a box or can, for example, is the amount of space inside. Volume can be used to describe the amount of juice in a pitcher or the amount of concrete needed to pour a foundation for a house.

The volume of a solid is the number of **cubic units** in the solid. A cubic centimeter and a cubic inch are illustrated.

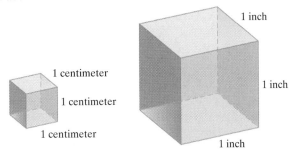

The **surface area** of a polyhedron is the sum of the areas of the faces of the polyhedron. For example, each face of the cube to the left above has an area of 1 square centimeter. Since there are 6 faces of the cube, the sum of the areas of the faces is 6 square centimeters. Surface area can be used to describe the amount of material needed to cover or form a solid. Surface area is measured in square units.

Formulas for finding the volumes, V, and surface areas, SA, of some common solids are given next.

Volume and Surface Area Formulas of Common Solids	
Solid	*Formulas*
RECTANGULAR SOLID	$V = lwh$ $SA = 2lh + 2wh + 2lw$ where h = height, w = width, l = length
CUBE	$V = s^3$ $SA = 6s^2$ where s = side

(*continued*)

Volume and Surface Area Formulas of Common Solids (continued)

Solid	Formulas
SPHERE 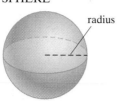 radius	$V = \dfrac{4}{3}\pi r^3$ $SA = 4\pi r^2$ where r = radius
CIRCULAR CYLINDER 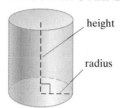 height, radius	$V = \pi r^2 h$ $SA = 2\pi rh + 2\pi r^2$ where h = height, r = radius
CONE 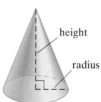 height, radius	$V = \dfrac{1}{3}\pi r^2 h$ $SA = \pi r\sqrt{r^2 + h^2} + \pi r^2$ where h = height, r = radius
SQUARE-BASED PYRAMID 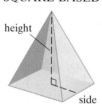 height, side	$V = \dfrac{1}{3}s^2 h$ $SA = B + \dfrac{1}{2}pl$ where B = area of base, p = perimeter of base, h = height, s = side, l = slant height

> ▶ **Helpful Hint**
> Volume is measured in cubic units. Surface area is measured in square units.

EXAMPLE 1   Find the volume and surface area of a rectangular box that is 12 inches long, 6 inches wide, and 3 inches high.

3 in.   6 in.   12 in.

Solution   Let $h = 3$ in., $l = 12$ in., and $w = 6$ in.

$$V = lwh$$
$$V = 12\text{ inches} \cdot 6\text{ inches} \cdot 3\text{ inches} = 216\text{ cubic inches}$$

The volume of the rectangular box is 216 cubic inches.

$$
\begin{aligned}
SA &= 2lh + 2wh + 2lw \\
&= 2(12\text{ in.})(3\text{ in.}) + 2(6\text{ in.})(3\text{ in.}) + 2(12\text{ in.})(6\text{ in.}) \\
&= 72\text{ sq in.} + 36\text{ sq in.} + 144\text{ sq in.} \\
&= 252\text{ sq in.}
\end{aligned}
$$

The surface area of the rectangular box is 252 square inches.

EXAMPLE 2 Find the volume and surface area of a ball of radius 2 inches. Give the exact volume and surface area and then use the approximation $\frac{22}{7}$ for π.

Solution

$$V = \frac{4}{3}\pi r^3 \qquad \text{Formula for volume of a sphere}$$

$$V = \frac{4}{3}\pi(2 \text{ in.})^3 \qquad \text{Let } r = 2 \text{ inches.}$$

$$= \frac{32}{3}\pi \text{ cu in.} \qquad \text{Simplify.}$$

$$\approx \frac{32}{3} \cdot \frac{22}{7} \text{ cu in.} \qquad \text{Approximate } \pi \text{ with } \frac{22}{7}.$$

$$= \frac{704}{21} \text{ or } 33\frac{11}{21} \text{ cu in.}$$

The volume of the sphere is exactly $\frac{32}{3}\pi$ cubic inches or approximately $33\frac{11}{21}$ cubic inches.

$$SA = 4\pi r^2 \qquad \text{Formula for surface area}$$

$$SA = 4\pi(2 \text{ in.})^2 \qquad \text{Let } r = 2 \text{ inches.}$$

$$= 16\pi \text{ sq in.} \qquad \text{Simplify.}$$

$$\approx 16 \cdot \frac{22}{7} \text{ sq in.} \qquad \text{Approximate } \pi \text{ with } \frac{22}{7}.$$

$$= \frac{352}{7} \text{ or } 50\frac{2}{7} \text{ sq in.}$$

The surface area of the sphere is exactly 16π square inches or approximately $50\frac{2}{7}$ square inches. □

APPENDIX B.3 EXERCISE SET

Find the volume and surface area of each solid. See Examples 1 and 2. For formulas that contain π, give an exact answer and then approximate using $\frac{22}{7}$ for π.

1.
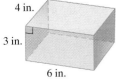
4 in.
3 in.
6 in.

2.

3 mi

5. (For surface area, use 3.14 for π and approximate to two decimal places.)

3 yd
2 yd

3.
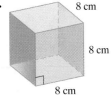
8 cm
8 cm
8 cm

4.
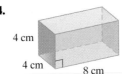
4 cm
4 cm
8 cm

6.

10 ft
6 ft

7.

10 in.

8. Find the volume only.

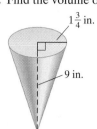

$1\frac{3}{4}$ in.

9 in.

9.

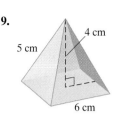

4 cm

5 cm

6 cm

10.

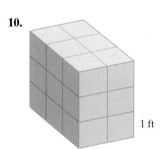

1 ft

Solve.

11. Find the volume of a cube with edges of $1\frac{1}{3}$ inches.

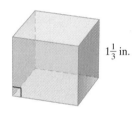

$1\frac{1}{3}$ in.

12. A water storage tank is in the shape of a cone with the pointed end down. If the radius is 14 ft and the depth of the tank is 15 ft, approximate the volume of the tank in cubic feet. Use $\frac{22}{7}$ for π.

14 ft

15 ft

13. Find the surface area of a rectangular box 2 ft by 1.4 ft by 3 ft.

14. Find the surface area of a box in the shape of a cube that is 5 ft on each side.

15. Find the volume of a pyramid with a square base 5 in. on a side and a height of 1.3 in.

16. Approximate to the nearest hundredth the volume of a sphere with a radius of 2 cm. Use 3.14 for π.

17. A paperweight is in the shape of a square-based pyramid 20 cm tall. If an edge of the base is 12 cm, find the volume of the paperweight.

18. A bird bath is made in the shape of a hemisphere (half-sphere). If its radius is 10 in., approximate the volume. Use $\frac{22}{7}$ for π.

10 in.

19. Find the exact surface area of a sphere with a radius of 7 in.

20. A tank is in the shape of a cylinder 8 ft tall and 3 ft in radius. Find the exact surface area of the tank.

21. Find the volume of a rectangular block of ice 2 ft by $2\frac{1}{2}$ ft by $1\frac{1}{2}$ ft.

22. Find the capacity (volume in cubic feet) of a rectangular ice chest with inside measurements of 3 ft by $1\frac{1}{2}$ ft by $1\frac{3}{4}$ ft.

23. An ice cream cone with a 4-cm diameter and 3-cm depth is filled exactly level with the top of the cone. Approximate how much ice cream (in cubic centimeters) is in the cone. Use $\frac{22}{7}$ for π.

24. A child's toy is in the shape of a square-based pyramid 10 in. tall. If an edge of the base is 7 in., find the volume of the toy.

Appendix C

Stretching and Compressing Graphs of Absolute Value Functions

In Section 2.7, we learned to shift and reflect graphs of common functions: $f(x) = x$, $f(x) = x^2$, $f(x) = |x|$, and $f(x) = \sqrt{x}$. Since other common functions are studied throughout this text, in this Appendix we concentrate on the absolute value function.

Recall that the graph of $h(x) = -|x - 1| + 2$, for example, is the same as the graph of $f(x) = |x|$ reflected about the x-axis, moved 1 unit to the right and 2 units upward. In other words,

$$h(x) = -|x - 1| + 2$$

opens
downward
(1, 2) location of vertex of
V-shape

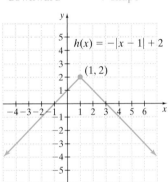

Let's now study the graphs of a few other absolute value functions.

EXAMPLE 1   Graph $h(x) = 2|x|$, and $g(x) = \dfrac{1}{2}|x|$.

Solution   Let's find and plot ordered pair solutions for the functions.

x	$h(x)$	$g(x)$
-2	4	1
-1	2	$\dfrac{1}{2}$
0	0	0
1	2	$\dfrac{1}{2}$
2	4	2

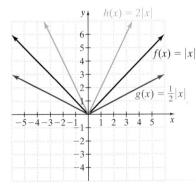

Notice that the graph of $h(x) = 2|x|$ is narrower than the graph of $f(x) = |x|$ and that the graph of $g(x) = \dfrac{1}{2}|x|$ is wider than the graph of $f(x) = |x|$.

In general, for the absolute function, we have the following:

> **The Graph of the Absolute Value Function**
> The graph of $f(x) = a|x - h| + k$
> - Has vertex (h, k) and is V-shaped.
> - Opens up if $a > 0$ and down if $a < 0$.
> - If $|a| < 1$, the graph is wider than the graph of $y = |x|$.
> - If $|a| > 1$, the graph is narrower than the graph of $y = |x|$.

EXAMPLE 2 Graph $f(x) = -\frac{1}{3}|x + 2| + 4$.

Solution Let's write this function in the form $f(x) = a|x - h| + k$. For our function, we have $f(x) = -\frac{1}{3}|x - (-2)| + 4$. Thus,

- vertex is $(-2, 4)$
- since $a < 0$, V-shape opens down
- since $|a| = \left|-\frac{1}{3}\right| = \frac{1}{3} < 1$, the graph is wider than $y = |x|$

We will also find and plot ordered pair solutions.

If $x = -5, f(-5) = -\frac{1}{3}|-5 + 2| + 4$, or 3

If $x = 1, f(1) = -\frac{1}{3}|1 + 2| + 4$, or 3

If $x = 3, f(3) = -\frac{1}{3}|3 + 2| + 4$, or $\frac{7}{3}$, or $2\frac{1}{3}$

x	$f(x)$
-5	3
1	3
3	$2\frac{1}{3}$

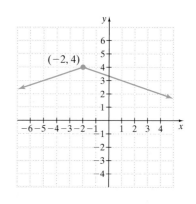

APPENDIX C | EXERCISE SET *MyMathLab*

Sketch the graph of each function. Label the vertex of the V-shape.

1. $f(x) = 3|x|$

2. $f(x) = 5|x|$

3. $f(x) = \frac{1}{4}|x|$

4. $f(x) = \frac{1}{3}|x|$

5. $g(x) = 2|x| + 3$

6. $g(x) = 3|x| + 2$

7. $h(x) = -\frac{1}{2}|x|$

8. $h(x) = -\frac{1}{3}|x|$

9. $f(x) = 4|x - 1|$

10. $f(x) = 3|x - 2|$

11. $g(x) = -\frac{1}{3}|x| - 2$

12. $g(x) = -\frac{1}{2}|x| - 3$

13. $f(x) = -2|x - 3| + 4$

14. $f(x) = -3|x - 1| + 5$

15. $f(x) = \frac{2}{3}|x + 2| - 5$

16. $f(x) = \frac{3}{4}|x + 1| - 4$

Appendix D

Solving Systems of Equations Using Determinants

OBJECTIVES

1 Define and Evaluate a 2 × 2 Determinant.

2 Use Cramer's Rule to Solve a System of Two Linear Equations in Two Variables.

3 Define and Evaluate a 3 × 3 Determinant.

4 Use Cramer's Rule to Solve a System of Three Linear Equations in Three Variables.

We have solved systems of two linear equations in two variables in four different ways: graphically, by substitution, by elimination, and by matrices. Now we analyze another method called **Cramer's rule.**

OBJECTIVE 1 ▶ Evaluating 2 × 2 determinants. Recall that a matrix is a rectangular array of numbers. If a matrix has the same number of rows and columns, it is called a **square matrix.** Examples of square matrices are

$$\begin{bmatrix} 1 & 6 \\ 5 & 2 \end{bmatrix} \qquad \begin{bmatrix} 2 & 4 & 1 \\ 0 & 5 & 2 \\ 3 & 6 & 9 \end{bmatrix}$$

A **determinant** is a real number associated with a square matrix. The determinant of a square matrix is denoted by placing vertical bars about the array of numbers. Thus,

The determinant of the square matrix $\begin{bmatrix} 1 & 6 \\ 5 & 2 \end{bmatrix}$ is $\begin{vmatrix} 1 & 6 \\ 5 & 2 \end{vmatrix}$.

The determinant of the square matrix $\begin{bmatrix} 2 & 4 & 1 \\ 0 & 5 & 2 \\ 3 & 6 & 9 \end{bmatrix}$ is $\begin{vmatrix} 2 & 4 & 1 \\ 0 & 5 & 2 \\ 3 & 6 & 9 \end{vmatrix}$.

We define the determinant of a 2 × 2 matrix first. (Recall that 2 × 2 is read "two by two." It means that the matrix has 2 rows and 2 columns.)

Determinant of a 2 × 2 Matrix

$$\begin{vmatrix} a & b \\ c & d \end{vmatrix} = ad - bc$$

EXAMPLE 1   Evaluate each determinant.

a. $\begin{vmatrix} -1 & 2 \\ 3 & -4 \end{vmatrix}$ **b.** $\begin{vmatrix} 2 & 0 \\ 7 & -5 \end{vmatrix}$

Solution   First we identify the values of $a, b, c,$ and d. Then we perform the evaluation.

a. Here $a = -1, b = 2, c = 3,$ and $d = -4$.

$$\begin{vmatrix} -1 & 2 \\ 3 & -4 \end{vmatrix} = ad - bc = (-1)(-4) - (2)(3) = -2$$

b. In this example, $a = 2, b = 0, c = 7,$ and $d = -5$.

$$\begin{vmatrix} 2 & 0 \\ 7 & -5 \end{vmatrix} = ad - bc = 2(-5) - (0)(7) = -10$$

A calculator may be used to find determinants. In Example 1a, we found that $\begin{vmatrix} -1 & 2 \\ 3 & -4 \end{vmatrix} = -2$.

```
[A]
     [[-1  2 ]
      [3  -4]]
det([A])
             -2
■
```

To check with a calculator, enter the elements of the determinant $\begin{vmatrix} -1 & 2 \\ 3 & -4 \end{vmatrix}$ in matrix A and then find the determinant of matrix A.

Calculators may be used to find determinants, and therefore, may also be used to solve systems of equations by Cramer's rule. □

OBJECTIVE 2 ▶ Using Cramer's rule to solve a system of two linear equations. To develop Cramer's rule, we solve the system $\begin{cases} ax + by = h \\ cx + dy = k \end{cases}$ using elimination. First, we eliminate y by multiplying both sides of the first equation by d and both sides of the second equation by $-b$ so that the coefficients of y are opposites. The result is that

$$\begin{cases} d(ax + by) = d \cdot h \\ -b(cx + dy) = -b \cdot k \end{cases} \quad \text{simplifies to} \quad \begin{cases} adx + bdy = hd \\ -bcx - bdy = -kb \end{cases}$$

We now add the two equations and solve for x.

$$
\begin{aligned}
adx + bdy &= hd \\
\underline{-bcx - bdy} &= \underline{-kb} \\
adx - bcx &= hd - kb \quad \text{Add the equations.} \\
(ad - bc)x &= hd - kb \\
x &= \frac{hd - kb}{ad - bc} \quad \text{Solve for } x.
\end{aligned}
$$

When we replace x with $\dfrac{hd - kb}{ad - bc}$ in the equation $ax + by = h$ and solve for y, we find that $y = \dfrac{ak - ch}{ad - bc}$.

Notice that the numerator of the value of x is the determinant of

$$\begin{vmatrix} h & b \\ k & d \end{vmatrix} = hd - kb$$

Also, the numerator of the value of y is the determinant of

$$\begin{vmatrix} a & h \\ c & k \end{vmatrix} = ak - hc$$

Finally, the denominators of the values of x and y are the same and are the determinant of

$$\begin{vmatrix} a & b \\ c & d \end{vmatrix} = ad - bc$$

This means that the values of x and y can be written in determinant notation:

$$x = \frac{\begin{vmatrix} h & b \\ k & d \end{vmatrix}}{\begin{vmatrix} a & b \\ c & d \end{vmatrix}} \quad \text{and} \quad y = \frac{\begin{vmatrix} a & h \\ c & k \end{vmatrix}}{\begin{vmatrix} a & b \\ c & d \end{vmatrix}}$$

For convenience, we label the determinants D, D_x, and D_y.

$$\begin{vmatrix} a & b \\ c & d \end{vmatrix} = D \qquad \begin{vmatrix} h & b \\ k & d \end{vmatrix} = D_x \qquad \begin{vmatrix} a & h \\ c & k \end{vmatrix} = D_y$$

x-coefficients, *y*-coefficients

x-column replaced by constants   *y*-column replaced by constants

These determinant formulas for the coordinates of the solution of a system are known as **Cramer's rule.**

Cramer's Rule for Two Linear Equations in Two Variables

The solution of the system $\begin{cases} ax + by = h \\ cx + dy = k \end{cases}$ is given by

$$x = \dfrac{\begin{vmatrix} h & b \\ k & d \end{vmatrix}}{\begin{vmatrix} a & b \\ c & d \end{vmatrix}} = \dfrac{D_x}{D} \qquad y = \dfrac{\begin{vmatrix} a & h \\ c & k \end{vmatrix}}{\begin{vmatrix} a & b \\ c & d \end{vmatrix}} = \dfrac{D_y}{D}$$

as long as $D = ad - bc$ is not 0.

When $D = 0$, the system is either inconsistent or the equations are dependent. When this happens, we need to use another method to see which is the case.

EXAMPLE 2   Use Cramer's rule to solve each system.

a. $\begin{cases} 3x + 4y = -7 \\ x - 2y = -9 \end{cases}$ **b.** $\begin{cases} 5x + y = 5 \\ -7x - 2y = -7 \end{cases}$

Solution

a. First we find D, D_x, and D_y.

$$\begin{cases} 3x + 4y = -7 \\ x - 2y = -9 \end{cases}$$

$a \quad b \quad h$

$c \quad d \quad k$

$$D = \begin{vmatrix} a & b \\ c & d \end{vmatrix} = \begin{vmatrix} 3 & 4 \\ 1 & -2 \end{vmatrix} = 3(-2) - 4(1) = -10$$

$$D_x = \begin{vmatrix} h & b \\ k & d \end{vmatrix} = \begin{vmatrix} -7 & 4 \\ -9 & -2 \end{vmatrix} = (-7)(-2) - 4(-9) = 50$$

$$D_y = \begin{vmatrix} a & h \\ c & k \end{vmatrix} = \begin{vmatrix} 3 & -7 \\ 1 & -9 \end{vmatrix} = 3(-9) - (-7)(1) = -20$$

Then $x = \dfrac{D_x}{D} = \dfrac{50}{-10} = -5$ and $y = \dfrac{D_y}{D} = \dfrac{-20}{-10} = 2$. The ordered pair solution is $(-5, 2)$.

As always, check the solution in both original equations.

b. Find D, D_x, and D_y for $\begin{cases} 5x + y = 5 \\ -7x - 2y = -7 \end{cases}$.

$$D = \begin{vmatrix} 5 & 1 \\ -7 & -2 \end{vmatrix} = 5(-2) - (1)(-7) = -3$$

$$D_x = \begin{vmatrix} 5 & 1 \\ -7 & -2 \end{vmatrix} = 5(-2) - (1)(-7) = -3$$

$$D_y = \begin{vmatrix} 5 & 5 \\ -7 & -7 \end{vmatrix} = 5(-7) - 5(-7) = 0$$

Then $x = \dfrac{D_x}{D} = \dfrac{-3}{-3} = 1$ and $y = \dfrac{D_y}{D} = \dfrac{0}{-3} = 0$.

The ordered pair solution is $(1, 0)$. Check this solution in both original equations. ☐

Let's solve the system in Example 2a using Cramer's rule and a calculator.

$$\begin{cases} 3x + 4y = -7 \\ x - 2y = -9 \end{cases}$$

Recall that,

$$D = \begin{vmatrix} 3 & 4 \\ 1 & -2 \end{vmatrix}, \quad D_x = \begin{vmatrix} -7 & 4 \\ -9 & -2 \end{vmatrix}, \quad D_y = \begin{vmatrix} 3 & -7 \\ 1 & -9 \end{vmatrix}$$

To evaluate D, D_x, and D_y, enter the elements of D in matrix A, D_x in matrix B, and D_y in matrix C and evaluate the determinant of each matrix. Then

$$x = \frac{D_x}{D} = \frac{\det[B]}{\det[A]} = -5 \quad \text{and} \quad y = \frac{D_y}{D} = \frac{\det[C]}{\det[A]} = 2$$

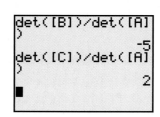

The solution is $(-5, 2)$, as expected.

EXAMPLE 3 Use Cramer's rule to solve the system

$$\begin{cases} 5x + y = 5 \\ -7x - 2y = -7 \end{cases}$$

Solution First we find D, D_x, and D_y.

$$D = \begin{vmatrix} 5 & 1 \\ -7 & -2 \end{vmatrix} = 5(-2) - (-7)(1) = -3$$

$$D_x = \begin{vmatrix} 5 & 1 \\ -7 & -2 \end{vmatrix} = 5(-2) - (-7)(1) = -3$$

$$D_y = \begin{vmatrix} 5 & 5 \\ -7 & -7 \end{vmatrix} = 5(-7) - 5(-7) = 0$$

Then

$$x = \frac{D_x}{D} = \frac{-3}{-3} = 1 \qquad y = \frac{D_y}{D} = \frac{0}{-3} = 0$$

The ordered pair solution is $(1, 0)$. ☐

OBJECTIVE 3 ▶ Evaluating 3 × 3 determinants. A 3 × 3 determinant can be used to solve a system of three equations in three variables. The determinant of a 3 × 3 matrix, however, is considerably more complex than a 2 × 2 one.

Determinant of a 3 × 3 Matrix

$$\begin{vmatrix} a_1 & b_1 & c_1 \\ a_2 & b_2 & c_2 \\ a_3 & b_3 & c_3 \end{vmatrix} = a_1 \cdot \begin{vmatrix} b_2 & c_2 \\ b_3 & c_3 \end{vmatrix} - a_2 \cdot \begin{vmatrix} b_1 & c_1 \\ b_3 & c_3 \end{vmatrix} + a_3 \cdot \begin{vmatrix} b_1 & c_1 \\ b_2 & c_2 \end{vmatrix}$$

Notice that the determinant of a 3 × 3 matrix is related to the determinants of three 2 × 2 matrices. Each determinant of these 2 × 2 matrices is called a **minor,** and every element of a 3 × 3 matrix has a minor associated with it. For example, the minor of c_2 is the determinant of the 2 × 2 matrix found by deleting the row and column containing c_2.

$$\begin{array}{ccc} a_1 & b_1 & c_1 \\ a_2 & b_2 & c_2 \\ a_3 & b_3 & c_3 \end{array} \qquad \text{The minor of } c_2 \text{ is} \qquad \begin{vmatrix} a_1 & b_1 \\ a_3 & b_3 \end{vmatrix}$$

Also, the minor of element a_1 is the determinant of the 2 × 2 matrix that has no row or column containing a_1.

$$\begin{array}{ccc} a_1 & b_1 & c_1 \\ a_2 & b_2 & c_2 \\ a_3 & b_3 & c_3 \end{array} \qquad \text{The minor of } a_1 \text{ is} \qquad \begin{vmatrix} b_2 & c_2 \\ b_3 & c_3 \end{vmatrix}$$

So the determinant of a 3 × 3 matrix can be written as

$$a_1 \cdot (\text{minor of } a_1) - a_2 \cdot (\text{minor of } a_2) + a_3 \cdot (\text{minor of } a_3)$$

Finding the determinant by using minors of elements in the first column is called **expanding** by the minors of the first column. *The value of a determinant can be found by expanding by the minors of any row or column.* The following **array of signs** is helpful in determining whether to add or subtract the product of an element and its minor.

$$\begin{array}{ccc} + & - & + \\ - & + & - \\ + & - & + \end{array}$$

If an element is in a position marked +, we add. If marked −, we subtract.   □

Answer to Concept Check:

$$\begin{array}{cccc} + & - & + & - \\ - & + & - & + \\ + & - & + & - \\ - & + & - & + \end{array}$$

Concept Check ☑

Suppose you are interested in finding the determinant of a 4 × 4 matrix. Study the pattern shown in the array of signs for a 3 × 3 matrix. Use the pattern to expand the array of signs for use with a 4 × 4 matrix.

EXAMPLE 4   Evaluate by expanding by the minors of the given row or column.

$$\begin{vmatrix} 0 & 5 & 1 \\ 1 & 3 & -1 \\ -2 & 2 & 4 \end{vmatrix}$$

a. First column   **b.** Second row

Solution

a. The elements of the first column are $0, 1,$ and $-2.$ The first column of the array of signs is $+, -, +.$

$$\begin{vmatrix} 0 & 5 & 1 \\ 1 & 3 & -1 \\ -2 & 2 & 4 \end{vmatrix} = 0 \cdot \begin{vmatrix} 3 & -1 \\ 2 & 4 \end{vmatrix} - 1 \cdot \begin{vmatrix} 5 & 1 \\ 2 & 4 \end{vmatrix} + (-2) \cdot \begin{vmatrix} 5 & 1 \\ 3 & -1 \end{vmatrix}$$

$$= 0(12 - (-2)) - 1(20 - 2) + (-2)(-5 - 3)$$
$$= 0 - 18 + 16 = -2$$

b. The elements of the second row are $1, 3,$ and $-1.$ This time, the signs begin with $-$ and again alternate.

$$\begin{vmatrix} 0 & 5 & 1 \\ 1 & 3 & -1 \\ -2 & 2 & 4 \end{vmatrix} = -1 \cdot \begin{vmatrix} 5 & 1 \\ 2 & 4 \end{vmatrix} + 3 \cdot \begin{vmatrix} 0 & 1 \\ -2 & 4 \end{vmatrix} - (-1) \cdot \begin{vmatrix} 0 & 5 \\ -2 & 2 \end{vmatrix}$$

$$= -1(20 - 2) + 3(0 - (-2)) - (-1)(0 - (-10))$$
$$= -18 + 6 + 10 = -2$$

Notice that the determinant of the 3×3 matrix is the same regardless of the row or column you select to expand by.   □

Concept Check ☑

Why would expanding by minors of the second row be a good choice for the determinant

$$\begin{vmatrix} 3 & 4 & -2 \\ 5 & 0 & 0 \\ 6 & -3 & 7 \end{vmatrix}?$$

OBJECTIVE 4 ▶ Using Cramer's rule to solve a system of three linear equations. A system of three equations in three variables may be solved with Cramer's rule also. Using the elimination process to solve a system with unknown constants as coefficients leads to the following.

Cramer's Rule for Three Equations in Three Variables

The solution of the system $\begin{cases} a_1x + b_1y + c_1z = k_1 \\ a_2x + b_2y + c_2z = k_2 \\ a_3x + b_3y + c_3z = k_3 \end{cases}$ is given by

$$x = \frac{D_x}{D} \qquad y = \frac{D_y}{D} \qquad \text{and} \qquad z = \frac{D_z}{D}$$

where

$$D = \begin{vmatrix} a_1 & b_1 & c_1 \\ a_2 & b_2 & c_2 \\ a_3 & b_3 & c_3 \end{vmatrix} \qquad D_x = \begin{vmatrix} k_1 & b_1 & c_1 \\ k_2 & b_2 & c_2 \\ k_3 & b_3 & c_3 \end{vmatrix}$$

$$D_y = \begin{vmatrix} a_1 & k_1 & c_1 \\ a_2 & k_2 & c_2 \\ a_3 & k_3 & c_3 \end{vmatrix} \qquad D_z = \begin{vmatrix} a_1 & b_1 & k_1 \\ a_2 & b_2 & k_2 \\ a_3 & b_3 & k_3 \end{vmatrix}$$

as long as D is not 0.

Answer to Concept Check:

Two elements of the second row are 0, which makes calculations easier.

EXAMPLE 5   Use Cramer's rule to solve the system

$$\begin{cases} x - 2y + z = 4 \\ 3x + y - 2z = 3 \\ 5x + 5y + 3z = -8 \end{cases}$$

Solution   First we find $D, D_x, D_y,$ and D_z. Beginning with D, we expand by the minors of the first column.

$$D = \begin{vmatrix} 1 & -2 & 1 \\ 3 & 1 & -2 \\ 5 & 5 & 3 \end{vmatrix} = 1 \cdot \begin{vmatrix} 1 & -2 \\ 5 & 3 \end{vmatrix} - 3 \cdot \begin{vmatrix} -2 & 1 \\ 5 & 3 \end{vmatrix} + 5 \cdot \begin{vmatrix} -2 & 1 \\ 1 & -2 \end{vmatrix}$$

$$= 1(3 - (-10)) - 3(-6 - 5) + 5(4 - 1)$$

$$= 13 + 33 + 15 = 61$$

$$D_x = \begin{vmatrix} 4 & -2 & 1 \\ 3 & 1 & -2 \\ -8 & 5 & 3 \end{vmatrix} = 4 \cdot \begin{vmatrix} 1 & -2 \\ 5 & 3 \end{vmatrix} - 3 \cdot \begin{vmatrix} -2 & 1 \\ 5 & 3 \end{vmatrix} + (-8) \cdot \begin{vmatrix} -2 & 1 \\ 1 & -2 \end{vmatrix}$$

$$= 4(3 - (-10)) - 3(-6 - 5) + (-8)(4 - 1)$$

$$= 52 + 33 - 24 = 61$$

$$D_y = \begin{vmatrix} 1 & 4 & 1 \\ 3 & 3 & -2 \\ 5 & -8 & 3 \end{vmatrix} = 1 \cdot \begin{vmatrix} 3 & -2 \\ -8 & 3 \end{vmatrix} - 3 \cdot \begin{vmatrix} 4 & 1 \\ -8 & 3 \end{vmatrix} + 5 \cdot \begin{vmatrix} 4 & 1 \\ 3 & -2 \end{vmatrix}$$

$$= 1(9 - 16) - 3(12 - (-8)) + 5(-8 - 3)$$

$$= -7 - 60 - 55 = -122$$

$$D_z = \begin{vmatrix} 1 & -2 & 4 \\ 3 & 1 & 3 \\ 5 & 5 & -8 \end{vmatrix} = 1 \cdot \begin{vmatrix} 1 & 3 \\ 5 & -8 \end{vmatrix} - 3 \cdot \begin{vmatrix} -2 & 4 \\ 5 & -8 \end{vmatrix} + 5 \cdot \begin{vmatrix} -2 & 4 \\ 1 & 3 \end{vmatrix}$$

$$= 1(-8 - 15) - 3(16 - 20) + 5(-6 - 4)$$

$$= -23 + 12 - 50 = -61$$

From these determinants, we calculate the solution:

$$x = \frac{D_x}{D} = \frac{61}{61} = 1 \quad y = \frac{D_y}{D} = \frac{-122}{61} = -2 \quad z = \frac{D_z}{D} = \frac{-61}{61} = -1$$

The ordered triple solution is $(1, -2, -1)$. Check this solution by verifying that it satisfies each equation of the system.   □

VOCABULARY & READINESS CHECK

Evaluate each determinant mentally.

1. $\begin{vmatrix} 7 & 2 \\ 0 & 8 \end{vmatrix}$

2. $\begin{vmatrix} 6 & 0 \\ 1 & 2 \end{vmatrix}$

3. $\begin{vmatrix} -4 & 2 \\ 0 & 8 \end{vmatrix}$

4. $\begin{vmatrix} 5 & 0 \\ 3 & -5 \end{vmatrix}$

5. $\begin{vmatrix} -2 & 0 \\ 3 & -10 \end{vmatrix}$

6. $\begin{vmatrix} -1 & 4 \\ 0 & -18 \end{vmatrix}$

Evaluate each determinant. See Example 1.

1. $\begin{vmatrix} 3 & 5 \\ -1 & 7 \end{vmatrix}$

2. $\begin{vmatrix} -5 & 1 \\ 1 & -4 \end{vmatrix}$

3. $\begin{vmatrix} 9 & -2 \\ 4 & -3 \end{vmatrix}$

4. $\begin{vmatrix} 4 & -1 \\ 9 & 8 \end{vmatrix}$

5. $\begin{vmatrix} -2 & 9 \\ 4 & -18 \end{vmatrix}$

6. $\begin{vmatrix} -40 & 8 \\ 70 & -14 \end{vmatrix}$

7. $\begin{vmatrix} \frac{3}{4} & \frac{5}{2} \\ -\frac{1}{6} & \frac{7}{3} \end{vmatrix}$

8. $\begin{vmatrix} \frac{5}{7} & \frac{1}{3} \\ \frac{6}{7} & \frac{2}{3} \end{vmatrix}$

Use Cramer's rule, if possible, to solve each system of linear equations. See Examples 2 and 3.

9. $\begin{cases} 2y - 4 = 0 \\ x + 2y = 5 \end{cases}$

10. $\begin{cases} 4x - y = 5 \\ 3x - 3 = 0 \end{cases}$

11. $\begin{cases} 3x + y = 1 \\ 2y = 2 - 6x \end{cases}$

12. $\begin{cases} y = 2x - 5 \\ 8x - 4y = 20 \end{cases}$

13. $\begin{cases} 5x - 2y = 27 \\ -3x + 5y = 18 \end{cases}$

14. $\begin{cases} 4x - y = 9 \\ 2x + 3y = -27 \end{cases}$

15. $\begin{cases} 2x - 5y = 4 \\ x + 2y = -7 \end{cases}$

16. $\begin{cases} 3x - y = 2 \\ -5x + 2y = 0 \end{cases}$

17. $\begin{cases} \frac{2}{3}x - \frac{3}{4}y = -1 \\ -\frac{1}{6}x + \frac{3}{4}y = \frac{5}{2} \end{cases}$

18. $\begin{cases} \frac{1}{2}x - \frac{1}{3}y = -3 \\ \frac{1}{8}x + \frac{1}{6}y = 0 \end{cases}$

Evaluate. See Example 4.

19. $\begin{vmatrix} 2 & 1 & 0 \\ 0 & 5 & -3 \\ 4 & 0 & 2 \end{vmatrix}$

20. $\begin{vmatrix} -6 & 4 & 2 \\ 1 & 0 & 5 \\ 0 & 3 & 1 \end{vmatrix}$

21. $\begin{vmatrix} 4 & -6 & 0 \\ -2 & 3 & 0 \\ 4 & -6 & 1 \end{vmatrix}$

22. $\begin{vmatrix} 5 & 2 & 1 \\ 3 & -6 & 0 \\ -2 & 8 & 0 \end{vmatrix}$

23. $\begin{vmatrix} 1 & 0 & 4 \\ 1 & -1 & 2 \\ 3 & 2 & 1 \end{vmatrix}$

24. $\begin{vmatrix} 0 & 1 & 2 \\ 3 & -1 & 2 \\ 3 & 2 & -2 \end{vmatrix}$

25. $\begin{vmatrix} 3 & 6 & -3 \\ -1 & -2 & 3 \\ 4 & -1 & 6 \end{vmatrix}$

26. $\begin{vmatrix} 2 & -2 & 1 \\ 4 & 1 & 3 \\ 3 & 1 & 2 \end{vmatrix}$

Use Cramer's rule, if possible, to solve each system of linear equations. See Example 5.

27. $\begin{cases} 3x + z = -1 \\ -x - 3y + z = 7 \\ 3y + z = 5 \end{cases}$

28. $\begin{cases} 4y - 3z = -2 \\ 8x - 4y = 4 \\ -8x + 4y + z = -2 \end{cases}$

29. $\begin{cases} x + y + z = 8 \\ 2x - y - z = 10 \\ x - 2y + 3z = 22 \end{cases}$

30. $\begin{cases} 5x + y + 3z = 1 \\ x - y - 3z = -7 \\ -x + y = 1 \end{cases}$

31. $\begin{cases} 2x + 2y + z = 1 \\ -x + y + 2z = 3 \\ x + 2y + 4z = 0 \end{cases}$

32. $\begin{cases} 2x - 3y + z = 5 \\ x + y + z = 0 \\ 4x + 2y + 4z = 4 \end{cases}$

33. $\begin{cases} x - 2y + z = -5 \\ 3y + 2z = 4 \\ 3x - y = -2 \end{cases}$

34. $\begin{cases} 4x + 5y = 10 \\ 3y + 2z = -6 \\ x + y + z = 3 \end{cases}$

CONCEPT EXTENSIONS

Find the value of x that will make each a true statement.

35. $\begin{vmatrix} 1 & x \\ 2 & 7 \end{vmatrix} = -3$

36. $\begin{vmatrix} 6 & 1 \\ -2 & x \end{vmatrix} = 26$

37. If all the elements in a single row of a determinant are zero, what is the value of the determinant? Explain your answer.

38. If all the elements in a single column of a determinant are 0, what is the value of the determinant? Explain your answer.

Appendix E

An Introduction to Using a Graphing Utility

THE VIEWING WINDOW AND INTERPRETING WINDOW SETTINGS

In this appendix, we will use the term **graphing utility** to mean a graphing calculator or a computer software graphing package. All graphing utilities graph equations by plotting points on a screen. While plotting several points can be slow and sometimes tedious for us, a graphing utility can quickly and accurately plot hundreds of points. How does a graphing utility show plotted points? A computer or calculator screen is made up of a grid of small rectangular areas called **pixels.** If a pixel contains a point to be plotted, the pixel is turned "on"; otherwise, the pixel remains "off." The graph of an equation is then a collection of pixels turned "on." The graph of $y = 3x + 1$ from a graphing calculator is shown in Figure A-1. Notice the irregular shape of the line caused by the rectangular pixels.

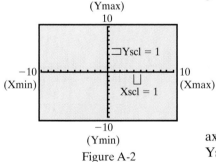
Figure A-1

The portion of the coordinate plane shown on the screen in Figure A-1 is called the **viewing window** or the **viewing rectangle.** Notice the x-axis and the y-axis on the graph. While tick marks are shown on the axes, they are not labeled. This means that from this screen alone, we do not know how many units each tick mark represents. To see what each tick mark represents and the minimum and maximum values on the axes, check the window setting of the graphing utility. It defines the viewing window. The window of the graph of $y = 3x + 1$ shown in Figure A-1 has the following setting (Figure A-2):

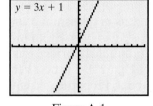

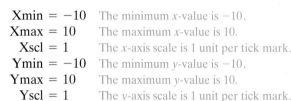

$$\begin{array}{ll} \text{Xmin} = -10 & \text{The minimum } x\text{-value is } -10. \\ \text{Xmax} = 10 & \text{The maximum } x\text{-value is } 10. \\ \text{Xscl} = 1 & \text{The } x\text{-axis scale is 1 unit per tick mark.} \\ \text{Ymin} = -10 & \text{The minimum } y\text{-value is } -10. \\ \text{Ymax} = 10 & \text{The maximum } y\text{-value is } 10. \\ \text{Yscl} = 1 & \text{The } y\text{-axis scale is 1 unit per tick mark.} \end{array}$$

Figure A-2

By knowing the scale, we can find the minimum and the maximum values on the axes simply by counting tick marks. For example, if both the Xscl (x-axis scale) and the Yscl are 1 unit per tick mark on the graph in Figure A-3, we can count the tick marks and find that the minimum x-value is -10 and the maximum x-value is 10. Also, the minimum y-value is -10 and the maximum y-value is 10. If the Xscl (x-axis scale) changes to 2 units per tick mark (shown in Figure A-4), by counting tick marks, we see that the minimum x-value is now -20 and the maximum x-value is now 20.

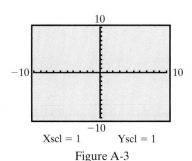

Xscl = 1   Yscl = 1

Figure A-3

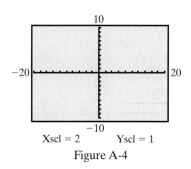

Xscl = 2   Yscl = 1

Figure A-4

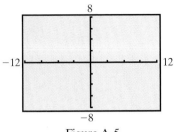

Figure A-5

It is also true that if we know the Xmin and the Xmax values, we can calculate the Xscl by the displayed axes. For example, the Xscl of the graph in Figure A-5 must be 3 units per tick mark for the maximum and minimum x-values to be as shown. Also, the Yscl of that graph must be 2 units per tick mark for the maximum and minimum y-values to be as shown.

We will call the viewing window in Figure A-3 a *standard* viewing window or rectangle. Although a standard viewing window is sufficient for much of this text, special care must be taken to ensure that all key features of a graph are shown. Figures A-6, A-7, and A-8 show the graph of $y = x^2 + 11x - 1$ on three different viewing windows. Note that certain viewing windows for this equation are misleading.

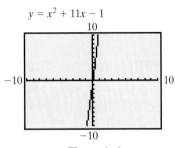

Figure A-6

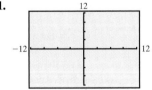

Figure A-7

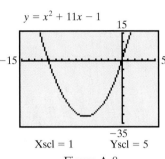

Figure A-8

How do we ensure that all distinguishing features of the graph of an equation are shown? It helps to know about the equation that is being graphed. For example, the equation $y = x^2 + 11x - 1$ is not a linear equation and its graph is not a line. This equation is a quadratic equation and, therefore, its graph is a parabola. By knowing this information, we know that the graph shown in Figure A-6, although correct, is misleading. Of the three viewing rectangles shown, the graph in Figure A-8 is best because it shows more of the distinguishing features of the parabola. Properties of equations needed for graphing will be studied in this text.

VIEWING WINDOW AND INTERPRETING WINDOW SETTINGS EXERCISE SET

In Exercises 1–4, determine whether all ordered pairs listed will lie within a standard viewing rectangle.

1. $(-9, 0), (5, 8), (1, -8)$
2. $(4, 7), (0, 0), (-8, 9)$
3. $(-11, 0), (2, 2), (7, -5)$
4. $(3, 5), (-3, -5), (15, 0)$

In Exercises 5–10, choose an Xmin, Xmax, Ymin, and Ymax so that all ordered pairs listed will lie within the viewing rectangle.

5. $(-90, 0), (55, 80), (0, -80)$
6. $(4, 70), (20, 20), (-18, 90)$
7. $(-11, 0), (2, 2), (7, -5)$
8. $(3, 5), (-3, -5), (15, 0)$
9. $(200, 200), (50, -50), (70, -50)$
10. $(40, 800), (-30, 500), (15, 0)$

Write the window setting for each viewing window shown. Use the following format:

Xmin =   Ymin =
Xmax =   Ymax =
Xscl =   Yscl =

11.

12.

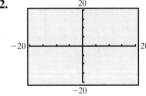

13.

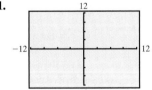

14.

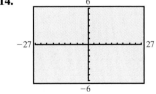

15.

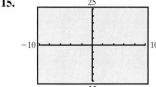

16.

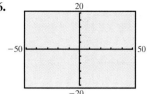

19.

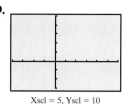

Xscl = 5, Yscl = 10

20.

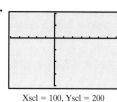

Xscl = 100, Yscl = 200

17.

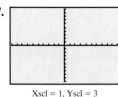

Xscl = 1, Yscl = 3

18.

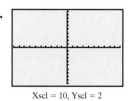

Xscl = 10, Yscl = 2

GRAPHING EQUATIONS AND SQUARE VIEWING WINDOW

In general, the following steps may be used to graph an equation on a standard viewing window.

> **Graphing an Equation in X and Y with a Graphing Utility on a Standard Viewing Window**
>
> **Step 1:** Solve the equation for y.
>
> **Step 2:** Using your graphing utility and enter the equation in the form
> Y = *expression involving x.*
>
> **Step 3:** Activate the graphing utility.

Special care must be taken when entering the *expression involving x* in Step 2. You must be sure that the graphing utility you are using interprets the expression as you want it to. For example, let's graph $3y = 4x$. To do so,

STEP 1: Solve the equation for y.

$$3y = 4x$$

$$\frac{3y}{3} = \frac{4x}{3}$$

$$y = \frac{4}{3}x$$

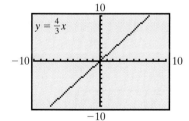

Figure A-9

STEP 2: Using your graphing utility, enter the expression $\frac{4}{3}x$ after the Y= prompt.

In order for your graphing utility to correctly interpret the expression, you may need to enter $(4/3)x$ or $(4 \div 3)x$.

STEP 3: Activate the graphing utility. The graph should appear as in Figure A-9.

Distinguishing features of the graph of a line include showing all the intercepts of the line. For example, the window of the graph of the line in Figure A-10 does not show both intercepts of the line, but the window of the graph of the same line in Figure A-11 does show both intercepts. Notice the notation below each graph. This is a shorthand notation of the range setting of the graph. This notation means [Xmin, Xmax] by [Ymin, Ymax].

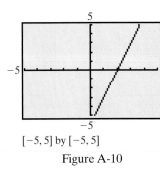

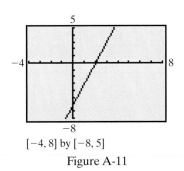

[−5, 5] by [−5, 5]
Figure A-10

[−4, 8] by [−8, 5]
Figure A-11

On a standard viewing window, the tick marks on the *y*-axis are closer together than the tick marks on the *x*-axis. This happens because the viewing window is a rectangle, and so 10 equally spaced tick marks on the positive *y*-axis will be closer together than 10 equally spaced tick marks on the positive *x*-axis. This causes the appearance of graphs to be distorted.

For example, notice the different appearances of the same line graphed using different viewing windows. The line in Figure A-12 is distorted because the tick marks along the *x*-axis are farther apart than the tick marks along the *y*-axis. The graph of the same line in Figure A-13 is not distorted because the viewing rectangle has been selected so that there is equal spacing between tick marks on both axes.

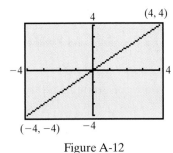

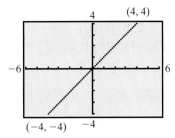

Figure A-12

Figure A-13

We say that the line in Figure A-13 is graphed on a *square* setting. Some graphing utilities have a built-in program that, if activated, will automatically provide a square setting. A square setting is especially helpful when we are graphing perpendicular lines, circles, or when a true geometric perspective is desired. Some examples of square screens are shown in Figures A-14 and A-15.

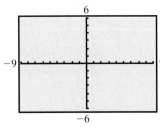

Figure A-14

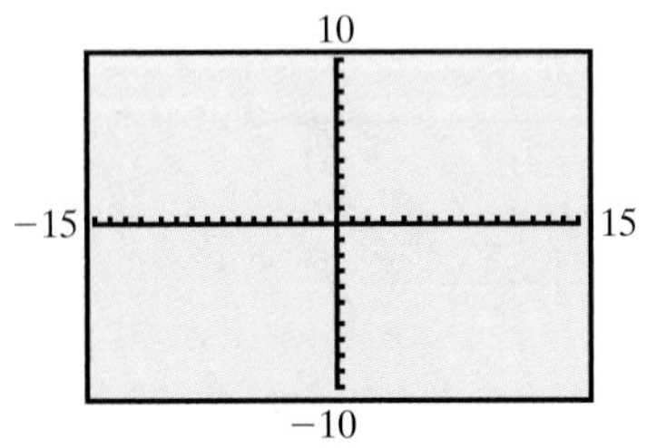

Figure A-15

Other features of a graphing utility such as Trace, Zoom, Intersect, and Table are discussed in appropriate Graphing Calculator Explorations in this text.

GRAPHING EQUATIONS AND SQUARE VIEWING WINDOW EXERCISE SET

Graph each linear equation in two variables, using the two different range settings given. Determine which setting shows all intercepts of a line.

1. $y = 2x + 12$
   Setting A: $[-10, 10]$ by $[-10, 10]$
   Setting B: $[-10, 10]$ by $[-10, 15]$

2. $y = -3x + 25$
   Setting A: $[-5, 5]$ by $[-30, 10]$
   Setting B: $[-10, 10]$ by $[-10, 30]$

3. $y = -x - 41$
   Setting A: $[-50, 10]$ by $[-10, 10]$
   Setting B: $[-50, 10]$ by $[-50, 15]$

4. $y = 6x - 18$
   Setting A: $[-10, 10]$ by $[-20, 10]$
   Setting B: $[-10, 10]$ by $[-10, 10]$

5. $y = \dfrac{1}{2}x - 15$
   Setting A: $[-10, 10]$ by $[-20, 10]$
   Setting B: $[-10, 35]$ by $[-20, 15]$

6. $y = -\dfrac{2}{3}x - \dfrac{29}{3}$
   Setting A: $[-10, 10]$ by $[-10, 10]$
   Setting B: $[-15, 5]$ by $[-15, 5]$

The graph of each equation is a line. Use a graphing utility and a standard viewing window to graph each equation.

7. $3x = 5y$ **8.** $7y = -3x$ **9.** $9x - 5y = 30$
10. $4x + 6y = 20$ **11.** $y = -7$ **12.** $y = 2$
13. $x + 10y = -5$ **14.** $x - 5y = 9$

Graph the following equations using the square setting given. Some keystrokes that may be helpful are given.

15. $y = \sqrt{x}$ $[-12, 12]$ by $[-8, 8]$
   Suggested keystrokes: $\sqrt{\ }\ x$
16. $y = \sqrt{2x}$ $[-12, 12]$ by $[-8, 8]$
   Suggested keystrokes: $\sqrt{\ }\ (2x)$
17. $y = x^2 + 2x + 1$ $[-15, 15]$ by $[-10, 10]$
   Suggested keystrokes: $x^\wedge 2 + 2x + 1$
18. $y = x^2 - 5$ $[-15, 15]$ by $[-10, 10]$
   Suggested keystrokes: $x^\wedge 2 - 5$
19. $y = |x|$ $[-9, 9]$ by $[-6, 6]$
   Suggested keystrokes: $ABS\,(x)$
20. $y = |x - 2|$ $[-9, 9]$ by $[-6, 6]$
   Suggested keystrokes: $ABS(x - 2)$

Graph each line. Use a standard viewing window; then, if necessary, change the viewing window so that all intercepts of each line show.

21. $x + 2y = 30$ **22.** $1.5x - 3.7y = 40.3$

Appendix F

Graphing Stat Plots and Regression Equations

To review graphing Stat Plots and Regressions Equations, we will use a linear model, but the procedure is the same if the data is modeled by a quadratic, cubic, exponential, or any other equation. (For any additional help, consult your graphing utility manual.)

1. **Collect the data.** The table below shows the number of two-income families, which has risen since 1970. For this table,

   $x = 0$ represents the number of years since 1970 and the y-values are the number of two-income families (in millions).

   Calculate a linear regression and use this equation to predict the number of two-income families in 2015.

Year (x)	0	5	10	15	20	30	35	37
Number of Couples (y)	15.3	16.5	20.1	21.5	24	26.4	30.6	31.2

▶ **Helpful Hint**

Before entering the data in Lists, know that Lists (L1, L2, L3, etc.) often get deleted and/or rearranged as you use them. To arrange back in numerical order, go to the Stat, Edit menu. Choose SetupEditor and press Enter twice. Make sure you see Done on the Home Screen. (See the below left screen.)

2. **Enter the data in Lists** by pressing Stat, Edit, Enter. Next to enter the data in two lists, we choose L1 and L2. (See the below middle and right screens.)

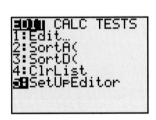 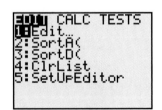 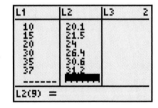

3. **To Create a Stat Plot of this data,** go to Plot1 by pressing 2nd, Y=, Enter. (See the below left screen.)

  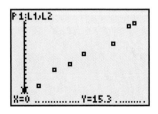

Use this screen (See the above middle screen.) to indicate that the Type graph you want is a scatter plot, the XList is in L1, the YList in L2, and we choose a square as the plotting mark. Next choose an appropriate window that will contain all the data points. (Pressing Zoom, ZoomStat will instruct the calculator to automatically choose a window to fit the data.) (See the above right screen.)

4. **Now calculate the linear regression equation** and paste it in Y1. Press Stat, Calc, LinReg(ax+b). We next indicate that the *x*-list is found in L1, the *y*-list is found in L2, and to paste the linear regression equation in Y1. (See the below left screen.) Press Enter and the regression equation is calculated and pasted in Y1. (See the below right screen.)

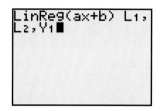
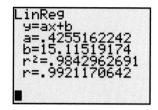

When you press the Y= editor, you can see that the equation is automatically entered into Y1. (See the below left screen.) Press graph and you will see the graph of Y1 along with the plotted data. (See the below middle screen.)

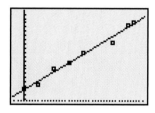

5. **Make the prediction.** If the trend continues to rise at the same rate, let's predict the number of two-income families by the year 2015. The regression equation is in Y1, so first return to the home screen. To paste Y1 on the home screen, press VARS, Y-VARS, FUNCTION, Y1. To calculate Y1 for the year 2015, remember that $x = 0$ represents the number of years since 1970, so we calculate Y1 for 45 (2015–1970.) (See the above right screen.)

If the trend continues, we can expect to see approximately 34.263 million two-income families in the year 2015.

Answers to Selected Exercises

CHAPTER 1 REAL NUMBERS, ALGEBRAIC EXPRESSIONS, AND EQUATIONS

Section 1.2
Practice Exercises

1. 14 sq cm **2.** 31 **3. a.** $\{6, 7, 8, 9\}$ **b.** $\{41, 42, 43, \ldots\}$ **4. a.** true **b.** true **5. a.** true **b.** false **c.** true **d.** false

6. a. 4 **b.** $\frac{1}{2}$ **c.** 1 **d.** -6.8 **e.** -4 **7. a.** -5.4 **b.** $\frac{3}{5}$ **c.** -18 **8. a.** $3x$ **b.** $2x - 5$ **c.** $3\frac{5}{8} + x$ **d.** $\frac{x}{2}$ **e.** $x - 14$ **f.** $5(x + 10)$

Vocabulary and Readiness Check 1.2

1. variables **3.** absolute value **5.** natural numbers **7.** integers **9.** rational number

Exercise Set 1.2

1. 35 **3.** 30.38 **5.** $\frac{3}{8}$ **7.** 22 **9.** 2000 mi **11.** 20.4 sq. ft **13.** \$10,612.80 **15.** $\{1, 2, 3, 4, 5\}$ **17.** $\{11, 12, 13, 14, 15, 16\}$ **19.** $\{0\}$

21. $\{0, 2, 4, 6, 8\}$ **23.** **25.** **27.**

29. answers may vary **31.** $\{3, 0, \sqrt{36}\}$ **33.** $\{3, \sqrt{36}\}$ **35.** $\{\sqrt{7}\}$ **37.** $\in$ **39.** $\notin$ **41.** $\notin$ **43.** $\notin$ **45.** true **47.** true

49. false **51.** false **53.** true **55.** false **57.** answers may vary **59.** -2 **61.** 4 **63.** 0 **65.** -3 **67.** answers may vary **69.** 6.2

71. $-\frac{4}{7}$ **73.** $\frac{2}{3}$ **75.** 0 **77.** $2x$ **79.** $2x + 5$ **81.** $x - 10$ **83.** $x + 2$ **85.** $\frac{x}{11}$ **87.** $12 - 3x$ **89.** $x + 2.3$ or $x + 2\frac{3}{10}$ **91.** $1\frac{1}{3} - x$

93. $\frac{5}{4 - x}$ **95.** $2(x + 3)$ **97.** 137 **99.** 69 **101.** answers may vary

Section 1.3
Practice Exercises

1. a. -8 **b.** -3 **c.** 5 **d.** -8.1 **e.** $\frac{1}{15}$ **2. a.** -8 **b.** -3 **c.** -12 **d.** 7.7 **e.** $\frac{8}{21}$ **f.** 1.8 **g.** -7 **3. a.** 12 **b.** -4

4. a. -15 **b.** $\frac{1}{2}$ **c.** -10.2 **d.** 0 **e.** $-\frac{2}{13}$ **f.** 36 **g.** -11.5 **5. a.** -2 **b.** 5 **c.** $-\frac{1}{6}$ **d.** -6 **e.** $\frac{1}{9}$ **f.** 0 **6. a.** 8 **b.** $\frac{1}{9}$

c. -36 **d.** 36 **e.** -64 **f.** -64 **7. a.** 7 **b.** $\frac{1}{4}$ **c.** -8 **d.** not a real number **e.** 10 **8. a.** 4 **b.** -1 **c.** 10 **9. a.** 2

b. 27 **c.** -29 **10.** 19 **11.** $-\frac{17}{44}$ **12. a.** 67 **b.** -100 **c.** $-\frac{39}{80}$ **13. a.** 46 **b.** 606 **c.** 119.098 **14.** 23; 50; 77

Vocabulary and Readiness Check 1.3

1. b, c **3.** b, d **5.** b **7.** 0 **9.** reciprocal **11.** exponent **13.** square root

Exercise Set 1.3

1. 5 **3.** -24 **5.** -11 **7.** -4 **9.** $\frac{4}{3}$ **11.** -2 **13.** $-\frac{1}{2}$ **15.** -6 **17.** -60 **19.** 0 **21.** 0 **23.** -3 **25.** 3 **27.** $-\frac{1}{6}$ **29.** 0.56

31. -7 **33.** -8 **35.** -49 **37.** 36 **39.** -8 **41.** $-\frac{1}{27}$ **43.** 7 **45.** $-\frac{2}{3}$ **47.** 4 **49.** 3 **51.** not a real number **53.** 48 **55.** -1

57. -9 **59.** 17 **61.** -4 **63.** -2 **65.** 11 **67.** $-\frac{3}{4}$ **69.** 7 **71.** -11 **73.** -2.1 **75.** $-\frac{1}{3}$ **77.** $-\frac{79}{15}$ **79.** $-\frac{4}{5}$ **81.** -81

83. $-\frac{20}{33}$ **85.** 93 **87.** -12 **89.** $-\frac{23}{18}$ **91.** 5 **93.** $-\frac{3}{19}$ **95.** 16.6 **97.** -0.61 **99.** -121.2 **101.** $x^2 + 2y$; $x = -2.7$, $y = 0.9$; 9.09

103. $A^2 + B^2$; $A = 4$, $B = -2.5$; 22.25 **105. a.** $\left(\frac{1}{2}\right)x = 1$; $\frac{1}{2}x = 1$ **b.** same; answers may vary **107.** 18; 22; 28; 208 **109.** 600; 150; 105

111. $\frac{5}{2}$ **113.** $\frac{13}{35}$ **115.** 4205 m **117.** $(2 + 7) \cdot (1 + 3)$ **119.** answers may vary **121.** 3.1623 **123.** 2.8107 **125.** -0.5876

Section 1.4
Practice Exercises

1. $-4x = 20$ **2.** $3(z - 3) = 9$ **3.** $x + 5 = 2x - 3$ **4.** $y + 2 = 4 + \frac{z}{8}$ **5. a.** $<$ **b.** $=$ **c.** $>$ **d.** $>$ **6. a.** $x - 3 \leq 5$ **b.** $y \neq -4$

c. $2 < 4 + \frac{1}{2}z$ **7. a.** 7 **b.** -4.7 **c.** $\frac{3}{8}$ **8. a.** $-\frac{3}{5}$ **b.** $\frac{1}{14}$ **c.** $-\frac{1}{2}$ **9.** $13x + 8$ **10.** $(3 \cdot 11)b = 33b$ **11. a.** $4x + 20y$ **b.** $-3 + 2z$

c. $0.3xy - 0.9x$ **12. a.** double negative property   **b.** distributive property   **c.** associative property of addition   **13. a.** $0.10x$ **b.** $26y$
c. $1.75z$ **d.** $0.15t$ **14. a.** $16 - x$ **b.** $180 - x$ **c.** $x + 2$ **d.** $x + 9$ **15. a.** $5ab$ **b.** $10x - 5$ **c.** $17p - 9$ **16. a.** $-pq + 7$
b. $x^2 + 19$ **c.** $5.8x + 3.8$ **d.** $-c - 8d + \dfrac{1}{4}$

Vocabulary and Readiness Check 1.4

1. $<$ **3.** $\neq$ **5.** $\geq$ **7.** additive inverse; $-a$ **9.** commutative   **11.** distributive   **13.** terms   **15.** x^{-1}

Exercise Set 1.4

1. $10 + x = -12$ **3.** $2x + 5 = -14$ **5.** $\dfrac{n}{5} = 4n$ **7.** $z - \dfrac{1}{2} = \dfrac{1}{2}z$ **9.** $7x \leq -21$ **11.** $2(x - 6) > \dfrac{1}{11}$ **13.** $2(x - 6) = -27$

15. $>$ **17.** $=$ **19.** $<$ **21.** $<$ **23.** $<$ **25.** $-5; \dfrac{1}{5}$ **27.** $-8; -\dfrac{1}{8}$ **29.** $\dfrac{1}{7}; -7$ **31.** 0; undefined   **33.** $\dfrac{7}{8}; -\dfrac{7}{8}$

35. Zero. For every real number $x, 0 \cdot x \neq 1$, so 0 has no reciprocal. It is the only real number that has no reciprocal because if $x \neq 0$, then $x \cdot \dfrac{1}{x} = 1$

by definition.   **37.** $y + 7x$ **39.** $w \cdot z$ **41.** $\dfrac{x}{5} \cdot \dfrac{1}{3}$ **43.** no; answers may vary   **45.** $(5 \cdot 7)x$ **47.** $x + (1.2 + y)$ **49.** $14(z \cdot y)$

51. 10 and 4. Subtraction is not associative.   **53.** $3x + 15$ **55.** $-2a - b$ **57.** $12x + 10y + 4z$ **59.** $-4x + 8y - 28$ **61.** $3xy - 1.5x$

63. $6 + 3x$ **65.** 0 **67.** 7 **69.** $(10 \cdot 2)y$ **71.** Associative property of addition   **73.** Commutative property of multiplication

75. $a(b + c) = ab + ac$ **77.** $0.1d$ **79.** $112 - x$ **81.** $90 - 5x$ **83.** $\$35.61y$ **85.** $2x + 2$ **87.** $-8y - 14$ **89.** $-9c - 4$ **91.** $4 - 8y$

93. $y^2 - 11yz - 11$ **95.** $3t - 14$ **97.** 0 **99.** $13n - 20$ **101.** $8.5y - 20.8$ **103.** $\dfrac{3}{8}a - \dfrac{1}{12}$ **105.** $20y + 48$ **107.** $-5x + \dfrac{5}{6}y - 1$

109. $-x - 6y - \dfrac{11}{24}$ **111.** $(5 \cdot 7)y$ **113.** $6.5y - 7.92x + 25.47$ **115.** no   **117.** 80 million   **119.** 35 million   **121.** 20.25%

Integrated Review

1. 16 **2.** -16 **3.** 0 **4.** -11 **5.** -5 **6.** $-\dfrac{1}{60}$ **7.** undefined   **8.** -2.97 **9.** 4 **10.** -50 **11.** 35 **12.** 92 **13.** $-15 - 2x$

14. $3x + 5$ **15.** 0 **16.** true   **17.** $-45x$ **18.** $-x - 8$ **19.** $7.6a + 7.2b$ **20.** $\dfrac{14}{9}y + \dfrac{7}{30}$

Section 1.5
Practice Exercises

1. 5 **2.** 0.2 **3.** -5 **4.** -4 **5.** $\dfrac{5}{6}$ **6.** $\dfrac{5}{4}$ **7.** -3 **8.** { } or $\varnothing$ **9.** all real numbers or $\{x | x$ is a real number$\}$

Vocabulary and Readiness Check 1.5

1. equivalent   **3.** addition   **5.** expression   **7.** equation   **9.** all real numbers   **11.** no solution

Exercise Set 1.5

1. 6 **3.** -22 **5.** 4.7 **7.** 10 **9.** -1.1 **11.** -5 **13.** -2 **15.** 0 **17.** 2 **19.** -9 **21.** $-\dfrac{10}{7}$ **23.** $\dfrac{9}{10}$ **25.** 4 **27.** 1 **29.** 5

31. $\dfrac{40}{3}$ **33.** 17 **35.** all real numbers   **37.** $\varnothing$ **39.** all real numbers   **41.** $\varnothing$ **43.** $6x - 5 = 4x - 21; x = -8$

45. $3(N - 3) + 2 = -1 + N; N = 3$ **47.** $\dfrac{1}{8}$ **49.** 0 **51.** all real numbers   **53.** 4 **55.** $\dfrac{4}{5}$ **57.** 8 **59.** $\varnothing$ **61.** -8 **63.** $-\dfrac{5}{4}$

65. -2 **67.** 23 **69.** $-\dfrac{2}{9}$ **71.** subtract 19 instead of adding; -3 **73.** $0.4 - 1.6 = -1.2$, not 1.2; -0.24 **75. a.** $4x + 5$ **b.** -3

c. answers may vary   **77.** answers may vary   **79.** $K = -11$ **81.** $K = -23$ **83.** answers may vary   **85.** 1 **87.** 3 **89.** -4.86 **91.** 1.53

The Bigger Picture

1. $-\dfrac{8}{3}$ **2.** 0 **3.** -1 **4.** 3 **5.** 6 **6.** $\varnothing$ **7.** $\dfrac{3}{2}$ **8.** all real numbers

Section 1.6
Practice Exercises

1. a. $3x + 6$ **b.** $6x - 1$ **2.** $3x + 17.3$ **3.** $14, 34, 70$ **4.** $\$450$ **5.** width: 32 in.; length: 48 in.   **6.** $25, 27, 29$

Vocabulary and Readiness Check 1.6

1. $>$ **3.** $=$ **5.** $31, 32, 33, 34$ **7.** $18, 20, 22$ **9.** $y, y + 1, y + 2$ **11.** $p, p + 1, p + 2, p + 3$

Exercise Set 1.6

1. $4y$ **3.** $3z + 3$ **5.** $(65x + 30)$ cents **7.** $10x + 3$ **9.** $2x + 14$ **11.** -5 **13.** $45, 145, 225$ **15.** approximately 1612.41 million acres
17. 2344 earthquakes **19.** 1275 shoppers **21.** 22% **23.** 417 employees **25.** 29.98 million **27.** $29°, 35°, 116°$ **29.** 28 m, 36 m, 38 m
31. 18 in., 18 in., 27 in., 36 in. **33.** 75, 76, 77 **35.** Fallon's zip code is 89406; Fernley's zip code is 89408; Gardnerville Ranchos' zip code is 89410
37. 317 thousand; 279 thousand; 184 thousand **39.** medical assistant: 215 thousand; postsecondary teacher jobs: 603 thousand; registered nurses:
623 thousand **41.** 757-200: 190 seats; 737-200: 113 seats; 737-300: 134 seats **43.** $430.00 **45.** 41.7 million **47.** $40°, 140°$ **49.** $64°, 32°, 84°$
51. square: 18 cm; triangle: 24 cm **53.** 76, 78, 80 **55.** 40.5 ft; 202.5 ft; 240 ft **57.** Los Angeles: 61.0 million, Atlanta: 74.3 million, Chicago: 62.3 million
59. incandescent: 1500 bulb hours; fluorescent: 100,000 bulb hours; halogen: 4000 bulb hours **61.** height: 48 in.; length: 108 in.
63. a. $4.9 billion **b.** $1.23 billion **65.** Russia: 5.6%; China: 19.3%; U.S.: 9.9% **67.** 309 pages **69.** Germany: 11; U.S.: 9; Canada: 7
71. answers may vary **73.** $50°$ **75. a.** 2032 **b.** 1015.8 **c.** 3 cigarettes per day; answers may vary
77. 500 boards; $30,000 **79.** company makes a profit

Section 1.7
Practice Exercises

1. $2000 + 0.06x$; 2060, 2180, 2300, 2420; $9000 **2.** Plumber 1: $160; Plumber 2: $175; Plumber 1 charges $15 less. **3.** $87.01; $92.67
4. Production of 800 items produces the break-even point. **5. a.** 2 seconds **b.** 5 seconds **c.** 7 seconds **d.** The maximum height, 196 feet,
occurs 3.5 seconds after the rocket was launched.

Vocabulary and Readiness Check 1.7

1. f **3.** b **5.** d

Exercise Set 1.7

1. 343; 512; 729; 1000; 1331 **a.** 729 cu cm **b.** 1331 cu ft **c.** 7 in. **3.** 298; 301.50; 305; 308.50 **a.** 44 hr **b.** 35 hr **c.** 37 hr
5. a. $y_1 = 15.25x$ **b.** $y_2 = 17x$ **c.** Rough draft: 15.25, 19.06, 22.88, 26.69, 30.50, 34.31, 38.13, 41.94, 45.75; Manuscript: 17, 21.25, 25.5, 29.75, 34,
38.25, 42.5, 46.75, 51 **d.** $19.06 **e.** $182.75 **7.** Circumference: 12.57, 31.42, 131.95, 591.88; Area: 12.57, 78.54, 1385.44, 27,877.36 **a.** 188.5 yd;
2827.43 sq yd **b.** 493.23 mm; 19,359.28 sq mm **9. a.** 5500, 8500, 10,500, 14,500, 16,500 **b.** $5500 and $16,500 **11. a.** $y_1 = 12.96 + 1.10x$
b. 12.96, 14.06, 15.16, 16.26, 17.36, 18.46, 19.56 **13. a.** $y_1 = 500 - 8x$ **b.** 500; 420; 340; 260; 180; 100; 20; -60 **c.** 30 min **d.** 62.5 min
15. a. First job: $18,500, $23,500, $28,500, $33,500; Second job: $22,300, $25,175, $28,050, $30,925 **b.** answers may vary **17. a.** $y_1 = 15 + 0.08924x$
b. 15; 59.62; 104.24; 148.86; 193.48 **19. a.** $y_1 = 1000 - 2.5x$ **b.** 1000, 850, 700, 550, 400, 250, 100 **b.** $6\frac{1}{3}$ hours later or 4:20 p.m.
21. a. $y_1 = 24 - \dfrac{x}{15}$ **b.** $y_2 = 24 - \dfrac{x}{25}$ **c.** 24; 17.3, 10.7, 4, -2.7, -9.3, -16; 24, 20, 16, 12, 8, 4, 0 **d.** 17.3 gal **e.** 20 gal **f.** 600 mi
g. 360 mi **h.** answers may vary **23. a.** $y_1 = 0.65x$ **b.** $19.47; $23.37; $12.51; $25.97; $11.67; $6.47; $16.74 **c.** $116.20 **d.** $124.33
e. $62.55 **25.** Cost: 9000, 10,375, 11,750, 13,125, 14,500; Revenue: 7500, 9375, 11,250, 13,125, 15,000 **27.** When 200 calculators are produced
and sold, the profit is $500. **29. a.** 4050; 4380; 4710; 5040; 5370 **b.** In 2025, the electricity demand is expected to be 5040 billion killowatt hours.
31. a. 1569 ft, 1606 ft, 1611 ft, 1584 ft, 1525 ft **b.** 1613 ft **33. a.** Anne: $49.00; Michelle: $50.00 **b.** distances less than $333\frac{1}{3}$ mi
35. a. $C = 90 + 75x$ **b.** 240; 390; 540; 690 **37. a.** 3.5 ft **b.** 5 ft **c.** 10 ft **39.** 23 ft by 46 ft; area: 1058 sq ft **41. a.** 6.73 or 6:44 a.m.
b. 4; July 1. **c.** 7:14 a.m. **d.** 6:17 a.m. **e.** answers may vary

Section 1.8
Practice Exercises

1. $t = \dfrac{I}{Pr}$ **2.** $y = \dfrac{7}{2}x - \dfrac{5}{2}$ **3.** $r = \dfrac{A - P}{Pt}$ **4.** $10,134.16 **5.** 25.6 hr; 25 hr 36 min **6.** $-40, -22, 50, 86$ **7.** The dimensions that give the
largest area are 22 feet by 44 feet with an area of 968 square feet.

Exercise Set 1.8

1. $t = \dfrac{D}{r}$ **3.** $R = \dfrac{I}{PT}$ **5.** $y = \dfrac{9x - 16}{4}$ **7.** $W = \dfrac{P - 2L}{2}$ **9.** $A = \dfrac{J + 3}{C}$ **11.** $g = \dfrac{W}{h - 3t^2}$ **13.** $B = \dfrac{T - 2C}{AC}$ **15.** $r = \dfrac{C}{2\pi}$
17. $r = \dfrac{E - IR}{I}$ **19.** $L = \dfrac{2s - an}{n}$ **21.** $v = \dfrac{3st^4 - N}{5s}$ **23.** $H = \dfrac{S - 2LW}{2L + 2W}$ **25.** $4703.71; $4713.99; $4719.22; $4722.74; $4724.45
27. a. $7313.97 **b.** $7321.14 **c.** $7325.98 **29.** $40°C$ **31.** 3.6 hr, or 3 hr and 36 min **33.** 171 packages **35.** 9 ft **37.** 2 gal
39. a. 1174.86 cu m **b.** 310.34 cu m **c.** 1485.20 cu m **41.** 128.3 mph **43.** 0.42 ft **45.** 41.125π ft ≈ 129.1325 ft
47. 13 ft by 26 ft; area: 338 sq ft **49.** $f = \dfrac{C - 4h - 4p}{9}$ **51.** 178 cal **53.** 1.5 g **55.** 0.388; 0.723; 1.00; 1.523; 5.202; 9.538; 19.193; 30.065; 39.505
57. $6.80 per person **59.** 4 times a year; answers may vary **61.** 0.25 sec **63.** $\dfrac{1}{4}$ **65.** $\dfrac{3}{8}$ **67.** $\dfrac{3}{8}$ **69.** $\dfrac{3}{4}$ **71.** 1 **73.** 1

Chapter 1 Vocabulary Check

1. algebraic expression **2.** opposite **3.** distributive **4.** absolute value **5.** exponent **6.** variable **7.** inequality **8.** reciprocals
9. commutative **10.** associative **11.** whole **12.** real **13.** contradiction **14.** identity **15.** formula **16.** solution
17. consecutive integers **18.** linear equation in one variable

Chapter 1 Review

1. 21   **3.** 324,000   **5.** $\{-2, 0, 2, 4, 6\}$ **7.** $\varnothing$ **9.** $\{\ldots, -1, 0, 1, 2\}$ **11.** false   **13.** true   **15.** true   **17.** true   **19.** true

21. true   **23.** true   **25.** true   **27.** true   **29.** $\left\{5, \dfrac{8}{2}, \sqrt{9}\right\}$ **31.** $\left\{\sqrt{7}, \pi\right\}$ **33.** $\left\{5, \dfrac{8}{2}, \sqrt{9}, -1\right\}$ **35.** -0.6 **37.** -1 **39.** $\dfrac{1}{0.6}$

41. 1   **43.** -35 **45.** 0.31   **47.** 13.3   **49.** 0   **51.** 0   **53.** -5 **55.** 4   **57.** 9   **59.** 3   **61.** $-\dfrac{32}{135}$ **63.** $-\dfrac{5}{4}$ **65.** $\dfrac{5}{8}$ **67.** -1

69. 1   **71.** -4 **73.** $\dfrac{5}{7}$ **75.** $\dfrac{1}{5}$ **77.** -5 **79.** 5   **81. a.** 6.28; 62.8; 628   **b.** increase   **83.** $-5x - 9$ **85.** $-15x^2 + 6$ **87.** $5.7x + 1.1$

89. $n + 2n = -15$ **91.** $6(t - 5) = 4$ **93.** $9x - 10 = 5$ **95.** $-4 < 7y$ **97.** $t + 6 \le -12$ **99.** distributive property

101. commutative property of addition   **103.** multiplicative inverse property   **105.** associative property of multiplication

107. multiplicative identity property   **109.** $(3 + x) + (7 + y)$ **111.** $2 \cdot \dfrac{1}{2}$, for example   **113.** $7 + 0$ **115.** $<$ **117.** $<$ **119.** 3   **121.** $-\dfrac{45}{14}$

123. 0   **125.** 6   **127.** all real numbers   **129.** $\varnothing$ **131.** -3 **133.** $\dfrac{96}{5}$ **135.** 32   **137.** 8   **139.** $\varnothing$ **141.** -7 **143.** 940.5 million

145. no such integers exist   **147.** 258 mi   **149.** 34, 39.36, 49, 55.36   **151.** Coast: 9.1, 10.2, 11.3, 12.4; Cross Gates: 12, 13.5, 15, 16.5   **153.** $w = \dfrac{V}{lh}$

155. $y = \dfrac{5x + 12}{4}$ **157.** $m = \dfrac{y - y_1}{x - x_1}$ **159.** $r = \dfrac{E}{I} - R$ or $r = \dfrac{E - IR}{I}$ **161.** $g = \dfrac{T}{r + vt}$ **163.** $B = \dfrac{2A - hb}{h}$ **165.** $r_1 = 2R - r_2$

167. a. \$3695.27   **b.** \$3700.81   **169. a.** $-40, 5, 50, 140$ **b.** 212°F   **c.** 32°F   **171.** 16 packages   **173.** $\dfrac{3}{4}; -\dfrac{4}{3}$ **175.** $-10x + 6.1$

177. $-\dfrac{6}{11}$ **179.** $-\dfrac{4}{15}$ **181.** 2   **183.** China: 137 million; USA: 102 million; France: 93 million   **185.** $h = \dfrac{3V}{\pi r^2}$

187. 23.56, 46.18, 79.19, 94.25, 128.28

Chapter 1 Test

1. true   **2.** false   **3.** false   **4.** false   **5.** true   **6.** false   **7.** -3 **8.** -56 **9.** -225 **10.** 3   **11.** 1   **12.** $-\dfrac{3}{2}$ **13.** 12   **14.** 1

15. a. 5.75; 17.25; 57.50; 115.00   **b.** increase   **16.** $2(x + 5) = 30$ **17.** $\dfrac{(6 - y)^2}{7} < -2$ **18.** $\dfrac{9z}{|-12|} \ne 10$ **19.** $3\left(\dfrac{n}{5}\right) = -n$

20. $20 = 2x - 6$ **21.** $-2 = \dfrac{x}{x + 5}$ **22.** distributive property   **23.** associative property of addition   **24.** additive inverse property

25. multiplication property of zero   **26.** $0.05n + 0.1d$ **27.** $-6x - 14$ **28.** $\dfrac{1}{2}a - \dfrac{9}{8}$ **29.** $2y - 10$ **30.** $-1.3x + 1.9$ **31.** 64.3118

32. Radius: 1, 1.9, 5, 7.45; Circumference: 6.28, 11.94, 31.42, 46.81; Area: 3.14, 11.34, 78.54, 174.37   **33.** 10   **34.** 1   **35.** $\varnothing$ **36.** all real numbers

37. $-\dfrac{80}{29}$ **38.** $y = \dfrac{3x - 8}{4}$ **39.** $g = \dfrac{s}{t^2 + vt}$ **40.** $C = \dfrac{5}{9}(F - 32)$ **41.** 230,323 people   **42.** approximately 8 dogs   **43.** \$3542.27

44. New York: 21.8 million; Seoul: 23.1 million; Tokyo: 33.4 million   **45. a.** 1900; 1950; 2000; 2050; 2100   **b.** \$24,600   **c.** \$14,000
46. a. at 2 and 3 seconds   **b.** 120 ft   **c.** 2.5 sec   **d.** 3.0 sec

CHAPTER 2 GRAPHS AND FUNCTIONS

Section 2.1

Practice Exercises

1. 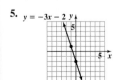   **2.** 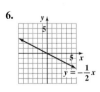   **a.** Quadrant IV   **b.** y-axis   **c.** Quadrant II   **d.** x-axis
e. Quadrant III   **f.** Quadrant I
3. yes, no, yes   **4. a.** \$4200   **b.** more than \$9000

5. 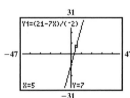   **6.**   **7.**   **8.**   **9. a.**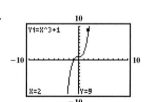

b. 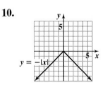   **10.**   **11. a.** 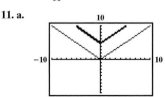   **b.**

Vocabulary and Readiness Check 2.1

1. $(5, 2)$ **3.** $(3, 0)$ **5.** $(-5, -2)$ **7.** $(-1, 0)$ **9.** QI **11.** QII **13.** QIII **15.** y-axis **17.** QIII **19.** x-axis

Exercise Set 2.1

1. Quadrant I **3.** Quadrant II **5.** Quadrant IV **7.** y-axis **9.** Quadrant III

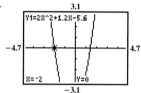   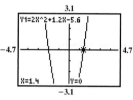 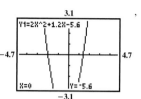

11. Quadrant IV **13.** x-axis **15.** Quadrant III **17.** $(2, 8)$; quadrant I **19.** $(1, -4)$; quadrant IV
21. possible answer: $[-10, 10, 1]$ by $[0, 20, 1]$ **23.** possible answer: $[-100, 100, 10]$ by $[-100, 100, 10]$
25. C **27.** D **29.** no; yes **31.** yes; yes **33.** yes; yes
35. linear, line; linear; line; nonlinear, parabola; linear, line; nonlinear, V-shaped; nonlinear, parabola; linear, line; nonlinear, V-shaped; nonlinear, cubic

37. $y = -2x + 10$ **39.** $y = \dfrac{(-7x - 4)}{3}$ **41.**

 , , ,

y-intercept: $(0, -5.6)$; x-intercepts: $(-2, 0)$ and $(1.4, 0)$

43. d **45.** C **47.** A **49.** D **51.** C **53.** A **55.** C **57.** **59.**

61. linear **63.** linear **65.** linear **67.** not linear **69.** linear **71.** not linear

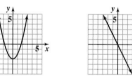

73. not linear **75.** linear **77.** linear **79.** not linear **81.** not linear **83.** not linear

85. **87. a.** parabola **b.** line **89. a.** line **b.** parabola **91. a.** 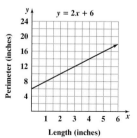 **b.** 14 in.

93. -5 **95.** $-\dfrac{1}{10}$ **97.** b **99.** b **101.** c **103.** 1991 **105.** answers may vary

107. $7000 **109.** $500 **111.** Depreciation is the same from year to year.

113. **115.** $y = x^2 + 2$

Section 2.2

Practice Exercises

1. a. Domain: {4, 5}; Range: {1, −3, −2, 6}   **b.** Domain: {3}; Range: {−4, −3, −2, −1, 0, 1, 2, 3, 4}   **c.** Domain: {Administrative Secretary, Game Developer, Engineer, Restaurant Manager, Marketing}; Range: {27, 50, 73, 35}   **2. a.** yes, a function   **b.** not a function   **c.** yes, a function
3. yes, a function   **4.** yes, a function   **5. a.** yes, a function   **b.** yes, a function   **c.** no, not a function   **d.** yes, a function   **e.** no, not a function
f. yes, a function   **6. a.** Domain: $\{x|-1 \le x \le 2\}$; Range: $\{y|-2 \le y \le 9\}$; yes, a function   **b.** Domain: $\{x|-1 \le x \le 1\}$; Range: $\{y|-4 \le y \le 4\}$;
not a function   **c.** Domain: all real numbers; Range: $\{y|y \le 4\}$; yes, a function   **d.** Domain: all real numbers; Range: all real numbers; yes, a function
7. a. 1   **b.** 6   **c.** −2   **d.** 15   **8. a.** 6.4   **b.** 37.6   **c.** −4   **d.** −2.726   **9. a.** −3   **b.** −2   **c.** 3   **d.** 1   **e.** −1 and 3   **f.** −3
10. $17.25, $26.16, $44.28, $60.95   **11.** $35 billion   **12.** $57.224 billion

Vocabulary and Readiness Check 2.2

1. origin   **3.** $x; y$ **5.** V-shaped   **7.** relation   **9.** domain   **11.** vertical

Exercise Set 2.2

1. domain: $\{-1, 0, -2, 5\}$; range: $\{7, 6, 2\}$; function   **3.** domain: $\{-2, 6, -7\}$; range: $\{4, -3, -8\}$; not a function

5. domain: $\{1\}$; range: $\{1, 2, 3, 4\}$; not a function   **7.** domain: $\left\{\dfrac{3}{2}, 0\right\}$; range: $\left\{\dfrac{1}{2}, -7, \dfrac{4}{5}\right\}$; not a function

9. domain: $\{-3, 0, 3\}$; range: $\{-3, 0, 3\}$; function   **11.** domain: $\{-1, 1, 2, 3\}$; range: $\{2, 1\}$; function   **13.** domain: {Iowa, Alaska, Delaware, Illinois, Connecticut, New York}; range: {5, 1, 19, 29}; function   **15.** domain: {32°, 104°, 212°, 50°}; range: {0°, 40°, 10°, 100°}; function
17. domain: $\{0\}$; range: $\{2, -1, 5, 100\}$; not a function   **19.** function   **21.** not a function   **23.** function   **25.** not a function   **27.** function
29. function   **31.** not a funcion   **33.** domain: $\{x|x \ge 0\}$; range: all real numbers; not a function   **35.** domain: $\{x|-1 \le x \le 1\}$; range: all real numbers; not a function   **37.** domain: all real numbers; range $\{y|y \le -3$ and $y \ge 3\}$; not a function   **39.** domain: $\{x|2 \le x \le 7\}$; range: $\{y|1 \le y \le 6\}$; not a function   **41.** domain: $\{-2\}$; range: all real numbers; not a function   **43.** domain: all real numbers; range: $\{y|y \le 3\}$; function
45. domain: all real numbers; range: all real numbers; function   **47.** answers may vary   **49.** yes   **51.** no   **53.** yes   **55.** yes   **57.** yes

59. no   **61.** 15   **63.** 38   **65.** 7   **67.** 3   **69. a.** 0   **b.** 1   **c.** −1   **71. a.** 246   **b.** 6   **c.** $\dfrac{9}{2}$ **73. a.** −5   **b.** −5   **c.** −5   **75. a.** 5.1
b. 15.5   **c.** 9.533   **77. a.** 23   **b.** 23   **c.** 33   **79.** 7   **81.** (1, −10)   **83.** (4, 56)   **85.** $f(-1) = -2$ **87.** $g(2) = 0$ **89.** −4, 0
91. −4   **93.** infinite number   **95.** 55, 90, 125, 160, 195   **97. a.** 36   **b.** 64   **c.** 84   **d.** 96   **e.** 100   **f.** 0   **99. a.** $17 billion
b. $15.592 billion   **101.** $15.54 billion   **103.** $f(x) = x + 7$ **105.** 25π sq cm   **107.** 2744 cu in.   **109.** 166.38 cm   **111.** 163.2 mg
113. a. 106.26; per capita consumption of poultry was 106.26 lb in 2006.   **b.** 108.54 lb   **115.** 5, −5, 6

117. $2, \dfrac{8}{7}, \dfrac{12}{7}$ **119.** 0, 0, −6   **121.** yes; 170 m   **123.** false
125. true   **127. a.** −3s + 12
b. −3r + 12
129. a. 132   **b.** $a^2 - 12$
131. answers may vary

Section 2.3

Practice Exercises

1. 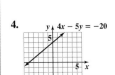   **2.** 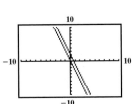 ; shifting the graph of $f(x)$ up 2 units   **3.** shifting the graph of $y = -x$ down 5 units

4. 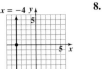   **5. a.** $\left(0, -\dfrac{2}{5}\right)$ **b.** (0, 4.1)   **6.**   **7.** 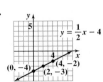   **8.**   **9.** 84; 96.50; 100; 124

Vocabulary and Readiness Check 2.3

1. linear   **3.** vertical; $(c, 0)$ **5.** $y; f(x); x$

Exercise Set 2.3

1.   **3.**   **5.**   **7.**   **9.** C   **11.** D   **13.**

15.   **17.**   **19.**   **21.** answers may vary   **23.**   **25.**

27.   **29.** C   **31.** A   **33.** The vertical line $x = 0$ has y-intercepts.   **35.** $x + 2y = 8$ **37.** $3x + 5y = 7$

39. $x + 8y = 8$ **41.** 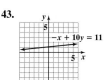 $y = 6x - 5$ **43.** $-x + 10y = 11$ **45.** $y = \dfrac{3}{2}$ **47.** $2x + 3y = 6$

49. $x = -3$ **51.** $y = \dfrac{3}{4}x + 2$ **53.** $y = x$ **55.** $y = \dfrac{1}{2}x$ **57.** $y = 4x - \dfrac{1}{3}$ **59.** $x = -3$

61. b, d   **63.** 677.25, 706.78, 807.19, 860.34   **65.** $\dfrac{3}{2}$ **67.** 6   **69.** $-\dfrac{6}{5}$

71. a. $(0, 500)$; if no tables are produced, 500 chairs can be produced   **b.** $(750, 0)$; if no chairs are produced, 750 tables can be produced   **c.** 466 chairs
73. a. \$64   **b.**   **c.** The line moves upward from left to right.   **75. a.** \$2855.12   **b.** 2012   **c.** answers may vary

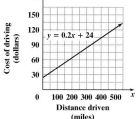

77. a. a line parallel to $y = -4x$ but with y-intercept $(0, 2)$ **b.** a line parallel to $y = -4x$ but with y-intercept $(0, -5)$ **79.** B   **81.** A

Section 2.4

Practice Exercises

1. $m = -\dfrac{1}{2}$;   **2.** $m = \dfrac{3}{5}$;   **3.** $m = \dfrac{2}{3}$; y-intercept: $(0, -3)$ **4.** \$81.84   **5.** undefined
6. $m = 0$ **7. a.** perpendicular   **b.** parallel

Vocabulary and Readiness Check 2.4

1. slope   **3.** m; $(0, b)$ **5.** horizontal   **7.** -1 **9.** upward   **11.** horizontal

Exercise Set 2.4

1. $\dfrac{9}{5}$ **3.** $-\dfrac{5}{6}$ **5.** $\dfrac{1}{3}$ **7.** 0   **9.** undefined   **11.** -1 **13.** $-\dfrac{2}{3}$ **15.** 1.5   **17.** 0.2   **19.** l_2 **21.** l_2 **23.** l_2 **25.** $m = 5$; $(0, -2)$

27. $m = -2$; $(0, 7)$ **29.** $m = \dfrac{2}{3}$; $\left(0, -\dfrac{10}{3}\right)$ **31.** $m = \dfrac{1}{2}$; $(0, 0)$ **33.** A   **35.** B   **37.** undefined   **39.** 0   **41.** undefined

43. answers may vary   **45.** $m = -1$; $(0, 5)$ **47.** $m = \dfrac{6}{5}$; $(0, 6)$ **49.** $m = 3$; $(0, 9)$ **51.** $m = 0$; $(0, 4)$ **53.** $m = 7$; $(0, 0)$

55. $m = 0$; $(0, -6)$ **57.** slope is undefined, no y-intercept   **59.** neither   **61.** parallel   **63.** perpendicular   **65.** answers may vary
67. $\dfrac{3}{2}$ **69.** $-\dfrac{1}{2}$ **71.** $\dfrac{2}{3}$ **73.** approximately -0.12 **75. a.** \$50,139   **b.** $m = 694.9$; The annual income increases \$694.90 every year.
c. y-intercept: $(0, 43,884.9)$; At year $x = 0$, or 2000, the annual average income was \$43,884.90.   **77. a.** $m = 33$; y-intercept: $(0, 42)$

b. The number of WiFi hotspots increases by 33 thousand for every 1 year. **c.** There were 42 thousand WiFi hotspots in 2003.
79. a. The yearly cost of tuition increases \$291.5 every 1 year. **b.** The yearly cost of tuition in 2000 was \$2944.05. **81.** $y = 5x + 32$

83. $y = 2x - 1$ **85.** $m = -\dfrac{20}{9}$ **87.** $-\dfrac{5}{3}$ **89.** $-\dfrac{7}{2}$ **91.** $\dfrac{2}{7}$ **93.** $\dfrac{5}{2}$ **95.** perpendicular **97.** parallel

99. a. $(6, 20)$ **b.** $(10, 13)$ **c.** $-\dfrac{7}{4}$ or -1.75 yd per sec **d.** $\dfrac{3}{2}$ or 1.5 yd per sec

Section 2.5

Practice Exercises

1. $y = -\dfrac{3}{4}x + 4$ **2.** 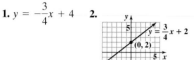 **3.** 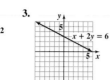 **4.** $y = -4x - 3$ **5.** $f(x) = -\dfrac{2}{3}x + \dfrac{4}{3}$ **6.** $2x + 3y = 5$ **7.** 12,568 house sales
8. $y = -2$ **9.** $x = 6$ **10.** $3x + 4y = 12$ **11.** $f(x) = \dfrac{4}{3}x - \dfrac{41}{3}$

Vocabulary and Readiness Check 2.5

1. $m = -4$, y-intercept: $(0, 12)$ **3.** $m = 5$, y-intercept: $(0, 0)$ **5.** $m = \dfrac{1}{2}$, y-intercept: $(0, 6)$ **7.** parallel **9.** neither

Exercise Set 2.5

1. $y = -x + 1$ **3.** $y = 2x + \dfrac{3}{4}$ **5.** $y = \dfrac{2}{7}x$ **7.** 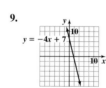 **9.** **11.** 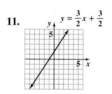 **13.** $y = 3x - 1$

15. $y = -2x - 1$ **17.** $y = \dfrac{1}{2}x + 5$ **19.** $y = -\dfrac{9}{10}x - \dfrac{27}{10}$ **21.** $f(x) = 3x - 6$ **23.** $f(x) = -2x + 1$ **25.** $f(x) = -\dfrac{1}{2}x - 5$

27. $f(x) = \dfrac{1}{3}x - 7$ **29.** $f(x) = -\dfrac{3}{8}x + \dfrac{5}{8}$ **31.** $2x + y = 3$ **33.** $2x - 3y = -7$ **35.** -2 **37.** 2

39. -2 **41.** $y = -4$ **43.** $x = 4$ **45.** $y = 5$ **47.** $f(x) = 4x - 4$ **49.** $f(x) = -3x + 1$ **51.** $f(x) = -\dfrac{3}{2}x - 6$ **53.** $2x - y = -7$

55. $f(x) = -x + 7$ **57.** $x + 2y = 22$ **59.** $2x + 7y = -42$ **61.** $4x + 3y = -20$ **63.** $x = -2$ **65.** $x + 2y = 2$ **67.** $y = 12$

69. $8x - y = 47$ **71.** $x = 5$ **73.** $f(x) = -\dfrac{3}{8}x - \dfrac{29}{4}$ **75. a.** $P(x) = 12,000x + 18,000$ **b.** \$102,000 **c.** end of the ninth yr

77. a. $y = -1000x + 13,000$ **b.** 9500 Fun Noodles **79. a.** $y = 14,220x + 150,900$ **b.** \$278,880 **c.** every year, the median price of a home

increases by \$14,220. **81. a.** $y = 20.2x + 387$ **b.** 568.8 thousand people **83.** 14 **85.** $\dfrac{7}{2}$ **87.** $-\dfrac{1}{4}$ **89.** true

91. $-4x + y = 4$ **93.** $2x + y = -23$ **95.** $3x - 2y = -13$ **97.** answers may vary

Integrated Review

1. **2.** **3.** **4.** **5.** 0 **6.** $-\dfrac{3}{5}$ **7.** $m = 3$; $(0, -5)$ **8.** $m = \dfrac{5}{2}$; $\left(0, -\dfrac{7}{2}\right)$
9. parallel **10.** perpendicular **11.** $y = -x + 7$ **12.** $x = -2$
13. $y = 0$ **14.** $f(x) = -\dfrac{1}{2}x - 8$ **15.** $f(x) = -5x - 6$

16. $f(x) = -4x + \dfrac{1}{3}$ **17.** $f(x) = \dfrac{1}{2}x - 1$ **18.** $y = 3x - \dfrac{3}{2}$

19. $y = 3x - 2$ **20.** $y = -\dfrac{5}{4}x + 4$ **21.** $y = \dfrac{1}{4}x - \dfrac{7}{2}$ **22.** $y = -\dfrac{5}{2}x - \dfrac{5}{2}$ **23.** $x = -1$ **24.** $y = 3$

Section 2.6

Practice Exercises

1. ≈ 9.82 billion **2.** \$27.29 per customer in 2016; The average increase in the amount spent for these services between 2001 and 2006 is \$0.54 per year.
3. b. $y = 0.25x + 2.29$; increasing line slope is positive **c.** The number of sick or injured workers is increasing at a rate of 0.25 per 100 workers per year.
d. 8.54 per 100 workers in the year 2015 **4. a.** $y = 195.91x + 38,438.48$ **b.** \$364,040.90 **5.** About 45%

Vocabulary and Readiness Check 2.6

1. Regression analysis **3.** Stat Plot; scatter plot **5.** linear regression equation

Exercise Set 2.6

1. 65.03 billion dollars **3.** 2000 **5. a.** $y = -27.457x + 704.6$ **b.** \$27.46 **c.** \$265.29 **d.** In the year 2000, the average amount spent for
residential phone services was \$704.60. **7. a.** $y = -0.703x + 52.248$ **b.** 15.0% **c.** 0.703% per year **9. a.** $y = 46.304x + 273.417$
b. 1848 thousand or 1,848,000 male prisoners **c.** 46 thousand or 46,000 per year

11. a. 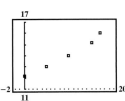 **b.** $y = 0.219x + 12.019$ **c.** The number of workers over 65 is increasing at the rate of 0.219% per year. **d.** We can expect to see about 17.5% of our workforce over 65 by the year 2015.

13. $y = 114.453x - 81,378.675$ **15.** $170,418 **17.** $y = 0.08x + 700$ **19. a.** $y = 0.262x + 48.756$ **b.** The life expectancy at birth of males is increasing at the rate of 0.262 years annually. **c.** 78.9 years old **21. a.** $y = 8.209x + 41.170$ **b.** 205.36 million members **c.** 2015 **d.** The number of PPO members is increasing at the rate of 8.209 million per year. **23. a.** $y = 3.933x - 11.025$ **b.** $158.09 per ticket **c.** $3.93 per year

25. a. 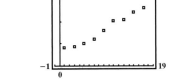 **b.** $y = 0.124x + 0.505$ **c.** The cost is increasing at the rate of 0.124 million, or $124,000 per year. **d.** $3.357 million or $3,357,000

27. 8 **29.** $\frac{1}{10}$ **31.** full service: $y = 7.157x + 62.484$; fast food: $y = 5.971x + 69.452$; 1996 **33.** answers may vary

Section 2.7

Practice Exercises

1. $f(4) = 5; f(-2) = 6; f(0) = -2$ **2.**   **3.**   **4.**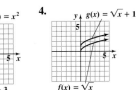

5. 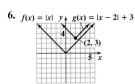 **6.**   **7.**

Vocabulary and Readiness Check 2.7

1. c **3.** d

Exercise Set 2.7

1.   **3.**   **5.**   **7.**   **9.**   domain: all real numbers; range: $\{y|y \geq 0\}$

11.   domain: all real numbers; range: $\{y|y < 5\}$ **13.**   domain: all real numbers; range: $\{y|y \leq 6\}$ **15.**   domain: $\{x|x \leq 0 \text{ or } x \geq 1\}$; range: $\{-4, -2\}$

17. A **19.** B **21.** A **23.** C **25.**   **27.**   **29.**

31.   **33.**   **35.** (graph) **37.** (graph) **39.** (graph)

41. **43.** **45.** **47.** **49.** A **51.** D **53.** answers may vary

55.

57. domain: $\{x|x \geq 2\}$; range: $\{y|y \geq 3\}$ **59.** domain: all real numbers; range: $\{y|y \leq 3\}$ **61.** $\{x|x \geq 20\}$ **63.** all real numbers **65.** $\{x|x \geq -103\}$

67. domain: all real numbers; range: $\{y|y \geq 0\}$ **69.** domain: all real numbers; range: $\{y|y \leq 0 \text{ or } y > 2\}$

Chapter 2 Vocabulary Check

1. relation **3.** point-slope **5.** range **7.** slope-intercept **9.** slope **11.** y **13.** linear function

Chapter 2 Review

1. **3.** no, yes **5.** yes, yes **7.** linear **9.** linear **11.** nonlinear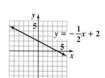

13. linear **15.** D **17.** C **19.** domain: $\left\{-\frac{1}{2}, 6, 0, 25\right\}$; range: $\left\{\frac{3}{4} \text{ or } 0.75, -12, 25\right\}$; function

21. domain: $\{2, 4, 6, 8\}$; range: $\{2, 4, 5, 6\}$; not a function **23.** domain: all real numbers; range: $\{y|y \leq -1 \text{ or } y \geq 1\}$; not a function

25. domain: all real numbers; range: $\{4\}$; function **27.** -3 **29.** 18

31. -3 **33.** 381 lb **35.** 0 **37.** $-2, 4$ **39.** **41.** **43.** A **45.** D **47.**

49. **51.** **53.** **55.** -3 **57.** $\frac{5}{2}$ **59.** $m = \frac{2}{5}, b = -\frac{4}{3}$ **61.** 0 **63.** l_2 **65.** l_2

67. a. $m = 0.3$; The cost increases by \$0.30 for each additional mile driven.

b. $b = 42$; The cost for 0 miles driven is \$42. **69.** parallel

71. **73.** **75.** $x = -2$ **77.** $y = 5$ **79.** $2x - y = 12$ **81.** $11x + y = -52$ **83.** $y = -5$

85. $f(x) = -x - 2$ **87.** $f(x) = -\frac{3}{2}x - 8$ **89.** $f(x) = -\frac{3}{2}x - 1$

91. a. $y = \frac{17}{22}x + 43$ **b.** 52 million

93. **95.** **97.** **99.** 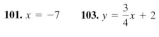 **101.** $x = -7$ **103.** $y = \frac{3}{4}x + 2$

105. $y = -\frac{3}{2}x - 8$ **107.** **109.**

Chapter 2 Test

1. A: quadrant IV   B: x-axis   C: quadrant II

2.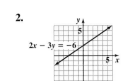
3.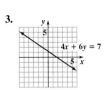
4.

5.

6. $-\dfrac{3}{2}$ **7.** $m = -\dfrac{1}{4}, b = \dfrac{2}{3}$ **8.** $y = -8$ **9.** $x = -4$ **10.** $y = -2$ **11.** $3x + y = 11$

12. $5x - y = 2$ **13.** $f(x) = -\dfrac{1}{2}x$ **14.** $f(x) = -\dfrac{1}{3}x + \dfrac{5}{3}$ **15.** $f(x) = -\dfrac{1}{2}x - \dfrac{1}{2}$ **16.** neither   **17.** B   **18.** A   **19.** D   **20.** C

21. domain: all real numbers; range: {5}; function   **22.** domain: {2}; range: all real numbers; not a function   **23.** domain: all real numbers; range: $\{y\,|\,y \geq 0\}$; function   **24.** domain: all real numbers: range: all real numbers: function   **25. a.** $25,193   **b.** $32,410   **c.** 2015   **d.** The average yearly earnings for high school graduates increases $1031 per year.   **e.** The average yearly earnings for a high school graduate in 2000 was $25,193.

26.   domain: all real numbers; range: $\{y\,|\,y > -3\}$

27.

28.

29.   domain: all real numbers; range: $\{y\,|\,y \leq -1\}$

30.

Chapter 2 Cumulative Review

1. 41; Sec. 1.2, Ex. 2   **3. a.** true   **b.** false   **c.** false   **d.** false; Sec. 1.2, Ex. 5   **5. a.** -6 **b.** -7 **c.** -16 **d.** 20.5
e. $\dfrac{1}{6}$ **f.** 0.94   **g.** -3; Sec. 1.3, Ex. 2   **7. a.** 9   **b.** $\dfrac{1}{16}$ **c.** -25 **d.** 25   **e.** -125 **f.** -125; Sec. 1.3, Ex. 6

9. a. >   **b.** =   **c.** <   **d.** <; Sec. 1.4, Ex. 5   **11. a.** $\dfrac{1}{11}$ **b.** $-\dfrac{1}{9}$ **c.** $\dfrac{4}{7}$; Sec. 1.4, Ex. 8   **13.** 0.4; Sec. 2.1, Ex. 2

15. { } or $\varnothing$; Sec. 2.1, Ex. 8   **17. a.** $3x + 3$ **b.** $12x - 3$; Sec. 2.2, Ex. 1   **19.** 86, 88 and 90; Sec. 2.2, Ex. 7   **21.** $\dfrac{V}{lw} = h$; Sec. 2.3, Ex. 1

23. a. quadrant IV   **b.** no quadrant   **c.** quadrant II   **d.** no quadrant   **e.** quadrant III   **f.** quadrant I; Sec. 2.1, Ex. 1

25. solutions: $(2, -6)$, $(0, -12)$; not a solution $(1, 9)$; Sec. 3.1, Ex. 2   **27.** yes; Sec. 3.2, Ex. 3   **29. a.** $\left(0, \dfrac{3}{7}\right)$ **b.** $(0, -3.2)$; Sec. 3.3, Ex. 3

31. $\dfrac{2}{3}$; Sec. 3.4, Ex. 3   **33.** $y = \dfrac{1}{4}x - 3$; Sec. 3.5 Ex. 1

CHAPTER 3   EQUATIONS AND INEQUALITIES

Section 3.1

Practice Exercises

1. 2   **2.** 6   **3.** -3.641 **4.** $275.60   **5.** cell phone: $876.46; residential phone: $375.12   **6.** all real numbers   **7.** { } or $\varnothing$ **8.** -3.8

Vocabulary and Readiness Check 3.1

1. intersection-of-graphs   **3.** x **5.** root, zero   **7.** -2 **9.** 1   **11.** 7   **13.** all real numbers

Exercise Set 3.1

1. 2   **3.** -1 **5.** -2 **7.** $3x - 13 = 2(x - 5) + 3$; 6   **9.** $-(2x + 3) + 2 = 5x - 1 - 7x$; all real numbers   **11.** $\varnothing$ **13.** all real numbers
15. $\varnothing$ **17.** all real numbers   **19.** 25.8   **21.** 3.5   **23.** -3.28 **25.** $\varnothing$ **27.** 4   **29.** 8   **31.** all real numbers   **33.** -0.32 **35.** 0.6
37. 17   **39.** 0.8   **41. a.** second consultant   **b.** first consultant   **c.** 6-hr job   **43. a.** first agency   **b.** second agency   **c.** 60 mi   **45.** $25,000
47. If the company produces 1667 calculators, the cost and the revenue will be approximately $250,050.   **49.** $\{-3, -2, -1\}$ **51.** $\{-3, -2, -1, 0, 1\}$
53. answers may vary   **55.** 22   **57.** greater than   **59.** 1.43   **61.** 1.51   **63.** 0.91   **65.** 22.87   **67.** The equation has no solution.

Section 3.2
Practice Exercises

1. a. ———→ $(-\infty, 3.5)$ **b.** ←——— $[-3, \infty)$ **c.** ←——→ $[-1, 4)$ **2.** ←——— $(4, \infty)$
      3.5   -3   -1 4   4

3. ←——┤ $(-\infty, -4]$ **4. a.** ←——→ $\left[\dfrac{2}{3}, \infty\right)$ **b.** ←——○ $(-4, \infty)$ **5.** $\left[-\dfrac{3}{2}, \infty\right)$
    -4 $\frac{2}{3}$ -4

6. $(-\infty, 13]$ **7. a.** $(-\infty, \infty)$ **b.** $\varnothing$ **8.** \$10,000 per month.   **9.** the entire year 2021 and after

Vocabulary and Readiness Check 3.2
1. d   **3.** b   **5.** $[-0.4, \infty)$ **7.** $(-\infty, -0.4]$ **9.** no   **11.** yes

Exercise Set 3.2

1. ←——→ $(-\infty, -3)$ **3.** ←——→ $[0.3, \infty)$ **5.** ←——→ $(-2, 5)$
      -3   0.3   -2 5

7. ←——→ $(-1, 5]$ **9.** ←——→ $[-2, \infty)$ **11.** ←——→ $(-\infty, 2]$
    -1 5   -2   2

13. ←——→ $[8, \infty)$ **15.** ←——→ $(-\infty, -4.7)$ **17.** $(-\infty, 4)$ **19.** $[-3, \infty)$ **21.** $\varnothing$ **23.** $(-\infty, -1]$ **25.** $(-\infty, 11]$
       8   -4.7

27. $(0, \infty)$ **29.** $(-13, \infty)$ **31.** $\left[-\dfrac{79}{3}, \infty\right)$ **33.** $\left(-\infty, -\dfrac{35}{6}\right)$ **35.** $(-\infty, -6)$ **37.** $(4, \infty)$ **39.** $[-0.5, \infty)$ **41.** $(-\infty, 7]$ **43.** $[0, \infty)$

45. $(-\infty, -29]$ **47.** $[3, \infty)$ **49.** $(-\infty, -1]$ **51.** $[-31, \infty)$ **53.** $(-\infty, -2]$ **55.** $(-\infty, -15)$ **57.** $\left[-\dfrac{37}{3}, \infty\right)$ **59.** $(-\infty, 5)$

61. $(-\infty, 9)$ **63.** $\left(-\infty, -\dfrac{11}{2}\right]$ **65.** $(-\infty, \infty)$ **67.** $\varnothing$ **69. a.** $\{x \mid x \geq 81\}$ **b.** A final exam grade of 81 or higher will result in an average of
77 or higher.   **71. a.** $\{x \mid x \leq 1040\}$ **b.** The luggage and cargo must weigh 1040 pounds or less.   **73. a.** $\{x \mid x \leq 20\}$
b. She can move at most 20 whole boxes at one time.   **75. a.** $\{x \mid x > 200\}$ **b.** If you make more than 200 calls, plan 1 is more economical.
77. $\{F \mid F \geq 932°\}$ **79. a.** 2011   **b.** answers may vary   **81.** decreasing; answers may vary   **83.** 5.7 gal   **85.** during 2008
87. answers may vary   **89.** 2, 3, 4   **91.** 2, 3, 4, …   **93.** 5   **95.** $\dfrac{13}{6}$ **97.** $\{x \mid x \geq 2\}$; $[2, \infty)$ **99.** ←——○ ; $(-\infty, 0)$
                                                                                                               0

101. $\{x \mid -2 < x \leq 1.5\}$; ←○——┤   **103.** $\{4\}$ **105.** $(4, \infty)$ **107.** answers may vary   **109.** answers may vary   **111.** answers may vary
                                    -2 1.5

The Bigger Picture
1. $-\dfrac{10}{3}$ **2.** 0   **3.** $(-1, \infty)$ **4.** 3   **5.** -6 **6.** $\varnothing$ **7.** $\left(-\infty, \dfrac{3}{2}\right]$ **8.** all real numbers or $(-\infty, \infty)$

Integrated Review
1. -5 **2.** $(-5, \infty)$ **3.** $\left[\dfrac{8}{3}, \infty\right)$ **4.** $[-1, \infty)$ **5.** 0   **6.** $\left[-\dfrac{1}{10}, \infty\right)$ **7.** $\left(-\infty, -\dfrac{1}{6}\right]$ **8.** 0   **9.** $\varnothing$ **10.** $\left[-\dfrac{3}{5}, \infty\right)$ **11.** 4.2   **12.** 6

13. -8 **14.** $(-\infty, -16)$ **15.** $\dfrac{20}{11}$ **16.** 1   **17.** $(38, \infty)$ **18.** -5.5 **19.** $\dfrac{3}{5}$ **20.** $(-\infty, \infty)$ **21.** 29   **22.** all real numbers

23. $(-\infty, 1)$ **24.** $\dfrac{9}{13}$ **25.** $(23, \infty)$ **26.** $(-\infty, 6]$ **27.** $\left(-\infty, \dfrac{3}{5}\right]$ **28.** $\left(-\infty, -\dfrac{19}{32}\right)$

Section 3.3
Practice Exercises
1. $\{1, 3\}$ **2.** $(-\infty, 2)$ **3.** $\{\ \}$ or $\varnothing$ **4.** $(-4, 2)$ **5.** $[-6, 8]$ **6.** $\{1, 2, 3, 4, 5, 6, 7, 9\}$ **7.** $\left(-\infty, \dfrac{3}{8}\right] \cup [3, \infty)$ **8.** $(-\infty, \infty)$

Vocabulary and Readiness Check 3.3
1. compound   **3.** or   **5.** $\cup$ **7.** and

Exercise Set 3.3
1. $\{2, 3, 4, 5, 6, 7\}$ **3.** $\{4, 6\}$ **5.** $\{\ldots, -2, -1, 0, 1, \ldots\}$ **7.** $\{5, 7\}$ **9.** $\{x \mid x$ is an odd integer or $x = 2$ or $x = 4\}$ **11.** $\{2, 4\}$
13. ←○——○ $(-3, 1)$ **15.** ←——┤ $\varnothing$ **17.** ←——○ $(-\infty, -1)$ **19.** $[6, \infty)$ **21.** $(-\infty, -3]$ **23.** $(4, 10)$
       -3 1   0   -1

25. $(11, 17)$ **27.** $[1, 4]$ **29.** $\left[-3, \dfrac{3}{2}\right]$ **31.** $\left[-\dfrac{7}{3}, 7\right]$ **33.** ←——○ $(-\infty, 5)$ **35.** ←┤——┤ $(-\infty, -4] \cup [1, \infty)$
                                                                                              5   -4 1

37. ←——→ $(-\infty, \infty)$ **39.** $[2, \infty)$ **41.** $(-\infty, -4) \cup (-2, \infty)$ **43.** $(-\infty, \infty)$ **45.** $\left(-\dfrac{1}{2}, \dfrac{2}{3}\right)$ **47.** $(-\infty, \infty)$ **49.** $\left[\dfrac{3}{2}, 6\right]$
       0

51. $\left(\dfrac{5}{4}, \dfrac{11}{4}\right)$ **53.** $\varnothing$ **55.** $\left(-\infty, -\dfrac{56}{5}\right) \cup \left[\dfrac{5}{3}, \infty\right)$ **57.** $\left(-5, \dfrac{5}{2}\right)$ **59.** $\left(0, \dfrac{14}{3}\right]$ **61.** $(-\infty, -3]$ **63.** $(-\infty, 1] \cup \left(\dfrac{29}{7}, \infty\right)$ **65.** $\varnothing$

67. $\left[-\dfrac{1}{2}, \dfrac{3}{2}\right)$ **69.** $\left(-\dfrac{4}{3}, \dfrac{7}{3}\right)$ **71.** $(6, 12)$ **73. a.** $(-5, 2.5)$ **b.** $(-\infty, -5) \cup (2.5, \infty)$ **75. a.** $[2, 9]$ **b.** $(-\infty, 2] \cup [9, \infty)$

77. -12 **79.** -4 **81.** $-7, 7$ **83.** 0 **85.** $2003, 2004, 2005$ **87.** $-20.2° \le F \le 95°$

89. $67 \le$ final score ≤ 94 **91.** $(6, \infty)$ **93.** $[3, 7]$ **95.** $(-\infty, -1)$

The Bigger Picture
1. $[-1, 3]$ **2.** $(-1, 6)$ **3.** 5.1 **4.** $(-\infty, -4)$ **5.** $(-\infty, -3]$ **6.** $(-\infty, -2) \cup (2, \infty)$ **7.** 4 **8.** $[19, \infty)$

Section 3.4
Practice Exercises
1. $-7, 7$ **2.** $-1, 4$ **3.** $-80, 70$ **4.** $-2, 2$ **5.** 0 **6.** $\{ \}$ or $\varnothing$ **7.** $\{ \}$ or $\varnothing$ **8.** $-\dfrac{3}{5}, 5$ **9.** 5

Vocabulary and Readiness Check 3.4
1. C **3.** B **5.** D

Exercise Set 3.4
1. $7, -7$ **3.** $4.2, -4.2$ **5.** $7, -2$ **7.** $8, 4$ **9.** $5, -5$ **11.** $3, -3$ **13.** 0 **15.** $\varnothing$ **17.** $\dfrac{1}{5}$ **19.** $|x| = 5$ **21.** $9, -\dfrac{1}{2}$

23. $-\dfrac{5}{2}$ **25.** answers may vary **27.** $\{-1, 4\}$ **29.** $\{2.5\}$ **31.** $4, -4$ **33.** $\varnothing$ **35.** $0, \dfrac{14}{3}$ **37.** $2, -2$ **39.** $7, -1$

41. $\varnothing$ **43.** $\varnothing$ **45.** $-\dfrac{1}{8}$ **47.** $\dfrac{1}{2}, -\dfrac{5}{6}$ **49.** $2, -\dfrac{12}{5}$ **51.** $3, -2$ **53.** $-8, \dfrac{2}{3}$ **55.** $\varnothing$ **57.** 4 **59.** $13, -8$ **61.** $3, -3$ **63.** $8, -7$ **65.** $2, 3$ **67.** $2, -\dfrac{10}{3}$ **69.** $\dfrac{3}{2}$ **71.** $\varnothing$

73. answers may vary **75.** 33% **77.** 38.4 lb **79.** answers may vary **81.** no solution **83.** $|x - 3| = 2$ **85.** $|2x - 1| = 4$ **87. a.** $c = 0$ **b.** c is a negative number **c.** c is a positive number **89.** $0.09, 0.59$ **91.** $-4, -1.37$

Section 3.5
Practice Exercises
1. $(-2, 2)$ **2.** $(-4, 2)$ **3.** $\left[-\dfrac{2}{3}, 2\right]$ **4.** $\{ \}$ or $\varnothing$ **5.** $(-\infty, -10] \cup [2, \infty)$

6. $(-\infty, \infty)$ **7.** $(-\infty, 0) \cup (12, \infty)$ **8.** 2

Vocabulary and Readiness Check 3.5
1. D **3.** C **5.** A

Exercise Set 3.5
1. $[-4, 4]$ **3.** $(1, 5)$ **5.** $(-5, -1)$ **7.** $[-10, 3]$ **9.** $[-5, 5]$

11. $\varnothing$ **13.** $[0, 12]$ **15.** $(-\infty, -3) \cup (3, \infty)$ **17.** $(-\infty, -24] \cup [4, \infty)$

19. $(-\infty, -4) \cup (4, \infty)$ **21.** $(-\infty, \infty)$ **23.** $\left(-\infty, \dfrac{2}{3}\right) \cup (2, \infty)$ **25.** $\{0\}$

27. $\left(-\infty, -\dfrac{3}{8}\right) \cup \left(-\dfrac{3}{8}, \infty\right)$ **29.** $[-2, 2]$ **31.** $(-\infty, -1) \cup (1, \infty)$

33. $(-5, 11)$ **35.** $(-\infty, 4) \cup (6, \infty)$ **37.** $\varnothing$ **39.** $(-\infty, \infty)$

41. $[-2, 9]$ **43.** $(-\infty, -11] \cup [1, \infty)$ **45.** $(-\infty, 0) \cup (0, \infty)$ **47.** $(-\infty, \infty)$

49. $\left[-\dfrac{1}{2}, 1\right]$ **51.** $(-\infty, -3) \cup (0, \infty)$ **53.** $\varnothing$ **55.** $\dfrac{3}{8}$

57. $\left(-\dfrac{2}{3}, 0\right)$ **59.** $(-\infty, -12) \cup (0, \infty)$ **61.** $[-1, 8]$ **63.** $\left[-\dfrac{23}{8}, \dfrac{17}{8}\right]$

65. $(-2, 5)$ **67.** $5, -2$ **69.** $(-\infty, -7] \cup [17, \infty)$ **71.** $-\dfrac{9}{4}$ **73.** $(-2, 1)$ **75.** $2, \dfrac{4}{3}$ **77.** $\varnothing$ **79.** $\dfrac{19}{2}, -\dfrac{17}{2}$ **81.** $\left(-\infty, -\dfrac{25}{3}\right) \cup \left(\dfrac{35}{3}, \infty\right)$

83. a. $\{-5, 11\}$ **b.** $(-5, 11)$ **c.** $(-\infty, -5] \cup [11, \infty)$ **85. a.** $\{-8, 4\}$ **b.** $[-8, 4]$ **c.** $(-\infty, -8) \cup (4, \infty)$ **87.** $\dfrac{1}{6}$ **89.** 0 **91.** $\dfrac{1}{3}$

93. -1.5 **95.** 0 **97.** $|x| < 7$ **99.** $|x| \le 5$ **101.** answers may vary **103.** $3.45 < x < 3.55$

The Bigger Picture

1. -8 **2.** $-13, 21$ **3.** $(-\infty, 6]$ **4.** $(-7, 5]$ **5.** $-\dfrac{13}{2}$ **6.** $\varnothing$ **7.** -50 **8.** -5 **9.** $\left[-\dfrac{9}{5}, 1\right]$ **10.** $(\infty, -13) \cup (-9, \infty)$ **11.** $-3, \dfrac{23}{9}$

12. $-11, -\dfrac{9}{7}$

Section 3.6
Practice Exercises

1. 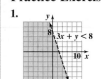 **2.**   **3.**   **4.**

Exercise Set 3.6

1.   **3.**   **5.**   **7.**   **9.** 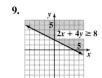 **11.** 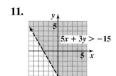 **13.** answers may vary

15.   **17.**   **19.**   **21.** 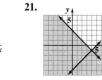 **23.** 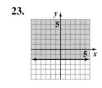 **25.**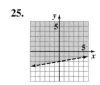

27.   **29.**   **31.**   **33.**   **35.**   **37.**

39.   **41.**   **43.**   **45.**   **47.** D **49.** A **51.** $x \geq 2$ **53.** $y \leq -3$ **55.** $y > 4$

57. $x < 1$ **59.** 8 **61.** -25 **63.** 16 **65.** $\dfrac{27}{125}$

67. domain: $[1, 5]$; range: $[1, 3]$; no

69. $x \leq 20$ and $y \geq 10$.   **71.**

Chapter 3 Vocabulary Check

1. compound inequality **2.** contradiction **3.** intersection **4.** union **5.** identity **6.** absolute value **7.** solution
8. linear inequality in one variable **9.** linear equation in one variable

Chapter 3 Review

1. 12 **3.** $\varnothing$ **5.** $\dfrac{17}{5}$ **7.** -1.51 **9.** $(3, \infty)$ **11.** $(-4, \infty)$ **13.** $(-\infty, 7]$ **15.** $(-\infty, 1)$

17. more economical to use housekeeper for more than 35 pounds per week **19.** 9.6 **21.** $\left[2, \dfrac{5}{2}\right]$ **23.** $\left(\dfrac{1}{8}, 2\right)$ **25.** $\left(\dfrac{7}{8}, \dfrac{27}{20}\right]$

27. $\left(\dfrac{11}{3}, \infty\right)$ **29.** 5, 11 **31.** $-1, \dfrac{11}{3}$ **33.** $-\dfrac{1}{6}$ **35.** $\varnothing$ **37.** $5, -\dfrac{1}{3}$ **39.** 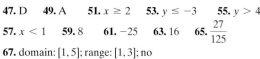 $\left(-\dfrac{8}{5}, 2\right)$

41. $\xleftarrow{}\underset{-3}{)}\underset{3}{(}\xrightarrow{}$ $(-\infty, -3) \cup (3, \infty)$ **43.** $\xleftarrow{}\underset{0}{|}\xrightarrow{}$ $\varnothing$ **45.** $\xleftarrow{}\underset{-27}{}\underset{-9}{}\xrightarrow{}$ $(-\infty, -27) \cup (-9, \infty)$

47. **49.** **51.** **53.**

55. 2   **57.** $(2, \infty)$ **59.** $(-\infty, \infty)$ **61.** 3, -3 **63.** $-10, -\dfrac{4}{3}$ **65.** $\left(-\dfrac{1}{2}, 2\right)$

Chapter 3 Test

1. $-\dfrac{29}{17}$ **2.** 0.15   **3.** 1, $\dfrac{2}{3}$ **4.** $\varnothing$ **5.** $(5, \infty)$ **6.** $(-\infty, 2]$ **7.** $\left(\dfrac{3}{2}, 5\right]$ **8.** $(-\infty, -2) \cup \left(\dfrac{4}{3}, \infty\right)$ **9.** $(-\infty, -5]$ **10.** $(-\infty, -2]$

11. $[-3, -1)$ **12.** $(-\infty, \infty)$ **13.** $(3, 7)$ **14.** 3, $-\dfrac{1}{5}$ **15.** $(-\infty, \infty)$ **16.** $\dfrac{3}{2}$ **17.**   **18.**   **19.**

20. $(-\infty, -3)$ **21.** $(-3, \infty)$ **22.** more than 850 sunglasses

Chapter 3 Cumulative Review

1. a. $\{101, 102, 103, \dots\}$ **b.** $\{2, 3, 4, 5\}$; Sec. 1.2, Ex. 3   **3. a.** 3   **b.** $\dfrac{1}{7}$ **c.** -2.7 **d.** -8 **e.** 0; Sec. 1.2, Ex. 6

5. a. -14 **b.** -4 **c.** 5   **d.** -10.2 **e.** $-\dfrac{5}{21}$; Sec. 1.3, Ex. 1   **7. a.** 3   **b.** 5   **c.** $\dfrac{1}{2}$ **d.** -6 **e.** not a real number; Sec. 1.3, Ex. 7

9. a. -2 **b.** -18 **c.** -1 **d.** -12; Sec. 1.3, Ex. 12   **11. a.** $x + 5 = 20$ **b.** $2(3 + y) = 4$ **c.** $x - 8 = 2x$ **d.** $\dfrac{z}{9} = 9 + z$; Sec. 1.4, Ex. 1–4

13. $5 + 7x$; Sec. 1.4, Ex. 9   **15.** 2; Sec. 1.5, Ex. 1   **17.** all real numbers; Sec. 1.5, Ex. 9   **19. a.** $3x + 3$ **b.** $12x - 3$; Sec. 1.6, Ex. 1

21. 23, 49, 92; Sec. 1.6, Ex. 3   **23.** $y = \dfrac{2x + 7}{3}$ or $y = \dfrac{2x}{3} + \dfrac{7}{3}$; Sec. 1.8, Ex. 2   **25.** $b = \dfrac{2A - Bh}{h}$; Sec. 1.8, Ex. 3

27. a. $[2, \infty)$ **b.** $(-\infty, -1)$ **c.** $(0.5, 3]$; Sec. 3.2, Ex. 1   **29.** $\left[\dfrac{5}{2}, \infty\right)$; Sec. 3.2, Ex. 5   **31. a.** $(-\infty, \infty)$ **b.** $\varnothing$; Sec. 3.2, Ex. 7   **33.** $\{4, 6\}$; Sec. 3.3, Ex. 1

35. $(-\infty, 4)$; Sec. 3.3, Ex. 2   **37.** $\{2, 3, 4, 5, 6, 8\}$; Sec. 3.3, Ex. 6   **39.** $(-\infty, \infty)$; Sec. 3.3, Ex. 8   **41.** 2, -2; Sec. 3.4, Ex. 1

43. 24, -20; Sec. 3.4, Ex. 3   **45.** 4; Sec. 3.4, Ex. 9   **47.** $[-3, 3]$; Sec. 3.5, Ex. 1   **49.** $(-\infty, \infty)$; Sec. 3.5, Ex. 6

CHAPTER 4   SYSTEMS OF LINEAR EQUATIONS AND INEQUALITIES

Section 4.1

Practice Exercises

1. a. yes   **b.** no   **2. a.** infinite number of solutions of the form $\{(x, y) \mid 3x - 2y = 4\}$ or $\{(x, y) \mid -9x + 6y = -12\}$

2. b. solution: $(1, 5)$ **c.** no solution   **3.** $(-5.23, 0.29)$

   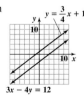

4. $\left(-\dfrac{1}{2}, 5\right)$ **5.** $\left(-\dfrac{9}{5}, -\dfrac{2}{5}\right)$ **6.** $(2, 1)$ **7.** $(-2, 0)$ **8.** $\varnothing$ or $\{\ \}$ **9.** $\{(x, y) \mid -3x + 2y = -1\}$ or $\{(x, y) \mid 9x - 6y = 3\}$ **10.** 730 units

Vocabulary and Readiness Check 4.1

1. B   **3.** A

Exercise Set 4.1

1. yes **3.** no **5.** no **7.** $(1.5, 1)$ **9.** $(-2, 0.75)$ **11.** $(2, -1)$ **13.** $(1, 2)$ **15.** $\varnothing$

17. No; answers may vary **19.** $(2, 8)$ **21.** $(0, -9)$ **23.** $(1, -1)$ **25.** $(-5, 3)$ **27.** $\left(\dfrac{5}{2}, \dfrac{5}{4}\right)$ **29.** $(1, -2)$ **31.** $(8, 2)$ **33.** $(7, 2)$

35. $\varnothing$ **37.** $\{(x, y) \mid 3x + y = 1\}$ **39.** $\left(\dfrac{3}{2}, 1\right)$ **41.** $(2, -1)$ **43.** $(-5, 3)$ **45.** $\{(x, y) \mid 3x + 9y = 12\}$ **47.** $\varnothing$ **49.** $\left(\dfrac{1}{2}, \dfrac{1}{5}\right)$ **51.** $(9, 9)$

53. $\{(x, y) \mid x = 3y + 2\}$ **55.** $\left(-\dfrac{1}{4}, \dfrac{1}{2}\right)$ **57.** $(3, 2)$ **59.** $(7, -3)$ **61.** $\varnothing$ **63.** $(3, 4)$ **65.** $(-2, 1)$ **67.** $(1.2, -3.6)$ **69.** true **71.** false

73. $6y - 4z = 25$ **75.** $x + 10y = 2$ **77.** 5000 DVDs; \$21 **79.** supply greater than demand **81.** $(1875, 4687.5)$ **83.** makes money

85. for x-values greater than 1875 **87.** answers may vary; One possibility: $\begin{cases} -2x + y = 1 \\ \quad x - 2y = -8 \end{cases}$

89. a. Consumption of red meat is decreasing while consumption of poultry is increasing. **b.** $(35, 103)$ **c.** In the year 2035, red meat and poultry consumption will each be about 103 pounds per person. **91.** $\left(\dfrac{1}{4}, 8\right)$ **93.** $\left(\dfrac{1}{3}, \dfrac{1}{2}\right)$ **95.** $\left(\dfrac{1}{4}, -\dfrac{1}{3}\right)$ **97.** $\varnothing$

Section 4.2

Practice Exercises

1. $(-1, 2, 1)$ **2.** $\{\ \}$ or $\varnothing$ **3.** $\left(\dfrac{2}{3}, -\dfrac{1}{2}, 0\right)$ **4.** $\{(x, y, z) \mid 2x + y - 3z = 6\}$ **5.** $(6, 15, -5)$

Exercise Set 4.2

1. a, b, d **3.** Yes; answers may vary **5.** $(-1, 5, 2)$ **7.** $(-2, 5, 1)$ **9.** $(-2, 3, -1)$ **11.** $\{(x, y, z) \mid x - 2y + z = -5\}$ **13.** $\varnothing$
15. $(0, 0, 0)$ **17.** $(-3, -35, -7)$ **19.** $(6, 22, -20)$ **21.** $\varnothing$ **23.** $(3, 2, 2)$ **25.** $\{(x, y, z) \mid x + 2y - 3z = 4\}$ **27.** $(-3, -4, -5)$
29. $\left(0, \dfrac{1}{2}, -4\right)$ **31.** $(12, 6, 4)$ **33.** 15 and 30 **35.** 5 **37.** $-\dfrac{5}{3}$ **39.** answers may vary **41.** answers may vary **43.** $(1, 1, -1)$
45. $(1, 1, 0, 2)$ **47.** $(1, -1, 2, 3)$ **49.** answers may vary

Section 4.3

Practice Exercises

1. a. 2037 **b.** yes; answers may vary **2.** 12 and 17 **3.** Atlantique: 500 kph; V150: 575 kph **4.** 0.95 liter of water; 0.05 liter of 99% HCL
5. 1500 packages **6.** $40°, 60°, 80°$

Exercise Set 4.3

1. 10 and 8 **3. a.** Enterprise class: 1101 ft; Nimitz class: 1092 ft **b.** 3.67 football fields **5.** plane: 520 mph; wind: 40 mph **7.** 20 qt of 4%; 40 qt
of 1% **9.** United Kingdom: 32,071 students; Italy: 24,858 students **11.** 9 large frames; 13 small frames **13.** -10 and -8 **15.** 2007
17. tablets: \$0.80; pens: \$0.20 **19.** B737: 450 mph; Piper: 90 mph **21. a.** answers may vary but notice the slope of each function **b.** 1991
23. 28 cm; 28 cm; 37 cm **25.** 600 mi **27.** $x = 75; y = 105$ **29.** 625 units **31.** 3000 units **33.** 1280 units **35. a.** $R(x) = 450x$
b. $C(x) = 200x + 6000$ **c.** 24 desks **37.** 2 units of Mix A; 3 units of Mix B; 1 unit of Mix C **39.** 5 in.; 7 in.; 10 in. **41.** 18, 13, and 9
43. 143 free throws; 177 two-point field goals; 121 three-point field goals **45.** $x = 60; y = 55; z = 65$ **47.** $5x + 5z = 10$ **49.** $-5y + 2z = 2$
51. 1996: 1,059,444; 2006: 1,085,209 **53. a.** $(112, 137)$ **b.** May 2016 **55.** $a = 1, b = -2, c = 3$
57. $a = 0.28, b = -3.71, c = 12.83$; 2.12 in. in Sept.

Integrated Review

1. C **2.** D **3.** A **4.** B **5.** $(1, 3)$ **6.** $\left(\dfrac{4}{3}, \dfrac{16}{3}\right)$ **7.** $(2, -1)$ **8.** $(5, 2)$ **9.** $\left(\dfrac{3}{2}, 1\right)$ **10.** $\left(-2, \dfrac{3}{4}\right)$ **11.** $\varnothing$

12. $\{(x, y) \mid 2x - 5y = 3\}$ **13.** $\left(1, \dfrac{1}{3}\right)$ **14.** $\left(3, \dfrac{3}{4}\right)$ **15.** $(-1, 3, 2)$ **16.** $(1, -3, 0)$ **17.** $\varnothing$ **18.** $\{(x, y, z) \mid x - y + 3z = 2\}$

19. $\left(2, 5, \dfrac{1}{2}\right)$ **20.** $\left(1, 1, \dfrac{1}{3}\right)$ **21.** 19 and 27 **22.** $70°; 70°; 100°; 120°$

Section 4.4

Practice Exercises

1. $(2, -1)$ **2.** $\varnothing$ **3.** $(-1, 1, 2)$ **4.** $(2, -1)$ **5.** $(-1, 1, 2)$

Vocabulary and Readiness Check 4.4

1. matrix **3.** row **5.** false **7.** true

Exercise Set 4.4

1. $(2, -1)$ **3.** $(-4, 2)$ **5.** $\varnothing$ **7.** $\{(x, y) \mid 3x - 3y = 9\}$ **9.** $(-2, 5, -2)$ **11.** $(1, -2, 3)$ **13.** $(4, -3)$ **15.** $(2, 1, -1)$ **17.** $(9, 9)$
19. $\varnothing$ **21.** $\varnothing$ **23.** $(1, -4, 3)$ **25.** function   **27.** not a function   **29.** -13 **31.** -36 **33.** 0 **35.** c   **37. a.** end of 1984
b. black-and-white sets; microwave ovens; The percent of households owning black-and-white television sets is decreasing and the percent of households owning microwave ovens is increasing; answers may vary   **c.** in 2002   **d.** no; answers may vary   **39.** answers may vary

Section 4.5

Practice Exercises

1.   **2.**   **3.**

Vocabulary and Readiness Check 4.5

1. system   **3.** corner

Exercise Set 4.5

1.   **3.**   **5.**   **7.**   **9.**   **11.**

13.   **15.**   **17.**   **19.**   **21.** C   **23.** D   **25.** 9   **27.** $\dfrac{4}{9}$ **29.** 5   **31.** 59
33. the line $y = 3$ **35.** answers may vary

Chapter 4 Vocabulary Check

1. system of equations   **2.** solution   **3.** consistent   **4.** square   **5.** inconsistent   **6.** matrix

Chapter 4 Review

1. $(-3, 1)$ **3.** $\varnothing$ **5.** $\left(3, \dfrac{8}{3}\right)$ **7.** $(2, 0, 2)$ **9.** $\left(-\dfrac{1}{2}, \dfrac{3}{4}, 1\right)$ **11.** $\varnothing$ **13.** $(1, 1, -2)$ **15.** 10, 40, and 48

17. 58 mph; 65 mph   **19.** 20 L of 10% solution; 30 L of 60% solution   **21.** 17 pennies; 20 nickels; 16 dimes

23. two sides: 22 cm each; third side; 29 cm   **25.** $(-3, 1)$ **27.** $\left(-\dfrac{2}{3}, 3\right)$ **29.** $\left(\dfrac{5}{4}, \dfrac{5}{8}\right)$ **31.** $(1, 3)$ **33.** $(1, 2, 3)$ **35.** $(3, -2, 5)$

37. $(1, 1, -2)$ **39.**   **41.**   **43.**   **45.**

47. $\left(\dfrac{7}{3}, -\dfrac{8}{3}\right)$ **49.** $\{(x, y) \mid 5x - 2y = 10\}$ **51.** $(-1, 3, 5)$ **53.** 28 units, 42 units, 56 units   **55.** 2000

Chapter 4 Test

1. $(1, 3)$ **2.** $\varnothing$ **3.** $(2, -3)$ **4.** $\{(x, y) \mid 10x + 4y = 10\}$ **5.** $(-1, -2, 4)$ **6.** $\varnothing$ **7.** $\left(\dfrac{7}{2}, -10\right)$

8. $\{(x, y) \mid x - y = -2\}$ **9.** $(5, -3)$ **10.** $(-1, -1, 0)$ **11.** 53 double rooms; 27 single rooms   **12.** 5 gal of 10%; 15 gal of 20%

13. 800 packages   **14.** $23°, 45°, 112°$ **15.**

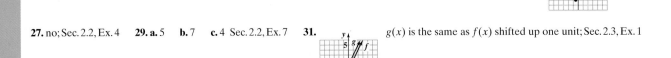

Chapter 4 Cumulative Review

1. a. true   **b.** true; Sec. 1.2, Ex. 4   **3. a.** 6   **b.** -7; Sec. 1.3, Ex. 3   **5. a.** -4 **b.** $-\dfrac{3}{7}$ **c.** 11.2; Sec. 1.4, Ex. 7   **7. a.** $6x + 3y$

b. $-3x + 1$ **c.** $0.7\,ab - 1.4a$; Sec. 1.4, Ex. 11   **9. a.** $-2x + 4$ **b.** $8yz$ **c.** $4z + 6.1$; Sec. 1.4, Ex. 15   **11.** -4; Sec. 1.5, Ex. 3

13. -4; Sec. 1.5, Ex. 7   **15.** 25 cm, 62 cm, 62 cm; Sec. 1.6, Ex. 5   **17.** $[-10, \infty)$ 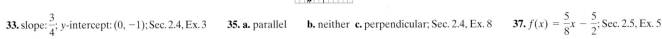 ; Sec. 3.2, Ex. 3   **19.** $(-3, 2)$; Sec. 3.3, Ex. 4

21. $1, -1$; Sec. 3.4, Ex. 4   **23.** $(4, 8)$; Sec. 3.5, Ex. 2   **25. a.** IV   **b.** y-axis   **c.** II   **d.** x-axis   **e.** III   **f.** I;   ; Sec. 2.1, Ex. 1

27. no; Sec. 2.2, Ex. 4   **29. a.** 5   **b.** 7   **c.** 4 Sec. 2.2, Ex. 7   **31.** $g(x)$ is the same as $f(x)$ shifted up one unit; Sec. 2.3, Ex. 1

33. slope: $\dfrac{3}{4}$; y-intercept: $(0, -1)$; Sec. 2.4, Ex. 3   **35. a.** parallel   **b.** neither **c.** perpendicular; Sec. 2.4, Ex. 8   **37.** $f(x) = \dfrac{5}{8}x - \dfrac{5}{2}$; Sec. 2.5, Ex. 5

39.   ; Sec. 3.6, Ex. 2   **41. a.** yes   **b.** no; Sec. 4.1, Ex. 1   **43.** $(-4, 2, -1)$; Sec. 4.2, Ex. 1   **45.** $(-1, 2)$; Sec. 4.4, Ex. 1

CHAPTER 5 EXPONENTS, POLYNOMIALS, AND POLYNOMIAL FUNCTIONS

Section 5.1
Practice Exercises

1. a. 3^6 **b.** x^7 **c.** y^9 **2. a.** $35z^4$ **b.** $-20.5t^6q^8$ **3. a.** 1   **b.** -1 **c.** 1   **d.** 3   **4. a.** z^5 **b.** 3^6 **c.** $9x^4$ **d.** $\dfrac{4a^7}{3}$ or $\dfrac{4}{3}a^7$

5. a. $\dfrac{1}{36}$ **b.** $\dfrac{1}{64}$ **c.** $\dfrac{3}{x^5}$ **d.** $\dfrac{1}{5y}$ **e.** $\dfrac{1}{k^7}$ **f.** $\dfrac{1}{25}$ **g.** $\dfrac{9}{20}$ **h.** z^8 **6. a.** $\dfrac{1}{z^{11}}$ **b.** $7t^8$ **c.** 9   **d.** $\dfrac{b^7}{3a^7}$ **e.** $\dfrac{2}{x^4}$ **7. a.** x^{3a+4} **b.** x^{2t+1}
8. a. 6.5×10^4 **b.** 3.8×10^{-5} **9. a.** 620,000   **b.** 0.03109

Vocabulary and Readiness Check 5.1

1. x **3.** 3   **5.** y^7 **7.** $\dfrac{5}{xy^2}$ **9.** $\dfrac{a^2}{bc^5}$ **11.** $\dfrac{x^4}{y^2}$

Exercise Set 5.1

1. 4^5 **3.** x^8 **5.** m^{14} **7.** $-20x^2y$ **9.** $-16x^6y^3p^2$ **11.** -1 **13.** 1   **15.** -1 **17.** 9   **19.** a^3 **21.** $-13z^4$ **23.** x **25.** $\dfrac{4}{3}x^3y^2$

27. $-6a^4b^4c^6$ **29.** $\dfrac{1}{16}$ **31.** $-\dfrac{1}{27}$ **33.** $\dfrac{1}{x^8}$ **35.** $\dfrac{5}{a^4}$ **37.** $\dfrac{y^2}{x^7}$ **39.** $\dfrac{1}{x^7}$ **41.** $4r^8$ **43.** 1   **45.** $\dfrac{b^7}{9a^7}$ **47.** $\dfrac{6x^{16}}{5}$ **49.** $-140x^{12}$

51. x^{16} **53.** $10x^{10}$ **55.** 6   **57.** $\dfrac{1}{z^3}$ **59.** -2 **61.** y^4 **63.** $\dfrac{13}{36}$ **65.** $\dfrac{3}{x}$ **67.** r^8 **69.** $\dfrac{1}{x^9y^4}$ **71.** $24x^7y^6$ **73.** $\dfrac{x}{16}$ **75.** 625

77. $\dfrac{1}{8}$ **79.** $\dfrac{a^5}{81}$ **81.** $\dfrac{7}{x^3z^5}$ **83.** x^{7a+5} **85.** x^{2t-1} **87.** x^{4a+7} **89.** z^{6x-7} **91.** x^{6t-1} **93.** 3.125×10^7 **95.** 1.6×10^{-2} **97.** 6.7413×10^4

99. 1.25×10^{-2} **101.** 5.3×10^{-5} **103.** 3.44992×10^{11} **105.** 3.5×10^6 **107.** 1.24×10^{11} **109.** 1.0×10^{-3} **111.** 0.0000000036

113. 93,000,000   **115.** 1,278,000   **117.** 7,350,000,000,000   **119.** 0.000000403   **121.** 300,000,000   **123.** \$153,000,000,000   **125.** 100   **127.** $\dfrac{27}{64}$

129. 64   **131.** answers may vary   **133.** answers may vary   **135. a.** x^{2a} **b.** $2x^a$ **c.** x^{a-b} **d.** x^{a+b} **e.** $x^a + x^b$ **137.** 7^{13} **139.** 7^{-11}

Section 5.2
Practice Exercises

1. a. z^{15} **b.** 625 **c.** $\dfrac{1}{27}$ **d.** x^{24} **2. a.** $32x^{15}$ **b.** $\dfrac{9}{25}$ **c.** $\dfrac{16a^{20}}{b^{28}}$ **d.** $9x$ **e.** $\dfrac{a^4b^{10}}{c^8}$ **3. a.** $\dfrac{b^{15}}{27a^3}$ **b.** y^{15} **c.** $\dfrac{64}{9}$ **d.** $\dfrac{b^8}{81a^6}$

4. a. $\dfrac{c^3}{125a^{36}b^3}$ **b.** $\dfrac{16x^{16}y^4}{25}$ **5. a.** $27x^a$ **b.** y^{5b+3} **6. a.** 1.7×10^{-2} **b.** 1.4×10^{10} **7.** 4.2×10^{-6}

Vocabulary and Readiness Check 5.2

1. x^{20} **3.** x^9 **5.** y^{42} **7.** z^{36} **9.** z^{18}

Exercise Set 5.2

1. $\dfrac{1}{9}$ **3.** $\dfrac{1}{x^{36}}$ **5.** $\dfrac{1}{y^5}$ **7.** $9x^4y^6$ **9.** $16x^{20}y^{12}$ **11.** $\dfrac{c^{18}}{a^{12}b^6}$ **13.** $\dfrac{y^{15}}{x^{35}z^{20}}$ **15.** $\dfrac{1}{125}$ **17.** x^{15} **19.** $\dfrac{8}{x^{12}y^{18}}$ **21.** $\dfrac{y^{16}}{64x^5}$ **23.** $\dfrac{64}{p^9}$

25. $-\dfrac{1}{x^9a^9}$ **27.** $\dfrac{x^5y^{10}}{5^{15}}$ **29.** $\dfrac{1}{x^{63}}$ **31.** $\dfrac{343}{512}$ **33.** $16x^4$ **35.** $-\dfrac{y^3}{64}$ **37.** $4^8x^2y^6$ **39.** 64 **41.** $\dfrac{x^4}{16}$ **43.** $\dfrac{1}{y^{15}}$ **45.** $\dfrac{x^9}{8y^3}$

47. $\dfrac{16a^2b^9}{9}$ **49.** $\dfrac{3}{8x^8y^7}$ **51.** $\dfrac{1}{x^{30}b^6c^6}$ **53.** $\dfrac{25}{8x^5y^4}$ **55.** $\dfrac{2}{x^4y^{10}}$ **57.** x^{9a+18} **59.** x^{12a+2} **61.** b^{10x-4} **63.** y^{15a+3} **65.** $16x^{4t+4}$

67. $5x^{-a}y^{-a+2}$ **69.** 1.45×10^9 **71.** 8×10^{15} **73.** 4×10^{-7} **75.** 3×10^{-1} **77.** 2×10^1 **79.** 1×10^1 **81.** 8×10^{-5} **83.** 1.1×10^7

85. 3.5×10^{22} **87.** 2×10^{-3} sec **89.** 6.232×10^{-11} cu m **91.** $-3m - 15$ **93.** $-3y - 5$ **95.** $-3x + 5$ **97.** $\dfrac{15y^3}{x^8}$ sq ft

99. 1.331928×10^{13} tons **101.** no **103.** 85 people per sq mi **105.** 6.4 times **107.** 31.7%

Section 5.3
Practice Exercises

1. a. 5 **b.** 3 **c.** 1 **d.** 11 **e.** 0 **2. a.** degree 4; trinomial **b.** degree 5; monomial **c.** degree 5; binomial **3.** 5
4. a. -15 **b.** -47 **5. a.** 290 feet; 226 feet **b.** 4 seconds **c.** 4.3 seconds **6. a.** $3x^4 - 5x$ **b.** $7ab - 3b$
7. a. $3a^4b + 4ab^2 - 5$ **b.** $2x^5 + y - x - 9$ **8.** $6x^3 + 6x^2 - 7x - 8$ **9.** $15a^4 - 15a^3 + 3$ **10.** $6x^2y^2 - 4xy^2 - 5y^3$ **11.** C
12. a. $(0, -12)$; U-shaped, opening upward **b.** $(0, 6)$; U-shaped, opening downward **c.** $(0, 4)$; S-shaped **d.** $(0, -1)$; S-shaped

Vocabulary and Readiness Check 5.3

1. coefficient **3.** binomial **5.** trinomial **7.** degree **9.** $6x$ **11.** $2y$ **13.** $7xy^2 - y^2$

Exercise Set 5.3

1. 0 **3.** 2 **5.** 3 **7.** 3 **9.** 9 **11.** degree 1; binomial **13.** degree 2; trinomial **15.** degree 3; monomial **17.** degree 4; none of these

19. 57 **21.** 499 **23.** $-\dfrac{11}{16}$ **25.** 1061 ft **27.** 549 ft **29.** $6y$ **31.** $11x - 3$ **33.** $xy + 2x - 1$ **35.** $6x^2 - xy + 16y^2$ **37.** $18y^2 - 17$

39. $3x^2 - 3xy + 6y^2$ **41.** $x^2 - 4x + 8$ **43.** $y^2 + 3$ **45.** $-2x^2 + 5x$ **47.** $-2x^2 - 4x + 15$ **49.** $4x - 13$ **51.** $x^2 + 2$ **53.** $12x^3 + 8x + 8$

55. $7x^3 + 4x^2 + 8x - 10$ **57.** $-18y^2 + 11yx + 14$ **59.** $-x^3 + 8a - 12$ **61.** $5x^2 - 9x - 3$ **63.** $-3x^2 + 3$ **65.** $8xy^2 + 2x^3 + 3x^2 - 3$

67. $7y^2 - 3$ **69.** $5x^2 + 22x + 16$ **71.** $\dfrac{3}{4}x^2 - \dfrac{1}{3}x^2y - \dfrac{8}{3}x^2y^2 + \dfrac{3}{2}y^3$ **73.** $-q^4 + q^2 - 3q + 5$ **75.** $15x^2 + 8x - 6$ **77.** $x^4 - 7x^2 + 5$

79. $\dfrac{1}{3}x^2 - x + 1$ **81.** 202 sq in. **83. a.** 284 ft **b.** 536 ft **c.** 756 ft **d.** 944 ft **e.** answers may vary **f.** 19 sec **85.** $80,000

87. $40,000 **89.** A **91.** D **93.** $15x - 10$ **95.** $-2x^2 + 10x - 12$ **97.** a and c
99. $(12x - 1.7) - (15x + 6.2) = 12x - 1.7 - 15x - 6.2 = -3x - 7.9$ **101.** answers may vary **103.** answers may vary
105. $3x^{2a} + 2x^a + 0.7$ **107.** $4x^{2y} + 2x^y - 11$ **109.** $(6x^2 + 14y)$ units **111.** $4x^2 - 3x + 6$ **113.** $-x^2 - 6x + 10$ **115.** $3x^2 - 12x + 13$
117. $15x^2 + 12x - 9$ **119. a.** $2a - 3$ **b.** $-2x - 3$ **c.** $2x + 2h - 3$ **121. a.** $4a$ **b.** $-4x$ **c.** $4x + 4h$
123. a. $4a - 1$ **b.** $-4x - 1$ **c.** $4x + 4h - 1$ **125. a.** 2.9 million **b.** 57.3 million **c.** 247.4 million **d.** answers may vary
127. a. 42.4 million **b.** 29.5 million **129. a.** 4.4 million **b.** 6.2 million

Section 5.4
Practice Exercises

1. a. $6x^6$ **b.** $40m^5n^2p^8$ **2. a.** $21x^2 - 3x$ **b.** $-15a^4 + 30a^3 - 25a^2$ **c.** $-5m^3n^5 - 2m^2n^4 + 5m^2n^3$
3. a. $2x^2 + 13x + 15$ **b.** $3x^3 - 19x^2 + 12x - 2$ **4.** $3x^4 - 12x^3 - 13x^2 - 8x - 10$ **5.** $x^2 - 2x - 15$
6. a. $6x^2 - 31x + 35$ **b.** $8x^4 - 10x^2y - 3y^2$ **7. a.** $x^2 + 12x + 36$ **b.** $x^2 - 4x + 4$ **c.** $9x^2 + 30xy + 25y^2$ **d.** $9x^4 - 48x^2b + 64b^2$
8. a. $x^2 - 49$ **b.** $4a^2 - 25$ **c.** $25x^4 - \dfrac{1}{16}$ **d.** $a^6 - 16b^4$ **9.** $4 + 12x - 4y + 9x^2 - 6xy + y^2$ **10.** $9x^2 - 6xy + y^2 - 25$
11. $x^4 - 32x^2 + 256$ **12.** $h^2 - h + 3$

Vocabulary and Readiness Check 5.4

1. b **3.** b **5.** d

Exercise Set 5.4

1. $-12x^5$ **3.** $12x^2 + 21x$ **5.** $-24x^2y - 6xy^2$ **7.** $-4a^3bx - 4a^3by + 12ab$ **9.** $2x^2 - 2x - 12$ **11.** $2x^4 + 3x^3 - 2x^2 + x + 6$
13. $15x^2 - 7x - 2$ **15.** $15m^3 + 16m^2 - m - 2$ **17.** $x^2 + x - 12$ **19.** $10x^2 + 11xy - 8y^2$ **21.** $3x^2 + 8x - 3$
23. $9x^2 - \dfrac{1}{4}$ **25.** $5x^4 - 17x^2y^2 + 6y^4$ **27.** $x^2 + 8x + 16$ **29.** $36y^2 - 1$ **31.** $9x^2 - 6xy + y^2$ **33.** $25b^2 - 36y^2$ **35.** $16b^2 + 32b + 16$
37. $4s^2 - 12s + 8$ **39.** $x^2y^2 - 4xy + 4$ **41.** answers may vary   **43.** $x^4 - 2x^2y^2 + y^4$ **45.** $x^4 - 8x^3 + 24x^2 - 32x + 16$ **47.** $x^4 - 625$
49. $9x^2 + 18x + 5$ **51.** $10x^5 + 8x^4 + 2x^3 + 25x^2 + 20x + 5$ **53.** $49x^2 - 9$ **55.** $9x^3 + 30x^2 + 12x - 24$ **57.** $16x^2 - \dfrac{2}{3}x - \dfrac{1}{6}$
59. $36x^2 + 12x + 1$ **61.** $x^4 - 4y^2$ **63.** $-30a^4b^4 + 36a^3b^2 + 36a^2b^3$ **65.** $2a^2 - 12a + 16$ **67.** $49a^2b^2 - 9c^2$ **69.** $m^2 - 8m + 16$
71. $9x^2 + 6x + 1$ **73.** $y^2 - 7y + 12$ **75.** $2x^3 + 2x^2y + x^2 + xy - x - y$ **77.** $9x^4 + 12x^3 - 2x^2 - 4x + 1$ **79.** $12x^3 - 2x^2 + 13x + 5$
81. $a^2 - 3a$ **83.** $a^2 + 2ah + h^2 - 3a - 3h$ **85.** $b^2 - 7b + 10$ **87.** -2 **89.** $\dfrac{3}{5}$ **91.** function   **93.** $7y(3z - 2) + 1 = 21yz - 14y + 1$
95. answers may vary   **97. a.** $a^2 + 2ah + h^2 + 3a + 3h + 2$ **b.** $a^2 + 3a + 2$ **c.** $2ah + h^2 + 3h$ **99.** $30x^2y^{2n+1} - 10x^2y^n$
101. $x^{3a} + 5x^{2a} - 3x^a - 15$ **103.** $\pi(25x^2 - 20x + 4)$ sq km   **105.** $(8x^2 - 12x + 4)$ sq in.   **107. a.** $6x + 12$
b. $9x^2 + 36x + 35$; one operation is addition, the other is multiplication.   **109.** $5x^2 + 25x$ **111.** $x^4 - 4x^2 + 4$ **113.** $x^3 + 5x^2 - 2x - 10$

Section 5.5

Practice Exercises

1. $8y$ **2. a.** $3(2x^2 + 3 + 5x)$ **b.** $3x - 8y^3$ **c.** $2a^3(4a - 1)$ **3.** $8x^3y^2(8x^2 - 1)$ **4.** $-xy^2(9x^3 - 5x - 7)$ **5.** $(3 + 5b)(x + 4)$
6. $(8b - 1)(a^3 + 2y)$ **7.** $(x + 2)(y - 5)$ **8.** $(a + 2)(a^2 + 5)$ **9.** $(x^2 + 3)(y^2 - 5)$ **10.** $(q + 3)(p - 1)$

Vocabulary and Readiness Check 5.5

1. factoring   **3.** least   **5.** false   **7.** false   **9.** 6   **11.** 5   **13.** x **15.** $7x$

Exercise Set 5.5

1. a^3 **3.** y^2z^2 **5.** $3x^2y$ **7.** $5xz^3$ **9.** $6(3x - 2)$ **11.** $4y^2(1 - 4xy)$ **13.** $2x^3(3x^2 - 4x + 1)$ **15.** $4ab(2a^2b^2 - ab + 1 + 4b)$
17. $(x + 3)(6 + 5a)$ **19.** $(z + 7)(2x + 1)$ **21.** $(x^2 + 5)(3x - 2)$ **23.** answers may vary   **25.** $(a + 2)(b + 3)$
27. $(a - 2)(c + 4)$ **29.** $(x - 2)(2y - 3)$ **31.** $(4x - 1)(3y - 2)$ **33.** $3(2x^3 + 3)$ **35.** $x^2(x + 3)$ **37.** $4a(2a^2 - 1)$
39. $-4xy(5x - 4y^2)$ or $4xy(-5x + 4y^2)$ **41.** $5ab^2(2ab + 1 - 3b)$ **43.** $3b(3ac^2 + 2a^2c - 2a + c)$ **45.** $(y - 2)(4x - 3)$
47. $(2x + 3)(3y + 5)$ **49.** $(x + 3)(y - 5)$ **51.** $(2a - 3)(3b - 1)$ **53.** $(6x + 1)(2y + 3)$ **55.** $(n - 8)(2m - 1)$ **57.** $3x^2y^2(5x - 6)$
59. $(2x + 3y)(x + 2)$ **61.** $(5x - 3)(x + y)$ **63.** $(x^2 + 4)(x + 3)$ **65.** $(x^2 - 2)(x - 1)$ **67.** $55x^7$ **69.** $125x^6$ **71.** $x^2 - 3x - 10$
73. $x^2 + 5x + 6$ **75.** $y^2 - 4y + 3$ **77.** d   **79.** $2\pi r(r + h)$ **81.** $A = 5600(1 + rt)$ **83.** answers may vary   **85.** none   **87.** a
89. $A = P(1 + RT)$ **91.** $y^n(3 + 3y^n + 5y^{7n})$ **93.** $3x^{2a}(x^{3a} - 2x^a + 3)$ **95. a.** $h(t) = -16(t^2 - 14)$ **b.** 160 ft   **c.** answers may vary

Section 5.6

Practice Exercises

1. $(x + 3)(x + 2)$ **2.** $(x - 3)(x - 8)$ **3.** $3x(x - 5)(x + 2)$ **4.** $2(b^2 - 9b - 11)$ **5.** $(2x + 1)(x + 6)$ **6.** $(4x - 3)(x + 2)$
7. $3b^2(2b - 5)(3b - 2)$ **8.** $(5x + 2y)^2$ **9.** $(5x + 2)(4x + 3)$ **10.** $(5x + 3)(3x - 1)$ **11.** $(x - 3)(3x + 8)$ **12.** $(3x^2 + 2)(2x^2 - 5)$

Vocabulary and Readiness Check 5.6

1. 5 and 2   **3.** 8 and 3

Exercise Set 5.6

1. $(x + 3)(x + 6)$ **3.** $(x - 8)(x - 4)$ **5.** $(x + 12)(x - 2)$ **7.** $(x - 6)(x + 4)$ **9.** $3(x - 2)(x - 4)$ **11.** $4z(x + 2)(x + 5)$
13. $(x + 5), (x - 3)$ **15.** $(x - 2)(x - 6)$ **17.** $(5x + 1)(x + 3)$ **19.** $(2x - 3)(x - 4)$ **21.** prime polynomial   **23.** $(2x - 3)^2$
25. $2(3x - 5)(2x + 5)$ **27.** $y^2(3y + 5)(y - 2)$ **29.** $(2x + y)(x - 3y)$ **31.** $2(7y + 2)(2y + 1)$ **33.** $(2x - 3)(x + 9)$
35. $(x^2 + 3)(x^2 - 2)$ **37.** $(5x + 8)(5x + 2)$ **39.** $(x^3 - 4)(x^3 - 3)$ **41.** $(a - 3)(a + 8)$ **43.** $(x - 27)(x + 3)$ **45.** $(x - 18)(x + 3)$
47. $3(x - 1)^2$ **49.** $(3x + 1)(x - 2)$ **51.** $(4x - 3)(2x - 5)$ **53.** $3x^2(2x + 1)(3x + 2)$ **55.** $(x + 7z)(x + z)$ **57.** $(x - 4)(x + 3)$
59. $3(a + 2b)^2$ **61.** prime polynomial   **63.** $(2x + 13)(x + 3)$ **65.** $(3x - 2)(2x - 15)$ **67.** $(x^2 - 6)(x^2 + 1)$ **69.** $x(3x + 1)(2x - 1)$
71. $(4a - 3b)(3a - 5b)$ **73.** $(3x + 5)^2$ **75.** $y(3x - 8)(x - 1)$ **77.** $2(x + 3)(x - 2)$ **79.** $(x + 2)(x - 7)$ **81.** $(2x^3 - 3)(x^3 + 3)$
83. $2x(6y^2 - z)^2$ **85.** $2xy(x + 3)(x - 2)$ **87.** $(x + 5y)(x + y)$ **89.** $x^2 - 9$ **91.** $4x^2 + 4x + 1$ **93.** $x^3 - 8$ **95.** $\pm5, \pm7$
97. $x(x + 4)(x - 2)$ **99. a.** 576 ft; 672 ft; 640 ft; 480 ft   **b.** answers may vary   **c.** $-16(t + 4)(t - 9)$ **101.** $(x^n + 2)(x^n + 8)$
103. $(x^n - 6)(x^n + 3)$ **105.** $(2x^n + 1)(x^n + 5)$ **107.** $(2x^n - 3)^2$

Section 5.7

Practice Exercises

1. $(b + 8)^2$ **2.** $5b(3x - 1)^2$ **3. a.** $(x + 4)(x - 4)$ **b.** $(5b - 7)(5b + 7)$ **c.** $5(3 - 2x)(3 + 2x)$ **d.** $\left(y - \dfrac{1}{9}\right)\left(y + \dfrac{1}{9}\right)$
4. a. $(x^2 + 100)(x + 10)(x - 10)$ **b.** $(x + 9)(x - 5)$ **5.** $(m + 3 + n)(m + 3 - n)$ **6.** $(x + 4)(x^2 - 4x + 16)$
7. $(a + 2b)(a^2 - 2ab + 4b^2)$ **8.** $(3 - y)(9 + 3y + y^2)$ **9.** $x^2(b - 2)(b^2 + 2b + 4)$

Vocabulary and Readiness Check 5.7

1. $(9y)^2$ **3.** $(8x^3)^2$ **5.** 5^3 **7.** $(2x)^3$ **9.** $(4x^2)^3$

Exercise Set 5.7

1. $(x + 3)^2$ **3.** $(2x - 3)^2$ **5.** $3(x - 4)^2$ **7.** $x^2(3y + 2)^2$ **9.** $(x + 5)(x - 5)$ **11.** $(3 + 2z)(3 - 2z)$ **13.** $(y + 9)(y - 5)$
15. $4(4x + 5)(4x - 5)$ **17.** $(x + 3)(x^2 - 3x + 9)$ **19.** $(z - 1)(z^2 + z + 1)$ **21.** $(m + n)(m^2 - mn + n^2)$ **23.** $y^2(x - 3)(x^2 + 3x + 9)$
25. $b(a + 2b)(a^2 - 2ab + 4b^2)$ **27.** $(5y - 2x)(25y^2 + 10yx + 4x^2)$ **29.** $(x + 3 + y)(x + 3 - y)$ **31.** $(x - 5 + y)(x - 5 - y)$
33. $(2x + 1 + z)(2x + 1 - z)$ **35.** $(3x + 7)(3x - 7)$ **37.** $(x - 6)^2$ **39.** $(x^2 + 9)(x + 3)(x - 3)$ **41.** $(x + 4 + 2y)(x + 4 - 2y)$
43. $(x + 2y + 3)(x + 2y - 3)$ **45.** $(x - 6)(x^2 + 6x + 36)$ **47.** $(x + 5)(x^2 - 5x + 25)$ **49.** prime polynomial **51.** $(2a + 3)^2$
53. $2y(3x + 1)(3x - 1)$ **55.** $(2x + y)(4x^2 - 2xy + y^2)$ **57.** $(x^2 - y)(x^4 + x^2y + y^2)$ **59.** $(x + 8 + x^2)(x + 8 - x^2)$

61. $3y^2(x^2 + 3)(x^4 - 3x^2 + 9)$ **63.** $(x + y + 5)(x^2 + 2xy + y^2 - 5x - 5y + 25)$ **65.** $(2x - 1)(4x^2 + 20x + 37)$ **67.** 5 **69.** $-\dfrac{1}{3}$ **71.** 0

73. 5 **75.** no; $x^2 - 4$ can be factored further **77.** yes **79.** $\pi R^2 - \pi r^2 = \pi(R + r)(R - r)$ **81.** $x^3 - y^2x$; $x(x + y)(x - y)$ **83.** $c = 9$
85. $c = 49$ **87.** $c = \pm 8$ **89. a.** $(x + 1)(x^2 - x + 1)(x - 1)(x^2 + x + 1)$ **b.** $(x + 1)(x - 1)(x^4 + x^2 + 1)$ **c.** answers may vary
91. $(x^n + 6)(x^n - 6)$ **93.** $(5x^n + 9)(5x^n - 9)$ **95.** $(x^{2n} + 25)(x^n + 5)(x^n - 5)$

Integrated Review Practice Exercises

1. a. $3xy(4x - 1)$ **b.** $(7x + 2)(7x - 2)$ **c.** $(5x - 3)(x + 1)$ **d.** $(3 + x)(x^2 + 2)$ **e.** $(2x + 5)^2$ **f.** cannot be factored
2. a. $(4x + y)(16x^2 - 4xy + y^2)$ **b.** $7y^2(x - 3y)(x + 3y)$ **c.** $3(x + 2 + b)(x + 2 - b)$ **d.** $x^2y(xy + 3)(x^2y^2 - 3xy + 9)$
e. $(x + 7 + 9y)(x + 7 - 9y)$

Integrated Review

1. $2y^2 + 2y - 11$ **2.** $-2z^4 - 6z^2 + 3z$ **3.** $x^2 - 7x + 7$ **4.** $7x^2 - 4x - 5$ **5.** $25x^2 - 30x + 9$ **6.** $x - 3$
7. $2x^3 - 4x^2 + 5x - 5 + \dfrac{8}{x + 2}$ **8.** $4x^3 - 13x^2 - 5x + 2$ **9.** $(x - 4 + y)(x - 4 - y)$ **10.** $2(3x + 2)(2x - 5)$ **11.** $x(x - 1)(x^2 + x + 1)$
12. $2x(2x - 1)$ **13.** $2xy(7x - 1)$ **14.** $6ab(4b - 1)$ **15.** $4(x + 2)(x - 2)$ **16.** $9(x + 3)(x - 3)$ **17.** $(3x - 11)(x + 1)$
18. $(5x + 3)(x - 1)$ **19.** $4(x + 3)(x - 1)$ **20.** $6(x + 1)(x - 2)$ **21.** $(2x + 9)^2$ **22.** $(5x + 4)^2$ **23.** $(2x + 5y)(4x^2 - 10xy + 25y^2)$
24. $(3x - 4y)(9x^2 + 12xy + 16y^2)$ **25.** $8x^2(2y - 1)(4y^2 + 2y + 1)$ **26.** $27x^2y(xy - 2)(x^2y^2 + 2xy + 4)$
27. $(x + 5 + y)(x^2 + 10x - xy - 5y + y^2 + 25)$ **28.** $(y - 1 + 3x)(y^2 - 2y + 1 - 3xy + 3x + 9x^2)$ **29.** $(5a - 6)^2$ **30.** $(4r + 5)^2$
31. $7x(x - 9)$ **32.** $(4x + 3)(5x + 2)$ **33.** $(a + 7)(b - 6)$ **34.** $20(x - 6)(x - 5)$ **35.** $(x^2 + 1)(x - 1)(x + 1)$ **36.** $5x(3x - 4)$
37. $(5x - 11)(2x + 3)$ **38.** $9m^2n^2(5mn - 3)$ **39.** $5a^3b(b^2 - 10)$ **40.** $x(x + 1)(x^2 - x + 1)$ **41.** prime **42.** $20(x + y)(x^2 - xy + y^2)$
43. $10x(x - 10)(x - 11)$ **44.** $(3y - 7)^2$ **45.** $a^3b(4b - 3)(16b^2 + 12b + 9)$ **46.** $(y^2 + 4)(y + 2)(y - 2)$ **47.** $2(x - 3)(x^2 + 3x + 9)$
48. $(2s - 1)(r + 5)$ **49.** $(y^4 + 2)(3y - 5)$ **50.** prime **51.** $100(z + 1)(z^2 - z + 1)$ **52.** $2x(5x - 2)(25x^2 + 10x + 4)$ **53.** $(2b - 9)^2$
54. $(a^4 + 3)(2a - 1)$ **55.** $(y - 4)(y - 5)$ **56.** $(c - 3)(c + 1)$ **57.** $A = 9 - 4x^2 = (3 + 2x)(3 - 2x)$

Section 5.8

Practice Exercises

1. $-8, 5$ **2.** $-4, \dfrac{2}{3}$ **3.** $-4, -\dfrac{2}{3}$ **4.** $-\dfrac{3}{4}$ **5.** $-\dfrac{1}{8}, 2$ **6.** $0, 3, -1$ **7.** $3, -3, -2$ **8.** 6 seconds **9.** $6, 8, 10$ units **10.** $f(x)$: C; $g(x)$: A; $h(x)$: B

Vocabulary and Readiness Check 5.8

1. $3, -5$ **3.** $3, -7$ **5.** $0, 9$

Exercise Set 5.8

1. $-3, \dfrac{4}{3}$ **3.** $\dfrac{5}{2}, -\dfrac{3}{4}$ **5.** $-3, -8$ **7.** $\dfrac{1}{4}, -\dfrac{2}{3}$ **9.** $1, 9$ **11.** $\dfrac{3}{5}, -1$ **13.** 0 **15.** $6, -3$ **17.** $\dfrac{2}{5}, -\dfrac{1}{2}$ **19.** $\dfrac{3}{4}, -\dfrac{1}{2}$ **21.** $-2, 7, \dfrac{8}{3}$

23. $0, 3, -3$ **25.** $2, 1, -1$ **27.** answers may vary **29.** $-\dfrac{7}{2}, 10$ **31.** $0, 5$ **33.** $-3, 5$ **35.** $-\dfrac{1}{2}, \dfrac{1}{3}$ **37.** $-4, 9$ **39.** $\dfrac{4}{5}$ **41.** $-5, 0, 2$

43. $-3, 0, \dfrac{4}{5}$ **45.** $\varnothing$ **47.** $-7, 4$ **49.** $4, 6$ **51.** $-\dfrac{1}{2}$ **53.** $-4, -3, 3$ **55.** $-5, 0, 5$ **57.** $-6, 5$ **59.** $-\dfrac{1}{3}, 0, 1$ **61.** $-\dfrac{1}{3}, 0$ **63.** $-\dfrac{7}{8}$

65. $\dfrac{31}{4}$ **67.** 1 **69. a.** incorrect **b.** correct **c.** correct **d.** incorrect **71.** -11 and -6 or 6 and 11 **73.** 75 ft **75.** 105 units

77. 12 cm and 9 cm **79.** 2 in. **81.** 10 sec **83.** width: $7\dfrac{1}{2}$ ft; length: 12 ft **85.** 10-in. sq tier **87.** 9 sec **89.** E **91.** F **93.** B
95. $(-3, 0), (0, 2)$; function **97.** $(-4, 0), (0, 2), (4, 0), (0, -2)$; not a function **99.** answers may vary

101. $x - 5 = 0$ or $x + 2 = 0$
$\quad\quad x = 5$ or $x = -2$

103.
$y(y - 5) = -6$
$y^2 - 5y + 6 = 0$
$(y - 2)(y - 3) = 0$
$y - 2 = 0$ or $y - 3 = 0$
$y = 2$ or $y = 3$

105. $-3, -\dfrac{1}{3}, 2, 5$ **107.** no; answers may vary **109.** answers may vary

111. answers may vary

The Bigger Picture

1. $-\dfrac{1}{2}, 6$ **2.** $(-7, 3)$ **3.** $-\dfrac{5}{3}$ **4.** $-\dfrac{3}{2}, 6$ **5.** $(-\infty, \infty)$ **6.** $-8, 3$ **7.** $-3, 10$ **8.** $(-\infty, 0]$

Chapter 5 Vocabulary Check

1. polynomial   **2.** factoring   **3.** exponents   **4.** degree of a term   **5.** monomial   **6.** 1   **7.** trinomial
8. quadratic equation   **9.** scientific notation   **10.** degree of a polynomial   **11.** binomial   **12.** 0

Chapter 5 Review

1. 4   **3.** -4 **5.** 1   **7.** $-\dfrac{1}{16}$ **9.** $-x^2y^7z$ **11.** $\dfrac{1}{a^9}$ **13.** $\dfrac{1}{x^{11}}$ **15.** $\dfrac{1}{y^5}$ **17.** -3.62×10^{-4} **19.** 410,000   **21.** $\dfrac{a^2}{16}$ **23.** $\dfrac{1}{16x^2}$

25. $\dfrac{1}{8^{18}}$ **27.** $-\dfrac{1}{8x^9}$ **29.** $\dfrac{-27y^6}{x^6}$ **31.** $\dfrac{xz}{4}$ **33.** $\dfrac{2}{27z^3}$ **35.** $2y^{x-7}$ **37.** -2.21×10^{-11} **39.** $\dfrac{x^3y^{10}}{3z^{12}}$ **41.** 5   **43.** $12x - 6x^2 - 6x^2y$

45. $4x^2 + 8y + 6$ **47.** $8x^2 + 2b - 22$ **49.** $12x^2y - 7xy + 3$ **51.** $x^3 + x - 2xy^2 - y - 7$ **53.** 58   **55.** $x^2 + 4x - 6$
57. $(6x^2y - 12x + 12)$ cm   **59.** $-12a^2b^5 - 28a^2b^3 - 4ab^2$ **61.** $9x^2a^2 - 24xab + 16b^2$ **63.** $15x^2 + 18xy - 81y^2$
65. $x^4 + 18x^3 + 83x^2 + 18x + 1$ **67.** $16x^2 + 72x + 81$ **69.** $16 - 9a^2 + 6ab - b^2$ **71.** $(9y^2 - 49z^2)$ sq units
73. $16x^2y^2z - 8xy^zb + b^2$ **75.** $8x^2(2x - 3)$ **77.** $2ab(3b + 4 - 2ab)$ **79.** $(a + 3b)(6a - 5)$ **81.** $(x - 6)(y + 3)$ **83.** $(p - 5)(q - 3)$
85. $x(2y - x)$ **87.** $(x - 4)(x + 20)$ **89.** $3(x + 2)(x + 9)$ **91.** $(3x + 8)(x - 2)$ **93.** $(15x - 1)(x - 6)$ **95.** $3(x - 2)(3x + 2)$
97. $(x + 7)(x + 9)$ **99.** $(x^2 - 2)(x^2 + 10)$ **101.** $(x + 9)(x - 9)$ **103.** $6(x + 3)(x - 3)$ **105.** $(4 + y^2)(2 + y)(2 - y)$ **107.** $(x - 7)(x + 1)$
109. $(y + 8)(y^2 - 8y + 64)$ **111.** $(1 - 4y)(1 + 4y + 16y^2)$ **113.** $2x^2(x + 2y)(x^2 - 2xy + 4y^2)$ **115.** $(x - 3 - 2y)(x - 3 + 2y)$
117. $(4a - 5b)^2$ **119.** $\dfrac{1}{3}, -7$ **121.** $0, 4, \dfrac{9}{2}$ **123.** $0, 6$ **125.** $-\dfrac{1}{3}, 2$ **127.** $-4, 1$ **129.** $0, 6, -3$ **131.** $0, -2, 1$ **133.** $-\dfrac{15}{2}, 7$ **135.** 5 sec
137. $3x^3 + 13x^2 - 9x + 5$ **139.** $8x^2 + 3x + 4.5$ **141.** -24 **143.** $6y^4(2y - 1)$ **145.** $2(3x + 1)(x - 6)$ **147.** $z^5(2z + 7)(2z - 7)$ **149.** $0, 3$

Chapter 5 Test

1. $\dfrac{1}{81x^2}$ **2.** $-12x^2z$ **3.** $\dfrac{3a^7}{2b^5}$ **4.** $-\dfrac{y^{40}}{z^5}$ **5.** 6.3×10^8 **6.** 1.2×10^{-2} **7.** 0.000005   **8.** 0.0009   **9.** $-5x^3y - 11x - 9$

10. $-12x^2y - 3xy^2$ **11.** $12x^2 - 5x - 28$ **12.** $25a^2 - 4b^2$ **13.** $36m^2 + 12mn + n^2$ **14.** $2x^3 - 13x^2 + 14x - 4$ **15.** $4x^2y(4x - 3y^3)$

16. $(x - 15)(x + 2)$ **17.** $(2y + 5)^2$ **18.** $3(2x + 1)(x - 3)$ **19.** $(2x + 5)(2x - 5)$ **20.** $(x + 4)(x^2 - 4x + 16)$

21. $3y(x + 3y)(x - 3y)$ **22.** $6(x^2 + 4)$ **23.** $2(2y - 1)(4y^2 + 2y + 1)$ **24.** $(x + 3)(x - 3)(y - 3)$ **25.** $4, -\dfrac{8}{7}$ **26.** $-3, 8$

27. $-\dfrac{5}{2}, -2, 2$ **28.** $(x + 2y)(x - 2y)$ **29. a.** 960 ft   **b.** 953.44 ft   **c.** 11 sec

Chapter 5 Cumulative Review

1. a. 3   **b.** 1   **c.** 2; Sec. 1.3, Ex. 8   **3.** 1; Sec. 1.5, Ex. 4   **5.** $11,607.55; Sec. 1.8, Ex. 4   **7. a.** $\left(-\infty, \dfrac{3}{2}\right]$

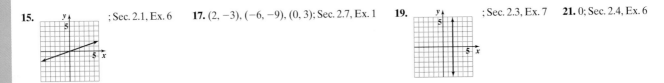

b. $(-3, \infty)$; Sec. 3.2, Ex. 4   **9.** $\left[-9, -\dfrac{9}{2}\right]$; Sec. 3.3, Ex. 5   **11.** 0; Sec. 3.4, Ex. 5   **13.** $\varnothing$; Sec. 3.5, Ex. 4

15. ; Sec. 2.1, Ex. 6   **17.** $(2, -3), (-6, -9), (0, 3)$; Sec. 2.7, Ex. 1   **19.** ; Sec. 2.3, Ex. 7   **21.** 0; Sec. 2.4, Ex. 6

23. $y = 3$; Sec. 2.5, Ex. 8   **25.** ; Sec. 3.6, Ex. 4   **27.** $\left(-4, \dfrac{1}{2}\right)$; Sec. 4.1, Ex. 4   **29.** $\left(\dfrac{1}{2}, 0, \dfrac{3}{4}\right)$; Sec. 4.2, Ex. 3

31. 7, 11; Sec. 4.3, Ex. 2   **33.** $\varnothing$; Sec. 4.4, Ex. 2   **35.** $30°, 40°, 110°$; Sec. 4.3, Ex. 6   **37. a.** 7.3×10^5 **b.** 1.04×10^{-6}; Sec. 5.1, Ex. 8

39. a. $\dfrac{y^6}{4}$ **b.** x^9 **c.** $\dfrac{49}{4}$ **d.** $\dfrac{y^{16}}{25x^5}$; Sec. 5.2, Ex. 3   **41.** 4; Sec. 5.3, Ex. 3   **43. a.** $10x^9$ **b.** $-7xy^{15}z^9$; Sec. 5.4, Ex. 1

45. $17x^3y^2(1 - 2x)$; Sec. 5.5, Ex. 3   **47.** $(x + 2)(x + 8)$; Sec. 5.6, Ex. 1   **49.** $-5, \dfrac{1}{2}$; Sec. 5.8, Ex. 2

CHAPTER 6 RATIONAL EXPRESSIONS

Section 6.1
Practice Exercises

1. a. $\{x|x \text{ is a real number}\}$ **b.** $\{x|x \text{ is a real number and } x \neq -3\}$ **c.** $\{x|x \text{ is a real number and } x \neq 2, x \neq 3\}$

2. a. $\{x|x \text{ is a real number and } x \neq -3\}$ **b.** $\left\{x \middle| x \text{ is a real number and } x \neq 1, x \neq \frac{5}{3}\right\}$

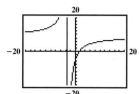 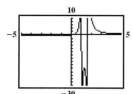

3. a. $\dfrac{1}{2z-1}$ **b.** $\dfrac{5x+3}{6x-5}$ **4. a.** 1   **b.** -1 **5.** $-\dfrac{5(2+x)}{x+3}$ **6. a.** $x^2-4x+16$ **b.** $\dfrac{5}{z-3}$ **7. a.** $\dfrac{2n+1}{n(n-1)}$ **b.** $-x^2$

8. a. $\dfrac{-y^3}{21(y+3)}$ **b.** $\dfrac{7x+2}{x+2}$ **9.** $\dfrac{5}{3(x-3)}$ **10. a.** \$7.20   **b.** \$3.60

Vocabulary and Readiness Check 6.1

1. rational   **3.** domain   **5.** 1   **7.** $\dfrac{-a}{b}; \dfrac{a}{-b}$ **9.** $\dfrac{xy}{10}$ **11.** $\dfrac{2y}{3x}$ **13.** $\dfrac{m^2}{36}$

Exercise Set 6.1

1. $\left\{x|x \text{ is a real number}\right\}$ **3.** $\{t|t \text{ is a real number and } t \neq 0\}$ **5.** $\{x|x \text{ is a real number and } x \neq 7\}$

7. $\left\{x \middle| x \text{ is a real number and } x \neq \frac{1}{3}\right\}$ **9.** $\{x|x \text{ is a real number and } x \neq -2, x \neq 0, x \neq 1\}$ **11.** $\{x|x \text{ is a real number and } x \neq 2, x \neq -2\}$

13. $1-2x$ **15.** $x-3$ **17.** $\dfrac{9}{7}$ **19.** $x-4$ **21.** -1 **23.** $-(x+7)$ **25.** $\dfrac{2x+1}{x-1}$ **27.** $\dfrac{x^2+5x+25}{2}$ **29.** $\dfrac{x-2}{2x^2+1}$

31. $\dfrac{1}{3x+5}$ **33.** $-\dfrac{4}{5}$ **35.** $-\dfrac{6a}{2a+1}$ **37.** $\dfrac{3}{2(x-1)}$ **39.** $\dfrac{x+2}{x+3}$ **41.** $\dfrac{3a}{5(a-b)}$ **43.** $\dfrac{1}{6}$ **45.** $\dfrac{x}{3}$ **47.** $\dfrac{4a^2}{a-b}$

49. $\dfrac{4}{(x+2)(x+3)}$ **51.** $\dfrac{1}{2}$ **53.** -1 **55.** $\dfrac{8(a-2)}{3(a+2)}$ **57.** $\dfrac{(x+2)(x+3)}{4}$ **59.** $\dfrac{2(x+3)(x-3)}{5(x^2-8x-15)}$ **61.** r^2-rs+s^2

63. $\dfrac{8}{x^2y}$ **65.** $\dfrac{(y+5)(2x-1)}{(y+2)(5x+1)}$ **67.** $\dfrac{5(3a+2)}{a}$ **69.** $\dfrac{5x^2-2}{(x-1)^2}$ **71.** $\dfrac{10}{3}, -8, -\dfrac{7}{3}$ **73.** $\dfrac{17}{48}, \dfrac{2}{7}, -\dfrac{3}{8}$

75. a. \$200 million   **b.** \$500 million   **c.** \$300 million   **d.** $\{x|x \text{ is a real number}\}$ **77.** $\dfrac{7}{5}$ **79.** $\dfrac{1}{12}$ **81.** $\dfrac{11}{16}$

83. b and d   **85.** no; answers may vary   **87.** $\dfrac{5}{x-2}$ sq m   **89.** $\dfrac{(x+2)(x-1)^2}{x^5}$ ft   **91.** answers may vary

93. a. 1   **b.** -1 **c.** neither   **d.** -1 **e.** -1 **f.** 1   **95.** $(x-5)(2x+7)$ **97.** $0, \dfrac{20}{9}, \dfrac{60}{7}, 20, \dfrac{140}{3}, 180, 380, 1980;$

99. $2x^2(x^n+2)$ **101.** $\dfrac{1}{10y(y^n+3)}$ **103.** $\dfrac{y^n+1}{2(y^n-1)}$

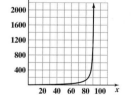

Section 6.2
Practice Exercises

1. a. $\dfrac{9+x}{11z^2}$ **b.** $\dfrac{3x}{4}$ **c.** $x-4$ **d.** $\dfrac{-3}{2a^2}$ **2. a.** $18x^3y^5$ **b.** $(x-2)(x+3)$ **c.** $(b-4)^2(b+4)(2b+3)$ **d.** $-4(y-3)(y+3)$

3. a. $\dfrac{20p+3}{5p^4q}$ **b.** $\dfrac{5y^2+19y-12}{(y+3)(y-3)}$ **c.** 3   **4.** $\dfrac{t^2-t-15}{(t+5)(t-5)(t+2)}$ **5.** $\dfrac{5x^2-12x-3}{(3x+1)(x-2)(2x-5)}$ **6.** $\dfrac{4}{x-2}$

Vocabulary and Readiness Check 6.2

1. a, b   **3.** c   **5.** $\dfrac{12}{y}$ **7.** $\dfrac{35}{y^2}$ **9.** $\dfrac{-x+4}{2x}$ **11.** $\dfrac{16}{y-2}$

Exercise Set 6.2

1. $-\dfrac{3}{xz^2}$ **3.** $\dfrac{x+2}{x-2}$ **5.** $x-2$ **7.** $\dfrac{-1}{x-2}$ or $\dfrac{1}{2-x}$ **9.** $-\dfrac{5}{x}$ **11.** $35x$ **13.** $x(x+1)$ **15.** $(x+7)(x-7)$ **17.** $6(x+2)(x-2)$

19. $(a + b)(a - b)^2$ **21.** $-4x(x + 3)(x - 3)$ **23.** $\dfrac{17}{6x}$ **25.** $\dfrac{35 - 4y}{14y^2}$ **27.** $\dfrac{-13x + 4}{(x + 4)(x - 4)}$ **29.** $\dfrac{3}{x + 4}$ **31.** 0 **33.** $-\dfrac{x}{x - 1}$

35. $\dfrac{-x + 1}{x - 2}$ **37.** $\dfrac{y^2 + 2y + 10}{(y + 4)(y - 4)(y - 2)}$ **39.** $\dfrac{5(x^2 + x - 4)}{(3x + 2)(x + 3)(2x - 5)}$ **41.** $\dfrac{-x^2 + 10x + 19}{(x - 2)(x + 1)(x + 3)}$ **43.** $\dfrac{x^2 + 4x + 2}{(2x + 5)(x - 7)(x - 1)}$

45. $\dfrac{5a + 1}{(a + 1)^2(a - 1)}$ **47.** $\dfrac{3}{x^2 y^3}$ **49.** $-\dfrac{5}{x}$ **51.** $\dfrac{25}{6(x + 5)}$ **53.** $\dfrac{-2x - 1}{x^2(x - 3)}$ **55.** $\dfrac{b(2a - b)}{(a + b)(a - b)}$ **57.** $\dfrac{2(x + 8)}{(x + 2)^2(x - 2)}$

59. $\dfrac{3x^2 + 23x - 7}{(2x - 1)(x - 5)(x + 3)}$ **61.** $\dfrac{5 - 2x}{2(x + 1)}$ **63.** $\dfrac{2(x^2 + x - 21)}{(x + 3)^2(x - 3)}$ **65.** $\dfrac{6x}{(x + 3)(x - 3)^2}$ **67.** $\dfrac{4}{3}$ **69.** $\dfrac{2x^2 + 9x - 18}{6x^2}$ or $\dfrac{(x + 6)(2x - 3)}{6x^2}$

71. $\dfrac{4a^2}{9(a - 1)}$ **73.** 4 **75.** $-\dfrac{4}{x - 1}$ **77.** $-\dfrac{32}{x(x + 2)(x - 2)}$ **79.** 10 **81.** $4 + x^2$ **83.** 10 **85.** 2 **87.** 3 **89.** $5\,\text{m}$

91. $\dfrac{2x - 3}{x^2 + 1} - \dfrac{x - 6}{x^2 + 1} = \dfrac{2x - 3 - x + 6}{x^2 + 1} = \dfrac{x + 3}{x^2 + 1}$ **93.** $\dfrac{4x}{x + 5}$ ft; $\dfrac{x^2}{(x + 5)^2}$ sq ft **95.** answers may vary **97.** answers may vary

99. answers may vary **101.** $\dfrac{3}{2x}$ **103.** $\dfrac{4 - 3x}{x^2}$ **105.**

Section 6.3
Practice Exercises

1. a. $\dfrac{1}{12m}$ **b.** $\dfrac{8x(x + 4)}{3(x - 4)}$ **c.** $\dfrac{b^2}{a^2}$ **2. a.** $\dfrac{8x(x + 4)}{3(x - 4)}$ **b.** $\dfrac{b^2}{a^2}$ **3.** $\dfrac{y(3xy + 1)}{x^2(1 + xy)}$ **4.** $\dfrac{1 - 6x}{15 + 6x}$

Vocabulary and Readiness Check 6.3

1. $\dfrac{7}{1 + z}$ **3.** $\dfrac{1}{x^2}$ **5.** $\dfrac{2}{x}$ **7.** $\dfrac{1}{9y}$

Exercise Set 6.3

1. 4 **3.** $\dfrac{7}{13}$ **5.** $\dfrac{4}{x}$ **7.** $\dfrac{9(x - 2)}{9x^2 + 4}$ **9.** $2x + y$ **11.** $\dfrac{2(x + 1)}{2x - 1}$ **13.** $\dfrac{2x + 3}{4 - 9x}$ **15.** $\dfrac{1}{x^2 - 2x + 4}$ **17.** $\dfrac{x}{5(x - 2)}$

19. $\dfrac{x - 2}{2x - 1}$ **21.** $\dfrac{x}{2 - 3x}$ **23.** $-\dfrac{y}{x + y}$ **25.** $-\dfrac{2x^3}{y(x - y)}$ **27.** $\dfrac{2x + 1}{y}$ **29.** $\dfrac{x - 3}{9}$ **31.** $\dfrac{1}{x + 2}$ **33.** 2

35. $\dfrac{xy^2}{x^2 + y^2}$ **37.** $\dfrac{2b^2 + 3a}{b(b - a)}$ **39.** $\dfrac{x}{(x + 1)(x - 1)}$ **41.** $\dfrac{1 + a}{1 - a}$ **43.** $\dfrac{x(x + 6y)}{2y}$ **45.** $\dfrac{5a}{2(a + 2)}$ **47.** $xy(5y + 2x)$

49. $\dfrac{xy}{2x + 5y}$ **51.** $\dfrac{x^2 y^2}{4}$ **53.** $-9x^3 y^4$ **55.** $-4, 14$ **57.** a and c **59.** $\dfrac{770a}{770 - s}$ **61.** a, b **63.** $\dfrac{1 + x}{2 + x}$ **65.** $x(x + 1)$

67. $\dfrac{x - 3y}{x + 3y}$ **69.** $3a^2 + 4a + 4$ **71. a.** $\dfrac{1}{a + h}$ **b.** $\dfrac{1}{a}$ **c.** $\dfrac{\dfrac{1}{a + h} - \dfrac{1}{a}}{h}$ **d.** $\dfrac{-1}{a(a + h)}$ **73. a.** $\dfrac{3}{a + h + 1}$

b. $\dfrac{3}{a + 1}$ **c.** $\dfrac{\dfrac{3}{a + h + 1} - \dfrac{3}{a + 1}}{h}$ **d.** $\dfrac{-3}{(a + h + 1)(a + 1)}$

Section 6.4
Practice Exercises

1. $3a^2 - 2a + 5$ **2.** $5a^2 b^2 - 8ab + 1 - \dfrac{8}{ab}$ **3.** $3x - 2$ **4.** $3x - 2$ **5.** $5x^2 - 6x + 8 + \dfrac{6}{x + 3}$ **6.** $2x^2 + 3x - 2 + \dfrac{-8x + 4}{x^2 + 1}$

7. $16x^2 + 20x + 25$ **8.** $4x^2 + x + 7 + \dfrac{12}{x - 1}$ **9.** $x^3 - 5x + 21 - \dfrac{51}{x + 3}$ **10. a.** -4 **b.** -4 **11.** 15

Exercise Set 6.4

1. $2a + 4$ **3.** $3ab + 4$ **5.** $2y + \dfrac{3y}{x} - \dfrac{2y}{x^2}$ **7.** $x + 1$ **9.** $2x - 8$ **11.** $x - \dfrac{1}{2}$ **13.** $2x^2 - \dfrac{1}{2}x + 5$ **15.** $2x^2 - 6$

17. $3x^3 + 5x + 4 - \dfrac{2x}{x^2 - 2}$ **19.** $2x^3 + \dfrac{9}{2}x^2 + 10x + 21 + \dfrac{42}{x - 2}$ **21.** $x + 8$ **23.** $x - 1$ **25.** $x^2 - 5x - 23 - \dfrac{41}{x - 2}$

27. $4x + 8 + \dfrac{7}{x-2}$ **29.** $x^6y + \dfrac{2}{y} + 1$ **31.** $5x^2 - 6 - \dfrac{5}{2x-1}$ **33.** $2x^2 + 2x + 8 + \dfrac{28}{x-4}$ **35.** $2x^3 - 3x^2 + x - 4$

37. $3x^2 + 4x - 8 + \dfrac{20}{x+1}$ **39.** $3x^2 + 3x - 3$ **41.** $x^2 + x + 1$ **43.** $-\dfrac{5y}{x} - \dfrac{15z}{x} - 25z$ **45.** $3x^4 - 2x$ **47.** 1 **49.** -133

51. 3 **53.** $-\dfrac{187}{81}$ **55.** $\dfrac{95}{32}$ **57.** $-\dfrac{5}{6}$ **59.** 2 **61.** 54 **63.** $(x-1)(x^2+x+1)$ **65.** $(5z+2)(25z^2-10z+4)$ **67.** $(y+2)(x+3)$

69. $x(x+3)(x-3)$ **71.** yes **73.** no **75.** a or d **77.** $(x^4 + 2x^2 - 6)$ m **79.** $(3x-7)$ in. **81.** $(x^3 - 5x^2 + 2x - 1)$ cm

83. $x^3 + \dfrac{5}{3}x^2 + \dfrac{5}{3}x + \dfrac{8}{3} + \dfrac{8}{3(x-1)}$ **85.** $\dfrac{3}{2}x^3 + \dfrac{1}{4}x^2 + \dfrac{1}{8}x - \dfrac{7}{16} + \dfrac{1}{16(2x-1)}$ **87.** $x^3 - \dfrac{2}{5}x$ **89.** $5x - 1 + \dfrac{6}{x}; x \neq 0$

91. $7x^3 + 14x^2 + 25x + 50 + \dfrac{102}{x-2}; x \neq 2$ **93.** answers may vary **95.** answers may vary

97. $(x+3)(x^2+4) = x^3 + 3x^2 + 4x + 12$ **99.** 0 **101.** $x^3 + 2x^2 + 7x + 28$

Section 6.5
Practice Exercises
1. 4 **2.** 7 **3.** $\{\}$ or $\varnothing$ **4.** -1 **5.** 1 **6.** $12, -1$

Vocabulary and Readiness Check 6.5
1. c **3.** a

Exercise Set 6.5
1. 72 **3.** 2 **5.** 6 **7.** $2, -2$ **9.** $\varnothing$ **11.** $-\dfrac{28}{3}$ **13.** 3 **15.** -8 **17.** 3 **19.** $\varnothing$ **21.** 1 **23.** 3 **25.** -1 **27.** 6 **29.** $\dfrac{1}{3}$

31. $-5, 5$ **33.** 3 **35.** 7 **37.** $\varnothing$ **39.** $\dfrac{4}{3}$ **41.** -12 **43.** $1, \dfrac{11}{4}$ **45.** $-5, -1$ **47.** $-\dfrac{7}{5}$ **49.** 5 **51.** length, 15 in.; width, 10 in.

53. 36% **55.** $12-19$ **57.** $40{,}000$ students **59.** answers may vary **61.** 800 pencil sharpeners **63.** $\dfrac{1}{9}, -\dfrac{1}{4}$ **65.** $3, 2$ **67.** 1.39

69. -0.08 **71.** $1, 2$ **73.** $-3, -\dfrac{3}{4}$ **75.** **77.**

The Bigger Picture
1. $\left(-2, \dfrac{16}{7}\right)$ **2.** $-2, \dfrac{16}{7}$ **3.** ± 11 **4.** 5 **5.** $-\dfrac{8}{5}$ **6.** $(-\infty, 2]$ **7.** $(-\infty, -5]$ **8.** $(7, 10]$ **9.** $(-\infty, -17) \cup (18, \infty)$ **10.** $0, -\dfrac{1}{3}, \dfrac{7}{5}$

Integrated Review
1. $\dfrac{1}{2}$ **2.** 10 **3.** $\dfrac{1+2x}{8}$ **4.** $\dfrac{15+x}{10}$ **5.** $\dfrac{2(x-4)}{(x+2)(x-1)}$ **6.** $-\dfrac{5(x-8)}{(x-2)(x+4)}$ **7.** 4 **8.** 8 **9.** -5 **10.** $-\dfrac{2}{3}$ **11.** $\dfrac{2x+5}{x(x-3)}$

12. $\dfrac{5}{2x}$ **13.** -2 **14.** $-\dfrac{y}{x}$ **15.** $\dfrac{(a+3)(a+1)}{a+2}$ **16.** $\dfrac{-a^2 + 31a + 10}{5(a-6)(a+1)}$ **17.** $-\dfrac{1}{5}$ **18.** $-\dfrac{3}{13}$ **19.** $\dfrac{4a+1}{(3a+1)(3a-1)}$

20. $\dfrac{-a-8}{4a(a-2)}$ or $-\dfrac{a+8}{4a(a-2)}$ **21.** $-1, \dfrac{3}{2}$ **22.** $\dfrac{x^2 - 3x + 10}{2(x+3)(x-3)}$ **23.** $\dfrac{3}{x+1}$ **24.** $\{x \mid x$ is a real number and $x \neq 2, x \neq -1\}$ **25.** -1

26. $\dfrac{22z - 45}{3z(z-3)}$ **27. a.** $\dfrac{x}{5} - \dfrac{x}{4} + \dfrac{1}{10}$ **b.** Write each rational expression term so that the denominator is the LCD, 20. **c.** $\dfrac{-x+2}{20}$

28. a. $\dfrac{x}{5} - \dfrac{x}{4} = \dfrac{1}{10}$ **b.** Clear the equation of fractions by multiplying each term by the LCD, 20. **c.** -2 **29.** b **30.** d **31.** d **32.** a **33.** d

Section 6.6
Practice Exercises
1. $a = \dfrac{bc}{b+c}$ **2.** 7 **3.** 3000 **4.** $1\dfrac{1}{5}$ hr **5.** 50 mph

Exercise Set 6.6

1. $C = \dfrac{5}{9}(F - 32)$ **3.** $I = A - QL$ **5.** $R = \dfrac{R_1 R_2}{R_1 + R_2}$ **7.** $n = \dfrac{2S}{a + L}$ **9.** $b = \dfrac{2A - ah}{h}$ **11.** $T_2 = \dfrac{P_2 V_2 T_1}{P_1 V_1}$ **13.** $f_2 = \dfrac{f_1 f}{f_1 - f}$

15. $L = \dfrac{n\lambda}{2}$ **17.** $c = \dfrac{2L\omega}{\theta}$ **19.** 1 and 5 **21.** 5 **23.** 4.5 gal **25.** 4470 women **27.** 15.6 hr **29.** 10 min **31.** 200 mph

33. 15 mph **35.** -8 and -7 **37.** 36 min **39.** 45 mph; 60 mph **41.** 5.9 hr **43.** 2 hr **45.** 135 mph **47.** 12 mi **49.** $\dfrac{7}{8}$

51. $1\dfrac{1}{2}$ min **53.** $2\dfrac{2}{9}$ hr **55.** 10 mph; 8 mph **57.** 2 hr **59.** by jet: 3 hr; by car: 4 hr **61.** 428 movies **63.** 6

65. 22 **67.** answers may vary; 60 in. or 5 ft **69.** 6 ohms **71.** $\dfrac{1}{R} = \dfrac{1}{R_1} + \dfrac{1}{R_2} + \dfrac{1}{R_3}; R = \dfrac{15}{13}$ ohms

Section 6.7

Practice Exercises

1. $k = \dfrac{4}{3}; y = \dfrac{4}{3}x$ **2.** $18\dfrac{3}{4}$ inches **3.** $k = 45; b = \dfrac{45}{a}$ **4.** $P = 653\dfrac{1}{3}$ kilopascals **5.** $A = kpa$ **6.** $k = 4; y = \dfrac{4}{x^3}$ **7.** $k = 81; y = \dfrac{81z}{x^3}$

Vocabulary and Readiness Check 6.7

1. direct **3.** joint **5.** inverse **7.** direct

Exercise Set 6.7

1. $k = \dfrac{1}{5}; y = \dfrac{1}{5}x$ **3.** $k = \dfrac{3}{2}; y = \dfrac{3}{2}x$ **5.** $k = 14; y = 14x$ **7.** $k = 0.25; y = 0.25x$ **9.** 4.05 lb **11.** 204,706 tons **13.** $k = 30; y = \dfrac{30}{x}$

15. $k = 700; y = \dfrac{700}{x}$ **17.** $k = 2; y = \dfrac{2}{x}$ **19.** $k = 0.14; y = \dfrac{0.14}{x}$ **21.** 54 mph **23.** 72 amps **25.** divided by 4 **27.** $x = kyz$

29. $r = kst^3$ **31.** $k = \dfrac{1}{3}; y = \dfrac{1}{3}x^3$ **33.** $k = 0.2; y = 0.2\sqrt{x}$ **35.** $k = 1.3; y = \dfrac{1.3}{x^2}$ **37.** $k = 3; y = 3xz^3$ **39.** 22.5 tons

41. 15π cu in. **43.** 8 ft **45.** $y = kx$ **47.** $a = \dfrac{k}{b}$ **49.** $y = kxz$ **51.** $y = \dfrac{k}{x^3}$ **53.** $y = \dfrac{kx}{p^2}$ **55.** $C = 8\pi$ in.; $A = 16\pi$ sq in.

57. $C = 18\pi$ cm; $A = 81\pi$ sq cm **59.** 9 **61.** 1 **63.** $\dfrac{1}{2}$ **65.** $\dfrac{2}{3}$ **67.** a **69.** c **71.** multiplied by 8 **73.** multiplied by 2

75. 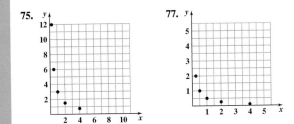 **77.**

Chapter 6 Vocabulary Check

1. complex fraction **2.** long division **3.** directly **4.** inversely **5.** least common denominator
6. synthetic division **7.** jointly **8.** opposites **9.** rational expression **10.** equation, expression

Chapter 6 Review

1. $\{x \mid x \text{ is a real number}\}$ **3.** $\{x \mid x \text{ is a real number and } x \neq 5\}$ **5.** $\{x \mid x \text{ is a real number and } x \neq 0, x \neq -8\}$ **7.** -1 **9.** $\dfrac{1}{x - 1}$

11. $\dfrac{2(x - 3)}{x - 4}$ **13.** $-\dfrac{3}{2}$ **15.** $\dfrac{a - b}{2a}$ **17.** $\dfrac{12}{5}$ **19.** $\dfrac{a - b}{5a}$ **21.** $-\dfrac{1}{x}$ **23.** $60x^2 y^5$ **25.** $5x(x - 5)$ **27.** $\dfrac{4 + x}{x - 4}$

29. $\dfrac{3}{2(x - 2)}$ **31.** $\dfrac{-7x - 6}{5(x - 3)(x + 3)}$ **33.** $\dfrac{5a - 1}{(a - 1)^2(a + 1)}$ **35.** $\dfrac{5 - 2x}{2(x - 1)}$ **37.** $\dfrac{4 - 3x}{8 + x}$ **39.** $\dfrac{5(4x - 3)}{2(5x^2 - 2)}$ **41.** $\dfrac{x(5y + 1)}{3y}$

43. $\dfrac{1 + x}{1 - x}$ **45.** $\dfrac{3}{a + h}$ **47.** $\dfrac{\dfrac{3}{a + h} - \dfrac{3}{a}}{h}$ **49.** $1 + \dfrac{x}{2y} - \dfrac{9}{4xy}$ **51.** $3x^3 + 9x^2 + 2x + 6 - \dfrac{2}{x - 3}$ **53.** $x^2 - 1 + \dfrac{5}{2x + 3}$

55. $3x^2 + 6x + 24 + \dfrac{44}{x - 2}$ **57.** $x^2 + 3x + 9 - \dfrac{54}{x - 3}$ **59.** 3043 **61.** $\dfrac{365}{32}$ **63.** 6 **65.** $\dfrac{3}{2}$ **67.** $\dfrac{2x + 5}{x(x - 7)}$ **69.** $\dfrac{-5(x + 6)}{2x(x - 3)}$

71. $R_2 = \dfrac{RR_1}{R_1 - R}$ **73.** $r = \dfrac{A - P}{Pt}$ **75.** 1, 2 **77.** $1\dfrac{23}{37}$ hr **79.** 8 mph **81.** 9 **83.** 2 **85.** $\dfrac{3}{5x}$ **87.** $\dfrac{5(a - 2)}{7}$ **89.** $\dfrac{13}{3x}$

91. $\dfrac{1}{x - 2}$ **93.** $\dfrac{2}{15 - 2x}$ **95.** $\dfrac{2(x - 1)}{x + 6}$ **97.** 2 **99.** $\dfrac{23}{25}$ **101.** 10 hr **103.** $63\dfrac{2}{3}$ mph; 45 mph **105.** 64π sq in.

Chapter 6 Test

1. $\{x|x$ is a real number and $x \neq 1\}$ **2.** $\{x|x$ is a real number and $x \neq -3, x \neq -1\}$ **3.** $-\dfrac{7}{8}$ **4.** $\dfrac{x}{x+9}$ **5.** $x^2 + 2x + 4$

6. $\dfrac{5}{3x}$ **7.** $\dfrac{3}{x^3}$ **8.** $\dfrac{x+2}{2(x+3)}$ **9.** $-\dfrac{4(2x+9)}{5}$ **10.** -1 **11.** $\dfrac{1}{x(x+3)}$ **12.** $\dfrac{5x-2}{(x-3)(x+2)(x-2)}$ **13.** $\dfrac{-x+30}{6(x-7)}$

14. $\dfrac{3}{2}$ **15.** $\dfrac{64}{3}$ **16.** $\dfrac{(x-3)^2}{x-2}$ **17.** $\dfrac{4xy}{3z} + \dfrac{3}{z} + 1$ **18.** $2x^2 - x - 2 + \dfrac{2}{2x+1}$ **19.** $4x^3 - 15x^2 + 45x - 136 + \dfrac{407}{x+3}$

20. 91 **21.** 8 **22.** $\dfrac{2}{7}$ **23.** 3 **24.** $x = \dfrac{7a^2 + b^2}{4a - b}$ **25.** 5 **26.** $\dfrac{6}{7}$ hr **27.** 16 **28.** 9 **29.** 256 ft

Chapter 6 Cumulative Review

1. a. $8x$ **b.** $8x + 3$ **c.** $x \div -7$ or $\dfrac{x}{-7}$ **d.** $2x - 1.6$ **e.** $x - 6$ **f.** $2(4 + x)$; Sec. 1.2, Ex. 8 **3.** 2; Sec. 1.5, Ex. 5

5. $-20, -12.22, 0, 21.11, 37.78$; Sec. 1.8, Ex. 6 **7.** $\varnothing$; Sec. 3.4, Ex. 7 **9.** -1; Sec. 3.5, Ex. 8 **11.** 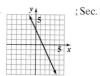 ; Sec. 2.1, Ex. 5

13. a. function **b.** not a function **c.** function; Sec. 2.2, Ex. 2 **15.** ; Sec. 2.3, Ex. 4 **17.** $y = -3x - 2$; Sec. 2.5, Ex. 4

19. ; Sec. 3.6, Ex. 3 **21.** $(0, -5)$; Sec. 4.1, Ex. 7 **23.** $\varnothing$; Sec. 4.2, Ex. 2 **25.** $30°, 110°, 40°$; Sec. 4.3, Ex. 6

27. $(1, -1, 3)$; Sec. 4.4, Ex. 3 **29. a.** 1 **b.** -1 **c.** 1 **d.** 2; Sec. 5.1, Ex. 3 **31. a.** $4x^b$ **b.** y^{5a+6}; Sec. 5.2, Ex. 5

33. a. 2 **b.** 5 **c.** 1 **d.** 6 **e.** 0; Sec. 5.3, Ex. 1 **35.** $9 + 12a + 6b + 4a^2 + 4ab + b^2$; Sec. 5.4, Ex. 9 **37.** $(b - 6)(a + 2)$; Sec. 5.5, Ex. 7

39. $2(n^2 - 19n + 40)$; Sec. 5.6, Ex. 4 **41.** $(x + 2 + y)(x + 2 - y)$; Sec. 5.7, Ex. 5 **43.** $-2, 6$; Sec. 5.8, Ex. 1 **45. a.** $\dfrac{1}{5x - 1}$

b. $\dfrac{9x + 4}{8x - 7}$; Sec. 6.1, Ex. 2 **47.** $\dfrac{5k^2 - 7k + 4}{(k + 2)(k - 2)(k - 1)}$; Sec. 6.2, Ex. 4 **49.** -2; Sec. 6.5, Ex. 2

CHAPTER 7 RATIONAL EXPONENTS, RADICALS, AND COMPLEX NUMBERS

Section 7.1
Practice Exercises

1. a. 7 **b.** 0 **c.** $\dfrac{4}{9}$ **d.** 0.8 **e.** z^4 **f.** $4b^2$ **g.** -6 **h.** not a real number **2.** 6.708 **3. a.** -1 **b.** 3 **c.** $\dfrac{3}{4}$ **d.** x^4 **e.** $-2x$

4. a. 10 **b.** -1 **c.** -9 **d.** not a real number **e.** $3x^3$ **5. a.** 4 **b.** $|x^7|$ **c.** $|x + 7|$ **d.** -7 **e.** $3x - 5$ **f.** $7|x|$ **g.** $|x + 2|$

6. a. 4 **b.** 2 **c.** 2 **d.** -2 **7.** 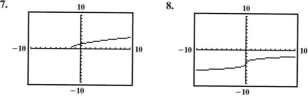 **8.**

Vocabulary and Readiness Check 7.1

1. index; radical sign; radicand **3.** is not **5.** $[0, \infty)$ **7.** $(16, 4)$ **9.** d **11.** d

Exercise Set 7.1

1. 10 **3.** $\dfrac{1}{2}$ **5.** 0.01 **7.** -6 **9.** x^5 **11.** $4y^3$ **13.** 2.646 **15.** 6.164 **17.** 14.142 **19.** 4 **21.** $\dfrac{1}{2}$

23. -1 **25.** x^4 **27.** $-3x^3$ **29.** -2 **31.** not a real number **33.** -2 **35.** x^4 **37.** $2x^2$ **39.** $9x^2$ **41.** $4x^2$

43. 8 **45.** -8 **47.** $2|x|$ **49.** x **51.** $|x - 5|$ **53.** $|x + 2|$ **55.** -11 **57.** $2x$ **59.** y^6 **61.** $5ab^{10}$

63. $-3x^4y^3$ **65.** a^4b **67.** $-2x^2y$ **69.** $\dfrac{5}{7}$ **71.** $\dfrac{x}{2y}$ **73.** $-\dfrac{z^7}{3x}$ **75.** $\dfrac{x}{2}$ **77.** $\sqrt{3}$ **79.** -1 **81.** -3 **83.** $\sqrt{7}$

85. C; $[0,\infty)$ **87.** D; $[3,\infty)$ **89.** A; $(-\infty,\infty)$ **91.** B; $(-\infty,\infty)$ **93.** $-32x^{15}y^{10}$ **95.** $-60x^7y^{10}z^5$ **97.** $\dfrac{x^9y^5}{2}$ **99.** not a real number

101. not a real number **103.** answers may vary **105.** 13 **107.** 18 **109.** 1.69 sq m **111.** answers may vary

Section 7.2
Practice Exercises

1. a. 6 **b.** 10 **c.** $\sqrt[5]{x}$ **d.** 1 **e.** -8 **f.** $5x^3$ **g.** $\sqrt[4]{3x}$ **2. a.** 64 **b.** -1 **c.** -27 **d.** $\dfrac{1}{125}$ **e.** $\sqrt[9]{(3x+2)^5}$

3. a. $\dfrac{1}{27}$ **b.** $\dfrac{1}{16}$ **4. a.** $y^{10/3}$ **b.** $x^{17/20}$ **c.** $\dfrac{1}{9}$ **d.** $b^{2/9}$ **e.** $\dfrac{81}{x^3y^{11/3}}$ **5. a.** $x^{14/15}-x^{13/5}$ **b.** $x+4x^{1/2}-12$ **6.** $x^{-1/5}(2-7x)$

7. a. $\sqrt[3]{x}$ **b.** $\sqrt{6}$ **c.** $\sqrt[4]{a^2b}$ **8. a.** $\sqrt[12]{x^7}$ **b.** $\sqrt[15]{y^2}$ **c.** $\sqrt[6]{675}$

Vocabulary and Readiness Check 7.2
1. true **3.** true **5.** multiply **7.** A **9.** C **11.** B

Exercise Set 7.2

1. 7 **3.** 3 **5.** $\dfrac{1}{2}$ **7.** 13 **9.** $2\sqrt[3]{m}$ **11.** $3x^2$ **13.** -3 **15.** -2 **17.** 8 **19.** 16 **21.** not a real number **23.** $\sqrt[5]{(2x)^3}$

25. $\sqrt[3]{(7x+2)^2}$ **27.** $\dfrac{64}{27}$ **29.** $\dfrac{1}{16}$ **31.** $\dfrac{1}{16}$ **33.** not a real number **35.** $\dfrac{1}{x^{1/4}}$ **37.** $a^{2/3}$ **39.** $\dfrac{5x^{3/4}}{7}$ **41.** $a^{7/3}$ **43.** x **45.** $3^{5/8}$

47. $y^{1/6}$ **49.** $8u^3$ **51.** $-b$ **53.** $\dfrac{1}{x^2}$ **55.** $27x^{2/3}$ **57.** $\dfrac{y}{z^{1/6}}$ **59.** $\dfrac{1}{x^{7/4}}$ **61.** $y-y^{7/6}$ **63.** $x^{5/3}-2x^{2/3}$ **65.** $4x^{2/3}-9$ **67.** $x^{8/3}(1+x^{2/3})$

69. $x^{1/5}(x^{1/5}-3)$ **71.** $x^{-1/3}(5+x)$ **73.** $\sqrt{x}$ **75.** $\sqrt[3]{2}$ **77.** $2\sqrt{x}$ **79.** $\sqrt{xy}$ **81.** $\sqrt[3]{a^2b}$ **83.** $\sqrt{x+3}$ **85.** $\sqrt[15]{y^{11}}$

87. $\sqrt[12]{b^5}$ **89.** $\sqrt[24]{x^{23}}$ **91.** $\sqrt{a}$ **93.** $\sqrt[6]{432}$ **95.** $\sqrt[15]{343y^5}$ **97.** $\sqrt[6]{125r^3s^2}$ **99.** $25\cdot3$ **101.** $16\cdot3$ or $4\cdot12$ **103.** $8\cdot2$ **105.** $27\cdot2$

107. 1509 calories **109.** 210.1 million **111.** $a^{1/3}$ **113.** $x^{1/5}$ **115.** 1.6818 **117.** 5.6645 **119.** $\dfrac{t^{1/2}}{u^{1/2}}$

Section 7.3
Practice Exercises

1. a. $\sqrt{35}$ **b.** $\sqrt{13z}$ **c.** 5 **d.** $\sqrt[3]{15x^2y}$ **e.** $\sqrt{\dfrac{5t}{2m}}$ **2. a.** $\dfrac{6}{7}$ **b.** $\dfrac{\sqrt{z}}{4}$ **c.** $\dfrac{5}{2}$ **d.** $\dfrac{\sqrt[4]{5}}{3x^2}$ **3. a.** $7\sqrt{2}$ **b.** $3\sqrt[3]{2}$ **c.** $\sqrt{35}$ **d.** $3\sqrt[4]{3}$

4. a. $6z^3\sqrt{z}$ **b.** $2pq^2\sqrt[3]{4pq}$ **c.** $2x^3\sqrt[4]{x^3}$ **5. a.** 4 **b.** $\dfrac{7}{3}\sqrt{z}$ **c.** $10xy^2\sqrt[3]{x^2}$ **d.** $6x^2y\sqrt[5]{2y}$ **6.** $\sqrt{17}\approx4.123$ **7.** $\left(\dfrac{13}{2},-4\right)$

Vocabulary and Readiness Check 7.3
1. midpoint; point **3.** midpoint **5.** false **7.** true **9.** false

Exercise Set 7.3

1. $\sqrt{14}$ **3.** 2 **5.** $\sqrt[3]{36}$ **7.** $\sqrt{6x}$ **9.** $\sqrt{\dfrac{14}{xy}}$ **11.** $\sqrt[4]{20x^3}$ **13.** $\dfrac{\sqrt{6}}{7}$ **15.** $\dfrac{\sqrt{2}}{7}$ **17.** $\dfrac{\sqrt[4]{x^3}}{2}$ **19.** $\dfrac{\sqrt[3]{4}}{3}$ **21.** $\dfrac{\sqrt[4]{8}}{x^2}$ **23.** $\dfrac{\sqrt[3]{2x}}{3y^4\sqrt{3}}$

25. $\dfrac{x\sqrt{y}}{10}$ **27.** $\dfrac{x\sqrt{5}}{2y}$ **29.** $-\dfrac{z^2\sqrt[3]{z}}{3x}$ **31.** $4\sqrt{2}$ **33.** $4\sqrt[3]{3}$ **35.** $25\sqrt{3}$ **37.** $2\sqrt{6}$ **39.** $10x^2\sqrt{x}$ **41.** $2y^2\sqrt[3]{2y}$ **43.** $a^2b\sqrt[4]{b^3}$

45. $y^2\sqrt{y}$ **47.** $5ab\sqrt{b}$ **49.** $-2x^2\sqrt[5]{y}$ **51.** $x^4\sqrt[3]{50x^2}$ **53.** $-4a^4b^3\sqrt{2b}$ **55.** $3x^3y^4\sqrt{xy}$ **57.** $5r^3s^4$ **59.** $\sqrt{2}$ **61.** 2 **63.** 10

65. x^2y **67.** $24m^2$ **69.** $\dfrac{15x\sqrt{2x}}{2}$ or $\dfrac{15x}{2}\sqrt{2x}$ **71.** $2a^2\sqrt[4]{2}$ **73.** 5 units **75.** $\sqrt{41}$ units ≈6.403 **77.** $\sqrt{10}$ units ≈3.162

79. $\sqrt{5}$ units ≈2.236 **81.** $\sqrt{192.58}$ units ≈13.877 **83.** $(4,-2)$ **85.** $\left(-5,\dfrac{5}{2}\right)$ **87.** $(3,0)$ **89.** $\left(-\dfrac{1}{2},\dfrac{1}{2}\right)$ **91.** $\left(\sqrt{2},\dfrac{\sqrt{5}}{2}\right)$

93. $(6.2,-6.65)$ **95.** $14x$ **97.** $2x^2-7x-15$ **99.** y^2 **101.** $-3x-15$ **103.** $x^2-8x+16$ **105.** $\dfrac{\sqrt[3]{64}}{\sqrt{64}}=\dfrac{4}{8}=\dfrac{1}{2}$ **107.** x^7 **109.** a^3bc^5

111. $z^{10}\sqrt[3]{z^2}$ **113.** $q^2r^5s\sqrt{q^3r^5}$ **115.** $r=1.6$ meters **117. a.** 3.8 times **b.** 2.9 times **c.** answers may vary

Section 7.4
Practice Exercises

1. a. $8\sqrt{17}$ **b.** $-5\sqrt[3]{5z}$ **c.** $3\sqrt{2}+5\sqrt[3]{2}$ **2. a.** $11\sqrt{6}$ **b.** $-9\sqrt[3]{3}$ **c.** $-2\sqrt{3x}$ **d.** $2\sqrt{10}+2\sqrt[3]{5}$ **e.** $4x\sqrt[3]{3x}$ **3. a.** $\dfrac{5\sqrt{7}}{12}$

b. $\dfrac{13\sqrt[3]{6y}}{4}$ **4. a.** $2\sqrt{5}+5\sqrt{3}$ **b.** $2\sqrt{3}+2\sqrt{2}-\sqrt{30}-2\sqrt{5}$ **c.** $6z+\sqrt{z}-12$ **d.** $-6\sqrt{6}+15$ **e.** $5x-9$ **f.** $6\sqrt{x+2}+x+11$

Vocabulary and Readiness Check 7.4

1. Unlike **3.** Like **5.** $6\sqrt{3}$ **7.** $7\sqrt{x}$ **9.** $8\sqrt[3]{x}$ **11.** $\sqrt{11}+\sqrt[3]{11}$ **13.** $10\sqrt[3]{2x}$

Exercise Set 7.4

1. $-2\sqrt{2}$ **3.** $10x\sqrt{2x}$ **5.** $17\sqrt{2}-15\sqrt{5}$ **7.** $-\sqrt[3]{2x}$ **9.** $5b\sqrt{b}$ **11.** $\dfrac{31\sqrt{2}}{15}$ **13.** $\dfrac{\sqrt[3]{11}}{3}$ **15.** $\dfrac{5\sqrt{5x}}{9}$ **17.** $14+\sqrt{3}$ **19.** $7-3y$

21. $6\sqrt{3}-6\sqrt{2}$ **23.** $-23\sqrt[3]{5}$ **25.** $2b\sqrt{b}$ **27.** $20y\sqrt{2y}$ **29.** $2y\sqrt[3]{2x}$ **31.** $6\sqrt[3]{11}-4\sqrt{11}$ **33.** $4x\sqrt[4]{x^3}$ **35.** $\dfrac{2\sqrt{3}}{3}$ **37.** $\dfrac{5x\sqrt[3]{x}}{7}$

39. $\dfrac{5\sqrt{7}}{2x}$ **41.** $\dfrac{\sqrt[3]{2}}{6}$ **43.** $\dfrac{14x\sqrt[3]{2x}}{9}$ **45.** $15\sqrt{3}$ in. **47.** $\sqrt{35}+\sqrt{21}$ **49.** $7-2\sqrt{10}$ **51.** $3\sqrt{x}-x\sqrt{3}$ **53.** $6x-13\sqrt{x}-5$

55. $\sqrt[3]{a^2}+\sqrt[3]{a}-20$ **57.** $6\sqrt{2}-12$ **59.** $2+2x\sqrt{3}$ **61.** $-16-\sqrt{35}$ **63.** $x-y^2$ **65.** $3+2x\sqrt{3}+x^2$ **67.** $23x-5x\sqrt{15}$

69. $2\sqrt[3]{2}-\sqrt[3]{4}$ **71.** $x+1$ **73.** $x+24+10\sqrt{x-1}$ **75.** $2x+6-2\sqrt{2x+5}$ **77.** $x-7$ **79.** $\dfrac{7}{x+y}$ **81.** $2a-3$ **83.** $\dfrac{-2+\sqrt{3}}{3}$

85. $22\sqrt{5}$ ft; 150 sq ft **87. a.** $2\sqrt{3}$ **b.** 3 **c.** answers may vary **89.** answers may vary

Section 7.5
Practice Exercises

1. a. $\dfrac{5\sqrt{3}}{3}$ **b.** $\dfrac{15\sqrt{x}}{2x}$ **c.** $\dfrac{\sqrt[3]{6}}{3}$ **2.** $\dfrac{\sqrt{15yz}}{5y}$ **3.** $\dfrac{\sqrt[3]{z^2x^2}}{3x^2}$ **4. a.** $\dfrac{5(3\sqrt{5}-2)}{41}$ **b.** $\dfrac{\sqrt{6}+5\sqrt{3}+\sqrt{10}+5\sqrt{5}}{-2}$ **c.** $\dfrac{6x-3\sqrt{xy}}{4x-y}$

5. $\dfrac{2}{\sqrt{10}}$ **6.** $\dfrac{5b}{\sqrt[3]{50ab^2}}$ **7.** $\dfrac{x-9}{4(\sqrt{x}+3)}$

Vocabulary and Readiness Check 7.5

1. conjugate **3.** rationalizing the numerator **5.** $\sqrt{2}-x$ **7.** $5+\sqrt{a}$ **9.** $-7\sqrt{5}-8\sqrt{x}$

Exercise Set 7.5

1. $\dfrac{\sqrt{14}}{7}$ **3.** $\dfrac{\sqrt{5}}{5}$ **5.** $\dfrac{2\sqrt{x}}{x}$ **7.** $\dfrac{4\sqrt[3]{9}}{3}$ **9.** $\dfrac{3\sqrt{2x}}{4x}$ **11.** $\dfrac{3\sqrt[3]{2x}}{2x}$ **13.** $\dfrac{3\sqrt{3a}}{a}$ **15.** $\dfrac{3\sqrt[4]{4}}{2}$ **17.** $\dfrac{2\sqrt{21}}{7}$ **19.** $\dfrac{\sqrt{10xy}}{5y}$ **21.** $\dfrac{\sqrt[3]{75}}{5}$

23. $\dfrac{\sqrt{6x}}{10}$ **25.** $\dfrac{\sqrt{3z}}{6z}$ **27.** $\dfrac{\sqrt[3]{6xy^2}}{3x}$ **29.** $\dfrac{3\sqrt[4]{2}}{2}$ **31.** $\dfrac{2\sqrt[4]{9x}}{3x^2}$ **33.** $\dfrac{5a\sqrt[5]{4ab^4}}{2a^2b^3}$ **35.** $-2(2+\sqrt{7})$ **37.** $\dfrac{7(3+\sqrt{x})}{9-x}$ **39.** $-5+2\sqrt{6}$

41. $\dfrac{2a+2\sqrt{a}+\sqrt{ab}+\sqrt{b}}{4a-b}$ **43.** $-\dfrac{8(1-\sqrt{10})}{9}$ **45.** $\dfrac{x-\sqrt{xy}}{x-y}$ **47.** $\dfrac{5+3\sqrt{2}}{7}$ **49.** $\dfrac{5}{\sqrt{15}}$ **51.** $\dfrac{6}{\sqrt{10}}$ **53.** $\dfrac{2x}{7\sqrt{x}}$ **55.** $\dfrac{5y}{\sqrt[3]{100xy}}$

57. $\dfrac{2}{\sqrt{10}}$ **59.** $\dfrac{2x}{11\sqrt{2x}}$ **61.** $\dfrac{7}{2\sqrt[3]{49}}$ **63.** $\dfrac{3x^2}{10\sqrt[3]{9x}}$ **65.** $\dfrac{6x^2y^3}{\sqrt{6z}}$ **67.** answers may vary **69.** $\dfrac{-7}{12+6\sqrt{11}}$ **71.** $\dfrac{3}{10+5\sqrt{7}}$

73. $\dfrac{x-9}{x-3\sqrt{x}}$ **75.** $\dfrac{1}{3+2\sqrt{2}}$ **77.** $\dfrac{x-1}{x-2\sqrt{x}+1}$ **79.** 5 **81.** $-\dfrac{1}{2}, 6$ **83.** 2, 6 **85.** $\sqrt[3]{25}$ **87.** $r=\dfrac{\sqrt{A\pi}}{2\pi}$ **89.** answers may vary

Integrated Review

1. 9 **2.** -2 **3.** $\dfrac{1}{2}$ **4.** x^3 **5.** y^3 **6.** $2y^5$ **7.** $-2y$ **8.** $3b^3$ **9.** 6 **10.** $\sqrt[4]{3y}$ **11.** $\dfrac{1}{16}$ **12.** $\sqrt[5]{(x+1)^3}$ **13.** y **14.** $16x^{1/2}$

15. $x^{5/4}$ **16.** $4^{11/15}$ **17.** $2x^2$ **18.** $\sqrt[4]{a^3b^2}$ **19.** $\sqrt[6]{x^3}$ **20.** $\sqrt[6]{500}$ **21.** $2\sqrt{10}$ **22.** $2xy^2\sqrt[4]{x^3y^2}$ **23.** $3x\sqrt[3]{2x}$ **24.** $-2b^2\sqrt[5]{2}$

25. $\sqrt{5x}$ **26.** $4x$ **27.** $7y^2\sqrt{y}$ **28.** $2a^2\sqrt[3]{3}$ **29.** $2\sqrt{5}-5\sqrt{3}+5\sqrt{7}$ **30.** $y\sqrt[3]{2y}$ **31.** $\sqrt{15}-\sqrt{6}$ **32.** $10+2\sqrt{21}$

33. $4x^2-5$ **34.** $x+2-2\sqrt{x+1}$ **35.** $\dfrac{\sqrt{21}}{3}$ **36.** $\dfrac{5\sqrt[3]{4x}}{2x}$ **37.** $\dfrac{13-3\sqrt{21}}{5}$ **38.** $\dfrac{7}{\sqrt{21}}$ **39.** $\dfrac{3y}{\sqrt[3]{33y^2}}$ **40.** $\dfrac{x-4}{x+2\sqrt{x}}$

Section 7.6
Practice Exercises

1. 18 **2.** $\dfrac{3}{8}, -\dfrac{1}{2}$ **3.** 10 **4.** 9 **5.** $\dfrac{3}{25}$ **6.** $6\sqrt{3}$ meters **7.** $\sqrt{193}$ in. ≈ 13.89

Vocabulary and Readiness Check 7.6

1. extraneous solution **3.** $x^2-10x+25$

Exercise Set 7.6

1. 8 **3.** 7 **5.** ∅ **7.** 7 **9.** 6 **11.** $-\dfrac{9}{2}$ **13.** 29 **15.** 4 **17.** -4 **19.** ∅ **21.** 7 **23.** 9 **25.** 50 **27.** ∅ **29.** $\dfrac{15}{4}$

31. 13 **33.** 5 **35.** -12 **37.** 9 **39.** -3 **41.** 1 **43.** 1 **45.** $\dfrac{1}{2}$ **47.** 0, 4 **49.** $\dfrac{37}{4}$ **51.** $3\sqrt{5}$ ft **53.** $2\sqrt{10}$ m

55. $2\sqrt{131}$ m ≈ 22.9 m **57.** $\sqrt{100.84}$ mm ≈ 10.0 mm **59.** 17 ft **61.** 13 ft **63.** 14,657,415 sq mi **65.** 100 ft **67.** 100

69. $\dfrac{\pi}{2}$ sec ≈ 1.57 sec **71.** 12.97 ft **73.** answers may vary **75.** $15\sqrt{3}$ sq mi ≈ 25.98 sq mi **77.** answers may vary **79.** 0.51 km

81. function **83.** function **85.** not a function **87.** $\dfrac{x}{4x+3}$ **89.** $-\dfrac{4z+2}{3z}$

91. $\sqrt{5x-1}+4=7$ **93.** 1 **95. a.–b.** answers may vary **97.** $-1, 2$ **99.** $-8, -6, 0, 2$
$$\sqrt{5x-1}=3$$
$$(\sqrt{5x-1})^2=3^2$$
$$5x-1=9$$
$$5x=10$$
$$x=2$$

The Bigger Picture

1. -19 **2.** $-\dfrac{5}{3}, 5$ **3.** $-\dfrac{9}{2}, 5$ **4.** $\left[\dfrac{-11}{5}, 1\right]$ **5.** $\left(-\dfrac{7}{5}, \infty\right)$ **6.** 25 **7.** $(-5, \infty)$ **8.** ∅ **9.** $(-\infty, -13)\cup(17, \infty)$ **10.** $\dfrac{17}{25}$

Section 7.7

Practice Exercises

1. a. $2i$ **b.** $i\sqrt{7}$ **c.** $-3i\sqrt{2}$ **2. a.** $-\sqrt{30}$ **b.** -3 **c.** $25i$ **d.** $3i$ **3. a.** $-1-4i$ **b.** $-3+5i$ **c.** $3-2i$ **4. a.** 20 **b.** $-5+10i$

c. $15+16i$ **d.** $8-6i$ **e.** 85 **5. a.** $\dfrac{11}{10}-\dfrac{7i}{10}$ **b.** $0-\dfrac{5i}{2}$ **6. a.** i **b.** 1 **c.** -1 **d.** 1

Vocabulary and Readiness Check 7.7

1. complex **3.** -1 **5.** real **7.** $9i$ **9.** $i\sqrt{7}$ **11.** -4 **13.** $8i$

Exercise Set 7.7

1. $2i\sqrt{6}$ **3.** $-6i$ **5.** $24i\sqrt{7}$ **7.** $-3\sqrt{6}$ **9.** $-\sqrt{14}$ **11.** $-5\sqrt{2}$ **13.** $4i$ **15.** $i\sqrt{3}$ **17.** $2\sqrt{2}$ **19.** $6-4i$ **21.** $-2+6i$

23. $-2-4i$ **25.** -40 **27.** $18+12i$ **29.** 7 **31.** $12-16i$ **33.** $-4i$ **35.** $\dfrac{28}{25}-\dfrac{21}{25}i$ **37.** $4+i$ **39.** $\dfrac{17}{13}+\dfrac{7}{13}i$ **41.** 63

43. $2-i$ **45.** $27+3i$ **47.** $-\dfrac{5}{2}-2i$ **49.** $18+13i$ **51.** 20 **53.** 10 **55.** 2 **57.** $-5+\dfrac{16}{3}i$ **59.** $17+144i$ **61.** $\dfrac{3}{5}-\dfrac{1}{5}i$

63. $5-10i$ **65.** $\dfrac{1}{5}-\dfrac{8}{5}i$ **67.** $8-i$ **69.** 7 **71.** $12-16i$ **73.** 1 **75.** i **77.** $-i$ **79.** -1 **81.** -64 **83.** $-243i$ **85.** $40°$

87. $x^2-5x-2-\dfrac{6}{x-1}$ **89.** 5 people **91.** 14 people **93.** 16.7% **95.** $-1-i$ **97.** 0 **99.** $2+3i$ **101.** $2+i\sqrt{2}$ **103.** $\dfrac{1}{2}-\dfrac{\sqrt{3}}{2}i$

105. answers may vary **107.** $6-6i$ **109.** yes

Chapter 7 Vocabulary Check

1. conjugate **2.** principal square root **3.** rationalizing **4.** imaginary unit **5.** cube root **6.** index, radicand **7.** like radicals
8. complex number **9.** distance **10.** midpoint

Chapter 7 Review

1. 9 **3.** -2 **5.** $-\dfrac{1}{7}$ **7.** -6 **9.** $-a^2b^3$ **11.** $2ab^2$ **13.** $\dfrac{x^6}{6y}$ **15.** $|-x|$ **17.** -27 **19.** $-x$ **21.** $5|(x-y)^5|$ **23.** $-x$

25. $(-\infty, \infty)$; $-2, -1, 0, 1, 2$ **27.** $-\dfrac{1}{3}$ **29.** $-\dfrac{1}{4}$ **31.** $\dfrac{1}{4}$ **33.** $\dfrac{343}{125}$ **35.** not a real number **37.** $5^{1/5}x^{2/5}y^{3/5}$ **39.** $5\sqrt[3]{xy^2z^5}$

41. $a^{13/6}$ **43.** $\dfrac{1}{a^{9/2}}$ **45.** a^4b^6 **47.** $\dfrac{b^{5/6}}{49a^{1/4}c^{5/3}}$ **49.** 4.472 **51.** 5.191 **53.** -26.246 **55.** $\sqrt[6]{1372}$

57. $2\sqrt{6}$ **59.** $2x$ **61.** $2\sqrt{15}$ **63.** $3\sqrt[3]{6}$ **65.** $6x^3\sqrt{x}$ **67.** $\dfrac{p^8\sqrt{p}}{11}$ **69.** $\dfrac{y\sqrt[4]{xy^2}}{3}$ **71. a.** $\dfrac{5}{\sqrt{\pi}}$ m or $\dfrac{5\sqrt{\pi}}{\pi}$ m

b. 5.75 in. **73.** $\sqrt{130}$ units ≈ 11.402 **75.** $7\sqrt{2}$ units ≈ 9.899 **77.** $\sqrt{275.6}$ units ≈ 16.601 **79.** $\left(-\dfrac{15}{2}, 1\right)$ **81.** $\left(\dfrac{1}{20}, -\dfrac{3}{16}\right)$

83. $\left(\sqrt{3}, -3\sqrt{6}\right)$ **85.** $2x\sqrt{3xy}$ **87.** $3a\sqrt[4]{2a}$ **89.** $\dfrac{3\sqrt{2}}{4x}$ **91.** $-4ab\sqrt[4]{2b}$ **93.** $x-6\sqrt{x}+9$ **95.** $4x-9y$ **97.** $\sqrt[3]{a^2}+4\sqrt[3]{a}+4$

99. $a+64$ **101.** $\dfrac{\sqrt{3x}}{6}$ **103.** $\dfrac{2x^2\sqrt{2x}}{y}$ **105.** $-\dfrac{10+5\sqrt{7}}{3}$ **107.** $-5+2\sqrt{6}$ **109.** $\dfrac{6}{\sqrt{2y}}$ **111.** $\dfrac{4x^3}{y\sqrt{2x}}$ **113.** $\dfrac{x-25}{-3\sqrt{x}+15}$

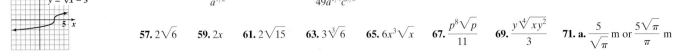

115. $\varnothing$ **117.** $\varnothing$ **119.** 16   **121.** $\sqrt{241}$ **123.** 4.24 ft   **125.** $-i\sqrt{6}$ **127.** $-\sqrt{10}$ **129.** $-13 - 3i$ **131.** $-12 - 18i$ **133.** $-5 - 12i$

135. $\dfrac{3}{2} - i$ **137.** x **139.** -10 **141.** $\dfrac{y^5}{2x^3}$ **143.** $\dfrac{1}{8}$ **145.** $\dfrac{1}{x^{13/2}}$ **147.** $\dfrac{n\sqrt{3n}}{11m^5}$ **149.** $4x - 20\sqrt{x} + 25$ **151.** (4, 16)   **153.** $\dfrac{2\sqrt{x} - 6}{x - 9}$

Chapter 7 Test

1. $6\sqrt{6}$ **2.** $-x^{16}$ **3.** $\dfrac{1}{5}$ **4.** 5   **5.** $\dfrac{4x^2}{9}$ **6.** $-a^6 b^3$ **7.** $\dfrac{8a^{1/3}c^{2/3}}{b^{5/12}}$ **8.** $a^{7/12} - a^{7/3}$ **9.** $|4xy|$ or $4|xy|$ **10.** -27 **11.** $\dfrac{3\sqrt{y}}{y}$

12. $\dfrac{8 - 6\sqrt{x} + x}{8 - 2x}$ **13.** $\dfrac{\sqrt[3]{b^2}}{b}$ **14.** $\dfrac{6 - x^2}{8(\sqrt{6} - x)}$ **15.** $-x\sqrt{5x}$ **16.** $4\sqrt{3} - \sqrt{6}$ **17.** $x + 2\sqrt{x} + 1$ **18.** $\sqrt{6} - 4\sqrt{3} + \sqrt{2} - 4$

19. -20 **20.** 23.685   **21.** 0.019   **22.** 2, 3   **23.** $\varnothing$ **24.** 6   **25.** $i\sqrt{2}$ **26.** $-2i\sqrt{2}$ **27.** $-3i$ **28.** 40   **29.** $7 + 24i$ **30.** $-\dfrac{3}{2} + \dfrac{5}{2}i$

31. $\dfrac{5\sqrt{2}}{2}$ **32.** $[-2, \infty)$; 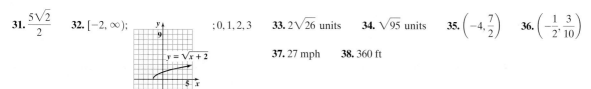 ; 0, 1, 2, 3   **33.** $2\sqrt{26}$ units   **34.** $\sqrt{95}$ units   **35.** $\left(-4, \dfrac{7}{2}\right)$ **36.** $\left(-\dfrac{1}{2}, \dfrac{3}{10}\right)$

37. 27 mph   **38.** 360 ft

Chapter 7 Cumulative Review

1. a. $2xy - 2$ **b.** $2x^2 + 23$ **c.** $3.1x - 0.3$ **d.** $-a - 7b + \dfrac{7}{12}$; Sec. 1.4, Ex. 16   **3.** $\dfrac{21}{11}$; Sec. 1.5, Ex. 6   **5.** \$4500 per month; Sec. 3.2, Ex. 8

7. $\varnothing$; Sec. 3.4, Ex. 6   **9.** $(-\infty, -3] \cup [9, \infty)$; Sec. 3.5, Ex. 7   **11.**  ; Sec. 2.1, Ex. 10   **13. a.** domain: $\{2, 0, 3\}$; range: $\{3, 4, -1\}$

b. domain: $\{-4, -3, -2, -1, 0, 1, 2, 3\}$; range: $\{1\}$ **c.** domain: {Lubbock, Colorado Springs, Omaha, Yonkers, Sacramento}; range: $\{307, 404, 445, 206, 197\}$;

Sec. 2.2, Ex. 1   **15.**  ; Sec. 2.3, Ex.8   **17.** undefined; Sec. 2.4, Ex. 5   **19.** $\left(-\dfrac{21}{10}, \dfrac{3}{10}\right)$; Sec. 4.1, Ex. 5   **21. a.** 2^7 **b.** x^{10}

c. y^7; Sec. 5.1, Ex. 1   **23.** 6×10^{-5}; Sec. 5.2, Ex. 7   **25. a.** -4 **b.** 11; Sec. 5.3, Ex. 4   **27. a.** $2x^2 + 11x + 15$ **b.** $10x^3 - 27x^2 + 32x - 21$;

Sec. 5.4, Ex. 3   **29.** $5x^2$; Sec. 5.5, Ex. 1   **31. a.** $x^2 - 2x + 4$ **b.** $\dfrac{2}{y - 5}$; Sec. 6.1, Ex. 6   **33. a.** $\dfrac{6x + 5}{3x^3y}$ **b.** $\dfrac{2x^2 + 7x - 6}{(x + 2)(x - 2)}$ **c.** 2; Sec. 6.2, Ex. 3

35. a. $\dfrac{x(x - 2)}{2(x + 2)}$ **b.** $\dfrac{x^2}{y^2}$; Sec. 6.3, Ex. 2   **37.** $2x^2 - x + 4$; Sec. 6.4, Ex. 1   **39.** $2x^2 + 5x + 2 + \dfrac{7}{x - 3}$; Sec. 6.4, Ex. 8   **41.** $\varnothing$; Sec. 6.5, Ex. 3

43. $x = \dfrac{yz}{y - z}$; Sec. 6.6, Ex. 1   **45.** constant of variation: 15; $u = \dfrac{15}{w}$; Sec. 6.7, Ex. 3   **47. a.** $\dfrac{1}{8}$ **b.** $\dfrac{1}{9}$; Sec. 7.2, Ex. 3

49. $\dfrac{x - 4}{5(\sqrt{x} - 2)}$; Sec. 7.5, Ex. 7

CHAPTER 8 QUADRATIC EQUATIONS AND FUNCTIONS

Section 8.1
Practice Exercises

1. $-3\sqrt{2}, 3\sqrt{2}$ **2.** $\pm\sqrt{10}$ **3.** $-3 \pm 2\sqrt{5}$ **4.** $\dfrac{2 + 3i}{5}, \dfrac{2 - 3i}{5}$ **5.** $-2 \pm \sqrt{7}$ **6.** $\dfrac{3 \pm \sqrt{5}}{2}$ **7.** $\dfrac{6 \pm \sqrt{33}}{3}$ **8.** $\dfrac{5 \pm i\sqrt{31}}{4}$ **9.** 6%

Vocabulary and Readiness Check 8.1
1. $\pm\sqrt{b}$ **3.** completing the square   **5.** 9   **7.** 1   **9.** 49

Exercise Set 8.1

1. $-4, 4$ **3.** $-\sqrt{7}, \sqrt{7}$ **5.** $-3\sqrt{2}, 3\sqrt{2}$ **7.** $-\sqrt{10}, \sqrt{10}$ **9.** $-8, -2$ **11.** $6 - 3\sqrt{2}, 6 + 3\sqrt{2}$ **13.** $\dfrac{3 - 2\sqrt{2}}{2}, \dfrac{3 + 2\sqrt{2}}{2}$

15. $-3i, 3i$ **17.** $-\sqrt{6}, \sqrt{6}$ **19.** $-2i\sqrt{2}, 2i\sqrt{2}$ **21.** $1 - 4i, 1 + 4i$ **23.** $-7 - \sqrt{5}, -7 + \sqrt{5}$ **25.** $-3 - 2i\sqrt{2}, -3 + 2i\sqrt{2}$

27. $x^2 + 16x + 64 = (x + 8)^2$ **29.** $z^2 - 12z + 36 = (z - 6)^2$ **31.** $p^2 + 9p + \dfrac{81}{4} = \left(p + \dfrac{9}{2}\right)^2$ **33.** $x^2 + x + \dfrac{1}{4} = \left(x + \dfrac{1}{2}\right)^2$ **35.** $-5, -3$

37. $-3 - \sqrt{7}, -3 + \sqrt{7}$ **39.** $\dfrac{-1 - \sqrt{5}}{2}, \dfrac{-1 + \sqrt{5}}{2}$ **41.** $-1 - \sqrt{6}, -1 + \sqrt{6}$ **43.** $\dfrac{6 - \sqrt{30}}{3}, \dfrac{6 + \sqrt{30}}{3}$ **45.** $\dfrac{3 - \sqrt{11}}{2}, \dfrac{3 + \sqrt{11}}{2}$

47. $-4, \dfrac{1}{2}$ **49.** $-1, 5$ **51.** $-4 - \sqrt{15}, -4 + \sqrt{15}$ **53.** $\dfrac{-3 - \sqrt{21}}{3}, \dfrac{-3 + \sqrt{21}}{3}$ **55.** $-1, \dfrac{5}{2}$ **57.** $-1 - i, -1 + i$ **59.** $3 - \sqrt{6}, 3 + \sqrt{6}$

61. $-2 - i\sqrt{2}, -2 + i\sqrt{2}$ **63.** $-5 - i\sqrt{3}, -5 + i\sqrt{3}$ **65.** $-4, 1$ **67.** $\dfrac{2 - i\sqrt{2}}{2}, \dfrac{2 + i\sqrt{2}}{2}$ **69.** $\dfrac{-3 - \sqrt{69}}{6}, \dfrac{-3 + \sqrt{69}}{6}$

71. 2 real number solutions **73.** no real number solutions **75.** 20% **77.** 4% **79.** answers may vary **81.** 8.11 sec **83.** 6.73 sec

85. simple; answers may vary **87.** $\dfrac{7}{5}$ **89.** $\dfrac{1}{5}$ **91.** $5 - 10\sqrt{3}$ **93.** $\dfrac{3 - 2\sqrt{7}}{4}$ **95.** $2\sqrt{7}$ **97.** $\sqrt{13}$ **99.** complex, but not real numbers

101. real solutions **103.** complex, but not real numbers **105.** $-6y, 6y$ **107.** $-x, x$ **109.** 6 in. **111.** 16.2 in. $\times$ 21.6 in.
113. 2.828 thousand units or 2828 units

Section 8.2

Practice Exercises

1. $2, -\dfrac{1}{3}$ **2.** $\dfrac{4 \pm \sqrt{22}}{3}$ **3.** $1 \pm \sqrt{17}$ **4.** $\dfrac{-1 \pm i\sqrt{15}}{4}$ **5. a.** one real solution **b.** two real solutions **c.** two complex, but not real

solutions **6.** 6 ft **7.** 2.4 sec

Vocabulary and Readiness Check 8.2

1. $x = \dfrac{-b \pm \sqrt{b^2 - 4ac}}{2a}$ **3.** $-5; -7$ **5.** $1; 0$

Exercise Set 8.2

1. $-6, 1$ **3.** $-\dfrac{3}{5}, 1$ **5.** 3 **7.** $\dfrac{-7 - \sqrt{33}}{2}, \dfrac{-7 + \sqrt{33}}{2}$ **9.** $\dfrac{1 - \sqrt{57}}{8}, \dfrac{1 + \sqrt{57}}{8}$ **11.** $\dfrac{7 - \sqrt{85}}{6}, \dfrac{7 + \sqrt{85}}{6}$ **13.** $1 - \sqrt{3}, 1 + \sqrt{3}$

15. $-\dfrac{3}{2}, 1$ **17.** $\dfrac{3 - \sqrt{11}}{2}, \dfrac{3 + \sqrt{11}}{2}$ **19.** $\dfrac{-5 - \sqrt{17}}{2}, \dfrac{-5 + \sqrt{17}}{2}$ **21.** $\dfrac{5}{2}, 1$ **23.** $-3 - 2i, -3 + 2i$ **25.** $-2 - \sqrt{11}, -2 + \sqrt{11}$

27. $\dfrac{3 - i\sqrt{87}}{8}, \dfrac{3 + i\sqrt{87}}{8}$ **29.** $\dfrac{3 - \sqrt{29}}{2}, \dfrac{3 + \sqrt{29}}{2}$ **31.** $\dfrac{-5 - i\sqrt{5}}{10}, \dfrac{-5 + i\sqrt{5}}{10}$ **33.** $\dfrac{-1 - \sqrt{19}}{6}, \dfrac{-1 + \sqrt{19}}{6}$

35. $\dfrac{-1 - i\sqrt{23}}{4}, \dfrac{-1 + i\sqrt{23}}{4}$ **37.** 1 **39.** $3 + \sqrt{5}, 3 - \sqrt{5}$ **41.** two real solutions **43.** one real solution **45.** two real solutions

47. two complex but not real solutions **49. a.** 2 real solutions **b.** There is one x-intercept.; 1 real solution **51.** 14 ft
53. $2 + 2\sqrt{2}$ cm, $2 + 2\sqrt{2}$ cm, $4 + 2\sqrt{2}$ cm **55.** width: $-5 + 5\sqrt{17}$ ft; length: $5 + 5\sqrt{17}$ ft **57. a.** $50\sqrt{2}$ m **b.** 5000 sq m
59. 37.4 ft by 38.5 ft **61.** base, $2 + 2\sqrt{43}$ cm; height, $-1 + \sqrt{43}$ cm **63.** 8.9 sec **65.** 2.8 sec **67.** $\dfrac{11}{5}$ **69.** 15

71. $(x^2 + 5)(x + 2)(x - 2)$ **73.** $(z + 3)(z - 3)(z + 2)(z - 2)$ **75.** b **77.** answers may vary **79.** 0.6, 2.4

81. Sunday to Monday **83.** Wednesday **85.** 32; yes **87. a.** 8630 stores **b.** 2011 **89.** answers may vary **91.** $\dfrac{\sqrt{3}}{3}$

93. $\dfrac{-\sqrt{2} - i\sqrt{2}}{2}, \dfrac{-\sqrt{2} + i\sqrt{2}}{2}$ **95.** $\dfrac{\sqrt{3} - \sqrt{11}}{4}, \dfrac{\sqrt{3} + \sqrt{11}}{4}$

Section 8.3

Practice Exercises

1. 8 **2.** $\dfrac{5 \pm \sqrt{137}}{8}$ **3.** $4, -4, 3i, -3i$ **4.** $1, -3$ **5.** $1, 64$ **6.** Katy: $\dfrac{7 + \sqrt{65}}{2} \approx 7.5$ hr; Steve: $\dfrac{9 + \sqrt{65}}{2} \approx 8.5$ hr

7. to Shanghai: 40 km/hr; to Ningbo: 90 km/hr

Exercise Set 8.3

1. 2 **3.** 16 **5.** 1, 4 **7.** $3 - \sqrt{7}, 3 + \sqrt{7}$ **9.** $\dfrac{3 - \sqrt{57}}{4}, \dfrac{3 + \sqrt{57}}{4}$ **11.** $\dfrac{1 - \sqrt{29}}{2}, \dfrac{1 + \sqrt{29}}{2}$ **13.** $-2, 2, -2i, 2i$

15. $-\dfrac{1}{2}, \dfrac{1}{2}, -i\sqrt{3}, i\sqrt{3}$ **17.** $-3, 3, -2, 2$ **19.** $125, -8$ **21.** $-\dfrac{4}{5}, 0$ **23.** $-\dfrac{1}{8}, 27$ **25.** $-\dfrac{2}{3}, \dfrac{4}{3}$ **27.** $-\dfrac{1}{125}, \dfrac{1}{8}$ **29.** $-\sqrt{2}, \sqrt{2}, -\sqrt{3}, \sqrt{3}$

31. $\dfrac{-9 - \sqrt{201}}{6}, \dfrac{-9 + \sqrt{201}}{6}$ **33.** $2, 3$ **35.** 3 **37.** $27, 125$ **39.** $1, -3i, 3i$ **41.** $\dfrac{1}{8}, -8$ **43.** $-\dfrac{1}{2}, \dfrac{1}{3}$ **45.** 4

47. -3 **49.** $-\sqrt{5}, \sqrt{5}, -2i, 2i$ **51.** $-3, \dfrac{3 - 3i\sqrt{3}}{2}, \dfrac{3 + 3i\sqrt{3}}{2}$ **53.** $6, 12$ **55.** $-\dfrac{1}{3}, \dfrac{1}{3}, -\dfrac{i\sqrt{6}}{3}, \dfrac{i\sqrt{6}}{3}$ **57.** 5 mph, then 4 mph

59. inlet pipe: 15.5 hr; hose: 16.5 hr   **61.** 55 mph, 66 mph   **63.** 8.5 hr   **65.** 12 or -8 **67. a.** $(x - 6)$ in.   **b.** $300 = (x - 6) \cdot (x - 6) \cdot 3$
c. 16 cm by 16 cm   **69.** 22 feet   **71.** $(-\infty, 3]$ **73.** $(-5, \infty)$ **75.** domain: $(-\infty, \infty)$; range: $(-\infty, \infty)$; function

77. domain: $(-\infty, \infty)$; range: $[-1, \infty)$; function   **79.** $1, -3i, 3i$ **81.** $-\dfrac{1}{2}, \dfrac{1}{3}$ **83.** $-3, \dfrac{3 - 3i\sqrt{3}}{2}, \dfrac{3 + 3i\sqrt{3}}{2}$ **85.** answers may vary

87. a. 150.94 ft/sec   **b.** 151.49 ft/sec   **c.** Bourdais: 102.9 mph; Pagenaud: 103.3 mph

Integrated Review

1. $-\sqrt{10}, \sqrt{10}$ **2.** $-\sqrt{14}, \sqrt{14}$ **3.** $1 - 2\sqrt{2}, 1 + 2\sqrt{2}$ **4.** $-5 - 2\sqrt{3}, -5 + 2\sqrt{3}$ **5.** $-1 - \sqrt{13}, -1 + \sqrt{13}$

6. $1, 11$ **7.** $\dfrac{-3 - \sqrt{69}}{6}, \dfrac{-3 + \sqrt{69}}{6}$ **8.** $\dfrac{-2 - \sqrt{5}}{4}, \dfrac{-2 + \sqrt{5}}{4}$ **9.** $\dfrac{2 - \sqrt{2}}{2}, \dfrac{2 + \sqrt{2}}{2}$ **10.** $-3 - \sqrt{5}, -3 + \sqrt{5}$

11. $-2 + i\sqrt{3}, -2 - i\sqrt{3}$ **12.** $\dfrac{-1 - i\sqrt{11}}{2}, \dfrac{-1 + i\sqrt{11}}{2}$ **13.** $\dfrac{-3 + i\sqrt{15}}{2}, \dfrac{-3 - i\sqrt{15}}{2}$ **14.** $3i, -3i$ **15.** $0, -17$

16. $\dfrac{1 + \sqrt{13}}{4}, \dfrac{1 - \sqrt{13}}{4}$ **17.** $2 + 3\sqrt{3}, 2 - 3\sqrt{3}$ **18.** $2 + \sqrt{3}, 2 - \sqrt{3}$ **19.** $-2, \dfrac{4}{3}$ **20.** $\dfrac{-5 + \sqrt{17}}{4}, \dfrac{-5 - \sqrt{17}}{4}$ **21.** $1 - \sqrt{6}, 1 + \sqrt{6}$

22. $-\sqrt{31}, \sqrt{31}$ **23.** $-\sqrt{11}, \sqrt{11}$ **24.** $-i\sqrt{11}, i\sqrt{11}$ **25.** $-11, 6$ **26.** $\dfrac{-3 + \sqrt{19}}{5}, \dfrac{-3 - \sqrt{19}}{5}$ **27.** $\dfrac{-3 + \sqrt{17}}{4}, \dfrac{-3 - \sqrt{17}}{4}$

28. $10\sqrt{2}\,\text{ft} \approx 14.1\,\text{ft}$ **29.** Jack: 9.1 hr; Lucy: 7.1 hr   **30.** 5 mph during the first part, then 6 mph

Section 8.4
Practice Exercises

1. $(-\infty, -3) \cup (4, \infty)$ **2.** $[0, 8]$ **3.** $(-\infty, 0] \cup [8, \infty)$ **4.** $(-\infty, -3] \cup [-1, 2]$ **5.** $\varnothing$ **6.** $(-4, 5]$ **7.** $\left(-\infty, -3\right) \cup \left(-\dfrac{8}{5}, \infty\right)$

Vocabulary and Readiness Check 8.4

1. $[-7, 3)$ **3.** $(-\infty, 0]$ **5.** $(-\infty, -12) \cup [-10, \infty)$

Exercise Set 8.4

1. $(-\infty, -5) \cup (-1, \infty)$ **3.** $[-4, 3]$ **5.** $[2, 5]$ **7.** $\left(-5, -\dfrac{1}{3}\right)$ **9.** $(2, 4) \cup (6, \infty)$ **11.** $(-\infty, -4] \cup [0, 1]$

13. $[-3, -2] \cup [2, 3]$ **15.** $(-7, 2)$ **17.** $(-1, \infty)$ **19.** $(-\infty, -1] \cup (4, \infty)$ **21.** $(-\infty, 2) \cup \left(\dfrac{11}{4}, \infty\right)$ **23.** $(0, 2] \cup [3, \infty)$

25. $(-\infty, -7) \cup (8, \infty)$ **27.** $\left[-\dfrac{5}{4}, \dfrac{3}{2}\right]$ **29.** $(-\infty, 0) \cup (1, \infty)$ **31.** $(-\infty, -4] \cup [4, 6]$ **33.** $\left(-\infty, -\dfrac{2}{3}\right) \cup \left[\dfrac{3}{2}, \infty\right)$

35. $\left(-4, -\dfrac{3}{2}\right) \cup \left(\dfrac{3}{2}, \infty\right)$ **37.** $(-\infty, -5] \cup [-1, 1] \cup [5, \infty)$ **39.** $\left(-\infty, -\dfrac{5}{3}\right) \cup \left(\dfrac{7}{2}, \infty\right)$ **41.** $(0, 10)$ **43.** $(-\infty, -4) \cup [5, \infty)$

45. $(-\infty, -6] \cup (-1, 0] \cup (7, \infty)$ **47.** $(-\infty, 1) \cup (2, \infty)$ **49.** $(-\infty, -8] \cup (-4, \infty)$ **51.** $(-\infty, 0] \cup \left(5, \dfrac{11}{2}\right]$ **53.** $(0, \infty)$

55.   **57.**   **59.**   **61.**   **63.** answers may vary

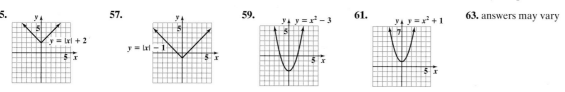

$y = |x| + 2$ $y = |x| - 1$ $y = x^2 - 3$ $y = x^2 + 1$

65. any number less than -1 or between 0 and 1   **67.** x is between 2 and 11   **69.**   **71.**

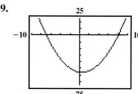

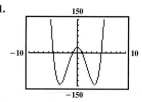

The Bigger Picture

1. $-9, \dfrac{7}{3}$ **2.** $(4, 7)$ **3.** $4, 5$ **4.** $\dfrac{-1 - \sqrt{13}}{6}, \dfrac{-1 + \sqrt{13}}{6}$ **5.** $[-2, 7)$ **6.** $\dfrac{5}{4}$ **7.** $\dfrac{1}{5}, 7$ **8.** $\left(-\infty, -\dfrac{1}{2}\right] \cup [4, \infty)$

9. $(-\infty, -8) \cup (22, \infty)$ **10.** $(7, \infty)$

Section 8.5
Practice Exercises

1.

2.

3.

4.

5.

Vocabulary and Readiness Check 8.5

1. quadratic   **3.** upward   **5.** lowest   **7.** $(0, 0)$ **9.** $(2, 0)$ **11.** $(0, 3)$ **13.** $(-1, 5)$

Exercise Set 8.5

1.   **3.**   **5.**   **7.**   **9.**   **11.**

13.   **15.**   **17.** 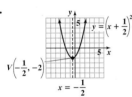   **19.**   **21.** 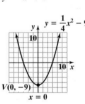   **23.**

25. 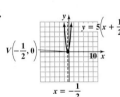   **27.**   **29.**   **31.** 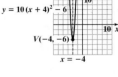   **33.** 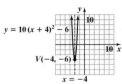   **35.**

37.   **39.**   **41.**   **43.**

45.   **47.**   **49.**   **51.**

53.   **55.** $x^2 + 8x + 16$ **57.** $z^2 - 16z + 64$ **59.** $y^2 + y + \dfrac{1}{4}$ **61.** $-6, 2$ **63.** $-5 - \sqrt{26}, -5 + \sqrt{26}$
65. $4 - 3\sqrt{2}, 4 + 3\sqrt{2}$ **67.** c   **69.** $f(x) = 5(x - 2)^2 + 3$ **71.** $f(x) = 5(x + 3)^2 + 6$

73.   **75.**   **77.**   **79. a.** 185,564 thousand   **b.** 233,088 thousand

Section 8.6
Practice Exercises

1.   **2.** vertex: $(0.5, -6.25)$; axis of symmetry: $x = \dfrac{1}{2}$ **3.**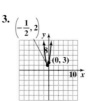

4. $(1, -4)$ **5.** vertex: $(2.5, 12.25)$; x-intercepts: $(-1, 0)$, $(6, 0)$; y-intercept: (0.6) 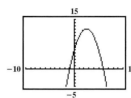   **6.** Maximum height 9 feet in $\dfrac{3}{4}$ second

Vocabulary and Readiness Check 8.6

1. (h, k) **3.** $0; 1$ **5.** $2; 1$ **7.** $1; 1$ **9.** down   **11.** up

Exercise Set 8.6

1. $(-4, -9)$ **3.** $(5, 30)$ **5.** $(1, -2)$ **7.** $\left(\dfrac{1}{2}, \dfrac{5}{4}\right)$ **9.** D   **11.** B

13.   **15.**   **17.**   **19.**   **21.**

23.   **25.**   **27.**   **29.**   **31.**

33. 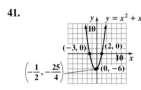   **35.**   **37.**   **39.**

41.   **43.**   **45.** 144 ft   **47. a.** 200 bicycles   **b.** $12,000   **49.** 30 and 30   **51.** 5, -5
53. length, 20 units; width, 20 units

55. 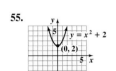   **57.**   **59.** 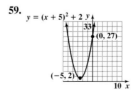   **61.** 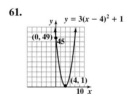   **63.**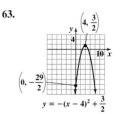

65. minimum value   **67.** maximum value   **69.**   **71.**   **73.** -0.84 **75.** 1.43

77. a. maximum; answers may vary   **b.** 2005   **c.** 31,523

Section 8.7

Practice Exercises

1. b. $y = 0.869x - 52.044$ **c.** $y = -0.006x^2 + 2.033x - 103.516$ **d.** quadratic regression equation; 48¢
2. a. $y = 0.064x + 22.933$ **b.** $y = -9.967x^2 + 0.071x + 22.926$ **c.** linear model: 23.8 MPG; quadratic model: 23.7 MPG   **d.** answers may vary
3. a. linear   **b.** $y = 7.767x + 316.6$ **c.** 511 truck sales in 2015
4. b. $y = -5.214x^2 + 62.314x + 471.286$ **c.** \$468.2 million in 2012   **5. a.** $y = 5.893x^2 - 17.679x + 266.429$ **b.** 785 films

Exercise Set 8.7

1. linear   **3.** neither   **5.** quadratic   **7.** quadratic   **9.** not linear or quadratic
11. $y = 8.252x + 472.775$; \$653.33; $y = -0.055x^2 + 9.025x + 471.107$; \$643.25
13. quadratic; $y = 0.75x^2 - 7.47x + 42.51$; \$51.09 billion   **15.** answers may vary.
17. a. $y = 14.607x + 1139.75$ **b.** $y = -1.042x^2 + 21.899x + 1132.458$ **c.** linear: 1315 million users in 2012; quadratic: 1245 million users in 2012
d. answers may vary   **19. a.** quadratic   **b.** $y = -13.010x^2 + 1727.626x - 37,489.948$ **c.** \$17,457 profit at \$80; \$19,332 profit at \$60
d. \$1500; no; answers may vary   **e.** \$66   **21.** $y = 82.509x - 51,908.333$ **23.** \$146,113

25.

Predicted MPG	Difference between Actual and Predicted MPG
23.1	0
23.2	0
23.3	-0.2
23.3	0.2
23.4	-0.1
23.4	0

27. $y = 0.75x^2 - 10.774x + 86.655$; answers may vary   **29.** quadratic   **31.** linear

33. quadratic   **35.** quadratic   **37.** $-3, 5$

39. $-\dfrac{3}{2}, 5$ **41.** 10   **43.** $(0, -8)$ **45.** $(0, 5)$

Chapter 8 Vocabulary Check

1. discriminant   **2.** $\pm\sqrt{b}$ **3.** $\dfrac{-b}{2a}$ **4.** quadratic inequality   **5.** completing the square   **6.** $(0, k)$ **7.** $(h, 0)$ **8.** (h, k)
9. quadratic formula   **10.** quadratic

Chapter 8 Review

1. $14, 1$ **3.** $-7, 7$ **5.** $\dfrac{-3 - \sqrt{5}}{2}, \dfrac{-3 + \sqrt{5}}{2}$ **7.** 4.25% **9.** two complex but not real solutions   **11.** two real solutions   **13.** 8   **15.** $-\dfrac{5}{2}, 1$

17. $\dfrac{5 - i\sqrt{143}}{12}, \dfrac{5 + i\sqrt{143}}{12}$ **19. a.** 20 ft   **b.** $\dfrac{15 + \sqrt{321}}{16}$ sec; 2.1 sec   **21.** $3, \dfrac{-3 + 3i\sqrt{3}}{2}, \dfrac{-3 - 3i\sqrt{3}}{2}$ **23.** $\dfrac{2}{3}, 5$ **25.** $1, 125$ **27.** $-1, 1, -i, i$

29. Jerome: 10.5 hr; Tim: 9.5 hr   **31.** $[-5, 5]$ **33.** $(5, 6)$ **35.** $(-\infty, -6) \cup \left(-\dfrac{3}{4}, 0\right) \cup (5, \infty)$ **37.** $(-5, -3) \cup (5, \infty)$ **39.** $\left(-\dfrac{6}{5}, 0\right) \cup \left(\dfrac{5}{6}, 3\right)$

41.   **43.**   **45.**   **47.**   **49.**   **51.**

53.   **55.** The numbers are both 210.

57. a. $y = 2.575x + 200.939$; 317 million   **b.** $y = 0.017x^2 + 1.919x + 203.991$; 325 million   **59.** $-5, 6$

61. $\dfrac{-1 - 3i\sqrt{3}}{2}, \dfrac{-1 + 3i\sqrt{3}}{2}$ **63.** $-i\sqrt{11}, i\sqrt{11}$ **65.** $-\dfrac{8\sqrt{7}}{7}, \dfrac{8\sqrt{7}}{7}$ **67.** $\left(-\infty, -\dfrac{5}{4}\right] \cup \left[\dfrac{3}{2}, \infty\right)$ **69.** $\left(2, \dfrac{7}{2}\right)$

Chapter 8 Test

1. $\dfrac{7}{5}, -1$ **2.** $-1 - \sqrt{10}, -1 + \sqrt{10}$ **3.** $\dfrac{1 + i\sqrt{31}}{2}, \dfrac{1 - i\sqrt{31}}{2}$ **4.** $3 - \sqrt{7}, 3 + \sqrt{7}$ **5.** $-\dfrac{1}{7}, -1$ **6.** $\dfrac{3 + \sqrt{29}}{2}, \dfrac{3 - \sqrt{29}}{2}$

7. $-2 - \sqrt{11}, -2 + \sqrt{11}$ **8.** $-1, 1, -i, i, -3$ **9.** $-1, 1, -i, i$ **10.** $6, 7$ **11.** $3 - \sqrt{7}, 3 + \sqrt{7}$ **12.** $\dfrac{2 - i\sqrt{6}}{2}, \dfrac{2 + i\sqrt{6}}{2}$

13. $\left(-\infty, -\dfrac{3}{2}\right) \cup (5, \infty)$ **14.** $(-\infty, -5] \cup [-4, 4] \cup [5, \infty)$ **15.** $(-\infty, -3) \cup (2, \infty)$ **16.** $(-\infty, -3) \cup [2, 3)$

17.   **18.** **19.** **20.**   **21.** $(5 + \sqrt{17})\,\text{hr} \approx 9.12\,\text{hr}$
22. a. 272 ft **b.** 5.12 sec **23.** 7 ft
24. a. quadratic **b.** $y = 0.019x^2 - 0.465x + 16.65$; 15.62 births per 1000 people

Chapter 8 Cumulative Review

1. a. $5 + y \geq 7$ **b.** $11 \neq z$ **c.** $20 < 5 - 2x$; Sec. 1.4, Ex. 6 **3.** slope: 1; 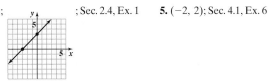 ; Sec. 2.4, Ex. 1 **5.** $(-2, 2)$; Sec. 4.1, Ex. 6

7. a. $6x^2 - 29x + 28$ **b.** $15x^4 - x^2y - 2y^2$; Sec. 5.4, Ex. 6 **9. a.** $4(2x^2 + 1)$ **b.** prime polynomial **c.** $3x^2(2 - x + 4x^2)$; Sec. 5.5, Ex. 2

11. $(x - 5)(x - 7)$; Sec. 5.6, Ex. 2 **13.** $3x(a - 2b)^2$; Sec. 5.7, Ex. 2 **15.** $-\dfrac{2}{3}$; Sec. 5.8, Ex. 4 **17.** $-2, 0, 2$; Sec. 5.8, Ex. 6

19. $\dfrac{1}{5x - 1}$; Sec. 6.1, Ex. 3a **21.** $\dfrac{7x^2 - 9x - 13}{(2x + 1)(x - 5)(3x - 2)}$; Sec. 6.2, Ex. 5 **23.** $\dfrac{xy + 2x^3}{y - 1}$; Sec. 6.3, Ex. 3 **25.** $3x^3y - 15x - 1 - \dfrac{6}{xy}$; Sec. 6.4, Ex. 2

27. a. 5 **b.** 5; Sec. 6.4, Ex. 10 **29.** -3; Sec. 6.5, Ex. 1 **31.** 2; Sec. 6.6, Ex. 2 **33.** $\dfrac{1}{6}$; $y = \dfrac{1}{6}x$; Sec. 6.7, Ex. 1 **35. a.** 3 **b.** $|x|$ **c.** $|x - 2|$
d. -5 **e.** $2x - 7$ **f.** $5|x|$ **g.** $|x + 1|$; Sec. 7.1, Ex. 5 **37. a.** $\sqrt{x}$ **b.** $\sqrt[3]{5}$ **c.** $\sqrt{rs^3}$; Sec. 7.2, Ex. 7 **39. a.** $5x\sqrt{x}$ **b.** $3x^2y^2\sqrt[3]{2y^2}$
c. $3z^2\sqrt[4]{z^3}$; Sec. 7.3, Ex. 4 **41. a.** $\dfrac{2\sqrt{5}}{5}$ **b.** $\dfrac{8\sqrt{x}}{3x}$ **c.** $\dfrac{\sqrt[3]{4}}{2}$; Sec. 7.5, Ex. 1 **43.** $\dfrac{2}{9}$; Sec. 7.6, Ex. 5 **45. a.** $\dfrac{1}{2} + \dfrac{3}{2}i$ **b.** $-\dfrac{7}{3}i$; Sec. 7.7, Ex. 5
47. $-1 + 2\sqrt{3}, -1 - 2\sqrt{3}$; Sec. 8.1, Ex. 3 **49.** 9; Sec. 8.3, Ex. 1

CHAPTER 9 EXPONENTIAL AND LOGARITHMIC FUNCTIONS

Section 9.1
Practice Exercises

1. a. $4x + 7$ **b.** $-2x - 3$ **c.** $3x^2 + 11x + 10$ **d.** $\dfrac{x + 2}{3x + 5}$, where $x \neq -\dfrac{5}{3}$ **2. a.** 50; 46 **b.** $9x^2 - 30x + 26$; $3x^2 - 2$
3. a. $x^2 + 6x + 14$ **b.** $x^2 + 8$ **4. a.** $(h \circ g)(x)$ **b.** $(g \circ f)(x)$

Vocabulary and Readiness Check 9.1
1. C **3.** F **5.** D

Exercise Set 9.1

1. a. $3x - 6$ **b.** $-x - 8$ **c.** $2x^2 - 13x - 7$ **d.** $\dfrac{x - 7}{2x + 1}$, where $x \neq -\dfrac{1}{2}$ **3. a.** $x^2 + 5x + 1$ **b.** $x^2 - 5x + 1$ **c.** $5x^3 + 5x$
d. $\dfrac{x^2 + 1}{5x}$, where $x \neq 0$ **5. a.** $\sqrt{x} + x + 5$ **b.** $\sqrt{x} - x - 5$ **c.** $x\sqrt{x} + 5\sqrt{x}$ **d.** $\dfrac{\sqrt{x}}{x + 5}$, where $x \neq -5$
7. a. $5x^2 - 3x$ **b.** $-5x^2 - 3x$ **c.** $-15x^3$ **d.** $-\dfrac{3}{5x}$, where $x \neq 0$ **9.** 42 **11.** -18 **13.** 0
15. $(f \circ g)(x) = 25x^2 + 1$; $(g \circ f)(x) = 5x^2 + 5$ **17.** $(f \circ g)(x) = 2x + 11$; $(g \circ f)(x) = 2x + 4$
19. $(f \circ g)(x) = -8x^3 - 2x - 2$; $(g \circ f)(x) = -2x^3 - 2x + 4$ **21.** $(f \circ g)(x) = |10x - 3|$; $(g \circ f)(x) = 10|x| - 3$
23. $(f \circ g)(x) = \sqrt{-5x + 2}$; $(g \circ f)(x) = -5\sqrt{x} + 2$ **25.** $H(x) = (g \circ h)(x)$ **27.** $F(x) = (h \circ f)(x)$ **29.** $G(x) = (f \circ g)(x)$
31. answers may vary; for example $g(x) = x + 2$ and $f(x) = x^2$ **33.** answers may vary; for example, $g(x) = x + 5$ and $f(x) = \sqrt{x} + 2$
35. answers may vary; for example, $g(x) = 2x - 3$ and $f(x) = \dfrac{1}{x}$ **37.** $y = x - 2$ **39.** $y = \dfrac{x}{3}$ **41.** $y = -\dfrac{x + 7}{2}$ **43.** 6 **45.** 4 **47.** 4
49. -1 **51.** answers may vary **53.** $P(x) = R(x) - C(x)$

Section 9.2
Practice Exercises

1. a. one-to-one **b.** not one-to-one **c.** not one-to-one **d.** not one-to-one **e.** not one-to-one **f.** not one-to-one
2. a. no, not one-to-one **b.** yes **c.** yes **d.** no, not a function **e.** no, not a function
3. $f^{-1}(x) = \{(4, 3), (0, -2), (8, 2), (6, 6)\}$ **4.** $f^{-1}(x) = 6 - x$

5. 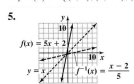 **6. a.** 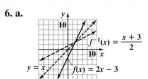 **b.** 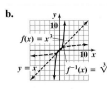 **7.** $f(f^{-1}(x)) = f\left(\dfrac{x + 1}{4}\right) = 4\left(\dfrac{x + 1}{4}\right) - 1 = x + 1 - 1 = x$

$f^{-1}(f(x)) = f^{-1}(4x - 1) = \dfrac{(4x - 1) + 1}{4} = \dfrac{4x}{4} = x$

Vocabulary and Readiness Check 9.2

1. $(2, 11)$ **3.** $(3, 7)$ **5.** horizontal   **7.** $x; x$

Exercise Set 9.2

1. one-to-one; $f^{-1} = \{(-1, -1), (1, 1), (2, 0), (0, 2)\}$ **3.** one-to-one; $h^{-1} = \{(10, 10)\}$ **5.** one-to-one; $f^{-1} = \{(12, 11), (3, 4), (4, 3), (6, 6)\}$
7. not one-to-one   **9.** one-to-one;
11. a. 3   **b.** 1   **13. a.** 1   **b.** -1

Rank in Population (Input)	1	19	35	4	48
State (Output)	CA	MD	NV	FL	ND

15. one-to-one   **17.** not one-to-one
19. one-to-one   **21.** not one-to-one

23. $f^{-1}(x) = x - 4$ **25.** $f^{-1}(x) = \dfrac{x + 3}{2}$ **27.** $f^{-1}(x) = 2x + 2$ **29.** $f^{-1}(x) = \sqrt[3]{x}$

31. $f^{-1}(x) = \dfrac{x - 2}{5}$ **33.** $f^{-1}(x) = 5x + 2$ **35.** $f^{-1}(x) = x^3$ **37.** $f^{-1}(x) = \dfrac{5 - x}{3x}$ **39.** $f^{-1}(x) = \sqrt[3]{x} - 2$

41. 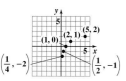   **43.**   **45.** 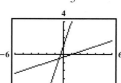   **47.** $(f \circ f^{-1})(x) = x; (f^{-1} \circ f)(x) = x$

49. $(f \circ f^{-1})(x) = x; (f^{-1} \circ f)(x) = x$ **51.** 5   **53.** 8   **55.** $\dfrac{1}{27}$ **57.** 9   **59.** $3^{1/2} \approx 1.73$ **61. a.** $(2, 9)$ **b.** $(9, 2)$

63. a. $\left(-2, \dfrac{1}{4}\right), \left(-1, \dfrac{1}{2}\right), (0, 1), (1, 2), (2, 5)$ **b.** $\left(\dfrac{1}{4}, -2\right), \left(\dfrac{1}{2}, -1\right), (1, 0), (2, 1), (5, 2)$

c.   **d.** 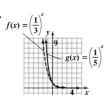   **65.** answers may vary   **67.** $f^{-1}(x) = \dfrac{x - 1}{3}$;   **69.** $f^{-1}(x) = x^3 - 1$;

 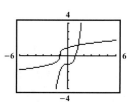

Section 9.3

Practice Exercises

1.   **2.**   **3.**   **4. a.** 2;   **b.** $\dfrac{4}{3}$;   **c.** -4 **5.** $3950.43   **6.** 60.86%
7. a. 67.21%   **b.** 3.09%

Vocabulary and Readiness Check 9.3

1. exponential   **3.** yes   **5.** yes; $(0, 1)$ **7.** $(0, \infty)$

Exercise Set 9.3

1.   **3.**   **5.**   **7.**   **9.**

11.   **13.**   **15.** 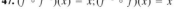   **17.** C   **19.** B   **21.** 3   **23.** $\dfrac{3}{4}$ **25.** $\dfrac{8}{5}$ **27.** $-\dfrac{2}{3}$ **29.** 4   **31.** $\dfrac{3}{2}$

33. $-\dfrac{1}{3}$ **35.** -2 **37.** 24.6 lb   **39.** 333 bison   **41.** 1.1 g   **43. a.** 658.1 pascals   **b.** 180.0 pascals   **45. a.** 134,342 students   **b.** 840, 276 students

47. \$7621.42   **49.** \$4065.59   **51.** 562 million cellular phone users   **53.** 4   **55.** $\varnothing$ **57.** 2, 3   **59.** 3   **61.** -1 **63.** answers may vary

65.   **67.**   **69.** The graphs are the same since $\left(\dfrac{1}{2}\right)^{-x} = 2x.$ **71.** 18.62 lb   **73.** 50.41 g

Section 9.4
Practice Exercises

1. a. $3^4 = 81$ **b.** $5^{-1} = \dfrac{1}{5}$ **c.** $7^{1/2} = \sqrt{7}$ **d.** $13^4 = y$ **2. a.** $\log_4 64 = 3$ **b.** $\log_6 \sqrt[3]{6} = \dfrac{1}{3}$ **c.** $\log_5 \dfrac{1}{125} = -3$ **d.** $\log_\pi z = 7$

3. a. 2   **b.** -3 **c.** $\dfrac{1}{2}$ **4. a.** -2 **b.** 2   **c.** 36   **d.** 0   **e.** 0   **5. a.** 4   **b.** -2 **c.** 5   **d.** 4   **6.**   **7.**

Vocabulary and Readiness Check 9.4

1. logarithmic   **3.** yes   **5.** no; none   **7.** $(-\infty, \infty)$

Exercise Set 9.4

1. $6^2 = 36$ **3.** $3^{-3} = \dfrac{1}{27}$ **5.** $10^3 = 1000$ **7.** $9^4 = x$ **9.** $\pi^{-2} = \dfrac{1}{\pi^2}$ **11.** $7^{1/2} = \sqrt{7}$ **13.** $0.7^3 = 0.343$

15. $3^{-4} = \dfrac{1}{81}$ **17.** $\log_2 16 = 4$ **19.** $\log_{10} 100 = 2$ **21.** $\log_\pi x = 3$ **23.** $\log_{10} \dfrac{1}{10} = -1$ **25.** $\log_4 \dfrac{1}{16} = -2$

27. $\log_5 \sqrt{5} = \dfrac{1}{2}$ **29.** 3   **31.** -2 **33.** $\dfrac{1}{2}$ **35.** -1 **37.** 0   **39.** 2   **41.** 4   **43.** -3 **45.** 2   **47.** 81

49. 7   **51.** -3 **53.** -3 **55.** 2   **57.** 2   **59.** $\dfrac{27}{64}$ **61.** 10   **63.** 4   **65.** 5   **67.** $\dfrac{1}{49}$ **69.** 3   **71.** 3   **73.** 1

75.   **77.**   **79.**   **81.**   **83.** 1

85. $\dfrac{x-4}{2}$ **87.** $\dfrac{2x+3}{x^2}$ **89.** $m-1$ **91. a.** $g(2) = 25$ **b.** $(25, 2)$ **c.** $f(25) = 2$ **93.** answers may vary   **95.** $\dfrac{9}{5}$ **97.** 1

99.   **101.** $y = \left(\dfrac{1}{3}\right)^x$ **103.** answers may vary   **105.** 0.0827

Section 9.5
Practice Exercises

1. a. $\log_8 15$ **b.** $\log_2 6$ **c.** $\log_5(x^2 - 1)$ **2. a.** $\log_5 3$ **b.** $\log_6 \dfrac{x}{3}$ **c.** $\log_4 \dfrac{x^2 + 1}{x^2 + 3}$ **3. a.** $8 \log_7 x$ **b.** $\dfrac{1}{4} \log_5 7$

4. a. $\log_5 512$ **b.** $\log_8 \dfrac{x^2}{x+3}$ **c.** $\log_7 15$ **5. a.** $\log_5 4 + \log_5 3 - \log_5 7$ **b.** $2 \log_4 a - 5 \log_4 b$ **6. a.** 1.39   **b.** 1.66   **c.** 0.28

Vocabulary and Readiness Check 9.5

1. 36   **3.** $\log_b 2^7$ **5.** x

Exercise Set 9.5

1. $\log_5 14$ **3.** $\log_4 9x$ **5.** $\log_6(x^2 + x)$ **7.** $\log_{10}(10x^2 + 20)$ **9.** $\log_5 3$ **11.** $\log_3 4$ **13.** $\log_2 \dfrac{x}{y}$ **15.** $\log_2 \dfrac{x^2 + 6}{x^2 + 1}$ **17.** $2 \log_3 x$

19. $-1 \log_4 5 = -\log_4 5$ **21.** $\dfrac{1}{2} \log_5 y$ **23.** $\log_2 5x^3$ **25.** $\log_4 48$ **27.** $\log_5 x^3 z^6$ **29.** $\log_4 4$, or 1   **31.** $\log_7 \dfrac{9}{2}$ **33.** $\log_{10} \dfrac{x^3 - 2x}{x+1}$

35. $\log_2 \dfrac{x^{7/2}}{(x+1)^2}$ **37.** $\log_8 x^{16/3}$ **39.** $\log_3 4 + \log_3 y - \log_3 5$ **41.** $\log_4 2 - \log_4 9 - \log_4 z$ **43.** $3 \log_2 x - \log_2 y$ **45.** $\dfrac{1}{2} \log_b 7 + \dfrac{1}{2} \log_b x$

47. $4 \log_6 x + 5 \log_6 y$ **49.** $3 \log_5 x + \log_5(x+1)$ **51.** $2 \log_6 x - \log_6(x+3)$ **53.** 1.2   **55.** 0.2   **57.** 0.35   **59.** 1.29   **61.** -0.68

63. −0.125   **65.**   **67.** −1   **69.** $\frac{1}{2}$ **71.** a and d   **73.** false   **75.** true   **77.** false

Integrated Review

1. $x^2 + x - 5$ **2.** $-x^2 + x - 7$ **3.** $x^3 - 6x^2 + x - 6$ **4.** $\dfrac{x - 6}{x^2 + 1}$ **5.** $\sqrt{3x - 1}$ **6.** $3\sqrt{x} - 1$

7. one-to-one; $\{(6, -2), (8, 4), (-6, 2), (3, 3)\}$ **8.** not one-to-one   **9.** not one-to-one   **10.** one-to-one

11. not one-to-one   **12.** $f^{-1}(x) = \dfrac{x}{3}$ **13.** $f^{-1}(x) = x - 4$ **14.** $f^{-1}(x) = \dfrac{x + 1}{5}$ **15.** $f^{-1}(x) = \dfrac{x - 2}{3}$

16.   **17.**   **18.**   **19.**   **20.** 3

21. 7   **22.** −8   **23.** 3   **24.** 2   **25.** $\frac{1}{2}$ **26.** 32   **27.** 4   **28.** 5   **29.** $\frac{1}{9}$ **30.** $\log_2 x^5$ **31.** $\log_2 5^x$ **32.** $\log_5 \dfrac{x^3}{y^5}$

33. $\log_5 x^9 y^3$ **34.** $\log_2 \dfrac{x^2 - 3x}{x^2 + 4}$ **35.** $\log_3 \dfrac{y^4 + 11y}{y + 2}$ **36.** $\log_7 9 + 2\log_7 x - \log_7 y$ **37.** $\log_6 5 + \log_6 y - 2\log_6 z$

Section 9.6

Practice Exercises

1. 1.1761   **2. a.** −2   **b.** 5   **c.** $\frac{1}{5}$ **d.** −3   **3.** $10^{3.4} \approx 2511.8864$ **4.** 5.6   **5.** 2.5649   **6. a.** 4   **b.** $\frac{1}{3}$

7. $\dfrac{e^8}{5} \approx 596.1916$ **8.** \$3051   **9.** 0.7740

Vocabulary and Readiness Check 9.6

1. 10   **3.** 7   **5.** 5   **7.** $\dfrac{\log 7}{\log 2}$ or $\dfrac{\ln 7}{\ln 2}$

Exercise Set 9.6

1. 0.9031   **3.** 0.3636   **5.** 0.6931   **7.** −2.6367   **9.** 1.1004   **11.** 1.6094   **13.** 1.6180   **15.** answers may vary   **17.** 2   **19.** −3   **21.** 2

23. $\frac{1}{4}$ **25.** 3   **27.** −7   **29.** −4   **31.** $\frac{1}{2}$ **33.** $\dfrac{e^7}{2} \approx 548.3166$ **35.** $10^{1.3} \approx 19.9526$ **37.** $\dfrac{10^{1.1}}{2} \approx 6.2946$ **39.** $e^{1.4} \approx 4.0552$

41. $\dfrac{4 + e^{2.3}}{3} \approx 4.6581$ **43.** $10^{2.3} \approx 199.5262$ **45.** $e^{-2.3} \approx 0.1003$ **47.** $\dfrac{10^{-0.5} - 1}{2} \approx -0.3419$ **49.** $\dfrac{e^{0.18}}{4} \approx 0.2993$ **51.** 1.5850   **53.** −2.3219

55. 1.5850   **57.** −1.6309   **59.** 0.8617   **61.** 4.2   **63.** 5.3   **65.** \$3656.38   **67.** \$2542.50   **69.** $\frac{4}{7}$ **71.** $x = \dfrac{3y}{4}$ **73.** −6, −1   **75.** $(2, -3)$

77. $\ln 50$; answers may vary   **79.**   **81.** 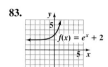   **83.** 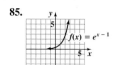   **85.**   **87.**

89.   **91.** 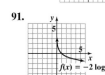   **93.** 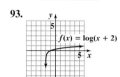   **95.** 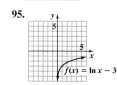   **97.** answers may vary

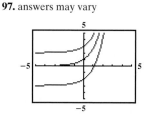

Section 9.7

Practice Exercises

1. $\dfrac{\log 9}{\log 5} \approx 1.3652$ **2.** 33   **3.** 1   **4.** $\frac{1}{3}$ **5.** 937 rabbits   **6.** 10 years   **7.** 3.67 years, or 3 years and 8 months

Exercise Set 9.7

1. $\dfrac{\log 6}{\log 3}$; 1.6309   **3.** $\dfrac{\log 3.8}{2 \log 3}$; 0.6076   **5.** $3 + \dfrac{\log 5}{\log 2}$; 5.3219   **7.** $\dfrac{\log 5}{\log 9}$; 0.7325   **9.** $\dfrac{\log 3}{\log 4} - 7$; -6.2075 **11.** $\dfrac{1}{3}\left(4 + \dfrac{\log 11}{\log 7}\right)$; 1.7441

13. $\dfrac{\ln 5}{6}$; 0.2682   **15.** 11   **17.** $9, -9$ **19.** $\dfrac{1}{2}$ **21.** $\dfrac{3}{4}$ **23.** 2   **25.** $\dfrac{1}{8}$ **27.** 11   **29.** $4, -1$ **31.** $\dfrac{1}{5}$ **33.** 100   **35.** $\dfrac{-5 + \sqrt{33}}{2}$

37. $\dfrac{192}{127}$ **39.** $\dfrac{2}{3}$ **41.** 103 wolves   **43.** 354,000 inhabitants   **45.** 14.7 yr   **47.** 9.9 yr   **49.** 1.7 yr   **51.** 8.8 yr   **53.** 24.5 lb   **55.** 55.7 in.

57. 11.9 lb/sq in.   **59.** 3.2 mi   **61.** 12 weeks   **63.** 18 weeks   **65.** $-\dfrac{5}{3}$ **67.** $\dfrac{17}{4}$ **69.** $f^{-1}(x) = \dfrac{x - 2}{5}$ **71.** 2.9%   **73.** answers may vary

75. a. $y = 25.275(1.261)^x$; **100**   **b.** \$31.88 billion   **c.** \$325.08 billion

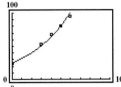

77. a. $y = 1634.766(1.331)^x$; **7500**   **b.** \$1228 million for 1999   **c.** \$21,470 million for 2009

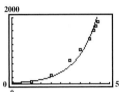

79. a. $y = 31.445(1.100)^x$; **2000**   **b.** \$1585.4 billion for 2001   **c.** \$3748.8 billion for 2010

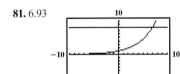

81. 6.93   **83.** $-3.68, 0.19$

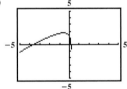

The Bigger Picture

1. $-\dfrac{3}{2}$ **2.** $\dfrac{\log 5}{\log 11} \approx 0.6712$ **3.** $[-5, \infty)$ **4.** $\left[-\dfrac{13}{3}, -2\right]$ **5.** $\left(-\dfrac{6}{5}, 0\right)$ **6.** $6, -\dfrac{1}{5}$ **7.** $\dfrac{21}{8}$ **8.** $-\dfrac{7}{3}, 3$ **9.** $(-\infty, \infty)$ **10.** $-2, 2$

11. $-5 + \sqrt{3}, -5 - \sqrt{3}$ **12.** $-\dfrac{1}{4}, 7$

Chapter 9 Vocabulary Check

1. inverse   **2.** composition   **3.** exponential   **4.** symmetric   **5.** Natural   **6.** Common   **7.** vertical; horizontal   **8.** logarithmic

Chapter 9 Review

1. $3x - 4$ **3.** $2x^2 - 9x - 5$ **5.** $x^2 + 2x - 1$ **7.** 18   **9.** -2 **11.** one-to-one;
$h^{-1} = \{(14, -9), (8, 6), (12, -11), (15, 15)\}$ **13.** one-to-one;
15. a. 3   **b.** 7   **17.** not one-to-one
19. not one-to-one   **21.** $f^{-1}(x) = x + 9$

Rank in Automobile Thefts (Input)	2	4	1	3
US Region (Output)	W	Midwest	S	NE

23. $f^{-1}(x) = \dfrac{x - 11}{6}$ **25.** $f^{-1}(x) = \sqrt[3]{x + 5}$

27. $g^{-1}(x) = \dfrac{6x + 7}{12}$ **29.**   **31.** $f^{-1}(x) = \dfrac{x + 3}{2}$;   **33.** -2 **35.** $\dfrac{3}{2}$ **37.** $\dfrac{8}{9}$ **39.**   **41.**

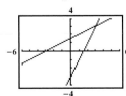

43. $1131.82   **45.** $\log_7 49 = 2$ **47.** $\left(\dfrac{1}{2}\right)^{-4} = 16$ **49.** $\dfrac{1}{64}$ **51.** 0   **53.** 8   **55.** 5   **57.** 4   **59.** $\dfrac{17}{3}$ **61.** $-1, 4$ **63.**

65. $\log_3 32$ **67.** $\log_7 \dfrac{3}{4}$ **69.** $\log_{11} 4$ **71.** $\log_5 \dfrac{x^3}{(x+1)^2}$ **73.** $3\log_3 x - \log_3(x+2)$ **75.** $\log_2 3 + 2\log_2 x + \log_2 y - \log_2 z$

77. 2.02   **79.** 0.5563   **81.** 0.2231   **83.** 3   **85.** -1 **87.** $\dfrac{e^2}{2}$ **89.** $\dfrac{e^{-1}+3}{2}$ **91.** 1.67 mm   **93.** 0.2920   **95.** $1957.30

97. $\dfrac{\log 7}{2 \log 3}$; 0.8856   **99.** $\dfrac{1}{2}\left(\dfrac{\log 6}{\log 3} - 1\right)$; 0.3155   **101.** $\dfrac{1}{3}\left(\dfrac{\log 4}{\log 5} + 5\right)$; 1.9538   **103.** $-\dfrac{\log 2}{\log 5} + 1$; 0.5693   **105.** $\dfrac{25}{2}$ **107.** $\varnothing$ **109.** $2\sqrt{2}$

111. 197,044 ducks   **113.** 20.9 yr   **115.** 38.5 yr   **117.** 8.5 yr   **119.** 2.82   **121.** $\dfrac{1}{2}$ **123.** 1   **125.** $-1, 5$ **127.** $e^{-3.2}$ **129.** $2e$ **131.** $\varnothing$

Chapter 9 Test

1. $2x^2 - 3x$ **2.** $3 - x$ **3.** 5   **4.** $x - 7$ **5.** $x^2 - 6x - 2$

6.   **7.** one-to-one   **8.** not one-to-one   **9.** one-to-one; $f^{-1}(x) = \dfrac{-x+6}{2}$ **10.** one-to-one; $f^{-1} = \{(0,0),(3,2)(5,-1)\}$

11. not one-to-one   **12.** $\log_3 24$ **13.** $\log_5 \dfrac{x^4}{x+1}$ **14.** $\log_6 2 + \log_6 x - 3\log_6 y$ **15.** -1.53 **16.** 1.0686   **17.** -1

18. $\dfrac{1}{2}\left(\dfrac{\log 4}{\log 3} - 5\right)$; -1.8691 **19.** $\dfrac{1}{9}$ **20.** $\dfrac{1}{2}$ **21.** 22   **22.** $\dfrac{25}{3}$ **23.** $\dfrac{43}{21}$ **24.** -1.0979

25.   **26.**   **27.** $5234.58   **28.** 6 yr   **29.** 64,913 prairie dogs   **30.** 15 yr   **31.** 1.2% per day

32. 3.95   **33.** $y = 247.71(1.138)^x$; $1971.1 billion in 2006

Chapter 9 Cumulative Review

1. a. 8   **b.** $-\dfrac{1}{3}$ **c.** -0.36 **d.** 0   **e.** $-\dfrac{2}{11}$ **f.** 42   **g.** 0; Sec. 1.3, Ex. 4   **3.** ; Sec. 2.1, Ex. 8

5. $\{(x, y, z)\mid x - 5y - 2z = 6\}$; Sec. 4.2, Ex. 4   **7. a.** x^3 **b.** 5^6 **c.** $5x$ **d.** $\dfrac{6y^2}{7}$; Sec. 5.1, Ex. 4   **9. a.** $102.60   **b.** $12.60; Sec. 6.1, Ex. 10

11. a. $\dfrac{3x}{2}$ **b.** $\dfrac{5+x}{7z^2}$ **c.** $x - 7$ **d.** $-\dfrac{1}{3y^2}$; Sec. 6.2, Ex. 1   **13.** $3x^2 + 2x + 3 + \dfrac{-6x+9}{x^2-1}$; Sec. 6.4, Ex. 6   **15.** -1; Sec. 6.5, Ex. 4

17. 6 mph; Sec. 6.6, Ex. 5   **19. a.** 3   **b.** -3 **c.** -5 **d.** not a real number   **e.** $4x$; Sec. 7.1, Ex. 4   **21. a.** $\sqrt[4]{x^3}$ **b.** $\sqrt[6]{x}$

c. $\sqrt[6]{72}$; Sec. 7.2, Ex. 8   **23. a.** $5\sqrt{3} + 3\sqrt{10}$ **b.** $\sqrt{35} + \sqrt{5} - \sqrt{42} - \sqrt{6}$ **c.** $21x - 7\sqrt{5x} + 15\sqrt{x} - 5\sqrt{5}$ **d.** $49 - 8\sqrt{3}$

e. $2x - 25$ **f.** $x + 22 + 10\sqrt{x-3}$; Sec. 7.4, Ex. 4   **25.** $\dfrac{\sqrt[4]{xy^3}}{3y^2}$; Sec. 7.5, Ex. 3   **27.** 3; Sec. 7.6, Ex. 4   **29.** $\dfrac{9 + i\sqrt{15}}{6}, \dfrac{9 - i\sqrt{15}}{6}$; Sec. 8.1, Ex. 8

31. $\dfrac{-1 + \sqrt{33}}{4}, \dfrac{-1 - \sqrt{33}}{4}$; Sec. 8.3, Ex. 2   **33.** $[0, 4]$; Sec. 8.4, Ex. 2   **35.** 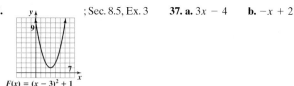; Sec. 8.5, Ex. 3   **37. a.** $3x - 4$ **b.** $-x + 2$

c. $2x^2 - 5x + 3$ **d.** $\dfrac{x-1}{2x-3}$, where $x \ne \dfrac{3}{2}$; Sec. 9.1, Ex. 1   **39.** $f^{-1}(x) = x - 3$; Sec. 9.2, Ex. 4   **41. a.** 2   **b.** -1 **c.** $\dfrac{1}{2}$; Sec. 9.4, Ex. 3

CHAPTER 10 CONIC SECTIONS

Section 10.1

Practice Exercises

1. $x = \dfrac{1}{2}y^2$ **2.** $x = -2(y+4)^2 - 1$ **3.** $y = -x^2 + 4x + 6$

4. $x = 3y^2 + 6y + 4$ **5.** 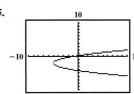 $x = y^2 + 4y - 1$ **6.** $x^2 + y^2 = 25$

7. 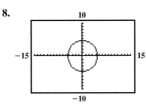 $(x - 3)^2 + (y + 2)^2 = 4$ **8.**

9. $(x + 2)^2 + (y + 5)^2 = 81$ **10.** 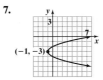 $x^2 + y^2 + 6x - 2y = 6$

Vocabulary and Readiness Check 10.1

1. conic sections   **3.** circle, center   **5.** radius   **7.** upward   **9.** to the left   **11.** downward

Exercise Set 10.1

1.   **3.**   **5.**   **7.**   **9.**

11.   **13.**   **15.**   **17.**   **19.**

21. 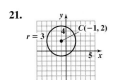   **23.** 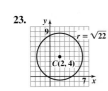   **25.** $(x - 2)^2 + (y - 3)^2 = 36$ **27.** $x^2 + y^2 = 3$ **29.** $(x + 5)^2 + (y - 4)^2 = 45$

31. The radius is $\sqrt{10}$.   **33.**   **35.** 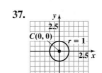   **37.**   **39.**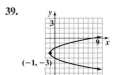

41.   **43.**   **45.** 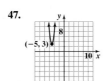   **47.**   **49.**

51. 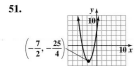   **53.** 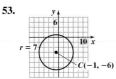   **55.** 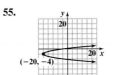   **57.** 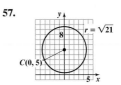   **59.**

61. **63.** **65.** **67.** **69.** $\dfrac{\sqrt{10}}{4}$ **71.** $2\sqrt{5}$

73. a. 67.5 meters **b.** ground level or 0 m **c.** 67.5 meters **d.** $(0, 67.5)$ **e.** $x^2 + (y - 67.5)^2 = (67.5)^2$

75. a. 76.5 m **b.** 7 m **c.** 83.5 m **d.** $(0, 83.5)$ **e.** $x^2 + (y - 83.5)^2 = (76.5)^2$

77. a. **b.** $x^2 + y^2 = 100$ **c.** $x^2 + y^2 = 25$ **79.** 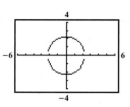 **81.**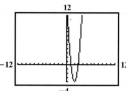

Section 10.2

Practice Exercises

1. $\dfrac{x^2}{25} + \dfrac{y^2}{4} = 1$ **2.** $9x^2 + 4y^2 = 36$ **3.** $\dfrac{(x - 4)^2}{49} + \dfrac{(y + 1)^2}{81} = 1$

4. $\dfrac{x^2}{9} - \dfrac{y^2}{16} = 1$ **5.** $9y^2 - 25x^2 = 225$

Vocabulary and Readiness Check 10.2

1. hyperbola **3.** focus **5.** hyperbola; $(0, 0)$; x; $(a, 0)$ and $(-a, 0)$ **7.** ellipse **9.** hyperbola **11.** hyperbola

Exercise Set 10.2

1. **3.** 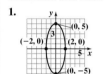 **5.** **7.** **9.**

11. **13.** **15.** **17.** **19.**

21. answers may vary **23.** parabola **25.** ellipse **27.** hyperbola

29. circle **31.** parabola **33.** ellipse **35.** hyperbola

37. parabola **39.** $(-\infty, 5)$ **41.** $[4, \infty)$ **43.** $-2x^3$ **45.** $-5x^4$ **47.** x-intercepts; 6 units **49.** x-intercepts; 6 units

51. ellipses: C, E, H; circles: B, F; hyperbolas: A, D, G   **53.** A: 49, 7; B: 0, 0; C: 9, 3; D: 64, 8; E: 64, 8; F: 0, 0; G: 81, 9; H: 4, 2

55. A: $\frac{7}{6}$; B: 0; C: $\frac{3}{5}$; D: $\frac{8}{5}$; E: $\frac{8}{9}$; F: 0; G: $\frac{9}{4}$; H: $\frac{1}{6}$ **57.** equal to zero   **59.** answers may vary   **61.** $\dfrac{x^2}{1.69 \cdot 10^{16}} + \dfrac{y^2}{1.5625 \cdot 10^{16}} = 1$

63. $9x^2 + 4y^2 = 36$ **65.**   **67.**   **69.**

Integrated Review

1. 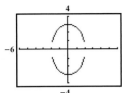   **2.**   **3.**   **4.**   **5.**

6. 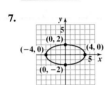   **7.**   **8.**   **9.**   **10.**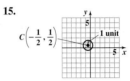

11.   **12.**   **13.**   **14.** 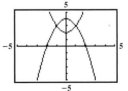   **15.**

Section 10.3

Practice Exercises

1. $(4, -2)$ **2.** $(-4, 3), (0, -1)$ **3.** $\varnothing$ **4.** $(2, \sqrt{3}); (2, -\sqrt{3}); (-2, \sqrt{3}); (-2, -\sqrt{3})$

Exercise Set 10.3

1. $(3, -4), (-3, 4)$ **3.** $(\sqrt{2}, \sqrt{2}), (-\sqrt{2}, -\sqrt{2})$ **5.** $(4, 0), (0, -2)$ **7.** $(-\sqrt{5}, -2), (-\sqrt{5}, 2), (\sqrt{5}, -2), (\sqrt{5}, 2)$ **9.** $\varnothing$ **11.** $(1, -2), (3, 6)$
13. $(2, 4), (-5, 25)$ **15.** $\varnothing$ **17.** $(1, -3)$ **19.** $(-1, -2), (-1, 2), (1, -2), (1, 2)$ **21.** $(0, -1)$ **23.** $(-1, 3), (1, 3)$ **25.** $(\sqrt{3}, 0), (-\sqrt{3}, 0)$
27. $\varnothing$ **29.** $(-6, 0), (6, 0), (0, -6)$ **31.** 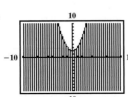   **33.**   **35.** $(8x - 25)$ in.   **37.** $(4x^2 + 6x + 2)$ m

39. answers may vary   **41.** 0, 1, 2, 3, or 4; answers may vary   **43.** 9 and 7; 9 and -7; -9 and 7; -9 and -7 **45.** 15 cm by 19 cm
47. 15 thousand compact discs; price: \$3.75   **49.**   **51.**

Section 10.4

Practice Exercises

1. $\dfrac{x^2}{36} + \dfrac{y^2}{16} \geq 1$ **2.** $y \leq x^2 + 2$ **3.** $16y^2 > 9x^2 + 144$

4. $\begin{cases} y \ge x^2 \\ y \le -3x + 2 \end{cases}$ **5.** $\begin{cases} x^2 + y^2 < 16 \\ \dfrac{x^2}{4} - \dfrac{y^2}{9} < 1 \\ y < x + 3 \end{cases}$

Exercise Set 10.4

1.   **3.**   **5.**   **7.** 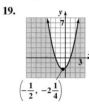   **9.**   **11.**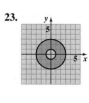

13.   **15.**   **17.**   **19.**   **21.**   **23.**

25.   **27.** 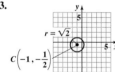   **29.**   **31.**   **33.**   **35.**

37. not a function   **39.** function   **41.** 1   **43.** $3a^2 - 2$ **45.** answers may vary   **47.**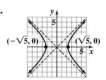

Chapter 10 Vocabulary Check

1. circle; center   **2.** nonlinear system of equations   **3.** ellipse   **4.** radius   **5.** hyperbola

Chapter 10 Review

1. $(x + 4)^2 + (y - 4)^2 = 9$ **3.** $(x + 7)^2 + (y + 9)^2 = 11$ **5.**   **7.**   **9.**

11.   **13.**   **15.**   **17.**   **19.**

21.   **23.**   **25.**   **27.** $(1, -2), (4, 4)$ **29.** $(-1, 1), (2, 4)$ **31.** $(2, 2\sqrt{2}), (2, -2\sqrt{2})$

33. (1, 4) **35.** 15 ft by 10 ft **37.**   **39.**   **41.**   **43.** $(x + 7)^2 + (y - 8)^2 = 25$

45.   **47.**   **49.**   **51.**   **53.**

55. (5, 1), (−1, 7) **57.**

Chapter 10 Test

1.   **2.**   **3.**   **4.**   **5.**

6.   **7.**   **8.**   **9.** (−5, −1), (−5, 1), (5, −1), (5, 1) **10.** (6, 12), (1, 2)

11.   **12.**   **13.**   **14.** height: 10 ft; width: 30 ft

Chapter 10 Cumulative Review

1. $(4 \cdot 9)y = 36y$; Sec. 1.4, Ex. 10 **3.** ; Sec. 2.3, Ex. 6 **5.** ∅; Sec. 4.1, Ex. 8

7. a. $125x^6$ **b.** $\dfrac{8}{27}$ **c.** $\dfrac{9p^8}{q^{10}}$ **d.** $64y^2$ **e.** $\dfrac{y^{14}}{x^{35}z^7}$; Sec. 5.2, Ex. 2 **9.** $-\dfrac{1}{6}$, 3; Sec. 5.8, Ex. 5 **11.** $\dfrac{12}{x - 1}$; Sec. 6.2, Ex. 6

13. a. $\dfrac{1}{9xy^2}$ **b.** $\dfrac{x(x - 2)}{2(x + 2)}$ **c.** $\dfrac{x^2}{y^2}$; Sec. 6.3, Ex. 1 **15.** $2x - 5$; Sec. 6.4, Ex. 3 **17.** 16; Sec. 6.4, Ex. 11 **19.** −6, −1; Sec. 6.5, Ex. 5

21. $2\dfrac{2}{9}$ hr; no; Sec. 6.6, Ex. 4 **23. a.** 1 **b.** −4 **c.** $\dfrac{2}{5}$ **d.** x^2 **e.** $-3x^3$; Sec. 7.1, Ex. 3 **25. a.** $z - z^{17/3}$ **b.** $x^{2/3} - 3x^{1/3} - 10$; Sec. 7.2, Ex. 5

27. a. 2 **b.** $\dfrac{5}{2}\sqrt{x}$ **c.** $14xy^2\sqrt[3]{x}$; **d.** $4a^2b\sqrt[4]{2a}$; Sec. 7.3, Ex. 5 **29. a.** $\dfrac{5\sqrt{5}}{12}$ **b.** $\dfrac{5\sqrt[3]{7x}}{2}$; Sec. 7.4, Ex. 3 **31.** $\dfrac{\sqrt{21xy}}{3y}$; Sec. 7.5, Ex. 2

33. 42; Sec. 7.6, Ex. 1 **35. a.** $-i$ **b.** 1 **c.** −1 **d.** 1; Sec. 7.7, Ex. 6 **37.** $-1 + \sqrt{5}$, $-1 - \sqrt{5}$; Sec. 8.1, Ex. 5

39. $2 + \sqrt{2}$, $2 - \sqrt{2}$; Sec. 8.2, Ex. 3 **41.** 2, −2, i, $-i$; Sec. 8.3, Ex. 3 **43.** $[-2, 3)$; Sec. 8.4, Ex. 6

45. ; Sec. 8.5, Ex. 5 **47.** (2, −16); Sec. 8.6, Ex. 4 **49.** $\sqrt{2} \approx 1.414$; Sec. 7.3, Ex. 6

CHAPTER 11 SEQUENCES, SERIES, AND THE BINOMIAL THEOREM

Section 11.1

Practice Exercises

1. $6, 9, 14, 21, 30$ **2. a.** $-\dfrac{1}{5}$ **b.** $\dfrac{1}{20}$ **c.** $\dfrac{1}{150}$ **d.** $-\dfrac{1}{95}$ **3. a.** $a_n = (2n - 1)$ **b.** $a_n = 3^n$ **c.** $a_n = \dfrac{n}{n+1}$ **d.** $a_n = -\dfrac{1}{n+1}$ **4.** $2022.40

Vocabulary and Readiness Check 11.1

1. general **3.** infinite **5.** -1

Exercise Set 11.1

1. $5, 6, 7, 8, 9$ **3.** $-1, 1, -1, 1, -1$ **5.** $\dfrac{1}{4}, \dfrac{1}{5}, \dfrac{1}{6}, \dfrac{1}{7}, \dfrac{1}{8}$ **7.** $2, 4, 6, 8, 10$ **9.** $-1, -4, -9, -16, -25$ **11.** $2, 4, 8, 16, 32$ **13.** $7, 9, 11, 13, 15$

15. $-1, 4, -9, 16, -25$ **17.** 75 **19.** 118 **21.** $\dfrac{6}{5}$ **23.** 729 **25.** $\dfrac{4}{7}$ **27.** $\dfrac{1}{8}$ **29.** -95 **31.** $-\dfrac{1}{25}$ **33.** $a_n = 4n - 1$ **35.** $a_n = -2^n$

37. $a_n = \dfrac{1}{3^n}$ **39.** 48 ft, 80 ft, and 112 ft **41.** $a_n = 0.10(2)^{n-1}$; $819.20 **43.** 2400 cases; 75 cases **45.** 50 sparrows in 2004: extinct in 2010

47. **49.** **51.** $\sqrt{13}$ units **53.** $\sqrt{41}$ units **55.** $1, 0.7071, 0.5774, 0.5, 0.4472$
57. $2, 2.25, 2.3704, 2.4414, 2.4883$

Section 11.2

Practice Exercises

1. $4, 9, 14, 19, 24$ **2. a.** $a_n = 5 - 3n$ **b.** -31 **3.** 51 **4.** 47 **5.** $a_n = 54{,}800 + 2200n$; $61{,}400 **6.** $8, -24, 72, -216$ **7.** $\dfrac{1}{64}$ **8.** -192
9. $a_1 = 3$; $r = \dfrac{3}{2}$ **10.** 75 units

Vocabulary and Readiness Check 11.2

1. geometric; ratio **3.** first; difference

Exercise Set 11.2

1. $4, 6, 8, 10, 12$ **3.** $6, 4, 2, 0, -2$ **5.** $1, 3, 9, 27, 81$ **7.** $48, 24, 12, 6, 3$ **9.** 33 **11.** -875 **13.** -60 **15.** 96 **17.** -28 **19.** 1250

21. 31 **23.** 20 **25.** $a_1 = \dfrac{2}{3}$; $r = -2$ **27.** answers may vary **29.** $a_1 = 2$; $d = 2$ **31.** $a_1 = 5$; $r = 2$ **33.** $a_1 = \dfrac{1}{2}$; $r = \dfrac{1}{5}$ **35.** $a_1 = x$; $r = 5$

37. $a_1 = p$; $d = 4$ **39.** 19 **41.** $-\dfrac{8}{9}$ **43.** $\dfrac{17}{2}$ **45.** $\dfrac{8}{81}$ **47.** -19 **49.** $a_n = 4n + 50$; 130 seats **51.** $a_n = 6(3)^{n-1}$

53. $486, 162, 54, 18, 6$; $a_n = \dfrac{486}{3^{n-1}}$; 6 bounces **55.** $a_n = 4000 + 125(n - 1)$ or $a_n = 3875 + 125n$; $5375 **57.** 25 g **59.** $\dfrac{11}{18}$ **61.** 40 **63.** $\dfrac{907}{495}$
65. $11{,}782.40, $5891.20, $2945.60, $1472.80 **67.** $19.652, 19.618, 19.584, 19.55$ **69.** answers may vary

Section 11.3

Practice Exercises

1. a. $-\dfrac{5}{4}$ **b.** 360 **2. a.** $\displaystyle\sum_{i=1}^{6} 5i$ **b.** $\displaystyle\sum_{i=1}^{4}\left(\dfrac{1}{5}\right)^i$ **3.** $\dfrac{655}{72}$ **4.** 95 plants

Vocabulary and Readiness Check 11.3

1. infinite **3.** summation **5.** partial sum

Exercise Set 11.3

1. -2 **3.** 60 **5.** 20 **7.** $\dfrac{73}{168}$ **9.** $\dfrac{11}{36}$ **11.** 60 **13.** 74 **15.** 62 **17.** $\dfrac{241}{35}$ **19.** $\displaystyle\sum_{i=1}^{5}(2i - 1)$ **21.** $\displaystyle\sum_{i=1}^{4} 4(3)^{i-1}$

23. $\displaystyle\sum_{i=1}^{6}(-3i + 15)$ **25.** $\displaystyle\sum_{i=1}^{4}\dfrac{4}{3^{i-2}}$ **27.** $\displaystyle\sum_{i=1}^{7} i^2$ **29.** -24 **31.** 0 **33.** 82 **35.** -20 **37.** -2 **39.** $1, 2, 3, \ldots, 10$; 55 trees

41. $a_n = 6(2)^{n-1}$; 96 units **43.** $a_n = 50(2)^n$; n represents the number of 12-hour periods; 800 bacteria **45.** 30 opossums; 68 opossums

47. 6.25 lb; 93.75 lb **49.** 16.4 in.; 134.5 in. **51.** 10 **53.** $\dfrac{10}{27}$ **55.** 45 **57.** 90 **59. a.** $2 + 6 + 12 + 20 + 30 + 42 + 56$
b. $1 + 2 + 3 + 4 + 5 + 6 + 7 + 1 + 4 + 9 + 16 + 25 + 36 + 49$ **c.** answers may vary **d.** true; answers may vary

Integrated Review

1. $-2, -1, 0, 1, 2$ **2.** $\dfrac{7}{2}, \dfrac{7}{3}, \dfrac{7}{4}, \dfrac{7}{5}, \dfrac{7}{6}$ **3.** $1, 3, 9, 27, 81$ **4.** $-4, -1, 4, 11, 20$ **5.** 64 **6.** -14 **7.** $\dfrac{1}{40}$ **8.** $-\dfrac{1}{82}$ **9.** $7, 4, 1, -2, -5$

10. $-3, -15, -75, -375, -1875$ **11.** $45, 15, 5, \dfrac{5}{3}, \dfrac{5}{9}$ **12.** $-12, -2, 8, 18, 28$ **13.** 101 **14.** $\dfrac{243}{16}$ **15.** 384 **16.** 185 **17.** -10 **18.** $\dfrac{1}{5}$

19. 50 **20.** 98 **21.** $\dfrac{31}{2}$ **22.** $\dfrac{61}{20}$ **23.** -10 **24.** 5

Section 11.4
Practice Exercises

1. 80 **2.** 1275 **3.** 105 blocks of ice **4.** $42\dfrac{5}{8}$ **5.** $\$987,856$ **6.** $9\dfrac{1}{3}$ **7.** 900 in.

Vocabulary and Readiness Check 11.4
1. arithmetic **3.** geometric **5.** arithmetic

Exercise Set 11.4
1. 36 **3.** 484 **5.** 63 **7.** 2.496 **9.** 55 **11.** 16 **13.** 24 **15.** $\dfrac{1}{9}$ **17.** -20 **19.** $\dfrac{16}{9}$ **21.** $\dfrac{4}{9}$ **23.** 185 **25.** $\dfrac{381}{64}$

27. $-\dfrac{33}{4}$, or -8.25 **29.** $-\dfrac{75}{2}$ **31.** $\dfrac{56}{9}$ **33.** $4000, 3950, 3900, 3850, 3800; 3450$ cars; $44,700$ cars **35.** Firm A (Firm A, $\$265,000$; Firm B, $\$254,000$)

37. $\$39,930; \$139,230$ **39.** 20 min; 123 min **41.** 180 ft **43.** Player A, 45 points; Player B, 75 points **45.** $\$3050$ **47.** $\$10,737,418.23$

49. 720 **51.** 3 **53.** $x^2 + 10x + 25$ **55.** $8x^3 - 12x^2 + 6x - 1$ **57.** $\dfrac{8}{10} + \dfrac{8}{100} + \dfrac{8}{1000} + \cdots; \dfrac{8}{9}$ **59.** answers may vary

Section 11.5
Practice Exercises

1. $p^7 + 7p^6r + 21p^5r^2 + 35p^4r^3 + 35p^3r^4 + 21p^2r^5 + 7pr^6 + r^7$ **2. a.** $\dfrac{1}{7}$ **b.** 840 **c.** 5 **d.** 1

3. $a^9 + 9a^8b + 36a^7b^2 + 84a^6b^3 + 126a^5b^4 + 126a^4b^5 + 84a^3b^6 + 36a^2b^7 + 9ab^8 + b^9$ **4.** $a^3 + 15a^2 + 75a + 125$

5. $27x^3 - 54x^2y + 36xy^2 - 8y^3$ **6.** $1,892,352x^5y^6$

Vocabulary and Readiness Check 11.5
1. 1 **3.** 24 **5.** 6

Exercise Set 11.5
1. $m^3 + 3m^2n + 3mn^2 + n^3$ **3.** $c^5 + 5c^4d + 10c^3d^2 + 10c^2d^3 + 5cd^4 + d^5$ **5.** $y^5 - 5y^4x + 10y^3x^2 - 10y^2x^3 + 5yx^4 - x^5$

7. answers may vary **9.** 8 **11.** 42 **13.** 360 **15.** 56 **17.** $a^7 + 7a^6b + 21a^5b^2 + 35a^4b^3 + 35a^3b^4 + 21a^2b^5 + 7ab^6 + b^7$

19. $a^5 + 10a^4b + 40a^3b^2 + 80a^2b^3 + 80ab^4 + 32b^5$ **21.** $q^9 + 9q^8r + 36q^7r^2 + 84q^6r^3 + 126q^5r^4 + 126q^4r^5 + 84q^3r^6 + 36q^2r^7 + 9qr^8 + r^9$

23. $1024a^5 + 1280a^4b + 640a^3b^2 + 160a^2b^3 + 20ab^4 + b^5$ **25.** $625a^4 - 1000a^3b + 600a^2b^2 - 160ab^3 + 16b^4$ **27.** $8a^3 + 36a^2b + 54ab^2 + 27b^3$

29. $x^5 + 10x^4 + 40x^3 + 80x^2 + 80x + 32$ **31.** $5cd^4$ **33.** d^7 **35.** $-40r^2s^3$ **37.** $6x^2y^2$ **39.** $30a^9b$

41. not one-to-one **43.** one-to-one **45.** not one-to-one

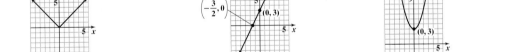

47. $x^2\sqrt{x} + 5\sqrt{3}x^2 + 30x\sqrt{x} + 30\sqrt{3}x + 45\sqrt{x} + 9\sqrt{3}$ **49.** 126 **51.** 28 **53.** answers may vary

Chapter 11 Vocabulary Check
1. finite sequence **2.** factorial of n **3.** infinite sequence **4.** geometric sequence, common ratio **5.** series **6.** general term
7. arithmetic sequence, common difference **8.** Pascal's triangle

Chapter 11 Review
1. $-3, -12, -27, -48, -75$ **3.** $\dfrac{1}{100}$ **5.** $a_n = \dfrac{1}{6n}$ **7.** 144 ft, 176 ft, 208 ft **9.** $660,000; 1,320,000; 2,640,000; 5,280,000; 10,560,000; 2010:$

$10,560,000$ infested acres **11.** $-2, -\dfrac{4}{3}, -\dfrac{8}{9}, -\dfrac{16}{27}, -\dfrac{32}{81}$ **13.** 111 **15.** -83 **17.** $a_1 = 3; d = 5$ **19.** $a_n = \dfrac{3}{10^n}$ **21.** $a_1 = \dfrac{8}{3}, r = \dfrac{3}{2}$

23. $a_1 = 7x, r = -2$ **25.** $8, 6, 4.5, 3.4, 2.5, 1.9;$ good **27.** $a_n = 2^{n-1}, \$512, \$536,870,912$ **29.** $a_n = 900 + (n - 1)150$ or $a_n = 150n + 750;$

$\$1650/\text{month}$ **31.** $1 + 3 + 5 + 7 + 9 = 25$ **33.** $\dfrac{1}{4} - \dfrac{1}{6} + \dfrac{1}{8} = \dfrac{5}{24}$ **35.** -4 **37.** -10 **39.** $\displaystyle\sum_{i=1}^{6} 3^{i-1}$ **41.** $\displaystyle\sum_{i=1}^{4} \dfrac{1}{4^i}$ **43.** $a_n = 20(2)^n;$

n represents the number of 8-hour periods; 1280 yeast **45.** Job A, $\$48,300$; Job B, $\$46,600$ **47.** 150 **49.** 900 **51.** -410 **53.** 936 **55.** 10

57. -25 **59.** $\$30,418; \$99,868$ **61.** $\$58; \$553$ **63.** 2696 mosquitoes **65.** $\dfrac{5}{9}$ **67.** $x^5 + 5x^4z + 10x^3z^2 + 10x^2z^3 + 5xz^4 + z^5$

69. $16x^4 + 32x^3y + 24x^2y^2 + 8xy^3 + y^4$ **71.** $b^8 + 8b^7c + 28b^6c^2 + 56b^5c^3 + 70b^4c^4 + 56b^3c^5 + 28b^2c^6 + 8bc^7 + c^8$

73. $256m^4 - 256m^3n + 96m^2n^2 - 16mn^3 + n^4$ **75.** $35a^4b^3$ **77.** 130 **79.** 40.5

Chapter 11 Test

1. $-\frac{1}{5}, \frac{1}{6}, -\frac{1}{7}, \frac{1}{8}, -\frac{1}{9}$ **2.** 247 **3.** $a_n = \frac{2}{5}\left(\frac{1}{5}\right)^{n-1}$ **4.** $a_n = (-1)^n 9n$ **5.** 155 **6.** -330 **7.** $\frac{144}{5}$ **8.** 1 **9.** 10 **10.** -60

11. $a^6 - 6a^5b + 15a^4b^2 - 20a^3b^3 + 15a^2b^4 - 6ab^5 + b^6$ **12.** $32x^5 + 80x^4y + 80x^3y^2 + 40x^2y^3 + 10xy^4 + y^5$ **13.** 925 people; 250 people initially

14. $1 + 3 + 5 + 7 + 9 + 11 + 13 + 15$; 64 shrubs **15.** 33.75 cm, 218.75 cm **16.** 320 cm **17.** 304 ft; 1600 ft **18.** $\frac{14}{33}$

Chapter 11 Cumulative Review

1. a. -5 **b.** 3 **c.** $-\frac{1}{8}$ **d.** -4 **e.** $\frac{1}{4}$ **f.** undefined; Sec. 1.3, Ex. 5 **3.** $\$2350$; Sec. 1.6, Ex. 4 **5.** $y = \frac{2x}{3} + \frac{7}{3}$; Sec. 1.8, Ex. 2

7. a. $15x^7$ **b.** $-9.6x^4p^{12}$; Sec. 5.1, Ex. 2 **9.** $x^3 - 4x^2 - 3x + 11 + \frac{12}{x+2}$; Sec. 6.4, Ex. 9 **11. a.** $5\sqrt{2}$ **b.** $2\sqrt[3]{3}$ **c.** $\sqrt{26}$

d. $2\sqrt[4]{2}$; Sec. 7.3, Ex. 3 **13.** 10%; Sec. 8.1, Ex. 9 **15.** $2, 7$; Sec. 8.3, Ex. 4 **17.** $\left(-\frac{7}{2}, -1\right)$; Sec. 8.4, Ex. 7 **19.** $\frac{25}{4}$ ft; $\frac{5}{8}$ sec; Sec. 8.6, Ex. 6

21. a. $25; 7$ **b.** $x^2 + 6x + 9$; $x^2 + 3$; Sec. 9.1, Ex. 2 **23.** $f^{-1} = \{(1,0), (7,-2), (-6,3), (4,4)\}$; Sec. 9.2, Ex. 3 **25. a.** 4 **b.** $\frac{3}{2}$

c. 6 **d.** 1.431; Sec. 9.3, Ex. 4 **27. a.** 2 **b.** -1 **c.** 3 **d.** 6; Sec. 9.4, Ex. 5 **29. a.** $\log_{11} 30$ **b.** $\log_3 6$ **c.** $\log_2(x^2 + 2x)$; Sec. 9.5, Ex. 1

31. $\$2509.30$; Sec. 9.6, Ex. 8 **33.** $\frac{\log 7}{\log 3} \approx 1.7712$; Sec. 9.7, Ex. 1 **35.** 18; Sec. 9.7, Ex. 2 **37.** 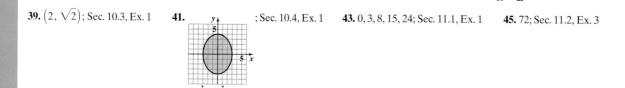 ; Sec. 10.2, Ex. 4

39. $\left(2, \sqrt{2}\right)$; Sec. 10.3, Ex. 1 **41.** 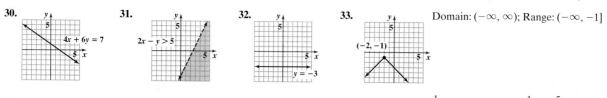 ; Sec. 10.4, Ex. 1 **43.** $0, 3, 8, 15, 24$; Sec. 11.1, Ex. 1 **45.** 72; Sec. 11.2, Ex. 3

47. a. $\frac{7}{2}$ **b.** 56; Sec. 11.3, Ex. 1 **49.** 465; Sec. 11.4, Ex. 2

APPENDIX A.2 PRACTICE FINAL EXAM

1. $6\sqrt{6}$ **2.** $-\frac{3}{2}$ **3.** 5 **4.** $\frac{1}{81x^2}$ **5.** $\frac{64}{3}$ **6.** $\frac{8a^{1/3}c^{2/3}}{b^{5/12}}$ **7.** $1900, 1950, 2000, 2050, 2100$ **a.** $\$24,600$ **b.** $\$14,000$ **8.** $3y(x + 3y)(x - 3y)$

9. $2(2y - 1)(4y^2 + 2y + 1)$ **10.** $(x + 3)(x - 3)(y - 3)$ **11.** $-5x^3y - 11x - 9$ **12.** $36m^2 + 12mn + n^2$ **13.** $2x^3 - 13x^2 + 14x - 4$

14. $\frac{x + 2}{2(x + 3)}$ **15.** $\frac{1}{x(x + 3)}$ **16.** $\frac{5x - 2}{(x - 3)(x + 2)(x - 2)}$ **17.** $-x\sqrt{5x}$ **18.** -20 **19.** $2x^2 - x - 2 + \frac{2}{2x + 1}$ **20.** $-\frac{29}{17}$ **21.** $1, \frac{2}{3}$

22. $4, -\frac{8}{7}$ **23.** $\left(\frac{3}{2}, 5\right]$ **24.** $(-\infty, -2) \cup \left(\frac{4}{3}, \infty\right)$ **25.** 3 **26.** $\frac{3 \pm \sqrt{29}}{2}$ **27.** $2, 3$ **28.** $\left(-\infty, -\frac{3}{2}\right) \cup (5, \infty)$ **29.** $\left(\frac{7}{2}, -10\right)$

30. 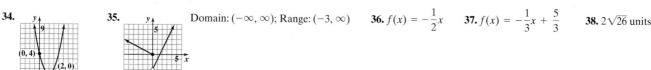 **31.** **32.** **33.** Domain: $(-\infty, \infty)$; Range: $(-\infty, -1]$

34. **35.** Domain: $(-\infty, \infty)$; Range: $(-3, \infty)$ **36.** $f(x) = -\frac{1}{2}x$ **37.** $f(x) = -\frac{1}{3}x + \frac{5}{3}$ **38.** $2\sqrt{26}$ units

39. $\left(-4, \dfrac{7}{2}\right)$ **40.** $\dfrac{3\sqrt{y}}{y}$ **41.** $\dfrac{8 - 6\sqrt{x} + x}{8 - 2x}$ **42.** $(-\infty, -3)$ **43.** New York: 21.8 million; Seoul: 23.1 million; Tokyo: 33.4 million **44.** 5

45. a. quadratic **b.** $y = 0.019x^2 - 0.465x + 16.65$; 15.62 births per 1000 people **46.** 7 ft

47. a. 272 ft **b.** 5.12 sec **48.** 5 gal of 10%; 15 gal of 20% **49.** $-2i\sqrt{2}$ **50.** $-3i$ **51.** $7 + 24i$ **52.** $-\dfrac{3}{2} + \dfrac{5}{2}i$

53. $(g \circ h)(x) = x^2 - 6x - 2$ **54.** $f^{-1}(x) = \dfrac{-x + 6}{2}$ **55.** $\log_5 \dfrac{x^4}{x + 1}$ **56.** -1 **57.** $\dfrac{1}{2}\left(\dfrac{\log 4}{\log 3} - 5\right)$; -1.8691 **58.** 22

59. $\dfrac{43}{21}$ **60.** 3.95 **61.** **62.** 64,913 prairie dogs **63.** **64.** **65.**

66. $(-5, -1)(-5, 1), (5, -1), (5, 1)$ **67.** $-\dfrac{1}{5}, \dfrac{1}{6}, -\dfrac{1}{7}, \dfrac{1}{8}, -\dfrac{1}{9}$ **68.** 155 **69.** 1 **70.** 10 **71.** $32x^5 + 80x^4y + 80x^3y^2 + 40x^2y^3 + 10xy^4 + y^5$

APPENDIX B.3 REVIEW OF VOLUME AND SURFACE AREA

1. $V = 72$ cu in.; $SA = 108$ sq in. **3.** $V = 512$ cu cm; $SA = 384$ sq cm

5. $V = 4\pi$ cu yd ≈ 12.56 cu yd; $SA = (2\sqrt{13}\pi + 4\pi)$ sq yd ≈ 35.20 sq yd **7.** $V = \dfrac{500}{3}\pi$ cu in. $\approx 523\dfrac{17}{21}$ cu in.; $SA = 100\pi$ sq in. $\approx 314\dfrac{2}{7}$ sq in.

9. $V = 48$ cu cm; $SA = 96$ sq cm **11.** $2\dfrac{10}{27}$ cu in. **13.** 26 sq ft **15.** $10\dfrac{5}{6}$ cu in. **17.** 960 cu cm **19.** 196π sq in.

21. $7\dfrac{1}{2}$ cu ft **23.** $12\dfrac{4}{7}$ cu cm

APPENDIX C
Exercise Set Appendix C

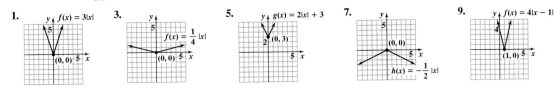

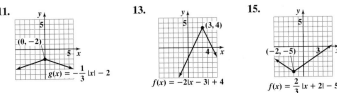

APPENDIX D
Vocabulary and Readiness Check
1. 56 **3.** -32 **5.** 20

Exercise Set Appendix D

1. 26 **3.** -19 **5.** 0 **7.** $\dfrac{13}{6}$ **9.** $(1, 2)$ **11.** $\{(x, y)\,|\,3x + y = 1\}$ **13.** $(9, 9)$ **15.** $(-3, -2)$ **17.** $(3, 4)$ **19.** 8 **21.** 0 **23.** 15

25. 54 **27.** $(-2, 0, 5)$ **29.** $(6, -2, 4)$ **31.** $(-2, 3, -1)$ **33.** $(0, 2, -1)$ **35.** 5 **37.** 0; answers may vary

APPENDIX E AN INTRODUCTION TO USING A GRAPHING UTILITY

Viewing Window and Interpreting Window Settings Exercise Set

1. yes **3.** no **5.** answers may vary **7.** answers may vary **9.** answers may vary

11. Xmin $= -12$ Ymin $= -12$ **13.** Xmin $= -9$ Ymin $= -12$ **15.** Xmin $= -10$ Ymin $= -25$ **17.** Xmin $= -10$ Ymin $= -30$
Xmax $= 12$ Ymax $= 12$ Xmax $= 9$ Ymax $= 12$ Xmax $= 10$ Ymax $= 25$ Xmax $= 10$ Ymax $= 30$
Xscl $= 3$ Yscl $= 3$ Xscl $= 1$ Yscl $= 2$ Xscl $= 2$ Yscl $= 5$ Xscl $= 1$ Yscl $= 3$

19. Xmin $= -20$ Ymin $= -30$
Xmax $= 30$ Ymax $= 50$
Xscl $= 5$ Yscl $= 10$

Graphing Equations and Square Viewing Window Exercise Set

1. Setting B   **3.** Setting B   **5.** Setting B

7.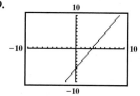

9.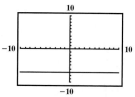

11.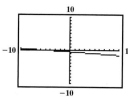

13.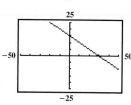

15.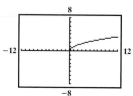

17.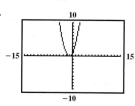

19.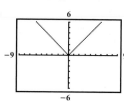

21.

Index

Photo Credits

Page 437 Kelly-Mooney Photography/Corbis/Bettmann

Page 454 Spike Mafford/Getty Images, Inc.—Photodisc

Page 459 Rob Melnychuk/Getty Images, Inc.—Photodisc

Page 460 Amy C. Etra/PhotoEdit Inc.

Page 460 John Serafin

Page 463 Michael Gadomski/Photo Researchers, Inc.

Page 467 Richard A. Cooke III/Getty Images Inc.—Stone Allstock

Page 471 Ken Welsh/AGE Fotostock America, Inc.—Royalty-free

Chapter 7

Page 484 William West/AFP/Getty Images

Page 500 © Susan Findlay/Masterfile

Page 529 Tom Carter/PhotoEdit Inc.

Chapter 8

Page 558 Tony Freeman/PhotoEdit Inc.

Page 561 David Parker/Science Photo Library/Photo Researchers, Inc.

Page 571 Dr. Chikaraishi/National Institute for Fusion Science

Page 572 © Shane Pedersen/Wildlight

Page 577 Zefa Collection/Corbis Zefa Collection

Page 580 Tim Flach/Getty Images, Inc.—Stone Allstock

Page 580 Arthur S. Aubry Photography/Getty Images, Inc.—Photodisc

Page 605 Robert Harding World Imagery

Page 620 AP Wide World Photos

Chapter 9

Page 624 Copyright © Andrew McClenaghan/Photo Researchers, Inc.

Page 649 © Alfred Pasieka/Peter Arnold

Page 673 © Weiss/Sunset/Animals Animals

Chapter 10

Page 688 Peter Adams/Getty Images—Creative Express Royalty Free

Page 697 Steve Vidler/ImageState/International Stock Photography Ltd.